Membrane Proteins: Structure, Function and Motion

Membrane Proteins: Structure, Function and Motion

Editor

Masoud Jelokhani-Niaraki

MDPI • Basel • Beijing • Wuhan • Barcelona • Belgrade • Manchester • Tokyo • Cluj • Tianjin

Editor
Masoud Jelokhani-Niaraki
Department of Chemistry & Biochemistry,
Wilfrid Laurier University,
Waterloo, ON, Canada

Editorial Office
MDPI
St. Alban-Anlage 66
4052 Basel, Switzerland

This is a reprint of articles from the Special Issue published online in the open access journal *International Journal of Molecular Sciences* (ISSN 1422-0067) (available at: https://www.mdpi.com/journal/ijms/special_issues/membrane_proteins).

For citation purposes, cite each article independently as indicated on the article page online and as indicated below:

LastName, A.A.; LastName, B.B.; LastName, C.C. Article Title. *Journal Name* **Year**, *Volume Number*, Page Range.

ISBN 978-3-0365-6537-8 (Hbk)
ISBN 978-3-0365-6538-5 (PDF)

Contents

About the Editor

Masoud Jelokhani-Niaraki

Masoud Jelokhani-Niaraki is a professor of biochemistry at Wilfrid Laurier University in Canada. His main research interests are in the biophysical chemistry of membrane proteins (ion channel, carrier proteins, and membrane surface proteins) and biologically active peptides (antimicrobial peptides, neuropeptides, and cell-penetrating peptides).

Preface to "Membrane Proteins: Structure, Function and Motion"

This book is a collection of reviews and original research articles on the physical biochemistry of membrane proteins. Different aspects of the complex and dynamic structure and functional behaviour of membrane proteins in cell membranes and their interaction with lipids, proteins, and other molecules and ions have been studied by some of the leading scientists in the field. This book is helpful to chemical and biological science students, instructors, and research scientists.

Masoud Jelokhani-Niaraki
Editor

International Journal of
Molecular Sciences

MDPI

Editorial

Membrane Proteins: Structure, Function and Motion

Masoud Jelokhani-Niaraki

Department of Chemistry and Biochemistry, Wilfrid Laurier University, Waterloo, ON N2L 3C5, Canada; mjelokhani@wlu.ca

Cell membranes are intricate multicomponent supramolecular structures, with a complex variable morphology and chemical composition. Membrane proteins are present in the cell membranes of all living organisms and contribute to their biological and physicochemical properties. The structure, function and mobility of membrane proteins are closely intertwined with the structure and composition of membranes and their surrounding environment. Cell membranes are essential for the steady-state homeostasis of cells and their reaction to environmental changes. Membrane proteins are involved in a variety of dynamic cellular processes, such as ionic and molecular transport, electron transport, signal transduction, enzymatic reactions and intercellular communication. Despite their relative abundance (25–30% of all proteins) and important roles in life processes, there is considerably less structural and functional information about membrane proteins in comparison to other types of proteins. The first high-resolution structure for a membrane protein was published in 1985 [1,2]. Currently, more than 1500 unique membrane protein structures have been determined in atomic resolution [3]. In comparison, by the end of 2022, close to 200,000 protein structures have been deposited in the Protein Data Bank [4].

The main challenge in studying the native structures of membrane proteins and their biological functions is in optimizing the experimental conditions for their extraction/purification from cell membranes and successive reconstitution in their native-like membrane milieus [5]. In their native environments, membrane proteins interact with membrane lipids, other membrane proteins, as well as extramembranous water, inorganic and small organic ions and other biomolecules. All these interactions, the everchanging chemical composition of the cell and the extracellular environmental factors influence the structure and biological function of membrane proteins in a time-dependent dynamic system. These variable interactions and the complexity of cell membrane structures are not easily reproducible (or impossible to reproduce) experimentally under in vitro or modified in vivo conditions. In addition to arduous technical problems in isolation and purification of membrane proteins in their intact native conformations, there are many questions about the molecular behaviour of membrane proteins that are not yet answered or are only partially answered. Some of these problems include the specific and non-specific interactions of membrane proteins with lipids and other membrane or membrane-interacting proteins, dynamic conformational changes in membrane proteins, membrane protein folding and oligomerization, modes of action of infectious structures (such as viruses) with membrane surface and membrane proteins, role of membrane proteins in morphology of cell and organelle membrane structures and the molecular action and adaptability of membrane proteins in their complex in vivo environment. As a result of these substantially challenging problems, in the past few decades, molecular studies of membrane proteins have become an attractive and flourishing research area in biochemistry, molecular biophysics, cell biology and systems biology. Visualizing/imagining the dynamic nature of membrane proteins and their various interconnections in the cell is becoming one of the key elements in understanding the colourful, intricate and mysterious molecular machinery of life.

In this compendium of reviews and research articles on the structure, function and motion of membrane proteins, we can see diverse experimental, theoretical and computational

Citation: Jelokhani-Niaraki, M. Membrane Proteins: Structure, Function and Motion. *Int. J. Mol. Sci.* **2023**, *24*, 468. https://doi.org/10.3390/ijms24010468

Received: 6 December 2022
Accepted: 16 December 2022
Published: 27 December 2022

approaches in studying different aspects of the biophysics and biochemistry of membrane proteins. Despite several important breakthroughs in deciphering the molecular aspects of membrane proteins and structure of membranes, this unusual diversity in approaches emphasizes the exploratory nature of the research in this field, which is still in its initial steps in a slightly illuminated but foggy territory.

At the beginning of this journey, the general reader can start from a brief but comprehensive review on the difficulties and uncertainties facing researchers in the current methods of extraction, purification and reconstitution of membrane proteins in native-like lipid surroundings, highlighting the advantages and disadvantages of each method [6]. Lipids, as the essential components in cell membranes, interact with embedded membrane proteins and influence their structure, function, association and conformational flexibility. Two insightful reviews emphasize the biophysical aspects of lipid–protein interacting systems and various computational and experimental methodologies currently used for studying these molecular systems [7,8].

The transport of ions, as well as small molecules and macromolecules, through membranes is closely connected to the membrane protein structure and function, structure of membranes and their chemical composition. A group of articles in this collection focuses on this aspect of membrane proteins. The mechanism of the transport of protons through the inner membrane of mitochondria by structurally connected carrier proteins adenine nucleotide translocase (ANT) and uncoupling protein (UCP) is considered in detail in a review and two articles. The detailed review introduces the current views on the connection between the structure and mechanism of proton transport in mitochondrial membrane proteins, UCP and ANT [9]. In a related article, it is hypothesized that the proton transport in ANT and UCPs has similar regulation patterns that can be explained by the fatty acid cycling concept [10]. A molecular dynamics (MD) simulation study, based on the homology modelling with ANT, shows that the UCP2 structure is impermeable to water and possesses additional functional elements, such as a specific fatty-acid-binding site, which is related to the proton transport mechanism across the inner-mitochondrial membranes [11]. Proton transport is also the subject of a review article on the connection and importance of cellular and subcellular protons on the mechanism of activity of some molecule transporters essential for the biological processes in plants [12]. In a biophysical study on reconstituted archaerhodopsin photoreceptor protein, a proton pump, the real-time pore-forming properties of this protein is studied in bilayers [13]. An inactivated non-conducting conformational state of KcsA (the bacterial potassium ion channel) with minimal affinity for potassium ions, while maintaining two binding sites for K^+ ions, was detected by using a long-chain quaternary ammonium channel blocker [14].

Molecular transport and targeting mediated by the membrane proteins in cell and organelle membranes are the subject of a review and two articles. Recent advances in the relatively unexplored area of protein targeting and protein/molecular transport in the chloroplast outer membrane are reviewed with a focus on the characterized chloroplast outer-membrane protein targeting pathways and new insights into novel targeting pathways using a bioinformatics approach [15]. Bacterial resistance-nodulation-cell division (RND) membrane protein transporters are a group of efflux pumps involved in the mechanisms of bacterial resistance against antibiotic drugs. In a comparative approach, these inner-membrane proteins are studied in two Gram-negative bacteria, with a special stress on the role of the inter-membrane peptidoglycan layer in stabilizing the RND protein complexes that surpass the inter-membrane space to reach the outer-bacterial membrane [16]. Transepithelial water flux at the tricellular tight junction for angulin-1 (lipolysis-stimulated lipoprotein receptor, LSR) is studied in epithelial cell lines to conclude that the transepithelial water permeability was affected only in tight cell lines [17].

The specific influence of lipid-interacting proteins and peptides on the membrane structure and its physicochemical properties is treated in three articles. The amyloid β (Aβ) cascade hypothesis has been proposed as a molecular mechanism for the neurodegenerative Alzheimer's disease (AD). Employing a combination of experimental and MD simulation methods, an alternative hypothesis is proposed for AD, based on the preservation of the mechanical balance in neuronal membranes, by investigating the effect of two Aβ peptides on membrane mechanical properties [18]. Tetraspanins are surface-active transmembrane proteins that form a network of protein–protein interactions within the plasma membrane, mediated by specific lipids. A single-molecule approach in mammary epithelial cells is used to study the membrane behavior and cell surface interconnected dynamics of tetraspanins, as well as the effect of gangliosides in the generation of tetraspanin-enriched areas [19]. It has been shown that small Heat-Shock Proteins (sHSPs) interact with lipids to modulate the physical state and integrity of cell membranes. Association with lipid membranes, strong preference of fluid membranes and the strong influence on the phase behaviour of plasma membranes are investigated for a less studied protein, HSPB1 [20].

Structure, function and dynamics of membrane and membrane-interacting proteins are the subject of one review and two computational articles. Membrane cytochrome b5 reductase is a single-electron oxidoreductase that facilitates the reduction of several biological acceptors in cellular membranes. Employing both computational and experimental methods of analysis of protein structures and dynamics, roles of amino acid and catalytic domains within cytochrome b5 reductase, as well as structural domains involved in cytochrome b5 interactions with other electron acceptors have been reviewed [21]. A kinetic model to be used for extensive simulations of the synaptic transmission process is challenging. In a systems biology study, a compartmentalized kinetic model for CA3-CA1 synaptic transmission is proposed to predict the functional impact caused by disease-associated variants of NMDA (N-Methyl-D-aspartic acid) receptors causing severe cognitive impairment [22]. In addition to focusing on the overall conformation and structural dynamics, microsecond MD simulations of UCP2 in the phospholipid bilayers show that ATP binding in the UCP2 cavity is tight and possesses a fatty-acid-binding site at the R60 region that is related to the proton transport mechanism across the inner-mitochondrial membrane [11].

The effects of interactions of infectious agents with cell membranes and membrane proteins, the cellular processes and cascades of events that follow, as well as their implications in the molecular mechanisms of infection and drug/vaccine development are the subject of one review and two articles. In a timely comprehensive survey of viral membrane proteins and their functional interactions and pathogenesis (with focus on SARS-CoV-2, or Severe Acute Respiratory Syndrome Coronavirus-2), the recent information on Coronavirus proteins and membrane proteins in the Coronaviridae family (including information on their structures, functions and participation in pathogenesis) is reviewed. In this review, some current CoV vaccine development strategies with purified proteins, attenuated viruses and DNA vaccines are also discussed [23]. Syk (Spleen tyrosine kinase) inhibitors can be considered as potential effective antimalarial drugs. This protein can be found in human erythrocytes targeting the membrane protein band 3. Tyr phosphorylation of band 3 occurs during the malaria parasite's growth, weakening the host cell membrane, causing easier reinfection. It is shown that the presence of Syk inhibitors decrease band 3 Tyr phosphorylation with the increase in the antimalarial drug's concentration [24]. Fusion of viral and host cell membranes is crucial in the life cycle of enveloped viruses. In influenza virus, this fusion is mediated by subunit 2 of hemagglutinin (HA) glycoprotein. MD simulations combined with experimental studies of three HAfp peptide variants are employed to characterize their free-energy landscape and interaction with the lipid bilayer. It is shown that the effect of deeply inserted peptides is significant in the membrane fusion process and correlates with the insertion depth of the N-terminal amino group [25].

Membrane proteins are involved in controlling/maintaining the integrity of cells, as well as defence and control mechanisms against programmed cell death. A review and two articles explore some examples of these functions of membrane proteins. FtsH proteins are membrane-bound ATP-dependent zinc metalloproteases essential in proteolysis of unneeded or damaged membrane proteins found in bacteria, animals and plants. In eukaryotic cells, their location is restricted to chloroplasts and mitochondria. Findings concerning the FtsHi pseudo-proteases and their involvement in protein import in the model organism

Arabidopsis thaliana are reviewed [26]. A critical step in apoptosis is the permeabilization of the outer-mitochondrial membrane, controlled by Bcl-2 family proteins. Binding and conformational dynamics of two main Bcl-2 family members, the pore-forming protein Bax and the truncated form of the activator protein Bid (tBid) are imaged and characterized at the single-particle level in a mitochondria-like planar lipid bilayer [27]. Comparable to apoptosis, ferroptosis is programmed cell death mediated by iron-dependent lipid peroxidation. The role of the NO• radical on suppression of ferroptosis has been suggested. To further explore the molecular mechanism of this suppression, a biochemical model, combined with lipidomics and structure-based modeling and simulations, has been utilized. The results of this study offer original insights into the molecular mechanism of repression of a ferroptotic peroxylated phosphatidylethanolamine (PE) production by NO• [28].

Involvement of membrane proteins in cell and organelle membranes can induce dynamic morphological changes that influence the overall cell function. These morphological modifications are detectable through high-resolution microscopy and connected to molecular interactions of membrane proteins in their native milieus. The membrane domain of eukaryotic HMG-CoA reductase (HMGR: 3-hydroxy-3-methylglutaryl CoA reductase) induces endoplasmic reticulum (ER) proliferation and membrane association into Organized Smooth Endoplasmic Reticulum (OSER) structures. OSER structures grow via the incorporation of ER membranes on their periphery and progressive compaction to the inside. The ER-HMGR domains are highly dynamic and can act as active components of the eukaryotic cells [29].

Limited mutations in membrane proteins may enhance their functional expression in experimental systems, without a significant effect on their functional sensitivity. An example is olfactory receptors (ORs) in vertebrate animals. ORs are members of the G protein-coupled receptor (GPCR) family. To overcome technical difficulties in studying these proteins, their functional expression has been improved by single amino acid substitution at one of the two sites in the OR-conserved residues without causing alterations in the odorant responsiveness, implying that specific sites within transmembrane domains in some ORs can regulate their membrane expression [30].

In conclusion, I would like to thank all the contributors to this collection and hope that it can contribute and add to the breadth and depth of biophysical and biochemical research on molecular and cellular aspects of Membrane Proteins.

Conflicts of Interest: The authors declare no conflict of interest.

References

1. Deisenhofer, J.; Epp, O.; Miki, K.; Huber, R.; Michel, H. Structure of the protein subunits in the photosynthetic reaction centre of *Rhodopseudomonas viridis* at 3Å resolution. *Nature* **1985**, *318*, 618–624. [CrossRef] [PubMed]
2. Finkelstein, J.M. Structures of membrane proteins. *Nature* **2014**, *511*, 21. [CrossRef]
3. MPSTRUC: Membrane Proteins of Known 3D Structure. Available online: https://blanco.biomol.uci.edu/mpstruc/ (accessed on 5 December 2022).
4. RSCB PDB: Protein Data Bank. Available online: https://www.rcsb.org/ (accessed on 5 December 2022).
5. Chorev, D.S.; Robinson, C.V. The Importance of the Membrane for Biophysical Measurements. *Nat. Chem. Biol.* **2020**, *16*, 1285–1292. [CrossRef] [PubMed]
6. Thoma, J.; Burmann, B.M. Fake it 'Till You Make It—The Pursuit of Suitable Membrane Mimetics for Membrane Protein Biophysics. *Int. J. Mol. Sci.* **2021**, *22*, 50. [CrossRef]
7. Jodaitis, L.; van Oene, T.; Martens, C. Assessing the Role of Lipids in the Molecular Mechanism of Membrane Proteins. *Int. J. Mol. Sci.* **2021**, *22*, 7267. [CrossRef]
8. Renard, K.; Byrne, B. Insights into the Role of Membrane Lipids in the Structure, Function and Regulation of Integral Membrane Proteins. *Int. J. Mol. Sci.* **2021**, *22*, 9026. [CrossRef]
9. Ardalan, A.; Smith, M.D.; Jelokhani-Niaraki, M. Uncoupling Proteins and Regulated Proton Leak in Mitochondria. *Int. J. Mol. Sci.* **2022**, *23*, 1528. [CrossRef]
10. Kreiter, J.; Rupprecht, A.; Škulj, S.; Brkljača, Z.; Žuna, K.; Knyazev, D.G.; Bardakji, S.; Vazdar, M.; Pohl, E.E. ANT1 Activation and Inhibition Patterns Support the Fatty Acid Cycling Mechanism for Proton Transport. *Int. J. Mol. Sci.* **2021**, *22*, 2490. [CrossRef]
11. Škulj, S.; Brkljača, Z.; Kreiter, J.; Pohl, E.E.; Vazdar, M. Molecular Dynamics Simulations of Mitochondrial Uncoupling Protein 2. *Int. J. Mol. Sci.* **2021**, *22*, 1214. [CrossRef]

12. Ku, Y.S.; Cheng, S.S.; Ng, M.S.; Chung, G.; Lam, H.M. The Tiny Companion Matters: The Important Role of Protons in Active Transports in Plants. *Int. J. Mol. Sci.* **2022**, *23*, 2824. [CrossRef]
13. Khangholi, N.; Finkler, M.; Seemann, R.; Ott, A.; Fleury, J.B. Photoactivation of Cell-Free Expressed Archaerhodopsin-3 in a Model Cell Membrane. *Int. J. Mol. Sci.* **2021**, *22*, 1981. [CrossRef]
14. Giudici, A.M.; Díaz-García, C.; Renart, M.L.; Coutinho, A.; Prieto, M.; González-Ros, J.M.; Poveda, J.A. Tetraoctylammonium, a Long Chain Quaternary Ammonium Blocker, Promotes a Noncollapsed, Resting-Like Inactivated State in KcsA. *Int. J. Mol. Sci.* **2021**, *22*, 490. [CrossRef] [PubMed]
15. Fish, M.; Nash, D.; German, A.; Overton, A.; Jelokhani-Niaraki, M.; Chuong, S.D.X.; Smith, M.D. New Insights into the Chloroplast Outer Membrane Proteome and Associated Targeting Pathways. *Int. J. Mol. Sci.* **2022**, *23*, 1571. [CrossRef] [PubMed]
16. Ma, M.; Lustig, M.; Salem, M.; Mengin-lecreulx, D.; Phan, G.; Broutin, I. MexAB-OprM Efflux Pump Interaction with the Peptidoglycan of *Escherichia coli* and *Pseudomonas aeruginosa*. *Int. J. Mol. Sci.* **2021**, *22*, 5328. [CrossRef]
17. Ayala-torres, C.; Krug, S.M.; Rosenthal, R.; Fromm, M. Angulin-1 (LSR) Affects Paracellular Water Transport, However Only in Tight Epithelial Cells. *Int. J. Mol. Sci.* **2021**, *22*, 7827. [CrossRef] [PubMed]
18. Drabik, D.; Chodaczek, G.; Kraszewski, S. Effect of Amyloid-β Monomers on Lipid Membrane Mechanical Parameters–Potential Implications for Mechanically Driven Neurodegeneration in Alzheimer's Disease. *Int. J. Mol. Sci.* **2021**, *22*, 18. [CrossRef] [PubMed]
19. Fernandez, L.; Malrieu, M.; Bénistant, C.; Dosset, P.; Rubinstein, E.; Odintsova, E.; Berditchevski, F.; Milhiet, P.E. CD82 and Gangliosides Tune CD81 Membrane Behavior. *Int. J. Mol. Sci.* **2021**, *22*, 8459. [CrossRef]
20. Csoboz, B.; Gombos, I.; Kóta, Z.; Dukic, B.; Klement, É.; Varga-Zsíros, V.; Lipinszki, Z.; Páli, T.; Vígh, L.; Török, Z. The Small Heat Shock Protein, HSPB1, Interacts with and Modulates the Physical Structure of Membranes. *Int. J. Mol. Sci.* **2022**, *23*, 7317. [CrossRef]
21. Guti, C.; Mart, O.H.; Monsalve, M.; Samhan-Arias, A.K. Structural Features of Cytochrome b5—Cytochrome b5 Reductase Complex Formation and Implications for the Intramolecular Dynamics of Cytochrome b5 Reductase. *Int. J. Mol. Sci.* **2022**, *23*, 118. [CrossRef]
22. Micheli, P.; Ribeiro, R.; Giorgetti, A. A Mechanistic Model of NMDA and AMPA Receptor-Mediated Synaptic Transmission in Individual Hippocampal CA3-CA1 Synapses: A Computational Multiscale Approach. *Int. J. Mol. Sci.* **2021**, *22*, 1536. [CrossRef]
23. Wong, N.A.; Saier, M.H. The SARS-Coronavirus Infection Cycle: A Survey of Viral Membrane Proteins, Their Functional Interactions and Pathogenesis. *Int. J. Mol. Sci.* **2021**, *22*, 1308. [CrossRef] [PubMed]
24. Marchetti, G.; Dessì, A.; Dallocchio, R.; Tsamesidis, I.; Pau, M.C.; Turrini, F.M.; Pantaleo, A. Syk Inhibitors: New Computational Insights into Their Intraerythrocytic Action in *Plasmodium falciparum* Malaria. *Int. J. Mol. Sci.* **2020**, *21*, 7009. [CrossRef] [PubMed]
25. Worch, R.; Dudek, A.; Borkowska, P.; Setny, P. Transient Excursions to Membrane Core as Determinants of Influenza Virus Fusion Peptide Activity. *Int. J. Mol. Sci.* **2021**, *22*, 5301. [CrossRef] [PubMed]
26. Mishra, L.S.; Funk, C. The FtsHi enzymes of *Arabidopsis thaliana*: Pseudo-Proteases with an Important Function. *Int. J. Mol. Sci.* **2021**, *22*, 5917. [CrossRef] [PubMed]
27. Rose, M.; Kurylowicz, M.; Mahmood, M.; Winkel, S.; Moran-Mirabal, J.M.; Fradin, C. Direct Measurement of the Affinity Between tBid and Bax in a Mitochondria-Like Membrane. *Int. J. Mol. Sci.* **2021**, *22*, 8240. [CrossRef]
28. Mikulska-Ruminska, K.; Anthonymuthu, T.S.; Levkina, A.; Shrivastava, I.H.; Kapralov, A.A.; Bayır, H.; Kagan, V.E.; Bahar, I. NO• Represses the Oxygenation of Arachidonoyl PE by 15LOX/PEBP1: Mechanism and Role in Ferroptosis. *Int. J. Mol. Sci.* **2021**, *22*, 5253. [CrossRef]
29. Grados-Torrez, R.E.; López-Iglesias, C.; Ferrer, J.C.; Campos, N. Loose Morphology and High Dynamism of OSER Structures Induced by the Membrane Domain of HMG-CoA Reductase. *Int. J. Mol. Sci.* **2021**, *22*, 9232. [CrossRef]
30. Fukutani, Y.; Nakamura, Y.; Muto, N.; Miyanaga, S.; Kanemaki, R.; Ikegami, K.; Noguchi, K.; Ohsawa, I.; Matsunami, H.; Yohda, M. Hot Spot Mutagenesis Improves the Functional Expression of Unique Mammalian Odorant Receptors. *Int. J. Mol. Sci.* **2022**, *23*, 277. [CrossRef]

International Journal of *Molecular Sciences*

MDPI

Review

Fake It 'Till You Make It—The Pursuit of Suitable Membrane Mimetics for Membrane Protein Biophysics

Johannes Thoma [1,2] and Björn M. Burmann [1,2,*]

[1] Wallenberg Centre for Molecular and Translational Medicine, University of Gothenburg, 405 30 Göteborg, Sweden; johannes.thoma@gu.se

[2] Department of Chemistry and Molecular Biology, University of Gothenburg, 405 30 Göteborg, Sweden

* Correspondence: bjorn.marcus.burmann@gu.se; Tel.: +46-317-863-937

Abstract: Membrane proteins evolved to reside in the hydrophobic lipid bilayers of cellular membranes. Therefore, membrane proteins bridge the different aqueous compartments separated by the membrane, and furthermore, dynamically interact with their surrounding lipid environment. The latter not only stabilizes membrane proteins, but directly impacts their folding, structure and function. In order to be characterized with biophysical and structural biological methods, membrane proteins are typically extracted and subsequently purified from their native lipid environment. This approach requires that lipid membranes are replaced by suitable surrogates, which ideally closely mimic the native bilayer, in order to maintain the membrane proteins structural and functional integrity. In this review, we survey the currently available membrane mimetic environments ranging from detergent micelles to bicelles, nanodiscs, lipidic-cubic phase (LCP), liposomes, and polymersomes. We discuss their respective advantages and disadvantages as well as their suitability for downstream biophysical and structural characterization. Finally, we take a look at ongoing methodological developments, which aim for direct in-situ characterization of membrane proteins within native membranes instead of relying on membrane mimetics.

Keywords: membrane protein; lipid bilayer; membrane mimetic

Citation: Thoma, J.; Burmann, B.M. Fake It 'Till You Make It—The Pursuit of Suitable Membrane Mimetics for Membrane Protein Biophysics. *Int. J. Mol. Sci.* **2021**, *22*, 50. https://dx.doi.org/10.3390/ijms22010050

Received: 9 December 2020
Accepted: 19 December 2020
Published: 23 December 2020

1. Introduction

The overwhelming majority of scientific articles on membrane proteins introduces this class of proteins by mentioning their contribution of roughly 30% to organisms' genomes, thus, highlighting their importance and resulting self-explanatory relevance. In the interest of avoiding a repetition of what has been written countless times before, let us instead spend a moment of consideration on the endangered species of the polar bear. The polar bear (*Ursus maritimus*) is perfectly adapted to life on the annual sea ice of the arctic circle, a very specific habitat, which it roams as wide-ranging predator hunting seals [1]. When polar bears are relocated from their natural environment to live in captivity in zoological gardens, they have a high tendency to develop abnormal repetitive behavior, such as stereotypical pacing and head nodding [2], and have severely increased infant mortality rates [3]. Similar alterations of behavior and infant mortality in captivity have been reported for several different species, whereupon the degree of these alterations is directly correlated to the extent of the environmental difference, but rarely as pronounced as in polar bears [3]. Much like this admittedly farfetched example, membrane proteins evolved to reside in the very specific amphipathic lipid bilayer environment of biological membranes and their removal from this environment often results in pronounced structural and functional ramifications [4–6]. It is, thus, one of the great challenges of membrane protein biophysics to characterize membrane proteins, while maintaining the specific nature of their lipid bilayer environment to be able study this class of proteins in a biologically meaningful context. The continuously progressing efforts to recreate this environment, in order to facilitate the biophysical characterization of membrane proteins are the subject of this review.

Lipid bilayers form the physical permeability barriers, which segregate cells and cellular compartments. Driven by the hydrophobic effect, amphiphilic lipid molecules self-assemble to form lamellar bilayers, with their hydrophobic moieties facing the core of the bilayer and their hydrophilic head groups facing the surrounding aqueous environment [7]. Cellular membranes are formed from a large variety of chemically very diverse lipids [8], ranging from hundreds of different lipid species in "simple" prokaryotic organisms like *Escherichia coli* to thousands in more complex eukaryotic organisms [9,10]. The diverse physiochemical properties of different cellular membranes are shaped by their lipid composition [9]. Biological membranes obtain their functionality only through the presence of specialized integral membrane proteins, which transmit molecules, energy and stimuli across these physical barriers. To fulfil these crucial functions membrane proteins depend on the properties of the surrounding membrane environment [11–13]. Factors, such as the lipid composition and bilayer asymmetry, membrane curvature, tension as well as the fluidity of the bilayer directly impact the structural and functional integrity of membrane proteins [13–15].

Unfortunately, in their native cellular form, membrane proteins immanently defy the requirements of biophysical experiments, which demand protein samples of high purity and high concentration, often in form of a solution. In addition to being insoluble, individual membrane protein species occur in rather low densities in cellular membranes. To bridge this gap and make membrane proteins experimentally accessible a multitude of membrane mimetics have been developed over the last decades. As implied by their name, membrane mimetics attempt to imitate the environment of lipid bilayers. In their most fundamental forms, as detergent micelles, this means simply to recreate the hydrophobic core environment of a lipid bilayer [16]. However, more complex forms such as bicelles and nanodiscs try to incorporate a certain number of lipid molecules (Figure 1). Moreover, purified membrane proteins can be reconstituted into bilayers of synthetic lipids or lipid extracts, attempting to closely resemble the original lipid bilayer a membrane protein was purified from.

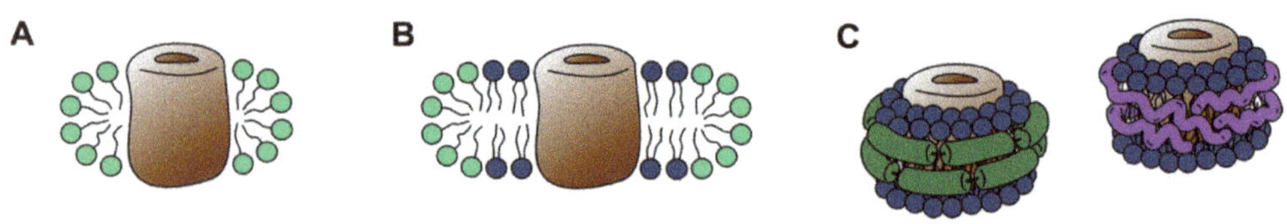

Figure 1. Solubilized states of membrane proteins. (**A**) Membrane protein in a detergent micelle (cyan). (**B**) Membrane protein in a disc-shaped bicelle containing lipids (blue). (**C**) Membrane proteins in nanodiscs stabilized by membrane scaffold protein (MSP, green) and amphipathic polymers (purple).

Naturally, all these methods have their pros and cons and not all of them are compatible with different biophysical methods. For example, solution NMR (Nuclear magnetic resonance) spectroscopy, single-particle cryo-EM (electron microscopy), and X-ra/neutron solution scattering methods demand membrane proteins in a solubilized form, typically in the form of micelles, bicelles or nanodiscs. Solid-state-NMR, negative stain EM, and AFM (atomic force microscopy) are suitable for larger membrane assemblies such as proteoliposomes, whereas X-Ray crystallography requires samples in a three-dimensional (3D) crystalline form. The hydrophobic contacts required to maintain the structural integrity of membrane proteins within these crystals can either be satisfied by detergents or in case of lipidic cubic phases (LCP) by lipid molecules. In this review, we survey the currently available membrane mimetic systems, weigh their advantages, as well as their disadvantages, evaluate how they impact the structural and functional states of membrane proteins and assess their suitability for various biophysical methods.

2. Detergent Micelles

The archetype of membrane mimetics are detergents, which are routinely used to solubilize membrane proteins. Detergents are amphipathic molecules, which self-assemble to form micelles in aqueous solutions. Based on their molecular structure and properties, namely the charge of their hydrophilic head groups, detergent molecules can be subdivided into ionic, non-ionic, and zwitterionic detergents. Thereby, ionic detergents such as sodium dodecyl sulfate (SDS) are considered "harsh", due to their (differently pronounced) denaturing effect on membrane proteins, ranging from minor structural alterations [17] to loss of function and complete unfolding [18]. Notably, some membrane proteins remain rather unaffected by ionic detergents. In particular, bacterial outer membrane proteins have shown to be very resistant to SDS denaturing, due to extensive hydrogen bonding networks stabilizing the transmembrane β-barrels of these proteins. In fact, the resulting altered migration behavior in SDS page is frequently used as an indicator of the folding state of outer membrane proteins [19,20].

For the solubilization of membrane proteins in a functional form more widely used are "mild" non-ionic detergents, such as Octyl-L-D-glucoside (OG) and Dodecyl-L-D-maltoside (DDM), which tend to retain the structural integrity of solubilized proteins and leave inter- and intra-molecular protein-protein interactions intact. The latter is particularly important for the solubilization of multimeric membrane protein complexes and especially DDM has proven quite useful for the purification of intact complexes [21–24]. The third group are zwitterionic detergents, such as Lauryldimethylamine-N-oxide (LDAO), the hydrophilic head groups of which have a positive, as well as a negative charge. While, the overall electrically neutral, zwitterionic detergents form an intermediate between ionic and non-ionic detergents, with a stronger solubilizing potential than non-ionic detergents and a less pronounced deactivating effect than ionic detergents [25].

The solubilization of biological membranes usually occurs *via* two stages with increasing detergent concentration (Figure 2). Initially, at low detergent concentrations, the detergent molecules insert into the lipid bilayer, resulting in destabilization and fragmentation. At high concentration, typically exceeding the detergents critical micellar concentration (CMC), the lipid bilayer is dissolved, resulting in binary lipid-detergent or protein-detergent, as well as ternary lipid-protein-detergent mixed micelles [26,27]. Thereby different detergents vary in their capacity to solubilize different cellular membranes. For example detergents like Triton X-100 and Sarkosyl have been shown to selectively solubilize inner membranes of Gram-negative bacteria and mitochondria, while leaving their outer membranes largely unaffected [28,29]. Lipid microdomains (sometimes referred to as lipid rafts), which typically contain cholesterol and saturated sphingo- and glycerophospholipids in a liquid ordered phase, are resistant to mild detergents, such as Triton X-100, and thus, remain as detergent-resistant membranes after solubilization [30,31]. Likewise, the amount of endogenous lipid molecules bound to membrane proteins varies greatly depending on the detergent used to solubilize a membrane [32]. In this context, mass spectrometry and in particular tandem mass spectrometry (MS/MS) is a powerful method to characterize not only the binding of lipids to membrane proteins, but also their effects on membrane protein oligomerization [33,34]. The choice of detergent depends ultimately on the planned downstream biophysical characterization and especially detergents with a low CMC, despite effectively solubilizing most membranes, can be difficult to remove and are, thus, of limited suitability for methods requiring detergent removal [35].

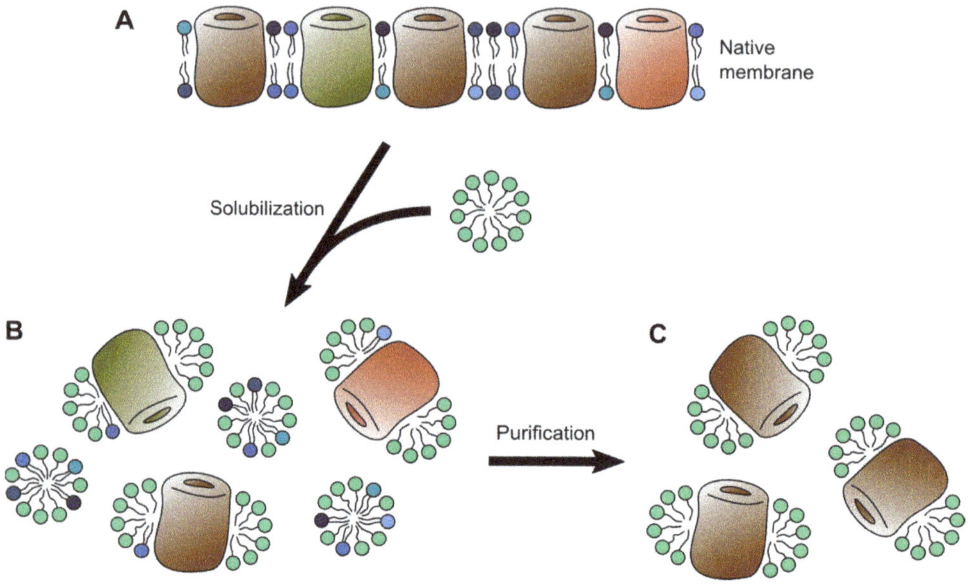

Figure 2. Solubilization and purification of membrane proteins. (**A**) Native membrane containing a membrane protein of interest over a background of other membrane proteins in a diverse lipid composition. (**B**) Mixed membrane protein-detergent and lipid-detergent micelles resulting from solubilization of the membrane through the addition of detergent. (**C**) Purified membrane protein in detergent micelles after removal of undesired lipid components and protein contaminations.

The purification of a membrane protein from a cellular membrane typically begins with the isolation and solubilization of the membrane of interest [36]. Due to the differences in the way detergents interact with cellular membranes, their efficient solubilization involves screening for a suitable detergent [37]. Finding a detergent that also stably maintains a membrane protein and is suitable for downstream applications can require substantial screening work [38,39]. Once solubilized, the purification of membrane proteins follows similar principles as the routine purification of soluble proteins, relying on chromatographic methods including affinity, gel filtration, and ion exchange chromatography [36]. An alternative route to obtaining solubilized membrane proteins in a pure and folded state is refolding into detergent micelles. To this end, membrane proteins are transferred from a fully unfolded state in concentrated solutions of chaotropic salts into a detergent containing refolding buffer to adopt a folded state within the detergent micelles [40,41]. Interestingly, it was shown that in a similar way the cellular chaperone machinery can be exploited and refolding into micelles results in exactly the same folded protein state, regardless whether folding was initiated from a chaperone or chaotropic reagents [42]. Combined with recombinant protein expression in inclusion bodies, this method can expedite protein purification and can prove particularly useful when large amounts of a protein are required. However, despite routinely used, membrane proteins do not necessarily fold into a native structure and refolding can result in non-native multimers and structural intermediates [43,44].

Using NMR spectroscopy membrane proteins can be directly characterized in a detergent solubilized form, without the need for additional downstream modifications [45]. Particularly useful when analyzing membrane proteins with NMR spectroscopy is the use of deuterated detergents, which eliminate interfering proton signals originating from the detergent [46]. While, a handful of membrane protein structures have been determined using NMR spectroscopy [47–50], the real strength of the method lies in its ability to probe dynamic processes [51]. In the micellar state NMR spectroscopy determines the residue-specific dynamics, can probe interactions with ligands and detect conformational changes in solution [45,52]. Moreover, unlike other structural techniques, NMR spectroscopy can yield detailed information on highly dynamic and unstructured regions of membrane proteins, such as loops [53,54]. Therefore, NMR spectroscopy could capture the subtle differences imposed on the structural conformation of membrane proteins by different detergents [54]. In this context, it should be stressed that, despite often resulting in high-resolution structural information, detergent stabilized states of membrane proteins, especially in Dodecylphosphorylcholine (DPC), might often be non-functional states [55].

Actuated by the recent resolution revolution [56], single-particle cryo-EM emerged as another powerful technique to characterize membrane proteins in detergent micelles. Unlike NMR spectroscopy, which excels at characterizing small proteins, single-particle cryo-EM is best suited for large proteins and protein complexes [57]. Several membrane protein structures stabilized in detergent micelles could be solved by cryo-EM, including bacterial β-barrel assembly machinery (BAM) complex, mitochondrial TOM core complex, and the spinach cytochrome b_6f complex [58–60]. However, despite efforts to streamline the preparation of membrane proteins for cryo-EM, the methodology is far from being routine work [61,62]. Detergent concentrations typically used in preparations of membrane proteins tend to interfere with the controlled formation of thin vitrified ice, often resulting in reduced image contrast [63]. Therefore, the preparation of membrane proteins for cryo-EM requires thorough removal of excess detergent [64]. Moreover, the surface to volume ratio of a sub-micrometer thin water film on an EM grid is much higher than in conventional liquid droplets and it is not yet fully clear how the consequentially altered fluid dynamics and the air-water interface impact solubilized macromolecules [61,65].

While, detergents arguably form a less than ideal environment for many membrane proteins their usage remains in most cases unavoidable. With very few exceptions, the solubilization of membrane proteins from the membranes of an expression host is commonly accomplished through the use of detergents [66,67]. The majority of membrane mimetics require the reconstitution of a membrane protein, which naturally requires the prior solubilization and purification of said membrane protein. Detergent micelles are therefore almost always the starting point for additional downstream applications involving more complex membrane mimetics.

3. 3D Crystals and Lipidic Cubic Phase

Despite a growing toolbox of alternative membrane mimetics, to date the majority of membrane protein structures have been solved by X-ray crystallography, utilizing 3D crystals of membrane proteins. The crystallization of detergent-stabilized solubilized membrane proteins follows similar methods and principles as the crystallization of soluble proteins based on vapor diffusion, microdialysis, and batch crystallization [68,69]. Crystals are, thereby, formed from a protein solution, supersaturated with precipitating agents such as salts or polyethylene glycol (PEG), which drive the aggregation of protein-detergent-complexes into ordered crystals [68]. Within these type II crystals the hydrophobic surfaces of membrane proteins remain satisfied through the co-crystallized detergent micelles, while the crystal lattice is preferentially formed by polar protein-protein interactions [70,71]. Successful crystallization often depends on the choice and nature of the detergent and high detergent concentrations or the use of detergents with a large micelle size can impair crystal formation [69].

Following the first high-resolution crystal structure of a membrane protein, the photosynthetic reaction center from *Rhodopseudomonas viridis* [72], a great variety of membrane protein structures could be solved using X-ray crystallography. These include groundbreaking structures such as bacterial potassium channel KcsA [73] and lactose permease LacY [74], which resulted in unprecedented insight into the molecular details of these proteins in particular and membrane proteins in general. Nevertheless, it should be noted that solving crystal structures of membrane proteins often only is possible due to substantial molecular engineering. This includes the introduction of mutations, which arrest proteins in a certain conformation [75]—an effect, which can also be achieved through the binding of ligands or antibody fragments [76,77]—deletions of parts of the proteins to eliminate unfavorable crystal contacts [78], or chimeric modifications of proteins [79]. In addition, ambiguities can inhere in crystal structures, imposed by the crystallization conditions, together with uncertainties in the identity and position of atoms and molecules within crystal structures [80]. Even similar sample preparations can lead to altered structures, as for example evidenced in the translocator protein (TSPO) that yielded two different α-helical bundles depending if the structure was determined in milder DM or DDM [81,82], compared to the harsher zwitterionic DPC [83] and it remains unclear whether the different structures represent simply alternative states of the protein. However, the structures, obtained in the milder detergents, are in better agreement with known functional mutations and the ligand binding site shows a larger degree of conservation [81].

A valuable alternative to detergent-micelle mediated 3D crystallization of membrane proteins is crystallization in the lipidic cubic phase (LCP) [84]. LCP takes advantage of the propensity of monounsaturated monoacylglycerols, such as monoolein, to form a bicontinuous cubic mesophase, a single lipid bilayer organized into a three-dimensional bilayer structure containing an aqueous channel system [85]. The cubic phase is formed spontaneously when the lipid is mixed with solubilized or dispersed protein solutions, while crystal formation is driven by the addition of a precipitant [86]. In addition to monoolein, LCP can contain various accessory lipids, which can either remain from co-purification with a membrane protein or can be specifically added during the crystallization process [87]. Through the stabilization in a lipid bilayer, LCP is thought to provide a more natural environment for membrane proteins and LCP has been successfully used, particularly with membrane proteins which have small polar surfaces, such as the seven-helix-bundles of rhodopsins [88–90] and G-protein-coupled receptors (GPCRs) [91,92].

One limitation of LCP crystallization is its propensity to result in microcrystals [85]. However, microcrystals grown in LCP are ideally suited to be studied using serial femtosecond crystallography (SFX) [93]. The latter utilizes ultrashort pulses generated by an X-ray free-electron laser (XFEL) [94] to sequentially collect data from a continuous stream of microcrystals [95]. By the short duration of highly intense X-ray pulses the crystals are vaporized before radiation damage can occur [94], thus facilitating the time-resolved characterization of dynamic structural transitions in crystallized proteins [96]. These studies have provided remarkable insight into the activation of different photoreceptors, by visualizing the so-called protein quake [97–99]. In addition to SFX, another method that has been show to provide high-resolution structural data of microcrystalline membrane proteins is so-called microcrystal electron diffraction (Micro ED) [100,101], which was recently applied to study the tetrameric sodium channel NaK in DDM based crystals [102] as well as the human adenosine A_{2A} receptor in LCP [103].

4. Bicelles and Nanodiscs

The first attempt to incorporate a substantial amount of lipids into solubilized membrane protein systems were bicelles. Bicelles (bilayered micelles; Figure 1) are formed by phospholipids, traditionally dimyristoyl-phosphatidylcholine (DMPC), in a planar discoidal bilayer assembly, which are surrounded by a scaffold of either a detergent, such as CHAPS, or short-chained lipids, such as dihexanoyl-phosphatidylcholine (DHPC) [104]. Thereby, bicelles can be designed to adopt a variety of shapes, ranging from small bilayer discs over wormlike structures to large perforated lamellar assemblies, dictated by the lipid-to-detergent, or longchain-to-shortchain lipid ratio, respectively [105,106]. In order to reconstitute membrane proteins into bicelles either a detergent-stabilized membrane protein is integrated into preformed bicelles or the bicelles can be formed through the addition of detergents to proteoliposomes that were assembled beforehand [106]. Through the incorporation of lipid molecules, bicelles have been shown to outperform micelles in their ability to maintain membrane proteins in a functional state [107,108]. However, it has also been shown that different lipid compositions of bicelles significantly influence membrane protein dynamics [109]. Moreover, molecular dynamics (MD) simulations indicate increased peptide solvation of transmembrane segments within small bicelles compared to larger bilayer systems [110].

While, originally developed as a membrane system for solid state NMR spectroscopy [111], bicelles found a broad audience in biophysics [106,112]. Due to the increased lipid content, bicelles are larger than most purely detergent-based micellar systems. Yet, their molecular tumbling permits detailed characterization of reconstituted membrane proteins by solution NMR spectroscopy [113,114]. By stabilizing the transmembrane region of the HIV-1 envelope spike (Env) in DMPC/DHPC bicelles the atomic resolution structure of the trimeric assembly could be determined [115–117], which remained elusive in the previous cryo-EM structure, possibly due to the detrimental influence of the DDM/sodium deoxycholate micelles [118]. Likewise, a direct comparison of the dimeric transmembrane domain of Glycophorin A in DPC micelles and DMPC/DHPC bicelles, respectively, showed reduced conformational fluctuation and enhanced stability of the transmembrane α-helixes in the bicellar lipid environment [119].

Bicelles have also been implemented as an alternative crystallization method trying to combine the incorporation of lipids, as used in LCP, with the facility of detergent based crystallization [120]. Bicelle crystallization exploits the temperature-dependent ability of lipid/amphiphile mixtures to exchange between different three-dimensional structural arrangements [121]. When this concept was initially introduced with studies of bacteriorhodopsin, the protein was found to be embedded in bicelles as a stable monomer instead of its usual trimeric assembly [120]. Bicelle-based 3D crystals have since these early studies successfully been used with multiple membrane proteins, including the human G-protein-coupled receptor (GPCR) β2 adrenergic receptor (β_2AR) [122], and the eukaryotic mitochondrial voltage-dependent anion channel (VDAC) [123].

Evolving from bicelles, the recent years have seen great development in the field of nanodiscs. Collectively, the term nanodiscs refers to the lipid bilayer particles similar to discoidal bicelles, which are surrounded by a scaffolding molecule (Figure 1). Yet, the size-range in which scaffolded nanodiscs can be prepared is somewhat limited, compared to bicelles. Following the first account of a lipid nanodisc surrounded by membrane scaffold protein (MSP) [124], several other types of scaffolds have been described, including saposin proteins (salipro) [125], as well as copolymer-scaffolded nanodiscs utilizing styrene-maleic acid (SMA) and diisobutylene/maleic acid (DIBMA) [126,127]. The formation of nanodiscs follows similar principles as the formation of bicelles, starting from a mixture of detergent-solubilized lipid, detergent-solubilized protein, and MSP or saposin, respectively, and is driven by subsequent detergent removal [125,128,129].

Unlike protein-based nanodiscs, which require the reconstitution of membrane proteins from a detergent-solubilized micellar state, co-polymers have shown certain detergent-like properties. When mixed with membrane preparations, these polymers can extract "native nanodiscs" containing membrane proteins together with a fraction of the lipid molecules surrounding the proteins [130,131]. While, nanodiscs certainly have the ability to preserve the local lipid composition around a membrane protein, they cannot maintain membrane asymmetry [132]. The latter is due to the dynamic nature of the different nanodisc systems. On the one hand equilibration between both leaflets can occur when lipid molecules flip around the edges of the nanodisc and on the other through diffusional as well as collisional transfer. Polymer-scaffolded nanodiscs have been shown to readily exchange proteins, lipids, and polymer components at much higher rates compared to MSP-scaffolded nanodiscs or unilamellar lipid vesicles [133,134]. In contrast, MSP-scaffolded nanodiscs have been found to exhibit internal lipid dynamics, which are comparable to lipids in liposomes [135].

Nanodiscs have been widely used with a wide range of biophysical methods, including NMR spectroscopy [136–138], electron microscopy [139,140] and atomic force microscopy [141]. Therefore, the use of nanodiscs has proven successful even with complex systems. For example, individually MSP-nanodisc-stabilized inner and outer membrane components of bacterial tripartite efflux pumps MexAB–OprM and AcrAB–TolC could be recombined to form stable intact complexes bearing two separate nanodiscs [142]. The open pore state of bacterial Tc toxin complex TcdA1, which could not be sufficiently stabilized in Tween-20 detergent micelles or liposomes [143,144], could be resolved in high detail embedded in nanodiscs [145], whereas the corresponding crystal structure could only reveal a closed pre-pore state [143]. Likewise, when characterized in nanodiscs, aforementioned HIV envelope protein Env showed substantial differences in its arrangement and orientation to the membrane surface, compared to micellar and bicellar environments [146].

Direct comparisons of membrane proteins in micellar, bicellar and nanodisc systems revealed distinct differences between these membrane mimetics. For example, the bacterial outer membrane protein OmpX has been shown to form a stable β-barrel in all three environments, however, the dynamics and molecular motion differed substantially between the mimetics [147,148]. Similarly, the comparison of outer membrane protein BamA in LDAO micelles, DMPC/DHPC bicelles, and MSP-bounded DMPC nanodiscs suggested altered dynamics between the three environments [149]. Likewise, the α-helical tetrameric potassium channel KcsA has been shown to have reduced stability in nanodiscs, indicating altered dynamic properties, compared to DDM micelles [150]. In contrast, the CC-chemokine receptor 5 (CCR5), a GPCR which is inherently instable in detergents, such as DDM, could be stabilized in nanodiscs for prolonged NMR studies [151]. The human anion channel VDAC showed nearly identical folds in micelles, bicelles, and nanodiscs, however, unlike the first two, nanodiscs incorporated not only monomeric VDAC, but multimeric states, similar to the ones observed in native membranes [152].

5. Liposomes (and Polymersomes)

Native lipid bilayers contain a diverse blend of membrane proteins, interweaved with additional components, such as lipoproteins and carbohydrates in the form of glycoproteins and glycolipids [7]. In most native membranes any membrane protein of interest is vastly outnumbered by these additional components, making it inaccessible to biophysical investigations. Few exceptions, such as the purple membranes from Halobacteria [153,154] or disc membranes from vertebral rod cells [155], in fact allow proteins to be studied in native membrane isolates, which are covered in the next section. For the vast majority of membrane proteins, the alternative lies in the bottom-up reconstruction of an artificial lipid membrane. To this end, a membrane protein of interest is solubilized and isolated from the membrane of a suitable expression system using detergents and, naturally, the same principles and limitations of detergent micelles apply as covered above [156,157]. Following purification, in a subsequent step, membrane proteins are reconstituted into an artificial lipid bilayer. Reconstitution is generally achieved *via* one of two alternative routes (Figure 3). In one method the self-assembly of a lipid bilayer is driven by detergent removal from a ternary mixture of micelles containing the protein and separately detergent-solubilized lipids [158]. Alternatively, the solubilized membrane protein can be inserted into preformed liposomes [159]. In both cases assembly and membrane insertion are driven by the removal of the detergent and can be achieved by several means, all aiming at reducing the detergent concentration (far) below the CMC. Thereby, the method of detergent removal can have a strong influence on the resulting proteoliposomes. Dialysis can result in homogeneous vesicles, but is time-consuming and limited to detergents with high CMC. More rapid methods, such as size exclusion chromatography, rapid dilution, and the use hydrophobic adsorbents often result in inhomogeneous protein distribution and in-complete detergent removal [35].

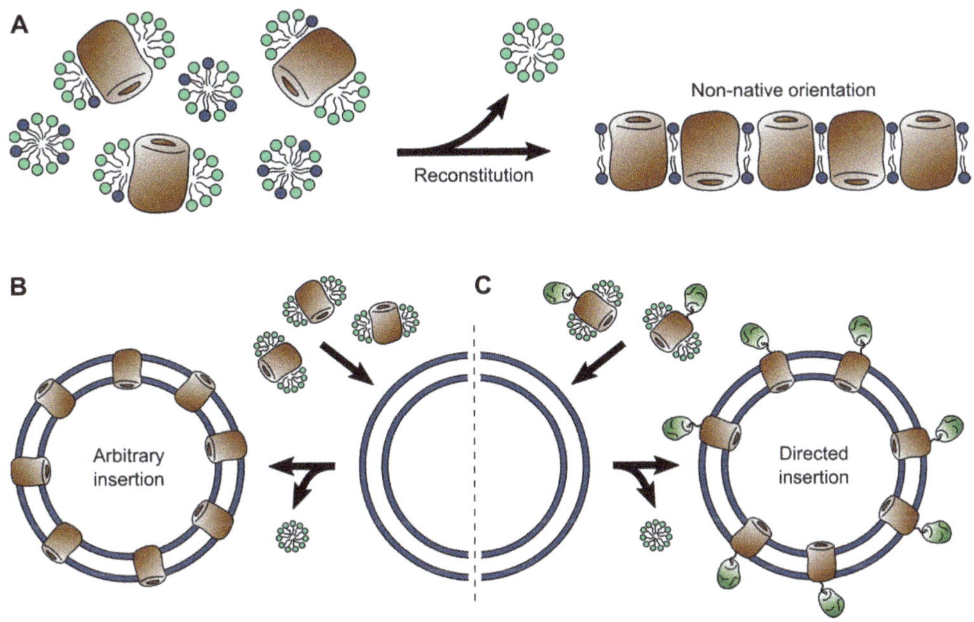

Figure 3. Reconstitution of membrane proteins into liposomes. (**A**) Bottom-up assembly from a ternary mixture (left), driven by detergent removal, typically results in an alternating up-down orientation of membrane proteins (right). (**B**) Undirected insertion of membrane proteins into preformed liposomes. (**C**) Directed insertion of membrane proteins into preformed liposomes with soluble domains facing outwards.

Self-assembled lipid bilayers allow better control over the lipid-to-protein ratio and thus can result in a very dense protein packing within the bilayer, culminating in a 2D crystalline assembly [160,161]. However, due to the rotational freedom during assembly, membranes formed this way typically contain membrane proteins in a non-native up-down configuration, with either half of the proteins inserted in the bilayer in opposite directions (Figure 3). While random orientation can occur as well during insertion into preformed liposomes, asymmetry between the two solvent accessible poles of a membrane protein, for example in membrane proteins containing a soluble domain, can bias insertion with the soluble domain facing outwards (Figure 3) [159,162,163]. In fact, by fusing a soluble domain to one side of a membrane protein, this phenomenon can be utilized to drive unidirectional insertion of the protein into liposomes [164].

Naturally, a reconstituted artificial lipid bilayer cannot fully reassemble the complexity of a native membrane and features like membrane asymmetry or local variations in the lipid composition, sometimes referred to as lipid rafts, are impossible to mimic. Nevertheless, since artificial bilayers allow precise control over the lipid composition they are ideal proxies for detailed studies on how different lipid composition of a membrane affect membrane proteins. The utilization of reconstituted lipid bilayers for example allowed detailed characterization of lipid-induced topological switches in proteins of the major facilitator family [165,166], which could be confirmed in vivo [167]. Similarly, liposomes of varying lipid composition are a valuable tool to understanding membrane protein folding at a molecular level. Therefore, not only the folding process from chemically denatured states is of interest [168,169], but much focus has been laid on chaperone-dependent protein folding [170,171].

Until challenged by the advent of single particle cryo-EM, two-dimensional (2D) crystalline membrane protein assemblies were the system of choice for structure determination by EM [172,173]. Unlike X-Ray crystallography, which requires 3D crystals that are sometimes difficult to obtain from membrane proteins, electron crystallography is ideally suited for the 2D crystals of membrane proteins. Electron crystallography revealed the first high resolution structures of Bacteriorhodopsin [174], Aquaporin [175], and light-harvesting complex [176]. A rather recent development is the use of protein-containing liposomes in cryo-electron tomography combined with subtomogram averaging for the high-resolution reconstruction of membrane proteins [177–179]. Moreover, liposomes have been increasingly used to investigate membrane proteins using solid-state NMR spectroscopy. Recent advances in fast and ultra-fast (>100 kHZ) magic angle spinning (MAS) solid-state NMR spectroscopy based on ^{1}H-detection resulted in enhanced sensitivity and resolution comparable to solution NMR, required to facilitate high resolution NMR studies of membrane proteins [180,181]. Solid-state NMR spectroscopy has, for example, been used to determine structural the details and conformational rearrangements of α-helical transmembrane proteins such as KcsA reconstituted in lipid bilayers [182,183], as well as bacterial β-barrel outer membrane proteins such as OmpA and OmpG [184,185]. Importantly, all of these structures revealed nuanced differences, such as structural alterations and altered dynamics compared to previously reported structures.

Another powerful tool for characterizing the membrane proteins reconstituted in lipid bilayers is atomic force microscopy (AFM), due to its ability to be operated in liquid environments at physiological temperatures [186,187]. Adsorbed to atomically flat surfaces, such as muscovite mica or highly oriented pyrolytic graphite (HOPG), AFM allows imaging of the membrane topographies of solid-supported planar lipid bilayers at molecular resolution [188–190]. AFM could reveal ligand-induced conformational changes of membrane proteins [191], details on the electrostatics of their accessible surfaces [192,193] and their dynamic behavior in lipid bilayers [194,195]. Moreover, AFM allows the direct physical manipulation of individual membrane proteins, facilitating detailed studies of their behavior under force [196–198], their folding behavior [199–201], as well as their interaction with tethered ligands [202].

In addition to being used for the structural characterization of membrane proteins, lipid bilayers allow the functional examination of membrane proteins, in particular molecular transport phenomena, which cannot be probed in solubilized states. Therefore, electrophysiological measurements, either based on fused liposomes or utilizing black lipid membranes, allow detailed characterization of diffusion through transmembrane pores as well as probing their electrochemical properties [203–206]. Similarly, molecular transport is routinely studied in intact liposomes based on osmotic swelling. The latter is particularly powerful when combined with fluorogenic reactions conducted inside the liposomes lumen in stopped-flow experiments [207,208].

The ability of artificial lipid bilayers to maintain membrane proteins in a functional state is showcased by their use as a bottom-up platform for the assembly of nanocells in synthetic biology, whereby membrane proteins are embedded in liposomes to create functional systems with potential applications as nano-sized reaction compartments, drug delivery vehicles and novel therapeutics [209–211]. In this context, a biomimetic alternative to lipid bilayers are membranes formed from amphiphilic block copolymers [212–214]. While the physical properties of these are very different from lipids, their bilayer forming abilities are driven by the same principles, and polymersomes can effectively maintain membrane proteins in a functional form [215,216]. In fact, in mixed polymer-lipid bilayers under certain conditions membrane proteins have been shown to preferentially reside within the polymer rather than the lipid fraction, depending on their relative fluidity [217,218].

6. Native Membranes

Whereas so far no mimetic entirely met the physiological requirements of membrane proteins, native membranes, in the majority of cases, cannot meet the experimental requirements of the modern biophysicist. Native membranes are notoriously difficult to handle. Typically, they contain a particular protein of interest only in small quantities over a much larger background of other membrane proteins. Nevertheless, native membrane preparations from *E. coli*, also known as Kabackosomes, were used to characterize molecular transport through membrane proteins long before the first membrane protein structures were uncovered [219,220]. Moreover, some specialized cellular membranes natively contain high densities of distinct membrane proteins, which are sufficiently pure to permit biophysical studies directly in the unaltered native membranes. The best example are the purple membranes from archaea such as *Halobacterium salinarum* and *Halobacterium halobium*, which naturally contain bacteriorhodopsin in a 2D crystalline form [154]. In fact, the first structural models of bacteriorhodopsin were determined using native purple membranes isolated from *H. halobium* [221]. While later preparations employing increased 2D crystals, obtained through detergent treatment of purple membranes, resulted in models with increased resolution [174,222], direct comparison of detergent treated crystals to native membranes showed that lipid boundaries separating individual trimers within the native membranes were removed [223]. A similar approach, based on partial de-lipidation was used to observe crystalline assemblies of porins in situ, in bacterial outer membrane sacculi [224].

While these early electron crystallographic studies required 2D crystalline assemblies of membrane proteins, advancements in cryo-electron tomography nowadays allow direct in situ analysis of membrane proteins in non-crystalline native membranes [225,226]. Although, data on integral membrane proteins so far remains sparse, the method has already proven its potential. Aided by membrane targeting nanoparticles, active preprotein-carrying TOM–TIM23 supercomplexes could be localized and subsequently visualized in situ in yeast mitochondrial membranes [227]. Intriguing examples showcasing the potential of cryo-electron tomography are the in situ structures of envelope spanning bacterial secretion systems, such as the structure of an intact primordial type III secretion system, which could be determined in *Chlamydia trachomatis* elementary bodies [228]. Other examples are the in situ structures of AcrAB–TolC efflux pump in intact *E. coli* cells [229], the type IV secretion system in intact *Legionella pneumophila* cells [230], as well as the membrane complex of a type VI secretion system in *E. coli* [231]. A very promising approach to study membrane proteins in situ is the enrichment of secreted extracellular vesicles with specific proteins, which could be exploited to study integral type I membrane proteins from *C. elegans* as well as Herpes simplex virus [232].

In addition to electron microscopy, solid-state NMR spectroscopy has been successfully employed to study membrane proteins embedded in native membranes. In this context, native *Escherichia coli* inner membranes were used as a proxy to characterize Anabaena sensory rhodopsin (ASR), which was found to form hexagonal packing in DMPC/DMPA liposomes but a square lattice assembly in the *E. coli* membrane [233]. Furthermore, solid-state NMR spectroscopy was used to characterize bacterial BAM complexes in native outer membranes, suggesting increased structural disorder in the native environment [234]. Recently, a combination of cryo-electron tomography and of ^{1}H-detected solid-state NMR spectroscopy was employed to reveal different conformations of the bacterial membrane protein YidC in native membranes [235].

One exceptional method to study native membranes is AFM, which does not depend on the preparation of highly homogeneous samples [236,237]. The ability of AFM to imaging native membranes at molecular resolution was initially demonstrated with native purple membranes. Following initial topographies, which allowed discrimination of individual proteins [238], AFM was optimized to reveal the submolecular details of single bacteriorhodopsin molecules in great detail [239]. Using AFM, two-dimensional arrays of aquaporin 0 (AQP0) could be observed in native lens core membranes, surrounded by densely packed gap junction channels, and AQP0 array formation could be followed using time-lapse AFM [240]. Similarly, AFM could reveal native supramolecular assembly of VDAC in dense and sparse domains in native yeast mitochondrial outer membranes [241] as well as the closely packed, paracrystalline dimeric arrangement of rhodopsin arrays in native mouse disc membranes [242]. More recently, AFM was used to distinguish small and large protruding proteins in dimeric photosystem II oxygen-evolving complexes within native spinach grana membranes [243]. In addition to high-resolution imaging, AFM-based single-molecule force spectroscopy (SMFS) was used to probe the force response of individual membrane proteins in native membranes [196]. In this context, protein-enriched outer membrane vesicles from *Escherichia coli* were recently employed to reveal subtle differences in the dynamics of outer membrane proteins between native membranes and reconstituted lipid bilayers [244,245].

7. Conclusions and Future Perspectives

The last several decades have seen an enormous increase in efforts to optimize membrane mimetics to facilitate the structural and functional characterization of membrane proteins using biophysical methods leading to a constantly growing toolbox of diverse options (Table 1). Multiple membrane proteins could be characterized in several different mimetics, allowing the direct comparison and revealing the influence of the different mimetics on the structure and dynamics of these membrane proteins. Whereas for some membrane proteins these different tools are all in good agreement, for others vast differences have been observed, imposed by the respective mimetics used. For some membrane proteins, only a small set of mimetics could stabilize their folded state sufficiently to permit biophysical characterization, whereas other mimetics let to destabilization, impairing detailed characterization. Only in very few cases could membrane proteins be characterized in their native bilayer environment, some of which only revealed subtle, possibly negligible differences to mimetic systems, whereas others were substantially impacted by the mimetic environment.

Table 1. Pros and cons of different membrane systems.

System	Pro	Contra	Suitable methods
Detergent micelles	Universally used; starting point for downstream applications	Can have denaturing effects; may disrupt complexes; de-lipidation of membrane proteins	Single-particle Cryo-EM; solution NMR; X-ray / neutron solution scattering; MS/MS
3D crystals	Most prevalent system for structure determination; can include lipids (LCP)	Non-native crystal contacts; often requires protein engineering; proteins are “locked” in one state; crystallization artifacts	X-ray crystallography; Micro-ED
Bicelles	Lipid system; can be used for 3D crystallization; variety of shapes and sizes	Limited lipid diversity; altered lipid dynamics	Solution NMR; solid-state NMR; (X-ray crystallography)
Nanodiscs	Lipid system; broad range of lipid compositions; possible to extract native lipid composition	Limited size range; altered lipid dynamics; membrane asymmetry is lost	Single particle Cryo-EM; solution NMR; Cryo-electron tomography; AFM
Liposomes	Lipid system; Broad range of lipid compositions; high protein density possible; facilitate transmembrane transport studies	Often non-native protein orientation; not possible to create asymmetric bilayers	Electron crystallography; Cryo-ET; solid-state NMR; AFM; electrophysiology; fluorometry
Native membranes	Native environment	Often difficult to handle; low content of protein of interest over “contaminants”	Cryo-ET; solid-state NMR; AFM

Unlike the case of the polar bear, which can in fact be observed in its natural environment, as well as in captivity, thus, allowing a direct comparison between the two environments, for the majority of membrane proteins observations embedded in the native lipid environment are until now inexistent. Despite decades of studying membrane proteins with exceptional effort and the development of a plethora of groundbreaking sophisticated methods, we have only caught the first glimpses providing snapshots of their structural details within artificial environments. For many membrane proteins, these studies yielded priceless insight into their molecular architecture, which are supported by a multitude of functional investigations in vitro, as well as in vivo, yet, their behavior under native cellular conditions remains elusive. Clearly, to close this gap much more research on membrane proteins embedded in their native environment is necessary in the future. The recent trend towards in situ membrane protein biophysics will certainly help to illuminate this blind spot and provide in-depth insight into the details underlying membrane protein function.

Author Contributions: Conceptualization: J.T. and B.M.B.; writing—original draft preparation: J.T.; writing—review and editing: J.T. and B.M.B.; visualization: J.T. All authors have read and agreed to the published version of the manuscript.

Funding: J.T. was supported by an EMBO Long-Term Fellowship (ALTF 413-2018). B.M.B. gratefully acknowledges funding from the Swedish Research Council (Vetenskapsrådet Starting Grant 2016-04721) and the Knut och Alice Wallenberg Foundation through a Wallenberg Academy Fellowship (2016.0163) as well as through the Wallenberg Centre for Molecular and Translational Medicine, University of Gothenburg, Sweden.

Conflicts of Interest: The authors declare no conflict of interest.

Abbreviations

AFM	Atomic force microscopy
AQP0	Aquaporin 0
ASR	Anabaena sensory rhodopsin
β_2AR	β2 adrenergic receptor
BAM	β-barrel assembly machinery
CCR5	CC-chemokine receptor 5
CMC	Critical micellar concentration
DM	Decyl-L-D-maltoside
DHPC	Dihexanoyl-phosphatidylcholine
DIBMA	Diisobutylene/maleic acid
DMPC	Dimyristoyl-phosphatidylcholine
DDM	Dodecyl-L-D-maltoside
DPC	Dodecylphosphorylcholine
EM	Electron microscopy
ET	Electron tomography
GPCRs	G-protein-coupled receptors
HOPG	Highly oriented pyrolytic graphite
LDAO	Lauryldimethylamine-N-oxide
LCP	Lipidic cubic phases
MAS	Magic angle spinning
MS	Mass spectrometry
MSP	Membrane scaffold protein
MicroED	Microcrystal electron diffraction
MD	Molecular dynamics
NMR	Nuclear magnetic resonance
OG	Octyl-L-D-glucoside
PEG	Polyethylene glycol
SDS	Sodium dodecyl sulfate
SFX	Serial femtosecond crystallography
SMA	Styrene-maleic acid
SMFS	Single-molecule force spectroscopy
Salipro	Saposin lipo-protein
XFEL	X-ray free-electron laser
VDAC	Voltage-dependent anion channel

References

1. Stirling, I.; Derocher, A.E. Possible impacts of climatic warming on polar bears. *Arctic* **1993**, *46*, 240–245. [CrossRef]
2. Cremers, P.W.F.H.; Geutjes, S.L. The cause of stereotypic behaviour in a male polar bear (*Ursus maritimus*). *Proc. Meas. Behav.* **2012**, *2012*, 28–31.

3. Clubb, R.; Mason, G. Captivity effects on wide-ranging carnivores. *Nature* **2003**, *425*, 473–474. [CrossRef] [PubMed]
4. Seddon, A.M.; Curnow, P.; Booth, P.J. Membrane proteins, lipids and detergents: Not just a soap opera. *Biochim. Biophys. Acta Biomembr.* **2004**, *1666*, 105–117. [CrossRef]
5. Chorev, D.S.; Robinson, C.V. The importance of the membrane for biophysical measurements. *Nat. Chem. Biol.* **2020**, *16*, 1285–1292. [CrossRef] [PubMed]
6. Zhou, H.-X.; Cross, T.A. Influences of membrane mimetic environments on membrane protein structures. *Annu. Rev. Biophys.* **2013**, *42*, 361–392. [CrossRef] [PubMed]
7. Singer, S.J.; Nicolson, G.L. The fluid mosaic model of the structure of cell membranes. *Science* **1972**, *175*, 720–731. [CrossRef] [PubMed]
8. Harayama, T.; Riezman, H. Understanding the diversity of membrane lipid composition. *Nat. Rev. Mol. Cell Biol.* **2018**, *19*, 281–296. [CrossRef]
9. Dowhan, W.; Bogdanov, M. Functional roles of lipids in membranes. In *Biochemistry of Lipids, Lipoproteins and Membranes*; Elsevier Science: Amsterdam, The Netherlands, 2002; pp. 1–35. ISBN 9780444634382.
10. Shevchenko, A.; Simons, K. Lipidomics: Coming to grips with lipid diversity. *Nat. Rev. Mol. Cell Biol.* **2010**, *11*, 593–598. [CrossRef]
11. Andersen, O.S.; Koeppe, R.E., II. Bilayer thickness and membrane protein function: An energetic perspective. *Annu. Rev. Biophys. Biomol. Struct.* **2007**, *36*, 107–130. [CrossRef]
12. Phillips, R.; Ursell, T.; Wiggins, P.; Sens, P. Emerging roles for lipids in shaping membrane-protein function. *Nature* **2009**, *459*, 379–385. [CrossRef] [PubMed]
13. Coskun, Ü.; Simons, K. Cell membranes: The lipid perspective. *Structure* **2011**, *19*, 1543–1548. [CrossRef]
14. Oates, J.; Watts, A. Uncovering the intimate relationship between lipids, cholesterol and GPCR activation. *Curr. Opin. Struct. Biol.* **2011**, *21*, 802–807. [CrossRef] [PubMed]
15. Van Klompenburg, W.; Nilsson, I.; Von Heijne, G.; De Kruijff, B. Anionic phospholipids are determinants of membrane protein topology. *EMBO J.* **1997**, *16*, 4261–4266. [CrossRef] [PubMed]
16. Garavito, R.M.; Ferguson-Miller, S. Detergents as tools in membrane biochemistry. *J. Biol. Chem.* **2001**, *276*, 32403–32406. [CrossRef] [PubMed]
17. Krishnamani, V.; Hegde, B.G.; Langen, R.; Lanyi, J.K. Secondary and tertiary structure of bacteriorhodopsin in the SDS denatured state. *Biochemistry* **2012**, *51*, 1051–1060. [CrossRef]
18. Lau, F.W.; Bowie, J.U. A method for assessing the stability of a membrane protein. *Biochemistry* **1997**, *36*, 5884–5892. [CrossRef]
19. Noinaj, N.; Kuszak, A.J.; Buchanan, S.K. Heat modifiability of outer membrane proteins from gram-negative bacteria. *Methods Mol. Biol.* **2015**, *1329*, 51–56. [CrossRef]
20. Kleinschmidt, J.H.; Wiener, M.C.; Tamm, L.K. Outer membrane protein A of *E. coli* folds into detergent micelles, but not in the presence of monomeric detergent. *Protein Sci.* **1999**, *8*, 2065–2071. [CrossRef]
21. Sachelaru, I.; Winter, L.; Knyazev, D.G.; Zimmermann, M.; Vogt, A.; Kuttner, R.; Ollinger, N.; Siligan, C.; Pohl, P.; Koch, H.G. YidC and SecYEG form a heterotetrameric protein translocation channel. *Sci. Rep.* **2017**, *7*, 1–15. [CrossRef]
22. Sverzhinsky, A.; Chung, J.W.; Deme, J.C.; Fabre, L.; Levey, K.T.; Plesa, M.; Carter, D.M.; Lypaczewski, P.; Coulton, J.W. Membrane protein complex ExbB4-ExbD1-TonB1 from *Escherichia coli* demonstrates conformational plasticity. *J. Bacteriol.* **2015**, *197*, 1873–1885. [CrossRef] [PubMed]
23. Gu, Y.; Li, H.; Dong, H.; Zeng, Y.; Zhang, Z.; Paterson, N.G.; Stansfeld, P.J.; Wang, Z.; Zhang, Y.; Wang, W.; et al. Structural basis of outer membrane protein insertion by the BAM complex. *Nature* **2016**, *531*, 64–69. [CrossRef] [PubMed]
24. Lau, W.C.Y.; Rubinstein, J.L. Structure of intact *Thermus thermophilus* V-ATPase by cryo-EM reveals organization of the membrane-bound VO motor. *Proc. Natl. Acad. Sci. USA* **2010**, *107*, 1367–1372. [CrossRef] [PubMed]
25. Luche, S.; Santoni, V.; Rabilloud, T. Evaluation of nonionic and zwitterionic detergents as membrane protein solubilizers in two-dimensional electrophoresis. *Proteomics* **2003**, *3*, 249–253. [CrossRef] [PubMed]
26. Almgren, M. Mixed micelles and other structures in the solubilization of bilayer lipid membranes by surfactants. *Biochim. Biophys. Acta Biomembr.* **2000**, *1508*, 146–163. [CrossRef]
27. Le Maire, M.; Champeil, P.; Møller, J.V. Interaction of membrane proteins and lipids with solubilizing detergents. *Biochim. Biophys. Acta Biomembr.* **2000**, *1508*, 86–111. [CrossRef]
28. Schnaitman, C. Solubilization of the cytoplasmic membrane of *Escherichia coli* by Triton X-100. *J. Bacteriol.* **1971**, *108*, 545–552. [CrossRef]
29. Filip, C.; Fletcher, G.; Wulff, J.L.; Earhart, C.F. Solubilization of the cytoplasmic membrane of *Escherichia coli* by the ionic detergent sodium-lauryl sarcosinate. *J. Bacteriol.* **1973**, *115*, 717–722. [CrossRef]
30. Schuck, S.; Honsho, M.; Ekroos, K.; Shevchenko, A.; Simons, K. Resistance of cell membranes to different detergents. *Proc. Natl. Acad. Sci. USA* **2003**, *100*, 5795–5800. [CrossRef]
31. London, E.; Brown, D.A. Insolubility of lipids in Triton X-100: Physical origin and relationship to sphingolipid/cholesterol membrane domains (rafts). *Biochim. Biophys. Acta Biomembr.* **2000**, *1508*, 182–195. [CrossRef]
32. Ilgü, H.; Jeckelmann, J.M.; Gachet, M.S.; Boggavarapu, R.; Ucurum, Z.; Gertsch, J.; Fotiadis, D. Variation of the detergent-binding capacity and phospholipid content of membrane proteins when purified in different detergents. *Biophys. J.* **2014**, *106*, 1660–1670. [CrossRef] [PubMed]

33. Gupta, K.; Donlan, J.A.C.; Hopper, J.T.S.; Uzdavinys, P.; Landreh, M.; Struwe, W.B.; Drew, D.; Baldwin, A.J.; Stansfeld, P.J.; Robinson, C.V. The role of interfacial lipids in stabilizing membrane protein oligomers. *Nature* **2017**, *541*, 421–424. [CrossRef] [PubMed]
34. Rimon, A.; Mondal, R.; Friedler, A.; Padan, E. Cardiolipin is an optimal phospholipid for the assembly, stability, and proper functionality of the dimeric form of NhaA Na^+/H^+ antiporter. *Sci. Rep.* **2019**, *9*, 1–11. [CrossRef] [PubMed]
35. Rigaud, J.; Levy, D.; Mosser, G.; Lambert, O. Detergent removal by non-polar polystyrene beads. *Eur. Biophys. J.* **1998**, *27*, 305–319. [CrossRef]
36. Lin, S.H.; Guidotti, G. *Chapter 35 Purification of Membrane Proteins*, 1st ed.; Elsevier Inc.: Amsterdam, The Netherlands, 2009; Volume 463.
37. Lantez, V.; Nikolaidis, I.; Rechenmann, M.; Vernet, T.; Noirclerc-Savoye, M. Rapid automated detergent screening for the solubilization and purification of membrane proteins and complexes. *Eng. Life Sci.* **2015**, *15*, 39–50. [CrossRef]
38. Kotov, V.; Bartels, K.; Veith, K.; Josts, I.; Subhramanyam, U.K.T.; Günther, C.; Labahn, J.; Marlovits, T.C.; Moraes, I.; Tidow, H.; et al. High-throughput stability screening for detergent-solubilized membrane proteins. *Sci. Rep.* **2019**, *9*, 1–19. [CrossRef]
39. Champeil, P.; Orlowski, S.; Babin, S.; Lund, S.; le Maire, M.; Møller, J.; Lenoir, G.; Montigny, C. A robust method to screen detergents for membrane protein stabilization, revisited. *Anal. Biochem.* **2016**, *511*, 31–35. [CrossRef]
40. Buchanan, S.K. β-barrel proteins from bacterial outer membranes: Structure, function and refolding. *Curr. Opin. Struct. Biol.* **1999**, *9*, 455–461. [CrossRef]
41. Fairman, J.W.; Noinaj, N.; Buchanan, S.K. The structural biology of β-barrel membrane proteins: A summary of recent reports. *Curr. Opin. Struct. Biol.* **2011**, *21*, 523–531. [CrossRef]
42. Burmann, B.M.; Hiller, S. Solution NMR studies of membrane-protein-chaperone complexes. *Chimia* **2012**, *66*, 759–763. [CrossRef]
43. Visudtiphole, V.; Thomas, M.B.; Chalton, D.A.; Lakey, J.H. Refolding of *Escherichia coli* outer membrane protein F in detergent creates LPS-free trimers and asymmetric dimers. *Biochem. J.* **2005**, *392*, 375–381. [CrossRef] [PubMed]
44. Wang, H.; Andersen, K.K.; Vad, B.S.; Otzen, D.E. OmpA can form folded and unfolded oligomers. *Biochim. Biophys. Acta Proteins Proteomics* **2013**, *1834*, 127–136. [CrossRef] [PubMed]
45. Liang, B.; Tamm, L.K. NMR as a tool to investigate the structure, dynamics and function of membrane proteins. *Nat. Struct. Mol. Biol.* **2016**, *23*, 468–474. [CrossRef] [PubMed]
46. Hiruma-Shimizu, K.; Shimizu, H.; Thompson, G.S.; Kalverda, A.P.; Patching, S.G. Deuterated detergents for structural and functional studies of membrane proteins: Properties, chemical synthesis and applications. *Mol. Membr. Biol.* **2015**, *32*, 139–155. [CrossRef]
47. Hiller, S.; Garces, R.G.; Malia, T.J.; Orekhov, V.Y.; Colombini, M.; Wagner, G. Solution structure of the integral human membrane protein VDAC-1 in detergent micelles. *Science* **2008**, *321*, 1206–1210. [CrossRef]
48. Gautier, A.; Mott, H.R.; Bostock, M.J.; Kirkpatrick, J.P.; Nietlispach, D. Structure determination of the seven-helix transmembrane receptor sensory rhodopsin II by solution NMR spectroscopy. *Nat. Struct. Mol. Biol.* **2010**, *17*, 768–774. [CrossRef]
49. Zhou, Y.; Cierpicki, T.; Jimenez, R.H.F.; Lukasik, S.M.; Ellena, J.F.; Cafiso, D.S.; Kadokura, H.; Beckwith, J.; Bushweller, J.H. NMR solution structure of the integral membrane enzyme DsbB: Functional insights into DsbB-catalyzed disulfide bond formation. *Mol. Cell* **2008**, *31*, 896–908. [CrossRef]
50. Arora, A.; Abildgaard, F.; Bushweller, J.H.; Tamm, L.K. Structure of outer membrane protein A transmembrane domain by NMR spectroscopy. *Nat. Struct. Biol.* **2001**, *8*, 334–338. [CrossRef]
51. Hiller, S.; Wagner, G. The role of solution NMR in the structure determinations of VDAC-1 and other membrane proteins. *Curr. Opin. Struct. Biol.* **2009**, *19*, 396–401. [CrossRef]
52. Opella, S.J.; Marassi, F.M. Applications of NMR to membrane proteins. *Arch. Biochem. Biophys.* **2017**, *628*, 92–101. [CrossRef]
53. Kucharska, I.; Seelheim, P.; Edrington, T.; Liang, B.; Tamm, L.K. OprG harnesses the dynamics of its extracellular loops to transport small amino acids across the outer membrane of *Pseudomonas aeruginosa*. *Structure* **2015**, *23*, 2234–2245. [CrossRef] [PubMed]
54. Renault, M.; Saurel, O.; Czaplicki, J.; Demange, P.; Gervais, V.; Löhr, F.; Réat, V.; Piotto, M.; Milon, A. Solution state NMR structure and dynamics of KpOmpA, a 210 residue transmembrane domain possessing a high potential for immunological applications. *J. Mol. Biol.* **2009**, *385*, 117–130. [CrossRef] [PubMed]
55. Kurauskas, V.; Hessel, A.; Ma, P.; Lunetti, P.; Weinhäupl, K.; Imbert, L.; Brutscher, B.; King, M.S.; Sounier, R.; Dolce, V.; et al. How detergent impacts membrane proteins: Atomic-level views of mitochondrial carriers in dodecylphosphocholine. *J. Phys. Chem. Lett.* **2018**, *9*, 933–938. [CrossRef] [PubMed]
56. Kühlbrandt, W. The resolution revolution. *Science* **2014**, *343*, 1443–1444. [CrossRef] [PubMed]
57. Cheng, Y. Single-particle Cryo-EM at crystallographic resolution. *Cell* **2015**, *161*, 450–457. [CrossRef] [PubMed]
58. Iadanza, M.G.; Higgins, A.J.; Schiffrin, B.; Calabrese, A.N.; Brockwell, D.J.; Ashcroft, A.E.; Radford, S.E.; Ranson, N.A. Lateral opening in the intact β-barrel assembly machinery captured by cryo-EM. *Nat. Commun.* **2016**, *7*, 12865. [CrossRef]
59. Bausewein, T.; Mills, D.J.; Langer, J.D.; Nitschke, B.; Nussberger, S.; Kühlbrandt, W. Cryo-EM Structure of the TOM Core Complex from *Neurospora crassa*. *Cell* **2017**, *170*, 693–700. [CrossRef]
60. Malone, L.A.; Qian, P.; Mayneord, G.E.; Hitchcock, A.; Farmer, D.A.; Thompson, R.F.; Swainsbury, D.J.K.; Ranson, N.A.; Hunter, C.N.; Johnson, M.P. Cryo-EM structure of the spinach cytochrome b 6 f complex at 3.6 Å resolution. *Nature* **2019**, *575*, 535–539. [CrossRef]

61. Sgro, G.G.; Costa, T.R.D. Cryo-EM grid preparation of membrane protein samples for single particle analysis. *Front. Mol. Biosci.* **2018**, *5*, 1–8. [CrossRef]
62. Carvalho, V.; Pronk, J.W.; Engel, A.H. Characterization of membrane proteins using cryo-electron microscopy. *Curr. Protoc. Protein Sci.* **2018**, *94*, 1–30. [CrossRef]
63. Schmidt-Krey, I.; Rubinstein, J.L. Electron cryomicroscopy of membrane proteins: Specimen preparation for two-dimensional crystals and single particles. *Micron* **2011**, *42*, 107–116. [CrossRef] [PubMed]
64. Hauer, F.; Gerle, C.; Fischer, N.; Oshima, A.; Shinzawa-Itoh, K.; Shimada, S.; Yokoyama, K.; Fujiyoshi, Y.; Stark, H. GraDeR: Membrane protein complex preparation for single-particle cryo-EM. *Structure* **2015**, *23*, 1769–1775. [CrossRef] [PubMed]
65. Glaeser, R.M.; Han, B.-G. Opinion: Hazards faced by macromolecules when confined to thin aqueous films. *Biophys. Rep.* **2017**, *3*, 1–7. [CrossRef] [PubMed]
66. Von Jagow, G.; Link, T.A.; Schägger, H. Purification Strategies for Membrane Proteins. In *A Practical Guide to Membrane Protein Purification;* Separation, Detection, and Characterization of Biological Macromolecules; Von Jagow, G., Schägger, H., Eds.; Academic Press: San Diego, CA, USA, 1994; Volume 2, pp. 3–21. ISBN 978-0-08-057172-0.
67. Schägger, H. Chromatographic Techniques and Basic Operations in Membrane Protein Purification. In *A Practical Guide to Membrane Protein Purification;* Separation, Detection, and Characterization of Biological Macromolecules; Von Jagow, G., Schägger, H., Eds.; Academic Press: San Diego, CA, USA, 1994; Volume 2, pp. 23–57. ISBN 978-0-08-057172-0.
68. Wiener, M.C. A pedestrian guide to membrane protein crystallization. *Methods* **2004**, *34*, 364–372. [CrossRef]
69. Birch, J.; Axford, D.; Foadi, J.; Meyer, A.; Eckhardt, A.; Thielmann, Y.; Moraes, I. The fine art of integral membrane protein crystallisation. *Methods* **2018**, *147*, 150–162. [CrossRef]
70. Michel, H. Crystallization of membrane proteins. *Trends Biochem. Sci.* **1983**, *8*, 56–59. [CrossRef]
71. Ostermeier, C.; Michel, H. Crystallization of membrane proteins. *Curr. Opin. Struct. Biol.* **1997**, *7*, 697–701. [CrossRef]
72. Deisenhofer, J.; Epp, O.; Miki, K.; Huber, R.; Michel, H. Structure of the protein subunits in the photosynthetic reaction centre of *Rhodopseudomonas viridis* at 3Å resolution. *Nature* **1985**, *318*, 618–624. [CrossRef]
73. Doyle, D.A.; Cabral, J.M.; Pfuetzner, R.A.; Kuo, A.; Gulbis, J.M.; Cohen, S.L.; Chait, B.T.; MacKinnon, R. The structure of the potassium channel: Molecular basis of K+ conduction and selectivity. *Science* **1998**, *280*, 69–77. [CrossRef]
74. Abramson, J.; Smirnova, I.; Kasho, V.; Verner, G.; Kaback, H.R.; Iwata, S. Structure and mechanism of the lactose permease of Escherichia coli. *Science* **2003**, *301*, 610–615. [CrossRef]
75. Vaidehi, N.; Grisshammer, R.; Tate, C.G. How can mutations thermostabilize G-protein-coupled receptors? *Trends Pharmacol. Sci.* **2016**, *37*, 37–46. [CrossRef] [PubMed]
76. Manglik, A.; Kobilka, B.K.; Steyaert, J. Nanobodies to study G protein-coupled receptor structure and function. *Annu. Rev. Pharmacol. Toxicol.* **2017**, *57*, 19–37. [CrossRef] [PubMed]
77. Wacker, D.; Stevens, R.C.; Roth, B.L. How ligands illuminate GPCR molecular pharmacology. *Cell* **2017**, *170*, 414–427. [CrossRef] [PubMed]
78. Rosenbaum, D.M.; Cherezov, V.; Hanson, M.A.; Rasmussen, S.G.F.; Foon, S.T.; Kobilka, T.S.; Choi, H.J.; Yao, X.J.; Weis, W.I.; Stevens, R.C.; et al. GPCR engineering yields high-resolution structural insights into β2-adrenergic receptor function. *Science* **2007**, *318*, 1266–1273. [CrossRef] [PubMed]
79. Thorsen, T.S.; Matt, R.; Weis, W.I.; Kobilka, B.K. Modified T4 lysozyme fusion proteins facilitate G protein-coupled receptor crystallogenesis. *Structure* **2014**, *22*, 1657–1664. [CrossRef] [PubMed]
80. Davis, A.M.; Teague, S.J.; Kleywegt, G.J. Application and limitations of x-ray crystallographic data in structure-based ligand and drug design. *Angew. Chem. Int. Ed.* **2003**, *42*, 2718–2736. [CrossRef] [PubMed]
81. Guo, Y.; Kalathur, R.C.; Liu, Q.; Kloss, B.; Bruni, R.; Ginter, C.; Kloppmann, E.; Rost, B.; Hendrickson, W.A. Structure and activity of tryptophan-rich TSPO proteins. *Science* **2015**, *347*, 551–555. [CrossRef]
82. Li, F.; Liu, J.; Zheng, Y.; Garavito, R.M.; Ferguson-Miller, S. Crystal structures of translocator protein (TSPO) and mutant mimic of a human polymorphism. *Science* **2015**, *347*, 555–558. [CrossRef]
83. Jaremko, Ł.; Jaremko, M.; Giller, K.; Becker, S.; Zweckstetter, M. Structure of the mitochondrial translocator protein in complex with a diagnostic ligand. *Science* **2014**, *343*, 1363–1366. [CrossRef]
84. Landau, E.M.; Rosenbusch, J.P. Lipidic cubic phases: A novel concept for the crystallization of membrane proteins. *Proc. Natl. Acad. Sci. USA* **1996**, *93*, 14532–14535. [CrossRef]
85. Caffrey, M. Crystallizing membrane proteins for structure determination: Use of lipidic mesophases. *Annu. Rev. Biophys.* **2009**, *38*, 29–51. [CrossRef]
86. Caffrey, M. Membrane protein crystallization. *J. Struct. Biol.* **2003**, *142*, 108–132. [CrossRef]
87. Cherezov, V.; Clogston, J.; Misquitta, Y.; Abdel-Gawad, W.; Caffrey, M. Membrane protein crystallization in meso: Lipid type-tailoring of the cubic phase. *Biophys. J.* **2002**, *83*, 3393–3407. [CrossRef]
88. Pebay-Peyroula, E.; Rummel, G.; Rosenbusch, J.P.; Landau, E.M. X-ray structure of bacteriorhodopsin at 2.5 Ångstroms from microcrystals grown in lipidic cubic phases. *Science* **1997**, *277*, 1676–1681. [CrossRef]
89. Kolbe, M.; Besir, H.; Essen, L.O.; Oesterhelt, D. Structure of the light-driven chloride pump halorhodopsin at 1.8 Å Resolution. *Science* **2000**, *288*, 1390–1396. [CrossRef] [PubMed]
90. Royant, A.; Nollert, P.; Edman, K.; Neutze, R.; Landau, E.M.; Pebay-Peyroula, E.; Navarro, J. X-ray structure of sensory rhodopsin II at 2.1-Å resolution. *Proc. Natl. Acad. Sci. USA* **2001**, *98*, 10131–10136. [CrossRef] [PubMed]

91. Cherezov, V.; Rosenbaum, D.M.; Hanson, M.A.; Rasmussen, S.G.F.; Foon, S.T.; Kobilka, T.S.; Choi, H.J.; Kuhn, P.; Weis, W.I.; Kobilka, B.K.; et al. High-resolution crystal structure of an engineered human β2-adrenergic G protein-coupled receptor. *Science* **2007**, *318*, 1258–1265. [CrossRef] [PubMed]
92. Chien, E.Y.T.; Liu, W.; Zhao, Q.; Katritch, V.; Han, G.W.; Hanson, M.A.; Shi, L.; Newman, A.H.; Javitch, J.A.; Cherezov, V.; et al. Structure of the human dopamine D3 receptor in complex with a D2/D3 selective antagonist. *Science* **2010**, *330*, 1091–1095. [CrossRef]
93. Weierstall, U.; James, D.; Wang, C.; White, T.A.; Wang, D.; Liu, W.; Spence, J.C.H.; Bruce Doak, R.; Nelson, G.; Fromme, P.; et al. Lipidic cubic phase injector facilitates membrane protein serial femtosecond crystallography. *Nat. Commun.* **2014**, *5*. [CrossRef]
94. Neutze, R.; Wouts, R.; Van Der Spoel, D.; Weckert, E.; Hajdu, J. Potential for biomolecular imaging with femtosecond X-ray pulses. *Nature* **2000**, *406*, 752–757. [CrossRef]
95. Chapman, H.N.; Fromme, P.; Barty, A.; White, T.A.; Kirian, R.A.; Aquila, A.; Hunter, M.S.; Schulz, J.; Deponte, D.P.; Weierstall, U.; et al. Femtosecond X-ray protein nanocrystallography. *Nature* **2011**, *470*, 73–78. [CrossRef]
96. Hajdu, J.; Neutze, R.; Sjögren, T.; Edman, K.; Szöke, A.; Wilmouth, R.C.; Wilmot, C.M. Analyzing protein functions in four dimensions. *Nat. Struct. Biol.* **2000**, *7*, 1006–1012. [CrossRef] [PubMed]
97. Arnlund, D.; Johansson, L.C.; Wickstrand, C.; Barty, A.; Williams, G.J.; Malmerberg, E.; Davidsson, J.; Milathianaki, D.; DePonte, D.P.; Shoeman, R.L.; et al. Visualizing a protein quake with time-resolved X-ray scattering at a free-electron laser. *Nat. Methods* **2014**, *11*, 923–926. [CrossRef] [PubMed]
98. Nango, E.; Royant, A.; Kubo, M.; Nakane, T.; Wickstrand, C.; Kimura, T.; Tanaka, T.; Tono, K.; Song, C.; Tanaka, R.; et al. A three-dimensional movie of structural changes in bacteriorhodopsin. *Science* **2016**, *354*, 1552–1557. [CrossRef] [PubMed]
99. Dods, R.; Båth, P.; Morozov, D.; Gagnér, V.A.; Arnlund, D.; Luk, H.L.; Kübel, J.; Maj, M.; Vallejos, A.; Wickstrand, C.; et al. Ultrafast structural changes within a photosynthetic reaction centre. *Nature* **2020**. [CrossRef] [PubMed]
100. Shi, D.; Nannenga, B.L.; Iadanza, M.G.; Gonen, T. Three-dimensional electron crystallography of protein microcrystals. *eLife* **2013**, *2*, 1–17. [CrossRef]
101. Nannenga, B.L.; Shi, D.; Leslie, A.G.W.; Gonen, T. High-resolution structure determination by continuous-rotation data collection in MicroED. *Nat. Methods* **2014**, *11*, 927–930. [CrossRef]
102. Liu, S.; Gonen, T. MicroED structure of the NaK ion channel reveals a Na^+ partition process into the selectivity filter. *Commun. Biol.* **2018**, *1*, 1–6. [CrossRef]
103. Zhu, L.; Bu, G.; Jing, L.; Shi, D.; Lee, M.Y.; Gonen, T.; Liu, W.; Nannenga, B.L. Structure determination from lipidic cubic phase embedded microcrystals by MicroED. *Structure* **2020**, *28*, 1149–1159. [CrossRef]
104. Sanders, C.R.; Prosser, R.S. Bicelles: A model membrane system for all seasons? *Structure* **1998**, *6*, 1227–1234. [CrossRef]
105. Harroun, T.A.; Koslowsky, M.; Nieh, M.P.; De Lannoy, C.F.; Raghunathan, V.A.; Katsaras, J. Comprehensive examination of mesophases formed by DMPC and DHPC mixtures. *Langmuir* **2005**, *21*, 5356–5361. [CrossRef] [PubMed]
106. Dürr, U.H.N.; Gildenberg, M.; Ramamoorthy, A. The magic of bicelles lights up membrane protein structure. *Chem. Rev.* **2012**, *112*, 6054–6074. [CrossRef] [PubMed]
107. Sanders, C.R.; Landis, G.C. Reconstitution of membrane proteins into lipid-rich bilayered mixed micelles for NMR studies. *Biochemistry* **1995**, *34*, 4030–4040. [CrossRef] [PubMed]
108. Fanucci, G.; Lee, J.; Cafiso, D. Membrane mimetic environments alter the conformation of the outer membrane protein BtuB. *J. Am. Chem. Soc.* **2003**, *125*, 13932–13933. [CrossRef] [PubMed]
109. Morrison, E.A.; Henzler-Wildman, K.A. Reconstitution of integral membrane proteins into isotropic bicelles with improved sample stability and expanded lipid composition profile. *Biochim. Biophys. Acta Biomembr.* **2012**, *1818*, 814–820. [CrossRef] [PubMed]
110. Vestergaard, M.; Kraft, J.F.; Vosegaard, T.; Thøgersen, L.; Schiøtt, B. Bicelles and other membrane mimics: Comparison of structure, properties, and dynamics from MD simulations. *J. Phys. Chem. B* **2015**, *119*, 15831–15843. [CrossRef]
111. Sanders, C.R.; Hare, B.J.; Howard, K.P.; Prestegard, J.H. Magnetically-oriented phospholipid micelles as a tool for the study of membrane-associated molecules. *Prog. Nucl. Magn. Reson. Spectrosc.* **1994**, *26*, 421–444. [CrossRef]
112. Diller, A.; Loudet, C.; Aussenac, F.; Raffard, G.; Fournier, S.; Laguerre, M.; Grélard, A.; Opella, S.J.; Marassi, F.M.; Dufourc, E.J. Bicelles: A natural "molecular goniometer" for structural, dynamical and topological studies of molecules in membranes. *Biochimie* **2009**, *91*, 744–751. [CrossRef]
113. Prosser, R.S.; Evanics, F.; Kitevski, J.L.; Al-Abdul-Wahid, M.S. Current applications of bicelles in NMR studies of membrane-associated amphiphiles and proteins. *Biochemistry* **2006**, *45*, 8453–8465. [CrossRef]
114. Piai, A.; Fu, Q.; Dev, J.; Chou, J.J. Optimal bicelle size q for solution NMR studies of the protein transmembrane partition. *Chem. A Eur. J.* **2017**, *23*, 1361–1367. [CrossRef]
115. Dev, J.; Park, D.; Fu, Q.; Chen, J.; Ha, H.J.; Ghantous, F.; Herrmann, T.; Chang, W.; Liu, Z.; Frey, G.; et al. Structural basis for membrane anchoring of HIV-1 envelope spike. *Science* **2016**, *353*, 172–175. [CrossRef] [PubMed]
116. Fu, Q.; Shaik, M.M.; Cai, Y.; Ghantous, F.; Piai, A.; Peng, H.; Rits-Volloch, S.; Liu, Z.; Harrison, S.C.; Seaman, M.S.; et al. Structure of the membrane proximal external region of HIV-1 envelope glycoprotein. *Proc. Natl. Acad. Sci. USA* **2018**, *115*, E8892–E8899. [CrossRef] [PubMed]
117. Chiliveri, S.C.; Louis, J.M.; Ghirlando, R.; Baber, J.L.; Bax, A. Tilted, Uninterrupted, monomeric HIV-1 gp41 transmembrane helix from residual dipolar couplings. *J. Am. Chem. Soc.* **2018**, *140*, 34–37. [CrossRef] [PubMed]

118. Lee, J.H.; Ozorowski, G.; Ward, A.B. Cryo-EM structure of a native, fully glycosylated, cleaved HIV-1 envelope trimer. *Science* **2016**, *351*, 1043–1048. [CrossRef] [PubMed]
119. Mineev, K.S.; Bocharov, E.V.; Volynsky, P.E.; Goncharuk, M.V.; Tkach, E.N.; Ermolyuk, Y.S.; Schulga, A.A.; Chupin, V.V.; Maslennikov, I.V.; Efremov, R.G.; et al. Dimeric structure of the transmembrane domain of Glycophorin A in lipidic and detergent environments. *Acta Nat.* **2011**, *3*, 90–98. [CrossRef]
120. Faham, S.; Bowie, J.U. Bicelle crystallization: A new method for crystallizing membrane proteins yields a monomeric bacteriorhodopsin structure. *J. Mol. Biol.* **2002**, *316*, 1–6. [CrossRef] [PubMed]
121. Ujwal, R.; Bowie, J.U. Crystallizing membrane proteins using lipidic bicelles. *Methods* **2011**, *55*, 337–341. [CrossRef]
122. Rasmussen, S.G.F.; Choi, H.J.; Rosenbaum, D.M.; Kobilka, T.S.; Thian, F.S.; Edwards, P.C.; Burghammer, M.; Ratnala, V.R.P.; Sanishvili, R.; Fischetti, R.F.; et al. Crystal structure of the human β2 adrenergic G-protein-coupled receptor. *Nature* **2007**, *450*, 383–387. [CrossRef]
123. Ujwal, R.; Cascio, D.; Colletier, J.P.; Faham, S.; Zhang, J.; Toro, L.; Ping, P.; Abramson, J. The crystal structure of mouse VDAC1 at 2.3 Å resolution reveals mechanistic insights into metabolite gating. *Proc. Natl. Acad. Sci. USA* **2008**, *105*, 17742–17747. [CrossRef]
124. Bayburt, T.H.; Sligar, S.G. Self-assembly of single integral membrane proteins into soluble nanoscale phospholipid bilayers. *Protein Sci.* **2003**, *12*, 2476–2481. [CrossRef]
125. Frauenfeld, J.; Löving, R.; Armache, J.P.; Sonnen, A.F.P.; Guettou, F.; Moberg, P.; Zhu, L.; Jegerschöld, C.; Flayhan, A.; Briggs, J.A.G.; et al. A saposin-lipoprotein nanoparticle system for membrane proteins. *Nat. Methods* **2016**, *13*, 345–351. [CrossRef] [PubMed]
126. Knowles, T.J.; Finka, R.; Smith, C.; Lin, Y.-P.P.; Dafforn, T.; Overduin, M. Membrane proteins solubilized intact in lipid containing nanoparticles bounded by styrene maleic acid copolymer. *J. Am. Chem. Soc.* **2009**, *131*, 7484–7485. [CrossRef] [PubMed]
127. Oluwole, A.O.; Danielczak, B.; Meister, A.; Babalola, J.O.; Vargas, C.; Keller, S. Solubilization of membrane proteins into functional lipid-bilayer nanodiscs using a diisobutylene/maleic acid copolymer. *Angew. Chem. Int. Ed.* **2017**, *56*, 1919–1924. [CrossRef] [PubMed]
128. Sligar, S.G.; Denisov, I.G. Nanodiscs: A toolkit for membrane protein science. *Protein Sci.* **2020**, 1–19. [CrossRef]
129. Denisov, I.G.; Sligar, S.G. Nanodiscs in membrane biochemistry and biophysics. *Chem. Rev.* **2017**, *117*, 4669–4713. [CrossRef]
130. Dörr, J.M.; Koorengevel, M.C.; Schäfer, M.; Prokofyev, A.V.; Scheidelaar, S.; van der Cruijsen, E.A.W.W.; Dafforn, T.R.; Baldus, M.; Killian, J.A. Detergent-free isolation, characterization, and functional reconstitution of a tetrameric K^+ channel: The power of native nanodiscs. *Proc. Natl. Acad. Sci. USA* **2014**, *111*, 18607–18612. [CrossRef]
131. Overduin, M.; Klumperman, B. Advancing membrane biology with poly(styrene-co-maleic acid)-based native nanodiscs. *Eur. Polym. J.* **2019**, *110*, 63–68. [CrossRef]
132. Xue, M.; Cheng, L.; Faustino, I.; Guo, W.; Marrink, S.J. Molecular mechanism of lipid nanodisc formation by styrene-maleic acid copolymers. *Biophys. J.* **2018**, *115*, 494–502. [CrossRef]
133. Cuevas Arenas, R.; Danielczak, B.; Martel, A.; Porcar, L.; Breyton, C.; Ebel, C.; Keller, S. Fast collisional lipid transfer among polymer-bounded nanodiscs. *Sci. Rep.* **2017**, *7*, 1–8. [CrossRef]
134. Danielczak, B.; Keller, S. Collisional lipid exchange among DIBMA-encapsulated nanodiscs (DIBMALPs). *Eur. Polym. J.* **2018**, *109*, 206–213. [CrossRef]
135. Martinez, D.; Decossas, M.; Kowal, J.; Frey, L.; Stahlberg, H.; Dufourc, E.J.; Riek, R.; Habenstein, B.; Bibow, S.; Loquet, A. Lipid internal dynamics probed in nanodiscs. *ChemPhysChem* **2017**, *18*, 2651–2657. [CrossRef] [PubMed]
136. Raschle, T.; Hiller, S.; Etzkorn, M.; Wagner, G. Nonmicellar systems for solution NMR spectroscopy of membrane proteins. *Curr. Opin. Struct. Biol.* **2010**, *20*, 471–479. [CrossRef] [PubMed]
137. Hagn, F.; Etzkorn, M.; Raschle, T.; Wagner, G. Optimized phospholipid bilayer nanodiscs facilitate high-resolution structure determination of membrane proteins. *J. Am. Chem. Soc.* **2013**, *135*, 1919–1925. [CrossRef] [PubMed]
138. Yokogawa, M.; Fukuda, M.; Osawa, M. Nanodiscs for structural biology in a membranous environment. *Chem. Pharm. Bull.* **2019**, *67*, 321–326. [CrossRef] [PubMed]
139. Efremov, R.G.; Gatsogiannis, C.; Raunser, S. *Lipid Nanodiscs as a Tool for High-Resolution Structure Determination of Membrane Proteins by Single-Particle Cryo-EM*, 1st ed.; Elsevier Inc.: Amsterdam, The Netherlands, 2017; Volume 594.
140. Kalienkova, V.; Alvadia, C.; Clerico Mosina, V.; Paulino, C. Single-particle cryo-EM of membrane proteins in lipid nanodiscs. *Methods Mol. Biol.* **2020**, *2127*, 245–273. [CrossRef] [PubMed]
141. Zocher, M.; Roos, C.; Wegmann, S.; Bosshart, P.D.; Dötsch, V.; Bernhard, F.; Müller, D.J. Single-molecule force spectroscopy from nanodiscs: An assay to quantify folding, stability, and interactions of native membrane proteins. *ACS Nano* **2012**, *6*, 961–971. [CrossRef]
142. Daury, L.; Orange, F.; Taveau, J.C.; Verchère, A.; Monlezun, L.; Gounou, C.; Marreddy, R.K.R.; Picard, M.; Broutin, I.; Pos, K.M.; et al. Tripartite assembly of RND multidrug efflux pumps. *Nat. Commun.* **2016**, *7*, 1–8. [CrossRef]
143. Meusch, D.; Gatsogiannis, C.; Efremov, R.G.; Lang, A.E.; Hofnagel, O.; Vetter, I.R.; Aktories, K.; Raunser, S. Mechanism of Tc toxin action revealed in molecular detail. *Nature* **2014**, *508*, 61–65. [CrossRef]
144. Gatsogiannis, C.; Lang, A.E.; Meusch, D.; Pfaumann, V.; Hofnagel, O.; Benz, R.; Aktories, K.; Raunser, S. A syringe-like injection mechanism in *Photorhabdus luminescens* toxins. *Nature* **2013**, *495*, 520–523. [CrossRef]
145. Gatsogiannis, C.; Merino, F.; Prumbaum, D.; Roderer, D.; Leidreiter, F.; Meusch, D.; Raunser, S. Membrane insertion of a Tc toxin in near-atomic detail. *Nat. Struct. Mol. Biol.* **2016**, *23*, 884–890. [CrossRef]

146. Rantalainen, K.; Berndsen, Z.T.; Antanasijevic, A.; Schiffner, T.; Zhang, X.; Lee, W.H.; Torres, J.L.; Zhang, L.; Irimia, A.; Copps, J.; et al. HIV-1 envelope and MPER antibody structures in lipid assemblies. *Cell Rep.* **2020**, *31*, 107583. [CrossRef] [PubMed]
147. Frey, L.; Lakomek, N.A.; Riek, R.; Bibow, S. Micelles, bicelles, and nanodiscs: Comparing the impact of membrane mimetics on membrane protein backbone dynamics. *Angew. Chemie Int. Ed.* **2017**, *56*, 380–383. [CrossRef] [PubMed]
148. Frey, L.; Hiller, S.; Riek, R.; Bibow, S. Lipid- and cholesterol-mediated time-scale-specific modulation of the outer membrane protein X dynamics in lipid bilayers. *J. Am. Chem. Soc.* **2018**, *140*, 15402–15411. [CrossRef]
149. Morgado, L.; Zeth, K.; Burmann, B.M.; Maier, T.; Hiller, S. Characterization of the insertase BamA in three different membrane mimetics by solution NMR spectroscopy. *J. Biomol. NMR* **2015**, *61*, 333–345. [CrossRef] [PubMed]
150. Imai, S.; Osawa, M.; Mita, K.; Toyonaga, S.; Machiyama, A.; Ueda, T.; Takeuchi, K.; Oiki, S.; Shimada, I. Functional equilibrium of the KcsA structure revealed by NMR. *J. Biol. Chem.* **2012**, *287*, 39634–39641. [CrossRef] [PubMed]
151. Yoshiura, C.; Kofuku, Y.; Ueda, T.; Mase, Y.; Yokogawa, M.; Osawa, M.; Terashima, Y.; Matsushima, K.; Shimada, I. NMR analyses of the interaction between CCR5 and its ligand using functional reconstitution of CCR5 in lipid bilayers. *J. Am. Chem. Soc.* **2010**, *132*, 6768–6777. [CrossRef] [PubMed]
152. Raschle, T.; Hiller, S.; Yu, T.Y.; Rice, A.J.; Walz, T.; Wagner, G. Structural and functional characterization of the integral membrane protein VDAC-1 in lipid bilayer nanodiscs. *J. Am. Chem. Soc.* **2009**, *131*, 17777–17779. [CrossRef]
153. Oesterhelt, D.; Stoeckenius, W. Rhodopsin-like protein from the purple membrane of *Halobacterium halobium*. *Nat. New Biol.* **1971**, *233*, 149–152. [CrossRef]
154. Blaurock, A.E.; Stoeckenius, W. Structure of the purple membrane. *Nat. New Biol.* **1971**, *233*, 152–155. [CrossRef]
155. Olive, J. The Structural Organization of Mammalian Retinal Disc Membrane. *Int Rev Cytol.* **1980**, *64*, 107–169. [CrossRef]
156. Arnold, T.; Linke, D. The use of detergents to purify membrane proteins. *Curr. Protoc. Protein Sci.* **2008**, 1–30. [CrossRef] [PubMed]
157. Linke, D. *Chapter 34 Detergents. An Overview*, 1st ed.; Elsevier Inc.: Amsterdam, The Netherlands, 2009; Volume 463.
158. Rémigy, H.W.; Caujolle-Bert, D.; Suda, K.; Schenk, A.; Chami, M.; Engel, A. Membrane protein reconstitution and crystallization by controlled dilution. *FEBS Lett.* **2003**, *555*, 160–169. [CrossRef]
159. Rigaud, J.L.; Pitard, B.; Levy, D. Reconstitution of membrane proteins into liposomes: Application to energy-transducing membrane proteins. *Biochim. Biophys. Acta Bioenerg.* **1995**, *1231*, 223–246. [CrossRef]
160. Hasler, L.; Heymann, J.; Engel, A. 2D crystallization of membrane proteins: Rationales and examples. *J. Struct. Biol.* **1998**, *121*, 162–171. [CrossRef] [PubMed]
161. Jap, B.; Zulauf, M.; Scheybani, T.; Hefti, A. 2D crystallization: From art to science. *Ultramicroscopy* **1992**, *46*, 45–84. [CrossRef]
162. Romsicki, Y.; Sharom, F.J. Phospholipid flippase activity of the reconstituted P-glycoprotein multidrug transporter. *Biochemistry* **2001**, *40*, 6937–6947. [CrossRef]
163. Tunuguntla, R.; Bangar, M.; Kim, K.; Stroeve, P.; Ajo-Franklin, C.M.; Noy, A. Lipid bilayer composition can influence the orientation of proteorhodopsin in artificial membranes. *Biophys. J.* **2013**, *105*, 1388–1396. [CrossRef]
164. Ritzmann, N.; Thoma, J.; Hirschi, S.; Kalbermatter, D.; Fotiadis, D.; Müller, D.J. Fusion domains guide the oriented insertion of light-driven proton pumps into liposomes. *Biophys. J.* **2017**, *113*, 1181–1186. [CrossRef]
165. Bogdanov, M.; Xie, J.; Heacock, P.; Dowhan, W. To flip or not to flip: Lipid-protein charge interactions are a determinant of final membrane protein topology. *J. Cell Biol.* **2008**, *182*, 925–935. [CrossRef]
166. Dowhan, W.; Bogdanov, M. Lipid-dependent membrane protein topogenesis. *Annu. Rev. Biochem.* **2009**, *78*, 515–540. [CrossRef]
167. Bogdanov, M.; Dowhan, W. Lipid-dependent generation of dual topology for a membrane protein. *J. Biol. Chem.* **2012**, *287*, 37939–37948. [CrossRef] [PubMed]
168. Kleinschmidt, J.H.; Den Blaauwen, T.; Driessen, A.J.M.; Tamm, L.K. Outer membrane protein A of *Escherichia coli* inserts and folds into lipid bilayers by a concerted mechanism. *Biochemistry* **1999**, *38*, 5006–5016. [CrossRef] [PubMed]
169. Plummer, A.M.; Fleming, K.G. BamA alone accelerates outer membrane protein folding *in vitro* through a catalytic mechanism. *Biochemistry* **2015**, *54*, 6009–6011. [CrossRef] [PubMed]
170. Gessmann, D.; Chung, Y.H.; Danoff, E.J.; Plummer, A.M.; Sandlin, C.W.; Zaccai, N.R.; Fleming, K.G. Outer membrane β-barrel protein folding is physically controlled by periplasmic lipid head groups and BamA. *Proc. Natl. Acad. Sci. USA* **2014**, *111*, 5878–5883. [CrossRef] [PubMed]
171. Schiffrin, B.; Calabrese, A.N.; Higgins, A.J.; Humes, J.R.; Ashcroft, A.E.; Kalli, A.C.; Brockwell, D.J.; Radford, S.E. Effects of periplasmic chaperones and membrane thickness on BamA-catalyzed outer-membrane Ppotein folding. *J. Mol. Biol.* **2017**, *429*, 3776–3792. [CrossRef] [PubMed]
172. Fujiyoshi, Y. The structural study of membrane proteins by electron crystallography. *Adv. Biophys.* **1998**, *35*, 25–80. [CrossRef]
173. Stahlberg, H.; Fotiadis, D.; Scheuring, S.; Rémigy, H.; Braun, T.; Mitsuoka, K.; Fujiyoshi, Y.; Engel, A. Two-dimensional crystals: A powerful approach to assess structure, function and dynamics of membrane proteins. *FEBS Lett.* **2001**, *504*, 166–172. [CrossRef]
174. Henderson, R.; Baldwin, J.M.; Ceska, T.A.; Zemlin, F.; Beckmann, E.; Downing, K.H. Model for the structure of bacteriorhodopsin based on high-resolution electron cryo-microscopy. *J. Mol. Biol.* **1990**, *213*, 899–929. [CrossRef]
175. Mitsuoka, K.; Murata, K.; Walz, T.; Hirai, T.; Agre, P.; Heymann, J.B.; Engel, A.; Fujiyoshi, Y. The structure of aquaporin-1 at 4.5-Å resolution reveals short α- helices in the center of the monomer. *J. Struct. Biol.* **1999**, *128*, 34–43. [CrossRef]
176. Kühlbrandt, W.; Wang, D.N.; Fujiyoshi, Y. Atomic model of plant light-harvesting complex by electron crystallography. *Nature* **1994**, *367*, 614–621. [CrossRef]

177. Hutchings, J.; Stancheva, V.; Miller, E.A.; Zanetti, G. Subtomogram averaging of COPII assemblies reveals how coat organization dictates membrane shape. *Nat. Commun.* **2018**, *9*, 4154. [CrossRef] [PubMed]
178. Kudryashev, M.; Castaño-Díez, D.; Deluz, C.; Hassaine, G.; Grasso, L.; Graf-Meyer, A.; Vogel, H.; Stahlberg, H. The Structure of the mouse serotonin 5-HT3 receptor in lipid vesicles. *Structure* **2016**, *24*, 165–170. [CrossRef] [PubMed]
179. Castaño-Díez, D.; Zanetti, G. *In situ* structure determination by subtomogram averaging. *Curr. Opin. Struct. Biol.* **2019**, *58*, 68–75. [CrossRef] [PubMed]
180. Opella, S.J. Solid-state NMR and membrane proteins. *J. Magn. Reson.* **2015**, *253*, 129–137. [CrossRef] [PubMed]
181. Schubeis, T.; Le Marchand, T.; Andreas, L.B.; Pintacuda, G. ^{1}H magic-angle spinning NMR evolves as a powerful new tool for membrane proteins. *J. Magn. Reson.* **2018**, *287*, 140–152. [CrossRef] [PubMed]
182. Lange, A.; Giller, K.; Hornig, S.; Martin-Eauclaire, M.F.; Pongs, O.; Becker, S.; Baldus, M. Toxin-induced conformational changes in a potassium channel revealed by solid-state NMR. *Nature* **2006**, *440*, 959–962. [CrossRef]
183. Jekhmane, S.; Medeiros-Silva, J.; Li, J.; Kümmerer, F.; Müller-Hermes, C.; Baldus, M.; Roux, B.; Weingarth, M. Shifts in the selectivity filter dynamics cause modal gating in K^+ channels. *Nat. Commun.* **2019**, *10*, 1–12. [CrossRef]
184. Saurel, O.; Iordanov, I.; Nars, G.; Demange, P.; Le Marchand, T.; Andreas, L.B.; Pintacuda, G.; Milon, A. Local and global dynamics in *Klebsiella pneumoniae* outer membrane protein a in lipid bilayers probed at atomic resolution. *J. Am. Chem. Soc.* **2017**, *139*, 1590–1597. [CrossRef]
185. Retel, J.S.; Nieuwkoop, A.J.; Hiller, M.; Higman, V.A.; Barbet-Massin, E.; Stanek, J.; Andreas, L.B.; Franks, W.T.; Van Rossum, B.J.; Vinothkumar, K.R.; et al. Structure of outer membrane protein G in lipid bilayers. *Nat. Commun.* **2017**, *8*, 1–10. [CrossRef]
186. Bippes, C.A.; Müller, D.J. High-resolution atomic force microscopy and spectroscopy of native membrane proteins. *Rep. Prog. Phys.* **2011**, *74*, 086601. [CrossRef]
187. Engel, A.; Gaub, H.E. Structure and mechanics of membrane proteins. *Annu. Rev. Biochem.* **2008**, *77*, 127–148. [CrossRef] [PubMed]
188. Engel, A.; Schoenenberger, C.-A.; Müller, D.J. High resolution imaging of native biological sample surfaces using scanning probe microscopy. *Curr. Opin. Struct. Biol.* **1997**, *7*, 279–284. [CrossRef]
189. Frederix, P.L.T.M.; Bosshart, P.D.; Engel, A. Atomic force microscopy of biological membranes. *Biophys. J.* **2009**, *96*, 329–338. [CrossRef] [PubMed]
190. Dufrêne, Y.F.; Ando, T.; Garcia, R.; Alsteens, D.; Martinez-Martin, D.; Engel, A.; Gerber, C.; Müller, D.J. Imaging modes of atomic force microscopy for application in molecular and cell biology. *Nat. Nanotechnol.* **2017**, *12*, 295–307. [CrossRef]
191. Mari, S.A.; Pessoa, J.; Altieri, S.; Hensen, U.; Thomas, L.; Morais-Cabral, J.H.; Müller, D.J. Gating of the MlotiK1 potassium channel involves large rearrangements of the cyclic nucleotide-binding domains. *Proc. Natl. Acad. Sci. USA* **2011**, *108*, 20802–20807. [CrossRef]
192. Philippsen, A.; Im, W.; Engel, A.; Schirmer, T.; Roux, B.; Müller, D.J. Imaging the electrostatic potential of transmembrane channels: Atomic probe microscopy of OmpF porin. *Biophys. J.* **2002**, *82*, 1667–1676. [CrossRef]
193. Pfreundschuh, M.; Hensen, U.; Müller, D. Quantitative imaging of the electrostatic field and potential generated by a transmembrane protein pore at subnanometer resolution. *Nano Lett.* **2013**, *13*, 5585–5593. [CrossRef]
194. Yamashita, H.; Voïtchovsky, K.; Uchihashi, T.; Contera, S.A.; Ryan, J.F.; Ando, T. Dynamics of bacteriorhodopsin 2D crystal observed by high-speed atomic force microscopy. *J. Struct. Biol.* **2009**, *167*, 153–158. [CrossRef]
195. Shibata, M.; Yamashita, H.; Uchihashi, T.; Kandori, H.; Ando, T. High-speed atomic force microscopy shows dynamic molecular processes in photoactivated bacteriorhodopsin. *Nat. Nanotechnol.* **2010**, *5*, 208–212. [CrossRef]
196. Oesterhelt, F.; Oesterhelt, D.; Pfeiffer, M.; Engel, A.; Gaub, H.E.; Müller, D.J. Unfolding pathways of individual bacteriorhodopsins. *Science* **2000**, *288*, 143–146. [CrossRef]
197. Kedrov, A.; Janovjak, H.; Sapra, K.T.; Müller, D.J. Deciphering molecular interactions of native membrane proteins by single-molecule force spectroscopy. *Annu. Rev. Biophys. Biomol. Struct.* **2007**, *36*, 233–260. [CrossRef] [PubMed]
198. Thoma, J.; Sapra, K.T.; Müller, D.J. Single-molecule force spectroscopy of transmembrane β-barrel proteins. *Annu. Rev. Anal. Chem.* **2018**, *11*, 375–395. [CrossRef] [PubMed]
199. Kessler, M.; Gottschalk, K.E.; Janovjak, H.; Müller, D.J.; Gaub, H.E. Bacteriorhodopsin folds into the membrane against an external force. *J. Mol. Biol.* **2006**, *357*, 644–654. [CrossRef] [PubMed]
200. Thoma, J.; Burmann, B.M.; Hiller, S.; Müller, D.J. Impact of holdase chaperones Skp and SurA on the folding of β-barrel outer-membrane proteins. *Nat. Struct. Mol. Biol.* **2015**, *22*, 795–802. [CrossRef]
201. Serdiuk, T.; Balasubramaniam, D.; Sugihara, J.; Mari, S.A.; Kaback, H.R.; Müller, D.J. YidC assists the stepwise and stochastic folding of membrane proteins. *Nat. Chem. Biol.* **2016**, *12*, 911–917. [CrossRef]
202. Alsteens, D.; Pfreundschuh, M.; Zhang, C.; Spoerri, P.M.; Coughlin, S.R.; Kobilka, B.K.; Müller, D.J. Imaging G protein-coupled receptors while quantifying their ligand-binding free-energy landscape. *Nat. Methods* **2015**, *1*, 12103–12108. [CrossRef]
203. Harsman, A.; Niemann, M.; Pusnik, M.; Schmidt, O.; Burmann, B.M.; Hiller, S.; Meisinger, C.; Schneider, A.; Wagner, R. Bacterial origin of a mitochondrial outer membrane protein translocase: New perspectives from comparative single channel electrophysiology. *J. Biol. Chem.* **2012**, *287*, 31437–31445. [CrossRef]
204. Hilty, C.; Winterhalter, M. Facilitated substrate transport through membrane proteins. *Phys. Rev. Lett.* **2001**, *86*, 5624–5627. [CrossRef]

205. Wang, J.; Terrasse, R.; Bafna, J.A.; Benier, L.; Winterhalter, M. Electrophysiological characterization of transport across outer-membrane channels from gram-negative bacteria in presence of lipopolysaccharides. *Angew. Chem. Int. Ed.* **2020**, *59*, 8517–8521. [CrossRef]
206. Kreir, M.; Farre, C.; Beckler, M.; George, M.; Fertig, N. Rapid screening of membrane protein activity: Electrophysiological analysis of OmpF reconstituted in proteoliposomes. *Lab Chip* **2008**, *8*, 587–595. [CrossRef]
207. Zeidel, M.L.; Ambudkar, S.V.; Smith, B.L.; Agre, P. Reconstitution of functional water channels in liposomes containing purified red cell CHIP28 protein. *Biochemistry* **1992**, *31*, 7436–7440. [CrossRef] [PubMed]
208. Posson, D.J.; Rusinova, R.; Andersen, O.S.; Nimigean, C.M. Stopped-flow fluorometric ion flux assay for ligand-gated ion channel studies. *Methods Mol. Biol.* **2018**, *1684*, 223–235. [CrossRef] [PubMed]
209. Nishimura, T.; Akiyoshi, K. Biotransporting biocatalytic reactors toward therapeutic nanofactories. *Adv. Sci.* **2018**, *5*. [CrossRef] [PubMed]
210. Karlsson, M.; Davidson, M.; Karlsson, R.; Karlsson, A.; Bergenholtz, J.; Konkoli, Z.; Jesorka, A.; Lobovkina, T.; Hurtig, J.; Voinova, M.; et al. Biomimetic nanoscale reactors and networks. *Annu. Rev. Phys. Chem.* **2004**, *55*, 613–649. [CrossRef] [PubMed]
211. Huang, Y.Y.; Sharma, S.K.; Dai, T.; Chung, H.; Yaroslavsky, A.; Garcia-Diaz, M.; Chang, J.; Chiang, L.Y.; Hamblin, M.R. Can nanotechnology potentiate photodynamic therapy? *Nanotechnol. Rev.* **2012**, *1*, 111–146. [CrossRef] [PubMed]
212. Discher, D.E.; Ahmed, F. Polymersomes. *Annu. Rev. Biomed. Eng.* **2006**, *8*, 323–341. [CrossRef]
213. Mecke, A.; Dittrich, C.; Meier, W. Biomimetic membranes designed from amphiphilic block copolymers. *Soft Matter* **2006**, *2*, 751–759. [CrossRef]
214. Zhang, X.; Tanner, P.; Graff, A.; Palivan, C.G.; Meier, W. Mimicking the cell membrane with block copolymer membranes. *J. Polym. Sci. Part A Polym. Chem.* **2012**, *50*, 2293–2318. [CrossRef]
215. Poschenrieder, S.T.; Schiebel, S.K.; Castiglione, K. Stability of polymersomes with focus on their use as nanoreactors. *Eng. Life Sci.* **2018**, *18*, 101–113. [CrossRef]
216. Palivan, C.G.; Goers, R.; Najer, A.; Zhang, X.; Car, A.; Meier, W. Bioinspired polymer vesicles and membranes for biological and medical applications. *Chem. Soc. Rev.* **2016**, *45*, 377–411. [CrossRef]
217. Kowal, J.; Wu, D.; Mikhalevich, V.; Palivan, C.G.; Meier, W. Hybrid polymer-lipid films as platforms for directed membrane protein insertion. *Langmuir* **2015**, *31*, 4868–4877. [CrossRef] [PubMed]
218. Thoma, J.; Belegrinou, S.; Rossbach, P.; Grzelakowski, M.; Kita-Tokarczyk, K.; Meier, W. Membrane protein distribution in composite polymer–lipid thin films. *Chem. Commun.* **2012**, *48*, 8811. [CrossRef] [PubMed]
219. Kaback, H.R. The role of the phosphoenolpyruvate-phosphotransferase system in the transport of sugars by isolated membrane preparations of *Escherichia coli*. *J. Biol. Chem.* **1968**, *243*, 3711–3724. [PubMed]
220. Kaback, H.R. Transport studies in bacterial membrane vesicles. *Science* **1974**, *186*, 882–892. [CrossRef] [PubMed]
221. Henderson, R.; Unwin, P.N.T. Three-dimensional model of purple membrane obtained by electron microscopy. *Nature* **1975**, *257*, 28–32. [CrossRef]
222. Baldwin, J.; Henderson, R. Measurement and evaluation of electron diffraction patterns from two-dimensional crystals. *Ultramicroscopy* **1984**, *14*, 319–335. [CrossRef]
223. Glaeser, R.M.; Jubb, J.S.; Henderson, R. Structural comparison of native and deoxycholate-treated purple membrane. *Biophys. J.* **1985**, *48*, 775–780. [CrossRef]
224. Hoenger, A.; Ghosh, R.; Schoenenberger, C.A.; Aebi, U.; Engel, A. Direct *in situ* structural analysis of recombinant outer membrane porins expressed in an OmpA-deficient mutant *Escherichia coli* strain. *J. Struct. Biol.* **1993**, *111*, 212–221. [CrossRef]
225. Briggs, J.A.G. Structural biology *in situ*-the potential of subtomogram averaging. *Curr. Opin. Struct. Biol.* **2013**, *23*, 261–267. [CrossRef]
226. Galaz-Montoya, J.G.; Ludtke, S.J. The advent of structural biology *in situ* by single particle cryo-electron tomography. *Biophys. Rep.* **2017**, *3*, 17–35. [CrossRef]
227. Gold, V.A.M.; Ieva, R.; Walter, A.; Pfanner, N.; Van Der Laan, M.; Kühlbrandt, W. Visualizing active membrane protein complexes by electron cryotomography. *Nat. Commun.* **2014**, *5*. [CrossRef] [PubMed]
228. Nans, A.; Kudryashev, M.; Saibil, H.R.; Hayward, R.D. Structure of a bacterial type III secretion system in contact with a host membrane *in situ*. *Nat. Commun.* **2015**, *6*. [CrossRef] [PubMed]
229. Shi, X.; Chen, M.; Yu, Z.; Bell, J.M.; Wang, H.; Forrester, I.; Villarreal, H.; Jakana, J.; Du, D.; Luisi, B.F.; et al. *In situ* structure and assembly of the multidrug efflux pump AcrAB-TolC. *Nat. Commun.* **2019**, *10*, 4–9. [CrossRef] [PubMed]
230. Ghosal, D.; Chang, Y.; Jeong, K.C.; Vogel, J.P.; Jensen, G.J. *In situ* structure of the *Legionella* Dot/Icm type IV secretion system by electron cryotomography. *EMBO Rep.* **2017**, *18*, 726–732. [CrossRef]
231. Rapisarda, C.; Cherrak, Y.; Kooger, R.; Schmidt, V.; Pellarin, R.; Logger, L.; Cascales, E.; Pilhofer, M.; Durand, E.; Fronzes, R. In situ and high-resolution cryo- EM structure of a bacterial type VI secretion system membrane complex. *EMBO J.* **2019**, *38*, 1–18. [CrossRef]
232. Zeev-Ben-Mordehai, T.; Vasishtan, D.; Siebert, C.A.; Whittle, C.; Grünewald, K. Extracellular vesicles: A platform for the structure determination of membrane proteins by cryo-EM. *Structure* **2014**, *22*, 1687–1692. [CrossRef]
233. Ward, M.E.; Wang, S.; Munro, R.; Ritz, E.; Hung, I.; Gor'Kov, P.L.; Jiang, Y.; Liang, H.; Brown, L.S.; Ladizhansky, V. *In situ* structural studies of anabaena sensory rhodopsin in the *E. coli* membrane. *Biophys. J.* **2015**, *108*, 1683–1696. [CrossRef]

234. Pinto, C.; Mance, D.; Julien, M.; Daniels, M.; Weingarth, M.; Baldus, M. Studying assembly of the BAM complex in native membranes by cellular solid-state NMR spectroscopy. *J. Struct. Biol.* **2019**, *206*, 1–11. [CrossRef]
235. Baker, L.A.; Sinnige, T.; Schellenberger, P.; de Keyzer, J.; Siebert, C.A.; Driessen, A.J.M.; Baldus, M.; Grünewald, K. Combined ^{1}H-detected solid-state NMR spectroscopy and electron cryotomography to study membrane proteins across resolutions in native environments. *Structure* **2018**, *26*, 161–170. [CrossRef]
236. Müller, D. AFM: A Nanotool in membrane biology. *Biochemistry* **2008**, *47*, 7986–7998. [CrossRef]
237. Liu, L.N.; Scheuring, S. High-resolution AFM imaging of native biological membranes. *Nanoscale Liq. Interfaces* **2013**, 657–681. [CrossRef]
238. Worcester, D.L.; Miller, R.G.; Bryant, P.J. Atomic force microscopy of purple membranes. *J. Microsc.* **1988**, *152*, 817–821. [CrossRef] [PubMed]
239. Müller, D.J.; Schabert, F.A.; Büldt, G.; Engel, A. Imaging purple membranes in aqueous solutions at sub-nanometer resolution by atomic force microscopy. *Biophys. J.* **1995**, *68*, 1681–1686. [CrossRef]
240. Buzhynskyy, N.; Hite, R.K.; Walz, T.; Scheuring, S. The supramolecular architecture of junctional microdomains in native lens membranes. *EMBO Rep.* **2007**, *8*, 51–55. [CrossRef] [PubMed]
241. Gonçalves, R.P.; Buzhynskyy, N.; Prima, V.; Sturgis, J.N.; Scheuring, S. Supramolecular assembly of VDAC in native mitochondrial outer membranes. *J. Mol. Biol.* **2007**, *369*, 413–418. [CrossRef]
242. Fotiadis, D.; Liang, Y.; Filipek, S.; Saperstein, D.A.; Engel, A.; Palczewski, K. Atomic-force microscopy: Rhodopsin dimers in native disc membranes. *Nature* **2003**, *421*, 127–128. [CrossRef]
243. Phuthong, W.; Huang, Z.; Wittkopp, T.M.; Sznee, K.; Heinnickel, M.L.; Dekker, J.P.; Frese, R.N.; Prinz, F.B.; Grossman, A.R. The use of contact mode atomic force microscopy in aqueous medium for structural analysis of spinach photosynthetic complexes. *Plant Physiol.* **2015**, *169*, 1318–1332. [CrossRef]
244. Thoma, J.; Manioglu, S.; Kalbermatter, D.; Bosshart, P.D.; Fotiadis, D.; Müller, D.J. Protein-enriched outer membrane vesicles as a native platform for outer membrane protein studies. *Commun. Biol.* **2018**, *1*, 23. [CrossRef]
245. Thoma, J.; Sun, Y.; Ritzmann, N.; Müller, D.J. POTRA domains, extracellular lid, and membrane composition modulate the conformational stability of the β-barrel assembly factor BamA. *Structure* **2018**, *26*, 987–996. [CrossRef]

International Journal of
Molecular Sciences

MDPI

Review

Assessing the Role of Lipids in the Molecular Mechanism of Membrane Proteins

Léni Jodaitis †, Thomas van Oene † and Chloé Martens *

Center for Structural Biology and Bioinformatics, Université Libre de Bruxelles, 1050 Brussels, Belgium; leni.jodaitis@ulb.be (L.J.); thomas.van.oene@ulb.be (T.v.O.)
* Correspondence: chloe.martens@ulb.be
† These authors contributed equally.

Abstract: Membrane proteins have evolved to work optimally within the complex environment of the biological membrane. Consequently, interactions with surrounding lipids are part of their molecular mechanism. Yet, the identification of lipid–protein interactions and the assessment of their molecular role is an experimental challenge. Recently, biophysical approaches have emerged that are compatible with the study of membrane proteins in an environment closer to the biological membrane. These novel approaches revealed specific mechanisms of regulation of membrane protein function. Lipids have been shown to play a role in oligomerization, conformational transitions or allosteric coupling. In this review, we summarize the recent biophysical approaches, or combination thereof, that allow to decipher the role of lipid–protein interactions in the mechanism of membrane proteins.

Keywords: membrane protein; lipid–protein interaction; cryo-electron microscopy; hydrogen–deuterium exchange mass spectrometry; native mass spectrometry; single-molecule Förster resonance energy transfer; double electron–electron resonance; native mass spectrometry

Citation: Jodaitis, L.; van Oene, T.; Martens, C. Assessing the Role of Lipids in the Molecular Mechanism of Membrane Proteins. *Int. J. Mol. Sci.* **2021**, *22*, 7267. https://doi.org/10.3390/ijms22147267

Academic Editor: Masoud Jelokhani-Niaraki

Received: 31 May 2021
Accepted: 1 July 2021
Published: 6 July 2021

1. Introduction

Integral membrane proteins (IMPs) comprise an important part of the known human proteome [1] and have long been under scrutiny because of their role in fundamental metabolic processes, making them attractive drug targets [2]. IMPs are responsible for transmembrane transport, signaling, cell homeostasis and energy transduction, and this enumeration is not exhaustive. They are challenging to study at the experimental level because of their location, embedded in the biological membrane. Detergent solubilization is widely used to extract IMPs from the membrane but this procedure usually strips IMPs from their molecular partners: lipids. The surrounding lipids are an integral part of the machinery that allows a membrane protein to perform its function. Functional modulation of membrane proteins by lipids was established more than thirty years ago [3,4]. Membrane bulk properties, such as fluidity and lateral pressure, have been shown to affect the function of transporters, channels and other IMPs [5–7]. Several methods are available to measure how a specific lipid species or lipid composition affects the function of IMPs. For example, the activity of channels can be determined with electrophysiology measurements in planar lipid bilayers, and that of transporters can be assessed using reconstitution in proteoliposomes with an ionic gradient [8]. However, the molecular details of this functional modulation are a challenge to capture. The progress achieved in structural biology of membrane proteins in the last 20 years [9] yielded snapshots of lipids within high-resolution structures. These observations have triggered a renewed interest into the role of lipid–protein interactions. Other methods are now meeting the challenge to pinpoint important lipid–protein interactions and decipher their role at the molecular level. Commonly used methods to study lipid–protein interactions at the molecular level are X-ray crystallography and cryogenic electron microscopy (cryo-EM) for structural resolution, nuclear magnetic resonance (NMR), electron paramagnetic resonance (EPR), single-molecule Förster Resonance Energy Transfer (sm-FRET) and hydrogen–deuterium exchange mass spectrometry (HDX-MS), to study their dynamics, and native MS for identification [10] (Table 1). These techniques often rely on the use of a membrane mimic to study the protein of interest within a lipid bilayer. Several reconstitution systems are available for this purpose, such as proteoliposomes, bicelles, nanodiscs [11] and, more recently, saposin discs [12] and peptidiscs [13,14]. The use of styrene maleic-acid lipid particles (SMALPs)—also called 'native nanodiscs' or lipodisqs—is also popular since it retains the native lipid composition of the IMP under study [15]. The pros and cons of each of these mimics have been reviewed elsewhere [16].

In this review, we give an overview of how lipid–protein interactions can shape the mechanisms of IMPs at the molecular level, and the experimental approaches to study these interactions. Such modulation of the molecular mechanism can happen in various ways, but specific mechanisms have now been observed and confirmed; lipids can bind at the interface between proteins and promote or hinder their oligomerization. They can stabilize specific structural intermediates and shift the conformational equilibrium between conformers. They can directly (competition) or indirectly (allostery) modulate ligand binding (Figure 1). A combination of all these mechanisms is likely happening in any living cell. The challenge lies in disentangling these effects experimentally. This review is structured in three sections that each presents a mechanism of lipid–protein interaction, namely, oligomerization, conformational regulation and modulation of ligand binding. Each section summarizes recent studies and details the methods used to identify the role of lipid–protein interactions in the mechanism of the protein(s) under study. This review is not extensive but aims to illustrate the complexity of lipid–protein interactions with select examples highlighting the pros and cons of different experimental approaches. The studies presented here feature mostly three biophysical techniques that have changed the landscape of structural biology in the last five years: cryo-EM, structural MS and single-molecule FRET. Research works using more established structural techniques (e.g., X-ray and NMR) will be mentioned rather than detailed. Other excellent reviews on lipid–protein interactions have covered these aspects [17–19]. Similarly, and for the sake of conciseness, this review focuses on experimental approaches and will not detail the advances enabled by molecular dynamics (MD) simulations. Select examples of studies combining MD simulations and biophysical approaches will be illustrated in the following sections. An excellent and comprehensive review by Corradi and colleagues summarizes how MD simulations have provided countless insights into the molecular role of lipids [20].

Table 1. Representative biophysical techniques used to study the lipid–protein interactions of IMPs.

Biophysical Techniques		Information Obtained
X-ray crystallography extrapolates the position and arrangement of atoms in single crystals from the diffraction pattern of X-ray.	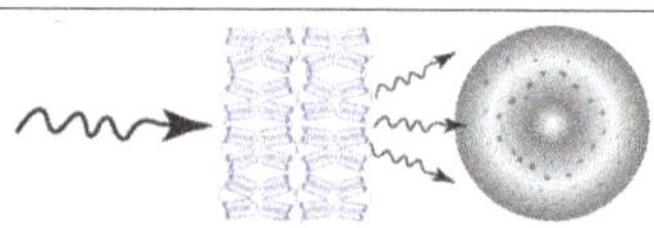	High-resolution structure.
Cryo-electron microscopy images flash frozen protein solutions with an electron beam. Single particles are aligned and classified in two-dimensional classes. The 3D structure is determined by reconstruction software.	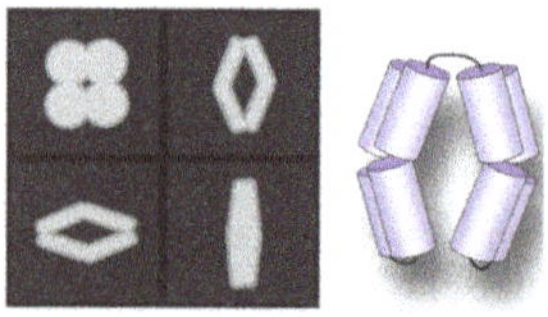	High-resolution structure.

Table 1. *Cont.*

Biophysical Techniques		Information Obtained
Native mass spectrometry (nMS) retains in the gas-phase non-covalent interactions pre-existing in solution. Protein-protein and protein-ligand complexes can be observed.	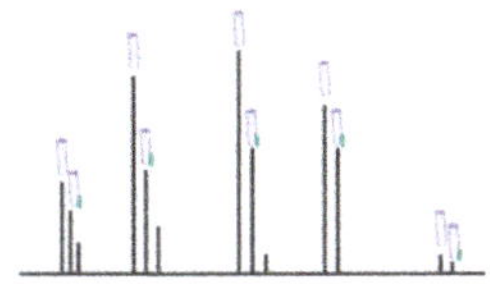	Protein–protein interactions, protein–ligand interactions.
Hydrogen–deuterium exchange (HDX) MS measures the rate of deuterium exchange of labile protons of the backbone amide. The exchange depends on stability of the H-bond and solvent accessibility.	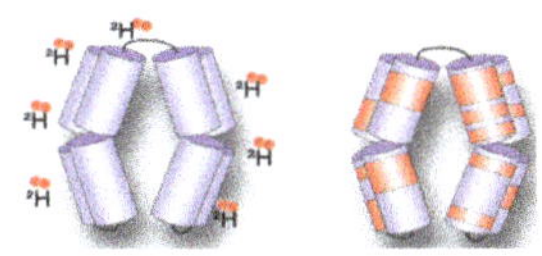	Structural dynamics, protein–protein interactions, protein–ligand interactions.
Single-molecule FRET measures the amount of energy transfer between a pair of fluorophores, a donor and an acceptor to perform small distances measurements (1–10 nanometers).	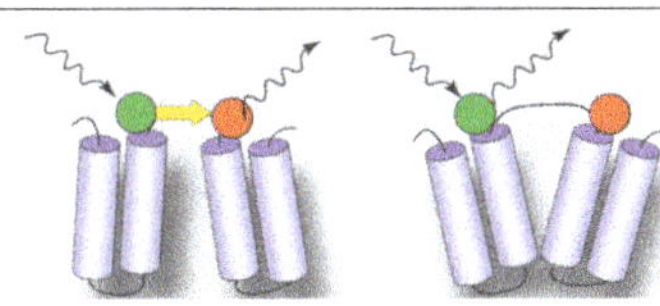	Conformational changes, kinetics.
Double electron–electron resonance measures the dipolar spin coupling between a pair of spin labels. The modulated spin echo contains distance information.	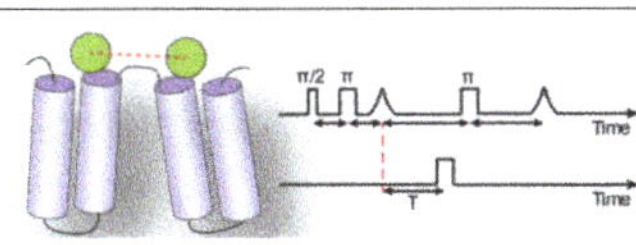	Conformational changes.
Electron paramagnetic resonance (cw-EPR) provides a read-out of the environment and the mobility of a paramagnetic probe within a protein, typically a nitroxide spin label.	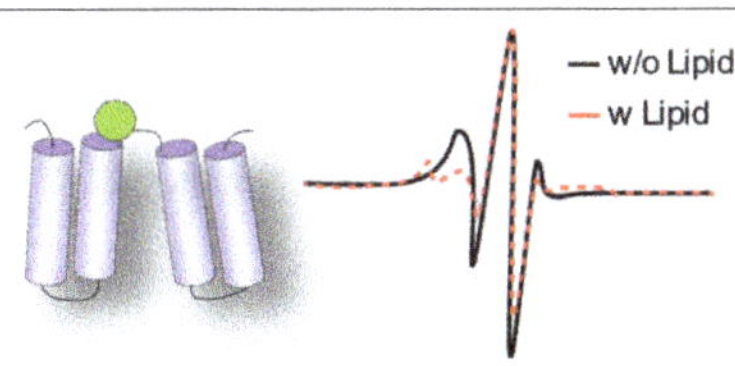	Mobility, environment, tertiary fold.
Multidimensional nuclear magnetic resonance (NMR) provides details about the local molecular environment of the nuclei of an isotopically labelled protein, allowing to deduce chemical bonds, distance and relative motions between nuclei.	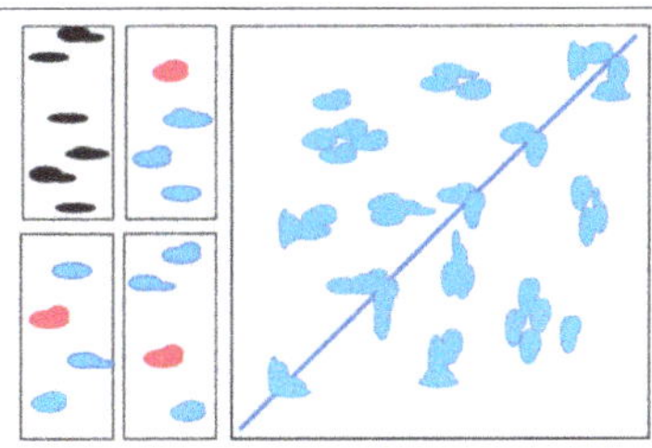	High-resolution structure, dynamics, protein–protein interactions, protein–ligand interactions.

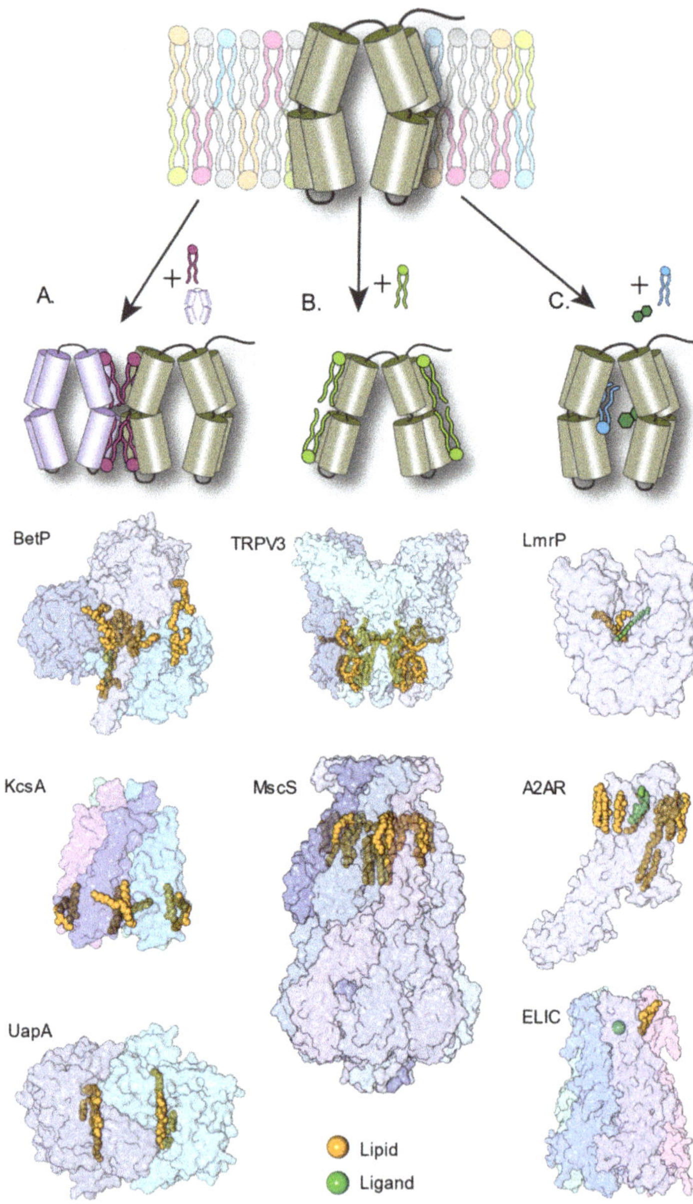

Figure 1. Schematic representation of known molecular mechanisms mediated by lipids and representative examples of lipid–protein complexes captured in high-resolution structures (**A**). Lipids as regulators of protein oligomerization and complex assembly. Cartoon representation of BetP (PDB ID 4C7R), KcsA (PDB ID 3IFX) and UapA (PDB ID 5I6C) proteins. (**B**). Lipid regulating conformational dynamics. Cartoon representation of TRPV3 (PDB ID 6LGP) and MscS (PDB ID 6PWN) proteins (**C**). Lipids modulating ligand binding directly (competition) or indirectly (allostery). Cartoon representation of LmrP (PDB ID 6T1Z), A2AR (PDB ID 5IUA) and ELIC (PDB ID 6HJX) proteins. PDB figures were generated using 'The Protein Imager' [21].

2. Lipids as Regulators of Protein Oligomerization and Complex Assembly

The role of lipids in the oligomerization of membrane protein complexes is well known [22,23] but the methods to observe such a mechanism vary widely. High-resolution structures of oligomeric IMPs occasionally capture lipids at the interface of the monomeric unit of oligomers [24–33]. In 1998, a crystal structure of bacteriorhodopsin revealed a lipid molecule between each monomer as well as a lipid patch in the central cavity [28]. In 2002, the structure of KcsA, a homotetrameric channel, was solved by X-ray crystallography and displayed the presence of a lipid between each monomer [34] (Figure 1). In 2013, BetP, an osmotic stress-regulated betaine transporter, was crystallized in complex with anionic lipids [27]. Central lipids bound within the trimer suggest that lipids are used as cofactors during trimeric assembly (Figure 1). These three IMPs belong to vastly different families and have different architectures, pointing towards the ubiquity of such phenomena. Most structures of the membrane proteins solved by X-ray crystallography are carried out in the presence of a detergent, which often disrupts oligomeric assemblies. Detergent molecules also compete for lipid-binding sites, hindering native lipid binding and potentially disrupting oligomers.

These limitations are now overcome with the advent of cryo-EM, that allows structure resolution in membrane mimics. For example, a recent study by Sun and colleagues illustrates the role of lipids in maintaining the multimeric assembly of ACIII [35], an actor of the photosynthetic electron transport chains of many bacteria. The whole complex was purified using SMALPs to retain the native lipids environment. The authors obtained a structure of the complex with eleven ordered phospholipid molecules. Most of the eleven lipids are found at protein–protein interfaces, between ActC and ActF subunits. The structure also revealed a triacylated cysteine residue at the N terminus of ActB, with two other phospholipids located just next to this lipid anchor. The authors hypothesize that the lipids are necessary to keep the complex together and in the membrane, a suggestion supported by MD simulations. Another multi-unit complex, the F_0/F_1 ATP synthase from *E. coli*, also presents lipids located at specific regions. Sobti and colleagues identify lipid densities bridging the rotor stator interface and speculate that the lipids have a role in maintaining an energetically optimal subunit arrangement [36]. The TRPV1 structure resolved in nanodiscs in the presence of a bivalent tarantula toxin reveals a tripartite complex wherein membrane lipids form bridging interactions between the toxin and channel. Hydrophobic fingers of the toxin insert through the outer leaflet of the bilayer [37]. Thus, lipids can really act as a glue that keeps a complex IMP together. As more and more high-resolution structures in membrane mimics are solved, it can reasonably be expected to observe more of such interactions.

Another technique that has helped revealing the role of lipids in oligomerization of IMPs is native mass spectrometry [38–41] (nMS). nMS retains non-covalent interactions of biological complexes pre-existing in solution in the gas phase. It is ideally suited to observe protein–lipid complexes, without the need for a high-resolution structure. Using this technique, Gupta and colleagues demonstrated elegantly the direct role of lipids in regulating oligomerization [38]. The authors ranked all known alpha-helical IMPs structures based on their computationally predicted oligomeric stability. They then performed nMS experiments on dimeric proteins predicted to have a weak oligomeric interface, LeuT and NhaA. The authors observed that the measured mass of the dimeric form was systematically higher than the summed mass of the two monomers. Blasting the complex apart in a mass spectrometer revealed that the mass excess was caused by cardiolipin (CL) sticking to the dimer. The authors then combined mutagenesis on the oligomeric interface, delipidation and relipidation experiments and the use of a CL-deficient expression strain to demonstrate that cardiolipin is required for dimerization of LeuT and NhaA. Another study that compared the role of lipids on the mechanism of sodium antiporters from different hosts confirmed the requirement of CL for NhaA dimerization [42]. A systematic nMS study by Reading et al. explored the effects of different lipid species on the oligomerization properties of the mechanosensitive channel MscL [43]. The equilibrium between pentameric and tetrameric oligomers was mostly dependent on the detergent used for nMS studies, but lipids fine-tuned such dependence. It is interesting to note that some lipids could act either as a stabilizer of the native pentameric form or destabilizer favoring lower oligomeric forms, depending on the temperature used for lipid incubation and the detergent used for transfer into the gas phase. This dual effect illustrates the exquisite sensitivity to the variables regulating oligomerization. In another nMS study, Pyle et al. established the requirement for specific lipids for the formation of a functional dimer of the eukaryotic purine transporter UapA [41]. nMS measurements of a delipidated UapA showed that the transporter was mainly present in its monomeric form. The restoration of the dimeric form was obtained by addition of the phospholipids phosphatidylethanolamine (PE) and phosphatidylinositol (PI) (Figure 2). The authors observed an additive effect of the two different phospholipids. In contrast, addition of phosphatidylcholine disrupts the dimeric interface, demonstrating the specificity of the mechanism. Another example of lipid-mediated dimerization is that of the transporter NHE9. The authors used a combination of cryo-EM structural resolution, nMS measurements, functional assays and MD simulations. They hypothesize that phosphoinositides bridge the monomer not at the dimer interface but just above, through interaction with a loop conserved through the same protein family [44].

A recent study discovered the role of negatively charged phospholipid phosphatidylinositol-4,5-biphosphate (PIP_2) into oligomerization of the serotonine transporter SERT without relying on nMS nor structural resolution [45]. Tissue-dependent oligomerization of SERT had been observed previously [46] and another study had identified a functional modulation of SERT by PIP_2 [47]. The authors used a specific technique of single-molecule fluorescence developed in their laboratory to look at the oligomerization kinetics of SERT in vivo, in the plasma membrane and the endoplasmic reticulum (ER). The plasma membrane contains about 1% of PIP_2 but the lipid species is virtually absent from the ER membrane. The authors disrupt putative PIP_2 binding sites by mutagenesis of SERT. In parallel, they co-express the receptor with a PIP_2-specific lipase. By combining these two approaches, they can show that the serotonin receptor oligomeric state is correlated with PIP_2 availability. These research works and others [48,49] demonstrated that lipids can be part of the subunit interface, and directly stabilize the complex assembly. These findings also explain a number of functional studies that had observed a lipid-dependent modulation of function [7,50].

While factors governing lipid-mediated oligomerization of IMPs are not yet understood, the research works presented here showcase both the versatility and importance of such interactions. Specifically, the use of nMS to determine lipid-induced oligomerization opens the door to studies of IMPs with an unknown structure. So far, nMS studies looking at the effect of lipids in oligomerization were carried out in mixed detergent–lipid micelles. Recently, the availability of MS instruments designed to bring bigger and more complex systems into the gas phase—coupled with dedicated software development for the analysis of IMPs embedded in nanodiscs (Unidec)—has facilitated nMS studies of proteins in nanodiscs [51,52] or SMALPs [53]. Such progress will lead to more research on lipid-mediated oligomerization in a detergent-free environment [54,55].

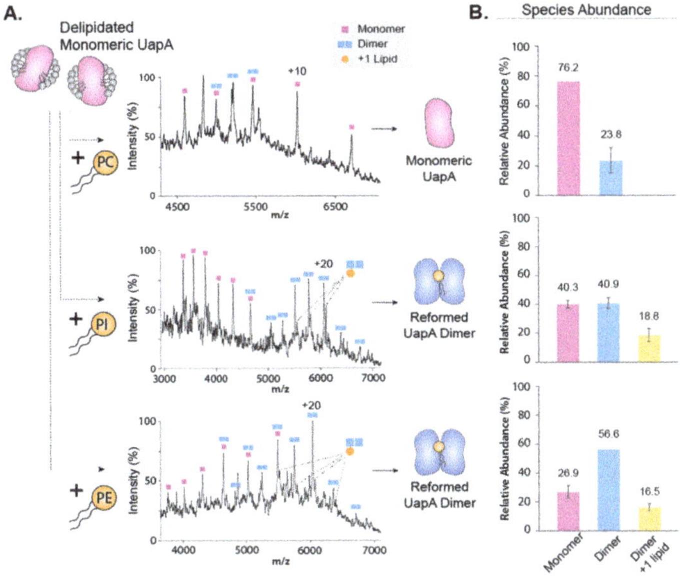

Figure 2. Addition of lipids to delipidated UapAG411V$_{\Delta 1\text{-}11}$ reforms the dimer. (**A**) Mass spectra showing the effects of PC (34:1, upper), PI (34:1, middle) and PE (34:1, lower) on the oligomerization of delipidated UapAG411V$_{\Delta 1\text{-}11}$. Peaks of the lipid-bound species are highlighted. (**B**) Relative abundances of monomer and dimer species in the presence of PC (upper), PI (middle) and PE (lower) were quantified using UniDec software [56]. Adapted with permission [41].

3. Lipids Stabilizing Conformational States

Membrane proteins are dynamic entities and have to adapt structurally during a functional cycle. For example, transporters will open to opposite sides of the membrane, whereas channels will open and close to provide a pathway for ions and GPCRs will transmit a transmembrane signal by switching between active and inactive states. These structural rearrangements happening within the membrane often imply an active role for the lipids, interacting directly or not with specific structural motifs to modulate the energy landscape. Different approaches have been used to experimentally determine the role of lipids in regulating conformational transitions. The most direct methods rely on distance measurements in membrane mimics or observation of high-resolution structures in different conformations. More indirect approaches measure the changes in solvent accessibility of the secondary structure elements or specific amino acids. Distance measurements depend on the introduction of probes, and such chemical modification of the protein under study may affect its function. In that regard, techniques aiming at looking at dynamics using accessibility are less invasive, and can be applied more broadly. Among these, solid-state NMR is now an established spectroscopic method to investigate the dynamics of membrane proteins in bilayers [57–59]. Its principles and applications are described in excellent reviews [60–62] and will not be discussed in the present work.

3.1. Measuring Distances

In the last five years, two techniques were mainly used to measure the nanometric distance changes that accompany conformational changes of IMPs: single-molecule Förster Resonance Energy Transfer (sm-FRET) and double electron–electron resonance (DEER also called PELDOR). Both rely on the introduction of a pair of probes on the protein, usually tethered chemically via a cysteine residue. Sm-FRET requires the introduction of a donor and an acceptor probe. The efficiency of energy transfer (E_{FRET}) to the acceptor after exciting the donor probe is distance-dependent. E_{FRET} is measured by detecting the fluorescence signal of both the donor and acceptor probes, using dedicated microscopes [63]. DEER uses a pair of identic spin labels, typically the thiol-specific probe MTSSL. DEER measurements extract the spin coupling between both labels, which is also distance-dependent [64]. Both techniques have pros and cons in terms of sensitivity, equipment needed, experimental setup and analysis of the signal, detailed in excellent reviews [65,66].

Among the diverse classes of IMPs, transporters are the target of choice for smFRET and DEER studies. The motions necessary to switch from inward open to outward open are large enough to be detectable (typically 15 to 60 angstroms). Furthermore, transporters often present structural conservation from prokaryotic to human homologs. Bacterial homologs can be used as proxies for structural studies of human counterparts, often harder to produce for biophysical studies. They are also often monomeric or dimeric, which facilitates the labeling strategy, compared to multimeric channels, for example. For these practical reasons, DEER [67–70] and smFRET [71–74] studies in membrane mimics are heavily biased towards transporters.

Liu and coworkers compared the conformational transitions of the lipid flippase MsbA in detergent micelles, nanodiscs and liposomes [74]. The authors looked at the dynamics of the nucleotide-binding domains (NBDs) and the transmembrane domains (TMDs) with smFRET, using a pair of probes on each side of the transporter. They could show that the dynamics of the TMDs are restricted in a lipid environment, while the dynamics of the NBDs are larger in proteoliposomes and detergent micelles compared to nanodiscs. The transition between different states is faster in liposomes compared to the other environments. A similar comparison between detergent micelles and the lipid environment of proteoliposomes was carried out for the peptide transporter McjD [71], using smFRET. The authors show that the trigger of the conformational transition is conserved regardless of the environment; both in detergent micelles and proteoliposomes, the addition of the peptide mccj25 opens the extracellular side, while the addition of ATP closes the NBDs. Similarly, the vitamin transporter BtuCD was studied by smFRET in nanodiscs and detergent micelles. The conformational dynamics of BtuCD confirmed the general trend: while the overall features of the transport cycle are conserved in detergent micelles and nanodiscs, the extent of the conformational changes differs [72]. A similar conclusion was reached in an extensive DEER study on the secondary multidrug transporter LmrP [68]. The authors performed distance measurements on the transporter in detergent micelles and nanodiscs and identified a specific structural motif that works as a protonation-dependent conformational switch [68]. They could show that the conformational states visited in the presence or absence of lipids are similar. However, the lipid environment modulates the pK_a of the conformational switch and facilitates a conformational transition at a physiological pH. To go further into their molecular characterization, they use synthetic phospholipids. They demonstrate that methylation of the phosphatidylethanolamine headgroup tips the conformational equilibrium towards the outward-open state. Such dependence on the chemistry of the membrane was also observed for the transporter DtpA [73]. The authors used the Salipro system to reconstitute the peptide transporter DtpA in soluble discoidal bilayers composed either of phosphatidylethanolamine POPE, phosphatidylserine POPS, phosphatidic acid POPA or brain lipids extract. They demonstrate that anionic lipids promote a 'very inward open state' not observed in crystal structures of DtpA homologs. It is worth noting that for both LmrP and DtpA, the synthetic lipids used for reconstitution have the same aliphatic tail. This suggests that the headgroup is the main driver of the change in the conformational equilibrium.

The importance of the lipid environment is well illustrated in a study by Jagessar et al. on the PfMATE multidrug exporter [75]. The isomerization between the outward-facing and inward-facing states is strictly dependent on the presence of a lipid bilayer. DEER distances measurements in detergent micelles show no isomerization upon ligand binding, but similar measurements in nanodiscs captured large-scale conformational changes upon protonation of PfMATE, indicative of the conformational transition (Figure 3). Another striking example of the importance of the lipid bilayer in stabilizing specific conformations comes from the DEER study of the multidrug transporter MdfA in nanodiscs. The lipid environment allowed to isolate a doubly occluded intermediate, where the periplasmic and cytoplasmic sides are both closed, compared to published crystals structures [70]. This observation led to the hypothesis that for this transporter, the state compatible with substrate binding is closed. Lipophilic substrates come from the membrane, in line with a model suggested for homologous transporter LmrP [76]. Another DEER study of the eukaryotic ABC transporter P-glycoprotein (P-gP) in nanodiscs shows that the high-energy post hydrolysis state of the transporter is stabilized by the lipid bilayer [69]. These DEER and smFRET studies on various transporters demonstrate that molecular level interaction between the membrane and the IMP can shape the conformational landscape.

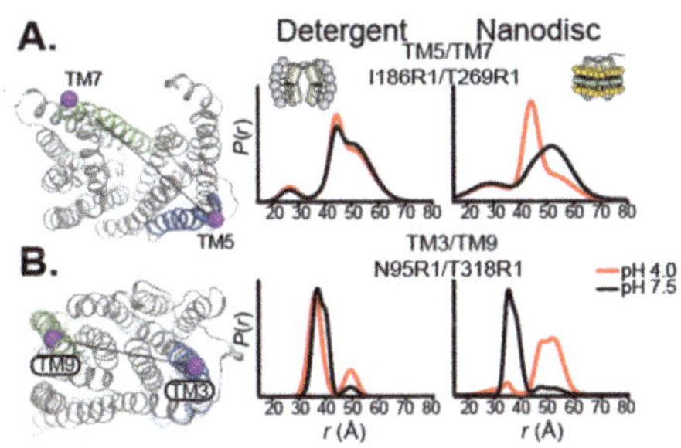

Figure 3. Ligand-dependent conformational dynamics of PfMATE require a lipid environment. (**A**) Representative spin label pair sampling distances between TM5 and TM7 on the extracellular side and (**B**) TM3 and TM9 on the intracellular side of PfMATE. The spin label locations are highlighted on the OF structure by purple spheres connected by a line. The helices targeted in the N-lobe and C-lobe are highlighted in blue and green, respectively. Distance distributions, representing the probability of a distance P(r) versus the distance (r) between spin labels, are shown in black traces at pH 7.5 and red traces at pH 4.0 in DDM micelles (**left**) and lipid nanodiscs (**right**). Adapted with permission from [75].

3.2. Looking at High-Resolution Structures

A direct way to look at how lipids modulate IMP conformations has become accessible thanks to cryo-EM. With a bit of luck, an IMP solved by cryo-EM in a lipid environment will reveal which lipids are bound where, and provide a direct read-out of their role in stabilizing a specific conformer. The technique also allows to solve different conformations present in a sample if the conformations are different enough to be sorted into different classes during the image processing workflow [77]. This has proven particularly fruitful for channels, whose important size (often >100 kDa) makes them easily amenable to cryo-EM structural resolution. In the last few years, an important number of channels has been solved in nanodiscs, such as thermo-sensitive channels (TRPM4 [26], TRPM1 [78], TRPV2 [79], TRPV3 [80,81], TRPV5 [82,83], TRPV6 [84] and TRPV1 [25]), mechanosensitive channels (MscS [85–87] and OSCA1.2 [88]), voltage-gated channels (TASK2 [89] and TPC1 [90]) and other channels, such as PKD2 [91] and TMEM16 [92]. Structures of a few reconstituted transporters have also been solved, such as the bacterial multidrug transporters AcrB [93] and TmrAB [94], and human multidrug transporter ABCG2 [95]. The majority of these structures have shown lipids bound to specific locations, leading to various hypotheses about their role. In the next paragraphs, we provide a few examples where the structural resolution has allowed to deduce a lipid-mediated mechanism of IMP function.

A telling example is that of the TRPV3 channel. Two research groups solved the structure of the channel in nanodiscs and their findings were published simultaneously [80,81]. Shimada and colleagues solved the structure of mouse TRPV3, while Deng and colleagues solved that of human TRPV3. Both teams captured the ligand-free, closed state. In the case of human TRPV3, a sensitization-prone mutant was used to obtain additional structures in the open and inactive states. In both studies, the structures deviate substantially from the previous ones obtained in the absence of a lipid environment [96–98]. The most noticeable change is that the pore in its closed state presents two constriction sites with a narrow selectivity filter, which is completely absent in the previous structures of the closed channel. In the mouse TRPV3, a phospholipid is observed in the pore. This lipid would be sterically clashing with the selectivity filter in the conducting pore, suggesting that the lipid stabilizes the narrow filter of the closed confirmation [81]. This lipid is not observed in the human TRPV3 structure, probably because the lipids used for nanodisc reconstitution are different. Both groups suggest that lipids are directly involved in the heat-sensing mechanism that activates the channel, as a sensor and relay of temperature changes.

Mechanosensitive channels were among the obvious targets for structure resolution in a lipid environment since their mechanism is directly related to lipid–protein interactions. These channels open in response to membrane tension, but whether the structural changes are caused by hydrophobic mismatch, changes in membrane curvature or anisotropic forces in the bilayer is debated [99]. Three groups solved the structure of the prototypical mechanosensitive channel MscS in nanodiscs by cryo-EM: Rasmussen et al. in 2019 [86], Reddy a few months later [85] and Zhang et al. in early 2021 [87]. The membrane-embedded MscS obtained by all three groups appears to be in the closed conformation, which is expected in the absence of tension. To test whether membrane thinning would cause the closed to open transition, Zhang and colleagues solved the structure of MscS in nanodiscs made of short-chains lipids (10-PC). The obtained structure is only partially open, demonstrating that hydrophobic mismatch is not sufficient to open the channel. The authors then incubated the reconstituted protein with cyclodextrin, to selectively remove lipids from the nanodiscs. They reasoned that the remaining lipids would have to stretch to cover the hydrophobic regions, therefore introducing tension in the membrane. The structure obtained in this bilayer under tension is that of a desensitized channel, where the transmembrane domain is collapsed, and the conducting pore is narrow. The authors then hypothesized that the completely open conformation is dynamic and transient, and hence difficult to capture. Zhang and colleagues favor it, using a mutant that does not desensitize. All the structures obtained present lipids in several key locations, which allow to deduce their role in the structural transitions. Lipids are observed inside the conducting pore in the closed state. These pore lipids act like a plug and prevent ion conduction. They move to the periphery of the pore and then out of it as the channel transitions to open and sub-conducting states, and as its transmembrane domain tilts and collapses. Additional lipids are located in grooves between the subunits and maintain the closed conformation by physically preventing sliding and tilting of the subunits to transition towards the conductive states. As the channel transitions to open and desensitized states, the pore lipids move to the periphery and then leave the pore, while the lipids leave the grooves, allowing the gradual collapse of the transmembrane domain. Thus, the motions of the lipids directly drive the structural changes and therefore function of the channel.

Another interesting effect of the lipid environment is observed for the channel LRRC8A. The structure of the channel was resolved in nanodiscs by Kern and colleagues [29]. The protein is a homohexameric channel that presents a six-fold symmetry. By comparing their structure with others obtained in the detergent digitonin, the authors observed that the lipid environment causes a change in the symmetry of the channel. LRRC8A in digitonin displays a three-fold symmetry and is formed of three pairs of asymmetric dimers. This difference in symmetry arises from the presence of lipids in a specific gap between the subunits. When the lipid environment is removed, the channel rearranges in a way that forms three smaller gaps and three larger gaps between the subunits, the large ones being filled with digitonin. Thus, the presence of lipids is required to lock the monomeric unit in a specific conformation that leads to a symmetric hexameric channel. These examples show unequivocally that IMPs work as lipoprotein complexes and their mechanism of action is dictated by lipids and lipids motions.

3.3. Measuring Accessibility Changes with H/D Exchange

A method that has shown a huge increase in its use to study membrane protein dynamics in the last five years is hydrogen–deuterium exchange coupled to mass spectrometry (HDX-MS) [100–103]. This method measures the rate of exchange of labile protons from the backbone amide upon incubation in a deuterated solvent, leading to mass shifts of the peptic peptides detectable by LC-MS [104]. The extent of H/D exchange is directly related to the stability of the H-bond and to its accessibility to the solvent. Deducing conformational intermediates from accessibility changes requires a structural framework, usually obtained through high-resolution structures of the IMP (or a homolog) in different conformations. Once this framework is established, changes in solvent accessibility can be correlated to specific conformational intermediates [105]. HDX-MS applied to IMPs has mostly been used to study the dynamics of GPCRs and transporters [106]. Many HDX-MS studies of GPCRs were carried out in bicelles [107,108], but these studies did not investigate the role of the lipid environment per se.

The use of HDX-MS to specifically interrogate the role of the lipid environment on the conformational dynamics was used to study the secondary transporters LeuT [109,110] as well as LacY and XylE [111], and the ABC transporters BmrA [112,113] and P-glycoprotein (P-gP). Most of these studies were carried out using the nanodisc system as a membrane mimic. The methodology was developed in 2010 by Hebling and colleagues on the model protein γ-Glutamyl carboxylase (GGCX) [114]. In 2017, this method was adapted by Adhikary et al. to study the neurotransmitter homolog LeuT [115,116]. The authors reconstituted the protein in the nanodiscs in either outward-favoring or inward-favoring conditions and mapped its overall conformational hallmarks. Aided by MD simulations in a bilayer, they pinpoint which secondary structure elements display a specific dynamic behavior depending on the conformation. Importantly, this research work illustrates how HDX-MS can offer a global view of IMP dynamics in a bilayer. Atkins and colleagues reconstituted the human multidrug transporter P-gP in nanodiscs and compared its HDX-MS profile with that of the protein in detergent micelles [117]. They found an overall similarly rough conformational landscape in both cases, but information is mostly limited to the NBDs because the sequence coverage of the TMDs is low. They nevertheless observed asymmetry in the exchange behavior between the NBDs, thereby confirming, with a non-invasive methodology, this observation, which was also made with DEER [69]. Another study on the secondary transporters from the Major Facilitator Superfamily (MFS) delves deeper into the molecular details of lipid–protein interactions. Martens and colleagues compared the conformational states of the secondary transporters XylE and LacY in nanodiscs of different lipid compositions comprising or not the main bacterial phospholipid PE [111]. They use MD simulations in bilayers matching the composition of the nanodiscs to identify a specific PE–protein interaction. They carry out HDX-MS experiments in nanodiscs on the WT and on proteins mutated at the site of the predicted interaction. They show that PE favors the inward open conformation trough electrostatic interaction with a conserved conformational switch (Figure 4). Another notable methodological improvement was the use of SMALPs for HDX-MS experiments, allowing to study IMPs in a native-like environment. Reading and colleagues prepared SMALPs of the rhomboid protease GlpG in three different native lipid environments, obtained through the use of either different bacterial strains or different growth conditions [118]. The difference in the lipid compositions (evaluated by lipidomics) could be linked with differences in local dynamics proving the versatility of the method.

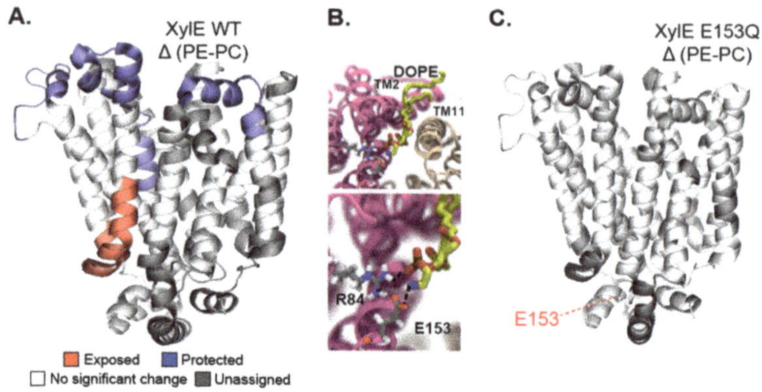

Figure 4. Lipid–protein interactions regulate the conformational equilibrium. (**A**) Differential deuterium uptake pattern (ΔHDX) of WT XylE in DOPE-PG-CL nanodiscs (native-like) minus DOPC-PG-CL (control) mapped onto the 3D structure of XylE (PDB: 4GBY). Red- and blue-colored regions indicate segments containing peptides with a positive ΔHDX (red—more deuteration) or negative ΔHDX (blue—less deuteration), respectively; white regions indicate that no significant ΔHDX is observed ($p \leq 0.01$), and gray indicates regions where peptides were not obtained for both the mutant and the WT conditions. (**B**) Representative MD snapshot of the close-up of the conserved, charged residues interacting with the PE headgroup of the phospholipid. Polar interactions with R84 and E153 prevent network formation and steric hindrance prevents contacts of the TM2 and TM11. (**C**) Mutagenesis of E153 abolishes a lipid-induced conformational shift. ΔHDX of XylE in PE:PG:CL nanodiscs (native-like) minus PC:PG:CL (control) nanodiscs mapped on the PDB structure. Adapted from [111].

These examples show that HDX-MS is a useful tool to follow changes in conformational dynamics of reconstituted IMPs. Since its use is facilitated by the availability of high-resolution structures, it is poised to become a standard tool for the structural biology of IMPs in native-like environments.

4. Lipids as Modulators of Ligand Binding

Lipid molecules can occupy specific sites within IMPs, either competing directly with ligand binding by modulating the drug-binding pocket [119] or causing long-range allosteric effects [120]. Detecting such effects typically requires a combination of high-resolution structural data and functional data on the mutants compromising lipid-binding sites [121–123]. A textbook example of allosteric coupling by a lipid molecule was provided by the structural resolution of the TRPV1 channel in nanodiscs [37]. The authors observed a shared binding pocket for phosphatidylinositol (PIP) lipid and the agonist resiniferatoxin (RTX). When the agonist is not present, the binding pocket is occupied by PIP. Binding of RTX kicks PIP from the pocket and by doing so changes the coordinating residues. This small structural rearrangement amplifies by pulling the S4–S5 linker away from the central axis thereby facilitating opening of the lower gate to activate the channel. Another recent example of allosteric modulation by a lipid molecule comes from a study of ELIC, a pentameric ligand-gated ion channel (pLGIC). Functional modulation of pLGICs by lipids is well documented [124,125]. The structure of the anionic channel was resolved by cryo-EM in nanodiscs and reveals a phospholipid molecule located near a conserved proline kink in transmembrane helix M4. The authors identify this kink as a conserved feature in anion selective channels and GABA receptors. The mobility of the helix couples ligand binding to channel opening and desensitization. A combination of mutagenesis and electrophysiology measurements show that occupation of this site either by a lipid molecule or a drug stabilizes a closed conformation [126]. Competitions between an exogenous drug an endogenous lipid appears to be an important mode of regulating channel opening, and is suggested for other IMPs (GABA receptor [127], serotonin receptor [119] and voltage-gated channel TPC1 [90]).

Another good example of the impact of lipid–protein interactions on allosteric modulation is provided by an interesting study on the A2AR receptor, which does not involve structure resolution. Radioligand-binding assays performed by Guixà-González et al. show that binding of cholesterol significantly reduces the binding of the antagonist to the receptor [128]. Subsequent MD analysis predicted direct entry of cholesterol from the membrane into the orthosteric binding site. The authors designed an elegant assay to confirm the MD predictions. The presence of a cholesterol molecule inside the receptor, clashing with the orthosteric site, would prevent labeling of cysteine residues with a reactive probe. Sequential addition of the probe and removal of cholesterol in the presence and absence of the antagonist strongly suggests that cholesterol is inside the receptor. The authors demonstrate that the observed inhibitory effect of cholesterol was not only due to allosteric changes (as previously shown for this GPCR and others [129]), but also to direct occupation of the orthosteric binding site. This research opens the way to the potential use of sterol and sterol-like compounds in GPCR therapeutics.

Another tool that facilitates the identification of coupled interactions between lipids and ligands or cofactors is native MS [130]. The order of incubation of the ligands/cofactors allow to determine whether drug binding correlates positively or negatively with lipid binding. One of the first studies showing such synergy was carried out by Marcoux et al., on the multidrug efflux pump P-glycoprotein [131]. Preincubation of the transporter with cardiolipin (CL) before addition of ligand cyclosporin A (CsA) had no effect on CsA binding. However, when CsA was added prior to CL incubation, an increase in lipid binding was observed. This suggests that CsA binding primes the transporter for subsequent lipid binding, either through a conformational change or adaptation of the ligand pocket. Such a method was used successfully on GPCRs [132], channels [133–136], transporters [131,137–139] and other IMPs [140], highlighting the role of specific lipid species in allosteric coupling.

A recent article combines both high-resolution structure determination, nMS and functional tests to demonstrate that the presence of a lipid can modulate polyspecificity of the bacterial multidrug transporter LmrP (Figure 5). The authors obtained the crystal structure of the protein in a ligand-bound state. An unknown density observed close to the ligand shows the features of a phospholipid molecule. Molecular dynamic simulations identify the phosphatidylglycerol (PG) molecule as the more likely candidate to stabilize the ligand inside the binding pocket. To confirm the presence of a PG molecule inside the binding pocket of LmrP, the authors performed nMS experiments in PG-containing nanodiscs. They compared the spectrum of WT LmrP and a mutant, designed to compromise lipid binding, at different voltages. They showed that a PG molecule remained attached to the WT protein at high voltages, but the mutant did not present such preferential binding. These measurements demonstrated a specific affinity for PG. Additional transport assays confirmed the importance of keeping the lipid-binding site intact for the efflux of some but not all ligands. The authors hypothesize that the lipid provides a malleable hydrophobic environment for diverse substrates, providing a rationale for the substrate polyspecificity of the multidrug transporter.

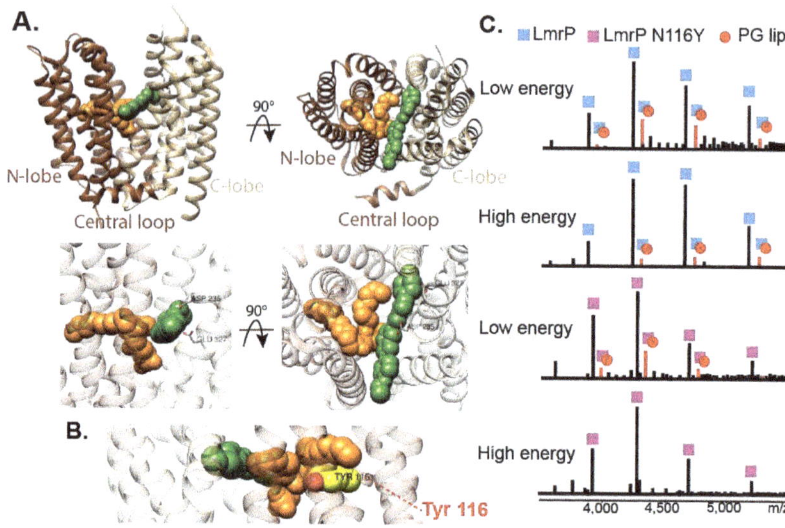

Figure 5. Structure of LmrP in complex with Hoechst 33342. (**A**) Cartoon representation of LmrP with the ligand Hoechst 33342 (green) and phosphatidic acid (orange) modeled from the densities observed (PDB ID 6T1Z). Hoechst 33342 forms polar interactions with D235 and E327 located in the C-lobe of LmrP. (**B**) Close up on the phospholipid. N116Y mutation perturbs lipid binding (**C**). nMS spectra showing compromised PG binding on the N116Y mutant. At a low activation energy (160 V), multiple lipids are bound to LmrP, including a peak corresponding to a single bound DOPG (in red). At high energy (200 V), this peak is still present. In the case of the N116Y mutant, although DOPG binds at low activation energy, it disappears at a high activation energy.

5. Conclusions and Perspectives

Membrane proteins work within a complex lipidic environment that participates in the molecular mechanism enabling function. It is becoming clear that membrane proteins have to be considered as lipoprotein complexes. The methods and approaches to study them are adapting to that complexity (Table 1). Recent work has shown that different lipid species can have different albeit interlinked roles in the mechanisms of IMPs [141,142]. For example, studies on the Na, K-ATPase have highlighted distinct roles for distinct lipid species. Two X-ray structures have identified two lipid-binding sites at different locations [143,144] and a third study has determined the role of lipids at these sites using native MS, transport assays and mutagenesis. Cholesterol and PS stabilize the protein without affecting its activity while PC or PE stimulate activity by accelerating the conformational transitions, but do not affect stability [145]. Integrative approaches that combine biochemical and biophysical characterization with high-resolution structures are becoming standard and provide updated structure–function relationships of the IMPs.

As our understanding of membrane proteins shifts towards describing lipoprotein complexes, other layers of complexity are now added into the equation. Eventually, obtaining molecular information in the cellular context is the next aim of structural biology. Techniques that report molecular interactions in vivo or in cellulo are fast developing. Cryo-electron tomography [146] and super resolution microscopy techniques [147] are making strides towards molecular-level resolution. Native mass spectrometry of whole cells has yielded novel insights into the organization of IMPs in their truly native membrane [148]. Not only the structural but also the temporal resolution of biophysical techniques are increasing [149–151], and the possibility to look at IMPs in action is getting closer. To this end, IMPs activated by ion gradients need the right environment. Yao and colleagues have devised a protocol to carry out cryo-EM experiments in proteoliposomes, using the prototypical AcrB MDR efflux pump as a model. Such a development paves the way for further studies of functioning IMPs [152]. The future looks bright and exciting for the field of structural biology of membrane proteins, with many methodological improvements making the mysteries of the cell visible to the human eye.

Author Contributions: All the authors curated the literature and participated in writing and editing the manuscript. All authors have read and agreed to the published version of the manuscript.

Funding: This research was funded by the Fonds National de la Recherche Scientifique FRS-FNRS. C.M. (Grant number 4.121.F.000517) and T.V.O. (Grant number 4.124.F.000193) benefit from a fellowship of the FRS-FNRS.

Institutional Review Board Statement: Not applicable.

Informed Consent Statement: Not applicable.

Data Availability Statement: Not applicable.

Acknowledgments: We thank Andy M. Lau for the graphical abstract figure.

Conflicts of Interest: Authors declare no conflict of interest.

References

1. Almén, M.S.; Nordström, K.J.; Fredriksson, R.; Schiöth, H.B. Mapping the Human Membrane Proteome: A Majority of the Human Membrane Proteins Can Be Classified According to Function and Evolutionary Origin. *BMC Biol.* **2009**, *7*, 50. [CrossRef] [PubMed]
2. Santos, R.; Ursu, O.; Gaulton, A.; Bento, A.P.; Donadi, R.S.; Bologa, C.G.; Karlsson, A.; Al-Lazikani, B.; Hersey, A.; Oprea, T.I.; et al. A Comprehensive Map of Molecular Drug Targets. *Nat. Rev. Drug Discov.* **2017**, *16*, 19–34. [CrossRef]
3. Criado, M.; Eibl, H.; Barrantes, F.J. Functional Properties of the Acetylcholine Receptor Incorporated in Model Lipid Membranes. Differential Effects of Chain Length and Head Group of Phospholipids on Receptor Affinity States and Receptor-Mediated Ion Translocation. *J. Biol. Chem.* **1984**, *259*, 9188–9198. [CrossRef]
4. Bogdanov, M.; Dowhan, W. Phosphatidylethanolamine Is Required for in Vivo Function of the Membrane-Associated Lactose Permease of Escherichia Coli. *J. Biol. Chem.* **1995**, *270*, 732–739. [CrossRef] [PubMed]
5. Lee, A.G. Lipid-Protein Interactions in Biological Membranes: A Structural Perspective. *Biochim. Biophys. Acta* **2003**, *1612*, 1–40. [CrossRef]
6. Marsh, D. Protein Modulation of Lipids, and Vice-Versa, in Membranes. *Biochim. Biophys. Acta BBA Biomembr.* **2008**, *1778*, 1545–1575. [CrossRef]
7. Lee, A.G. How Lipids Affect the Activities of Integral Membrane Proteins. *Biochim. Biophys. Acta BBA Biomembr.* **2004**, *1666*, 62–87. [CrossRef]
8. Findlay, H.E.; Booth, P.J. The Folding, Stability and Function of Lactose Permease Differ in Their Dependence on Bilayer Lipid Composition. *Sci. Rep.* **2017**, *7*, 13056. [CrossRef]
9. Choy, B.C.; Cater, R.J.; Mancia, F.; Pryor, E.E. A 10-Year Meta-Analysis of Membrane Protein Structural Biology: Detergents, Membrane Mimetics, and Structure Determination Techniques. *Biochim. Biophys. Acta BBA Biomembr.* **2021**, *1863*, 183533. [CrossRef]
10. Laganowsky, A.; Reading, E.; Allison, T.M.; Ulmschneider, M.B.; Degiacomi, M.T.; Baldwin, A.J.; Robinson, C.V. Membrane Proteins Bind Lipids Selectively to Modulate Their Structure and Function. *Nature* **2014**, *510*, 172–175. [CrossRef]
11. Bayburt, T.H.; Sligar, S.G. Membrane Protein Assembly into Nanodiscs. *FEBS Lett.* **2010**, *584*, 1721–1727. [CrossRef] [PubMed]
12. Flayhan, A.; Mertens, H.D.T.; Ural-Blimke, Y.; Molledo, M.M.; Svergun, D.I.; Löw, C. Saposin Lipid Nanoparticles: A Highly Versatile and Modular Tool for Membrane Protein Research. *Structure* **2018**, *26*, 345–355.e5. [CrossRef] [PubMed]
13. Angiulli, G.; Dhupar, H.S.; Suzuki, H.; Wason, I.S.; Duong Van Hoa, F.; Walz, T. New Approach for Membrane Protein Reconstitution into Peptidiscs and Basis for Their Adaptability to Different Proteins. *eLife* **2020**, *9*, e53530. [CrossRef] [PubMed]
14. Carlson, M.L.; Young, J.W.; Zhao, Z.; Fabre, L.; Jun, D.; Li, J.; Li, J.; Dhupar, H.S.; Wason, I.; Mills, A.T.; et al. The Peptidisc, a Simple Method for Stabilizing Membrane Proteins in Detergent-Free Solution. *eLife* **2018**, *7*, e34085. [CrossRef]
15. Bada Juarez, J.F.; Harper, A.J.; Judge, P.J.; Tonge, S.R.; Watts, A. From Polymer Chemistry to Structural Biology: The Development of SMA and Related Amphipathic Polymers for Membrane Protein Extraction and Solubilisation. *Chem. Phys. Lipids* **2019**, *221*, 167–175. [CrossRef]
16. Thoma, J.; Burmann, B.M. Fake It 'Till You Make It—The Pursuit of Suitable Membrane Mimetics for Membrane Protein Biophysics. *Int. J. Mol. Sci.* **2021**, *22*, 50. [CrossRef]
17. Saliba, A.-E.; Vonkova, I.; Gavin, A.-C. The Systematic Analysis of Protein–Lipid Interactions Comes of Age. *Nat. Rev. Mol. Cell Biol.* **2015**, *16*, 753–761. [CrossRef]
18. Hsia, C.-Y.; Richards, M.J.; Daniel, S. A Review of Traditional and Emerging Methods to Characterize Lipid–Protein Interactions in Biological Membranes. *Anal. Methods* **2015**, *7*, 7076–7094. [CrossRef]

19. Bagheri, Y.; Ali, A.A.; You, M. Current Methods for Detecting Cell Membrane Transient Interactions. *Front. Chem.* **2020**, *8*. [CrossRef]
20. Corradi, V.; Sejdiu, B.I.; Mesa-Galloso, H.; Abdizadeh, H.; Noskov, S.Y.; Marrink, S.J.; Tieleman, D.P. Emerging Diversity in Lipid–Protein Interactions. *Chem. Rev.* **2019**, *119*, 5775–5848. [CrossRef] [PubMed]
21. Tomasello, G.; Armenia, I.; Molla, G. The Protein Imager: A Full-Featured Online Molecular Viewer Interface with Server-Side HQ-Rendering Capabilities. *Bioinformatics* **2020**, *36*, 2909–2911. [CrossRef]
22. Popot, J.L.; Engelman, D.M. Membrane Protein Folding and Oligomerization: The Two-Stage Model. Available online: https://pubs.acs.org/doi/pdf/10.1021/bi00469a001 (accessed on 31 May 2021).
23. Mohole, M.; Kumar, G.A.; Sengupta, D.; Chattopadhyay, A. Chapter 6—Molecular Determinants of GPCR Oligomerization. In *GPCRs*; Jastrzebska, B., Park, P.S.-H., Eds.; Academic Press: Cambridge, MA, USA, 2020; pp. 97–108. [CrossRef]
24. Gao, S.; Valinsky, W.C.; On, N.C.; Houlihan, P.R.; Qu, Q.; Liu, L.; Pan, X.; Clapham, D.E.; Yan, N. Employing NaChBac for Cryo-EM Analysis of Toxin Action on Voltage-Gated Na+ Channels in Nanodisc. *Proc. Natl. Acad. Sci. USA* **2020**, *117*, 14187–14193. [CrossRef]
25. McGoldrick, L.L.; Singh, A.K.; Demirkhanyan, L.; Lin, T.-Y.; Casner, R.G.; Zakharian, E.; Sobolevsky, A.I. Structure of the Thermo-Sensitive TRP Channel TRP1 from the Alga Chlamydomonas Reinhardtii. *Nat. Commun.* **2019**, *10*, 4180. [CrossRef]
26. Autzen, H.E.; Myasnikov, A.G.; Campbell, M.G.; Asarnow, D.; Julius, D.; Cheng, Y. Structure of the Human TRPM4 Ion Channel in a Lipid Nanodisc. *Science* **2018**, *359*, 228–232. [CrossRef] [PubMed]
27. Koshy, C.; Schweikhard, E.S.; Gärtner, R.M.; Perez, C.; Yildiz, Ö.; Ziegler, C. Structural Evidence for Functional Lipid Interactions in the Betaine Transporter BetP. *EMBO J.* **2013**, *32*, 3096–3105. [CrossRef]
28. Essen, L.-O.; Siegert, R.; Lehmann, W.D.; Oesterhelt, D. Lipid Patches in Membrane Protein Oligomers: Crystal Structure of the Bacteriorhodopsin-Lipid Complex. *Proc. Natl. Acad. Sci. USA* **1998**, *95*, 11673–11678. [CrossRef]
29. Kern, D.M.; Oh, S.; Hite, R.K.; Brohawn, S.G. Cryo-EM Structures of the DCPIB-Inhibited Volume-Regulated Anion Channel LRRC8A in Lipid Nanodiscs. *eLife* **2019**, *8*, e42636. [CrossRef] [PubMed]
30. Song, D.; Jiao, H.; Liu, Z. Phospholipid Translocation Captured in a Bifunctional Membrane Protein MprF. *Nat. Commun.* **2021**, *12*, 2927. [CrossRef]
31. Periole, X.; Huber, T.; Marrink, S.-J.; Sakmar, T.P. G Protein-Coupled Receptors Self-Assemble in Dynamics Simulations of Model Bilayers. *J. Am. Chem. Soc.* **2007**, *129*, 10126–10132. [CrossRef] [PubMed]
32. Fotiadis, D.; Liang, Y.; Filipek, S.; Saperstein, D.A.; Engel, A.; Palczewski, K. Rhodopsin Dimers in Native Disc Membranes. *Nature* **2003**, *421*, 127–128. [CrossRef] [PubMed]
33. Prasanna, X.; Chattopadhyay, A.; Sengupta, D. Cholesterol Modulates the Dimer Interface of the B2-Adrenergic Receptor via Cholesterol Occupancy Sites. *Biophys. J.* **2014**, *106*, 1290–1300. [CrossRef] [PubMed]
34. Valiyaveetil, F.I.; Zhou, Y.; MacKinnon, R. Lipids in the Structure, Folding, and Function of the KcsA K^+ Channel. *Biochemistry* **2002**, *41*, 10771–10777. [CrossRef] [PubMed]
35. Sun, C.; Benlekbir, S.; Venkatakrishnan, P.; Wang, Y.; Hong, S.; Hosler, J.; Tajkhorshid, E.; Rubinstein, J.L.; Gennis, R.B. Structure of the Alternative Complex III in a Supercomplex with Cytochrome Oxidase. *Nature* **2018**, *557*, 123–126. [CrossRef] [PubMed]
36. Sobti, M.; Walshe, J.L.; Wu, D.; Ishmukhametov, R.; Zeng, Y.C.; Robinson, C.V.; Berry, R.M.; Stewart, A.G. Cryo-EM Structures Provide Insight into How *E. coli* F_1F_o ATP Synthase Accommodates Symmetry Mismatch. *Nat. Commun.* **2020**, *11*, 2615. [CrossRef]
37. Gao, Y.; Cao, E.; Julius, D.; Cheng, Y. TRPV1 Structures in Nanodiscs Reveal Mechanisms of Ligand and Lipid Action. *Nature* **2016**, *534*, 347–351. [CrossRef]
38. Gupta, K.; Donlan, J.A.C.; Hopper, J.T.S.; Uzdavinys, P.; Landreh, M.; Struwe, W.B.; Drew, D.; Baldwin, A.J.; Stansfeld, P.J.; Robinson, C.V. The Role of Interfacial Lipids in Stabilizing Membrane Protein Oligomers. *Nature* **2017**, *541*, 421–424. [CrossRef]
39. Gupta, K.; Li, J.; Liko, I.; Gault, J.; Bechara, C.; Wu, D.; Hopper, J.T.S.; Giles, K.; Benesch, J.L.P.; Robinson, C.V. Identifying Key Membrane Protein Lipid Interactions Using Mass Spectrometry. *Nat. Protoc.* **2018**, *13*, 1106–1120. [CrossRef]
40. Liko, I.; Degiacomi, M.T.; Mohammed, S.; Yoshikawa, S.; Schmidt, C.; Robinson, C.V. Dimer Interface of Bovine Cytochrome c Oxidase Is Influenced by Local Posttranslational Modifications and Lipid Binding. *Proc. Natl. Acad. Sci. USA* **2016**, *113*, 8230–8235. [CrossRef]
41. Pyle, E.; Kalli, A.C.; Amillis, S.; Hall, Z.; Lau, A.M.; Hanyaloglu, A.C.; Diallinas, G.; Byrne, B.; Politis, A. Structural Lipids Enable the Formation of Functional Oligomers of the Eukaryotic Purine Symporter UapA. *Cell Chem. Biol.* **2018**, *25*, 840–848.e4. [CrossRef]
42. Landreh, M.; Marklund, E.G.; Uzdavinys, P.; Degiacomi, M.T.; Coincon, M.; Gault, J.; Gupta, K.; Liko, I.; Benesch, J.L.P.; Drew, D.; et al. Integrating Mass Spectrometry with MD Simulations Reveals the Role of Lipids in Na^+/H^+ Antiporters. *Nat. Commun.* **2017**, *8*, 13993. [CrossRef]
43. Reading, E.; Walton, T.A.; Liko, I.; Marty, M.T.; Laganowsky, A.; Rees, D.C.; Robinson, C.V. The Effect of Detergent, Temperature, and Lipid on the Oligomeric State of MscL Constructs: Insights from Mass Spectrometry. *Chem. Biol.* **2015**, *22*, 593–603. [CrossRef]
44. Winkelmann, I.; Matsuoka, R.; Meier, P.F.; Shutin, D.; Zhang, C.; Orellana, L.; Sexton, R.; Landreh, M.; Robinson, C.V.; Beckstein, O.; et al. Structure and Elevator Mechanism of the Mammalian Sodium/Proton Exchanger NHE9. *EMBO J.* **2020**, *39*, e105908. [CrossRef]
45. Anderluh, A.; Hofmaier, T.; Klotzsch, E.; Kudlacek, O.; Stockner, T.; Sitte, H.H.; Schütz, G.J. Direct PIP 2 Binding Mediates Stable Oligomer Formation of the Serotonin Transporter. *Nat. Commun.* **2017**, *8*, 14089. [CrossRef]

46. Anderluh, A.; Klotzsch, E.; Reismann, A.W.A.F.; Brameshuber, M.; Kudlacek, O.; Newman, A.H.; Sitte, H.H.; Schütz, G.J. Single Molecule Analysis Reveals Coexistence of Stable Serotonin Transporter Monomers and Oligomers in the Live Cell Plasma Membrane. *J. Biol. Chem.* **2014**, *289*, 4387–4394. [CrossRef] [PubMed]
47. Buchmayer, F.; Schicker, K.; Steinkellner, T.; Geier, P.; Stübiger, G.; Hamilton, P.J.; Jurik, A.; Stockner, T.; Yang, J.-W.; Montgomery, T.; et al. Amphetamine Actions at the Serotonin Transporter Rely on the Availability of Phosphatidylinositol-4,5-Bisphosphate. *Proc. Natl. Acad. Sci. USA* **2013**, *110*, 11642–11647. [CrossRef] [PubMed]
48. Ardalan, A.; Sowlati-Hashjin, S.; Uwumarenogie, S.O.; Fish, M.; Mitchell, J.; Karttunen, M.; Smith, M.D.; Jelokhani-Niaraki, M. Functional Oligomeric Forms of Uncoupling Protein 2: Strong Evidence for Asymmetry in Protein and Lipid Bilayer Systems. *J. Phys. Chem. B* **2021**, *125*, 169–183. [CrossRef] [PubMed]
49. Kourkoulou, A.; Grevias, P.; Lambrinidis, G.; Pyle, E.; Dionysopoulou, M.; Politis, A.; Mikros, E.; Byrne, B.; Diallinas, G. Specific Residues in a Purine Transporter Are Critical for Dimerization, ER Exit, and Function. *Genetics* **2019**, *213*, 1357–1372. [CrossRef] [PubMed]
50. Perozo, E.; Kloda, A.; Cortes, D.M.; Martinac, B. Physical Principles Underlying the Transduction of Bilayer Deformation Forces during Mechanosensitive Channel Gating. *Nat. Struct. Biol.* **2002**, *9*, 696–703. [CrossRef] [PubMed]
51. Marty, M.T. Nanodiscs and Mass Spectrometry: Making Membranes Fly. *Int. J. Mass Spectrom.* **2020**, *458*, 116436. [CrossRef]
52. Ro, S.Y.; Schachner, L.F.; Koo, C.W.; Purohit, R.; Remis, J.P.; Kenney, G.E.; Liauw, B.W.; Thomas, P.M.; Patrie, S.M.; Kelleher, N.L.; et al. Native Top-down Mass Spectrometry Provides Insights into the Copper Centers of Membrane-Bound Methane Monooxygenase. *Nat. Commun.* **2019**, *10*, 2675. [CrossRef]
53. Keener, J.E.; Zambrano, D.E.; Zhang, G.; Zak, C.K.; Reid, D.J.; Deodhar, B.S.; Pemberton, J.E.; Prell, J.S.; Marty, M.T. Chemical Additives Enable Native Mass Spectrometry Measurement of Membrane Protein Oligomeric State within Intact Nanodiscs. *J. Am. Chem. Soc.* **2019**, *141*, 1054–1061. [CrossRef]
54. Townsend, J.A.; Sanders, H.M.; Rolland, A.D.; Prell, J.S.; Wang, J.; Marty, M.T. Influenza A M2 Channel Oligomerization is Sensitive to its Chemical Environment. *bioRxiv* **2021**. [CrossRef]
55. Hoi, K.K.; Bada Juarez, J.F.; Judge, P.J.; Yen, H.-Y.; Wu, D.; Vinals, J.; Taylor, G.F.; Watts, A.; Robinson, C.V. Detergent-Free Lipodisq Nanoparticles Facilitate High-Resolution Mass Spectrometry of Folded Integral Membrane Proteins. *Nano Lett.* **2021**, *21*, 2824–2831. [CrossRef]
56. Marty, M.T.; Baldwin, A.J.; Marklund, E.G.; Hochberg, G.K.A.; Benesch, J.L.P.; Robinson, C.V. Bayesian Deconvolution of Mass and Ion Mobility Spectra: From Binary Interactions to Polydisperse Ensembles. *Anal. Chem.* **2015**, *87*, 4370–4376. [CrossRef]
57. Cho, M.-K.; Gayen, A.; Banigan, J.R.; Leninger, M.; Traaseth, N.J. Intrinsic Conformational Plasticity of Native EmrE Provides a Pathway for Multidrug Resistance. *J. Am. Chem. Soc.* **2014**, *136*, 8072–8080. [CrossRef] [PubMed]
58. Gayen, A.; Banigan, J.R.; Traaseth, N.J. Ligand-Induced Conformational Changes of the Multidrug Resistance Transporter EmrE Probed by Oriented Solid-State NMR Spectroscopy. *Angew. Chem.* **2013**, *125*, 10511–10514. [CrossRef]
59. Van der Cruijsen, E.A.W.; Nand, D.; Weingarth, M.; Prokofyev, A.; Hornig, S.; Cukkemane, A.A.; Bonvin, A.M.J.J.; Becker, S.; Hulse, R.E.; Perozo, E.; et al. Importance of Lipid–Pore Loop Interface for Potassium Channel Structure and Function. *Proc. Natl. Acad. Sci. USA* **2013**, *110*, 13008–13013. [CrossRef]
60. Mandala, V.S.; Williams, J.K.; Hong, M. Structure and Dynamics of Membrane Proteins from Solid-State NMR. *Annu. Rev. Biophys.* **2018**, *47*, 201–222. [CrossRef]
61. Ladizhansky, V. Applications of Solid-State NMR to Membrane Proteins. *Biochim. Biophys. Acta BBA Proteins Proteom.* **2017**, *1865* (11, Part B), 1577–1586. [CrossRef]
62. Ullrich, S.J.; Glaubitz, C. Perspectives in Enzymology of Membrane Proteins by Solid-State NMR. *Acc. Chem. Res.* **2013**, *46*, 2164–2171. [CrossRef]
63. Lerner, E.; Cordes, T.; Ingargiola, A.; Alhadid, Y.; Chung, S.; Michalet, X.; Weiss, S. Toward Dynamic Structural Biology: Two Decades of Single-Molecule Förster Resonance Energy Transfer. *Science* **2018**, *359*, eaan1133. [CrossRef]
64. Jeschke, G. DEER Distance Measurements on Proteins. *Annu. Rev. Phys. Chem.* **2012**, *63*, 419–446. [CrossRef] [PubMed]
65. Mchaourab, H.S.; Steed, P.R.; Kazmier, K. Toward the Fourth Dimension of Membrane Protein Structure: Insight into Dynamics from Spin-Labeling EPR Spectroscopy. *Structure* **2011**, *19*, 1549–1561. [CrossRef]
66. Roy, R.; Hohng, S.; Ha, T. A Practical Guide to Single-Molecule FRET. *Nat. Methods* **2008**, *5*, 507–516. [CrossRef] [PubMed]
67. Chang, Y.-N.; Jaumann, E.A.; Reichel, K.; Hartmann, J.; Oliver, D.; Hummer, G.; Joseph, B.; Geertsma, E.R. Structural Basis for Functional Interactions in Dimers of SLC26 Transporters. *Nat. Commun.* **2019**, *10*, 2032. [CrossRef]
68. Martens, C.; Stein, R.A.; Masureel, M.; Roth, A.; Mishra, S.; Dawaliby, R.; Konijnenberg, A.; Sobott, F.; Govaerts, C.; Mchaourab, H.S. Lipids Modulate the Conformational Dynamics of a Secondary Multidrug Transporter. *Nat. Struct. Mol. Biol.* **2016**, *23*, 744–751. [CrossRef] [PubMed]
69. Dastvan, R.; Mishra, S.; Peskova, Y.B.; Nakamoto, R.K.; Mchaourab, H.S. Mechanism of Allosteric Modulation of P-Glycoprotein by Transport Substrates and Inhibitors. *Science* **2019**, *364*, 689–692. [CrossRef] [PubMed]
70. Yardeni, E.H.; Mishra, S.; Stein, R.A.; Bibi, E.; Mchaourab, H.S. The Multidrug Transporter MdfA Deviates from the Canonical Model of Alternating Access of MFS Transporters. *J. Mol. Biol.* **2020**, *432*, 5665–5680. [CrossRef]
71. Husada, F.; Bountra, K.; Tassis, K.; de Boer, M.; Romano, M.; Rebuffat, S.; Beis, K.; Cordes, T. Conformational Dynamics of the ABC Transporter McjD Seen by Single-Molecule FRET. *EMBO J.* **2018**, *37*, e100056. [CrossRef]

72. Yang, M.; Livnat Levanon, N.; Acar, B.; Aykac Fas, B.; Masrati, G.; Rose, J.; Ben-Tal, N.; Haliloglu, T.; Zhao, Y.; Lewinson, O. Single-Molecule Probing of the Conformational Homogeneity of the ABC Transporter BtuCD. *Nat. Chem. Biol.* **2018**, *14*, 715–722. [CrossRef]
73. Lasitza-Male, T.; Bartels, K.; Jungwirth, J.; Wiggers, F.; Rosenblum, G.; Hofmann, H.; Löw, C. Membrane Chemistry Tunes the Structure of a Peptide Transporter. *Angew. Chem. Int. Ed.* **2020**, *59*, 19121–19128. [CrossRef] [PubMed]
74. Liu, Y.; Liu, Y.; He, L.; Zhao, Y.; Zhang, X.C. Single-Molecule Fluorescence Studies on the Conformational Change of the ABC Transporter MsbA. *Biophys. Rep.* **2018**, *4*, 153–165. [CrossRef]
75. Jagessar, K.L.; Claxton, D.P.; Stein, R.A.; Mchaourab, H.S. Sequence and Structural Determinants of Ligand-Dependent Alternating Access of a MATE Transporter. *Proc. Natl. Acad. Sci. USA* **2020**, *117*, 4732–4740. [CrossRef]
76. Masureel, M.; Martens, C.; Stein, R.A.; Mishra, S.; Ruysschaert, J.-M.; Mchaourab, H.S.; Govaerts, C. Protonation Drives the Conformational Switch in the Multidrug Transporter LmrP. *Nat. Chem. Biol.* **2014**, *10*, 149–155. [CrossRef]
77. Sigworth, F.J. Principles of Cryo-EM Single-Particle Image Processing. *Microscopy* **2016**, *65*, 57–67. [CrossRef] [PubMed]
78. Chen, Q.; She, J.; Zeng, W.; Guo, J.; Xu, H.; Bai, X.; Jiang, Y. Structure of Mammalian Endolysosomal TRPML1 Channel in Nanodiscs. *Nature* **2017**, *550*, 415–418. [CrossRef] [PubMed]
79. Pumroy, R.A.; Samanta, A.; Liu, Y.; Hughes, T.E.; Zhao, S.; Yudin, Y.; Rohacs, T.; Han, S.; Moiseenkova-Bell, V.Y. Molecular Mechanism of TRPV2 Channel Modulation by Cannabidiol. *eLife* **2019**, *8*, e48792. [CrossRef] [PubMed]
80. Deng, Z.; Maksaev, G.; Rau, M.; Xie, Z.; Hu, H.; Fitzpatrick, J.A.J.; Yuan, P. Gating of Human TRPV3 in a Lipid Bilayer. *Nat. Struct. Mol. Biol.* **2020**, *27*, 635–644. [CrossRef]
81. Shimada, H.; Kusakizako, T.; Dung Nguyen, T.H.; Nishizawa, T.; Hino, T.; Tominaga, M.; Nureki, O. The Structure of Lipid Nanodisc-Reconstituted TRPV3 Reveals the Gating Mechanism. *Nat. Struct. Mol. Biol.* **2020**, *27*, 645–652. [CrossRef] [PubMed]
82. Hughes, T.E.; Del Rosario, J.S.; Kapoor, A.; Yazici, A.T.; Yudin, Y.; Fluck, E.C., III; Filizola, M.; Rohacs, T.; Moiseenkova-Bell, V.Y. Structure-Based Characterization of Novel TRPV5 Inhibitors. *eLife* **2019**, *8*, e49572. [CrossRef] [PubMed]
83. Dang, S.; van Goor, M.K.; Asarnow, D.; Wang, Y.; Julius, D.; Cheng, Y.; van der Wijst, J. Structural Insight into TRPV5 Channel Function and Modulation. *Proc. Natl. Acad. Sci. USA* **2019**, *116*, 8869–8878. [CrossRef]
84. McGoldrick, L.L.; Singh, A.K.; Saotome, K.; Yelshanskaya, M.V.; Twomey, E.C.; Grassucci, R.A.; Sobolevsky, A.I. Opening of the Human Epithelial Calcium Channel TRPV6. *Nature* **2018**, *553*, 233–237. [CrossRef]
85. Reddy, B.; Bavi, N.; Lu, A.; Park, Y.; Perozo, E. Molecular Basis of Force-from-Lipids Gating in the Mechanosensitive Channel MscS. *eLife* **2019**, *8*, e50486. [CrossRef] [PubMed]
86. Rasmussen, T.; Flegler, V.J.; Rasmussen, A.; Böttcher, B. Structure of the Mechanosensitive Channel MscS Embedded in the Membrane Bilayer. *J. Mol. Biol.* **2019**, *431*, 3081–3090. [CrossRef] [PubMed]
87. Zhang, Y.; Daday, C.; Gu, R.-X.; Cox, C.D.; Martinac, B.; de Groot, B.L.; Walz, T. Visualization of the Mechanosensitive Ion Channel MscS under Membrane Tension. *Nature* **2021**, *590*, 509–514. [CrossRef] [PubMed]
88. Jojoa-Cruz, S.; Saotome, K.; Murthy, S.E.; Tsui, C.C.A.; Sansom, M.S.; Patapoutian, A.; Ward, A.B. Cryo-EM Structure of the Mechanically Activated Ion Channel OSCA1.2. *eLife* **2018**, *7*, e41845. [CrossRef]
89. Matthies, D.; Bae, C.; Toombes, G.E.; Fox, T.; Bartesaghi, A.; Subramaniam, S.; Swartz, K.J. Single-Particle Cryo-EM Structure of a Voltage-Activated Potassium Channel in Lipid Nanodiscs. *eLife* **2018**, *7*, e37558. [CrossRef]
90. Kintzer, A.F.; Green, E.M.; Dominik, P.K.; Bridges, M.; Armache, J.-P.; Deneka, D.; Kim, S.S.; Hubbell, W.; Kossiakoff, A.A.; Cheng, Y.; et al. Structural Basis for Activation of Voltage Sensor Domains in an Ion Channel TPC1. *Proc. Natl. Acad. Sci. USA* **2018**, *115*, E9095–E9104. [CrossRef]
91. Shen, P.S.; Yang, X.; DeCaen, P.G.; Liu, X.; Bulkley, D.; Clapham, D.E.; Cao, E. The Structure of the Polycystic Kidney Disease Channel PKD2 in Lipid Nanodiscs. *Cell* **2016**, *167*, 763–773.e11. [CrossRef] [PubMed]
92. Feng, S.; Dang, S.; Han, T.W.; Ye, W.; Jin, P.; Cheng, T.; Li, J.; Jan, Y.N.; Jan, L.Y.; Cheng, Y. Cryo-EM Studies of TMEM16F Calcium-Activated Ion Channel Suggest Features Important for Lipid Scrambling. *Cell Rep.* **2019**, *28*, 567–579.e4. [CrossRef]
93. Qiu, W.; Fu, Z.; Xu, G.G.; Grassucci, R.A.; Zhang, Y.; Frank, J.; Hendrickson, W.A.; Guo, Y. Structure and Activity of Lipid Bilayer within a Membrane-Protein Transporter. *Proc. Natl. Acad. Sci. USA* **2018**, *115*, 12985–12990. [CrossRef]
94. Hofmann, S.; Januliene, D.; Mehdipour, A.R.; Thomas, C.; Stefan, E.; Brüchert, S.; Kuhn, B.T.; Geertsma, E.R.; Hummer, G.; Tampé, R.; et al. Conformation Space of a Heterodimeric ABC Exporter under Turnover Conditions. *Nature* **2019**, *571*, 580–583. [CrossRef]
95. Taylor, N.M.I.; Manolaridis, I.; Jackson, S.M.; Kowal, J.; Stahlberg, H.; Locher, K.P. Structure of the Human Multidrug Transporter ABCG2. *Nature* **2017**, *546*, 504–509. [CrossRef]
96. Singh, A.K.; McGoldrick, L.L.; Demirkhanyan, L.; Leslie, M.; Zakharian, E.; Sobolevsky, A.I. Structural Basis of Temperature Sensation by the TRP Channel TRPV3. *Nat. Struct. Mol. Biol.* **2019**, *26*, 994–998. [CrossRef] [PubMed]
97. Zubcevic, L.; Herzik, M.A.; Wu, M.; Borschel, W.F.; Hirschi, M.; Song, A.S.; Lander, G.C.; Lee, S.-Y. Conformational Ensemble of the Human TRPV3 Ion Channel. *Nat. Commun.* **2018**, *9*, 4773. [CrossRef]
98. Zubcevic, L.; Borschel, W.F.; Hsu, A.L.; Borgnia, M.J.; Lee, S.-Y. Regulatory Switch at the Cytoplasmic Interface Controls TRPV Channel Gating. *eLife* **2019**, *8*, e47746. [CrossRef] [PubMed]
99. Flegler, V.J.; Rasmussen, T.; Böttcher, B. More Than Just Closed and Open: Unraveling a Mechanosensor. *Trends Biochem. Sci.* **2021**, *0*. [CrossRef] [PubMed]
100. Englander, S.W.; Kallenbach, N.R. Hydrogen Exchange and Structural Dynamics of Proteins and Nucleic Acids. *Q. Rev. Biophys.* **1983**, *16*, 521–655. [CrossRef] [PubMed]

101. Konermann, L.; Pan, J.; Liu, Y.-H. Hydrogen Exchange Mass Spectrometry for Studying Protein Structure and Dynamics. *Chem. Soc. Rev.* **2011**, *40*, 1224–1234. [CrossRef]
102. Wales, T.E.; Eggertson, M.J.; Engen, J.R. Considerations in the Analysis of Hydrogen Exchange Mass Spectrometry Data. In *Mass Spectrometry Data Analysis in Proteomics*; Matthiesen, R., Ed.; Methods in Molecular Biology; Humana Press: Totowa, NJ, USA, 2013; pp. 263–288. [CrossRef]
103. Smith, D.L.; Deng, Y.; Zhang, Z. Probing the Non-Covalent Structure of Proteins by Amide Hydrogen Exchange and Mass Spectrometry. *J. Mass Spectrom.* **1997**, *32*, 135–146. [CrossRef]
104. Engen, J.R. Analysis of Protein Conformation and Dynamics by Hydrogen/Deuterium Exchange MS. *Anal. Chem.* **2009**, *81*, 7870–7875. [CrossRef]
105. Martens, C.; Shekhar, M.; Lau, A.M.; Tajkhorshid, E.; Politis, A. Integrating Hydrogen–Deuterium Exchange Mass Spectrometry with Molecular Dynamics Simulations to Probe Lipid-Modulated Conformational Changes in Membrane Proteins. *Nat. Protoc.* **2019**, *14*, 3183–3204. [CrossRef] [PubMed]
106. Martens, C.; Politis, A. A Glimpse into the Molecular Mechanism of Integral Membrane Proteins through Hydrogen–Deuterium Exchange Mass Spectrometry. *Protein Sci.* **2020**, *29*, 1285–1301. [CrossRef]
107. Komolov, K.E.; Du, Y.; Duc, N.M.; Betz, R.M.; Rodrigues, J.P.G.L.M.; Leib, R.D.; Patra, D.; Skiniotis, G.; Adams, C.M.; Dror, R.O.; et al. Structural and Functional Analysis of a B2-Adrenergic Receptor Complex with GRK5. *Cell* **2017**, *169*, 407–421.e16. [CrossRef] [PubMed]
108. Liu, H.; Kim, H.R.; Deepak, R.N.V.K.; Wang, L.; Chung, K.Y.; Fan, H.; Wei, Z.; Zhang, C. Orthosteric and Allosteric Action of the C5a Receptor Antagonists. *Nat. Struct. Mol. Biol.* **2018**, *25*, 472–481. [CrossRef] [PubMed]
109. Merkle, P.S.; Gotfryd, K.; Cuendet, M.A.; Leth-Espensen, K.Z.; Gether, U.; Loland, C.J.; Rand, K.D. Substrate-Modulated Unwinding of Transmembrane Helices in the NSS Transporter LeuT. *Sci. Adv.* **2018**, *4*, eaar6179. [CrossRef]
110. Adhikary, S.; Deredge, D.J.; Nagarajan, A.; Forrest, L.R.; Wintrode, P.L.; Singh, S.K. Conformational Dynamics of a Neurotransmitter:Sodium Symporter in a Lipid Bilayer. *Proc. Natl. Acad. Sci. USA* **2017**, *114*, E1786–E1795. [CrossRef]
111. Martens, C.; Shekhar, M.; Borysik, A.J.; Lau, A.M.; Reading, E.; Tajkhorshid, E.; Booth, P.J.; Politis, A. Direct Protein-Lipid Interactions Shape the Conformational Landscape of Secondary Transporters. *Nat. Commun.* **2018**, *9*, 4151. [CrossRef]
112. Mehmood, S.; Domene, C.; Forest, E.; Jault, J.-M. Dynamics of a Bacterial Multidrug ABC Transporter in the Inward- and Outward-Facing Conformations. *Proc. Natl. Acad. Sci. USA* **2012**, *109*, 10832–10836. [CrossRef]
113. Chaptal, V.; Zampieri, V.; Wiseman, B.; Orelle, C.; Martin, J.; Nguyen, K.-A.; Magnard, S.; Gobet, A.; Cesare, M.D.; Javed, W.; et al. Drug-Bound and -Free Outward-Facing Structures of a Multidrug ABC Exporter Point to a Swing Mechanism. *bioRxiv* **2021**. [CrossRef]
114. Hebling, C.M.; Morgan, C.R.; Stafford, D.W.; Jorgenson, J.W.; Rand, K.D.; Engen, J.R. Conformational Analysis of Membrane Proteins in Phospholipid Bilayer Nanodiscs by Hydrogen Exchange Mass Spectrometry. *Anal. Chem.* **2010**, *82*, 5415–5419. [CrossRef] [PubMed]
115. Yamashita, A.; Singh, S.K.; Kawate, T.; Jin, Y.; Gouaux, E. Crystal Structure of a Bacterial Homologue of Na^+/Cl^--Dependent Neurotransmitter Transporters. *Nature* **2005**, *437*, 215–223. [CrossRef] [PubMed]
116. Penmatsa, A.; Gouaux, E. How LeuT Shapes Our Understanding of the Mechanisms of Sodium-Coupled Neurotransmitter Transporters. *J. Physiol.* **2014**, *592*, 863–869. [CrossRef] [PubMed]
117. Li, M.J.; Guttman, M.; Atkins, W.M. Conformational Dynamics of P-Glycoprotein in Lipid Nanodiscs and Detergent Micelles Reveal Complex Motions on a Wide Time Scale. *J. Biol. Chem.* **2018**, *293*, 6297–6307. [CrossRef]
118. Reading, E.; Hall, Z.; Martens, C.; Haghighi, T.; Findlay, H.; Ahdash, Z.; Politis, A.; Booth, P.J. Interrogating Membrane Protein Conformational Dynamics within Native Lipid Compositions. *Angew. Chem. Int. Ed.* **2017**, *56*, 15654–15657. [CrossRef] [PubMed]
119. Xu, P.; Huang, S.; Zhang, H.; Mao, C.; Zhou, X.E.; Cheng, X.; Simon, I.A.; Shen, D.-D.; Yen, H.-Y.; Robinson, C.V.; et al. Structural Insights into the Lipid and Ligand Regulation of Serotonin Receptors. *Nature* **2021**, *592*, 469–473. [CrossRef] [PubMed]
120. Dawaliby, R.; Trubbia, C.; Delporte, C.; Masureel, M.; Van Antwerpen, P.; Kobilka, B.K.; Govaerts, C. ALLOSTERIC REGULATION OF GPCR ACTIVITY BY PHOSPHOLIPIDS. *Nat. Chem. Biol.* **2016**, *12*, 35–39. [CrossRef]
121. Takahashi, N.; Hamada-Nakahara, S.; Itoh, Y.; Takemura, K.; Shimada, A.; Ueda, Y.; Kitamata, M.; Matsuoka, R.; Hanawa-Suetsugu, K.; Senju, Y.; et al. TRPV4 Channel Activity Is Modulated by Direct Interaction of the Ankyrin Domain to $PI(4,5)P_2$. *Nat. Commun.* **2014**, *5*, 4994. [CrossRef]
122. Zhang, M.; Gui, M.; Wang, Z.-F.; Gorgulla, C.; Yu, J.J.; Wu, H.; Sun, Z.J.; Klenk, C.; Merklinger, L.; Morstein, L.; et al. Cryo-EM Structure of an Activated GPCR–G Protein Complex in Lipid Nanodiscs. *Nat. Struct. Mol. Biol.* **2021**, *28*, 258–267. [CrossRef]
123. Staus, D.P.; Hu, H.; Robertson, M.J.; Kleinhenz, A.L.W.; Wingler, L.M.; Capel, W.D.; Latorraca, N.R.; Lefkowitz, R.J.; Skiniotis, G. Structure of the M2 Muscarinic Receptor–β-Arrestin Complex in a Lipid Nanodisc. *Nature* **2020**, *579*, 297–302. [CrossRef]
124. Thompson, M.J.; Baenziger, J.E. Structural Basis for the Modulation of Pentameric Ligand-Gated Ion Channel Function by Lipids. *Biochim. Biophys. Acta BBA Biomembr.* **2020**, *1862*, 183304. [CrossRef]
125. Bocquet, N.; Nury, H.; Baaden, M.; Le Poupon, C.; Changeux, J.-P.; Delarue, M.; Corringer, P.-J. X-Ray Structure of a Pentameric Ligand-Gated Ion Channel in an Apparently Open Conformation. *Nature* **2009**, *457*, 111–114. [CrossRef] [PubMed]
126. Kumar, P.; Wang, Y.; Zhang, Z.; Zhao, Z.; Cymes, G.D.; Tajkhorshid, E.; Grosman, C. Cryo-EM Structures of a Lipid-Sensitive Pentameric Ligand-Gated Ion Channel Embedded in a Phosphatidylcholine-Only Bilayer. *Proc. Natl. Acad. Sci. USA* **2020**, *117*, 1788–1798. [CrossRef] [PubMed]

127. Kim, J.J.; Gharpure, A.; Teng, J.; Zhuang, Y.; Howard, R.J.; Zhu, S.; Noviello, C.M.; Walsh, R.M.; Lindahl, E.; Hibbs, R.E. Shared Structural Mechanisms of General Anaesthetics and Benzodiazepines. *Nature* **2020**, *585*, 303–308. [CrossRef]
128. Guixà-González, R.; Albasanz, J.L.; Rodriguez-Espigares, I.; Pastor, M.; Sanz, F.; Martí-Solano, M.; Manna, M.; Martinez-Seara, H.; Hildebrand, P.W.; Martín, M.; et al. Membrane Cholesterol Access into a G-Protein-Coupled Receptor. *Nat. Commun.* **2017**, *8*, 14505. [CrossRef]
129. Paila, Y.D.; Chattopadhyay, A. The Function of G-Protein Coupled Receptors and Membrane Cholesterol: Specific or General Interaction? *Glycoconj. J.* **2008**, *26*, 711. [CrossRef]
130. Gault, J.; Donlan, J.A.C.; Liko, I.; Hopper, J.T.S.; Gupta, K.; Housden, N.G.; Struwe, W.B.; Marty, M.T.; Mize, T.; Bechara, C.; et al. High-Resolution Mass Spectrometry of Small Molecules Bound to Membrane Proteins. *Nat. Methods* **2016**, *13*, 333–336. [CrossRef]
131. Marcoux, J.; Wang, S.C.; Politis, A.; Reading, E.; Ma, J.; Biggin, P.C.; Zhou, M.; Tao, H.; Zhang, Q.; Chang, G.; et al. Mass Spectrometry Reveals Synergistic Effects of Nucleotides, Lipids, and Drugs Binding to a Multidrug Resistance Efflux Pump. *Proc. Natl. Acad. Sci. USA* **2013**, *110*, 9704–9709. [CrossRef]
132. Yen, H.-Y.; Hoi, K.K.; Liko, I.; Hedger, G.; Horrell, M.R.; Song, W.; Wu, D.; Heine, P.; Warne, T.; Lee, Y.; et al. PIP2 Stabilises Active States of GPCRs and Enhances the Selectivity of G-Protein Coupling. *Nature* **2018**, *559*, 423–427. [CrossRef] [PubMed]
133. Cong, X.; Liu, Y.; Liu, W.; Liang, X.; Laganowsky, A. Allosteric Modulation of Protein-Protein Interactions by Individual Lipid Binding Events. *Nat. Commun.* **2017**, *8*, 2203. [CrossRef]
134. Schrecke, S.; Zhu, Y.; McCabe, J.W.; Bartz, M.; Packianathan, C.; Zhao, M.; Zhou, M.; Russell, D.; Laganowsky, A. Selective Regulation of Human TRAAK Channels by Biologically Active Phospholipids. *Nat. Chem. Biol.* **2021**, *17*, 89–95. [CrossRef] [PubMed]
135. Patrick, J.W.; Boone, C.D.; Liu, W.; Conover, G.M.; Liu, Y.; Cong, X.; Laganowsky, A. Allostery Revealed within Lipid Binding Events to Membrane Proteins. *Proc. Natl. Acad. Sci. USA* **2018**, *115*, 2976–2981. [CrossRef] [PubMed]
136. Liu, Y.; LoCaste, C.E.; Liu, W.; Poltash, M.L.; Russell, D.H.; Laganowsky, A. Selective Binding of a Toxin and Phosphatidylinositides to a Mammalian Potassium Channel. *Nat. Commun.* **2019**, *10*, 1352. [CrossRef]
137. Bolla, J.R.; Sauer, J.B.; Wu, D.; Mehmood, S.; Allison, T.M.; Robinson, C.V. Direct Observation of the Influence of Cardiolipin and Antibiotics on Lipid II Binding to MurJ. *Nat. Chem.* **2018**, *10*, 363–371. [CrossRef] [PubMed]
138. Su, C.-C.; Klenotic, P.A.; Bolla, J.R.; Purdy, G.E.; Robinson, C.V.; Yu, E.W. MmpL3 Is a Lipid Transporter That Binds Trehalose Monomycolate and Phosphatidylethanolamine. *Proc. Natl. Acad. Sci. USA* **2019**, *116*, 11241–11246. [CrossRef]
139. Bechara, C.; Nöll, A.; Morgner, N.; Degiacomi, M.T.; Tampé, R.; Robinson, C.V. A Subset of Annular Lipids Is Linked to the Flippase Activity of an ABC Transporter. *Nat. Chem.* **2015**, *7*, 255–262. [CrossRef]
140. Liko, I.; Degiacomi, M.T.; Lee, S.; Newport, T.D.; Gault, J.; Reading, E.; Hopper, J.T.S.; Housden, N.G.; White, P.; Colledge, M.; et al. Lipid Binding Attenuates Channel Closure of the Outer Membrane Protein OmpF. *Proc. Natl. Acad. Sci. USA* **2018**, *115*, 6691–6696. [CrossRef]
141. Mehmood, S.; Corradi, V.; Choudhury, H.G.; Hussain, R.; Becker, P.; Axford, D.; Zirah, S.; Rebuffat, S.; Tieleman, D.P.; Robinson, C.V.; et al. Structural and Functional Basis for Lipid Synergy on the Activity of the Antibacterial Peptide ABC Transporter McjD. *J. Biol. Chem.* **2016**, *291*, 21656–21668. [CrossRef]
142. Hariharan, P.; Tikhonova, E.; Medeiros-Silva, J.; Jeucken, A.; Bogdanov, M.V.; Dowhan, W.; Brouwers, J.F.; Weingarth, M.; Guan, L. Structural and Functional Characterization of Protein–Lipid Interactions of the Salmonella Typhimurium Melibiose Transporter MelB. *BMC Biol.* **2018**, *16*, 85. [CrossRef]
143. Cornelius, F.; Habeck, M.; Kanai, R.; Toyoshima, C.; Karlish, S.J.D. General and Specific Lipid–Protein Interactions in Na,K-ATPase. *Biochim. Biophys. Acta BBA Biomembr.* **2015**, *1848*, 1729–1743. [CrossRef]
144. Shinoda, T.; Ogawa, H.; Cornelius, F.; Toyoshima, C. Crystal Structure of the Sodium–Potassium Pump at 2.4 Å Resolution. *Nature* **2009**, *459*, 446–450. [CrossRef]
145. Habeck, M.; Kapri-Pardes, E.; Sharon, M.; Karlish, S.J.D. Specific Phospholipid Binding to Na,K-ATPase at Two Distinct Sites. *Proc. Natl. Acad. Sci. USA* **2017**, *114*, 2904–2909. [CrossRef]
146. Schur, F.K. Toward High-Resolution in Situ Structural Biology with Cryo-Electron Tomography and Subtomogram Averaging. *Curr. Opin. Struct. Biol.* **2019**, *58*, 1–9. [CrossRef]
147. Sigal, Y.M.; Zhou, R.; Zhuang, X. Visualizing and Discovering Cellular Structures with Super-Resolution Microscopy. *Science* **2018**, *361*, 880–887. [CrossRef]
148. Chorev, D.S.; Baker, L.A.; Wu, D.; Beilsten-Edmands, V.; Rouse, S.L.; Zeev-Ben-Mordehai, T.; Jiko, C.; Samsudin, F.; Gerle, C.; Khalid, S.; et al. Protein Assemblies Ejected Directly from Native Membranes Yield Complexes for Mass Spectrometry. *Science* **2018**, *362*, 829–834. [CrossRef] [PubMed]
149. Kaledhonkar, S.; Fu, Z.; White, H.; Frank, J. Time-Resolved Cryo-Electron Microscopy Using a Microfluidic Chip. In *Protein Complex Assembly: Methods and Protocols*; Marsh, J.A., Ed.; Methods in Molecular Biology; Springer: New York, NY, USA, 2018; pp. 59–71. [CrossRef]
150. Kaledhonkar, S.; Fu, Z.; Caban, K.; Li, W.; Chen, B.; Sun, M.; Gonzalez, R.L.; Frank, J. Late Steps in Bacterial Translation Initiation Visualized Using Time-Resolved Cryo-EM. *Nature* **2019**, *570*, 400–404. [CrossRef] [PubMed]

151. Mäeots, M.-E.; Lee, B.; Nans, A.; Jeong, S.-G.; Esfahani, M.M.N.; Ding, S.; Smith, D.J.; Lee, C.-S.; Lee, S.S.; Peter, M.; et al. Modular Microfluidics Enables Kinetic Insight from Time-Resolved Cryo-EM. *Nat. Commun.* **2020**, *11*, 3465. [CrossRef]
152. Yao, X.; Fan, X.; Yan, N. Cryo-EM Analysis of a Membrane Protein Embedded in the Liposome. *Proc. Natl. Acad. Sci. USA* **2020**, *117*, 18497–18503. [CrossRef]

International Journal of *Molecular Sciences*

MDPI

Review

Insights into the Role of Membrane Lipids in the Structure, Function and Regulation of Integral Membrane Proteins

Kenta Renard and Bernadette Byrne *

Department of Life Sciences, Imperial College London, Exhibition Road, London SW7 2AZ, UK; kenta.renard18@imperial.ac.uk
* Correspondence: b.byrne@imperial.ac.uk

Abstract: Membrane proteins exist within the highly hydrophobic membranes surrounding cells and organelles, playing key roles in cellular function. It is becoming increasingly clear that the membrane does not just act as an appropriate environment for these proteins, but that the lipids that make up these membranes are essential for membrane protein structure and function. Recent technological advances in cryogenic electron microscopy and in advanced mass spectrometry methods, as well as the development of alternative membrane mimetic systems, have allowed experimental study of membrane protein–lipid complexes. These have been complemented by computational approaches, exploiting the ability of Molecular Dynamics simulations to allow exploration of membrane protein conformational changes in membranes with a defined lipid content. These studies have revealed the importance of lipids in stabilising the oligomeric forms of membrane proteins, mediating protein–protein interactions, maintaining a specific conformational state of a membrane protein and activity. Here we review some of the key recent advances in the field of membrane protein–lipid studies, with major emphasis on respiratory complexes, transporters, channels and G-protein coupled receptors.

Keywords: integral membrane protein; membrane lipid; structure; function; oligomeric state; cryo-EM; advanced mass spectrometry; molecular dynamics; membrane mimetic systems

Citation: Renard, K.; Byrne, B. Insights into the Role of Membrane Lipids in the Structure, Function and Regulation of Integral Membrane Proteins. *Int. J. Mol. Sci.* **2021**, *22*, 9026. https://doi.org/10.3390/ijms22169026

Academic Editor: Masoud Jelokhani-Niaraki

Received: 20 July 2021
Accepted: 17 August 2021
Published: 21 August 2021

Publisher's Note: MDPI stays neutral with regard to jurisdictional claims in published maps and institutional affiliations.

1. Introduction

Biological membranes form barriers separating cellular or organellar contents from the external environment. These are comprised of complex mixtures of polar lipids and membrane proteins. The long-standing Fluid-Mosaic model described the structure of the membrane as a bilayer of freely laterally diffusing polar lipids forming a highly hydrophobic core and acting as solvent for the membrane proteins [1]. Whilst some features of this model still hold true, it is becoming increasingly clear that the membrane is more organised than this model suggested [2], with the presence of discrete domains or rafts in the plasma membrane acting as signalling hubs [3]. A further key feature of biological membranes is their asymmetry, with the individual monolayers that make up the bilayer having distinct lipid compositions and associated functional implications [4]. It is also becoming increasingly evident that the membrane lipids do not just act as a solvent for membrane proteins but have critical roles in their structure and function [5]. Indeed, it seems in many cases that the functional unit is a complex of membrane protein and associated lipids. An interesting study using the rhomboid protease GlpG as a model protein suggested that the presence of cavities and pockets on the external membrane-facing surfaces of a protein induce instability key for the membrane protein conformational changes. Lipid interactions in these regions do not limit the conformational flexibility of the protein but do reduce the instability associated with the presence of the cavities [6].

Until recently it has been challenging to obtain definitive information on the precise nature and role of the interactions between individual membrane proteins and membrane lipids. This has been due in part to the limitations in technologies and the fact that membrane proteins are typically solubilised from the membrane for structural and other biophysical analysis, a process using detergents that is designed to remove most if not all the interacting lipids. This has led to some controversy regarding some membrane protein structures [7]. Despite this, it has been possible to explore many important membrane protein–lipid interactions. In this review, we summarise recent advances in our understanding of the roles of lipids in a range of membrane proteins and membrane protein complexes (Table 1).

Table 1. Summary of the lipid interactions covered in this manuscript.

		Membrane Protein					
		Respiratory Complexes	**2° Active Transporters**	**ATP-Dependent Pumps/ Transporters**	**Channels**	**GPCRs**	**Other**
Lipid entity	Cholesterol		Activity regulation [8,9]	Activity and changes in conformation [10], Modulation of substrate binding affinity [11]	Channel inhibition [12]	Activation [13], Signalling [14,15], Stability [16,17], Allosteric regulation [18], Lower affinity ligand binding [17], Oligomerisation [19–23]	
	PA				Activity [24]	Stabilisation [25]	
	PE		Dimer formation and function [26], Dimer formation [27], conformer stability [28]		Conformer stabilisation and channel desensitisation [29]	Agonist and antagonist binding affinities [30], Increase G protein coupling [31,32]	Protein assembly [33]
	PC					Dimerisation [34], β-arrestin interaction and function [35], Increase G protein coupling [31]	
	PG		Oligomerisation and function [36]		Conformer stabilisation and channel desensitisation [37]	Increase G protein coupling [30,38], Active conformer stability [39], β-arrestin interaction and function [35], Decrease G protein coupling [31]	
	PS		Dimer formation [27],			Stabilisation [25], Dimerisation [34], Decrease G protein coupling [31]	
	PI		Dimer formation and function [26], Possible stabilisation [27]				

Table 1. *Cont.*

	Membrane Protein					
	Respiratory Complexes	**2° Active Transporters**	**ATP-Dependent Pumps/ Transporters**	**Channels**	**GPCRs**	**Other**
$PtdIns(4,5)P_2$		Possible dimer stabilisation [40,41],		Activation [42,43], Inactive conformer stability [44]	Active conformer stability [25,45], Increase G protein interaction [25], β-arrestin interaction and complex stability [46], Important for G protein recruitment [47]	
Glycolipids and sphingolipids					Ligand binding [48,49], Signalling [48], Allostery [47]	
Cardiolipin	Oligomer stabilisation [50]	Oligomerisation [51], Allosteric regulation [52]				
DHA and unsaturated tails					Active conformer stability [53,54], Increase G protein coupling [53], Partitioning to lipid rafts [55], Oligomerisation [34,56]	
Saturated tails					Dimerisation and function [57]	

2. Lipids and Respiratory Complexes

High-resolution structural studies have provided a number of insights into the role that tightly associated lipids play in the structure of membrane proteins. In virtually all cases the proteins are extracted from the membrane using detergent. However, in many cases, tightly associated lipids remain in complex with the protein even after detergent extraction and purification. If the lipids are sufficiently ordered, then they can be observed in X-ray crystal structures. One early example of lipids being clearly visible was in the crystal structure of Formate Dehydrogenase-N (Fdh-N) [50]. Each protomer of Fdh-N contains three subunits with the γ-subunit and a single transmembrane (TM) domain of the β-subunit forming the integral membrane region. FdhN crystallised as a physiological trimer with the interactions between the individual membrane domains mediated by molecules of cardiolipin (CL) forming essential interactions between the integral membrane regions of the individual protomers and clearly stabilising the oligomer. The integral membrane regions of the individual protomers have little role in this interaction, with the lipid almost entirely responsible for mediating trimer formation within the membrane. Although Fdh-N functions as a monomer, it is highly likely that trimer formation is critical for stability and thus the CL molecules are essential for formation of the physiological oligomeric state of the protein [50].

More recently, and in an exciting development, researchers have exploited a detergent-free approach using styrene maleic acid copolymer (SMA) to extract the Alternative Complex III (ACIII) from the bacterium *Flavobacterium johnsoniae* to produce SMA lipid particles

(SMALPs) containing the ACIII along with native lipids [58]. The SMA approach punches holes in the membrane and surrounds the lipid and protein particles rather than disrupting the hydrophobic interactions between the membrane protein and the membrane lipids as detergents do [59]. One very appealing consequence of this mode of extraction from the membrane is that hitherto uncharacterised membrane complexes can be isolated. The researchers in this case effectively isolated a super complex of ACIII and cytochrome c oxidase. In addition, electron density assigned to 11 phospholipid (PL) molecules was also discernible in the structure in two key regions. The first region is between two of the ACIII subunits, suggesting a role for the PLs in the stability of the protein, and the second region flanks a triacylated cysteine residue in the ActB subunit, close to the site of menaquinol entry into the protein, suggesting a role for the PLs in the function of the protein [58]. Reports of lipids binding to other respiratory complexes are covered in an earlier review [60]. These include the yeast bc_1 complex with an initial crystal structure obtained in complex with five closely associated lipid molecules [61]. Interestingly, in this case the researchers were able to alter the amount of bound lipid by changing the purification protocol, limiting the amount of time the protein spent on an ion exchange chromatography column. Protein produced with this optimised purification protocol was more active and yielded a crystal structure with an additional bound lipid [62].

A cryo-EM structure revealed lipid bound to the complete F-type ATPase from sheep [63]. Two lipid molecules are bound into the c-ring, part of the integral membrane, Fo domain, central to proton translocation. The lipids are bound into both the matrix and the intramembrane space sides of the c-ring. The e subunit of Fo, which forms part of the hook apparatus, interacts with the lipid, possibly a lysolipid, bound to the intra-membrane side of the c-ring via the C-terminal Lys residue. This lipid-mediated connection between different regions of the Fo domain is thought to increase the stability of the complex but is also likely to play a key role in ATPase function. Confirmational changes in the protein would cause movement of subunit e away from the c-ring, removing the associated lipid. It is suggested that this is an early step in full opening of the proton translocation channel [63].

3. Secondary Active Transporters

3.1. CitS

The structure of the CitS transporter from *Salmonella enterica* reveals the presence of a lipid, assigned as phosphatidylethanolamine (PE) on the edge of the protein [64]. It is not clear if this lipid has any specific role in CitS structure and function, but the lipid molecule is only detectable on one of the two protomers in the CitS dimer. Given that the two protomers are in different conformational states it is possible that binding of the lipid is conformationally specific.

3.2. *UapA and ScBOR1p*

In some cases, lipids might be associated with a membrane protein but too disordered to be visible in the crystal structure. This was the case with UapA, a xanthine/uric acid transporter from *Aspergillus nidulans*, whose structure was determined as a closely associated, functionally relevant dimer with no visible lipids [65]. Native mass spectrometry (MS) analysis is proving to be a very powerful method for exploring membrane protein–lipid interactions [51,66], and this approach revealed that the UapA dimer isolates in the detergent dodecylmaltoside in complex with membrane lipids [26]. Further lipidomics analysis identified that UapA co-purifies with the membrane lipids, phosphatidylinositol (PI), phosphatidylcholine (PC) and PE. Loss of these PLs results in dissociation of the UapA dimer into the monomeric state, with the dimer recoverable through the addition of exogenous PE and/or PI. Molecular dynamics simulations predicted the location of a lipid binding site made up of three Arg residues (R287, R478, R479) at the dimer interface and on the intracellular membrane leaflet. Subsequent mutagenic and MS analyses indicated that substitution of these three Arg residues caused loss of function and resulted in protein that was almost exclusively in the monomeric form. Addition of exogenous lipid to the mutant lacking the binding site was unable to recover the dimer form, strongly indicating that binding of lipids to this site in the wild-type protein is key in the formation and maintenance of the physiological dimeric state [26]. This research highlighted that the PLs were essential for functional dimer formation and since the crystal structure of UapA is a dimer, PLs must be present in the crystals of UapA, albeit too disordered to be detectable in the final structure [65].

UapA is structurally and mechanistically related to other transporters from the solute carrier (SLC) 4 and SLC26 families, including the BOR proteins, boron transporters. The BOR protein from *Saccharomyces cerevisiae*, ScBOR1p, isolates as a monomer in both dodecyl-β-D-maltoside (DDM) and Triton X-100 [27], but lipidomics analysis reveals that it co-purifies in the presence of PI, PE, PC and phosphatidylserine (PS). As with UapA, addition of exogenous lipid to delipidated ScBOR1p causes the monomers to associate into dimers as revealed by native MS. A similar lipid binding site was predicted at the intracellular side of the dimer interface of ScBOR1p, from a model of the protein based on the crystal structure of *Arabidopsis thaliana* BOR1 [67]. Mutagenesis of the lipid binding site in ScBOR1p prevents lipid dependent dimer formation but does not abolish transport function, indicating that the dimer is not critical for function and highlighting that although there are clear similarities in the interactions between UapA and ScBOR1p and membrane lipids, the precise functional outcome of that interaction differs between the two proteins [27].

3.3. Monoamine Transporters, hSERT and hDAT, and the Homologue LeuT

Differences in relatively related transporters and their interactions with lipids are also seen in the serotonin (hSERT) and dopamine (hDAT) transporters, both monoamine transporters. There is strong support for both transporters being oligomeric in the native membrane, with the hSERT oligomers stabilised by phosphatidylinositol-4,5-bisphosphate (PtdIns(4,5)P_2) binding [40]. However, whilst there is clear evidence of binding of PtdIns(4,5)P_2 to the N-terminus of hDAT and a suggested role in amphetamine-induced dopamine efflux through the hDAT [41], stability of the hDAT dimers is independent of PtdIns(4,5)P_2 as assessed by single molecule fluorescence microscopy before and after enzymatic depletion of PtdIns(4,5)P_2 in the membrane [68]. Crystal structures of both the drosophila DAT [69,70] and the hSERT transporter [71] identify putative cholesterol-binding sites. Cholesterol is important for the functioning of hSERT. Recent analysis indicates that mutation of a cholesterol-binding site in hSERT or depletion of membrane cholesterol results in the transporter preferentially adopting the inward facing conformation which in turn reduces transporter activity [8,9]. Conversely, mutations in the cholesterol binding site that favour cholesterol binding cause the transporter to preferentially adopt an outward facing conformation [9]. Thus, association and dissociation of cholesterol at a key site in the protein may play an essential role in regulating hSERT activity. Given the fact that similar effects of cholesterol binding on transporter conformation are reported for hDAT [72,73], it is possible that cholesterol plays a similar regulatory role in hDAT.

A recent publication from the groups of Michael Landreh and Carol Robinson explored the concept of annular lipids that provide the hydrophobic environment essential for maintaining the overall structure of the protein [74] versus specific lipids in more detail [75]. They developed a method which exploits the fact that loosely bound, annular lipids on the surface of a protein are more prone to exchange for detergent molecules than closely associated lipid molecules. Following detergent exchange, the protein is analysed by native MS which detects changes in the abundance of bound lipid. The results of this study revealed that the bacterial presenilin homologue forms only weak interactions with annular lipids whereas LeuT, a bacterial homologue of the monoamine transporters, binds both specific lipids at a dimer interface and annular lipids on the periphery [75]. Additional analysis has indicated a role for CL in the oligomerisation of LeuT [51], a lipid also implicated in allosteric regulation of the bacterial lipid II exporter, MurJ [52].

3.4. The Betaine Transporter, BetP

Although it can be difficult to obtain structural insights into transporter-lipid interactions as a result of both the loss of lipids and the poor structural resolution of the lipid molecules, it is possible. One example is the bacterial transporter, BetP, involved in osmoregulation, which crystallised in complex with lipids whose density was clearly discernible in the high resolution (2.7 Å) structure. The lipid density was assigned as eight palmitoyl-oleoyl phosphatidyl glycerol (PG) molecules [36], with seven of the PG molecules located on the intracellular side of the protein and one located on the extracellular side. Five of these lipids mediate protomer–protomer interactions within the BetP trimer (Figure 1A) playing a fundamental role in oligomer formation. The remaining three lipid molecules, including the lipid on the extracellular side of the protein, are more loosely associated with the periphery of the protein (Figure 1B,C) and are likely to be annular lipids. The lipids bound to BetP are associated with regions known to be involved in conformational changes associated with transport activity and transport regulation, strongly indicating that in addition to the quaternary structure of the protein, lipid binding is also critical for function [36].

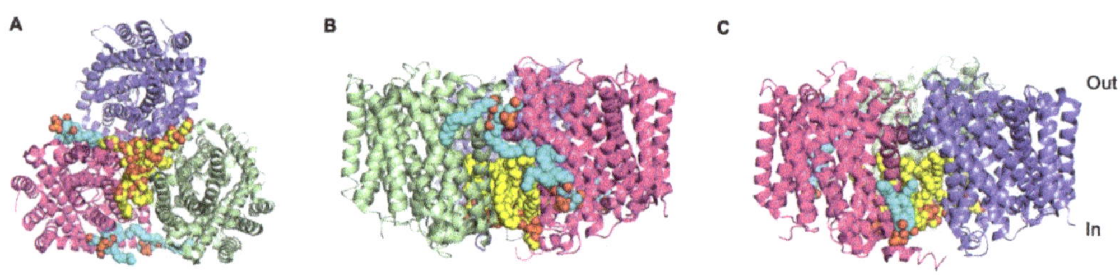

Figure 1. Crystal structure of the trimer of BetP in complex with the lipid (PDB: 4C7R [36]). The individual protomers are shown as green, bright pink and blue ribbons. The lipids key in trimer formation are shown in sphere representation with yellow carbon atoms. The annular lipids bound to the periphery of the protein are shown in sphere representation with cyan carbon atoms. (**A**) The protein–lipid complex is shown from the intracellular side of the membrane. (**B**,**C**) The protein–lipid complex in two different views looking through the membrane to illustrate the different locations of the peripherally bound lipids.

3.5. The Cationic Amino Acid Transporter, GkApcT

The presence of lipid, in this case cholesterol, was essential for crystallisation of a cationic amino acid transporter homologue from the thermophilic bacterium, *Geobacillus kaustophilus*, GkApcT [76]. A cholesterol molecule sits in a pocket formed by interactions between the GkApcT and another single transmembrane domain protein, MgtS, stabilising this interaction. Clearly this is a very interesting lipid dependency since bacteria do not produce cholesterol. The authors of the study suggest that the cholesterol may be a functional replacement for a group of chemically similar lipids, called the hopanoids, found in some bacteria [77]. Whilst the precise role of the potential hopanoid interaction in vivo is not clear, it is possible that the lipid plays a role in transporter regulation as described above for cholesterol and the monoamine transporters.

3.6. The Major Facilitator Superfamily Sugar Transporters, LacY and XylE

An exciting new development in understanding the structural and functional implications of membrane protein–lipid interactions is hydrogen–deuterium exchange mass spectrometry (HDX-MS). HDX-MS defines the solvent accessibility of different regions of a protein by monitoring the exchange of hydrogen to deuterium; the exchange reaction depends on intrinsic protein motions [78]. H-bonding networks greatly reduce the rate and efficiency of HDX, whereas highly dynamic regions typically undergo a higher level of HDX [79]. In a recent study, Martens et al., [28] explored the conformational state of two bacterial sugar transporters, LacY and XylE, that had been purified in detergent and then reconstituted into nanodiscs incorporating a mix of PC, PG and CL or PE, PG and CL. Nanodiscs are membrane mimetic systems formed from a membrane scaffold protein, exogenous lipids and the target protein, with the scaffold protein wrapping around the complex of lipids and protein. This arrangement shields the protein from aqueous solution and produces a more native-like environment than detergent micelles whilst also allowing ready variation of the lipid composition [80,81]. The results of the study on LacY and XylE revealed that in the presence of PE both transporters were preferentially in the inward-facing conformational state. The ability to control the lipid composition surrounding isolated protein through nanodisc reconstitution is a powerful means of exploring the role of individual lipids; however, it does suffer from the disadvantage that the protein is isolated initially in detergent and then lipids added. An alternative approach is to use the SMALPs together with HDX-MS, allowing analysis of the conformational dynamics of membrane proteins encapsulated in their native membranes as described for the bacterial rhomboid protease, GlpG [82]. Here, the authors were able to alter the lipid composition at the level of the bacterial membrane by varying the expression strain or the expression temperature. They found that the conformational flexibility of the protein differed depending on the lipid composition. Such approaches have major potential for exploring the protein–lipid relationships of other classes of membrane proteins. Additionally, researchers are developing direct methods, performing MS on proteins removed directly from the membrane without the need for any extraction agents [83]. This technology is still in the early stages, but it has potential to provide information on the direct physiological interactions of membrane proteins with membrane lipids as well as protein–protein interactions.

4. ATP Dependent Pumps and Transporters

Lipids are also crucial for ATP dependent pumps and transporters. The Na^+, K^+ ATPase, for example, was crystallised in complex with a molecule of cholesterol bound between the α and β subunits [84]. The precise role of the cholesterol bound to this protein has been subject to debate, possibly due to the non-physiological conditions used for study of the effect of the lipid. However, recent research uses the addition of methyl-β-cyclodextrin to deplete cholesterol from membrane fragments while keeping all the other membrane components intact [10]. This study revealed that Na^+, K^+ ATPase is less active when there is a lower amount of cholesterol in the membrane and suggests that this is due to less efficient transition between different conformational states critical for the transport cycle.

Cholesterol is also important for the ATP binding cassette (ABC) transporter, PgP. PgP is an extremely important example from this family responsible for multi-drug resistance in many types of cancers [85]. The biochemical relationship between PgP and cholesterol appears to be multi-faceted, involving possible roles for cholesterol in ATPase activity, in modulating binding affinity of some transported substrates, as well as a possible direct function for PgP in cholesterol trafficking [11]. Although there is no direct structural information available for cholesterol binding to the PgP, recent MD simulations have identified several putative cholesterol binding sites and have suggested that the cholesterol cluster specifically on one side of the protein interacts with TM domain 1 [86]. The study also revealed that cholesterol flipping from one leaflet of the membrane can occur along the surface of the protein [86].

5. Channels

Lipids play fundamental roles in the regulation of ion channels [87]. A recent study on the two-protein domain potassium (K_{2P}) channel, TRAAK, using a combination of native MS and liposome-based potassium efflux assay, allowed a detailed analysis of the lipids involved in the regulation of channel activity. The results revealed that TRAAK was activated by high affinity binding of phosphatidic acid (PA) [24].

Ligand–gated ion channels have also been analysed using MS methods revealing that the pentameric, Erwinia ligand–gated ion channel (ELIC) co-purifies in DDM with PG and

PE, with a preference for PG binding [37]. The authors of this study demonstrated that addition of exogenous POPG increases the thermostability of ELIC, and they identified a likely Arg-rich lipid binding site between two subunits on the intracellular side of the TM domains. This site involves residues from TMs 1 and 4. An additional potential site on the extracellular side of the membrane was also suggested. Importantly, the researchers showed that PG binding stabilises the channel in the open, active conformation, with mutations that reduce lipid binding, increasing channel desensitization [37]. A recent structure of ELIC has conclusively identified a lipid binding site on the intracellular side of the membrane (Figure 2A), which in the structure contained a bound molecule of PE [29]. The site is similar to that predicted in the Tong et al., (2019) study and involves one of the Arg residues (Figure 2B) identified by that earlier study [37]. The structure together with MD simulations indicated that the binding of the lipid was critical for stabilising the kinked structure of the TM4 (Figure 2B), and in the absence of the stabilising influence of the lipid, this region of the protein was much more conformationally dynamic, and it is this that was suggested to be the molecular basis for the increased desensitization seen in the lipid-binding site mutants [29]. Taken together, all these findings strongly indicate that lipid binding is key to regulating ELIC gating, a feature that may also be important in some eukaryotic pentameric LGICs. Clearly, further studies are required to confirm precisely which lipid plays this role in the physiological membrane.

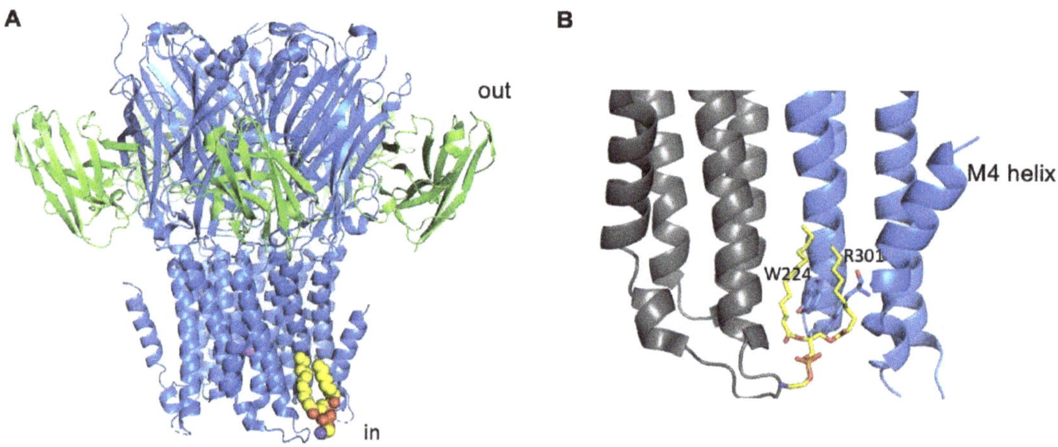

Figure 2. Structure of the Erwinia ligand–gated ion channel (ELIC) in complex with the lipid (PDB: 6HJX [29]). (**A**) The channel is shown in blue, and the nanobody used to facilitate crystallization is shown in green, both in ribbon representation. A bound Na^+ ion is shown in purple, and the bound lipid is shown in yellow. (**B**) Zoomed-in view of the lipid-binding site with only two channel subunits shown (one grey and one blue) revealing the key interacting Arg and Trp residues shown in blue stick representation with blue carbon atoms and illustrating the kinked M4 helix (blue). The lipid is shown in stick representation with yellow carbon atoms.

Duncan et al., [42] used MD simulations to build on earlier biochemical and structural studies [88–90] exploring lipid binding to the inward rectifier (Kir2) potassium channels. Their study confirmed the importance of PtdIns(4,5)P_2 binding in activation of the Kir2 channels and suggested that there is cross talk between PtdIns(4,5)P_2 binding and binding of a further phospholipid, most likely PS, to a second distinct lipid binding site.

In addition to extensive data showing that cholesterol has a key role in channel inhibition [12], these findings reveal that membrane protein–lipid interactions can be very complex indeed.

Lipids have been revealed in the structures of Transient Receptor Potential (TRP) channels, nonselective cation channels sensitive to a range of environmental changes including alterations in temperature and pressure [91]. A recent example is the TRP from the alga, *Chlamydomonas reinhardtii*, whose tetrameric structure revealed the presence of three lipid molecules per protomer [43]. These were assigned as two PtdIns(4,5)P_2 molecules and one PC molecule binding to three distinct sites. PtdIns(4,5)P_2 binding at one site, site 2, is key for channel opening, and removal of that binding site results in a loss of PtdIns(4,5)P_2-induced channel activation. The second PtdIns(4,5)P_2 binding site is similar to the vanilloid binding site, occupied by a PtdIns(4,5)P_2 molecule in the structure of the mammalian TRPV1 channel. In the case of the TRPV1 channel, it is thought that the bound PtdIns(4,5)P_2 stabilises the inactive state of the channel, which is activated upon binding of a specific ligand into the vanilloid binding site, displacing the bound PtdIns(4,5)P_2 molecule [44]. Whilst it is not clear precisely what role the equivalent site plays in the *C. reinhardtii* TRP channel, it seems probable that membrane lipids can act as allosteric modulators of these proteins. This is supported by structures of the temperature sensitive, mouse TRPV3 channel which revealed a lipid, likely to be a phospholipid, bound to the vanilloid binding site in the closed state but not in the open state of the protein. Loss of the lipid at the phospholipid binding site is postulated to be key to transition from the closed to the sensitized and ultimately the open state of the channel upon heat-induced activation [92].

6. G-Protein Coupled Receptors

G-protein coupled receptors (GPCRs) are crucial for cellular responses to a range of bioactive molecules including hormones, neurotransmitters and many drugs. As a result of their biological and pharmacological importance, they have been extensively studied. A vast body of research has accumulated on the roles of lipids in GPCR structure and function. Figure 3 illustrates some of the membrane lipids and their roles. Given the nature of the current state of the art with respect to GPCR-lipid interactions, the following sections have been organised mainly according to lipid rather than protein.

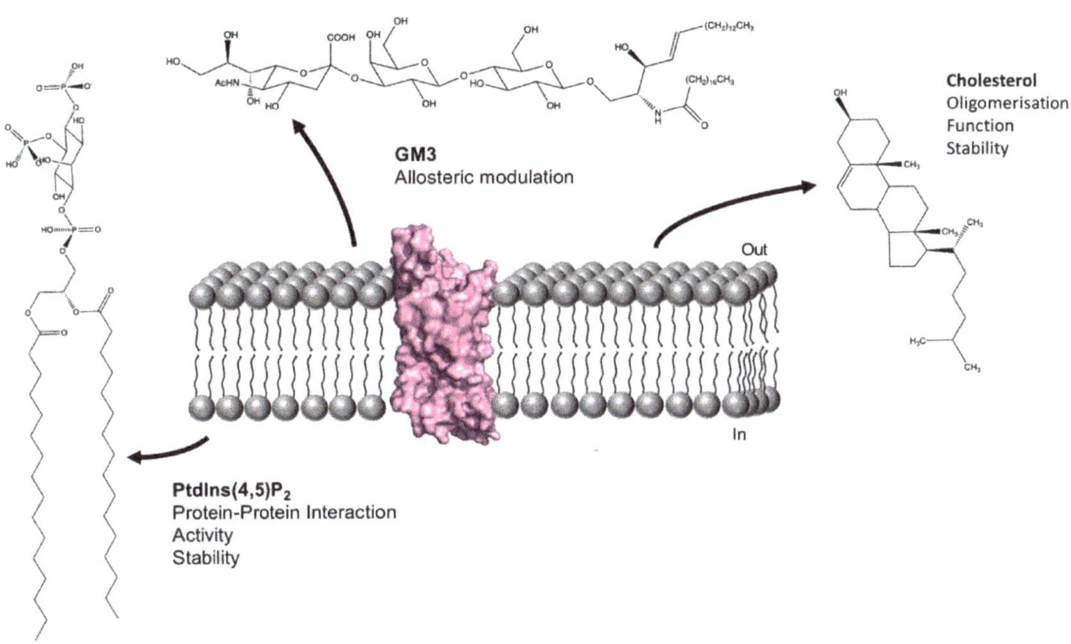

Figure 3. Schematic illustrating some key GPCR-lipid interactions. The structure of the $A_{2A}R$ (PDB: 5UEN [93]) is shown in space filling representation embedded in the lipid membrane. The chemical structures of GM3, cholesterol and PtdIns(4,5)P_2 are shown together with details of their known effects on GPCR structure and function.

6.1. GPCRs and Cholesterol

It has long been known that cholesterol has a key role in GPCR structure and function [94]. Cholesterol directly affects the ligand–binding ability of several GPCRs, including the subtype 2 galanin receptor and the serotonin 1A receptor [95], and there is evidence that cholesterol also plays a role in GPCR signalling, for example, increasing basal activity of the cannabinoid 2 receptor [14]. In the recent study of the class F GPCR, Smoothened, cholesterol is revealed to traffic through a channel in the receptor and play a fundamental role in receptor activation [13].

In other cases, the more indirect effects of cholesterol on the biophysical properties of the membrane appear to be important [96]. Cholesterol is important for the stability of receptors, as supported by a raft of different GPCR structures reviewed in Gimpl [16]. Cholesterol binding was observed in a groove created by TMs 1-4 in a high-resolution structure of the β_2-adrenergic receptor [97], leading to the identification of the Cholesterol Consensus Motif (CCM) found in multiple receptors. Interestingly many subsequent GPCR structures exhibit cholesterol binding at other sites and not the CCM, even when a CCM–binding site is present [16]. In addition to the CCM, a range of other cholesterol binding motifs are found in GPCRs which may accommodate these lipid molecules. The wide variety of cholesterol–binding sites seen across the different GPCR structures indicates that cholesterol binds promiscuously across the surface of the proteins, with specificity conferred by individual conformational states and the individual requirements of a given receptor [16]. Further computational analysis of a range of X-ray and cryo-EM structures indicates that cholesterol binds to a number of regions of GPCRs and that these sites are not characterised by specific motifs [98]. This is supported by a recent MD simulation study on 28 individual GPCR structures, including some active and inactive states of the same receptor [18]. In this case, the study revealed that the numbers and sites of the binding of cholesterol molecules differ between both different receptors and alternate conformational states of the same receptor.

A nice example of receptor–specific interactions with cholesterol is provided by the recent structure of the Oxytocin receptor (OTR) [17] which was crystallised in complex with a molecule of cholesterol bound to a site between helices 4 and 5. This study also revealed that mutation of residues involved in cholesterol binding reduced the stability of the OTR in the presence of exogenous cholesterol hemisuccinate (CHS), compared to a receptor construct with the cholesterol binding site intact. Furthermore, mutation of these cholesterol binding residues substantially reduced agonist and antagonist binding compared to the WT OTR. Given the proximity between the cholesterol–binding site and the ligand binding site, it is suggested that cholesterol binding is crucial for maintaining the optimal arrangement of amino acid residues within the ligand–binding site [17]. Further research on the OTR supports the fact that cholesterol is key for high affinity ligand binding but also that the act of ligand binding stabilises the interaction between the receptor and the bound cholesterol [99]. It is postulated that ligand binding may induce dimer formation, thus burying one or more cholesterol molecules at the dimer interface.

There is much evidence that cholesterol plays a role in GPCR oligomerisation. Initial indications of this came from the first structure of the β_2-adrenergic receptor β_2AR, which revealed a role for cholesterol in mediating dimer formation through a TM1 and TM7 interface [19]. A range of subsequent studies have provided supporting evidence of cholesterol having a role in both receptor homo-oligomerisation [20–22] and hetero-oligomerisation [23] of GPCRs. In the case of the β_2AR receptor, the cholesterol interacts with the palmitoyl group post-translationally added to a Cys residues in the C-terminal region of the protein [19]. Such an interaction has also been suggested for the μ-opioid receptor. In this case, removing the palmitoylation site reduced cholesterol association with the receptor and this decreased receptor signalling. Cholesterol depletion also reduced receptor signalling [15]. However, subsequent MD simulation analysis indicates that this cholesterol–palmitoyl interaction seems to occur preferentially in the inactive form of the receptor, and in the case of μ-opioid receptor, cholesterol does not appear to have a clear role in dimerization [100].

A structure of the yeast GPCR, Ste2, in the dimeric form and in complex with 2 cognate heterotrimeric G-proteins, has recently been reported [101]. In this structure, density assigned to 6 cholesteryl hemisuccinate (CHS) molecules was observed close to the dimer interface. These were assigned as CHS, since this sterol was added to the buffers during isolation of the receptor. However, it is possible that some if not all of these are native ergosterol molecules with a role in stabilising the dimer interface and carried through the solubilisation and purification of the receptor.

There is also some evidence from MD simulations that cholesterol and phospholipid compete for binding at some receptor sites, with phospholipids shown [102] and suggested to [39] displace cholesterol bound to the adenosine 2A receptor ($A_{2A}R$). Given that lipid binding is stronger when the receptor is in the active state and in complex with G-protein, a combination of specific bound lipids at defined sites is likely to play a role in regulating receptor activity [102].

6.2. GPCRs and Phospholipids

Many studies have revealed the contribution of phospholipids in modulating the stability and activity of GPCRs, as well as the selectivity of G-protein coupling. Dawaliby and colleagues demonstrated that DOPE induced a significantly reduced affinity for agonist binding to the β_2AR reconstituted into high–density lipoparticles compared to DOPG [30]. In contrast, β_2AR reconstituted in DOPE lipoparticles exhibited higher binding affinity for the antagonist compared to DOPG and DOPI. Further experiments revealed that β_2AR preferentially co-purifies with PG, and that PG provides the most favourable environment for binding to a G-protein mimetic [30], indicating that in the case of this receptor, negatively charged lipids are important for receptor activation. These findings indicated that PLs modulate receptor activity by stabilising different specific receptor conformations, and this is further supported by MD simulations on the $A_{2A}R$ which indicate that PG together with ligand binding induces the active form of the receptor, while a combination of ligand and PC is unable to induce the active from of the receptor [39]. An additional MD–based survey of 28 GPCR structures, from different classes, identified PIP lipids as forming the closest interactions with the receptors, although the precise molecular basis of the interactions seems to differ for individual receptors [18]. The important role of PIP lipids is underlined by a study that utilised a combination of mass spectrometry analyses and MD simulations revealing that PtdIns(4,5)P_2 binds to positively charged residues on the intracellular side of class A GPCRs, stabilising the active states of the receptors [25]. Similar results have been obtained for the GSHR, ghrelin receptor, with FRET analysis using labelled PtdIns(4,5)P_2 and labelled ghrelin receptor revealing that PtdIns(4,5)P_2 binds preferentially to the active form of the receptor [45].

MD studies on the neurotensin receptor (NTSR1), revealed that POPC promoted much greater dimer formation than physiological–simulated membranes based on brain polar lipids. The dimer interfaces adopted in POPC involved TMs 1, 5 and 6 in both symmetrical and asymmetrical protomer arrangements [34]. In contrast, in the brain polar lipid membrane, the NTS1 dimers form with a range of different interfaces involving TMs 1-6, in agreement with experimental studies on the same receptor [103]. This MD study also highlighted that different lipids stabilise different dimer conformations, with, for example, PS stabilising a symmetrical dimer involving TMs 3 and 4 of each protomer [34]. This dimer interaction interface leaves TMs 5 and 6 free to interact with the G-protein, suggesting that PS binding at the dimerization interface may be more favourable for the active forms of the receptor than PC. Since dimerization/oligomerisation interfaces are suggested to be partially dependent on protomer conformation [100,104,105], phospholipids can favour receptor–receptor contacts at particular interfaces by binding favourably to certain conformations. This suggests that by stabilising a certain receptor oligomeric state, phospholipids may modulate receptor activity.

Whilst it is clear that phospholipid head groups are important, there is also support for the fact that the acyl tails of phospholipids play a role in GPCR function and organisation. A recent study explored the effect that lipids with long (22 C), polyunsaturated tails derived from docosahexaenoic acid (DHA) have on the $A_{2A}R$. The findings revealed that the DHA–derived lipids resulted in increased populations of $A_{2A}R$ in the active conformation and greater G-protein coupling compared to lipids with shorter acyl tails but the same head group [53]. A number of MD studies have supported a role for DHA-containing unsaturated phospholipids as these order around the NTS1 [34] and drive $A_{2A}R$ to partition to lipid rafts [55]. A very recent MD study indicated that solvation of $A_{2A}R$ by unsaturated acyl chains is thermodynamically more favourable than saturated acyl chains, shifting the equilibrium towards active conformers [54]. In contrast, saturated acyl tails, which form part of the lipid raft domains from which DHA was excluded, allow formation of functional dimeric rhodopsin [57].

Phospholipids also exert an influence over GPCRs by changing the bulk membrane properties [106]. For example, unsaturated chains are known to cause hydrophobic mismatch between receptors and the membrane [107]. This can drive non-specific receptor oligomerisation [34,56], as a means of combatting the free energy penalty caused by the mismatch [108]. Mismatch–driven oligomerisation may also partially be a result of receptor activation [96]. However, this does not necessarily mean higher-order structures driven by mismatch are not functionally important; mismatch is suggested to aid partitioning of rhodopsin to lipid domains in central regions of the disc membrane, thus allowing efficient coupling to G-proteins [57].

6.3. GPCR Complexes and Phospholipids

There is increasing evidence that the lipid bilayer plays a key role in interactions between GPCRs and key binding partners. β-arrestin binding is responsible for both desensitization and internalisation of GPCRs and G-protein independent intracellular signalling [109,110]. The recent structure of an engineered form of muscarinic M2 receptor in complex with β-arrestin 1 obtained in nanodiscs (comprised of POPC, POPG and the membrane scaffold protein, MSP1D1E3) revealed that β-arrestin 1 interacted with the nanodisc encapsulated lipids as well as the receptor [35]. Additional data suggested that this β-arrestin 1-lipid interaction might be crucial for physiological receptor-β-arrestin 1 affinity by providing an additional source of complex stabilisation. The β-arrestin 1-lipid interaction is also important for β-arrestin 1 function in terms of modulating agonist binding to the receptor and receptor desensitization and internalisation [35]. Further support for lipids playing a role in receptor-β-arrestin binding has come from an additional cryo-EM complex structure; in this case, the NTSR1 in complex with a modified form of β-arrestin 1 [46]. The structure revealed that a molecule of PtdIns(4,5)P_2 mediates interactions between the receptor and the β-arrestin (Figure 4). Mutating the PtdIns(4,5)P_2 binding site in β–arrestin results in reduced β-arrestin binding to the receptor. These findings strongly suggest that the lipid has a role in the recruitment of β-arrestin and subsequent stability of the receptor-β-arrestin complex [46]. It is possible that there is receptor–dependent variability in the precise nature of the interactions with β-arrestin but the variability described in just these two examples may also reflect differences in sample preparation prior to structural analysis. However, it is clear that lipids have the potential to modulate receptor function at the level of the GPCR itself as well as through direct interaction with GPCR effector molecules.

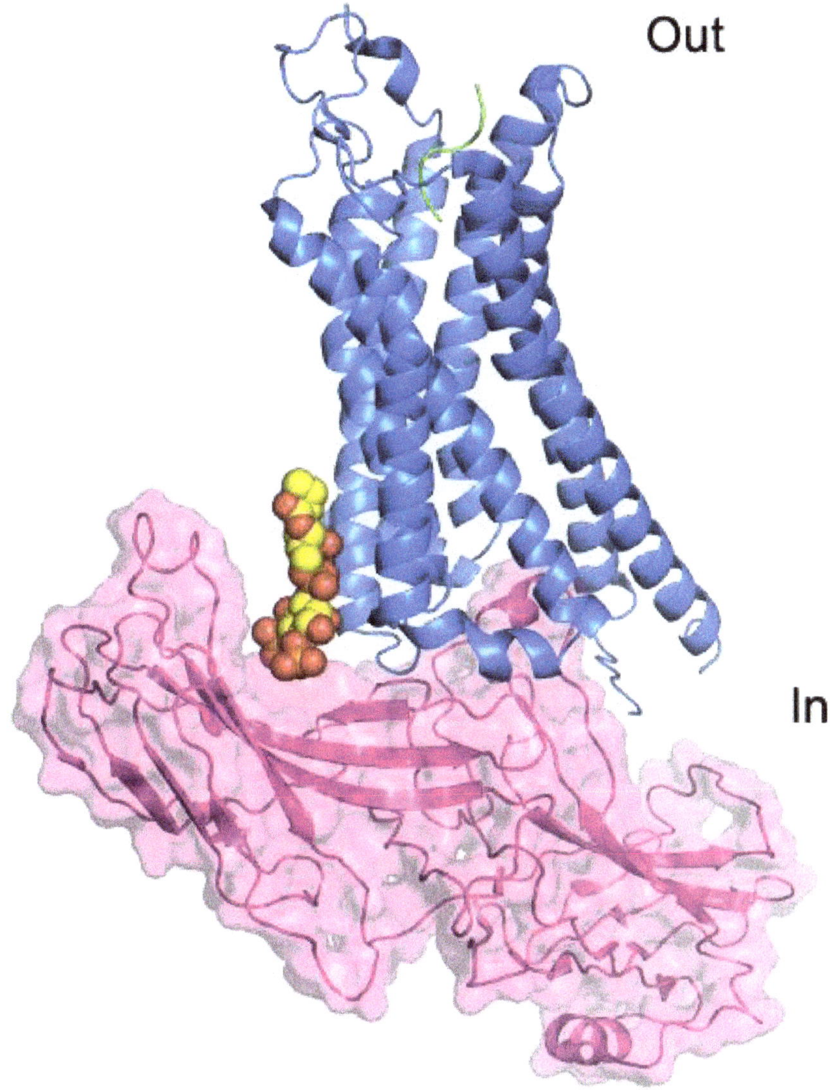

Figure 4. Structure of the NTSR + β-arrestin complex (PDB: 6UP7 [46]). The NTSR is shown in blue ribbon representation with the bound shown in green. The β-arrestin is shown in pink transparent surface representation. The PtdIns(4,5)P_2 in contact with both the receptor and the β-arrestin is shown in space filling representation with yellow C atoms.

Lipids are also suggested to play a role in interactions between GPCRs and G-proteins. The MS analysis study by Yen et al., revealed that PtdIns(4,5)P_2 bound to the intracellular regions of GPCRs stabilises the active conformation of the receptor and increases interactions with the G-protein [25]. In contrast, fluorescence spectroscopy and mutational analysis determined that PG/PS lipids diminish coupling of G_{i1} and G_{i3} to $β_2$AR under conditions of low Ca^{2+}, likely as a result of repulsion between the negative charges of PG/PS and G_i [31]. Higher levels of $β_2$AR-G_{i3} coupling were observed in the presence of PE/PC lipids, with Ca^{2+} mediating the G_{i3}-PE/PC interaction [31]. This suggests that phospholipids not only aid discrimination between G protein types by increasing the population of a conformer that couples to a specific G protein but also by directly promoting GPCR-G-protein interactions by mediating key electrostatic interactions.

Using microscale thermophoresis, Zhang et al., [38] suggested that NTSR1 coupling to $G\alpha_i$ was mediated by PG at the interaction interface, in contrast with earlier studies which suggested NTSR1 coupling to $G\alpha_i$ is enhanced by PE-rich membranes [32]. Yet further studies using native MS revealed that the NTSR1 purifies in the presence of PS, PA and PIP species and shows preferential binding to these lipids when added exogenously compared to PC [25]. This study revealed no detectable PG binding, with the authors suggesting that any effects of PG could be a result of alterations in the local membrane charge around the receptor. Among the lipids tested, PtdIns(4,5)P_2 bound most effectively to the NTSR1, as well as the β_1AR and the A_{2A}R, and mutagenesis of a predicted PtdIns(4,5)P_2 binding site formed on the intracellular side of the protein and involving principally positively charged residues in TM4 resulted in loss of PtdIns(4,5)P_2 binding [25]. Further analysis revealed that NTSR1-G-protein coupling was increased in the presence of PtdIns(4,5)P_2 and given the location of PtdIns(4,5)P_2 binding to the receptor, this is likely to be the result of the lipid mediating interactions between the receptor and the G-protein. Since similar results were obtained with the β_1AR and the A_{2A}R, this suggests that the role of PtdIns(4,5)P_2 in G-protein coupling is common to other Class A GPCRs [25,102].

MD simulations indicate that PtdIns(4,5)P_2 interacts with the glucagon receptor, GCGR, a class B receptor, in some sites conserved with those in class A GPCRs [47]. However, no PtdIns(4,5)P_2 binding was detectable at the TM3/ICL2 site shown to be important for class A receptor G-protein recruitment. These findings suggest that the roles PtdIns(4,5)P_2 plays in class B receptor function are distinct from those of class A receptors [47].

6.4. GPCRs and Sphingolipids and Glycolipids

Use of a mycotoxin known to reduce sphingolipid content reduces the amount of cell surface localised 5HT-1A receptor [111], in line with earlier results indicating the same treatment reduced specific ligand binding and associated downstream cAMP signalling for this receptor [48]. In contrast, similar studies on the angiotensin II type 1A receptor and the bitter taste receptor, T2R14, indicated no change in receptor signalling as a result of sphingomyelin depletion [112].

Gangliosides (GMs) are a type of glycosphingolipid found mostly on the membrane outer leaflet [113]. Coarse-grain MD simulations proposed that GM1 binds to an identified and conserved "sphingolipid binding domain" on extracellular loop (ECL) 1 of the 5HT-1A receptor and modulates ligand binding [49]. MD simulations also indicate that GM3 binds to ECL1-3 and extracellular portions of TMs of both the class B glucagon receptor (GCGR) [47] and the class A A_{2A}R [102] through basic and aromatic residues. In the case of GCGR, GM3 binding to the extracellular domain (ECD, also responsible for ligand binding) affects the conformational dynamics of this region of the receptor, thus potentially acting as an allosteric modulator affecting the ability of the receptor to bind ligands [47].

Given the high percentage of sphingolipids and glycosphingolipids in lipid rafts, these and other findings support an important relationship between lipid rafts and GPCRs [114]. However, other MD simulations have indicated that while GMs were enriched around GPCRs, sphingomyelin was depleted around the 28 GPCR structures they probed, relative to the bulk membrane, suggesting that some sphingolipid species play little role in stability, function or organisation of these receptors [18]. However, it is also possible that the differences in the simulation methodology used are responsible for some of the different results obtained. The lack of high–resolution GPCR structures in complex with GMs and SMs currently limit our overall understanding of the precise nature of these interactions.

7. Membrane Lipids and Other Membrane Proteins

Lysosome–associated protein transmembrane 4B (LAPTMB) is responsible for mediating traffic of amino acid transporters to lysosomes under conditions of high nutrient availability. Experimental and MD simulation data indicated that a lipid binding site in TM3 specific for ceramide is crucial for correct dimerization of LAPTMB and the amino acid transporter proteins [115]. EPR and mutagenesis-based analysis of Annexin B12 indicates that oligomerisation of the protein is highly dependent on membrane lipids [116], in addition to protein–protein interactions. However, the precise lipids that mediate the oligomer formation have yet to be identified. Connexins are integral membrane proteins that associate to form gap junctions between cells and allow the passage of information and small molecules from one cell to another. A likely lipid binding site was recently identified in the cryo-EM structure of the Cx31.3 connexin hemi-channel obtained at a resolution of 2.4 Å. This binding site located within the pore cavity is suggested to have a role in connexin hemi-channel assembly [33]. Density close to this binding site was assigned as a PE molecule, which had been extracted from the membrane and copurified with the hemi-channel [33]. Further work on the Connexin-46/50 full cell–cell junction focused on the protein obtained in DMPC-containing nanodiscs. The integral membrane domains on the extracellular side of the two hemi-channels are stabilised by extensive clusters of lipid molecules. These ordered lipid molecules extend further out from the protein than is typical for lipids forming specific interactions, a finding further supported by MD simulations [117]. This study raises the possibility that formation of the full cell–cell junction induces local order in the membrane environment [117].

8. Conclusions

As summarised in the current review, insights into membrane protein–lipid interactions are increasing rapidly. The development of novel, non-detergent-based membrane mimetic systems and the use of lipid reconstitution approaches such as nanodiscs for mass spectrometry and electron microscopy sample preparation pave the way for a more physiologically relevant understanding of the interactions between membrane proteins and membrane lipids. In addition, it is clear that the ability to probe potential membrane protein–lipid interactions in silico using Molecular Dynamics simulations where the researcher has unparalleled control of the membrane lipid composition provides an excellent means of providing context to experimental findings and forms the basis for further studies. We are just starting to unpick these complex and important molecular relationships and there is much more to do.

Author Contributions: Conceptualization, K.R. and B.B.; writing—original draft preparation, K.R. and B.B. writing—review and editing, K.R. and B.B. Both authors have read and agreed to the published version of the manuscript.

Funding: This research was funded by Biotechnology and Biological Sciences Research Council (BBSRC) grant BB/N016467/1 awarded to Bernadette Byrne.

Acknowledgments: The authors also wish to thank Nicole Deacon-Smith for help with the figure preparation.

Conflicts of Interest: The authors declare no conflict of interest.

References

1. Singer, S.J.; Nicolson, G.L. The Fluid Mosaic Model of the Structure of Cell Membranes. *Science* **1972**, *175*, 720–731. [CrossRef]
2. Kalappurakkal, J.M.; Sil, P.; Mayor, S. Towards a New Picture of the Living Plasma Membrane. *Protein Sci.* **2020**, *29*, 1355–1365. [CrossRef] [PubMed]
3. Simons, K.; Gerl, M.J. Revitalizing Membrane Rafts: New Tools and Insights. *Nat. Rev. Mol. Cell Biol.* **2010**, *11*, 688–699. [CrossRef]
4. Devaux, P.F.; Morris, R. Transmembrane Asymmetry and Lateral Domains in Biological Membranes. *Traffic* **2004**, *5*, 241–246. [CrossRef]
5. Laganowsky, A.; Reading, E.; Allison, T.M.; Ulmschneider, M.B.; Degiacomi, M.T.; Baldwin, A.J.; Robinson, C.V. Membrane Proteins Bind Lipids Selectively to Modulate Their Structure and Function. *Nature* **2014**, *510*, 172–175. [CrossRef] [PubMed]

6. Guo, R.; Cang, Z.; Yao, J.; Kim, M.; Deans, E.; Wei, G.; Kang, S.; Hong, H. Structural Cavities Are Critical to Balancing Stability and Activity of a Membrane-Integral Enzyme. *Proc. Natl. Acad. Sci. USA* **2020**, *117*, 22146–22156. [CrossRef] [PubMed]
7. Guo, Y. Be Cautious with Crystal Structures of Membrane Proteins or Complexes Prepared in Detergents. *Crystals* **2020**, *10*, 86. [CrossRef] [PubMed]
8. Bjerregaard, H.; Severinsen, K.; Said, S.; Wiborg, O.; Sinning, S. A Dualistic Conformational Response to Substrate Binding in the Human Serotonin Transporter Reveals a High Affinity State for Serotonin. *J. Biol. Chem.* **2015**, *290*, 7747–7755. [CrossRef]
9. Laursen, L.; Severinsen, K.; Kristensen, K.B.; Periole, X.; Overby, M.; Müller, H.K.; Schiøtt, B.; Sinning, S. Cholesterol Binding to a Conserved Site Modulates the Conformation, Pharmacology, and Transport Kinetics of the Human Serotonin Transporter. *J. Biol. Chem.* **2018**, *293*, 3510–3523. [CrossRef]
10. Garcia, A.; Lev, B.; Hossain, K.R.; Gorman, A.; Diaz, D.; Pham, T.H.N.; Cornelius, F.; Allen, T.W.; Clarke, R.J. Cholesterol Depletion Inhibits Na+, K+-ATPase Activity in a near-Native Membrane Environment. *J. Biol. Chem.* **2019**, *294*, 5956–5969. [CrossRef]
11. Sharom, F.J. Complex Interplay between the P-Glycoprotein Multidrug Efflux Pump and the Membrane: Its Role in Modulating Protein Function. *Front. Oncol.* **2014**, *4*, 41. [CrossRef] [PubMed]
12. Singh, D.K.; Rosenhouse-Dantsker, A.; Nichols, C.G.; Enkvetchakul, D.; Levitan, I. Direct Regulation of Prokaryotic Kir Channel by Cholesterol. *J. Biol. Chem.* **2009**, *284*, 30727–30736. [CrossRef] [PubMed]
13. Qi, X.; Friedberg, L.; Bose-Boyd, R.D.; Long, T.; Li, X. Sterols in an Intramolecular Channel of Smoothened Mediate Hedgehog Signaling. *Nat. Chem. Biol.* **2020**, *16*, 1368–1375. [CrossRef]
14. Yeliseev, A.; Iyer, M.R.; Joseph, T.T.; Coffey, N.J.; Cinar, R.; Zoubak, L.; Kunos, G.; Gawrisch, K. Cholesterol as a Modulator of Cannabinoid Receptor CB2 Signaling. *Sci. Rep.* **2021**, *11*, 3706–3718. [CrossRef] [PubMed]
15. Zheng, H.; Pearsall, E.A.; Hurst, D.P.; Zhang, Y.; Chu, J.; Zhou, Y.; Reggio, P.H.; Loh, H.H.; Law, P.-Y. Palmitoylation and Membrane Cholesterol Stabilize μ-Opioid Receptor Homodimerization and G Protein Coupling. *BMC Cell Biol.* **2012**, *13*, 6–18. [CrossRef]
16. Gimpl, G. Interaction of G Protein Coupled Receptors and Cholesterol. *Chem. Phys. Lipids* **2016**, *199*, 61–73. [CrossRef]
17. Waltenspühl, Y.; Schöppe, J.; Ehrenmann, J.; Kummer, L.; Plückthun, A. Crystal Structure of the Human Oxytocin Receptor. *Sci. Adv.* **2020**, *6*, eabb5419. [CrossRef]
18. Sejdiu, B.I.; Tieleman, D.P. Lipid-Protein Interactions Are a Unique Property and Defining Feature of G Protein-Coupled Receptors. *Biophys. J.* **2020**, *118*, 1887–1900. [CrossRef]
19. Cherezov, V.; Rosenbaum, D.M.; Hanson, M.A.; Rasmussen, S.G.F.; Thian, F.S.; Kobilka, T.S.; Choi, H.J.; Kuhn, P.; Weis, W.I.; Kobilka, B.K.; et al. High-Resolution Crystal Structure of an Engineered Human 2-Adrenergic G Protein Coupled Receptor. *Science* **2007**, *318*, 1258–1265. [CrossRef]
20. Manglik, A.; Kruse, A.C.; Kobilka, T.S.; Thian, F.S.; Mathiesen, J.M.; Sunahara, R.K.; Pardo, L.; Weis, W.I.; Kobilka, B.K.; Granier, S. Crystal Structure of the M-Opioid Receptor Bound to a Morphinan Antagonist. *Nature* **2012**, *485*, 321–326. [CrossRef]
21. Huang, W.; Manglik, A.; Venkatakrishnan, A.J.; Laeremans, T.; Feinberg, E.N.; Sanborn, A.L.; Kato, H.E.; Livingston, K.E.; Thorsen, T.S.; Kling, R.C.; et al. Structural Insights into M-Opioid Receptor Activation. *Nature* **2015**, *524*, 315–321. [CrossRef]
22. Gahbauer, S.; Böckmann, R.A. Membrane-Mediated Oligomerization of G Protein Coupled Receptors and Its Implications for GPCR Function. *Front. Physiol.* **2016**, *7*, 494. [CrossRef]
23. Prasanna, X.; Mohole, M.; Chattopadhyay, A.; Sengupta, D. Role of Cholesterol-Mediated Effects in GPCR Heterodimers. *Chem. Phys. Lipids* **2020**, *227*, 104852. [CrossRef]
24. Schrecke, S.; Zhu, Y.; McCabe, J.W.; Bartz, M.; Packianathan, C.; Zhao, M.; Zhou, M.; Russell, D.; Laganowsky, A. Selective Regulation of Human TRAAK Channels by Biologically Active Phospholipids. *Nat. Chem. Biol.* **2021**, *17*, 89–95. [CrossRef]
25. Yen, H.-Y.; Hoi, K.K.; Liko, I.; Hedger, G.; Horrell, M.R.; Song, W.; Wu, D.; Heine, P.; Warne, T.; Lee, Y.; et al. PtdIns(4,5)P2 Stabilizes Active States of GPCRs and Enhances Selectivity of G-Protein Coupling. *Nature* **2018**, *559*, 423–427. [CrossRef] [PubMed]
26. Pyle, E.; Kalli, A.C.; Amillis, S.; Hall, Z.; Lau, A.M.; Hanyaloglu, A.C.; Diallinas, G.; Byrne, B.; Politis, A. Structural Lipids Enable the Formation of Functional Oligomers of the Eukaryotic Purine Symporter UapA. *Cell Chem.* **2018**, *25*, 840–848. [CrossRef] [PubMed]
27. Pyle, E.; Guo, C.; Hofmann, T.; Schmidt, C.; Ribiero, O.; Politis, A.; Byrne, B. Protein–lipid Interactions Stabilize the Oligomeric State of BOR1p from Saccharomyces Cerevisiae. *Anal. Chem.* **2019**, *91*, 13071–13079. [CrossRef] [PubMed]
28. Martens, C.; Shekhar, M.; Borysik, A.J.; Lau, A.M.; Reading, E.; Tajkhorshid, E.; Booth, P.J.; Politis, A. Direct Protein–lipid Interactions Shape the Conformational Landscape of Secondary Transporters. *Nat. Commun.* **2018**, *9*, 4151. [CrossRef] [PubMed]
29. Hénault, C.M.; Govaerts, C.; Spurny, R.; Brams, M.; Estrada-Mondragon, A.; Lynch, J.; Bertrand, D.; Pardon, E.; Evans, G.L.; Woods, K.; et al. A Lipid Site Shapes the Agonist Response of a Pentameric Ligand-Gated Ion Channel. *Nat. Chem. Biol.* **2019**, *15*, 1156–1164. [CrossRef]
30. Dawaliby, R.; Trubbia, C.; Delporte, C.; Masureel, M.; Antwerpen, P.V.; Kobilka, B.K.; Govaerts, C. Allosteric Regulation of G Protein–Coupled Receptor Activity by Phospholipids. *Nat. Chem. Biol.* **2016**, *12*, 35–39. [CrossRef]
31. Strohman, M.J.; Maeda, S.; Hilger, D.; Masureel, M.; Du, Y.; Kobilka, B.K. Local Membrane Charge Regulates β_2 Adrenergic Receptor Coupling to Gi3. *Nat. Commun.* **2019**, 1–10. [CrossRef] [PubMed]
32. Dijkman, P.M.; Watts, A. Lipid Modulation of Early G Protein-Coupled Receptor Signalling Events. *BBA Biomembr.* **2015**, *1848*, 2889–2897. [CrossRef] [PubMed]

33. Lee, H.-J.; Jeong, H.; Hyun, J.; Ryu, B.; Park, K.; Lim, H.-H.; Yoo, J.; Woo, J.-S. Cryo-EM Structure of Human Cx31.3/GJC3 Connexin Hemichannel. *Sci. Adv.* **2020**, *6*, eaba4996. [CrossRef] [PubMed]
34. Gahbauer, S.; Böckmann, R.A. Comprehensive Characterization of Lipid-Guided G Protein-Coupled Receptor Dimerization. *J. Phys. Chem. B* **2020**, *124*, 2823–2834. [CrossRef]
35. Staus, D.P.; Hu, H.; Robertson, M.J.; Kleinhenz, A.L.W.; Wingler, L.M.; Capel, W.D.; Latorraca, N.R.; Lefkowitz, R.J.; Skiniotis, G. Structure of the M2 Muscarinic Receptor-β-Arrestin Complex in a Lipid Nanodisc. *Nature* **2020**, *579*, 297–302. [CrossRef]
36. Koshy, C.; Schweikhard, E.S.; Gärtner, R.M.; Perez, C.; Yildiz, Ö.; Ziegler, C. Structural Evidence for Functional Lipid Interactions in the Betaine Transporter BetP. *EMBO J.* **2013**, *32*, 3096–3105. [CrossRef]
37. Tong, A.; II, J.T.P.; Hsu, F.-F.; Schmidpeter, P.A.; Nimigean, C.M.; Sharp, L.; Brannigan, G.; Cheng, W.W. Direct Binding of Phosphatidylglycerol at Specific Sites Modulates Desensitization of a Ligand-Gated Ion Channel. *ELife* **2019**, *8*, e50766. [CrossRef]
38. Zhang, M.; Gui, M.; Wang, Z.-F.; Gorgulla, C.; Yu, J.J.; Wu, H.; Sun, Z.-Y.J.; Klenk, C.; Merklinger, L.; Morstein, L.; et al. Cryo-EM Structure of an Activated GPCR-G Protein Complex in Lipid Nanodiscs. *Nat. Struct. Mol. Biol.* **2021**, *28*, 258–267. [CrossRef]
39. Bruzzese, A.; Dalton, J.A.R.; Giraldo, J. Insights into Adenosine A2A Receptor Activation through Cooperative Modulation of Agonist and Allosteric Lipid Interactions. *PLoS Comput. Biol.* **2020**, *16*, e1007818. [CrossRef] [PubMed]
40. Anderluh, A.; Hofmaier, T.; Klotzsch, E.; Kudlacek, O.; Stockner, T.; Sitte, H.H.; Schütz, G.J. Direct PIP2 Binding Mediates Stable Oligomer Formation of the Serotonin Transporter. *Nat. Commun.* **2017**, *8*, 14089. [CrossRef]
41. Hamilton, P.J.; Belovich, A.N.; Khelashvili, G.; Saunders, C.; Erreger, K.; Javitch, J.A.; Sitte, H.H.; Weinstein, H.; Matthies, H.J.G.; Galli, A. PIP2 Regulates Psychostimulant Behaviors through Its Interaction with a Membrane Protein. *Nat. Chem. Biol.* **2014**, *10*, 582–589. [CrossRef] [PubMed]
42. Duncan, A.L.; Corey, R.A.; Sansom, M.S.P. Defining How Multiple Lipid Species Interact with Inward Rectifier Potassium (Kir2) Channels. *Proc. Natl. Acad. Sci. USA* **2020**, *117*, 7803–7813. [CrossRef] [PubMed]
43. McGoldrick, L.L.; Singh, A.K.; Demirkhanyan, L.; Lin, T.-Y.; Casner, R.G.; Zakharian, E.; Sobolevsky, A.I. Structure of the Thermo-Sensitive TRP Channel TRP1 from the Alga Chlamydomonas Reinhardtii. *Nat. Commun.* **2019**, *10*, 4180. [CrossRef]
44. Gao, Y.; Cao, E.; Julius, D.; Cheng, Y. TRPV1 Structures in Nanodiscs Reveal Mechanisms of Ligand and Lipid Action. *Nature* **2016**, *534*, 347–351. [CrossRef] [PubMed]
45. Damian, M.; Louet, M.; Gomes, A.A.S.; M'Kadmi, C.; Denoyelle, S.; Cantel, S.; Mary, S.; Bisch, P.M.; Fehrentz, J.-A.; Catoire, L.J.; et al. Allosteric Modulation of Ghrelin Receptor Signaling by Lipids. *Nat. Commun.* **2021**, *12*, 3938. [CrossRef] [PubMed]
46. Huang, W.; Masureel, M.; Qu, Q.; Janetzko, J.; Inoue, A.; Kato, H.E.; Robertson, M.J.; Nguyen, K.C.; Glenn, J.S.; Skiniotis, G.; et al. Structure of the Neurotensin Receptor 1 in Complex with β-Arrestin 1. *Nature* **2020**, *579*, 303–308. [CrossRef]
47. Ansell, T.B.; Song, W.; Sansom, M.S.P. The Glycosphingolipid GM3 Modulates Conformational Dynamics of the Glucagon Receptor. *Biophys. J.* **2020**, *119*, 300–313. [CrossRef]
48. Paila, Y.D.; Ganguly, S.; Chattopadhyay, A. Metabolic Depletion of Sphingolipids Impairs Ligand Binding and Signaling of Human Serotonin 1A Receptors. *Biochemistry* **2010**, *49*, 2389–2397. [CrossRef]
49. Prasanna, X.; Jafurulla, M.; Sengupta, D.; Chattopadhyay, A. The Ganglioside GM1 Interacts with the Serotonin1A Receptor via the Sphingolipid Binding Domain. *BBA Biomembr.* **2016**, *1858*, 2818–2826. [CrossRef]
50. Jormakka, M.; Törnroth, S.; Byrne, B.; Iwata, S. Molecular Basis of Proton Motive Force Generation: Structure of Formate Dehydrogenase-N. *Science* **2002**, *295*, 1863–1868. [CrossRef] [PubMed]
51. Gupta, K.; Donlan, J.A.C.; Hopper, J.T.S.; Uzdavinys, P.; Landreh, M.; Struwe, W.B.; Drew, D.; Baldwin, A.J.; Stansfeld, P.J.; Robinson, C.V. The Role of Interfacial Lipids in Stabilizing Membrane Protein Oligomers. *Nature* **2017**, 1–15. [CrossRef] [PubMed]
52. Bolla, J.R.; Sauer, J.B.; Wu, D.; Mehmood, S.; Allison, T.M.; Robinson, C.V. Direct Observation of the Influence of Cardiolipin and Antibiotics on Lipid II Binding to MurJ. *Nat. Chem.* **2018**, *10*, 363–371. [CrossRef] [PubMed]
53. Mizumura, T.; Kondo, K.; Kurita, M.; Kofuku, Y.; Natsume, M.; Imai, S.; Shiraishi, Y.; Ueda, T.; Shimada, I. Activation of Adenosine A2A Receptor by Lipids from Docosahexaenoic Acid Revealed by NMR. *Sci. Adv.* **2020**, *6*, eaay8544. [CrossRef]
54. Leonard, A.N.; Lyman, E. Activation of G-Protein-Coupled Receptors Is Thermodynamically Linked to Lipid Solvation. *Biophys. J.* **2021**, *120*, 1777–1787. [CrossRef]
55. Javanainen, M.; Enkavi, G.; Guixà-Gonzaléz, R.; Kulig, W.; Martinez-Seara, H.; Levental, I.; Vattulainen, I. Reduced Level of Docosahexaenoic Acid Shifts GPCR Neuroreceptors to Less Ordered Membrane Regions. *PLoS Comput. Biol.* **2019**, *15*, e1007033. [CrossRef]
56. Mondal, S.; Johnston, J.M.; Wang, H.; Khelashvili, G.; Filizola, M.; Weinstein, H. Membrane Driven Spatial Organization of GPCRs. *Sci. Rep.* **2013**, *3*, 2909. [CrossRef]
57. Hayashi, F.; Saito, N.; Tanimoto, Y.; Okada, K.; Morigaki, K.; Seno, K.; Maekawa, S. Raftophilic Rhodopsin-Clusters Offer Stochastic Platforms for G Protein Signalling in Retinal Discs. *Commun. Biol.* **2019**, *2*, 209. [CrossRef] [PubMed]
58. Sun, C.; Benlekbir, S.; Venkatakrishnan, P.; Wang, Y.; Hong, S.; Hosler, J.; Tajkhorshid, E.; Rubinstein, J.L.; Gennis, R.B. Structure of the Alternative Complex III in a Supercomplex with Cytochrome Oxidase. *Nature* **2018**, *557*, 123–126. [CrossRef] [PubMed]
59. Knowles, T.J.; Finka, R.; Finka, R.; Smith, C.; Smith, C.; Lin, Y.-P.; Dafforn, T.; Dafforn, T.; Overduin, M. Membrane Proteins Solubilized Intact in Lipid Containing Nanoparticles Bounded by Styrene Maleic Acid Copolymer. *J. Am. Chem. Soc.* **2009**, *131*, 7484–7485. [CrossRef]
60. Palsdottir, H.; Hunte, C. Lipids in Membrane Protein Structures. *Biochim. Biophys. Acta Biomembr.* **2004**, *1666*, 2–18. [CrossRef]

61. Lange, C.; Nett, J.H.; Trumpower, B.L.; Hunte, C. Specific Roles of Protein—Phospholipid Interactions in the Yeast Cytochrome Bc1 Complex Structure. *EMBO J.* **2001**, *20*, 6591–6600. [CrossRef]
62. Palsdottir, H.; Lojero, C.G.; Trumpower, B.L.; Hunte, C. Structure of the Yeast Cytochrome Bc1 Complex with a Hydroxyquinone Anion Qo Site Inhibitor Bound. *J. Biol. Chem.* **2003**, *278*, 31303–31311. [CrossRef]
63. Pinke, G.; Zhou, L.; Sazanov, L.A. Cryo-EM Structure of the Entire Mammalian F-Type ATP Synthase. *Nat. Struct. Mol. Biol.* **2020**, *27*, 1077–1085. [CrossRef]
64. Wöhlert, D.; Grötzinger, M.J.; Kühlbrandt, W.; Yildiz, Ö. Mechanism of Na(+)—Dependent Citrate Transport from the Structure of an Asymmetrical CitS Dimer. *ELife* **2015**, *4*, e09375. [CrossRef]
65. Alguel, Y.; Amillis, S.; Leung, J.; Lambrinidis, G.; Capaldi, S.; Scull, N.J.; Craven, G.; Iwata, S.; Armstrong, A.; Mikros, E.; et al. Structure of Eukaryotic Purine/H(+) Symporter UapA Suggests a Role for Homodimerization in Transport Activity. *Nat. Commun.* **2016**, *7*, 11336. [CrossRef] [PubMed]
66. Bolla, J.R.; Agasid, M.T.; Mehmood, S.; Robinson, C.V. Membrane Protein–Lipid Interactions Probed Using Mass Spectrometry. *Annu. Rev. Biochem.* **2019**, *88*, 85–111. [CrossRef] [PubMed]
67. Thurtle-Schmidt, B.H.; Stroud, R.M. Structure of Bor1 Supports an Elevator Transport Mechanism for SLC4 Anion Exchangers. *Proc. Natl. Acad. Sci. USA* **2016**, *113*, 10542–10546. [CrossRef] [PubMed]
68. Das, A.K.; Kudlacek, O.; Baumgart, F.; Jaentsch, K.; Stockner, T.; Sitte, H.H.; Schütz, G.J. Dopamine Transporter Forms Stable Dimers in the Live Cell Plasma Membrane in a Phosphatidylinositol 4,5-Bisphosphate-Independent Manner. *J. Biol. Chem.* **2019**, *294*, 5632–5642. [CrossRef]
69. Penmatsa, A.; Wang, K.H.; Gouaux, E. X-Ray Structure of Dopamine Transporter Elucidates Antidepressant Mechanism. *Nature* **2013**, *503*, 85–90. [CrossRef]
70. Wang, K.H.; Penmatsa, A.; Gouaux, E. Neurotransmitter and Psychostimulant Recognition by the Dopamine Transporter. *Nature* **2015**, *521*, 322–327. [CrossRef]
71. Coleman, J.A.; Green, E.M.; Gouaux, E. X-Ray Structures and Mechanism of the Human Serotonin Transporter. *Nature* **2016**, *532*, 334–339. [CrossRef]
72. Hong, W.C.; Amara, S.G. Membrane Cholesterol Modulates the Outward Facing Conformation of the Dopamine Transporter and Alters Cocaine Binding. *J. Biol. Chem.* **2010**, *285*, 32616–32626. [CrossRef]
73. Jones, K.T.; Zhen, J.; Reith, M.E.A. Importance of Cholesterol in Dopamine Transporter Function. *J. Neurochem.* **2012**, *123*, 700–715. [CrossRef] [PubMed]
74. Bechara, C.; Robinson, C.V. Different Modes of Lipid Binding to Membrane Proteins Probed by Mass Spectrometry. *J. Am. Chem. Soc.* **2015**, *137*, 5240–5247. [CrossRef] [PubMed]
75. Bolla, J.R.; Corey, R.A.; Sahin, C.; Gault, J.; Hummer, A.; Hopper, J.T.S.; Lane, D.P.; Drew, D.; Allison, T.M.; Stansfeld, P.J.; et al. A Mass-Spectrometry-Based Approach to Distinguish Annular and Specific Lipid Binding to Membrane Proteins. *Angew. Chem. Int. Ed.* **2020**, *59*, 3523–3528. [CrossRef] [PubMed]
76. Jungnickel, K.E.J.; Parker, J.L.; Newstead, S. Structural Basis for Amino Acid Transport by the CAT Family of SLC7 Transporters. *Nat. Commun.* **2018**, 1–12. [CrossRef]
77. Pearson, A.; Page, S.R.F.; Jorgenson, T.L.; Fischer, W.W.; Higgins, M.B. Novel Hopanoid Cyclases from the Environment. *Environ. Microbiol.* **2007**, *9*, 2175–2188. [CrossRef]
78. Mistarz, U.H.; Brown, J.M.; Haselmann, K.F.; Rand, K.D. Probing the Binding Interfaces of Protein Complexes Using Gas-Phase H/D Exchange Mass Spectrometry. *Struct. Fold. Des.* **2016**, *24*, 310–318. [CrossRef]
79. Konermann, L.; Pan, J.; Liu, Y.-H. Hydrogen Exchange Mass Spectrometry for Studying Protein Structure and Dynamics. *Chem. Soc. Rev.* **2011**, *40*, 1224–1234. [CrossRef] [PubMed]
80. Esmaili, M.; Eldeeb, M.A.; Moosavi-Movahedi, A.A. Current Developments in Native Nanometric Discoidal Membrane Bilayer Formed by Amphipathic Polymers. *Nanomaterials* **2021**, *11*, 1771. [CrossRef]
81. Farrelly, M.D.; Martin, L.L.; Thang, S.H. Polymer Nanodiscs and Their Bioanalytical Potential. *Chem. Eur. J.* **2021**. [CrossRef] [PubMed]
82. Reading, E.; Hall, Z.; Martens, C.; Haghighi, T.; Findlay, H.; Ahdash, Z.; Politis, A.; Booth, P.J. Interrogating Membrane Protein Conformational Dynamics within Native Lipid Compositions. *Angew. Chem. Int. Ed.* **2017**, *129*, 15860–15863. [CrossRef]
83. Chorev, D.S.; Baker, L.A.; Wu, D.; Beilsten-Edmands, V.; Rouse, S.L.; Zeev-Ben-Mordehai, T.; Jiko, C.; Samsudin, F.; Gerle, C.; Khalid, S.; et al. Protein Assemblies Ejected Directly from Native Membranes Yield Complexes for Mass Spectrometry. *Science* **2018**, *362*, 829–834. [CrossRef] [PubMed]
84. Shinoda, T.; Ogawa, H.; Cornelius, F.; Toyoshima, C. Crystal Structure of the Sodium-Potassium Pump at 2.4 A Resolution. *Nature* **2009**, *459*, 446–450. [CrossRef]
85. Eckford, P.; Sharom, F.J. ABC Efflux Pump-Based Resistance to Chemotherapy Drugs. *Chem. Rev.* **2009**, *109*, 2989–3011. [CrossRef] [PubMed]
86. Thangapandian, S.; Kapoor, K.; Tajkhorshid, E. Probing Cholesterol Binding and Translocation in P-Glycoprotein. *BBA Biomembr.* **2020**, *1862*, 183090. [CrossRef]
87. Duncan, A.L.; Song, W.; Sansom, M.S.P. Lipid-Dependent Regulation of Ion Channels and G Protein–Coupled Receptors: Insights from Structures and Simulations. *Annu. Rev. Pharmacol. Toxicol.* **2020**, *60*, 31–50. [CrossRef]

88. Hansen, S.B.; Tao, X.; MacKinnon, R. Structural Basis of PIP2 Activation of the Classical Inward Rectifier K+ Channel Kir2.2. *Nature* **2011**, *477*, 495–498. [CrossRef] [PubMed]
89. Cheng, W.W.L.; D'Avanzo, N.; Doyle, D.A.; Nichols, C.G. Dual-Mode Phospholipid Regulation of Human Inward Rectifying Potassium Channels. *Biophys. J.* **2011**, *100*, 620–628. [CrossRef]
90. Lee, S.-J.; Wang, S.; Borschel, W.; Heyman, S.; Gyore, J.; Nichols, C.G. Secondary Anionic Phospholipid Binding Site and Gating Mechanism in Kir2.1 Inward Rectifier Channels. *Nat. Commun.* **2013**, *4*, 2786. [CrossRef] [PubMed]
91. Zhang, X.; Hu, M.; Yang, Y.; Xu, H. Organellar TRP Channels. *Nat. Struct. Mol. Biol.* **2018**, *25*, 1009–1018. [CrossRef] [PubMed]
92. Nadezhdin, K.D.; Neuberger, A.; Trofimov, Y.A.; Krylov, N.A.; Sinica, V.; Kupko, N.; Vlachova, V.; Zakharian, E.; Efremov, R.G.; Sobolevsky, A.I. Structural Mechanism of Heat-Induced Opening of a Temperature-Sensitive TRP Channel. *Nat. Struct. Mol. Biol.* **2021**, *28*, 564–572. [CrossRef] [PubMed]
93. Glukhova, A.; Thal, D.M.; Nguyen, A.T.; Vecchio, E.A.; Jörg, M.; Scammells, P.J.; May, L.T.; Sexton, P.M.; Christopoulos, A. Structure of the Adenosine A1 Receptor Reveals the Basis for Subtype Selectivity. *Cell* **2017**, *168*, 867–877. [CrossRef] [PubMed]
94. Pucadyil, T.J.; Chattopadhyay, A. Role of Cholesterol in the Function and Organization of G-Protein Coupled Receptors. *Prog. Lipid Res.* **2006**, *45*, 295–333. [CrossRef]
95. Paila, Y.D.; Chattopadhyay, A. The Function of G-Protein Coupled Receptors and Membrane Cholesterol: Specific or General Interaction? *Glycoconj. J.* **2008**, *26*, 711–720. [CrossRef] [PubMed]
96. Salas-Estrada, L.A.; Leioatts, N.; Romo, T.D.; Grossfield, A. Lipids Alter Rhodopsin Function via Ligand-like and Solvent-like Interactions. *Biophys. J.* **2018**, *114*, 355–367. [CrossRef] [PubMed]
97. Hanson, M.A.; Cherezov, V.; Griffith, M.T.; Roth, C.B.; Jaakola, V.-P.; Chien, E.Y.T.; Velasquez, J.; Kuhn, P.; Stevens, R.C. A Specific Cholesterol Binding Site Is Established by the 2.8 Å Structure of the Human B2-Adrenergic Receptor. *Structure* **2008**, *16*, 897–905. [CrossRef]
98. Taghon, G.J.; Rowe, J.B.; Kapolka, N.J.; Isom, D.G. Predictable Cholesterol Binding Sites in GPCRs Lack Consensus Motifs. *Structure* **2021**, *29*, 499–506. [CrossRef] [PubMed]
99. Lemel, L.; Nieścierowicz, K.; García-Fernández, M.D.; Darré, L.; Durroux, T.; Busnelli, M.; Pezet, M.; Rébeillé, F.; Jouhet, J.; Mouillac, B.; et al. The Ligand-Bound State of a G Protein-Coupled Receptor Stabilizes the Interaction of Functional Cholesterol Molecules. *J. Lipid Res.* **2021**, *62*, 100059. [CrossRef]
100. Marino, K.A.; Wade, R.C.; Prada-Gracia, D.; Provasi, D.; Filizola, M. Impact of Lipid Composition and Receptor Conformation on the Spatio-Temporal Organization of μ-Opioid Receptors in a Multi-Component Plasma Membrane Model. *PLoS Comput. Biol.* **2016**, *12*, e1005240. [CrossRef]
101. Velazhahan, V.; Ma, N.; Pándy-Szekeres, G.; Kooistra, A.J.; Lee, Y.; Gloriam, D.E.; Vaidehi, N.; Tate, C.G. Structure of the Class D GPCR Ste2 Dimer Coupled to Two G Proteins. *Nature* **2021**, *589*, 148–153. [CrossRef] [PubMed]
102. Song, W.; Yen, H.-Y.; Robinson, C.V.; Sansom, M.S.P. State-Dependent Lipid Interactions with the A2a Receptor Revealed by MD Simulations Using In Vivo-Mimetic Membranes. *Structure* **2019**, *27*, 392–403. [CrossRef]
103. Dijkman, P.M.; Castell, O.K.; Goddard, A.D.; Munoz-Garcia, J.C.; de Graaf, C.; Wallace, M.I.; Watts, A. Dynamic Tuneable G Protein-Coupled Receptor Monomer-Dimer Populations. *Nat. Commun.* **2018**, *9*, 1710. [CrossRef] [PubMed]
104. Ferré, S.; Casadó, V.; Devi, L.A.; Filizola, M.; Jockers, R.; Lohse, M.J.; Milligan, G.; Pin, J.-P.; Guitart, X. G Protein–Coupled Receptor Oligomerization Revisited: Functional and Pharmacological Perspectives. *Pharmacol. Rev.* **2014**, *66*, 413–434. [CrossRef]
105. Møller, T.C.; Hottin, J.; Clerté, C.; Zwier, J.M.; Durroux, T.; Rondard, P.; Prézeau, L.; Royer, C.A.; Pin, J.-P.; Margeat, E.; et al. Oligomerization of a G Protein-Coupled Receptor in Neurons Controlled by Its Structural Dynamics. *Sci. Rep.* **2018**, *8*, 10414. [CrossRef] [PubMed]
106. Jones, A.J.Y.; Gabriel, F.; Tandale, A.; Nietlispach, D. Structure and Dynamics of GPCRs in Lipid Membranes: Physical Principles and Experimental Approaches. *Molecules* **2020**, *25*, 4729. [CrossRef]
107. Soubias, O.; Teague, W.E.; Hines, K.G.; Gawrisch, K. The Role of Membrane Curvature Elastic Stress for Function of Rhodopsin-like G Protein-Coupled Receptors. *Biochimie* **2014**, *107*, 28–32. [CrossRef]
108. Cheng, X.; Smith, J.C. Biological Membrane Organization and Cellular Signaling. *Chem. Rev.* **2019**, *119*, 5849–5880. [CrossRef]
109. Rajagopal, S.; Shenoy, S.K. GPCR Desensitization: Acute and Prolonged Phases. *Cell. Signal.* **2018**, *41*, 9–16. [CrossRef]
110. Luttrell, L.M.; Wang, J.; Plouffe, B.; Smith, J.S.; Yamani, L.; Kaur, S.; Jean-Charles, P.-Y.; Gauthier, C.; Lee, M.-H.; Pani, B.; et al. Manifold Roles of β-Arrestins in GPCR Signaling Elucidated with SiRNA and CRISPR/Cas9. *Sci. Signal.* **2018**, *11*, eaat7650. [CrossRef]
111. Kumar, A.; Sarkar, P.; Chattopadhyay, A. Metabolic Depletion of Sphingolipids Reduces Cell Surface Population of the Human Serotonin 1A Receptor Due to Impaired Trafficking. *ACS Chem. Neurosci.* **2021**, *12*, 1189–1196. [CrossRef]
112. Shaik, F.A.; Chelikani, P. Differential Effects of Membrane Sphingomyelin and Cholesterol on Agonist-Induced Bitter Taste Receptor T2R14 Signaling. *Mol. Cell. Biochem.* **2020**, *463*, 57–66. [CrossRef]
113. Ingólfsson, H.I.; Melo, M.N.; van Eerden, F.J.; Arnarez, C.; Lopez, C.A.; Wassenaar, T.A.; Periole, X.; de Vries, A.H.; Tieleman, D.P.; Marrink, S.J. Lipid Organization of the Plasma Membrane. *J. Am. Chem. Soc.* **2014**, *136*, 14554–14559. [CrossRef]
114. Sengupta, D.; Prasanna, X.; Mole, M.; Chattopadhyay, A. Exploring GPCR–Lipid Interactions by Molecular Dynamics Simulations: Excitements, Challenges, and the Way Forward. *J. Phys. Chem. B* **2018**, *122*, 5727–5737. [CrossRef] [PubMed]

115. Zhou, K.; Dichlberger, A.; Martinez-Seara, H.; Nyholm, T.K.M.; Li, S.; Kim, Y.A.; Vattulainen, I.; Ikonen, E.; Blom, T. A Ceramide-Regulated Element in the Late Endosomal Protein LAPTM4B Controls Amino Acid Transporter Interaction. *ACS Central Sci.* **2018**, *4*, 548–558. [CrossRef] [PubMed]
116. Tao, M.; Isas, J.M.; Langen, R. Annexin B12 Trimer Formation Is Governed by a Network of Protein–protein and Protein–lipid Interactions. *Sci. Rep.* **2020**, *10*, 1–11. [CrossRef] [PubMed]
117. Flores, J.A.; Haddad, B.G.; Dolan, K.A.; Myers, J.B.; Yoshioka, C.C.; Copperman, J.; Zuckerman, D.M.; Reichow, S.L. Connexin-46/50 in a Dynamic Lipid Environment Resolved by CryoEM at 1.9 Å. *Nat. Commun.* **2020**, *11*, 4331. [CrossRef] [PubMed]

International Journal of
Molecular Sciences

MDPI

Review

Uncoupling Proteins and Regulated Proton Leak in Mitochondria

Afshan Ardalan [1,†], Matthew D. Smith [2] and Masoud Jelokhani-Niaraki [1,*]

1 Department of Chemistry and Biochemistry, Wilfrid Laurier University, Waterloo, ON N2L 3C5, Canada; afshan.ardalan@utoronto.ca
2 Department of Biology, Wilfrid Laurier University, Waterloo, ON N2L 3C5, Canada; msmith@wlu.ca
* Correspondence: mjelokhani@wlu.ca
† Present Address: Department of Biochemistry, University of Toronto, Toronto, ON M5S 1A8, Canada.

Abstract: Higher concentration of protons in the mitochondrial intermembrane space compared to the matrix results in an electrochemical potential causing the back flux of protons to the matrix. This proton transport can take place through ATP synthase complex (leading to formation of ATP) or can occur via proton transporters of the mitochondrial carrier superfamily and/or membrane lipids. Some mitochondrial proton transporters, such as uncoupling proteins (UCPs), transport protons as their general regulating function; while others are symporters or antiporters, which use the proton gradient as a driving force to co-transport other substrates across the mitochondrial inner membrane (such as phosphate carrier, a symporter; or aspartate/glutamate transporter, an antiporter). Passage (or leakage) of protons across the inner membrane to matrix from any route other than ATP synthase negatively impacts ATP synthesis. The focus of this review is on regulated proton transport by UCPs. Recent findings on the structure and function of UCPs, and the related research methodologies, are also critically reviewed. Due to structural similarity of members of the mitochondrial carrier superfamily, several of the known structural features are potentially expandable to all members. Overall, this report provides a brief, yet comprehensive, overview of the current knowledge in the field.

Keywords: mitochondrial carriers; uncoupling proteins; ADP/ATP carrier; membrane protein structure and function; regulation and mechanism of proton transport; membrane protein oligomerization; ATP synthesis; biphasic proton transport model; alternating access mechanism; reactive oxygen species control

Citation: Ardalan, A.; Smith, M.D.; Jelokhani-Niaraki, M. Uncoupling Proteins and Regulated Proton Leak in Mitochondria. *Int. J. Mol. Sci.* **2022**, 23, 1528. https://doi.org/10.3390/ijms23031528

Academic Editor: Giovanni Natile

Received: 24 December 2021
Accepted: 26 January 2022
Published: 28 January 2022

1. Introduction

Solute carrier proteins (SLCs) are integral Membrane proteins (MP), responsible for transport of solutes across intracellular membranes [1]. The largest family of SLCs in the human body is the mitochondrial carrier superfamily (MCF) or SLC25 with 53 members [2]. MCF proteins are mostly antiporters, transporting different solutes across the inner membrane of mitochondria (IMM) in opposite directions. Some MCF proteins are uniporters, transporting solute molecules in one direction, whereas others are thought to be substrate-proton symporters [2]. Because of amino acid sequence similarities among the MCF members it has been suggested that these proteins have comparable secondary and tertiary structures [3,4]. This review elaborates on recent advances about the structure and function of members of the MCF that are capable of regulated transport of protons across the IMM. We will discuss the common structural and functional features of these proteins, then review the concepts of proton motive force, coupling and uncoupling, and finally discuss the suggested mechanistic models for proton transport.

2. Structure of MCF Members

Most of the information about the structure of MCF members is based on the knowledge of ADP/ATP carrier protein (AAC) as the most characterized member of the MCF. AAC is the only member of the MCF for which high resolution X-ray structures are available [5,6] (Figure 1A–F). The first high-resolution X-ray structure of this carrier protein (bovine AAC1 or BtAAC) was reported in 2003 (PDB ID: 1OKC) [5]. AAC imports ADP from the intermembrane space (IMS) into the matrix and exports ATP from the matrix to the IMS [6]. The structure of AAC reveals a trilateral pseudosymmetric structure that includes three tandem repeats, that is thought to be typical of all MCF members. Each repeat is composed of ~100 amino acids distributed over two transmembrane (TM) α-helices and a connecting loop containing a hydrophilic short helix parallel to the membrane surface on the matrix side [5] (Figure 1A,D).

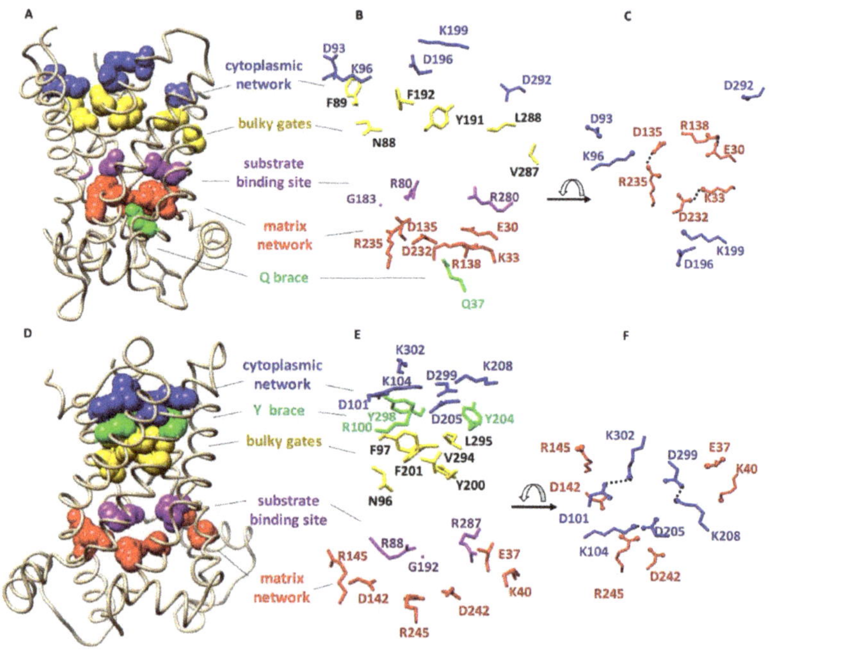

Figure 1. Overall structure and conserved functional residues of AAC. (**A**) X-ray structure of bovine AAC in the cytoplasmic state (1OKC). The conserved functional residues are shown as spheres; (**B**) Detailed representation of the side chains and directions of conserved residues of bovine AAC; (**C**) Top view of broken cytoplasmic and formed matrix networks of bovine AAC in the cytoplasmic state; (**D**) X-ray structure of TtAAC in the matrix state (6GCI). The conserved functional residues are shown as spheres; (**E**) Detailed representation of the side chains of TtAAC; (**F**) Top view of the broken matrix and formed cytoplasmic networks of Tt AAC in the matrix state. Figure generated by USCF Chimera 1.13rc.

Two conserved signature motifs are present at the water-membrane interface of mitochondrial carrier proteins. The first signature motif, PX[DE]XX[RK], is located on odd-numbered helices (helices 1, 3 and 5) close to the matrix side [5,7]. The matrix salt-bridge network (or matrix network) is formed as a result of electrostatic interactions between positive and negative residues of the matrix side signature motifs [5,7] (Figure 1). Conserved glutamine residues, known as the Q-brace/glutamine-brace and located in close proximity to the matrix signature motif (Px[DE]xx[KR]xxxQ) on helices 1 and/or 3 and/or 5, provide extra structural support by forming hydrogen bonds with the salt-bridge network residues

2 (Figure 1). In the past few years computational and experimental evidence have suggested larger (composed of more amino acid residues) and more complex matrix networks for some MCF members, such as AAC1 and uncoupling protein 2 (UCP2) [8]. The second signature motif, [FY][DE]XX[RK], is located on even numbered helices (helices 2, 4 and 6) close to the IMS. Similar to the matrix network, an intramolecular salt-bridge network is also formed from electrostatic interactions of positively and negatively charged residues of the signature motifs close to the IMS, which is known as the cytoplasmic salt-bridge network (cytoplasmic network) [7]. The cytoplasmic network is supported by hydrogen bond formation between tyrosine residues that are in the motif with the residues of the cytoplasmic network (Y brace/tyrosine-brace). MCF members have one to three of each brace (Q or Y) [2] (Figure 1).

Based on the structure of AAC1 it has been proposed that a single conserved common substrate/inhibitor/solute binding site (SBS) is present at the center of the transporter cavity in all MCF members (Figure 1). This binding site resides at the bottom of a solvent accessible cavity and includes three protein-substrate specific contact points at even numbered helices [9,10]. It has been suggested that contact points I (located on helix 2) and II (located on helix 4) can confer substrate specificity to the carrier; however, it seems that the positively charged contact point III (located on helix 6) does not contribute to substrate specificity [2] (Figure 1).

Besides signature motifs, other sequences are also conserved in the members of MCF, such as the $\pi G\pi X\pi G$ motif on helices 1, 3 and 5 and the $\pi XXX\pi$ motif on helices 2, 4 and 6 [6]. The symbol π refers to amino acids with short side chains; however, occasionally amino acids with bulky side chains are found at the π positions 2.

Formation of homo- and/or hetero-oligomers in the membrane to optimize/regulate protein functions is common among MPs [11–13]. Therefore, understanding oligomerization and interconversion between different oligomeric states can provide insights about a protein's mechanism of function [13]. Lipids can influence and control protein oligomerization in the membrane [13]. The role of lipids in controlling oligomerization have been shown to be more important in MPs with weakly interacting subunits and small oligomerization interface(s) (small, buried surface area and no salt-bridge formation between the subunits) [12]. It has also been suggested that cells can alter lipid composition of the membranes to regulate the presence and abundance of oligomeric species [12].

The presence of a GXXXG motif on TM helices has been shown to be involved in homo-oligomerization of MPs as a result of helix-helix interactions [14]. This motif was first shown to play an important role in dimerization of glycophorin A [15]. NMR structure of the glycophorin A showed that two glycine residues, that are separated from one another by three amino acid residues, form a groove that allows the helices to pack closely [16]. Furthermore, the GXXXG motif is known to be the most abundant sequence involved in helix–helix association of E. coli inner MPs (~80%) [14,17]. In addition to GXXXG, $\pi XXX\pi$ (π = small residues such as G, A, S or T) and GXXXAXXG have also been shown to play an important role in oligomerization of proteins [14]. On the other hand, such oligomerization motifs are very abundant in MPs and cannot be used as the only evidence to predict a dimerization interface [14].

The quaternary structure and protein-lipid interactions of members of the MCF are not fully understood and are subjects of debate between researchers [11,18–20]. Some reports suggest that all mitochondrial carriers function as monomers [19,21], while others propose oligomeric structures for many of them [11,20]. More specifically, among proton transporters of the MCF, there has not been a consensus on the functional molecular forms of AAC [21–23], UCP [11,18,21,24] and phosphate carrier (PIC) [20,21,25]. It has been suggested that AAC forms either a monomer [21] or a dimer [22], or both [26] in the membrane. Out of six molecules of cardiolipin (CL) that are bound to subunits of dimeric AAC, two are thought to sit at the dimer interface and act as a "glue" between the monomeric subunits [22]. Furthermore, it has been suggested that CL might have a role in cross-talk between AAC subunits in the dimer [22]. Moiseeva et al. [26] showed that external factors (such as tonicity of the mitochondria or incubation at 4 °C) can influence the oligomerization of AAC [26]. PICs are also proposed to function either as a monomer [21] or dimer [25]. The case of UCPs is more complicated, since monomeric [27], dimeric [24], and tetrameric [11,28,29] structures have been all reported to be functional.

3. Mechanism of Function in MCF Members

Comparable amino acid sequences and the presence of conserved motifs led to suggestions that all members of the MCF transport their respective solutes (substrates) via an alternating access mechanism [2,30]. In this hypothesized mechanism, mitochondrial anion carriers can adopt two different conformational states: a cytoplasmic state (V), in which the solvent-accessible cavity of the carrier is open towards the IMS, the cytoplasmic network is broken and the matrix network is intact (like the letter "V"); and the matrix state (Λ), in which the solvent-accessible cavity of the carrier is open towards the matrix side (like the symbol "Λ"), the cytoplasmic network is formed and the matrix network is broken [9]. The model predicts that members of the MCF can alternate between matrix and cytoplasmic states by movement of the helices towards or away from the central Z axis (normal to the bilayer) as a consequence of substrate binding [6] (Figure 2).

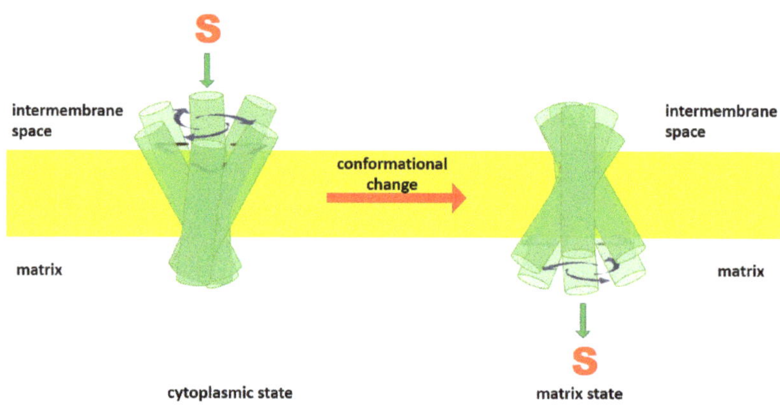

Figure 2. Alternating access mechanism is proposed to be a common mechanism for substrate translocation by members of the MCF. In this model the protein's cavity opens alternately towards the cytoplasmic and matrix sides alternatively.

4. Proton Motive Force and Coupling

Being the powerhouse of the cell, mitochondria transform energy to produce consumable energy molecules (ATP) that can be used to drive metabolic processes in the cell. For this purpose, the tricarboxylic acid (TCA) cycle [31] (also known as the citric acid cycle or Krebs cycle) and oxidative phosphorylation take place [32] inside the matrix and across the inner membrane (IMM). The final product of lipid, sugar, or amino acid catabolism, acetyl-CoA enters the Krebs cycle when it is combined with oxaloacetate (a four-carbon molecule) in the mitochondria matrix to form a six-carbon citrate molecule. After seven subsequent enzymatic steps, oxaloacetate is regenerated and can receive another molecule of acetyl-CoA [33]. During the Krebs cycle, the two carbons from the acetyl group of acetyl-CoA are released in the form of two CO_2 molecules, and the extra electrons are used to generate reduced nicotinamide adenine dinucleotide (NADH) and flavin adenine dinucleotide ($FADH_2$) that carry the electrons to the electron transport chain (ETC), which is comprised of a series of protein complexes located in the IMM [8].

The movement of electrons from NADH and $FADH_2$ through the ETC components results in the efflux of protons from the matrix to the IMS, thus generating an electrochemical gradient across the IMM widely known as the proton motive force [34] that eventually results in formation of ATP through the activity of the ATP synthase complex [34].

The consequence of translocation of protons from the matrix to the IMS in different steps of the ETC is the generation of a ~200 mV electric potential across the IMM [32]. After the Krebs cycle and ETC, the resulting electrochemical gradient across the IMM is transformed into a more readily consumable form of energy for the cell through a "coupling" process. Coupling of the electrochemical gradient to ATP synthesis was first proposed by Peter Mitchell in 1961 [34]. ATP synthase (also known as complex V and F0F1 ATPase) provides a route for protons to flow in accordance with their concentration gradient from the IMS to the matrix. The outcome of this passage is conformational changes in ATP Synthase that result in ATP synthesis from ADP and Pi [8]. The mechanism of ATP synthesis was hypothesized and developed by Paul Boyer and John Walker between 1964 to 1994 [35]. It is estimated that the transport of between two to five protons are needed for the synthesis of one molecule of ATP. However, relating these numbers to the number of required NADH or $FADH_2$ is not straightforward as protons can leak across the IMM through routes other than the ATP synthase complex. Three ATP per one NADH or two ATP molecules per one $FADH_2$ are common estimations [32].

5. Uncoupling in the Mitochondria

Translocation/leak of protons that are pumped into the IMS during the electron transport process back to the matrix by any paths other than the ATP synthase complex results in dissipation of the electrochemical gradient across the IMM, and thus "uncouples" electron transport from ATP synthesis [36]. Proton leak across the IMM is the outcome of two leakage processes: Inducible and basal proton leak [37]. Inducible proton leak is mainly controlled by a specific family within the MCF known as UCPs and can be regulated by fatty acids, superoxides, lipid peroxidation products and adenine nucleotides [38]. Basal proton leak is non-regulated leakage of protons across the IMM, which depends on the fatty acyl composition of the membrane and the presence of mitochondrial carriers that are able to translocate protons [37,38]. Basal proton leak has been extensively discussed in several reviews [37,38]; our focus will be more on inducible and regulated proton leak in mitochondria.

Inducible proton leak in mitochondria can be closely related to the proton transport function of UCPs—proteins that are not universally present in all mitochondria. UCP1 (thermogenin, the prototypic member of the UCP family) was discovered in 1978 by Heaton et al. in hamsters [39]. This protein was purified from brown adipose tissue of rat and hamster mitochondria in high yields by Lin and Klingenberg within two years of its discovery [40]. Almost 20 years after the discovery of UCP1, four other homologues of UCPs were discovered: UCP2 (1997) [41,42], UCP3 (1997) [43,44], UCP4 (1999) [45], and UCP5 (1998) [46]. UCPs have different expression patterns in the human body. UCP1 is mainly found in brown fat adipose tissue [36]. UCP2 is expressed in several tissues such as skeletal muscles, heart, liver, kidney, lung, pancreas, spleen, and macrophages [36]. UCP3 is mostly expressed in brown adipose tissue and muscles [36]. UCPs 4 and 5 are specifically neuronal and mainly expressed in brain tissues [36]. UCPs 2, 4 and 5 are considered neuronal UCPs, as they are predominantly found in the central nervous system (CNS) [28].

Another candidate for contributing to inducible proton leak in mitochondria is AAC. Even though AAC has long been known as a basal proton leaker, it has recently been suggested to be involved in regulated proton transport [47,48]. The next two sections of this review will discuss the regulated proton leakers of the IMM.

6. Regulated Proton Transport across the IMM Is Facilitated by Uncoupling Proteins

As mentioned above, UCPs are the only subfamily of the MCF known to be responsible for regulated proton leak across the IMM [38]. Generally, due to low purification yield and intricate preparation steps (such as extraction from membrane, purification in native form, and reconstitution in a bilayer-mimicking environment) analyses of structure and function of MPs have been difficult, cumbersome, and controversial [49]. UCPs are no exception, as there is not universal agreement on UCPs' tertiary and quaternary structures [11,18,21,24], or their mechanism of function and inhibition [50–53].

Inconsistencies regarding UCPs' structure arise from various sources including the organism in which recombinant UCPs are expressed in or purified from, the method of expression, choice of detergent for extraction, as well as the method of purification and reconstitution. As a continuing challenge in obtaining membrane proteins in their native structure and function and regardless of the type of organism used for expression, as long as a single or a mix of detergents is used for isolation of the protein from cell membranes and its reconstitution in lipid membrane models, reproducing an exact native-like structure for proteins would not be possible. This is due to the inherently denaturing character of detergents and the fundamental differences in their physicochemical properties from those of lipids of cell membranes.

The traditionally understood structure for UCPs, based on methods such as ultracentrifugation [24], cross-linking [18] and nucleotide binding studies [54], is dimeric. In 2013, a tetrameric (dimer of dimers) structure of UCPs was reported for the first time for UCP1 [11]. In this study, UCP1 was directly expressed in and purified from *E. coli* membranes under native conditions and reconstituted in POPC liposomes for structural and functional analyses [11]. Observation of a high molecular weight band at ~132 kDa on polyacrylamide gel (in a semi-native PAGE analysis) as well as evaluations of the protein structure by circular dichroism (CD), ultracentrifugation, and mass spectrometry led Hoang et al. to report a tetrameric structure for UCP1 [11]. Their data showed that gradual addition of the potent denaturing agent SDS to reconstituted tetrameric UCP1 resulted in a gradual decrease in the abundance of the high molecular weight band (at 132 kDa) and corresponding appearance of bands consistent with the size of dimers (~66 kDa) and monomers (~33 kDa) [11]. Disappearance of the tetrameric UCP1 band occurred at lower concentrations of SDS compared to that of dimeric UCP (the kd values for (tetramer)/(dimer) and (dimer)/(monomer) were 1.3 mM and 23 mM, respectively) [11]. Reconstituted UCPs 2, 4 and 5 in POPC (+CL) liposomes have also been reported to be tetrameric showing a behavior similar to UCP1 upon addition of SDS [28]. These observations led to the proposition that the UCP tetramer is in fact a dimer of dimers, in which there is a loose binding interface between the dimers and a tight binding interface between the monomers within each pair of dimers (Figure 3) [11,28]. Consistent with SDS titration data, computational analysis of a UCP2 tetramer demonstrated that the dissociation energy of tetramer to dimer was ~20 kcal/mol lower than dissociation of dimer to monomer (tetramer to dimer: 60 kcal/mol; and dimer to monomer: 80 kcal/mol) [55]. Molecular dynamics (MD) simulations of UCP2 in POPC bilayer, also demonstrated a pseudosymmetrical structure for the tetramer in which the transport state of each dimeric pair open towards one side of the membrane was opposed by the other dimeric pair open on the other side of the membrane (either VVΛΛ or ΛΛVV) [55]. Furthermore, two salt-bridges were observed between the monomers within a dimer while no salt-bridge was observed between the dimers (Figure 3) [55]. Based on the original assumption that protons (or other substrates) were transported through monomeric subunits, these results suggested a possibility of co-existence of all three monomeric, dimeric and tetrameric UCPs as functional forms of the protein(s).

In a detergent-based study using size exclusion chromatography, CD, electrophoretic assays, and isothermal titration calorimetry (ITC), Lee et al. (2015) suggested that the only functional state of UCP1 is monomeric [27]. These researchers performed their structural evaluations of the protein in a non-native environment (detergent in buffer), used reconstituted proteins in detergent and applied an unclear binding model (for stoichiometry) in their ITC experiment, and had an ambiguous interpretation of their data regarding the molecular form of the protein by ignoring the possibility of oligomerization. These shortcomings undermine the certainty of their conclusions.

It is noteworthy that, in addition to the possibility of self-association, UCPs might be able to interact with other members of the MCF, such as AAC, to form heterodimers [56].

In addition, different homologues of UCPs are proposed to be able to interact with one another and form heterodimers, such as UCP2 + UCP3 [57]. It has been also suggested that homo-tetrameric MPs are most probably formed by dimerization of dimers [11,28,58].

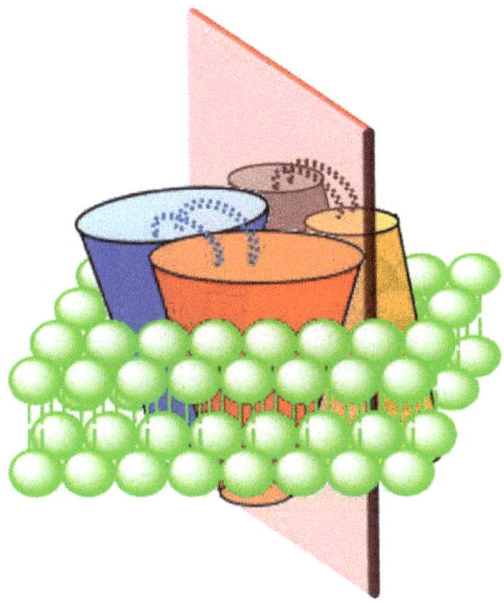

Figure 3. Schematic representation of a pseudosymmetric UCP tetramer. This tetramer is in fact a dimer of dimers with two tight-binding interfaces (between red and blue subunits and between yellow and gray subunits) and a loose-binding interface (shown with a light red plate). There are two salt-bridges between the monomers within each dimer shown with blue and grey dotted lines.

Even though there is no available high resolution X-ray structure for UCP homologues, structures of human UCP1 [59] and mouse UCP2 (PDB ID: 2LCK) (Figure 4) [60] have been obtained by NMR spectroscopy. Proteins used for both studies were expressed in and purified from E. coli cells and reconstituted in dodecylphosphocholine (DPC) detergent micelles (UCP1), or detergent (DPC) lipid-mixed micelles (UCP2) for structural analyses [59,60]. Use of DPC detergent micelles as the reconstitution environment has been argued to destabilize UCP structure [49]. The reported NMR structures of UCPs 1 and 2 possess the common structural features of MCF proteins; however, the internal cavity in UCP monomers is larger compared to what is commonly observed in other members of the MCF [49].

It is widely accepted that proton flux by UCPs is activated by fatty acids [28,61]. Evaluation of the effect of fatty acid features (length, structural rigidity, hydrophobicity) on proton transport rate of neuronal UCPs reveals that unsaturated and long chain fatty acids are generally more potent proton transport activators [28,62]. Alterations of circular dichroism (CD) spectra of neuronal UCPs reconstituted in phosphatidylcholine (PC) vesicles upon addition of fatty acids imply that fatty acids are able to induce conformational changes in UCPs [28]. NMR spectroscopy of UCP1 and 2 suggests that there is a groove between helices 1 and 6, which might be a fatty acid binding site; the hydrophobic chain of the fatty acid lays in the groove and its headgroup faces the matrix side [53,59]. Hoang et al. [62] proposed the presence of a specific geometrically optimized binding site for fatty acids in UCPs 2, 4 and 5. Furthermore, mutagenetic and functional analysis of UCPs by different researchers and different methods provide important insights about fatty acid-UCP interactions, as well as residues that are involved in UCP-mediated transport. Table 1 shows the residues that are involved in proton transport regulation along with the major findings of related studies [4,28,53,59,60,62–68].

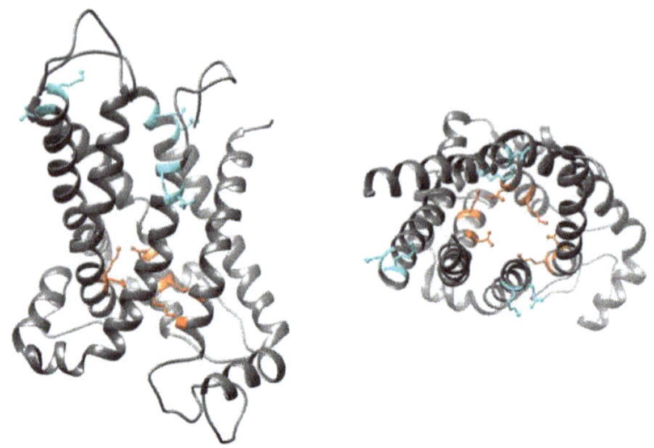

Figure 4. 350 ns MD simulated structure of UCP2 in cytoplasmic state. Left: sideview, and right: top view, residues involved in the cytoplasmic and matrix salt-bridge networks are shown in cyan and orange, respectively. Figure was generated by USCF Chimera 1.13rc. Adapted, with modifications, with permission from Ardalan et al. *J. Phys. Chem. B* **2021**, *125*, 9130–9144. Copyright© 2021 American Chemical Society.

Table 1. Amino acid residues involved in proton transport regulation.

Protein	H^+ Transport (Activated by Fatty Acids)	Inhibition (by Purine Nucleotides)	Methods	Findings Related to H^+ Transport Mechanism	Ref.
UCP2	R279 R60 R241 R88 K141 K16 K271	R185 K141 R88	-NMR -mutagenesis -fluorescence	1-The fatty acid chain probably partitions into a groove between helices 1 and 6 which is likely its binding site and the carboxylate head group interacts with R60 and K271. 2-Fatty acid flipping is required for H^+ transport activity of UCP2. 3-GDP can allosterically dislocate fatty acid from its binding.	[53,60]
UCP1	K269 K56		-NMR -mutagenesis -fluorescence -MD simul.	There is a specific binding site between helices 1 and 6 close to the matrix for fatty acid.	[59]
UCP1	D28 R84 R183 R277		-patch clamp of whole mitoplast	1-Proposed induced fit mechanism (shuttling model + alternating access) 2-Long chain low pKa fatty acid anions can inhibit proton transport thus transport of proton and fatty acid is from the same translocation path.	[63,64]
UCP1		R84 R183 R277	-mutagenesis -fluorescence	Single point mutations of R84Q, R183T and R277L resulted in more than 93% decrease in inhibition.	[66]
UCP1	D210 D28 H146 H148	R84 R183 R277	-mutagenesis -fluorescence	E168Q, R84I and R92T inhibited Cl^- transport but did not affect proton transport.	[67]
UCP2	R96 k104		-mutagenesis -fluorescence	1-Longer fatty acids showed better transport rate: PA > MA > LA 2-Sensitivity of mutants to fatty acid length was different: K104Q > R96Q > >UCP2 3-R76Q and R88Q decreased the chloride transport rate up to 82–83%. 4-Succinic acid (C4) did not activate proton transport of UCP2.	[62]

Table 1. *Cont.*

Protein	H^+ Transport (Activated by Fatty Acids)	Inhibition (by Purine Nucleotides)	Methods	Findings Related to H^+ Transport Mechanism	Ref.
UCP2			-fluorescence	Addition of fatty acid inhibited UCP2 inherent chloride transport up to ~50%, suggesting that fatty acids and chloride ions (anions) share the same path.	[28]
UCP2	Salt-bridges: D35-K141 D138-K239 D138-R88 D236-R185 D236-R279	K38 K141 K239 R88 R185 R279	-fluorescence -CD -MD simulation	UCP2 tetramer can transport protons via a **biphasic two-state molecular model:** -All four subunits are functional. -Monomers of each dimer -are in similar transport state. -The function of one dimeric unit could be regulated by the other dimeric unit. ATP can occlude the translocon channel and prevent conformational switching.	[68]
UCP1	Single binding site: D28, R84, R183, S184, I187, S230, R277 (Found based on sequence alignment with AAC, Figure S1)				[9]

Four potential models have been proposed to describe the role of fatty acids in activation of proton transport in monomeric subunits of UCP/UCP1: (i) fatty acid cycling model [51]; (ii) buffering/cofactor model [52]; (iii) proton shuttling model [64] and (iv) alternating access/induced transition fit (ITF) model [63] (Figure 5).

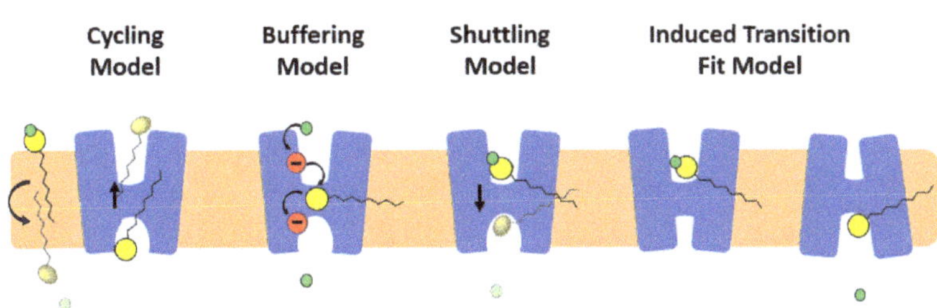

Figure 5. Proposed mechanisms of proton transport in UCPs. In the fatty acid cycling model, protonated fatty acid (yellow head group and a black tail) flip-flops across the lipid bilayer, releases its proton (green circle) and is transported from to the IMS as an anion by UCP. In the buffering model, fatty acid anion binds to UCP and accepts/donates protons via its carboxylate group from and to titratable amino acids of the translocon channel. In the shuttling model long chain fatty acids remain bound to the protein while their head moves back and forth across the IMM (getting protonated in the IMS and releasing the proton into the matrix). In the ITF, UCP changes conformation from cytoplasmic to matrix state and vice versa upon movement of fatty acid across the translocon channel.

In the fatty acid cycling model, a (protonated) fatty acid that resides in the IMS is transferred to the matrix through a "flip-flop" process across the lipid bilayer and releases its proton to the matrix. Transport of the fatty acid anion back from the matrix to the IMS is then facilitated by UCP (Figure 5) [51]. NMR results from Zhao et al. [59] and MD simulations by Škulj et al. [69] are compatible with the cycling model. In the buffering model, fatty acid anions bind to UCP and accept/donate protons via their carboxylate group from and to titratable amino acids of the translocon channel (the central channel of the carrier), all the way to the matrix side, where the proton is released (Figure 5) [50,52]. The shuttling model proposes that the proton transport mechanism depends on the length of the fatty acid; short-chain protonated fatty acids can be transported from the IMS to the matrix (where they release their proton) via UCPs' translocon channel, whereas long chain fatty acids remain in the translocation channel. Long-chain fatty acids remain bound to the protein while their head group moves back and forth across the IMM (getting protonated in the IMS and releasing the proton into the matrix) (Figure 5) [64]. In 2017, an updated version of the shuttling model "ITF" was proposed as the fourth proton transport model of UCPs [63]. In this model, the elements of alternating access have been combined with the shuttling mechanism. Based on this ITF model, the arginine residues of the single binding site (R84, R13, R277) of UCP1 attract long chain anionic fatty acids towards the cavity of the protein where D28 is present [63]. Proximity of the anionic fatty acid's head and D28 will result in increasing the pK_a of both fatty acid and aspartate residue leading to attraction of a proton from the IMS. Co-presence of fatty acid and proton in the cavity will lead to further conformational changes from the cytoplasmic to matrix state at which point the proton gets released into the matrix (Figure 5) [63].

All four proposed fatty-acid-mediated transport models involve mechanisms of proton transport through monomeric units of UCPs. Recently, a more complex Biphasic Two-State molecular model has been proposed for proton transport through oligomeric states of UCPs [55,68]. This model is based on a thorough comparative structural and functional analysis of tetrameric UCP2 and several mutants in model lipid bilayers [55,68].

In the Biphasic Two-State model each monomer can separately transport protons via the alternating access mechanism. There are two interconnected dimeric units in the tetrameric structure of UCP (tetramer is considered as a dimer of dimers). At any moment, monomers within a dimeric unit are at the same stage/phase of transport. In other words, both monomers of a dimeric unit are either in the cytoplasmic state (VV) or in the matrix state (ΛΛ). There is a binary functional and structural phase difference between the two dimeric units of the tetramer, which means that when one dimeric unit is in the cytoplasmic state and ready to absorb protons in the cytoplasm the other dimeric unit is in the matrix state and release protons into the matrix (pair one: VV, pair 2: ΛΛ, overall tetramer: VVΛΛ). Concurrently, conformational changes of one dimeric pair can influence the active/inactive conformation of the other pair. Considering the loose interface between the two dimeric pairs [55], it has been suggested that dimerization of dimers might be an indication of a higher level of regulation of proton transport by UCPs, as compared with proton transport through monomers [68] (Figure 6). None of these models are universally accepted and continue to be debated [50,69].

It is noteworthy that long-chain fatty acids are not the only activators of proton transport in UCPs; superoxides and alkenals (reactive oxygen-derivatized products), such as HNE (4-hydroxy 2-nonenal), can also activate UCPs [70]. Based on experimental results, it has been suggested that superoxides are activators of proton transport in UCPs from the matrix side [71].

Di- and tri-phosphate purine nucleotides (ADP, ATP, GDP, and GTP) inhibit UCPs' proton flux activity [72]. The degree of inhibition of UCPs by purine nucleotides is different; for example, UCP2 is more inhibited by ATP compared to ADP [29]. The mechanism of inhibition by purine nucleotides is not fully understood; but the current working model for inhibition is based on mutational studies on UCP1. These studies indicated that three arginine residues are essential in binding of GDP by UCP1: R84, R183, and R277 [65]. Single mutations R183T and R277L resulted in loss of more than 90% of proton transport inhibition without any changes in the uninhibited rate of transport. A single mutation of R88Q caused 99% loss of inhibition, while the proton transport rate doubled [65]. The three R residues are conserved among all human UCPs and make up part of the single SBS (based on homology/sequence alignment with other members of MCF (Table 1)) [9]. It has been demonstrated that addition of purine nucleotides to UCPs reconstituted in lipid vesicles induces minor changes in the protein's overall conformation [29,73]. One NMR study of UCP2 showed that GDP could broaden NMR resonances on helices 1, 2, 3 and 4 [60], which implies that GDP interacts with these helices [74]. This study suggested that R185 and K141 of UCP2 can interact with GDP [60]. A complementary study by the same group added R88 to the list of GDP-interacting residues [53]. Interestingly, two out of the three amino acids were among those that have been shown to also interact with fatty acids (Table 1) [53]. A suggested inhibitory mechanism of GDP is that it can bind inside the translocation channel of the protein and allosterically displace fatty acid from its binding site [53]. A recent study on the structure and function of tetrameric UCP2 provided evidence that purine nucleotides such as ATP can bind to positive residue at/close to the matrix network (K38, K141, K239, R88, R185, R279) in each subunit and occlude the internal channel thus inhibiting the protein from conformational changes required for proton transport [68].

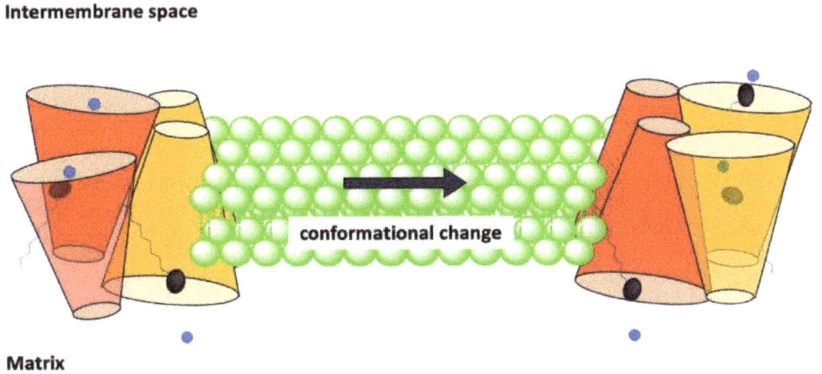

Figure 6. The Biphasic Two-State model for proton transport and inhibition of tetrameric UCP2. The monomeric subunits within a dimeric unit are functional and conformationally correlated. Initially the first dimeric pair (red) is in the cytoplasmic state (open to the IMS) where the fatty acid and proton can be absorbed. Upon movement of the protonated head group of the fatty acid from the IMS to the matrix, the red dimeric pair's conformation changes to the matrix state (open towards the matrix). The second dimeric pair (yellow), which was initially in the matrix state, transforms to the cytoplasmic state in parallel to the movement of fatty acid's head group towards the IMS. The four subunits of the tetramer do not have identical conformations at any time; however, the monomers within one dimeric unit are always at comparable conformational states (compare the red subunits conformation with yellow). Deprotonated fatty acids are shown in black, and protons are shown as blue spheres. Fatty acids are protonated at the IMS and deprotonated in the matrix. Changes in conformation of one dimeric pair can induce conformational changes in the other dimeric pair.

There are different reports about the stoichiometry of UCP inhibition by purine nucleotides; some have concluded that inhibition is achieved by one nucleotide per monomer [27], while others suggest the stoichiometry is one nucleotide per dimer [75]. The number of nucleotide inhibitors per tetrameric form of UCPs could conform with either suggestion, as only one dimeric unit is in the open state on each side of the membrane (Figure 6).

In addition to protons, all UCP homologues are capable of transporting chloride [29]. However, the rate at which UCPs transport chloride ions is lower compared to that of protons and no fatty acid is required to initiate/activate chloride transport [28,29]. In fact, fatty acids can inhibit the chloride transport activity of UCPs, suggesting that both species share the same path [76].

In addition to protons, UCPs transport a wide spectrum of ions including (but not limited to) monovalent anions such as chloride, bromide, nitrate [62,76], and alkyl sulfonates [77]. It has also been reported that UCP2 is able to transport anionic C4 metabolites such as aspartate, malate, and oxaloacetate [78].

In our experience [11], and as reported elsewhere [22], UCP has no tightly bound CLs; however, the presence of CL in UCPs 2,4 and 5′s lipidic environment affected their proton and chloride transport function and enhanced their helical conformation [29].

7. Proton Transport in ADP/ATP Carrier

AAC is an antiporter of ADP and ATP across the IMM. As a result of the conversion between the cytoplasmic and matrix states, AAC is responsible for transporting an amount of ADP and ATP that is approximately equivalent to the average weight of the human body per day [6]. Two inhibitors of AAC, carboxyatractyloside (CATR) [79] and bongkrekic acid (BKA) [80,81], can deactivate the transport and lock the protein structure in cytoplasmic and matrix states, respectively.

The structural elements that are involved in the alternative access mechanism can all be observed in the crystal structures of the carrier (Figure 1) [2,6]: (i) The three salt-bridges of the matrix network (K33-D232, R235-D135 and R235-E30) and the Q-brace (Q37) from BtAAC, V; (ii) The cytoplasmic network salt-bridges (D299-K208, D205-K104 and D101-Q302) and the Y brace Y298, Y204 (and R100) from TtAAC, Λ; and (iii) The SBS residues R88, G192 and R287 for TtAAC and R80, G183 and R 280 for BtAAC.

After the matrix state structure was suggested by Ruprecht et al. (Figure 1) [6], the same researchers proposed a detailed mechanism of transport, which is in agreement with the previously proposed alternating access mechanism [6]. In this mechanism, matrix and cytoplasmic conformational states can be interconverted as a result of interconnected movement of the helices (between 12–17 Å) towards the central z-axis of the carrier or away from it [82]. It has also been proposed that the SBS has minimal movement during the conversion of the two conformational states and is accessible from either side [82]. In fact, the contact points appear to act as a fulcrum during conformational changes [6,82].

A comparison of CATR-locked (V) and BKA-locked (Λ) structures reveals that only 20% of the carrier structure changes during the interconversion of the two conformational states [6]. One notable change is the movement of cytoplasmic ends of the even-numbered helices resulting in formation or disruption of the cytoplasmic network [6]. Surface analyses of the cytoplasmic and matrix states agree with the alternating access mechanism, as in each state the exposed site provides a positively charged accessible surface available for ATP or ADP binding; conversely, the opposite side is closed at the same time with no/very low positively charged surface available for interaction [6].

The proposed transport mechanism of AAC is thorough and detailed [6], however, it suggests that non-solvent accessible gates in both cytoplasmic and matrix states prevent proton leak through the protein. On the other hand, AAC has long been known to cause basal proton leak from the IMS to the matrix [38,83,84]. Furthermore, it has been shown recently that AAC might in fact be involved in regulated proton transport [47,48]. A patch-clamp study of mouse AAC in mitoplasts (mitochondria with the OMM removed) provides valuable insights into this carrier's proton transport mechanism [47], which is consistent with, yet complementary to, the mechanism proposed by Ruprecht et al. [6]. In this study, AAC was shown to be capable of transporting protons only in the presence of micromolar concentrations of fatty acids on the cytosolic/IMS side [47]. In the proposed mechanism, fatty acid functions as a co-factor bound to the carrier, and its carboxylate head moves through the translocation pathway. Therefore, inhibitors of the translocation pathway (CATR and BKA) inhibit the AAC-mediated proton transport [47]. Fatty acid can bind to the carrier regardless of its initial conformational state (matrix state or cytoplasmic state), and proton transport cannot induce any changes in the carriers' conformational state [47]. Moreover, translocation of ADP and ATP by AAC decreases the proton transport activity. Based on these observations, it has been hypothesized that AAC functions via two transport modes both sharing the same translocation path: (i) ADP/ATP antiporter mode based on the alternating access mechanism (cytoplasmic state and matrix state interconversions [2]), and (ii) proton transport mode, which relies on presence of/activation by fatty acids [47].

8. Importance of Regulated Proton Leak in the Mitochondria

Mitochondria contribute significantly to the production of cellular ROS [85]. Approximately 1 to 2% of the oxygen utilized by mitochondria results in the production of superoxide anion ($O2^{\bullet-}$) [85]. In cases where excess ROS generation leads to oxidative stress, mitochondria overexpress UCPs [86]. The evidence for the preventive role of UCPs against oxidative stress has been discussed in many studies [87–90]. UCP activity may also decrease the rate of apoptosis (programmed cell death) induced by ROS species [87]. These proteins can be activated by superoxide anions, resulting in the dissipation of the electrochemical gradient across the IMM [85]. Overall, UCPs are considered to contribute to a feedback loop which is activated by ROS, and leads to decrease in superoxide concentration [85,87].

Among UCPs, UCP1 is known to have a thermogenic physiological function in brown adipose tissues. It has also been shown that neuronal UCPs can increase the temperature in neuronal microenvironments during uncoupling activity [86]. As neuronal UCPs mostly accumulate in axon terminals, their thermogenesis can lead to changes in neurotransmission mechanisms [86]. It has been suggested that temperature differences between neuronal cells in the absence and presence of UCPs result in a temperature gradient that facilitates the diffusion of neurochemicals toward their postsynaptic targets [91]. Furthermore, although UCPs decrease the amount of synthesized ATP per mitochondrion, these proteins trigger mitochondrial biogenesis, thereby raising overall cellular ATP levels. Higher cellular ATP concentrations in presynaptic nerve terminals, where UCP is present, assist active processes (such as formation, transportation, and exocytosis of vesicles), and consequently facilitates neurotransmission [86,91].

Mitochondria can modulate the Ca^{2+} content of intact cells via uptake and release mechanisms, which are carried out separately. Mitochondrial ATP production and metabolism can be regulated by Ca^{2+} uptake activity in the IMM, which is mostly performed by the mitochondrial Ca^{2+} uniporter (MCU) [92]. The fundamental role of UCPs in regulating MCU activity has been reported in intact cells and in isolated liver mitochondria [57]. However, the involvement of UCPs in regulation of Ca^{2+} transporters remain controversial [93,94]. Using gene silencing in human and mice cell lines, Trenker et al. [93] showed that UCPs 2 and 3 could affect both the amount and the rate of Ca^{2+} uptake [93]. Additionally, it has been proposed that UCPs can act as conductive ion channels for Ca^{2+} [92,93].

Among members of the MCF, a few are known to transport protons across the IMM (Figure S1). Other than UCPs, of which proton transport is the common regulated function [38], most other transporters of protons are in fact anion carriers, which use the proton (gradient) as a driving force to co-transport other substrates across the IMM [95]. Substrate-proton co-transport by some MCF proteins is accomplished mainly by symport or antiport mechanisms [9,95]. Figure S1 aligns the amino acid sequences and shows the main structural and functional features of these proton transporters.

9. Conclusions

Overall, proton leak plays a vital role in mitochondrial bioenergetics as well as ROS generation and oxidative stress. Direct and indirect evidence have shown increased rate of mitochondrial proton leak upon aging [96], Thus proton transporters (basal and regulated) can be considered as potential drug targets for therapy of numerous diseases and understanding the underlying mechanism of proton transport is extremely valuable.

In this report we reviewed the current developments regarding the structure and function of proton transporters with special focus on UCPs. UCPs have been linked to thermogenesis as well as several metabolic diseases such as diabetes, hypertension, obesity, and cancer, making them fascinating targets for drug development [36]. Despite their biological importance there is no agreement on their mechanism of activity or their functional structure. For example, there is evidence that UCPs can exist in the IMM as functional monomers, dimers, tetramers, or a mixture of all. Protons are transported from the internal channel of the monomer or monomeric subunits of the dimer or tetramer. Regardless of what has been observed in experimental models, oligomeric state of UCPs (and other MCF members) might be interchangeable and differ in response to environmental factors or stress (in cell and organism), which can be also observed in mitochondrial structure (morphology) and its dynamic lipid and protein composition. In studying and interpreting dynamic processes in mitochondria or cell, it is essential to always consider the limitations applied to studying the model systems with current technologies when relating the results to a living, independent and ever-changing (non-rigid) system.

Supplementary Materials: The following supporting information can be downloaded at: https://www.mdpi.com/article/10.3390/ijms23031528/s1.

Author Contributions: Conceptualization, A.A. and M.J.-N.; investigation, A.A.; writing—original draft preparation, A.A.; writing—review and editing, A.A., M.D.S. and M.J.-N.; supervision, M.J.-N. All authors have read and agreed to the published version of the manuscript.

Funding: This research was funded by Natural Sciences and Engineering Research Council of Canada (NSERC) Discovery and Canada Foundation for Innovation (CFI) grants to M.J.-N. (05900 and 6786, respectively), and M.D.S. (NSERC Discovery 05437). A.A. was a recipient of the Ontario Trillium Scholarship for the duration of her doctoral studies at Wilfrid Laurier University.

Institutional Review Board Statement: Not applicable.

Informed Consent Statement: Not applicable.

Data Availability Statement: Not applicable.

Acknowledgments: The authors would like to acknowledge the technical contributions of Mikko Karttunen and Shahin Sowlati-Hashjin (both at Western University, Canada) in the computational component of this research, through SHARCNET (www.sharcnet.ca (last accessed on 23 December 2021)) and Compute Canada (www.computecanada.ca (last accessed on 23 December 2021)).

Conflicts of Interest: The authors declare no conflict of interest. The funders had no role in the design of the study; in the collection, analyses, or interpretation of data; in the writing of the manuscript, or in the decision to publish the results.

References

1. Schumann, T.; König, J.; Henke, C.; Willmes, D.M.; Bornstein, S.R.; Jordan, J.; Fromm, M.F.; Birkenfeld, A.L. Solute carrier transporters as potential targets for the treatment of metabolic disease. *Pharmacol. Rev.* **2020**, *72*, 343–379. [CrossRef] [PubMed]
2. Ruprecht, J.J.; Kunji, E.R. The SLC25 mitochondrial carrier family: Structure and mechanism. *Trends Biochem. Sci.* **2020**, *45*, 244–258. [CrossRef] [PubMed]
3. Aquila, H.; Link, T.A.; Klingenberg, M. The uncoupling protein from brown fat mitochondria is related to the mitochondrial ADP/ATP carrier. Analysis of sequence homologies and of folding of the protein in the membrane. *EMBO J.* **1985**, *4*, 2369–2376. [CrossRef] [PubMed]
4. Kunji, E.R.; Robinson, A.J. The conserved substrate binding site of mitochondrial carriers. *Biochim. Biophys. Acta Bioenerg. Bioenerg.* **2006**, *1757*, 1237–1248. [CrossRef]
5. Pebay-Peyroula, E.; Dahout-Gonzalez, C.; Kahn, R.; Trézéguet, V.; Lauquin, G.J.-M.; Brandolin, G. Structure of mitochondrial ADP/ATP carrier in complex with carboxyatractyloside. *Nature* **2003**, *426*, 39–44. [CrossRef]
6. Ruprecht, J.J.; King, M.S.; Zögg, T.; Aleksandrova, A.A.; Pardon, E.; Crichton, P.G.; Steyaert, J.; Kunji, E.R. The molecular mechanism of transport by the mitochondrial ADP/ATP carrier. *Cell* **2019**, *176*, 435–447. [CrossRef]
7. Robinson, A.J.; Overy, C.; Kunji, E.R. The mechanism of transport by mitochondrial carriers based on analysis of symmetry. *Proc. Natl. Acad. Sci. USA* **2008**, *105*, 17766–17771. [CrossRef]
8. Yi, Q.; Li, Q.; Yao, S.; Chen, Y.; Guan, M.-X.; Cang, X. Molecular dynamics simulations on apo ADP/ATP carrier shed new lights on the featured motif of the mitochondrial carriers. *Mitochondrion* **2019**, *47*, 94–102. [CrossRef]
9. Robinson, A.J.; Kunji, E.R. Mitochondrial carriers in the cytoplasmic state have a common substrate binding site. *Proc. Natl. Acad. Sci. USA* **2006**, *103*, 2617–2622. [CrossRef]
10. Klingenberg, M. The ADP, ATP shuttle of the mitochondrion. *Trends Biochem. Sci.* **1979**, *4*, 249–252. [CrossRef]
11. Hoang, T.; Smith, M.D.; Jelokhani-Niaraki, M. Expression, folding, and proton transport activity of human uncoupling protein-1 (UCP1) in lipid membranes: Evidence for associated functional forms. *J. Biol. Chem.* **2013**, *288*, 36244–36258. [CrossRef] [PubMed]
12. Gupta, K.; Donlan, J.A.; Hopper, J.T.; Uzdavinys, P.; Landreh, M.; Struwe, W.B.; Drew, D.; Baldwin, A.J.; Stansfeld, P.J.; Robinson, C.V. The role of interfacial lipids in stabilizing membrane protein oligomers. *Nature* **2017**, *541*, 421–424. [CrossRef] [PubMed]
13. Bolla, J.R.; Agasid, M.T.; Mehmood, S.; Robinson, C.V. Membrane protein–lipid interactions probed using mass spectrometry. *Annu. Rev. Biochem.* **2019**, *88*, 85–111. [CrossRef] [PubMed]
14. Li, E.; Wimley, W.C.; Hristova, K. Transmembrane helix dimerization: Beyond the search for sequence motifs. *Biochim. Biophys. Acta Biomembranes* **2012**, *1818*, 183–193. [CrossRef] [PubMed]

15. Lemmon, M.A.; Flanagan, J.M.; Treutlein, H.R.; Zhang, J.; Engelman, D.M. Sequence specificity in the dimerization of transmembrane.alpha.-helixes. *Biochem.* **1992**, *31*, 12719–12725. [CrossRef]
16. MacKenzie, K.R.; Prestegard, J.H.; Engelman, D.M. A transmembrane helix dimer: Structure and implications. *Science* **1997**, *276*, 131–133. [CrossRef]
17. Russ, W.P.; Engelman, D.M. The GxxxG motif: A framework for transmembrane helix-helix association. *J. Mol. Biol.* **2000**, *296*, 911–919. [CrossRef]
18. Klingenberg, M.; Appel, M. The uncoupling protein dimer can form a disulfide cross-link between the mobile C-terminal SH groups. *Eur. J. Biochem.* **1989**, *180*, 123–131. [CrossRef]
19. Kunji, E.R.; Crichton, P.G. Mitochondrial carriers function as monomers. *Biochim. Biophys. Acta Bioenerg.* **2010**, *1797*, 817–831. [CrossRef]
20. Wohlrab, H. Homodimeric intrinsic membrane proteins. Identification and modulation of interactions between mitochondrial transporter (carrier) subunits. *Biochem. Biophys. Res. Commun.* **2010**, *393*, 746–750. [CrossRef]
21. Crichton, P.G.; Harding, M.; Ruprecht, J.J.; Lee, Y.; Kunji, E.R. Lipid, detergent, and Coomassie Blue G-250 affect the migration of small membrane proteins in blue native gels: Mitochondrial carriers migrate as monomers not dimers. *J. Biol. Chem.* **2013**, *288*, 22163–22173. [CrossRef] [PubMed]
22. Klingenberg, M. Cardiolipin and mitochondrial carriers. *Biochim. Biophys. Acta Biomembr.* **2009**, *1788*, 2048–2058. [CrossRef] [PubMed]
23. Trézéguet, V.; Le Saux, A.; David, C.; Gourdet, C.; Fiore, C.; Dianoux, A.-C.; Brandolin, G.; Lauquin, G.J.-M. A covalent tandem dimer of the mitochondrial ADP/ATP carrier is functional in vivo. *Biochim. Biophys. Acta Bioenerg.* **2000**, *1457*, 81–93. [CrossRef]
24. Lin, C.; Hackenberg, H.; Klingenberg, E. The uncoupling protein from brown adipose tissue mitochondria is a dimer. A hydrodynamic study. *FEBS Lett.* **1980**, *113*, 304–306. [CrossRef]
25. Wohlrab, H. Novel inter-and intrasubunit contacts between transport-relevant residues of the homodimeric mitochondrial phosphate transport protein. *Biochem. Biophys. Res. Commun.* **2004**, *320*, 685–688. [CrossRef] [PubMed]
26. Moiseeva, V.; Murugova, T.; Vangeli, I.; Byvshev, I.; Ravaud, S.; Simonyan, R.; Gordeliy, V.; Pebay-Peyroula, E.; Yaguzhinsky, L. On the mechanism and functional significance of the ADP/ATP carrier (AAC) dimerization. *Biochem. Suppl. Ser. A Membr. Cell Biol.* **2017**, *11*, 321–329. [CrossRef]
27. Lee, Y.; Willers, C.; Kunji, E.R.; Crichton, P.G. Uncoupling protein 1 binds one nucleotide per monomer and is stabilized by tightly bound cardiolipin. *Proc. Natl. Acad. Sci. USA* **2015**, *112*, 6973–6978. [CrossRef]
28. Hoang, T.; Kuljanin, M.; Smith, M.D.; Jelokhani-Niaraki, M. A biophysical study on molecular physiology of the uncoupling proteins of the central nervous system. *Biosci. Rep.* **2015**, *35*, e00226. [CrossRef]
29. Hoang, T.; Smith, M.D.; Jelokhani-Niaraki, M. Toward understanding the mechanism of ion transport activity of neuronal uncoupling proteins UCP2, UCP4, and UCP5. *Biochemistry* **2012**, *51*, 4004–4014. [CrossRef]
30. Jardetzky, O. Simple allosteric model for membrane pumps. *Nature* **1966**, *211*, 969–970. [CrossRef]
31. Krebs, H.A.; Johnson, W.A. The role of citric acid in intermediate metabolism in animal tissues. *FEBS Lett.* **1980**, *117*, K2–K10. [CrossRef]
32. Osellame, L.D.; Blacker, T.S.; Duchen, M.R. Cellular and molecular mechanisms of mitochondrial function. *Best Pract. Res. Clin. Endocrinol. Metab.* **2012**, *26*, 711–723. [CrossRef] [PubMed]
33. Krebs, H. The tricarboxylic acid cycle. In *Chemical Pathways of Metabolism*; Elsevier: Amsterdam, The Netherlands, 1954; pp. 109–171.
34. Reid, R.; Moyle, J.; Mitchell, P. Synthesis of adenosine triphosphate by a protonmotive force in rat liver mitochondria. *Nature* **1966**, *212*, 257–258. [CrossRef]
35. Boyer, P.D. Energy, life, and ATP (Nobel lecture). *Angew. Chem. Int. Ed.* **1998**, *37*, 2296–2307. [CrossRef]
36. Echtay, K.S.; Bienengraeber, M.; Mayinger, P.; Heimpel, S.; Winkler, E.; Druhmann, D.; Frischmuth, K.; Kamp, F.; Huang, S.-G. Uncoupling proteins: Martin Klingenberg's contributions for 40 years. *Arch. Biochem. Biophys.* **2018**, *657*, 41–55. [CrossRef]
37. Busiello, R.A.; Savarese, S.; Lombardi, A. Mitochondrial uncoupling proteins and energy metabolism. *Front. Physiol.* **2015**, *6*, 36. [CrossRef]
38. Brown, G.C.; Murphy, M.P.; Jastroch, M.; Divakaruni, A.S.; Mookerjee, S.; Treberg, J.R.; Brand, M.D. Mitochondrial proton and electron leaks. *Essays Biochem.* **2010**, *47*, 53–67. [CrossRef]
39. Heaton, G.M.; Wagenvoord, R.J.; Kemp Jr, A.; Nicholls, D.G. Brown-adipose-tissue mitochondria: Photoaffinity labelling of the regulatory site of energy dissipation. *Eur. J. Biochem.* **1978**, *82*, 515–521. [CrossRef]
40. Lin, C.; Klingenberg, M. Isolation of the uncoupling protein from brown adipose tissue mitochondria. *FEBS Lett.* **1980**, *113*, 299–303. [CrossRef]
41. Fleury, C.; Neverova, M.; Collins, S.; Raimbault, S.; Champigny, O.; Levi-Meyrueis, C.; Bouillaud, F.; Seldin, M.F.; Surwit, R.S.; Ricquier, D. Uncoupling protein-2: A novel gene linked to obesity and hyperinsulinemia. *Nat. Genet.* **1997**, *15*, 269–272. [CrossRef]
42. Gimeno, R.E.; Dembski, M.; Weng, X.; Deng, N.; Shyjan, A.W.; Gimeno, C.J.; Iris, F.; Ellis, S.J.; Woolf, E.A.; Tartaglia, L.A. Cloning and characterization of an uncoupling protein homolog: A potential molecular mediator of human thermogenesis. *Diabetes* **1997**, *46*, 900–906. [CrossRef]
43. Boss, O.; Samec, S.; Paoloni-Giacobino, A.; Rossier, C.; Dulloo, A.; Seydoux, J.; Muzzin, P.; Giacobino, J.-P. Uncoupling protein-3: A new member of the mitochondrial carrier family with tissue-specific expression. *FEBS Lett.* **1997**, *408*, 39–42. [CrossRef]

44. Vidal-Puig, A.; Solanes, G.; Grujic, D.; Flier, J.S.; Lowell, B.B. UCP3: An uncoupling protein homologue expressed preferentially and abundantly in skeletal muscle and brown adipose tissue. *Biochem. Biophys. Res. Commun.* **1997**, *235*, 79–82. [CrossRef]
45. Mao, W.; Yu, X.X.; Zhong, A.; Li, W.; Brush, J.; Sherwood, S.W.; Adams, S.H.; Pan, G. UCP4, a novel brain-specific mitochondrial protein that reduces membrane potential in mammalian cells. *FEBS Lett.* **1999**, *443*, 326–330. [CrossRef]
46. Sanchis, D.; Fleury, C.; Chomiki, N.; Goubern, M.; Huang, Q.; Neverova, M.; Grégoire, F.; Easlick, J.; Raimbault, S.; Lévi-Meyrueis, C. BMCP1, a novel mitochondrial carrier with high expression in the central nervous system of humans and rodents, and respiration uncoupling activity in recombinant yeast. *J. Biol. Chem.* **1998**, *273*, 34611–34615. [CrossRef]
47. Bertholet, A.M.; Chouchani, E.T.; Kazak, L.; Angelin, A.; Fedorenko, A.; Long, J.Z.; Vidoni, S.; Garrity, R.; Cho, J.; Terada, N. H+ transport is an integral function of the mitochondrial ADP/ATP carrier. *Nature* **2019**, *571*, 515–520. [CrossRef]
48. Kreiter, J.; Rupprecht, A.; Škulj, S.; Brkljača, Z.; Žuna, K.; Knyazev, D.G.; Bardakji, S.; Vazdar, M.; Pohl, E.E. Ant1 activation and inhibition patterns support the fatty acid cycling mechanism for proton transport. *Int. J. Mol. Sci.* **2021**, *22*, 2490. [CrossRef]
49. Chipot, C.; Dehez, F.; Schnell, J.R.; Zitzmann, N.; Pebay-Peyroula, E.; Catoire, L.J.; Miroux, B.; Kunji, E.R.; Veglia, G.; Cross, T.A. Perturbations of native membrane protein structure in alkyl phosphocholine detergents: A critical assessment of NMR and biophysical studies. *Chem. Rev.* **2018**, *118*, 3559–3607. [CrossRef]
50. Krauss, S.; Zhang, C.-Y.; Lowell, B.B. The mitochondrial uncoupling-protein homologues. *Nat. Rev. Mol. Cell Biol.* **2005**, *6*, 248–261. [CrossRef]
51. Ježek, P.; Engstová, H.; Žáčková, M.; Vercesi, A.E.; Costa, A.D.; Arruda, P.; Garlid, K.D. Fatty acid cycling mechanism and mitochondrial uncoupling proteins. *Biochim. Biophys. Acta Bioenerg.* **1998**, *1365*, 319–327. [CrossRef]
52. Winkler, E.; Klingenberg, M. Effect of fatty acids on H+ transport activity of the reconstituted uncoupling protein. *J. Biol. Chem.* **1994**, *269*, 2508–2515. [CrossRef]
53. Berardi, M.J.; Chou, J.J. Fatty acid flippase activity of UCP2 is essential for its proton transport in mitochondria. *Cell Metab.* **2014**, *20*, 541–552. [CrossRef]
54. Lin, C.S.; Klingenberg, M. Characteristics of the isolated purine nucleotide binding protein from brown fat mitochondria. *Biochemistry* **1982**, *21*, 2950–2956. [CrossRef]
55. Ardalan, A.; Sowlati-Hashjin, S.; Uwumarenogie, S.O.; Fish, M.; Mitchell, J.; Karttunen, M.; Smith, M.D.; Jelokhani-Niaraki, M. Functional Oligomeric Forms of Uncoupling Protein 2: Strong Evidence for Asymmetry in Protein and Lipid Bilayer Systems. *J. Phys. Chem. B* **2020**, *125*, 169–183. [CrossRef] [PubMed]
56. Echtay, K.S.; Esteves, T.C.; Pakay, J.L.; Jekabsons, M.B.; Lambert, A.J.; Portero-Otín, M.; Pamplona, R.; Vidal-Puig, A.J.; Wang, S.; Roebuck, S.J. A signalling role for 4-hydroxy-2-nonenal in regulation of mitochondrial uncoupling. *EMBO J.* **2003**, *22*, 4103–4110. [CrossRef]
57. Trenker, M.; Malli, R.; Fertschai, I.; Levak-Frank, S.; Graier, W.F. Uncoupling proteins 2 and 3 are fundamental for mitochondrial Ca^{2+} uniport. *Nat. Cell Biol.* **2007**, *9*, 445–452. [CrossRef]
58. Ahnert, S.E.; Marsh, J.A.; Hernández, H.; Robinson, C.V.; Teichmann, S.A. Principles of assembly reveal a periodic table of protein complexes. *Science* **2015**, *350*, 6266. [CrossRef]
59. Zhao, L.; Wang, S.; Zhu, Q.; Wu, B.; Liu, Z.; OuYang, B.; Chou, J.J. Specific interaction of the human mitochondrial uncoupling protein 1 with free long-chain fatty acid. *Structure* **2017**, *25*, 1371–1379. [CrossRef]
60. Berardi, M.J.; Shih, W.M.; Harrison, S.C.; Chou, J.J. Mitochondrial uncoupling protein 2 structure determined by NMR molecular fragment searching. *Nature* **2011**, *476*, 109–113. [CrossRef]
61. Jabůrek, M.; Vařecha, M.; Gimeno, R.E.; Dembski, M.; Ježek, P.; Zhang, M.; Burn, P.; Tartaglia, L.A.; Garlid, K.D. Transport function and regulation of mitochondrial uncoupling proteins 2 and 3. *J. Biol. Chem.* **1999**, *274*, 26003–26007. [CrossRef]
62. Hoang, T.; Matovic, T.; Parker, J.; Smith, M.D.; Jelokhani-Niaraki, M. Role of positively charged residues of the second transmembrane domain in the ion transport activity and conformation of human uncoupling protein-2. *Biochemistry* **2015**, *54*, 2303–2313. [CrossRef] [PubMed]
63. Bertholet, A.M.; Kirichok, Y. UCP1: A transporter for H+ and fatty acid anions. *Biochimie* **2017**, *134*, 28–34. [CrossRef] [PubMed]
64. Fedorenko, A.; Lishko, P.V.; Kirichok, Y. Mechanism of fatty-acid-dependent UCP1 uncoupling in brown fat mitochondria. *Cell* **2012**, *151*, 400–413. [CrossRef] [PubMed]
65. Modrianský, M.; Murdza-Inglis, D.L.; Patel, H.V.; Freeman, K.B.; Garlid, K.D. Identification by site-directed mutagenesis of three arginines in uncoupling protein that are essential for nucleotide binding and inhibition. *J. Biol. Chem.* **1997**, *272*, 24759–24762. [CrossRef] [PubMed]
66. Echtay, K.S.; Winkler, E.; Bienengraeber, M.; Klingenberg, M. Site-directed mutagenesis identifies residues in uncoupling protein (UCP1) involved in three different functions. *Biochemistry* **2000**, *39*, 3311–3317. [CrossRef] [PubMed]
67. Klingenberg, M.; Echtay, K.S. Uncoupling proteins: The issues from a biochemist point of view. *Biochim. Biophys. Acta Bioenerg.* **2001**, *1504*, 128–143. [CrossRef]
68. Ardalan, A.; Sowlati-Hashjin, S.; Oduwoye, H.; Uwumarenogie, S.O.; Karttunen, M.; Smith, M.D.; Jelokhani-Niaraki, M. Biphasic Proton Transport Mechanism for Uncoupling Proteins. *J. Phys. Chem. B* **2021**, *125*, 9130–9144. [CrossRef]
69. Škulj, S.; Brkljača, Z.; Kreiter, J.; Pohl, E.E.; Vazdar, M. Molecular Dynamics Simulations of Mitochondrial Uncoupling Protein 2. *Int. J. Mol. Sci.* **2021**, *22*, 1214. [CrossRef]
70. Esteves, T.C.; Brand, M.D. The reactions catalysed by the mitochondrial uncoupling proteins UCP2 and UCP3. *Biochim. Biophys. Acta Bioenerg.* **2005**, *1709*, 35–44. [CrossRef]

71. Echtay, K.S. Mitochondrial uncoupling proteins—What is their physiological role? *Free Radic. Biol. Med.* **2007**, *43*, 1351–1371. [CrossRef]
72. Rial, E.; Poustie, A.; Nicholls, D.G. Brown-adipose-tissue mitochondria: The regulation of the 32,000-Mr uncoupling protein by fatty acids and purine nucleotides. *Eur. J. Biochem.* **1983**, *137*, 197–203. [CrossRef] [PubMed]
73. Rebuffet, E.; Frick, A.; Järvå, M.; Törnroth-Horsefield, S. Cell-free production and characterisation of human uncoupling protein 1–3. *Biochem. Biophys. Rep.* **2017**, *10*, 276–281. [CrossRef] [PubMed]
74. Ye, Y.; Liu, X.; Chen, Y.; Xu, G.; Wu, Q.; Zhang, Z.; Yao, C.; Liu, M.; Li, C. Labeling strategy and signal broadening mechanism of protein NMR spectroscopy in Xenopus laevis oocytes. *Chemistry* **2015**, *21*, 8686–8690. [CrossRef] [PubMed]
75. Ricquier, D. UCP1, the mitochondrial uncoupling protein of brown adipocyte: A personal contribution and a historical perspective. *Biochimie* **2017**, *134*, 3–8. [CrossRef] [PubMed]
76. Nicholls, D.G. Brown adipose tissue mitochondria. *Biochim. Biophys. Acta Rev. Bioenerg.* **1979**, *549*, 1–29. [CrossRef]
77. Jezek, P.; Garlid, K. New substrates and competitive inhibitors of the Cl-translocating pathway of the uncoupling protein of brown adipose tissue mitochondria. *J. Biol. Chem.* **1990**, *265*, 19303–19311. [CrossRef]
78. Vozza, A.; Parisi, G.; De Leonardis, F.; Lasorsa, F.M.; Castegna, A.; Amorese, D.; Marmo, R.; Calcagnile, V.M.; Palmieri, L.; Ricquier, D. UCP2 transports C4 metabolites out of mitochondria, regulating glucose and glutamine oxidation. *Proc. Natl. Acad. Sci. USA* **2014**, *111*, 960–965. [CrossRef]
79. Vignais, P.V.; Vignais, P.M.; Defaye, G. Adenosine diphosphate translocation in mitochondria. Nature of the receptor site for carboxyatractyloside (*gummiferin*). *Biochemistry* **1973**, *12*, 1508–1519. [CrossRef]
80. Henderson, P.J.; Lardy, H.A. Bongkrekic acid: An inhibitor of the adenine nucleotide translocase of mitochondria. *J. Biol. Chem.* **1970**, *245*, 1319–1326. [CrossRef]
81. Erdelt, H.; Weidemann, M.J.; Buchholz, M.; Klingenberg, M. Some principle effects of bongkrekic acid on the binding of adenine nucleotides to mitochondrial membranes. *Eur. J. Biochem.* **1972**, *30*, 107–122. [CrossRef]
82. Ruprecht, J.J.; Kunji, E.R. Structural changes in the transport cycle of the mitochondrial ADP/ATP carrier. *Curr. Opin. Struct. Biol.* **2019**, *57*, 135–144. [CrossRef] [PubMed]
83. Brustovetsky, N.; Klingenberg, M. The reconstituted ADP/ATP carrier can mediate H+ transport by free fatty acids, which is further stimulated by mersalyl. *J. Biol. Chem.* **1994**, *269*, 27329–27336. [CrossRef]
84. Skulachev, V.P. Fatty acid circuit as a physiological mechanism of uncoupling of oxidative phosphorylation. *FEBS Lett.* **1991**, *294*, 158–162. [CrossRef]
85. Baffy, G. Mitochondrial uncoupling in cancer cells: Liabilities and opportunities. *Biochim. Biophys. Acta Bioenerg.* **2017**, *1858*, 655–664. [CrossRef] [PubMed]
86. Andrews, Z.B.; Diano, S.; Horvath, T.L. Mitochondrial uncoupling proteins in the CNS: In support of function and survival. *Nat. Rev. Neurosci.* **2005**, *6*, 829–840. [CrossRef] [PubMed]
87. Ruiz-Ramírez, A.; López-Acosta, O.; Barrios-Maya, M.A.; El-Hafidi, M. Cell death and heart failure in obesity: Role of uncoupling proteins. *Oxid. Med. Cell. Longev.* **2016**, 9340654. [CrossRef]
88. Jia, P.; Wu, X.; Pan, T.; Xu, S.; Hu, J.; Ding, X. Uncoupling protein 1 inhibits mitochondrial reactive oxygen species generation and alleviates acute kidney injury. *EBioMedicine* **2019**, *49*, 331–340. [CrossRef]
89. Cadenas, S. Mitochondrial uncoupling, ROS generation and cardioprotection. *Biochim. Biophys. Acta Bioenerg.* **2018**, *1859*, 940–950. [CrossRef]
90. Hass, D.T.; Barnstable, C.J. Uncoupling proteins in the mitochondrial defense against oxidative stress. *Prog. Retin. Eye Res.* **2021**, *83*, 100941. [CrossRef]
91. Horvath, T.L.; Warden, C.H.; Hajos, M.; Lombardi, A.; Goglia, F.; Diano, S. Brain uncoupling protein 2: Uncoupled neuronal mitochondria predict thermal synapses in homeostatic centers. *J. Neurosci.* **1999**, *19*, 10417–10427. [CrossRef]
92. Waldeck-Weiermair, M.; Malli, R.; Naghdi, S.; Trenker, M.; Kahn, M.J.; Graier, W.F. The contribution of UCP2 and UCP3 to mitochondrial Ca^{2+} uptake is differentially determined by the source of supplied Ca^{2+}. *Cell Calcium* **2010**, *47*, 433–440. [CrossRef]
93. Trenker, M.; Fertschai, I.; Malli, R.; Graier, W.F. UCP2/3—Likely to be fundamental for mitochondrial Ca^{2+} uniport. *Nat. Cell Biol.* **2008**, *10*, 1237–1240. [CrossRef]
94. Brookes, P.S.; Parker, N.; Buckingham, J.A.; Vidal-Puig, A.; Halestrap, A.P.; Gunter, T.E.; Nicholls, D.G.; Bernardi, P.; Lemasters, J.J.; Brand, M.D. UCPs—Unlikely calcium porters. *Nat. Cell Biol.* **2008**, *10*, 1235–1237. [CrossRef] [PubMed]
95. Kunji, E.R.; Robinson, A.J. Coupling of proton and substrate translocation in the transport cycle of mitochondrial carriers. *Curr. Opin. Struct. Biol.* **2010**, *20*, 440–447. [CrossRef] [PubMed]
96. Zhang, H.; Alder, N.N.; Wang, W.; Szeto, H.; Marcinek, D.J.; Rabinovitch, P.S. Reduction of elevated proton leak rejuvenates mitochondria in the aged cardiomyocyte. *eLife* **2020**, *9*, e60827. [CrossRef]

International Journal of
Molecular Sciences

MDPI

Article

ANT1 Activation and Inhibition Patterns Support the Fatty Acid Cycling Mechanism for Proton Transport

Jürgen Kreiter [1], Anne Rupprecht [1,2], Sanja Škulj [3], Zlatko Brkljača [3], Kristina Žuna [1], Denis G. Knyazev [4], Sarah Bardakji [1], Mario Vazdar [3,5] and Elena E. Pohl [1,*]

1 Institute of Physiology, Pathophysiology and Biophysics, Department of Biomedical Sciences, University of Veterinary Medicine, 1210 Vienna, Austria; juergen.kreiter@vetmeduni.ac.at (J.K.); anne.rupprecht@med.uni-rostock.de (A.R.); kristina.zuna@vetmeduni.ac.at (K.Ž.); sarah.bardakji@vetmeduni.ac.at (S.B.)
2 Institute of Pharmacology and Toxicology, Rostock University Medical Center, 18057 Rostock, Germany
3 Division of Organic Chemistry and Biochemistry, Rudjer Bošković Institute,10000 Zagreb, Croatia; Sanja.Skulj@irb.hr (S.Š.); Zlatko.Brkljaca@irb.hr (Z.B.); Mario.Vazdar@irb.hr (M.V.)
4 Institute of Biophysics, Johannes Kepler University, 4020 Linz, Austria; denis.knyazev@jku.at
5 Institute of Organic Chemistry and Biochemistry, Czech Academy of Sciences, 16610 Prague 6, Czech Republic
* Correspondence: elena.pohl@vetmeduni.ac.at

Abstract: Adenine nucleotide translocase (ANT) is a well-known mitochondrial exchanger of ATP against ADP. In contrast, few studies have shown that ANT also mediates proton transport across the inner mitochondrial membrane. The results of these studies are controversial and lead to different hypotheses about molecular transport mechanisms. We hypothesized that the H^+-transport mediated by ANT and uncoupling proteins (UCP) has a similar regulation pattern and can be explained by the fatty acid cycling concept. The reconstitution of purified recombinant ANT1 in the planar lipid bilayers allowed us to measure the membrane current after the direct application of transmembrane potential $\Delta\Psi$, which would correspond to the mitochondrial states III and IV. Experimental results reveal that ANT1 does not contribute to a basal proton leak. Instead, it mediates H^+ transport only in the presence of long-chain fatty acids (FA), as already known for UCPs. It depends on FA chain length and saturation, implying that FA's transport is confined to the lipid-protein interface. Purine nucleotides with the preference for ATP and ADP inhibited H^+ transport. Specific inhibitors of ATP/ADP transport, carboxyatractyloside or bongkrekic acid, also decreased proton transport. The H^+ turnover number was calculated based on ANT1 concentration determined by fluorescence correlation spectroscopy and is equal to $14.6 \pm 2.5\ s^{-1}$. Molecular dynamic simulations revealed a large positively charged area at the protein/lipid interface that might facilitate FA anion's transport across the membrane. ANT's dual function—ADP/ATP and H^+ transport in the presence of FA—may be important for the regulation of mitochondrial membrane potential and thus for potential-dependent processes in mitochondria. Moreover, the expansion of proton-transport modulating drug targets to ANT1 may improve the therapy of obesity, cancer, steatosis, cardiovascular and neurodegenerative diseases.

Keywords: fatty acid anion transport; proton transport; ADP/ATP carrier protein; mitochondrial transporter; arachidonic acid; long-chain fatty acids

Citation: Kreiter, J.; Rupprecht, A.; Škulj, S.; Brkljača, Z.; Žuna, K.; Knyazev, D.G.; Bardakji, S.; Vazdar, M.; Pohl, E.E. ANT1 Activation and Inhibition Patterns Support the Fatty Acid Cycling Mechanism for Proton Transport. *Int. J. Mol. Sci.* **2021**, *22*, 2490. https://doi.org/10.3390/ijms22052490

Academic Editor: Masoud Jelokhani-Niaraki

Received: 8 January 2021
Accepted: 24 February 2021
Published: 2 March 2021

1. Introduction

In mitochondria, oxidative phosphorylation accounts for ATP production by phosphorylating ADP using proton (H^+) gradient generated by the respiratory chain proteins (coupling). H^+ can return to the matrix by alternative pathways (uncoupling): (i) inhibitor-non sensitive basal H^+ leak (J_B) and (ii) protein-mediated inhibitor-sensitive proton transport (J_H) [1–3]. J_B is sensitive to the membrane potential, mitochondrial inner membrane surface area, the composition of phospholipids and free fatty acids (FA) and was observed in mitochondria of all tissues [1]. Uncoupling proteins (UCP) are implicated in the mediation of J_H [4–9]. As several tissues such as liver, kidney, skin, and others lack any UCPs under physiological conditions (for review, see [10]), mitochondrial adenine nucleotide translocase (ANT, also cited in the literature as AAC or ADP/ATP carrier) was proposed to provide an alternative pathway for proton transport alongside its well-known function to exchange ADP for ATP [11–15].

The H^+ transporting function of ANT in the presence of palmitate has been first observed in experiments with isolated mitochondria [16,17]. The addition of free FA to the proteoliposomes reconstituted with purified ANT caused the transmembrane potential ($\Delta\Phi$) decrease, which was restored by carboxyatractyloside (CATR) and bongkrekic acid (BA) [18]. In brown-fat mitochondria from mice knockout for UCP1, fatty-acid-induced uncoupling could also be inhibited by CATR [19]. The H^+ conductance of muscle mitochondria from mice knockout for ANT1 was reported to be half that of wild-type controls [20]. Recently, ANT-mediated H^+ transport was observed in patched mitoplasts [21]. Although the H^+ transporting function of ANT1 seems to be accepted, discrepancies in results obtained in various experimental systems led to different views on the proton transport mechanism. In the 1980–1990s, several groups recognized that the proton transport could occur by the flip-flop of the protonated form of long-chain fatty acid (FA) without membrane proteins' participation [22,23]. However, a fatty acid anion (FA^-) transport is a rate-limiting step in FA circulation and has to be accelerated by proteins. In 1991, Skulachev proposed the "fatty acid circuit hypothesis", claiming that proteins such as ANT1 and UCP1 mediate the return of the FA^- to the cytosolic side of the membrane, resulting in net proton transport catalyzed by the protein [24]. Our previous results obtained for UCP1-UCP3 can be well described based on the FA cycling model and are consistent with the translocation of FA^- at the protein/lipid interface.

In contrast, Bertholet et al. proposed FA to be co-factors in H^+ transport by ANT based on patch-clamp experiments [21]. In this model, FA is not translocated but stays in one place as a part of the protein translocating pathway, where it is (de-)protonated. This mechanism differed from the mechanism suggested by the same group for UCP1, in which UCP1 was regarded as a FA^-/H^+ symporter [25]. Moreover, this model does not necessarily assume direct binding of H^+ to the FA anion and allows the proton transport in both directions.

Here, we hypothesized that the H^+-transport mediated by ANT has a regulation pattern similar to UCPs and can be explained by the FA cycling concept. The goals of this study were (i) to investigate the dependence of ANT-mediated H^+ transport on FA structure, (ii) to estimate ANT-specific H^+ turnover number, and (iii) to examine whether the specific ANT substrates inhibit H^+ transport.

2. Results

2.1. ANT1-Mediated Substrate Transport

To evaluate whether recombinant murine ANT1 was correctly refolded in proteoliposomes, we performed ADP/ATP transport measurements using proteoliposomes initially filled with radioactively labeled ^{3}H-ATP [26]. After adding ADP to the bulk solution, we measured the release of ^{3}H-ATP with time (Supplementary Figure S2). The determined ADP/ATP exchange rate depends on ANT1 content (k_{ANT} = 5.53 ± 0.74 mmol/min/g) and corresponds well to the reported results for ANT reconstituted into liposomes (Supplementary Table S1) [26–29].

2.2. Basal Proton Leak

We further investigated the controversially discussed ANT1 involvement in the basal leak [20,21]. For this, we formed planar bilayer lipid membranes from proteoliposomes reconstituted with recombinant ANT1 [26]. Figure 1 demonstrates that the total specific conductances, G_m and G_0, of bilayer membranes made from DOPC, DOPE, and cardiolipin were similar in the presence and absence of ANT1 (G_m =10.0 ± 2.0 nS/cm^2 and G_0 =8.4

$\pm$ 2.3 nS/cm^2) if no purine nucleotides (PN) were added. The addition of ATP and ADP on both sides of the membrane led to a substantial G_m increase, which was directly proportional to the applied membrane potentials, reaching G_m = 39.5 $\pm$ 4.9 nS/cm^2 at 190 mV (Figure 1, Supplementary Figure S3 and Supplementary Table S2. This increase vanished after adding the specific inhibitor of ADP/ATP transport—CATR (Figure 1) and can be explained by the electrogenic shift due to ATP/ADP exchange by ANT1. This experiment showed that ANT1 has no measurable impact on the proton leak without FAs.

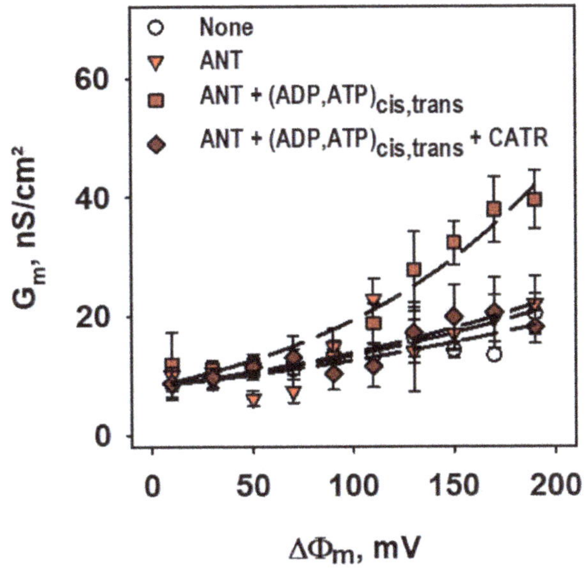

Figure 1. ANT1 does not contribute to the basal proton leak. Total membrane conductance (G_m) was measured at different membrane potentials ($\Delta\Phi$) and membrane compositions (s. legend). Planar bilayer membranes were made of 45:45:10 mol % PC:PE:CL. Lipid concentration was 1.5 mg/(mL of buffer solution). Protein concentration measured by BCA assay was 4 µg/(mg of lipid). Buffer contained 50 mM Na_2SO_4, 10 mM Tris, 10 mM MES and 0.6 mM EGTA at pH = 7.34 and T = 306 K. ADP, ATP and CATR were added at concentrations 2 mM, 2 mM and 100 µM. Lines represent the least square regression fit of an exponential function to the data. Data are the mean $\pm$ SD of at least three independent experiments.

2.3. *ANT1-Mediated Proton Transport in the Presence of FA*

The addition of polyunsaturated arachidonic acid (AA) to the membrane in the absence of ANT1 led to a potential-dependent increase in G_m. It confirms FA's importance as weak uncouplers, especially at high potentials (G_m was one order of magnitude higher at 190 mV) relevant for mitochondrial membranes [30]. The reconstitution of ANT1 in the membrane increased G_m in the presence of AA 4-fold ($G_m^{ANT, AA}/G_m^{AA}$) (Figure 2a). At 190 mV $G_m^{ANT, AA}$, G_m^{AA} and G_0 were equal to 1750 $\pm$ 220 nS/cm^2, 440 $\pm$ 135 nS/cm^2 and 20.4 $\pm$ 3.4 nS/cm^2, respectively (Supplementary Figure S4 and Supplementary Table S2). Notably, ANT1-mediated G_m depended on the structure of FAs. It increased with the elongation of FA chain length in order palmitic (PA, 16:0) $\rightarrow$ arachidic (ArA, 20:0) acid and was the highest by unsaturated AA (20:4) (Figure 2b).

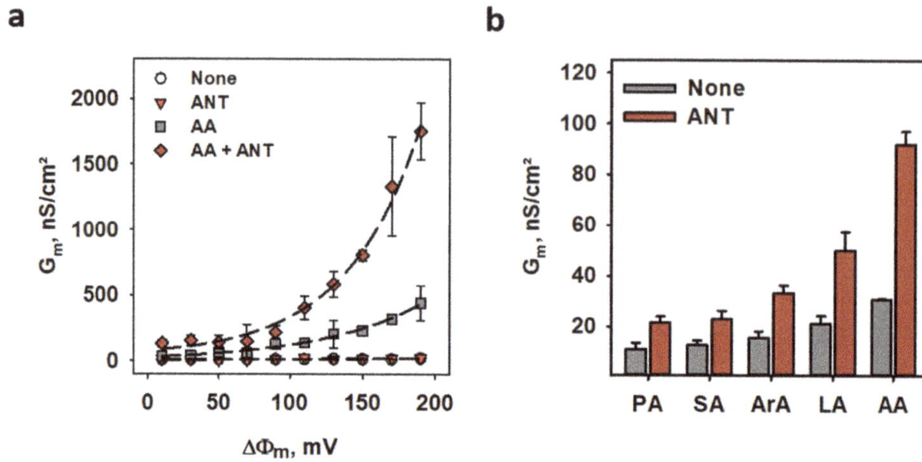

Figure 2. Fatty acids are required to activate ANT1-mediated proton transport. (**a**) Total membrane conductance (G_m) of lipid bilayers in the presence of AA (gray squares), ANT1 (red triangles), ANT1 and AA (dark red diamonds) and in the absence of AA and ANT1 (white circles) at different membrane potentials ($\Delta\Phi_m$). Lines represent the least square regression fit of an exponential function to the data. (**b**) Dependence of total membrane conductance (G_m) on fatty acid chain length and unsaturation in the presence (red) and absence (gray) of ANT1. PA, SA, ArA, LA, and AA indicate palmitic, stearic, arachidic, linoleic, and arachidonic acids. In all measurements, planar bilayer membranes were made of 45:45:10 mol % PC:PE:CL reconstituted with 15 mol % FA, except indicated otherwise. Lipid concentration was 1.5 mg/(mL of buffer solution). Protein concentration measured by BCA assay was 4 μg/(mg of lipid). The buffer solution contained 50 mM Na_2SO_4, 10 mM Tris, 10 mM MES and 0.6 mM EGTA at pH = 7.34 and T = 306 K. Data are the mean ± SD of at least three independent experiments.

2.4. Proton Turnover Number of ANT1

To determine the H^+ turnover number of ANT, we recorded current-voltage characteristics in the presence and absence of a transmembrane pH gradient (Figure 3, insert) [31].

To estimate a protein to lipid ratio, we measured the number of fluorescently-labeled ANT per liposome using fluorescence correlation spectroscopy (FCS) [32] (s. Methods and Supplementary Figure S5). By comparing the number of the fluorescent particles in proteoliposomes before ($N_{ANT, none}$ = 1.60 ± 0.01) and after ($N_{ANT, SDS}$ = 13.83 ± 0.04) the addition of 2 % (*v*/*v*) SDS, and assuming one ANT protein per detergent micelle after micellization, we calculated 8.67 ± 0.74 ANT molecules per liposome. The protein to lipid ratio estimated according to Equation (4) was 1:12,000.

From the potential shift and proton/lipid ratio, we then estimated that ANT has a turnover rate of 14.6 ± 2.5 H^+/s (Figure 3), being similar to those of uncoupling proteins (Supplementary Table S3) [6,7,9,33–35].

2.5. Inhibition of ANT1-Mediated Proton Transport

Specific inhibitors of nucleotide transport lock ANT either in its cytosolic-opened c-side (CATR) or in its matrix-opened m-side (BA) [15] and inhibit both ADP/ATP exchange and FA-mediated proton leak. The comparison of CATR and BA effect on G_m (Figure 4a) showed that inhibition by CATR was more effective than by BA. It is displayed by the EC50 values of 18.9 ± 1.8 μM for CATR and 32.3 ± 11.4 μM for BA, respectively (Figure 4b and Supplementary Table S4). Maximum inhibition values (I_{max} = 64.2 ± 2.8% and I_{max} = 44.3 ± 5.7% in the presence of CATR or BA) indicate that ANT conformation in the bilayer is approximately 60% in the c-state and 40% in the m-state in our system (Figure 4c and Supplementary Table S4).

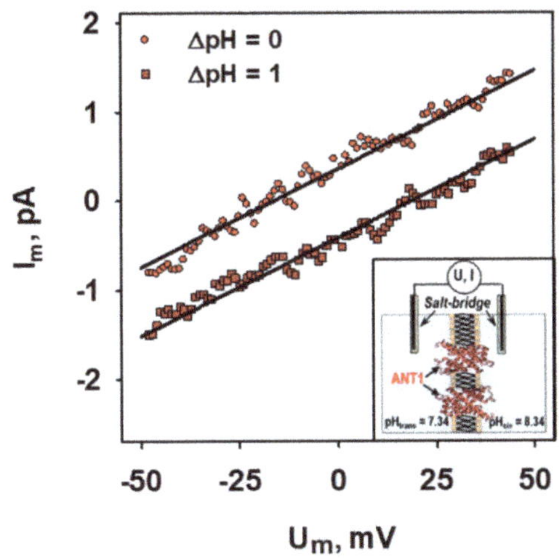

Figure 3. The proton turnover number of ANT is similar to UCPs. Representative current-voltage recordings of lipid bilayer membranes reconstituted with ANT1 in the presence (squares) and absence (circles) of ΔpH = 1.0 across the membrane. Lines represent a linear fit to the data. Planar bilayer membranes were made of 45:45:10 mol % PC:PE:CL reconstituted with 15 mol % AA. Buffer contained 50 mM Na_2SO_4, 10 mM Tris, 10 mM MES and 0.6 mM EGTA at pH = 7.34 and T = 306 K. Lipid concentration was 1.5 mg/(mL of buffer solution). Protein concentration measured by BCA assay was 4 µg/(mg of lipid). Insert: Experimental setup of the measurements to establish a transmembrane pH gradient.

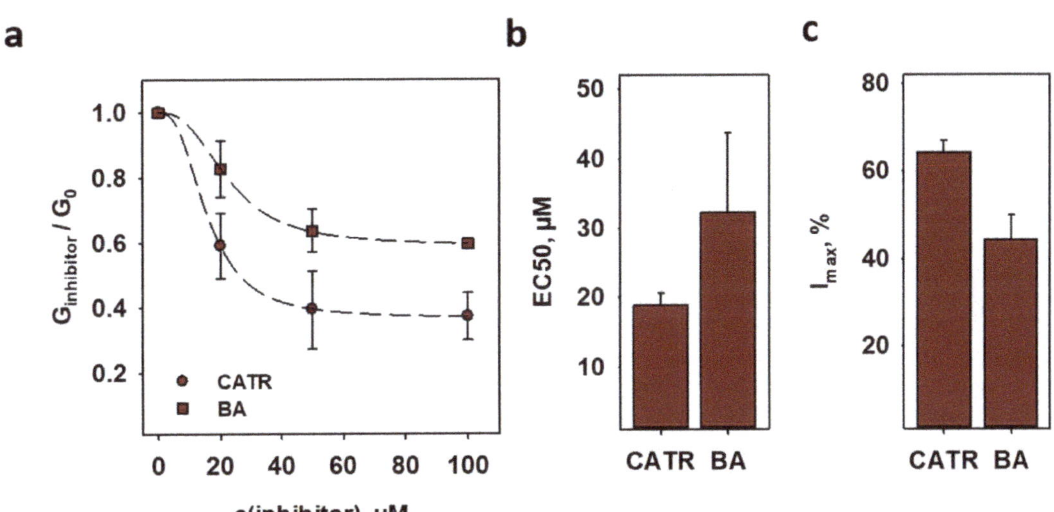

Figure 4. The ADP/ATP exchange inhibitors inhibit ANT-mediated proton leak. (**a**) Dose-dependent inhibition of ANT-mediated proton leak by the inhibitors CATR (circles) and bongkrekic acid (BA, triangles) on proton leak (dark red). Lines are a least square regression fit of a sigmoidal function to the data. (**b**) EC50 and (**c**) maximum inhibition I_{max} as fit parameters of (**a**). Planar bilayer membranes were made of 45:45:10 mol % PC:PE:CL reconstituted with 15 mol % AA. Buffer contained 50 mM Na_2SO_4, 10 mM Tris, 10 mM MES and 0.6 mM EGTA at pH = 7.34 and T = 306 K. Lipid concentration was 1.5 mg/(mL of buffer solution). Protein concentration measured by BCA assay was 4 µg/(mg of lipid). Data are the mean ± SD of at least three independent experiments.

To investigate the interdependence of proton transport, activated by FA, and ADP/ATP transport, we measured G_m of ANT1-containing lipid bilayers reconstituted with AA in the presence and absence of purine nucleotides (PN). Adenine nucleotides inhibited H^+ transport much more effectively than guanosine nucleotides (Figure 5a). The EC50 values correlated well with the known narrow substrate specificity of ANT (Figure 5b and Supplementary Table S4) [36]. ATP and ADP fully inhibited G_m, whereas all other PN decreased G_m by 50% (Figure 5c and Supplementary Table S4).

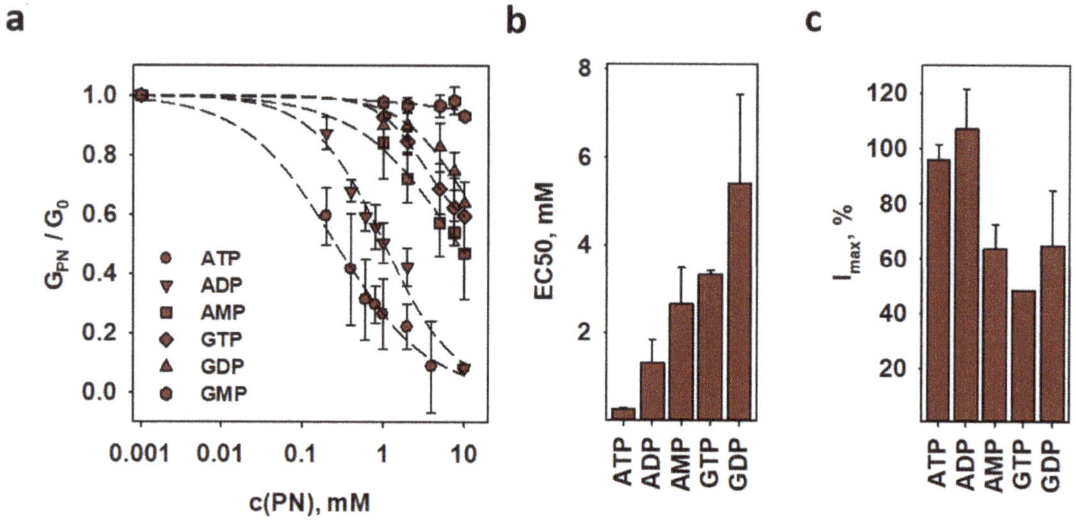

Figure 5. FA activated proton leak is preferably inhibited by ADP and ATP and maintains the substrate specificity of ANT. (**a**) Membrane conductance of lipid bilayers reconstituted with AA and ANT1 in the presence of different purine nucleotides. Lines are a least square regression fit of a sigmoidal function to the data. (**b**) EC50 and (**c**) maximum inhibition values as fit function parameters in (**a**). Values for GMP were dropped due to the low effect. For experimental conditions, see Figure 4. Data are the mean ± SD of at least three independent experiments.

2.6. Analysis of the ANT´s Surface Electrostatic Potential using Molecular Dynamic Simulations

The high similarity of the ANT activation pattern to those of UCP1, UCP2 and UCP3 [6,7,35] leads to the hypothesis that the H^+ transport can be explained by the FA cycling mechanism. We analyzed the ANT's surface electrostatic potential in the DOPC bilayers to test whether a possible FA translocation pathway may be localized at the lipid-protein interface. The calculation revealed a large positively charged patch (Figure 6a) that might facilitate FA anion's sliding alongside the protein. The ATP binding significantly decreased the positive electrostatic potential (Figure 6b) due to its strong screening at the bottom of the cavity [37]. The existence of such a patch would explain the inhibition of H^+ transport by ATP observed in electrophysiological experiments. GTP has less effect on the positive electrostatic potential (Figure 6c) because of its different orientation in the cavity (Figure 6d) and its "weaker interaction with the hydrophobic pocket that binds the adenine moiety" [36].

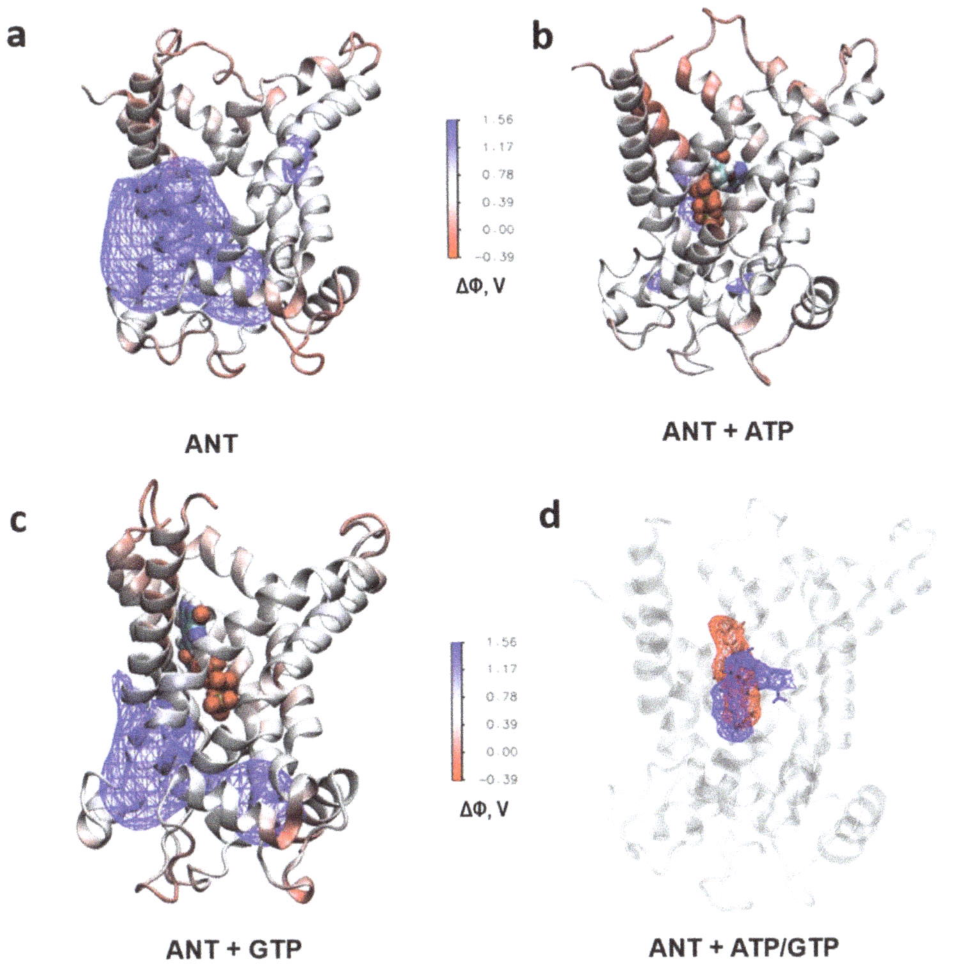

Figure 6. The purine nucleotides ATP and GTP differently modulate the electrostatic surface potential of ANT upon binding. (**a**–**c**) Electrostatic potential (Δ of ANT1 in the absence (**a**) and presence of bound ATP (**b**) and GTP (**c**) calculated by molecular dynamics simulations. The isosurface of the potential of 0.9 V is shown with the wireframe. (**d**) Different average binding location of ATP (blue wireframe) and GTP (red wireframe) in ANT1.

3. Discussion

We investigated the regulation of H^+ transport using planar bilayer membranes reconstituted with the recombinant mouse ANT1. This model allowed us (i) to measure membrane conductance at precisely defined lipid and buffer composition, (ii) to apply mitochondria-relevant potentials directly, and (iii) to separate the ANT1-originated effects from the simultaneous effects of other proteins. The latter is a main disadvantage by the interpretation of experiments on (isolated) mitochondria or mitoplasts representing swollen mitochondria lacking an intact outer membrane.

We confirmed that ANT1 has a dual function performing H^+ transport additionally to the substrate transport. Proton transport occurs only in the presence of the long-chain FAs and reveals high sensitivity to the FA chain length and saturation (Figure 7). The data on ANT1 activation showed remarkable similarity to the activation pattern of uncoupling proteins (UCP1-UCP3) [6,7,35]. The dependence of proton transport rate on the FA structure can be explained by the FA cycling model, assuming that the transport of FA anions, which is the rate-limiting step, occurs at the lipid-protein interface as proposed for UCP2 [6].

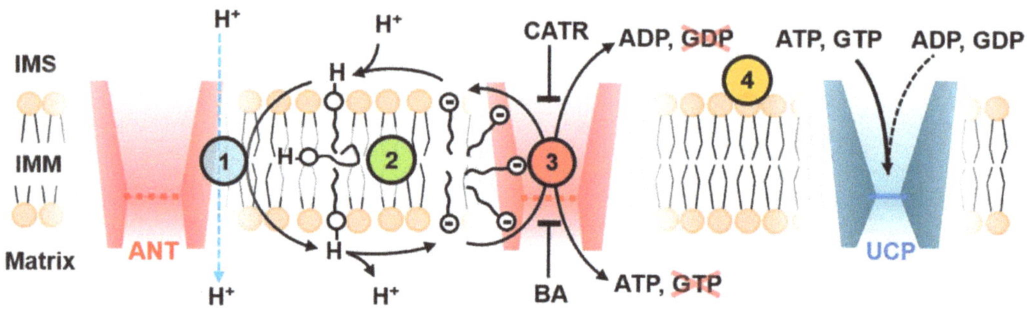

Figure 7. ANT1 transport features point to the fatty acid cycling mechanism. The proton transport rate of ANT1 (Table S3) in the presence of fatty acids (FA) is similar to that proposed for UCPs and depends on the FA structure (Circle 1, blue). ANT1 facilitates the FA anion's transport at the protein-lipid interface, which is supported by the membrane conductance dependence on the FA structure (Circle 2, green). The FA anions slide alongside the electrostatic surface potential of ANT1; its modulation by binding ADP/ATP specific substrates inhibits the FA anion transport (Circle 3, red). The inhibition of ANT1-mediated proton leak is the strongest for the ANT1 substrates—ADP and ATP. It is in contrast to UCPs, in which the triphosphate nucleotides ATP and GTP are the most potent inhibitors (Circle 4, yellow).

Although the FA hydrophobicity increases with both chain length and unsaturation, the FA anions occupy similar positions at the lipid-water interface [38]. FA anions should also not further penetrate the ANT1 structure unless a hydrophobic pocket would pull the FA into the ANT1 interior. However, such a membrane-spanning pocket is not found in the ANT structure [39].

Whereas the FA activation pattern of the proton transport seems to be similar in ANT and UCPs, the inhibition pattern is not (Figure 7). We explain it by the fact that these proteins transport different substrates. However, all ANT-specific substrates bind at the substrate-binding site in the ANT cavity [39]. As shown by molecular dynamics simulation, after the ATP binding, the electrostatic potential is diminished, and FAs are potentially less attracted to the ANT surface (Figure 6a,b). Since all substrates (ATP^{4-}, ADP^{3-}, $CATR^{4-}$ and BA^{3-}) are similarly charged and bind to the same region, the electrostatic potential of ANT will be significantly altered upon their binding [36,39,40]. GTP has a different orientation when bound to ANT, which has less impact on ANT's electrostatic potential (Figure 6c,d and Figure 7). The surface electric charge directly correlated with the inhibition potency of PN. This observation strongly supports the data on the binding site competition between PNs and FAs.

Our model is in strong contrast to the model, which proposed that FAs bind within the ANT cavity and act as a co-factor of H^+ transport [21]. The authors based their model on the experiments showing that non-protonable sulfonated FA failed to induce any transmembrane current in isolated mitoplasts. The use of sulfonated FA is very controversial, as chemical and geometrical properties of the crucial head group are altered compared to the carboxylic head group of native FA. However, the absence of any transient current by non-protonable FA acids is well described by the FA cycling model, in which the net charge transport of H^+ is impaired by the inability of the sulfonated FA to transport a proton across the membrane [41,42]. Upon FA addition to the mitoplast matrix, Bertholet et al. [21] observed no current in contrast to the FA addition to the cytosolic side, showing that the H^+ transport is independent of ANT conformation. The authors claimed that FA reaches its putative binding site inside ANT only from the cytosolic side. However, it is questionable how FAs should activate H^+ transport inside ANT, as protons have to cross at least one salt-bridge network [29], independent of ANT conformation. The FA binding to

ANT and subsequently H^+ binding to the FA in the protein cavity will most probably not provide the energy of roughly 10 kcal/mol to break the strong salt-bridge network [42–46].

The inhibition of H^+ transport in the FA co-factor model can be clearly described as a competition between FA and purine nucleotides for the binding site that we also observed for UCP1 and UCP3 [7]. However, the FA binding site is not further characterized, and it is not clear if there is a common binding target for the FA anion and the adenine nucleotides and/or specific transport inhibitors. Nevertheless, the dependency on FA chain length and unsaturation would imply a loose binding of FAs inside the protein to account for the different structures, which contrasts the high energy required to break the salt-bridge network. Thus, the model proposed by Bertholet et al. [21] seems to fail in describing our experimental results.

We determined the ANT-specific H^+ turnover number of 14.6 $\pm$ 2.5 H^+/s, which is similar to turnover numbers determined earlier for UCPs. Besides ANT and UCPs, FA-activated H^+ leak was also shown for other mitochondrial carriers, including the aspartate/glutamate carrier, dicarboxylate carrier, 2-oxoglutarate carrier, and the phosphate carrier [47–50]. Thus, it is reasonable to conclude a dual transport function for the before mentioned proteins: (i) the substrate transport to maintain mitochondrial respiration and (ii) the proton transport, which may affect the inner mitochondrial membrane potential. We speculate that these carriers have a similar mechanism of FA-mediated activation of H^+ transport due to their high homology. Simultaneous activation of several proton transporters could ensure an essential drop in potential. The latter is crucial for regulating potential-dependent processes in mitochondria, such as reactive oxygen species production, cell death, autophagy, protein secretion, metabolic adaptations, and cell signaling [51]. The controlling of the mitochondrial uncoupling can be used to treat several human diseases, such as obesity, cardiovascular diseases, or neurological disorders.

4. Methods

4.1. Chemicals

Agarose (#3810), KCl (#6781), Na_2SO_4 (#8560), 2-(N-morpholino)ethanesulfonic acid (MES, #4256), sodium dodecyl sulfate (SDS, #0183), tris(hydroxymethyl)-aminomethane (Tris, #AE15), chloroform (#AE54) and ethylene glycol-bis(β-aminoethyl ether)-N,N,N′,N′-tetraacetic acid (EGTA, #3054) were purchased from Carl Roth GmbH & Co. KG (Karlsruhe, Germany). Hexane (#296090), hexadecane (#296317), palmitic acid (#P0500), stearic acid (#S4751), arachidic acid (#A3631), linoleic acid (#L1376) and arachidonic acid (#A3611), dimethyl sulfoxide (DMSO, #472301), the purine nucleotides adenine and guanine tri-, di-, and mono-phosphate (ATP, #A2383; ADP, #A2754; AMP, #01930; GTP #G8877; GDP, #G7127; and GMP, #G8377), carboxyatractyloside (CATR, #C4992) and bongkrekic acid (BA, #B6179) were purchased from Sigma-Aldrich (Vienna, Austria). 1,2-dioleoyl-sn-glycero-3-phosphocholine (DOPC, #850375P), 1,2-dioleoyl-sn-glycero-3-phosphoethanolamine (DOPE, #850725P) and cardiolipin (CL, #710335P) came from Avanti Polar Lipids Inc. (Alabaster, AL, USA).

4.2. Cloning, Purification and Reconstitution of Murine ANT1

Cloning, purification and reconstitution of murine ANT1 followed a previously established protocol [26]. The protein concentration in proteoliposomes was measured with the Micro BCA™ Protein Assay Kit (Thermo Fisher Scientific, Prod. #23235, Waltham, MA, USA). Protein purity was verified by SDS-PAGE and silver staining (Supplementary Figure S1).

4.3. Exchange Rate Measurements of mANT1

ANT-mediated exchange of ADP/ATP was measured radioactively using ^{3}H-ATP (Prod. #NET420250UC, Perkin Elmer, Waltham, MA, USA) following the protocol as described elsewhere [26] (Supplementary Figure S2).

4.4. Electrophysiological Measurements of mANT1

Planar lipid bilayers were formed from proteoliposomes as described previously [31,52]. FAs were added to the lipid phase before membrane formation. Proper membrane formation was verified by measuring membrane capacitance ($C = 0.72 \pm 0.05\ \mu F/cm^2$), which is independent of the presence of protein, FA and inhibitor. Current-voltage (I-U) measurements were performed with a patch-clamp amplifier (EPC 10USB, Werner Instruments, Holliston, MA, USA). The specific total membrane conductance (G_m) at 0 mV was obtained as the slope of a linear fit of the experimental data at applied voltages from −50 mV to + 50 mV and normalized to the membrane area in cm^2. Purine nucleotides (solved in distilled water, pH = 7.34) and ANT-specific inhibitors BA and CATR (solved in DMSO) were added to the buffer solution before forming bilayer membranes. The concentrations of each substrate are indicated in the figure legends. Membrane conductance expressed in relative units was calculated according to [7].

Measurement and calculation of H^+ turnover rate of ANT followed the established protocol [31]. The addition of Tris increased the pH value of the buffer solution on the cis side of the membrane to a value of pH = 8.34.

4.5. Fluorescence Correlation Spectroscopy (FCS)

The average number of ANT1 per liposome was measured using FCS [32,53]. In brief, proteoliposomes obtained after reconstitution of the ANT1 were extruded using a Mini-Extruder system (Avanti Polar Lipids Inc., Alabaster, AL, USA) with a membrane nanopore filter with a pore diameter of 100 nm (Avestin Europe GmbH, Mannheim, Germany, LFM-100). ANT1 was labeled with ATTO 488-maleimide [54] (Sigma-Aldrich, Vienna, Austria; 28562-1MG-F). We used size exclusion chromatography (Sephadex® G-50 Superfine, Merck, Vienna, Austria; GE17-0041-01) to remove the unbound dye. The average residence time, τ_D, and ANT1- containing proteoliposomes number in the focal volume were derived from the autocorrelation function ($G(\tau)$) of the temporal fluorescence signal (Supplementary Figure S5). To measure the signal, a commercial laser scanning microscope (LSM 510 META/ConfoCor 3, Carl Zeiss, Jena, Germany) equipped with avalanche diodes and a 40× water immersion objective was used. The standard model for two-component free 3D diffusion was applied [55]:

$$G(\tau) = 1 + 1/(n\ (1+ \tau/\tau_D)) \tag{1}$$

where the number of fluorescent particles, n, in the detection volume, V_{eff}, was determined as $n = V_{eff}\ C$, where C is the particle concentration. The diffusion coefficient (D) was determined as $D = \omega^2/4\tau_D$, where $\omega = 0.16\ \mu m$ is the diameter of the confocal volume cross-section as determined from the calibration experiments.

Dissolving the liposomes with 2% (v/v) SDS was expected to increase the particle number if liposomes contained more than one ANT1. The average number of ANT1 per liposome, $<N_{ANT1}>$, was obtained from the ratio of the particle number per confocal volume after and before the addition of SDS.

The protein per lipid ratio (ρ) was estimated by:

$$\rho = \frac{N_{ANT1}}{N_{Lipids}} = \left.\frac{< N_{ANT1} >}{< N_{Lipids} >}\right|_{Liposomes} \tag{2}$$

The average number of lipids per liposomes ($<N_{Lipids}>$) is calculated by the ratio of the surface area of the liposome with radius r and the average area per lipid ($A_L \approx 0.6\ nm^2$) of a membrane containing DOPC, DOPE and CL [56,57]:

$$< N_{Lipids} >= 2\frac{4\pi r^2}{A_L} \tag{3}$$

Thus, Equation (2) gives:

$$\rho = < N_{ANT1} > \frac{1}{2}\frac{A_L}{4\pi r^2} \tag{4}$$

4.6. Molecular Dynamics Simulations

We performed all-atom molecular dynamics (MD) simulations of ANT1 protein in a 1,2-dioleoyl-sn-glycero-3-phosphocholine (DOPC) bilayer. Residues (residue 1 and residues 294–297) missing from the crystal structure of ANT1 (PDB code: 1okc) [40] without CATR were added using Modeller 9 [58] and implemented into the DOPC bilayer using CHARMM-GUI (http://www.charmm-gui.org/ (accessed on 7 January 2021)) [59–61]. Three system setups were prepared—the wild-type ANT1, the wild-type ANT1 with ATP^{4-} bound in the cytosolic-open state (c-state) [43], as well as GTP^{4-} bound in the same position. All simulation boxes contained ANT1 protein (with a total charge of +19), 73 DOPC molecules per leaflet (146 per system), ~11,500 water molecules, and the necessary number of Cl^- anions to neutralize the net charge, depending on whether ATP^{4-} or GTP^{4-} are added to the system. All systems were first minimized and equilibrated in six steps using the CHARMM-GUI protocol [62] and then simulated for a further 100 ns without any restraints with a 2 fs time step in a periodic rectangular box of 7.9 nm × 7.9 nm × 9.4 nm using the isobaric-isothermal ensemble (NPT) and periodic boundary conditions in all directions at T = 310 K, maintained via Nosé–Hoover thermostat [63] independently for the DOPC, water/ions and protein subsystems with a coupling constant of 1.0 ps^{-1}. The pressure was set to 1.013 bar and controlled with a semi-isotropic Parrinello-Rahman barostat [64] with a time constant for pressure coupling of 5 ps^{-1}. Long-range electrostatics were calculated using the particle-mesh Ewald (PME) method [65] with real space Coulomb interactions cut off at 1.2 nm using a Fourier spacing of 0.12 nm and a Verlet cut-off scheme. All simulated systems were described by the CHARMM36m force field [66]. The electrostatic potential maps of all systems were calculated using VMD's PMEPOT plugin [67]. All simulations were run with the GROMACS 5.1.4 software package [68] and visualized with the VMD molecular graphics program [69].

4.7. Statistics

Data analysis and fitting of electrophysiological measurements were performed using Sigma Plot 12.5 (Systat Software GmbH, Erkrath, Germany) and displayed as mean ± SD of at least three independent measurements.

Supplementary Materials: The following are available online at https://www.mdpi.com/1422-0067/22/5/2490/s1.

Author Contributions: Conceptualization, E.E.P., A.R., J.K. and M.V.; funding acquisition, E.E.P. and M.V.; investigation, J.K., D.G.K., K.Ž., S.Š., Z.B., S.B. and M.V.; project administration, E.E.P.; resources, E.E.P. and M.V.; supervision, E.E.P. and M.V.; writing—original draft, J.K., M.V. and E.E.P.; writing—review and editing, J.K., A.R., S.Š., Z.B., D.G.K., K.Ž., S.B., E.E.P. and M.V. All authors have read and agreed to the published version of the manuscript.

Funding: We acknowledge support from the Austrian Research Fund (FWF, P31559 to E.E.P.) and Croatian Science Foundation (Project No. IP-2019-04-3804 to M.V.). We thank the computer cluster Isabella based in SRCE—University of Zagreb, University Computing Centre for computational resources. M. V. thanks UOCHB Sabbatical visit program for support.

Institutional Review Board Statement: Not applicable.

Informed Consent Statement: Not applicable.

Data Availability Statement: The datasets generated and/or analyzed during this study are available from the corresponding authors on reasonable request.

Acknowledgments: We thank Y.N. Antonenko (Belozersky Institute of Physico-Chemical Biology, Lomonosov Moscow State University, Moscow, Russia) for the valuable comments on the manuscript.

Conflicts of Interest: The authors declare no conflict of interest.

References

1. Brand, M.D. The proton leak across the mitochondrial inner membrane. *Biochim. Biophys. Acta* **1990**, *1018*, 128–133. [CrossRef]
2. Rolfe, D.F.; Brand, M.D. Contribution of mitochondrial proton leak to skeletal muscle respiration and to standard metabolic rate. *Am. J. Physiol.* **1996**, *271 Pt 1*, C1380–C1389. [CrossRef]
3. Rolfe, D.F.; Brand, M.D. The physiological significance of mitochondrial proton leak in animal cells and tissues. *Biosci. Rep.* **1997**, *17*, 9–16. [CrossRef] [PubMed]
4. Krauss, S.; Zhang, C.Y.; Lowell, B.B. The mitochondrial uncoupling-protein homologues. *Nat. Rev. Mol. Cell Biol.* **2005**, *6*, 248–261. [CrossRef] [PubMed]
5. Skulachev, V.P. Uncoupling: New approaches to an old problem of bioenergetics. *Biochim. Biophys. Acta* **1998**, *1363*, 100–124. [CrossRef]
6. Beck, V.; Jaburek, M.; Demina, T.; Rupprecht, A.; Porter, R.K.; Jezek, P.; Pohl, E.E. Polyunsaturated fatty acids activate human uncoupling proteins 1 and 2 in planar lipid bilayers. *FASEB J.* **2007**, *21*, 1137–1144. [CrossRef]
7. Macher, G.; Koehler, M.; Rupprecht, A.; Kreiter, J.; Hinterdorfer, P.; Pohl, E.E. Inhibition of mitochondrial UCP1 and UCP3 by purine nucleotides and phosphate. *Biochim. Biophys. Acta Biomembr.* **2018**, *1860*, 664–672. [CrossRef] [PubMed]
8. Zackova, M.; Jezek, P. Reconstitution of novel mitochondrial uncoupling proteins UCP2 and UCP3. *Biosci. Rep.* **2002**, *22*, 33–46. [CrossRef]
9. Hoang, T.; Smith, M.D.; Jelokhani-Niaraki, M. Toward understanding the mechanism of ion transport activity of neuronal uncoupling proteins UCP2, UCP4, and UCP5. *Biochemistry* **2012**, *51*, 4004–4014. [CrossRef]
10. Pohl, E.E.; Rupprecht, A.; Macher, G.; Hilse, K.E. Important Trends in UCP3 Investigation. *Front. Physiol.* **2019**, *10*, 470. [CrossRef] [PubMed]
11. Heldt, H.W.; Jacobs, H.; Klingenberg, M. Endogenous Adp of Mitochondria, an Early Phosphate Acceptor of Oxidative Phosphorylation as Disclosed by Kinetic Studies with C14 Labelled Adp and Atp and with Atractyloside. *Biochem. Biophys. Res. Commun.* **1965**, *18*, 174–179. [CrossRef]
12. Pfaff, E.; Klingenberg, M.; Heldt, H.W. Unspecific permeation and specific exchange of adenine nucleotides in liver mitochondria. *Biochim. Biophys. Acta* **1965**, *104*, 312–315. [CrossRef]
13. Vignais, P.V. The mitochondrial adenine nucleotide translocator. *J. Bioenerg.* **1976**, *8*, 9–17. [CrossRef] [PubMed]
14. Kunji, E.R.; Aleksandrova, A.; King, M.S.; Majd, H.; Ashton, V.L.; Cerson, E.; Springett, R.; Kibalchenko, M.; Tavoulari, S.; Crichton, P.G.; et al. The transport mechanism of the mitochondrial ADP/ATP carrier. *Biochim. Biophys. Acta* **2016**, *1863*, 2379–2393. [CrossRef]
15. Klingenberg, M. The ADP and ATP transport in mitochondria and its carrier. *Biochim. Biophys. Acta* **2008**, *1778*, 1978–2021. [CrossRef] [PubMed]
16. Andreyev, A.Y.; Bondareva, T.O.; Dedukhova, V.I.; Mokhova, E.N.; Skulachev, V.P.; Volkov, N.I. Carboxyatractylate inhibits the uncoupling effect of free fatty acids. *FEBS Lett.* **1988**, *226*, 265–269. [CrossRef]
17. Andreyev, A.Y.; Bondareva, T.O.; Dedukhova, V.I.; Mokhova, E.N.; Skulachev, V.P.; Tsofina, L.M.; Volkov, N.I.; Vygodina, T.V. The ATP/ADP-antiporter is involved in the uncoupling effect of fatty acids on mitochondria. *Eur. J. Biochem.* **1989**, *182*, 585–592. [CrossRef]
18. Brustovetsky, N.; Klingenberg, M. The reconstituted ADP/ATP carrier can mediate H+ transport by free fatty acids, which is further stimulated by mersalyl. *J. Biol. Chem.* **1994**, *269*, 27329–27336. [CrossRef]
19. Shabalina, I.G.; Kramarova, T.V.; Nedergaard, J.; Cannon, B. Carboxyatractyloside effects on brown-fat mitochondria imply that the adenine nucleotide translocator isoforms Ant1 and Ant2 may be responsible for basal and fatty acid-induced uncoupling, respectively. *Biochem. J.* **2006**, *399*, 405–414. [CrossRef]
20. Brand, M.D.; Pakay, J.L.; Ocloo, A.; Kokoszka, J.; Wallace, D.C.; Brookes, P.S.; Cornwall, E.J. The basal proton conductance of mitochondria depends on adenine nucleotide translocase content. *Biochem. J.* **2005**, *392 Pt 2*, 353–362. [CrossRef]
21. Bertholet, A.M.; Chouchani, E.T.; Kazak, L.; Angelin, A.; Fedorenko, A.; Long, J.Z.; Vidoni, S.; Garrity, R.; Cho, J.; Terada, N.; et al. H(+) transport is an integral function of the mitochondrial ADP/ATP carrier. *Nature* **2019**, *571*, 515–520. [CrossRef] [PubMed]
22. Kamp, F.; Hamilton, J.A. pH gradients across phospholipid membranes caused by fast flip-flop of un-ionized fatty acids. *Proc. Natl. Acad. Sci. USA* **1992**, *89*, 11367–11370. [CrossRef] [PubMed]
23. Pohl, E.E.; Peterson, U.; Sun, J.; Pohl, P. Changes of intrinsic membrane potentials induced by flip-flop of long-chain fatty acids. *Biochemistry* **2000**, *39*, 1834–1839. [CrossRef] [PubMed]
24. Skulachev, V.P. Fatty acid circuit as a physiological mechanism of uncoupling of oxidative phosphorylation. *FEBS Lett.* **1991**, *294*, 158–162. [CrossRef]
25. Fedorenko, A.; Lishko, P.V.; Kirichok, Y. Mechanism of fatty-acid-dependent UCP1 uncoupling in brown fat mitochondria. *Cell* **2012**, *151*, 400–413. [CrossRef] [PubMed]
26. Kreiter, J.; Beitz, E.; Pohl, E.E. A Fluorescence-Based Method to Measure ADP/ATP Exchange of Recombinant Adenine Nucleotide Translocase in Liposomes. *Biomolecules* **2020**, *10*, 685. [CrossRef] [PubMed]
27. Kramer, R.; Klingenberg, M. Electrophoretic control of reconstituted adenine nucleotide translocation. *Biochemistry* **1982**, *21*, 1082–1089. [CrossRef] [PubMed]

28. Heidkamper, D.; Muller, V.; Nelson, D.R.; Klingenberg, M. Probing the role of positive residues in the ADP/ATP carrier from yeast. The effect of six arginine mutations on transport and the four ATP versus ADP exchange modes. *Biochemistry* **1996**, *35*, 16144–16152. [CrossRef] [PubMed]
29. King, M.S.; Kerr, M.; Crichton, P.G.; Springett, R.; Kunji, E.R.S. Formation of a cytoplasmic salt bridge network in the matrix state is a fundamental step in the transport mechanism of the mitochondrial ADP/ATP carrier. *Biochim. Biophys. Acta* **2016**, *1857*, 14–22. [CrossRef] [PubMed]
30. Rupprecht, A.; Sokolenko, E.A.; Beck, V.; Ninnemann, O.; Jaburek, M.; Trimbuch, T.; Klishin, S.S.; Jezek, P.; Skulachev, V.P.; Pohl, E.E. Role of the transmembrane potential in the membrane proton leak. *Biophys. J* **2010**, *98*, 1503–1511. [CrossRef] [PubMed]
31. Kreiter, J.; Pohl, E.E. A Micro-agar Salt Bridge Electrode for Analyzing the Proton Turnover Rate of Recombinant Membrane Proteins. *J. Vis. Exp.* **2019**, *143*, e58552. [CrossRef]
32. Erokhova, L.; Horner, A.; Kugler, P.; Pohl, P. Monitoring single-channel water permeability in polarized cells. *J. Biol. Chem.* **2011**, *286*, 39926–39932. [CrossRef]
33. Nicholls, D.G. The effective proton conductance of the inner membrane of mitochondria from brown adipose tissue. Dependency on proton electrochemical potential gradient. *Eur. J. Biochem.* **1977**, *77*, 349–356. [CrossRef] [PubMed]
34. Klingenberg, M.; Winkler, E. The reconstituted isolated uncoupling protein is a membrane potential driven H+ translocator. *EMBO J.* **1985**, *4*, 3087–3092. [CrossRef]
35. Urbankova, E.; Voltchenko, A.; Pohl, P.; Jezek, P.; Pohl, E.E. Transport kinetics of uncoupling proteins. Analysis of UCP1 reconstituted in planar lipid bilayers. *J. Biol. Chem.* **2003**, *278*, 32497–32500. [CrossRef] [PubMed]
36. Mifsud, J.; Ravaud, S.; Krammer, E.M.; Chipot, C.; Kunji, E.R.; Pebay-Peyroula, E.; Dehez, F. The substrate specificity of the human ADP/ATP carrier AAC1. *Mol. Membr. Biol.* **2013**, *30*, 160–168. [CrossRef] [PubMed]
37. Krammer, E.M.; Ravaud, S.; Dehez, F.; Frelet-Barrand, A.; Pebay-Peyroula, E.; Chipot, C. High-chloride concentrations abolish the binding of adenine nucleotides in the mitochondrial ADP/ATP carrier family. *Biophys. J.* **2009**, *97*, L25–L27. [CrossRef]
38. Pashkovskaya, A.A.; Vazdar, M.; Zimmermann, L.; Jovanovic, O.; Pohl, P.; Pohl, E.E. Mechanism of Long-Chain Free Fatty Acid Protonation at the Membrane-Water Interface. *Biophys. J.* **2018**, *114*, 2142–2151. [CrossRef] [PubMed]
39. Ruprecht, J.J.; Kunji, E.R. Structural changes in the transport cycle of the mitochondrial ADP/ATP carrier. *Curr. Opin. Struct. Biol.* **2019**, *57*, 135–144. [CrossRef] [PubMed]
40. Pebay-Peyroula, E.; Dahout-Gonzalez, C.; Kahn, R.; Trezeguet, V.; Lauquin, G.J.; Brandolin, G. Structure of mitochondrial ADP/ATP carrier in complex with carboxyatractyloside. *Nature* **2003**, *426*, 39–44. [CrossRef] [PubMed]
41. Jaburek, M.; Varecha, M.; Jezek, P.; Garlid, K.D. Alkylsulfonates as probes of uncoupling protein transport mechanism. Ion pair transport demonstrates that direct H(+) translocation by UCP1 is not necessary for uncoupling. *J. Biol. Chem.* **2001**, *276*, 31897–31905.
42. Jezek, P.; Holendova, B.; Garlid, K.D.; Jaburek, M. Mitochondrial Uncoupling Proteins: Subtle Regulators of Cellular Redox Signaling. *Antioxid. Redox Signal.* **2018**, *29*, 667–714. [PubMed]
43. Wang, Y.; Tajkhorshid, E. Electrostatic funneling of substrate in mitochondrial inner membrane carriers. *Proc. Natl. Acad. Sci. USA* **2008**, *105*, 9598–9603. [CrossRef] [PubMed]
44. Dehez, F.; Pebay-Peyroula, E.; Chipot, C. Binding of ADP in the mitochondrial ADP/ATP carrier is driven by an electrostatic funnel. *J. Am. Chem. Soc.* **2008**, *130*, 12725–12733. [CrossRef]
45. Pietropaolo, A.; Pierri, C.L.; Palmieri, F.; Klingenberg, M. The switching mechanism of the mitochondrial ADP/ATP carrier explored by free-energy landscapes. *Biochim. Biophys. Acta* **2016**, *1857*, 772–781. [CrossRef]
46. Škulj, S.; Brkljača, Z.; Vazdar, M. Molecular Dynamics Simulations of the Elusive Matrix-Open State of Mitochondrial ADP/ATP Carrier. *Isr. J. Chem.* **2020**, *60*, 735–743. [CrossRef]
47. Samartsev, V.N.; Smirnov, A.V.; Zeldi, I.P.; Markova, O.V.; Mokhova, E.N.; Skulachev, V.P. Involvement of aspartate/glutamate antiporter in fatty acid-induced uncoupling of liver mitochondria. *Biochim. Biophys. Acta* **1997**, *1319*, 251–257. [CrossRef]
48. Wieckowski, M.R.; Wojtczak, L. Involvement of the dicarboxylate carrier in the protonophoric action of long-chain fatty acids in mitochondria. *Biochem. Biophys. Res. Commun.* **1997**, *232*, 414–417. [PubMed]
49. Engstova, H.; Zackova, M.; Ruzicka, M.; Meinhardt, A.; Hanus, J.; Kramer, R.; Jezek, P. Natural and azido fatty acids inhibit phosphate transport and activate fatty acid anion uniport mediated by the mitochondrial phosphate carrier. *J. Biol. Chem.* **2001**, *276*, 4683–4691. [CrossRef]
50. Yu, X.X.; Lewin, D.A.; Zhong, A.; Brush, J.; Schow, P.W.; Sherwood, S.W.; Pan, G.; Adams, S.H. Overexpression of the human 2-oxoglutarate carrier lowers mitochondrial membrane potential in HEK-293 cells: Contrast with the unique cold-induced mitochondrial carrier CGI-69. *Biochem. J.* **2001**, *353 Pt 2*, 369–375. [CrossRef]
51. Demine, S.; Renard, P.; Arnould, T. Mitochondrial Uncoupling: A Key Controller of Biological Processes in Physiology and Diseases. *Cells* **2019**, *8*, 795. [CrossRef]
52. Beck, V.; Jaburek, M.; Breen, E.P.; Porter, R.K.; Jezek, P.; Pohl, E.E. A new automated technique for the reconstitution of hydrophobic proteins into planar bilayer membranes. Studies of human recombinant uncoupling protein 1. *Biochim. Biophys. Acta* **2006**, *1757*, 474–479. [CrossRef] [PubMed]
53. Knyazev, D.G.; Lents, A.; Krause, E.; Ollinger, N.; Siligan, C.; Papinski, D.; Winter, L.; Horner, A.; Pohl, P. The bacterial translocon SecYEG opens upon ribosome binding. *J. Biol. Chem.* **2013**, *288*, 17941–17946. [CrossRef] [PubMed]

54. Majima, H.J.; Oberley, T.D.; Furukawa, K.; Mattson, M.P.; Yen, H.C.; Szweda, L.I.; St Clair, D.K. Prevention of mitochondrial injury by manganese superoxide dismutase reveals a primary mechanism for alkaline-induced cell death. *J. Biol. Chem.* **1998**, *273*, 8217–8224. [CrossRef] [PubMed]
55. Magde, D.; Elson, E.L.; Webb, W.W. Fluorescence correlation spectroscopy. II. An experimental realization. *Biopolymers* **1974**, *13*, 29–61. [CrossRef]
56. Boscia, A.L.; Treece, B.W.; Mohammadyani, D.; Klein-Seetharaman, J.; Braun, A.R.; Wassenaar, T.A.; Klosgen, B.; Tristram-Nagle, S. X-ray structure, thermodynamics, elastic properties and MD simulations of cardiolipin/dimyristoylphosphatidylcholine mixed membranes. *Chem. Phys. Lipids* **2014**, *178*, 1–10. [CrossRef] [PubMed]
57. De Vries, A.H.; Mark, A.E.; Marrink, S.J. The Binary Mixing Behavior of Phospholipids in a Bilayer: A Molecular Dynamics Study. *J. Phys. Chem. B* **2004**, *108*, 2454–2463. [CrossRef]
58. Webb, B.; Sali, A. Comparative Protein Structure Modeling Using MODELLER. *Curr. Protoc. Bioinform.* **2014**, *47*, 5.6.1–5.6.37. [CrossRef]
59. Jo, S.; Lim, J.B.; Klauda, J.B.; Im, W. CHARMM-GUI Membrane Builder for mixed bilayers and its application to yeast membranes. *Biophys. J.* **2009**, *97*, 50–58. [CrossRef]
60. Wu, E.L.; Cheng, X.; Jo, S.; Rui, H.; Song, K.C.; Davila-Contreras, E.M.; Qi, Y.; Lee, J.; Monje-Galvan, V.; Venable, R.M.; et al. CHARMM-GUI Membrane Builder toward realistic biological membrane simulations. *J. Comput. Chem.* **2014**, *35*, 1997–2004. [CrossRef] [PubMed]
61. Lee, J.; Cheng, X.; Swails, J.M.; Yeom, M.S.; Eastman, P.K.; Lemkul, J.A.; Wei, S.; Buckner, J.; Jeong, J.C.; Qi, Y.; et al. CHARMM-GUI Input Generator for NAMD, GROMACS, AMBER, OpenMM, and CHARMM/OpenMM Simulations Using the CHARMM36 Additive Force Field. *J. Chem. Theory Comput.* **2016**, *12*, 405–413. [CrossRef] [PubMed]
62. Jo, S.; Kim, T.; Im, W. Automated Builder and Database of Protein/Membrane Complexes for Molecular Dynamics Simulations. *PLoS ONE* **2007**, *2*, e880. [CrossRef]
63. Nosé, S. A molecular dynamics method for simulations in the canonical ensemble. *Mol. Phys.* **1984**, *52*, 255–268. [CrossRef]
64. Parrinello, M.; Rahman, A. Polymorphic transitions in single crystals: A new molecular dynamics method. *J. Appl. Phys.* **1981**, *52*, 7182–7190. [CrossRef]
65. Essmann, U.; Perera, L.; Berkowitz, M.L.; Darden, T.; Lee, H.; Pedersen, L.G. A smooth particle mesh Ewald method. *J. Chem. Phys.* **1995**, *103*, 8577–8593. [CrossRef]
66. Huang, J.; Rauscher, S.; Nawrocki, G.; Ran, T.; Feig, M.; de Groot, B.L.; Grubmuller, H.; MacKerell, A.D., Jr. CHARMM36m: An improved force field for folded and intrinsically disordered proteins. *Nat. Methods* **2017**, *14*, 71–73. [CrossRef]
67. Aksimentiev, A.; Schulten, K. Imaging alpha-hemolysin with molecular dynamics: Ionic conductance, osmotic permeability, and the electrostatic potential map. *Biophys. J.* **2005**, *88*, 3745–3761. [CrossRef]
68. Abraham, M.J.; Murtola, T.; Schulz, R.; Páll, S.; Smith, J.C.; Hess, B.; Lindahl, E. GROMACS: High performance molecular simulations through multi-level parallelism from laptops to supercomputers. *SoftwareX* **2015**, *1–2*, 19–25. [CrossRef]
69. Humphrey, W.; Dalke, A.; Schulten, K. VMD: Visual molecular dynamics. *J. Mol. Graph.* **1996**, *14*, 33–38. [CrossRef]

International Journal of *Molecular Sciences*

MDPI

Article

Molecular Dynamics Simulations of Mitochondrial Uncoupling Protein 2

Sanja Škulj [1], Zlatko Brkljača [1], Jürgen Kreiter [2], Elena E. Pohl [2,*] and Mario Vazdar [1,3,*]

1 Division of Organic Chemistry and Biochemistry, Ruđer Bošković Institute, Bijenička 54, 10000 Zagreb, Croatia; Sanja.Skulj@irb.hr (S.Š.); Zlatko.Brkljaca@irb.hr (Z.B.)
2 Department of Biomedical Sciences, Institute of Physiology, Pathophysiology and Biophysics, University of Veterinary Medicine, 1210 Vienna, Austria; juergen.kreiter@vetmeduni.ac.at
3 Institute of Organic Chemistry and Biochemistry, Czech Academy of Sciences, Flemingovo nám. 2, 16610 Prague, Czech Republic
* Correspondence: elena.pohl@vetmeduni.ac.at (E.E.P.); Mario.Vazdar@irb.hr (M.V.)

Abstract: Molecular dynamics (MD) simulations of uncoupling proteins (UCP), a class of transmembrane proteins relevant for proton transport across inner mitochondrial membranes, represent a complicated task due to the lack of available structural data. In this work, we use a combination of homology modelling and subsequent microsecond molecular dynamics simulations of UCP2 in the DOPC phospholipid bilayer, starting from the structure of the mitochondrial ATP/ADP carrier (ANT) as a template. We show that this protocol leads to a structure that is impermeable to water, in contrast to MD simulations of UCP2 structures based on the experimental NMR structure. We also show that ATP binding in the UCP2 cavity is tight in the homology modelled structure of UCP2 in agreement with experimental observations. Finally, we corroborate our results with conductance measurements in model membranes, which further suggest that the UCP2 structure modeled from ANT protein possesses additional key functional elements, such as a fatty acid-binding site at the R60 region of the protein, directly related to the proton transport mechanism across inner mitochondrial membranes.

Keywords: membrane protein; long-chain fatty acid; proton transfer; purine nucleotide; conductance measurements in model membranes; uncoupling

Citation: Škulj, S.; Brkljača, Z.; Kreiter, J.; Pohl, E.E; Vazdar, M. Molecular Dynamics Simulations of Mitochondrial Uncoupling Protein 2. *Int. J. Mol. Sci.* **2021**, *22*, 1214. https://doi.org/10.3390/ijms22031214

Academic Editor: Masoud Jelokhani-Niaraki

Received: 18 December 2020
Accepted: 22 January 2021
Published: 26 January 2021

1. Introduction

Uncoupling protein 2 (UCP2) belongs to the mitochondrial SLC25 superfamily of anion transporters. It was implicated in the pathogenesis of multiple physiological and pathological processes, such as diabetes, ischemia, metabolic disorders, (neuro) inflammation, cancer, and aging. Based on its proton transporting function, UCP2 was first suggested to act as a mild uncoupler to reduce oxidative stress [1–3]. Later, it was shown to transport C4 metabolites out of mitochondria [4], facilitating the tricarboxylic acid (TCA) cycle. A recently proposed dual transport function for UCP2 (proton and substrate) increases the similarity of UCP2 to the ANT (also abbreviated as AAC in literature), which transports protons [5–7], additionally to ATP/ADP exchange.

The mechanism of how UCP2 controls proton transport across mitochondrial membranes is still not understood. So far, it is established that long-chain fatty acids (FAs) are an integral part of the mechanism and are crucial for proton transfer [8–10]. Currently, several mechanistic models exist that explain the proton transfer mechanism. In the first one, so-called the "FA cycling" model, FAs act as protonophores. Due to the excess of protons in the mitochondrial intermembrane space, FA carboxyl anions are easily protonated and they can flip-flop across the membrane very fast in the neutral form to the matrix [11–13] where a proton is subsequently released. After that, UCP2 facilitates the otherwise very slow transfer of the negatively charged fatty acid by a still unknown mechanism back to the intermembrane space and the cycle starts again [1,8,14]. The dependence of H^+ transport rate on FA saturation, FA chain length [9] and fluidity of the membrane [15] indicates that FA^- transport likely occurs at the protein–lipid interface.

The second group of models does not involve flip-flop of FAs. Instead, it proposes that carboxyl groups of negatively charged amino acids of the UCPs can accept a proton from a FA and transport it through the hypothetic channel in the UCP ("FA proton buffering" model) [16,17]. Alternatively, the FA anion binds in the cavity inside the UCP interior ("FA shuttle" model). Upon proton binding to the FA anion, a conformational change occurs which shuttles the FA together with a proton, which is subsequently released in the mitochondrion matrix and the cycle is repeated [6,18].

Currently, the consensus on the exact mechanism of how UCP2 works is far from being reached, mainly due to the shortage of reliable structural information. A potential breakthrough in the UCP2 investigation occurred in 2011 when an NMR structure of UCP2 was published [19]. In theory, the structure should have served as an ideal starting point for all potential molecular simulations and detailed structural and mechanistic analyses. Unfortunately, it turned out that the UCP2 structure extracted from commonly used detergent dodecyl phosphocholine (DPC) is not functionally relevant [20]. Moreover, it is now quite established that alkyl phosphocholine detergents destabilize and denature α-helical membrane proteins, leading to a distorted protein secondary structure. It raises important questions on the appropriateness of alkyl phosphocholine detergents as the extraction media for the determination of membrane protein structure by solution NMR. A lively debate is currently still taking place whether the disturbance of the protein structure by these types of detergents is prohibitive for further understanding of the protein function [21–24] or if it can still be used for capturing the most important functional aspects [25–27]. Despite recent developments in membrane protein structure determination, such as detergent-free solubilization of membrane proteins using styrene-maleic acid lipid particles [28,29] and cryo-EM keeping the lipid environment intact [30,31], the handling of small mitochondrial carriers, and in particular uncoupling proteins, remains challenging because of their size and low abundance in mitochondrial membranes.

Molecular dynamics (MD) simulations are an attractive complementary option for studying membrane proteins, provided that sampling times are sufficiently long to sample their dynamics in membranes adequately [28–30]. Membrane proteins are encoded by ca. 30% of the human genome and their total number is predicted to be significantly higher [31]. Only about 1000 unique membrane protein structures are determined today [32] representing a small fraction of the total number of membrane proteins found in humans. However, since an increasing number of membrane protein structures are determined by solution NMR using contentious alkyl phosphonates as the extraction media [33], long MD simulations (in the microsecond time range) in combination with homology modeling [34,35] often represent the only available option for studying membrane protein structure and dynamics.

Currently, MD simulations of monomeric UCP proteins reported in the literature are primarily based on the available UCP2 NMR structure [20,36]. MD simulations of UCP2 protein made by Zoonens and coworkers indicated that DPC detergent induced large structural deformations of UCP2 protein helices, which in turn created a large water channel, thus facilitating continuous water leakage across the protein [20]. Since the proton conductance is unattainable under these conditions, its experimental measurements have additionally confirmed that UCP2 protein, extracted with the help of DPC detergent, is not functionally relevant. In contrast, the closely related UCP1 protein extracted using DPC and the structurally different detergent TX-100 remained physiologically active [20]. Interestingly, a recent study showing the oligomerization of UCP2 monomers did not describe differences between NMR and homology model structures [37].

Motivated by the lack of relevant MD simulations for the monomeric structures which would help to decipher the function of UCP2 protein in membranes, we turned to homology modeling using the structure of mitochondrial ADP/ATP carrier (ANT,

PDB code 1OKC) [38] as a template for UCP2 MD simulations. ANT, a member of the mitochondrial carrier protein family SLC25 [39], is also found in inner mitochondrial membranes. Its primary function is to exchange ADP against ATP across mitochondrial membranes [40–42]. However, it has been reported that ANT also works as a proton carrier, similar to UCP proteins, by a mechanism still undetermined at the molecular level [5–7]. Taking into account the sequence identity between ANT and UCP2 of 24% [43], a similarity in the overall shape containing six membrane domains, as well as sharing of the proton transporting function in mitochondria, we chose 1OKC structure as a starting template for homology modeling and subsequent microsecond MD simulations. Finally, we compared the simulation results with the data obtained in model membranes reconstituted with UCP2 to validate our MD model.

2. Results

2.1. Structural Properties of Modeled Membrane Proteins

Firstly, we aligned the primary sequences of two proteins, murine UCP2 protein and bovine ANT protein (Figure 1). Although the homology of both proteins is 24%, the crucial amino acid positions and motifs characteristic for even and odd-numbered transmembrane helices, namely $\pi G\pi x\pi G$ (helices 1, 3 and 5) and $\pi xxx\pi$ (helices 2, 4 and 6), are conserved. Furthermore, the positions of amino acids corresponding to matrix or cytosolic salt bridge network, as well as proline kinks at odd-numbered helices, also remain preserved (Figure 1).

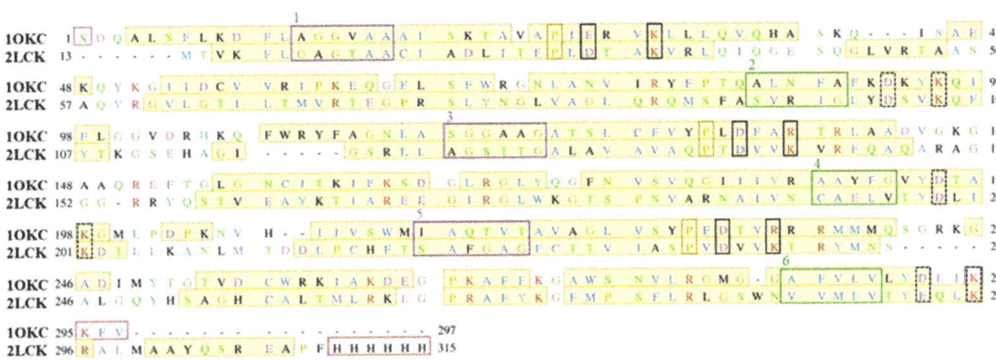

Figure 1. Alignment of the primary sequences of UCP2 protein (PDB code: 2LCK) and bovine ANT protein (PDB code: 1OKC). Secondary structure alpha helices are denoted in the yellow shade, and missing residues from the crystallographic structure are presented with red boxes. Residues that constitute the salt bridge network at the cytoplasmic side are shown in dashed black boxes, while potential residues that could constitute the salt bridge network at the matrix side are enclosed in solid black boxes. $\pi G\pi x\pi G$ (helices 1, 3 and 5) and $\pi xxx\pi$ (helices 2, 4 and 6) motifs are depicted in violet and green boxes, respectively. Residues responsible for proline kinks are enclosed in yellow boxes.

Taking into account that strategic pillars, including cytosolic and matrix salt bridges as well as shape-forming proline kinks in both structures were conserved, we felt that it was safe to take the ANT structure as a starting point for subsequent MD simulations. In our previous MD simulations of two different crystallographic structures belonging to differently open ANT states towards the cytosolic [38] or matrix [42] side of the inner mitochondrial membrane, we have shown that it is possible to capture important conformational changes of the protein embedded in the membrane within microsecond MD simulations [44]. Moreover, we have also revealed that an order shorter timescales of ca. 200 ns, used in previous MD simulations of UCP2 NMR structures in DOPC bilayers [20], were not sufficiently long for a more relevant description of critical regions in the protein, such as reversible salt bridge breaking and forming [44]. In addition to MD simulations of the $UCP2_h$, we also repeated simulations of $UCP2_{NMR}$ based on the NMR structure of UCP2 [20] and compared them to referent ANT simulations at relevant microsecond timescales. A schematic representation of the UCP2 structure is depicted in Figure S1.

Figure 2 shows the time evolution of root mean square deviation (RMSD), which indicates the three studied protein structures' stability in time. RMSD deviation was the largest for $UCP2_{NMR}$, which was not surprising, given the fact that this structure obtained in alkylphosphocholine detergent was determined in a non-optimal environment [20,22,33]. It was visible that the extension of simulations by Zoonens et al. [20] showed even larger deformations of the structure (especially after 1 μs), which will be analyzed in more detail in later sections. MD simulations of $UCP2_h$ were more stable, although a small increase in the RMSD occurred at the end of simulation time. However, these instabilities were not as severe as in the $UCP2_{NMR}$ simulations and were of a similar order of magnitude as the RMSD oscillations observed for the referent ANT structure. However, although very useful for the general description of protein stability, the RMSD analysis was a simple "one-number" analysis and did not contain information on the conformational changes of specific residues [45]. For this reason, we turned to root mean square fluctuation (RMSF) analysis (Figure 3), which provided important (but not time-resolved) data on the flexibility of particular residues. Importantly, we saw that the $UCP2_{NMR}$ structure was more flexible (and less stable) around residues found in the water phase, oriented to the matrix side (especially around residues 250–270 and C-terminus) compared to $UCP2_h$ and referent ANT structures. Figure S2 shows the time evolution of the $UCP2_{NMR}$ and $UCP2_h$ secondary structures which remained preserved for both structures in the simulation time.

Finally, another very useful analysis of general structural parameters was obtained by the principal component analysis (PCA) of backbone carbon atoms of the protein. PCA is a procedure that reduces a multidimensional complex set of all possible conformational degrees of freedom to lower dimensions along which the main conformational changes of protein are identified. The PCA analysis of $UCP2_{NMR}$ and $UCP2_h$ is shown in the Figure 4. We can see that the area spanned by the first two principal components (PC1 and PC2) was much larger in the case of $UCP2_{NMR}$ structure in comparison to $UCP2_h$ structure. This further supports the above analysis, showing that $UCP2_{NMR}$ structure was more flexible and less stable in DOPC phospholipid bilayer. The analysis indicates that the protein structure tried to find its optimal position and a proper fold in the membrane, which was not attainable at a microsecond time scale and probably orders of magnitude longer simulation times were needed. In contrast, the area spanned by PC1/PC2 in the $UCP2_h$ structure was much smaller and relatively compact, demonstrating that it was stable in the bilayer within our simulation time. It was in line with RMSD and RMSF analyses and with reported microsecond MD simulations of ANT protein [44].

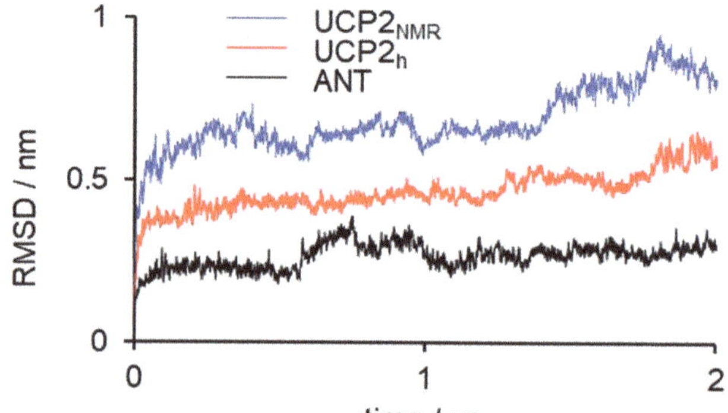

Figure 2. Time propagation of the RMSD values for $UCP2_{NMR}$, $UCP2_h$ and the referent structure ANT. The RMSD is calculated for backbone carbon atoms (C_α) of protein.

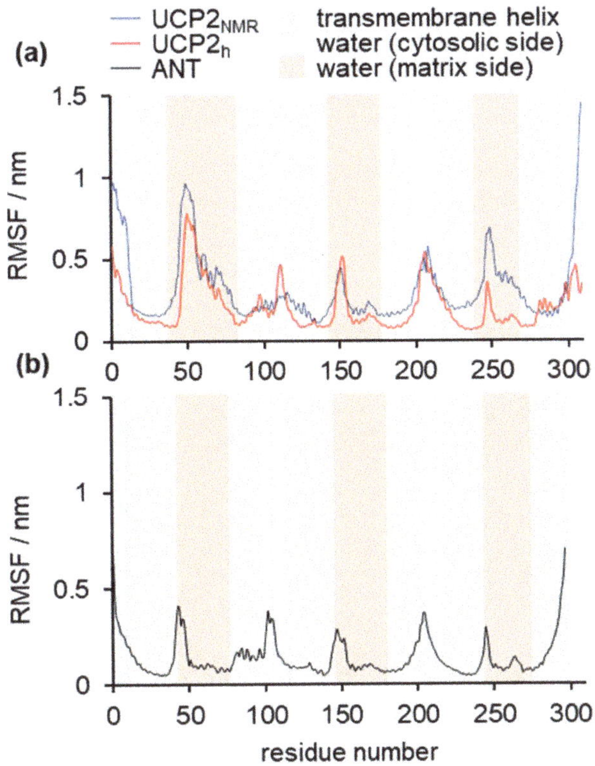

Figure 3. RMSF analysis of (**a**) $UCP2_{NMR}$ and $UCP2_h$ structures and (**b**) referent ANT structure. The RMSF is calculated for backbone carbon atoms (C_α) of protein. Light red color corresponds to protein residues inside the phospholipid bilayer, light blue represents protein residues immersed in water at the cytosolic side, whereas dark pink corresponds to the protein residues immersed in water at the matrix side.

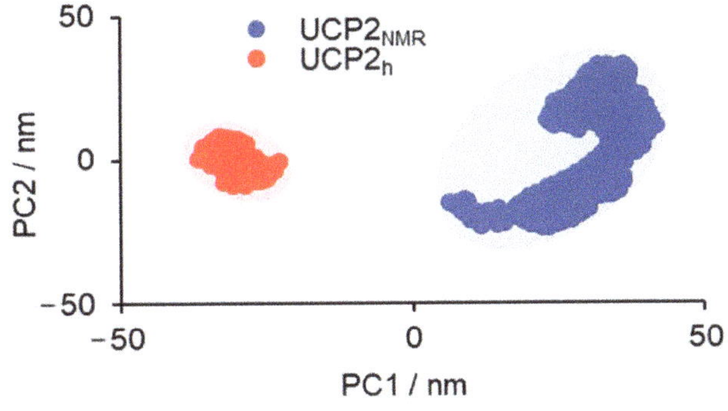

Figure 4. PCA analysis-2D projection of $UCP2_{NMR}$ and $UCP2_h$ protein conformations onto common first and second principal components (PC1 and PC2) are presented in blue and red color, respectively.

2.2. Stability of Salt Bridges Exposed to the Cytosolic and Matrix Side of the Inner Mitochondrial Membrane

As a next step, we now focused on the stability of the salt bridge networks formed at the cytosolic and matrix sides of the modeled UCP2 structure (Figure S1). Opening and closing the cytosolic and matrix side of the ANT protein via salt bridges, which are connected to the transport of ADP and ATP nucleotides across inner mitochondrial membranes, involves at least 10 kcal mol^{-1} for breaking the salt bridge network [40–42]. However, it is essential that water does not leak through the protein interior since it would abolish strictly controlled proton transport due to water-mediated ion exchange, as had been shown by functional leakage assays [20]. Therefore, as an initial prerequisite for controlled proton transfer, the UCP2 protein should be impermeable to water in order not to allow short-circuiting of the system, which is possible only if the salt bridge network is closed and constricts the protein at the matrix [38] or cytosolic side [42] as found in the corresponding crystallographic structures of ANT and subsequent MD simulations [44]. These experimental results further indicated that the proton transport mechanism, either via UCPs or ANT [6], was not controlled by direct transport of proton through the protein interior, but involved the transport of FA anion (and in turn proton) alongside the protein/lipid interface [1,8,9,46].

The salt bridge networks analysis showed that in the case of the $UCP2_{NMR}$ structure (Figure 5a,b), only one residue pair (Asp35-Lys141), located at the matrix side of the protein, permanently formed a salt bridge within our simulation time. In contrast, two other salt bridges located at the matrix side (Asp236-Lys38 and Asp138-Lys239) were not making a salt bridge, as well as three other salt bridge pairs at the cytosolic side (Asp198-Lys104, Asp101-Lys295, and Glu292-Lys201). On the other hand, salt bridges at the matrix side formed in the case of $UCP2_h$ structure (Asp35-Lys141, Asp236-Lys38, and Asp138-Lys239) were stable and persistent (Figure 5c) just as in the case of the analogous salt bridges in the referent ANT structure (Glu29-Arg137, Asp231-Lys32, Asp134-Arg234) presented in Figure 5e. Cytosolic salt bridges were partially closed in $UCP2_h$ (Figure 5d). In contrast, they were fully opened in the case of ANT (Figure 5f). These results imply that water leakage should be largely suppressed in the case of the $UCP2_h$ structure due to the closed matrix side of the protein, which is the pivotal condition for the protein structure to have a relevant functional role in the proton transfer mechanism. We reached similar conclusions by analysis of the referent ANT structure. However, in the case of $UCP2_{NMR}$ structure, we showed that due to the opened matrix side of the protein, water leakage was possible across the protein interior (more details are found in the next section). It was similar to the observations by Zoonens et al. from their shorter analogous MD simulations [20]. We should also mention that in the case of $UCP2_h$ and ANT structures, the distances between pairs of negatively charged residues at the matrix side (i.e., EG-motif), which were highly conserved across mitochondrial ADP/ATP carriers [47], kept three-fold pseudosymmetry in contrast to the $UCP2_{NMR}$ structure where this motif was not conserved, and distances between the negatively charged residues were larger (Figure S3). In this way, we further showed that the $UCP2_{NMR}$ structure was unstable and functionally irrelevant when embedded in phospholipid bilayers, which were structurally significantly different compared to the alkyl phosphocholine environment serving as an extracting agent [20,22,33].

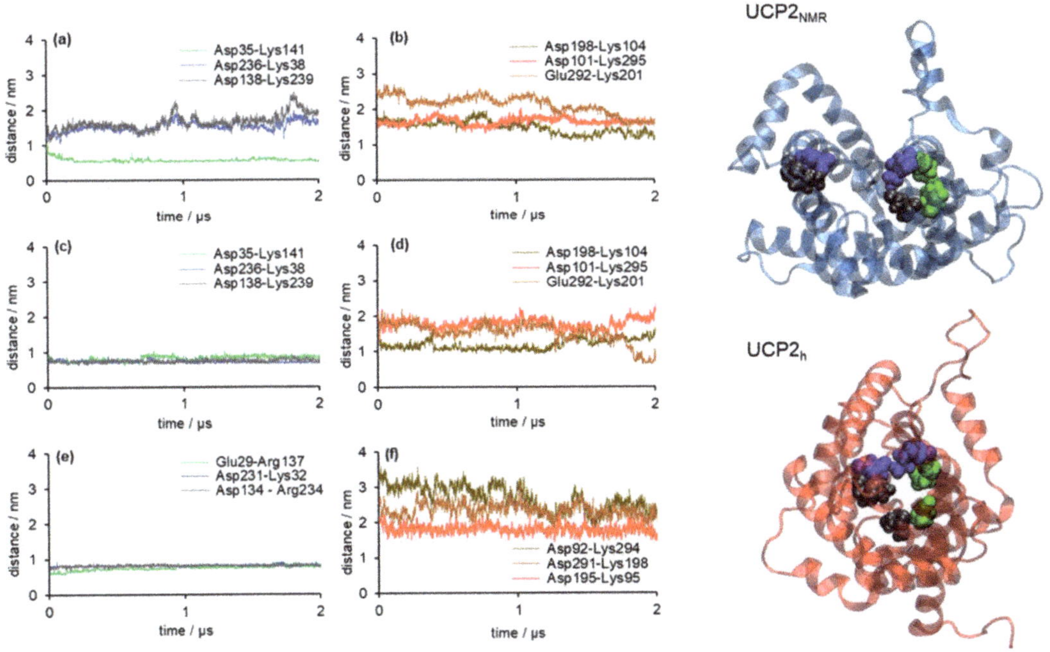

Figure 5. Analysis of the salt bridge network for UCP2 model based on (**a**,**b**) the UCP2 NMR structure ($UCP2_{NMR}$), (**c**,**d**) the UCP2 model based on the crystallographic structure of ANT ($UCP2_h$), and (**e**,**f**) the referent ANT structure. Distances between residues that can form a salt bridge network at the matrix side are shown in panels (**a**,**c**,**e**). Distances between residues that can form a salt bridge network at the cytoplasmic side are shown in panels (**b**,**d**,**f**). Distances are calculated between centers of mass of the corresponding residues. A top-down view on the matrix exposed side of selected protein snapshots of $UCP2_{NMR}$ and $UCP2_h$ structures is shown on the right.

2.3. *Water Leakage across the Protein and Permeability Calculations*

The analysis of the salt bridge networks in the previous section suggests that $UCP2_{NMR}$ structure should be more water permeable due to the simultaneously open matrix and cytosolic sides of the protein in contrast to the partially closed $UCP2_h$ structure (Figure 5). To quantitatively analyze this assumption, we performed a detailed analysis of the water density inside the protein for both structures and calculated corresponding water osmotic permeability coefficients P_f using the method described in Zoonens et al. [20]. The analysis of averaged water density inside the protein along the z-axis for $UCP2_{NMR}$, $UCP2_h$ and the referent ANT structure is shown in Figure 6a, with the time evolution shown in Figure S4. Interestingly, although the averaged number density of water was averaged across z-coordinate and did not include the differences in the x- and y- directions, the minimal value of the number density was similar for all structures, being less than a half of the water molecule per nm^3. Thus, it was not very informative of the possible formation of a continuous water channel, which would possibly enable water-mediated direct proton transfer leading to the inactive UCP protein [44]. However, we should mention here that the presence of a continuous water channel is not a key prerequisite for efficient proton transfer across the membrane protein and that electrostatic effects resulting in a high energy barrier for proton transfer predominate, such as in a case of aquaporins [48,49].

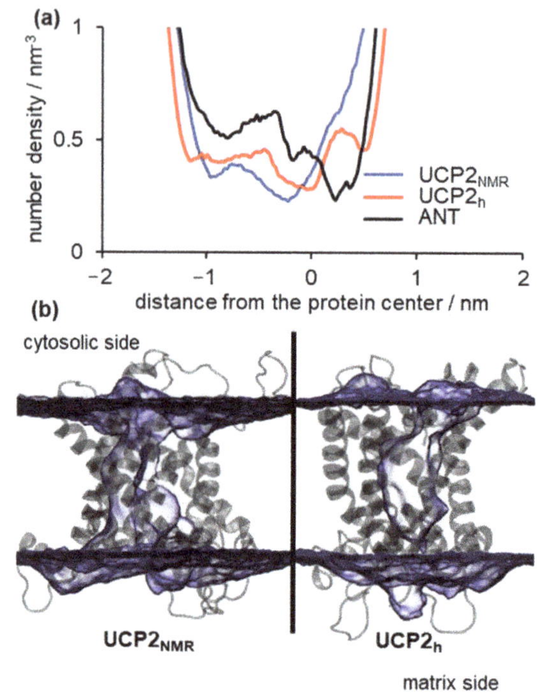

Figure 6. (**a**) *z*-averaged water number density based on 2 μs simulations for UCP2$_h$, UCP2$_{NMR}$ and referent ANT structures, (**b**) snapshots presenting volume map of water in transparent blue with surface isovalue set to 0.2 for UCP2$_{NMR}$ structure (left side) and homology modeled UCP2$_h$ structure (right side). Cytosolic and matrix sides of UCP2 protein are indicated.

A better look at Figure 6a revealed that the area, and in turn the total volume of water, was largest in the case of the UCP2$_{NMR}$ structure (blue curve), in contrast to UCP2$_h$ and ANT number density profiles, which were considerably wider (red and black curve, respectively). This is better visualized in Figure 6b, where we saw that the average volume map of water was continuous along the protein interior (left panel) in the case of UCP2$_{NMR}$ structure. In contrast, two disjointed volume maps existed in the case of the UCP2$_h$ structure, indicating that the water channel was not formed (right panel), similar to ANT protein.

However, these analyses are still not fully quantitative, and therefore we turned to water osmotic permeability calculations P_f, to compare the data to other systems. The results of the calculations are shown in Table 1.

Table 1. Water osmotic permeability coefficients calculated for four distinct membrane protein structures in different simulation times.

Structure	Permeability (cm^3 s^{-1})
UCP2$_{NMR}$ after equilibration	$(5.7 \pm 0.4) \times 10^{-13}$
UCP2$_{NMR}$ after 200 ns	$(3.2 \pm 0.2) \times 10^{-13}$
UCP2$_{NMR}$ after 2 μs	$(1.3 \pm 0.1) \times 10^{-13}$
UCP2$_h$ after 2 μs	$(2.0 \pm 0.5) \times 10^{-16}$

Osmotic permeability coefficients were calculated for four different membrane structures. First, we calculated water osmotic permeability for the equilibrated UCP2$_{NMR}$ structure (i.e., only after short initial equilibration), which closely corresponded to the experimental NMR structure. The calculated P_f is $(5.7 \pm 0.4) \times 10^{-13}$ cm^3 s^{-1}, which is comparable to the value of 5.3×10^{-13} cm^3 s^{-1} obtained by Zoonens et al. for an analogous system [20]. Similarly to their observations, the P_f decreased after 200 ns to $(3.2 \pm 0.2) \times 10^{-13}$ cm^3 s^{-1}) and finally after 2 µs it assumed the value of $(1.3 \pm 0.1) \times 10^{-13}$ cm^3 s^{-1} which showed a certain collapse of the water pore in the protein. However, this number was still comparable to the water osmotic permeability of the α-hemolysine, where this value was calculated to be 1.9×10^{-12} cm^3 s^{-1} [50]. Since these values were comparable, it was clear that the $UCP2_{NMR}$ structure was behaving quite similarly to the water channel, which was physiologically irrelevant for UCP2 function [20]. On the other hand, the water permeability coefficient of $UCP2_h$ structure, calculated after 2 µs, was by three orders of magnitude lower being $P_f = (2.0 \pm 0.5) \times 10^{-16}$ cm^3 s^{-1}. This was additionally confirmed by simple counting of water molecules that crossed across the protein, where we saw that $UCP2_h$ and ANT protein were virtually impermeable in contrast to the $UCP2_{NMR}$ structure (Table S1). These results are in accordance with the general structure analysis described in the previous sections, and promote $UCP2_h$ structure as a potentially relevant structure for further MD simulation and mechanistic studies. We should mention here that the MD simulations of UCP2 based on ANT homology structure presented in Reference [37] agreed with the presented MD simulations. In particular, the constriction at the matrix side and opening of the cytosolic side of UCP2 protein had been observed as well, together with the low number density of water inside the protein cavity similar to the $UCP2_h$ structure (Figures 5 and 6).

2.4. Binding of ATP in the UCP2 Cavity

To further verify key functional elements of the $UCP2_h$ structure, we performed a series of MD simulations (see Simulation Details) to evaluate the binding properties of ATP nucleotide in the UCP2 cavity which are known to inhibit proton transport in UCPs. It has been suggested in the literature that three positively charged arginine residues in the UCP1 cavity bind a negatively charged phosphate group, which leads to the conformational change of the protein and inhibition of proton transport [2,51–53]. This mechanism can be extended to other UCP proteins as well since the arginine residues responsible for nucleotide binding are conserved in other homologs as illustrated in Figure S5. MD simulations of ATP binding in the $UCP2_h$ cavity show that the ATP phosphate group binds tightly in the protein cavity, having all three phosphate groups bound to arginines R88, R185, and R279 (Figure 7). This can be inspected by analyzing average distances between phosphorous atoms present in ATP and the center of mass of arginine residues in the UCP2 structure (Figure S6). In particular, we observed that in the case of $UCP2_h$ structure, the binding of phosphate to arginine residues was tight, with all three phosphate groups bound to arginine residues R88, R185 and R279. This was in striking contrast to $UCP2_{NMR}$ structures after 20 ns and also after 2 µs where simultaneous binding of ATP phosphate groups to arginine residues did not occur and average distances between the groups were significantly larger (Figure S6), implying in turn weaker binding of ATP. This is in line with previously suggested binding motifs of ATP in UCP1 and UCP3 [54] and very tight binding of GDP in the UCP1 cavity determined by titration calorimetry experiments [52]. Moreover, it has been found that in the case of AAC3 protein, binding of carboxyatractyloside (CATR) inhibitor at the analogous location in the protein cavity as ATP was by several orders of magnitude weaker if the AAC protein structure was obtained by extraction with DPC detergent in contrast to the native crystallographic structure [24]. Therefore, we believe that the molecular description of ATP binding in the $UCP2_h$ cavity further promotes the relevance of the homology modelled structure for further MD simulation studies.

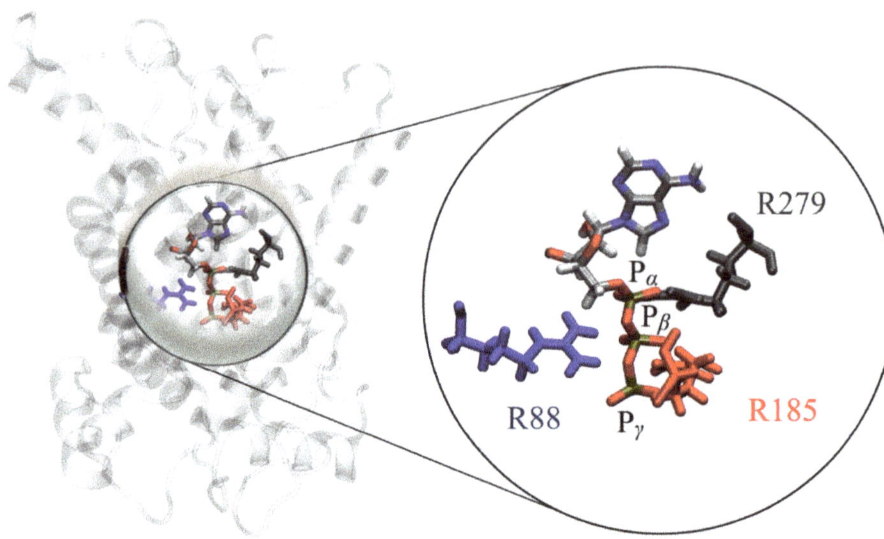

Figure 7. Simultaneous ATP nucleotide binding to three arginines (R88, R185, R279) in the case of $UCP2_h$ protein, with R279 (depicted in gray) being found to primarily bind to P_α. R88 (shown in blue) binds to P_β (occasionally to P_α), while R185 (depicted in red) binds predominantly to P_γ. Water molecules are omitted for the sake of clarity.

2.5. Binding of Fatty Acid to UCP2

In previous sections, we focused on the general structural parameters of the UCP2 protein obtained by MD simulations. Finally, we turned to additional experimental verification of MD results by complementary experiments using model membranes to prove whether the suggested $UCP2_h$ structure might be physiologically relevant.

NMR titration experiments, as well as proton flux assay measurements, suggest that a patch of positively charged residues around R60 residue in UCP2 (consisted of K271, R267, R40, and R71 residues) is relevant for the proton transport mechanism since it serves as a binding site for FA anion [25]. We should stress here that the proton flux assay measurements presented in Reference [25] were performed in liposomes with correctly folded protein. It suggests that the active site in UCP2 is preserved regardless of the protein extraction medium. We performed experiments with the recombinant mutant UCP2 to check whether the mutation from R60 to S60 affected the proton conductance (see details in Materials and Methods). We also performed 500 ns of complementary MD simulations with the AA anion (AA^-) to visualize the binding process at the molecular level. Figure 8c shows the comparison of the experimental total membrane conductance (G_m) in two systems: the wild type UCP2-WT protein and the UCP2-R60S mutant protein. Both proteins were measured in the presence and absence of AA. We observed two effects. Firstly, the addition of AA^- was essential for the increase of G_m as shown in our previous works [9,10], as the conductance increased by order of magnitude compared to the neat WT protein. Secondly, the mutation from R60 to S60 had a significant effect on the G_m, thus further supporting the hypothesis that R60 is a possible binding site for AA^-. It agreed with NMR titration experiments and proton flux assays [25]. Moreover, the addition of ATP, which is an efficient inhibitor of the UCP2 proton-transporting function [3,10,43,55], also showed that it was effective only in the UCP2-WT and less effective in the UCP2-R60S (Figure S7), confirming that R60 had an important role in the proton transport mechanism. However, based on the present MD simulation results, it is still questionable whether other positively charged residues along the outer protein ring at the matrix side could also serve as potential binding sites of AA^-.

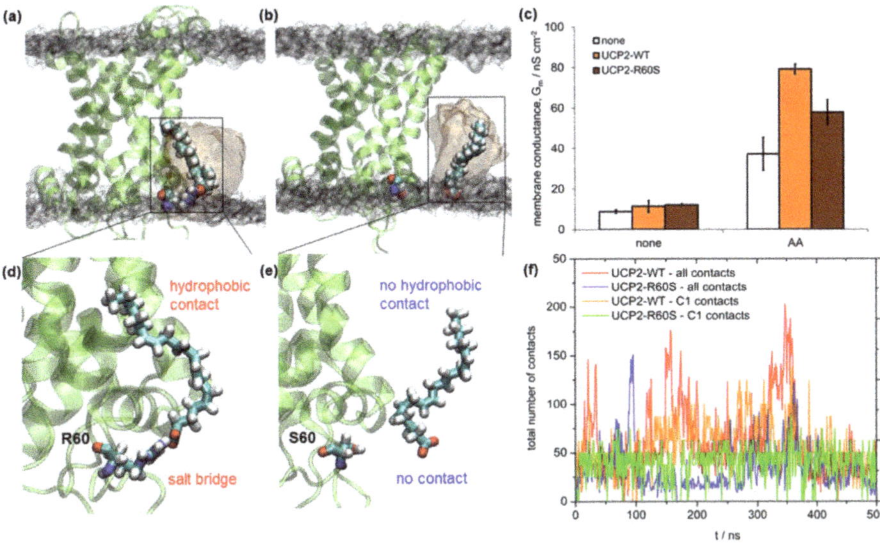

Figure 8. Snapshots of 500 ns simulations of UCP2 homology modeled structures ($UCP2_h$) with added arachidonic acid anion (AA^-). A starting structure of protein used for binding calculations is taken from UCP2h MD simulations after 2 µs. $UCP2_h$ structure with AA^- (**a**) bound to R60 and (**b**) not bound to the mutated binding site S60. A volume map of phosphorus atoms of DOPC lipid is presented in transparent gray color (surface isovalue set to 0.0064) and volume map of AA^- is presented in transparent orange color (surface isovalue set to 0.02). (**c**) Specific membrane conductance of lipid bilayers in the absence of protein (white), and presence of UCP2-WT (orange) or UCP2-R60S (brown). Membranes were made of 45:45:10 mol.% DOPC:DOPE:CL reconstituted with 15 mol.% AA where indicated. Lipid and protein concentrations were 1.5 mg/mL and 4 µg per mg of lipid, respectively. The buffer solution contained 50 mM Na_2SO_4, 10 mM TRIS, 10 mM MES and 0.6 mM EGTA at pH = 7.34 and T = 306 K. Data are represented as the mean and standard deviation from three independent experiments. (**d**) Zoomed region of AA^- binding to $UCP2_h$ structure. (**e**) Zoomed region of AA^- binding to UCP2-R60S mutant structure. (**f**) Total number of contacts within 0.35 nm between all AA^- atoms and UCP2-WT and (red color) and UCP2-R60S mutant structures (blue color), respectively. Total number of contacts within 0.35 nm between AA^- carboxyl atom towards $UCP2_h$ (orange color) and UCP2-R60S structures (green color).

A detailed analysis of MD simulations was in full agreement with conductance measurements using model membranes. In Figure 8a,d (which is a zoomed region of the binding site) we can see that AA^- fitted nicely to the binding site, at the same time having two distinct modes of interaction, i.e., salt bridge formation between R60 and AA^- and stabilizing hydrophobic interactions between AA^- and protein α-helices. This was also clearly visible in the analysis of total contacts between AA^- and protein (Figure 8f, red and orange curves). Conversely, interaction between AA^- and S60 was severely diminished (Figure 8b,e), resulting in no formation of a salt bridge between S60 and AA^-, as well as a greatly reduced number of total contacts between AA^- and protein α-helices within 0.35 nm (blue and green curve, Figure 8c). Interestingly, the number of contacts within 0.35 nm between carboxylic carbon atom C1 of AA^- was similar in both cases. The present analysis implies that the unsaturated 20:4 AA anion, with its four cis double bonds, which are conformationally quite restricted, fits much better to the $UCP2_h$-WT structure than the UCP2-R60S mutant.

Interestingly, the activation of UCP2-WT with saturated 20:0 arachidic acid (ArA) showed lower G_m than with its unsaturated 20:4 counterpart [10], thus further pointing out the importance of hydrophobic contacts for proper binding of FA anion to R60, which were reduced and entropically unfavored in a far more flexible fully saturated ArA.

Finally, we also analyzed how AA^- binds to the $UCP2_{NMR}$ structure in two cases, after equilibration (to mimic the NMR experimental structure) and after 2 µs of MD simulations (to see the effect of relaxation in the bilayer). In the case of the equilibrated NMR structure, AA^- bound very poorly to R60, showing no permanent salt bridge formation (evidenced by a smaller number of contacts between C1 atom and protein in comparison to the $UCP2_h$ structure shown in Figure 8) and very few total contacts with the protein (Figure 9a,c,d). In addition to the increased permeability of the protein to water, which is a fundamental structural problem, we also saw that binding of AA^- to R60 was not established, which even further disqualified $UCP2_{NMR}$ structure in regard to new mechanistic studies. After 2 µs of MD simulations, the total number of contacts remained low and the situation was actually even worse as AA^- could not achieve proper hydrophobic interactions with one of the α-helices which was displaced from the rest of the UCP2 protein structure (Figure 9b,e,f). This implies a possible denaturation or even disintegration of the experimental $UCP2_{NMR}$ structure when transferred from the alkyl phosphonate detergent environment to the phospholipid bilayer milieu.

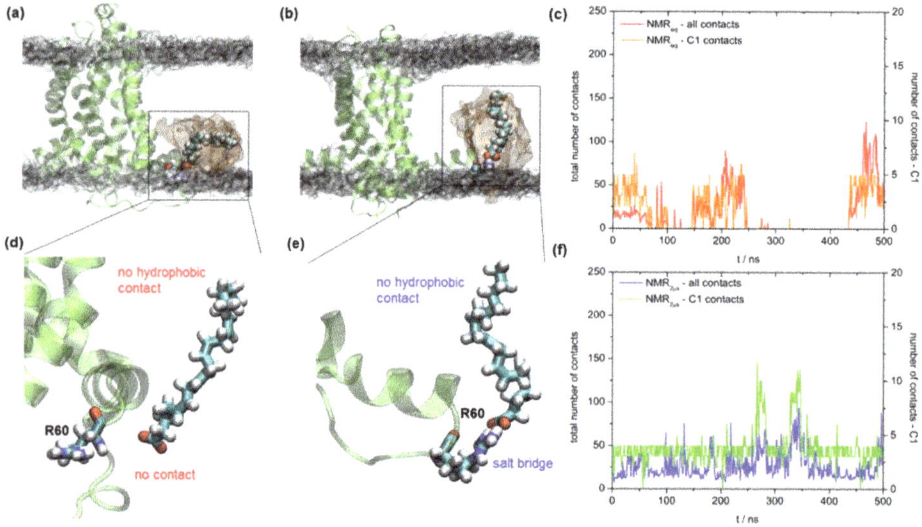

Figure 9. Snapshots of 500 ns simulations of $UCP2_{NMR}$ structures with added arachidonic acid anion (AA^-). The starting structures of protein used for binding calculations are taken from $UCP2_{NMR}$ MD simulations (**a**) after equilibration and (**b**) after 2 µs. A volume map of phosphorus atoms of DOPC lipid is presented in transparent gray color (surface isovalue set to 0.0064) and volume map of AA^- is presented in transparent orange color (surface isovalue set to 0.02). (**c**) Total number of contacts within 0.35 nm between all AA^- atoms and $UCP2_{NMR}$ after equilibration. (**d**) Zoomed region of AA^- binding to $UCP2_{NMR}$ structure after equilibration. (**e**) Zoomed region of AA^- binding to $UCP2_{NMR}$ structure after 2 µs. (**f**) Total number of contacts within 0.35 nm between all AA^- atoms and $UCP2_{NMR}$ after 2 µs.

3. Discussion

The absence of the relevant UCP structure hinders the understanding of UCP2 transport mechanisms and biological functions. To find a possible remedy for the current situation, we turned to homology modeling using ANT crystallographic structure and experimental UCP2 NMR structure as starting structures for long microsecond MD simulations of UCP2 proteins ($UCP2_h$ and $UCP2_{NMR}$). We showed by microsecond MD simulations that $UCP2_h$ structure is almost water impermeable ($P_f = 2.0 \times 10^{-16}$ cm^3 s^{-1}), with a water osmotic permeability coefficient lower by three orders of magnitude than the corresponding $UCP2_{NMR}$ structure after 2 µs of MD simulation time ($P_f = 1.3 \times 10^{-13}$ cm^3 s^{-1}). This is a consequence of the salt bridge network formation at the matrix side of the protein in $UCP2_h$, formed by arginine/lysine and aspartate/glutamate residues, as well as conservation of the threefold symmetry between negatively charged residues at the matrix side (EG motif), similar to water-impermeable ANT proteins [44,56,57]. In contrast, the stable formation of the identical salt bridges and the EG motif at the matrix side of $UCP2_{NMR}$ structure were not observed, thus enabling water to almost freely diffuse across the protein. Finally, we showed that ATP binds to all three arginine residues in the UCP2 cavity only in the case of $UCP2_h$ structure, in agreement with previous experimental observations [53,54]. We should mention here that MD simulations of UCP2 proteins are not fully converged (which is almost impossible with the current computational power) and that $UCP2_h$ structure is possibly not the global minimum. Still, we believe that the presented $UCP2_h$ structure is relevant for future studies since it corresponds well to available experimental data and that its behavior is described sufficiently well.

The described $UCP2_h$ structure can be easily transferred to membranes formed by DOPC, DOPE and cardiolipin, which would then more realistically describe the lipid composition of inner mitochondrial membranes and our model membranes. However, at this stage, we believe that further complication of a model system is not necessary, as we have shown previously in MD simulations of ANT proteins in different, more complex environments [44]. Due to a high complexity of membrane protein/lipid bilayer systems and a necessity for exceedingly long simulation times, pitfalls resulting from incomplete coverage of whole phase space in heterogeneous systems, even at the microsecond timescales and relatively small bilayers, are most likely.

To test the validity of MD simulation results regarding UCP2 functionality, we compared the binding of UCP2 activator AA to both models, $UCP2_h$ and $UCP2_{NMR}$. Additionally, we measured the activation of the recombinant UCP2 due to the AA binding. A larger membrane conductance of the UCP2 in the presence of AA compared to the UCP2-R60S, in which the putative AA^- binding site was modified, is in full agreement with MD simulation results. It convincingly shows that the suggested AA^- binding region around R60 is available only in the $UCP2_h$ protein structure and that in UCP2-R60S the binding of AA^- is prevented. This shows that the $UCP2_h$ structure correctly predicts the binding of AA^- and is sensitive to selective mutations, such as R60S. In contrast, the similar binding of FA to the $UCP2_{NMR}$ structure, both the experimental one and after 2 μs of MD simulation time is not optimal, thus further questioning the stability and physiological relevance of experimental NMR structure.

The proposed MD model of $UCP2_h$ can largely contribute to the investigation of the UCP2-mediated proton transport mechanism in general and the protein binding site for FA in particular. Currently, two basic mechanisms are proposed for UCP1-mediated proton transport (see Introduction). A crucial difference between the "FA cycling" and "FA shuttle" hypotheses is a localization of the binding site for the activating FA. The "FA shuttle" model states that FA^- binds inside the pore from the cytosolic side of UCP1 and transfers protons by shuttling from the cytosolic to the matrix side of mitochondria [58]. In contrast, the "FA cycling" hypothesis suggests that FA^- binds to protein from the matrix side and is transported to the intermembrane space in its deprotonated form. Our initial results that show that FA can bind from the matrix site contradict the results of patch-clamp experiments on mitoplasts, showing the inability of UCP1 to bind FAs on the matrix side [18]. Further developments of the UCP2 model will pave the way towards clarifying this question and the complete transport mechanism of UCP2.

In conclusion, we propose the protocol involving homology modelling of UCP2 protein based on ANT structure and subsequent microsecond molecular dynamics simulations. It yields a functionally relevant structure, which can be used for future mechanistic studies of proton/FA transfer in mitochondria.

4. Materials and Methods

4.1. Simulation Details

To analyze the UCP2 protein structure, schematically represented in Figure S1, we simulated and compared two different structures of UCP2 protein and bovine ANT protein as a reference. The simulated structures include (a) the published NMR structure of mUCP2

(PDB code 2LCK, organism *Mus musculus*) [19] determined by NMR molecular fragment replacement, (b) the homology modeled structure of the primary sequence of UCP2 protein to crystallographic structure of ANT protein (PDB code 1OKC, 2.2 Å, high resolution, organism *Bos taurus*) [38] whose homology to mUCP2 is 24%, and (c) a crystallographic structure of ANT protein, which served as a template for the homology modeling of UCP2. Moreover, we used it as a reference structure for investigated UCP2 structures.

Three different systems were prepared using the membrane builder module of CHARMM-GUI (http://www.charmm-gui.org/) [59–61] from three UCP2/ANT starting protein structures immersed in 1,2-dioleoyl-sn-glycero-3-phosphocholine (DOPC) membrane containing 230 lipid molecules, 28,750 water molecules, and 15/19 chloride ions to neutralize the net charge of UCP2/ANT protein. DOPC was selected as it represents one of the main lipids of inner mitochondrial membranes. The homology modeled structure of UCP2 and missing residues (compared to the crystallographically determined structures of ANT)—residue 1 and residues 294–297 for 1OKC and residues 1–13 for 2LCK—were added using USCF Chimera program [62]. All arginines and lysines were prepared in their protonated forms, histidines and cysteines in their neutral forms, with glutamates and aspartates being in their deprotonated forms. CHARMM-GUI membrane builder minimization and equilibration procedure was used for all systems [59]. After equilibration, we simulated each system for 2.0 μs using unbiased all-atom molecular dynamics (MD) simulations in a periodic rectangular box of 9.5 nm $\times$ 9.5 nm $\times$ 13.5 nm with a time step of 2 fs with CHARMM36m force field [63] and TIP3P water model [64].

In order to check the potential binding site of the fatty acid anion in UCP2, we also simulated an additional four systems of UCP2 protein structures with added arachidonic acid anion (AA^-). AA was used since it shows a substantial effect on the proton conductance across mitochondrial membranes [10]. AA^- was added to the binding site arginine R60 in three systems: (a) the homology modelled structure of UCP2 ($UCP2_h$) simulated after 2 μs, (b) the NMR structure after initial equilibration which closely resembles experimental NMR structure, and (c) the NMR experimental structure ($UCP2_{NMR}$) simulated after 2 μs. In the fourth system, AA^- was added to the mutated binding site (R60 to S60) in $UCP2_h$ simulated after 2 μs ($UCP2\text{-}R60_h$). Each of the four AA^- containing systems mentioned above was simulated for 500 ns and analyzed further.

All production MD simulations were performed in the isobaric-isothermal ensemble (NPT) at T = 310 K, which was maintained via Nosé–Hoover thermostat [65,66] independently for three groups: DOPC, water/ions, and protein subsystem with a coupling constant of 1.0 ps^{-1}. The pressure was set to 1.013 bar and controlled with semi-isotropic Parrinello–Rahman barostat [67] with a time constant for pressure coupling of 5 ps^{-1}. Periodic boundary conditions (PBC) were imposed in all three directions, with long range electrostatic interactions calculated by the particle-mesh Ewald (PME) method [68] with real space Coulomb interactions cut-off at 1.2 nm using a Fourier spacing of 0.12 nm and Verlet cut-off scheme. All simulations were propagated with GROMACS 2018.4 software package [69] and visualized with VMD (Visualize Molecular Dynamics) program [70].

4.2. Homology Modeling

For the homology modeled structure, we used a sequence from 2LCK protein. Using Blast protein comparative structure modelling [71] in UniProt Protein Knowledgebase [72], we obtained the best matching tertiary structures: 1OKC, 2C3E, and 4C9G. The best matching tertiary structure 1OKC (lowest significance value—4.7 $\times$ 10^{-20} and highest alignment score—88) serves as a structure template. From the target-template sequence alignment, we generated 10 models and chose the model with the lowest discrete optimized protein energy (DOPE) score of 0.85 was chosen as the representative structure to carry out simulations. Homology models were constructed using programs MODELLER 9 [73] and Chimera 1.13.1 [62].

4.3. Permeability Calculations

We calculated osmotic permeabilities for four distinct UCP2 structures obtained from unbiased MD simulations, closely following the procedure illustrated by Zoonens et al. [20] and based on the algorithm described by Aksementiev and Schulten [50]. In this respect, the osmotic permeability of homologically modeled UCP2 was calculated using structures obtained after 2 μs of unbiased MD simulation, while the same property was determined for three distinct UCP2 NMR structures, i.e., structures obtained immediately after equilibration, after 0.2 μs and 2 μs of unbiased MD simulation. The four chosen systems were propagated in the NVT ensemble, where position restraints were applied on C_α atoms of the protein (500 kcal mol^{-1} nm^{-2}). The simulations were propagated for 10 ns with a 2 fs time step, with the last 5 ns used in the subsequent analysis. The simulation snapshots were saved every 250 steps, i.e., every 0.5 ps. PBC conditions were applied in all three directions and treated the long-range electrostatics using the PME method (see above for details).

The pore formed in the structure of both $UCP2_{NMR}$ and $UCP2_h$ possesses a complex topology; thus, careful choice of the region for the osmotic permeability calculations is necessary. Following the procedure of Zoonens et al. [20], we calculated the permeability taking into account only the central region of the pore. This region was rather well-preserved during simulations, and its topology relatively simple, i.e., it could be accurately described as being roughly cylindrical in nature. In this respect, the chosen region for the $UCP2_{NMR}$ structures was defined by two roughly coplanar rings, each consisting of six C_α atoms. Each C_α atom was chosen to belong to a different transmembrane helix present in the protein. The chosen C_α atoms form a bottom and a top ring, and belong to residues 34, 85, 137, 181, 239, and 274, and residues 20, 101, 120, 194, 227, 288, respectively. Due to a different topology of the $UCP2_h$ protein, we used a similar yet somewhat different choice of C_α atoms to delineate its central pore. In this respect, a bottom and a top ring were described using C_α atoms belonging to residues 34, 82, 137, 181, and 274, and residues 20, 101, 120, 192 and 288, respectively (see Figure S8).

Thus, the region encapsulated between the two chosen rings has the form of a cylinder with bases at the centers of mass of the bottom (R_0) and the top ring (R_1), respectively. The axis of this cylinder lies along the vector R_1–R_0. The radius of the cylinder (r = 2 nm) was chosen so that it is large enough to enclose all water molecules found in the analyzed pore of UCP2 protein.

Water molecules collective displacement within the protein pore at time $t + \Delta t$ of the MD simulation trajectory can now be calculated using the approach developed by Zhu et al. [74], i.e., via

$$n(t+\Delta t) = n(t) + \sum_{i \in S(t,t+\Delta t)} \left(\frac{\Delta \mathbf{r}_i \Delta \mathbf{e}}{L} \right)$$

where the union of all subsets of water molecules that are found inside the cylinder at time t and $t + \Delta t$ is denoted by $S(t + \Delta t)$, $\Delta \mathbf{r}_i$ represents the displacement of i-th water molecule in the time window t to $t + \Delta t$, $\mathbf{e}$ represents the unit vector along R_1–R_0. L denotes the length of a cylindrical region and is approximately equal to 1.95 nm in all considered cases. Importantly, displacements of water molecules that enter or exit the cylindrical region between the two consecutive frames were cut at the boundaries of the region in such a way that only the displacement of such water molecules inside the region is taken into account.

The collective diffusion coefficient of water inside the protein, D_n, was calculated using $\langle n(t) \rangle^2 = 2D_n t$, with the average being obtained over 100 subtrajectories, each being 50 ps in length (5 ns of overall post-equilibration simulation time/100), Figure S9. The osmotic permeability was estimated using $P_f = v_w D_n$, where the average volume of a single water molecule is denoted by v_w. Finally, thus obtained osmotic permeabilities were scaled by a factor of 1/2.87, since real water possesses larger viscosity compared to the used TIP3P water model [50].

4.4. Binding of ATP in the UCP2 Cavity

To inspect the geometry of the binding site of ATP in the UCP2 protein cavity, we performed a set of simulations (50 per inspected UCP2 structure). We initially placed the ATP molecule inside the UCP2 cavity for three distinct protein structures, $UCP2_h$, and two $UCP2_{NMR}$ structures. More precisely, to represent $UCP2_h$, we chose the structure obtained after 2 µs of its respective free MD simulation, while two distinct $UCP2_{NMR}$ structures, namely the structures obtained after 20 ns and after 2 µs of their respective free MD simulation, were utilized to inspect the behavior of ATP in the cavity of $UCP2_{NMR}$. For each UCP2 structure, 50 different MD simulations in the duration of 20 ns each were performed (overall 1 µs per UCP2 structure). The ATP starting position was maintained for each of the 50 simulations, with the initial velocities being randomly generated, following the Boltzmann distribution. Thus, while the starting structure in each 50 simulation sets (for each investigated UCP2 structure) is represented by the same point in the conformational phase space, its position in the momentum phase space is different, representing overall distinct starting structures. To inspect whether the initial conditions biased the obtained results, we performed additional simulations (again 20 ns each) for the $UCP2_h$ structure, where we placed ATP molecule in five different spots in the cavity of $UCP2_h$ and performed 10 simulations for each starting configuration. In all simulations mentioned above, only the last 10 ns were used in the analysis. The first 10 ns were omitted (equilibration time). CHARMM36m force field parameters were used to describe ATP moiety together with the aforementioned parameters of UCP2 and DOPC lipids. The simulations were performed in the NVT ensemble, with all other MD simulation parameters being identical to the ones applied in the long 2 µs simulations (see Simulation Details).

4.5. Chemicals

The 1,2-dioleoyl-sn-glycero-3-phosphocholine(DOPC), 1,2-dioleoyl-sn-glycero-3-phosphoetha-nolamine (DOPE), cardiolipin (CL) from bovine heart, arachidonic acid (AA), Triton X-114 octylpolyoxyethylene, dithiothreitol (DTT), bovine serum albumin (BSA), adenosine and guanosine triphosphate (ATP and GTP), sodium sulfate (Na_2SO_4), diammonium hydrogen phosphate ($(NH_4)_2HPO_4$), 2-(N-morpholino)ethanesulfonic acid (MES), 2-Amino-2-(hydroxymethyl)propane-1,3-diol (Tris), ethylene glycol-bis(β-aminoethyl ether)-N,N,N′,N′-tetraacetic acid (EGTA), hexane, hexadecane and sodium dodecyl sulfate (SDS) were obtained from Sigma-Aldrich (Munich, Germany). Chloroform was from Merck KGaA (Darmstadt, Germany).

4.6. Cloning, Mutation and Expression of mUCP2 and Reconstitution into Liposomes

Mouse UCP2 (mUCP2) was cloned and expressed, as described previously [75,76]. In brief, the ORF of mUCP2 was inserted into the pET24a- expression plasmid. For expression of the protein the plasmid of wild type mUCP2 was transferred into *E. coli* cells (strain Rosetta) and grown to reach OD_{600nm} between 0.3 and 0.5. The protein expression was then induced by adding 1 mM isopropyl-β-D-thiogalactopyranoside (IPTG). *E. coli* cells were incubated for 3 h before harvesting by centrifugation. Inclusion bodies (IB) containing the expressed proteins were collected by disruption of cells using a French press following centrifugation at 14,000× *g* [77].

In vitro site-directed mutagenesis was carried out on expression plasmids containing the cDNA of mUCP2 as templates. The mutation was introduced with a designed oligonucleotide to alter the R60 (CGT) to S (AGT) using Q5 site-directed mutagenesis kit (New England Biolabs GmbH, Frankfurt am Main, Germany) and confirmed by sequencing.

Recombinant mUCP2WT and mUCP2R60S were purified and refolded from inclusion bodies, and reconstituted into liposomes according to the previously described protocol [54]. In brief, mUCP3 were solubilized from IB using 2% sarcosyl and 1 mM DTT. 100 mg *E. coli* polar lipid (Avanti polar lipids, Alabaster, AL, USA), 300 µg Triton X-114, 75 µg octylpolyoxyethylene and 2 mM GTP were added to the solubilized mUCP2. Sarcosyl and GTP were removed by dialysis with the assay buffer (50 mM Na_2SO_4, 10 mM Tris, 10 mM MES and 0.6 mM EGTA, pH 7.34). The sample was passed through a hydroxyapatite column (Bio-Rad, Laboratories, Inc., Feldkirchen, Germany) to remove decomposed proteins. Nonionic detergents were eliminated using Bio Beads (Bio-Rad, Germany). The purity of the recombinant proteins was verified by silver staining (Figure S10). The correct folding was proved by the activity assay—protein activation or inhibition.

4.7. Measurements of Electrical Parameters of Membranes Reconstituted with mUCP2

Planar lipid bilayers were formed from (proteo-) liposomes [78,79] made of 45:45:10 mol.% DOPC:DOPE:CL. Lipid concentration was 1.5 mg/mL and protein to lipid ratio—4 μg per mg of lipid. Arachidonic acid (AA) at a concentration of 15 mol.% was directly added to the lipid phase before membrane formation. Buffer contained 50 mM Na_2SO_4, 10 mM Tris, 10 mM MES and 0.6 mM EGTA at pH = 7.34 and T = 306 K. Proper membrane formation was verified by measuring specific capacitance (C = 718 ± 34 nF/cm^2) that was independent of protein, AA and ATP content. Current−voltage (I-U) measurements were performed with a patch-clamp amplifier (EPC 10 USB, HEKA Elektronik Dr Schulze GmbH, Lambrecht, Germany). Total membrane conductance at 0 mV was obtained from the slope of a linear fit of experimental data at applied voltages from −50 mV to +50 mV (Figure S7). ATP was dissolved in a buffer solution to a concentration of 400 mM and the stock solution pH was adjusted to 7.34. The volume of 3.75 μL of the stock solution was added to 750 μL buffer solution for a final concentration of 2 mM ATP. Incubation time was 30 min at T = 306 K. Data were analyzed using Sigma Plot (Systat Software GmbH, Erkrath, Germany).

Supplementary Materials: The following are available online at https://www.mdpi.com/1422-0067/22/3/1214/s1. Figure S1: A schematic representation of UCP2 protein in the inner mitochondrial membrane, Figure S2: Time propagation of the secondary structure in simulations of UCP2 protein, Figure S3: Analysis of the EG motif in UCP2 and ANT structures, Figure S4: Time evolution of z-averaged water number density in simulations of UCP2 protein, Figure S5: Alignment of the primary sequences of UCP1, UCP2 and UCP3 proteins, Figure S6: Analysis of ATP binding in $UCP2_h$, $UCP2_{NMR}$ and ANT structures, Figure S7: Experimental measurements of UCP2 conductance, Figure S8: Cylindrical region of UCP2 used for permeability calculations, Figure S9: Mean square displacements (MSDs) for diffusion coefficient calculations, Figure S10: Representative silverstaining of murine UCP2WT and UCP2R60S, Table S1: Number of water molecules passing through the membrane for $UCP2_h$, $UCP2_{NMR}$ and ANT.

Author Contributions: Conceptualization, M.V.; Funding acquisition, E.E.P. and M.V.; Investigation, S.Š., Z.B., J.K. and M.V.; Project administration, M.V.; Resources, E.E.P. and M.V.; Supervision, E.E.P. and M.V.; Writing—original draft, S.Š. and M.V.; Writing—review & editing, S.Š., Z.B., J.K., E.E.P. and M.V. All authors have read and agreed to the published version of the manuscript.

Funding: This research was supported by the Croatian Science Foundation (Project No. IP-2019-04-3804 to M.V.) and the Austrian Research Fund (FWF, P31559 to E.E.P.). We thank the computer cluster Isabella based in SRCE—the University of Zagreb, University Computing Centre for computational resources. M.V. was supported by the UOCHB Sabbatical visit program. Open Access was funded by the Austrian Science Fund (FWF).

Institutional Review Board Statement: Not Applicable.

Informed Consent Statement: Not Applicable.

Data Availability Statement: The datasets generated and/or analyzed during this study are available from the corresponding authors on reasonable request.

Acknowledgments: The authors thank the reviewers for their useful comments. We thank Sarah Bardakji (Vetmeduni Vienna, Austria) for excellent technical assistance.

Conflicts of Interest: The authors declare no conflict of interest.

References

1. Skulachev, V.P. Fatty acid circuit as a physiological mechanism of uncoupling of oxidative phosphorylation. *FEBS Lett.* **1991**, *294*, 158–162. [CrossRef]
2. Krauss, S.; Zhang, C.Y.; Lowell, B.B. The mitochondrial uncoupling-protein homologues. *Nat. Rev. Mol. Cell Biol.* **2005**, *6*, 248–261. [CrossRef] [PubMed]
3. Ježek, P.; Holendová, B.; Garlid, K.D.; Jabůrek, M. Mitochondrial Uncoupling Proteins: Subtle Regulators of Cellular Redox Signaling. *Antioxid. Redox Signal.* **2018**, *29*, 667–714. [CrossRef] [PubMed]
4. Vozza, A.; Parisi, G.; De Leonardis, F.; Lasorsa, F.M.; Castegna, A.; Amorese, D.; Marmo, R.; Calcagnile, V.M.; Palmieri, L.; Ricquier, D.; et al. UCP2 transports C4 metabolites out of mitochondria, regulating glucose and glutamine oxidation. *Proc. Natl. Acad. Sci. USA* **2014**, *111*, 960–965. [CrossRef] [PubMed]
5. Andreyev, A.Y.; Bondareva, T.O.; Dedukhova, V.I.; Mokhova, E.N.; Skulachev, V.P.; Tsofina, L.M.; Volkov, N.I.; Vygodina, T.V. The ATP/ADP-antiporter is involved in the uncoupling effect of fatty acids on mitochondria. *Eur. J. Biochem.* **1989**, *182*, 585–592. [CrossRef]
6. Bertholet, A.M.; Chouchani, E.T.; Kazak, L.; Angelin, A.; Fedorenko, A.; Long, J.Z.; Vidoni, S.; Garrity, R.; Cho, J.; Terada, N.; et al. H+ transport is an integral function of the mitochondrial ADP/ATP carrier. *Nature* **2019**, *571*, 515–520. [CrossRef]
7. Brustovetsky, N.; Klingenberg, M. The reconstituted ADP/ATP carrier can mediate H+ transport by free fatty acids, which is further stimulated by mersalyl. *J. Biol. Chem.* **1994**, *269*, 27329–27336. [CrossRef]
8. Garlid, K.D.; Orosz, D.E.; Modrianska, M.; Vassanelli, S.; Jezek, P. On the mechanism of fatty acid-induced proton transport by mitochondrial uncoupling protein. *J. Biol. Chem.* **1996**, *271*, 2615–2620. [CrossRef]
9. Beck, V.; Jaburek, M.; Demina, T.; Rupprecht, A.; Porter, R.K.; Jezek, P.; Pohl, E.E. Polyunsaturated fatty acids activate human uncoupling proteins 1 and 2 in planar lipid bilayers. *FASEB J.* **2007**, *21*, 1137–1144. [CrossRef]
10. Malingriaux, E.A.; Rupprecht, A.; Gille, L.; Jovanovic, O.; Jezek, P.; Jaburek, M.; Pohl, E.E. Fatty Acids are Key in 4-Hydroxy-2-Nonenal-Mediated Activation of Uncoupling Proteins 1 and 2. *PLoS ONE* **2013**, *8*, e77786. [CrossRef]
11. Kamp, F.; Hamilton, J.A. pH Gradients across Phospholipid Membranes Caused by Fast Flip-flop of Un-ionized Fatty Acids. *Proc. Natl. Acad. Sci. USA* **1992**, *89*, 11367–11370. [CrossRef] [PubMed]
12. Kamp, F.; Zakim, D.; Zhang, F.; Noy, N.; Hamilton, J.A. Fatty acid flip-flop in phospholipid bilayers is extremely fast. *Biochemistry* **1995**, *34*, 11928–11937. [CrossRef] [PubMed]
13. Škulj, S.; Vazdar, M. Calculation of Apparent p K a Values of Saturated Fatty Acids with Different Lengths in DOPC Phospholipid Bilayers. *Phys. Chem. Chem. Phys.* **2019**, *21*, 10052–10060. [CrossRef] [PubMed]
14. Wu, X.; Gale, P.A. Small-Molecule Uncoupling Protein Mimics: Synthetic Anion Receptors as Fatty Acid-Activated Proton Transporters. *J. Am. Chem. Soc.* **2016**, *138*, 16508–16514. [CrossRef]
15. Jovanović, O.; Pashkovskaya, A.A.; Annibal, A.; Vazdar, M.; Burchardt, N.; Sansone, A.; Gille, L.; Fedorova, M.; Ferreri, C.; Pohl, E.E. The molecular mechanism behind reactive aldehyde action on transmembrane translocations of proton and potassium ions. *Free Radic. Biol. Med.* **2015**, *89*, 1067–1076. [CrossRef]
16. Winkler, E.; Klingenberg, M. Effect of fatty acids on H+ transport activity of the reconstituted uncoupling protein. *J. Biol. Chem.* **1994**, *269*, 2508–2515. [CrossRef]
17. Klingenberg, M. UCP1-A sophisticated energy valve. *Biochimie* **2017**, *134*, 19–27. [CrossRef]
18. Fedorenko, A.; Lishko, P.V.; Kirichok, Y. Mechanism of fatty-acid-dependent UCP1 uncoupling in brown fat mitochondria. *Cell* **2012**, *151*, 400–413. [CrossRef]
19. Berardi, M.J.; Shih, W.M.; Harrison, S.C.; Chou, J.J. Mitochondrial uncoupling protein 2 structure determined by NMR molecular fragment searching. *Nature* **2011**, *476*, 109–114. [CrossRef]
20. Zoonens, M.; Comer, J.; Masscheleyn, S.; Pebay-Peyroula, E.; Chipot, C.; Miroux, B.; Dehez, F. Dangerous Liaisons between Detergents and Membrane Proteins. The Case of Mitochondrial Uncoupling Protein 2. *J. Am. Chem. Soc.* **2013**, *135*, 15174–15182. [CrossRef]
21. Dehez, F.; Schanda, P.; King, M.S.; Kunji, E.R.S.; Chipot, C. Mitochondrial ADP/ATP Carrier in Dodecylphosphocholine Binds Cardiolipins with Non-native Affinity. *Biophys. J.* **2017**, *113*, 2311–2315. [CrossRef] [PubMed]
22. Kurauskas, V.; Hessel, A.; Ma, P.; Lunetti, P.; Weinhäupl, K.; Imbert, L.; Brutscher, B.; King, M.S.; Sounier, R.; Dolce, V.; et al. How Detergent Impacts Membrane Proteins: Atomic-Level Views of Mitochondrial Carriers in Dodecylphosphocholine. *J. Phys. Chem. Lett.* **2018**, *9*, 933–938. [CrossRef] [PubMed]
23. Kurauskas, V.; Hessel, A.; Dehez, F.; Chipot, C.; Bersch, B.; Schanda, P. Dynamics and interactions of AAC3 in DPC are not functionally relevant. *Nat. Struct. Mol. Biol.* **2018**, *25*, 745–747. [CrossRef] [PubMed]
24. King, M.S.; Crichton, P.G.; Ruprecht, J.J.; Kunji, E.R.S. Concerns with yeast mitochondrial ADP/ATP carrier's integrity in DPC. *Nat. Struct. Mol. Biol.* **2018**, *25*, 747–749. [CrossRef] [PubMed]
25. Berardi, M.J.; Chou, J.J. Fatty acid flippase activity of UCP2 is essential for its proton transport in mitochondria. *Cell Metab.* **2014**, *20*, 541–552. [CrossRef] [PubMed]
26. Brüschweiler, S.; Yang, Q.; Run, C.; Chou, J.J. Substrate-modulated ADP/ATP-transporter dynamics revealed by NMR relaxation dispersion. *Nat. Struct. Mol. Biol.* **2015**, *22*, 636–641. [CrossRef] [PubMed]

27. Yang, Q.; Brüschweiler, S.; Zhao, L.; Chou, J.J. Reply to 'Concerns with yeast mitochondrial ADP/ATP carrier's integrity in DPC' and 'Dynamics and interactions of AAC3 in DPC are not functionally relevant'. *Nat. Struct. Mol. Biol.* **2018**, *25*, 749–750. [CrossRef]
28. Lindahl, E.; Sansom, M.S. Membrane proteins: Molecular dynamics simulations. *Curr. Opin. Struct. Biol.* **2008**, *18*, 425–431. [CrossRef]
29. Weng, J.; Wang, W. Molecular dynamics simulation of membrane proteins. *Adv. Exp. Med. Biol.* **2014**, *805*, 305–329. [CrossRef]
30. Dutagaci, B.; Heo, L.; Feig, M. Structure refinement of membrane proteins via molecular dynamics simulations. *Proteins Struct. Funct. Bioinform.* **2018**, *86*, 738–750. [CrossRef]
31. Almén, M.S.; Nordström, K.J.V.; Fredriksson, R.; Schiöth, H.B. Mapping the human membrane proteome: A majority of the human membrane proteins can be classified according to function and evolutionary origin. *BMC Biol.* **2009**, *7*, 50. [CrossRef] [PubMed]
32. Newport, T.D.; Sansom, M.S.P.; Stansfeld, P.J. The MemProtMD database: A resource for membrane-embedded protein structures and their lipid interactions. *Nucleic Acids Res.* **2019**, *47*, D390–D397. [CrossRef] [PubMed]
33. Chipot, C.; Dehez, F.; Schnell, J.R.; Zitzmann, N.; Pebay-Peyroula, E.; Catoire, L.J.; Miroux, B.; Kunji, E.R.S.; Veglia, G.; Cross, T.A.; et al. Perturbations of Native Membrane Protein Structure in Alkyl Phosphocholine Detergents: A Critical Assessment of NMR and Biophysical Studies. *Chem. Rev.* **2018**, *118*, 3559–3607. [CrossRef] [PubMed]
34. Gromiha, M.M.; Nagarajan, R.; Selvaraj, S. Protein Structural Bioinformatics: An Overview. In *Encyclopedia of Bioinformatics and Computational Biology*; Elsevier: Amsterdam, The Netherlands, 2019; pp. 445–459.
35. Abeln, S.; Feenstra, K.A.; Heringa, J. Protein Three-Dimensional Structure Prediction. In *Encyclopedia of Bioinformatics and Computational Biology*; Elsevier: Amsterdam, The Netherlands, 2019; pp. 497–511.
36. Zhao, L.; Wang, S.; Zhu, Q.; Wu, B.; Liu, Z.; OuYang, B.; Chou, J.J. Specific Interaction of the Human Mitochondrial Uncoupling Protein 1 with Free Long-Chain Fatty Acid. *Structure* **2017**, *25*, 1371–1379.e3. [CrossRef] [PubMed]
37. Ardalan, A.; Sowlati-Hashjin, S.; Uwumarenogie, S.O.; Fish, M.; Mitchell, J.; Karttunen, M.; Smith, M.D.; Jelokhani-Niaraki, M. Functional Oligomeric Forms of Uncoupling Protein 2: Strong Evidence for Asymmetry in Protein and Lipid Bilayer Systems. *J. Phys. Chem. B* **2021**, *125*, 169–183. [CrossRef] [PubMed]
38. Pebay-Peyroula, E.; Dahout-Gonzalez, C.; Kahn, R.; Trézéguet, V.; Lauquin, G.J.-M.; Brandolin, G. Structure of mitochondrial ADP/ATP carrier in complex with carboxyatractyloside. *Nature* **2003**, *426*, 39–44. [CrossRef]
39. Monné, M.; Palmieri, F. Antiporters of the mitochondrial carrier family. In *Current Topics in Membranes*; Academic Press Inc.: Cambridge, MA, USA, 2014; Volume 73, pp. 289–320.
40. Pietropaolo, A.; Pierri, C.L.; Palmieri, F.; Klingenberg, M. The switching mechanism of the mitochondrial ADP/ATP carrier explored by free-energy landscapes. *Biochim. Biophys. Acta-Bioenerg.* **2016**, *1857*, 772–781. [CrossRef]
41. Ruprecht, J.J.; Kunji, E.R. Structural changes in the transport cycle of the mitochondrial ADP/ATP carrier. *Curr. Opin. Struct. Biol.* **2019**, *57*, 135–144. [CrossRef]
42. Ruprecht, J.J.; King, M.S.; Zögg, T.; Aleksandrova, A.A.; Pardon, E.; Crichton, P.G.; Steyaert, J.; Kunji, E.R.S. The Molecular Mechanism of Transport by the Mitochondrial ADP/ATP Carrier. *Cell* **2019**, *176*, 435–447.e15. [CrossRef]
43. Pohl, E.E.; Rupprecht, A.; Macher, G.; Hilse, K.E. Important trends in UCP3 investigation. *Front. Physiol.* **2019**, *10*, 470. [CrossRef]
44. Škulj, S.; Brkljača, Z.; Vazdar, M. Molecular Dynamics Simulations of the Elusive Matrix-Open State of Mitochondrial ADP/ATP Carrier. *Isr. J. Chem.* **2020**, *60*, 735–743. [CrossRef]
45. Bahar, I.; Lezon, T.R.; Yang, L.-W.; Eyal, E. Global Dynamics of Proteins: Bridging Between Structure and Function. *Annu. Rev. Biophys.* **2010**, *39*, 23–42. [CrossRef] [PubMed]
46. Skulachev, V.P. Anion carriers in fatty acid-mediated physiological uncoupling. *J. Bioenerg. Biomembr.* **1999**, *31*, 431–445. [CrossRef] [PubMed]
47. Ruprecht, J.J.; Hellawell, A.M.; Harding, M.; Crichton, P.G.; McCoy, A.J.; Kunji, E.R.S. Structures of yeast mitochondrial ADP/ATP carriers support a domain-based alternating-access transport mechanism. *Proc. Natl. Acad. Sci. USA* **2014**, *111*, E426–E434. [CrossRef] [PubMed]
48. Burykin, A.; Warshel, A. What Really Prevents Proton Transport through Aquaporin? Charge Self-Energy versus Proton Wire Proposals. *Biophys. J.* **2003**, *85*, 3696–3706. [CrossRef]
49. Eisenberg, B. Why Can't Protons Move through Water Channels? *Biophys. J.* **2003**, *85*, 3427–3428. [CrossRef]
50. Aksimentiev, A.; Schulten, K. Imaging α-hemolysin with molecular dynamics: Ionic conductance, osmotic permeability, and the electrostatic potential map. *Biophys. J.* **2005**, *88*, 3745–3761. [CrossRef]
51. Ježek, P.; Modrianský, M.; Garlid, K.D. Inactive fatty acids are unable to flip-flop across the lipid bilayer. *FEBS Lett.* **1997**, *408*, 161–165. [CrossRef]
52. Lee, Y.; Willers, C.; Kunji, E.R.S.; Crichton, P.G. Uncoupling protein 1 binds one nucleotide per monomer and is stabilized by tightly bound cardiolipin. *Proc. Natl. Acad. Sci. USA* **2015**, *112*, 6973–6978. [CrossRef]
53. Modrianský, M.; Murdza-Inglis, D.L.; Patel, H.V.; Freeman, K.B.; Garlid, K.D. Identification by site-directed mutagenesis of three arginines in uncoupling protein that are essential for nucleotide binding and inhibition. *J. Biol. Chem.* **1997**, *272*, 24759–24762. [CrossRef]
54. Macher, G.; Koehler, M.; Rupprecht, A.; Kreiter, J.; Hinterdorfer, P.; Pohl, E.E. Inhibition of mitochondrial UCP1 and UCP3 by purine nucleotides and phosphate. *Biochim. Biophys. Acta-Biomembr.* **2018**, *1860*, 664–672. [CrossRef] [PubMed]

55. Garlid, K.D.; Jabůrek, M.; Ježek, P. The mechanism of proton transport mediated by mitochondrial uncoupling proteins. *FEBS Lett.* **1998**, *438*, 10–14. [CrossRef]
56. Wang, Y.; Tajkhorshid, E. Electrostatic funneling of substrate in mitochondrial inner membrane carriers. *Proc. Natl. Acad. Sci. USA* **2008**, *105*, 9598–9603. [CrossRef] [PubMed]
57. Dehez, F.; Pebay-Peyroula, E.; Chipot, C. Binding of ADP in the mitochondrial ADP/ATP carrier is driven by an electrostatic funnel. *J. Am. Chem. Soc.* **2008**, *130*, 12725–12733. [CrossRef]
58. Bertholet, A.M.; Kirichok, Y. UCP1: A transporter for H+ and fatty acid anions. *Biochimie* **2017**, *134*, 28–34. [CrossRef]
59. Jo, S.; Kim, T.; Im, W. Automated builder and database of protein/membrane complexes for molecular dynamics simulations. *PLoS ONE* **2007**, *2*, e880. [CrossRef]
60. Wu, E.L.; Cheng, X.; Jo, S.; Rui, H.; Song, K.C.; Dávila-Contreras, E.M.; Qi, Y.; Lee, J.; Monje-Galvan, V.; Venable, R.M.; et al. CHARMM-GUI membrane builder toward realistic biological membrane simulations. *J. Comput. Chem.* **2014**, *35*, 1997–2004. [CrossRef]
61. Lee, J.; Cheng, X.; Swails, J.M.; Yeom, M.S.; Eastman, P.K.; Lemkul, J.A.; Wei, S.; Buckner, J.; Jeong, J.C.; Qi, Y.; et al. CHARMM-GUI Input Generator for NAMD, GROMACS, AMBER, OpenMM, and CHARMM/OpenMM Simulations Using the CHARMM36 Additive Force Field. *J. Chem. Theory Comput.* **2016**, *12*, 405–413. [CrossRef]
62. Pettersen, E.F.; Goddard, T.D.; Huang, C.C.; Couch, G.S.; Greenblatt, D.M.; Meng, E.C.; Ferrin, T.E. UCSF Chimera-A visualization system for exploratory research and analysis. *J. Comput. Chem.* **2004**, *25*, 1605–1612. [CrossRef]
63. Huang, J.; Rauscher, S.; Nawrocki, G.; Ran, T.; Feig, M.; De Groot, B.L.; Grubmüller, H.; MacKerell, A.D. CHARMM36m: An improved force field for folded and intrinsically disordered proteins. *Nat. Methods* **2016**, *14*, 71–73. [CrossRef]
64. Jorgensen, W.L.; Chandrasekhar, J.; Madura, J.D.; Impey, R.W.; Klein, M.L. Comparison of simple potential functions for simulating liquid water. *J. Chem. Phys.* **1983**, *79*, 926–935. [CrossRef]
65. Nosé, S. A molecular dynamics method for simulations in the canonical ensemble. *Mol. Phys.* **1984**, *52*, 255–268. [CrossRef]
66. Hoover, W.G. Canonical dynamics: Equilibrium phase-space distributions. *Phys. Rev. A* **1985**, *31*, 1695–1697. [CrossRef] [PubMed]
67. Parrinello, M.; Rahman, A. Polymorphic transitions in single crystals: A new molecular dynamics method. *J. Appl. Phys.* **1981**, *52*, 7182–7190. [CrossRef]
68. Essmann, U.; Perera, L.; Berkowitz, M.L.; Darden, T.; Lee, H.; Pedersen, L.G. A smooth particle mesh Ewald method. *J. Chem. Phys.* **1995**, *103*, 8577–8593. [CrossRef]
69. Abraham, M.J.; Murtola, T.; Schulz, R.; Pall, S.; Smith, J.C.; Hess, B.; Lindah, E. Gromacs: High performance molecular simulations through multi-level parallelism from laptops to supercomputers. *SoftwareX* **2015**, *1–2*, 19–25. [CrossRef]
70. Humphrey, W.; Dalke, A.; Schulten, K. VMD: Visual molecular dynamics. *J. Mol. Graph.* **1996**, *14*, 33–38. [CrossRef]
71. Altschul, S.F.; Gish, W.; Miller, W.; Myers, E.W.; Lipman, D.J. Basic local alignment search tool. *J. Mol. Biol.* **1990**, *215*, 403–410. [CrossRef]
72. Pundir, S.; Martin, M.J.; O'Donovan, C. UniProt protein knowledgebase. In *Methods in Molecular Biology*; Humana Press Inc.: Totowa, NJ, USA, 2017; Volume 1558, pp. 41–55.
73. Webb, B.; Sali, A. Comparative protein structure modeling using MODELLER. *Curr. Protoc. Bioinform.* **2014**, *47*, 5.6.1–5.6.32. [CrossRef]
74. Zhu, F.; Tajkhorshid, E.; Schulten, K. Collective diffusion model for water permeation through microscopic channels. *Phys. Rev. Lett.* **2004**, *93*, 224501. [CrossRef]
75. Rupprecht, A.; Sokolenko, E.A.; Beck, V.; Ninnemann, O.; Jaburek, M.; Trimbuch, T.; Klishin, S.S.; Jezek, P.; Skulachev, V.P.; Pohl, E.E. Role of the transmembrane potential in the membrane proton leak. *Biophys. J.* **2010**, *98*, 1503–1511. [CrossRef] [PubMed]
76. Hilse, K.E.; Rupprecht, A.; Kalinovich, A.; Shabalina, I.G.; Pohl, E.E. Quantification of Mitochondrial UCP3 Expression in Mouse Tissues. *Biophys. J.* **2014**, *106*, 592a. [CrossRef]
77. Hilse, K.E.; Kalinovich, A.V.; Rupprecht, A.; Smorodchenko, A.; Zeitz, U.; Staniek, K.; Erben, R.G.; Pohl, E.E. The expression of UCP3 directly correlates to UCP1 abundance in brown adipose tissue. *Biochim. Biophys. Acta-Bioenerg.* **2016**, *1857*, 72–78. [CrossRef]
78. Beck, V.; Jabůrek, M.; Breen, E.P.; Porter, R.K.; Ježek, P.; Pohl, E.E. A new automated technique for the reconstitution of hydrophobic proteins into planar bilayer membranes. Studies of human recombinant uncoupling protein 1. *Biochim. Biophys. Acta-Bioenerg.* **2006**, *1757*, 474–479. [CrossRef] [PubMed]
79. Kreiter, J.; Pohl, E.E. A Micro-agar Salt Bridge Electrode for Analyzing the Proton Turnover Rate of Recombinant Membrane Proteins. *J. Vis. Exp.* **2019**, *143*. [CrossRef]

International Journal of
Molecular Sciences

MDPI

Review

The Tiny Companion Matters: The Important Role of Protons in Active Transports in Plants

Yee-Shan Ku [1,*], Sau-Shan Cheng [1], Ming-Sin Ng [1], Gyuhwa Chung [2] and Hon-Ming Lam [1,*]

1 Centre for Soybean Research of the State Key Laboratory of Agrobiotechnology and School of Life Sciences, The Chinese University of Hong Kong, Hong Kong, China; chengsaushan@yahoo.com (S.-S.C.); sammingsin0212@gmail.com (M.-S.N.)
2 Department of Biotechnology, Chonnam National University, Yeosu 59626, Korea; chung@chonnam.ac.kr
* Correspondence: ysamyku@cuhk.edu.hk (Y.-S.K.); honming@cuhk.edu.hk (H.-M.L.); Tel.: +852-3943-8132 (Y.-S.K.); +852-3943-6336 (H.-M.L.)

Abstract: In plants, the translocation of molecules, such as ions, metabolites, and hormones, between different subcellular compartments or different cells is achieved by transmembrane transporters, which play important roles in growth, development, and adaptation to the environment. To facilitate transport in a specific direction, active transporters that can translocate their substrates against the concentration gradient are needed. Examples of major active transporters in plants include ATP-binding cassette (ABC) transporters, multidrug and toxic compound extrusion (MATE) transporters, monosaccharide transporters (MSTs), sucrose transporters (SUTs), and amino acid transporters. Transport via ABC transporters is driven by ATP. The electrochemical gradient across the membrane energizes these secondary transporters. The pH in each cell and subcellular compartment is tightly regulated and yet highly dynamic, especially when under stress. Here, the effects of cellular and subcellular pH on the activities of ABC transporters, MATE transporters, MSTs, SUTs, and amino acid transporters will be discussed to enhance our understanding of their mechanics. The relation of the altered transporter activities to various biological processes of plants will also be addressed. Although most molecular transport research has focused on the substrate, the role of protons, the tiny counterparts of the substrate, should also not be ignored.

Keywords: ATP-binding cassette (ABC) transporter; multidrug and toxic compound extrusion (MATE) transporter; monosaccharide transporter (MST); sucrose transporter (SUT); amino acid transporter; detoxification; nutrient transport; stress adaptation; proton gradient; cellular pH

Citation: Ku, Y.-S.; Cheng, S.-S.; Ng, M.-S.; Chung, G.; Lam, H.-M. The Tiny Companion Matters: The Important Role of Protons in Active Transports in Plants. *Int. J. Mol. Sci.* **2022**, *23*, 2824. https://doi.org/10.3390/ijms23052824

Academic Editor: Masoud Jelokhani-Niaraki

Received: 29 November 2021
Accepted: 2 March 2022
Published: 4 March 2022

1. Introduction

Transporters play important roles in the transport of nutrients, hormones, and metabolites for the purposes of growth, development, and adaptation to stresses. To deliver the substrates in a specific direction, active transporters, which can transport substrates against their concentration gradients, are required. Transmembrane transporters mediate the transport of substrates across biological membranes. The polypeptides of transmembrane transporters contain transmembrane segments (TMSs), which are integrated in the membrane [1]. The segments of the integral membrane proteins embedded in the membrane are called transmembrane domains (TMDs) [2]. Such active transport is often driven by the proton gradient across the cellular/subcellular membrane. Therefore, changes in the cellular/subcellular pH could regulate the transport activities and determine the direction of transport. In each subcellular compartment, the pH is highly regulated and yet dynamic. It has been reported that stresses result in changes in cellular pH, which can form part of the stress signal [3]. For example, it was reported that flooding and drought stress induced an increase in the pH of xylem sap [3]; ionic stress induced an increase in cytosolic pH and a decrease in vacuolar pH [4]; fungal infection induced an increase in the pH of apoplastic sap; and pattern-triggered immunity (PTI) induced a decrease in cytosolic pH [5]. The changes in pH in different cellular compartments may imply the change in pH difference as well as the change in electrical potential difference across the biological membrane. Furthermore, the pH itself could also affect the activity of the transporter and the protonation state of the substrates to be transported. Altogether, these factors influence the transport efficiency and, thus, bring forth the physiological regulations.

In plants, cellular pH has been reported to be mainly regulated by proton pumps including the plasma membrane ATPase (PM-ATPase), the vacuolar-type ATPase (V-ATPase), the vacuolar pyrophosphatase (V-PPase) [6,7] and Na^+/H^+ antiporter (NHX) [8]. PM-ATPase hydrolyzes ATP to release H^+, which is then exported out of the cell by PM-ATPase [9]. Such transport of H^+ results in the proton gradient and electrical potential gradient across the membrane [9,10]. Under salt stress, it has been suggested that the PM-ATPase creates the electrochemical proton gradient to enable the extrusion of Na^+ out of the cell or the intrusion of Na^+ inside the vacuole by NHX [11]. V-ATPase has been known to be responsible for vacuolar acidification and create electrochemical proton gradient across the vacuolar membrane to energize substrate transport in and out of the vacuole [12,13]. In addition, V-PPase was suggested to have a higher vacuolar acidification potential compared to V-ATPase and also contribute to energize various transporters, such as Ca^{2+}/H^+, Na^+/H^+ and Zn^{2+}/H^+ exchangers and phosphate transporter, at the vacuolar membrane [14]. The pH of various cellular compartments is associated with the activities of proton pumps and transporters that transport substrates in exchange of H^+.

Phytohormone transporters, alkaloid transporters, ion and ion chelator transporters, sugar transporters, and amino acid transporters are common active transporters in plants. Examples of phytohormone transporters, alkaloid transporters, and ion and ion chelator transporters include ATP-binding cassette (ABC) transporters and Multidrug Additionally, Toxic compound Extrusion (MATE) transporters, while examples of sugar transporters include MonoSaccharide Transporters (MSTs) and SUcrose Transporters (SUTs). These transporters have been reported to play important roles in various biological processes including cellular detoxification, nutrient transport, and stress adaptation. Previous reports have largely focused on the transport of substrates such as metabolites, sugars, and amino acids by these transporters. However, considering the mechanics of the transport activities, the protons required in exchange of these substrates also deserve the same attention. The awareness of the role of protons, and therefore pH, in transport activities will bring forth a more comprehensive understanding of these transporters.

2. Phytohormone Transporters

ABC transporters and MATE transporters are the two major types of multidrug transporters that transport various substrates including phytohormones, alkaloids, ion, and ion chelators [15].

2.1. Overview of ABC Transporters

The ATP-binding cassette (ABC) transporters belong to one of the largest superfamilies of transporters and are present in many organisms ranging from prokaryotes to human [16]. Driven by the free energy change associated with ATP hydrolysis, ABC transporters have been reported to drive the export or import of various substrates across biological membranes against the electrochemical gradient [17]. Based on the protein structure, ABC transporters can be classified into three structural types: the full transporter having two TMDs and two nucleotide-binding cytosolic domains (NBDs), the half transporter having one TMD and one NBD, and the third type of transporter having no TMD and two NBDs [18]. Half transporters could dimerize to form virtual full transporters [18]. The NBD of ABC transporters consists of the signature LSGGQ motif, which enables them to be distinguished from other ATPases [18]. Besides the signature LSGGQ motif, the NBD also contains other conserved motifs including Walker A, Q-loop, Walker B, D-loop, and switch H-loop [18]. The Walker A and B motifs are important for ATP binding by forming a P-loop, while the residues of Q-loop and H-loop interact with the γ-phosphate of the ATP [18].

In plants, ABC transporters were first identified as being involved in cellular detoxification by mediating the storage of xenobiotics inside the vacuole [19]. They have subsequently been reported to be involved in other biological processes such as the regulation of growth and development, tolerance to abiotic and biotic stresses, and nutrient uptake [20]. Terrestrial plants contain more ABC transporter genes in their genomes than other living organisms [21]. These transporters can facilitate the transportation of diverse substrates including hormones such as auxins and abscisic acid (ABA), and secondary metabolites such as flavonoids, reactive oxygen species (ROS), and lipid molecules [20,22,23]. Thus, ABC transporters play crucial roles in the development and survival of plants by translocating different molecules under diverse conditions. The subcellular localizations of ABC transporters in various membranes, including the membrane of peroxisome, mitochondrion, vacuole, chloroplast, and plasma membrane, have been reported [24–28].

2.1.1. The Deprotonation of Auxin Molecules in Cytoplasm Implies the Need for Active Transporters

Auxin molecules can migrate by mass flow in the vascular system or being transported from cell to cell [29]. In the cytoplasm where the pH is near neutral, most of the indoleacetic acid (IAA) molecules exist in the anionic form, which is not favorable for diffusion across the plasma membrane [29]. Therefore, the transport of ionized auxin molecules across the membrane would need active auxin transporters [29].

It was reported that ABCB transporters are associated with the active transport of auxin driven by ATP hydrolysis [30]. *ABCB*s are mainly expressed in the leaf and root meristem and are related to the transportation of auxin in the apoplast [15]. AtABCB1 and AtABCB19 are ABC transporters mediating intracellular auxin transport and were reported to be involved in anther development [31]. The Arabidopsis double knockout mutant of *abcb1abcb19* showed reduced apical dominance, dwarfism, short hypocotyl, and poor fertility [32]. It was also reported that the direction of transport of IAA mediated by AtABCB4 could be reversed depending on the relative IAA concentration [33]. Although the substrate, IAA, was transported down the concentration gradient, such transport still had to be mediated by AtABCB4 [33]. However, the structural basis for such reverse of substate transport direction by the transport has remained unclear. In *Oryza sativa*, under salt stress, it was found that the *OsABCB* genes in leaf and root had differential expression responses towards the stress [34]. Under salt stress, a finetune of auxin levels in various plant tissues is needed to archive the balance of different biological processes such as biosynthesis and signal perception [35]. From these examples, it could be deduced that the deprotonation of auxin under stress is coupled with the regulated expression of ABC transporters to archive the finetune of auxin transports.

2.1.2. The Deprotonation of ABA Molecules in the Apoplast Implies the Need for Active Transporters

ABA is involved in the signaling of both abiotic and biotic stresses [36]. One of the major roles of ABA is the regulation of stomatal opening [36]. The possible active transport of ABA between cells and between tissues has been suggested [37–39]. The ABA molecule is negatively charged in the cytosol where the pH is near neutral. During drought stress, the pH in the apoplast increases [40]. At a higher pH, ABA molecules undergo further proton dissociation. The negative charge of the ABA molecules implies the need for active ABA transport across membranes. AtABCG25 is an ABA exporter involved in the intercellular signaling pathway [41]. It is mainly expressed in vascular tissues and localized in the plasma membrane, suggesting the role of AtABCG25 in ABA export [41]. *atabcg25* knockout mutants exhibited enhanced ABA-sensitive phenotypes at the early growth stage compared to the wild-type [41]. On the other hand, other ABC transporters could also be ABA importers. For example, AtABCG40 is an ABA uptake transporter in the plasma membrane. Such function of AtABCG40 was demonstrated in both yeast and tobacco BY2 cells ectopically expressing *AtABCG40* [42]. Compared to the wild-type, *atabcg40* mutants had a slower uptake of ABA into the protoplast and a slower rate of stomatal closure under ABA treatment [42]. Furthermore, in wheat, Lr34, a PDR-type ABCG transporter, was found to be an ABA transporter. The ectopic expression of the *Lr34res* allele from a rice blast-resistant wheat cultivar improved the accumulation of ABA and enhanced the tolerance of rice to rice blast [43]. These examples show the important roles of ABC transporters in transporting ABA between cells, especially when ABA molecules are further deprotonated under stress.

2.2. *Overview of MATE Transporters*

Multidrug and toxic compound extrusion (MATE) transporters are antiporters that transport various molecules across a membrane in exchange for sodium ion (Na^+) or proton (H^+) [44]. The movement of Na^+ or H^+ across the membrane results in an electrochemical gradient that drives the transport of the substrates, which could be macromolecules or ions, in the opposite direction [44]. Most eukaryotic MATE transporters utilize H^+ while prokaryotic MATE transporters can use either Na^+ or H^+ for this purpose [44]. In plants, MATE transporters typically consist of 12 transmembrane domains and have been reported to be located at membrane structures including the plasma membrane, vacuolar membrane, mitochondrial membrane, chloroplast envelope, and the surface of small vesicles [45].

The Phytohormone Transport Activities of MATE Transporters Could Be Dependent on pH or Electrochemical Proton Gradient

AtDTX50 (detoxification efflux carrier) was reported to be a plasma membrane-localized MATE-type protein for ABA efflux [46]. *AtDTX50* is mainly expressed in vascular tissues and guard cells where ABA is synthesized [46]. Under drought stress, the pH in xylem sap was found to increase from 6.1 to 6.7 [47]. Among pH 6, 7, and 8, AtDTX50 was found to be most active at pH 7 [46]. It therefore appears reasonable that, during drought, the xylem sap pH becomes closer to pH 7 to maximize the ABA-export activity of AtDTX50. However, from the mechanical perspective, it is not clear why the ABA-exporting function of AtDTX50 was highest at pH 7. Considering the near neutral cytosolic pH, when the pH of the medium was lower than 7, the proton gradient as well as the electrical potential difference across the cell membrane increased. However, such increases did not appear to enable an increased ABA export in exchange of the influx of H^+ from the medium. It is possible that the pH itself, besides the pH gradient across the membrane, has effects on the protein function. Since MATE is not the sole transporter of ABA, it is possible that other transporters could complement the ABA transport when one transporter is not working at its optimal capacity.

Besides ABA, MATE transporter has also been reported to transport salicylic acid (SA). ENHANCED DISEASE SUSCEPTIBILITY5 (EDS5) was reported to be a chloroplast envelope-localized MATE-type protein that transports SA from chloroplast to cytoplasm [48]. The *eds5* mutant was formerly known as the *sid* (SA induction deficient) mutant [49,50]. It was found that when the proton gradient was destroyed by nigericin or m-chlorophenyl hydrazone (CCCP), the SA-exporting activity of EDS5 was disrupted [48]. Therefore, it was suggested that the SA-exporting activity of EDS5 is dependent on the electrochemical gradient generated by protons [48]. The expression of *EDS5* was induced by *Pseudomonas syringae* [51]. Upon pathogen infection, in PTI, cytosolic acidification occurs [5]. It would therefore be reasonable to speculate that, upon pathogen infection, the SA-exporting activity of EDS5 is enhanced. Altogether, the induced expression of *EDS5* and the enhanced exporting activity of the protein promote the resistance to the pathogen.

3. Alkaloid Transporters

3.1. *ABC Transporter for the Storage of Alkaloids*

Besides phytohormones, ABC transporter is also able to transport alkaloids. ABC transporters play an important role in vacuolar sequestration, which is an important process for minimizing the effects of toxic compounds and heavy metals [52,53]. Various secondary metabolites, including terpenoids, alkaloids and phenolics, are employed for defense purposes in plants [54]. However, the accumulation of these compounds can be toxic. Therefore, in plant cells, the secondary metabolites are either stored as the non-toxic precursors in the vacuole or exported out of the cell [55]. Berberine, a member of the alkaloid family, which is one of the largest groups of secondary metabolites in plants, was reported to accumulate in the rhizome of *Coptis japonica*, as an antimicrobial compound [56]. It was reported that CjMDR1 (*Coptis japonica* multidrug resistance 1) mediates the transport of berberine. *CjMDR1* is expressed in the xylem tissue and encodes a multidrug-resistance protein (MDR)-type ABC transporter to facilitate the transport of berberine from the root to the rhizome [56].

3.2. MATE Transporter for the Storage of Alkaloids

Besides exporting substrates from the cell, MATE transporters have also been shown to mediate substrate influx into organelles such as vacuoles. In *Nicotiana tabacum*, Nt-JAT1 (*Nicotiana tabacum* jasmonate-inducible alkaloid transporter 1) was reported to be a MATE-type antiporter that mediates the influx of nicotine into the vacuole in exchange of H^+ [57]. The uptake function of Nt-JAT1 was found to be facilitated by F0F1-ATPase (F-ATPase), the ΔpH generator in the membrane, and was reduced by the dissipation of the pH gradient [57]. The lower pH inside the vacuole compared to the cytosol enabled the export of H^+ from the vacuole. In this case, the nicotine influx into the vacuole resulted in the increase in cytosolic H^+.

4. Ion and Ion Chelator Transporters

4.1. ABC Transporters for the Detoxification of Heavy Metal

ABC transporters have also been reported to be involved in the detoxification of various toxic heavy metals such as cadmium (Cd), mercury (Hg), and aluminum (Al). In Arabidopsis, compared to the wild-type, the double knock-out mutant *abcc1abcc2* was hypersensitive to arsenic (As) and had reduced vacuolar uptake of As(III)–PC_2 and apoPC_2 [26]. However, the overexpression of *AtABCC1* alone could not promote the tolerance to As [26], unless it was co-overexpressed with *AtPCS1* (*PhytoChelatin* (*PC*) *Synthase*) [26]. Using yeast as the model, it was shown that the ectopic expression of *AtABCC1* or *AtABCC2* mediated the microsomal uptake of PC_2-As more efficiently than that of apo-PC_2 [26]. It was therefore concluded that AtABCC1 and AtABCC2 are the transporters of PC and PC conjugates and mediate the uptake of PC-conjugated As into the vacuole for detoxification [26]. In another report, *AtABCC1* and *AtABCC2* were shown to mediate the tolerance to Cd and Hg [58]. In Arabidopsis, the double knock-out mutant *abcc1abcc2* was also hypersensitive to Cd(II) and Hg(II), and was demonstrated to be impaired in the vacuolar sequestration of Cd [58]. In addition, AtABCC1 and AtABCC2 enhanced the tolerance of PC-producing yeast to Hg(III) [58]. In rice, the ortholog of AtABCC1 and AtABCC2, OsABCC1, was also demonstrated to be a transporter that mediates the uptake of PC-As into the vacuole [59]. ABC transporters involved in mediating the tolerance to other heavy metals such as Al have also been reported and reviewed [60].

4.2. The Ion/Ion Chelator Transport Activities of MATE Transporters Are pH Dependent

Using *Xenopus* oocytes as the model, it was shown that a decrease in the pH of the medium favored the SbMATE (*Sorghum bicolor* MATE)-mediated efflux of citrate from the oocytes [61]. Similarly, the Arabidopsis MATE transporters, TT12 [62], FRD3 [63], and AtDTX1 [64], were also found to be substrate/H^+ antiporters. H^+-ATPase was suggested as the provider of H^+ for MATE transporters in exchange for their substrates [65]. When under aluminum (Al) toxicity stress, the concerted activities of plasma membrane-bound H^+-ATPase and a MATE transporter mediated the efflux of citrate and played a role in maintaining a stable cytosolic pH [65]. The Al-induced extrusion of citrate has been known as a strategy to enhance tolerance to Al toxicity. Under neutral pH in the cytosol, H^+ is dissociated from citric acid from the TCA cycle [65]. The H^+ is then transported out of the cell by H^+-ATPase [65]. The remaining citrate is then exported from the cell by MATE, possibly with the transport of H^+ back into the cytosol [65]. The extruded citrate molecules chelate Al cations to prevent Al cations from entering the cell [66]. This process is likely to result in the increase in cytosolic H^+ level. It was speculated that this drop in cytosolic pH may be related to the pH regulation in the cytosol and the balancing of charges for secondary ion transports [65].

The MATE transporters DTX33 and DTX35 in Arabidopsis were reported to be tonoplast-localized MATE-type proteins that mediate Cl^- influx into the vacuole and regulate the vacuolar turgor [67]. The *dtx33/dtx35* double mutant is impaired in stomatal opening and is more tolerant to drought. Although DTX33 and DTX35 were demonstrated to function as ion channels with the measured reversal potentials in agreement with the theoretical Nernst potentials at different pH and dependent on the Cl^- concentration, it was also demonstrated that their activities were dependent on vacuolar pH [67]. Among vacuolar pH 5, 6, and 7, DTX33 and DTX35 had the highest activities at pH 5 [67]. However, whether the change in transport activity was due to the pH itself or the change in pH gradient, which means the change in proton gradient as well as electrical potential gradient, remains unclear.

The above examples show that the export of substrates out of the cell and the import of substrate into the vacuole by MATE transporters could mediate the upregulation of the cytosolic H^+ concentration [57,65]. On the other hand, the differential H^+ concentrations in different cellular compartments, which result in the pH gradient across the biological membrane where the MATE transporters are localized, regulate the activities of the transporters. Under stress, the fluctuation of cellular pH possibly plays a role to regulate the activities of MATE transporters.

5. Sugar Transporters

5.1. Classification and Structural Properties

In plants, MonoSaccharide Transporters (MSTs), SUcrose Transporters (SUTs; or SUCs, SUcrose Carriers), and Sugars Will Eventually be Exported Transporters (SWEETs) are the three major types of sugar transporters [68,69]. SUTs and MSTs belong to the major facilitator superfamily (MFS), which has the characteristic 12 transmembrane domains (TMDs), in the form of 6 N-terminal TMDs connected to 6 C-terminal TMDs via a cytosolic loop [68,70,71]. Despite having similar architectures, SUTs and MSTs could be differentiated from each other by the different domain structures [68]. On the other hand, SWEETs have seven TMDs and are characterized by an MtN3/saliva domain [68]. In terms of the transport mechanism, MSTs and SUTs are proton/sucrose symporters [72–74], while SWEETs are reported to be uniporters of sugars [72,73]. In the following sections, only MSTs and SUTs, whose functions are H^+-dependent, will be discussed.

5.2. The SUT Family

In photosynthetic cells, sucrose is derived from the photosynthetically fixed carbon and is the major form of sugar transported via the phloem to other tissues [75,76]. SUTs are H^+/sucrose symporters involved in loading sucrose into the phloem against the concentration gradient and are driven by the proton motive force across the plasma membrane of the sieve element–companion cell complex (SE-CCC) [77].

5.2.1. The Activities of SUTs Are pH-Dependent

The pH dependence of SUT activities has been reported in various species. For example, in Arabidopsis, AtSUC4 was reported to be a proton/sucrose symporter localized in the vacuole membrane [78]. The proton motive force was suggested to be provided by the pumping of cytosolic protons into the vacuole by V-type ATPases and V-PPases [78]. Using yeast as the model, the proton-coupled and pH-dependent uptake of sucrose mediated by AtSUT4 was demonstrated [79]. Further experiments showed that AtSUC4 could act as a H^+/sucrose antiporter or symporter depending on the pH difference between vacuole lumen and the medium outside [78]. When the vacuole lumen was more acidic than the medium outside, sucrose was imported into the vacuole with the export of H^+ from the vacuole; when the medium outside the vacuole was more acidic, sucrose was transported together with proton into the vacuole [78]. Such switch of antiporter/symporter activity is illustrated in Figure 1. Similar bidirectional transport of sucrose was demonstrated using the phloem-localized ZmSUT1 [80]. The sucrose/H^+ symporter activity of ZmSUT1 was demonstrated to be pH-dependent [80]. It was suggested that both the sucrose gradient and the proton motive force determine the sucrose/proton symport direction [80].

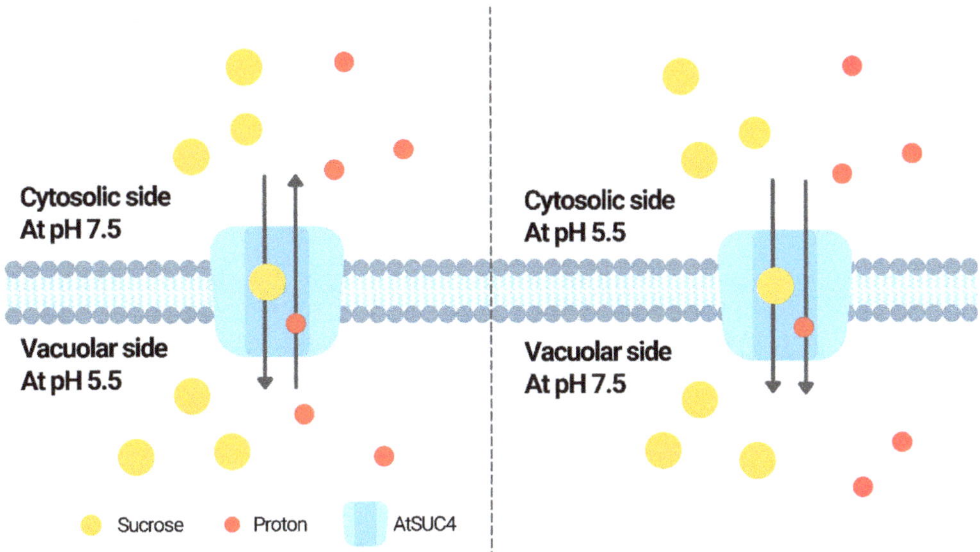

Figure 1. The switch of antiporter/symporter activity of AtSUC4. When the vacuole lumen was more acidic than the medium outside, AtSUC4 mediates the import of sucrose into the vacuole with the export of H^+ from the vacuole; when the medium outside the vacuole was more acidic, AtSUC4 mediates the transport of sucrose together with proton into the vacuole [78].

5.2.2. The Role of Sucrose Transport during Stress-Induced Cellular pH Fluctuations

Atsuc4 mutants were found to be more sensitive to stresses including salt, osmotic, cold, and ABA treatments compared to the wild-type [81]. Probably due to the impaired sugar distribution, these mutants were found to have higher and lower sucrose, fructose and glucose in the shoots and roots, respectively, compared to the wild-type [81]. The ABA-induced expressions of stress-responsive genes, including ABA-responsive element binding factors (ABFs) and the genes upstream or downstream to ABFs, were inhibited in *Atsuc4* mutants [81]. Although the detailed mechanism remained unclear, it was suggested that AtSUC4 is involved in the crosstalk between ABA signaling and sucrose signaling [81]. The accumulation of sucrose in the root is agriculturally important for those crops that are harvested for their edible roots. In sweet potato (*Ipomoea batatas*), similar to *AtSUC4*, *IbSUT4* was found to regulate the accumulation of sucrose in the root and ABA signaling under stress [82]. The overexpression of *IbSUT4* in Arabidopsis reduced the sucrose level in the leaf but improved the sucrose level in the root under salt stress [82]. This is possibly the result of a drop in cytosolic pH induced by salt stress [83], which reduced the sucrose export from the vacuole by SUT4.

The *Phaseolus vulgaris* SUT1.1, when expressed in transgenic yeast, was found to have a higher transport activity when the pH of the medium was more acidic [84]. The tonoplast- and plasma membrane-localized PvSUT1.1 was reported to be a sucrose-proton co-transporter that is probably involved in the export of sucrose from the leaf through the phloem [84]. In the same study, the expression of *PvSUT1.1* was found to be repressed in the leaf upon heat stress [84]. Since the export of sucrose from the leaf is important for heat tolerance, the heat stress susceptibility of *P. vulgaris* was therefore hypothesized to be associated with the repression of *PvSUT1.1* under high temperature [84]. However, under high temperature in the leaf, H^+ can leak through membranes [85]. Although not discussed in this study, such proton leaks in the cytosol may influence the transport activity of SUT in the tonoplast and plasma membrane. The net effect on the export of sucrose from the cell is unknown. To explain the SUT activity under stress, more mechanistic considerations, other than transcriptional controls, will be needed.

It has been widely accepted that SUTs are responsible for phloem loading. However, the analyses of expression data, including microarray, RNAseq, and quantitative PCR, from 167 experiments on various plant species showed that the effects on *SUT* expressions by factors such as photosynthetic rate, light level, and CO_2 concentration are limited [86]. It was speculated that when the photosynthetic rate is increased, the existing SUT proteins may already have enough capacity to increase phloem loading without the need to increase their transcript levels [86]. Other regulatory mechanisms, such as differential intracellular localization and protein dimerization, were suggested [86]. Mechanistically, the changes in proton distribution under different environmental conditions could also potentially alter the SUT activity. Using *Amaranthus caudatus* and *Vicia faba* as the models for C_4 and C_3 plants, respectively, despite different kinetics and extents, the C_4 and C_3 plants both exhibited light-dependent cytosolic alkalization and vacuolar acidification [87]. Such alkalization and acidification peaked when the photosynthetic apparatus was maximally energized under high energy flux rates in the absence of CO_2 [87]. Under such conditions, it is likely that the SUT activity will be affected. Therefore, the expression levels of *SUT*s under different environmental conditions could not sufficiently explain the regulation of the SUT function.

5.3. The MST Family

5.3.1. Classification and Structural Properties

Based on the protein sequences, MSTs can be further classified into several clades, including the STP, HXT, PLT-VGT, SBG-GLT-GLUT1, ERD6-like, and TMT clades [74]. Monosaccharides including glucose, fructose, galactose, mannose, and xylose are possible substrates of MSTs [88].

Based on the crystal structure of STP10 from *A. thaliana* that has recently been resolved [89,90], it was suggested that the protein's affinity for sugar is mainly due to the N-terminal domain and the Lid domain of the protein, while the substrate specificity is mediated by the C-terminal domain, which interacts with specific hydroxyl groups of the substrate [89]. At the apoplast, which is acidic, protonation of the Asp42 residue occurs and finetunes the protein structure to enhance the binding affinity to the substrate [89]. The subsequent unloading of glucose then enables a modification in the protein structure, which results in the release of the proton from Asp42 [90]. Since the important functional domains for the proton-coupled substrate transport are conserved among STPs, it was suggested that this transport model explains the general mechanism of action of the transporter [89,90].

5.3.2. The Activities of MSTs Are pH-Dependent

In *A. thaliana*, AtPLT5 (polyol transporter 5) was reported to be a plasma membrane-localized MST-like protein that mediates the transport of sorbitol, glucose, galactose, ribose, xylose, mannitol, glycerol, and inositol [91]. Using transgenic *Xenopus* oocytes as the model and glucose and glycerol as substrate examples, the transport activity of AtPLT5 was demonstrated to be pH-dependent [91]. Maximal transport activity was demonstrated at pH 5.5 [91]. At pH 6.5, the transport activity was reduced; at pH 7.5, there was no transport activity [91]. When the extracellular pH was brought back to 5.5, the transport activity resumed [91]. AtPLT5 has a broad spectrum of substrates and is found to be expressed in various tissues [91]. Therefore, it is proposed that AtPLT5 is possibly involved in the retrieval of sugars from the apoplast [91]. The pH-dependent sugar/proton symporter activity of STPs was also reported in apple (*Malus domestica*) [92]. MdSTP13a was reported to be the transporter of both hexose and sucrose competitively to provide the sugars for pollen tube growth [92]. Using transgenic yeast as the model, optimal glucose or sucrose uptake by MdSTP13a was established at pH 6 [92]. An increase or decrease in the pH resulted in declined transport activity [92]. Different STPs have different optimal pH for their transport activities. For example, DgSTP1 from *Datisca glomerata* has the optimal pH for transport activity at pH 4.5 [93]. In plants, the transport of sugars is a major strategy to distribute or store nutrients [77]. Since different cellular compartments have different pH, understanding the pH dependence of the activities of sugar transporters is essential for the interpretation of the biological functions.

5.3.3. Paralogs of MSTs Have Differential Expression Patterns to Serve Different Functions

In *A. thaliana*, *AtSTP1* was found to have a consistently high expression level in both the root and leaf among all 14 of the *STP*s identified, under normal conditions [94]. However, in the root, the expression of *AtSTP13* was highly inducible by salinity and ABA treatment [94]. Although the expression of *ATSTP1* in the root was also induced by salinity, the fold change is much less than that of *AtSTP13* [94]. Both *stp1* and *stp13* mutants had reduced abilities to uptake glucose and fructose, while *stp1* also had reduced galactose uptake [94]. After salt treatment, the leak of glucose from the *stp13* mutant was enhanced [94]. Based on the expression data and the sugar flux data, it was suggested that AtSTP13 mediates the reabsorption of monosaccharides leaked from damaged cells under salt stress while AtSTP1 is the major contributor of monosaccharide uptake under normal conditions [94]. Such differential functions of STPs in the same species are in line with another expression study on *STP*s in *O. sativa*. In this study, the expressions of *STP*s were found to be responsive to stresses including cold, high temperature, and submergence [95]. However, the patterns of expression upon the same stress were diverse among various *STP*s, which also had different expression patterns in different tissues [95].

6. Amino Acid Transporters

Plants can absorb inorganic and organic nitrogen from the environment via the root system [96]. In soil, inorganic nitrogen is usually found in the forms of nitrate and ammonium, while organic nitrogen usually exists in the forms of free amino acids, urea, and short peptides [97]. The uptake of nitrogen-containing molecules by plants is mediated by the respective transporters of these molecules, especially amino acid transporters, which have been known to play a major role in distributing nitrogen throughout the whole plant [98]. Amino acid accumulation and signaling have been suggested to play important roles in stress responses.

6.1. Classification and Structural Properties

Amino acid transporters are found in diverse plant species. They are categorized into three major families: the Amino acid Transporter Family (ATF) (which is also known as the Amino Acid/Auxin Permease (AAAP) family), the Amino acid-Polyamine-organoCation (APC) family and the Usually Multiple Acids Move In and out Transporter family (UMAMIT) [99–102]. The ATF family consists of eight subfamilies: general Amino Acid Permeases (AAPs), Lysine and Histidine Transporters (LHTs), Proline Transporters (ProTs), γ-Aminobutyric acid Transporters (GATs), aromatic and neutral amino acid transporters, indole-3-acetic acid transporters (AUXs), amino acid transporter-like proteins and Vesicular Aminergic-Associated Transporters (VAATs). The APC family consists of three subfamilies: Cationic Amino acid Transporters (CATs), Amino acid/Choline Transporters (ACTs), and Polyamine H^+-Symporters (PHSs). Members of the ATF family and the APC family usually share similar transport activities and protein structures. These transporters usually function by one of these mechanisms: solute-cation symport, solute-solute antiport, or facilitated diffusion at the plasma membrane [103], and they have a common protein structure with 10–14 TMDs [103,104]. UMAMITs were more recently identified compared to the ATF and APC families. These transporters belong to the nodulin-like gene family, functioning as the bidirectional facilitator for amino acid transport [102,105].

6.2. Amino Acid Transporters Are Driven by Proton Motive Force

Amino acid transporters act as secondary active transporters, with the specific amino acid being transported coupled to the proton motive force generated by the primary active H^+-pumping complex featuring a proton-pumping ATPase at the membrane [9,100,106–108]. The majority of amino acid transporters characterized in plants are proton-amino acid symporters [9,100,106–108]. During amino acid import, a transient alkalization of the extracellular medium was observed [106]. In transgenic yeasts expressing Arabidopsis amino acid transporter genes such as *AAP3*, *AAP4*, or *AAP5*, the proline uptake rate was increased when the external pH was made to decrease. In other words, the higher external H^+ concentration leads to a stronger transportation driving force for the uptake of the amino acid [108]. The requirement for the driving force provided by the proton gradient was further evidenced by the amino acid transport being abolished upon the addition of the protonophore, CCCP, a compound used for disrupting the proton gradient across the mitochondrial membrane by increasing its permeability to protons [106,108–111].

6.3. The Expressions of Amino Acid Transporters Are Stress-Responsive

Changes in the expressions of amino acid transporters were reported under abiotic stresses such as salt and water stress in Arabidopsis, rice, wheat, and barley [112–116]. Spatiotemporal differences in the induction patterns of different amino acid transporters upon stress were often observed within the same plant species or among their functional homologs. The *Hordeum vulgare* (barley) proline transporter, HvProT, was suggested to be crucial in transporting proline to the root tip region upon salt stress [116]. The *Triticum aestivum* (wheat) proline transporter 3, TaProT3, was upregulated in the root under salt stress whereas TaProT1 and TaProT2 were downregulated [115]. Through genome-wide identification and evolutionary expression analyses on wheat amino acid transporters, numerous stress- and hormone-responsive *cis*-regulatory elements were found within the promoter regions of the amino acid transporter genes [117]. Similar *cis*-regulatory element analyses were conducted between *Brassica napus* (rapeseed) and Arabidopsis, and the results were consistent with previous research [117], where multiple possible transcription factor recognition sites were discovered in the promoter regions of these transporter genes [118]. Based on the gene expression data, both TaAATs and BnAAPs responded differently to different abiotic stresses, suggesting the involvement of different interconnected regulatory networks of transporters in response to specific stresses [117,118].

6.4. Amino Acids as Osmolytes and Their Involvement in Ion Transport Mechanisms during Stress Responses

The accumulation of proline is widely known to protect plants from water-related stresses such as salinity, drought, and freezing [119]. The high solubility of proline and low inhibitory effect on seed germination make it a good candidate as a non-toxic osmolyte. One of the toxic effects of salt stress is the mineral nutrient imbalance brought forth by the over-accumulation of Na^+ and Cl^-, and the reduced levels of other essential ions such as K^+ and Ca^{2+} [120]. In different plant species, including *Olea europaea* [121] and *Cucumis sativus* [122], it was reported that the application of proline promoted salt tolerance. In *O. europaea*, it was shown that proline treatment reduced the level of Na^+ but enhanced the level of Ca^{2+} in the leaf under salt treatment [121]. In *C. sativus*, it was shown that proline treatment reduced the level of Cl^- in the leaf under salt treatment [122]. Exogenous proline and glycine betaine application were also reported to relieve the inhibition on both root and shoot growth of barley seedlings due to NaCl [123]. These observations suggest a possible involvement of amino acids, such as proline and glycine betaine, in the ion transportation mechanism.

6.5. Amino Acid Accumulation and Salicylic Acid (SA) Signaling

Plants carry a readily utilizable form of nitrogen. It is conceivable that the amino acid composition in plants was one of the determinants of plant–pathogen interactions. The importance of amino acid transporters in sustaining pathogen growth is not surprising. For example, there are a large number of transporter protein-encoding genes in the *Psuedomonas syringae* genome [124]. The cellular concentrations of amino acids, the uptake of inorganic nitrogen, and the relocation of amino acids might contribute to the plant's susceptibility towards the pathogen [125,126]. It has been suggested that amino acid-derived signaling can regulate SA accumulation and signaling [127,128]. A broad-specificity, high-affinity AAP homolog, *Lysine Histidine Transporter 1* (*LHT1*), expressed in the rhizodermis and mesophyll in Arabidopsis [111] was related to the pathogen susceptibility exhibited by the plant. The knockout Arabidopsis mutant *lht1* had reduced susceptibility to the hemibiotrophic pathogen, *P. syringae* [125]. It was hypothesized that the cellular redox status is modulated by nitrogen metabolism, where glutamine deficiency plays an important role in enhancing the defense responses of *lht1* plants. By studying the accumulation patterns of H_2O_2 and NO in the mesophyll and the spatial expression patterns of the pathogen-induced *LHT1*, it was suggested that LHT1 helped to maintain a low reactive oxygen species (ROS) level by keeping the glutamine level high. As a result, the low ROS level hindered the activation of SA-mediated defense responses [125].

In another study, Arabidopsis overexpressing *AtCAT1* (*Cation Amino acid Transporter 1*) incorporated lysine at a higher rate and was more resistant to *P. syringae* compared to the wild-type and the *cat1* mutant [126]. In the *AtCAT1* overexpressor, the resistance to the bacterial pathogen *P. syringae* was enhanced with an increased SA level in the leaves [126]. Therefore, the expression levels of amino acid transporters might affect biotic stress responses by manipulating the cellular amino acid concentrations [125,126].

6.6. Amino Acid Transporters Are Involved in the Regulation of Cellular pH and Rhizospheric pH

Although amino acids usually carry charges and their transportation across membranes is often coupled with protons, there was no significant direct correlation among amino acid concentrations, transport, and cellular pH [97,129]. Nevertheless, there have been reports of amino acid assimilation and transportation altering pH homeostasis in plants. Ammonium, nitrogen, and nitrate assimilation produce protons that affect the cellular pH [130]. Aspartic acid and glutamic acid are suggested to be involved in balancing the excessive H^+ production during nitrogen assimilation [131]. Moreover, it was reported that the aluminum-activated malate transporter in wheat, TaALMT1, promoted the acidification of an alkaline rhizosphere by facilitating the exudation of both malate and the zwitterionic buffer, gamma-aminobutyric acid (GABA) from the plant root [132]. The expression level of *TaALMT1* was positively correlated with the growth performance of the wheat plant in the alkaline environment. The expression pattern of the amino acid transporter *SlCAT9* in *Solanum lycopersicum* was linked to the exchange of GABA for glutamate and aspartate during fruit ripening, and eventually led to changes in the amino acid composition of the developing fruit [131]. The result is a change in the flavor and the nutritional value of the fruit [131]. The correlation between GABA concentration and low cytosolic pH during the early stages of fruit development was demonstrated using nuclear magnetic resonance (NMR) [133]. It was suggested that the export of GABA out of the vacuole in exchange for the import of glutamate or aspartate could serve to counterbalance the proton charges through the 'reverse' GABA shunt pathway [131].

6.7. Protons Are the Unneglectable Regulators of Active Transporters under Stresses

The activities of active transporters have been reported to be pH-dependent. Such property implies the regulation of the transporter activities at biological membranes between various cellular compartments with different pH. An obvious example is the increase in proline uptake by amino acid transporters when the extracellular pH was made to decrease [108]. In addition, the different pH of the vacuole, cytosol, and apoplast further implies the finetune of the transport direction to achieve different purposes. For example, the more acidic vacuole compared to the cytosol favors the storage of toxic nicotine in the vacuole [57] and the influx of Cl^- into the vacuole for turgor regulation [67], while the acidification of the rhizosphere favors the exudation of malate and GABA from plant cells to the rhizosphere by TaALMT1 [132]. The pH dependence properties of active transporters even allow bidirectional transport H^+. For example, AtSUC4 was demonstrated to have reversible direction of H^+ transport when the H^+ gradient between the membrane was made to reverse, although the structural basis of such reverse transport direction by the transporter has remained unclear [78]. Reverse transport direction has been observed in other transporters. The possible mechanisms underlying the reverse transport direction include the existence of two discrete conformation states of the transporter protein, which allows the binding of the substrate to either side of the transporter, and the existence of the substrate binding site, which is close to both sides of the transporter protein [134]. For example, based on the protein crystal structure, different conformations of BetP, a betaine transporter, were observed [135]. The existence of both the outward-facing conformation and the inward-facing conformation underlies the possibility of alternating access of the substrate [135]. In another study, it was reported that Glt_{ph}H7 has the substrate binding site close to both sides of the transporter to allow alternating access of glutamate from either side of the transporter [136]. However, the structural basis of the revere transport direction by the transporters discussed in this review has remained unknown. More mechanical and structural studies on the transporters will be needed to address this phenomenon.

The pH of different cellular compartments is highly regulated and yet highly dynamic, especially when under stress. The alteration in pH of different cellular compartments under stress implies the regulation of transporter activities. For example, under drought stress, the pH in xylem sap was found to be increased from 6.1 to 6.7 [47], which is closer to pH 7, the optimal pH for the transport activity of AtDTX50 [46]. The significance of such pH-dependent activity lies in the possibility of improving ABA transport activity under stress. In addition, under stress, the deprotonation of phytohormones such as auxin and ABA molecules highlights the necessity of active transporters when the diffusion across membranes is made more difficult. Such reliance on active transporters for phytohormone transportation allows a highly controlled transport of the phytohormones under stresses.

Considering the mechanics of the active transporters under stress, it could be deduced that the altered expression of the transporters under stresses could not fully explain the regulation on the transport capacities. In some cases, the altered transporter activity under stress could be coupled with the altered expression of the gene encoding the transporter to improve the transport efficiency. For example, the pathogen-induced expression of *EDS5* is coupled with possible pathogen-induced cytosolic acidification [5,48,51], which favors the SA-exporting activity of EDS5. However, in some cases, the expression of transporters could not sufficiently explain the altered substrate transport under stress. For example, it was found that photosynthetic rate, light level, and CO_2 concentration had limited effects on the expressions of *SUT*s [86]. However, an improved sucrose transport upon the increased photosynthetic rate is expected. In this case, the possible alteration of the sucrose transport efficiency could be the explanation. It is therefore important to understand the mechanics of the active transporters under different situations when the cellular pH fluctuates.

The modulation of the expression of active transporters in plants has been considered a common approach to modulate the accumulation of desired metabolites [137]. Since protons play important roles in regulating active transports in plants, attempts to modulate the transporter activities could be broadened to consider the manipulation of genes that regulate the proton levels in various cellular compartments. Such manipulation allows the finetune of transporter activities under stress when cellular pH fluctuates.

The movement of H^+ across the biological membranes results in electrochemical proton gradient, which means the proton gradient as well as the electrical potential gradient across the membrane, to energize the substrate transport. In the above discussion, several proteins, such as AtEDS5 [48,51], AtSUC4 [78,79,81], and AtLHT1 [111,125], were reported to have their activities dependent on the proton gradient. Such observations are in line with the suggestion that the active transporters are energized by the electrochemical proton gradient across the biological membranes. In several examples, such as AtDTX50 [46], AtDTX33 [67], AtDTX35 [67], PvSUT1.1 [84], AtPLT5 [91], MdSTP13a [92], and HvProT [116], the transport activities were shown to be different under different pH of the vacuole/medium. However, it is not clear whether such differences of activities are merely a result of the altered electrochemical proton gradient across the membrane or a result of the structural change of the protein under different pH. Detailed studies on the conformation of the proteins under different pH will be needed to address the question.

Examples of the activities, pH dependence, and the biological significance of ABC transporters, MATE transporters, MSTs, SUTs, and amino acid transporters are summarized in Table 1.

Table 1. Examples of the activities, pH dependence, and the biological significance of ATP-binding cassette (ABC) transporters, multidrug and toxic compound extrusion (MATE) transporters, monosaccharide transporters (MSTs), sucrose transporters (SUTs), and amino acid transporters.

Transporter Type	Transporter Name	Transport Activity	pH Dependence	Biological Significance	References
MATE	Nt-JAT1	Mediates nicotine influx into the vacuole	Makes use of pH gradient across vacuolar membrane, with pH inside vacuole lower than in cytosol	Storage of toxic compound	[57]
	AtDTX50	Mediates ABA efflux under drought stress	Most active at pH 7 when compared among pH 6, 7, and 8	Promotes ABA efflux under drought stress when the pH of the xylem sap becomes closer to neutral	[46]
	AtDTX33	Mediates Cl^- influx into vacuoles; mutant impaired in stomatal opening	Most active at vacuolar pH 5 when compared among vacuolar pH 5, 6, and 7	Promotes Cl^- influx into the vacuole, which is more acid than the cytosol, for turgor regulation	[67]
	AtDTX35	Mediates Cl^- influx into vacuoles; mutant impaired in stomatal opening	Most active at vacuolar pH 5 when compared among vacuolar pH 5, 6, and 7	Promotes Cl^- influx into the vacuole, which is more acid than the cytosol, for turgor regulation	[67]
	AtEDS5	Mediates SA efflux from chloroplast to cytoplasm	The transport activity is driven by the proton gradient across the biological membrane	The efflux of SA from chloroplast to cytosol promotes stress tolerance	[48,51]
SUC	AtSUC4	Mediates vacuolar sucrose storage	Acts as a H^+/sucrose antiporter or symporter depending on the pH difference between vacuole lumen and the medium outside	Facilitates sugar distribution under stress; compared to the wild-type, mutants have higher and lower sucrose, fructose and glucose in shoots and roots, respectively, and are more sensitive to salt, osmotic, cold and ABA treatments	[78,79,81]

Table 1. *Cont.*

Transporter Type	Transporter Name	Transport Activity	pH Dependence	Biological Significance	References
	PvSUT1.1	Exports sucrose from leaf through phloem	Higher activity at lower pH in medium	Involved in sucrose translocation between different tissues of plant, the downregulated expression is possibly associated with the heat susceptibility of the plant	[84]
MST	AtPLT5	Mediates transport of a large spectrum of polyols	Maximal transport activity at pH 5.5; activity reduced at pH 6.5 and no activity at pH 7	Proposed to be involved in the retrieval of sugars from the apoplast	[91]
	MdSTP13a	Mediates transport of a hexose and sucrose for pollen tube growth	Optimal uptake at pH 6 in yeast model	Growth and development	[92]
	AtSTP1	Inducible by salinity but mainly involved in the distribution of monosaccharides under normal conditions; mutant with reduced ability to uptake glucose, fructose and galactose	unknown	Growth and development, adaptation to the environment	[94]
	AtSTP13	Inducible by salinity and ABA treatments; involved in the reabsorption of monosaccharides leaked from damaged cells; mutant with reduced ability to uptake glucose, fructose and galactose	unknown	Growth and development, adaptation to the environment	[94]
ATF	HvProT	Proline transportation during salt stress	pH-dependent; the proline uptake activity o yeast mutant complemented with HvProT was the highest at pH 4.5 among pH 4.5, 5.5, and 6.5	Adaptation to the environment	[116]
AAP	AtLHT1	Transports a broad spectrum of amino acids; knockout mutant with reduced susceptibility to *P. syringae*; suspected to be involved in SA pathway	pH gradient dependent	Resistance to biotic stress	[111,125]
APC	AtCAT1	Lysine incorporation; overexpressor more resistant to *P. syringae* with increased SA level in leaves	unknown	Resistance to biotic stress	[126]
	SlCAT9	Exchange of GABA for glutamate and aspartate during fruit ripening	The transport of GABA has been suggested to play a role in regulating cytosolic pH	Growth and development	[131]

Table 1. *Cont.*

Transporter Type	Transporter Name	Transport Activity	pH Dependence	Biological Significance	References
ABC	AtABCB1	Transports auxin; double mutant with *Atabcb19* resulted in developmental problem with poor fertility	unknown	Growth and development	[31]
	AtABCB19	Transports auxin; double mutant with *Atabcb1* resulted in developmental problem with poor fertility	unknown	Growth and development	[31]
	AtABCG25	Exports ABA; mutant with ABA-sensitive phenotype at early growth stage	unknown	Adaptation to the environment	[41]
	AtABCG40	Uptakes ABA; mutant with slow uptake of ABA and insensitivity towards ABA	unknown	Adaptation to the environment	[42]
	CjMDR1	Transports berberine from root to rhizome	unknown	Adaptation to the environment	[56]
	Lr34	Transports ABA; ectopic expression in wheat enhanced the tolerance to rice blast	unknown	Adaptation to the environment	[43]
	AtABCC1	Mediates microsomal uptake of PC and PC conjugates for heavy metal detoxification in vacuole	unknown	Adaptation to the environment	[26]
	AtABCC2	Mediates microsomal uptake of PC and PC conjugates for heavy metal detoxification in vacuole	unknown	Adaptation to the environment	[26]

7. Conclusions

ABC transporters, MATE transporters, MSTs, SUTs, and amino acid transporters are involved in the transport of ions, toxic compounds, sugars, hormones, and amino acids. The transport of these molecules is important in the growth, development, and stress adaptations of plants. Since these transporters are driven by proton motive force, the activities of these transporters are highly regulated by cellular pH, which is usually influenced by both abiotic and biotic stresses. Besides directly participating in the transport activity by providing the proton motive force, pH also influences the charges of the substrates to be transported. In addition, the pH itself could possibly affect the transporter function. These factors together affect the transport efficiency. The role of the proton in molecular transports deserves more attention than it has received thus far. In-depth structural and mechanical studies will be needed to delineate the effects of pH, proton gradient, and electrical potential gradient on the transporter functions.

Author Contributions: Y.-S.K. put together the first complete draft and prepared the final version. H.-M.L. planned and coordinated the writing and prepared the final version. Y.-S.K., S.-S.C., M.-S.N., G.C. and H.-M.L. contributed to the search of the literature and writing. All authors have read and agreed to the published version of the manuscript.

Funding: This work was supported by grants from the Hong Kong Research Grants Council: General Research Fund (14143916) and Area of Excellence Scheme (AoE/M-403/16).

Institutional Review Board Statement: Not applicable.

Informed Consent Statement: Not applicable.

Data Availability Statement: Not applicable.

Acknowledgments: Jee-Yan Chu copy-edited the manuscript. Any opinions, findings, conclusions or recommendations expressed in this publication do not reflect the views of the Government of the Hong Kong Special Administrative Region or the Innovation and Technology Commission.

Conflicts of Interest: Authors declare no conflict of interest.

References

1. Bañó-Polo, M.; Baeza-Delgado, C.; Orzáez, M.; Marti-Renom, M.A.; Abad, C.; Mingarro, I. Polar/ionizable residues in transmembrane segments: Effects on helix-helix packing. *PLoS ONE* **2012**, *7*, e44263. [CrossRef] [PubMed]
2. Van Lehn, R.C.; Zhang, B.; Miller, T.F. Regulation of multispanning membrane protein topology via post-translational annealing. *Elife* **2015**, *4*, e08697. [CrossRef] [PubMed]
3. Wilkinson, S. PH as a stress signal. *Plant Growth Regul.* **1999**, *29*, 87–99. [CrossRef]
4. Kader, M.A.; Lindberg, S. Cytosolic calcium and pH signaling in plants under salinity stress. *Plant Signal. Behav.* **2010**, *5*, 233–238. [CrossRef]
5. Li, J.; Staiger, C.J. Understanding cytoskeletal dynamics during the plant immune response. *Annu. Rev. Phytopathol.* **2018**, *56*, 513–533. [CrossRef]
6. Cosse, M.; Seidel, T. Plant proton pumps and cytosolic pH-homeostasis. *Front. Plant Sci.* **2021**, *12*, 672873. [CrossRef]
7. Morsomme, P.; Boutry, M. The plant plasma membrane H^+-ATPase: Structure, function and regulation. *Biochim. Biophys. Acta (BBA)-Biomembr.* **2000**, *1465*, 1–16. [CrossRef]
8. Bassil, E.; Coku, A.; Blumwald, E. Cellular ion homeostasis: Emerging roles of intracellular NHX Na^+/H^+ antiporters in plant growth and development. *J. Exp. Bot.* **2012**, *63*, 5727–5740. [CrossRef]
9. Sze, H.; Li, X.; Palmgren, M.G.; Sze, H.; Li, X.; Palmgrenb, M.G. Energization of plant cell membranes by H^+-pumping ATPases: Regulation and biosynthesis. *Plant Cell* **1999**, *11*, 677–689.
10. Haruta, M.; Gray, W.M.; Sussman, M.R. Regulation of the plasma membrane proton pump (H^+-ATPase) by phosphorylation. *Curr. Opin. Plant Biol.* **2016**, *28*, 68–75. [CrossRef]
11. Janicka-Russak, M. Plant plasma membrane H^+-ATPase in adaptation of plants to abiotic stresses. In *Abiotic Stress Response in Plants-Physiological, Biochemical and Genetic Perspectives*; Shanker, A., Venkatteswarlu, B., Eds.; Intech Open: London, UK, 2011.
12. Davies, J.M. Vacuolar energization: Pumps, shunts and stress. *J. Exp. Bot.* **1997**, *48*, 633–641. [CrossRef]
13. Ratajczak, R. Structure, function and regulation of the plant vacuolar H^+-translocating ATPase. *Biochim. Biophys. Acta (BBA)-Biomembr.* **2000**, *1465*, 17–36. [CrossRef]
14. Segami, S.; Asaoka, M.; Kinoshita, S.; Fukuda, M.; Nakanishi, Y.; Maeshima, M. Biochemical, structural and physiological characteristics of vacuolar H^+-pyrophosphatase. *Plant Cell Physiol.* **2018**, *59*, 1300–1308. [CrossRef] [PubMed]
15. Park, J.; Lee, Y.; Martinoia, E.; Geisler, M. Plant hormone transporters: What we know and what we would like to know. *BMC Biol.* **2017**, *15*, 93. [CrossRef] [PubMed]
16. Henikoff, S.; Greene, E.A.; Pietrokovski, S.; Bork, P.; Attwood, T.K.; Hood, L. Gene families: The taxonomy of protein paralogs and chimeras. *Science* **1997**, *278*, 5338. [CrossRef] [PubMed]
17. Wilkens, S. Structure and mechanism of ABC transporters. *F1000Prime Rep.* **2015**, *7*, 14. [CrossRef] [PubMed]
18. Lane, T.S.; Rempe, C.S.; Davitt, J.; Staton, M.E.; Peng, Y.; Soltis, D.E.; Melkonian, M.; Deyholos, M.; Leebens-Mack, J.H.; Chase, M.; et al. Diversity of ABC transporter genes across the plant kingdom and their potential utility in biotechnology. *BMC Biotechnol.* **2016**, *16*, 47. [CrossRef]
19. Martinola, E.; Grill, E.; Tommaslnl, R.; Kreuz, K.; Amrhein, N. ATP-dependent glutathione S-conjugate "export" pump in the vacuolar membrane of plants. *Nature* **1993**, *364*, 247–249. [CrossRef]
20. Do, T.H.T.; Martinoia, E.; Lee, Y. Functions of ABC transporters in plant growth and development. *Curr. Opin. Plant Biol.* **2018**, *41*, 32–38. [CrossRef]
21. Hwang, J.; Song, W.; Hong, D.; Ko, D.; Yamaoka, Y.; Jang, S.; Yim, S.; Lee, E.; Khare, D.; Kim, K.; et al. Plant ABC transporters enable many unique aspects of a terrestrial plant's lifestyle. *Mol. Plant* **2016**, *9*, 338–355. [CrossRef]
22. Ha, T.; Do, T.; Choi, H.; Palmgren, M.; Martinoia, E.; Hwang, J.-U.; Lee, Y. *Arabidopsis* ABCG28 is required for the apical accumulation of reactive oxygen species in growing pollen tubes. *Proc. Natl. Acad. Sci. USA* **2019**, *116*, 25.
23. Footitt, S.; Slocombe, S.P.; Larner, V.; Kurup, S.; Wu, Y.; Larson, T.; Graham, I.; Baker, A.; Holdsworth, M. Control of germination and lipid mobilization by *COMATOSE*, the *Arabidopsis* homologue of human ALDP. *EMBO J.* **2002**, *21*, 2912–2922. [CrossRef] [PubMed]
24. Kunz, H.-H.; Scharnewski, M.; Feussner, K.; Feussner, I.; Flügge, U.-I.; Fulda, M.; Gierth, M. The ABC transporter PXA1 and peroxisomal β-oxidation are vital for metabolism in mature leaves of *Arabidopsis* during extended darkness. *Plant Cell* **2009**, *21*, 2733–2749. [CrossRef]

25. Dahuja, A.; Kumar, R.R.; Sachdev, A.; Watts, A.; Singh, B.; Goswami, S.; Sachdev, A.; Praveen, S. Role of ATP-binding cassette transporters in maintaining plant homeostasis under abiotic and biotic stresses. *Physiol. Plant.* **2021**, 785–801. [CrossRef] [PubMed]
26. Song, W.-Y.; Park, J.; Mendoza-Cózatl, D.G.; Suter-Grotemeyer, M.; Shim, D.; Hörtensteiner, S.; Geisler, M.; Weder, B.; Rea, P.A.; Rentsch, D.; et al. Arsenic tolerance in *Arabidopsis* is mediated by two ABCC-type phytochelatin transporters. *Proc. Natl. Acad. Sci. USA* **2010**, *49*, 21187–21192. [CrossRef] [PubMed]
27. von Voithenberg, L.V.; Park, J.; Stübe, R.; Lux, C.; Lee, Y.; Philippar, K. A novel prokaryote-type ECF/ABC transporter module in chloroplast metal hoemostasis. *Front. Plant Sci.* **2019**, *10*, 1264. [CrossRef] [PubMed]
28. Crouzet, J.; Roland, J.; Peeters, E.; Trombik, T.; Ducos, E.; Nader, J.; Boutry, M. NtPDR1, a plasma membrane ABC transporter from *Nicotiana tabacum*, is involved in diterpene transport. *Plant Mol. Biol.* **2013**, *82*, 181–192. [CrossRef] [PubMed]
29. Zažímalová, E.; Murphy, A.S.; Yang, H.; Hoyerová, K.; Hošek, P. Auxin transporters-Why so many? *Cold Spring Harb. Perspect. Biol.* **2010**, *2*, a001552.
30. Geisler, M. Getting to the right side. *Plant Physiol.* **2016**, *172*, 2081. [CrossRef] [PubMed]
31. Cecchetti, V.; Brunetti, P.; Napoli, N.; Fattorini, L.; Altamura, M.M.; Costantino, P.; Cardarelli, M. ABCB1 and ABCB19 auxin transporters have synergistic effects on early and late *Arabidopsis* anther development. *J. Integr. Plant Biol.* **2015**, *57*, 1089–1098. [CrossRef] [PubMed]
32. Noh, B.; Murphy, A.S.; Spalding, E.P. *Multidrug Resistance*-like genes of Arabidopsis required for auxin transport and auxin-mediated development. *Plant Cell* **2001**, *13*, 2441–2454.
33. Yang, H.; Murphy, A.S. Functional expression and characterization of Arabidopsis ABCB, AUX 1 and PIN auxin transporters in *Schizosaccharomyces pombe*. *Plant J.* **2010**, *59*, 179–191. [CrossRef] [PubMed]
34. Chai, C.; Subudhi, P.K. Comprehensive analysis and expression profiling of the *OsLAX* and *OsABCB* auxin transporter gene families in rice (*Oryza sativa*) under phytohormone stimuli and abiotic stresses. *Front. Plant Sci.* **2016**, *7*, 593. [CrossRef] [PubMed]
35. Ribba, T.; Garrido-Vargas, F.; O'Brien, J.A. Auxin-mediated responses under salt stress: From developmental regulation to biotechnological applications. *J. Exp. Biol.* **2020**, *71*, 3843–3853. [CrossRef] [PubMed]
36. Ku, Y.-S.; Sintaha, M.; Cheung, M.-Y.; Lam, H.-M. Plant hormone signaling crosstalks between biotic and abiotic stress responses. *Int. J. Mol. Sci.* **2018**, *19*, 3206. [CrossRef] [PubMed]
37. Nambara, E.; Marion-Poll, A. Abscisic acid biosynthesis and catabolism. *Annu. Rev. Plant Biol.* **2005**, *56*, 165–185. [CrossRef]
38. Kuromori, T.; Sugimoto, E.; Shinozaki, K. Intertissue signal transfer of abscisic acid from vascular. *Plant Physiol.* **2014**, *164*, 1587–1592. [CrossRef] [PubMed]
39. Seo, M.; Koshiba, T. Transport of ABA from the site of biosynthesis to the site of action. *J. Plant Res.* **2011**, *124*, 501–507. [CrossRef] [PubMed]
40. Geilfus, C.-M. The pH of the apoplast: Dynamic factor with functional impact under stress. *Mol. Plant* **2017**, *10*, 1371–1386. [CrossRef] [PubMed]
41. Kuromori, T.; Miyaji, T.; Yabuuchi, H.; Shimizu, H.; Sugimoto, E.; Kamiya, A.; Moriyama, Y.; Shinozaki, K. ABC transporter AtABCG25 is involved in abscisic acid transport and responses. *Proc. Natl. Acad. Sci. USA* **2010**, *107*, 2361–2366. [CrossRef] [PubMed]
42. Kang, J.; Hwang, J.; Lee, M.; Kim, Y.-Y.; Assmann, S.M.; Martinoia, E.; Lee, Y. PDR-type ABC transporter mediates cellular uptake of the phytohormone abscisic acid. *Proc. Natl. Acad. Sci. USA* **2010**, *107*, 2355–2360. [CrossRef] [PubMed]
43. Krattinger, S.G.; Kang, J.; Bräunlich, S.; Boni, R.; Chauhan, H.; Selter, L.L.; Robinson, M.D.; Schmid, M.W.; Wiederhold, E.; Hensel, G.; et al. Abscisic acid is a substrate of the ABC transporter encoded by the durable wheat disease resistance gene Lr34. *New Phytol.* **2019**, *223*, 853–866. [CrossRef] [PubMed]
44. Omote, H.; Hiasa, M.; Matsumoto, T.; Otsuka, M.; Moriyama, Y. The MATE proteins as fundamental transporters of metabolic and xenobiotic organic cations. *Trends Pharmacol. Sci.* **2006**, *27*, 11. [CrossRef] [PubMed]
45. Takanashi, K.; Shitan, N.; Yazaki, K. The multidrug and toxic compound extrusion (MATE) family in plants. *Plant Biotechnol.* **2014**, *31*, 417–430. [CrossRef]
46. Zhang, H.; Zhu, H.; Pan, Y.; Yu, Y.; Luan, S.; Li, L. A DTX/MATE-type transporter facilitates abscisic acid efflux and modulates ABA sensitivity and drought tolerance in *Arabidopsis*. *Mol. Plant* **2014**, *7*, 1522–1532. [CrossRef] [PubMed]
47. Wilkinson, S.; Davies, W.J. Xylem sap pH increase: A drought signal received at the apoplastic face of the guard cell that involves the suppression of saturable abscisic acid uptake by the epidermal Symplast. *Plant Physiol.* **1997**, *113*, 559–573. [CrossRef] [PubMed]
48. Serrano, M.; Wang, B.; Aryal, B.; Garcion, C.; Abou-Mansour, E.; Heck, S.; Geisler, M.; Mauch, F.; Nawrath, C.; Métraux, J.-P. Export of salicylic acid from the chloroplast requires the multidrug and toxin extrusion-like transporter EDS5. *Plant Physiol.* **2013**, *162*, 1815–1821. [CrossRef]
49. Nawrath, C.; Métraux, J.-P. Salicylic acid induction-deficient mutants of *Arabidopsis* express *PR-2* and *PR-5* and accumulate high levels of camalexin after pathogen inoculation. *Plant Cell* **1999**, *11*, 1393–1404. [PubMed]
50. Lefevere, H.; Bauters, L.; Gheysen, G. Salicylic acid biosynthesis in plants. *Front. Plant Sci.* **2020**, *11*, 338. [CrossRef] [PubMed]
51. Nawrath, C.; Heck, S.; Parinthawong, N.; Métraux, J.-P. EDS5, an essential component of salicylic acid-dependent signaling for disease resistance in Arabidopsis, is a member of the MATE transporter family. *Plant Cell* **2002**, *14*, 275–286. [CrossRef]

52. Coleman, J.; Blake-Kalff, M.; Davies, T.G.E. Detoxification of xenobiotics by plants: Chemical modification and vacuolar compartmentation. *Trends Plant Sci.* **1997**, *2*, 144–151. [CrossRef]
53. Hall, J.L. Cellular mechanisms for heavy metal detoxification and tolerance. *J. Exp. Biol.* **2002**, *53*, 1–11. [CrossRef]
54. Bennett, B.Y.R.N.; Wallsgrove, R.M. Secondary metabolites in plant defence mechanisms. *New Phytol.* **1994**, *127*, 617–633. [CrossRef]
55. Shitan, N. Secondary metabolites in plants: Transport and self-tolerance mechanisms. *Biosci. Biotechnol. Biochem.* **2016**, *80*, 1283–1293. [CrossRef]
56. Shitan, N.; Bazin, I.; Dan, K.; Obata, K.; Kigawa, K.; Ueda, K.; Sato, F.; Forestier, C.; Yazaki, K. Involvement of CjMDR1, a plant multidrug-resistance-type ATP-binding cassette protein, in alkaloid transport in Coptis japonica. *Proc. Natl. Acad. Sci. USA* **2003**, *100*, 751–756. [CrossRef]
57. Morita, M.; Shitan, N.; Sawada, K.; Van Mongtagu, M.C.E.; Inzéc, D.; Rischer, H.; Goossens, A.; Oksman-caldentey, K.; Moriyama, Y.; Yazaki, K. Vacuolar transport of nicotine is mediated by a multidrug and toxic compound extrusion (MATE) transporter in *Nicotiana tabacum*. *Proc. Natl. Acad. Sci. USA* **2009**, *106*, 2447–2452. [CrossRef]
58. Park, J.; Song, W.-Y.; Ko, D.; Eom, Y.; Hansen, T.H.; Schiller, M.; Lee, T.G.; Martinoia, E.; Lee, Y. The phytochelatin transporters AtABCC1 and AtABCC2 mediate tolerance to cadmium and mercury. *Plant J.* **2012**, *69*, 278–288. [CrossRef]
59. Song, W.-Y.; Yamaki, T.; Yamaji, N.; Ko, D.; Jung, K.H.; Fujii-Kashino, M.; An, G.; Martinoia, E.; Lee, Y.; Ma, J.F. A rice ABC transporter, OsABCC1, reduces arsenic accumulation in the grain. *Proc. Natl. Acad. Sci. USA* **2014**, *111*, 15699–15704. [CrossRef]
60. Song, W.-Y.; Park, J.; Eisenach, C.; Maeshima, M.; Lee, Y.; Martinoia, E. ABC transporters and heavy metals. In *Plant ABC Transporters*; Geisler, M., Ed.; Springer International Publishing: Cham, Switzerland, 2014; pp. 1–17, ISBN 9783319065113.
61. Doshi, R.; Mcgrath, A.P.; Piñeros, M.; Szewczyk, P.; Garza, D.M.; Kochian, L.V.; Chang, G. Functional characterization and discovery of modulators of SbMATE, the agronomically important aluminium tolerance transporter from *Sorghum bicolor*. *Sci. Rep.* **2017**, *7*, 17996. [CrossRef]
62. Marinova, K.; Pourcel, L.; Weder, B.; Schwarz, M.; Barron, D.; Routaboul, J.-M.; Debeaujon, I.; Klein, M. The *Arabidopsis* MATE transporter TT12 acts as a vacuolar flavonoid/H^+- antiporter active in proanthocyanidin-accumulating cells of the seed coat. *Plant Cell* **2007**, *19*, 2023–2038. [CrossRef]
63. Durrett, T.P.; Gassmann, W.; Rogers, E.E. The FRD3-mediated efflux of citrate into the root vasculature is necessary for efficient iron translocation. *Plant Physiol.* **2007**, *144*, 197–205. [CrossRef] [PubMed]
64. Li, L.; He, Z.; Pandey, G.K.; Tsuchiya, T.; Luan, S. Functional cloning and characterization of a plant efflux carrier for multidrug and heavy metal detoxification. *J. Biol. Chem.* **2002**, *277*, 5360–5368. [CrossRef] [PubMed]
65. Zhang, J.; Wei, J.; Li, D.; Kong, X.; Rengel, Z.; Chen, L.; Yang, Y.; Cui, X.; Chen, Q. The role of the plasma membrane H^+-ATPase in plant responses to aluminum toxicity. *Front. Plant Sci.* **2017**, *8*, 1757. [CrossRef] [PubMed]
66. Zhang, X.; Long, Y.; Huang, J.; Xia, J. Molecular mechanisms for coping with Al toxicity in plants. *Int. J. Mol. Sci.* **2019**, *20*, 1551. [CrossRef]
67. Zhang, H.; Zhao, F.; Tang, R.; Yu, Y.; Song, J.; Wang, Y.; Li, L. Two tonoplast MATE proteins function as turgor-regulating chloride channels in *Arabidopsis*. *Proc. Natl. Acad. Sci. USA* **2017**, *144*, E2036–E2045. [CrossRef]
68. Misra, V.A.; Wafula, E.K.; Wang, Y.; DePamphilis, C.W.; Timko, M.P. Genome-wide identification of MST, SUT and SWEET family sugar transporters in root parasitic angiosperms and analysis of their expression during host parasitism. *BMC Plant Biol.* **2019**, *19*, 196. [CrossRef]
69. Doidy, J.; Vidal, U.; Lemoine, R. Sugar transporters in Fabaceae, featuring SUT MST and SWEET families of the model plant *Medicago truncatula* and the agricultural crop *Pisum sativum*. *PLoS ONE* **2019**, *14*, e0223173. [CrossRef]
70. Wang, Y.; Chen, Y.; Wei, Q.; Wan, H.; Sun, C. Phylogenetic relationships of sucrose transporters (SUTs) in plants and genome-wide characterization of *SUT* genes in Orchidaceae reveal roles in floral organ development. *PeerJ* **2021**, *9*, e11961. [CrossRef]
71. Drew, D.; North, R.A.; Nagarathinam, K.; Tanabe, M. Structures and general transport mechanisms by the major facilitator superfamily (MFS). *Chem. Rev.* **2021**, *121*, 5289–5335. [CrossRef]
72. Jeena, G.S.; Kumar, S.; Shukla, R.K. Structure, evolution and diverse physiological roles of SWEET sugar transporters in plants. *Plant Mol. Biol.* **2019**, *100*, 351–365. [CrossRef]
73. Eom, J.; Chen, L.; Sosso, D.; Julius, B.T.; Lin, I.W.; Qu, X.; Braun, D.M.; Frommer, W.B. SWEETs, transporters for intracellular and intercellular sugar translocation. *Curr. Opin. Plant Biol.* **2015**, *25*, 53–62. [CrossRef] [PubMed]
74. Lalonde, S.; Frommer, W.B. SUT sucrose and MST monosaccharide transporter inventory of the Selaginella genome. *Front. Plant Sci.* **2012**, *3*, 24. [CrossRef] [PubMed]
75. Lemoine, R.; La Camera, S.; Atanassova, R.; Dédaldéchamp, F.; Allario, T.; Pourtau, N.; Bonnemain, J.-L.; Laloi, M.; Coutos-Thévenot, P.; Maurousset, L.; et al. Source-to-sink transport of sugar and regulation by environmental factors. *Front. Plant Sci.* **2013**, *4*, 272. [CrossRef] [PubMed]
76. Nookaraju, A.; Upadhyaya, C.P.; Pandey, S.K.; Young, K.E.; Hong, S.J.; Park, S.K.; Park, S.W. Molecular approaches for enhancing sweetness in fruits and vegetables. *Sci. Hortic.* **2010**, *127*, 1–15. [CrossRef]
77. Julius, B.T.; Leach, K.A.; Tran, T.M.; Mertz, R.A.; Braun, D.M. Sugar transporters in plants: New insights and discoveries. *Plant Cell Physiol.* **2017**, *58*, 1442–1460. [CrossRef]

78. Schulz, A.; Beyhl, D.; Marten, I.; Wormit, A.; Neuhaus, E.; Poschet, G.; Büttner, M.; Schneider, S.; Sauer, N.; Hedrich, R. Proton-driven sucrose symport and antiport are provided by the vacuolar transporters SUC4 and TMT1/2. *Plant J.* **2011**, *68*, 129–136. [CrossRef]
79. Weise, A.; Barker, L.; Kühn, C.; Lalonde, S.; Buschmann, H.; Frommer, W.B.; Ward, J.M. A new subfamily of sucrose transporters, SUT4, with low affinity/high capacity localized in enucleate sieve elements of plants. *Plant Cell* **2000**, *12*, 1345–1355. [CrossRef]
80. Carpaneto, A.; Geiger, D.; Bamberg, E.; Sauer, N.; Fromm, J.; Hedrich, R. Phloem-localized, proton-coupled sucrose carrier ZmSUT1 mediates sucrose efflux under the control of the sucrose gradient and the proton motive force. *J. Biol. Chem.* **2005**, *280*, 21437–21443. [CrossRef]
81. Gong, X.; Liu, M.; Zhang, L.; Ruan, Y.; Ding, R.; Ji, Y.; Zhang, N.; Zhang, S.; Farmer, J.; Wang, C. Arabidopsis *AtSUC2* and *AtSUC4*, encoding sucrose transporters, are required for abiotic stress tolerance in an ABA-dependent pathway. *Physiol. Plant.* **2015**, *153*, 119–136. [CrossRef]
82. Wang, D.; Liu, H.; Wang, H.; Zhang, P.; Shi, C. A novel sucrose transporter gene *IbSUT4* involves in plant growth and response to abiotic stress through the ABF-dependent ABA signaling pathway in Sweetpotato. *BMC Plant Biol.* **2020**, *20*, 157. [CrossRef]
83. Gao, D.; Knight, M.R.; Trewavas, A.J.; Sattelmacher, B.; Plieth, C. Self-reporting Arabidopsis expressing pH and [Ca^{2+}] indicators unveil ion dynamics in the cytoplasm and in the apoplast under abiotic stress. *Plant Physiol.* **2004**, *134*, 898–908. [CrossRef]
84. Santiago, J.P.; Ward, J.M.; Sharkey, T.D. *Phaseolus vulgaris* SUT1.1 is a high affinity sucrose-proton co-transporter. *Plant Direct* **2020**, *4*, e00260. [CrossRef]
85. Sharkey, T.D.; Zhang, R. High temperature effects on electron and proton circuits of photosynthesis. *J. Integr. Plant Biol.* **2010**, *52*, 712–722. [CrossRef]
86. Xu, Q.; Chen, S.; Yunjuan, R.; Chen, S.; Liesche, J. Regulation of sucrose transporters and phloem loading in response to environmental cues. *Plant Physiol.* **2018**, *176*, 930–945. [CrossRef]
87. Raghavendra, A.S.; Yin, Z.-H.; Heber, U. Light-dependent pH changes in leaves of C4 plants Comparison of the pH response to carbon dioxide and oxygen with that of C3 plants. *Planta* **1993**, *189*, 278–287. [CrossRef]
88. Slewinski, T.L. Diverse functional roles of monosaccharide transporters and their homologs in vascular plants: A physiological perspective. *Mol. Plant* **2011**, *4*, 641–662. [CrossRef]
89. Paulsen, P.A.; Custódio, T.F.; Pedersen, B.P. Crystal structure of the plant symporter STP10 illuminates sugar uptake mechanism in monosaccharide transporter superfamily. *Nat. Commun.* **2019**, *10*, 407. [CrossRef]
90. Bavnhøj, L.; Paulsen, P.A.; Flores-Canales, J.C.; Schiøtt, B.; Pedersen, B.P. Molecular mechanism of sugar transport in plants unveiled by structures of glucose/H^+ symporter STP10. *Nat. Plants* **2021**, *7*, 1409–1419. [CrossRef]
91. Klepek, Y.; Geiger, D.; Stadler, R.; Klebl, F.; Landouar-arsivaud, L. Arabidopsis POLYOL TRANSPORTER5, a new member of the monosaccharide transporter-like superfamily, mediates H^+-symport of numerous substrates, including *myo*-inositol, glycerol, and ribose. *Plant Cell* **2005**, *17*, 204–218. [CrossRef]
92. Li, C.; Meng, D.; Piñeros, M.A.; Mao, Y.; Dandekar, A.M.; Cheng, L. A sugar transporter takes up both hexose and sucrose for sorbitol-modulated in vitro pollen tube growth in apple. *Plant Cell* **2020**, *32*, 449–469. [CrossRef]
93. Schubert, M.; Koteyeva, N.K.; Wabnitz, P.W.; Santos, P.; Büttner, M.; Sauner, N.; Demchenko, K.; Pawlowski, K. Plasmodesmata distribution and sugar partitioning in nitrogen-fixing root nodules of *Datisca glomerata*. *Planta* **2011**, *233*, 139–152. [CrossRef]
94. Yamada, K.; Kanai, M.; Osakabe, Y.; Ohiraki, H.; Shinozaki, K.; Yamaguchi-Shinozaki, K. Monosaccharide absorption activity of Arabidopsis roots depends on expression profiles of transporter genes under high salinity conditions. *J. Biol. Chem.* **2011**, *286*, 43577–43586. [CrossRef]
95. Kong, W.; An, B.; Zhang, Y.; Yang, J.; Li, S.; Sun, T.; Li, Y. Sugar transporter proteins (STPs) in Gramineae crops: Comparative analysis, phylogeny, evolution, and expression profiling. *Cells* **2019**, *8*, 560. [CrossRef]
96. Muratore, C.; Espen, L.; Prinsi, B. Nitrogen uptake in plants: The plasma membrane root transport systems from a physiological and proteomic perspective. *Plants* **2021**, *10*, 681. [CrossRef]
97. Feng, H.; Fan, X.; Miller, A.J.; Xu, G. Plant nitrogen uptake and assimilation: Regulation of cellular pH homeostasis. *J. Exp. Bot.* **2020**, *71*, 4380–4392. [CrossRef]
98. Yao, X.; Nie, J.; Bai, R.; Sui, X. Amino acid transporters in plants: Identification and function. *Plants* **2020**, *9*, 972. [CrossRef]
99. Yang, G.; Wei, Q.; Huang, H.; Xia, J. Amino acid transporters in plant cells: A brief review. *Plants* **2020**, *9*, 967. [CrossRef]
100. Ortiz-Lopez, A.; Chang, H.-C.; Bush, D.R. Amino acid transporters in plants. *Biochim. Biophys. Acta* **2000**, *1465*, 275–280. [CrossRef]
101. Okumoto, S.; Pilot, G. Amino acid export in plants: A missing link in nitrogen cycling. *Mol. Plant* **2011**, *4*, 453–463. [CrossRef]
102. Dinkeloo, K.; Boyd, S.; Pilot, G. Update on amino acid transporter functions and on possible amino acid sensing mechanisms in plants. *Semin. Cell Dev. Biol.* **2018**, *74*, 105–113. [CrossRef]
103. Jack, D.L.; Paulsen, I.T.; Saier, J. The amino acid/polyamine/organocation (APC) superfamily of transporters specific for amino acids, polyamines and organocations. *Microbiology* **2000**, *146*, 1797–1814. [CrossRef]
104. Young, G.B.; Jack, D.L.; Smith, D.W.; Saier, M.H., Jr. The amino acid/auxin: Proton symport permease family. *Biochim. Biophys. Acta (BBA)-Biomembr.* **1999**, *1415*, 306–322. [CrossRef]
105. Zhao, C.; Pratelli, R.; Yu, S.; Shelley, B.; Collakova, E.; Pilot, G. Detailed characterization of the UMAMIT proteins provides insight into their evolution, amino acid transport properties, and role in the plant. *J. Exp. Bot.* **2021**, *72*, 6400–6417. [CrossRef]

106. Williams, L.E.; Nelson, S.J.; Hall, J.L. Characterization of solute transport in plasma membrane vesicles isolated from cotyledons of *Ricinus communis* L.-II. Evidence for a proton-coupled mechanism for sucrose and amino acid uptake. *Planta* **1990**, *182*, 540–545. [CrossRef]
107. Li, Z.C.; Bush, D.R. Structural determinants in substrate recognition by proton-amino acid symports in plasma membrane vesicles isolated from sugar beet leaves. *Arch. Biochem. Biophys.* **1992**, *294*, 519–526. [CrossRef]
108. Fischer, W.N.; Kwart, M.; Hummel, S.; Frommer, W.B. Substrate specificity and expression profile of amino acid transporters (AAPs) in *Arabidopsis*. *J. Biol. Chem.* **1995**, *270*, 16315–16320. [CrossRef]
109. Okumoto, S.; Koch, W.; Tegeder, M.; Fischer, W.N.; Biehl, A.; Leister, D.; Stierhof, Y.D.; Frommer, W.B. Root phloem-specific expression of the plasma membrane amino acid proton co-transporter AAP3. *J. Exp. Bot.* **2004**, *55*, 2155–2168. [CrossRef]
110. Montamat, F.; Maurousset, L.; Tegeder, M.; Frommer, W.; Delrot, S. Cloning and expression of amino acid transporters from broad bean. *Plant Mol. Biol.* **1999**, *41*, 259–268. [CrossRef]
111. Hirner, A.; Ladwig, F.; Stransky, H.; Okumoto, S.; Keinath, M.; Harms, A.; Frommer, W.B.; Koch, W. *Arabidopsis* LHT1 is a high-affinity transporter for cellular amino acid uptake in both root epidermis and leaf mesophyll. *Plant Cell* **2006**, *18*, 1931–1946. [CrossRef]
112. Rentsch, D.; Hirner, B.; Schmelzer, E.; Frommer, W.B. Salt stress-induced proline transporters and salt stress-repressed broad specificity amino acid permeases identified by suppression of a yeast amino acid permease-targeting mutant. *Plant Cell* **1996**, *8*, 1437–1446.
113. Wang, J.J.; Hou, Q.Q.; Li, P.H.; Yang, L.; Sun, X.C.; Benedito, V.A.; Wen, J.Q.; Chen, B.B.; Mysore, K.S.; Zhao, J. Diverse functions of multidrug and toxin extrusion (MATE) transporters in citric acid efflux and metal homeostasis in *Medicago truncatula*. *Plant J.* **2017**, *90*, 79–95. [CrossRef]
114. Zhao, H.; Ma, H.; Yu, L.; Wang, X.; Zhao, J. Genome-wide survey and expression analysis of amino acid transporter gene family in rice (*Oryza sativa* L.). *PLoS ONE* **2012**, *7*, e49210. [CrossRef]
115. Wan, Y.; King, R.; Mitchell, R.A.C.; Hassani-Pak, K.; Hawkesford, M.J. Spatiotemporal expression patterns of wheat amino acid transporters reveal their putative roles in nitrogen transport and responses to abiotic stress. *Sci. Rep.* **2017**, *7*, 5461. [CrossRef]
116. Ueda, A.; Shi, W.; Sanmiya, K.; Shono, M.; Takabe, T. Functional analysis of salt-inducible proline transporter of barley roots. *Plant Cell Physiol.* **2001**, *42*, 1282–1289. [CrossRef]
117. Tian, R.; Yang, Y.; Chen, M. Genome-wide survey of the amino acid transporter gene family in wheat (*Triticum aestivum* L.): Identification, expression analysis and response to abiotic stress. *Int. J. Biol. Macromol.* **2020**, *162*, 1372–1387. [CrossRef]
118. Zheng, L.; Ma, S.; Zhou, T.; Yue, C.; Hua, Y.; Huang, J. Genome-wide identification of Brassicaceae *B-BOX* genes and molecular characterization of their transcriptional responses to various nutrient stresses in allotetraploid rapeseed. *BMC Plant Biol.* **2021**, *21*, 288. [CrossRef]
119. Yancey, P.H. Organic osmolytes as compatible, metabolic and counteracting cytoprotectants in high osmolarity and other stresses. *J. Exp. Biol.* **2005**, *208*, 2819–2830. [CrossRef]
120. El Moukhtari, A.; Cabassa-Hourton, C.; Farissi, M.; Savouré, A. How does proline treatment promote salt stress tolerance during crop plant development? *Front. Plant Sci.* **2020**, *11*, 1127. [CrossRef]
121. Ben Ahmed, C.; Magdich, S.; Ben Rouina, B.; Sensoy, S.; Boukhris, M.; Ben Abdullah, F. Exogenous proline effects on water relations and ions contents in leaves and roots of young olive. *Amino Acids* **2011**, *40*, 565–573. [CrossRef]
122. Huang, Y.; Bie, Z.; Liu, Z.; Zhen, A.; Wang, W. Protective role of proline against salt stress is partially related to the improvement of water status and peroxidase enzyme activity in cucumber. *Soil Sci. Plant Nutr.* **2009**, *55*, 698–704. [CrossRef]
123. Lone, M.I.; Kueh, J.S.H.; Wyn Jones, R.G.; Bright, S.W.J. Influence of proline and glycinebetaine on salt tolerance of cultured barley embryos. *J. Exp. Bot.* **1987**, *38*, 479–490. [CrossRef]
124. Buell, C.R.; Joardar, V.; Lindeberg, M.; Selengut, J.; Paulsen, I.T.; Gwinn, M.L.; Dodson, R.J.; Deboy, R.T.; Durkin, A.S.; Kolonay, J.F.; et al. The complete genome sequence of the *Arabidopsis* and tomato pathogen *Pseudomonas syringae* pv. *tomato* DC3000. *Proc. Natl. Acad. Sci. USA* **2003**, *100*, 10181–10186. [CrossRef]
125. Liu, G.; Ji, Y.; Bhuiyan, N.H.; Pilot, G.; Selvaraj, G.; Zou, J.; Wei, Y. Amino acid homeostasis modulates salicylic acid-associated redox status and defense responses in *Arabidopsis*. *Plant Cell* **2010**, *22*, 3845–3863. [CrossRef]
126. Yang, H.; Postel, S.; Kemmerling, B.; Ludewig, U. Altered growth and improved resistance of *Arabidopsis* against *Pseudomonas syringae* by overexpression of the basic amino acid transporter *AtCAT1*. *Plant Cell Environ.* **2014**, *37*, 1404–1414. [CrossRef]
127. Song, J.T.; Lu, H.; Mcdowell, J.M.; Greenberg, J.T. A key role for ALD1 in activation of local and systemic defenses in *Arabidopsis*. *Plant J.* **2004**, *40*, 200–212. [CrossRef]
128. Song, J.T.; Lu, H.; Greenberg, J.T. Divergent roles in *Arabidopsis thaliana* development and defense of two homologous genes, *Aberrant Growth and Death2* and *Agd2-Like Defense Response Protein1*, encoding novel aminotransferases. *Plant Cell* **2004**, *16*, 353–366. [CrossRef]
129. Näsholm, T.; Kielland, K.; Ganeteg, U. Uptake of organic nitrogen by plants. *New Phytol.* **2008**, *182*, 31–48. [CrossRef]
130. Raven, J.A.; Smith, F.A. Nitrogen assimilation and transport in vacuolar land plants in relation to intraceuular pH regulation. *New Phytol.* **1976**, *76*, 415–431. [CrossRef]
131. Snowden, C.J.; Thomas, B.; Baxter, C.J.; Smith, J.A.C.; Sweetlove, L.J. A tonoplast Glu/Asp/GABA exchanger that affects tomato fruit amino acid composition. *Plant J.* **2015**, *81*, 651–660. [CrossRef]

132. Kamran, M.; Ramesh, S.A.; Gilliham, M.; Tyerman, S.D.; Bose, J. Role of TaALMT1 malate-GABA transporter in alkaline pH tolerance of wheat. *Plant Cell Environ.* **2020**, *43*, 2443–2459. [CrossRef]
133. Rolin, D.; Baldet, P.; Just, D.; Chevalier, C.; Biran, M.; Raymond, P. NMR study of low subcellular pH during the development of cherry tomato fruit. *Aust. J. Plant Physiol.* **2000**, *27*, 61–69.
134. Grewer, C.; Gameiro, A.; Zhang, Z.; Tao, Z.; Braams, S.; Rauen, T. Glutamate forward and reverse transport: From molecular mechanism to transporter-mediated release after ischemia. *IUBMB Life* **2008**, *60*, 609–619. [CrossRef]
135. Perez, C.; Koshy, C.; Yildiz, Ö.; Ziegler, C. Alternating-access mechanism in conformationally asymmetric trimers of the betaine transporter BetP. *Nature* **2012**, *490*, 126–130. [CrossRef]
136. Yernool, D.; Boudker, O.; Jin, Y.; Gouaux, E. Structure of a glutamate transporter homologue from *Pyrococcus horikoshii*. *Nature* **2004**, *431*, 811–818. [CrossRef]
137. Nogia, P.; Pati, P.K. Plant secondary metabolite transporters: Diversity, functionality, and their modulation. *Front. Plant Sci.* **2021**, *12*, 758202. [CrossRef]

International Journal of
Molecular Sciences

MDPI

Article

Photoactivation of Cell-Free Expressed Archaerhodopsin-3 in a Model Cell Membrane

Navid Khangholi †, Marc Finkler †, Ralf Seemann, Albrecht Ott * and Jean-Baptiste Fleury *

Experimental Physics and Center for Biophysics, Saarland University, 66123 Saarbrücken, Germany; navid.khangholi@physik.uni-saarland.de (N.K.); marc.finkler@uni-saarland.de (M.F.); r.seemann@physik.uni-saarland.de (R.S.)

* Correspondence: albrecht.ott@physik.uni-saarland.de (A.O.); jean-baptiste.fleury@physik.uni-saarland.de (J.-B.F.)

† Navid Khangholi and Marc Finkler contributed equally to this paper.

Abstract: Transmembrane receptor proteins are located in the plasma membranes of biological cells where they exert important functions. Archaerhodopsin (Arch) proteins belong to a class of transmembrane receptor proteins called photoreceptors that react to light. Although the light sensitivity of proteins has been intensely investigated in recent decades, the electrophysiological properties of pore-forming Archaerhodopsin (Arch), as studied in vitro, have remained largely unknown. Here, we formed unsupported bilayers between two channels of a microfluidic chip which enabled the simultaneous optical and electrical assessment of the bilayer in real time. Using a cell-free expression system, we recombinantly produced a GFP (green fluorescent protein) labelled as a variant of Arch-3. The label enabled us to follow the synthesis of Arch-3 and its incorporation into the bilayer by fluorescence microscopy when excited by blue light. Applying a green laser for excitation, we studied the electrophysiological properties of Arch-3 in the bilayer. The current signal obtained during excitation revealed distinct steps upwards and downwards, which we interpreted as the opening or closing of Arch-3 pores. From these steps, we estimated the pore radius to be 0.3 nm. In the cell-free extract, proteins can be modified simply by changing the DNA. In the future, this will enable us to study the photoelectrical properties of modified transmembrane protein constructs with ease. Our work, thus, represents a first step in studying signaling cascades in conjunction with coupled receptor proteins.

Keywords: Archaerhodopsin-3; lipid bilayer; microfluidics; cell-free gene expression

Citation: Khangholi, N.; Finkler, M.; Seemann, R.; Ott, A.; Fleury, J.-B. Photoactivation of Cell-Free Expressed Archaerhodopsin-3 in a Model Cell Membrane. *Int. J. Mol. Sci.* **2021**, *22*, 11981. https://doi.org/10.3390/ijms222111981

Academic Editor: Masoud Jelokhani-Niaraki

Received: 14 October 2021
Accepted: 3 November 2021
Published: 5 November 2021

1. Introduction

In biological cells, transmembrane proteins are embedded in the plasma membrane, composed of a lipid bilayer [1–3]. Certain transmembrane proteins act as receptors for external stimuli [4,5] such as light, mechanical stress, or the presence of specific molecular compounds [5–9]. Transmembrane receptor proteins transform the stimulus into signals that can be processed further downstream by the molecular machinery of the cell [2,4,10]. The stimulus produces structural or conformational changes in the transmembrane protein that will, for instance, result in an activation or deactivation of a specific ion channel, the release of a G protein (G protein-coupled receptors) or the activation of certain enzymes (enzyme-coupled receptors) [1,2,6,9,11–13].

Archaerhodopsin proteins (Arch) are among a class of transmembrane receptor proteins called photoreceptors that react to light [14,15]. Upon illumination with green light, Arch will undergo deprotonation of a Schiff base, resulting in proton pumping [16–19]. In this way, Arch actively transports protons through the membrane and out of the cell [14,18,19]. In vivo, the resulting proton gradient enables ATP synthases to produce ATP [18,20]. Photoreceptors typically consist of seven transmembrane helices and the chromophore retinal [16,18,21]. The structure of Arch corresponds to G protein-coupled receptors that include rhodopsin. Because bacteriorhodopsin and Arch share similar structures and functions, their activation is assumed to be the same (see Figure 1c). Rhodopsin acts as a photoreceptor as well [1,15,22]. However, Arch acts as an ion channel, not as a G protein-coupled receptor. There are different kinds of Arch, which have slightly different properties [21]. Besides being photo active, Arch-3 fluorescence is also voltage sensitive [16,23], and it is often used in optogenetics as a voltage sensor [16,17]. To our knowledge, an electrophysiological characterization of Arch-3 has not yet been reported in the literature [18].

In this work, we present a new and simple microfluidic approach [24] to study the conducting properties of ion channels. We incorporated recombinantly produced Arch-3-EGFP into a free-standing lipid bilayer that mimics a biological membrane. A cell-free expression system was used for the production and reconstitution of Arch-3-EGFP. Such in vitro systems enable better control of various biochemical parameters and processes than in vivo systems [25–28]. They are regularly used to study gene circuits or reaction cascades [29–31]. In vitro, proteins, in the form of wild-type proteins or with modifications, can easily be produced recombinantly by the simple addition of the corresponding coding DNA [32–34]. Moreover, in vitro systems can mimic in vivo systems, allowing the transfer of results to the in vivo situation [35–37]. Using this approach, we report the first electrophysiological characterization of single Arch-3 channels in a model cell membrane. The method presented here (sketched in Figure 1) can be understood as a first step towards the further investigation of different signaling cascades.

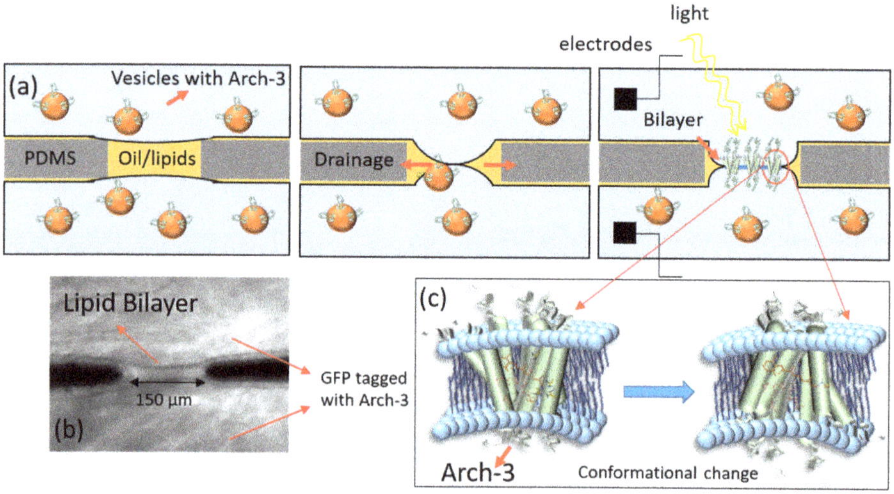

Figure 1. (**a**) A schematic of the formation of a lipid bilayer at the intersection of two microfluidic channels. First, the channel is filled with oil–lipid solution. Then, a cell-free expression reaction solution containing synthesized Arch-3-EGFP fused to the membrane of vesicles is injected into both channels. A remaining oil–filled lipid bilayer separates the two channels (left). Lipids from the oil phase and lipid vesicles with the inserted Arch-3-EGFP from the aqueous phase form a monolayer at both oil–water interfaces. As oil drains into the PDMS, the two interfaces gently meet to form a lipid bilayer containing Arch-3-EGFP (middle). The visualization of the inserted Arch-3-EGFP is achieved by exciting EGFP with blue light. AgCl electrodes inserted into the microfluidic device are used for electrical measurements (right). (**b**) The image shows the fluorescent signal of the EGFP tag from Arch-3 at the site of the suspended bilayer; PDMS elements remain dark. (**c**) The image shows the conformational changes of Arch-3 upon light exposure mediated by the deprotonation of retinal. This results in proton pumping which can be detected using electrophysiological measurements (scheme based on channel rhodopsin, adapted from [22]).

2. Results and Discussion

A DOPC/DPhPC bilayer containing Arch-3-EGFP was formed in a microfluidic device as described in the method section and as sketched in Figure 1. To verify the formation of a lipid bilayer, electrophysiological measurements were performed by applying a potential difference of 20 mV between both channels containing the reaction solution. The lipid bilayer separated the two ion-conducting water reservoirs, whereas the water–oil–water sandwich acted as a capacitor. Measuring the capacitance of such a sandwich in real time enabled the detection of bilayer formation; the graph in Figure 2a shows the related data from our experiments. The initial signal fluctuating around 10 pF corresponds to the situation with two monolayers separated by a macroscopic oil layer. The jump in the capacitance signal corresponds to the formation of a bilayer, a so-called zipping process. The following gradual increase in the capacitance demonstrates the growth of the bilayer area. This is due to the drainage of the oil to the PDMS at the plateau border.

Hydrostatic control of the flows enabled us to keep the bilayer area fairly constant for 1 h. The fluorescent image of EGFP tagged to Arch-3 under blue illumination, as shown in Figure 1b, confirmed the presence of Arch-3-EGFP in the vicinity of the lipid bilayer. After switching the laser illumination from blue to green, Arch-3 was activated, and an ion current was detected across the suspended lipid bilayer in real time upon light stimulation. The graph in Figure 2b shows the current intensity as a function of time, measured in the absence of light and after a short (~10 s) exposure to green laser light (532 nm) at ~40 s. Before light exposure, the current stayed almost constant. During the light pulse, the electrical current peaked, which means that ions were passing through the protein. After a few minutes, the signal decayed to its initial value. We interpret this observation as the signature of light-induced activation and subsequent deactivation of Arch-3. According to the literature, in any type of a rhodopsin photocycle, the deprotonation of the Schiff base after light excitation occurs within a range of picoseconds [38,39]. The deprotonation opens the channel; however, the subsequent recovery of the Schiff base requires milliseconds to several seconds [38–41]. Until recovery, the channel remains open for protons to pass through. Fitting an exponential decay to the curve from the time it begins to fall gives a recovery time τ of about 84 s.

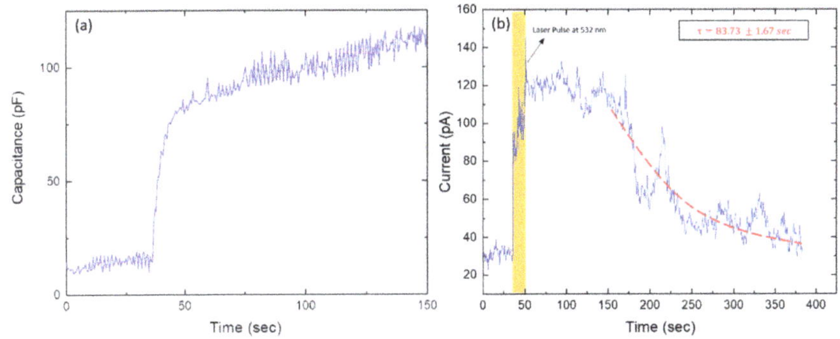

Figure 2. (**a**) Electrical capacitance as measured between the two microfluidic channels, separated by an oil phase (see Figure 1a) as a function of time. The signature of bilayer formation is the sudden increase in capacitance. (**b**) A real-time current recording of the bilayer containing Arch-3-EGFP. In the absence of light, we observe a current signal fluctuating around a constant value. The sudden jump in the current signal demonstrates the activation of Arch-3 caused by a green laser pulse applied to the bilayer at $\sim 40 \text{ s} \leq t \leq 50 \text{ s}$. The signal decays, with a time constant of ~84 s, to its initial dark value. The recovery time corresponds to the re-protonation of the Schiff base as reported in [38–40].

To characterize Arch-3 activation more deeply, we continuously exposed an Arch-3-containing bilayer to monochromatic light with a wavelength of 532 nm. During light exposure, we observed an overall increase in the conductance, which was composed of a stepwise increase and decrease in the current signal, as shown in Figure 3a. This signal is caused by a combination of opening and closing channels and the simultaneous fusion of Arch-3-EGFP containing vesicles from the aqueous phase with the bilayer. If Arch-3-EGFP is added, the current recording will expose a positive jump. The individual current steps ΔI, as indicated by the red line in Figure 3a, were obtained using the "change point analysis" algorithm as part of the software Origin (Origin 2021b; OriginLab Corporation, Northampton, MA, USA). In total, over 300 steps were analyzed. The histogram shown in Figure 3b displays the frequency of certain current steps ΔI grouped within intervals of 2.5 pA. From the histogram, it can be seen that most of the jumps were in the range of (16.25 ± 6.25) pA, where the full width at half maximum was used to determine the experimental uncertainty. The obtained value is in excellent agreement with the value found in the literature for bacteriorhodopsin (BR) [42]. The large current steps ΔI causing the asymmetric distribution towards larger current steps in the histogram could be due to multiple channels in the bilayer simultaneously being activated or deactivated.

Although Arch-3 is a proton pump, protons can passively pass while the channel is open. Considering the current step with the highest probability, i.e., the maximum peak of the histogram at ΔI = 16.25 pA, and assuming that Arch-3 has a passive opening in the bilayer, we can determine the radius of a single Arch-3 channel from the relation [43]

$$r = \sqrt{\frac{l\Delta I}{\pi GCV}} \tag{1}$$

where $l \approx 3$ nm is the length of the protein assumed to be close to the thickness of the bilayer, ΔI = 16.25 pA is the amplitude of the typical current jump for one channel opening or closing, and $V = 20$ mV is the applied voltage. The molarity C of the reaction solution was converted from the osmolarity, which was measured as 1.02 osmol/kg. Because of the complexity of the reaction solution, the number of dissociable particles per molecule (n) was assumed to be two, as is the case for NaCl. Thus, the molarity was obtained as $C = 0.5$ M. Using this value, the molar conductivity results as $G = 18.4$ S·M^{-1}·m^{-1}. The corresponding prediction of the pore radius of an Arch-3 channel, $r \approx (0.31 \pm 0.02)$ nm, is in excellent agreement with the radius of the pore of BR (0.4–1.1) nm [42] and of rhodopsin (0.45–0.7) nm [41], which also consist of seven transmembrane helices forming the same structure.

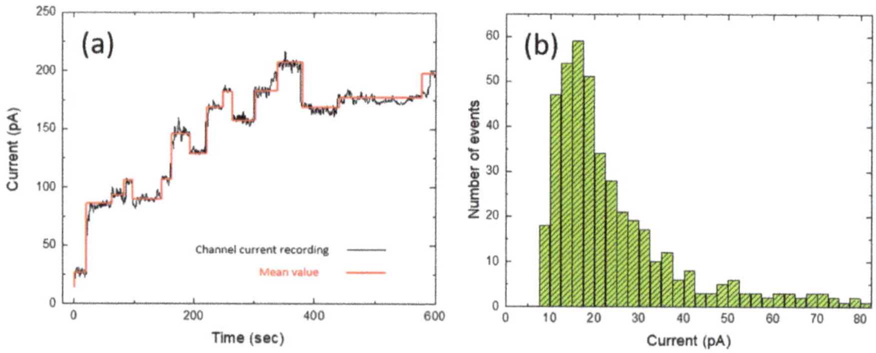

Figure 3. (**a**) Current signal as a function of time for an Arch-3-containing bilayer under continuous 532 nm laser illumination (black line). The red line was obtained by "change point analysis" from the software Origin (Origin 2021b; OriginLab Corporation, Northampton, Massachusetts, USA). It shows the mean value of each current step. (**b**) The histogram shows the frequency of certain current steps grouped in intervals of 2.5 pA.

3. Conclusions

In this paper, we presented a simple and new microfluidic approach to investigate the electrophysiological properties of the recombinantly produced transmembrane protein Arch-3 inserted in a free-standing DOPC/DPhPC bilayer. By applying a voltage across such an Arch-3-containing DOPC/DPhPC bilayer, the light-induced opening of individual Arch-3 ion channels could be observed. The corresponding pore radius of the Arch-3 ion channel was determined to be (0.31 ± 0.02) nm, which is in excellent agreement with values found for similar protein pores.

The in vitro system described here presents the advantage of quick testing and prototyping of modified Arch-3, since only the DNA needs to be adapted. Moreover, even non-canonical amino acids can be incorporated [28,44,45]. Because G protein-coupled receptors involved in many signaling cascades exhibit a similar structure, we expect our work to be helpful for in vitro studies focusing on this kind of protein. This may pave the way for the creation of artificial signaling cascades.

4. Materials and Methods

4.1. Gene Expression

For gene expression, a commercially available cell-free expression system (*E. coli* T7 S30 Extract System for Circular DNA; Promega, Madison, Wisconsin, USA) was used. The plasmid in our experiments was VV020: WT Arch-3-EGFP in pET28b, a gift from Adam Cohen (Addgene plasmid # 58488; http://n2t.net/addgene:58488 (last accessed on 4 September 2021); RRID: Addgene_58488; Addgene, Watertown, Massachusetts, USA) [16]. The gene expression was performed in the presence of SUV (small unilamellar vesicles). For this, first a solution containing vesicles was prepared as follows: 400 µL of S30 Premix without Amino Acids, 50 µL of Amino Acid Mixture Minus Cysteine at 1 mM, 50 µL of Amino Acid Mixture Minus Leucine at 1 mM and 500 µL of ultrapure water were mixed. To this solution, 1 mg of DPhPC was added and sonicated three times, applying the continuous cycle of 4 s pulse, 4 s break for 4 min, and 2 min pause between each cycle. After that, the solution was put into the fridge to rest for 24 h.

Gene expression reactions were performed as follows: The components of the cell-free expression system were combined to obtain a "mastermix" containing 40 µL of S30 Premix without Amino Acids, 5 µL of Amino Acid Mixture Minus Cysteine at 1 mM, 5 µL of Amino Acid Mixture Minus Leucine at 1 mM and 30 µL of T7 S30 Extract System for Circular DNA. To this mastermix, 10 nM of plasmid DNA was added and filled up with vesicles solution to achieve a total reaction volume of 100 µL. The expression reaction was then incubated at 37 °C for 48 h. The synthesized Arch-3-EGFP was incorporated into the vesicles. Each reaction solution was directly injected into the microfluidic device after expression and could no longer be used after 24 h.

4.2. Lipid Preparation, and Device and Bilayer Fabrication

DOPC (1,2-dioleoyl-sn-glycero-3-phosphocholine), and DPhPC (1,2-diphytanoyl-sn-glycero-3-phosphocholine) were used for bilayer preparation. The lipids were from Avanti Polar Lipids (Avanti Polar Lipids, Birmingham, Alabama, USA). To prepare the oil–lipid solution, 5 mg of lipids (1:1 DOPC/DPhPC) were dissolved in 1 mL of pure squalene oil (Sigma-Aldrich, St. Louis, Missouri, USA) at 45 °C while undergoing continuous stirring for 3 h.

The microfluidic chip was produced by standard soft lithographic protocols and consisted of Sylgard 184 bonded to a glass slide, see e.g., ref. [24] for fabrication details. The chip was designed with two side-to-side rectangular channels with a width of 500 µm and height of 100 µm, forming an X geometry (see Figure 4).

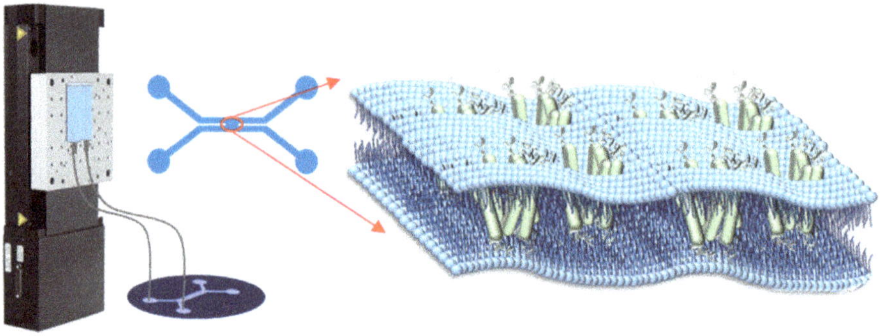

Figure 4. Design and structure of the microfluidic setup including the hydrostatic pressure system.

The membrane was formed across an orifice with a width of about 150 μm that connected the two parallel channels, as sketched in Figure 4. A hydrostatic pressure system was used to control the flow of the aqueous solution in the microfluidic chip [24]. The two inlets were connected to two syringes, which were fixed on a motorized stage, and the two outlets were left open. By adjusting the height of the motorized stage, positive or negative pressures could be applied to the channels causing the aqueous solution to move forward or backward.

For bilayer formation, the whole chip was first filled with the squalene oil containing dissolved lipids. Subsequently, the cell-free expression reaction solution containing synthesized Arch-3-EGFP proteins that were fused to vesicles was injected gently into both microfluidic channels, displacing the oil but leaving behind an oil inclusion at the orifice connecting the microfluidic channels. During this process the two oil–water interfaces were being decorated with a monolayer of lipids and Arch-3-EGFP. Due to the drainage of oil into the PDMS, the two lipid monolayers came into contact with each other, leading to the formation of a bilayer. While bilayer formation, the Arch-3-EGFP at the interface of the two monolayers fuses into the bilayer (sketched in Figure 1).

4.3. Microscope Setup and Electrical Measurements

An inverted epifluorescence microscope (Axio Observer Z1; Zeiss, Oberkochen, Germany) with 473 nm (blue) and 532 nm (green) laser illumination was used. As Arch-3 was tagged with enhanced green fluorescent protein (EGFP), we used the wavelength of 473 nm to excite the EGFP and monitor Arch-3 production (see Figure 2b). The electrical properties of the Arch-3-EGFP-containing bilayer were analyzed by electrophysiological measurements using a patch-clamp amplifier, EPC 10 USB (Heka Electronics, Reutlingen, Germany). For that purpose, Ag/AgCl electrodes were prepared by inserting a 5 cm-long silver wire into a borosilicate glass pipet containing 150 mM of NaCl electrolyte solution while applying 5 V for 30 min. The prepared electrodes were inserted into the inlets of the microfluidic device. The current passing through the bilayer was measured over time using an excitation signal with an amplitude of 20 mV and a time resolution of 100 ms.

Author Contributions: Conceptualization, N.K., M.F., R.S., A.O. and J.-B.F.; methodology, N.K., M.F. and J.-B.F.; software, N.K. and M.F.; validation, N.K. and M.F.; formal analysis, N.K. and M.F.; investigation, N.K. and M.F.; resources, R.S. and A.O.; data curation, N.K. and M.F.; writing—original draft preparation, N.K. and M.F. with help from A.O.; writing—review and editing, N.K., M.F., R.S., A.O. and J.-B.F.; visualization, N.K. and M.F.; supervision, R.S., A.O. and J.-B.F.; project administration, R.S., A.O. and J.-B.F.; funding acquisition, R.S., A.O. and J.-B.F. All authors have read and agreed to the published version of the manuscript.

Funding: This work was supported by the German Research Foundation (Projects B4, and C1 of CRC 1027) and, in part, by the Human Frontier Science Program (HFSP, RGP0037/2015).

Conflicts of Interest: The authors declare no conflict of interest.

References

1. Lefkowitz, R.J. Seven transmembrane receptors: Something old, something new. *Acta Physiol.* **2007**, *190*, 9–19. [CrossRef] [PubMed]
2. Gudermann, T.; Nürnberg, B.; Schultz, G. Receptors and G proteins as primary components of transmembrane signal transduction-Part 1. G-protein-coupled receptors: Structure and function. *J. Mol. Med.* **1995**, *73*, 51–63. [CrossRef] [PubMed]
3. Orwick-Rydmark, M.; Lovett, J.E.; Graziadei, A.; Lindholm, L.; Hicks, M.R.; Watts, A. Detergent-free incorporation of a seven-transmembrane receptor protein into nanosized bilayer lipodisq particles for functional and biophysical studies. *Nano Lett.* **2012**, *12*, 4687–4692. [CrossRef]
4. Zhulin, I.B.; Nikolskaya, A.N.; Galperin, M.Y. Common extracellular sensory domains in transmembrane receptors for diverse signal transduction pathways in Bacteria and Archaea. *J. Bacteriol.* **2013**, *185*, 285–294. [CrossRef] [PubMed]
5. Stock, J.B.; Stock, A.M.; Mottonen, J.M. Signal transduction in bacteria. *Nature* **1990**, *344*, 395–400. [CrossRef]
6. Árnadóttir, J.; Chalfie, M. Eukaryotic mechanosensitive channels. *Annu. Rev. Biophys.* **2010**, *39*, 111–137. [CrossRef]
7. Nagel, G.; Szellas, T.; Huhn, W.; Kateriya, S.; Adeishvili, N.; Berthold, P.; Ollig, D.; Hegemann, P.; Bamberg, E. Channelrhodopsin-2, a directly light-gated Cation-Selective Membrane Channel. *Proc. Natl. Acad. Sci. USA* **2003**, *100*, 13940–13945. [CrossRef]
8. Sandoval, P.J.; Santiago, J. In Vitro Analytical Approaches to Study Plant Ligand-Receptor Interactions. *Plant Physiol.* **2020**, *182*, 1697–1712. [CrossRef]
9. Stora, T.; Lakey, J.H.; Vogel, H. Ion-channel gating in transmembrane receptor proteins: Functional activity in tethered lipid membranes. *Angew. Chem. Int. Ed.* **1999**, *38*, 389–392. [CrossRef]
10. Simon, M.I.; Strathmann, M.P.; Gautam, N. Diversity of G proteins in Signal Transduction. *Science* **1991**, *252*, 802–808. [CrossRef]
11. Pierce, K.L.; Premont, R.T.; Lefkowitz, R.J. Seven-transmembrane receptors. *Nat. Rev. Mol. Cell Biol.* **2002**, *3*, 639–650. [CrossRef] [PubMed]
12. Bethani, I.; Skånland, S.S.; Dikic, I.; Acker-Palmer, A. Spatial organization of transmembrane receptor signalling. *EMBO J.* **2010**, *9*, 2677–2688. [CrossRef]
13. Haswell, E.S.; Phillips, R.; Rees, D.C. Mechanosensitive channels: What can they do and how do they do it? *Structure* **2011**, *19*, 1356–1369. [CrossRef] [PubMed]
14. Mukohata, Y.; Sugiyama, Y.; Ihara, K.; Yoshida, M. An Australian Halobacterium contains a novel Proton Pump Retinal Protein: Archaerhodopsin. *Biochem. Biophys. Res. Commun.* **1988**, *151*, 1339–1345. [CrossRef]
15. Ganapathy, S.; Kratz, S.; Chen, Q.; Hellingwerf, K.J.; de Groot, H.J.; Rothschild, K.J.; de Grip, W.J. Redshifted and Near-infrared Active Analog Pigments Based upon Archaerhodopsin-3. *Photochem. Photobiol.* **2019**, *95*, 959–968. [CrossRef]
16. Maclaurin, D.; Venkatachalam, V.; Lee, H.; Cohen, A.E. Mechanism of voltage-sensitive fluorescence in a microbial rhodopsin. *Proc. Natl. Acad. Sci. USA* **2013**, *110*, 5939–5944. [CrossRef]
17. Flytzanis, N.C.; Bedbrook, C.N.; Chiu, H.; Engqvist, M.K.; Xiao, C.; Chan, K.Y.; Sternberg, P.W.; Arnold, F.H.; Gradinaru, V. Archaerhodopsin variants with enhanced voltage-sensitive fluorescence in mammalian and *Caenorhabditis elegans* neurons. *Nat. Commun.* **2014**, *5*, 4894. [CrossRef] [PubMed]
18. Juarez, J.F.B.; Judge, P.J.; Adam, S.; Axford, D.; Vinals, J.; Birch, J.; Kwan, T.O.; Hoi, K.K.; Yen, H.-Y.; Vial, A. Structures of the archaerhodopsin-3 transporter reveal that disordering of internal water networks underpins receptor sensitization. *Nat. Commun.* **2021**, *12*, 629. [CrossRef]
19. El-Gaby, M.; Zhang, Y.; Wolf, K.; Schwiening, C.J.; Paulsen, O.; Shipton, O.A. Archaerhodopsin Selectively and Reversibly Silences Synaptic Transmission through Altered pH. *Cell Rep.* **2016**, *16*, 2259–2268. [CrossRef]
20. Mukohata, Y.; Sugiyama, Y.; Ihara, K. Photophosphorylation Elements in Halobacteria: An A-type ATP Synthase and Bacterial Rhodopsins. *J. Bioenerg. Biomembr.* **1992**, *24*, 547–553. [CrossRef]
21. Cao, Z.; Ding, X.; Peng, B.; Zhao, Y.; Ding, J.; Watts, A.; Zhao, X. Novel expression and characterization of a light driven proton pump archaerhodopsin 4 in a *Halobacterium salinarum* strain. *Biochim. Biophys. Acta* **2015**, *1847*, 390–398. [CrossRef]
22. Deisseroth, K.; Hegemann, P. The form and function of channelrhodopsin. *Science* **2017**, *357*, eaan5544. [CrossRef] [PubMed]
23. Gong, Y.; Li, J.Z.; Schnitzer, M.J. Enhanced Archaerhodopsin Fluorescent Protein Voltage Indicators. *PLoS ONE* **2013**, *8*, e66959. [CrossRef]
24. Khangholi, N.; Seemann, R.; Fleury, J.-B. Simultaneous measurement of surface and bilayer tension in a microfluidic chip. *Biomicrofluidics* **2020**, *14*, 024117. [CrossRef] [PubMed]
25. Shin, J.; Noireaux, V. An *E. coli* Cell-Free Expression Toolbox: Application to Synthetic Gene Circuits and Artificial Cells. *ACS Synth. Biol.* **2012**, *1*, 29–41. [CrossRef]
26. Finkler, M.; Ott, A. Bead-based assay for spatiotemporal gene expression control in cell-free transcription–translation systems. *Biotechniques* **2019**, *66*, 29–33. [CrossRef] [PubMed]
27. Finkler, M.; Kurt, Ö.; Grimm, F.; Hartz, P.; Ott, A. A bead-based method for the removal of the amino acid lysine from cell-free transcription-translation systems. *J. Biotechnol.* **2020**, *324S*, 100024. [CrossRef]
28. Finkler, M.; Ravanbodshirazi, S.; Grimm, F.; Hartz, P.; Ott, A. Full incorporation of the noncanonical amino acid hydroxylysine as a surrogate for lysine in green fluorescent protein. *Bioorg. Med. Chem.* **2021**, *41*, 116207. [CrossRef]
29. Sunami, T.; Hosoda, K.; Suzuki, H.; Matsuura, T.; Yomo, T. Cellular compartment model for exploring the effect of the lipidic membrane on the kinetics of encapsulated biochemical reactions. *Langmuir* **2010**, *26*, 8544–8551. [CrossRef]

30. Izri, Z.; Garenne, D.; Noireaux, V.; Maeda, Y.T. Gene Expression in on-Chip Membrane-Bound Artificial Cells. *ACS Synth. Biol.* **2019**, *8*, 1705–1712. [CrossRef]
31. Nevenzal, H.; Noach-Hirsh, M.; Skornik-Bustan, O.; Brio, L.; Barbiro-Michaely, E.; Glick, Y.; Avrahami, D.; Lahmi, R.; Tzur, A.; Gerber, D. A high-throughput integrated microfluidics method enables tyrosine autophosphorylation discovery. *Commun. Biol.* **2019**, *2*, 42. [CrossRef]
32. Silverman, A.D.; Karim, A.S.; Jewett, M.C. Cell-free gene expression: An expanded repertoire of applications. *Nat. Rev. Genet.* **2020**, *21*, 151–170. [CrossRef] [PubMed]
33. Katzen, F.; Chang, G.; Kudlicki, W. The past, present and future of cell-free protein synthesis. *Trends Biotechnol.* **2005**, *23*, 150–156. [CrossRef] [PubMed]
34. Shin, J.; Noireaux, V. Efficient cell-free expression with the endogenous *E. Coli* RNA polymerase and sigma factor 70. *J. Biol. Eng.* **2010**, *4*, 8. [CrossRef]
35. McSweeney, M.A.; Styczynski, M.P. Effective Use of Linear DNA in Cell-Free Expression Systems. *Front. Bioeng. Biotechnol.* **2021**, *9*, 715328. [CrossRef] [PubMed]
36. Senoussi, A.; Lee Tin Wah, J.; Shimizu, Y.; Robert, J.; Jaramillo, A.; Findeiss, S.; Axmann, I.M.; Estevez-Torres, A. Quantitative Characterization of Translational Riboregulators Using an in Vitro Transcription-Translation System. *ACS Synth. Biol.* **2018**, *7*, 1269–1278. [CrossRef] [PubMed]
37. Damiati, S.; Mhanna, R.; Kodzius, R.; Ehmoser, E.-K. Cell-Free Approaches in Synthetic Biology Utilizing Microfluidics. *Genes* **2018**, *9*, 144. [CrossRef]
38. Koch, M.; Dencher, N.; Oesterhelt, D.; Plöhn, H.; Rapp, G.; Büldt, G. Time-resolved X-ray diffraction study of structural changes associated with the photocycle of bacteriorhodopsin. *EMBO J.* **1991**, *10*, 521–526. [CrossRef]
39. Nakasako, M.; Kataoka, M.; Amemiya, Y.; Tokunaga, F. Crystallographic characterization by X-ray diffraction of the M-intermediate from the photo-cycle of bacteriorhodopsin at room temperature. *FEBS Lett.* **1991**, *292*, 73–75.
40. Dencher, N.A.; Dresselhaus, D.; Zaccai, G.; Büldt, G. Structural changes in bacteriorhodopsin during proton translocation revealed by neutron diffraction. *Proc. Natl. Acad. Sci. USA* **1989**, *86*, 7876–7879. [CrossRef]
41. Montal, M.; Darszon, A.; Trissl, H.W. Transmembrane channel formation in rhodopsin-containing bilayer membranes. *Nature* **1977**, *267*, 221–225. [CrossRef] [PubMed]
42. Radzwill, N.; Gerwert, K.; Steinhoff, H.J. Time-resolved detection of transient movement of helices F and G in doubly spin-labeled bacteriorhodopsin. *Biophys. J.* **2001**, *80*, 2856–2866. [CrossRef]
43. Heo, P.; Ramakrishnan, S.; Coleman, J.; Rothman, J.E.; Fleury, J.B.; Pincet, F. Highly Reproducible Physiological Asymmetric Membrane with Freely Diffusing Embedded Proteins in a 3D-Printed Microfluidic Setup. *Small* **2019**, *15*, 1900725. [CrossRef] [PubMed]
44. Worst, E.G.; Exner, M.P.; De Simone, A.; Schenkelberger, M.; Noireaux, V.; Budisa, N.; Ott, A. Cell-free expression with the toxic amino acid canavanine. *Bioorg. Med. Chem. Lett.* **2015**, *25*, 3658–3660. [CrossRef] [PubMed]
45. Worst, E.G.; Exner, M.P.; De Simone, A.; Schenkelberger, M.; Noireaux, V.; Budisa, N.; Ott, A. Residue-specific Incorporation of Noncanonical Amino Acids into Model Proteins Using an *Escherichia coli* Cell-free Transcription-translation System. *J. Vis. Exp.* **2016**, *114*, 54273. [CrossRef]

International Journal of *Molecular Sciences*

MDPI

Article

Tetraoctylammonium, a Long Chain Quaternary Ammonium Blocker, Promotes a Noncollapsed, Resting-Like Inactivated State in KcsA

Ana Marcela Giudici [1,†], Clara Díaz-García [2,†], Maria Lourdes Renart [1,†], Ana Coutinho [2,3], Manuel Prieto [2], José M. González-Ros [1,*] and José Antonio Poveda [1,*]

1 Instituto de Investigación, Desarrollo e Innovación en Biotecnología Sanitaria de Elche (IDiBE), Instituto de Biología Molecular y Celular (IBMC), Universidad Miguel Hernández, E-03202 Elche, Spain; marcela@umh.es (A.M.G.); lrenart@umh.es (M.L.R.)

2 Institute for Bioengineering and Bioscience (IBB), Instituto Superior Técnico, Universidade de Lisboa, 1049-001 Lisboa, Portugal; clara.dg93@gmail.com (C.D.-G.); ana.coutinho@tecnico.ulisboa.pt (A.C.); manuel.prieto@tecnico.ulisboa.pt (M.P.)

3 Departamento de Química e Bioquímica, Faculty of Sciences, Universidade de Lisboa, 1749-016 Lisboa, Portugal

* Correspondence: gonzalez.ros@umh.es (J.M.G.-R.); ja.poveda@umh.es (J.A.P.); Tel.: +34-966-658-757 (J.M.G.-R.); +34-966-658-466 (J.A.P.)

† These three authors contributed equally to this work.

Citation: Giudici, A.M.; Díaz-García, C.; Renart, M.L.; Coutinho, A.; Prieto, M.; González-Ros, J.M.; Poveda, J.A. Tetraoctylammonium, a Long Chain Quaternary Ammonium Blocker, Promotes a Noncollapsed, Resting-Like Inactivated State in KcsA. *Int. J. Mol. Sci.* **2021**, *22*, 490. https://doi.org/10.3390/ijms22020490

Received: 10 December 2020
Accepted: 1 January 2021
Published: 6 January 2021

Abstract: Alkylammonium salts have been used extensively to study the structure and function of potassium channels. Here, we use the hydrophobic tetraoctylammonium (TOA^+) to shed light on the structure of the inactivated state of KcsA, a tetrameric prokaryotic potassium channel that serves as a model to its homologous eukaryotic counterparts. By the combined use of a thermal denaturation assay and the analysis of homo-Förster resonance energy transfer in a mutant channel containing a single tryptophan (W67) per subunit, we found that TOA^+ binds the channel cavity with high affinity, either with the inner gate open or closed. Moreover, TOA^+ bound at the cavity allosterically shifts the equilibrium of the channel's selectivity filter conformation from conductive to an inactivated-like form. The inactivated TOA^+–KcsA complex exhibits a loss in the affinity towards permeant K^+ at pH 7.0, when the channel is in its closed state, but maintains the two sets of K^+ binding sites and the W67–W67 intersubunit distances characteristic of the selectivity filter in the channel resting state. Thus, the TOA^+–bound state differs clearly from the collapsed channel state described by X-ray crystallography and claimed to represent the inactivated form of KcsA.

Keywords: potassium channels; tetraalkylammonium salts; protein thermal stability; homo-FRET; C-type inactivation; binding affinity; selectivity filter conformation; steady-state and time-resolved fluorescence anisotropy

1. Introduction

Quaternary ammonium salts (QAs) have long been used as channel blockers in the characterization of K^+ channels [1,2]. Initial studies were performed on the giant squid axon and *D. melanogaster* Shaker channels [3–9], and it has already been suggested that QAs bind to the conduction pore, hence impeding the K^+ current. Nonetheless, the length of the alkyl chains and the hydrophobicity of the different QAs determine differences in their molecular mechanisms of blockade. Thus, while the shorter-chain QAs seem to block the channel by simply obstructing the conduction pathway, the longer-chain, more hydrophobic derivatives are also believed to induce a slow channel inactivation process [6].

The elucidation of molecular details on the interaction between QAs and K^+ channels have advanced significantly with the resolution of the X-ray structures of QAs bound to KcsA, a prokaryotic K^+ channel from *Streptomyces lividans*. This channel is a homotetrameric membrane protein, where each monomer includes two α-helical transmembrane segments (TM1 and TM2) and N- and C-terminal cytoplasmic ends (Scheme 1). The four C-terminal ends are arranged as a helical bundle, with a conformation that is sensitive to pH, acting as a channel gate (inner gate) [10]. On the other hand, the channel pore includes an aqueous cavity, a short tilted helix, and a selectivity filter (SF) with the sequence TVGYG, homologous to that of the eukaryotic K^+ channels [11], that constitutes a second channel gate (outer gate). The backbone carbonyls of the SF residues conform to four K^+ binding sites (sites S1–S4, from the extracellular to the intracellular side) [11,12], which can adopt different conformations at high or low K^+ concentrations [12–22]. The conformation at a low K^+ concentration shows no ions at the center of the SF (sites S2 and S3), thus adopting a "collapsed" structure that impedes ion flow through it. K^+ binds to S1 and S4 in this conformation, with an average occupancy of just one ion distributed between those two sites. However, at high K^+ concentrations, a conformational change is induced by a second K^+ entering the filter, with a final average occupancy of two K^+ ions per channel, either at the S1–S3 or the S2–S4 positions, thus enabling ion conduction [12,13,17,22]. The nonpermeant Na^+, on the other hand, does not induce such a conformational change and shows an average occupancy of one ion per channel at the S1 and S4 sites [23].

In terms of functional activity, KcsA undergoes a cycle that includes four main different states, which reveals the concerted action of the two channel gates. At neutral pH and in the presence of K^+, the channel is in a closed/conductive resting state, whereby the cytoplasmic inner gate impedes ion flow, while the SF (the outer gate) displays a conductive form. At acidic pH, the inner gate opens, allowing ion flow in this open/conductive state and making KcsA a proton-activated channel [24–26]. However, this is not a stable state, and the outer gate evolves to a conformation reminiscent of C-type inactivation in eukaryotic K^+ channels [27–29], which impedes ion flow in this open/inactivated form [27,28,30]. Such an inactivation process is modulated by a network of interactions that includes the so-called inactivation triad, i.e., residues E71, D80, and W67 from each subunit [17]. The cycle is completed when the pH returns to neutrality, which closes the inner gate, causing the transient closed/inactivated state to evolve to the initial closed/conductive resting state [31].

The crystallographic studies on KcsA complexed to different QAs reveal that the QA binding site for the hydrophilic ammonium head group is located in the internal water-filled cavity of the channel, directly underneath the innermost cation binding site (S4) of the SF (see Scheme 1). The QAs are further stabilized in the cavity through the insertion of their alkyl chains of varying lengths into the hydrophobic channel wall so that this hydrophobic component becomes an important source of binding stability. Thus, hydrophobic compounds such as TBA^+ (tetrabutylammonium), THA^+ (tetrahexylammonium) and TOA^+ (tetrabutylammonium) bind to KcsA with very high affinity (nM range) [32,33]. In the particular case of TOA^+–KcsA complexes, the X-ray structure also reveals that the channel SF is in a collapsed conformation at pH 7.0 and high K^+ concentrations, i.e., the S2 and S3 K^+ binding sites are absent, similar to that previously seen in the collapsed X-ray structure of the channel alone in the presence of very low K^+ concentrations. This led those authors to conclude that such a collapsed structure is an inactivated state induced by the binding of TOA^+. However, those X-ray studies used an L90C KcsA mutant without the C-terminal domain, in which the additional presence of a Fab fragment bound to the extracellular channel loop is suspected to restrict the conformational plasticity of the SF [17,34]. Indeed, previous studies from our laboratory on several inactivated models of KcsA, some of which were also predicted as collapsed from X-ray studies, have shown that the stack of K^+ binding sites in the inactivated filters remains accessible to cations, as in the resting channel. Therefore, rather than being collapsed, the inactivated SF seems "resting-like" [35]. In apparent agreement with our observations, other authors found only modest conformational changes in the G77 residue during inactivation compared to the resting state [36], further supporting the tenet that the SF remains "resting-like" upon inactivation, with all four K^+ binding sites accessible to cations. Our goal in this paper is to contribute to the elucidation of this controversy on the conformation of the SF in the inactivated channel state by studying in detail the effects of TOA^+ binding to the KcsA channel, which, up to now, was believed to result in an inactivated channel with a collapsed SF [33].

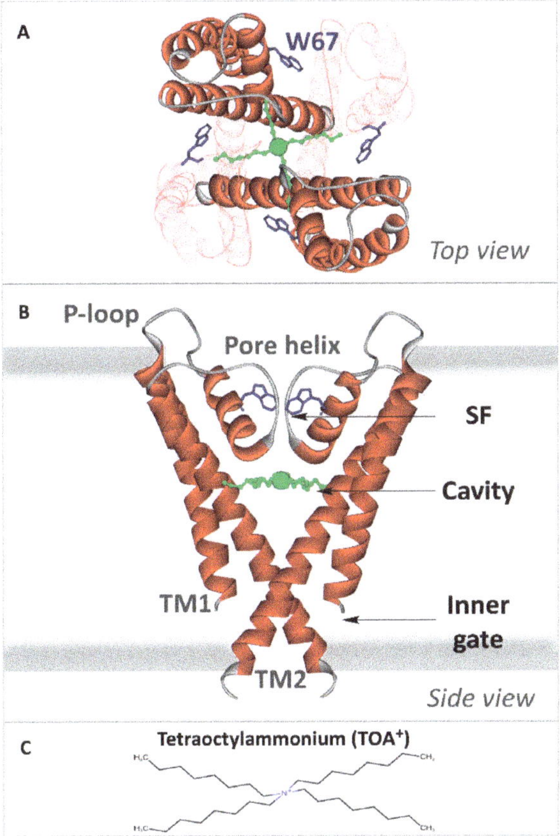

Scheme 1. Transmembrane portion of KcsA. Crystallographic structure of the C-terminal truncated KcsA in the closed state with TOA^+ (green balls and sticks, with a larger ball representing the quaternary nitrogen) bound at the cavity (PDB: 2W0F). In the top view (panel (**A**)), two of the four subunits appear faintly drawn to facilitate the observation. The four W67 residues are depicted as blue sticks. In the side view (panel (**B**)), only two of the four monomers have been drawn as a solid red ribbon for the sake of clarity. Each monomer consists of two transmembrane helices (TM1 and TM2) connected by the P-loop region, a short, tilted pore helix, and the selectivity filter (SF). The thick grey lines indicate the membrane limits. Both views illustrate how the TOA^+ alkyl chains traverse the TM2 helix. Panel (**C**) represents the tetraoctylammonium ion in perspective.

2. Results

2.1. TOA^+ Binds to Wild-Type KcsA with High Affinity and Diminishes the Thermal Stability of the Channel

An assay based on the thermal denaturation of the KcsA protein was previously developed to study the binding of different channel ligands to KcsA, including permeant or nonpermeant cations, membrane lipids, and others [35,37–40]. Figure 1A,B shows examples of thermal denaturation curves to illustrate that the addition of TOA^+ at micromolar concentrations to the detergent-solubilized channel leads to a concentration-dependent decrease in the thermal stability of the wild-type protein, both at pH 7.0 and pH 4.0. Because of the destabilizing effects of TOA^+ on KcsA, these experiments were always conducted at a constant 10-mM concentration of Na^+ to ensure that the tetrameric channel does not dissociate into monomers at the different TOA^+ concentrations tested (Figure 1A,B insets).

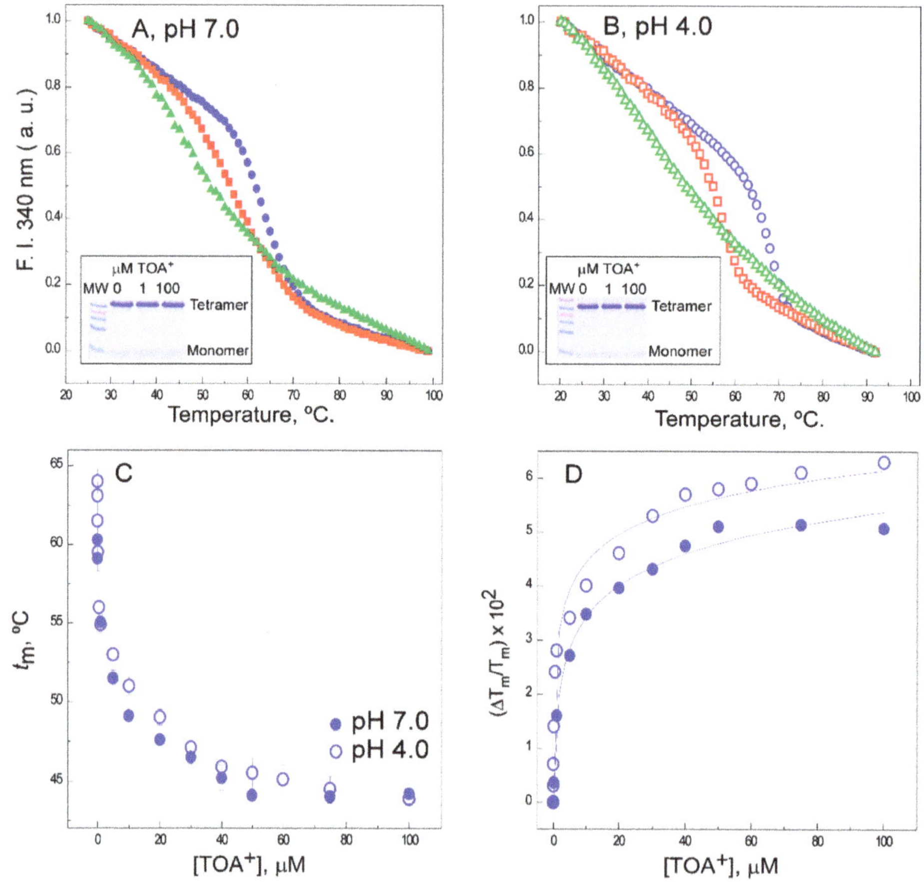

Figure 1. Effects of tetraoctylammonium (TOA^+) on the thermal denaturation of wild-type (WT) KcsA. The intrinsic fluorescence at pH 7.0 (panel (**A**)) and pH 4.0 (panel (**B**)) and in the absence of TOA^+ (full and empty circles, respectively) or upon the addition of 1 μM (full and empty squares, respectively) or 100 μM (full and empty triangles, respectively) of TOA^+. The insets show SDS-PAGE analysis of the different samples to illustrate the integrity of tetrameric WT KcsA at room temperature. Panel (**C**) shows the dependence of the midpoint temperature of the protein denaturation process (t_m) of WT KcsA with increasing concentrations of TOA^+, either at pH 7.0 (•) or pH 4.0 (○). The results are the average t_m (in Celsius) ± S.D. from three independent titrations. Panel (**D**) illustrates the fitting of Equation (1) to the experimental data from Panel (**C**) (see Methods). The apparent dissociation constants for the TOA^+-channel complexes and their 95% confidence intervals estimated at pH 7.0 and pH 4.0 were 1.8 (0.54–6.02) $\times 10^{-7}$ M and 3.6 (1.89–6.98) $\times 10^{-8}$ M, respectively. The observed differences between such dissociation constant (K_D) values were not statistically significant.

The midpoint temperatures (t_ms) from the thermal denaturation curves at different TOA^+ concentrations were used to build titration binding curves, such as those shown in Figure 1C. The fitting of such curves to a simple two-state binding equilibrium (see Methods) allows us to estimate the apparent dissociation constants (K_Ds) for the binding of TOA^+ to WT KcsA, which were in the 10^{-7} and 10^{-8} M range at pH 7.0 and pH 4.0, respectively (Figure 1D). These indicate that TOA^+ has a very high affinity for binding to the channel under either experimental condition. This is in contrast with previous observations on the binding to KcsA of tetrabutylammonium (TBA^+), a shorter acyl chain QA, for which the binding affinity at pH 7.0 (K_D ~5×10^{-9} M) is similar to that reported here for TOA^+ but drops five orders of magnitude at pH 4.0 (K_D~3.5×10^{-4} M) [35]. Such a dramatic difference in binding affinities could be rationalized based on the crystallographic information on the binding site for these compounds in the channel protein [12,32,33,41] As indicated in our introduction, the channel-bound QAs are further stabilized at the cavity through the interaction of their alkyl chains with the hydrophobic channel wall. In fact, the long alkyl chains of TOA^+ completely traverse the channel protein wall [33], so that hydrophobic interactions become particularly important to stabilize the TOA^+–KcsA complex.

2.2. The Presence of TOA^+ in the Cavity Diminishes the Affinity of WT KcsA to Bind K^+ in the Closed Channel State

The availability of cation binding sites in the SF of the TOA^+–KcsA complex has been explored by thermal denaturation experiments of such complexes in the presence of increasing concentrations of either permeant K^+ or nonpermeant Na^+. In these experiments, an excess TOA^+ concentration of 100 μM, identical to that used in the crystallographic studies referenced above, was maintained constant throughout the titrations with the cations. Figure 2A,C shows representative binding curves for Na^+, at both pH 7.0 and pH 4.0, respectively. Binding curves for Na^+ to WT KcsA in the absence of TOA^+ [37] are also included in all panels to facilitate comparison.

In these conditions, it is known that Na^+ binds to a single set of sites provided by the S1 and S4 crystallographic sites in a nonconductive pore conformation [23] (PDB 2ITC). This binding process has a K_D in the millimolar range, slightly higher at pH 4.0 than at pH 7.0 (Table 1), which should correspond to the overall K_D for Na^+ binding to the alluded S1 and S4 sites. In the TOA^+–KcsA complex, access to the S4 site is blocked by the presence of TOA^+; however, there is still Na^+ binding to the available S1 site (Figure 2A,C), although the extent of thermal stabilization (on top of the thermal destabilization caused by TOA^+) is lower and its K_D decreases an order of magnitude compared to the samples in the absence of TOA^+ (Figure 2B,D and Table 1). These observations on Na^+ binding to the available S1 site in the QA–KcsA complex are quite similar at pH 7.0 and pH 4.0 and analogous to the results previously seen in the presence of TBA^+ [35,39].

Table 1. Apparent dissociation constants (K_Ds) for the binding of Na^+ and K^+ to the WT KcsA channel at pH 7.0 and pH 4.0 in the absence or presence of 100 μM TOA^+. Mean K_D values given here come from the experiments reported in Figure 2, Figure 3 and Figure 5. Since the estimated K_D values are derived from a logarithmic function (Equation (1)), we used the 95% confidence intervals of these values for statistical comparisons instead of giving mean ± S.D. values since only parametric analysis is appropriate on the logarithmic scale for such data. [a] Significant difference with respect to the same sample in the absence of TOA^+ ($p < 0.05$). [b] Significant difference with respect to the same sample in the absence of TOA^+ at pH 4.0 ($p < 0.05$).

Tested Cation	Sets of Binding Sites Detected	WT KcsA, pH 7.0				WT KcsA, pH 4.0			
		No Added TOA^+		+100 μM TOA^+		No Added TOA^+		+100 μM TOA^+	
		K_D (M) from Tm	95% CI	K_D (M) from Tm	95% CI	K_D (M) from Tm	95% CI	K_D (M) from Tm	95% CI
Na^+	1	3.3×10^{-3}	$(2.5–4.3) \times 10^{-3\,b}$	1.4×10^{-2}	$(0.9–2.1) \times 10^{-2\,a,b}$	7.7×10^{-3}	$(6.6–8.9) \times 10^{-3}$	1.3×10^{-2}	$(0.9–1.8) \times 10^{-2\,a,b}$
K^+	2	2.8×10^{-7}	$(2.0–3.9) \times 10^{-7\,b}$	5.0×10^{-5}	$(3.3–7.5) \times 10^{-5\,a,b}$	3.8×10^{-4}	$(2.1–6.9) \times 10^{-4}$	1.9×10^{-4}	$(0.9–4.9) \times 10^{-4}$
		4.2×10^{-3}	$(3.2–5.4) \times 10^{-3\,b}$	1.6×10^{-3}	$(1.2–2.1) \times 10^{-2\,a,b}$	1.1×10^{-1}	$(0.7–1.5) \times 10^{-1}$	1.0×10^{-1}	$(0.8–1.4) \times 10^{-1}$
		W67 KcsA, pH 7.0				W67 KcsA, pH 4.0			
		K_D(M) from Tm	95% CI	K_D(M) from Tm	95% CI	K_D(M) from Tm	95% CI	K_D(M) from Tm	95% CI
K^+	2	1.0×10^{-5}	$(0.7–1.5) \times 10^{-5\,b}$	3.2×10^{-5}	$(1.2–8.6) \times 10^{-5\,b}$	4.8×10^{-4}	$(2.7–8.7) \times 10^{-4}$	1.2×10^{-4}	$(1.0–1.4) \times 10^{-4\,a,b}$
		9.4×10^{-4}	$(5.8–14.0) \times 10^{-4\,b}$	5.2×10^{-3}	$(3.3–8.1) \times 10^{-3\,a,b}$	1.4×10^{-2}	$(1.1–1.7) \times 10^{-2}$	8.6×10^{-3}	$(7.1–10.4) \times 10^{-3\,a,b}$
		K_D(M) from $\langle r \rangle_{SS}$	95% CI	K_D(M) from $\langle r \rangle_{SS}$	95% CI	K_D(M) from $\langle r \rangle_{SS}$	95% CI	K_D(M) from $\langle r \rangle_{SS}$	95% CI
K^+	1	4.5×10^{-4}	$(3.6–5.4) \times 10^{-4\,b}$	4.3×10^{-3}	$(3.4–5.2) \times 10^{-3\,a,b}$	4.4×10^{-2}	$(4.3–4.6) \times 10^{-2}$	1.9×10^{-2}	$(1.2–2.5) \times 10^{-2\,a,b}$

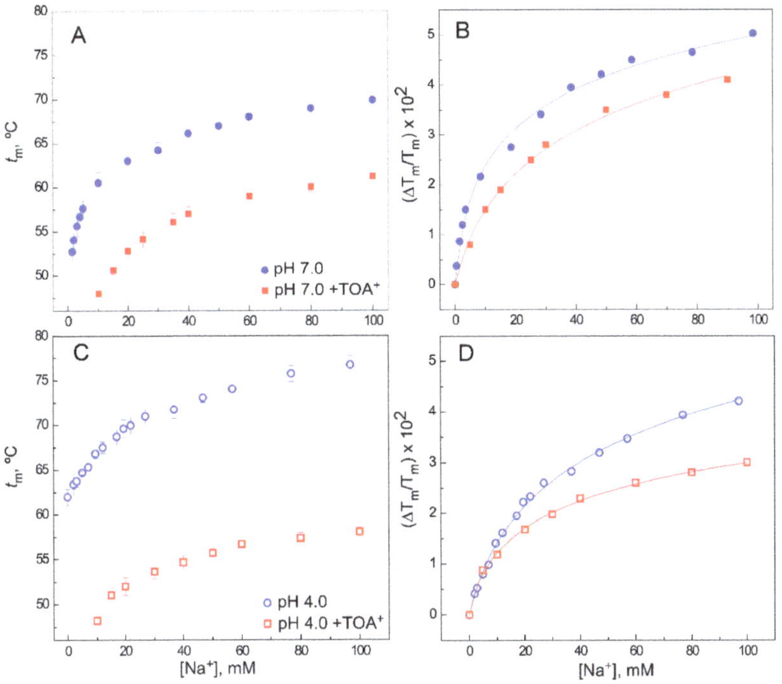

Figure 2. Effect of TOA^+ on Na^+ binding to WT KcsA. Panels (**A**,**C**) illustrate Na^+ binding to the WT KcsA channel at pH 7.0 (panel (**A**)) or pH 4.0 (panel (**C**)), in the absence (circles) and presence (squares) of 100 μM TOA^+, monitored through the Na^+-concentration-dependence of the midpoint temperature of the protein denaturation process (t_m). Each experimental point is the average t_m (in Celsius) ± S.D. (n = 3). Panels (**B**,**D**) show the fitting of Equation (1) to the experimental data from panels (**A**,**C**), respectively (see Methods). The apparent K_D values estimated for the above binding events are given in Table 1.

Experiments similar to the above were also conducted at increasing K^+ concentrations instead of Na^+. As reported previously in the absence of QAs [37], K^+ binds to two different sets of binding sites in WT KcsA at pH 7.0 (Figure 3A), which is consistent with crystallographic evidence on the ability of permeant cations to induce concentration-dependent transitions between nonconductive and conductive conformations of the SF [12,13]. The first set of such sites, assigned to the crystal S1 and S4 sites, shows a high affinity for K^+ (micromolar K_D), thus securing displacement of potentially competing nonpermeant cations. The second set of sites results from the contribution of all S1 to S4 crystallographic sites, is available only to permeant cations when the filter is in the conductive conformation and shows low affinity (millimolar K_D) to favor cation dissociation and permeation. Figure 3B,C also shows that in the presence of a saturating concentration of TOA^+, two different sets of sites are still available for K^+ binding in the TOA^+–KcsA complex, which, therefore, is clearly noncollapsed. This indicates that the TOA^+-bound channel retains the ability to undergo K^+ concentration-dependent transitions between different conformations of its SF. Despite such similarity, it is observed that the binding curve, mainly in the lower K^+ concentration range, shows a lower slope than in the absence of TOA^+, indicating a loss in K^+ binding affinity. Indeed, Table 1 shows that the K_D values for the two K^+ binding events in the TOA^+–KcsA complex are increased with respect to those in the absence of the QA. This is particularly noticeable in the first, high affinity K^+ binding event in which the K_D values differed by approximately two orders of magnitude. It should be noted that in the presence of TOA^+, the entrance to the channel's SF through the S4 site is permanently blocked by the bound TOA^+. In the first K^+ binding event, which takes place in the low

K^+ concentration range, the SF is in a nonconductive state, collapsed at the S2/S3 sites. This, along with the TOA^+ blockade of the S4 site, indicates that the decreased affinity for K^+ under the TOA^+ blockade should be attributed to the binding of K^+ to the S1 site in the nonconductive channel state. Nonetheless, as the K^+ concentration increases, a second K^+ binding event takes place, causing the filter to undergo a conformational transition in which the permeant cation reaches internal binding sites within the pore to provide the characteristic increase in thermal stability to the protein. Figure 3D shows the results from experiments similar to those described in the previous paragraph but now conducted at pH 4.0 to induce channel inactivation. These experiments are further complicated because pH 4.0 causes an additional thermal stabilization of the channel protein compared to pH 7.0 [35]. Still, as described above for the resting channel at pH 7.0, both in the WT KcsA alone or when complexed to TOA^+, the two sets of K^+ binding sites are also detected in the K^+ titrations of all samples at pH 4.0, when the channel is inactivated (Figure 3E,F). This was previously reported as a common feature in three different models of inactivated KcsA channels [35]. As expected for an inactivated state, all WT KcsA samples at pH 4.0 show a decrease in the affinity for K^+. Indeed, Table 1 shows that when compared to the pH 7.0 samples, the K_Ds for K^+ binding at pH 4.0 increase approximately three and two orders of magnitude, respectively, for the first and second K^+ binding events. Such effects on the binding affinity caused by pH-induced channel inactivation are comparable to those caused by the presence of TOA^+ bound to the resting state of the channel at pH 7.0. Interestingly, in contrast to TOA^+, the shorter chain-length TBA^+ does not critically change the interaction between the SF and the permeant cations [32,33,39].

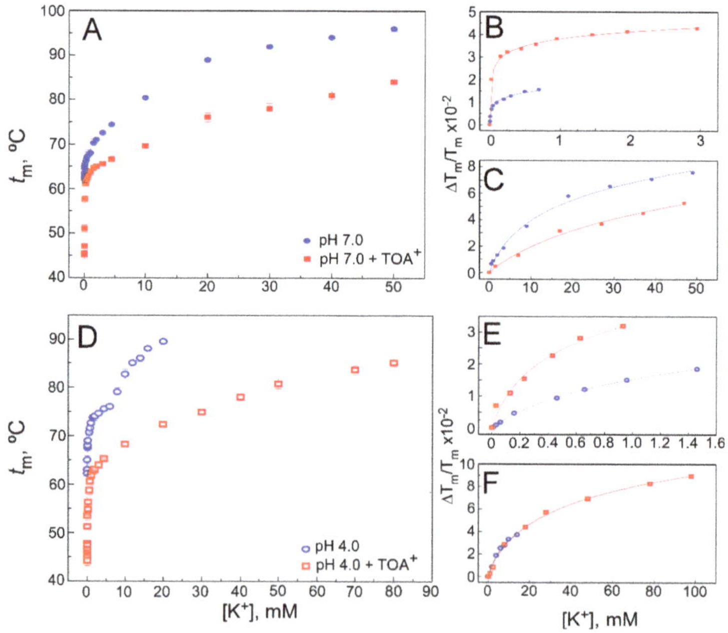

Figure 3. Effect of TOA^+ on K^+ binding to WT KcsA. Panels (**A**,**D**) illustrate K^+ binding to the WT KcsA channel in the absence (circles) and presence of 100 μM TOA^+ (squares) at pH 7.0 (panel (**A**)) or pH 4.0 (panel (**D**)), monitored through the K^+-concentration-dependence of the midpoint temperature of the protein denaturation process (t_m). Each experimental point is the average *tm* (in Celsius) ± S.D. (n = 3). When the low (panels (**B**,**E**)) and high (panels (**C**,**F**)) K^+ concentration ranges are analyzed separately, Equation (1) adequately fits the data from panels (**A**,**D**), indicating that two different sets of K^+ binding sites are present in the WT KcsA. The apparent K_D values estimated for the above binding events are given in Table 1.

2.3. The Homo-FRET Process in the W67 KcsA Mutant Reports Differences between the Nonconductive, Conductive, and Inactivated Conformations of the Selectivity Filter

We recently reported an analytical framework to analyze the homo-Förster resonance energy transfer (homo-FRET) within a quadruple mutant W26, 68, 87, 113F KcsA channel bearing a single tryptophan residue (W67) per subunit (hereby called W67 KcsA) to characterize the interplay between the pore helix conformation and the cation occupancy at the SF [34] (Figure 4A). The ion channel activity of such W67 KcsA mutant was previously shown to be very similar to that of the WT KcsA channel [34].

The time-resolved anisotropy decays of this mutant protein at pH 7.0 (Figure 4B) show that the rate of the homo-FRET process (k_1) increases in the presence of increasing K^+ concentrations and, consequently, the steady-state anisotropy ($<r>_{SS}$) and the W67–W67 intersubunit distance decrease along with it, further defining the characteristic two consecutive K^+ binding events in the mutant channel (Figure 4C,D).

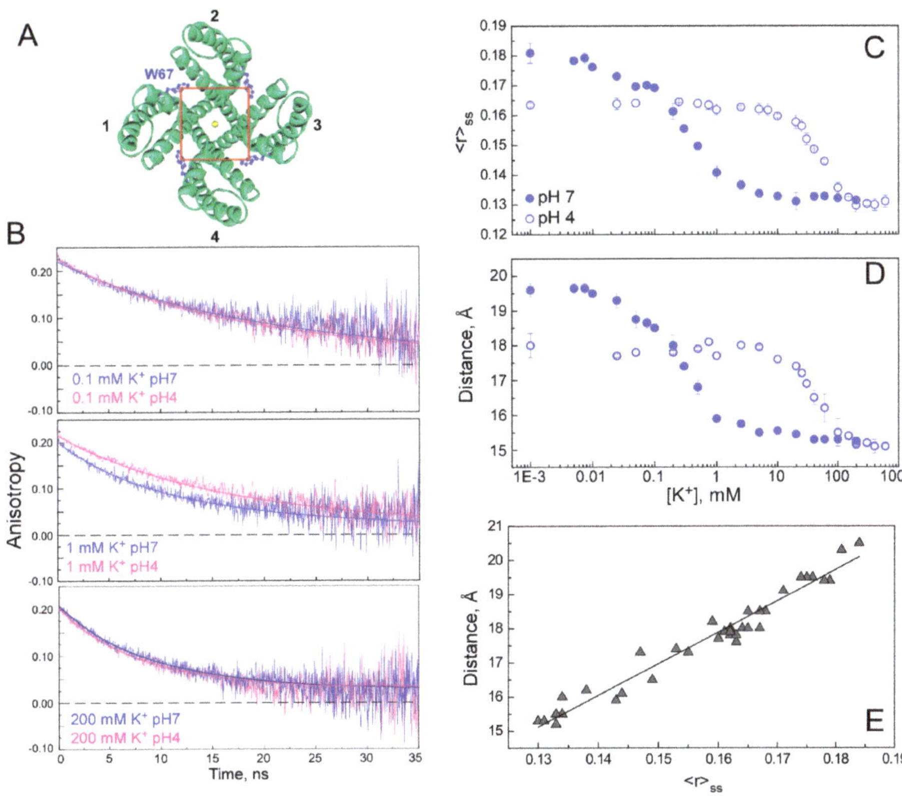

Figure 4. Influence of pH and K^+ concentration on the homo-Förster resonance energy transfer (homo-FRET) process among the tryptophan residues in the W67 KcsA channel. Panel (**A**) shows a schematic top view of the tetrameric structure of KcsA (PDB: 1K4C; each monomer numbered 1 to 4) and the location of the W67 residues. Panel (**B**) illustrates representative fluorescence anisotropy decays obtained for W67 KcsA in the presence of 0.1, 1, and 200 mM K^+ at both pH 7.0 (inner gate closed) and pH 4.0 (inner gate open). Equation (3) was then fitted to each data set, and the W67–W67 intersubunit distances were calculated from Equation (4). Panels (**C**,**D**) show the changes in the steady-state anisotropy $<r>_{SS}$ (panel (**C**)) and in the W67–W67 intersubunit lateral distances calculated from the time-resolved anisotropy decays (panel (**D**)) at different K^+ concentrations at both pH 7.0 and pH 4.0. Each titration curve is built from duplicate measurements from independent experiments (average ± S.D.) at each of the 20 to 25 different K^+ concentrations covering the indicated concentration range. Panel (**E**) shows the linear correlation between the $<r>_{SS}$ and W67–W67 lateral distances (distance = 92.165 × $<r>_{SS}$ + 3.015; r^2 = 0.960).

Different from the samples at pH 7.0, we found that the first K^+ binding event at $K^+ < 0.1$ mM was no longer detected at pH 4.0, where the channel's inner gate is open and the SF is inactivated. To test whether or not this first K^+ binding event is present at pH 4.0, we ran complementary thermal denaturation experiments on the mutant W67 channel. In contrast to the observations from steady-state or time-resolved anisotropy fluorescence measurements, the thermal denaturation experiments showed that the W67 KcsA mutant behaves just like the WT-like channel as both K^+ binding events were detected at either pH 7.0 or pH 4.0 (Figure 5).

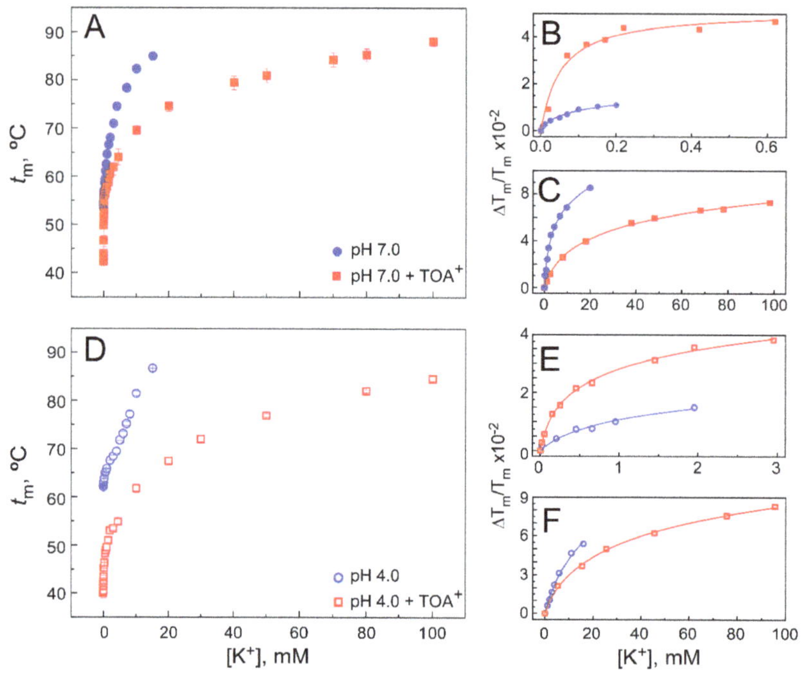

Figure 5. Effect of TOA^+ on K^+ binding to W67 KcsA. Panels (**A**,**D**) illustrate K^+ binding to the W67 KcsA channel in the absence (circles) and presence of 100 μM TOA^+ (squares) at pH 7.0 (panel (**A**)) or pH 4.0 (panel (**D**)), monitored through the K^+-concentration-dependence of the midpoint temperature of the protein denaturation process (t_m). Each experimental point is the average t_m (in Celsius) ± S.D. ($n = 3$). As to the WT KcsA from Figure 3, when the low (panels (**B**,**E**)) and high (panels (**C**,**F**)) K^+ concentration ranges are analyzed separately, Equation (1) adequately fits the data from panels (**A**,**D**), which is evidence that the two different sets of K^+ binding sites are present in W67 KcsA. The apparent K_D values estimated for the above binding events are given in Table 1.

Such apparent discrepancy between the two techniques could be simply explained by assuming that when the inner gate is opened by the acidic pH, the conformational changes involved in thermal stabilization at the low μM K^+ range take place without significant modifications at the pore helix level, where the W67 reporter is located, as reflected by the maintenance of the steady-state anisotropy $<r>_{SS}$ (~0.165) or the intersubunit distances (~18 Å), characteristics of the collapsed SF at low K^+ concentrations and pH 7.0.

The comparison of anisotropy results from W67 KcsA at pH 7.0 and pH 4.0 at intermediate cation concentrations (the second K^+ binding event) shows a clear decrease in K^+ binding affinity in the inactivated, pH 4.0 state (Figure 4C,D). This is confirmed by parallel observations from thermal denaturation experiments using either WT or W67 KcsA channels (see Figures 3 and 5 and Table 1). Additionally, when the K^+ concentration is raised above 0.1 mM at pH 7.0 (channel in the closed-conductive conformation) or

5–10 mM at pH 4.0 (channel in the open-inactivated form), the <r>$_{SS}$ and W67–W67 intersubunit distances decrease progressively to similar values of ~ 0.130 and 15 Å, respectively, in both cases. This indicates that the final conformations adopted by the SF at high K^+ are indistinguishable in terms of intersubunit W67 distances (see Table 2), regardless of pH.

Table 2. Comparison of the W67–W67 intersubunit distances calculated from the available X-ray crystallography data and from the homo-FRET analysis of the W67 KcsA mutant channel.

Inner Gate Conformation	Ionic Condition	*Intersubunit Distances* (Å)		
		C(δ2)-C(ε2) Inter-Tryptophan Lateral Distances from X-ray Structures (a)		Homo-FRET Analysis (b)
		Extracellular Fab, 1–125 KcsA	Intracellular Fab, full-length KcsA	
Closed	Low K^+	17.8 (1K4D)	n.a. (c)	18.5 ± 0.2
	High K^+	17.3 (1K4C)	15.9 (3EFF)	15.4 ± 0.2
	Low K^+ + TOA^+	n.a.	n.a.	19.1 ± 0.3
	High K^+ + TOA^+	17.6 (2W0F)	n.a.	16.1 ± 0.2
Open	Low K^+	n.a.	n.a.	17.9 ± 0.2 (e)
	High K^+	17.7 (d) (3F5W)	15.2 (d) (3PJS)	15.3 ± 0.1 (e)
	Low K^+ + TOA^+	n.a.	n.a.	18.4 ± 0.2 (e)
	High K^+ + TOA^+	n.a.	n.a.	16.2 ± 0.3 (e)

(a) PDB accession numbers of each crystallographic structure are given in parentheses. (b) The calculated distances from the homo-FRET analysis in the W67 mutant channel represent the mean ± S.D. of at least two independent experiments performed at pH 7 (closed state) or pH 4.0 (open state). (c) n.a., not available. (d) The X-ray data from the open states were collected at pH 7.0 using constitutively open mutant channels. (e) The opening of the inner gate was induced by acidic pH.

Interestingly, at both pH conditions, the K^+-dependent changes in the steady-state anisotropy values and the W67–W67 intersubunit distances calculated from the time-resolved anisotropy decays (Figure 4C,D) show an excellent linear correlation. Therefore, the intersubunit distance for a given experimental condition can be easily estimated by simply interpolating the steady-state anisotropy value in the correlation curve shown in Figure 4E. Such a correlation curve was built with the data presented in this work, as well as data previously determined for saturating concentrations of different cations [34].

2.4. The Presence of TOA^+ Bound at the Cavity Promotes the Inactivated State of the Selectivity Filter and Allosterically Changes the Pore Helix Conformation

Once the characterization of the K^+ binding behavior of the W67 KcsA mutant channel at pH 7.0 and 4.0 was completed, we ran similar experiments but now in the presence of TOA^+ in order to test the effect of this channel blocker on the SF and pore helix conformation and dynamics. Prior to that, control thermal denaturation assays showed that TOA^+ binds to the W67 mutant channel in a manner similar to that observed when WT KcsA was used in the experiments. Indeed, the apparent dissociation constants and the corresponding confidence intervals for the TOA^+–W67 KcsA complex, estimated at pH 7.0 and pH 4.0, were 10.7 (10.1–11.4) $\times$ 10^{-8} M and 5.2 (3.6–7.4) $\times$ 10^{-8} M, respectively. Furthermore, thermal denaturation assays of K^+ binding to the W67 mutant channel in the presence of 100 μM TOA^+ showed that binding of K^+ to two different sets of sites was detected at either pH 7.0 or pH 4.0 (Figure 5). This is also similar to that previously observed in the WT protein and indicates that the stack of K^+ binding sites at the SF remains available upon binding of TOA^+ to the W67 mutant channel cavity.

In the homo-FRET experiments, two main effects of TOA^+ binding to the W67 KcsA channel were detected: first, on the $\langle r\rangle_{SS}$ (and, thus, on the W67–W67 intersubunit distance) and, second, on the affinity of the SF to bind K^+. As to the former, Figure 6A,C shows that the presence of TOA^+ promotes an increase in the $\langle r\rangle_{SS}$ values at all the K^+ concentrations tested compared to samples prepared in the absence of TOA^+. Such an effect is much more noticeable at pH 7.0 than at pH 4.0 when the channel is already inactivated by the acidic pH. Indeed, the titration curve from the TOA^+–containing samples at pH 7.0 closely resembles those obtained at pH 4.0, with or without TOA^+, suggesting that similarly to the acidic pH, TOA^+ by itself causes inactivation of the channel SF at pH 7.0. The $\langle r\rangle_{SS}$ values from above were interpolated from the correlation curve shown in Figure 4E to yield the corresponding W67–W67 intersubunit distances. As concluded from the $\langle r\rangle_{SS}$ values, it is observed that the intersubunit distances are increased in the TOA^+–containing samples throughout the K^+ titration curve and that such effects are more noticeable at pH 7.0 than at pH 4.0 (Figure 6B,D). This increase in the intersubunit distance at the SF and the resulting loosening of its structure seems consistent with the decreased thermal stability induced by TOA^+ in these samples (Figures 3 and 5). Finally, the intersubunit distances of several samples prepared in the presence of TOA^+ were also determined experimentally from the time-resolved anisotropy decays. Such experimental values showed an excellent agreement with those calculated by interpolation from the steady-state anisotropy (black symbols in Figure 6B,D), attesting to the reliability of the simpler interpolation approach.

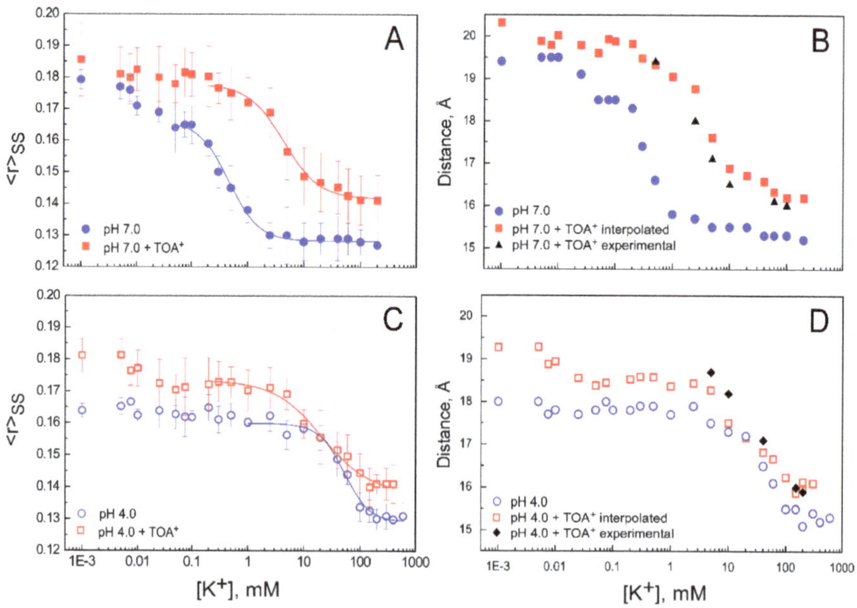

Figure 6. Effect of TOA^+ on the conformation of the W67 KcsA channel. Panels (**A**) and (**C**) show a representative experiment of the dependence of the W67 $\langle r\rangle_{SS}$ on K^+ concentration at both pH 7.0 and pH 4.0, in the absence or presence of 100 μM TOA^+. The anisotropy values correspond to the average of ten measurements ± S.D. The solid lines represent the best fit of Equation (10) to the data. The apparent K_Ds for the above binding events were calculated from two independent experiments, and their average values, along with the confidence intervals, are given in Table 1. Panels (**B**,**D**) illustrate the dependence of the W67–W67 intersubunit distances on K^+ concentration. The results from the samples without TOA^+ were obtained experimentally from time-resolved anisotropy decays, as explained in Figure 4. In the samples containing 100 μM TOA^+, such distances were obtained by simply interpolating the $\langle r\rangle_{SS}$ values using the correlation plot from Figure 4E. Nonetheless, the intersubunit distances in several samples containing 100 μM TOA^+ were also determined from time-resolved anisotropy decays (black symbols) to test the goodness of the interpolation procedure.

As to the effects of TOA^+ on the affinity of K^+ to bind to the SF, the titration curves in Figure 6A,B show that at pH 7.0, the sigmoidal decrease in $<r>_{SS}$ or in the intersubunit distance in the TOA^+-containing samples occurs at higher K^+ concentrations than in its absence. This attests to a TOA^+-induced loss in the affinity of the SF to bind K^+. Again, this is not as noticeable in the samples at pH 4.0, in which the channel is already inactivated.

As indicated previously, the homo-FRET experiments monitor quite precisely the second K^+ binding event to the KcsA channel, which corresponds to the K^+-induced transitions from the nonconductive to either the conductive (pH 7.0) or the inactivated (pH 4.0) states. Therefore, in an attempt to quantitate the observed loss in affinity from the anisotropy changes, an analysis was performed by fitting the model described in Equation (10) (see Methods) to the data corresponding to the second K^+ binding event in Figure 6A,C. At pH 7.0, the K_D value increases by an order of magnitude when TOA^+ is present (0.45 to 4.3 mM), whereas at pH 4.0, it remains fairly constant (44 to 19 mM). Such K_D values derived from homo-FRET are in qualitative agreement with the K_D values estimated from thermal denaturation (Table 1 and Figure 5). Again, these observations further suggest that the binding of TOA^+ causes inactivation of the channel SF at pH 7.0.

In addition to the main observations on the K_Ds for K^+ binding, it was observed that the fit of Equation (10) to the data required a cooperativity parameter (h) higher than 1 only in the case of the pH 4.0 samples in the absence of TOA^+ (a "h" parameter of 2, with a 95% confidence interval of 1.8 to 2.2, was used in those samples). This finding adds an additional feature to the pH-induced inactivated state of the channel since a cooperative binding process seems involved in facilitating the final inactivated conformation. A similar cooperative effect on the K^+ binding to the open channel was previously described by NMR [21]. Such a presumed cooperative process, however, occurs at fairly high K^+ concentrations and disappears when TOA^+ is bound to the channel cavity.

3. Discussion

In this paper, a combined approach using thermal denaturation and homo-FRET assays was used to characterize the effects of TOA^+ binding to KcsA. The addition of TOA^+ to the channel protein induces a concentration-dependent decrease in protein thermal stability, opposite to the stabilizing effect observed of a shorter-chain QA, TBA^+ [35,39]. When comparing the binding of these two QAs at pH 7.0 and pH 4.0, we observed that the closed-channel state at pH 7.0 exhibits a similar K_D for both TOA^+ and TBA^+. On the other hand, when the inner gate is open by pH 4.0, the affinity for TOA^+ remains unaffected, while that for TBA^+ decreases five orders of magnitude [35]. Since the only difference between these QA blockers is the length of their alkyl chains, it is concluded that the four extra carbon atoms in TOA^+ are critical for better hydrophobic interaction with the protein channel wall. There is no crystallographic information on QA–KcsA complexes at pH 4.0, but based on related evidence [42–44], it seems reasonable to expect that the widening of the channel cavity accompanies the acidic-pH-induced untangling of the cytoplasmic α-helical bundle and the bending of the TM2 segment away from the symmetry axes of the channel. Such a widening of the cavity could diminish the hydrophobic component in the binding of the shorter TBA^+, making it prompt to dissociate from the complex. However, the longer acyl chains of TOA^+ extend further so as to traverse the channel protein wall completely [33]. This should keep it firmly associated with the hydrophobic residues even in the open conformation at pH 4.0, thus explaining why it maintains a high binding affinity. The X-ray crystallographic data at pH 7.0 also revealed that while TBA^+ establishes van der Waals interactions with I100 and F103 residues from the cavity wall (TM2 helix), TOA^+ adds interactions with L36 (from the TM1 helix), T74 (near the SF), and G99 and S102 (from the TM2 helix) [33]. These additional interactions between the protein and the QA should be involved in providing a higher affinity for the binding of TOA^+ to the open state of the channel compared to TBA^+, thus preventing its dissociation from the complex, as suggested by the earlier electrophysiological studies [5,6]. Another difference in the interaction between these two QAs with the channel consists of a change on the side-chain rotamer of the F103 residue in the TOA^+–bound complex. The possible relevance of such observation is discussed below.

A thermal denaturation assay was also used to characterize the interaction of WT KcsA with Na^+ and K^+ in the presence of TOA^+, in terms of both the number of binding events detected and their respective affinities. In the case of Na^+, a single binding event with a slightly lower affinity than the control, in the absence of TOA^+ (see Table 1), was observed, which corresponds to Na^+ binding to its only available site, the extracellular S1 binding site. These observations on Na^+ binding were quite similar at pH 7.0 and pH 4.0, or in the presence of TBA^+, as indicated by the results. In contrast, the K^+ binding studies indicated that the presence of bound TOA^+ specifically affects the binding of the permeant species and is sensitive to the inner gate opening by the acidic pH. Thus, it is concluded that the effect of TOA^+ on the ion–protein interactions specifically affects the binding to the channel of the permeant K^+, and, therefore, monitoring of K^+ binding becomes a useful tool to detect both the acidic pH-induced conformational change of the SF to an inactivated state and the changes induced by TOA^+. As mentioned in the Introduction section, binding of K^+ to WT KcsA is described by two consecutive binding events, with dissociation constants in the μM and mM range, respectively. Here, it is shown that in the presence of TOA^+ bound to the channel cavity, the two K^+ binding events still remain, suggesting that the SF, rather than collapsing, retains the ability to accommodate K^+ at the stack of K^+ binding sites and to undergo the K^+-concentration-dependent conformational transition. Nonetheless, bound TOA^+ induces a decrease in the affinity for K^+ in both binding events at pH 7.0 when the inner gate is in the closed conformation. This is similar to that observed in the absence of TOA^+ upon acidic-pH-induced channel inactivation. Therefore, this suggests that TOA^+ binding by itself causes inactivation at pH 7.0 when the inner gate is closed. Indeed, a similar decrease in the affinity of the channel for K^+ at pH 7.0 was also detected in mutant channels where the inactivation process is favored [35,45]. Furthermore, in apparent agreement with such conclusion, the presence of bound TOA^+ in the pH 4.0 samples has only modest effects on the affinity of the two binding events for K^+, likely because the channel is already inactivated. We have no evidence to propose a molecular mechanism to explain how TOA^+ induces channel inactivation at neutral pH, but it could be speculated that the change on the side-chain rotation of the F103 residue in the TOA^+–bound complex indicated above could be involved. The reason to speculate on such a possibility is that F103 is believed to be an essential residue in the allosteric crosstalk between the inner and outer channel gates [46–48] involved in the regulation of the channel's functional cycle.

In order to gain structural information on the KcsA–TOA^+–K^+ complex, we use the quadruple mutant KcsA W26, 68, 87, 113F, which carries a single tryptophan (W67) as a fluorescent reporter of the SF conformation and dynamics. In this WT–like mutant channel, the homo-FRET process between the W67 residues from each subunit allows us to estimate the changes in steady-state anisotropy and the intersubunit lateral distances according to the type and concentration of cations within the SF [34]. Here, we first characterize in detail the acidic-pH-induced inactivated state in the absence of QAs. Even though the thermal denaturation assay from above detected two consecutive binding events for K^+ at pH 4.0, the homo-FRET process is only sensitive to the transition from the nonconductive to the inactivated state. The analysis of this latter event shows a clear decrease in K^+ binding affinity at intermediate concentrations of the cation, although the final conformation at saturating amounts of K^+ is almost identical to that observed in the closed-conductive state.

As to the effects of TOA^+ binding on the W67 mutant channel, we observed that TOA^+ bound at the cavity allosterically modifies the conformation of the pore helices, leading to longer W67–W67 intersubunit distances at any K^+ concentration at both pH 7.0 and pH 4.0. This loosening in the outer mouth packing seems consistent with the observed decrease in the thermal stability of the protein. The changes in the pore helix conformation, along with the decreased affinity for K^+ at pH 7.0 caused by TOA^+, seen in both homo-FRET and thermal denaturation experiments, are very similar to those effects caused by inactivation at pH 4.0. Therefore, as in the WT channel, it is concluded that TOA^+ binding at pH 7.0 also causes channel inactivation in the W67 KcsA mutant.

The intersubunit distances determined from the time-resolved anisotropy decays can be compared to those calculated from the published X-ray data, which are usually obtained in the presence of Fab fragments bound to the channel to improve crystal resolution. Even though there are no X-ray data on WT KcsA at pH 4.0, some constitutively open mutant channels were successfully crystallized in different conditions [24,47,49]. Table 2 summarizes the W67–W67 distances calculated from the time-resolved anisotropy decays and compares them to the W67–W67 Cδ2-Cε2 lateral distances derived from X-ray data.

It is observed that the W67–W67 lateral distances calculated from the anisotropy decays of the open/inactivated W67 KcsA channel at pH 4.0 and high K^+ concentrations (15.3 ± 0.1 Å) are almost identical to those W67–W67 Cδ2-Cε2 distances determine from X-ray data obtained in the presence of an intracellular Fab fragment (15.2 Å; PDB 3PJS), but not when an extracellular Fab was used to form the crystals (17.7 Å; PDB 3F5W). These results highlight how the binding of the extracellular Fab fragment alters the conformation of the extracellular loop and the SF dynamics. In fact, the same Fab fragment has been described to have a profound effect on KcsA inactivation [17]. In this respect, it should be noted that the available X-ray data on the TOA^+–KcsA complex was obtained in the presence of the extracellular Fab fragment [33]. Based on such data, the authors concluded that the SF conformation in the TOA^+–inactivated channel is a collapsed structure, similar to that detected in KcsA alone at low K^+ concentrations [50] (PDB 1K4D), where the inner S2 and S3 K^+ binding sites are absent. In contrast to such a conclusion, we find that the stack of K^+ binding sites, although with a lower affinity, remains accessible in the TOA^+–KcsA complex and that the W67–W67 intersubunit distances are very much like those found in the resting channel in the absence of TOA^+. We attribute such discrepancies to the perturbing effects of the extracellular Fab fragment and/or to the C-terminal deletion on the X-ray data and conclude that rather than being collapsed, the inactivated TOA^+–bound state of the SF at pH 7.0 and high K^+ has a "resting-like" conformation. This is not unique to the inactivated TOA^+–induced state as it is shared by several inactivated models of KcsA [35,51,52], and it seems, therefore, a general feature of the inactivated SF of KcsA.

In summary, the results obtained by the combination of the thermal denaturation assay and the analysis of the homo-FRET process among the W67 residues of each subunit in KcsA reinforce the argument that the long chain QA TOA^+ stabilizes an inactivated conformation of the SF, as suggested by earlier electrophysiological studies [5,6], which is characterized by a lower affinity for K^+ without affecting the interaction with Na^+. However, in contrast to the conclusion from the X-ray studies, this inactivated state is not collapsed but, rather, in a "resting-like" conformation, where the differences between the conductive and inactivated SFs are more subtle. However, if the inactivated SF is "resting-like", what makes it nonconductive? Earlier electrophysiological work concluded that inactivation is associated with a loss of K^+ from the selectivity filter in potassium channels [53,54] and that the presence of the cations inside the selectivity filter is fundamental to stabilizing it in the conductive conformation [55–57]. In this respect, the drop in K^+ affinity detected in our thermal denaturation and homo-FRET experiments would increase the probability of partial K^+ depletion from the filter, thus hampering ion conduction.

4. Materials and Methods

4.1. Materials

N-Dodecyl-β-D-maltoside (DDM) ULTROL® Grade was from Merck (Madrid, Spain). Hepes, succinic acid, N-methyl-D-glucamine (NMG), NaCl, KCl, tetraoctylammonium (TOA^+) chloride, and dimethylsulfoxide (DMSO) were from Sigma-Aldrich (Madrid, Spain). Ni^{2+}-Sepharose Fast Flow resin was from GE Healthcare (Madrid, Spain).

4.2. KcsA Heterologous Expression and Purification

The quadruple mutant of KcsA bearing a single Trp per monomer (W67 KcsA) was generated by mutating the rest of the native Trp residues 26, 68, 87, and 113 to Phe (W26, 68, 87, 113F). Wild-type (WT) and W67 channels were expressed in *E.coli* M15 (pRep4) and purified by Ni^{2+}/His-tag affinity chromatography according to previous reports [34,58]. Proteins were purified in 20 mM HEPES buffer, pH 7.0, containing 5 mM DDM, 5 mM NMG, and different concentrations of KCl or NaCl. To prepare samples at pH 4.0, aliquots of the above were dialyzed against 10 mM succinic acid buffer, pH 4.0, containing 5 mM DDM, 5 mM NMG, and the corresponding amounts of NaCl or KCl. The tetrameric state of the protein was routinely checked by SDS-PAGE (12%) [59].

4.3. Thermal Denaturation Assay and Cation Binding Analysis

Thermal denaturation of WT and W67 KcsA channels was performed in a Varian Cary Eclipse spectrofluorometer by recording the temperature dependence of the protein intrinsic emission fluorescence at 340 nm after excitation at 280 nm, as previously described [37]. In these experiments, the protein was diluted to 1 μM protein concentration (in terms of monomers of KcsA) in either 20 mM Hepes buffer, pH 7.0, 5 mM DDM, and 10 mM NaCl (pH 7.0 buffer) or 10 mM succinic acid buffer, pH 4.0, 5 mM DDM, and 10 mM NaCl (pH 4.0 buffer), with additional amounts of NaCl, KCl, or TOA^+, as required. The presence of 10 mM NaCl to start all titration experiments was required to maintain the tetrameric structure of the protein channels and to provide minimal stability to the proteins to start the thermal denaturation recordings.

When using TOA^+, a concentrated stock (22 mM) was prepared in DMSO, and then aliquots from this stock were added to the samples and incubated for 30 min at room temperature. For titrations of either Na^+ or K^+ in the presence of TOA^+, the same procedure as before was done, adding Na^+ or K^+ from a concentrated stock in the final step. The final amount of DMSO in the sample was always less than 1%, which, per se, has no effects on protein thermal stability.

The midpoint temperature of dissociation and unfolding of the tetramer (t_m, in Celsius) was calculated from the thermal denaturation curves by fitting a two-state unfolding model to the data [38]. The dissociation constants of the KcsA–cation complexes (K_Ds) can be estimated from:

$$\frac{|\Delta T_m|}{T_m} = \frac{|T_m - (T_m)_0|}{T_m} = \frac{R(T_m)_0}{\Delta H_0} \ln\left(1 + \frac{[L]}{K_D}\right) \quad (1)$$

where T_m and $(T_m)_0$ refer to the denaturation temperature (in Kelvin) for the protein in the presence and absence of ligands, respectively, R is the gas constant, and ΔH_0 is the enthalpy change upon protein denaturation in the absence of ligands. The change in T_m in the absence and presence of ligands (ΔT_m) is expressed in absolute terms ($|\Delta T_m|$) since the presence of a given ligand can induce either an increase or a decrease of the observable value.

4.4. Steady-State Fluorescence Measurements

The steady-state fluorescence anisotropy of W67 KcsA was measured on a Horiba Jobin Yvon Fluorolog-3-21 or SLM 8000 spectrofluorometer and calculated as:

$$< r >_{SS} = \frac{I_{VV} - G \cdot I_{VH}}{I_{VV} + 2\,G \cdot I_{VH}} \quad (2)$$

where I_{VV} and I_{VH} are the fluorescence intensities (blank subtracted) of the vertically and horizontally polarized emission when the sample is excited with vertically polarized light, respectively, and the G factor ($G = I_{HV}/I_{HH}$) is the instrument correction factor. The samples were measured at 340 nm using an excitation wavelength of 300 nm [34]. A final protein concentration of 5 μM in either the pH 7.0 or pH 4.0 buffer, with or without TOA^+, was used throughout. Ten measurements were done for each sample to calculate the average steady-state anisotropy values (±standard deviation).

4.5. Time-Resolved Fluorescence Measurements and Intersubunit Distance Calculations

Time-resolved fluorescence and anisotropy measurements with picosecond resolution were obtained using the time-correlated single-photon timing (SPT) technique. The fluorescence decays were measured at 345 nm (λ_{exc} = 300 nm) using an emission polarizer set at the magic angle (54.7°) relative to the vertically polarized excitation beam produced by a frequency-doubled Rhodamine 6 G laser [60].

The fluorescence and anisotropy decays were analyzed as described [34]. The W67–W67 intersubunit lateral distances were calculated from the time-resolved anisotropy measurements according to:

$$r(t) = \frac{r(0)}{4}\left[1 + e^{-4k_1 t} + 2e^{\left(-\frac{9}{4}k_1 t\right)}\right] \cdot e^{-\frac{t}{\varphi_g}} \tag{3}$$

where r(0) is the initial anisotropy, k_1 is the rate constant for homo-FRET between neighboring tryptophan residues, and φ_g is the rotational correlation time of the KcsA–DDM complex (43 ns) [34]. The intertryptophan lateral distance R can be directly calculated via k_1:

$$k_1 = \frac{1}{\tau}\left(\frac{R_0}{R}\right)^6 \tag{4}$$

where τ is the intensity-weighted mean fluorescence lifetime, and R_0 is the critical radius computed with an orientation factor $\kappa^2 = 2/3$. The goodness of fit was evaluated by statistical criteria (random distribution of weighted residuals and autocorrelation plots and a reduced $\chi^2 < 1.2$).

4.6. Calculation of the K^+ Binding Affinity to KcsA from Changes in the Steady-State Anisotropy Values

The binding of K^+ to KcsA W67, monitored by the changes in the $\langle r \rangle_{SS}$, presented a sigmoidal behavior that was also used to calculate the binding affinity of the channel for this permeant cation. Since we were mostly interested in the K^+ binding event related to the equilibrium between the nonconductive ($KcsA_{(NC)}$) and conductive (pH 7.0) or inactivated (pH 4.0) SF conformations of the channel ($KcsA{\bullet}K^+{}_{(C/I)}$), we analyzed this second binding event, which can be described by a two-state equilibrium:

$$KcsA_{(NC)} + K^+ \leftrightarrow KcsA{\bullet}K^+{}_{(C/I)}$$

The total concentration of protein corresponds to:

$$[KcsA]_{total} = [KcsA]_{NC} + \left[KcsA{\bullet}K^+\right]_{C/I} \tag{5}$$

and the dissociation constant (K_D) is therefore defined by:

$$K_D = \frac{[KcsA]_{NC} \cdot \left[K^+\right]}{\left[KcsA{\bullet}K^+\right]_{C/I}} \tag{6}$$

It could be assumed that the binding of a cation X^+ to the protein is a cooperative process that is described by an empiric Hill function (sigmoid curve), so the free and bound molar fractions of KcsA as a function of cation concentration can be expressed as:

$$x_{free} = x_{NC} = \frac{[KcsA]_{NC}}{[KcsA]_{total}} = \frac{{K_d}^h}{{K_d}^h + \left[X^+\right]^h} \tag{7}$$

$$x_{bound} = x_{C/I} = \frac{[KcsA]_{C/I}}{[KcsA]_{total}} = \frac{\left[X^+\right]^h}{{K_d}^h + \left[X^+\right]^h} \tag{8}$$

where h is the Hill coefficient (cooperativity index).

The steady-state fluorescence anisotropy of a sample, $\langle r \rangle_{SS}$, prepared at a given X^+ concentration is a weighted average of its limiting values, i.e., the steady-state fluorescence anisotropy of the nonconductive and conductive/inactivated conformations of KcsA, $\langle r \rangle_{NC}$, and $\langle r \rangle_{C/I}$, respectively. However, at variance with the usual case, the weighting factors here are the relative fluorescence intensities emitted by each species and not their molar fractions directly [61]. Mathematically, this is described as:

$$\langle r \rangle = \frac{x_{NC} \cdot \Phi_{NC}}{x_{NC} \cdot \Phi_{NC} + x_{C/I} \cdot \Phi_{C/I}} \cdot \langle r \rangle_{NC} + \frac{x_{C/I} \cdot \Phi_{C/I}}{x_{NC} \cdot \Phi_{NC} + x_{C/I} \cdot \Phi_{C/I}} \cdot \langle r \rangle_{C/I} \quad (9)$$

where Φ_{NC} and $\Phi_{C/I}$ are the quantum yields of the nonconductive and conductive (pH 7.0) or inactivated states (pH 4.0), respectively.

If we define a parameter Q as the relative change in KcsA quantum yield upon K^+ binding ($\Phi_{C/I}/\Phi_{NC}$) and combine the equations from above, the general expression that can be used to fit $\langle r \rangle_{SS}$ data obtained in equilibrium binding studies is, therefore:

$$\langle r \rangle = \langle r \rangle_{NC} + (Q \langle r \rangle_{C/I} - \langle r \rangle_{NC}) \cdot \frac{[X^+]^h}{K_d{}^h + [X^+]^h} \quad (10)$$

During the fitting procedures, the Q parameter was fixed at a constant value of 0.95 and 0.8 for the experiments carried out at pH 7.0 and pH 4.0, respectively.

Author Contributions: Conceptualization: J.A.P., M.P., and J.M.G.-R.; formal analysis: A.M.G., M.L.R., A.C., and C.D.-G.; investigation: A.M.G., C.D.-G., and M.L.R.; writing—original draft: J.M.G.-R.; writing—review and editing: M.L.R., C.D.-G., A.C., M.P., J.A.P., and J.M.G.-R. All authors have read and agreed to the published version of the manuscript.

Funding: This work was partly supported by grants PGC2018-093505-B-I00 from the Spanish "Ministerio de Ciencia e Innovación"/FEDER, UE, and UIDB/04565/2020 from FCT, Portugal. C.D.-G. acknowledges support from the Medical Biochemistry and Biophysics Doctoral Programme (M2B-PhD) FCT (reference: SFRH/PD/BD/135154/2017).

Institutional Review Board Statement: Not applicable.

Informed Consent Statement: Not applicable.

Data Availability Statement: Data is contained in the article.

Acknowledgments: The time-resolved fluorescence intensity and anisotropy measurements were performed by Aleksander Fedorov at the IBB in Lisbon (Portugal). Eva Martinez provided excellent technical help throughout this work.

Conflicts of Interest: The authors declare no conflict of interest.

References

1. Yellen, G. The moving parts of voltage-gated ion channels. *Q. Rev. Biophys.* **1998**, *31*, 239–295. [CrossRef] [PubMed]
2. Hille, B. *Ion Channels of Excitable Membranes*; Mass Sinauer Assoc. Inc.: Sunderland, UK, 2001; p. 5. ISBN 0-87893-321-2.
3. Armstrong, C.M.; Binstock, L. Anomalous rectification in the squid giant axon injected with tetraethylammonium chloride. *J. Gen. Physiol.* **1965**, *48*, 859–872. [CrossRef] [PubMed]
4. Hille, B. The selective inhibition of delayed potassium currents in nerve by tetraethylammonium ion. *J. Gen. Physiol.* **1967**, *50*, 1287–1302. [CrossRef] [PubMed]
5. Armstrong, C.M. Inactivation of the potassium conductance and related phenomena caused by quaternary ammonium ion injection in squid axons. *J. Gen. Physiol.* **1969**, *54*, 553–575. [CrossRef] [PubMed]
6. Armstrong, C.M. Interaction of tetraethylammonium ion derivatives with the potassium channels of giant axons. *J. Gen. Physiol.* **1971**, *58*, 413–437. [CrossRef] [PubMed]
7. Choi, K.L.; Aldrich, R.W.; Yellen, G. Tetraethylammonium blockade distinguishes two inactivation mechanisms in voltage-activated K^+ channels. *Proc. Natl. Acad. Sci. USA* **1991**, *88*, 5092–5095. [CrossRef]
8. Heginbotham, L.; MacKinnon, R. The aromatic binding site for tetraethylammonium ion on potassium channels. *Neuron* **1992**, *8*, 483–491. [CrossRef]

9. Choi, K.L.; Mossman, C.; Aubé, J.; Yellen, G. The internal quaternary ammonium receptor site of Shaker potassium channels. *Neuron* **1993**, *10*, 533–541. [CrossRef]
10. Uysal, S.; Vasquez, V.; Tereshko, V.; Esaki, K.; Fellouse, F.A.; Sidhu, S.S.; Koide, S.; Perozo, E.; Kossiakoff, A. Crystal structure of full-length KcsA in its closed conformation. *Proc. Natl. Acad. Sci. USA* **2009**, *106*, 6644–6649. [CrossRef]
11. Doyle, D.A.; Morais, C.J.; Pfuetzner, R.A.; Kuo, A.; Gulbis, J.M.; Cohen, S.L.; Chait, B.T.; MacKinnon, R. The structure of the potassium channel: Molecular basis of K^+ conduction and selectivity. *Science* **1998**, *280*, 69–77. [CrossRef]
12. Zhou, Y.; Morais-Cabral, J.H.; Kaufman, A.; MacKinnon, R. Chemistry of ion coordination and hydration revealed by a K^+ channel-Fab complex at 2.0 Å resolution. *Nature* **2001**, *414*, 43–48. [CrossRef] [PubMed]
13. Morais-Cabral, J.H.; Zhou, Y.; MacKinnon, R. Energetic optimization of ion conduction rate by the K^+ selectivity filter. *Nature* **2001**, *414*, 37–42. [CrossRef]
14. Wylie, B.J.; Bhate, M.P.; McDermott, A.E. Transmembrane allosteric coupling of the gates in a potassium channel. *Proc. Natl. Acad. Sci. USA* **2014**, *111*, 185–190. [CrossRef] [PubMed]
15. Zhou, Y.; MacKinnon, R. The occupancy of ions in the K^+ selectivity filter: Charge balance and coupling of ion binding to a protein conformational change underlie high conduction rates. *J.Mol.Biol.* **2003**, *333*, 965–975. [CrossRef] [PubMed]
16. Chill, J.H. NMR study of the tetrameric KcsA potassium channel in detergent micelles. *Protein Sci.* **2006**, *15*, 684–698. [CrossRef] [PubMed]
17. Cordero-Morales, J.F.; Cuello, L.G.; Zhao, Y.; Jogini, V.; Cortes, D.M.; Roux, B.; Perozo, E. Molecular determinants of gating at the potassium-channel selectivity filter. *Nat. Struct. Mol. Biol.* **2006**, *13*, 311–318. [CrossRef] [PubMed]
18. Baker, K.A.; Tzitzilonis, C.; Kwiatkowski, W.; Choe, S.; Riek, R. Conformational dynamics of the KcsA potassium channel governs gating properties. *Nat. Struct. Mol. Biol.* **2007**, *14*, 1089–1095. [CrossRef]
19. Ader, C.; Schneider, R.; Hornig, S.; Velisetty, P.; Vardanyan, V.; Giller, K.; Ohmert, I.; Becker, S.; Pongs, O.; Baldus, M. Coupling of activation and inactivation gate in a K-channel: Potassium and ligand sensitivity. *EMBO J.* **2009**, *28*, 2825–2834. [CrossRef]
20. Bhate, M.P.; Wylie, B.J.; Tian, L.; McDermott, A.E. Conformational dynamics in the selectivity filter of KcsA in response to potassium ion concentration. *J. Mol. Biol.* **2010**, *401*, 155–166. [CrossRef]
21. Imai, S.; Osawa, M.; Takeuchi, K.; Shimada, I. Structural basis underlying the dual gate properties of KcsA. *Proc. Natl. Acad. Sci. USA* **2010**, *107*, 6216–6221. [CrossRef]
22. Cheng, W.W.L.; McCoy, J.G.; Thompson, A.N.; Nichols, C.G.; Nimigean, C.M. Mechanism for selectivity-inactivation coupling in KcsA potassium channels. *Proc. Natl. Acad. Sci. USA* **2011**, *108*, 5272–5277. [CrossRef] [PubMed]
23. Lockless, S.W.; Zhou, M.; MacKinnon, R. Structural and thermodynamic properties of selective ion binding in a K^+ channel. *PLoS Biol.* **2007**, *5*, e121. [CrossRef] [PubMed]
24. Cuello, L.G.; Cortes, D.M.; Jogini, V.; Sompornpisut, A.; Perozo, E. A molecular mechanism for proton-dependent gating in KcsA. *FEBS Lett.* **2010**, *584*, 1126–1132. [CrossRef] [PubMed]
25. Hirano, M.; Onishi, Y.; Yanagida, T.; Ide, T. Role of the KcsA channel cytoplasmic domain in pH-dependent gating. *Biophys. J.* **2011**, *101*, 2157–2162. [CrossRef] [PubMed]
26. Posson, D.J.; Thompson, A.N.; McCoy, J.G.; Nimigean, C.M. Molecular interactions involved in proton-dependent gating in KcsA potassium channels. *J. Gen. Physiol.* **2013**, *142*, 613–624. [CrossRef] [PubMed]
27. Hoshi, T.; Zagotta, W.N.; Aldrich, R.W. Two types of inactivation in Shaker K^+ channels: Effects of alterations in the carboxy-terminal region. *Neuron* **1991**, *7*, 547–556. [CrossRef]
28. Liu, Y.; Jurman, M.E.; Yellen, G. Dynamic rearrangement of the outer mouth of a K^+ channel during gating. *Neuron* **1996**, *16*, 859–867. [CrossRef]
29. Gao, L.; Mi, X.; Paajanen, V.; Wang, K.; Fan, Z. Activation-coupled inactivation in the bacterial potassium channel KcsA. *Proc. Natl. Acad. Sci. USA* **2005**, *102*, 17630–17635. [CrossRef]
30. Li, J.; Ostmeyer, J.; Cuello, L.G.; Perozo, E.; Roux, B. Rapid constriction of the selectivity filter underlies C-type inactivation in the KcsA potassium channel. *J. Gen. Physiol.* **2018**, *150*, 1408–1420. [CrossRef]
31. Cuello, L.G.; Cortes, D.M.; Perozo, E. The gating cycle of a K^+ channel at atomic resolution. *Elife* **2017**, *6*, e28032. [CrossRef]
32. Yohannan, S.; Hu, Y.; Zhou, Y. Crystallographic Study of the Tetrabutylammonium Block to the KcsA K^+ Channel. *J. Mol. Biol.* **2007**, *366*, 806–814. [CrossRef] [PubMed]
33. Lenaeus, M.J.; Burdette, D.; Wagner, T.; Focia, P.J.; Gross, A. Structures of KcsA in complex with symmetrical quaternary ammonium compounds reveal a hydrophobic binding site. *Biochemistry* **2014**, *53*, 5365–5373. [CrossRef] [PubMed]
34. Renart, M.L.; Giudici, A.M.; Poveda, J.A.; Fedorov, A.; Berberan-Santos, M.N.; Prieto, M.; Díaz-García, C.; González-Ros, J.M.; Coutinho, A. Conformational plasticity in the KcsA potassium channel pore helix revealed by homo-FRET studies. *Sci. Rep.* **2019**, *9*, 6215–6228. [CrossRef] [PubMed]
35. Giudici, A.M.; Renart, M.L.; Díaz-García, C.; Morales, A.; Poveda, J.A.; González-Ros, J.M. Accessibility of cations to the selectivity filter of KcsA in the inactivated state: An equilibrium binding study. *Int. J. Mol. Sci.* **2019**, *20*, 689. [CrossRef]
36. Devaraneni, P.K.; Komarov, A.G.; Costantino, C.A.; Devereaux, J.J.; Matulef, K.; Valiyaveetil, F.I. Semisynthetic K^+ channels show that the constricted conformation of the selectivity filter is not the C-type inactivated state. *Proc. Natl. Acad. Sci. USA* **2013**, *110*, 15698–15703. [CrossRef]
37. Renart, M.L.; Triano, I.; Poveda, J.A.; Encinar, J.A.; Fernández, A.M.; Ferrer-Montiel, A.V.; Gómez, J.; González Ros, J.M. Ion binding to KcsA: Implications in ion selectivity and channel gating. *Biochemistry* **2010**, *49*, 9480–9487. [CrossRef]

38. Triano, I.; Barrera, F.N.; Renart, M.L.; Molina, M.L.; Fernandez-Ballester, G.; Poveda, J.A.; Fernandez, A.M.; Encinar, J.A.; Ferrer-Montiel, A.V.; Otzen, D.; et al. Occupancy of nonannular lipid binding sites on KcsA greatly increases the stability of the tetrameric protein. *Biochemistry* **2010**, *49*, 5397–5404. [CrossRef]
39. Renart, M.L.; Montoya, E.; Giudici, A.M.; Poveda, J.A.; Fernández, A.M.; Morales, A.; González-Ros, J.M. Selective exclusion and selective binding both contribute to ion selectivity in KcsA, a model potassium channel. *J. Biol. Chem.* **2017**, *292*, 15552–15560. [CrossRef]
40. Montoya, E.; Lourdes Renart, M.; Marcela Giudici, A.; Poveda, J.A.; Fernández, A.M.; Morales, A.; González-Ros, J.M. Differential binding of monovalent cations to KcsA: Deciphering the mechanisms of potassium channel selectivity. *Biochim. Biophys. Acta Biomembr.* **2017**, *1859*, 779–788. [CrossRef]
41. Faraldo-Gómez, J.D.; Kutluay, E.; Jogini, V.; Zhao, Y.; Heginbotham, L.; Roux, B. Mechanism of Intracellular Block of the KcsA K^+ Channel by Tetrabutylammonium: Insights from X-ray Crystallography, Electrophysiology and Replica-exchange Molecular Dynamics Simulations. *J. Mol. Biol.* **2007**, *365*, 649–662. [CrossRef]
42. Perozo, E.; Marien Cortes, D.; Cuello, L.G. Three-dimensional architecture and gating mechanism of a K^+ channel studied by EPR spectroscopy. *Nat. Struct. Biol.* **1998**, *5*, 459–469. [CrossRef] [PubMed]
43. Perozo, E.; Cortes, D.M.; Cuello, L.G. Structural rearrangements underlying K^+-channel activation gating. *Science* **1999**, *285*, 73–78. [CrossRef] [PubMed]
44. Iwamoto, M.; Shimizu, H.; Inoue, F.; Konno, T.; Sasaki, Y.C.; Oiki, S. Surface structure and its dynamic rearrangements of the KcsA potassium channel upon gating and tetrabutylammonium blocking. *J. Biol. Chem.* **2006**, *281*, 28379–28386. [CrossRef] [PubMed]
45. Renart, M.L.; Montoya, E.; Fernandez, A.M.; Molina, M.L.; Poveda, J.A.; Encinar, J.A.; Ayala, J.L.; Ferrer-Montiel, A.V.; Gomez, J.; Morales, A.; et al. Contribution of ion binding affinity to ion selectivity and permeation in KcsA, a model potassium channel. *Biochemistry* **2012**, *51*, 3891–3900. [CrossRef]
46. Cuello, L.G.; Jogini, V.; Cortes, D.M.; Pan, A.C.; Gagnon, D.G.; Dalmas, O.; Cordero-Morales, J.F.; Chakrapani, S.; Roux, B.; Perozo, E. Structural basis for the coupling between activation and inactivation gates in K^+ channels. *Nature* **2010**, *466*, 272–275. [CrossRef]
47. Cuello, L.G.; Jogini, V.; Cortes, D.M.; Perozo, E. Structural mechanism of C-type inactivation in K^+ channels. *Nature* **2010**, *466*, 203–208. [CrossRef]
48. Pan, A.C.; Cuello, L.G.; Perozo, E.; Roux, B. Thermodynamic coupling between activation and inactivation gating in potassium channels revealed by free energy molecular dynamics simulations. *J. Gen. Physiol.* **2011**, *138*, 571–580. [CrossRef]
49. Uysal, S.; Cuello, L.G.; Cortes, D.M.; Koide, S.; Kossiakoff, A.A.; Perozo, E. Mechanism of activation gating in the full-length KcsA K^+ channel. *Proc. Natl. Acad. Sci. USA* **2011**, *108*, 11896–11899. [CrossRef]
50. Zhou, M.; Morais-Cabral, J.H.; Mann, S.; MacKinnon, R. Potassium channel receptor site for the inactivation gate and quaternary amine inhibitors. *Nature* **2001**, *411*, 657–661. [CrossRef]
51. Matulef, K.; Komarov, A.G.; Costantino, C.A.; Valiyaveetil, F.I. Using protein backbone mutagenesis to dissect the link between ion occupancy and C-Type inactivation in K^+ channels. *Proc. Natl. Acad. Sci. USA* **2013**, *110*, 17886–17891. [CrossRef]
52. Matulef, K.; Annen, A.W.; Nix, J.C.; Valiyaveetil, F.I. Individual Ion Binding Sites in the K^+ Channel Play Distinct Roles in C-type Inactivation and in Recovery from Inactivation. *Structure* **2016**, *24*, 750–761. [CrossRef] [PubMed]
53. Baukrowitz, T.; Yellen, G. Modulation of K^+ current by frequency and external [K^+]: A tale of two inactivation mechanisms. *Neuron* **1995**, *15*, 951–960. [CrossRef]
54. Xu, Y.; Bhate, M.P.; McDermott, A.E. Transmembrane allosteric energetics characterization for strong coupling between proton and potassium ion binding in the KcsA channel. *Proc. Natl. Acad. Sci. USA* **2017**, *114*, 8788–8793. [CrossRef] [PubMed]
55. Swenson, R.P.; Armstrong, C.M. K^+ channels close more slowly in the presence of external K^+ and Rb^+. *Nature* **1981**, 427–429. [CrossRef] [PubMed]
56. Rasmusson, R.L.; Morales, M.J.; Wang, S.; Liu, S.; Campbell, D.L.; Brahmajothi, M.V.; Strauss, H.C. Inactivation of voltage-gated cardiac K^+ channels. *Circ. Res.* **1998**, 739–750. [CrossRef]
57. Santos, J.S.; Syeda, R.; Montal, M. Stabilization of the conductive conformation of a voltage-gated K^+ (Kv) channel: The lid mechanism. *J. Biol. Chem.* **2013**, *288*, 16619–16628. [CrossRef]
58. Barrera, F.N.; Renart, M.L.; Molina, M.L.; Poveda, J.A.; Encinar, J.A.; Fernández, A.M.; Neira, J.L.; González-Ros, J.M. Unfolding and refolding in vitro of a tetrameric, alpha-helical membrane protein: The prokaryotic potassium channel KcsA. *Biochemistry* **2005**, *44*, 14344–14352. [CrossRef]
59. Molina, M.L.; Encinar, J.A.; Barrera, F.N.; Fernández-Ballester, G.; Riquelme, G.; Gonzalez-Ros, J.M. Influence of C-terminal protein domains and protein-lipid interactions on tetramerization and stability of the potassium channel KcsA. *Biochemistry* **2004**, *43*, 14924–14931. [CrossRef]
60. Poveda, J.A.; Prieto, M.; Encinar, J.A.; González-Ros, J.M.; Mateo, C.R. Intrinsic tyrosine fluorescence as a tool to study the interaction of the Shaker B "ball" peptide with anionic membranes. *Biochemistry* **2003**, *42*, 7124–7132. [CrossRef]
61. Lakowicz, J.R. Principles of Fluorescence Spectroscopy. In *Principles of Fluorescence Spectroscopy*, 3rd ed.; Springer: New York, NY, USA, 2006; pp. 372–374.

International Journal of
Molecular Sciences

Review

New Insights into the Chloroplast Outer Membrane Proteome and Associated Targeting Pathways

Michael Fish [1,†], Delaney Nash [2,†], Alexandru German [1], Alyssa Overton [2], Masoud Jelokhani-Niaraki [3], Simon D. X. Chuong [2] and Matthew D. Smith [1,*]

1 Department of Biology, Wilfrid Laurier University, Waterloo, ON N2L 3C5, Canada; fish1960@mylaurier.ca (M.F.); germ2740@mylaurier.ca (A.G.)
2 Department of Biology, University of Waterloo, Waterloo, ON N2L 3G1, Canada; d2nash@uwaterloo.ca (D.N.); akoverton@uwaterloo.ca (A.O.); schuong@uwaterloo.ca (S.D.X.C.)
3 Department of Chemistry & Biochemistry, Wilfrid Laurier University, Waterloo, ON N2L 3C5, Canada; mjelokhani@wlu.ca
* Correspondence: msmith@wlu.ca
† These authors contributed equally to the work.

Abstract: Plastids are a dynamic class of organelle in plant cells that arose from an ancient cyanobacterial endosymbiont. Over the course of evolution, most genes encoding plastid proteins were transferred to the nuclear genome. In parallel, eukaryotic cells evolved a series of targeting pathways and complex proteinaceous machinery at the plastid surface to direct these proteins back to their target organelle. Chloroplasts are the most well-characterized plastids, responsible for photosynthesis and other important metabolic functions. The biogenesis and function of chloroplasts rely heavily on the fidelity of intracellular protein trafficking pathways. Therefore, understanding these pathways and their regulation is essential. Furthermore, the chloroplast outer membrane proteome remains relatively uncharted territory in our understanding of protein targeting. Many key players in the cytosol, receptors at the organelle surface, and insertases that facilitate insertion into the chloroplast outer membrane remain elusive for this group of proteins. In this review, we summarize recent advances in the understanding of well-characterized chloroplast outer membrane protein targeting pathways as well as provide new insights into novel targeting signals and pathways more recently identified using a bioinformatic approach. As a result of our analyses, we expand the known number of chloroplast outer membrane proteins from 117 to 138.

Keywords: chloroplast-targeting pathways; chloroplast outer membrane proteome; signal anchored protein; tail anchored protein; β-barrel protein; β-signal; chloroplast transit peptide; TOC complex; AKR2; OEP80

Citation: Fish, M.; Nash, D.; German, A.; Overton, A.; Jelokhani-Niaraki, M.; Chuong, S.D.X.; Smith, M.D. New Insights into the Chloroplast Outer Membrane Proteome and Associated Targeting Pathways. *Int. J. Mol. Sci.* **2022**, *23*, 1571. https://doi.org/10.3390/ijms23031571

Academic Editor: Koichi Kobayashi

Received: 15 December 2021
Accepted: 27 January 2022
Published: 29 January 2022

1. Introduction

The compartmentalization of eukaryotic cells by membrane-bound organelles provides diverse environments for a wide variety of biochemical pathways essential to cell function and survival. Plastids are a dynamic class of organelle that evolved from an ancient cyanobacterial endosymbiont [1–3]. As they evolved, plastids adopted various central functions in cell metabolism and biosynthesis as well as in signalling, embryogenesis, leaf development, gravitropism, temperature response, and plant–microbe interactions [4]. Unlike other organelles, plastids can differentiate into various types from a common precursor, known as the proplastid, which serve different metabolic needs throughout plant tissues [5–8]. Plastids can also transition between types in response to different developmental [9,10] or environmental cues [11–15]. One such example is the process of photomorphogenesis, where, in the presence of light, the etioplasts of leaf cells grown in the dark are converted to green photosynthetic chloroplasts, the most notable and well-characterized type of plastids [16]. The ability of plants to trigger these types of transitions and maintain a variety of plastid types in different tissues and at different stages of life is what allowed for the radiant expansion of the plant kingdom [6]. The interconversion of plastids is made possible by the coordinated remodelling of the plastid proteome [17–19].

Over the course of evolution, a large majority of plastid genes were transferred to the nuclear genome by horizontal gene transfer [1,20–23]. As a result, plastid biogenesis and function rely on the fidelity of intracellular protein trafficking pathways to deliver the corresponding proteins to plastids [24]. To recognize and import chloroplast precursor proteins synthesized in the cytosol, plant cells evolved complex proteinaceous machinery at the outer and inner membranes of the chloroplast known as the general import apparatus, composed of the translocon at the outer membrane of the chloroplast (TOC complex) and the translocon at the inner membrane of the chloroplast (TIC complex) [1,25]. Chloroplast outer membrane proteins, including components of the TOC complex, use alternative targeting pathways to reach the chloroplast outer membrane [26]. A unique set of challenges exist for membrane proteins that are targeted to their resident membranes post-translationally. As the hydrophobic segments of their amino acid chains are synthesized on cytosolic ribosomes, chaperones are often required to maintain stability, prevent misfolding and avoid aggregation before the proteins reach their target membranes [27]. In some cases, the chaperones themselves serve as targeting elements whereas in other cases, a targeting sequence may engage a receptor at the target membrane surface or local secondary and tertiary structures may engage the membrane directly, inducing self-insertion [28].

Classifying common targeting pathways is challenging, as the specific mechanisms used by many outer membrane proteins remain uncharacterized. This is, in large part, due to the limited number of known and confirmed chloroplast outer membrane proteins, as well as the difficulties associated with studying membrane proteins in vitro [29]. It is even the case that some proteins use a combination of redundant strategies [26], increasing the complexity of distinguishing between these mechanisms further. Advances in proteomics and the development of more powerful bioinformatic tools have led to the identification and characterization of an increasing number of chloroplast outer membrane proteins in recent years [29,30]. In this review, we discuss current advances in our understanding of the targeting signals and pathways used by chloroplast outer membrane proteins during their biogenesis. Further, we put forth a proteome-wide bioinformatic approach for identifying novel chloroplast outer membrane protein-targeting signals and pathways. This analysis allowed us to expand the current list of chloroplast outer membrane proteins from 117 to 138.

2. The Chloroplast Outer Membrane

2.1. Composition and Function of the Chloroplast Outer Membrane

Like their bacterial ancestors and mitochondria, the other endosymbiotic organelles found in eukaryotes, chloroplasts are surrounded by two membranes that differ in function and composition [31]. The inner membrane is studded with transport proteins and tightly regulates the flux of ions and metabolites between the intermembrane space (IMS) and the stroma, the interior compartment of the chloroplast [32]. The inner membrane is composed primarily of galactolipids with some phospholipids and sulfolipids [33]. In contrast, the chloroplast outer membrane is permeable to ions and metabolites and controls the recognition and import/insertion of chloroplast proteins [32]. It also serves as the site for galactolipid biosynthesis [34] and is composed primarily of phospholipids and galactolipids in equal proportions with some sulfolipids [33]. The entire chloroplast outer membrane proteome is encoded by genes in the nucleus, synthesized on cytosolic ribosomes, and targeted post-translationally [20–22]. Therefore, these proteins must contain signals that direct them to the chloroplast outer membrane. Inoue (2015) published a comprehensive list of chloroplast outer membrane proteins in *Arabidopsis thaliana*, of which there are 117, and categorized them based on their function [30]. This represents a diverse proteome with functions including: solute and ion transport, protein import, protein turnover and modification, lipid metabolism, carbohydrate metabolism and regulation, other metabolism and regulation and intracellular communication, as well as proteins of unknown function [30]. More proteins still have been identified that associate with the chloroplast outer membrane surface but are not inserted in the membrane [35]. This functional diversity highlights the crucial role that the chloroplast outer membrane plays in plastid biogenesis, cell metabolism and intracellular signalling between plastids and the rest of the cell. Despite the advances made in our understanding of chloroplast precursor protein import into the stroma, the targeting and insertion mechanisms for chloroplast outer membrane proteins remains elusive. This gap is further exaggerated when compared to our understanding of mitochondrial outer membrane proteins [36].

2.2. Topologies of Chloroplast Outer Membrane Proteins

Based on their structure and topology, chloroplast outer membrane proteins fit into several categories. α-helical proteins are categorized by the number and location of their transmembrane domain(s) (TMD(s)). Signal anchored (SA) proteins contain single membrane-spanning α-helices located at their N-terminus and tail anchored (TA) proteins contain single membrane-spanning α-helices located at their C-terminus [37]. Both SA and TA proteins adopt similar topologies in which their soluble domains are exposed to the cytosol, with some containing short extensions into the IMS. Although very few have been characterized, some α-helical proteins are anchored by a single membrane-spanning α-helix in the middle of their sequences, containing both N- and C-terminal soluble domains exposed at opposite sides of the membrane. These are not classified as SA or TA proteins. Further, some α-helical proteins in the chloroplast outer membrane contain two or more membrane-spanning α-helices. Finally, β-barrel proteins span the membrane through the formation of a cylindrical barrel composed of β-strands [27,38]. Beyond these well-established categories of integral membrane proteins are proteins that contain either or both α-helices and β-strands, which form uncharacterized membrane anchors, proteins that are anchored to the membrane by the covalent attachment to lipid molecules, and peripheral membrane proteins that rely on electrostatic or hydrophobic interactions at the membrane surface, as well as on interactions with integral membrane proteins [27,39].

3. Protein Entry into Chloroplasts: Structure and Function of the TOC Complex

The mechanisms by which chloroplast precursor proteins are targeted to the chloroplast, recognized, and imported to the stroma by the general import apparatus have been extensively reviewed [25,40–43]. In brief, chloroplast precursor proteins are synthesized on cytosolic ribosomes and targeted to the chloroplast post-translationally via a cleavable N-terminal transit peptide (TP) [25,44–46]. TP sequences are moderately hydrophobic, containing an amphipathic α-helix. They are typically rich in hydroxylated and basic residues, void of acidic residues and often contain multiple proline residues [46–50]. A lack of arginine residues in TPs differentiate them from mitochondrial presequences and preclude mitochondrial targeting [51]. Molecular chaperones in the cytosol guide precursor proteins to the chloroplast surface in an import competent state [52]. Heat shock proteins (Hsp) 70 and 90 are involved in the transport of most chloroplast precursor proteins containing TPs [53] and may use TOC64 as a receptor [54,55]. Hsp70 forms a guidance complex with the 14-3-3 protein, which interacts directly with the TOC complex [52,53]. The TOC complex mediates recognition and import of nuclear-encoded precursor proteins into the chloroplast through the TOC-TIC supercomplex, after which the TP is cleaved in the stroma by a stromal processing peptidase [56,57].

The components of the core TOC complex were originally identified in pea (*Pisum sativum*) and *Arabidopsis thaliana* [58], although its composition and function appear to be highly conserved across plant species [59,60]. TOC34 and TOC159 GTPase receptors act to recognize the cleavable N-terminal TPs of chloroplast precursor proteins before they are translocated across the outer membrane via the TOC75 translocation channel [61]. Cryo-electron microscopy was used to shed light on the organization of the TOC complex and its subunits [62]. This, in combination with affinity purification and electrophoretic techniques, suggest the TOC complex exists in a 4:4:1 (TOC75:TOC34:TOC159) arrangement [62–65].

Further, paralogous TOC complexes have been identified in *Arabidopsis thaliana*, which may be responsible for the recognition and import of different subsets of chloroplast precursor proteins (housekeeping vs. photosynthetic) [5,17–19]. Functionally distinct TOC complexes are defined by the presence of TOC34 (TOC33) and TOC159 (TOC90, TOC120, TOC132) receptor homologs, where TOC75 is central to all TOC complexes [66]. The abundance and ratio of various paralogous TOC complexes in the plastid membrane are likely to play a role in plastid biogenesis and the transition of plastids between various types based on developmental and environmental signals [9], such as in the process of photomorphogenesis described previously.

Less understood is the proteinaceous machinery present in the chloroplast outer membrane that is responsible for the recognition and insertion of α-helical and β-barrel chloroplast outer membrane proteins. In mitochondria, the translocon at the outer membrane of the mitochondrion (TOM) complex imports β-barrel proteins, and they are then transferred to the sorting and assembly machinery (SAM) complex by IMS chaperones before being integrated into the membrane [67,68]. Most, if not all, α-helical outer membrane proteins of mitochondria are inserted directly via the mitochondrial import (MIM) complex [67]. Such distinct pathways have not been identified for chloroplast outer membrane proteins to date.

4. Biogenesis of α-Helical Chloroplast Outer Membrane Proteins

4.1. Signal Anchored (SA) Proteins

SA and TA proteins represent a diverse array of functions, acting as receptors in pathways of protein translocation, membrane fusion, vesicle trafficking, electron transport, apoptosis and protein quality control [39,69]. SA proteins lack a cleavable TP and are anchored in the membrane by a single hydrophobic α-helix of approximately 20 amino acid residues in length at their N-terminus [30]. This TMD is flanked by a C-terminal positively charged region (CPR) and, together, the TMD and CPR act as an intrinsic targeting signal [70]. At the sequence level, it is difficult to identify a conserved sequence motif that may direct targeting. SA proteins do not share sequence similarity at their N-termini the way they do at their C-termini. Instead, proteinaceous factors in the cytosol may recognize SA proteins directed to chloroplasts and mitochondria based on their physicochemical characteristics, which are highly conserved [27]. Specifically, the majority of TMDs within SA proteins of chloroplasts and mitochondria are only moderately hydrophobic relative to SA proteins directed to the endoplasmic reticulum, with hydrophobicity values below 0.4 on the Wimley and White scale [70]. Increasing the hydrophobicity of the TMD will redirect a chloroplast SA protein to the plasma membrane [71]. Additionally, the CPR contains three or more positively charged amino acid residues that assist in the evasion of the signal recognition particle (SRP) that directs SA proteins to the ER [27]. Like the TMD, substitutions that reduce the positive charge of the CPR redirect targeting of chloroplast SA proteins to the plasma membrane [70].

It is still an open question how the plant cell achieves selective targeting of SA proteins to chloroplasts and mitochondria. Despite the ambiguity in the targeting signals of chloroplast and mitochondrion SA proteins, a cytosolic factor has been identified that is responsible for targeting chloroplast SA proteins to the membrane surface. Ankyrin repeat-containing protein 2 (AKR2) interacts with the SA targeting signal during translation and acts as a chaperone to shield the hydrophobic TMD and prevent aggregation to keep the SA protein in a membrane insertion competent state [39,72,73]. AKR2 achieves selective targeting through its lipid binding domain, which recognizes monogalactosyldiacylglycerol, a lipid unique to plastid membranes, and phosphatidylglycerol headgroups [74]. AKR2 may recognize subtle differences in the density of positively charged residues and amino acid residue composition of the CPR as well as its distance from the TMD [27]. The mechanism by which SA proteins are inserted in the chloroplast outer membrane is not well understood, and there is some evidence that it may vary between SA proteins [27]. Currently, it is not clear whether SA proteins can insert spontaneously into the membrane or if a, yet to be discovered, insertase is involved; regardless, targeting and insertion seem to be dependent on the presence of the TOC complex [27]. The targeting mechanisms of SA, TA and β-barrel chloroplast outer membrane proteins are depicted in Figure 1.

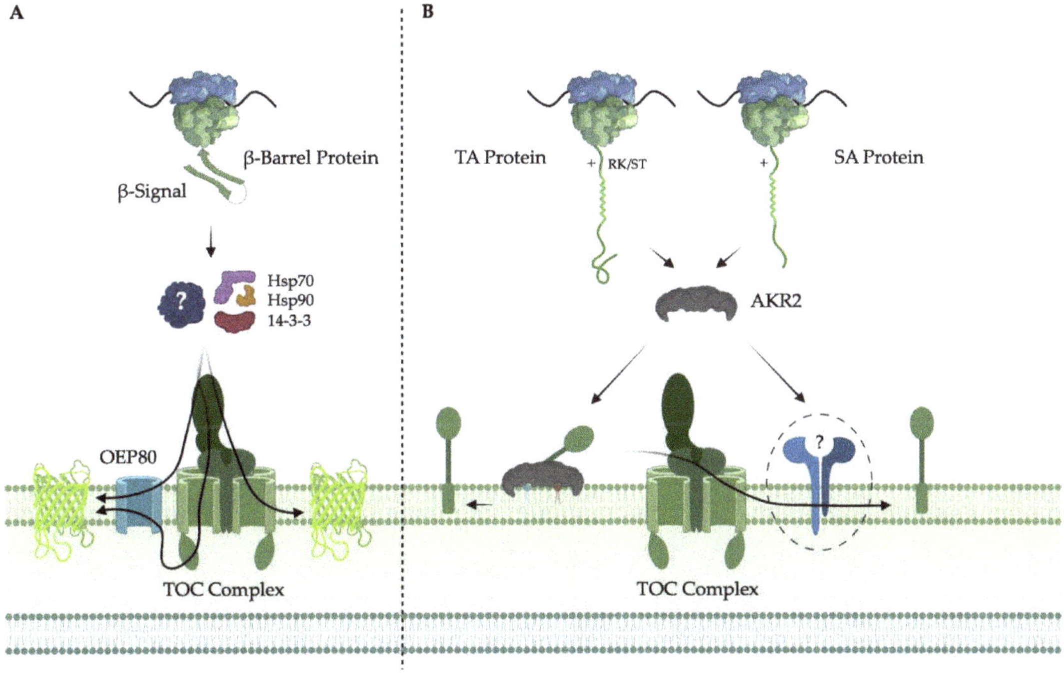

Figure 1. Signal anchor (SA), tail anchor (TA) and β-Barrel Mediated Targeting Pathways to the Chloroplast Outer Membrane. (**A**) β-barrel proteins are targeted to the chloroplast outer membrane by a variety of signals, like the β-signal. Although some may use Hsp70, Hsp90 and 14-3-3 proteins, a specific cytosolic chaperone that aids in their targeting is yet to be discovered. OEP80 likely plays a role in the insertion of β-barrels, although the translocon at the outer membrane of the chloroplast (TOC complex) is also involved in their targeting; (**B**) SA and TA proteins are targeted to the chloroplast outer membrane by their moderately hydrophobic N-terminal and C-terminal transmembrane α-helix, respectively, and a C-terminal positively charged region, sometimes accompanied by an RK/ST motif for TA proteins. Both are guided by ankyrin repeat-containing protein 2 (AKR2), which interacts with monogalactosyldiacylglycerol (MGDG) and phosphatidylglycerol (PG) in the chloroplast outer membrane. Whether they are inserted by the TOC complex or an undiscovered insertase is unknown, but interaction with the TOC complex is an essential step in their targeting. Chloroplast outer membrane (COM); intermembrane space (IMS); chloroplast inner membrane (CIM). Created using BioRender.com (accessed on 1 December 2021).

4.2. Tail Anchored (TA) Proteins

Like SA proteins, TA proteins lack a cleavable TP, but are anchored in the membrane by a single α-helix at their C-terminus [75]. The TMD is flanked by a C-terminal sequence (CTS) and, together, the TMD and CTS act as an intrinsic targeting signal [76]. Interestingly, the physicochemical characteristics of the TA are not as important to targeting. The hydrophobicity of the TMD varies widely across TA proteins and the importance of the CTS varies from protein to protein, although a net positive charge seems to contribute to chloroplast-targeting specificity [39]. Like the CPR of SA proteins, this assists in the evasion of the SRP. Eliminating the net positive charge redirects TA proteins to the mitochondrion [37]. A subset of TA proteins contains an RK/ST motif in their CTS. This motif is interchangeable among TA proteins in the subset. The charge distribution within the RK/ST motif appears to play a larger role than net charge [69,77]. For some TA proteins, the CTS is both necessary and sufficient for targeting, where for others, it is necessary but not sufficient [37]. The latter seems to be the case for those TA proteins that contain GTPase (G) domains, such as the TOC34 receptors discussed in detail below.

Like SA proteins, AKR2 also interacts with TA proteins in the cytosol to guide them to the chloroplast outer membrane, although there is evidence that other cytosolic factors are involved [37,69]. The targeting of TA proteins is well characterized in mammalian cells and yeast, and involves the guided entry of TA proteins (GET) and transmembrane recognition complex (TRC) pathways, respectively [78,79]. GET homologs have been identified in *Arabidopsis thaliana* [77] and have been implicated in the targeting of TOC34 receptors [80]. GET3B has been shown to target TA proteins to the thylakoid membrane [81]. Although cytosolic factors contribute to the efficiency of TA protein targeting, specificity for the chloroplast outer membrane seems to rely on lipid composition of the membrane, which may be important for insertion [82]. It was previously thought that SA and TA protein insertion in the chloroplast outer membrane occurred exclusively through interactions with the naked chloroplast outer membrane. More recently, it has been shown that targeting of both SA and TA proteins to the chloroplast outer membrane requires TOC75 and competes with precursor proteins for insertion, suggesting they do use the TOC complex, at least as a first step in their insertion [83].

5. Biogenesis of β-Barrel Chloroplast Outer Membrane Proteins

β-barrel proteins are characterized by their distinct topology from the more common α-helical TMDs of many membrane proteins. They are defined as proteins composed of 8–24 β-strands, where individual strands are usually 9–11 amino acids in length and are tilted approximately 45 degrees from the plane of the membrane. Alternating patterns of amino acid side chains result in amphiphilic segments with a hydrophilic face lining the interior of the pore and a hydrophobic face exposed to the lipid bilayer. The structure is stabilized by hydrogen bonding networks between the peptide backbone of neighbouring β-strands. These pores are found exclusively in the outer membrane of plastids and mitochondria of eukaryotes, as well as the outer membrane of Gram-negative bacteria from which they originated [84,85]. They act as membrane anchors or channels that recognize and transport a wide variety of substrates (ions, small molecules, peptides, nucleic acids, and proteins) with varying levels of specificity [84]. Their shared ancestry, along with their homologous structure and function would suggest easily identifiable targeting elements; but their primary sequences are highly divergent [27], complicating attempts to identify conserved targeting sequences. Their targeting information is thought to, instead, be dispersed among the primary sequence and displayed in the form of secondary and/or tertiary structures [38,86].

Bacterial and mitochondrial β-barrels contain β-signals, which are conserved motifs in the C-terminal β-strand(s) of the barrel [87] that initiate interactions with the β-barrel assembly machinery (BAM) and SAM complexes in bacterial and mitochondrial outer membranes, respectively. It was previously thought that membrane-embedded β-barrels were highly rigid and unlikely to open laterally. This is not the case. In Gram-negative bacteria and mitochondria, the β-signal acts as an insertion signal, triggering the lateral opening of BAM and SAM pores [88]. It has been shown that the targeting of β-barrel proteins to mitochondria relies on the hydrophobicity of the C-terminal β-hairpin. Specifically, a hydrophilic amino acid residue positioned at the C-terminus of the penultimate β-strand determines mitochondrial targeting. Interestingly, deletion of the C-terminal β-hairpin of chloroplast β-barrel proteins disrupts their targeting and redirects them to mitochondria [85]. Mislocalization of β-barrels between chloroplasts and mitochondria does not occur in plant cells but does occur in vitro [38], suggesting cytosolic factors yet to be discovered play a crucial role in targeting fidelity. Although mutagenesis can alter the localization of a chloroplast β-barrel to mitochondria [38], the reverse has not been demonstrated. This would suggest that chloroplast β-barrel proteins have gained additional targeting elements that both enable chloroplast localization and prevent localization to mitochondria. In fact, there is evidence that chloroplast β-barrels contain distinct groups of signals and may reach the chloroplast outer membrane using a variety of targeting pathways. Some may even be capable of self-insertion [89,90]. Unlike SA and TA proteins, the unassisted targeting of a β-barrel protein to the chloroplast outer membrane has not been demonstrated experimentally to date. For some chloroplast outer membrane β-barrels, targeting information is contained in a set of N-terminal β-strands that engage the TOC complex and trigger import into the IMS [91]. They are then integrated into the outer membrane by OEP80, a process dependent on the C-terminal β-strand [92]. OEP80 is required for the accumulation of other β-barrel proteins in the chloroplast outer membrane and may represent part of the machinery that recognizes and inserts β-barrels in the chloroplast outer membrane [93,94].

6. Targeting and Assembly of the TOC Complex

6.1. TOC75 Translocation Channel Targeting

Although the components of the TOC complex fit well into the above-mentioned chloroplast outer membrane protein categories based on their structure and topology, their targeting signals and pathways seem to diverge from those described for β-barrel and TA proteins (Figure 2). TOC75 is the first component of the TOC complex to be inserted in the membrane and facilitates the targeting and insertion of TOC34 and TOC159 receptors [26,95,96]. TOC75 is a β-barrel protein, related to the OMP85 family of proteins that include the BAM and SAM β-barrels [97,98], although no other homologs of the Bam and SAM complex machinery have been detected in the chloroplast outer membrane. Interestingly, and unlike other β-barrel proteins, TOC75 is synthesized as a precursor protein in the cytosol and contains a cleavable bipartite N-terminal chloroplast TP [95,99]. The TP may help to ensure TOC75 targeting, and insertion is coupled to the formation of new TOC complexes, as well as to maintain its reverse topology [26]. Although there is evidence that existing TOC complexes facilitate the biogenesis of new TOC complexes in the chloroplast outer membrane [99], the bipartite signal could act as a temporary anchor to couple insertion by a β-barrel assembly machinery such as OEP80 with TOC complex integration [26]. A poly-glycine stretch downstream of the TP arrests import through the TOC-TIC super-complex and the TP and poly-glycine stretch are cleaved by the stromal processing peptidase and a type I signal peptidase in the IMS, respectively, yielding mature TOC75 that is inserted into the membrane by an unknown mechanism [99]. The poly-glycine stretch may also act to avoid proteinaceous factors in the IMS that drive import [100], instead releasing the protein to be inserted into the membrane laterally by the TOC complex, or via OEP80. Interestingly, OEP80 has been shown to contain a cleavable N-terminal TP, but not a poly-glycine stretch [101,102]. It is not isolated from mature TOC complexes [103] and its targeting is not in competition with chloroplast precursor proteins [101], suggesting it is inserted in a different manner than TOC75.

6.2. TOC34 GTPase Receptor Family Targeting

"Free" TOC75, or TOC75 not associated with mature TOC complexes, serves as the site for TOC34 receptor integration in the chloroplast outer membrane and its assembly into maturing TOC complexes, simultaneously [96]. TOC34 receptors are TA proteins and are guided to the chloroplast outer membrane by interactions between their C-terminal α-helical TMD and AKR2 [37,104]. Arsenite/tail-anchored protein-transporting ATPase (ARSA1) has also been implicated in the targeting of TOC34 to the chloroplast outer membrane and TOM7, a mitochondrial TA protein, to mitochondria [78]. ARSA1 is similar to the GET proteins that target TA proteins to the ER in eukaryotes. Several ARSA homologs have been identified in plants and may be responsible for the targeting of TA proteins in plant cells [79]. There is evidence to support that TA proteins, like TOC34, engage the TOC complex thereafter [37]. Whether this occurs after insertion by an undiscovered insertase is unclear. Alternatively, there is evidence that interactions with galactolipids unique to the plastid membrane play an important role in TOC34 targeting [37]. This selectivity occurs at the membrane surface, independent of proteinaceous factors in the cytosol [82], and may facilitate self-insertion before or in addition to interaction with TOC75. Although there is evidence to suggest that TOC34 may self-insert into the chloroplast outer membrane [105–107], there is additional evidence to suggest that insertion is enhanced in the presence of nucleotides and inhibited when exposed to chloroplasts after proteolytic treatment [108], strengthening the theory that a yet-to-be-discovered insertase is involved. Additionally, the G domain of TOC34 receptors and the C-terminal tail exposed to the IMS have been implicated in targeting and integration into TOC complexes [106,109]. GTP hydrolysis could induce interactions between TOC34 monomers to produce TOC34 dimers that promote membrane insertion and/or TOC complex assembly [110]. Alternatively, GTP hydrolysis could induce an active conformation for monomeric TOC34 interaction with TOC75.

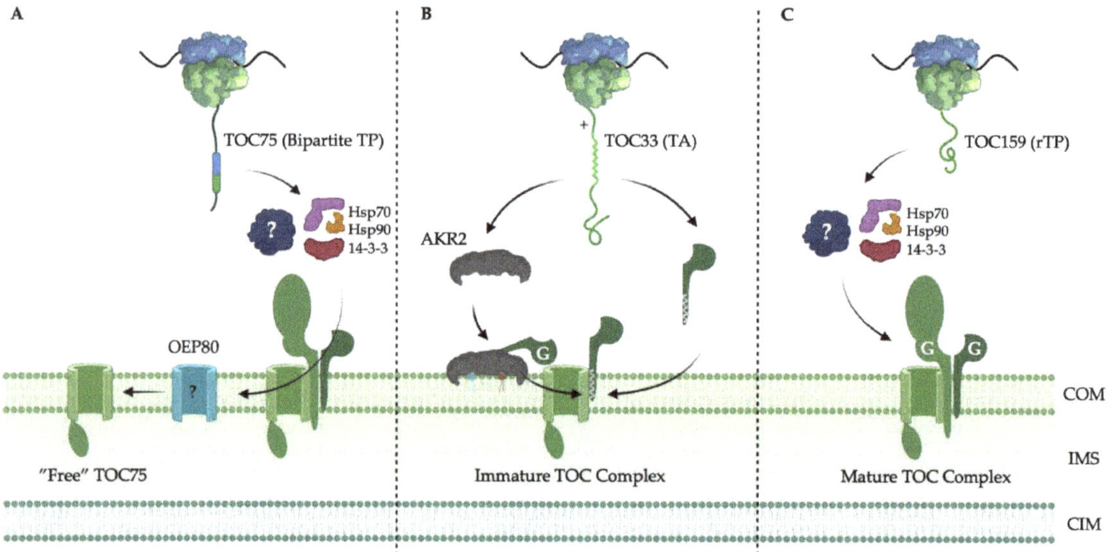

Figure 2. Targeting and assembly of the TOC complex: (**A**) TOC75 is targeted to the chloroplast outer membrane by a bipartite transit peptide (TP). The recognition and insertion of "free" TOC75 requires mature translocons at the outer membrane of the chloroplast (TOC complexes) and may be aided by OEP80. The presence of a TP suggests the use of chaperones such as Hsp70, Hsp90 and 14-3-3 employed by the canonical TP-mediated targeting pathway, although the existence of an uncharacterized cytosolic factor is possible; (**B**) TOC33/34 is a tail anchored (TA) protein and is targeted to the chloroplast outer membrane by both the TA and the GTPase (G) domain to form immature TOC complexes. Like other TA proteins, ankyrin repeat-containing protein 2 (AKR2) acts as a cytosolic chaperone; (**C**) TOC159/132/120 is targeted to the chloroplast outer membrane by a reverse TP-like sequence at the C-terminus (rTP), which may also engage chaperones employed by the canonical TP-mediated targeting pathway. Both TOC GTPase receptors rely on their G domains for successful targeting and TOC complex integration. The targeting of each component in sequence is coupled to the formation of mature TOC complexes. Chloroplast outer membrane (COM); intermembrane space (IMS); chloroplast inner membrane (CIM). Created using BioRender.com (accessed on 1 December 2021).

6.3. TOC159 GTPase Receptor Family Targeting

Relative to TOC75 and TOC34, we know very little about how TOC159 is targeted to and inserted in the chloroplast outer membrane [111]. Like TOC34, TOC159 targeting and insertion relies on TOC75, and there is evidence that TOC34 also supports the targeting and integration of TOC159 [107,112,113]. Whether the binding of TOC34 to TOC75 induces a conformation favourable to TOC159 binding and insertion and/or the interactions between the homologous G domains of the TOC34 and TOC159 receptors enhances these interactions, the membrane (M) domain clearly plays a critical role [5]. Several lines of evidence exist to support the targeting capability of the M domain [107,114,115]. The C-terminus of the M domain has demonstrated the ability to target fluorescent fusion proteins to the chloroplast outer membrane, likely due to a reverse TP-like sequence that shares physicochemical characteristics with canonical N-terminal chloroplast TPs [116,117]. This reverse TP may engage the other TOC complex components in the same way that chloroplast TPs do. Until recently, it was thought that TOC159 receptors used an atypical membrane anchor [118,119]. Recent structural prediction by AlphaFold would suggest that TOC159 receptors are anchored in the membrane by a β-barrel such as TOC75 [120]. The G domain also demonstrates intrinsic targeting capabilities [113], suggesting it may play an important role in targeting specificity and/or assembly of TOC159 into premature TOC complexes that contain TOC75 and TOC34 receptors [107]. In fact, it may even be required for the integration of TOC159 into mature TOC complexes [112,113]. It is important to distinguish between targeting, insertion, and TOC complex integration as they may represent distinct but interdependent processes. The targeting of TOC complex proteins likely evolved redundant targeting measures in this way to couple their targeting with the assembly of TOC complexes, further ensuring fidelity.

7. A Bioinformatic Approach to Identifying Novel Chloroplast Outer Membrane Targeting Signals and Pathways

Much remains unknown about how most chloroplast outer membrane proteins make their way to and are inserted in the chloroplast outer membrane. The targeting of TOC complex components highlights the complexity of chloroplast outer membrane targeting pathways and their dependence on not only protein structure but coupled assembly into protein complexes. The fact that there are 50 proteins known to be dually targeted to chloroplasts and mitochondria further illustrates the subtle yet powerful role physicochemical characteristics within targeting signals play in targeting fidelity [121–124]. A better understanding of the variety of targeting signals within the chloroplast outer membrane proteome, the structures of their membrane anchors, physicochemical characteristics, and role in targeting and insertion is important in expanding our understanding of already characterized pathways and essential to identifying novel ones. We are the first to develop a bioinformatic approach (Figure 3) to categorize the chloroplast outer membrane proteome based on these properties. The goal of this approach is to select candidates for experimental studies that will be useful in the validation of additional targeting signals and pathways.

Briefly, we scanned the literature to produce an updated version of the chloroplast outer membrane proteome, generated by Inoue (2015) [30]. We identified 21 additional proteins with potential to localize to the chloroplast outer membrane [22,29], bringing the total number of proteins in the chloroplast outer membrane proteome to 138 (Table 1). We sorted these proteins into the functional categories provided by Inoue (2015) for the previous list of 117 proteins. Five proteins (At2g25660 [125], At3g49560 [126], At4g26670 [126], At5g24650 [126] and At5g55510 [127]) are involved in protein import; three proteins (At1g54150 [128], At1g59560 [128] and At5g13530 [129]) are involved in protein turnover and modification; five proteins (At2g40690 [130], At3g63520 [131], At4g12470 [132], At4g13550 [133] and At5g16010 [127]) are involved in lipid metabolism; one protein (At2g32290 [127]) is involved in carbohydrate metabolism and regulation; two proteins (At1g26340 [127] and At5g02580 [127]) are involved in other metabolism and regulation; two proteins (At2g34585 and At3g03870) have unknown functions; and three proteins (At3g07430 [134], At3g19720 [135] and At3g57090 [136]) are involved in organellar fission, a function not described by Inoue (2015). Next, we compiled a database that allowed us to group proteins based on structural and physicochemical characteristics (such as secondary structure elements, amino acid composition and pI at each terminus), experimentally validated to be unique to SA, TA and β-barrel proteins. The predictive aspects of the database were tested against experimentally determined structures and targeting function of well-studied chloroplast outer membrane proteins. Finally, we added elements to the database, which allowed us to identify potential chloroplast TPs at the N-terminus and reverse TP-like sequences at the C-terminus independent of categorized targeting pathways. At each stage, thresholds were established, and a small number of conflicts were investigated further by careful analysis of secondary structure predictions. Together, this approach allowed us to positively identify TOC75 and OEP80 as β-barrel proteins containing an N-terminal TP and TOC34 as a TA protein. The results showed that 30% of proteins were categorized as single pass α-helical proteins, with 12% being SA, 14% being TA and 4% being "other"; 25% of proteins were categorized as multi pass α-helical proteins; 9% of proteins were categorized as β-barrel proteins; and 36% of proteins were categorized as "other", containing no predictable transmembrane elements. The 21 newly identified proteins are equally dispersed among the categories, except for the β-barrel category, into which none of the 21 newly identified proteins were sorted. Among the last group, 35% are predicted to contain a cleavable TP at their N-terminus and 39% are predicted to have a reverse TP-like sequence at their C-terminus. Interestingly, cleavable N-terminal TPs and reverse TP-like sequences at the C-terminus were also predicted, in small quantities, across all other categories, which suggests that more proteins than just TOC75 could use a bipartite signal sequence. This was also the case for the 21 newly identified chloroplast outer membrane proteins. It is also important to note that no protein was predicted to contain both an N-terminal TP and a reverse TP-like sequence at the C-Terminus. In summary, at least 65% of chloroplast outer membrane proteins were identified as having the potential to use novel targeting signals and pathways. Clearly, there is a disproportionate focus on traditional SA, TA and β-barrel-mediated targeting pathways in the literature.

Table 1. An updated list of the chloroplast outer membrane proteome and associated targeting pathways. Identified and predicted proteins of the chloroplast outer membrane proteome were categorized by potential targeting pathway and signal according to bioinformatic analyses.

AGI [1]	Name [2]	N TP [3]	C TP-Like [4]
Signal Anchored (SA) Proteins			
At1g67690	M3 Protease		✓
At2g19860	HXK2		
At2g34585 *	-		
At2g38670	PECT1		
At3g17970	TOC64-III		✓
At3g21865	PEX22		✓
At3g52420	OEP7		
At3g63170	FAP1	✓	
At4g12470 *	pEARLI1-Like Lipid Transfer Protein	✓	
At4g27680	NTPase		
At4g29130	HXK1		
At5g17770	CBR		
At5g20520	WAV2		✓
At5g25900	KO1/GA3		
At5g51020	CRL		
At5g64816	-		✓
Tail Anchored (TA) Proteins			
At1g02280	TOC33		
At1g09920	-		
At1g13900	PAP2		
At1g26340 *	Cytochrome b5		
At1g27300	-		
At1g27390	Tom20-2		
At2g16070	PDV2		

Table 1. *Cont.*

AGI [1]	Name [2]	N TP [3]	C TP-Like [4]
At2g32240	DUF869		
At3g03870 *	F20H23.8 Protein		✓
At3g27820	MDAR4		
At3g57090 *	Mitochondrial Fission 1 Protein		
At3g63150	MIRO2		
At4g32250	Tyrosine Kinase		
At4g35000	APX3		
At5g05000	TOC34		✓
At5g11560	-		✓
At5g21990	OEP61-TPR		
At5g27330	-		
At5g27540	MIRO1		
At5g56730	Peptidase M16		
Other Single Passα-Helical Proteins			
At1g16000	OEP9		
At1g80890	OEP9.2		
At2g25660 *	TIC236	✓	✓
At3g52230	OMP24 Homolog		
At3g63160	OEP6		
Multi-Passα-Helical Proteins			
At1g12230	Transaldolase	✓	
At1g34430	PDC E2	✓	
At1g44170	ALDH3H1		✓
At1g54150 *	E3 Ubiquitin-Protein Ligase SPL2		✓
At1g59560 *	E3 Ubiquitin-Protein Ligase SPL1		
At1g63900	SP1		✓
At1g64850	-		✓
At1g68680	-		
At1g77590	LACS9		✓
At2g01320	WBC7		
At2g11810	MGD3		✓
At2g28900	OEP16-1		
At2g34590	PDC E1 Beta	✓	
At2g40690 *	G3P Dehydrogenase	✓	
At2g44640	-		
At2g47770	TSPO		
At3g07430 *	YlmG Homolog	✓	
At3g49560 *	TIM Protein		
At3g51870	PAPST1 Homolog		
At3g62880	OEP16-4		
At4g15440	HPL Homolog		
At4g15810	NTPase		
At4g16160	OEP16-2		✓
At4g16450	Complex I Subunit		✓
At4g26670 *	TIM Protein		
At4g27990	YGGT-B Protein	✓	
At4g31780	MGD1	✓	
At4g38920	Vacuolar ATPase Subunit		
At5g06290	Prx B	✓	
At5g13530 *	E3 Ubiquitin-Protein Ligase KEG		
At5g16010 *	Dehydrogenase		
At5g21920	YGGT-A Protein		
At5g24650 *	TIM Protein		
At5g35210	PTM		
At5g55510 *	TIM Protein		

Table 1. *Cont.*

AGI [1]	Name [2]	N TP [3]	C TP-Like [4]
β-Barrel Proteins			
At1g20816	OEP21-1		
At1g45170	OEP24-1		
At1g76405	OEP21-2		
At2g06010	-		
At2g43950	OEP37	✓	
At3g01280	VDAC1		
At3g44160	P39/OEP80 TR1		
At3g46740	TOC75-III	✓	
At3g48620	P36/OEP80 TR2		
At4g09080	TOC75-IV		
At5g15090	VDAC3		
At5g19620	OEP80/TOC75-V	✓	
At5g42960	OEP24-2		
Other Proteins			
At1g02560	ClpP5	✓	
At1g07930	E-Tu		
At1g09340	CRB		
At1g70480	DUF220		✓
At2g16640	TOC132		✓
At2g17390	AKR2B		✓
At2g17695	OEP23		
At2g20890	THF1/PSB29	✓	
At2g24440	-		✓
At2g27490	ATCOAE		
At2g32290 *	Beta-Amylase 6	✓	
At2g32650	PTAC18-Like	✓	
At3g01500	Beta CA1	✓	
At3g06510	SFR2/GGGT		✓
At3g06960	TGD4		
At3g11670	DGD1	✓	
At3g12580	Hsc70-4		✓
At3g16620	TOC120		✓
At3g16950	PDC E3	✓	
At3g19720 *	ARC5		
At3g25690	CHUP1		
At3g25860	PDC E2	✓	
At3g26070	PAP/FBN3a	✓	
At3g26740	CCL	✓	
At3g46030	Histone H2B		✓
At3g46780	pTAC16	✓	
At3g49350	-	✓	
At3g53560	TPR Protein	✓	
At3g63520 *	Carotenoid Cleaving Protein		
At4g00550	DGD2		✓
At4g02482	Putative GTPase	✓	
At4g02510	TOC159		✓
At4g05050	UBQ11		
At4g13550 *	Putative Triglyceride Lipase	✓	
At4g14430	Enoyl-CoA Isomerase		✓
At4g17170	RAB2		✓
At4g36650	pBrP		
At5g02500	Hsc70-1		✓
At5g02580 *	Argininosuccinate Lyase		
At5g16870	PTH2 Family Protein		
At5g20300	TOC90		✓
At5g20410	MGD2		✓

Table 1. *Cont.*

AGI [1]	Name [2]	N TP [3]	C TP-Like [4]
At5g23190	CYP86B1		✓
At5g35360	CAC2/BC	✓	
At5g42070	-	✓	
At5g43070	WPP1		✓
At5g53280	PDV1		
At5g58140	PHOT2		✓
At5g59840	RAB8A-Like Protein		✓

[1] *Arabidopsis* gene identifier (AGI) number, [2] Proteins without a name are marked with -, [3] N-Terminal transit peptide (N TP), [4] TC-Terminal transit peptide-like sequence (C TP-Like), * Proteins not included in the list published by Inoue (2015) [30].

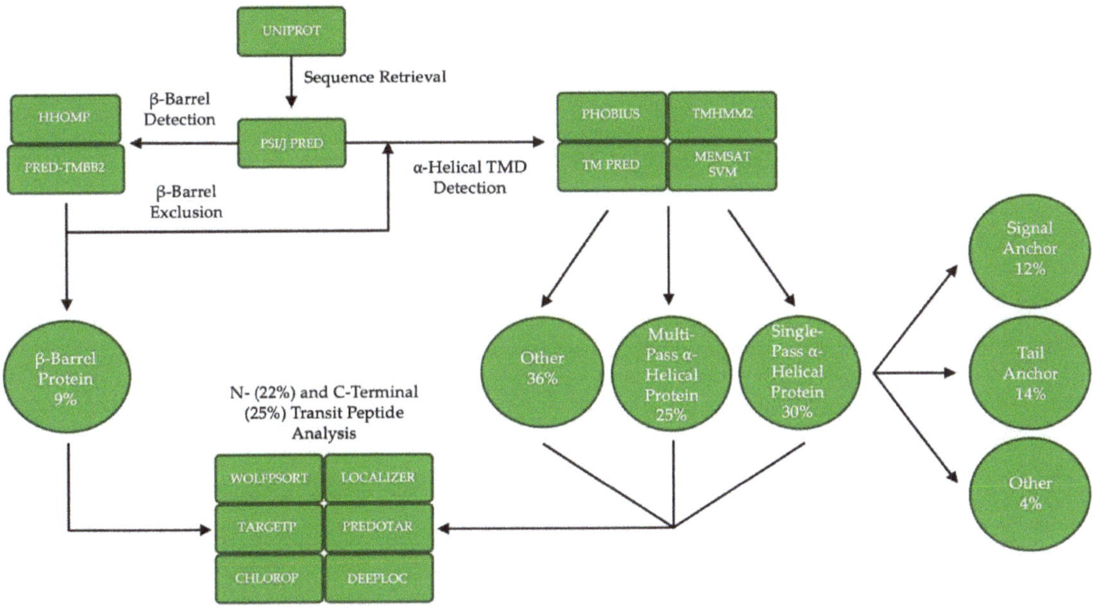

Figure 3. Bioinformatic approach to categorizing the chloroplast outer membrane proteome by targeting pathway. Bioinformatic tools used for sequence retrieval, β-barrel detection, β-barrel exclusion, α-helical transmembrane domain (TMD) detection and N- and C-terminal transit peptide detection [137–148] are provided in green squares, where resultant targeting pathway categories and percent composition of the proteome sorted into each category are given in green circles.

8. Conclusions and Future Directions

In summary, plastids, such as the chloroplast, play a central role in a variety of metabolic and signalling processes within plant cells. The biogenesis and function of chloroplasts rely heavily on the fidelity of intracellular protein targeting pathways. Like mitochondria, chloroplasts evolved from an ancient bacterial endosymbiont and the two organelles share many common characteristics in the post-translational targeting of their nuclear-encoded proteomes [3]. Similar to mitochondrial presequences, chloroplast transit peptide sequences are highly divergent, but conserved physicochemical and structural properties govern their interactions with proteinaceous factors in the cytosol, recognition by import complexes at the membrane surface and even direct interactions with specific lipids [47], all of which contribute to targeting specificity. Despite recent advances in our understanding of the targeting of chloroplast precursor proteins and their recognition and import by the TOC and TIC complexes, much remains unknown about the targeting of chloroplast outer membrane proteins, such as TOC159. This gap is further emphasized when compared to our knowledge of outer membrane proteins of mitochondria. Here, we have reviewed the current understanding of how SA, TA and β-barrel chloroplast outer membrane proteins are targeted to the organelle. Further, we used a novel bioinformatic approach to expand the current list of known chloroplast outer membrane proteins from 117 to 138 and provided new insight into novel targeting signals and pathways that could be used by a significant portion of these proteins, yet to be explored experimentally.

In mitochondria, two distinct complexes, SAM and MIM, ensure insertion and assembly of β-barrel and α-helical outer membrane proteins. These processes are mediated by the TOM complex [67]. Such complexes have not been identified in chloroplasts and so our understanding of the molecular mechanisms by which β-barrel and α-helical chloroplast outer membrane proteins are recognized, inserted and assembled at the chloroplast outer membrane is severely limited. Whether the TOC complex assumes these roles or, like the TOM complex, merely mediates interactions between membrane proteins and their insertases is unclear. It is crucial that future studies focus on identifying the targeting signals, cytosolic factors and integration complexes involved in chloroplast outer membrane protein biogenesis. Specifically, experiments should focus on clarifying the role of the TOC complex in the insertion of different types of chloroplast outer membrane proteins; identifying β-signals within the chloroplast outer membrane proteome and the cytosolic chaperone involved in β-barrel targeting; and further characterizing OEP80′s interaction with the TOC complex and its role in β-barrel insertion and assembly. Beyond this, we hope to further define bipartite signals that incorporate TPs and reverse TP-like sequences at the C-terminus, and how these sequences contribute to the modularity of targeting signals. AKR2 shuttles SA and TA proteins in the cytosol to the chloroplast outer membrane through specific interactions with lipids unique to the plastid membrane [74]. This emphasizes the interplay between the proteins of the outer membrane and the lipid molecules in which they are embedded; specifically, the role chloroplast outer membrane lipid composition plays in targeting specificity, another area that requires further exploration.

Author Contributions: Conceptualization, M.F., D.N., S.D.X.C. and M.D.S.; bioinformatic analyses, D.N., A.G. and M.F.; writing—original draft preparation, M.F.; writing—review and editing, M.F., D.N., A.G., A.O., M.J.-N., S.D.X.C. and M.D.S. All authors have read and agreed to the published version of the manuscript.

Funding: This research was funded by Natural Sciences and Engineering Research Council of Canada (NSERC); postgraduate doctoral scholarship awarded to M.F. (PGSD3-2021-559542) and Discovery Grants awarded to M.D.S. (RGPIN-2017-05437), S.D.X.C. (RGPIN-2017-04416) and M.J.-N. (RGPIN-2019-05900).

Conflicts of Interest: The authors declare no conflict of interest. The funders had no role in the collection, analyses, or interpretation of data; in the writing of the manuscript, or in the decision to publish the results.

References

1. Gould, S.B.; Waller, R.F.; McFadden, G.I. Plastid evolution. *Annu. Rev. Plant Biol.* **2008**, *59*, 491–517. [CrossRef]
2. Bölter, B. En route into chloroplasts: Preproteins' way home. *Photosynth. Res.* **2018**, *138*, 263–275. [CrossRef] [PubMed]
3. Day, P.M.; Theg, S.M. Evolution of protein transport to the chloroplast envelope membranes. *Photosynth. Res.* **2018**, *138*, 315–326. [CrossRef] [PubMed]
4. Kessler, F.; Schnell, D. Chloroplast biogenesis: Diversity and regulation of the protein import apparatus. *Curr. Opin. Cell Biol.* **2009**, *21*, 494–500. [CrossRef] [PubMed]
5. Paila, Y.D.; Richardson, L.G.L.; Schnell, D.J. New insights into the mechanism of chloroplast protein import and its integration with protein quality control, organelle biogenesis and development. *J. Mol. Biol.* **2015**, *427*, 1038–1060. [CrossRef]
6. Choi, H.; Yi, T.; Ha, S.H. Diversity of plastid types and their interconversions. *Front. Plant Sci.* **2021**, *12*, 692024. [CrossRef] [PubMed]
7. Chu, C.C.; Swamy, K.; Li, H.M. Tissue-specific regulation of plastid protein import via transit-peptide motifs. *Plant Cell* **2020**, *32*, 1204–1217. [CrossRef]

8. Pogson, B.J.; Ganguly, D.; Albrecht-Borth, V. Insights into chloroplast biogenesis and development. *Biochim. Biophys. Acta-Bioenerg.* **2015**, *1847*, 1017–1024. [CrossRef]
9. Chu, C.C.; Li, H.M. Developmental regulation of protein import into plastids. *Photosynth. Res.* **2018**, *138*, 327–334. [CrossRef]
10. Shanmugabalaji, V.; Chahtane, H.; Accossato, S.; Rahire, M.; Gouzerh, G.; Lopez-Molina, L.; Kessler, F. Chloroplast biogenesis controlled by DELLA-TOC159 interaction in early plant development. *Curr. Biol.* **2018**, *28*, 2616–2623. [CrossRef]
11. Yan, J.; Smith, M.D.; Glick, B.R.; Liang, Y. Effects of ACC deaminase containing rhizobacteria on plant growth and expression of Toc GTPases in tomato (*Solanum lycopersicum*) under salt stress. *Botany* **2014**, *92*, 775–781. [CrossRef]
12. Doroodian, P.; Hua, Z. The ubiquitin switch in plant stress response. *Plants* **2021**, *10*, 246. [CrossRef] [PubMed]
13. Richardson, L.G.L.; Singhal, R.; Schnell, D.J. The integration of chloroplast protein targeting with plant developmental and stress responses. *BMC Biol.* **2017**, *15*, 118. [CrossRef] [PubMed]
14. Jarvis, P.; López-Juez, E. Biogenesis and homeostasis of chloroplasts and other plastids. *Nat. Rev. Mol. Cell Biol.* **2013**, *14*, 787–802. [CrossRef] [PubMed]
15. Yang, X.; Li, Y.; Qi, M.; Liu, Y.; Li, T. Targeted control of chloroplast quality to improve plant acclimation: From protein import to degradation. *Front. Plant Sci.* **2019**, *10*, 958. [CrossRef] [PubMed]
16. Fukazawa, H.; Tada, A.; Richardson, L.G.L.; Kakizaki, T.; Uehara, S.; Ito-Inaba, Y.; Inaba, T. Induction of *TOC* and *TIC* genes during photomorphogenesis is mediated primarily by cryptochrome 1 in Arabidopsis. *Sci. Rep.* **2020**, *10*, 20255. [CrossRef] [PubMed]
17. Inoue, H.; Rounds, C.; Schnell, D.J. The molecular basis for distinct pathways for protein import into arabidopsis chloroplasts. *Plant Cell* **2010**, *22*, 1947–1960. [CrossRef]
18. Li, H.; Teng, Y.S. Transit peptide design and plastid import regulation. *Trends Plant Sci.* **2013**, *18*, 360–366. [CrossRef]
19. Chu, C.C.; Li, H.M. Protein import into isolated pea root leucoplasts. *Front. Plant Sci.* **2015**, *6*, 690. [CrossRef] [PubMed]
20. Leister, D. Chloroplast research in the genomic age. *Trends Genet.* **2003**, *19*, 47–56. [CrossRef]
21. Kleffmann, T.; Russenberger, D.; Von Zychlinski, A.; Christopher, W.; Sjölander, K.; Gruissem, W.; Baginsky, S. The *Arabidopsis thaliana* chloroplast proteome reveals pathway abundance and novel protein functions. *Curr. Biol.* **2004**, *14*, 354–362. [CrossRef] [PubMed]
22. Zybailov, B.; Rutschow, H.; Friso, G.; Rudella, A.; Emanuelsson, O.; Sun, Q.; Van Wijk, K.J. Sorting signals, N-terminal modifications and abundance of the chloroplast proteome. *PLoS ONE* **2008**, *3*, e1994. [CrossRef] [PubMed]
23. Zimorski, V.; Ku, C.; Martin, W.F.; Gould, S.B. Endosymbiotic theory for organelle origins. *Curr. Opin. Microbiol.* **2014**, *22*, 38–48. [CrossRef] [PubMed]
24. Jarvis, P.; Paul, J. Targeting of nucleus-encoded proteins to chloroplasts in plants. *New Phytol.* **2008**, *179*, 257–285. [CrossRef]
25. Schnell, D.J. The TOC GTPase receptors: Regulators of the fidelity, specificity and substrate profiles of the general protein import machinery of chloroplasts. *Protein J.* **2019**, *38*, 343–350. [CrossRef]
26. Richardson, L.G.L.; Paila, Y.D.; Siman, S.R.; Chen, Y.; Smith, M.D.; Schnell, D.J. Targeting and assembly of components of the TOC protein import complex at the chloroplast outer envelope membrane. *Front. Plant Sci.* **2014**, *5*, 269. [CrossRef]
27. Kim, J.; Na, Y.J.; Park, S.J.; Baek, S.H.; Kim, D.H. Biogenesis of chloroplast outer envelope membrane proteins. *Plant Cell Rep.* **2019**, *38*, 783–792. [CrossRef]
28. Sjuts, I.; Soll, J.; Bölter, B. Import of soluble proteins into chloroplasts and potential regulatory mechanisms. *Front. Plant Sci.* **2017**, *8*, 168. [CrossRef]
29. Bouchnak, I.; Brugière, S.; Moyet, L.; Le Gall, S.; Salvi, D.; Kuntz, M.; Tardif, M.; Rolland, N. Unraveling hidden components of the chloroplast envelope proteome: Opportunities and limits of better MS sensitivity. *Mol. Cell. Proteom.* **2019**, *18*, 1285–1306. [CrossRef]
30. Inoue, K. Emerging knowledge of the organelle outer membranes—Research snapshots and an updated list of the chloroplast outer envelope proteins. *Front. Plant Sci.* **2015**, *6*, 278. [CrossRef]
31. Inoue, K. Emerging roles of the chloroplast outer envelope membrane. *Trends Plant Sci.* **2011**, *16*, 550–557. [CrossRef] [PubMed]
32. Breuers, F.K.H.; Bräutigam, A.; Weber, A.P.M. The plastid outer envelope—A highly dynamic interface between plastid and cytoplasm. *Front. Plant Sci.* **2011**, *2*, 97. [CrossRef]
33. Schleiff, E.; Tien, R.; Salomon, M.; Soll, J. Lipid composition of outer leaflet of chloroplast outer envelope determines topology of OEP7. *Mol. Biol. Cell* **2001**, *12*, 4090–4102. [CrossRef] [PubMed]
34. Cline, K.; Keegstra, K. Galactosyltransferases involved in galactolipid biosynthesis are located in the outer membrane of pea chloroplast envelopes. *Plant Physiol.* **1983**, *71*, 366–372. [CrossRef]
35. Sahin, C.; Reid, D.J.; Marty, M.T.; Landreh, M. Scratching the surface: Native mass spectrometry of peripheral membrane protein complexes. *Biochem. Soc. Trans.* **2020**, *48*, 547–558. [CrossRef]
36. Murcha, M.W.; Kmiec, B.; Kubiszewski-Jakubiak, S.; Teixeira, P.F.; Glaser, E.; Whelan, J. Protein import into plant mitochondria: Signals, machinery, processing, and regulation. *J. Exp. Bot.* **2014**, *65*, 6301–6335. [CrossRef]
37. Dhanoa, P.K.; Richardson, L.G.L.; Smith, M.D.; Gidda, S.K.; Henderson, M.P.A.; Andrews, D.W.; Mullen, R.T. Distinct pathways mediate the sorting of tail-anchored proteins to the plastid outer envelope. *PLoS ONE* **2010**, *5*, e10098. [CrossRef] [PubMed]
38. Lee, D.W.; Lee, J.; Hwang, I. Sorting of nuclear-encoded chloroplast membrane proteins. *Curr. Opin. Plant Biol.* **2017**, *40*, 1–7. [CrossRef]

39. Lee, J.; Kim, D.H.; Hwang, I. Specific targeting of proteins to outer envelope membranes of endosymbiotic organelles, chloroplasts, and mitochondria. *Front. Plant Sci.* **2014**, *5*, 173. [CrossRef]
40. Richardson, L.G.L.; Schnell, D.J. Origins, function, and regulation of the TOC-TIC general protein import machinery of plastids. *J. Exp. Bot.* **2020**, *71*, 1226–1238. [CrossRef]
41. Thomson, S.M.; Pulido, P.; Jarvis, R.P. Protein import into chloroplasts and its regulation by the ubiquitin-proteasome system. *Biochem. Soc. Trans.* **2020**, *48*, 71–82. [CrossRef]
42. Andrès, C.; Agne, B.; Kessler, F. The TOC complex: Preprotein gateway to the chloroplast. *Biochim. Biophys. Acta-Mol. Cell Res.* **2010**, *1803*, 715–723. [CrossRef] [PubMed]
43. Nakai, M. New perspectives on chloroplast protein import. *Plant Cell Physiol.* **2018**, *59*, 1111–1119. [CrossRef] [PubMed]
44. Smith, M.D. Protein import into chloroplasts: An ever-evolving story. *Can. J. Bot.* **2006**, *84*, 531–542. [CrossRef]
45. Lee, D.W.; Jung, C.; Hwang, I. Cytosolic events involved in chloroplast protein targeting. *Biochim. Biophys. Acta-Mol. Cell Res.* **2013**, *1833*, 245–252. [CrossRef]
46. Patron, N.J.; Waller, R.F. Transit peptide diversity and divergence: A global analysis of plastid targeting signals. *BioEssays* **2007**, *29*, 1048–1058. [CrossRef]
47. Lee, D.W.; Hwang, I. Evolution and design principles of the diverse chloroplast transit peptides. *Mol. Cells* **2018**, *41*, 161–167. [CrossRef] [PubMed]
48. Lee, D.W.; Yoo, Y.J.; Razzak, M.A.; Hwang, I. Prolines in transit peptides are crucial for efficient preprotein translocation into chloroplasts. *Plant Physiol.* **2018**, *176*, 663–677. [CrossRef] [PubMed]
49. Bruce, B.D. The paradox of plastid transit peptides: Conservation of function despite divergence in primary structure. *Biochim. Biophys. Acta-Mol. Cell Res.* **2001**, *1541*, 2–21. [CrossRef]
50. Dong, W.L.; Jong, K.K.; Lee, S.; Choi, S.; Kim, S.; Hwang, I. Arabidopsis nuclear-encoded plastid transit peptides contain multiple sequence subgroups with distinctive chloroplast-targeting sequence motifs. *Plant Cell* **2008**, *20*, 1603–1622. [CrossRef]
51. Lee, D.W.; Lee, S.; Lee, J.; Woo, S.; Razzak, M.A.; Vitale, A.; Hwang, I. Molecular mechanism of the specificity of protein import into chloroplasts and mitochondria in plant cells. *Mol. Plant* **2019**, *12*, 951–966. [CrossRef]
52. Flores-Pérez, Ú.; Jarvis, P. Molecular chaperone involvement in chloroplast protein import. *Biochim. Biophys. Acta-Mol. Cell Res.* **2013**, *1833*, 332–340. [CrossRef] [PubMed]
53. Hristou, A.; Grimmer, J.; Baginsky, S. The secret life of chloroplast precursor proteins in the cytosol. *Mol. Plant* **2020**, *13*, 1111–1113. [CrossRef] [PubMed]
54. Panigrahi, R.; Whelan, J.; Vrielink, A. Exploring ligand recognition, selectivity and dynamics of TPR domains of chloroplast Toc64 and mitochondria Om64 from *Arabidopsis thaliana*. *J. Mol. Recognit.* **2014**, *27*, 402–414. [CrossRef]
55. Qbadou, S.; Becker, T.; Mirus, O.; Tews, I.; Soll, J.; Schleiff, E. The molecular chaperone Hsp90 delivers precursor proteins to the chloroplast import receptor Toc64. *EMBO J.* **2006**, *25*, 1836–1847. [CrossRef] [PubMed]
56. Kouranov, A.; Schnell, D.J. Analysis of the interactions of preproteins with the import machinery over the course of protein import into chloroplasts. *J. Cell Biol.* **1997**, *139*, 1677–1685. [CrossRef] [PubMed]
57. Smith, M.D.; Rounds, C.M.; Wang, F.; Chen, K.; Afitlhile, M.; Schnell, D.J. atToc159 is a selective transit peptide receptor for the import of nucleus-encoded chloroplast proteins. *J. Cell Biol.* **2004**, *165*, 323–334. [CrossRef]
58. Kessler, F.; Blobel, G.; Patel, H.A.; Schnell, D.J. Identification of two GTP-binding proteins in the chloroplast protein import machinery. *Science* **1994**, *266*, 1035–1039. [CrossRef]
59. Yan, J.; Campbell, J.H.; Glick, B.R.; Smith, M.D.; Liang, Y. Molecular characterization and expression analysis of chloroplast protein import components in tomato (*Solanum lycopersicum*). *PLoS ONE* **2014**, *9*, e95088. [CrossRef]
60. Ramundo, S.; Asakura, Y.; Salomé, P.A.; Strenkert, D.; Boone, M.; Mackinder, L.C.M.; Takafuji, K.; Dinc, E.; Rahire, M.; Crèvecoeur, M.; et al. Coexpressed subunits of dual genetic origin define a conserved supercomplex mediating essential protein import into chloroplasts. *Proc. Natl. Acad. Sci. USA* **2020**, *117*, 32739–32749. [CrossRef]
61. Agne, B.; Kessler, F. Protein transport in organelles: The Toc complex way of preprotein import. *FEBS J.* **2009**, *276*, 1156–1165. [CrossRef] [PubMed]
62. Schleiff, E.; Soll, J.; Küchler, M.; Kühlbrandt, W.; Harrer, R. Characterization of the translocon of the outer envelope of chloroplasts. *J. Cell Biol.* **2003**, *160*, 541–551. [CrossRef] [PubMed]
63. Chen, L.J.; Li, H.M. Stable megadalton TOC–TIC supercomplexes as major mediators of protein import into chloroplasts. *Plant J.* **2017**, *92*, 178–188. [CrossRef] [PubMed]
64. Kikuchi, S.; Hirohashi, T.; Nakai, M. Characterization of the preprotein translocon at the outer envelope membrane of chloroplasts by blue native PAGE. *Plant Cell Physiol.* **2006**, *47*, 363–371. [CrossRef]
65. Chen, K.Y.; Li, H.M. Precursor binding to an 880-kDa Toc complex as an early step during active import of protein into chloroplasts. *Plant J.* **2007**, *49*, 149–158. [CrossRef]
66. Jackson-Constan, D.; Keegstra, K. Arabidopsis genes encoding components of the chloroplastic protein import apparatus. *Plant Physiol.* **2001**, *125*, 1567–1576. [CrossRef] [PubMed]
67. Gupta, A.; Becker, T. Mechanisms and pathways of mitochondrial outer membrane protein biogenesis. *Biochim. Biophys. Acta-Bioenerg.* **2021**, *1862*, 148323. [CrossRef]
68. Lundquist, K.; Billings, E.; Bi, M.; Wellnitz, J.; Noinaj, N. The assembly of β-barrel membrane proteins by BAM and SAM. *Mol. Microbiol.* **2021**, *115*, 425–435. [CrossRef]

69. Teresinski, H.J.; Gidda, S.K.; Nguyen, T.N.D.; Howard, N.J.M.; Porter, B.K.; Grimberg, N.; Smith, M.D.; Andrews, D.W.; Dyer, J.M.; Mullen, R.T. An RK/ST C-terminal motif is required for Ttrgeting of OEP7.2 and a subset of other Arabidopsis tail-anchored proteins to the plastid outer envelope membrane. *Plant Cell Physiol.* **2019**, *60*, 516–537. [CrossRef] [PubMed]
70. Lee, J.; Lee, H.; Kim, J.; Lee, S.; Kim, D.H.; Kim, S.; Hwang, I. Both the hydrophobicity and a positively charged region flanking the c-terminal region of the transmembrane domain of signal-anchored proteins play critical roles in determining their targeting specificity to the endoplasmic reticulum or endosymbiotic org. *Plant Cell* **2011**, *23*, 1588–1607. [CrossRef]
71. Lee, Y.J.; Sohn, E.J.; Lee, K.H.; Lee, D.W.; Hwang, I. The transmembrane domain of AtTco64 and its C-terminal lysine-rich flanking region are targeting signals to the chloroplast outer envelope membrane. *Mol. Cells* **2004**, *17*, 281–291.
72. Bae, W.; Lee, Y.J.; Kim, D.H.; Lee, J.; Kim, S.; Sohn, E.J.; Hwang, I. AKR2A-mediated import of chloroplast outer membrane proteins is essential for chloroplast biogenesis. *Nat. Cell Biol.* **2008**, *10*, 220–227. [CrossRef] [PubMed]
73. Kim, D.H.; Lee, J.E.; Xu, Z.Y.; Geem, K.R.; Kwon, Y.; Park, J.W.; Hwang, I. Cytosolic targeting factor AKR2A captures chloroplast outer membrane-localized client proteins at the ribosome during translation. *Nat. Commun.* **2015**, *6*, 6843. [CrossRef] [PubMed]
74. Kim, D.H.; Park, M.J.; Gwon, G.H.; Silkov, A.; Xu, Z.Y.; Yang, E.C.; Song, S.; Song, K.; Kim, Y.; Yoon, H.S.; et al. An ankyrin repeat domain of AKR2 drives chloroplast targeting through coincident binding of two chloroplast lipids. *Dev. Cell* **2014**, *30*, 598–609. [CrossRef] [PubMed]
75. Zhuang, X.; Chung, K.P.; Jiang, L. Targeting tail-anchored proteins into plant organelles. *Proc. Natl. Acad. Sci. USA* **2017**, *114*, 1762–1764. [CrossRef] [PubMed]
76. Moog, D. Higher complexity requires higher accuracy: Tail-anchored protein targeting to the outer envelope membrane of plant plastids via a specific C-terminal motif. *Plant Cell Physiol.* **2019**, *60*, 489–491. [CrossRef]
77. Marty, N.J.; Teresinski, H.J.; Hwang, Y.T.; Clendening, E.A.; Gidda, S.K.; Sliwinska, E.; Zhang, D.; Miernyk, J.A.; Brito, G.C.; Andrews, D.W.; et al. New insights into the targeting of a subset of tail-anchored proteins to the outer mitochondrial membrane. *Front. Plant Sci.* **2014**, *5*, 426. [CrossRef] [PubMed]
78. Formighieri, C.; Cazzaniga, S.; Kuras, R.; Bassi, R. Biogenesis of photosynthetic complexes in the chloroplast of Chlamydomonas reinhardtii requires ARSA1, a homolog of prokaryotic arsenite transporter and eukaryotic TRC40 for guided entry of tail-anchored proteins. *Plant J.* **2013**, *73*, 850–861. [CrossRef] [PubMed]
79. Lin, T.W.; Chen, C.C.; Wu, S.M.; Chang, Y.C.; Li, Y.C.; Su, Y.W.; Hsiao, C.D.; Chang, H.Y. Structural analysis of chloroplast tail-anchored membrane protein recognition by ArsA1. *Plant J.* **2019**, *99*, 128–143. [CrossRef]
80. Maestre-Reyna, M.; Wu, S.M.; Chang, Y.C.; Chen, C.C.; Maestre-Reyna, A.; Wang, A.H.J.; Chang, H.Y. In search of tail-anchored protein machinery in plants: Reevaluating the role of arsenite transporters. *Sci. Rep.* **2017**, *7*, 46022. [CrossRef]
81. Anderson, S.A.; Satyanarayan, M.B.; Wessendorf, R.L.; Lu, Y.; Fernandez, D.E. A homolog of GuidedEntry of Tail-anchored proteins3 functions in membrane-specific protein targeting in chloroplasts of Arabidopsis. *Plant Cell* **2021**, *33*, 2812–2833. [CrossRef] [PubMed]
82. Kriechbaumer, V.; Abell, B.M. Chloroplast envelope protein targeting fidelity is independent of cytosolic components in dual organelle assays. *Front. Plant Sci.* **2012**, *3*, 148. [CrossRef] [PubMed]
83. Tu, S.L.; Chen, L.J.; Smith, M.D.; Su, Y.S.; Schnell, D.J.; Li, H.M. Import pathways of chloroplast interior proteins and the outer-membrane protein OEP14 converge at Toc75. *Plant Cell* **2004**, *16*, 2078–2088. [CrossRef] [PubMed]
84. Jores, T.; Rapaport, D. Early stages in the biogenesis of eukaryotic β-barrel proteins. *FEBS Lett.* **2017**, *591*, 2671–2681. [CrossRef]
85. Klinger, A.; Gosch, V.; Bodensohn, U.; Ladig, R.; Schleiff, E. The signal distinguishing between targeting of outer membrane β-barrel protein to plastids and mitochondria in plants. *Biochim. Biophys. Acta-Mol. Cell Res.* **2019**, *1866*, 663–672. [CrossRef] [PubMed]
86. Walther, D.M.; Rapaport, D. Biogenesis of mitochondrial outer membrane proteins. *Biochim. Biophys. Acta-Mol. Cell Res.* **2009**, *1793*, 42–51. [CrossRef] [PubMed]
87. Jores, T.; Klinger, A.; Groß, L.E.; Kawano, S.; Flinner, N.; Duchardt-Ferner, E.; Wöhnert, J.; Kalbacher, H.; Endo, T.; Schleiff, E.; et al. Characterization of the targeting signal in mitochondrial β-barrel proteins. *Nat. Commun.* **2016**, *7*, 12036. [CrossRef] [PubMed]
88. Höhr, A.I.C.; Lindau, C.; Wirth, C.; Qiu, J.; Stroud, D.A.; Kutik, S.; Guiard, B.; Hunte, C.; Becker, T.; Pfanner, N.; et al. Membrane protein insertion through a mitochondrial β-barrel gate. *Science* **2018**, *359*, eaah6834. [CrossRef] [PubMed]
89. Horne, J.E.; Brockwell, D.J.; Radford, S.E. Role of the lipid bilayer in outer membrane protein folding in Gram-negative bacteria. *J. Biol. Chem.* **2020**, *295*, 10340–10367. [CrossRef]
90. Gessmann, D.; Chung, Y.H.; Danoff, E.J.; Plummer, A.M.; Sandlin, C.W.; Zaccai, N.R.; Fleming, K.G. Outer membrane β-barrel protein folding is physically controlled by periplasmic lipid head groups and BamA. *Proc. Natl. Acad. Sci. USA* **2014**, *111*, 5878–5883. [CrossRef]
91. Day, P.M.; Inoue, K.; Theg, S.M. Chloroplast outer membrane b-barrel proteins use components of the general import apparatus. *Plant Cell* **2019**, *31*, 1845–1855. [CrossRef]
92. Gross, L.E.; Klinger, A.; Spies, N.; Ernst, T.; Flinner, N.; Simm, S.; Ladig, R.; Bodensohn, U.; Schleiff, E. Insertion of plastidic b-barrel proteins into the outer envelopes of plastids involves an intermembrane space intermediate formed with Toc75-V/OEP80. *Plant Cell* **2021**, *33*, 1657–1681. [CrossRef]
93. Huang, W.; Ling, Q.; Bédard, J.; Lilley, K.; Jarvis, P. In vivo analyses of the roles of essential Omp85-related proteins in the chloroplast outer envelope membrane. *Plant Physiol.* **2011**, *157*, 147–159. [CrossRef] [PubMed]

94. Patel, R.; Hsu, S.C.; Bédard, J.; Inoue, K.; Jarvis, P. The Omp85-related chloroplast outer envelope protein OEP80 is essential for viability in Arabidopsis. *Plant Physiol.* **2008**, *148*, 235–245. [CrossRef] [PubMed]
95. Tranel, P.J.; Keegstra, K. A novel, bipartite transit peptide targets OEP75 to the outer membrane of the chloroplastic envelope. *Plant Cell* **1996**, *8*, 2093–2104. [CrossRef]
96. Kouranov, A.; Chen, X.; Fuks, B.; Schnell, D.J. Tic20 and Tic22 are new components of the protein import apparatus at the chloroplast inner envelope membrane. *J. Cell Biol.* **1998**, *143*, 991–1002. [CrossRef] [PubMed]
97. Day, P.M.; Potter, D.; Inoue, K. Evolution and targeting of omp85 homologs in the chloroplast outer envelope membrane. *Front. Plant Sci.* **2014**, *5*, 535. [CrossRef]
98. Schleiff, E.; Becker, T. Common ground for protein translocation: Access control for mitochondria and chloroplasts. *Nat. Rev. Mol. Cell Biol.* **2011**, *12*, 48–59. [CrossRef]
99. Inoue, K.; Baldwin, A.J.; Shipman, R.L.; Matsui, K.; Theg, S.M.; Ohme-Takagi, M. Complete maturation of the plastid protein translocation channel requires a type I signal peptidase. *J. Cell Biol.* **2005**, *171*, 425–430. [CrossRef]
100. Endow, J.K.; Rocha, A.G.; Baldwin, A.J.; Roston, R.L.; Yamaguchi, T.; Kamikubo, H.; Inoue, K. Polyglycine acts as a rejection signal for protein transport at the chloroplast envelope. *PLoS ONE* **2016**, *11*, e0167802. [CrossRef]
101. Inoue, K.; Potter, D. The chloroplastic protein translocation channel Toc75 and its paralog OEP80 represent two distinct protein families and are targeted to the chloroplastic outer envelope by different mechanisms. *Plant J.* **2004**, *39*, 354–365. [CrossRef]
102. Gross, L.E.; Spies, N.; Simm, S.; Schleiff, E. Toc75-V/OEP80 is processed during translocation into chloroplasts, and the membrane-embedded form exposes its POTRA domain to the intermembrane space. *FEBS Open Bio* **2020**, *10*, 444–454. [CrossRef] [PubMed]
103. Eckart, K.; Eichacker, L.; Sohrt, K.; Schleiff, E.; Heins, L.; Soll, J. A Toc75-like protein import channel is abundant in chloroplasts. *EMBO Rep.* **2002**, *3*, 557–562. [CrossRef] [PubMed]
104. Kim, D.H.; Hwang, I. Direct targeting of proteins from the cytosol to organelles: The ER versus endosymbiotic organelles. *Traffic* **2013**, *14*, 613–621. [CrossRef] [PubMed]
105. Schleiff, E.; Klösgen, R.B. Without a little help from "my" friends: Direct insertion of proteins into chloroplast membranes? *Biochim. Biophys. Acta-Mol. Cell Res.* **2001**, *1541*, 22–33. [CrossRef]
106. Qbadou, S.; Tien, R.; Soll, J.; Schleiff, E. Membrane insertion of the chloroplast outer envelope protein, Toc34: Constrains for insertion and topology. *J. Cell Sci.* **2003**, *116*, 837–846. [CrossRef] [PubMed]
107. Wallas, T.R.; Smith, M.D.; Sanchez-Nieto, S.; Schnell, D.J. The roles of Toc34 and Toc75 in targeting the Toc159 preprotein receptor to chloroplasts. *J. Biol. Chem.* **2003**, *278*, 44289–44297. [CrossRef] [PubMed]
108. Hofmann, N.R.; Theg, S.M. Chloroplast outer membrane protein targeting and insertion. *Trends Plant Sci.* **2005**, *10*, 450–457. [CrossRef] [PubMed]
109. Sun, Y.J.; Forouhar, F.; Li, H.M.; Tu, S.L.; Yeh, Y.H.; Kao, S.; Shr, H.L.; Chou, C.C.; Chen, C.; Hsiao, C.D. Crystal structure of pea toc34, a novel gtpase of the chloroplast protein translocon. *Nat. Struct. Biol.* **2002**, *9*, 95–100. [CrossRef] [PubMed]
110. Lee, J.; Wang, F.; Schnell, D.J. Toc receptor dimerization participates in the initiation of membrane translocation during protein import into chloroplasts. *J. Biol. Chem.* **2009**, *284*, 31130–31141. [CrossRef]
111. Ivanova, Y.; Smith, M.D.; Chen, K.; Schnell, D.J. Members of the Toc159 import receptor family represent distinct pathways for protein targeting to plastids. *Mol. Biol. Cell* **2004**, *15*, 3379–3392. [CrossRef]
112. Smith, M.D.; Hiltbrunner, A.; Kessler, F.; Schnell, D.J. The targeting of the atToc159 preprotein receptor to the chloroplast outer membrane is mediated by its GTPase domain and is regulated by GTP. *J. Cell Biol.* **2002**, *159*, 833–843. [CrossRef]
113. Bauer, J.; Hiltbrunner, A.; Weibel, P.; Vidi, P.A.; Alvarez-Huerta, M.; Smith, M.D.; Schnell, D.J.; Kessler, F. Essential role of the G-domain in targeting of the protein import receptor atToc159 to the chloroplast outer membrane. *J. Cell Biol.* **2002**, *159*, 845–854. [CrossRef]
114. Muckel, E.; Soll, J. A protein import receptor of chloroplasts is inserted into the outer envelope membrane by a novel pathway. *J. Biol. Chem.* **1996**, *271*, 23846–23852. [CrossRef] [PubMed]
115. Lee, K.H.; Kim, S.J.; Lee, Y.J.; Jin, J.B.; Hwang, I. The M domain of atToc159 plays an essential role in the import of proteins into chloroplasts and chloroplast biogenesis. *J. Biol. Chem.* **2003**, *278*, 36794–36805. [CrossRef]
116. Lung, S.C.; Smith, M.D.; Weston, J.K.; Gwynne, W.; Secord, N.; Chuong, S.D.X. The C-terminus of Bienertia sinuspersici Toc159 contains essential elements for its targeting and anchorage to the chloroplast outer membrane. *Front. Plant Sci.* **2014**, *5*, 722. [CrossRef]
117. Lung, S.C.; Chuong, S.D.X. A transit peptide-like sorting signal at the C terminus directs the bienertia sinuspersici preprotein receptor toc159 to the chloroplast outer membrane. *Plant Cell* **2012**, *24*, 1560–1578. [CrossRef]
118. Bauer, J.; Chen, K.; Hiltbunner, A.; Wehrli, E.; Eugster, M.; Schnell, D.; Kessler, F. The major protein import receptor of plastids is essential for chloroplast biogenesis. *Nature* **2000**, *403*, 203–207. [CrossRef] [PubMed]
119. Chen, K.; Chen, X.; Schnell, D.J. Initial binding of preproteins involving the Toc159 receptor can be bypassed during protein import into chloroplasts. *Plant Physiol.* **2000**, *122*, 813–822. [CrossRef]
120. Tunyasuvunakool, K.; Adler, J.; Wu, Z.; Green, T.; Zielinski, M.; Žídek, A.; Bridgland, A.; Cowie, A.; Meyer, C.; Laydon, A.; et al. Highly accurate protein structure prediction for the human proteome. *Nature* **2021**, *596*, 590–596. [CrossRef] [PubMed]
121. Sharma, M.; Bennewitz, B.; Klösgen, R.B. Dual or not dual?—comparative analysis of fluorescence microscopy-based approaches to study organelle targeting specificity of nuclear-encoded plant proteins. *Front. Plant Sci.* **2018**, *9*, 1350. [CrossRef]
122. Garg, S.G.; Gould, S.B. The role of charge in protein targeting evolution. *Trends Cell Biol.* **2016**, *26*, 894–905. [CrossRef] [PubMed]

123. Ge, C.; Spånning, E.; Glaser, E.; Wieslander, Å. Import determinants of organelle-specific and dual targeting peptides of mitochondria and chloroplasts in *Arabidopsis thaliana*. *Mol. Plant* **2014**, *7*, 121–136. [CrossRef] [PubMed]
124. Carrie, C.; Small, I. A reevaluation of dual-targeting of proteins to mitochondria and chloroplasts. *Biochim. Biophys. Acta-Mol. Cell Res.* **2013**, *1833*, 253–259. [CrossRef]
125. Chen, Y.L.; Chen, L.J.; Chu, C.C.; Huang, P.K.; Wen, J.R.; Li, H. min TIC236 links the outer and inner membrane translocons of the chloroplast. *Nature* **2018**, *564*, 125–129. [CrossRef] [PubMed]
126. Rossig, C.; Reinbothe, C.; Gray, J.; Valdes, O.; Von Wettstein, D.; Reinbothe, S. Three proteins mediate import of transit sequence-less precursors into the inner envelope of chloroplasts in *Arabidopsis thaliana*. *Proc. Natl. Acad. Sci. USA* **2013**, *110*, 19962–19967. [CrossRef]
127. Gaudet, P.; Livstone, M.S.; Lewis, S.E.; Thomas, P.D. Phylogenetic-based propagation of functional annotations within the Gene Ontology consortium. *Brief. Bioinform.* **2011**, *12*, 449–462. [CrossRef]
128. Ling, Q.; Jarvis, P. Regulation of chloroplast protein import by the ubiquitin E3 ligase SP1 is important for stress tolerance in plants. *Curr. Biol.* **2015**, *25*, 2527–2534. [CrossRef]
129. Liu, H.; Stone, S.L. Abscisic acid increases Arabidopsis ABI5 transcription factor levels by promoting KEG E3 ligase self-ubiquitination and proteasomal degradation. *Plant Cell* **2010**, *22*, 2630–2641. [CrossRef]
130. Kachroo, A.; Venugopal, S.C.; Lapchyk, L.; Falcone, D.; Hildebrand, D.; Kachroo, P. Oleic acid levels regulated by glycerolipid metabolism modulate defense gene expression in Arabidopsis. *Proc. Natl. Acad. Sci. USA* **2004**, *101*, 5152–5157. [CrossRef] [PubMed]
131. Schwartz, S.H.; Qin, X.; Zeevaart, J.A.D. Characterization of a Novel Carotenoid Cleavage Dioxygenase from Plants. *J. Biol. Chem.* **2001**, *276*, 25208–25211. [CrossRef]
132. Xu, Z.Y.; Zhang, X.; Schläppi, M.; Xu, Z.Q. Cold-inducible expression of AZI1 and its function in improvement of freezing tolerance of *Arabidopsis thaliana* and Saccharomyces cerevisiae. *J. Plant Physiol.* **2011**, *168*, 1576–1587. [CrossRef]
133. Higashi, Y.; Okazaki, Y.; Takano, K.; Myouga, F.; Shinozaki, K.; Knoch, E.; Fukushima, A.; Saito, K. HEAT INDUCIBLE LIPASE1 remodels chloroplastic monogalactosyldiacylglycerol by liberating α-linolenic acid in arabidopsis leaves under heat stress. *Plant Cell* **2018**, *30*, 1887–1905. [CrossRef] [PubMed]
134. Kabeya, Y.; Nakanishi, H.; Suzuki, K.; Ichikawa, T.; Kondou, Y.; Matsui, M.; Miyagishima, S. ya The YlmG protein has a conserved function related to the distribution of nucleoids in chloroplasts and cyanobacteria. *BMC Plant Biol.* **2010**, *10*, 57. [CrossRef]
135. Gao, H.; Kadirjan-Kalbach, D.; Froehlicht, J.E.; Osteryoung, K.W. ARC5, a cytosolic dynamin-like protein from plants, is part of the chloroplast division machinery. *Proc. Natl. Acad. Sci. USA* **2003**, *100*, 4328–4333. [CrossRef]
136. Zhang, X.; Hu, J. Two small protein families, DYNAMIN-RELATED PROTEIN3 and FISSION1, are required for peroxisome fission in Arabidopsis. *Plant J.* **2009**, *57*, 146–159. [CrossRef]
137. Drozdetskiy, A.; Cole, C.; Procter, J.; Barton, G.J. JPred4: A protein secondary structure prediction server. *Nucleic Acids Res.* **2015**, *43*, W389–W394. [CrossRef] [PubMed]
138. Emanuelsson, O.; Nielsen, H.; Heijne, G. Von ChloroP, a neural network-based method for predicting chloroplast transit peptides and their cleavage sites. *Protein Sci.* **1999**, *8*, 978–984. [CrossRef] [PubMed]
139. Krogh, A.; Larsson, B.; Von Heijne, G.; Sonnhammer, E.L.L. Predicting transmembrane protein topology with a hidden Markov model: Application to complete genomes. *J. Mol. Biol.* **2001**, *305*, 567–580. [CrossRef]
140. Nugent, T.; Jones, D.T. Detecting pore-lining regions in transmembrane protein sequences. *BMC Bioinform.* **2012**, *13*, 169. [CrossRef]
141. Bagos, P.G.; Liakopoulos, T.D.; Spyropoulos, I.C.; Hamodrakas, S.J. PRED-TMBB: A web server for predicting the topology of β-barrel outer membrane proteins. *Nucleic Acids Res.* **2004**, *32*, 400–404. [CrossRef]
142. Käll, L.; Krogh, A.; Sonnhammer, E.L.L. Advantages of combined transmembrane topology and signal peptide prediction-the Phobius web server. *Nucleic Acids Res.* **2007**, *35*, 429–432. [CrossRef] [PubMed]
143. Bateman, A.; Martin, M.J.; Orchard, S.; Magrane, M.; Agivetova, R.; Ahmad, S.; Alpi, E.; Bowler-Barnett, E.H.; Britto, R.; Bursteinas, B.; et al. UniProt: The universal protein knowledgebase in 2021. *Nucleic Acids Res.* **2021**, *49*, D480–D489. [CrossRef]
144. Horton, P.; Park, K.J.; Obayashi, T.; Fujita, N.; Harada, H.; Adams-Collier, C.J.; Nakai, K. WoLF PSORT: Protein·localization predictor. *Nucleic Acids Res.* **2007**, *35*, 585–587. [CrossRef]
145. Sperschneider, J.; Catanzariti, A.M.; Deboer, K.; Petre, B.; Gardiner, D.M.; Singh, K.B.; Dodds, P.N.; Taylor, J.M. LOCALIZER: Subcellular localization prediction of both plant and effector proteins in the plant cell. *Sci. Rep.* **2017**, *7*, 44598. [CrossRef]
146. Armenteros, J.J.A.; Salvatore, M.; Emanuelsson, O.; Winther, O.; Von Heijne, G.; Elofsson, A.; Nielsen, H. Detecting sequence signals in targeting peptides using deep learning. *Life Sci. Alliance* **2019**, *2*, e201900429. [CrossRef] [PubMed]
147. Small, I.; Peeters, N.; Legeai, F.; Lurin, C. Predotar: A tool for rapidly screening proteomes for N-terminal targeting sequences. *Proteomics* **2004**, *4*, 1581–1590. [CrossRef] [PubMed]
148. Almagro Armenteros, J.J.; Sønderby, C.K.; Sønderby, S.K.; Nielsen, H.; Winther, O. DeepLoc: Prediction of protein subcellular localization using deep learning. *Bioinformatics* **2017**, *33*, 4049. [CrossRef]

International Journal of
Molecular Sciences

MDPI

Article

MexAB-OprM Efflux Pump Interaction with the Peptidoglycan of *Escherichia coli* and *Pseudomonas aeruginosa*

Miao Ma [1,2,†], Margaux Lustig [1,†], Michèle Salem [1], Dominique Mengin-Lecreulx [3], Gilles Phan [1] and Isabelle Broutin [1,*]

1 Laboratoire CiTCoM, Université de Paris, CNRS, 75006 Paris, France; mm2303@cam.ac.uk (M.M.); margaux.lustig@parisdescartes.fr (M.L.); michele.salem@u-paris.fr (M.S.); gilles.phan@u-paris.fr (G.P.)
2 Department of Biochemistry, University of Cambridge, Cambridge CB2 1GA, UK
3 Institute for Integrative Biology of the Cell (I2BC), Université Paris-Saclay, CEA, CNRS, 91198 Gif-sur-Yvette, France; dominique.mengin-lecreulx@i2bc.paris-saclay.fr
* Correspondence: isabelle.broutin@u-paris.fr
† These authors contributed equally to this work.

Abstract: One of the major families of membrane proteins found in prokaryote genome corresponds to the transporters. Among them, the resistance-nodulation-cell division (RND) transporters are highly studied, as being responsible for one of the most problematic mechanisms used by bacteria to resist to antibiotics, i.e., the active efflux of drugs. In Gram-negative bacteria, these proteins are inserted in the inner membrane and form a tripartite assembly with an outer membrane factor and a periplasmic linker in order to cross the two membranes to expulse molecules outside of the cell. A lot of information has been collected to understand the functional mechanism of these pumps, especially with AcrAB-TolC from *Escherichia coli*, but one missing piece from all the suggested models is the role of peptidoglycan in the assembly. Here, by pull-down experiments with purified peptidoglycans, we precise the MexAB-OprM interaction with the peptidoglycan from *Escherichia coli* and *Pseudomonas aeruginosa*, highlighting a role of the peptidoglycan in stabilizing the MexA-OprM complex and also differences between the two Gram-negative bacteria peptidoglycans.

Keywords: membrane; peptidoglycan; efflux pump assembly; resistance; *Pseudomonas*

Citation: Ma, M.; Lustig, M.; Salem, M.; Mengin-Lecreulx, D.; Phan, G.; Broutin, I. MexAB-OprM Efflux Pump Interaction with the Peptidoglycan of *Escherichia coli* and *Pseudomonas aeruginosa*. *Int. J. Mol. Sci.* **2021**, *22*, 5328. https://doi.org/10.3390/ijms22105328

Academic Editor: Masoud Jelokhani-Niaraki

Received: 14 April 2021
Accepted: 12 May 2021
Published: 18 May 2021

1. Introduction

The cell wall of Gram-negative bacteria comprises three main layers: the outer membrane lipid bilayer coated by lipopolysaccharides at the exterior; the inner membrane that meets the cytoplasm; and in between the periplasmic matrix where peptidoglycan (PG) confers the mechanical rigidity of the cell wall. Membrane transporters such as efflux pumps span these layers and expulse a wide variety of drugs leading to natural and acquired antimicrobial resistances. Resistance Nodulation-cell Division (RND) family is one of the six families of efflux pumps [1,2], acting as a three-component complex made of an inner-membrane transporter (RND), a periplasmic membrane fusion protein (MFP), and an outer-membrane factor channel (OMF). We focus our interest on the efflux pump MexAB-OprM from *Pseudomonas aeruginosa (P. aeruginosa)*, an opportunistic Gram-negative bacterium responsible of severe nosocomial infections and having developed resistance to several families of antibiotics. It has been recently classified as high priority by WHO in antimicrobial research [3] and belongs to the group ESKAPE named by the initials of the pathogens for which there is an urgent need for new treatments. The constitutive RND efflux pumps MexAB-OprM from *P. aeruginosa* and AcrAB-TolC from *Escherichia coli* (*E. coli*) have been extensively studied by different approaches making them archetypal models for the structural and functional comprehension of the efflux pump mechanism. The structure of each protein forming the pump has been solved by X-ray crystallography [4–12] and the whole assembly was recently determined by cryo-electron microscopy (cryo-EM) [13–16].

The comparison of these structures from several homologous pumps, together with in vitro measurements of transport through proteoliposomes [17,18], in cellulo determination of the minimal inhibitory concentration (MIC) of antibiotics for mutant strains [19,20], and the extensive modeling and molecular dynamic calculations [21,22], have led to a more precise understanding of the efflux pathways [23–28] and the conformational modifications [14,29,30] necessary to expulse the drugs out of the cell. We know now that the RND protein works as homotrimer with a proton motive force via proton relay amino acids localized in the membranous domain. This proton transfer is combined with the movement of the RND periplasmic domains where each monomer adopts alternatively Loose, Tight, and Close or Open conformations [31], to capture the drug at different entries and push it toward the RND funnel domain [16,32]. The drug then crosses the periplasm by a tunnel formed by the hexameric MFP proteins and is finally expulsed out of the cell through the OMF channel. Even if some hypotheses have been given on the final opening mechanism of the OMF [15,33,34], it is still not clear how and when the MFP and OMF interact. In vitro and in absence of its outer membrane partner, the MFP seems to be unstructured or too flexible to adopt a defined shape around the RND as illustrated by the recent structure of TriABC solved by cryo-EM [35]. In cellulo the tripartite pump must traverse the PG layer to convey the molecules, however, little is known about the relationship between the PG and the pumps. Is the PG a passive mesh made of long glycan chains crosslinked by short peptides or does it make specific interactions with the efflux pump components? If yes, which components are involved in PG recognition? In order to answer these questions, we purified the PG from *E. coli* and *P. aeruginosa* and performed pull-down assays to analyze their interaction with purified OprM and MexA proteins. Here we show that the presence of PG stabilizes the MexA-OprM binary complex formation, highlighting the PG as an important new actor of the assembly of the RND efflux pumps.

2. Results

2.1. Estimation of the Expected PG Interacting Partners

We recently solved the structure of the whole MexAB-OprM pump from *P. aeruginosa* by cryo-EM [16] showing a structure of around 230 Å long between the two membranous domains, which is compatible with the size of the periplasm estimated by cryo-transmission electron microscopy (Cryo-TEM) [36]. Matias et al. measured frozen-hydrated sections of *E. coli* and *P. aeruginosa* showing some differences between the two bacteria. The distances between the inner membrane (IM) and the outer membrane (OM) for *E. coli* and *P. aeruginosa* were estimated as 210 Å and 239 Å, respectively, the distances between the PG and the OM were 53 Å and 61 Å, and the thicknesses of the PG were 63 Å and 24 Å. These measurements are not necessarily physiological values as the sample preparation with the used technique resulted in a compression of the bacterial sections that, even if taken into account, increases the uncertainty on the measure. Nevertheless, the large difference of PG width between the two bacteria is remarkable. When adding this width difference to the PG-OM distance, this could reach 116 Å and 85 Å in *E. coli* and *P. aeruginosa*, respectively. If we report these values on the AcrAB-TolC and MexAB-OprM structures (Figure 1), in *P. aeruginosa* the PG would surround the ending part of the periplasmic α-helical coiled-coil domain of OprM and would sweep MexA tips, whereas in *E. coli* the PG would cover both α-helical coiled-coil domains of TolC and AcrA. Anyhow, it appeared clearly that the PG do not interact with the inner membrane RND protein. Consequently, in the present study only OprM and MexA were analyzed for their possible interaction with PG.

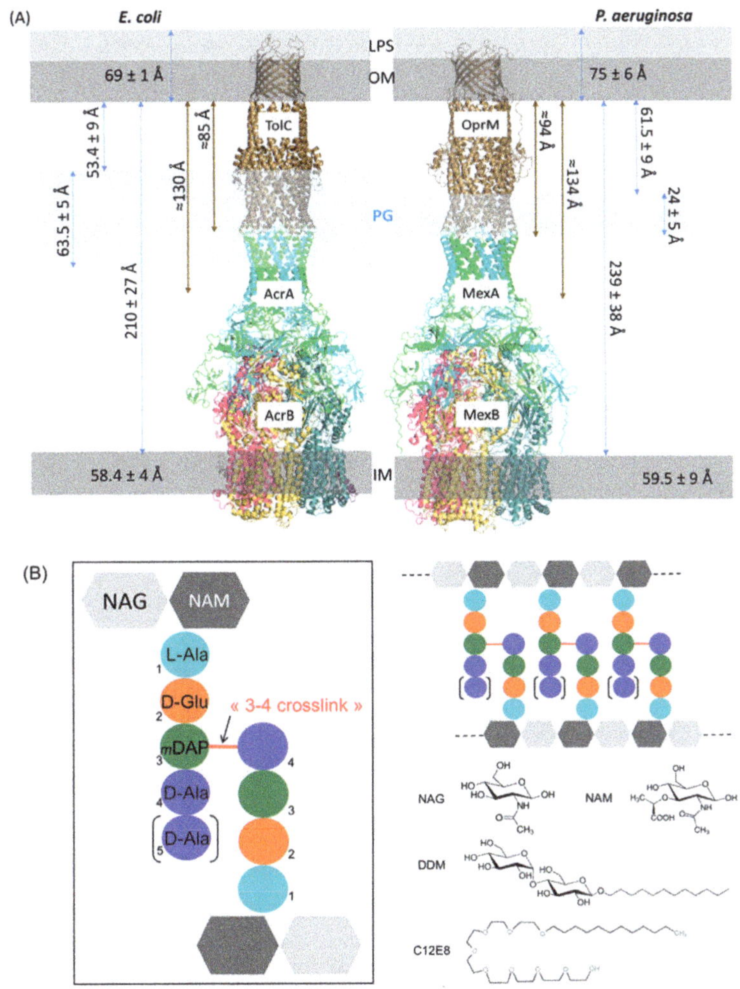

Figure 1. Schematic representation and localization of the PG. (**A**) Structures of the efflux pumps AcrAB-TolC (left) and MexAB-OprM (right) from *E. coli* and *P. aeruginosa*, respectively, presented in their membranous environment. Sizes illustrated by blue arrows come from Cryo-TEM measures performed on high-pressure freezing bacteria sections [36]. Sizes illustrated by brown arrows have been measured on the 3D structures (PDB codes 5ng5 [14] for AcrAB-TolC and 6ta6 [16] for MexAB-OprM). The trimeric OMFs are colored in brown, the MFPs' trimers of dimers are colored in green and cyan to highlight the different role of the two MFPs forming the dimer, the three monomers forming the RNDs are colored in magenta, yellow and blue depending on their respective functioning states (LTO) [4]. Scale is approximate. Lipopolysaccharides (LPS), outer membrane (OM), peptidoglycan (PG), inner membrane (IM). (**B**) General structure of peptidoglycan organization in Gram-negative bacteria: PG is mainly made of repeating units of disaccharide *N*-acetylglucosamine-β(1-4)-*N*-acetylmuramic acid (NAG-NAM) interconnected by tetra- or pentapeptides cross-linked at amino acid positions 3 and 4 (3-4 crosslinkage). The peptide composition is L-Ala (cyan), D-Ala (blue), D-Glu (orange) and *m*DAP (*meso*-diaminopimelic acid, green). The schematic structures of the two detergents used to solubilize the membrane proteins OprM and MexA are presented for comparison: dodecyl-β-D-maltopyranoside (DDM), which has a similar glucidic component to NAG and NAM, and Octaethylene glycol-monododecyl ether ($C_{12}E_8$).

2.2. Composition Analysis of the PG from *E. coli* and *P. aeruginosa*

The peptidoglycan layer is made of linear chains of two alternating amino sugars: *N*-acetylglucosamine (GlcNAc or NAG) and *N*-acetylmuramic acid (MurNAc or NAM), which are attached to a short (3- to 5-residue) amino acids chain. The repeating disaccharide-peptide units are cross-linked via peptide bonds to form a tight mesh barrier (Figure 1B). Even if it is admitted that the PG of *E. coli* and *P. aeruginosa*, both Gram-negative bacteria, present similar composition [37], we decided to quantify the PG composition of the specific *P. aeruginosa* strain used for the binding analysis of this study, PAO1, and to compare it to the PG of *E. coli* that has been extensively studied (Table 1). Besides it is necessary to grow PAO1 in Mueller–Hinton (MH) medium supplemented with salts instead of Luria or Lysogeny Broth (LB) medium to perform in cellulo anti-bioresistance experiments [38]. Therefore, PAO1 strain was grown in both MH and LB media for comparison. For each culture, the PG was purified and submitted or not to enzymatic treatments with α-amylase (E1), to remove high-molecular-weight glycogen, or with both E1 and pronase E (E2), to remove peptidoglycan associated proteins. Each of the six resulting samples were analyzed for their respective composition in the characteristic major components of Gram-negative bacteria, namely NAG, NAM, Ala, Glu, and diaminopimelic acid (DAP) (Figure 1B). The content of Gly is analyzed to verify the absence of protein contaminants although it occurs that a few amount of Gly can be found at position 4 or 5 of the peptide instead of the D-Ala [39]. We show that the treatment with the proteases is necessary to eliminate residual proteins or peptides co-purified with the PG. Nevertheless, in our study the first enzymatic treatment with α-amylase seems to be sufficient for the PG purification. No significant difference was observed between the cultures performed in LB or in MH medium. The relative ratio of NAG/NAM/Ala/Glu/DAP is close to 1/1/1.5/1/1 as expected for *P. aeruginosa*. The small enrichment in Ala is normal as the peptides linking the sugars could vary as tri, tetra and pentapeptides depending on the presence of D-Ala at positions 4 and 5 (Figure 1B) [40,41]. In view of these results, no difference was expected between pull-down experiments performed with PG purified from PAO1 cultures grown in LB and treated with α-amylase only or with the two enzymes E1 and E2. As this hypothesis was verified with our proteins (data not shown), we chose to present only the experiments performed with the PG treated with the two enzymes for clarity.

Table 1. PG composition of *E. coli* and *P. aeruginosa* bacterial strains normalized to the NAG concentration.

	NAG	NAM	Ala	Glu	DAP	Gly
E. coli						
crude PG extract	1	0.93	6.35	5.81	0.98	3.19
purified PG	1	0.98	1.82	1.13	1.03	0.19
P. aeruginosa						
LB medium crude PG extract	1	1.02	2.38	1.78	1.11	0.31
LB medium PG treated by E1	1	0.98	1.57	1.06	0.99	0.04
LB medium PG treated by E1+E2	1	0.98	1.60	1.16	1.00	0.08
MH medium crude PG extract	1	0.94	2.70	2.21	1.11	0.63
MH medium PG treated by E1	1	1.03	1.58	1.09	1.02	0.03
MH medium PG treated by E1+E2	1	0.98	1.60	1.25	1.05	0.19

2.3. *Pull-Down Analysis of MexA and OprM with Purified PG*

2.3.1. Comparison of the PG from *E. coli* and *P. aeruginosa*

To analyze the possible interaction between the upper part of the MexAB-OprM efflux pump and the PG, pull-down experiments were performed. The MexA and OprM proteins have been purified from recombinant expression in *E. coli* C43 strain driving our choice to first analyze their interaction with the PG of *E. coli*. The PG has been purified classically [39] from a culture performed on an untransformed C43 strain stopped during the exponential phase of growth. After PG incubation with MexA, OprM, or the complex MexA-OprM at a 2:1 ratio, each sample was extensively washed and the pull-down pellet of PG analyzed by SDS-PAGE, as well as the supernatant of each washing step of the pull-down experiment. As shown in Figure 2, although the PG does not retain most of the proteins, there is still a small proportion of MexA present in the PG pellet, which is not the case for OprM. Concerning the MexA-OprM complex, it clearly appears a supplementary band of OprM, suggesting the complex MexA-OprM-PG affinity to be strong enough to resist to the extensive washing procedure after incubation with the PG, or a stabilization of the MexA-OprM complex in presence of the PG.

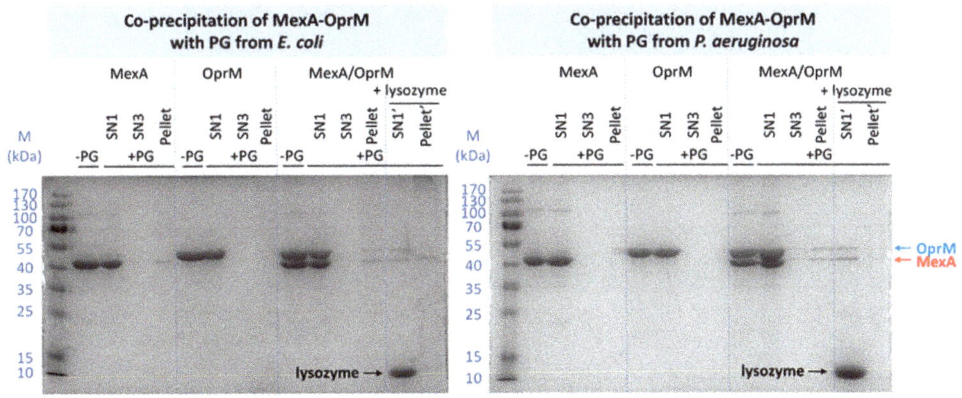

Figure 2. Pull-down experiments performed with the PG extracted from *E. coli* (**left**) and from *P. aeruginosa* (**right**), and with the MexA and OprM proteins in DDM detergent. M: marker of size in kDa; SNn: supernatant of the n washing and centrifugation step; Pellet: final pellet after the 3rd centrifugation; -PG: purified proteins without PG; +PG: pull-down with the PG; + lysozyme: extraction by lysozyme treatment of the remaining proteins from the final pull-down pellet after the three washing steps. SN1′: supernatant after lysozyme treatment after the first centrifugation; Pellet′: pellet after two washing and centrifugation steps.

The same experiment was performed with the PG purified from PAO1 in exactly the same conditions. No interaction can be observed between OprM and the PG nor between MexA and PG. Nevertheless, when a pre-mixing of the proteins was performed before the pull-down assay, the complex was retained by the PG of *P. aeruginosa* as observed previously with the PG of *E. coli*. For both experiments, the specificity of the pull-down has been verified by releasing MexA and OprM after PG hydrolysis with lysozyme (Figure 2).

2.3.2. Effects of Different Parameters on the Co-Precipitation of MexA-OprM with PG from *P. aeruginosa*

In order to verify the specificity of the stabilized complex, several controls were performed. As MexA and OprM are proteins from *P. aeruginosa*, these experiments were only performed with the PG extracted from this bacterium.

To verify that the presence of PG did not modify the MexA:OprM complex ratio, the experiment was repeated with a 3:1 ratio. The excess of MexA did not modify the relative intensity of the bands in the retained complex (Figure S1), even if more complexes were trapped, indicating that only the 2:1 ratio was stabilized by the PG if referring to our previous intensities calibration [42].

The C-terminus of OprM, corresponding to the 13 last amino acids, was not visible in any structure of OprM resolved by crystallography or cryo-EM. As no functional role of this fragment has been identified so far, it has been hypothesized it could play a role in the PG recognition. The pull-down experiment was repeated with this OprMΔCter protein. No significant difference was detectable in the quantity of MexA-OprMΔCter complex retained by PG compared to the MexA-OprM complex analyzed in the same conditions (see Figure S2).

Another parameter that can modify the result is the nature of the detergent. The chosen detergent, DDM, is the one that has been used to purify and to solve the structure of the isolated proteins and that of the whole pump reconstituted in nanodiscs [11,12,16]. Nevertheless, as the maltoside head group of this detergent can mimic the NAG-NAM disaccharide structure of the PG (Figure 1B), it can be conceived that the presence of the detergent can perturb the interaction of the proteins with the PG. Consequently, a change of detergent was performed for OprM and MexA, replacing DDM by $C_{12}E_8$, a PEG alkyl ether detergent presenting no similarity with the PG structure (Figure 1B). The co-precipitation with PG was performed in the same conditions as used for the proteins in DDM. As shown on Figure 3, a faint interaction of OprM($C_{12}E_8$) and of MexA($C_{12}E_8$) with PG is now observed. However, unexpectedly, when the proteins are purified in this detergent the quantity of the retained complex is largely increased, confirming a stabilization of the complex by PG but also revealing an effect of the detergent on this interaction. From the final evaluation of the relative intensity of the bands corresponding to the retained proteins, even though a large quantity of MexA is released after the first washing step (Figure 3), the MexA:OprM pull-down ratio in $C_{12}E_8$ seems to be comparable to the one observed with DDM detergent. It is not clear if the detergent is directly involved in the interaction with the PG, but the behavior of the protein obviously depends on it. For instance, OprM is less stable in $C_{12}E_8$ as an important quantity of protein was lost during the detergent exchange. Because the final quantity of retained complex is important, two additional washing steps were added showing further release of MexA compared to OprM. Thus, instability of the membrane proteins in this detergent could be a bias in our interpretation of PG pull-down.

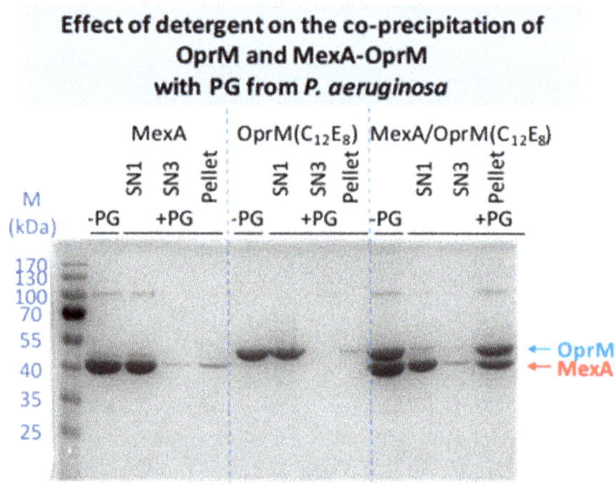

Figure 3. Pull-down experiments performed with the PG extracted from *P. aeruginosa* and with the MexA and OprM proteins after exchange of the DDM detergent for the $C_{12}E_8$. The same labeling as for Figure 2 have been used.

3. Discussion

The RND efflux pumps form a tripartite assembly which go through the two membranes of Gram-negative bacteria and the PG. Even if the distances between these three elements are known to fluctuate in spite of direct bonding between OM and PG via lipoproteins, position of PG has been restricted for long at the equatorial region of the OMF protein (see [2] for an example). This is mainly due to the fact the first models of the assembly were built with a direct interaction between the RND transporter and the OMF, suggesting that the PG could not reach the periplasmic coiled-coil helices of the OMF proteins (TolC or OprM) already involved in the binding of the MFP (AcrA or MexA). Since 2015, all the solved cryo-EM structures of AcrAB-TolC [13,14] and MexAB-OprM [15,16] revealed an elongated pump with a tip-to-tip interaction between the OMF and the MFP changing our mind about the position and the possible role of PG in the assembly. Nevertheless, as the PG of *E. coli* is three times thicker than the one of *P. aeruginosa*, it is supposed to cover a large portion of the tripartite assembly, including the coiled-coil domains of TolC and the α-hairpin loops of AcrA together with part of the equatorial domain of TolC, unlike the PG of *P. aeruginosa* that seems to contact the extremities of MexA only (Figure 1A). A recent structure of AcrAB-TolC solved by cryo-tomography in cellulo [43] confirmed the suggested interacting domains of the efflux pump with PG.

Concerning their composition, the two purified PGs are very similar as reported in Table 1. It can be noticed that after the α-amylase action the PG of PAO1 seems to be already free of residual amino-acids contrarily to the PG of *E. coli* that is known to be poorly purified by the sole action of this enzyme. The main difference in crude PG composition is a clear excess of Glu and Ala, especially with PAO1 grown in MH medium. This highlights the importance to get rid of residual interacting proteins as they could alter the binding experiments.

For proteins purified in DDM, when mixed with OprM, no interaction was detected whatever the PG as shown by the absence of the protein in the pellet after extensive washing process, even if a faint band was still detected in the supernatant after the second round of washing, justifying an additional washing step. Concerning MexA, it has been retained with the PG only when using the *E. coli* one. This can be due to the thickness difference between the two PGs. This result indicates a weak interaction that might involve a specific repartition of polar or basic residues on MexA in favor of PG binding. In the in cellulo structure of the *E. coli* efflux pump solved by cryo-tomography, Shi et al. [43] were able to reconstitute images corresponding only to the bottom part of the pump, AcrAB. This bipartite assembly corresponded to 38% of the reconstituted images and presented AcrA in a favorable position to be embedded in the PG. On the contrary they did not mention images corresponding to isolated TolC even if this can also be due to the low resolution of the method. In order to identify the binding sites with PG on the two proteins, they did crosslinking with 3,3′-dithiobis(sulfosuccinimidyl propionate) (DTSSP) followed by mass-spectrometry analysis. They identified several peptides on the two proteins localized in the equatorial and coiled-coil domains of TolC and in the α-hairpin and lipoyl domains of AcrA. When comparing the identified peptides with the equivalent ones in OprM and MexA after sequences alignment, it appears that there is neither conserved residues nor PG binding-motif. The absence of interaction of OprM with the PGs was not intuitive as it is postulated that the OMF is positioned in a waiting state until an RND-MFP complex diffusing in the inner membrane arrives. Nevertheless, it was not excluded that in vitro the detergent dodecyl-β-D-maltopyranoside (DDM) used for proteins purification could prevent the proper interaction of PG by competition, as DDM presents sugar moiety similar to the one in PG unit (see Figure 1B). For comparison, an exchange of detergent to $C_{12}E_8$ was performed for OprM and MexA, revealing a weak interaction of individual proteins with PG, of the same magnitude as the one observed between MexA (in DDM) and the PG of *E. coli* revealing a significant role of the detergent used to purify the proteins.

Concerning the pull-down experiments performed on the pre-mixed MexA-OprM complex, the two proteins are both retained, suggesting a stabilization of the complex in presence of the PG, whatever the PG origin (*E. coli* or *P. aeruginosa*, Figure 2). Despite the lack of structural information, it was very tempting to consider the last 13 C-terminus residues of OprM to interact with the PG as this part of the protein is rich in polar (Thr, Gln) and charged (Lys) amino acids able to interact with the NAG-NAM or the DAP as observed by Boags et al. [44] in a molecular dynamic study on the interaction of TolR and OmpA with PG. Their calculation resulted in the prediction of a predominant role of the flexible C-terminus of TolR involved in first anchoring of the protein to the PG before making more specific interaction. To explore this hypothesis, the pull-down experiment was repeated with the OprM protein deleted of its 13 last amino acids, not visible in the crystallographic nor the cryo-EM structures, but the use of this truncated OprMΔCter did not modify the result of the pull-down experiment, the quantity of MexA-OprMΔCter complex stabilized by PG being conserved (see Figure S2), thus invalidating a role of OprM C-terminus in the stabilization by PG. Concerning the experiments performed with the MexA-OprM complex after an exchange of detergent, the stabilization of the complex seems largely increased, even if a large amount of MexA is present in the supernatant after the first washing step (Figure 3). This stabilization can be due to the nature and the length of the $C_{12}E_8$ detergent which is longer than DDM. Nevertheless, a synergistic interaction is also observed when using $C_{12}E_8$ detergent and the PG seems to rebalance the quantity of protein forming the complex.

This synergistic interaction with PG was previously suggested by Xu et al. [45] when studying the binding of TolC and AcrA with the PG of *E. coli*. Nevertheless, the results were not easy to compare as only the final PG pellet after mixing with TolC and AcrA was analyzed on an SDS-PAGE revealed by Western blot. This stabilization can be interpreted by two different hypotheses. It can be due to specific electrostatic interactions as the helical domains of the two proteins are submitted to large conformation changes during the assembly and opening process that could make particular regions accessible during the process. Nevertheless, it is difficult to know at which step the proteins will be in position to interact with the PG, preventing from a rigorous model building. The second hypothesis could be a geometrical restriction. The structure of a synthetic PG has been solved by NMR [46] and the structure of TolC was docked in the PG holes showing that the protein can enter in the PG without enlargement of the holes by enzymes as the PG pores diameter have been measured to be ≈70 Å. Nevertheless, at that time the PG interacting zone was suggested to be the equatorial domain of TolC measured to be as large as the PG hole. The coiled-coil domain diameter is smaller favoring the insertion of the protein. On the MFP side, the helical tips of the hexamer form a circle larger than the TolC or OprM extremity but still slightly smaller than 70 Å, with an enlargement when in complex with the OMF proteins. So, it can be suggested that MexA do not interact with the PG when uncomplexed as its hexameric tunnel is too small. In presence of OprM, the slightly enlarged tunnel of MexA would be sufficient to contact the PG, stabilizing the complex. It will be difficult to decide between the two hypotheses, but in both cases the PG clearly appears as an important player in the mechanism of the pump assembly.

4. Materials and Methods

4.1. Purification of the Peptidoglycan

PGs from *E. coli* and *P. aeruginosa* were prepared according to the methods of Mengin-Lecreulx and van Heijenoort [47] with some modifications. Cultures of *E. coli* were performed in LB and cultures of *P. aeruginosa* were performed in LB or in MH media. They were grown at 37 °C under 180 rpm agitation until reaching the middle of the exponential phase at an optical density OD_{600nm} of 0.8. It is important to stop all the cultures at the same growing step, here the exponential phase, as the composition of PG is growth-phase dependent [47–49]. A similar protocol was used to purify the PG from *E. coli* or *P. aeruginosa* cultures as follows. Cells are pelleted by centrifugation at 4 °C, 5000× *g* for 30 min then washed with 25 mL of 0.9% NaCl. Cells are pelleted once again by centrifugation at 4 °C, 5000× *g* for 30 min and resuspended in 10 mL of 0.9% NaCl, then added drop by drop in 10 mL of boiling SDS 8% with strong stirring. The mixture is left under boiling conditions for 2 h with stirring then cooled down at room temperature overnight without stirring and finally centrifuged at 200,000× *g*, 20 °C for 30 min to collect the PG sacculus. The pellet is washed at least three times to remove SDS with 6 volumes of warm sterile MilliQ water (30–40 °C), centrifuged at 200,000× *g*, 20 °C for 30 min and the final pellet is resuspended in distilled water or in 10 mM Tris-HCl pH 7.5 if an enzymatic treatment is performed. The pellet might change from white to transparent during the wash. The final PG suspension can be used directly or submitted to digestion with proteases in order to remove the possibly bound proteins. It can be treated with 100 µg/mL α-amylase for 1 h at 37 °C then with 200 µg/mL of preheated pronase E overnight at 37 °C. The digestions are stopped by boiling 30 min in 1 volume of SDS at 4% final concentration. The resulting PG, either collected after the α-amylase digestion or after the second step is then washed as previously described and the final pellet is resuspended in 1 volume of distilled water for further use.

For quantification and composition analysis, a small portion (50 µL) of the sample is hydrolyzed in HCl 6M at 95 °C for 16 h, dried, and solubilized in 500 µL of citrate buffer pH 2.2 before analysis with a Hitachi model 8800 amino acid analyzer (ScienceTec) as described in Barreteau et al. [50]. The characteristic constituents of the PG are *N*-acetylmuramic acid (MurNAc or NAM), diaminopimelic acid (DAP) and *N*-acetylglucosamine (GlcNAc or NAG) for Gram-negative bacteria, with an expected ratio of approximatively 1:1:1 [47].

4.2. Cloning and Purification of the Analyzed Proteins

Wild type OprM and MexA were cloned and purified following the protocol described in [11,42] with minor modifications. OprM deleted of the flexible C-terminus residues 473–485 (OprMΔCter) was generated as 5′-NdeI and 3′-XmaI fragment using polymerase chain reaction (PCR) with PAO1 genome as template. A 6-His tag was included at the C-terminus for Ni-NTA affinity purification. Forward and reverse primers are GGAATTCCATATGAAACGGTCCTTCCTTTCC and TCCCCCCGGGTCATGATGATGATGATGGTCAGGTCTGCTGGTTCCAGCCGCCGCCGA, respectively. The PCR fragment was then inserted into pBAD33 expression vector as described in [11]. Heterologous expression of OprM full-length or OprMΔCter (deletion of the last 13 C-terminus residues 473–485) inserted in a pBAD33 plasmid is performed in a *E. coli* C43 strain deleted of *acrB* gene. The preculture is performed at 37 °C in LB medium under 200 rpm agitation and inoculated at OD_{600nm} = 0.05 in LB medium containing 25 µg/mL of chloramphenicol. Cell were grown to OD_{600nm} = 0.1 at 37 °C, 200 rpm, then cooled down to 20 °C. Cells were induced at OD_{600nm} = 0.7 with arabinose (0.02% final, *w*/*v*) and grown overnight at 20 °C before centrifugation. The cell pellet was resuspended in TBS buffer (20 mM Tris-HCl pH 8, NaCl 150 mM) and broken by the use of a cell disrupter (CellD) at 30,000 psi (2400 Bar) before centrifugation at 10,000× *g* for 20 min. The supernatant was diluted down to 1 mg/mL and solubilized into TBS with DDM 2%, imidazole 10 mM during 1 h at room temperature. The insoluble fraction is pelleted by ultracentrifugation at 100,000× *g* for 1 h at 4 °C and the solubilized fraction is loaded on a Ni-NTA column pre-equilibrated in TBS with 0.05% DDM (*w*/*v*). The column is washed with the same buffer supplemented with 20 mM imidazole pH 8. The protein is eluted between 100 to 250 mM imidazole, concentrated and injected on a Superdex 200 gel-filtration column equilibrated with the same buffer without imidazole. OprM is eluted as a trimer and concentrated at 5 mg/mL before use. MexA is purified following the protocol described in [17] which is similar to the one used for OprM at the exception of the culture, which is performed in TB medium, and grown at 30 °C during 2.5 h after induction.

The exchange of DDM detergent for $C_{12}E_8$ has been performed on Ni-NTA by reloading the pure proteins on the affinity column. After extensive washing with TBS with 0.025% $C_{12}E_8$ (*w*/*v*) and 20 mM imidazole pH 8, the samples were eluted in the same buffer at 500 mM imidazole. The samples were not further purified but several cycles of concentration and dilution on amicon®10K were performed to get rid of the imidazole.

Samples are finally resuspended in TBS with 0.025% $C_{12}E_8$ (*w*/*v*) and 300 mM NaCl before use for the pull-down experiments.

The purify of the proteins can be evaluated by analyzing the well without PG (-PG) on Figures 2, 3 and S2.

4.3. Pull Down with PG

The purified PG is insoluble in water. It is vortexed before mixing with the proteins. The pull-down experiment is performed with 50 nmoles of PG (in terms of DAP content, see paragraph 4.1) in 100 μL of the following buffer: 20 mM Tris-HCl, pH 8.0, 300 mM NaCl, 0.05% DDM. The PG:protein ratio is 100:1 for OprM, 100:2 for MexA, 100:2:1 for MexA-OprM to account for the 2:1 MexA:OprM ratio in the efflux pump assembly. The mix is incubated on a rolling wheel for 1 h at room temperature before centrifugation at 18,000× *g* for 5 min. The supernatant is kept for further analysis on an SDS-PAGE. The pellet is washed with 2 volumes of buffer before centrifugation. This washing–centrifugation cycle is repeated 3 times. The final pellet is resuspended in 1 volume of buffer. The pure protein, the three supernatants and the final pellet are analyzed on a 12% SDS-PAGE after 5 min boiling. The experiment was repeated 5 times. For 2 experiments the final pellet was treated with lysozyme to release the pulled-down proteins. The pellet is resuspended in 1 volume of buffer with 1 mg/mL final concentration of lysozyme and incubated on a rolling wheel for 1.5 h at room temperature before centrifugation at 18,000× *g* for 5 min. The pellet is washed with 2 volumes of buffer and solubilized in 1 volume. The first supernatant and the final solubilized pellet after the second washing step are analyzed on a 12% SDS-PAGE.

5. Conclusions

The MexAB-OprM efflux pump spans the complete cell wall of *P. aeruginosa*, crossing the peptidoglycan in the periplasmic compartment. Here we analyzed the possible specific interaction of the MexA and OprM proteins with the PG purified from *E. coli* or *P. aeruginosa* revealing, when proteins are purified in DDM, no interaction for OprM and an interaction of MexA only with the PG of *E. coli* that is three times thicker in vivo than the one of *P. aeruginosa*. Nevertheless, previous association of the two proteins before analysis reveals a stabilization of the complex–PG interaction in both cases. This is further reinforced with the PG of *P. aeruginosa* after the exchange of DDM detergent for $C_{12}E_8$ for the two proteins. Even if no covalent binding is expected, this synergistic interaction could correspond to a required step of the assembly mechanism, suggesting the PG to be an important actor of the process. It could help to stabilize MexA to keep it in the proper quaternary structure in order to increase the speed of the pump formation with OprM present in the opposite membrane.

Supplementary Materials: Supplementary Materials can be found at https://www.mdpi.com/1422-0067/22/10/5328/s1. Figure S1: Pull-down experiments performed with the PG extracted from *P. aeruginosa* and with the MexA and OprM proteins mixed at two different ratios, 2:1 and 3:1., Figure S2: Pull-down experiments performed with the PG extracted from *P. aeruginosa* and with the MexA and OprM modified proteins, OprMΔCter corresponding to a truncation of 13 residues at its C-terminus.

Author Contributions: Conceptualization, G.P. and I.B.; Funding acquisition, I.B.; Investigation, M.M., M.L., M.S., D.M.-L. and G.P.; Methodology, D.M.-L., G.P. and I.B.; Project administration, D.M.-L., G.P. and I.B.; Supervision, G.P. and I.B.; Validation, D.M.-L., G.P. and I.B.; Writing—original draft, I.B.; Writing—review and editing, D.M.-L. and G.P. All authors have read and agreed to the published version of the manuscript.

Funding: This research was funded by Agence Nationale de la Recherche, grant numbers ANR-17-CE11-0028-02 and ANR-12-BSV8-0010. MM was supported by ANR-12-BSV8-0010. M.L. was supported by a ministerial doctoral funding.

Acknowledgments: We thank Maria Antónia Sousa Correia da Silva for her participation in the pull-down experiments during her mobility training in IB laboratory.

Conflicts of Interest: The authors declare no conflict of interest.

Abbreviations

Abbreviations

PG	peptidoglycan
NAM	*N*-acetylmuramic acid
NAG	*N*-acetylglucosamine
DAP	diaminopimelic acid

References

1. Delmar, J.A.; Su, C.C.; Yu, E.W. Bacterial Multidrug Efflux Transporters. *Annu. Rev. Biophys.* **2014**, *43*, 93–117. [CrossRef] [PubMed]
2. Du, D.J.; van Veen, H.W.; Luisi, B.F. Assembly and operation of bacterial tripartite multidrug efflux pumps. *Trends Microbiol.* **2015**, *23*, 311–319. [CrossRef] [PubMed]
3. Tacconelli, E.; Carrara, E.; Savoldi, A.; Harbarth, S.; Mendelson, M.; Monnet, D.L.; Pulcini, C.; Kahlmeter, G.; Kluytmans, J.; Carmeli, Y.; et al. Discovery, research, and development of new antibiotics: The WHO priority list of antibiotic-resistant bacteria and tuberculosis. *Lancet Infect Dis.* **2018**, *18*, 318–327. [CrossRef]
4. Murakami, S.; Nakashima, R.; Yamashita, E.; Matsumoto, T.; Yamaguchi, A. Crystal structures of a multidrug transporter reveal a functionally rotating mechanism. *Nature* **2006**, *443*, 173–179. [CrossRef] [PubMed]
5. Seeger, M.A.; Schiefner, A.; Eicher, T.; Verrey, F.; Diederichs, K.; Pos, K.M. Structural asymmetry of AcrB trimer suggests a peristaltic pump mechanism. *Science* **2006**, *313*, 1295–1298. [CrossRef] [PubMed]
6. Mikolosko, J.; Bobyk, K.; Zgurskaya, H.I.; Ghosh, P. Conformational flexibility in the multidrug efflux system protein AcrA. *Structure* **2006**, *14*, 577–587. [CrossRef] [PubMed]
7. Koronakis, V.; Sharff, A.; Koronakis, E.; Luisi, B.; Hughes, C. Crystal structure of the bacterial membrane protein TolC central to multidrug efflux and protein export. *Nature* **2000**, *405*, 914–919. [CrossRef] [PubMed]
8. Sennhauser, G.; Bukowska, M.A.; Briand, C.; Grutter, M.G. Crystal Structure of the Multidrug Exporter MexB from *Pseudomonas aeruginosa*. *J. Mol. Biol.* **2009**, *389*, 134–145. [CrossRef]
9. Symmons, M.F.; Bokma, E.; Koronakis, E.; Hughes, C.; Koronakis, V. The assembled structure of a complete tripartite bacterial multidrug efflux pump. *Proc. Natl. Acad. Sci. USA* **2009**, *106*, 7173–7178. [CrossRef] [PubMed]
10. Akama, H.; Kanemaki, M.; Yoshimura, M.; Tsukihara, T.; Kashiwagi, T.; Yoneyama, H.; Narita, S.; Nakagawa, A.; Nakae, T. Crystal structure of the drug discharge outer membrane protein, OprM, of *Pseudomonas aeruginosa*—Dual modes of membrane anchoring and occluded cavity end. *J. Biol. Chem.* **2004**, *279*, 52816–52819. [CrossRef]
11. Phan, G.; Benabdelhak, H.; Lascombe, M.B.; Benas, P.; Rety, S.; Picard, M.; Ducruix, A.; Etchebest, C.; Broutin, I. Structural and Dynamical Insights into the Opening Mechanism of *P. aeruginosa* OprM Channel. *Structure* **2010**, *18*, 507–517. [CrossRef] [PubMed]
12. Monlezun, L.; Phan, G.; Benabdelhak, H.; Lascombe, M.B.; Enguene, V.Y.N.; Picard, M.; Broutin, I. New OprM structure highlighting the nature of the N-terminal anchor. *Front. Microbiol.* **2015**, *6*, 667. [CrossRef]
13. Du, D.J.; Wang, Z.; James, N.R.; Voss, J.E.; Klimont, E.; Ohene-Agyei, T.; Venter, H.; Chiu, W.; Luisi, B.F. Structure of the AcrAB-TolC multidrug efflux pump. *Nature* **2014**, *509*, 512–515. [CrossRef] [PubMed]
14. Wang, Z.; Fan, G.Z.; Hryc, C.F.; Blaza, J.N.; Serysheva, I.I.; Schmid, M.F.; Chiu, W.; Luisi, B.; Du, D.J. An allosteric transport mechanism for the AcrAB-TolC multidrug efflux pump. *eLife* **2017**, *6*, e24905. [CrossRef] [PubMed]
15. Tsutsumi, K.; Yonehara, R.; Ishizaka-Ikeda, E.; Miyazaki, N.; Maeda, S.; Iwasaki, K.; Nakagawa, A.; Yamashita, E. Structures of the wild-type MexAB-OprM tripartite pump reveal its complex formation and drug efflux mechanism. *Nat. Commun.* **2019**, *10*, 1520. [CrossRef] [PubMed]
16. Glavier, M.; Puvanendran, D.; Salvador, D.; Decossas, M.; Phan, G.; Garnier, C.; Frezza, E.; Cece, Q.; Schoehn, G.; Picard, M.; et al. Antibiotic export by MexB multidrug efflux transporter is allosterically controlled by a MexA-OprM chaperone-like complex. *Nat. Commun.* **2020**, *11*, 4948. [CrossRef] [PubMed]
17. Enguene, V.Y.N.; Verchere, A.; Phan, G.; Broutin, I.; Picard, M. Catch me if you can: A biotinylated proteoliposome affinity assay for the investigation of assembly of the MexA-MexB-OprM efflux pump from *Pseudomonas aeruginosa*. *Front. Microbiol.* **2015**, *6*, 541.
18. Verchere, A.; Dezi, M.; Adrien, V.; Broutin, I.; Picard, M. In vitro transport activity of the fully assembled MexAB-OprM efflux pump from *Pseudomonas aeruginosa*. *Nat. Commun.* **2015**, *6*, 6890. [CrossRef]
19. Sakurai, K.; Yamasaki, S.; Nakao, K.; Nishino, K.; Yamaguchi, A.; Nakashima, R. Crystal structures of multidrug efflux pump MexB bound with high-molecular-mass compounds. *Sci. Rep.* **2019**, *9*, 4359. [CrossRef]
20. Kobylka, J.; Kuth, M.S.; Muller, R.T.; Geertsma, E.R.; Pos, K.M. AcrB: A mean, keen, drug efflux machine. *Ann. N. Y. Acad. Sci.* **2020**, *1459*, 38–68. [CrossRef]
21. Reading, E.; Ahdash, Z.; Fais, C.; Ricci, V.; Xuan, W.K.; Grimsey, E.; Stone, J.; Malloci, G.; Lau, A.M.; Findlay, H.; et al. Perturbed structural dynamics underlie inhibition and altered efflux of the multidrug resistance pump AcrB. *Nat. Commun.* **2020**, *11*, 5565. [CrossRef]

22. Simsir, M.; Broutin, I.; Mus-Veteau, I.; Cazals, F. Studying dynamics without explicit dynamics: A structure-based study of the export mechanism by AcrB. *Proteins* **2021**, *89*, 259–275. [CrossRef]
23. Yamaguchi, A.; Nakashima, R.; Sakurai, K. Structural basis of RND-type multidrug exporters. *Front. Microbiol.* **2015**, *6*, 327. [CrossRef] [PubMed]
24. Zwama, M.; Yamaguchi, A. Molecular mechanisms of AcrB-mediated multidrug export. *Res. Microbiol.* **2018**, *169*, 372–383. [CrossRef] [PubMed]
25. Zwama, M.; Yamasaki, S.; Nakashima, R.; Sakurai, K.; Nishino, K.; Yamaguchi, A. Multiple entry pathways within the efflux transporter AcrB contribute to multidrug recognition. *Nat. Commun.* **2018**, *9*, 124. [CrossRef] [PubMed]
26. Atzori, A.; Malviya, V.N.; Malloci, G.; Dreier, J.; Pos, K.M.; Vargiu, A.V.; Ruggerone, P. Identification and characterization of carbapenem binding sites within the RND-transporter AcrB. *Biochim. Biophys. Acta Biomembr.* **2019**, *1861*, 62–74. [CrossRef] [PubMed]
27. Tam, H.K.; Malviya, V.N.; Foong, W.E.; Herrmann, A.; Malloci, G.; Ruggerone, P.; Vargiu, A.V.; Pos, K.M. Binding and Transport of Carboxylated Drugs by the Multidrug Transporter AcrB. *J. Mol. Biol.* **2020**, *432*, 861–877. [CrossRef] [PubMed]
28. Atzori, A.; Malloci, G.; Cardamone, F.; Bosin, A.; Vargiu, A.V.; Ruggerone, P. Molecular Interactions of Carbapenem Antibiotics with the Multidrug Efflux Transporter AcrB of *Escherichia coli*. *Int. J. Mol. Sci.* **2020**, *21*, 860. [CrossRef] [PubMed]
29. Vargiu, A.V.; Ramaswamy, V.K.; Malvacio, I.; Malloci, G.; Kleinekathofer, U.; Ruggerone, P. Water-mediated interactions enable smooth substrate transport in a bacterial efflux pump. *Biochim. Biophys. Acta Gen. Subj.* **2018**, *1862*, 836–845. [CrossRef] [PubMed]
30. Hazel, A.J.; Abdali, N.; Leus, I.V.; Parks, J.M.; Smith, J.C.; Zgurskaya, H.I.; Gumbart, J.C. Conformational Dynamics of AcrA Govern Multidrug Efflux Pump Assembly. *ACS Infect. Dis.* **2019**, *5*, 1926–1935. [CrossRef]
31. Schulz, R.; Vargiu, A.V.; Collu, F.; Kleinekathofer, U.; Ruggerone, P. Functional Rotation of the Transporter AcrB: Insights into Drug Extrusion from Simulations. *PLoS Comput. Biol.* **2010**, *6*, e1000806. [CrossRef]
32. Vargiu, A.V.; Ramaswamy, V.K.; Malloci, G.; Malvacio, I.; Atzori, A.; Ruggerone, P. Computer simulations of the activity of RND efflux pumps. *Res. Microbiol.* **2018**, *169*, 384–392. [CrossRef] [PubMed]
33. Bavro, V.N.; Pietras, Z.; Furnham, N.; Perez-Cano, L.; Fernandez-Recio, J.; Pei, X.Y.; Misra, R.; Luisi, B. Assembly and channel opening in a bacterial drug efflux machine. *Mol. Cell* **2008**, *30*, 114–121. [CrossRef] [PubMed]
34. Lopez, C.A.; Travers, T.; Pos, K.M.; Zgurskaya, H.I.; Gnanakaran, S. Dynamics of Intact MexAB-OprM Efflux Pump: Focusing on the MexA-OprM Interface. *Sci. Rep.* **2017**, *7*, 16521. [CrossRef]
35. Fabre, L.; Ntreh, A.T.; Yazidi, A.; Leus, I.V.; Weeks, J.W.; Bhattacharyya, S.; Ruickoldt, J.; Rouiller, I.; Zgurskaya, H.I.; Sygusch, J.A. "Drug Sweeping" State of the TriABC Triclosan Efflux Pump from *Pseudomonas aeruginosa*. *Structure* **2021**, *29*, 261–274.E6. [CrossRef]
36. Matias, V.R.F.; Al-Amoudi, A.; Dubochet, J.; Beveridge, T.J. Cryo-transmission electron Microscopy of frozen-hydrated sections of *Escherichia coli* and *Pseudomonas aeruginosa*. *J. Bacteriol.* **2003**, *185*, 6112–6118. [CrossRef] [PubMed]
37. Quintela, J.C.; Caparros, M.; De Pedro, M.A. Variability of Peptidoglycan Structural Parameters in Gram-Negative Bacteria. *FEMS Microbiol. Lett.* **1995**, *125*, 95–100. [CrossRef]
38. Barry, A.L.; Reller, L.B.; Miller, G.H.; Washington, J.A.; Schoenknect, F.D.; Peterson, L.R.; Hare, R.S.; Knapp, C. Revision of Standards for Adjusting the Cation Content of Mueller-Hinton Broth for Testing Susceptibility of *Pseudomonas aeruginosa* to Aminoglycosides. *J. Clin. Microbiol.* **1992**, *30*, 585–589. [CrossRef]
39. Glauner, B. Separation and Quantification of Muropeptides with High-Performance Liquid-Chromatography. *Anal. Biochem.* **1988**, *172*, 451–464. [CrossRef]
40. Yanai, A.; Kato, K.; Beppu, T.; Arima, K. Peptidoglycan of *Pseudomonas aeruginosa*. *Agr. Biol. Chem.* **1976**, *40*, 1505–1508. [CrossRef]
41. Crump, G.M.; Zhou, J.H.; Mashayekh, S.; Grimes, C.L. Revisiting peptidoglycan sensing: Interactions with host immunity and beyond. *Chem. Commun.* **2020**, *56*, 13313–13322. [CrossRef] [PubMed]
42. Ferrandez, Y.; Monlezun, L.; Phan, G.; Benabdelhak, H.; Benas, P.; Ulryck, N.; Falson, P.; Ducruix, A.; Picard, M.; Broutin, I. Stoichiometry of the MexA-OprM binding, as investigated by blue native gel electrophoresis. *Electrophoresis* **2012**, *33*, 1282–1287. [CrossRef] [PubMed]
43. Shi, X.D.; Chen, M.Y.; Yu, Z.L.; Bell, J.M.; Wang, H.; Forrester, I.; Villarreal, H.; Jakana, J.; Du, D.J.; Luisi, B.; et al. In situ structure and assembly of the multidrug efflux pump AcrAB-TolC. *Nat. Commun.* **2019**, *10*, 2635. [CrossRef] [PubMed]
44. Boags, A.T.; Samsudin, F.; Khalid, S. Binding from Both Sides: TolR and Full-Length OmpA Bind and Maintain the Local Structure of the *E-coli* Cell Wall. *Structure* **2019**, *27*, 713–724. [CrossRef]
45. Xu, Y.B.; Moeller, A.; Jun, S.Y.; Le, M.; Yoon, B.Y.; Kim, J.S.; Lee, K.; Ha, N.C. Assembly and Channel Opening of Outer Membrane Protein in Tripartite Drug Efflux Pumps of Gram-negative Bacteria. *J. Biol. Chem.* **2012**, *287*, 11740–11750. [CrossRef]
46. Meroueh, S.O.; Bencze, K.Z.; Hesek, D.; Lee, M.; Fisher, J.F.; Stemmler, T.L.; Mobashery, S. Three-dimensional structure of the bacterial cell wall peptidoglycan. *Proc. Natl. Acad. Sci. USA* **2006**, *103*, 4404–4409. [CrossRef]
47. Mengin-Lecreulx, D.; van Heijenoort, J. Effect of growth conditions on peptidoglycan content and cytoplasmic steps of its biosynthesis in *Escherichia coli*. *J. Bacteriol.* **1985**, *163*, 208–212. [CrossRef]
48. Pisabarro, A.G.; De Pedro, M.A.; Vazquez, D. Structural Modifications in the Peptidoglycan of *Escherichia coli* Associated with Changes in the State of Growth of the Culture. *J. Bacteriol.* **1985**, *161*, 238–242. [CrossRef]
49. Vollmer, W.; Bertsche, U. Murein (peptidoglycan) structure, architecture and biosynthesis in *Escherichia coli*. *Biochim. Biophys. Acta Biomembr.* **2008**, *1778*, 1714–1734. [CrossRef]

50. Barreteau, H.; Patin, D.; Bouhss, A.; Blanot, D.; Mengin-Lecreulx, D.; Touzé, T. CbrA mediates colicin M resistance in *Escherichia coli* through modification of undecaprenyl-phosphate-linked peptidoglycan precursors. *J. Bacteriol.* **2020**, *202*, e00436-20. [CrossRef]

International Journal of *Molecular Sciences*

Article

Angulin-1 (LSR) Affects Paracellular Water Transport, However Only in Tight Epithelial Cells

Carlos Ayala-Torres, Susanne M. Krug, Rita Rosenthal † and Michael Fromm *,†

Clinical Physiology/Nutritional Medicine, Medical Department, Division of Gastroenterology, Infectiology and Rheumatology, Charité—Universitätsmedizin Berlin, 12203 Berlin, Germany; carlos-mario.ayala-torres@charite.de (C.A.-T.); susanne.m.krug@charite.de (S.M.K.); rita.rosenthal@charite.de (R.R.)

* Correspondence: michael.fromm@charite.de

† These authors contributed equally to this work.

Abstract: Water transport in epithelia occurs transcellularly (aquaporins) and paracellularly (claudin-2, claudin-15). Recently, we showed that downregulated tricellulin, a protein of the tricellular tight junction (tTJ, the site where three epithelial cells meet), increased transepithelial water flux. We now check the hypothesis that another tTJ-associated protein, angulin-1 (*alias* lipolysis-stimulated lipoprotein receptor, LSR) is a direct negative actuator of tTJ water permeability depending on the tightness of the epithelium. For this, a tight and an intermediate-tight epithelial cell line, MDCK C7 and HT-29/B6, were stably transfected with CRISPR/Cas9 and single-guide RNA targeting angulin-1 and morphologically and functionally characterized. Water flux induced by an osmotic gradient using 4-kDa dextran caused water flux to increase in angulin-1 KO clones in MDCK C7 cells, but not in HT-29/B6 cells. In addition, we found that water permeability in HT-29/B6 cells was not modified after either angulin-1 knockout or tricellulin knockdown, which may be related to the presence of other pathways, which reduce the impact of the tTJ pathway. In conclusion, modulation of the tTJ by knockout or knockdown of tTJ proteins affects ion and macromolecule permeability in tight and intermediate-tight epithelial cell lines, while the transepithelial water permeability was affected only in tight cell lines.

Keywords: angulin-1; LSR; tricellulin; tricellular tight junction; paracellular water transport; tight epithelium; MDCK C7 cells; intermediate-tight epithelium; HT-29/B6 cells

Citation: Ayala-Torres, C.; Krug, S.M.; Rosenthal, R.; Fromm, M. Angulin-1 (LSR) Affects Paracellular Water Transport, However Only in Tight Epithelial Cells. *Int. J. Mol. Sci.* **2021**, *22*, 7827. https://doi.org/10.3390/ijms22157827

Academic Editor: Masoud Jelokhani-Niaraki

Received: 31 May 2021
Accepted: 20 July 2021
Published: 22 July 2021

1. Introduction

Adequate transport of solutes and water across epithelial barriers is indispensable for maintaining normal physiological homeostasis in all animals [1,2]. Fluid is moved either across the plasma membranes of the cells that comprise the epithelial layer (transcellular transport) or between these cells (paracellular transport). The discovery of aquaporin water channels provided a first molecular basis for transcellular water movement [3].

Paracellular transport involves specialized structures called tight junctions (TJ) that regulate the flow of solutes through paracellular pathways and maintain cell polarity, thereby functioning as a barrier of epithelial and endothelial cellular sheets [4]. Tricellular tight junctions (tTJs) form at the convergence of bicellular tight junctions (bTJs) where three epithelial cells meet in polarized epithelia [5,6]. Claudin family proteins and occludin are main components of bicellular TJs, which are important for the barrier function and permselectivity [7–10]. To date, 27 members of the claudin family have been identified in humans. Many of these have sealing functions (claudins 1, 3, 5, 11, 14, 19) while some claudins form channels across TJs, which are permeable either for cations (claudins 2, 10b, 15) or for anions (claudins 10a, 17). For several claudins, their effects on epithelial barriers are inconsistent and the function is unclear (claudins 4, 7, 8, 16) [11]. The claudin composition of the TJ mainly determines the tightness of an epithelium or an epithelial cell line, respectively.

In addition to their role as channels for small cations, Rosenthal et al. demonstrated that claudin-2 and claudin-15 also transport water [12–14].

The mechanisms that determine the use of paracellular versus transcellular routes for water transport are multifaceted. For example, in the kidney, transport of water is achieved in part through the paracellular pathway in the proximal nephron [15], whereas in the distal nephron and collecting duct, it takes place through transcellular routes [16]. Mutations in claudins have been identified to cause hereditary diseases, demonstrating that regulation of the paracellular permeability is crucial to the normal functions of various organs [11].

The tricellular TJ is more complex than bTJ. In the center of tTJ, bTJ elements from each of the three cells come together to form a central tube of about 1 μm in length and 10 nm in diameter [17]. To date, two types of integral membrane proteins, tricellulin and angulin family proteins, are identified as molecular components of tTJ [18]. Tricellulin, which was the first tTJ protein to be identified [5], plays a critical role in sealing the tTJ against the passage of solutes of up to 10 kDa [19,20], and it was recently found that it also plays a role in the paracellular water transport in the tight epithelial cell line, MDCK C7 [21].

As said, angulins contribute to forming the central element of the tTJ. Angulin family proteins, comprising lipolysis-stimulated lipoprotein receptor (LSR), immunoglobulin-like domain containing receptor-1 (ILDR1) and -2 (ILDR2), are type-I transmembrane proteins with an extracellular immunoglobulin-like domain. Due to their common structure and function as tTJs-associated membrane proteins, LSR, ILDR1, and ILDR2 were designated as angulin-1, angulin-2, and angulin-3, respectively [22].

The original role of LSR/angulin-1 was described in lipid metabolism studies, as LSR was first discovered to be expressed in hepatocytes, where it plays a role in the clearance of triglyceride-rich lipoproteins and low-density lipoproteins, and also acts as an apolipoprotein B/E–containing lipoprotein receptor [23]. At tricellular contacts, angulin-1 recruits tricellulin, and the interaction between the cytoplasmic domain of angulin-1 and the C-terminal cytoplasmic domain of tricellulin is required for this recruitment [24,25]. Angulin-1 together with tricellulin are required for full barrier function of epithelial cells with high transepithelial electrical resistance [25]. Loss or downregulation of angulin-1 disrupts the barriers with relocalization of tTJ molecule tricellulin in various cell types [24,26]. In addition, loss of angulin-1 enhances cancer cell motility [27,28] and the LSR knockout mice die before embryonic day 15.5 (E15.5), but the cause of death remains unclear [29].

The LSR-related proteins immunoglobulin-like domain containing receptors angulin-2 and angulin-3 are also expressed complementarily in many epithelial cell types at tricellular contacts of many epithelial cells and recruit tricellulin [22], for instance, it is known that angulin-1 and angulin-2 are co-expressed in the large intestine and the kidney [22].

As for the tTJ, Gong et al. provided evidence that water transport is increased in isolated kidney tubules in the absence of angulin-2, but is normally inhibited in the presence of angulin-2 [30]. In contrast to this, Hempstock et al. found in the colon and kidney of angulin-2 KO mice no detectable abnormalities in water transport and maintained barrier function of the epithelia. They concluded that angulin-1 changes its expression pattern and location, which compensates for the loss of angulin-2 [31].

These findings led us to hypothesize that angulin-1 is a direct negative actuator of tTJ water permeability depending on the tightness of the epithelium, in this respect similar to tricellulin. Therefore, we aimed to characterize the effect of angulin-1 expression on water transport in a tight and an intermediate-tight epithelial cell line, MDCK C7 and HT-29/B6, respectively. In order to define the tightness of these cell lines, we previously performed a two-path impedance spectroscopy study where we analyzed paracellular, transcellular, and transepithelial resistance in MDCK C7, HT-29/B6, and MDCK C11 cells. As a result, MDCK C7 cells were defined as a tight and HT-29/B6 cells as a moderate tight epithelial cell line, as defined by the ratio R^{para}/R^{trans} and the absolute resistance values of the two pathways [32]. In our present study, we found that the water passage through the tTJ was changed in the absence of angulin-1 only in MDCK C7 cells, thus in a tight epithelium.

2. Results

2.1. Establishment of Angulin-1 Knockout in MDCK C7 and HT-29/B6 Cells

After transfection of MDCK C7 and HT-29/B6 cells with CRISPR/Cas9, HDR plasmids and sgRNA targeting angulin-1, puromycin-resistant cell clones were screened for angulin-1 knockout by Western blotting. In this study, two angulin-1 knockout clones and their controls were investigated in both cell lines (KO 18 and KO 36 and their respective control clones 14 and 18 in MDCK C7 cell line; and KO 12 and KO 32 and their respective controls 15 and 29 in HT-29/B6 cell line). The angulin-1 KO clones in both cell lines were mono-allelic knockouts (Figure 1a,b, Supplementary Tables S1 and S2).

The localization of angulin-1 in the controls and its removal from the TJ in the knockouts was confirmed for all clones as dots at the tTJ by immunofluorescence confocal laser-scanning microscopy. ZO-1 [33] in HT-29/B6 cells and occludin [34] in MDCK C7 cells served as a TJ marker (Figure 1c,d).

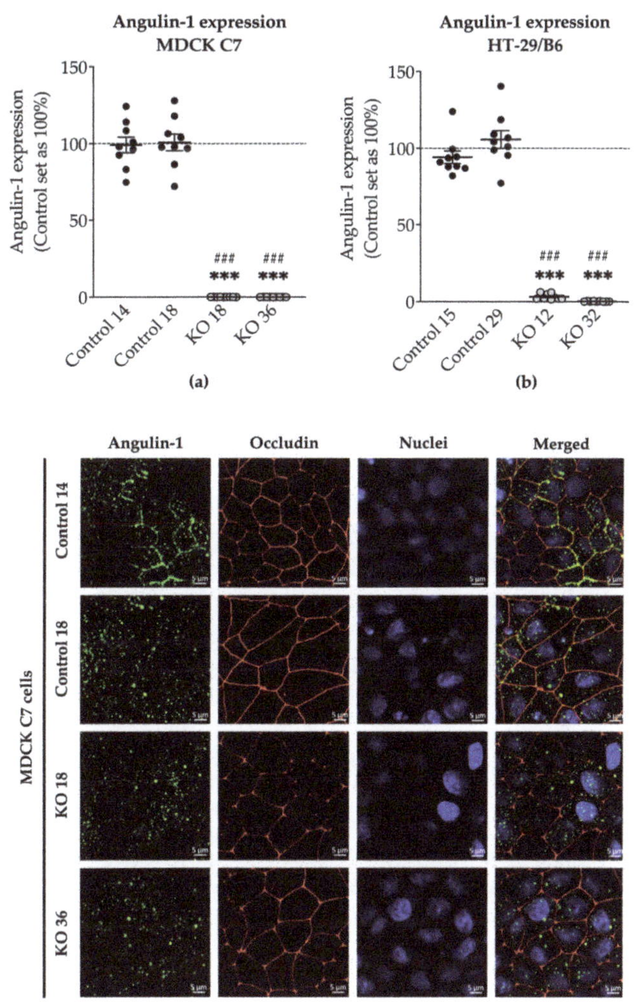

Figure 1. *Cont.*

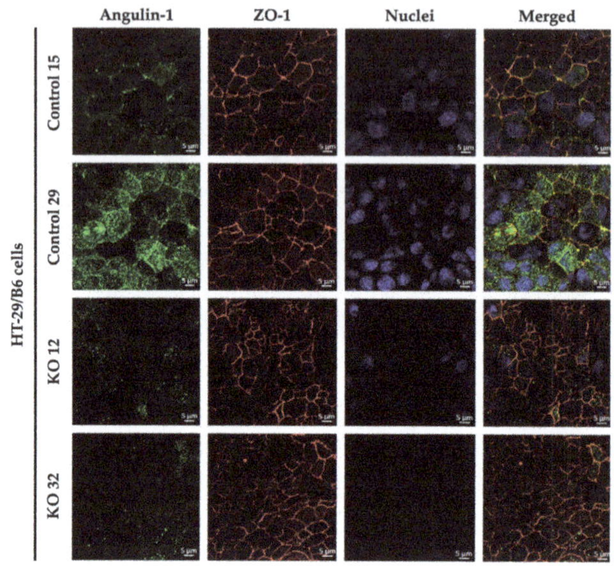

(d)

Figure 1. Expression and localization of angulin-1 in MDCK C7 and HT-29/B6 cells. Densitometric analysis of angulin-1 protein expression levels in control and angulin-1 knockout (**a**) MDCK C7 and (**b**) HT-29/B6 cells. sgRNA targeting angulin-1 led to a high decrease in angulin-1 expression in both cell lines (n = 9, MDCK C7: *** $p \leq 0.001$ with regard to control 14 and ### $p \leq 0.001$ with regard to control 18 and HT-29/B6: *** $p \leq 0.001$ with regard to control 15 and ### $p \leq 0.001$ with regard to control 29). Immunostaining of angulin-1 in the clones used throughout this study in (**c**) MDCK C7 and (**d**) HT-29/B6 cells. In the KO clones of both cell lines, angulin-1 disappeared from the tTJ in comparison with their controls. Angulin-1: green; Occludin or ZO-1: red; DAPI (nuclei): blue.

2.2. *Effects of Angulin-1 Knockout on Endogenous Proteins of MDCK C7 and HT-29/B6 Cells*

Knockout of one TJ protein may cause relevant variation in other proteins potentially involved in transepithelial water transport. Therefore, we examined levels of tricellulin, the three angulins, occludin, several claudins, aquaporin (AQP) water channels, AQP-1, -3, -4 and -7, SGLT1, and LI-cadherin (Figure 2a,b). As known from previous studies, claudin-2 is not expressed in MDCK C7 cells [12,35]. The densitometric analysis revealed some clonal variability in TJ protein expression between the knockout clones and the controls.

In detail, occludin, claudin-1, -3, -4, -5, -7, -8, and AQP-7 were reduced in the MDCK C7 angulin-1 KO 36 clone (Figure 2c). The MDCK C7 angulin-1 KO 18 clone showed a reduction in occludin, claudin-1, -5, and -7 and a slight increase in AQP-1 compared with the controls (Figure 2c). On the other hand, angulin-1 knockout resulted in increased levels of tricellulin, claudin-1, -5, -7 and -8 and LI-cadherin and decreased levels of claudin-2, AQP-4 and SGLT1 in HT-29/B6 angulin-1 KO 12 clone (Figure 2d). The HT-29/B6 angulin-1 KO 32 clone showed an upregulation of tricellulin, claudin-1, -2, -3, -5, -7, -8, and LI-cadherin (Figure 2d).

Clonal variability in protein expression was also observed between both control clones and both KO clones. None of the other two angulins were significantly altered in the knockout clones of both cell lines.

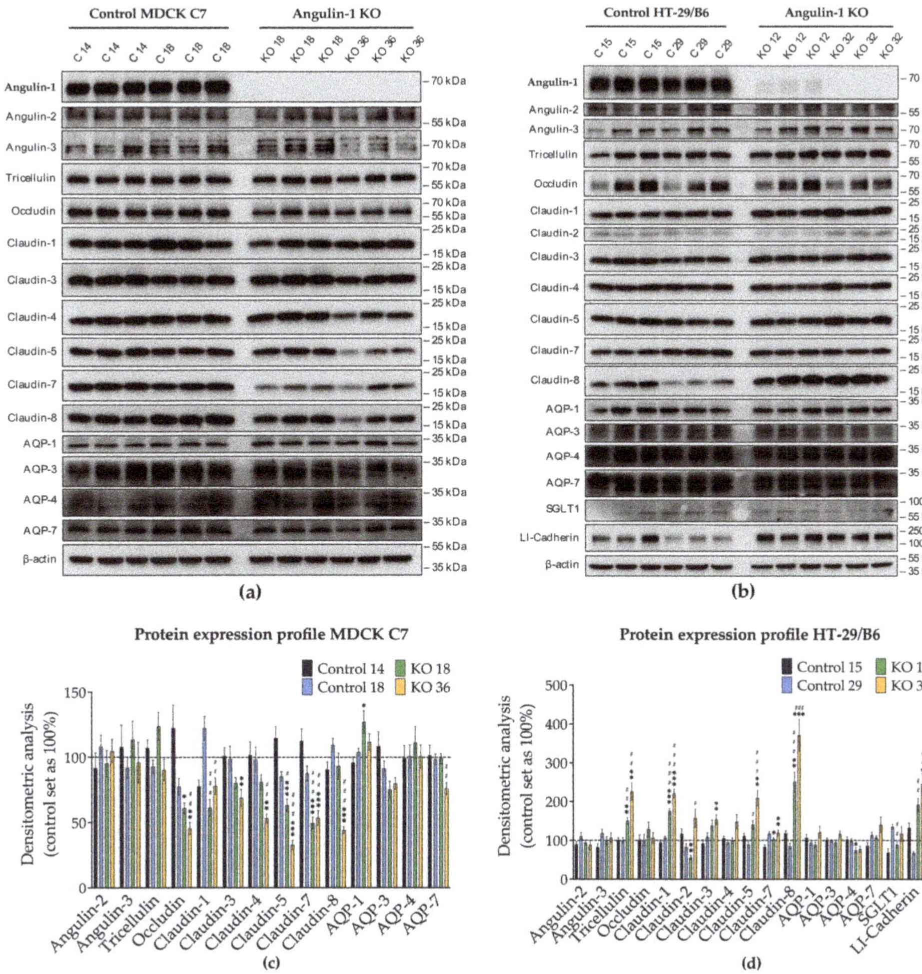

Figure 2. Angulin, tricellulin, occludin, claudin, and aquaporin (AQP) expression in control and angulin-1 knockout cells. Representative Western blots of angulin-1 knockout in (**a**) MDCK C7 and (**b**) HT-29/B6 cells. Densitometric analysis of protein expression levels in (**c**) MDCK C7 and (**d**) HT-29/B6 cells after angulin-1 knockout in comparison to the corresponding vector-transfected controls ($n = 9$, N = 3). β-Actin was used as an internal control for normalization to protein content. Statistical analysis was performed using one-way ANOVA test to compare between the four clones in both cell lines (MDCK C7: * $p \leq 0.05$, ** $p \leq 0.01$, *** $p \leq 0.001$ with regard to control 14 and # $p \leq 0.05$, ## $p \leq 0.01$, ### $p \leq 0.001$ with regard to control 18 and HT-29/B6: * $p \leq 0.05$, ** $p \leq 0.01$, *** $p \leq 0.001$ with regard to control 15 and # $p \leq 0.05$, ## $p \leq 0.01$, ### $p \leq 0.001$ with regard to control 29).

2.3. *Effects of Angulin-1 Knockout on Tricellulin Localization in MDCK C7 and HT-29/B6 Cells*

It is known that angulin-1 knockout could have different side effects on the localization of other tight junction proteins, specifically tricellulin. Immunofluorescence studies in combination with confocal microscopy revealed a slight delocalization of tricellulin from the tTJs to the bTJ after angulin-1 knockout in both cell lines, whereas occludin was concentrated at the corners of tTJs (Figure 3). These data suggest that angulin-1 has a role in the localization of tricellulin in both cell lines, nevertheless it does not play an essential role in the localization of other TJ proteins (data not shown).

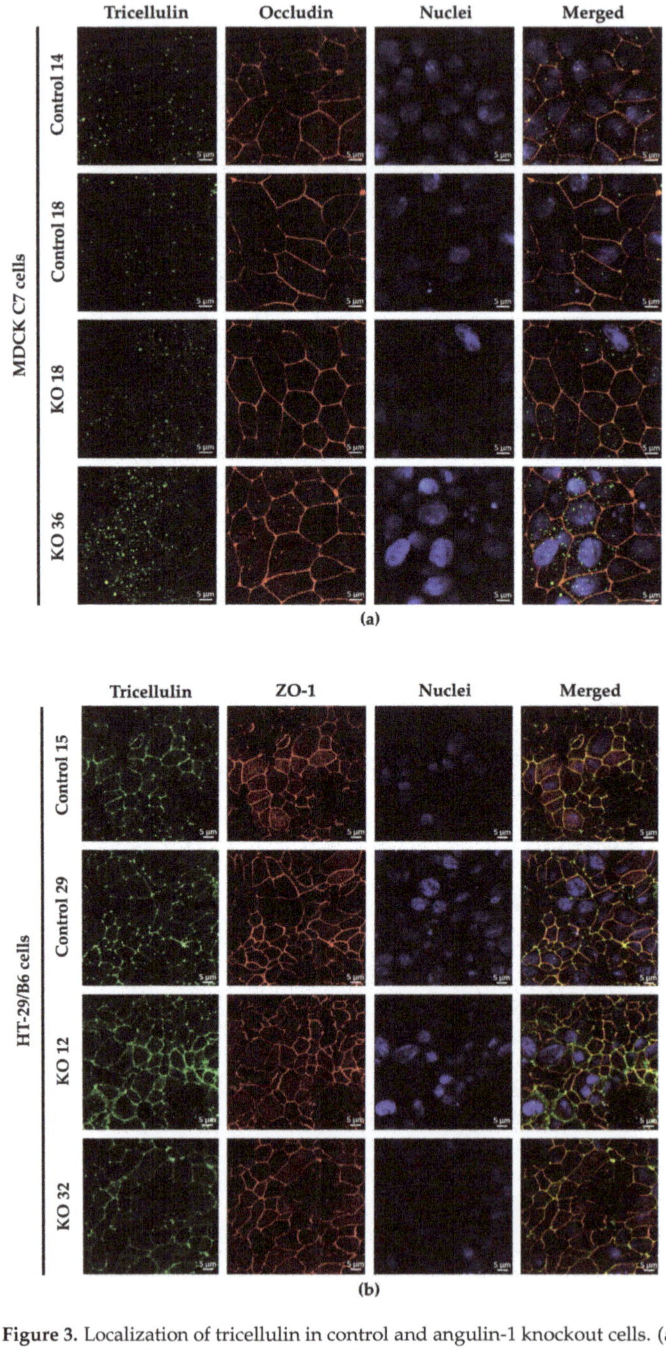

Figure 3. Localization of tricellulin in control and angulin-1 knockout cells. (**a**) Immunostaining of tricellulin in MDCK C7 angulin-1 KO cells. After angulin-1 knockout, tricellulin was still located in the tTJ, but additionally in the bTJ. (**b**) Immunostaining of tricellulin in HT-29/B6 angulin-1 KO cells. After angulin-1 knockout, tricellulin was found in both tTJ and bTJ, similar to control cells.

2.4. *Effect of Angulin-1 Knockout on the TJ Ultrastructural Level in MDCK C7 and HT-29/B6 Cells*

To obtain insight of whether and to what extent angulin-1 influences the barrier properties of the TJ in MDCK C7 and HT-29/B6 cells, the ultrastructure of the TJs was analyzed by freeze-fracture electron microscopy (Figure 4). Although it would be interesting to investigate the tTJ structure in the angulin-1 KO cells, we could not find a sufficient number of tTJs in order to discuss the changes of the structure because the tTJs are quite rare and difficult to find in the freeze fracture replica samples.

Comparison of the bTJs (Figure 4a,b) of controls and angulin-1 KO clones showed no alteration in the ultrastructure and revealed a regular meshwork in both cell lines.

Regarding bTJs (Table 1), no alteration was found in the horizontally oriented filaments arranged perpendicular to the paracellular diffusion pathway between the controls and the angulin-1 knockouts in both cell lines. Neither the numbers of strands nor the meshwork depth was changed after angulin-1 KO (Table 1). The network density, which is calculated from the ratio of the number of strands to the network depth, did not differ between controls and angulin-1 knockout clones (Table 1). The number of breaks (>20 nm) per µm horizontal length of single-strands in bTJ was not significantly different between controls and the angulin-1 KO clones (Table 1). More importantly, analysis of strand appearance as being either of continuous- or particle- (pearl string) type revealed no changes that correlated with the observations reported above. Continuous strands appeared in all examined microscopy fields in control clones as well as in angulin-1 KO clones (Table 1). In addition, the TJs of the MDCK C7 controls and the angulin-1 KO clones were composed exclusively of linear strands; however, the HT-29/B6 control 29 and the angulin-1 KO 32 clones showed a slightly curved pattern (Table 1).

Table 1. Morphometric analysis of TJ ultrastructure of MDCK C7 and HT-29/B6 angulin-1 KO cells. No significant differences in TJ ultrastructure of control cells and angulin-1 KO cells could be detected using freeze-fracture electron microscopic analysis in MDCK C7 and HT-29/B6 cells.

		MDCK C7 Cells				HT-29/B6 Cells			
		Control 14 (*n* = 29)	**Control 18** (*n* = 26)	**KO 18** (*n* = 18)	**KO 36** (*n* = 28)	**Control 15** (*n* = 17)	**Control 29** (*n* = 20)	**KO 12** (*n* = 11)	**KO 32** (*n* = 27)
Number of strands		3.62 ± 0.23	3.42 ± 0.20	4.17 ± 0.35	3.46 ± 0.23	4.53 ± 0.39	3.61 ± 0.34	3.91 ± 0.31	4.59 ± 0.41
Meshwork depth [a] (nm)		151.03 ± 11.00	165.00 ± 22.42	171.11 ± 24.78	157.50 ± 16.22	276.47 ± 37.42	228.89 ± 36.71	219.09 ± 23.06	310.93 ± 39.08
Strand density [b] (1/pm)		23.97 ± 2.32	20.75 ± 3.07	24.35 ± 4.09	22.00 ± 2.70	16.38 ± 2.64	15.78 ± 2.94	17.84 ± 2.36	14.77 ± 2.27
Number of breaks [c] per µm		0.0 ± 0.0	0.01 ± 0.009	0.0 ± 0.0	0.0 ± 0.0	0.02 ± 0.015	0.0 ± 0.0	0.03 ± 0.029	0.0 ± 0.0
Strand appearance	Continuous (%)	100	92	100	100	94	100	100	93
	Particle (%)	0	8	0	0	6	0	0	7
Strand pattern	Straight (%)	100	100	100	100	100	89	100	85
	Curved (%)	0	0	0	0	0	11	0	15

The meshwork depth [a] is defined as the distance between the apical and the contra-apical strand. Breaks [c] are defined as strand discontinuities >20 nm within the compact TJ meshwork; their number is given per µm length of horizontally oriented strands. The density [b] of bTJ (bicellular tight junction) strands is the ratio of strand number and meshwork depth and given in number per pm meshwork depth.

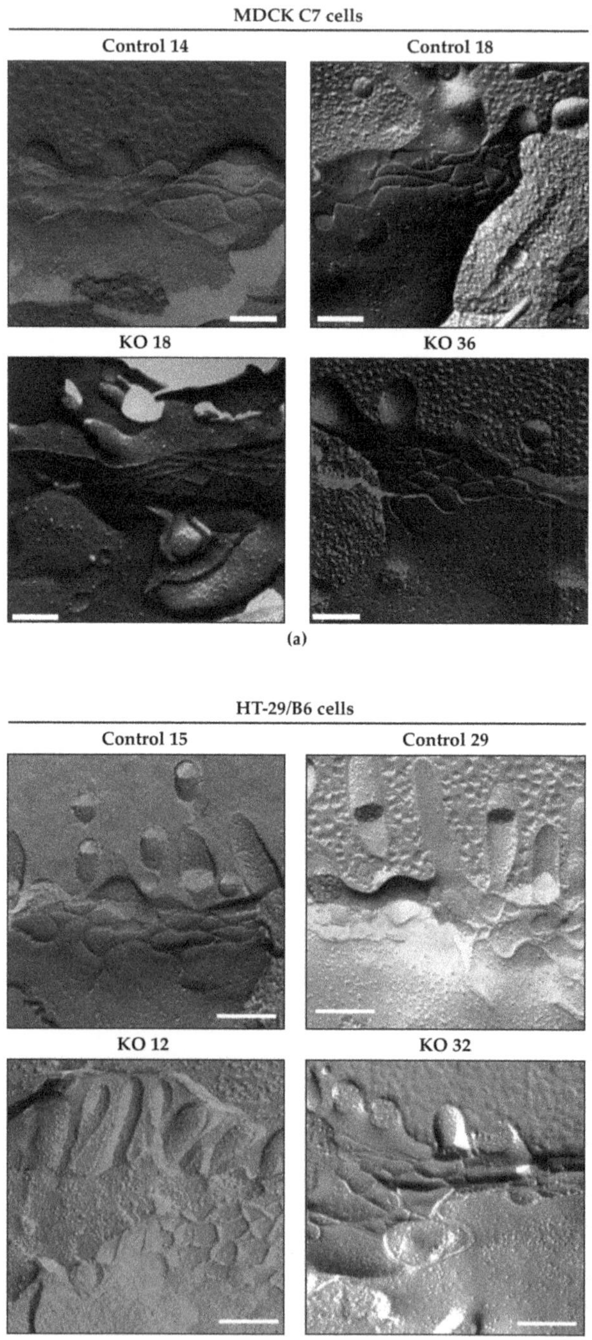

Figure 4. Freeze-fracture electron microscopy of angulin-1 knockout in (**a**) MDCK C7 and (**b**) HT-29/B6 cells. Photos were taken at ×51,000; Bars: 200 nm. The bTJ strands of the control cells and angulin-1 KO clones revealed a regular meshwork, characterized by continuous-type areas. bTJs of angulin-1 KO cells showed no ultrastructural difference compared with control clones in both cell lines.

2.5. Effects of Angulin-1 Knockout on Ion Permeability in MDCK C7 and HT-29/B6 Cells

To examine whether the loss of angulin-1 affects the epithelial barrier development, the transepithelial resistance (TER), reflecting inverse overall ion permeability, was measured on controls and knockout clones.

TER was reduced in angulin-1 knockout clones in both cell lines (Figure 5a,b, Supplementary Tables S1 and S2). Angulin-1 KO caused a stronger decrease of TER in MDCK C7 cells than in HT-29/B6 cells (7 to 14 times in MDCK C7 cells and 2 to 7 times in HT-29/B6 cells).

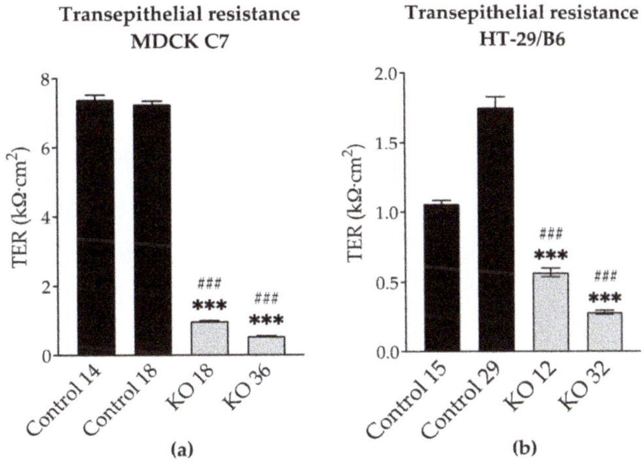

Figure 5. Functional analysis of angulin-1 knockout in MDCK C7 and HT-29/B6 cells. Effect of angulin-1 knockout on transepithelial resistance (TER) in (**a**) MDCK C7 ($n = 50$) and (**b**) HT-29/B6 cells ($n = 43$). Angulin-1 knockout strongly reduced TER in both cell lines. Statistical analysis was performed using one-way ANOVA test was used to compare between the four clones in both cell lines. (MDCK C7: *** $p \leq 0.001$ with regard to control 14 and ### $p \leq 0.001$ with regard to control 18 and HT-29/B6: *** $p \leq 0.001$ with regard to control 15 and ### $p \leq 0.001$ with regard to control 29).

2.6. Effects of Angulin-1 Knockout on Macromolecule Permeability in MDCK C7 and HT-29/B6 Cells

The next question was whether angulin-1 knockout also affected the permeability to macromolecules. Therefore, we examined the permeability of the different clones to FITC-dextran 4 kDa (FD4) in two different conditions: (1) under an osmotic gradient or (2) under isosmotic conditions. As a result, FD4 permeability exhibited a strong increase in the angulin-1 knockout in HT-29/B6 cells specifically under isosmotic conditions (Figure 6c,d, Supplementary Table S2). In contrast, angulin-1 knockout did not increase the permeability to FD4 in MDCK C7 cells under isosmotic conditions (Figure 6b, Supplementary Table S1). Interestingly, under an osmotic gradient, only the KO 18 clone in MDCK C7 cells showed a slightly increase in FD4 permeability (Figure 6a, Supplementary Table S1). In MDCK C7 cells, there was no difference between FD4 permeability measured under osmotic and isosmotic conditions, whereas in HT-29/B6 cells, the FD4 permeability measured under isosmotic conditions was higher than measured under osmotic conditions.

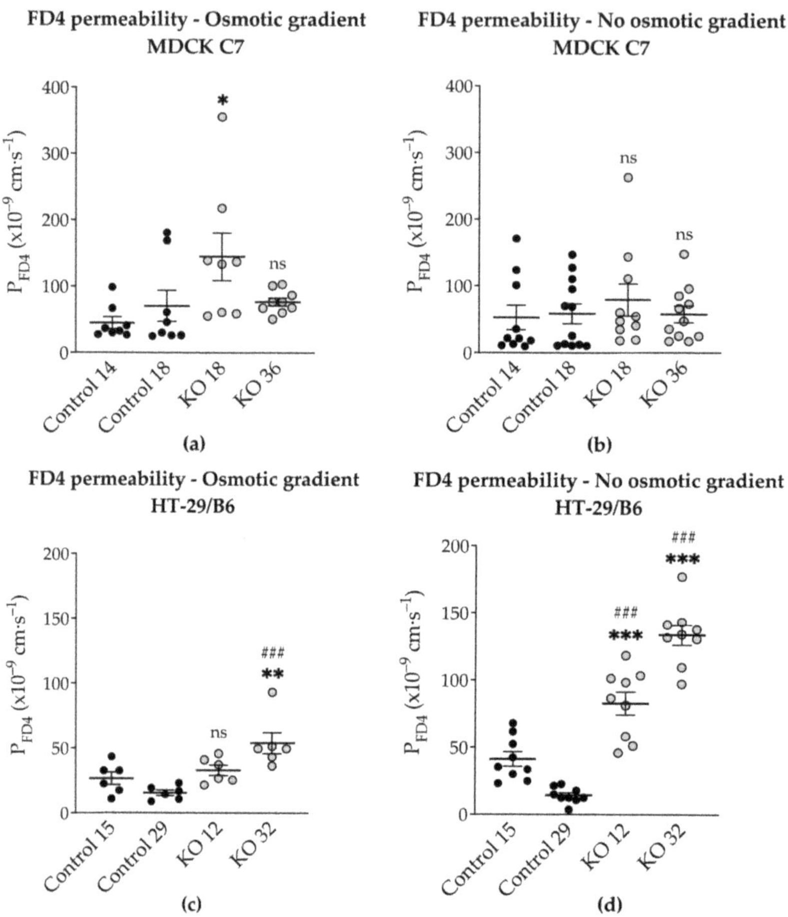

Figure 6. Functional analysis of angulin-1 knockout in MDCK C7 and HT-29/B6 cells. Effect of angulin-1 knockout on permeability for 4-kDa FITC-dextran (FD4) in MDCK C7 cells under an (**a**) osmotic gradient and (**b**) isosmotic condition. Permeability was increased only in the KO 18 clone under an osmotic gradient (n = 8–10). Effect of angulin-1 knockout on permeability for FD4 in HT-29/B6 cells under an (**c**) osmotic gradient and (**d**) isosmotic condition. Permeability was increased in both knockout clones under isosmotic conditions (n = 6–9), under osmotic gradient only the KO 32 clone showed an increase in permeability. Statistical analysis was performed using one-way ANOVA test was used to compare between the four clones in both cell lines (ns: not significant, MDCK C7: * $p \leq 0.05$ with regard to control 14 and HT-29/B6: ** $p \leq 0.01$, *** $p \leq 0.001$ with regard to control 15 and ### $p \leq 0.001$ with regard to control 29).

2.7. *Effect of Angulin-1 Knockout on Transepithelial Water Transport in MDCK C7 and HT-29/B6 Cells*

To analyze the effect of angulin-1 knockout on water permeability of MDCK C7 and HT-29/B6 cells, water flux was measured after induction with an osmotic gradient produced by 100 mM mannitol either in the apical or the basolateral side of the cell layers (Figure 7, Supplementary Tables S1 and S2). This concentration produced measured osmolality gradients of 100 mOsm for 100 mM mannitol.

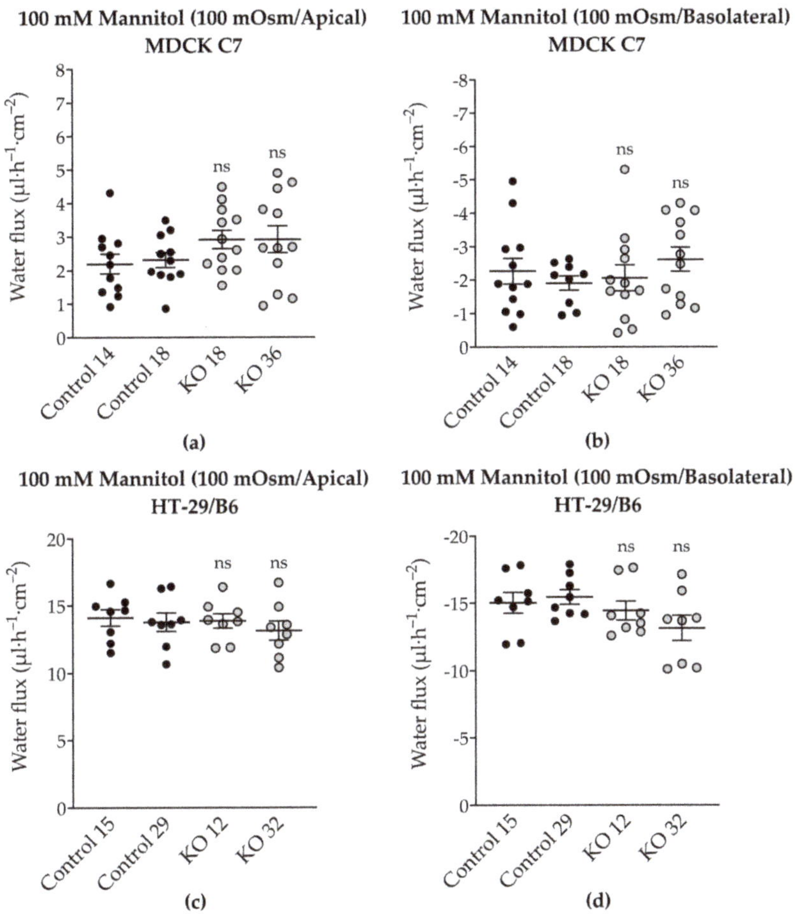

Figure 7. Water flux in control and angulin-1 knockout cells stimulated by an osmotic gradient on the apical or the basolateral side. Water flux induced by a gradient of 100 mM mannitol in (**a**,**b**) MDCK C7 and in (**c**,**d**) HT-29/B6 cells was unchanged. Statistical analysis was performed using one-way ANOVA test was used to compare between the four clones in both cell lines (ns: not significant).

Surprisingly, and in contrast to what was previously found after tricellulin knockdown in MDCK C7 cells, angulin-1 knockout did not significantly change the water flux under the osmotic gradient induced by 100 mOsm mannitol (Figure 7a,b) compared with their respective controls. In the same way, in HT-29/B6 angulin-1 KO cells, water flux did not significantly change under the osmotic gradient induced by mannitol (Figure 7c,d) in comparison with their controls.

In a parallel series of experiments, the water flux in the MDCK C7 angulin-1-depleted clones was measured after induction with an osmotic gradient produced by 37 mM 4-kDa dextran on the apical side because as we pointed out before, mannitol could go through the tTJ and interact with the pore or with the water molecules blocking their movement in the other direction. The water flux increased in both KO clones if 37 mM 4-kDa dextran (100 mOsm/Apical side) was used (Figure 8). This indicates that removal of angulin-1 from the tTJ altered the flux of water in a tight epithelium and that it is dependent on the chemical nature of the osmotic gradient.

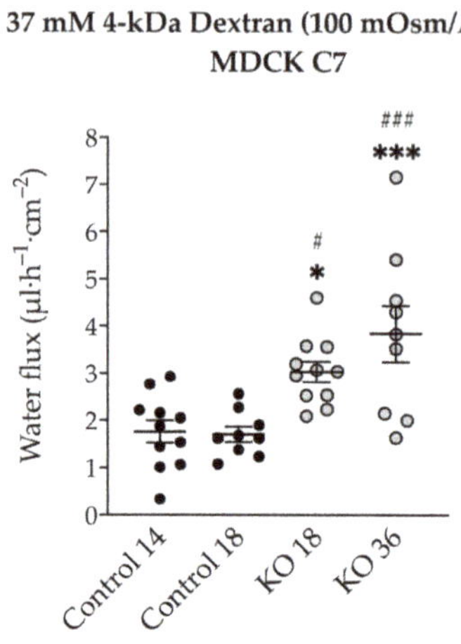

Figure 8. Water flux in angulin-1 knockout MDCK C7 cells induced by a gradient of 37 mM 4-kDa dextran on the apical side (n = 10–11). The transepithelial water transport increased in both angulin-1 knockout clones. (* $p \leq 0.05$, *** $p \leq 0.001$ with regard to control 14 and # $p \leq 0.05$, ### $p \leq 0.001$ with regard to control 18).

As shown in Figure 9, there were differences in FD4 permeability under isosmotic and osmotic gradient conditions between both angulin-1 KO cell types, but there was no difference in FD4 permeability in MDCK C7 control and angulin-1 KO cells under both osmotic conditions. In contrast, in HT-29/B6 cells, no difference in control cells was found, except for an increase in FD4 permeability without any osmotic gradient (Figure 9). A possible explanation for this puzzling finding would be an interaction between the movement of water in the opposite direction to the movement of dextran under osmotic conditions in the tTJ of angulin-1 KO HT-29/B6 cells. This interaction appears not to exist in MDCK C7 angulin-1 KO cells.

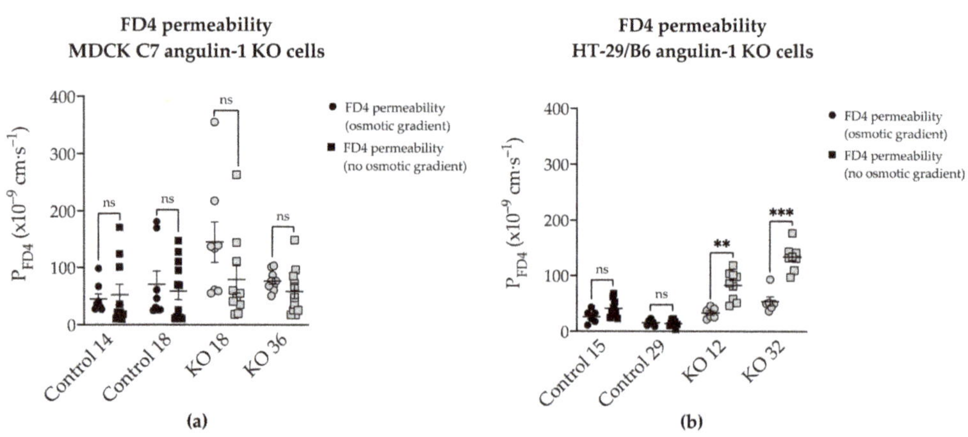

Figure 9. Comparison of angulin-1 knockout effect on permeability for 4-kDa FITC-dextran (FD4) under isosmotic and osmotic gradient conditions in (**a**) MDCK C7 and (**b**) HT-29/B6 cells. The FD4 permeability was higher under an isosmotic condition than under an osmotic gradient in angulin-1 KO HT-29/B6 cells, whereas no differences were observed in angulin-1 KO MDCK C7 cells. (ns: not significant, ** $p \leq 0.01$, *** $p \leq 0.001$).

2.8. Effect of Tricellulin KD on Transepithelial Water Transport in HT-29/B6 Cells

Previously, we described that in MDCK C7 cells tricellulin knockdown increased the transepithelial water flux compared to controls under different osmotic gradients [21]. In order to find out whether or not tricellulin has a similar effect in HT-29/B6 cells, tricellulin KD clones were generated and characterized in a similar way as we did for the MDCK C7 tricellulin KD clones (Table 2, Supplementary Figures S1–S4). Tricellulin KD in HT-29/B6 cells reduced the TER and increased the macromolecule permeability (Supplementary Figure S2). Interestingly, in tricellulin KD clones, the cation and water channel claudin-2 was upregulated (Supplementary Figure S3); however, it was not exclusively localized in the TJ, but also subjunctionally and intracellularly (Supplementary Figure S4a), and therefore this change did not affect ion and water permeability. In concordance with this, the ratio P^{Na}/P^{Cl} did not change (Supplementary Figure S4b,c), which means that tricellulin increases the movement of ions without charge selectivity.

Table 2. Characteristics of HT-29/B6 tricellulin knockdown clones and the corresponding control. Two tricellulin knockdown clones and its corresponding control were analyzed in this study (Control 12, KD 11 and KD 17). Data of tricellulin expression have been obtained by densitometric analysis of Western blots using β-actin for normalization. Paracellular permeability measurements for FD4 were carried out in the Ussing chamber. Data of P^{Na}/P^{Cl} permeability and absolute permeabilities for Na^+ and Cl^- (P^{Na}, P^{Cl}) were obtained from dilution potential measurements in the Ussing chamber. Water flux measurements were performed in a modified Ussing chamber with water flux induced by different osmotic gradients.

		Control 12	**KD 11**	**KD 17**
Tricellulin expression (%)		100.0 ± 8.9 (n = 11)	64.6 ± 4.7 *** (n = 10)	55.7 ± 6.3 *** (n = 10)
TER (kΩ·cm^2)		1.44 ± 0.07 (n = 12)	0.51 ± 0.03 *** (n = 12)	0.54 ± 0.02 *** (n = 12)
P_{FD4} ($\times 10^{-9}$ cm·s^{-1}) Osmotic gradient		21.34 ± 3.80 (n = 7)	42.96 ± 2.56 *** (n = 7)	67.55 ± 6.54 *** (n = 7)
P^{Na} ($\times 10^{-6}$ cm·s^{-1})		0.61 ± 0.10 (n = 10)	1.50 ± 0.11 *** (n = 10)	2.46 ± 0.30 *** (n = 11)
P^{Cl} ($\times 10^{-6}$ cm·s^{-1})		0.80 ± 0.18 (n = 10)	1.84 ± 0.17 ** (n = 10)	2.98 ± 0.33 ** (n = 11)
P^{Na}/P^{Cl}		0.91 ± 0.08 (n = 10)	0.92 ± 0.06 (n = 10)	0.90 ± 0.08 (n = 11)
Water flux—apical side (μL·h^{-1}·cm^{-2})	100 mM mannitol (100 mOsm)	14.9 ± 1.0 (n = 8)	14.4 ± 1.7 (n = 8)	15.3 ± 0.6 (n = 8)
	37 mM 4-kDa dextran (100 mOsm)	19.6 ± 1.2 (n = 8)	16.2 ± 1.0 (n = 8)	19.1 ± 1.4 (n = 8)
	100 mM 4-kDa dextran (900 mOsm)	58.6 ± 2.0 (n = 10)	54.0 ± 2.4 (n = 9)	56.9 ± 1.6 (n = 10)
Water flux—basolateral side (μL·h^{-1}·cm^{-2})	37 mM 4-kDa dextran (100 mOsm)	−16.4 ± 1.4 (n = 10)	−10.5 ± 1.0 * (n = 7)	−16.7 ± 0.9 (n = 6)

Significances refer to the control. n number of experiments, * $p \leq 0.05$, ** $p \leq 0.01$, *** $p \leq 0.001$.

Regarding tricellulin, its knockdown did not regulate the transepithelial water transport through the tTJ in HT-29/B6 cells (Figure 10). In HT-29/B6 cells claudin-2 is genuinely expressed and a large component of water transport may have travelled through the bicellular TJ so that a possible contribution of the tTJ may not have reached significant levels.

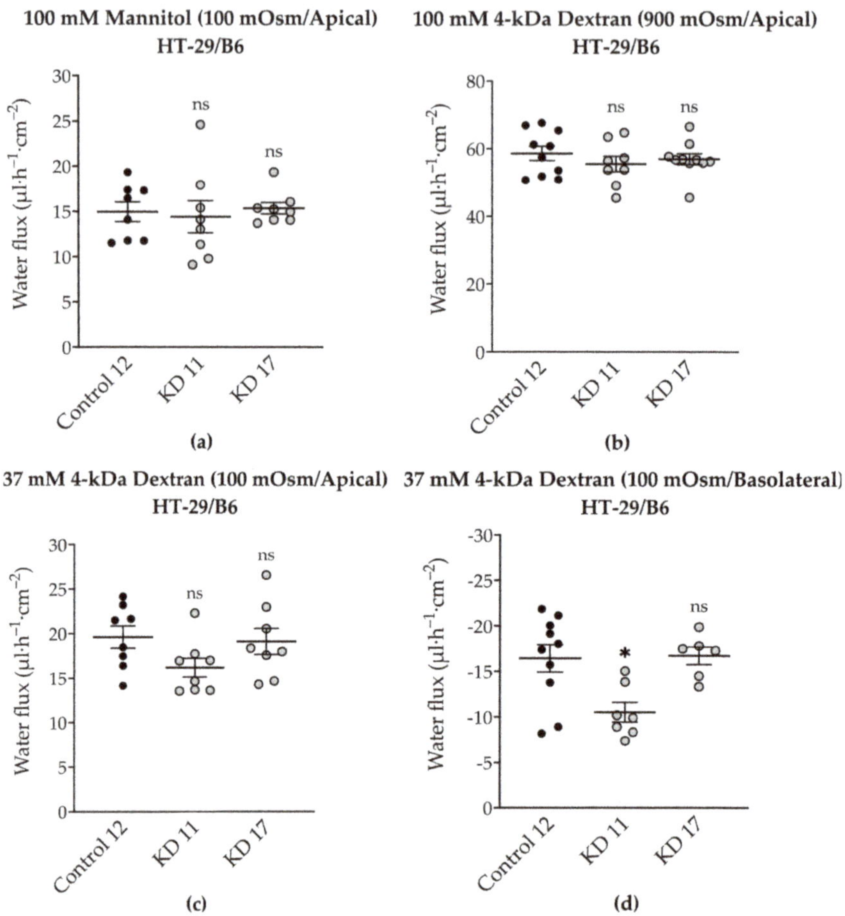

Figure 10. Water flux in control and tricellulin knockdown HT-29/B6 cells stimulated by an osmotic gradient. (**a**) Water flux induced by a gradient of 100 mM mannitol on the apical side. (**b**) Water flux induced by a gradient of 100 mM 4-kDa dextran on the apical side. (**c**) Water flux induced by a gradient of 37 mM 4-kDa dextran gradient on the apical side. (**d**) Water flux induced by a gradient of 37 mM 4-kDa dextran gradient on the basolateral side of the cell layer. (n = 8–10, ns: not significant, * $p \leq 0.05$).

3. Discussion

Tricellulin and angulin family proteins have been identified as molecular constituents of tTJs [5,22,25]. Tricellulin and angulins localize along the central elements of tTJs [5,25]. As angulins recruit tricellulin to TJs, the angulin–tricellulin axis is proposed to be of importance in tTJ formation [22,25,36]. Therefore, a dysregulation of angulins may critically affect the permeability for molecules up to 10 kDa in different epithelial organs.

In the present study, our focus was placed on elucidating the role of angulin-1 on paracellular water permeability. For this, the effect of angulin-1 knockout was analyzed in two different cell lines, MDCK C7, a tight epithelial cell line, and HT-29/B6, an intermediate-tight epithelial cell line. It turned out that angulin-1 is involved in water permeability only in the tight epithelial cell line (similar to what we previously reported on tricellulin [21]), while both angulin-1 and tricellulin have no effect on water permeability in an intermediate-tight epithelium. These results indicate that in tight epithelia, the tTJ is also essential for maintaining a barrier for water.

3.1. Angulin-1 Knockout Alters the Expression and Localization of Other Proteins in MDCK C7 and HT-29/B6 Cells

The expression of several tightening claudins in MDCK C7 cells was reduced in angulin-1 knockout cells, suggesting these claudins were partly removed from the TJ strands in the absence of angulin-1, in contrast, claudins and tricellulin expression were increased after angulin-1 knockout in HT-29/B6 cells, suggesting that these claudins were additionally incorporated into TJ strands in the absence of angulin-1. Most important, in both angulin-1 KO cell lines, tricellulin was partly removed from the tTJ and was also detectable in the bTJ [25].

On one hand, angulin-1 knockout in MDCK C7 cells induced a decrease in the expression level of occludin, claudin-1 and -4 (Figure 2c). This might affect the barrier function and formation as evidenced by Kirschner et al. through different knockdowns of claudin-1 and -4, occludin, and ZO-1, causing increased paracellular permeability for ions and larger molecules [37]. Furthermore, the absence of occludin and claudin-1 in both KO clones prevents the specific localization of tricellulin at tricellular contacts and promotes its localization at the bTJ [38,39]. These observations suggested that the proteins functionally influence each other.

On the other hand, angulin-1 knockout in HT-29/B6 cells increased the expression of barrier-forming claudins [39] and tricellulin (Figure 2d). Therefore, it can be concluded that the observed clonal variation of the claudins and tricellulin in sum is balanced and provides a constant barrier function of the bTJ and tTJ. These results suggested that the effects of angulin-1 and tricellulin relocalization from tTJs to bTJs on epithelial barrier function are controversial [39]. Interestingly, claudin-2 was downregulated in the KO 12 clone and upregulated in the KO 32 clone and appeared not to affect the water permeability. In relation to transcellular water transport, the major aquaporins of human colon cell line were found unchanged or downregulated, this suggests that changes in transepithelial water permeability could be attributed to the removal of angulin-1 and unlikely due to aquaporins. Lastly, the expression of LI-cadherin was increased in both KO clones, this could affect the water permeability under our experimental conditions. It was shown that water transport was impaired under hypertonic conditions when LI-cadherin is reduced [40,41]. Therefore, LI-cadherin might be important for water reabsorption of the gut at intraluminal hypertonic conditions [40,41].

Interestingly, the expression of angulin-2 and -3 did not change after the knockout of angulin-1 as a form of compensation in both cell lines. It was shown that compensatory functions exist between the angulin family members, e.g., in the organ of Corti [42], the large intestine and kidney [31], but this seems to be not relevant in the cell lines used for this study.

3.2. Angulin-1 Knockout Did Not Alter the Ultrastructure of the Bicellular Tight Junction in MDCK C7 and HT-29/B6 Cells

In previous RNAi experiments, the loss of tricellulin led to an unstable ultrastructure of tight junction, which ultimately caused the junctional complex to collapse completely [5]. In both cell lines, the loss of angulin-1 did not modify or alter the bTJ compared to the vector control clones, the TJ ultrastructure remains unchanged after angulin-1 knockout in MDCK C7 and HT-29/B6 cells (Figure 4).

It is established that, depending on the presence of claudin-2, strand discontinuities appear or disappear [43]. In our study, no discontinuities appear even if claudin-2 is increased in the HT-29/B6 angulin-1 KO 36 clone, possibly due to the upregulation of tricellulin.

Thus, an increase in ion permeability in both cell lines and an increase in macromolecule permeability in HT-29/B6 cells could possibly be caused by the opening of the tTJ because the only difference between the controls and the angulin-1 KO clones was the width of the central tube. However, the morphometric parameters could not be analyzed because of the low abundance of the same structures in both cell lines. After considering all possibilities, it is concluded that the knockout of angulin-1 did not induce resolvable changes in the ultrastructure of bTJs and tTJs that could affect water transport measurements.

3.3. Angulin-1 Knockout Increases the Ion Permeability in MDCK C7 and HT-29/B6 Cells

By the knockout of angulin-1, also the pore pathway was addressed (Figure 5). Similar to what was observed in tricellulin knockdown cells, the reduction in the expression of angulin-1 resulted in a lowered transepithelial resistance (TER). It is important to note that in MDCK C7 cells the removal of angulin-1 resulted in the reduction of the expression of most claudins while in HT-29/B6 the effect was the opposite. Therefore, it could be said that the reduction in TER on HT-29/B6 cells is due only to the reduction in angulin-1; however, in MDCK C7 cells the effect could be intensified due to the downregulation of other tightening TJ proteins. In conclusion, the decrease of angulin-1 within the tTJ also leads in both cell lines to a drop in paracellular resistance, which reciprocally reflects the permeability for ions.

3.4. Angulin-1 Knockout Increases the Macromolecule Permeability Only in HT-29/B6 Cells

As it is known, knockdown of angulin-1 in the mouse mammary gland Eph4 epithelial cells increases the permeability to macromolecules between 332 Da and 40 kDa [22]. Considering the above, the leak pathway was assessed (Figure 6). The present data showed that in HT-29/B6, a removal of angulin-1 strongly regulates the permeability for 4-kDa macromolecules. Nevertheless, it might be surprising that the same phenotype, in MDCK C7 cells, only slightly increased the permeability for macromolecules in one of the KO clones (KO 18), probably because in this KO clone, the tricellulin delocalization from the tTJ was larger than in the KO 36 clone.

Considering that HT-29/B6 is an intermediate-tight epithelial cell line and that its tTJ contributes at least 38% to the total paracellular conductance, it can be said that, not only did it affect the paracellular passage of the small solutes, including inorganic ions, but also of the macromolecules [20]. However, the contribution of tTJ in MDCK C7 cells is not yet known, but it would be expected to be higher than in HT-29/B6 cells because this cell line is highly tight [20], and therefore the permeability to FD4 would be higher after angulin-1 knockout, but this was not the case. This is perhaps due to tricellulin which has an affinity for claudin-based TJ strands within the plasma membrane if not directed to tricellular contacts by angulins [44] and also because occludin was concentrated in the corners of tTJ after angulin-1 KO (Figure 1c) probably stabilizing this structure. One could assume that due to redistribution of tricellulin into the bTJ a slight enhancement of bTJ barrier properties could occur [44].

Additionally, the permeability to FD4 was measured in two different conditions after angulin-1 KO and, as can be seen, in MDCK C7 cells there were no differences between isosmotic and osmotic conditions (Figure 9a). Conversely, in HT-29/B6 cells it was observed that the permeability to FD4 is higher under isosmotic conditions than under an osmotic gradient. A possible explanation for this might be that under osmotic conditions water moves in the opposite direction and this inhibits the movement of 4-kDa FITC-dextran, therefore a lower FD4 permeability was measured (Figure 9b). This water movement through the tTJ is too small to be detected in the overall transepithelial water flux measured in HT-29/B6 cells.

3.5. Angulin-1 Knockout Increases Transepithelial Water Transport Only in MDCK C7 Cells

To complete the tTJ permeability pattern, the transepithelial water transport was measured in the angulin-1 KO clones of MDCK C7 and HT-29/B6 cells. As described earlier, while the KO clones caused a decrease in TER in both cell lines, only in HT-29/B6 cells did angulin-1 KO increase the permeability to a 4-kDa macromolecule. Interestingly, the angulin-1 KO did not increase the water flux driven by 100 mM mannitol in both cell lines (Figure 7) and the total amount of water was independent of the direction of the osmotic gradient. The decision to use 4-kDa dextran was taken due to the dispersion of the values found in MDCK C7 cells when mannitol was used as an osmotic gradient; however, the values obtained with HT-29/B6 cells were more reproducible and therefore their study with 4-kDa dextran was obviated. As result, the water flux increased in the angulin-1 KO clones compared with their controls in MDCK C7 cells when 37 mM 4-kDa dextran was used as osmotic gradient (Figure 8).

Remarkably, in the case of MDCK C7 cells, an increase in water transport was expected like that found after reducing the expression of tricellulin in tTJ [21]; however, this was not the case when mannitol was used possibly due to the small diameter of this molecule with respect to the 4-kDa dextran which would facilitate its movement through tTJ thereby inhibiting the water flux in the converse direction. It is important to keep in mind that the reduction in the expression of most barrier-forming claudins could have an influence on the water transport after angulin-1 KO in MDCK C7 cells. As demonstrated by immunofluorescence confocal laser-scanning microscopy, tricellulin was redistributed from tTJ to bTJ, which could maintain and strengthen the epithelial barrier after the downregulation of several claudins. It is also important to note that the water fluxes measured in angulin-1 KO cells were lower than those measured in our previous work on tricellulin KD clones [21], demonstrating once again the importance of tricellulin in tTJ (for instance, angulin-1 KO 36 showed higher tricellulin displacement leading to higher water flux). In addition, it can be concluded that the reduction of tightening claudins did not increase the water flux in the KO clones with respect to their controls.

This phenomenon was also observed in HT-29/B6 cells; however, here the regulation of TJ expression was the opposite to what we observe in MDCK C7 cells, several claudins were upregulated, nevertheless, and no change in water transport could be measured. As shown in Figure 9, FD4 permeability in HT-29/B6 angulin-1 KO cells (not in control cells) measured under osmotic conditions is lower than under isosmotic conditions, potentially due to an opposite water flow across the tTJ. This means that the water flow across the tTJ, which was induced by a gradient using mannitol or dextran, could also be inhibited by the opposite passage of the osmotically active substances across the tTJ in these cells. This could also be the case in MDCK C7 angulin-1 KO cells when using mannitol for producing the osmotic gradient. In addition, the strong upregulation of tightening claudins in the bTJ in HT-29/B6 angulin-KO cells could diminish the water flux across the bTJ and thus mask a slight increase in water flux across the tTJ. Thus, we cannot completely exclude an increased water flow across the tTJ after downregulation of tricellulin or knockout of angulin-1, but this does not significantly contribute to the overall large water flow across an intermediate tight epithelial cell line like the HT-29/B6 cells.

In summary, the present study suggests a functional difference between angulin-2 and angulin-1 on the paracellular water transport via tTJs, which is an interesting issue to be pursued for the understanding of variations of tTJs. The effects of angulin-1 knockout on the epithelial barrier function vary by cell type and according to the interdependence between different TJ species affecting cell expression, localization and renewal. The reason could be the differences in angulin-1 interaction partners in bTJs and tTJ [22,39]. Therefore, combinations of proteins influenced by angulin-1 knockout in bTJs and tTJ vary by cell type and tissue, and the resulting epithelial barrier function of the mammalian epithelial cell sheet causes cell specific changes.

3.6. Tricellulin Knockdown Does Not Significantly Affect Paracellular Water Transport in the HT-29/B6 Cell Line

As was previously done for tricellulin knockdown clones in MDCK C7 cells [21], two tricellulin clones were selected in HT-29/B6 cells that differ in the grade of tricellulin reduction (Supplementary Figure S1). While both KD clones caused a decrease in TER and an increase in permeability to a 4-kDa macromolecule, none of them increased the water flux driven by different osmotic gradients. This may be due, as already described for the HT-29/B6 angulin knockout cells, to an inhibition of the water flux across the tTJ by a backflow of the osmotically active substances like mannitol or dextran. In addition, a large component of water transport may have traveled through the bTJ and a possible contribution from the tTJ may not have reached significant levels (Figure 10). On the other hand, there may be a threshold in the expression of tricellulin, that must be overcome to find significant differences in water transport, that were not reached with the knockdowns generated in this research. In addition, there is a possibility that in general the paracellular water flow is lower than the transcellular water flux or that there is a claudin-2-independent paracellular pathway for water.

In conclusion, in HT-29/B6 cells (intermediate-tight) the contribution of tricellulin to transepithelial water permeability is negligible, in contrast to MDCK C7 cells (tight) where tricellulin was able to regulate osmotically induced transepithelial water flux.

4. Materials and Methods

4.1. Cell Culture, Transfection and TER Measurement

The kidney cell line MDCK C7 (RRID: CVCL_0423) exhibits a high transepithelial resistance and other basic properties (no claudin-2 expression) making it an excellent model of a tight epithelium, and the intestinal cell line HT-29/B6 (RRID: CVCL_LJ30) exhibits a basic properties (moderate transepithelial resistance, low claudin-2 expression) making it an excellent model of an intermediate-tight epithelium. For stable angulin-1 knockout, MDCK C7 and HT-29/B6 cells were transfected with three sgRNA-CRISPR/Cas9 vectors containing sequences targeting different exons of angulin-1 and corresponding HDR plasmids for direct homologous repair or sgRNA-CRISPR/Cas9 negative control (Santa Cruz, Heidelberg, Germany). For stable tricellulin knockdown, HT-29/B6 cells were transfected with pLKO.1-puro vector containing a sequence for shRNA targeting tricellulin (tricellulin shRNA; Sigma-Aldrich, Schnelldorf, Germany) or pLKO.1-puro empty vector as a negative control (Sigma-Aldrich, Schnelldorf, Germany).

The transfected cells were incubated at 37 °C and 5% CO_2 in sterilization incubators held. For the cultivation of the cells sterile culture vessels made of plastic were used. For MDCK C7 knockout cells, a nutrient medium Earl's salts MEM (minimal essential medium) supplemented with 10% FBS as well as 100 U/mL penicillin/100 µg/mL streptomycin and 1.5 µg/mL of puromycin was used. For HT-29/B6 knockdown and knockout cells, RPMI-1640 with stable L-glutamine medium supplemented with 10% FBS as well as 100 U/mL penicillin/100 µg/mL streptomycin and 1.5 µg/mL of puromycin was used. Every second to third day the medium was changed. Consumption of nutrients was also visible due to a color change of the medium contained a detecting pH indicator.

For protein quantification, water flux measurements and electrophysiological studies, cell monolayers were cultured on porous culture plate inserts (Millicell PCF filters, pore size 0.4 µm, effective area 0.6 cm^2, Millipore GmbH, Schwalbach, Germany) for 7–10 days before they were used for experiments.

Transepithelial resistance (TER) was measured at 37 °C using chopstick electrodes (STX2, World Precision Instruments, Friedberg, Germany). Electrodes were reproducibly positioned by a semi-automatic motor-driven device and signals were processed by a low-frequency clamp (both own design). The resistances of the bathing solution and the blank filter support were subtracted from measured values, which were finally converted to $\Omega \cdot cm^2$ [45].

4.2. Western Blot Analysis

Western Blot analysis was carried out as described before [21]. Cells grown on culture-plate inserts were scraped and homogenized in total lysis buffer containing 10 mM of Tris, 150 mM of NaCl, 0.5% of Triton X-100, and 0.1% of sodium dodecyl sulphate (SDS), and protease inhibitors (cOmplete™ EDTA free; Roche, Basel, Switzerland). The samples were incubated for 2 h at 4 °C (vortexed every 20 min), and then centrifuged at 11,000× *g* during 20 min. The pellet was discarded and the protein concentration in the supernatant was determined by the BCA (bicinchoninic acid) method (reagents were purchased from Pierce (Perbio Science, Bonn, Germany)) and quantified with a plate reader (Tecan Deutschland, Crailsheim, Germany). After this, samples were prepared with a loading buffer containing

100 mM Tris-HCl (pH 6.8), 2% SDS, 10% glycerol, 100 mM DTT and 0.001% bromophenol blue. Each sample was then boiled at 95 °C for 10 min before loading onto the gel.

Aliquots between 10 and 15 µg protein samples were separated by 12% SDS-polyacrylamide gel electrophoresis and then transferred to a PVDF membrane (Perkin Elmer, Rodgau, Germany) for detection of angulins, TAMPs, claudins, AQPs, SGLT-1 and LI-cadherin. After blocking for 2 h in 1% PVP-40 and 0.05% Tween-20, membranes were incubated overnight with primary antibodies specific for angulin-1 (Sigma-Aldrich, Taufkirchen, Germany), angulin-2 (Aviva Systems Biology, San Diego, CA, USA), and angulin-3 (Sigma-Aldrich, Taufkirchen, Germany), tricellulin (Thermo Fisher Scientific, Invitrogen, Darmstadt, Germany), occludin (Thermo Fisher Scientific, Invitrogen, Darmstadt, Germany), claudin-1, -2, -3, -4, -5, -7, and -8 (Thermo Fisher Scientific, Invitrogen, Darmstadt, Germany), AQP-1 (OriGene Technologies, Herford, Germany), AQP-3, -4 and -7 (Santa Cruz, Heidelberg, Germany), SGLT1 (LSBio, Corston, UK) and LI-cadherin (Santa Cruz, Heidelberg, Germany).

After removing the first antibody and three washing steps, the membranes were incubated for 2 h with the second peroxidase-conjugated antibody (anti-mouse or anti-rabbit) in 1.5% of milk powder prepared in TBST 1X. The primary antibodies were used in dilutions of 1:1000 with the exception of AQP antibodies (1:1500), tricellulin (1:2000), and angulin-1 (1:3000). The secondary antibodies were used in dilutions of 1:1000. For detection of the chemiluminescence signal induced by addition of Lumi-LightPLUS Western blotting kit (Roche, Mannheim, Germany) a Fusion FX7 (Vilber Lourmat, Eberhardzell, Germany) was used. Densitometric analysis was performed with a quantification software (Image Studio™ Lite, LI-COR Biosciences, Lincoln, NE, USA). Equal protein loading in each lane was verified by comparison with signals for β-actin (Sigma-Aldrich, Taufkirchen, Germany). For Western blot analysis, lysates of at least three individual cell cultures were used and one representative experiment is shown.

4.3. Immunofluorescent Staining

Immunofluorescence analysis was carried out as described before [21]. Immunofluorescence studies were performed on culture-plate inserts. Confluent monolayers were rinsed with PBS, fixed with 4% paraformaldehyde for 20 min, washed three times with PBS, and permeabilized for 10 min with PBS containing 0.5% (*v*/*v*) Triton X-100. To block non-specific binding sites, cells were then incubated in PBS containing 1% (*w*/*v*) BSA and 5% (*v*/*v*) goat serum (blocking solution; Biochrom) for 60 min. All subsequent washing procedures were performed with this blocking solution.

After blocking, cells were incubated overnight at 4 °C with primary antibodies for angulin-1 (1:1000; Sigma-Aldrich), tricellulin (1:500; Thermo Fisher Scientific, Invitrogen), occludin (1:250; Thermo Fisher Scientific, Invitrogen) and ZO-1 (1:250; Thermo Fisher Scientific, Invitrogen), followed by washing steps and incubation during 60 min at room temperature with the respective secondary antibodies (Alexa Fluor 488 goat anti-rabbit and Alexa Fluor 594 goat anti-mouse, each 1:500; Thermo Fisher Scientific, Waltham MA, USA) and 4′,6-diamidino-2-phenylindole (DAPI, 1:1000; Roche, Mannheim, Germany). Cell culture inserts were mounted on microscope slides using ProTag MountFluor (Biocyc, Luckenwalde, Germany). Images were obtained with a confocal laser-scanning microscope (LSM 780, Zeiss, Jena, Germany) and processed using ZEN software (Carl Zeiss, Oberkochen, Germany, ZEN black edition 2012 SP1, ver. 8.1).

4.4. Freeze Fracture Electron Microscopy

At a confluency of 100%, cells were washed with PBS (containing Ca^{2+}/Mg^{2+}) and fixed with 2.5% glutaraldehyde at RT for 2 h. After washing twice with PBS (containing Ca^{2+}/Mg^{2+}), the cells were stored at 4 °C in 0.25% glutaraldehyde. Small rectangles of the bottom of the cell culture filters were cut out, the attached cells were cryoprotected in 30% glycerol for 30 min, placed between two gold specimen holders and shock frozen in R422D (TEGA GmbH, Würzburg, Germany) cooled by liquid nitrogen (−210 °C).

The samples were fractured using the freeze-fracture device Denton DV-502 (Denton Vacuum, Moorestown, NJ, USA) at −100 °C and 2×10^{-10} Torr. The samples were vaporized at −150 °C (2×10^{-10} Torr) with a layer of platinum and then a layer of carbon. This results in a thin metal film on the broken sample. The replicas were cleaned with 12% sodium hypochlorite, washed several times in ddH_2O and mounted on a copper mesh grid. The Zeiss 902A electron microscope was used to examine the replicas at 80 kV. Magnifications between 20,000 and 50,000 was used.

4.5. Measurement of 4-kDa FITC-Dextran Flux

Flux studies were performed in conventional Ussing chambers for cell-culture inserts [45] under voltage clamp conditions. Dextran flux was measured in 5 mL circulating 111-Ringer's ((in mM) 119.7 NaCl, 21.4 $NaHCO_3$, 2.5 Na_2HPO_4, 0.6 NaH_2PO_4, 5.7 KCl, 1.3 $MgCl_2$, 1.2 $CaCl_2$, and 10.0 D(+)-glucose) containing also 37 mM unlabeled 4-kDa dextran (SERVA, Heidelberg, Germany) on both sides of the cells for isosmotic conditions and only on the apical side for osmotic conditions. After addition of 0.2 mM 4-kDa FITC-labeled dialyzed dextran (Sigma-Aldrich) to the apical chamber (final concentration), basolateral samples (200 µL) were collected at 0-, 20-, 40-, 60-, 80-, 100-, 120- and 140-min. Tracer fluxes were determined from FITC-dextran samples, which were measured with a fluorometer at 520 nm (Spectramax Gemini, Molecular Devices, Ismaning, Germany). Dextran permeability was calculated from $p = J/\Delta c$ with p = permeability ($cm \cdot s^{-1}$), J = flux ($mol \cdot h^{-1} \cdot cm^{-2}$) and c = concentration (mol/L).

4.6. Dilution Potential Measurements

Dilution potential measurements for the determination of ion permeabilities were performed in Ussing chambers for cell-culture inserts [45]. Water-jacketed gas lifts kept at 37 °C were filled with 10 mL circulating fluid on each side. The bathing solution contained (in mM) 119.7 NaCl, 21.4 $NaHCO_3$, 5.7 KCl, 1.3 $MgCl_2$, 1.2 $CaCl_2$, 3.0 HEPES, and 10.0 D(+)-glucose, and was gassed with 95% O_2 and 5% CO_2 to ensure a pH value of 7.4. All experimental data were corrected for the resistance of the empty filter and the bathing solution. Dilution potentials were measured with modified bathing solution on the apical or basolateral side of the epithelial monolayer. In the modified bathing solution, NaCl was iso-osmotically replaced by mannitol. The ratio of P^{Na} and P^{Cl} and the absolute permeabilities for Na^+ and Cl^- were calculated as described before [46].

4.7. Measurement of Transepithelial Water Transport

Water flux measurements were performed using a modified Ussing chamber developed by our group [12–14,21]. The gas lifts of the common Ussing chamber were replaced by two glass tubules with a small diameter, where the menisci of the perfusion solution are clearly visible. From changes in the menisci, the water flux is calculated. Throughout these experiments, transepithelial voltage (mV) was clamped to 0 mV and transepithelial resistance (TER, $\Omega \cdot cm^2$) and short-circuit current (I_{SC}, $\mu A \cdot cm^{-2}$) were recorded. Resistances of bathing solution and blank filter support were measured prior to each experiment and subtracted. The stability of TER served as an indicator of cell viability.

Cell filters were mounted in Ussing chambers and perfused with HEPES-buffered solution with the following composition (in mM): 144.8 NaCl, 2.4 Na_2HPO_4, 0.6 NaH_2PO_4, 5.4 KCl, 1.2 $MgCl_2$, 1.2 $CaCl_2$, 10.6 HEPES, and 10.0 D(+)-glucose. The pH value of the perfusion solution was pH 7.4. A rotary pump ensured constant circulation of the perfusion solution (4.0 $mL \cdot min^{-1}$) and thus a fast fluid exchange in both hemichambers (volume 500 µL) to avoid effects of unstirred layers on water permeability. Water flux was induced by a transepithelial osmotic gradient: (i) 100 mM mannitol, (ii), 37 mM 4 kDa-dextran or (iii) 100 mM 4 kDa-dextran. The solution was added in the apical or basolateral compartment of the Ussing chamber. The osmolality of the perfusion solutions (mosmol/kg, abbreviated mOsm) was determined using a freezing point depression osmometer (Osmomat 3000, Gonotec, Berlin, Germany). The osmolality of the HEPES-buffered solutions is annotated in Table 3.

Table 3. Osmolality of the HEPES-buffered solutions using a freezing point depression osmometer.

		n	Concentration (mM)	Osmolality (mOsm/kg of Water) Mean ± SEM
Angulin-1 KO	HEPES only	17	-	289.0 ± 13.7
	+Mannitol	17	100 mM	391.4 ± 14.6
	+4-kDa dextran	5	37 mM	396.0 ± 4.5
Tricellulin KD	HEPES only	10	-	308.0 ± 4.4
	+Mannitol	10	100 mM	410.3 ± 1.9
	+4-kDa dextran	10	37 mM	396.2 ± 6.8
	+4-kDa dextran	10	100 mM	877.8 ± 13.7

The fluid level in both glass tubes was monitored by a visual system ColorView XS (Olympus Soft Imaging Solutions GmbH, Munster, Germany) at time 0 min and with intervals of 10 min over a period of 120 min. Transepithelial water flux, given as flux per square centimeter and hour, was calculated after special calibration from the difference between the menisci at the registration times. Fluxes directed from the basolateral to the apical compartment were defined as positive flux.

4.8. Statistical Analysis

Data are expressed as mean values ± SEM, indicating n as the number of single measurements, and N is the number of independent experiments, which means independent seeding of cells. Statistical analysis was performed using Student's t-test between the tricellulin KD clones and their corresponding control (KD 11 and KD 17 versus control 12). In case of angulin-1 KO clones, statistical significance of the difference between means was evaluated using one-way ANOVA and the Tukey and Dunnett adjustment for multiple testing.

MDCK C7 cells: $^{*}\,p \leq 0.05$, $^{**}\,p \leq 0.01$, $^{***}\,p \leq 0.001$ with regard to control 14 and $^{\#}\,p \leq 0.05$, $^{\#\#}\,p \leq 0.01$, $^{\#\#\#}\,p \leq 0.001$ with regard to control 18. HT-29/B6 cells: $^{*}\,p \leq 0.05$, $^{**}\,p \leq 0.01$, $^{***}\,p \leq 0.001$ with regard to control 15 and $^{\#}\,p \leq 0.05$, $^{\#\#}\,p \leq 0.01$, $^{\#\#\#}\,p \leq 0.001$ with regard to control 29. Statistical analysis was performed using GraphPad Prism 8 software (GraphPad; San Diego, CA, USA). A probability value of $p \leq 0.05$ was considered statistically significant.

5. Conclusions

The major goal of this research was to clarify whether or not the expression of the tricellular tight junction (tTJ) protein, angulin-1, is involved in controlling paracellular water transport in cell lines with different levels of tightness (HT-29/B6 and MDCK C7 cells). This work supports the previous hypothesis that tTJs function differentially in distinct epithelial cell lines.

In brief, we provide insights into tTJ permeability by showing that angulin-1 can function as regulator of the macromolecular barrier in the tricellular tight junction regardless of the tightness of the epithelium (Supplementary Table S3). Manipulating the tTJ permeability (via tricellulin or via angulin-1) represents a mechanism to regulate epithelial water homeostasis only in tight epithelial cell lines (MDCK C7 cells); nevertheless, in intermediate-tight epithelial cell lines (HT-29/B6 cells) the contribution of tTJ seems to be negligible. Thus, the contribution of tTJ to transepithelial water transport depends on the tightness of the epithelium. Probably, then, the regulation of water paracellular transport by osmolytes in each type of epithelial cell has physiological significance depending on the organ to which it belongs.

Supplementary Materials: Supplementary Materials can be found at https://www.mdpi.com/1422-0067/22/15/7827/s1.

Author Contributions: Conceptualization, C.A.-T., S.M.K., R.R. and M.F.; funding acquisition, M.F.; investigation, C.A.-T.; methodology, R.R. and M.F.; supervision, S.M.K., R.R. and M.F.; validation, S.M.K., R.R. and M.F.; visualization, C.A.-T.; writing—original draft, C.A.-T.; writing—review and editing, S.M.K., R.R. and M.F. All authors have read and agreed to the published version of the manuscript.

Funding: This research was funded by the Deutsche Forschungsgemeinschaft (DFG), Research Training Group GRK 2318, and the Open Access Publication Fund of the Charité—Universitätsmedizin Berlin.

Institutional Review Board Statement: Not applicable.

Informed Consent Statement: Not applicable.

Data Availability Statement: Data is contained within the article or Supplementary Materials.

Acknowledgments: We would like to thank Britta Jebautzke, In-Fah M. Lee, and Anja Fromm (Clinical Physiology/Nutritional Medicine, Medical Department, Division of Gastroenterology, Infectiology and Rheumatology, Charité—Universitätsmedizin Berlin, 12203 Berlin, Germany) for their expert technical assistance.

Conflicts of Interest: The authors declare no conflict of interest.

Abbreviations

Abbreviations

angulin-1	lipolysis-stimulated lipoprotein receptor (LSR)
angulin-2	immunoglobulin-like domain containing receptor-1 (ILDR1)
angulin-3	immunoglobulin-like domain containing receptor-2 (ILDR2)
bTJ	bicellular tight junction, located between two cells
CRISPR/Cas9	Clustered Regularly Interspaced Short Palindromic Repeats/Cas9
d(Å)	diameter (Å)
FD4	4-kDa FITC-dextran
HDR	homology-directed repair
KD	knockdown
KO	knockout
SEM	standard error of the mean
TJ	tight junction
TER	transepithelial resistance ($\Omega{\cdot}cm^2$)
tTJ	tricellular tight junction, located at corners of three cells
P	Apparent permeability ($cm{\cdot}s^{-1}$)

References

1. Zeuthen, T. General models for water transport across leaky epithelia. *Int. Rev. Cytol.* **2002**, *215*, 285–317. [CrossRef]
2. Marchiando, A.M.; Graham, W.V.; Turner, J.R. Epithelial barriers in homeostasis and disease. *Annu. Rev. Pathol.* **2010**, *5*, 119–144. [CrossRef]
3. Agre, P.; Preston, G.M.; Smith, B.L.; Jung, J.S.; Raina, S.; Moon, C.; Guggino, W.B.; Nielsen, S. Aquaporin CHIP: The archetypal molecular water channel. *Am. J. Physiol.* **1993**, *265*, F463–F476. [CrossRef]
4. Angelow, S.; Ahlstrom, R.; Yu, A.S. Biology of claudins. *Am. J. Physiol. Renal Physiol.* **2008**, *295*, F867–F876. [CrossRef] [PubMed]
5. Ikenouchi, J.; Furuse, M.; Furuse, K.; Sasaki, H.; Tsukita, S.; Tsukita, S. Tricellulin constitutes a novel barrier at tricellular contacts of epithelial cells. *J. Cell Biol.* **2005**, *171*, 939–945. [CrossRef] [PubMed]
6. Furuse, M.; Izumi, Y.; Oda, Y.; Higashi, T.; Iwamoto, N. Molecular organization of tricellular tight junctions. *Tissue Barriers* **2014**, *2*, e28960. [CrossRef] [PubMed]
7. Varadarajan, S.; Stephenson, R.E.; Miller, A.L. Multiscale dynamics of tight junction remodeling. *J. Cell Sci.* **2019**, *132*. [CrossRef] [PubMed]
8. Furuse, M. Molecular basis of the core structure of tight junctions. *Cold Spring Harb. Perspect. Biol.* **2010**, *2*, a002907. [CrossRef] [PubMed]
9. Van Itallie, C.M.; Lidman, K.F.; Tietgens, A.J.; Anderson, J.M. Newly synthesized claudins but not occludin are added to the basal side of the tight junction. *Mol. Biol. Cell* **2019**, *30*, 1406–1424. [CrossRef]
10. Schulzke, J.D.; Gitter, A.H.; Mankertz, J.; Spiegel, S.; Seidler, U.; Amasheh, S.; Saitou, M.; Tsukita, S.; Fromm, M. Epithelial transport and barrier function in occludin-deficient mice. *Biochim. Biophys. Acta* **2005**, *1669*, 34–42. [CrossRef]
11. Günzel, D.; Fromm, M. Claudins and other tight junction proteins. *Compr. Physiol.* **2012**, *2*, 1819–1852. [CrossRef]

12. Rosenthal, R.; Milatz, S.; Krug, S.M.; Oelrich, B.; Schulzke, J.D.; Amasheh, S.; Günzel, D.; Fromm, M. Claudin-2, a component of the tight junction, forms a paracellular water channel. *J. Cell Sci.* **2010**, *123*, 1913–1921. [CrossRef]
13. Rosenthal, R.; Günzel, D.; Krug, S.M.; Schulzke, J.D.; Fromm, M.; Yu, A.S. Claudin-2-mediated cation and water transport share a common pore. *Acta Physiol.* **2017**, *219*, 521–536. [CrossRef] [PubMed]
14. Rosenthal, R.; Günzel, D.; Piontek, J.; Krug, S.M.; Ayala-Torres, C.; Hempel, C.; Theune, D.; Fromm, M. Claudin-15 forms a water channel through the tight junction with distinct function compared to claudin-2. *Acta Physiol.* **2020**, *228*, e13334. [CrossRef]
15. Schnermann, J.; Huang, Y.; Mizel, D. Fluid reabsorption in proximal convoluted tubules of mice with gene deletions of claudin-2 and/or aquaporin1. *Am. J. Physiol. Renal Physiol.* **2013**, *305*, F1352–F1364. [CrossRef]
16. Robben, J.H.; Knoers, N.V.; Deen, P.M. Cell biological aspects of the vasopressin type-2 receptor and aquaporin 2 water channel in nephrogenic diabetes insipidus. *Am. J. Physiol. Renal Physiol.* **2006**, *291*, F257–F270. [CrossRef]
17. Staehelin, L.A. Further observations on the fine structure of freeze-cleaved tight junctions. *J. Cell Sci.* **1973**, *13*, 763–786. [CrossRef]
18. Higashi, T.; Miller, A.L. Tricellular junctions: How to build junctions at the TRICkiest points of epithelial cells. *Mol. Biol. Cell* **2017**, *28*, 2023–2034. [CrossRef] [PubMed]
19. Krug, S.M.; Amasheh, S.; Richter, J.F.; Milatz, S.; Günzel, D.; Westphal, J.K.; Huber, O.; Schulzke, J.D.; Fromm, M. Tricellulin Forms a Barrier to Macromolecules in Tricellular Tight Junctions without Affecting Ion Permeability. *Mol. Biol. Cell* **2009**, *20*, 3713–3724. [CrossRef] [PubMed]
20. Krug, S.M. Contribution of the tricellular tight junction to paracellular permeability in leaky and tight epithelia. *Ann. N. Y. Acad. Sci.* **2017**, *1397*, 219–230. [CrossRef] [PubMed]
21. Ayala-Torres, C.; Krug, S.M.; Schulzke, J.D.; Rosenthal, R.; Fromm, M. Tricellulin Effect on Paracellular Water Transport. *Int. J. Mol. Sci.* **2019**, *20*, 5700. [CrossRef]
22. Higashi, T.; Tokuda, S.; Kitajiri, S.; Masuda, S.; Nakamura, H.; Oda, Y.; Furuse, M. Analysis of the 'angulin' proteins LSR, ILDR1 and ILDR2–tricellulin recruitment, epithelial barrier function and implication in deafness pathogenesis. *J. Cell Sci.* **2013**, *126*, 966–977. [CrossRef] [PubMed]
23. Stenger, C.; Pincon, A.; Hanse, M.; Royer, L.; Comte, A.; Koziel, V.; Olivier, J.L.; Pillot, T.; Yen, F.T. Brain region-specific immunolocalization of the lipolysis-stimulated lipoprotein receptor (LSR) and altered cholesterol distribution in aged LSR+/− mice. *J. Neurochem.* **2012**, *123*, 467–476. [CrossRef]
24. Nakatsu, D.; Kano, F.; Taguchi, Y.; Sugawara, T.; Nishizono, T.; Nishikawa, K.; Oda, Y.; Furuse, M.; Murata, M. JNK1/2-dependent phosphorylation of angulin-1/LSR is required for the exclusive localization of angulin-1/LSR and tricellulin at tricellular contacts in EpH4 epithelial sheet. *Genes Cells* **2014**, *19*, 565–581. [CrossRef]
25. Masuda, S.; Oda, Y.; Sasaki, H.; Ikenouchi, J.; Higashi, T.; Akashi, M.; Nishi, E.; Furuse, M. LSR defines cell corners for tricellular tight junction formation in epithelial cells. *J. Cell Sci.* **2011**, *124*, 548–555. [CrossRef]
26. Czulkies, B.A.; Mastroianni, J.; Lutz, L.; Lang, S.; Schwan, C.; Schmidt, G.; Lassmann, S.; Zeiser, R.; Aktories, K.; Papatheodorou, P. Loss of LSR affects epithelial barrier integrity and tumor xenograft growth of CaCo-2 cells. *Oncotarget* **2017**, *8*, 37009–37022. [CrossRef] [PubMed]
27. Kohno, T.; Konno, T.; Kojima, T. Role of Tricellular Tight Junction Protein Lipolysis-Stimulated Lipoprotein Receptor (LSR) in Cancer Cells. *Int. J. Mol. Sci.* **2019**, *20*, 3555. [CrossRef] [PubMed]
28. Shimada, H.; Satohisa, S.; Kohno, T.; Konno, T.; Takano, K.I.; Takahashi, S.; Hatakeyama, T.; Arimoto, C.; Saito, T.; Kojima, T. Downregulation of lipolysis-stimulated lipoprotein receptor promotes cell invasion via claudin-1-mediated matrix metalloproteinases in human endometrial cancer. *Oncol. Lett.* **2017**, *14*, 6776–6782. [CrossRef]
29. Mesli, S.; Javorschi, S.; Berard, A.M.; Landry, M.; Priddle, H.; Kivlichan, D.; Smith, A.J.; Yen, F.T.; Bihain, B.E.; Darmon, M. Distribution of the lipolysis stimulated receptor in adult and embryonic murine tissues and lethality of LSR-/- embryos at 12.5 to 14.5 days of gestation. *Eur. J. Biochem.* **2004**, *271*, 3103–3114. [CrossRef] [PubMed]
30. Gong, Y.; Himmerkus, N.; Sunq, A.; Milatz, S.; Merkel, C.; Bleich, M.; Hou, J. ILDR1 is important for paracellular water transport and urine concentration mechanism. *Proc. Natl. Acad. Sci. USA* **2017**, *114*, 5271–5276. [CrossRef]
31. Hempstock, W.; Sugioka, S.; Ishizuka, N.; Sugawara, T.; Furuse, M.; Hayashi, H. Angulin-2/ILDR1, a tricellular tight junction protein, does not affect water transport in the mouse large intestine. *Sci. Rep.* **2020**, *10*, 10374. [CrossRef]
32. Krug, S.M.; Fromm, M.; Günzel, D. Two-path impedance spectroscopy for measuring paracellular and transcellular epithelial resistance. *Biophys. J.* **2009**, *97*, 2202–2211. [CrossRef]
33. Stevenson, B.R.; Siliciano, J.D.; Mooseker, M.S.; Goodenough, D.A. Identification of ZO-1: A high molecular weight polypeptide associated with the tight junction (zonula occludens) in a variety of epithelia. *J. Cell Biol.* **1986**, *103*, 755–766. [CrossRef]
34. Tsukita, S.; Furuse, M. Occludin and claudins in tight-junction strands: Leading or supporting players? *Trends Cell Biol.* **1999**, *9*, 268–273. [CrossRef]
35. Amasheh, S.; Meiri, N.; Gitter, A.H.; Schöneberg, T.; Mankertz, J.; Schulzke, J.D.; Fromm, M. Claudin-2 expression induces cation-selective channels in tight junctions of epithelial cells. *J. Cell Sci.* **2002**, *115*, 4969–4976. [CrossRef] [PubMed]
36. Furuse, M.; Oda, Y.; Higashi, T.; Iwamoto, N.; Masuda, S. Lipolysis-stimulated lipoprotein receptor: A novel membrane protein of tricellular tight junctions. *Ann. N. Y. Acad. Sci.* **2012**, *1257*, 54–58. [CrossRef] [PubMed]
37. Kirschner, N.; Rosenthal, R.; Furuse, M.; Moll, I.; Fromm, M.; Brandner, J.M. Contribution of tight junction proteins to ion, macromolecule, and water barrier in keratinocytes. *J. Investig. Dermatol.* **2013**, *133*, 1161–1169. [CrossRef]

38. Ikenouchi, J.; Sasaki, H.; Tsukita, S.; Furuse, M.; Tsukita, S. Loss of occludin affects tricellular localization of tricellulin. *Mol. Biol. Cell* **2008**, *19*, 4687–4693. [CrossRef]
39. Nakatsu, D.; Kano, F.; Shinozaki-Narikawa, N.; Murata, M. Pyk2-dependent phosphorylation of LSR enhances localization of LSR and tricellulin at tricellular tight junctions. *PLoS ONE* **2019**, *14*, e0223300. [CrossRef]
40. Weth, A.; Dippl, C.; Striedner, Y.; Tiemann-Boege, I.; Vereshchaga, Y.; Golenhofen, N.; Bartelt-Kirbach, B.; Baumgartner, W. Water transport through the intestinal epithelial barrier under different osmotic conditions is dependent on LI-cadherin trans-interaction. *Tissue Barriers* **2017**, *5*, e1285390. [CrossRef]
41. Vereshchaga, Y.; Arnold, N.; Baumgartner, W. Physiological relevance of epithelial geometry: New insights into the standing gradient model and the role of LI cadherin. *PLoS ONE* **2018**, *13*, e0208791. [CrossRef] [PubMed]
42. Higashi, T.; Katsuno, T.; Kitajiri, S.; Furuse, M. Deficiency of angulin-2/ILDR1, a tricellular tight junction-associated membrane protein, causes deafness with cochlear hair cell degeneration in mice. *PLoS ONE* **2015**, *10*, e0120674. [CrossRef] [PubMed]
43. Furuse, M.; Sasaki, H.; Tsukita, S. Manner of interaction of heterogeneous claudin species within and between tight junction strands. *J. Cell Biol.* **1999**, *147*, 891–903. [CrossRef] [PubMed]
44. Krug, S.M.; Hayaishi, T.; Iguchi, D.; Watari, A.; Takahashi, A.; Fromm, M.; Nagahama, M.; Takeda, H.; Okada, Y.; Sawasaki, T.; et al. Angubindin-1, a novel paracellular absorption enhancer acting at the tricellular tight junction. *J. Control. Release* **2017**, *260*, 1–11. [CrossRef]
45. Kreusel, K.M.; Fromm, M.; Schulzke, J.D.; Hegel, U. Cl- secretion in epithelial monolayers of mucus-forming human colon cells (HT-29/B6). *Am. J. Physiol.* **1991**, *261*, C574–C582. [CrossRef] [PubMed]
46. Günzel, D.; Stuiver, M.; Kausalya, P.J.; Haisch, L.; Krug, S.M.; Rosenthal, R.; Meij, I.C.; Hunziker, W.; Fromm, M.; Müller, D. Claudin-10 exists in six alternatively spliced isoforms that exhibit distinct localization and function. *J. Cell Sci.* **2009**, *122*, 1507–1517. [CrossRef] [PubMed]

International Journal of
Molecular Sciences

MDPI

Article

Effect of Amyloid-β Monomers on Lipid Membrane Mechanical Parameters–Potential Implications for Mechanically Driven Neurodegeneration in Alzheimer's Disease

Dominik Drabik [1,*], Grzegorz Chodaczek [2] and Sebastian Kraszewski [3]

1 Laboratory of Cytobiochemistry, Faculty of Biotechnology, University of Wrocław, F. Joliot-Curie 14a, 50-383 Wrocław, Poland
2 Łukasiewicz Research Network-PORT Polish Center for Technology Development, Stabłowicka 147, 54-066 Wrocław, Poland; Grzegorz.Chodaczek@port.lukasiewicz.gov.pl
3 Department of Biomedical Engineering, Faculty of Fundamental Problems of Technology, Wrocław University of Science and Technology, Pl. Grunwaldzki 13, 50-377 Wrocław, Poland; Sebastian.Kraszewski@PWr.edu.pl
* Correspondence: Dominik.Drabik@UWr.edu.pl

Abstract: Alzheimer's disease (AD) is a neurodegenerative disease that results in memory loss and the impairment of cognitive skills. Several mechanisms of AD's pathogenesis were proposed, such as the progressive accumulation of amyloid-β (Aβ) and τ pathology. Nevertheless, the exact neurodegenerative mechanism of the Aβ remains complex and not fully understood. This paper proposes an alternative hypothesis of the mechanism based on maintaining the neuron membrane's mechanical balance. The incorporation of Aβ decreases the lipid membrane's elastic properties, which eventually leads to the impairment of membrane clustering, disruption of mechanical wave propagation, and change in gamma oscillations. The first two disrupt the neuron's ability to function correctly while the last one decreases sensory encoding and perception enabling. To begin discussing this mechanical-balance hypothesis, we measured the effect of two selected peptides, Aβ-40 and Aβ-42, as well as their fluorescently labeled modification, on membrane mechanical properties. The decrease of bending rigidity, consistent for all investigated peptides, was observed using molecular dynamic studies and experimental flicker-noise techniques. Additionally, wave propagation was investigated with molecular dynamic studies in membranes with and without incorporated neurodegenerative peptides. A change in membrane behavior was observed in the membrane system with incorporated Aβ.

Keywords: membrane mechanics; molecular dynamics; flicker-noise spectroscopy; neurodegeneration; amyloid-beta peptides; pressure wave; giant unilamellar vesicles

Citation: Drabik, D.; Chodaczek, G.; Kraszewski, S. Effect of Amyloid-β Monomers on Lipid Membrane Mechanical Parameters–Potential Implications for Mechanically Driven Neurodegeneration in Alzheimer's Disease. *Int. J. Mol. Sci.* **2021**, *22*, 18. https://dx.doi.org/10.3390/ijms22010018

Received: 30 November 2020
Accepted: 17 December 2020
Published: 22 December 2020

Publisher's Note: MDPI stays neutral with regard to jurisdictional claims in published maps and institutional affiliations.

1. Introduction

Alzheimer's disease (AD) is an irreversible neuropathological disease with a progressive loss of the structure and function of neurons that slowly leads to memory and cognitive skills impairment [1]. AD is considered the most common cause of dementia, with increasing age being the most significant factor for AD occurrence. Many aspects of AD's pathophysiology have been investigated and understood. However, several knowledge gaps still exist [2]. At the microscale, the AD's brain is characterized by the presence of both amyloid plaques and neurofibrillary tangles [3]. Amyloid plaques display a broad range of morphological and biochemical characteristics and contain numerous proteins, mostly amyloid-β (Aβ). It is generally believed that AD's pathogenesis is related to alternations in APP (Aβ precursor protein) processing, which results in the progressive accumulation of Aβ protein [4]. The most common form of this protein is 40 amino acids long and is called Aβ-40. Less common, yet believed to be associated with the disease, is the most hydrophobic and toxic peptide isoform, Aβ-42 [2]. Due to its physical characteristics,

Aβ often acquires the configuration of β-pleated sheets and shows a greater tendency to aggregate, forming the core of the amyloid plaque composed of Aβ oligomers and fibrils. It is known that clearance abnormality leads to the accumulation of Aβ in the brain and central nervous system. Despite several studies showing the neurotoxicity of various forms of Aβ, the mechanism through which Aβ monomers, oligomers and other APP metabolites might lead to synaptic damage and neurodegeneration is not completely clear [5]. Several possibilities are under investigation, including alternation in signaling pathways related to synaptic plasticity, neuronal cell death, neurogenesis, and ion homeostasis disruption [6,7]. On the other hand, recent clinical trials showed that the elimination of Aβ does not affect the progression of AD [8]. To this end, the focus has shifted to the tau (τ) protein as a secondary pathogenic event that causes neurodegeneration [9]. While this is important for developing therapeutic strategies, it does not change the fact that Aβ proteins are mainly responsible for the further development of AD.

As stated previously, plaques' major components are the Aβ peptides derived from the APP proteolytic processing at lipid membrane domains [10]. The membrane's ability to dynamically cluster its components regulates the spatial and temporal assembly of signaling and trafficking molecules. The formation of short-lived signaling platforms can be vital in this case [11,12]. These platforms can be classical sphingolipid-cholesterol ordered domains, ceramide-rich platforms, or other areas with different biophysical properties [13]. For instance, gangliosides' presence was reported to increase the incorporation of Aβ-42 [14]. This is especially important for cell signaling, axon sorting and guidance, neural development, and synaptic plasticity [15]. Furthermore, microdomains in neurons are required to maintain dendritic spines and healthy synapses making them essential for neural communication, memory, and learning [16]. A membrane, to preserve the ability to form such microdomains (signaling platforms) spontaneously, would require a certain mechanical balance. The disruption of such a delicate balance could, in this case, lead to the slow and progressive loss of membrane functions [17]. Interestingly, the composition of detergent-resistant membranes purified from AD brains is abnormal–they are more ordered and more viscous [18]. It was also shown that neurons in AD have a significantly different membrane composition, including a lower level of sphingomyelin and a higher level of ceramides [19,20], which are known to significantly alter mechanical properties. This implies a potential disturbance of mechanical balance. Furthermore, it was recently suggested that the Hodgkin Huxley model of nerve propagation, based on local ion current flow, does not fully explain how membrane potential cause the opening and closing of the ionic channels. Based on the propagation of pressure waves and membrane mechanical properties, a complement model hypothesis was proposed [21,22].

In this paper, we decided to investigate whether Aβ monomers' accumulation in the membrane disrupts the membrane's mechanical balance. We focused on measuring the mechanical changes of POPC membranes as they can be used as a starting point for discussion or more complex approaches with neuron-mimicking membranes. We investigated the effect of Aβ monomers on mechanical properties, mainly focusing on the bending rigidity coefficient. These parameters were measured using flicker-noise spectroscopy, which links spontaneous bilayer fluctuations with its mechanical properties–and Molecular Dynamics simulations of whole lipid vesicles. Furthermore, we have simulated the pressure wave propagation on membranes with and without Aβ peptide to investigate the latter's effect on membrane behavior. Even if the proposed study is not thoroughgoing research, we aimed to draw attention to another vital aspect, namely mechanically-driven molecular phenomena during neurodegeneration in AD.

2. Results & Discussion

2.1. *Effect of Aβ Peptides on Structural Parameters*

Table 1 presents the calculated basic structural properties of a membrane with incorporated different Aβ monomers. Snapshots of simulated systems are presented in Figure 1. Both membrane thickness and area per lipid (APL) of vesicles with incorporated peptides differed with statistical significance from other populations. Specifically, one-way ANOVA reported the difference between the means of membrane thickness and APL as statistically significant. The following post-hoc Tukey test reported the difference to be significant between all investigated populations. It also should be noted that the difference of the POPC population's parameters from other Aβ populations' parameters is especially conspicuous. The membrane thickness was higher in vesicles with Aβ, suggesting that their presence contributed to the elevation of either whole lipid molecules or just phosphorus atoms in the bilayer due to bilayer remodeling. While APL was lower when compared to the base system, this result is not surprising as additional particles in the bilayer contributed to the more tightly packed conformation. These results somewhat contradict the literature, as it was suggested that binding of Aβ peptide to the membrane might result in both compression of the bilayer (higher APL) and making it thinner (lower membrane thickness) [7]. On the other hand, it was reported that both membrane properties (i.e., lipid bilayer thickness) regulate the generation and surface-induced aggregation of Aβ peptides and the incorporation of Aβ peptides induces membrane remodeling, such as elevation of lipids in the vicinity of the peptides [23]. Figure 2 shows the bilayer profiles of systems with different Aβ peptides. Interestingly, the peptides are localized in the whole interphase region. This positioning in the lipid bilayer was in agreement with the literature, as Aβ peptides in monomeric form have been reported to localize in the intermediate region between the lipid head group and acyl chains. Only when the dimer was formed did it slowly submerge deeper into the bilayer, eventually reaching the transmembrane state [24]. All of the investigated vesicles sustained their quasi-spherical geometry.

Table 1. Summary of Calculated Parameters from a Molecular Dynamics (MD) Study and Flicker Noise Spectroscopy Measurements for POPC membrane with incorporated Aβ (amyloid-β) monomers.

System Description	Membrane Thickness [nm]	APL of Vesicle [Å^2]	κ (MD) [J]	κ (Flicker, SA) [J]	κ (Flicker, AVB) [J]
POPC [25]	(3.43 ± 0.01)	(61.4 ± 0.1)	$(7.3) \times 10^{-20}$	$(11.7 \pm 6.9) \times 10^{-20}$	$(10.5 \pm 2.2) \times 10^{-20}$
POPC with 10 m% Aβ-40	(3.63 ± 0.01)	(56.4 ± 0.1)	$(4.3) \times 10^{-20}$	$(3.7 \pm 1.4) \times 10^{-20}$	$(3.6 \pm 2.0) \times 10^{-20}$
POPC with 10 m% Aβ-42	(3.65 ± 0.01)	(56.3 ± 0.1)	$(4.8) \times 10^{-20}$	$(3.1 \pm 1.9) \times 10^{-20}$	$(3.3 \pm 2.1) \times 10^{-20}$
POPC with 10 m% Aβ-40-TAMRA	(3.66 ± 0.01)	(56.0 ± 0.1)	$(4.0) \times 10^{-20}$	$(3.4 \pm 1.9) \times 10^{-20}$	$(3.1 \pm 1.3) \times 10^{-20}$

APL, area per lipid; κ, bending rigidity coefficient; MD, molecular dynamics; AVB, average-based approach; SA, statistical approach.

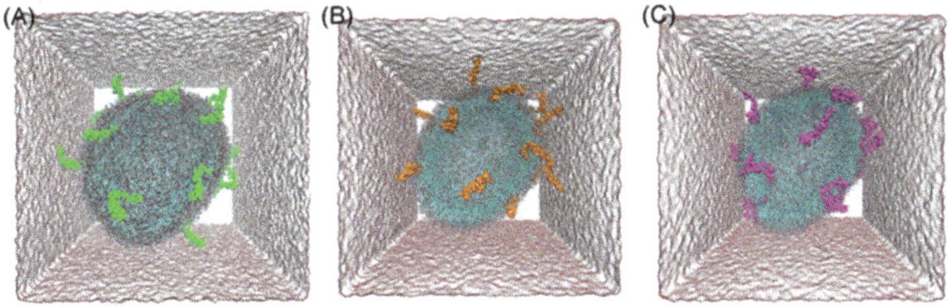

Figure 1. Snapshots of POPC vesicle systems with incorporated 10 m% Aβ monomers: (**A**) Aβ-40, (**B**) Aβ-42 and (**C**) Aβ-40-TAMRA.

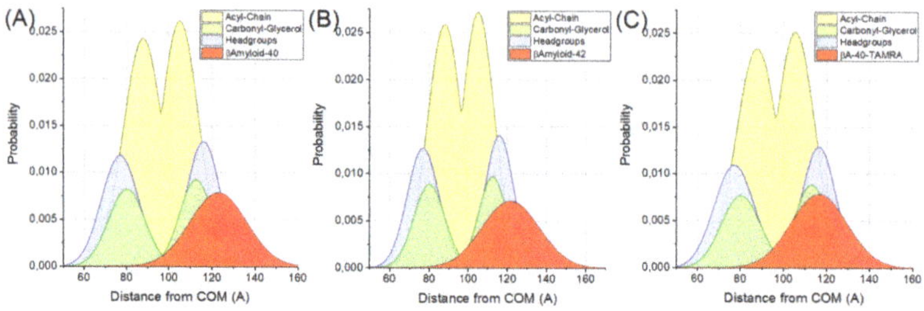

Figure 2. Probabilities of specified membrane regions for POPC vesicles with (**A**) Aβ-40, (**B**) Aβ-42 and (**C**) Aβ-40-TAMRA peptides incorporated in function of distance from center of mass. Specified regions are headgroups (blue), carbonyl-glycerol (green), acyl-chain (yellow) and peptides (red).

2.2. Effect of Aβ Peptides on Bending Rigidity Coefficient

The bending rigidity coefficient was determined using both computational and experimental techniques. These results are presented in Table 1. Power spectra of investigated MD, which were used for bending rigidity determination, are presented in Figure 3. The analysis of these systems showed that the bending rigidity of vesicles with Aβ peptides was almost twice as low as in the reference POPC system. However, there was no difference in mechanical properties between the systems with incorporated peptides. These results are in agreement with our experimental results. When measured using flickernoise spectroscopy, the bending rigidity differed with statistical significance between the populations using both tests: ANOVA and Kruskal–Wallis. The post-hoc Tukey test showed that the difference only occurs between the reference POPC vesicle and the systems with Aβ peptides. There was no statistically significant difference between systems with Aβ monomers. Results were similar for both statistical (SA) and average-based (AVB) approaches. More details are presented in Section 1 Figure S1 of Supporting Information. This decrease of membrane bending rigidity due to the presence of Aβ peptides is in agreement with the literature [26]. This effect was reported to be more robust in the case of oligomers and fibrils, but was still present in the case of the peptides. Aβ-42 was also reported to decrease Young's modulus in neural cells [27]. Furthermore, it was reported that Aβ oligomers are inducing neural elasticity changes [28]. These results and references are consistent with our hypothesis stating that Aβ peptides do influence the membrane's mechanical properties. Moreover, it was recently shown that remodeling of the membranes occurs after the incorporation of Aβ peptides [23]. However, this phenomenon was only observed in bilayers with low or medium bending rigidity (such as POPC) and not in membranes with higher bending rigidity (DLPC-1,2-dilauroyl-sn-glycero-3-phosphocholine). Combining the above with our results suggests that incorporating Aβ, which decreases the bending rigidity, could lead to further progression of membrane modeling and local disruption of membrane topology. This could result in the disruption of mechanical balance that could halve the ability to form spontaneous signaling platforms. While it was argued that Aβ incorporation is mostly electrostatically driven [7,29], it should also be noted that remodeling of membrane occurred in membranes with lower bending rigidity. It could somewhat explain why the likelihood of neurodegenerative disease occurrence increases with age. Aging of neural cells–aging of cells in general–does not tend to change their electrochemical potential but is known to change their mechanical properties [30]. Additionally, the progressing decrease of the bending rigidity (caused by aging) might influence the curvature of the lipid bilayers and/or induce packing defects, resulting in the exposure of hydrophobic clefts in the vesicle surface. Both factors were reported to promote the interaction of Aβ with the bilayer and its aggregation properties as well [31,32]. It should be noted that such a change in lipid membrane mechanical properties is especially relevant in neurons. It can modify both elasticity and viscosity, which influences signal transduction, leading to the loss of synaptic plasticity and impairment of neuronal signal propagation [33]. They can directly affect transmembrane proteins' functioning, sever metabolic pathways, and disrupt transport between the membrane itself.

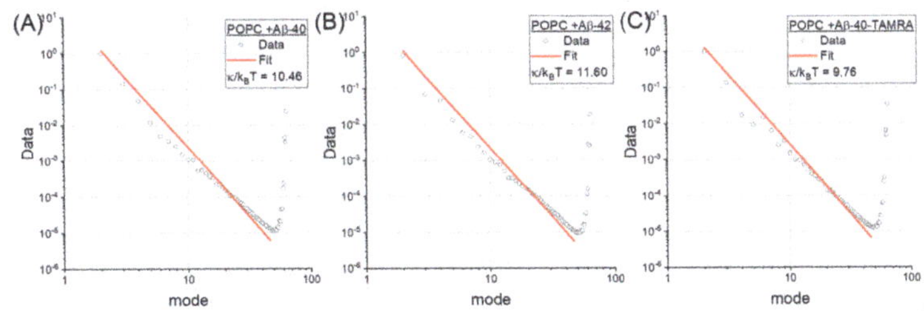

Figure 3. Power spectra along with model fits of investigated Molecular Dynamics systems with (**A**) Aβ-40, (**B**) Aβ-42 and (**C**) Aβ-40-TAMRA monomers, which were used for κ (bending rigidity) determination.

2.3. *Effect of Aβ Peptides Pressure Wave Propagation*

Finally, we investigated the effect of Aβ monomers' presence in the membrane on propagation of the mechanical wave. According to Barz et al. [21] a mechanical (pressure) wave is necessary for the neuronal membrane to trigger ion pumps. To this end, two systems were subjected to the effect of pressure wave induced by water slab velocity change. One of the systems had Aβ monomers incorporated. The second did not. A snapshot and evolution in time of the system with incorporated peptides is presented in Figure 4. The evolution of amplitude of the wave observed on the membrane is presented in Figure 5B. It can be clearly seen that the amplitude of the wave is oscillating, reaching local maximum and minimum values alternately. However, a significant difference between the system with incorporated peptides and the control system can be observed. The maximum values of amplitude occur significantly later in the control system. Furthermore, in the system with Aβ peptides, additional increases in the amplitude were observed that were not seen in the control system (see Figure 5C). This strongly suggests that Aβ peptides' presence influences the wave propagation and the membrane's response to pressure waves. If a sufficient number of peptides were incorporated, this could differ the membrane's response to disrupt nerve impulse propagation by inducing pump response too fast. Additionally, it could alter the membrane fluctuations to a point where a shift in membrane resonance frequency would occur. It was recently hypothesized that gamma neural oscillations (30–55 Hz) are responsible for sensory encoding and perception enabling [34]. Changes in gamma oscillations can be observed in several neurodegenerative diseases [35,36]. Interestingly, membrane properties strongly determine the characteristics of emergent gamma oscillations [37] and brain stimulation with gamma oscillations was reported to improve spatial and recognition memory during AD [38]. Our results suggest a very strict dependency between the lipid membrane's mechanical properties and the resonance frequency and behavior of the membrane. Moreover, the results strongly support the alternative hypothesis of nerve propagation based on membrane mechanical properties and pressure wave propagation.

Figure 4. Snapshots of POPC system with incorporated Aβ peptides throughout pressure wave propagation. In panel (**A**) water particles with increased velocity are marked blue. In panel (**B**) a snapshot of system after 13 ps is presented. In panel (**C**) a final snapshot of simulation after 3 ns is presented.

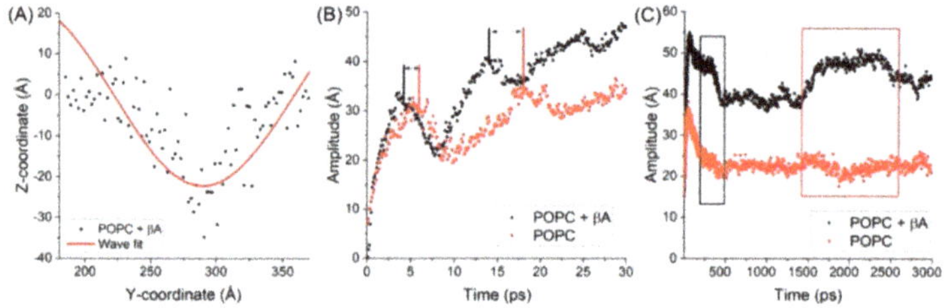

Figure 5. In panel (**A**) the fitting of sin function to histogrammed phosphorus atoms' positions to determine the characteristics of the wave. In panel (**B**) and (**C**) evolution of the amplitude is presented in shorter and longer times of the simulations, respectively.

3. Materials and Methods

3.1. Materials

Lipid POPC (1-palmitoyl-2-oleoyl-glycero-3-phosphocholine) was purchased from Avanti Polar Lipids (Alabaster, AL, USA). Fluorescent probe Atto488-DOPE was purchased from Atto-Tech (Siegen, Germany). Peptides β-Amyloid 1–40 and β-Amyloid 1–42 (Human) were purchased from Sigma-Aldrich (Poznań, Poland). Labelled peptide β-Amyloid 1-40-TAMRA was purchased from Eurogentec (Seraing, Belgium). Detergent DOTM (Decyl-B-D-1-Thiomaltopyranoside) was purchased from Sigma Aldrich. Bio-Beads SM-2 were purchased from Bio-Rad (Warszawa, Poland). PBS tablets were purchased from VWR (Gdańsk, Poland).

3.2. Preparation of Giant Unilamellar Vesicles (GUVs)

A modified electroformation method was used to enable the incorporation of Aβ peptides [39]. Briefly, 20 µL of chosen lipid and detergent mixture in chloroform were deposited in small quantities (as 2 µL droplets) onto platinum electrodes. The concentration of lipid POPC was 1 mg/mL, while the concentration of DOTM detergent was calculated so that its final concentration in the electroformation chamber was equal to 75 µM. Two electrodes were set parallel to one another at a distance of 5 mm. The electrodes were kept for 1 h under reduced pressure to remove traces of organic solvents. Next, the electrodes were immersed in 400 mM sucrose solution. This was followed by applying AC voltage to electrodes with 1 Hz frequency and 1 V amplitude. The voltage was increased by 1 V every hour up to 4 V [40]. After the electroformation, chambers were left for 1 h without an electrical field applied to allow the descent of vesicles from electrodes. This was followed by buffer exchange and peptide incorporation. Drops of GUVs solution (50 µL) were transferred to 100 µL of 1.5 × PBS solution with 75 µM DOTM. The peptides were dissolved in 1% NH_4OH solution and sonicated for 30 s. This was followed by the addition of 5× PBS buffer to obtain 1× PBS. The peptides were added to GUVs/PBS solution in such quantity to obtain 10 m% with respect to lipids and incubated for 12 h at room temperature. After that, DOTM removal was carried out using BioBeads in two batches. More detailed verification of Aβ monomers incorporation in GUVs can be found in Supporting Information in Section 2 Figures S2–S9.

3.3. Confocal Microscopy Imaging and Acquisition

A Cell Observer SD spinning disk confocal microscope (Zeiss, Jena, Germany) equipped with a Plan-Apochromat 100×/1.46 oil immersion objective (Zeiss) was used for vesicle recording. 512 × 512 pixels images were recorded with an EMCCD camera (Rolera, QImaging, British Columbia, Canada) using 2 × 2 binning with 0.133 μm pixel size with a video integration time of 30 ms. At least 5000 images were recorded for each vesicle. Samples were illuminated with 488 nm laser and emitted light passed through 527/54 filter. All samples were measured at 23 ± 1 °C. All measurements have been performed in a dedicated PTFE observation chamber with 300 μm height to reduce the effect of uncontrolled vesicle movements. The value of the depth of focus was equal to 0.85 μm. To enhance the quality of analysis the radius of a vesicle was calculated for each image. In case the change of radius was considered as an outliner, the image in the series was discarded from further analysis. It occurred due to misdetection caused by noise or other reasons described in previous work [41].

3.4. Flicker-Noise Spectroscopy Analysis

The flicker-noise spectroscopy technique is based on the analysis of a vesicle shape fluctuations over time. It is used to determine the bending rigidity coefficient from those fluctuations using the Helfrich's theory. Measurements were performed following our established protocol [41]. Briefly, the membrane fluctuation spectrum was extracted from every single recorded image of the same lipid vesicle using custom software. To calculate the bending rigidity coefficient from a set of time-lapsed two-dimensional images a correlation with three-dimensional membrane elasticity model was established. This was achieved by means of the angular autocorrelation function. The bending rigidity coefficient κ and membrane tension σ can be determined using two approaches, namely the statistical [41,42] and the average-based approach [41,43]. While the average-based approach might seem more straightforward, the main advantage of the statistical one is the histograms that show the characteristics of vesicle fluctuations and are in agreement with the model.

3.5. Molecular Dynamics Simulations

Molecular dynamics (MD) has been effectively used as a tool for studies of Aβ structures [44]. However, due to a high computational cost, many of these studies are performed on shorter Aβ segments in order to decrease the system size. Nevertheless, the MD simulations are used to study Aβ structures such as peptides [45], dimers [46] and oligomers [47]. In our work, we adopted a structure of Aβ peptides presented by Crescenzi et al. [48]. The full-atomistic MD simulation was performed using NAMD 2.9 [49] software with CHARMM36 united-atom force field [50] under NPT conditions (constant: Number of particles, Pressure and Temperature). Two different types of simulation were carried out, namely peptide incorporation on planar bilayer simulations and lipid vesicle simulation.

For peptides on planar bilayer simulations: a planar POPC membrane system consisting of 200 lipid molecules (100 on each of leaflets) was used. To determine peptide docking in the bilayer, several simulations were performed with various placement of the investigated peptide on the bilayer. The simulation was run till the peptide was incorporated into the bilayer and remained incorporated for at least 30 ns or it was not incorporated into the bilayer. Successful simulations for Aβ-40, Aβ-42, and Aβ-40 TAMRA were carried out for 62, 46, and 51 ns, respectively. The obtained stable position of the investigated peptide in the bilayer was later used for its manual incorporation in the vesicle system. More detailed information about systems setup and properties is presented in Supporting Information in Section 3 Figures S10–S15.

For lipid vesicle simulations: a POPC vesicle was modelled as a liposome of 20 nm radius both sides hydrated with TIP3P water molecules, giving a final simulation box of 30 nm^3. The vesicle system was selected over the planar system, as it differs in curvature, which can significantly change attached peptide's behavior [51]. Furthermore, we showed that, in some cases, results from vesicles systems agreed with the experimental data, while for the planar system results were completely different [25]. Three-dimensional periodic boundary conditions were applied to deal with potential energy disruption due to the origin cell discontinuity. The vesicle system was created using a custom script in Matlab. Starting APL was set as 68.1 on average [52], but was corrected accounting for the effect of vesicle's curvature. The APL value was multiplied by 0.95 for the inner and by 1.05 for the outer leaflets, respectively. The vesicle system was equilibrated prior to the addition of Aβ peptides for 100 ns. This was followed by peptide incorporation, which was done using a custom script. Peptides were equally distributed on the vesicle, merged with the liposome system and solvated. After peptide incorporation vesicles were equilibrated for additional 20 ns, followed by running for at least 10 ns and then analyzed. To determine the stable equilibration time-point, six selected parameters (vesicle radius, the thickness of lipid bilayer, mean values and standard deviations of both inner and outer leaflets) were continuously monitored. More detailed information about systems setup and properties is presented in Supporting Information in Section 4 Figures S16–S18.

For pressure wave propagation simulations: simulations were performed according to the established procedure [53,54]. Specifically, the planar system with incorporated β peptides was multiplied 4 times with an additional water slab in Z-axis. After equilibration of the membrane system was switched from NPT to NVE (constant: Number of particles, Volume and Energy) conditions. The pressure wave was modelled as the momentum change of water particles in Z-axis by averaged velocity Δvz defined by Equation (1), where I is pressure impulse, A denotes the area of changed water particles, m the mass of water, and N the number of changed water particles.

$$\Delta v_z = \frac{I \cdot A}{m \cdot N} \quad (1)$$

The simulated pressure wave was equal to 10 $\mu N/m^2 \cdot s$ (1 mPa·s). An evolution of bilayer bending and return to equilibrium was investigated. The position of phosphorus atoms was used to bin membrane position in the OX plane. The obtained bending characteristic of membrane was fitted with the sine function to parametrize the system's behavior and evolution in time.

3.6. Determination of Bending Rigidity Coefficient in MD

To determine the bending rigidity of model lipid vesicles, we adopted an algorithm originally developed by Braun & Sachs [25,55]. It has an advantage over other approaches [56–58] as it determines mechanical properties based on fluctuations of the bilayer within the vesicle, which can be different than for planar lipids [25]. In short, each lipid is described by a vector spreading from the head (phosphorus atom) up to tail position (midpoint of both 16th carbon atoms in each of tails). This is followed by the discrete surface representation θ, φ using a grid. For each time-point, the surface of fluctuations is established by detecting of fitted sphere's origin point, converting bilayer fluctuations into spherical coordinates and subtracting the radius value. Finally, the average of both inner and outer leaflets fluctuations is calculated. This is followed by spectral harmonics analysis (SPHA) for calculated fluctuations. Eventually, the Helfrich's approach is used by establishing spherical harmonic coefficients a_{lm}. The obtained a_{lm} undulation power spectrum can be interpreted according to the Helfrich continuum model for undulations on a sphere with vanishing spontaneous curvature.

3.7. Determination of Basic Structural Parameters

Additionally, basic structural parameters were determined from performed MD simulations to establish the effect of Aβ peptides incorporation. These include membrane thickness, area per lipid and vesicle density profiles. For each frame position a sphere fit to phosphorus atoms in inner leaflet, in outer leaflet and to both was done in order to obtain radius for inner leaflet, for outer leaflet and for whole vesicle, respectively. Membrane thickness is calculated as a difference between the radius of outer and inner layers. The area per lipid for the whole vesicle was calculated according to Equation (2) using Braun and Sachs approach [55].

$$ALP_{vesicle} = \frac{4\pi r^2_{vesicle}}{\frac{1}{2}(n_{L,inner} + n_{L,outer})} \tag{2}$$

To determine density vesicle profiles, three crucial zones of each vesicle area were selected: Head-groups, Carbonyl-Glycerol, and Acyl-Chain, respectively. The distance from radius was calculated for each particle and then histogrammed. Obtained results were followed by the normal distribution fit.

3.8. Statistics

To test for the significant difference between the parameters, one-way ANOVA with the significance level set at 0.05 was used for presenting the normal distribution of data. The Tukey test was used as a post hoc test. The Kruskal–Wallis test with the significance level set at 0.05 was additionally used for data with the risk of not the following a normal distribution. All statistical analysis were performed using the OriginPro 2015 (OriginLabs) software. Averaged values are presented with standard deviation values.

4. Conclusions

In this paper, we investigated the effect of Aβ monomers on the mechanical properties of the POPC vesicle. Additionally, we have investigated the change in basic membrane parameters, such as membrane thickness and APL. We showed that membrane thickness increases and APL decreases after the incorporation of Aβ. We postulate that this results from the bilayer's mechanical remodeling after Aβ monomer incorporation reported previously in the literature. We observed a decrease in the bending rigidity coefficient after incorporating Aβ in both MD simulation and the flicker-noise experiment. Since it was reported in the literature that Aβ-induced membrane remodeling is more likely when bending rigidity is smaller, we believe that the progressing decrease of bending rigidity might be considered a driving factor of neurodegeneration progression. We managed to prove that incorporating Aβ peptides influences the mechanical properties of lipid membranes, suggesting that membrane's mechanical properties may play a more important role in neurodegenerative disorders. Finally, we have investigated the difference in wave propagation on membranes with and without incorporated Aβ peptides. We showed that the presence of the peptides did change the behavior of the system. This could lead to an impediment of nerve impulse propagation and/or shift in membrane resonance frequency and, as a result, disrupt gamma oscillations. In summary, this paper aims to draw attention to an important hypothesis that mechanically driven molecular phenomena that originate from the membrane could be a critical factor in the pathogenesis of AD. We believe that we presented satisfactory results to support this claim. We acknowledge that this study was limited to the effect on POPC lipid membrane only. Further studies should be focused on more biologically relevant membranes, specifically on the disruption of their ability to cluster dynamically. The effect of mechanical changes on lipid metabolism/signaling should also be addressed in future studies.

Supplementary Materials: Supplementary materials can be found at https://www.mdpi.com/1422-0067/22/1/18/s1, Supporting Information: (1) Peptide incorporation into planar POPC bilayer, (2) System setup for POPC vesicles with incorporated peptides, (3) Verification of Aβ incorporation in GUVs, (4) Bending rigidity measurements using flicker-noise spectroscopy.

Author Contributions: Conceptualization, D.D. and S.K.; methodology, D.D. and G.C.; software, D.D.; validation, D.D.; formal analysis, D.D.; investigation, D.D.; resources, D.D.; data curation, D.D.; writing—original draft preparation, D.D.; writing—review and editing, G.C. and S.K.; visualization, D.D.; supervision, S.K.; project administration, D.D.; funding acquisition, D.D. All authors have read and agreed to the published version of the manuscript.

Funding: This work was possible thanks to the financial support from the National Science Centre (Poland) grants Nos. 2016/21/N/NZ1/02767 and 2018/30/E/NZ1/00099.

Data Availability Statement: Most of the data is available in the manuscript supplementary information. The simulation data presented in this study are available on request from the corresponding author. This data are not publicly available due to GBs files sizes.

Acknowledgments: Numerical resources for Molecular Dynamics simulations were granted by Wroclaw Centre of Networking and Supercomputing, grant No. 274.

Conflicts of Interest: The authors declare no conflict of interest.

Abbreviations

AD	Alzheimer's disease
Aβ	amyloid-β
SA	statistical approach
AVB	average-based approach
MD	molecular dynamics
APL	area per lipid

References

1. Irvine, G.B.; El-Agnaf, O.M.; Shankar, G.M.; Walsh, D.M. Protein Aggregation in the Brain: The Molecular Basis for Alzheimer's and Parkinson's Diseases. *Mol. Med.* **2008**, *14*, 451–461. [CrossRef] [PubMed]
2. Sanabria-Castro, A.; Alvarado-Echeverria, I.; Monge-Bonilla, C. Molecular Pathogenesis of Alzheimer's Disease: An Update. *Ann. Neurosci.* **2017**, *24*, 46–54. [CrossRef] [PubMed]
3. Masters, C.; Simms, G.; Weinman, N.; Multhaup, G.; McDonald, B.; Beyreuther, K. Amyloid plaque core protein in Alzhemier disease and Down syndrome. *Proc. Natl. Acad. Sci. USA* **1985**, *82*, 4245–4249. [CrossRef] [PubMed]
4. Crews, L.; Masllah, E. Molecular mechanisms of neurodegeneration in Alzheimer's disease. *Hum. Mol. Genet.* **2010**, *19*, R12–R20. [CrossRef] [PubMed]
5. Peña, F.; Gutiérrez-Lerma, A.; Quiroz-Baez, R.; Arias, C. The role of beta-amyloid protein in synaptic function: Implications for Alzheimer's disease therapy. *Curr. Neuropharmacol.* **2006**, *4*, 149–163. [CrossRef]
6. Maccioni, R.B.; Cambiazo, V. Role of microtubule-associated proteins in the control of microtubule assembly. *Physiol. Rev.* **1995**, *75*, 835–864. [CrossRef]
7. Niu, Z.Z.; Zhang, Z.; Zhao, W.; Yang, J. Interactions between amyloid β peptide and lipid membranes. *Biochim. Biophys. Acta (BBA) Biomembr.* **2018**, *1860*, 1663–1669. [CrossRef]
8. Holmes, C.C.; Boche, D.; Wilkinson, D.; Yadegarfar, G.; Hopkins, V.; Bayer, A.; Jones, R.W.; Bullrock, R.; Love, S.; Neal, J.W. Long-term effects of Aβ42 immunisation in Alzheimer's disease: Follow-up of a randomised, placebo-controlled phase I trial. *Lancet* **2008**, *372*, 216–223. [CrossRef]
9. Roberson, E.D.; Scearce-Levie, K.; Palop, J.J.; Yan, F.; Cheng, I.H.; Wu, T.; Gerstein, H.; Yu, G.Q.; Mucke, L. Reducing endogenous tau ameliorates amyloid beta-induced deficits in an Alzheimer's disease mouse model. *Science* **2007**, *316*, 750–754. [CrossRef]
10. Vardy, E.R.L.C.; Catto, A.J.; Hooper, N.M. Proteolytic mechanisms in amyloid-β peptide concentration as a predictor of synaptic change in Alzheimer's disease. *Am. J. Pathol.* **2005**, *155*, 853–862.
11. Rushworth, J.V.; Hooper, N.M. Lipid Rafts: Linking Alzheimer's Amyloid-β Production, Aggregation, and Toxicity at Neuronal Membranes. *Int. J. Alzheimer's Dis.* **2011**, *2011*, 603052. [CrossRef] [PubMed]
12. Allen, J.A.; Halverson-Tamboli, R.A.; Rasenick, M.M. Lipid raft microdomains and neurotransmitter signalling. *Nat. Rev. Neurosci.* **2007**, *8*, 128–140. [CrossRef] [PubMed]
13. Bieberich, E. Sphingolipids and lipid rafts: Novel concepts and methods of analysis. *Chem. Phys. Lipids* **2018**, *216*, 114–131. [CrossRef] [PubMed]

14. Ahyayauch, H.; de la Arada, I.; Masserini, M.E.; Arrondo, J.L.R.; Goni, F.M.; Alonso, A. The Binding of Abeta42 Peptide Monomers to Sphingomyelin/Cholesterol/Ganglioside Bilayers Assayed by Density Gradient Ultracentrifugation. *Int. J. Mol. Sci.* **2020**, *21*, 1674. [CrossRef] [PubMed]
15. Guirland, C.; Zheng, J. Membrane lipid rafts and their role in axon guidance. *Adv. Exp. Med. Biol.* **2007**, *621*, 144–155. [PubMed]
16. Hering, H.; Lin, C.C.; Sheng, M. Lipid rafts in the maintenance of synapses, dendritic spines, and surface AMPA receptor stability. *J. Neurosci.* **2003**, *23*, 3262–3271. [CrossRef]
17. Marchesi, V.T.A. alternative interpretation of the amyloid Abeta hypothesis with regard to the pathogenesis of Alzheimer's disease. *Proc. Natl. Acad. Sci. USA* **2005**, *102*, 9093–9098. [CrossRef]
18. Martín, B.; Fabelo, N.; Santpere, G.; Puig, B.; Marín, R.; Ferrer, I.; Diaz, M. Lipid alterations in lipid rafts from Alzheimer's disease human brain cortex. *J. Alzheimer's Dis.* **2010**, *19*, 489–502. [CrossRef]
19. Hicks, D.A.; Nalivaeva, N.N.; Turner, A.J. Lipid rafts and Alzheimer's disease: Protein-lipid interactions and perturbation of signaling. *Front. Physiol.* **2012**, *3*, 189. [CrossRef]
20. Zhang, X.; Liu, W.; Zan, J.; Wu, C.; Tan, W. Untargeted lipidomics reveals progression of early Alzheimer's disease in APP/PS1 transgenic mice. *Sci. Rep.* **2020**, *10*, 14509. [CrossRef]
21. Barz, H.; Schreiber, A.; Barz, U. Impulses and pressure waves cause excitement and conduction in the nervous system. *Med. Hypotheses* **2013**, *81*, 768–772. [CrossRef] [PubMed]
22. Barz, H.; Schreiber, A.; Barz, U. Nerve impulse propagation: Mechanical wave model and HH model. *Med. Hypotheses* **2020**, *137*, 109540. [CrossRef] [PubMed]
23. Meker, S.; Chin, H.; Sut, T.N.; Cho, N.J. Amyloid-beta Peptide Triggers Membrane Remodeling in Supported Lipid Bilayers Depending on Their Hydrophobic Thickness. *Langmuir* **2018**, *34*, 9548–9560. [CrossRef] [PubMed]
24. Chang, C.C.; Edwald, E.; Veatch, S.; Steel, D.G.; Gafni, A. Interactions of amyloid-beta peptides on lipid bilayer studied by single molecule imaging and tracking. *Biochim. Biophys. Acta (BBA) Biomembr.* **2018**, *1860*, 1616–1624. [CrossRef] [PubMed]
25. Drabik, D.; Chodaczek, G.; Kraszewski, S.; Langner, M. Mechanical Properties Determination of DMPC, DPPC, DSPC, and HSPC Solid-Ordered Bilayers. *Langmuir* **2020**, *36*, 3826–3835. [CrossRef] [PubMed]
26. Grasso, G.; Lionello, C.; Stojceski, F. Highlighting the effect of amyloid beta assemblies on the mechanical properties and conformational stability of cell membrane. *J. Mol. Graph. Model.* **2020**, *100*, 107670. [CrossRef]
27. Gao, Q.; Fang, Y.; Zhang, S.; Wong, H.S.H.; Chan, Y.E.; Wong, S.S.M.; Yung, K.K.L.; Lai, K.W.C. Dynamic effect of beta-amyloid 42 on cell mechanics. *J. Biomech.* **2019**, *86*, 79–88. [CrossRef]
28. Ungureanu, A.A.; Benilova, I.; Krylychkina, O.; Braeken, D.; De Strooper, B.; Van Haesendonck, C.; Dotti, C.G.; Bartic, C. Amyloid beta oligomers induce neuronal elasticity changes in age-dependent manner: A force spectroscopy study on living hippocampal neurons. *Sci. Rep.* **2016**, *6*, 25841. [CrossRef]
29. Tofoleanu, F.; Buchete, N.V. Alzheimer Abeta peptide interactions with lipid membranes: Fibrils, oligomers and polymorphic amyloid channels. *Prion* **2012**, *6*, 339–345. [CrossRef]
30. Phillip, J.M.; Aifuwa, I.; Walston, J.; Wirtz, D. The Mechanobiology of Aging. *Annu. Rev. Biomed. Eng.* **2015**, *17*, 113–141. [CrossRef]
31. Terakawa, M.S.; Lin, Y.; Kinoshita, M.; Kanemura, S.; Itoh, D.; Sugiki, T.; Okumura, M.; Ramamoorthy, A.; Lee, Y.H. Impact of membrane curvature on amyloid aggregation. *Biochim. Biophys. Acta (BBA) Biomembr.* **2018**, *1860*, 1741–1764. [CrossRef] [PubMed]
32. Kinoshita, M.; Kakimoto, E.; Terakawa, M.S.; Lin, Y.; Ikenoue, T.; So, M.; Sugiki, T.; Ramamoorthy, A.; Goto, Y.; Lee, Y.H. Model membrane size-dependent amyloidogenesis of Alzheimer's amyloid-beta peptides. *Phys. Chem. Chem. Phys.* **2017**, *19*, 16257–16266. [CrossRef] [PubMed]
33. Al-Rekabi, Z.; Contera, S. Multifrequency AFM reveals lipid membrane mechanical properties and the effect of cholesterol in modulating viscoelasticity. *Proc. Natl. Acad. Sci. USA* **2018**, *115*, 2658–2663. [CrossRef] [PubMed]
34. Shin, H.; Moore, C.I. Persistent Gamma Spiking in SI Nonsensory Fast Spiking Cells Predicts Perceptual Success. *Neuron* **2019**, *103*, 1150–1163.e5. [CrossRef] [PubMed]
35. Iaccarino, H.F.; Singer, A.C.; Martorell, A.J.; Rudenko, A.; Gao, F.; Gillingham, T.Z.; Mathys, H.; Seo, J.; Kritskiy, O.; Abdurrob, F.; et al. Gamma frequency entrainment attenuates amyloid load and modifies microglia. *Nature* **2016**, *540*, 230–235. [CrossRef]
36. Iaccarino, H.F.; Singer, A.C.; Martorell, A.J.; Rudenko, A.; Gao, F.; Gillingham, T.Z.; Mathys, H.; Seo, J.; Kritskiy, O.; Abdurrob, F.; et al. Author Correction: Gamma frequency entrainment attenuates amyloid load and modifies microglia. *Nature* **2018**, *562*, E1. [CrossRef]
37. Moca, V.V.; Nikolic, D.; Singer, W.; Muresan, R.C. Membrane resonance enables stable and robust gamma oscillations. *Cereb. Cortex* **2014**, *24*, 119–142. [CrossRef]
38. Martorell, A.J.; Paulson, A.L.; Suk, H.J.; Abdurrob, F.; Drummond, G.T.; Guan, W.; Young, J.Z.; Kim, D.N.; Kritskiy, O.; Barker, S.J.; et al. Multi-sensory Gamma Stimulation Ameliorates Alzheimer's-Associated Pathology and Improves Cognition. *Cell* **2019**, *177*, 256–271.e22. [CrossRef]
39. Dezi, M.; Di Cicco, A.; Bassereau, P.; Levy, D. Detergent-mediated incorporation of transmembrane proteins in giant unilamellar vesicles with controlled physiological contents. *Proc. Natl. Acad. Sci. USA* **2013**, *110*, 7276–7281. [CrossRef]
40. Drabik, D.; Doskocz, J.; Przybyło, M. Effects of electroformation protocol parameters on quality of homogeneous GUV populations. *Chem. Phys. Lipids* **2018**, *212*, 88–95. [CrossRef]

41. Drabik, D.; Przybyło, M.; Chodaczek, G.; Iglič, A.; Langner, M. The modified fluorescence based vesicle fluctuation spectroscopy technique for determination of lipid bilayer bending properties. *Biochim. Biophys. Acta (BBA) Biomembr.* **2016**, *1858*, 244–252. [CrossRef] [PubMed]
42. Meleard, P.; Pott, T.; Bouvrais, H.; Ipsen, J.H. Advantages of statistical analysis of giant vesicle flickering for bending elasticity measurements. *Eur. Phys. J. E* **2011**, *34*, 1–14. [CrossRef] [PubMed]
43. Pecreaux, J.J.; Döbereiner, H.-G.; Prost, J.; Joanny, J.-F.; Bassereau, P. Refined contour analysis of giant unilamellar vesicles. *Eur. Phys. J. E* **2004**, *13*, 277–290. [CrossRef] [PubMed]
44. Tran, L.; Ha-Duong, T. Exploring the Alzheimer amyloid-beta peptide conformational ensemble: A review of molecular dynamics approaches. *Peptides* **2015**, *69*, 86–91. [CrossRef] [PubMed]
45. Ntarakas, N.; Ermilova, I.; Lyubartsev, A.P. Effect of lipid saturation on amyloid-beta peptide partitioning and aggregation in neuronal membranes: Molecular dynamics simulations. *Eur. Biophys. J. Biophys. Lett.* **2019**, *48*, 813–824. [CrossRef] [PubMed]
46. Urbanc, B.; Cruz, L.; Ding, F.; Sammond, D.; Khare, S.; Buldyrev, S.V.; Stanley, H.E.; Dokholyan, N.V. Molecular dynamics simulation of amyloid beta dimer formation. *Biophys. J.* **2004**, *87*, 2310–2321. [CrossRef]
47. Grasso, G.; Danani, A. Molecular simulations of amyloid beta assemblies. *Adv. Phys. X* **2020**, *5*. [CrossRef]
48. Crescenzi, O.; Tomaselli, S.; Guerrini, R.; Salvadori, S.; D'Ursi, A.M.; Temussi, P.A.; Picone, D. Solution structure of the Alzheimer amyloid beta-peptide (1–42) in an apolar microenvironment: Similarity with a virus fusion domain. *Eur. J. Biochem.* **2002**, *269*, 5642–5648. [CrossRef]
49. Philips, J.C.; Braun, R.; Wang, W.; Gumbart, J.; Tajkhorsid, E.; Villa, E.; Chipot, C.; Skeel, R.D.; Kalé, L.; Schulten, K. Scalable molecular dynamics with NAMD. *J. Comput. Chem.* **2005**, *26*, 1781–1802. [CrossRef]
50. Lee, S.; Tran, A.; Allsopp, M.; Lim, J.B.; Hénin, J.; Klauda, J.B. CHARMM36 United Atom. Chain Model. for Lipids and Surfactants. *J. Phys. Chem. B* **2014**, *118*, 547–556. [CrossRef]
51. Koller, D.; Lohner, K. The role of spontaneous lipid curvature in the interaction of interfacially active peptides with membranes. *Biochim. Biophys. Acta (BBA) Biomembr.* **2014**, *1838*, 2250–2259. [CrossRef] [PubMed]
52. Kučerka, N.; Nieh, M.-P.; Katsaras, J. Fluid phase lipid areas and bilayer thicknesses of commonly used phosphatidylcholines as a function of temperature. *Biochim. Biophys. Acta (BBA) Biomembr.* **2011**, *1808*, 2761–2771. [CrossRef] [PubMed]
53. Kfoury, R.; Marzban, B.; Makki, E.; Greenfield, M.L.; Yuan, H. Effect of pressure profile of shock waves on lipid membrane deformation. *PLoS ONE* **2019**, *14*, e0212566. [CrossRef] [PubMed]
54. Koshiyama, K.; Kodama, T.; Yano, T.; Fujikawa, S. Structural change in lipid bilayers and water penetration induced by shock waves: Molecular dynamics simulations. *Biophys. J.* **2006**, *91*, 2198–2205. [CrossRef]
55. Braun, A.R.; Sachs, J.N. Determining Structural and Mechanical Properties from Molecular Dynamics Simulations of Lipid Vesicles. *J. Chem. Theory Comput.* **2014**, *10*, 4160–4168. [CrossRef]
56. Khelashvili, G.; Johner, N.; Zhao, G.; Harries, D.; Scott, H.L. Molecular origins of bending rigidity in lipids with isolated and conjugated double bonds: The effect of cholesterol. *Chem. Phys. Lipids* **2014**, *178*, 18–26. [CrossRef]
57. Levine, Z.A.; Venable, R.M.; Watson, M.C.; Lerner, M.G.; Shea, J.-E.; Pastor, R.W.; Brown, F.L.H. Determination of Biomembrane Bending Moduli in Fully Atomistic Simulations. *J. Am. Chem. Soc.* **2014**, *136*, 13582–13585. [CrossRef]
58. Kawamoto, S.; Nakamura, T.; Nielson, S.O.; Shinoda, W. A guiding potential method for evaluating the bending rigidity of tensionless lipid membranes from molecular simulation. *J. Chem. Phys.* **2013**, *139*, 034108. [CrossRef]

International Journal of *Molecular Sciences*

MDPI

Article

CD82 and Gangliosides Tune CD81 Membrane Behavior

Laurent Fernandez [1,†], Morgane Malrieu [1], Christine Bénistant [1], Patrice Dosset [1], Eric Rubinstein [2,3], Elena Odintsova [4], Fedor Berditchevski [4] and Pierre-Emmanuel Milhiet [1,5,*]

1 Centre de Biologie Structurale (CBS), INSERM (National Institute of Health and Medical Research), CNRS (Centre National de la Recherche Scientifique), Université de Montpellier, 34090 Montpellier, France; laurentfernandez281288@gmail.com (L.F.); morgane.malrieu@gmail.com (M.M.); christine.benistant@cbs.cnrs.fr (C.B.); pat@cbs.cnrs.fr (P.D.)
2 INSERM (National Institute of Health and Medical Research), U602, 94807 Villejuif, France; eric.rubinstein@inserm.fr
3 Institut André Lwoff, Université Paris 11, 94807 Villejuif, France
4 Institute of Cancer and Genomic Sciences, University of Birmingham, Edgbaston, Birmingham B15 2TT, UK; e.odintsova@bham.ac.uk (E.O.); F.BERDITCHEVSKI@bham.ac.uk (F.B.)
5 Centre de Biologie Structurale, 29, rue de Navacelles, 34090 Montpellier, France
* Correspondence: pem@cbs.cnrs.fr
† Present address: ARNA (Nucleic Acids: Natural and Artificial Regulations), INSERM (National Institute of Health and Medical Research), CNRS (Centre National de la Recherche Scientifique), Université dé Bordeaux, UMR 5320, U1212, IECB, 33000 Bordeaux, France.

Abstract: Tetraspanins are a family of transmembrane proteins that form a network of protein–protein interactions within the plasma membrane. Within this network, tetraspanin are thought to control the lateral segregation of their partners at the plasma membrane through mechanisms involving specific lipids. Here, we used a single molecule tracking approach to study the membrane behavior of tetraspanins in mammary epithelial cells and demonstrate that despite a common overall behavior, each tetraspanin (CD9, CD81 and CD82) has a specific signature in terms of dynamics. Furthermore, we demonstrated that tetraspanin dynamics on the cell surface are dependent on gangliosides. More specifically, we found that CD82 expression increases the dynamics of CD81 and alters its localization at the plasma membrane, this has no effect on the behavior of CD9. Our results provide new information on the ability of CD82 and gangliosides to differentially modulate the dynamics and organization of tetraspanins at the plasma membrane and highlight that its lipid and protein composition is involved in the dynamical architecture of the tetraspanin web. We predict that CD82 may act as a regulator of the lateral segregation of specific tetraspanins at the plasma membrane while gangliosides could play a crucial role in establishing tetraspanin-enriched areas.

Keywords: tetraspanins; CD81; CD82; gangliosides; single-molecule tracking; microdomain; membrane diffusion; fluorescence microscopy

Citation: Fernandez, L.; Malrieu, M.; Bénistant, C.; Dosset, P.; Rubinstein, E.; Odintsova, E.; Berditchevski, F.; Milhiet, P.-E. CD82 and Gangliosides Tune CD81 Membrane Behavior. *Int. J. Mol. Sci.* **2021**, *22*, 8459. https://doi.org/10.3390/ijms22168459

Academic Editor: Masoud Jelokhani-Niaraki

Received: 21 June 2021
Accepted: 3 August 2021
Published: 6 August 2021

1. Introduction

Tetraspanins are transmembrane proteins characterized by four membrane spanning hydrophobic domains, delineating two extracellular domains [1]. The large extracellular loop contains conserved residues, which contribute to the formation of a fold specific for tetraspanins [2]. Tetraspanins have been shown to modulate adhesion strengthening, cell migration, receptor signaling, cell–cell fusion [3] and trafficking of associated proteins [4]. Some tetraspanins, including CD82, have been associated with cancer progression [5–7]. This involvement in a wide spectrum of cellular processes can be explained by the ability of tetraspanins to form a network of protein–protein interactions within membranes, the so-called tetraspanin webs (or tetraspanin-enriched microdomains (TEM or TERM)) that rely on the homo- or hetero-multimerization of tetraspanins and stable interactions with other membrane proteins [8]. Importantly, using advanced fluorescence microscopy techniques, we and others have demonstrated that several tetraspanins (e.g., CD9, CD81 and CD151) are dynamic molecules that can diffuse within the plasma membrane [9,10]. Using Single Molecule Tracking (SMT), two main modes of motion were identified: Brownian and confined with a combination of these two modes corresponding to the transient confinement of proteins [11,12]. Specifically, we demonstrated that tetraspanins are often confined within specific tetraspanins-enriched areas (TEAs) and that these areas form stable platforms in permanent exchange with the rest of the membrane [13]. Moreover, the functional relevance of this dynamics has been highlighted in the context of viral infection [14,15] and cell migration [16,17].

The tetraspanin CD82 (also known as KAI-1) has been identified as a metastasis suppressor in prostate cancer [18]. Subsequent studies demonstrated that the expression of CD82 protein is frequently lost or downregulated during tumor progression in various epithelial cancers [7]. While the role of CD82 in the modulation of cancer-/metastasis-related pathways is not well understood, several studies have demonstrated that CD82 influences tumor cell migration and invasion by affecting integrin function or through association with EWI-2 and CD81 [19–22]. We also found that CD82 attenuates EGF-induced signaling [22] and regulates compartmentalization and dimerization of ErbB receptors [23]. Importantly, some of the CD82-dependent functions appear to be linked to gangliosides, glycosphingolipids found in the outer leaflet of the plasma membrane. Gangliosides have been proposed to form molecular assemblies containing tetraspanins and their partners (also referred to as glycosynapses [24]). In this regard, it has been demonstrated that various gangliosides modulate interactions of CD82 with its molecular partners [23]. Specifically, ganglioside depletion has a negative effect on the interaction between CD82 and EGFR, CD9, $\alpha_3\beta_1$ integrin and CD151. In another study CD82 was shown to interact directly with the gangliosides GM2 and GM3, impairing cell migration by decreasing EGFR expression and phosphorylation and cMet activation [25]. Taken together, these observations support the idea that gangliosides could regulate, at least in part, the function of CD82-contaning complexes.

In this work, we examined how CD82 and gangliosides can affect the lateral organization of the tetraspanins CD9 and CD81 at the plasma membrane of non-malignant human mammary epithelial cell line (HB2 cells) using SMT, a technique pioneered by Schindler's group in biological membranes using dye molecules [26], allowing the observation of the motion of individual particles. This technique has been applied to investigate the motion of lipids and proteins within membranes and largely used in the context of raft microdomains (see the review [27]). As mentioned above, we already implemented in our group such a technique to study tetraspanin lateral segregation based on the use of monovalent Fab fragments directly labelled with a chemical dye to prevent the potential steric hindrance sometimes observed with quantum dots [28]. We used here SMT, combined with an automatic classification of membrane behavior of single transmembrane proteins based on neural network [29], to compare the membrane behavior of CD82 with that of two other tetraspanins, CD9 and CD81. Our experiments revealed that despite overlapping molecular partnerships, all three tetraspanin proteins display significant differences in their behavior at the plasma membrane. Specifically, we demonstrate that while CD82 expression increases the dynamics of CD81 and modifies its localization at the plasma membrane, the behavior and cellular distribution of CD9 is not modified. Finally, our results suggest that gangliosides are not involved in CD82-dependent changes of CD81 dynamics but rather modulate the overall organization of tetraspanins into TEAs.

2. Results

2.1. Tetraspanins Display Different Dynamics at the Plasma Membrane of HB2 Cells

We have previously demonstrated that the tetraspanins CD9 and CD81, which are structurally related, are highly dynamic molecules at the plasma membrane of epithelial cells (PC3, HeLa, HuH7 and HepG2) [13,30,31]. Despite their high mobility, a significant number of proteins can be temporarily confined in TEAs. In this study, we wanted to extend these observations and investigate the behavior of CD82, a diverged member of the tetraspanin family. In these experiments we used HB2 cells [22], which constitutively express CD9 and CD81. We first performed SMT experiments on HB2/CD82 cells plated on coverslips and labelled with Fab fragments of antibody coupled to Atto647N and raised against the tetraspanin of interest. The advantages of Fab fragments are their small size and that they are monovalent but we cannot completely exclude that their binding could perturb the tetraspanin network. We are however very confident about our results since we used a very low concentration of Fab fragments (see Methods for details) in order to reach single-molecule imaging conditions and similar results about CD9 and CD81 partitioning were obtained in HeLa cells in which the membrane partnership is probably different from that of HB2 cells [15]. The recorded movies (see Supplemental Movies 1, 2 and 3 for representative examples and Figure S1) were then analyzed using our in-house software Patrack based on MSD (Mean Square displacement) analysis [29] (see details in the Materials and Methods section).

Although all three tetraspanins display similar diffusion modes already described for CD9 [13], namely pure Brownian, confined, or a combination of Brownian and confined diffusion (referred to as mixed diffusion), we observed notable differences between the dynamics of the proteins at the plasma membrane of HB2/CD82 cells (Figure 1A,B). Specifically, we found that CD9 is more dynamic than CD81, which, in turn, is more dynamic than CD82 (Figure 1A,B). Indeed, the median diffusion coefficients of CD9, CD81 and CD82 are respectively 0.17, 0.06 and 0.03 $\mu m^2/s$ (median values of apparent diffusion coefficients are calculated from 1000 trajectories, Figure 1A). These differences are due in part to the fact that a higher proportion of CD9 molecules undergoes pure Brownian trajectories, the diffusion of which is faster than that of mixed and confined trajectories (68% as compared to 55% and 45% for CD81 and CD82). Furthermore, we also observed that the rate of diffusion of CD9 molecules exhibiting Brownian motion is faster than that of CD81 and CD82 molecules (median = 0.25 $\mu m^2/s$ vs. 0.15 $\mu m^2/s$ and 0.10 $\mu m^2/s$, respectively). We also tracked CD82 proteins in HB2/Zeo and observed the same behavior with a median diffusion coefficient in the same range (0.02 $\mu m^2/s$).

2.2. CD82 Specifically Increases CD81 Dynamics at the Plasma Membrane of HB2 Cells

The differential dynamics of CD9, CD81 and CD82 described above could be due to tetraspanin-specific differences in forming molecular partnerships with other proteins on the plasma membrane. Firstly, we investigated whether the dynamics of CD9 and CD81, well established CD82 partners, were affected by this tetraspanin. Specifically, we compared the behavior of CD81 and CD9 proteins in HB2/CD82 and HB2/Zeo cells (the expression level of CD82 in HB2/Zeo cells was approximately ten times lower than that seen in HB2/CD82 cells, Figure S2) [29]. These experiments demonstrated that CD82 expression had no impact on CD9 behavior but strongly modified the behavior of CD81. Specifically, the median diffusion coefficient of CD81 in HB2/Zeo cells was three times lower compared to HB2/CD82 cells (Figure 2A). This was partly due to a lower proportion of Brownian CD81 molecules in these cells on one hand (40% in HB2/zeo cells as compared to 55% in HB2/CD82) and a higher proportion of confined molecules on the other hand (39% in HB2/Zeo cells vs. 25% in HB2/CD82 cells). The specific effect of CD82 on CD81 dynamics was further supported in reverse experiments in which CD82 knockdown using RNAi resulted in a decrease of CD81 diffusion coefficient at the plasma membrane (Figure 2A) to a value close to a value comparable to that seen in HB2/Zeo cells. Interestingly, CD82 did not influence the diffusion coefficients of CD81 molecules displaying Brownian motion, but led to an increase of the diffusion coefficients of both mixed and confined CD81 molecules (Figure S3A). These results suggest that CD82 reduces the interaction of CD81 with membrane or juxtamembrane component(s), which may restrict its diffusion. Importantly, expression levels of CD9 and CD81 were comparable in HB2/Zeo and HB2/CD82 cells (Figure S2C).

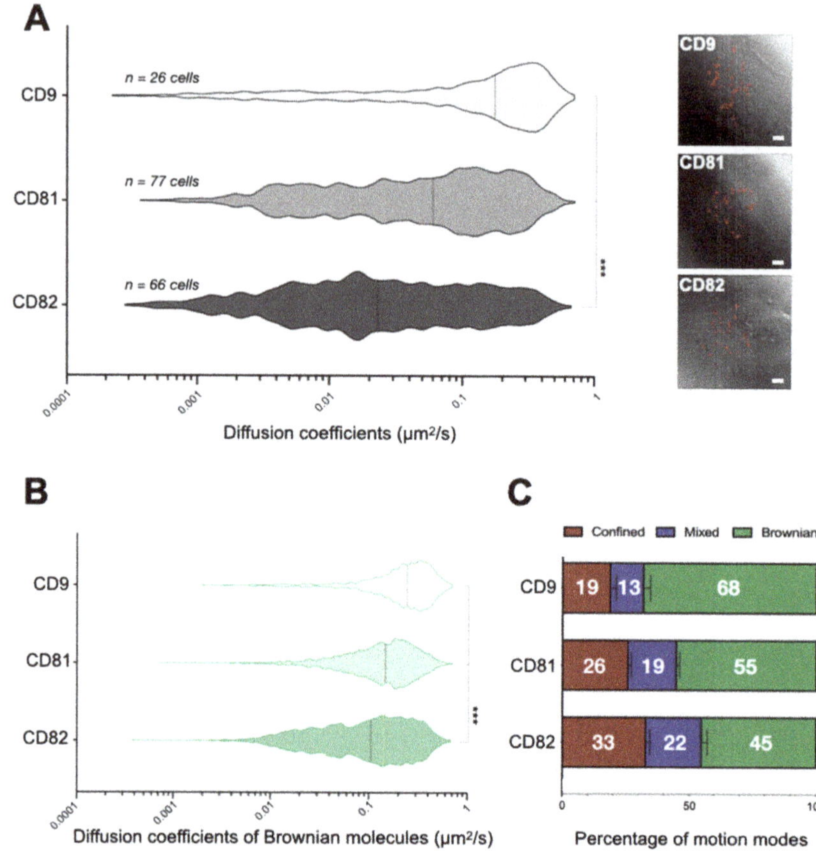

Figure 1. Membrane dynamics of tetraspanins CD82, CD81, and CD9 at the plasma membrane of HB2/CD82 cells. (**A**) Distribution of the apparent diffusion coefficients calculated for all individual tetraspanins molecules analyzed in HB2/CD82 cells. The violin plots were built with 1000 trajectories for each tetraspanin. The dotted lines indicate the median of the populations. The images on the right are examples of DIC images of cells taken after single molecule tracking and some analyzed trajectories are shown in red (scale bars, 5 µm). (**B**) Distribution of the apparent diffusion coefficients of Brownian tetraspanin molecules analyzed in HB2/CD82 cells. The dotted lines indicate the median of the populations. *** indicate that the difference between the populations is significant with a *p* value below 0.0001 as determined by the Mann–Whitney U test. (**C**) Histograms representing the percentage of tetraspanin molecules exhibiting Brownian, confined and mixed motion relative to the total number of trajectories. The error bars represent the standard deviation of at least three independent experiments.

In other experiments, we observed that the dynamics of the $\alpha_3\beta_1$ integrin, which interacts indirectly with CD82 though another tetraspanin, CD151, was not affected by the expression of CD82 (Figure S2A).

2.3. CD82 Modulates the Localization of CD81 at the Plasma Membrane

The increase of CD81 confinement in cells expressing high levels of CD82 strengthened the idea that CD82 could affect the partnership of tetraspanins within the tetraspanin web. To test this hypothesis, we examined co-localization of CD82, CD9 and CD81 at the plasma membrane of HB2/Zeo and HB2/CD82 cells using ensemble labeling (term used to describe a conventional labeling with saturation of the antigenic sites as compared to single molecule detection) combined with TIRF microscopy (Figure 3A). As previously reported for other cell lines [13,32], these tetraspanins were enriched in dot-like structures or larger patches at the basal membrane (referred to as TEAs). Many of these structures contained at least two of the 3 tetraspanins, albeit at different relative ratios. Quantification of the colocalization by Pearson correlation coefficient (PCC) analysis revealed that the expression of CD82 did not affect the high degree of colocalization of CD81 and CD9 (Figure 3A). Interestingly, CD82 expression led to an enrichment of CD9 and CD81 at the cell periphery in most of the imaged cells.

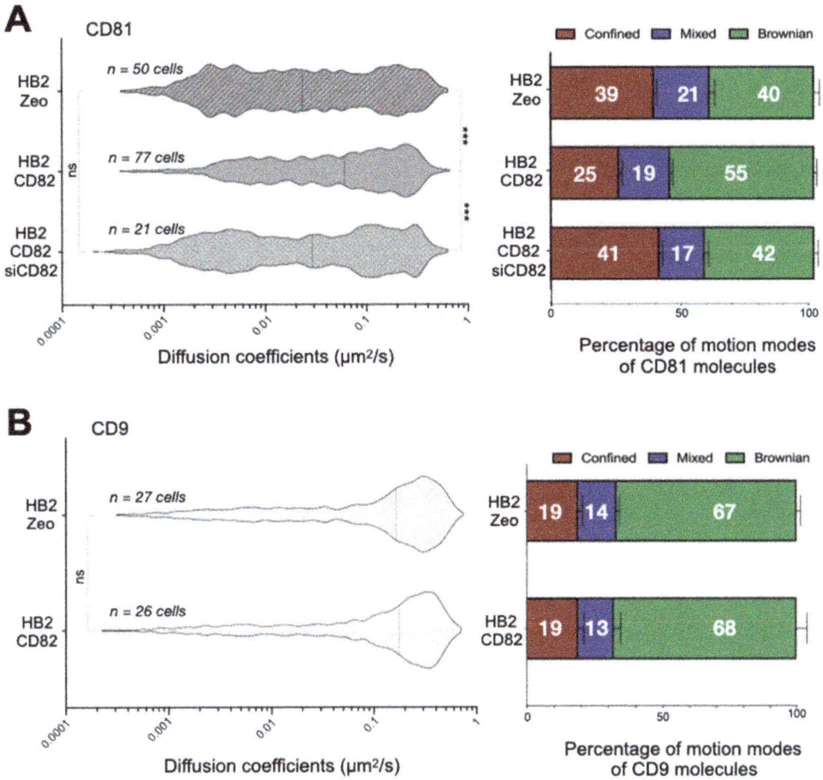

Figure 2. Left Panels: Distribution of the apparent diffusion coefficients calculated for all individual CD81 (**A**) or CD9 (**B**) molecules at the plasma membrane of HB2/CD82, HB2/zeo cells or HB2/CD82 transfected with siRNA targeting CD82. The violin plots were built with 1000 trajectories for each tetraspanin. The dotted lines indicate the median of the populations. *** indicate that the difference between the populations is significant with a p value below 0.0001 as determined by the Mann–Whitney U test («ns» for non-significant). Right Panels: Histograms representing the percentage of CD81 (**A**) or CD9 (**B**) molecules exhibiting Brownian, confined and mixed motion relative to the total number of trajectories. The error bars represent the standard deviation of at least three independent experiments.

We also evaluated CD82-CD81 and CD82-CD9 colocalization in HB2/CD82 cells. As for CD9 and CD81, CD82 was found in dot-like structures and in larger patches at the cell membrane of HB2/CD82 cells (Figure 3B). The calculated PCC showed that both CD9 and CD81 are well co-localized with CD82 at the basal membrane of HB2/CD82 cells: the PCC was 73% for CD82 and CD81 and 61% for CD82 and CD9 (Figure 3B).

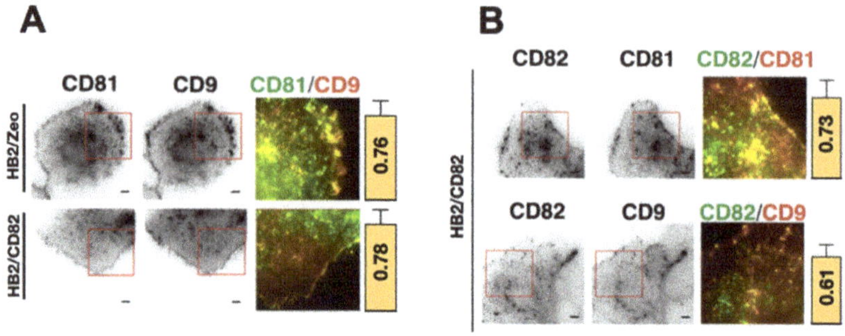

Figure 3. Ensemble labelling of CD9, CD81 and CD82 in HB2/Zeo and HB2/CD82 cells. (**A**) CD9-CD81 colocalization in both cell lines; (**B**) CD82-CD9 or CD81 in HB2/CD82 cells. The scale bar represents 5 μm. Images on the right part of both panels correspond to zooms delineated by the red boxes and represent the merge of the two channels. The yellow pixels represent the colocalization between tetraspanins. The histograms represent the calculated Pearson's correlation coefficients between the two signals in HB2/Zeo and HB2/CD82 cells using the «colocalizer studio» plugin of the open-source community image processing software Icy (Release 1.9.4.0). Calculations were done on at least 10 cells and the error bars represent the standard deviation.

2.4. *Effect of Gangliosides on the Dynamics and Organization of Tetraspanins*

A functional link between CD9 and CD82 and gangliosides has been previously reported [33]. Notably, CD82 was shown to upregulate the expression of the gangliosides GM1, GD1a [23] and GM3 [24]. To extend this finding, we analyzed the lipid composition of HB2/Zeo and HB2/CD82 cells using a lipidomic approach (of note this approach does not allow quantification of GM1 due to technical limitations). Cellular levels of GD1a and GM3 were almost doubled up in CD82-expressing HB2 cells with only a slight increase in the amount of GM2. By contrast, there were no significant changes in the quantity of phosphatidylcholine (PC), phosphatidyl ethanolamine (PE), sphingomyelin (SM) and cholesterol in HB2/CD82 cells compared to HB2/zeo cells (Figure S4A). Thus, this first lipidomic approach on tetraspanins demonstrates that CD82 expression specifically affects the expression of a subset of gangliosides.

Ganglioside expression can be lowered by treating the cells with the ceramide analogue D-*threo*-1-phenyl-2-hexadecanoylamino-3-morpholino-1-propanol·HCl (PPMP). This molecule acts very early in the ganglioside synthesis pathway by inhibiting the glucosyl ceramide (GlcCer) synthase (GlcCer is the first precursor of gangliosides) [34]. To verify the specificity of this treatment in our experimental models, we analyzed the lipid composition of cells treated with this reagent. As expected, treatment with PPMP strongly decreased the cellular levels of all the three gangliosides GD1a, GM2 and GM3 (Figure 4A), and had no effect on the levels of cholesterol and PC (Figure S4A). Interestingly, treatment with PPMP also resulted in a slight decrease in the level of PE and an increase of SM. Importantly, expression of tetraspanins [35] and interaction between them (see immunoprecipitation in Figure S5) were preserved after PPMP treatment.

First, we studied the effect of gangliosides on CD82 dynamics in HB2/CD82 cells. The overall median diffusion coefficient of CD82 doubled from 0.03 $\mu m^2/s$ in HB2/CD82 to 0.06 $\mu m^2/s$ in PPMP-treated HB2/CD82 cells (Figure 4B). This was due in part to a higher fraction of CD82 molecules exhibiting Brownian motion upon PPMP treatment. Importantly, PPMP also impacted the diffusion coefficients of CD82 molecules displaying Brownian behavior (Figure 4C,D): the median diffusion coefficient of Brownian CD82 molecules increased from 0.10 $\mu m^2/s$ in HB2/CD82 cells to 0.16 $\mu m^2/s$ in PPMP-treated HB2/CD82 cells (Figure 4D). This increase was mainly due to the loss of a population of Brownian CD82 molecules diffusing slowly.

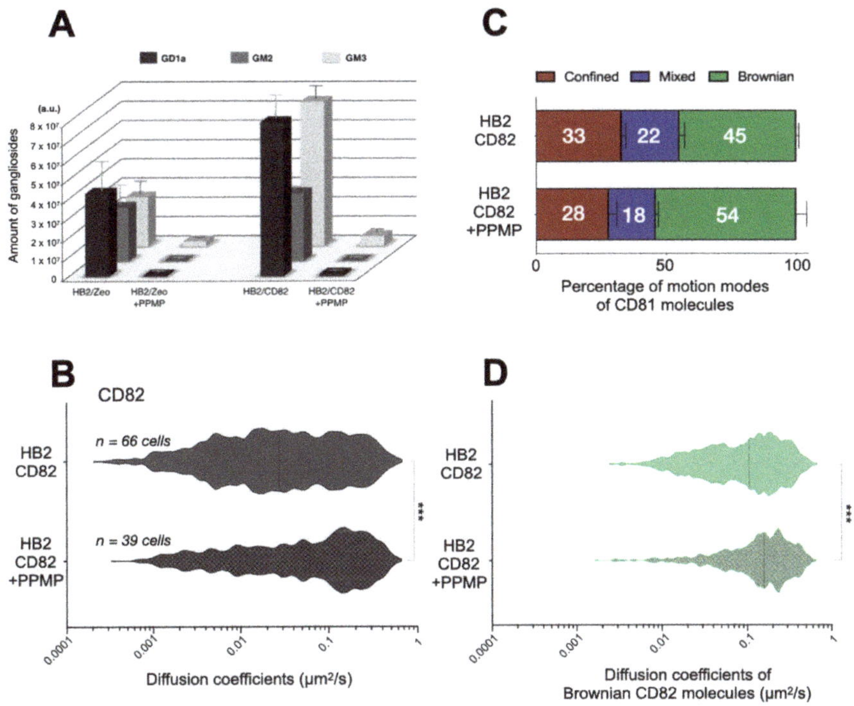

Figure 4. (**A**) Mass spectrometry analysis of ganglioside expression in HB2/Zeo or HB2/CD82 treated or not with PPMP. The analyses have been done in triplicate with samples containing 10^6 cells. The error bars represent the SEM of three independent experiments. (**B**) Distribution of the apparent diffusion coefficients calculated for all the individual CD82 molecules analyzed in HB2/CD82 cells treated or not with PPMP. The violin plots were built with 1000 trajectories for each condition. (**C**) Histograms representing the percentage of CD82 molecules exhibiting Brownian, confined and mixed mode relative to the total number of trajectories in HB2/CD82 cells treated or not with PPMP. The error bars represent the standard deviation of at least three independent experiments. (**D**) Distribution of the apparent diffusion coefficients of Brownian CD82 molecules. *** indicate that the difference between the two populations is significant with a p value below 0.0001 as determined by the Mann–Whitney U test.

PPMP treatment of HB2/CD82 cells did not modify the lateral diffusion of CD9 at the plasma membrane (Figure 5B) but increased the diffusion of CD81. This effect of PPMP on CD81 dynamics did not require CD82 expression since PPMP had an even more pronounced effect on the diffusion of CD81 in HB2/Zeo cells (0.02 µm^2/s versus 0.07 µm^2/s upon PPMP treatment). By contrast, the increase in CD81 median diffusion in PPMP-treated cells appears to be a consequence of an increase in the proportion of Brownian trajectories since we observed no impact on the diffusion coefficient of CD81 molecules displaying Brownian motion after the treatment.

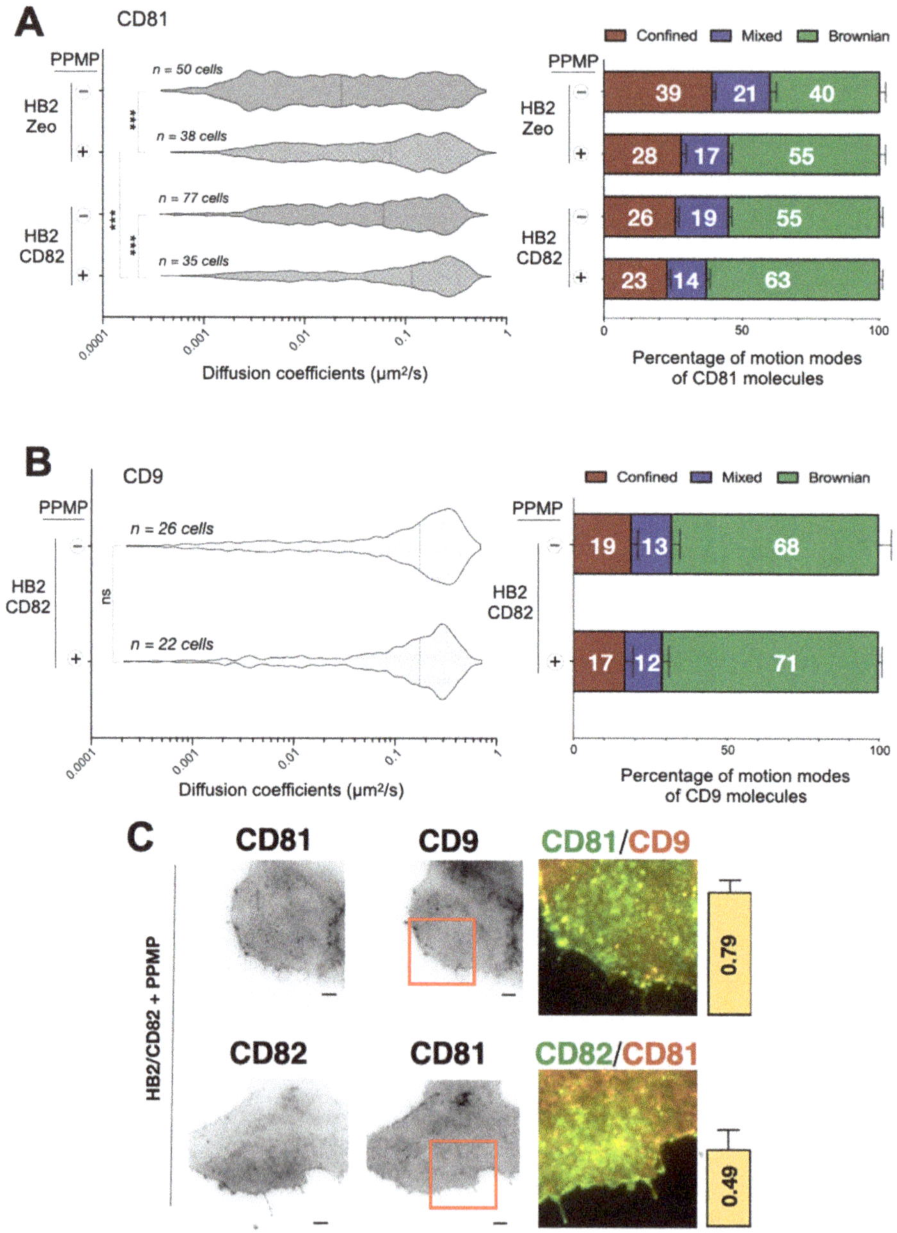

Figure 5. (**A**) Left: Distribution of the apparent diffusion coefficients calculated for all individual CD81 molecules analyzed in HB2/CD82 and HB2/Zeo cells treated or not with PPMP. The violin plots were built with 1000 trajectories for each condition. The dotted lines indicate the median of the populations. *** indicate that the difference between the populations is significant with a p value below 0.0001 as determined by the Mann–Whitney U test. Right: histograms representing the percentage of tetraspanin molecules exhibiting Brownian, confined and mixed motion relative to the total number of trajectories. The error bars represent the standard deviation of at least three independent experiments. (**B**) Left: Distribution of the apparent diffusion coefficients calculated for all individual CD9 molecules at the plasma membrane of HB2/CD82 treated or not with PPMP. The violin plots were built with 1000 trajectories for each condition. The dotted lines indicate the

median of the populations. «ns» indicate that the difference between the populations is not significant as determined by the Mann–Whitney U test. Right: Histograms representing the percentage of CD9 molecules exhibiting Brownian, confined and mixed motion relative to the total number of trajectories. Error bars represent the standard deviation of at least three independent experiments. (**C**) Ensemble labelling of CD81 and CD9 (Top panel), CD82 and CD81 (Bottom panel) in HB2/CD82 cells treated with PPMP. The scale bar represents 5 μm. Images on the right part correspond to zooms delineated by the red boxes and represent the merge of the two channels. The yellow pixels represent the colocalization between tetraspanins. The histograms represent the calculated Pearson's correlation coefficients between the two signals in HB2/Zeo and HB2/CD82 cells using the «colocalizer studio» plugin of Icy software. Calculations were done on at least 10 cells and the error bars represent the standard deviation.

We also wanted to assess whether ganglioside depletion after the PPMP treatment affects cellular distribution of tetraspanins. TIRF microscopy analysis of ensemble labeling revealed that the distribution patterns of CD82, CD81 and CD9 are different in PPMP-treated HB2 cells (compare Figures 3 and 5C). Specifically, the dot-like structures and patches observed in untreated cells are less prominent in the PPMP-treated cells with all three tetraspanins distributing more homogeneously at the basal membrane of HB2 cells. In addition, the accumulation of tetraspanins at the cell periphery was lost in these cells. upon ganglioside depletion. Interestingly, while the level of CD9 and CD81 colocalization in HB2/Zeo and HB2/CD82 cells was not affected upon PPMP treatment, co-localization between CD82 and CD81 in HB2/CD82 cells was diminished (indicated by the decrease of the PCC from 0.76 in untreated cells to 0.45 in PPMP-treated cells). These results suggest that there is specific and differential contribution of gangliosides in co-clustering of tetraspanins on the plasma membrane (Figure 5C). However, co-immunoprecipitation showed minor alterations if any of the association of CD82, CD9 and CD81 with one another, confirming that gangliosides are not essential for the interaction between these tetraspanins (Figure S5).

3. Discussion

It is now well established that tetraspanins form a dynamic network of protein–protein and protein–lipid interactions that likely defines their ability to diffuse in the plane of the plasma membrane. Previous studies have separately addressed the dynamics of CD9, CD81 and CD151 diffusion within the membrane of fibroblastic (CHO) [30], endothelial (HUVEC) [11] and epithelial cells (HeLa, PC3, HuH7) [13,14,30]. While clear differences in the behavior of these tetraspanins have been observed, it remained unknown whether this was due to intrinsic differences between tetraspanin proteins themselves or cell type specific differences in molecular composition of the plasma membrane. To address this question we compared the membrane dynamics of CD9 and CD81 (two closely related members of the tetraspanin family [36]), with that of CD82, a more distant tetraspanin (reviewed in [9]). We have also examined for the first time the contribution of gangliosides to the surface dynamics of tetraspanin proteins. Importantly, these experiments were performed in the context of one cellular model, which allowed us to draw a definitive conclusion on the specific contribution of gangliosides to the surface behavior of different tetraspanins. Our results demonstrated that all three tetraspanins display different membrane dynamics. Furthermore, we also showed that one tetraspanin can specifically affect the behavior of another one. Finally, we discovered that gangliosides could differentially contribute to the dynamics of tetraspanins on the cell surface.

Like many other membrane proteins, CD9, CD81 and CD82 display a combination of different diffusion modes, with a fraction of the molecules diffusing in a Brownian mode, while others are locally confined, either transiently or permanently. Interestingly, both apparent diffusion coefficients and motion types varied between the 3 tetraspanins in HB2 cells. Specifically, CD9 dynamics were higher than that of CD81 dynamics, which were themselves greater than that of CD82. As our previous work had already shown that CD9 was much more dynamic than CD81 at the plasma membrane [15], the higher dynamics of CD9 as compared to CD81 is likely to be due to different intrinsic properties between the two molecules. The lower diffusion coefficients calculated for CD81 and CD82, as compared with CD9, are due to a large extent to a higher fraction of confined molecules, which diffuse slowly. We have previously shown using a combination of single molecule tracking and ensemble labeling that CD9 molecules are confined in areas enriched in these molecules, which we referred to TEAs, suggesting that these TEAs correspond to the confinement of several molecules at the same place and at the same time [13]. By analogy, we can hypothesize that more CD82 molecules than CD81 molecules and more CD81 molecules than CD9 molecules are trapped in TEAs or that they are trapped for a longer time. In this regard, analysis of the distribution of these tetraspanins shows that there are more CD82- and CD81-enriched areas. The higher confinement of CD82 and CD81 may be due to a more favorable interaction with discrete proteins or lipids that serve as nucleation factors within the TEAs or stabilize them (see below). It cannot be explained by a differential expression of CD9 and CD81, which were comparable in HB2/Zeo and HB2/CD82 cells (Figure S2C). Importantly, the diffusion coefficients of molecules with Brownian behavior were also different between different tetraspanins, with CD9 Brownian molecules diffusing more than twice as fast as CD82 Brownian molecules (CD81 displayed an intermediate speed). A possible explanation is that CD82 (and to a lesser extent CD81) makes more contacts with membrane or submembrane components that restrict its diffusion. In this regard, a strong link between tetraspanins and the actin cytoskeleton has been previously reported [37]. Since the link between tetraspanins and the actin cytoskeleton is likely to be indirect, differential behavior of CD9, CD81 and CD82 may be explained by their specific preferences in "choosing" their molecular partners that connect them to the actin network. Consequently, membrane dynamics of tetraspanins may be determined by the size of the clusters formed with their specific partner proteins. Indeed, larger nanostructures diffuse slower according to the hydrodynamic model developed by Saffman and Delbrück that predicts a logarithmic dependence of the diffusion coefficient with the radius R of the diffusant [38]. CD82 may be embedded in larger membrane assemblies than CD81, which itself could be in larger assemblies than CD9. Alternatively, the lipid environment within or around the tetraspanin assemblies could explain the differential behavior of the tetraspanins. Indeed, tetraspanins were shown to directly interact [1,39,40] and the influence of cholesterol content into the membrane on the diffusion of transmembrane proteins has also been described for a few transmembrane proteins including CD9 and CD81 [15], Patched1, the receptor of the secreted Hedgehog ligand Sonic Hedgehog [41] or NrCAM, a cell adhesion molecule of the L1 family [42]. Moreover, our experiments involving ganglioside depletion strongly further support the idea that changes in the lipid composition of the plasma membrane may have a differential effect on the membrane dynamics of various tetraspanins (see below).

In line with the importance of the composition and the size of TEAs on tetraspanin membrane dynamics and partitioning, we demonstrate here that the expression of a particular tetraspanin protein could specifically influence the dynamics of another tetraspanin family member. Indeed, increased expression of CD82 specifically decreased the proportion of "confined" CD81 molecules at the basal membrane, thus leading to the increased number of CD81 molecules exhibiting Brownian motion. Given that the diffusion coefficient of Brownian CD81 molecules was not affected by CD82 expression, these results suggest that CD82 may be involved in displacement of CD81 from the TEAs. Importantly, the modulation of CD81 dynamics by CD82 expression was specific since the behavior of CD9 and $\alpha_3\beta_1$ integrin was not affected. Our results also suggest that while CD82 does not seem to affect the co-localization between CD9 and CD81, it may modify cellular distribution of CD81 and CD9 by directing the proteins to the periphery of the cells. Similarly, several tetraspanins have been described to be enriched at the periphery of breast cancer cells [43] and we have shown that expression of TSPAN5 in U2OS cells led to an enrichment of its partner ADAM10 at the cell periphery [44]. While molecular pathways linking CD82 expression with redistribution of other tetraspanins require further investigation, it is tempting to speculate that the underlying mechanisms may involve other membrane tetraspanin partners. For example, we found that EWI proteins, which have been described as primary partners of CD81 and CD9, also interact with CD82 in HB2 cells as shown with immunoprecipitation experiments (Figure S5).

Gangliosides are key components of the plasma membrane in eukaryotic cells and have been associated with a large variety of cellular processes, especially in the formation and function of microdomains. We investigated the effects of ganglioside expression on the behaviors of the tetraspanins CD82, CD9 and CD81 by tuning the expression of these lipids with PPMP. PPMP treatment was associated with a strong increase in CD82 and CD81 dynamics but not in CD9 dynamics, showing that the effects of PPMP are not due to the general change in membrane properties. This increase is mainly due to a decrease in the number of confined CD82 and CD81 molecules, indicating that confinement of these molecules, possibly in TEA, is highly dependent on the presence of gangliosides. However, the intrinsic link between gangliosides and these tetraspanins may be different as the diffusion coefficient of Brownian CD82 molecules (but not that of Brownian CD81 molecules) was increased upon ganglioside depletion. These observations are consistent with our previous work showing that ganglioside depletion induced CD82 partitioning into the light fraction of the sucrose density gradient, indicating a change in membrane environment [35]. The co-localization experiments performed here further support the role of gangliosides in surface distribution of tetraspanins. Indeed, the size of tetraspanin assemblies observed in HB2/Zeo and HB2/CD82 cells were reduced in cells treated with PPMP and tetraspanins were more homogeneously distributed. Interestingly we discovered that ganglioside depletion only affected co-localization between CD9 and CD81, thus highlighting the role of gangliosides in the structural heterogeneity of tetraspanin clusters on the cell surface [45]. In addition, a specific increase in the diffusion coefficient of Brownian CD82 molecules was observed upon ganglioside depletion (no modifications in the Brownian diffusion coefficients were observed for other tetraspanins or for $\alpha 3$ integrin). Since CD82 has been proposed to directly interact with gangliosides GM2, GM3 and GD1a [46,47], this interaction could explain why only Brownian CD82 molecules are sensitive to ganglioside depletion. Importantly, despite these modifications of membrane compartmentalization, lowering the level of gangliosides yields only minor [35] or no change (this study) in the interaction of CD82 with CD9 and CD81. Thus, gangliosides are dispensable for these interactions and are probably more involved in the dynamic behavior of diffusing CD82 proteins (in agreement with the study of Aikihiro Kusumi's group describing gangliosides as lipids that are very dynamic molecules moving in and out of membrane microdomains in an extremely dynamic manner [48]). It is therefore possible that transient interaction of gangliosides with CD82 could increase the Brownian diffusion coefficient of this tetraspanin.

In conclusion, we emphasized in this work that both lipid and protein compositions of the plasma membrane are involved in the dynamical architecture of the tetraspanin web. Importantly, each tetraspanin appears to have a specific signature in terms of dynamics, which is partly based on the ganglioside composition of the plasma membrane. These observations lay a solid foundation for further analysis focused on the role of tetraspanins in regulation of the membrane dynamics and, ultimately, functionalities of tetraspanin-based protein complexes.

4. Materials and Methods

4.1. Antibodies

Fab fragments or full length mAbs raised against CD81 (TS81), CD9 (SYB-1), and CD82 (TS82) were produced, purified and labeled with Atto647N or Cy3B as previously described [13,14]. The three antibodies recognize the extracellular loops of the corresponding tetraspanin [49,50].

4.2. Cell Culture and Treatments

HB2 mammary epithelial cells isolated from human breast milk were provided by Dr. Fedor Berditchevski and Dr. Elena Odintsova. Different cell lines were used: HB2/Zeo cells weakly expressing CD82 and HB2/CD82 cells overexpressing CD82. All these lines were cultured in a complete DMEM medium containing 10% heat inactivated fetal bovine serum (FBS), 1 mM pyruvate, 10 μg/mL hydrocortisone and 10 μg/mL insulin. Cells in culture were tested for mycoplasma contamination using MycoAlert Mycoplasma Detection kit from Lonza according to manufacturer instructions.

For single molecule tracking experiments, around 10^5 cells were plated 24 h before the experiment in 6-well culture dishes containing 25 mm diameter coverslips that had been previously cleaned using plasma etcher.

For PPMP treatment, cells were grown to a confluence around 50% and D-threo-1-phenyl-2-hexadecanoylamino-3-morpholino-1-propanol·HCl (PPMP) purchased from Matreya was added to reach a final concentration of 2 μM (this concentration was tuned in order to prevent cell death). After one-day treatment, the medium was removed and a fresh medium containing 5 μM of PPMP was added. After two days of culture, the medium was again exchanged for medium containing 10 μM and then routinely cultured as described above. Treatment efficiency was either assessed by flow cytometry using anti-GD1a antibody or by mass spectrometry.

4.3. SiRNA Experiments

Cells were transfected with Lipofectamine 2000 purchased from ThermoFisher. For CD82 expression knockdown, the following duplex was used: GCTGGGTCAGCTTCTA-CAAdTdT and TTGTAGAAGCTGACCCAGCdCdG. Briefly, cells were plated on 6-well dishes at 70–80% confluence the day prior to the transfection. For each well, 5 μL of Lipofectamine 2000 were added to 250 μL of Opti-MEM and the CD82 siRNA was added to a final concentration of 100 pM in 250 μL of Opti-MEM. The two samples were incubated at room temperature for 20 min and then mixed together. The complete medium of each well was then exchanged for 2 mL of Opti-MEM and the transfection mixture was added to each well. After 8 h, the Opti-MEM was removed and 2 mL of complete medium was added to each well. The following day, cells were detached and plated on 25 mm coverslips for further single molecule tracking experiments. After the experiments, the efficiency of CD82 siRNA was assessed by flow cytometry on cells used for the tracking.

4.4. Single-Molecule Tracking

SMT experiments were carried out as previously described [13]. Briefly, cells plated on coverslips were incubated in red phenol-free DMEM at 37 °C for 10 min with Atto647N-labeled Fab fragments of mAbs raised against CD81 (TS81), CD9 (SYB-1) or CD82 (TS82) at concentrations in the range of 1 to 10 ng/mL. For single molecule experiments, ~ one probe per Fab is required. Ensemble labeling was performed with full antibody. Homemade objective-type TIRF setup allowing multicolor single-molecule imaging and equipped with a Plan Fluor 100×/1.45 NA objective (Zeiss, Le Pecq, France Brattleboro, VT) was used. All the experiments were performed with a 100 ms integration time. The localization of each fluorescence peak was determined with subpixel resolution by fitting a two-dimensional elliptical Gaussian function. The accuracy of the position measurement in living cells was estimated to be 50 nm by fitting a 2D Gaussian to the emission intensity distribution of an immobile single molecule conjugated with Atto647N.

Movies were analyzed using homemade software (named 'PaTrack') implemented in visual C++ ([29], freely available using the link). Trajectories were constructed using the individual diffraction limited signal of each molecule. The center of each fluorescence peak was determined with subpixel resolution by fitting a two-dimensional elliptical Gaussian function. The two-dimensional trajectories of single molecules were constructed frame per frame. Only trajectories containing at least 41 points were retained. Diffusion coefficient values were determined from a linear fit to the MSD (mean square displacement)-τ plots between the first and the fourth points (D1–4) according to the equation MSD(t) = 4Dt. The determination of the motion modes was performed using homemade algorithm based on a neural network that has been trained using synthetic trajectories to detect pure Brownian, confined and directed motion modes. Thanks to a sliding window, the trajectory is analyzed and the different modes can be confidently detected within a trajectory for segments larger than 10 frames. Once the motion mode is identified, the different segments are analyzed by plotting the MSD versus time lag. The MSD curve was linearly fitted (Brownian) or adjusted with a quadratic curve (4Dt + v2t2) (directed diffusion) or exponential curve L2/3(1−exp(−12Dt/L2) (confined diffusion), where L is the side of a square domain, the confinement diameter being related to L by Ø conf = (2/$\sqrt{}$L) [51]. The apparent diffusion coefficient values were determined from a linear fit between the first and fourth points (D1–4). The algorithm was tested with simulated trajectories displaying pure Brownian, confined or directed behavior or a combination of these 3 modes and successfully applied to a set of single molecule experiments previously recorded for tetraspanins diffusing into plasma membranes.

4.5. Flow Cytometry Experiments

Cells were grown in 25 cm^2 flasks until reaching exponential growth. Then, cells were harvested using an enzyme-free cell dissociation reagent (purchased from Gibco) and centrifuged at 100× *g* for 5 min at 4 °C. The cell pellet was thoroughly suspended in cold PBS. The cells were then centrifuged again and the pellet was suspended in PBS/1% FBS buffer containing 2 µg/mL of mouse IgG raised against the protein of interest. After 30 min of incubation, cells were centrifuged at 100× *g* for 5 min. The pellet was then suspended in PBS/1% FBS. This washing step was repeated 3 times. The pellet was suspended in PBS/1% FBS containing 1 µg/mL of secondary antibodies raised against mouse IgG and coupled to Alexa568. As a control, one sample was incubated only with the secondary antibody. After one hour of incubation at 4 °C in the dark, the washing steps were repeated 3 times. After the last wash, the cell pellet was suspended in 500 µL of cold PBS with 2% PFA. After 10 min, cells were washed to remove PFA and were analyzed using a Bio-Rad flow cytometer. Post-acquisition analyses were performed using the software FlowJo.

4.6. Dual-Color Immunofluorescence Using TIRF Microscopy

One day prior to the experiment, 2×10^5 cells were plated in 6-well culture plates containing 25 mm diameter coverslips that had been previously plasma-cleaned. Cells were then fixed with PBS buffer containing 4% PFA for 10 min. Cells were then washed 3 times with PBS. PBS/1% FBS buffer containing 5 µg/mL of antibodies raised against the proteins of interest and coupled to Cy3b or Atto647N was then added to the coverslips for 30 min. Cells were then washed 3 times with PBS/1% FBS. Coverslips were mounted on a glass slide using Prolong Diamond and incubated overnight at 4 °C. Cells were imaged using a TIRF microscope at 10 images/s. One hundred frames were acquired and then stacked. Pearson's correlation coefficients were calculated using the colocalizer studio plugin of the software Icy.

4.7. Mass Spectrometry Analysis of Lipids

Total lipids were extracted overnight at 4 °C from the samples dissolved in 0.5 mL water with 10 volumes (5 mL) of $CHCl_3/CH_3OH$ (1:1, *v*/*v*). The residual pellet obtained after centrifugation (1500× *g*, 5 min) was extracted twice with 2 mL of the same solvent. The three lipid extracts were pooled, dried under a stream of nitrogen, and solubilized in 3 mL $CHCl_3/CH_3OH$ (1:1, *v*/*v*). Gangliosides were then separated from other lipids using phase partition by adding 1 mL water. After centrifugation, the upper aqueous phase was collected while the lower organic phase was reextracted twice with 2 mL CH_3OH/water (1:1, *v*/*v*). The three upper phases containing gangliosides were combined, dried under a stream of nitrogen, and solubilized in 2 mL CH_3OH/PBS 10 mM (1:1, *v*/*v*). This ganglisoside extract was desalted on a C18 silica gel column (Sep-Pak Vac 6 cc, 500 mg; Waters), washed with 7 mL CH3OH and pre-equilibrated with 7 mL CH_3OH/PBS 10 mM (1:1, *v*/*v*) before injection. After washing with 10 mL water, purified gangliosides were eluted with 6 mL CH_3OH and 4 mL $CHCl_3/CH_3OH$ (2:1, *v*/*v*). Liquid chromatography (LC) was performed at 30 °C using a Dionex UltiMateTM 3000 LC system from ThermoScientific equipped with an autosampler. Separation of GM3, GM2, GD3, GD1a, GD1b, GT1b, and GQ1b standards was achieved under hydrophilic interaction liquid chromatography (HILIC) conditions using a silica Kinetex column (GM1 measurement was not available when we have performed the experiments). The mobile phase was composed of acetonitrile/water (90:10, *v*/*v*) containing 10 mM ammonium acetate and acetonitrile/water (50:50, *v*/*v*) containing 10 mM ammonium acetate. The solvent-gradient system was as follows: 0–1 min 100%, 4 min 79%, 9 min 78%, 14–18 min 50%, and 19–48 min 100%. The flow rate was 400 µL/min and the injection volume was 10 µL. Eluates from the HPLC system were then analyzed by mass spectrometry (LC-MS/MS).

The cholesterol levels were analyzed using a Cholesterol Quantitation Kit from Sigma following the manufacturer's protocol.

4.8. Immunoprecipitations of Tetraspanins and Partners

Cells were grown in 75 cm^2 flasks until reaching 80% confluence. Cells were then washed three times with PBS and incubated with 10 mL of biotin at 0.5 mg/mL for 45 min at 4 °C. The cells were then washed three times with PBS and lysed in a buffer containing 30 mM Tris pH 7.4, 150 mM NaCl, protease inhibitors and 1% of Brij97. After 30 min of incubation at 4 °C, the insoluble material was removed by centrifugation at 10,000× *g* for 15 min. The cell lysate was then incubated with inactivated goat serum and G-sepharose protein beads for 2 h. The beads were removed by centrifugation and the lysate was incubated with 1 µg of primary antibody and 10 µL of G-sepharose protein beads for 400 µL of lysate for 2 h at 4 °C on a rotating wheel. The beads were washed four times using the lysis buffer with 0.5% Brij97 and Laemmli buffer was added. Immunoprecipitated proteins were separated by SDS-PAGE under non-reducing conditions and transferred to a low fluorescence PVDF membrane. Immunoprecipitates were analyzed by Western blot using streptavidin coupled to Alexa680 and blots revealed using the Odyssey Infrared Imaging System (LI-COR Biosciences GmbH, Bad Homburg, Germany). The antibody TS82 was coupled to biotin and revealed using streptavidin coupled to Alexa680.

Supplementary Materials: The following are available online at https://www.mdpi.com/1422-0067/22/16/8459/s1.

Author Contributions: Conceptualization, E.O., F.B., E.R. and P.-E.M.; methodology, L.F., M.M. and P.-E.M.; software, P.D.; validation, L.F., F.B., E.O., E.R. and P.-E.M.; formal analysis, L.F., P.-E.M.; investigation, L.F., M.M., C.B., E.R., P.D.; resources, P.-E.M.; data curation, P.-E.M.; writing—original draft preparation, L.F., P.-E.M.; writing—review and editing, L.F., E.R., C.B., E.O., F.B., P.-E.M.; supervision, P.-E.M., C.B., E.R., F.B.; project administration, P.-E.M.; funding acquisition, P.-E.M. All authors have read and agreed to the published version of the manuscript.

Funding: This research was supported by France BioImaging (FBI, ANR10INSB04) and the GIS IBISA (Infrastructures en Biologie Santé et Agronomie).

Institutional Review Board Statement: Not applicable.

Informed Consent Statement: Not applicable.

Data Availability Statement: Data is contained within the article or Supplementary Material.

Acknowledgments: L.F. was a recipient of the french ministry of research and a FRM fellow. We thank Tomonori Saotome, Anthony Lozano and Thibaud Dieudonné for technical help. We thank Emmanuel Margeat for his help in single-molecule microscopy experiments. Mass spectrometry analysis was performed by the Chemosens platform in Dijon (France).

Conflicts of Interest: The authors declare no conflict of interest. The funders had no role in the design of the study; in the collection, analyses, or interpretation of data; in the writing of the manuscript, or in the decision to publish the results.

References

1. Zimmerman, B.; Kelly, B.; McMillan, B.J.; Seegar, T.C.; Dror, R.O.; Kruse, A.C.; Blacklow, S.C. Crystal Structure of a Full-Length Human Tetraspanin Reveals a Cholesterol-Binding Pocket. *Cell* **2016**, *167*, 1041–1051.e11. [CrossRef] [PubMed]
2. Seigneuret, M.; Delaguillaumie, A.; Lagaudriere-Gesbert, C.; Conjeaud, H. Structure of the Tetraspanin Main Extracellular Domain: A Partially Conserved Fold with a Structurally Variable Domain Insertion. *J. Biol. Chem.* **2001**, *276*, 40055–40064. [CrossRef]
3. Hemler, M.E. Tetraspanin Proteins Mediate Cellular Penetration, Invasion, and Fusion Events and Define a Novel Type of Membrane Microdomain. *Annu. Rev. Cell Dev. Biol.* **2003**, *19*, 397–422. [CrossRef] [PubMed]
4. Charrin, S.; Jouannet, S.; Boucheix, C.; Rubinstein, E. Tetraspanins at a glance. *J. Cell Sci.* **2014**, *127*, 3641–3648. [CrossRef] [PubMed]
5. Hemler, M.E. Tetraspanin proteins promote multiple cancer stages. *Nat. Rev. Cancer* **2014**, *14*, 49–60. [CrossRef]
6. Feng, J.; Huang, C.; Wren, J.; Wang, D.-W.; Yan, J.; Zhang, J.; Sun, Y.; Han, X.; Zhang, X.A. Tetraspanin CD82: A suppressor of solid tumors and a modulator of membrane heterogeneity. *Cancer Metastasis Rev.* **2015**, *34*, 619–633. [CrossRef]
7. Tsai, Y.C.; Weissman, A.M. Dissecting the diverse functions of the metastasis suppressor CD82/KAI1. *FEBS Lett.* **2011**, *585*, 3166–3173. [CrossRef]
8. Claas, C.; Stipp, C.S.; Hemler, M.E. Evaluation of Prototype Transmembrane 4 Superfamily Protein Complexes and Their Relation to Lipid Rafts. *J. Biol. Chem.* **2001**, *276*, 7974–7984. [CrossRef] [PubMed]
9. Charrin, S.; Le Naour, F.; Silvie, O.; Milhiet, P.-E.; Boucheix, C.; Rubinstein, E. Lateral organization of membrane proteins: Tetraspanins spin their web. *Biochem. J.* **2009**, *420*, 133–154. [CrossRef] [PubMed]
10. Yáñez-Mó, M.; Barreiro, O.; Gordon, M.; Sala-Valdés, M.; Sánchez-Madrid, F. Tetraspanin-enriched microdomains: A functional unit in cell plasma membranes. *Trends Cell Biol.* **2009**, *19*, 434–446. [CrossRef]
11. Barreiro, O.; Zamai, M.; Yáñez-Mó, M.; Tejera, E.; López-Romero, P.; Monk, P.; Gratton, E.; Caiolfa, V.R.; Sánchez-Madrid, F. Endothelial adhesion receptors are recruited to adherent leukocytes by inclusion in preformed tetraspanin nanoplatforms. *J. Cell Biol.* **2008**, *183*, 527–542. [CrossRef]
12. Yang, X.H.; Mirchev, R.; Deng, X.; Yacono, P.; Yang, H.L.; Golan, D.E.; Hemler, M.E. CD151 restricts α6 integrin diffusion mode. *J. Cell Sci.* **2012**, *125*, 1478–1487. [CrossRef]
13. Espenel, C.; Margeat, E.; Dosset, P.; Arduise, C.; Le Grimellec, C.; Royer, C.; Boucheix, C.; Rubinstein, E.; Milhiet, P.-E. Single-molecule analysis of CD9 dynamics and partitioning reveals multiple modes of interaction in the tetraspanin web. *J. Cell Biol.* **2008**, *182*, 765–776. [CrossRef] [PubMed]
14. Krementsov, D.; Rassam, P.; Margeat, E.; Roy, N.H.; Schneider-Schaulies, J.; Milhiet, P.-E.; Thali, M. HIV-1 Assembly Differentially Alters Dynamics and Partitioning of Tetraspanins and Raft Components. *Traffic* **2010**, *11*, 1401–1414. [CrossRef]
15. Dahmane, S.; Rubinstein, E.; Milhiet, P.-E. Viruses and Tetraspanins: Lessons from Single Molecule Approaches. *Viruses* **2014**, *6*, 1992–2011. [CrossRef] [PubMed]
16. Danglot, L.; Chaineau, M.; Dahan, M.; Gendron, M.-C.; Boggetto, N.; Perez, F.; Galli, T. Role of TI-VAMP and CD82 in EGFR cell-surface dynamics and signaling. *J. Cell Sci.* **2010**, *123*, 723–735. [CrossRef] [PubMed]
17. Yang, X.; Claas, C.; Kraeft, S.-K.; Chen, L.B.; Wang, Z.; Kreidberg, J.A.; Hemler, M.E. Palmitoylation of Tetraspanin Proteins: Modulation of CD151 Lateral Interactions, Subcellular Distribution, and Integrin-dependent Cell Morphology. *Mol. Biol. Cell* **2002**, *13*, 767–781. [CrossRef]
18. Miranti, C. Controlling cell surface dynamics and signaling: How CD82/KAI1 suppresses metastasis. *Cell. Signal.* **2009**, *21*, 196–211. [CrossRef] [PubMed]
19. Ono, M.; Handa, K.; Withers, D.A.; Hakomori, S.-I. Glycosylation Effect on Membrane Domain (GEM) Involved in Cell Adhesion and Motility: A Preliminary Note on Functional α3, α5-CD82 Glycosylation Complex in ldlD 14 Cells. *Biochem. Biophys. Res. Commun.* **2000**, *279*, 744–750. [CrossRef]
20. Lee, J.H.; Seo, Y.-W.; Park, S.R.; Kim, Y.J.; Kim, K.K. Expression of a splice variant of KAI1, a tumor metastasis suppressor gene, influences tumor invasion and progression. *Cancer Res.* **2003**, *63*, 7247–7255.
21. Ruseva, Z.; Geiger, P.X.C.; Hutzler, P.; Kotzsch, M.; Luber, B.; Schmitt, M.; Gross, E.; Reuning, U. Tumor suppressor KAI1 affects integrin αvβ3-mediated ovarian cancer cell adhesion, motility, and proliferation. *Exp. Cell Res.* **2009**, *315*, 1759–1771. [CrossRef] [PubMed]
22. Odintsova, E.; Sugiura, T.; Berditchevski, F. Attenuation of EGF receptor signaling by a metastasis suppressor, the tetraspanin CD82/KAI-1. *Curr. Biol.* **2000**, *10*, 1009–1012. [CrossRef]
23. Odintsova, E.; Voortman, J.; Gilbert, E.; Berditchevski, F. Tetraspanin CD82 regulates compartmentalisation and ligand-induced dimerization of EGFR. *J. Cell Sci.* **2003**, *116*, 4557–4566. [CrossRef]
24. Hakomori, S.-I. The glycosynapse. *Proc. Natl. Acad. Sci. USA* **2002**, *99*, 225–232. [CrossRef]
25. Li, Y.; Huang, X.; Zhang, J.; Li, Y.; Ma, K. Synergistic inhibition of cell migration by tetraspanin CD82 and gangliosides occurs via the EGFR or cMet-activated Pl3K/Akt signalling pathway. *Int. J. Biochem. Cell Biol.* **2013**, *45*, 2349–2358. [CrossRef]

26. Schmidt, T.; Schutz, G.J.; Baumgartner, W.; Gruber, H.J.; Schindler, H. Imaging of single molecule diffusion. *Proc. Natl. Acad. Sci. USA* **1996**, *93*, 2926–2929. [CrossRef]
27. Kusumi, A.; Fujiwara, T.; Tsunoyama, T.A.; Kasai, R.; Liu, A.; Hirosawa, K.M.; Kinoshita, M.; Matsumori, N.; Komura, N.; Ando, H.; et al. Defining raft domains in the plasma membrane. *Traffic* **2019**, *21*, 106–137. [CrossRef]
28. Abraham, L.; Lu, H.Y.; Falcão, R.C.; Scurll, J.; Jou, T.; Irwin, B.; Tafteh, R.; Gold, M.R.; Coombs, D. Limitations of Qdot labelling compared to directly-conjugated probes for single particle tracking of B cell receptor mobility. *Sci. Rep.* **2017**, *7*, 11379. [CrossRef] [PubMed]
29. Dosset, P.; Rassam, P.; Fernandez, L.; Espenel, C.; Rubinstein, E.; Margeat, E.; Milhiet, P.-E. Automatic detection of diffusion modes within biological membranes using back-propagation neural network. *BMC Bioinform.* **2016**, *17*, 197. [CrossRef]
30. Potel, J.; Rassam, P.; Montpellier, C.; Kaestner, L.; Werkmeister, E.; Tews, B.A.; Couturier, C.; Popescu, C.-I.; Baumert, T.F.; Rubinstein, E.; et al. EWI-2wint promotes CD81 clustering that abrogates Hepatitis C Virus entry. *Cell. Microbiol.* **2013**, *15*, 1234–1252. [CrossRef]
31. Harris, H.J.; Clerte, C.; Farquhar, M.J.; Goodall, M.; Hu, K.; Rassam, P.; Dosset, P.; Wilson, G.K.; Balfe, P.; Ijzendoorn, S.C.; et al. Hepatoma polarization limits CD 81 and hepatitis C virus dynamics. *Cell. Microbiol.* **2012**, *15*, 430–445. [CrossRef] [PubMed]
32. Berditchevski, F. Complexes of tetraspanins with integrins: More than meets the eye. *J. Cell Sci.* **2001**, *114*, 4143–4151. [CrossRef]
33. Park, S.-Y.; Yoon, S.-J.; Freire-De-Lima, L.; Kim, J.-H.; Hakomori, S.-I. Control of cell motility by interaction of gangliosides, tetraspanins, and epidermal growth factor receptor in A431 versus KB epidermoid tumor cells. *Carbohydr. Res.* **2009**, *344*, 1479–1486. [CrossRef] [PubMed]
34. Abe, A.; Inokuchi, J.-I.; Jimbo, M.; Shimeno, H.; Nagamatsu, A.; Shayman, J.A.; Shukla, G.S.; Radin, N.S. Improved Inhibitors of Glucosylceramide Synthase1. *J. Biochem.* **1992**, *111*, 191–196. [CrossRef] [PubMed]
35. Odintsova, E.; Butters, T.D.; Monti, E.; Sprong, H.; van Meer, G.; Berditchevski, F. Gangliosides play an important role in the organization of CD82-enriched microdomains. *Biochem. J.* **2006**, *400*, 315–325. [CrossRef]
36. Delaguillaumie, A.; Harriague, J.; Kohanna, S.; Bismuth, G.; Rubinstein, E.; Seigneuret, M.; Conjeaud, H. Tetraspanin CD82 controls the association of cholesterol-dependent microdomains with the actin cytoskeleton in T lymphocytes: Relevance to co-stimulation. *J. Cell Sci.* **2004**, *117*, 5269–5282. [CrossRef]
37. Delaguillaumie, A.; Lagaudrière-Gesbert, C.; Popoff, M.R.; Conjeaud, H. Rho GTPases Link Cytoskeletal Rearrangements and Activation Processes Induced via the Tetraspanin CD82 in T Lymphocytes. *J. Cell Sci.* **2002**, *115*, 433–443. [CrossRef] [PubMed]
38. Saffman, P.G.; Delbruck, M. Brownian motion in biological membranes. *Proc. Natl. Acad. Sci. USA* **1975**, *72*, 3111–3113. [CrossRef]
39. Charrin, S.; Manié, S.; Thiele, C.; Billard, M.; Gerlier, D.; Boucheix, C.; Rubinstein, E. A physical and functional link between cholesterol and tetraspanins. *Eur. J. Immunol.* **2003**, *33*, 2479–2489. [CrossRef]
40. Umeda, R.; Satouh, Y.; Takemoto, M.; Nakada-Nakura, Y.; Liu, K.; Yokoyama, T.; Shirouzu, M.; Iwata, S.; Nomura, N.; Sato, K.; et al. Structural insights into tetraspanin CD9 function. *Nat. Commun.* **2020**, *11*, 1606. [CrossRef]
41. Weiss, L.E.; Milenkovic, L.; Yoon, J.; Stearns, T.; Moerner, W.E. Motional dynamics of single Patched1 molecules in cilia are controlled by Hedgehog and cholesterol. *Proc. Natl. Acad. Sci. USA* **2019**, *116*, 5550–5557. [CrossRef] [PubMed]
42. Falk, J.; Thoumine, O.; Dequidt, C.; Choquet, D.; Faivre-Sarrailh, C. NrCAM Coupling to the Cytoskeleton Depends on Multiple Protein Domains and Partitioning into Lipid Rafts. *Mol. Biol. Cell* **2004**, *15*, 4695–4709. [CrossRef] [PubMed]
43. Berditchevski, F.; Odintsova, E. Characterization of Integrin–Tetraspanin Adhesion Complexes: Role of Tetraspanins in Integrin Signaling. *J. Cell Biol.* **1999**, *146*, 477–492. [CrossRef] [PubMed]
44. Jouannet, S.; Saint-Pol, J.; Fernandez, L.; Nguyen, V.; Charrin, S.; Boucheix, C.; Brou, C.; Milhiet, P.-E.; Rubinstein, E. TspanC8 tetraspanins differentially regulate the cleavage of ADAM10 substrates, Notch activation and ADAM10 membrane compartmentalization. *Cell. Mol. Life Sci.* **2016**, *73*, 1895–1915. [CrossRef] [PubMed]
45. Zuidscherwoude, M.; Göttfert, F.; Dunlock, V.-M.; Figdor, C.; Bogaart, G.V.D.; Van Spriel, A.B. The tetraspanin web revisited by super-resolution microscopy. *Sci. Rep.* **2015**, *5*, 12201. [CrossRef] [PubMed]
46. Todeschini, A.R.; Dos Santos, J.N.; Handa, K.; Hakomori, S.-I. Ganglioside GM2/GM3 complex affixed on silica nanospheres strongly inhibits cell motility through CD82/cMet-mediated pathway. *Proc. Natl. Acad. Sci. USA* **2008**, *105*, 1925–1930. [CrossRef]
47. Todeschini, A.R.; Dos Santos, J.N.; Handa, K.; Hakomori, S.-I. Ganglioside GM2-Tetraspanin CD82 Complex Inhibits Met and Its Cross-talk with Integrins, Providing a Basis for Control of Cell Motility through Glycosynapse. *J. Biol. Chem.* **2007**, *282*, 8123–8133. [CrossRef]
48. Suzuki, K.G.; Ando, H.; Komura, N.; Konishi, M.; Imamura, A.; Ishida, H.; Kiso, M.; Fujiwara, T.; Kusumi, A. Revealing the Raft Domain Organization in the Plasma Membrane by Single-Molecule Imaging of Fluorescent Ganglioside Analogs. *Methods Enzymol.* **2018**, *598*, 267–282. [CrossRef]
49. Charrin, S.; Le Naour, F.; Labas, V.; Billard, M.; Le Caer, J.-P.; Emile, J.-F.; Petit, M.-A.; Boucheix, C.; Rubinstein, E. EWI-2 is a new component of the tetraspanin web in hepatocytes and lymphoid cells. *Biochem. J.* **2003**, *373*, 409–421. [CrossRef] [PubMed]
50. Yalaoui, S.; Zougbédé, S.; Charrin, S.; Silvie, O.; Arduise, C.; Farhati, K.; Boucheix, C.; Mazier, D.; Rubinstein, E.; Froissard, P. Hepatocyte Permissiveness to Plasmodium Infection Is Conveyed by a Short and Structurally Conserved Region of the CD81 Large Extracellular Domain. *PLoS Pathog.* **2008**, *4*, e1000010. [CrossRef]
51. Qian, H.; Sheetz, M.; Elson, E. Single particle tracking. Analysis of diffusion and flow in two-dimensional systems. *Biophys. J.* **1991**, *60*, 910–921. [CrossRef]

International Journal of
Molecular Sciences

MDPI

Communication

The Small Heat Shock Protein, HSPB1, Interacts with and Modulates the Physical Structure of Membranes

Balint Csoboz [1,2,†], Imre Gombos [1,†], Zoltán Kóta [3,4], Barbara Dukic [1], Éva Klement [1,4], Vanda Varga-Zsíros [1], Zoltán Lipinszki [1], Tibor Páli [3], László Vígh [1] and Zsolt Török [1,*]

1 Institute of Biochemistry, Biological Research Centre, 6726 Szeged, Hungary; bcsoboz@gmail.com (B.C.); gombos.imre@brc.hu (I.G.); dukic.barbara@brc.hu (B.D.); klement.eva@brc.hu (É.K.); zsiros.vanda@brc.hu (V.V.-Z.); lipinszki.zoltan@brc.hu (Z.L.); vigh.laszlo@brc.hu (L.V.)
2 Institute of Medical Biology, University of Tromsø, 9008 Tromsø, Norway
3 Institute of Biophysics, Biological Research Centre, 6726 Szeged, Hungary; kota.zoltan@brc.hu (Z.K.); pali.tibor@brc.hu (T.P.)
4 Single Cell Omics Advanced Core Facility, Hungarian Centre of Excellence for Molecular Medicine, 6726 Szeged, Hungary
* Correspondence: torok.zsolt@brc.hu
† These authors contributed equally to this work.

Citation: Csoboz, B.; Gombos, I.; Kóta, Z.; Dukic, B.; Klement, É.; Varga-Zsíros, V.; Lipinszki, Z.; Páli, T.; Vígh, L.; Török, Z. The Small Heat Shock Protein, HSPB1, Interacts with and Modulates the Physical Structure of Membranes. *Int. J. Mol. Sci.* **2022**, 23, 7317. https://doi.org/10.3390/ijms23137317

Academic Editor: Masoud Jelokhani-Niaraki

Received: 30 May 2022
Accepted: 29 June 2022
Published: 30 June 2022

Publisher's Note: MDPI stays neutral with regard to jurisdictional claims in published maps and institutional affiliations.

Abstract: Small heat shock proteins (sHSPs) have been demonstrated to interact with lipids and modulate the physical state of membranes across species. Through these interactions, sHSPs contribute to the maintenance of membrane integrity. HSPB1 is a major sHSP in mammals, but its lipid interaction profile has so far been unexplored. In this study, we characterized the interaction between HSPB1 and phospholipids. HSPB1 not only associated with membranes via membrane-forming lipids, but also showed a strong affinity towards highly fluid membranes. It participated in the modulation of the physical properties of the interacting membranes by altering rotational and lateral lipid mobility. In addition, the in vivo expression of HSPB1 greatly affected the phase behavior of the plasma membrane under membrane fluidizing stress conditions. In light of our current findings, we propose a new function for HSPB1 as a membrane chaperone.

Keywords: small HSP; lipid–protein interaction; membrane chaperone; membrane fluidity; stress response

1. Introduction

The heat shock response is a highly conserved response of a cell to challenging environmental stress conditions [1], and the conserved set of proteins termed heat shock proteins (HSPs) is critical to the maintenance of cellular integrity under stress conditions. HSPs are primarily responsible for sustaining cellular protein homeostasis via their chaperone activity, by aiding the assembly and folding of proteins, and by inducing their degradation after irreversible damage [2].

Aside from these well-characterized functions, HSPs were also found to interact with biological membranes through binding to membrane lipids [3]. Based on previous observations, the interaction between HSPs and lipids could alleviate the deleterious consequences of stress on membranes and their integral proteins by stabilizing membrane structure [4]. Experimental evidence accumulating over the past several decades suggests that among HSPs, the family of small heat shock proteins (sHSPs) could specifically associate with lipids and membranes [5]. The sHSP family is considered the most functionally diverse of the HSPs, consisting of proteins homologous to each other mostly in their alpha-crystallin domain, forming large hetero/homo-oligomeric complexes [6]. The initial characterization of sHSP interaction with lipids and their subsequent role in preserving the integrity of membranes via these interactions has been established by studying the HSP17 protein of the blue-green algae *Synechocystis* PCC 6803. The greater proportion of HSP17 was found to be associated with thylakoid membranes [7,8]. Its interaction with lipids increased the microviscosity of large unilamellar vesicles, consisting of synthetic or cyanobacterial lipids [9] and stabilized the lamellar liquid crystalline phase at the expense of the non-lamellar phase in membranes composed of the non-bilayer lipid, di-elaidoyl-phosphatidylethanolamine [10]. These studies [8–10] were among the first to demonstrate that HSP17 not only had an affinity towards specific membrane lipids, but also a previously unrecognized ability to stabilize membranes by modulating their lipid phase behavior. Several other examples, such as the sHSP proteins, IbpA and IbpB, of *Escherichia coli* [11] and HSP16 of *Mycobacterium tuberculosis* [12] reinforced the idea of the association of sHSPs with membranes. Moreover, the sHSP, Lo18, of the lactic acid bacterium, *Oenococcus oeni*, was upregulated and localized to the membrane fraction upon administration of the membrane fluidizer, benzyl alcohol (BA) [13]. Lo18 was also shown to interact with liposomes formed from lipids of *Oenococcus oeni* membranes and to reduce membrane fluidity of these vesicles at elevated temperatures [13]. Examples of the lipid interaction of sHSPs in other organisms are sparse. So far, only a few mammalian sHSPs were found to be associated with membranes. While HSP11 [14] and the muscle-specific HSPB2 [15] were linked to mitochondrial membranes, another mammalian sHSP, α-crystallin, was shown to associate with the plasma membranes of fiber cells in the lens of the eye [16] and to have a stabilizing effect on model membranes formed from synthetic lipids [10].

While it is clear that members of the sHSP family interact with membrane lipid vesicles in vitro, it is unclear whether this is a general property of sHSPs among different species and of different members of the sHSP family within a species. Moreover, detailed knowledge of how sHSPs affect membrane physical properties and whether they restore membrane functionality during/after heat shock is still lacking with only few reports. The existing studies suggest that the functional consequences of the association of sHSPs with membranes may include a reduced level of fluidity [9], elevated bilayer stability [10], and the overall restoration of membrane functionality during heat stress. Based on the above studies, we hypothesized that sHSPs may have a membrane-protective role across species to keep the membrane structure intact during elevated temperatures or other membrane-perturbing conditions. The association between sHSPs and membranes may constitute a general mechanism that preserves membrane integrity when the lipid order is compromised. One of the most ubiquitously expressed mammalian sHSPs, HSPB1, appears to be a potential candidate that could serve as a general membrane-stabilizing protein in mammalian cells. HSPB1's membrane association was initially described in the context of its interaction with the membrane-connected cytoskeletal network [17–19], but recently, it has been reported to associate with synthetic liposome membranes [20]. Other physiological observations further substantiated the possible association between HSPB1 and the plasma membrane by demonstrating that HSPB1 was among the few HSPs that were upregulated upon the treatment of mouse cells with the membrane fluidizing compound, BA [21,22].

In this study, we aimed to describe the potential lipid/membrane interaction of HSPB1. We were particularly interested in determining if HSPB1 could affect the physical state of membranes and the mobility of lipids to a similar degree as described for other sHSPs. Overall, we aimed to find out if HSPB1 contributed to membrane integrity during and after stress conditions.

2. Results

2.1. HSPB1 Interacts with Lipids: A Monolayer Study

Initially, we investigated the interaction of HSPB1 with lipids by the Langmuir monolayer method. This technique allowed us to measure the surface pressure increase in the lipid monolayer caused by insertion of a chosen molecule between the lipid molecules spread on the air–water interface. HSPB1 showed a strong preference for a distinct group of phospholipids modeling biological membranes with different fluidity and phase behavior. The increase in surface pressure ($\Delta\pi$), which is proportional to the number and level of proteins inserted into the monolayer, is concentration-dependent (Figure 1a). A protein concentration of 1 μM giving a significant surface pressure increase in all lipid mixtures was chosen for subsequent experiments. To test the effect of membrane fluidity, mono-component lipid monolayers of dioleoyl phosphatidylcholine (DOPC) or 1-palmitoyl-2-oleoylphosphatidylcholine (POPC) were applied. We observed stronger interaction with DOPC, which forms more fluid monolayers than POPC at a given initial surface pressure (Figure 1b). The elevation of surface pressure by protein insertion was reversed after proteinase K treatment (Figure 1d). Biomembranes contain raft microdomains enriched in phospholipids, sphingomyelin (SM), and cholesterol. To mimic the phase behavior of these microdomains, a ternary lipid mixture of POPC, SM, and dihydrocholesterol (DChol) (1:1:1) was applied, which contained both liquid-disordered fluid (Ld) and liquid-ordered (Lo) raft phases [23]. HSPB1 showed a weaker interaction with this raft mixture compared to POPC alone, while $\Delta\pi$ increased significantly when the amount of POPC was doubled in the mixture. Since ternary mixtures with higher PC ratios contain more liquid-disordered phases and higher overall lateral diffusion [23,24] (POPC/SM/DChol = 2:1:1) (Figure 1c), the significantly increased interaction of the protein with the lipids suggests a higher preference of HSPB1 for the liquid-disordered phase or a more fluid lipid phase, in general. This explanation was confirmed by the experiment where the interaction could be decreased by increasing the amount of the Lo phase by increasing the percentage of DChol in the mixture (Figure 1c); however, above a certain cholesterol level (2:1:3), the lipid interaction of HSPB1 was elevated again. The formation of free hydrophobic cholesterol patches in the monolayer at this high cholesterol concentration could account for this phenomenon [25]. These data suggest that HSPB1 tends to interact with lipids or lipid mixtures having higher fluidity, and the presence of free cholesterol could also be crucial for its binding to lipids. As the hydrophobic core of the bilayer becomes more accessible in the presence of free cholesterol [26], it is plausible that it can contribute to the binding of HSPB1 by creating a hydrophobic docking surface on the membrane.

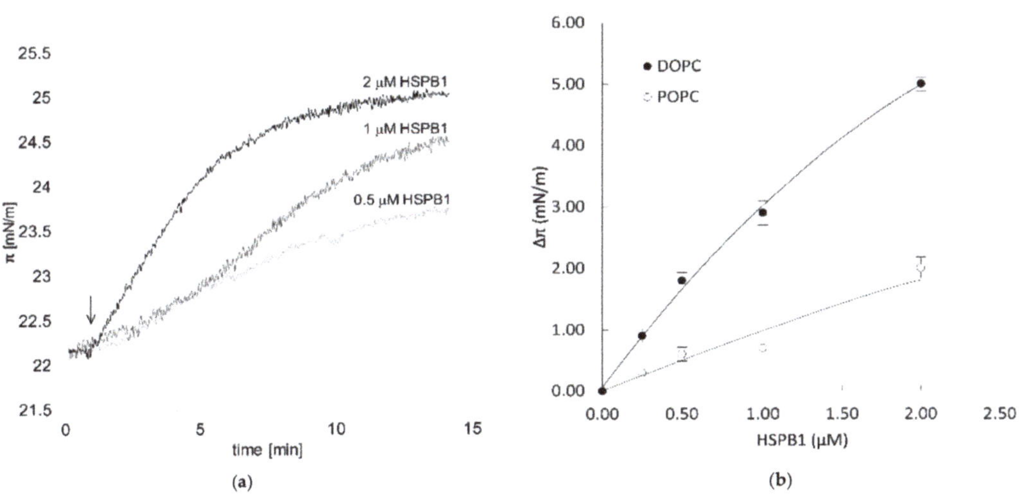

Figure 1. *Cont.*

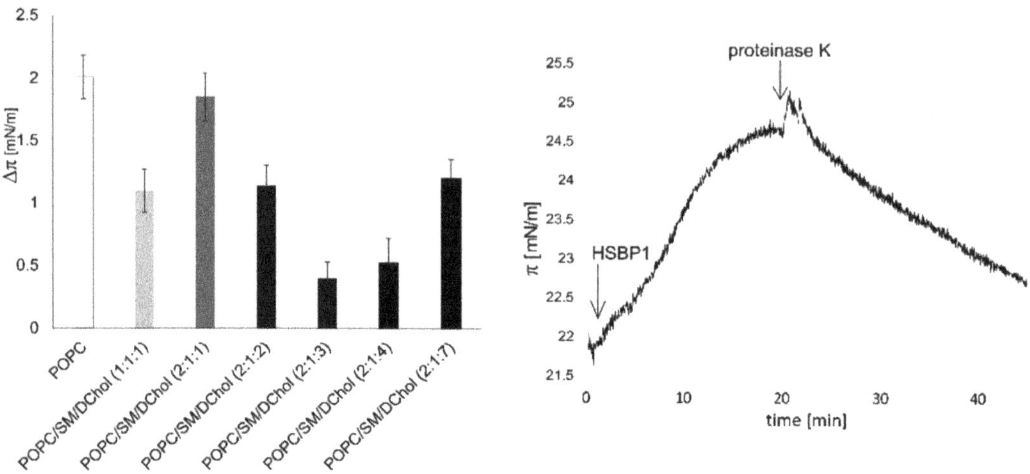

(c) (d)

Figure 1. Lipid preference of HSPB1 measured by monolayer surface pressure. (**a**) HSPB1 was injected underneath the lipid monolayer of POPC and the surface pressure (π_i) was measured after pressure equilibration. The black arrow indicates the time of injection. (**b**) Surface pressure change in DOPC or POPC monolayers five minutes following the addition of different amounts of HSPB1 underneath monolayers formed at an initial surface pressure of 22 mN/m. (**c**) Comparison of 1 μM HSPB1-induced maximum surface pressure increase in monolayers of pure POPC and different ternary lipid mixtures of POPC, SM, and DChol (π_I = 22 mN/m). (**d**) Effect of proteinase K treatment on the interaction of HSPB1 with POPC monolayers. Arrows indicate the time of HSPB1 (1 μM) and proteinase K (1 μg/mL) injection underneath the monolayer. Solid and dashed lines represent polynomial fitted curves.

2.2. HSPB1 Interaction Decreases Both Rotational and Lateral Fluidity in Model Membranes

In recent decades, spin-label EPR spectroscopy has become a useful technique for studying lipids, biological membranes, and lipid–protein interactions. We utilized this method to determine the effect of HSPB1 on the rotational mobility of lipids in model membranes.

The EPR spectra of 5-doxyl-stearic acid (5-SASL) and the corresponding outer splitting values are shown in Figure 2a,b, respectively, in the membranes of DOPC and POPC and in the absence and presence of HSPB1. The spectral parameter outer splitting ($2A_{max}$) represents the mobility of the lipid acyl chain segments bearing the nitroxyl group. The spectra of the pure membranes are qualitatively similar, but their outer splitting reflect the expected difference in chain dynamics of phospholipids with one (POPC) vs. two (DOPC) double bonds; POPC has more restricted rotational dynamics (as evidenced by larger outer splitting) than DOPC, because of the disordering/fluidizing effect of lipid chain unsaturation in membranes. Adding HSPB1 to DOPC causes a reduction in membrane fluidity, indicating an interaction between the lipids and the protein, whereas it has a negligible effect on POPC membranes (Figure 2a,b). We also measured the same samples using 16-SASL spin labeling, in which the nitroxyl label came from the more hydrophobic central region of the membrane. Those spectra did not show any difference in the presence of HSPB1, suggesting that the interaction between lipids and HSPB1 occurs in the head group region of the membrane (data not shown).

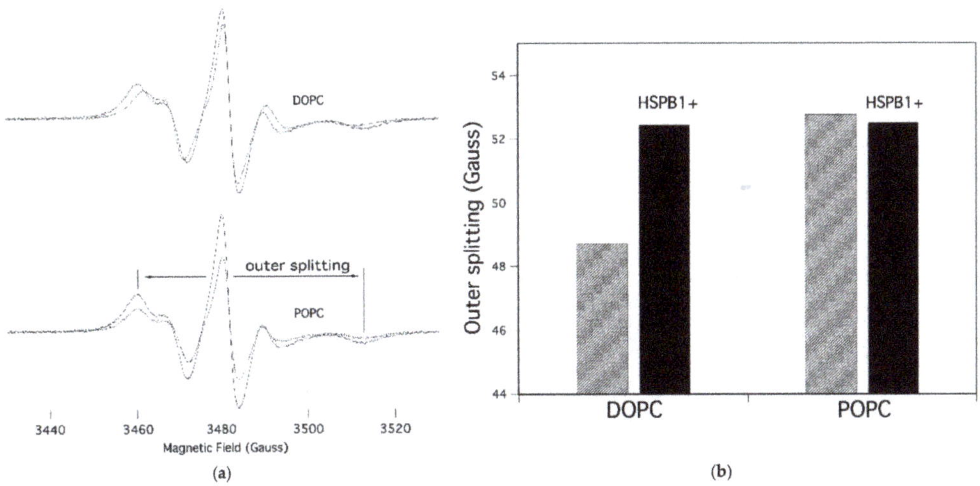

Figure 2. EPR spectra of 5-SASL in different lipid vesicles. (**a**) Spectra measured in the absence (dashed line) and in the presence (solid line) of HSPB1 (lipid to protein ratio, 100:1) are shown together. Spectral parameter outer splitting ($2A_{max}$) is indicated. (**b**) Outer splitting ($2A_{max}$) values of the control (striped) and HSPB1-containing (black) samples in different lipid membranes.

The lateral diffusion of STAR488-PEG-cholesterol in the supported bilayer of DOPC was measured by image-based total internal reflection–fluorescence correlation spectroscopy (ITIR-FCS). The diffusion coefficient (D) reflects the lateral diffusion of the probe and can be calculated from the diffusion law graph [27]. The D value of the fluorescent probe in DOPC was 4 $\mu m/s^2$, which is similar to values found in other fluid phase studies [28]. HSPB1 administration resulted in a 75% drop in D, suggesting drastic structural alterations in the DOPC bilayer induced by the HSPB1 interaction (Figure 3a). The reduced lateral mobility was almost completely restored by subsequent proteinase K treatment. Interestingly, the same HSPB1 treatment affected the mobility of the fluorescent cholesterol probe only slightly and in the opposite direction if the model membrane was made of DOPC/SM/DChol (1:1:1), modeling a liquid-ordered, membrane raft composition (Figure 3b). This is in line with our observation that heat shock-induced HSPB1 is partially present in the membranes of B16-F10 cells, and within the total membrane fraction, it predominantly accumulates in the non-raft membrane regions (Figures S1 and S2). Proteinase K treatment reduced the effect of HSPB1 on the D value in both supported bilayers.

2.3. The Presence of HSPB1 Preserves Membrane Order in Mammalian Cells

After we characterized the fluidity-sensitive interaction of HSPB1 in model systems, we tested its effect on the membranes of living cells. We used *E. coli* bacteria and B16-F10 murine melanoma cells, which both overexpress HSPB1. Environment-sensitive, membrane-intercalating fluorescent probes are widely available and can be used to follow the alterations in membrane properties such as fluidity and lipid packing. One of these probes, di-4-ANEPPDHQ, shows an emission shift according to its localization in a liquid-ordered or liquid-disordered lipid phase. Using this probe allowed us to quantitate membrane order by determining the general polarization (GP). This ratiometric value was calculated from fluorescence intensities recorded in two spectral channels. As a normalized ratio, GP provides a measure of membrane order, in the range between −1 (liquid crystalline) and +1 (gel), where a lower GP value means higher membrane fluidity. First, we tested whether the overexpression of HSPB1 altered the GP of di-4-ANEPPDHQ, and found only slightly more ordered membranes in the bacteria used for HSPB1 production and purification, but no spectral shift in B16-F10 murine cells (Figure 4). This difference could be due to variations in the efficacy of HSPB1 expression in the two cell types (Figure S3). To perturb the membrane structure, we used the membrane-fluidizing agent, benzyl alcohol (BA). Exposure to BA significantly decreased the GP in B16-F10 cells, suggesting a lower membrane order (higher fluidity). However, the fluidizing effect of BA was completely blocked by prior HSPB1 overexpression in both bacterial and mammalian cells (Figure 4). The artificially increased expression of HSPB1 in B16-F10/HSPB1 cells was comparable to the levels induced by heat or BA treatment in wild-type cells (Figure S3B).

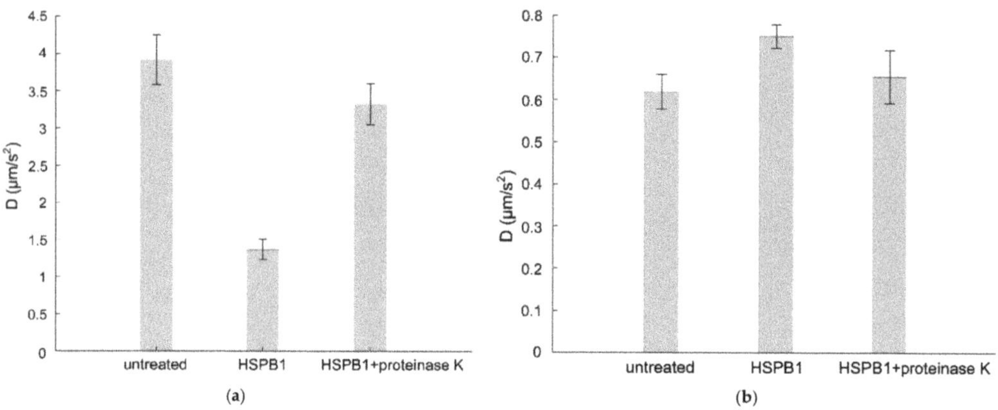

Figure 3. Effect of HSPB1 on the diffusion constant of STAR488-PEG-Chol fluorescent probe in supported bilayers. ITIR-FCS measurement was performed on a supported lipid bilayer made from (**a**) DOPC or (**b**) a 1:1:1 mixture of DOPC/SM/Chol before (first column) and after (second column) 15 min of 50 μg HSPB1 administration. The third column gives diffusion constants of HSPB1-treated samples after incubation with 1 μg/mL proteinase K for 15 min.

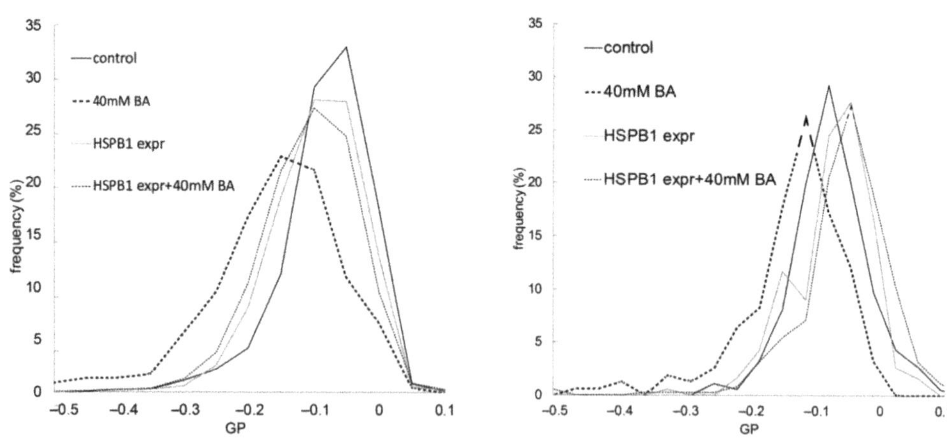

Figure 4. Effect of HSPB1 on membrane order impaired by benzyl alcohol (BA). Membrane order was monitored by calculating the general polarization (GP) of the di-4-ANEPPDHQ environment-sensitive probe. Control and HSPB1-overexpressing (**a**) B16-F10 cells and (**b**) *E. coli* cells were pretreated with 40 mM BA for 15 min after HSPB1 expression. GP values were read out from processed and segmented fluorescence images and the distribution was plotted as a histogram. The Kolmogorov–Smirnov test was performed to analyze the equality of GP distributions in sample pairs. Control and BA-treated samples differed from each other significantly ($p < 0.05$), but GP distribution was equal in cells overexpressing HSPB1 with and without BA administration.

3. Discussion

Biological membranes present a unique barrier, critical for the compartmentalization of a living cell from its environment. Factors compromising the structure of membranes could result in serious consequences for the integrity of a cell. Thus, characterization of the mechanisms for maintaining and repairing cell membrane structures is crucial for a deeper understanding of cellular physiology, stress responses, and the development of membrane-associated pathologies. It is presumed that membrane-interacting proteins, potentially early proto-HSPs, played an important role in the formation and stabilization of early membranes during evolution [29]. It is likely that this could be the origin of the documented membrane interactions for various HSPs [29]. HSPs are part of a cellular stress intervention pathway, and one of the positive results of these interactions could be the stabilization of membranes during stress [4,30].

In our study, we tried to widen the scope of understanding of the action of HSPs at the membrane level by describing the interaction between the stress protein, HSPB1, and lipid membranes in a fluidity-dependent manner. Our findings are consistent with the reported actions of other membrane-associated small HSPs. However, in contrast to the few reports on membrane-interacting mammalian sHSPs, the expression of HSPB1 is not tissue-specific. Thus, the action of this protein on membranes could be considered more as a ubiquitous and robust cellular tool for repair and maintenance of membranes. Based on our data, the range of intracellular functions of HSPB1 can now be extended to include membrane stabilization, through the lipid–protein interaction delineated here. The functional results of this interaction can potentially be manifold. For example, by directly stabilizing the membrane structure, HSPB1 could alleviate the deleterious effects of membrane over-fluidization, which can lead to the aggregation of membrane proteins even in the absence of heat stress.

The bilayer properties of membranes have an influence not only on the function of proteins embedded in them but can also be a defining factor in their capacity for aggregation within the membrane [31]. The fluid state of a membrane can affect the aggregation of normally soluble membrane-resident proteins. This has been indicated by a coarse-grained molecular dynamics simulation where the fluidity of the membrane was shown to have a determining effect on the rate of amyloid nucleation and subsequent aggregation, which was mainly caused by increased exposure to the hydrophobic core of the bilayer [32]. An indication that the size and curvature of a membrane are influencing factors in the amyloid fibrillation process has also been described in the highly fluid DOPC model membranes [33]. Another molecular simulation study demonstrated that increased membrane fluidity could be an important factor for α-synuclein membrane binding [34]. The overexpression of HSPB1 in a transgenic mouse model has also been found to decrease neuronal apoptosis upon membrane fluidizing ethanol treatment [35]. Overall, the stabilization of hyper-fluidic membranes by HSPB1 could have a major impact on the aggregation of membrane proteins. The direct interaction between HSPB1 and membrane proteins could represent an unorthodox chaperone activity separate from its classical function. The induction of HSPB1 has been connected with membrane lipid rafts and caveolae-1 signaling. HSPB1 levels were found to be lower in caveolin-1-deficient breast cancer cells [36], and the treatment of keratinocytes with agents such as filipin or methyl-β-cyclodextrin that disrupt lipid raft-caveolae suppressed sulfur mustard-induced HSPB1 mRNA and protein expression [37]. The alterations in the membrane cholesterol pool have been reported to have a fine-tuning effect on HSPB1 expression under stress conditions [38], suggesting that stress induction of HSPB1 can be partially regulated at the membrane level. For example, the surface coalescence of lipid rafts, which is a key event in lipid raft-associated signal initiation, is highly dependent on the fluid state of the membrane [39]. Thus, it is plausible that the readjustment of the membrane fluid state by HSPB1 could act as a re-setting mechanism, decreasing its expression as part of a negative feedback loop.

In our study, HSPB1 showed strong preference for DOPC as an interaction partner compared to POPC, and was more active in fluid ternary lipid mixtures. Our EPR measurements confirmed that HSPB1 operated in the lipid head group region, suggesting that, unlike HSPA1, it does not incorporate into the bilayer, but the interaction only occurs at the membrane surface. Our data show, however, that this interaction is sufficient for the protein to exert its effect on the physical structure of the bilayer by modulating the rotational mobility of the lipids. The active membrane modulatory nature of this interaction was verified by fluorescence correlation spectroscopy measurements. The presence of HSPB1 decreased the mobility of lipids forming the fluid phase in a supported bilayer, suggesting a general counteraction against imposed membrane fluidity. This effect was greatly reduced when cholesterol was present in the bilayer, which is in agreement with our observation (Figure S1) and with those in the literature [40] that described HSPB1 as only being present in the non-raft fraction of the membrane. HSPB1 was recently reported to interact with POPS, POPG, and POPC vesicles with higher affinity towards lipids with electrostatic charge (POPS and POPG). These experiments, which were performed with lipids containing the same acyl chains, are suggestive of the possibility that HSPB1 interaction with membranes depends on charged lipid head groups [20]. However, our biophysical measurements shed a different light on how critical the fluidity and packing order of lipids is in the lipid interaction of HSPB1. Based on our results, we hypothesize that HSPB1 interaction strongly depends on the phase properties of the interacting membrane, aside from the electrostatic charges of the lipid head groups, thus extending the former model. According to our observations, HSPB1 acts as a general membrane-stabilizing agent when the membrane fluidity increases and the hydrophobic regions of the membrane are increasingly exposed. In this regard, HSPB1 could be considered as a membrane-related stress-responsive actor that balances out the deleterious effect of sudden increases in membrane fluidity. HSPB1 modulates the membrane towards a more ordered state, as revealed by our experiments in which fluidization by BA was prevented by overexpression of the small heat shock protein in mammalian and bacterial cells. This feature of HSPB1 likely evolved under heat stress in order to compensate for the increased membrane fluidizing effect of heat. One could argue that the immediate effect of heat-induced membrane fluidization did not overlap with the stress-induced expression of this protein, as the former is immediate and the latter is a consequent event. We suggest that the membrane stabilizing function of HSPB1 represents an adaptive response of cells, allowing them to build up an acquired resistance to the next stress event. Therefore, we propose that stress-induced HSPB1 is a part of a toolkit for acquired stress resistance against membrane over-fluidization, and possibly against membrane-level stress conditions, in general. This hypothesis is in line with the widely documented literature on the ubiquitous role of HSPB1 in acquired stress resistance.

4. Materials and Methods

4.1. Expression and Purification of HSPB1

Recombinant human HSPB1 was expressed in *Escherichia coli* BL21 (DE3) (Thermo Fisher Scientific, Waltham, MA, USA) cells by using the pAK3038Hsp27 plasmid [41]. The subsequent purification of the recombinant protein was carried out as described in Buchner et al. [42]. In brief, *E. coli* BL21 (DE3) cells harboring the pAK3038Hsp27 plasmid were grown in the presence of ampicillin (Sigma-Aldrich, St. Louis, MO, USA) to the desired optical density, then induced with 0.5 mM isopropylthiogalactoside (IPTG) (Sigma-Aldrich). The expression of HSPB1 upon IPTG induction was assessed by Western blotting (Figure S3A). Cells were incubated for 3 h upon IPTG induction, then harvested by centrifugation for 10 min at 2600× g and at 4 °C and lysed as described in [42]. After lysis, the proteins were precipitated with 35% ammonium sulfate and purified by ion exchange chromatography on a Fractogel EMD DEAE column (Merck Millipore, Burlington, MA, USA) using a 50 to 600 mM NaCl linear gradient. Eluted fractions were characterized for the presence of HSPB1 by SDS-PAGE and Western blot analysis (anti-HSPB1, SMC-161, StressMarq, Victoria, BC, Canada).

4.2. Langmuir Monolayer Experiments

Monolayer experiments were carried out essentially as described in [43] using a KSV3000 Langmuir-Blodgett instrument (KSV Instruments, Helsinki, Finland) and a Teflon dish containing 6.5 mL of PBS buffer with a surface area of 9 cm^2 at 23 °C. Surface pressure was measured by the Wilhelmy method, using a platinum plate. Monomolecular lipid layers were spread on a buffer–air interface to give the desired initial surface pressure. The sub-phase was continuously stirred with a magnetic bar. Different concentrations of purified HSPB1, and subsequently, 1 µg/mL proteinase K (Thermo Fisher Scientific) was added underneath the monolayer with constant stirring.

4.3. EPR Spectroscopy

For spin-label EPR measurements, 5 µL of 5-SASL or 16-SASL spin label solution (Sigma-Aldrich) in ethanol (2 mg/mL) was added to 1 mg of lipid in chloroform solution (resulting in a ca. 50:1 lipid/spin label molar ratio). After vortexing, the solution was dried under nitrogen gas and incubated under vacuum overnight at room temperature. The dried samples were hydrated with PBS buffer (pH 7) above the melting temperature of the lipid. HSPB1 protein was added to the lipid suspension from an 11 mg/mL stock solution in PBS, resulting in a lipid:protein molar ratio of 100:1 in the samples.

A glass capillary with an internal diameter of 1 mm was then filled with 10 µL of spin-labeled sample. The EPR spectra were recorded using either a Bruker (Rheinstetten, Germany) ECS-106 or a Bruker ELEXSYS-II E580 X-band spectrometer at room temperature, with the following instrument settings: microwave frequency of 9.4 GHz, microwave power of 5 mW, field modulation of 0.7 G, scan range of 100 G, and conversion time of 40.96 s. Final spectra were the mean of four scans. Data analysis was performed using Igor Pro (Wavemetrics, Inc.; Portland, OR, USA).

4.4. ITIR-FCS

Giant unilamellar vesicles of DOPC or ternary lipid mixtures of DOPC/egg-sphingom yelin/cholesterol (Sigma-Aldrich) were prepared in 100 mM sucrose solution by electroswelling using a Nanion Vesicle Prep Pro device (Nanion, Munich, Germany). Vesicle suspensions in 250 µL of ultra-pure water were pipetted onto coverslips. After 20 min, the supported lipid bilayers were formed and the solution was replaced with 250 µL of ultra-pure water. Objective-type TIRF illumination was used to achieve the thinnest excited sample volume, with a high numerical aperture objective (alphaPlan-FLUAR100; Zeiss, Oberkochen, Germany). The data were acquired using a ProEM512 EMCCD camera (Princeton Instruments, Trenton, NJ, USA) with a 3 ms effective exposure time and 20 × 40 pixel acquisition area per measurement (pixel size 0.16 µm). The ImFCS plugin (https://www.dbs.nus.edu.sg/lab/BFL/imfcs_image_j_plugin.html, accessed on 16 May 2022) for ImageJ software was used for data evaluation (Rasband, W.S., ImageJ, U.S. National Institutes of Health, Bethesda, MD, USA, https://imagej.nih.gov/ij/, 1997–2018, accessed on 16 May 2022). The data analysis was performed as described earlier [44].

4.5. Imaging of di-4-ANEPPDHQ and Calculation of General Polarization (GP)

E. coli BL21 (DE3) (Thermo Fisher Scientific) and B16-F10 mouse melanoma cells (ATCC) were used in these experiments. For inducible protein expression, the *E. coli* BL21 (DE3) cells were induced with IPTG as described earlier in Section 4 (Figure S3A). In the case of the B16-F10 cells, HSPB1 was inserted into the pcDNA 4/TO (Thermo Fisher Scientific) expression vector, and the pcDNA6/TR vector (Thermo Fisher Scientific) was used as the source of the tetracycline repressor. The cells were co-transfected with both vectors using ExGen 500 (Thermo Fisher Scientific). Colonies were selected by the simultaneous addition of zeocin and blasticidine (both from Thermo Fisher Scientific). The expression of HSPB1 was induced by the addition of doxycycline hyclate (Sigma-Aldrich, D9891) to the cell culture media (2 µg/mL) for 24 h before the experiment. The expression of HSPB1 upon doxycycline induction, heat shock, and membrane fluidizing BA treatment was assessed by Western blot (Figure S3B).

Bacterial and mammalian cells were labeled with the environment-sensitive, membrane-incorporating dye, di-4-ANEPPDHQ (Thermo Fisher Scientific), added to the growth medium at final concentrations of 5 μM and 1.5 μM, respectively. Following a 5 min incubation at RT, image acquisition was carried out on a Leica TCS SP5 confocal system. An argon ion laser at 488 nm was used for excitation, and detection ranges of PMTs were set to 500–580 nm and 620–720 nm for the two emission channels, respectively. The di-4-ANEPPDHQ data were typically displayed as pseudo-colored GP images. The GP values were calculated according to the following equation: GP = $(I\_(500\text{-}580) - G\, I\_(620\text{-}750))/(I\_(500\text{-}580) + G\, I\_(620\text{-}750))$ using an ImageJ macro. *I* represents the intensity in each pixel in the indicated spectral channel (numbers are in nm) and *G* represents the calibration factor, which compensates for the differences in the efficiency of collection in the two channels. Further image processing of segmenting cells and cell membranes was performed with CellProfiler [45] in combination with ilastik [46]. More than 1000 ROIs of membranes were segmented on at least 5 images per sample. GP values were read out and sorted into classes to calculate distribution. The Kolmogorov–Smirnov test was performed to analyze the equality of GP distributions in sample pairs using the MATLAB software (MathWorks, Natick, MA, USA).

HSPB1 expression in *E. coli* cells was induced with 0.5 μM IPTG (Sigma-Aldrich) 4 h before the measurements. Doxycycline hyclate (Sigma-Aldrich) was used to induce the expression of HSPB1 in B16F10 cells 24 h before the measurement. Both bacterial and mammalian cells were incubated with BA (Sigma-Aldrich) for 15 min prior to the measurement.

Supplementary Materials: The following supporting information is available at: https://www.mdpi.com/article/10.3390/ijms23137317/s1. References [41,47] are cited in the Supplementary Materials.

Author Contributions: Conceptualization: L.V. and Z.T.; funding acquisition: L.V. and Z.T.; investigation: B.C., I.G., Z.K., B.D., É.K., V.V.-Z., Z.L. and T.P.; writing—original draft preparation: B.C. and I.G.; writing—review and editing: B.C., I.G. and Z.T.; visualization: I.G., B.C. and Z.K.; project administration: Z.T. All authors have read and agreed to the published version of the manuscript.

Funding: This research was funded by the Hungarian Basic Research Fund (OTKA ANN 132280), by the Hungarian Academy of Sciences (Lendület Grant LP2017-7/2017), and by the EU's Horizon 2020 Research and Innovation Program under grant agreement no. 739593.

Institutional Review Board Statement: Not applicable.

Informed Consent Statement: Not applicable.

Data Availability Statement: The datasets generated and/or analyzed during this study are available from the corresponding authors on request.

Acknowledgments: We are truly grateful for the HSPB1 expression plasmids provided by Mathias Gaestel.

Conflicts of Interest: The authors declare no competing financial interest.

Abbreviations

DOPC: 1,2-dioleoyl-sn-glycero-3-phosphocholine; POPC: 1-palmitoyl-2-oleoyl-sn-glycero-3-phosphocholine; SM: sphingomyelin; DChol: dansyl-cholesterol.

References

1. Lindquist, S. The Heat-Shock Response. *Annu. Rev. Biochem.* **1986**, *55*, 1151–1191. [CrossRef]
2. Kampinga, H. Chaperones in Preventing Protein Denaturation in Living Cells and Protecting Against Cellular Stress. *Handb. Exp. Pharmacol.* **2006**, *172*, 1–42. [CrossRef]

3. Török, Z.; Crul, T.; Maresca, B.; Schütz, G.J.; Viana, F.; Dindia, L.; Piotto, S.; Brameshuber, M.; Balogh, G.; Péter, M.; et al. Plasma membranes as heat stress sensors: From lipid-controlled molecular switches to therapeutic applications. *Biochim. Biophys. Acta* **2014**, *1838*, 1594–1618. [CrossRef]
4. Horváth, I.; Multhoff, G.; Sonnleitner, A.; Vígh, L. Membrane-associated stress proteins: More than simply chaperones. *Biochim. Biophys. Acta* **2008**, *1778*, 1653–1664. [CrossRef]
5. Nakamoto, H.; Vígh, L. The small heat shock proteins and their clients. *Cell. Mol. Life Sci.* **2006**, *64*, 294–306. [CrossRef]
6. Carra, S.; Alberti, S.; Arrigo, P.A.; Benesch, J.L.; Benjamin, I.J.; Boelens, W.; Bartelt-Kirbach, B.; Brundel, B.J.J.M.; Buchner, J.; Bukau, B.; et al. The growing world of small heat shock proteins: From structure to functions. *Cell Stress Chaperones* **2017**, *22*, 601–611. [CrossRef]
7. Horváth, I.; Glatz, A.; Varvasovszki, V.; Török, Z.; Páli, T.; Balogh, G.; Kovács, E.; Nádasdi, L.; Benkö, S.; Joó, F.; et al. Membrane physical state controls the signaling mechanism of the heat shock response in *Synechocystis* PCC 6803: Identification of *hsp17* as a "fluidity gene". *Proc. Natl. Acad. Sci. USA* **1998**, *95*, 3513–3518. [CrossRef]
8. Balogi, Z.; Cheregi, O.; Giese, K.C.; Juhász, K.; Vierling, E.; Vass, I.; Vígh, L.; Horváth, I. A Mutant Small Heat Shock Protein with Increased Thylakoid Association Provides an Elevated Resistance Against UV-B Damage in Synechocystis 6803. *J. Biol. Chem.* **2008**, *283*, 22983–22991. [CrossRef]
9. Török, Z.; Goloubinoff, P.; Horváth, I.; Tsvetkova, N.M.; Glatz, A.; Balogh, G.; Varvasovszki, V.; Los, D.A.; Vierling, E.; Crowe, J.H.; et al. *Synechocystis* HSP17 is an amphitropic protein that stabilizes heat-stressed membranes and binds denatured proteins for subsequent chaperone-mediated refolding. *Proc. Natl. Acad. Sci. USA* **2001**, *98*, 3098–3103. [CrossRef]
10. Tsvetkova, N.M.; Horváth, I.; Török, Z.; Wolkers, W.F.; Balogi, Z.; Shigapova, N.; Crowe, L.M.; Tablin, F.; Vierling, E.; Crowe, J.H.; et al. Small heat-shock proteins regulate membrane lipid polymorphism. *Proc. Natl. Acad. Sci. USA* **2002**, *99*, 13504–13509. [CrossRef]
11. Miyake, T.; Araki, S.; Tsuchido, T. Synthesis and Sedimentation of a Subset of 15-kDa Heat Shock Proteins in *Escherichia coli* Cells Recovering from Sublethal Heat Stress. *Biosci. Biotechnol. Biochem.* **1993**, *57*, 578–583. [CrossRef]
12. Lee, B.Y.; A Hefta, S.; Brennan, P.J. Characterization of the major membrane protein of virulent Mycobacterium tuberculosis. *Infect. Immun.* **1992**, *60*, 2066–2074. [CrossRef]
13. Coucheney, F.; Gal, L.; Beney, L.; Lherminier, J.; Gervais, P.; Guzzo, J. A small HSP, Lo18, interacts with the cell membrane and modulates lipid physical state under heat shock conditions in a lactic acid bacterium. *Biochim. Biophys. Acta-Biomembr.* **2005**, *1720*, 92–98. [CrossRef]
14. Bellyei, S.; Szigeti, A.; Pozsgai, E.; Boronkai, A.; Gomori, E.; Hocsak, E.; Farkas, R.; Sumegi, B.; Gallyas, F. Preventing apoptotic cell death by a novel small heat shock protein. *Eur. J. Cell Biol.* **2007**, *86*, 161–171. [CrossRef]
15. Nakagawa, M.; Tsujimoto, N.; Nakagawa, H.; Iwaki, T.; Fukumaki, Y.; Iwaki, A. Association of HSPB2, a Member of the Small Heat Shock Protein Family, with Mitochondria. *Exp. Cell Res.* **2001**, *271*, 161–168. [CrossRef]
16. Cobb, B.A.; Petrash, J.M. Characterization of α-Crystallin-Plasma Membrane Binding. *J. Biol. Chem.* **2000**, *275*, 6664–6672. [CrossRef]
17. Guay, J.; Lambert, H.; Gingras-Breton, G.; Lavoie, J.N.; Huot, J.; Landry, J. Regulation of actin filament dynamics by p38 map kinase-mediated phosphorylation of heat shock protein 27. *J. Cell Sci.* **1997**, *110*, 357–368. [CrossRef]
18. Clarke, J.P.; Mearow, K.M. Cell Stress Promotes the Association of Phosphorylated HspB1 with F-Actin. *PLoS ONE* **2013**, *8*, e68978. [CrossRef]
19. Lavoie, J.N.; Lambert, H.; Hickey, E.; Weber, L.A.; Landry, J. Modulation of cellular thermoresistance and actin filament stability accompanies phosphorylation-induced changes in the oligomeric structure of heat shock protein 27. *Mol. Cell. Biol.* **1995**, *15*, 505–516. [CrossRef]
20. De Maio, A.; Cauvi, D.M.; Capone, R.; Bello, I.; Egberts, W.V.; Arispe, N.; Boelens, W. The small heat shock proteins, HSPB1 and HSPB5, interact differently with lipid membranes. *Cell Stress Chaperones* **2019**, *24*, 947–956. [CrossRef]
21. Nagy, E.; Balogi, Z.; Gombos, I.; Åkerfelt, M.; Björkbom, A.; Balogh, G.; Török, Z.; Maslyanko, A.; Fiszer-Kierzkowska, A.; Lisowska, K.; et al. Hyperfluidization-coupled membrane microdomain reorganization is linked to activation of the heat shock response in a murine melanoma cell line. *Proc. Natl. Acad. Sci. USA* **2007**, *104*, 7945–7950. [CrossRef]
22. Csoboz, B.; Balogh, G.; Kusz, E.; Gombos, I.; Peter, M.; Crul, T.; Gungor, B.; Haracska, L.; Bogdanovics, G.; Torok, Z.; et al. Membrane fluidity matters: Hyperthermia from the aspects of lipids and membranes. *Int. J. Hyperth.* **2013**, *29*, 491–499. [CrossRef]
23. Ionova, I.V.; Livshits, V.A.; Marsh, D. Phase Diagram of Ternary Cholesterol/Palmitoylsphingomyelin/Palmitoyloleoyl-Phosphatidylcholine Mixtures: Spin-Label EPR Study of Lipid-Raft Formation. *Biophys. J.* **2012**, *102*, 1856–1865. [CrossRef]
24. Eckstein, J.; Berndt, N.; Holzhütter, H.-G. Computer Simulations Suggest a Key Role of Membranous Nanodomains in Biliary Lipid Secretion. *PLoS Comput. Biol.* **2015**, *11*, e1004033. [CrossRef]
25. Török, Z.; Pilbat, A.-M.; Gombos, I.; Hocsák, E.; Sümegi, B.; Horváth, I.; Vígh, L. Evidence on Cholesterol-Controlled Lipid Raft Interaction of the Small Heat Shock Protein HSPB11. *Heat Shock Proteins Cell. Traffick. Cell Stress Proteins Health Dis.* **2012**, *6*, 75–85. [CrossRef]
26. Kessel, A.; Ben-Tal, N.; May, S. Interactions of Cholesterol with Lipid Bilayers: The Preferred Configuration and Fluctuations. *Biophys. J.* **2001**, *81*, 643–658. [CrossRef]
27. Ng, X.W.; Bag, N.; Wohland, T. Characterization of Lipid and Cell Membrane Organization by the Fluorescence Correlation Spectroscopy Diffusion Law. *Chimia (Aarau)* **2015**, *69*, 112–119. [CrossRef]

28. Bag, N.; Sankaran, J.; Paul, A.; Kraut, R.S.; Wohland, T. Calibration and Limits of Camera-Based Fluorescence Correlation Spectroscopy: A Supported Lipid Bilayer Study. *ChemPhysChem* **2012**, *13*, 2784–2794. [CrossRef]
29. De Maio, A.; Hightower, L. The interaction of heat shock proteins with cellular membranes: A historical perspective. *Cell Stress Chaperones* **2021**, *26*, 769–783. [CrossRef]
30. Vigh, L.; Nakamoto, H.; Landry, J.; Muñoz, A.G.; Harwood, J.; Horvath, I. Membrane Regulation of the Stress Response from Prokaryotic Models to Mammalian Cells. *Ann. N. Y. Acad. Sci.* **2007**, *1113*, 40–51. [CrossRef]
31. Burke, K.A.; Yates, E.A.; Legleiter, J. Biophysical Insights into How Surfaces, Including Lipid Membranes, Modulate Protein Aggregation Related to Neurodegeneration. *Front. Neurol.* **2013**, *4*, 17. [CrossRef]
32. Krausser, J.; Knowles, T.P.J.; Šarić, A. Physical mechanisms of amyloid nucleation on fluid membranes. *Proc. Natl. Acad. Sci. USA* **2020**, *117*, 33090–33098. [CrossRef]
33. Caillon, L.; Lequin, O.; Khemtémourian, L. Evaluation of membrane models and their composition for islet amyloid polypeptide-membrane aggregation. *Biochim. Biophys. Acta-Biomembr.* **2013**, *1828*, 2091–2098. [CrossRef]
34. Viennet, T.; Wördehoff, M.M.; Uluca, B.; Poojari, C.; Shaykhalishahi, H.; Willbold, D.; Strodel, B.; Heise, H.; Buell, A.K.; Hoyer, W.; et al. Structural insights from lipid-bilayer nanodiscs link α-Synuclein membrane-binding modes to amyloid fibril formation. *Commun. Biol.* **2018**, *1*, 44. [CrossRef]
35. Dukay, B.; Walter, F.R.; Vigh, J.P.; Barabási, B.; Hajdu, P.; Balassa, T.; Migh, E.; Kincses, A.; Hoyk, Z.; Szögi, T.; et al. Neuroinflammatory processes are augmented in mice overexpressing human heat-shock protein B1 following ethanol-induced brain injury. *J. Neuroinflammation* **2021**, *18*, 22. [CrossRef]
36. Ciocca, D.R.; Cuello-Carrión, F.D.; Natoli, A.L.; Restall, C.; Anderson, R.L. Absence of caveolin-1 alters heat shock protein expression in spontaneous mammary tumors driven by Her-2/neu expression. Histochem. *Cell Biol.* **2012**, *137*, 187–194.
37. Black, A.T.; Hayden, P.J.; Casillas, R.P.; Heck, D.E.; Gerecke, D.R.; Sinko, P.J.; Laskin, D.L.; Laskin, J.D. Regulation of Hsp27 and Hsp70 expression in human and mouse skin construct models by caveolae following exposure to the model sulfur mustard vesicant, 2-chloroethyl ethyl sulfide. *Toxicol. Appl. Pharmacol.* **2011**, *253*, 112–120. [CrossRef]
38. Crul, T.; Csoboz, B.; Gombos, I.; Marton, A.; Peter, M.; Balogh, G.; Vizler, C.; Szente, L.; Vigh, L. Modulation of Plasma Membrane Composition and Microdomain Organization Impairs Heat Shock Protein Expression in B16-F10 Mouse Melanoma Cells. *Cells* **2020**, *9*, 951. [CrossRef]
39. Balogh, G.; Maulucci, G.; Gombos, I.; Horváth, I.; Török, Z.; Péter, M.; Fodor, E.; Páli, T.; Benkő, S.; Parasassi, T.; et al. Heat Stress Causes Spatially-Distinct Membrane Re-Modelling in K562 Leukemia Cells. *PLoS ONE* **2011**, *6*, e21182. [CrossRef]
40. Vega, V.L.; Rodríguez-Silva, M.; Frey, T.; Gehrmann, M.; Diaz, J.C.; Steinem, C.; Multhoff, G.; Arispe, N.; De Maio, A. Hsp70 Translocates into the Plasma Membrane after Stress and Is Released into the Extracellular Environment in a Membrane-Associated Form that Activates Macrophages. *J. Immunol.* **2008**, *180*, 4299–4307. [CrossRef]
41. Jakob, U.; Gaestel, M.; Engel, K.; Buchner, J. Small heat shock proteins are molecular chaperones. *J. Biol. Chem.* **1993**, *268*, 1517–1520. [CrossRef]
42. Buchner, J.; Ehrnsperger, M.; Gaestel, M.; Walke, S. Purification and characterization of small heat shock proteins. *Methods Enzymol.* **1998**, *290*, 339–349. [CrossRef]
43. Glatz, A.; Pilbat, A.-M.; Németh, G.L.; Vince-Kontár, K.; Jósvay, K.; Hunya, Á.; Udvardy, A.; Gombos, I.; Péter, M.; Balogh, G.; et al. Involvement of small heat shock proteins, trehalose, and lipids in the thermal stress management in Schizosaccharomyces pombe. *Cell Stress Chaperones* **2015**, *21*, 327–338. [CrossRef]
44. Peksel, B.; Gombos, I.; Péter, M.; Vigh, L.; Tiszlavicz, Á; Brameshuber, M.; Balogh, G.; Schütz, G.J.; Horváth, I.; Török, Z. Mild heat induces a distinct "eustress" response in Chinese Hamster Ovary cells but does not induce heat shock protein synthesis. *Sci. Rep.* **2017**, *7*, 15643. [CrossRef]
45. Stirling, D.R.; Swain-Bowden, M.J.; Lucas, A.M.; Carpenter, A.E.; Cimini, B.A.; Goodman, A. CellProfiler 4: Improvements in speed, utility and usability. *BMC Bioinform.* **2021**, *22*, 433. [CrossRef]
46. Berg, S.; Kutra, D.; Kroeger, T.; Straehle, C.N.; Kausler, B.X.; Haubold, C.; Schiegg, M.; Ales, J.; Beier, T.; Rudy, M.; et al. ilastik: Interactive machine learning for (bio)image analysis. *Nat. Methods* **2019**, *16*, 1226–1232. [CrossRef]
47. Csoboz, B.; Gombos, I.; Tatrai, E.; Tovari, J.; Kiss, A.L.; Horvath, I.; Vigh, L. Chemotherapy induced PRL3 expression promotes cancer growth via plasma membrane remodeling and specific alterations of caveolae-associated signaling. *Cell Commun. Signal.* **2018**, *16*, 51. [CrossRef]

International Journal of *Molecular Sciences*

MDPI

Review

Structural Features of Cytochrome b_5–Cytochrome b_5 Reductase Complex Formation and Implications for the Intramolecular Dynamics of Cytochrome b_5 Reductase

Carlos Gutiérrez-Merino [1,*], Oscar H. Martínez-Costa [2,3], Maria Monsalve [2] and Alejandro K. Samhan-Arias [2,3,*]

1 Department of Biochemistry and Molecular Biology, Faculty of Sciences and Instituto de Biomarcadores de Patologías Moleculares, Universidad de Extremadura, Av. Elvas S/N, 06006 Badajoz, Spain
2 Instituto de Investigaciones Biomédicas 'Alberto Sols' (CSIC-UAM), Arturo Duperier, 4, 28029 Madrid, Spain; oscar.martinez@uam.es (O.H.M.-C.); mpmonsalve@iib.uam.es (M.M.)
3 Department of Biochemistry, Faculty of Medicine, Universidad Autónoma de Madrid (UAM), Arzobispo Morcillo, 4, 28029 Madrid, Spain
* Correspondence: carlosgm@unex.es (C.G.-M.); alejandro.samhan@uam.es (A.K.S.-A.)

Citation: Gutiérrez-Merino, C.; Martínez-Costa, O.H.; Monsalve, M.; Samhan-Arias, A.K. Structural Features of Cytochrome b_5–Cytochrome b_5 Reductase Complex Formation and Implications for the Intramolecular Dynamics of Cytochrome b_5 Reductase. *Int. J. Mol. Sci.* **2022**, *23*, 118. https://doi.org/10.3390/ijms23010118

Academic Editors: Vitaliy Borisov and Gianfranco Gilardi

Received: 19 November 2021
Accepted: 21 December 2021
Published: 23 December 2021

Abstract: Membrane cytochrome b_5 reductase is a pleiotropic oxidoreductase that uses primarily soluble reduced nicotinamide adenine dinucleotide (NADH) as an electron donor to reduce multiple biological acceptors localized in cellular membranes. Some of the biological acceptors of the reductase and coupled redox proteins might eventually transfer electrons to oxygen to form reactive oxygen species. Additionally, an inefficient electron transfer to redox acceptors can lead to electron uncoupling and superoxide anion formation by the reductase. Many efforts have been made to characterize the involved catalytic domains in the electron transfer from the reduced flavoprotein to its electron acceptors, such as cytochrome b_5, through a detailed description of the flavin and NADH-binding sites. This information might help to understand better the processes and modifications involved in reactive oxygen formation by the cytochrome b_5 reductase. Nevertheless, more than half a century since this enzyme was first purified, the one-electron transfer process toward potential electron acceptors of the reductase is still only partially understood. New advances in computational analysis of protein structures allow predicting the intramolecular protein dynamics, identifying potential functional sites, or evaluating the effects of microenvironment changes in protein structure and dynamics. We applied this approach to characterize further the roles of amino acid domains within cytochrome b_5 reductase structure, part of the catalytic domain, and several sensors and structural domains involved in the interactions with cytochrome b_5 and other electron acceptors. The computational analysis results allowed us to rationalize some of the available spectroscopic data regarding ligand-induced conformational changes leading to an increase in the flavin adenine dinucleotide (FAD) solvent-exposed surface, which has been previously correlated with the formation of complexes with electron acceptors.

Keywords: cytochrome b_5 reductase; cytochrome b_5; superoxide anion radical; electron transfer; protein intrinsic dynamics

1. The Electron Transfer from the Cytochrome b_5 Reductase to Multiple Acceptors and Implication in Reactive Oxygen Species Formation

Flavoproteins can be categorized into three main types: oxidases, dehydrogenases, and monooxygenases. The distinction between them has been made based on the use of a semiquinone flavin radical as part of the catalytic mechanism, which is present in dehydrogenases but absent in monooxygenases and oxidases [1]. Early on, based on kinetic studies using stopped-flow techniques, Prof. Philipp Sttritmatter suggested the electron transfer mechanism of the cytochrome b_5 reductase (Cb_5R) [2]. This electron transfer process depends on the rapid formation of a complex with NADH, which is rapidly oxidized and meets the requirements for an intermediate generation in the catalytic reaction. In addition, Prof. Sttritmatter [2] showed evidence of the existence of, at least, one intermediate involved in flavin reduction and a second implicated in flavin reoxidation, both part of a complete oxidative cycle. The reaction of NADH with the oxidized enzyme is fast and occurs in less than 2 ms, prompting the formation of a stable complex. Since the oxidation of the reduced enzyme complex is faster than the reduction in the flavoprotein, the oxidized flavoprotein and the reduced pyridine nucleotide complex are the predominant forms during the catalytic turnover. In this reaction, one-electron oxidation can be measured, leading to the formation of a flavin semiquinone radical which can be spectroscopically measured [3]. The formation of the flavin semiquinone radical was reported to be independent of the substrate used: cytochrome b_5 (Cb_5) or ferricyanide. Later, it was shown that Cb_5 presence increases the binding constant of oxidized nicotinamide adenine dinucleotide (NAD^+) of the quaternary complex [4]. The pyridine nucleotide remains bound to the enzyme throughout the entire oxidation process. The final oxidation step is the rate-limiting step in the catalytic cycle. NAD^+ slowly autoxidizes and disproportionates the FAD semiquinone [4]. As shown later by pulse radiolysis techniques, one electron can react with NAD^+, yielding $NAD^\bullet$ [5]. The electron can be transferred to the NAD^+-bound oxidized enzyme to form the blue and red semiquinone or a mixture of the two forms of the enzyme, where pK_a value of this flavin radical was approximately 6.3 [5]. Very few changes in the electron transfer mechanism have been shown since this mechanism was suggested. The involvement of one tyrosine residue and SH groups was suggested early on to be implicated in the catalytic processes, although no specific amino acid residues were identified [2].

At that time, the role of free radicals produced by enzymes in diseases had not been explored. Notably, the transient formation of a flavin radical in the electron transfer mechanism suggested that reactive oxygen species could be formed within the catalytical mechanism of the Cb_5R. When oxygen is present in the media, reactive oxygen species production can be catalyzed by flavins and flavoproteins [1,6]. In the case of flavoproteins, oxidases, hydroxylases, and dehydrogenases, they all rapidly or slowly react with oxygen in a flavin-semiquinone-dependent or -independent way leading to the formation of free radicals [1]. However, this might be modulated by endogenous electron acceptors or biological molecules present in the subcellular compartments where Cb_5R localizes.

Cb_5R is a pleiotropic oxidoreductase enzyme that donates electrons to many redox acceptors aside from oxygen. This enzyme modulates many metabolic pathways by electron transfer to biological electron acceptors. This enzyme has different isoforms, but the catalytic domain is highly conserved between them [7]. The presence of a soluble isoform Cb_5R was described in erythrocytes as the main responsible for the enzymatic recycling of methemoglobin [8]. The membrane isoform of Cb_5R has a domain deeply inserted within the lipid bilayer that anchors this enzyme to different subcellular membranes. The membrane isoform is located in the outer leaflet of the endoplasmic reticulum (ER), the mitochondrial outer membrane (MOM), and the plasma membrane [9]. The Cb_5R membrane isoform is made up of 300 amino acid residues comprising a soluble domain, which is formed by 275 residues oriented toward the cytosol, and the N-terminal tail of 24 residues anchoring the protein to the membrane [10]. This soluble domain is similar to the erythrocyte soluble isoform that only presents the C-terminal soluble domain [11]. The membrane isoform is present in almost all mammalian cells as a membrane-bound protein, including erythrocytes. The molecular mechanism for the distribution of the reductase across different subcellular locations is not completely clear. The topography of the membrane domain allows the membrane isoform to penetrate deeply into the lipid bilayer. Notably, the mechanism for spreading along the membrane and how the enzyme confines to the cytoplasmic leaflet in a hairpin conformation have been questioned [11,12].

1.1. The Endoplasmic Reticulum Cb_5R

Cb_5R membrane isoforms have N-terminal domains that anchor the soluble domain into the microsomal membrane [10,11], and the MOM [13]. A single point mutation demonstrated that the same gene product localizes the protein in two different subcellular compartments [14]. The primary function of the Cb_5R located at the ER is the electron transfer to desaturases and participation in detoxification pathways in conjunction with Cb_5 and cytochrome P450s [15]. Direct mutagenesis experiments showed that myristylation was key for the reductase targeting the MOM, while the non-myristoylated mutant was only found in the ER [16]. The first seven residues of the membrane Cb_5R isoform translation product, which includes the methionine, constitute the myristylation signal [17]. The codons that target the protein for myristylation can be excluded or included in the transcript in a tissue-specific manner [11]. The biological function of Cb_5R in membranes is also facilitated by the membrane isoform of cytochrome b_5 (Cb_5), which also possesses a membrane-binding domain in the protein structure and a water-soluble domain that interacts with the cytosol. This is particularly relevant for many metabolic functions of Cb_5R because Cb_5 is an electron carrier of many membrane-bound enzymes of the lipid metabolism, such as cytochrome P450 monooxygenases and membrane redox chains (reviewed in [15]). It is noteworthy that both soluble and membrane-bound Cb_5 isoforms modulate the activity of membrane cytochrome P450 monooxygenases involved in xenobiotic metabolism [18]. Additionally, soluble Cb_5 can act as an electron acceptor of the reductase present in membranes [19]. Since both cytochrome P450 monooxygenases and Cb_5R can produce reactive oxygen species (ROS) in the absence of electron acceptors [15,20], Cb_5 can be regarded as an antioxidant protein that prevents excessive intracellular ROS production during drug detoxification [18]. This effect can also be considered as derived from microsomal cytochromes P450 requirement of two electrons and two protons for the oxidation of substrates. Although the two electrons can be provided by cytochrome P450 reductase, the second electron can also be donated by Cb_5, leading to a more rapid protonation of the anionic ferric hydroperoxy–heme intermediate of P450 [21]. This fosters the Cb_5 stimulatory effect through a more efficient coupling of the system components.

1.2. The MOM Cb_5R

Interestingly, in *Saccharomyces cerevisiae*, mitochondrial Cb_5R sorts into different mitochondrial compartments, due to 40 amino acid residues at the N-terminal end of the membrane isoform Cb_5R not conserved in other flavoenzymes. Two forms of 34 kDa and 32 kDa have been detected in this yeast locating at the MOM and in the intermembrane space, respectively [22]. The 34 kDa form has a putative 21 amino-terminal matrix terminal signal, followed by 21 uncharged hydrophobic residues that are transported into the intermembrane space, which is cut by the inner membrane protease 1 to generate the 32 kDa form of the reductase. The function of the intermembrane space isoform present in yeast is not clearly defined, although it was early suggested to be involved in the electron transfer from external NADH to cytochrome c, thereby mediating antimycin-insensitive, energy-coupled oxidation of external NADH by yeast mitochondria [23].

The function assigned to the mammalian Cb_5R located at the MOM depends on its coupling with other proteins. Some studies have associated the reductase present at this location with the formation of a complex with the mitochondrial amidoxime reducing component and the outer mitochondrial Cb_5. This complex is in charge of the activation of prodrugs containing an amidoxime structure and detoxification pathways [24]. A very recent role of Cb_5R located at the MOM has been suggested in endothelial cells expressing NADPH oxidase 4 (NOX4). As a mitigator of inflammatory activation, Cb_5R would reduce the NOX4-dependent production of H_2O_2 via the reduction of ubiquinone (CoQ), a substrate of Cb_5R. [25]. Yuan et al. (2021) suggest that the electron transfer through CoQ plays a role in shifting the outer membrane NOX4 superoxide anion radical ($O_2^{\bullet-}$) production to H_2O_2 [25]. The presence of a transmembrane domain in charge of H_2O_2 production associated with the E-loop [26] suggests that indeed this enzyme NOX4 might work as a redox-signaling transducer. Additionally, NOX4–Cb_5R complexes seem necessary for proper coupling between both enzymes, because excessive ubiquinol oxidation has been reported to generate H_2O_2 and $O_2^{\bullet -}$ [27–30]. Notably, the importance of correct clustering of Cb_5R with other proteins or associated with a correct membrane environment has been pointed out in the plasma membrane of neurons [19,31–33].

1.3. The Plasma Membrane Cb_5R

A proposal has been made to rationalize the location of the membrane isoform at the plasma membrane using erythrocyte membranes—namely, an N-terminal-extended erythroid polypeptide with an N-terminus 12 uncharged reticulocyte-specific residues, in addition to 17 residues of the membrane that anchors the myristoylated reductase [11,34,35]. The membrane isoform of Cb_5R is the major NADH consuming enzyme at the plasma membrane of vesicles derived from rat brain synaptosomes (SPMV) and the neuronal plasma membranes [19,31]. We have reported that Cb_5R accounts for 80% of the NADH oxidase activity, by analyzing the activation data in the presence of cytochrome activation and soluble Cb_5 (3–4 μM) [19,31]. Cb_5 regulates this enzyme function [19]. The amount of Cb_5 present in SPMVs is not at saturation since the NADH oxidase activity is stimulated by supplementation with soluble Cb_5, which elicits a three-fold activation on this activity [19]. Cb_5R location at the plasma membrane lipid rafts suggests a role of this enzyme in cholesterol metabolism. A role regarding an in situ cholesterol formation at this location should not be discarded based on Cb_5 functions in the last steps of cholesterol synthesis [9,15]. We have reported that Cb_5R/Cb_5 can be a source of reactive oxygen species using the purified enzyme, biological membranes, and culture cells [28,31,33]. Cholesterol-rich plasma membrane sub microdomains have been suggested to be a major extramitochondrial $O_2^{\bullet -}$ source in cerebellar granule cells cultures [32].

In addition, our laboratory has described that the NADH oxidase activity of SPMVs is similarly stimulated in the presence of horse heart Cyt *c* with an IC_{50} of 6 μM [36]. The Cyt *c* stimulated NADH oxidase activity of SPMV can be inhibited by the addition of antibodies against Cb_5R to the assay [19]. This fact correlates with the ability of recombinant Cb_5R to use also Cyt *c* as an electron acceptor and a role of Cb_5R performing this function in membranes [33]. The activity of the membrane Cb_5R bound to the neuronal plasma membrane can be modulated by changes in cytosolic levels of Cb_5 and Cyt *c* in the low micromolar concentration range. However, this activity is about three times more sensitive to Cb_5 than to Cyt *c* [19,36]. The similar dissociation constant of the Cb_5:Cb_5R complex with respect to that of the Cyt *c*:Cb_5R complex (0.4–0.5 μM) [33,37] correlates with other possible endogenous ligands of Cyt *c* that might be present in these membranes.

We postulated a protective role of Cb_5R by its ability to reduce oxidized Cyt *c* [33], a widely recognized pro-apoptotic factor that is needed for caspases activation [38,39] and cardiolipin-induced Cyt *c* peroxidase activity [40], early events in apoptotic cell death [33].

1.4. Antioxidants Recycling Activity of Cb_5R and the Importance in Cellular ROS Balance

Cb_5R function was early revealed as an enzyme in charge of antioxidants recycling [39,41–43]. Tocopherol is a lipophilic antioxidant important for the stability of lipids and phospholipids placed within the (sub)cellular membranes [44]. α-tocopherol has a leading role against lipid peroxidation, and the redox reaction with other cellular antioxidants such as GSH, CoQ, and ascorbate have been directly related to membrane protection and cell function maintenance [45,46]. Tocopheroxyl radical is generated upon the reaction of α-tocopherol with lipid hydroperoxides. Antioxidants can react with the tocopheroxyl radical to recycle it [45,46]. Additionally, those cellular elements capable of reducing radicals are key to restoring tocopherol's antioxidant power [47]. Cb_5R and other proteins with quinone reductases activity, such as the plasma membrane NADH quinone oxidoreductase (NQO1) and the mitochondrial NADH-quinone oxidoreductase of mitochondrial electron transfer complex I, have been postulated to present this function. Cb_5R reduces CoQ at the expense of soluble NADH through a one-electron reaction mechanism. NQO1 also reduces

CoQ through a two-electron reaction mechanism using both NADH and NADPH [48]. This difference between both systems may be significant under oxidative stress conditions since the formation of the semiquinone radical could lead to the production of $O_2^{\bullet-}$ upon by its reaction with molecular oxygen, as reported, by the mitochondrial redox chain and, also found by Cb_5R reacting with CoQ mimetics [28]. The semiquinone radical and possible deleterious side reactions can be avoided by Cb_5R recycling the semiquinone radical, reducing the radical rather than reducing it after generation of the oxidized quinone. Semiquinone generation has been proposed to be key for the formation of ROS [49]. This is also supported by experiments showing that the formation of CoQ semiquinone radical and recycling of vitamin E homologs are $O_2^{\bullet-}$-dependent reactions, which agree with the ability of Cb_5R to produce $O_2^{\bullet-}$ [20,28,31,33]. In addition, the formation of ROS in biological systems depends on the reaction between ubiquinol and electron partners accepting two electrons leading to the formation of oxidized CoQ. The proximity of systems recycling CoQ in two electron-reducing pathways and the slower reaction rate between Cb_5R and the semiquinone radical vs. reaction with other molecules leads to radical reactions in which Cb_5R cannot participate. Additionally, a key molecule in semiquinone reduction is soluble ascorbate [50]. Ascorbate radical and ubiquinol are products of this reaction. Ascorbate radical can be recycled by the NADH: ascorbate radical reductase activity of Cb_5R [41,42,50]. The reaction rates of $O_2^{\bullet-}$/ascorbate and $O_2^{\bullet-}$/ascorbate radical are reported: 5×10^4 M^{-1} s^{-1} and 2.6×10^8 M^{-1} s^{-1}, respectively [51]. These values are close to the ones reported for the reaction rate of $O_2^{\bullet-}$ with superoxide dismutase [52].

2. Structural Features of Soluble Cb_5R Helps to Rationalize the Electron Transfer Processes Using NADH as a Substrate

2.1. FAD-Binding Domain of Cb_5R

Several conserved motifs have been observed in Cb_5R's primary structure in comparison with other FAD-containing proteins, which have been implicated in the binding of the flavin group, such as "$RxY^T{}_Sxx^SN$" [53]. In rat Cb_5R, the FAD-binding domain is comprised of amino acid residues 33–147 located at the amino-terminal side (Figure 1, underlined sequence) [54]. The loop formed by residues 110–125 (Figure 1, yellow background) contributes to most of the interactions between the FAD-binding domain of the reductase and the adenine dinucleotide moiety of the FAD group in which water mediates the interaction [54]. The conserved motif "$^{91}RxY^T{}_Sxx^S{}_N{}^{97}$" has been proposed to rule flavin binding to the apoprotein in the flavin transhydrogenase superfamily of oxidoreductases, in which the NADH:*Cb_5R* is included (Figure 1, light blue) [55]. Nishida and Kimura proposed that R63, Y65, and S99 residues of the motif of the pig soluble enzyme (R91, Y93, and S127 residues in Figure 1, dark blue) are important in the flavin coordination through the hydrogen bonding of the phosphate and ribityl moieties of the cofactor [56]. Later, Barber's research group showed that indeed R91 was not essential for flavin binding but participated in tethering the ADP moiety of the FAD cofactor by H bonding to the protein [57]. Moreover, a role of T66 of the pig enzyme (T94 in Figure 1, purple) in the formation and stability of the NAD^+–FAD semiquinone complex during NADH turnover was shown [58]. Later, the role of P92 and Y93 residues of the rat Cb_5R was reevaluated, concluding that these residues are not important for FAD incorporation into the apoprotein. Notably, mutagenesis experiments showed that Y93 residue contacts with the FAD group and modulates spectroscopical, catalytic, and thermodynamic properties of the FAD cofactor [55]. This suggests that these residues are important for the redox modulation of the enzyme by Cb_5 [37]. Moreover, methemoglobinemia is reported when mutations in Cb_5R's residues, V105M and M126V, are present. These mutations induce perturbations of the FAD-binding domain [54] since the side-chain atoms of M126 are housed by the hydrophobic pocket formed by residues: L80, T82, V89, Y93, Y129, I139, and the hydrophobic atoms of R91 [54].

```
        1       10        20        30        40        50        60        70        80
PIG     ............................STPAITLENPDIKYPLRLIDKEVVNHDTRRFRFALPSPEHILGLPVGQHIYL
HUMAN   GAQLSTLGHMVLFPVWFLYSLLMKLFQRSTPAITLESPDIKYPLRLIDREIISHDTRRFRFALPSPQHILGLPVGQHIYL
BOVINE  GAQLSTLGHVVLSPVWFLYSLIMKLFQRSTPAITLENPDIKYPLRLIDKEVISHDTRRFRFALPSPEHILGLPVGQHIYL
MOUSE   GAQLSTLSHVVLSPVWFIYSLFMKLFQRSTPAITLENPDIKYPLRLIDKEVISPDTRRFRFALPSPQHILGLPIGQHIYL
RAT     GAQLSTLSRVVLSPVWFVYSLFMKLFQRSSPAITLENPDIKYPLRLIDKEIISHDTRRFRFALPSPQHILGLPIGQHIYL

        81       90        100       110       120       130       140       150      160
PIG     SARIDGNLVIRPYTPVSSDDDKGFVDLVIKVYFKDTHPKFPAGGKMSQYLESMKIGDTIEFRGPNGLLVYQGKGKFAIRP
HUMAN   SARIDGNLVVRPYTPISSDDDKGFVDLVIKVYFKDTHPKFPAGGKMSQYLESMQIGDTIEFRGPSGLLVYQGKGKFAIRP
BOVINE  SARIDGNLVIRPYTPVSSDDDKGFVDLVIKVYFKDTHPKFPAGGKMSQYLESMKIGDTIEFRGPNGLLVYQGKGKFAIRP
MOUSE   STRIDGNLVIRPYTPVSSDDDKGFVDLVVKVYFKDTHPKFPAGGKMSQYLENMKIGDTIEFRGPNGLLVYQGKGKFAIRA
RAT     STRIDGNLVIRPYTPVSSDDDKGFVDLVVKVYFKDTHPKFPAGGKMSQYLENMNIGDTIEFRGPNGLLVYQGKGKFAIRA

        161     170       180       190       200       210       220       230      240
PIG     DKKSSPVIKTVKSVGMIAGGTGITPMLQVIRAIMKDPDDHTVCHLLFANQTEKDILLRPELEELRNEHSARFKLWYTVDR
HUMAN   DKKSNPIIRTVKSVGMIAGGTGITPMLQVIRAIMKDPDDHTVCHLLFANQTEKDILLRPELEELRNKHSARFKLWYTLDR
BOVINE  DKKSDPVIKTVKSVGMIAGGTGITPMLQVIRAIMKDPDDHTVCHLLFANQTEKDILLRPELEELRNEHSARFKLWYTVDK
MOUSE   DKKSNPVVRTVKSVGMIAGGTGITPMLQVIRAVLKDPNDHTVCYLLFANQSEKDILLRPELEELRNEHSARFKLWYTVDK
RAT     DKKSNPVVRTVKSVGMIAGGTGITPMLQVIRAVLKDPNDHTVCYLLFANQSEKDILLRPELEELRNEHSSRFKLWYTVDK

        241     250       260       270       280       290       300
PIG     APEAWDYSQGFVNEEMIRDHLPPPEEEPLVLMCGPPPMIQYACLPNLERVGHPKERCFAF
HUMAN   APEAWDYGQGFVNEEMIRDHLPPPEEEPLVLMCGPPPMIQYACLPNLDHVGHPTERCFVF
BOVINE  APEAWDYSQGFVNEEMIRDHLPPPEEEPLVLMCGPPPMIQYACLPNLDRVGHPKERCFAF
MOUSE   APDAWDYSQGFVNEEMIRDHLPTPGEEPLILMCGPPPMIQFACLPNLERVGHPKERCFTF
RAT     APDAWDYSQGFVNEEMIRDHLPPPGEETLILMCGPPPMIQFACLPNLERVGHPKERCFTF
```

Figure 1. Alignment of Cb_5R isoforms from different sources. Pig (Unitprot: P83686), human (Unitprot: P00387), bovine (Unitprot: P20070) and rat (Unitprot: P20070). Amino acid residues 33–147 located at the amino-terminal side of the FAD-binding domain (underlined sequence); residues 110–125 forming a loop in the FAD-binding domain (yellow background); the conserved motif "91RxYTSxxSN97" (light blue background); location of the R91, Y93, and S127 (dark blue background); location of the T66 of the pig enzyme implicated in the formation and stability of the NAD+–FAD semiquinone complex during NADH turnover which correlate with T94 in the human isoform (purple background); location of the residues forming the plateau where the NAD^+ group sets and in which the adenine ring and the diphosphate group packed between the side chain residues of these amino acid residues (golden background); Proposed residues that could alter the hydrophobic environment formed by the residues I177, L206, A208, T237, M256, M278, and C283 affecting the binding of NAD^+ (red background); suggested sequence ^{273}CGPPPM^{278} to be critical for the correct orientation of the nicotinamide moiety with the flavin for efficient hydride transfer (grey background).

2.2. The NADH-Binding Pocket and Hydride Transfer to the Flavin Group of the Reductase

Comparative analysis with other FAD-binding proteins allowed the identification of some motifs in the primary sequence of Cb_5R that are involved in the interaction with reduced pyridine nucleotide, and, also, in the selectivity of the interaction between NADH/NADPH and the enzyme such as "GxGxxP" and "CGxxxxM" [53]. In addition, an NADH-binding lobe was proposed by Barber's research group, with a function in setting NAD^+ to the plateau formed by proline residues 275–277, in which the adenine ring is packed between the side chain residues, in parallel to F251 and P277 (Figure 1, golden background). Moreover, the NADH diphosphate group packs against P275 and the oxygen atoms from the phosphate group from H^+-bonds with Q210, T181, Y112, and K110 [54]

(Figure 1, golden background). Experiments with the rat isoform indicate that mutations in V253 residue affect the NAD^+ binding by altering the hydrophobic environment formed by the residues I177, L206, A208, T237, M256, M278, and C283 [59] (Figure 1, red background). The sulfhydryl group of C245 (C283 of the rat Cb_5R) also forms Van der Waals contact with the nicotinamide C3 atom at the *si*-face [60]. This interaction is considered to alter the redox potential of the NAD^+ upon binding to the enzyme. C273 was observed to form part of a sequence ^{273}CGPPPM^{278}, which was suggested to be critical for the correct orientation of the nicotinamide moiety with the flavin for efficient hydride transfer [61] (Figure 1, grey background).

The second conserved motif among pyridine nucleotide-binding FAD proteins is "GxGxxP", which is located in residues between the G180-P185 residues of the carboxyl-terminal lobe of rat Cb_5R and a recognized function in the binding of reduced pyridine nucleotides [62]. The importance of the G179 residue preceding this motif was shown by direct mutagenesis since its mutations induce changes in both the adequate NADH/NADPH selectivity and NADH binding and efficient hydride transfer [62]. Some catalytic properties are shared between dehydrogenases that use NADPH as a substrate. Enzymes that present a dehydrogenase activity accept two electrons from the substrate to carry electrons to a final metalloprotein, normally a heme or iron-dependent protein that acts as an electron acceptor [63]. Iyanagi suggested that for the case of Cb_5R, its dehydrogenase activity is associated with a direct hydride ion transfer to the flavin group [63]. The two-electron reduced enzyme−NAD^+ complex (E-$FADH^-$−NAD^+) then transfers two electrons to two one-electron acceptors one by one, after which the reduced enzyme returns to the oxidized state.

3. Cb_5-Interacting Domain of Cb_5R

The topography of the human Cb_5/Cb_5R-interacting domain has been reported in [37]. The results obtained in NMR experimental studies and docking simulation allowed the identification of the Cb_5R amino acids more directly involved in the interaction with Cb_5—namely, K41, Y79, N87, L88, V89, V90, R91, P92, F120, G123, K125, R142, S145, L147, P160, D161, K162, K163, M272, P276, P277, R279, Q280, Y281, L284, T294, F298, V299, and F300 [37]. In addition, the results reported in Samhan-Arias et al. [37] allowed the prediction of two salt bridges and four H-bond pairs formation in the Cb_5/Cb_5R interacting. The predicted salt bridges are K162(Cb_5R):E42(Cb_5) and K125(Cb_5R):E48(Cb_5), and the predicted H bonds are V89/V90(Cb_5R):E47(Cb_5), Y281(Cb_5R):D64(Cb_5), Y79(Cb_5R):P44/H43/Q53(Cb_5) and L88(Cb_5R):E48/R51(Cb_5).

Electron Transfer from the Flavoprotein to Cb_5

Several research groups have studied the electron transfer kinetics to Cb_5 from the reduced Cb_5R. In general, the electron transfer reactions between flavins/flavoproteins and electron acceptors can follow one or two electron-dependent pathways [64]. Both electron transfer mechanisms and can be distinguished spectroscopically by analyzing the UV–visible spectrum of the flavin group [1,65]. The oxidized flavin is yellow, with maximum intensity at 370 and 450 nm, and the hydroquinone form (fully reduced) has very weak broad spectra [65]. The unprotonated form of red anionic semiquinone has a maximum band of 370 nm. However, if N(5) of FAD is protonated, a strong band at 570 nm resulting from the blue neutral semiquinone formation is observed [65]. The formation of one or the other redox state in flavoproteins depends on the protein microenvironment because hydrophobicity and proximal charged amino acids might be key variables regulating the flavin redox state. The Cb_5R's flavin semiquinone stabilization is affordable when a more positive reduction potential for the first single-electron transfer step is found over the second single electron transfer step [65].

Based on crystallographic data, the electron pathways from the reductase to Cb_5 have been suggested to directly proceed following the hydrogen bond paths (FAD-N5$\cdots$Y65/T66 $\cdots$H49$\cdots$$Cb_5$) [66]. These residues are labeled as Y93, T94, and H77 in Figure 2, panel b

(Y65, T66, and H49 in the manuscript). Additionally, this is supported by the observations showing a stabilization of the N5 atom of the isoalloxazine ring of FAD by hydrogen bonding to Y93 and T94 and the highly conserved H77. The importance of H77 has been shown by mutagenesis experiments of this residue. Y93, T94, and H77 residues are in the backside of the crevice, where the FAD binds to Cb_5R (Figure 2). Notably, the Cb_5-binding site that we have described by docking analysis indicates that the heme group is approximately at a 12 Å distance from the H77 (Figure 3a,b). It might suggest a possible role of these residues in the bifurcation of electrons to different acceptors or to increase the half-life of the flavin radical species if they are generated in an oxidative environment.

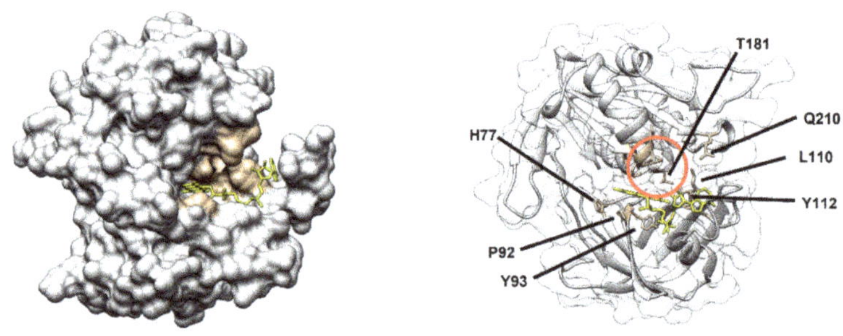

(**a**) (**b**)

Figure 2. FAD protruding site of Cb_5R (PDB: 1UMK) and residues involved in NADH binding. Surface representation of the human Cb_5R structure obtained by crystallography (PDB:1UMK) and location of the FAD (represented as yellow sticks) protruding site is shown in panel (**a**). Residues interacting with the diphosphate groups and the adenine ring of NAD^+, as reported, are labeled in brown. The same representation is shown in panel (**b**), where the surface is transparent, and the residues interacting with NAD^+ (brown), and the backbone can be viewed (grey). The red circle labels the ^{273}CGPPPM^{278} motif, which forms the NAD^+-binding surface.

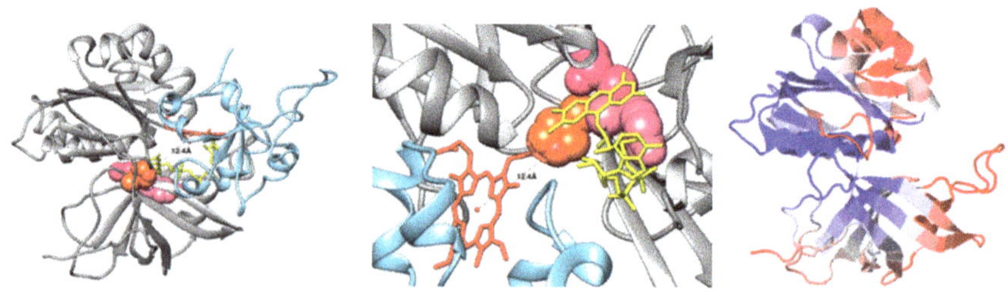

(a) (**b**) (c)

Figure 3. Representation of the Cb_5:Cb_5R complex model obtained by docking and representation of Cb_5R areas that are susceptible for mobility through intraprotein dynamics. The Cb_5:Cb_5R complex model has been previously published [37]. Cb_5R's backbone is shown in grey color, with side-chain residues of Y93, T94, and H77, which are represented as light pink, dark pink, and red-colored balloons, respectively, the flavin group depicted as sticks in yellow interact with the heme group (represented as red sticks) of Cb_5 backbone (represented in light blue color). Also, a distance of 12.4 Å from the iron heme group of Cb_5 to NE2 of H77 was measured and labeled, as shown in panel (**a**). The zoomed area where a distance of 12.4 Å was found between the iron heme group of Cb_5 and NE2 of H77 is shown in panel (**b**). The chain corresponding to Cb_5R (1UMK) from our model for the Cb_5:Cb_5R complex was submitted to the dynOmics portal, as indicated in the text, and a figure representing Cb_5R was obtained, where the size of fluctuations by the lowest frequency (slowest) two Gaussian network model (GNM) modes (blue: almost rigid; and red: highly mobile), as indicated by the color bar under the diagram window) (panel (**c**)).

4. Flavin Fluorescence and Structural Alterations by Complexes Formation with Electron Acceptors

Cb_5 addition to the reductase induced an increase in the Cb_5R flavin autofluorescence, compared with the free FAD cofactor [37]. This effect was not associated with the release of the flavin group from the protein. The increase in the reported fluorescence intensity by the complex formation with Cb_5 suggests that Cb_5R undergoes a conformational change induced by the presence of electron acceptors of this enzyme. Therefore, we used these changes to characterize the complex formation between the soluble isoforms of Cb_5R and Cb_5 [37]. From the dependence upon Cb_5 of the increase in Cb_5R's flavin autofluorescence, we calculated a K_d value of 0.5 μM Cb_5. The stoichiometry for the complex formation between these proteins is 1:1 [37]. As indicated above, these changes in fluorescence monitor structural alterations in the binding site of FAD and correlate with changes in the redox potential of Cb_5R upon binding of Cb_5. The reported redox potential of Cb_5R was to -196 ± 8 mV and shifted to -239 ± 4 mV upon complex formation with Cb_5 [37]. Indeed, this result is in good agreement with the results obtained with other flavoproteins, which have led to the conclusion that flavoproteins can modulate their redox potential upon interaction with redox partners [67] due to changes in the local pH, polarity of the environment, and interactions with protein amino acid residues and ligands [68]. This can be seen as a particular case of the relevance of changes in the microenvironment of protein prosthetic groups due to local hydrophobicity and pH changes [37,69]. In the bibliography, some Cb_5R mutants that retain the FAD group have been reported to present an increase in FAD fluorescence [55]. The tyrosine variants Y93A, D, F, H, and S exhibit alterations in the flavin visible spectra associated with blue shifts in the spectra.

Moreover, in the wild type, P92S, and A, the intrinsic flavin fluorescence was quenched, while some Y93 mutants (substitutions of Y by H, W and A, D, F, and S) exhibited a fluorescent increase, compared with free FAD [55]. Changes in the FAD spectral properties also correlated with shifts up to 20–30 mV in the midpoint potential. The described fluorescence increase in these mutants was lower than that of free FAD or FMN, as we observed for the changes found in the Cb_5R autofluorescence and redox potential upon complexing with Cb_5 [37]. An explanation for this behavior can be correlated with the existence of two conformational states for FAD: (1) a closed FAD conformation in which the π–π stacking interactions between the adenine and isoalloxazine rings largely quench FAD fluorescence, and (2) a FAD open conformation or extended conformation induced upon binding to nearby peptide side chains. A shift from the close to the open conformation of FAD should produce an increase in flavin fluorescence. A change in the Cb_5R's FAD exposition to the solvent could also help to understand the observed changes in spectroscopic and redox properties [37]. This has also been documented, concerning the structural models obtained by crystallography of the fully reduced and the oxidized form of porcine liver Cb_5R that was recently resolved [60]. Local conformational changes in the NADH- and FAD-binding domains were found in Cb_5R in the different redox states [60]. This led to the authors suggesting a new role of T66 (T94 in Figure 1) interaction in the release of a proton from the N5 atom of the isoalloxazine ring of FAD, increasing the solvent-accessible surface area of FAD [60]. These results also support that the N5 atom of FAD in Cb_5R is stabilized by hydrogen bonding with CαH of Y65 and amide-H of T66 (Y93 and T94 in Figure 1) [66]. The biological function of these changes would prevent reoxidation backflow of the catalytic cycle and the acceleration of the electron transfer to one electron acceptor such as Cb_5. It is noteworthy that Y93, T94, and H77 are located at the backside of the FAD group of Cb_5R structure when the enzyme is observable in an orientation where the FAD-protruding site is observable (Figures 2 and 3). By computational methods, we found that the FAD-protruding site was associated with the domain where Cb_5 docks, and interact with Cb_5R (Figure 3a), which contains a significant number of hydrophobic residues; this also helps to rationalize the observed changes on Cb_5R's FAD fluorescence. The distance between the residues implicated in the reduction in electron acceptors is still far from the heme iron group, i.e., the distance from Cb_5 heme iron to the NE2 atom of H77 was estimated to be approximately 12 Å (Figure 3a,b). This challenges the role of this residue in the one-electron reduction to the Cb_5, although its role and implication in the electron transfer to other redox partners of the Cb_5R remain to be experimentally demonstrated.

To obtain answers to some of the opened questions regarding the interaction of Cb_5R with ligands, we evaluated the structural features of Cb_5R via molecular dynamics by submitting the PDB file of the protein (1UMK) to the dynOmics portal 1.0 server (http://dynomics.pitt.edu/ (accessed on 21 June 2021)) [70].

5. Intramolecular Dynamics of Cb_5R

Proteins in physiological conditions show a number of motions that help them adapt to intermolecular interactions or accomplish their biological functions [70]. Thus, the intrinsic dynamics are unique and characteristic for each protein. First, we identified the residues with a higher degree of mobility based on the elastic network model 1.0. The image obtained after submission shows a colored image in which the Cb_5R structure is colored based on the size of fluctuations driven by the slowest two Gaussian network model (GNM) modes (blue: almost rigid; and red: highly mobile), where the low-frequency modes are highly relative to the biological functions (Figure 3c). We found that the Cb_5R areas with a higher degree of mobility define hot motifs centered on the following amino acid residues: Q67 to L80, from V90 to I109, from Lys110 to Q136, T211, A244, H289. In contrast, Cb_5R areas with a lower mobility rate are V74, T94, I139, and L205, as indicated by blue-colored motifs (lower mobility), as shown in Figure 3c versus the red-colored motifs that correspond to higher mobility areas.

Regarding the interaction of Cb_5R with NADH, crystallographic data obtained with the enzyme in the presence of NADH show differences between Cb_5R's structure in the reduced and the oxidized state and can be correlated with an increase in the solvent-accessible surface area of FAD [69]. These data suggest that Cb_5R likely undergoes conformational changes dependent on the interaction with ligands and highlight the existence of protein areas that could sense a structural perturbation such as ligand binding. These areas strongly respond to perturbations through a significant change in their local conformation. In addition, effector areas efficiently communicate perturbations or associated "information" to other sites [70]. We analyzed the presence of sensing and effector areas that could respond to perturbation in certain domains on the Cb_5R structure (Figure 4). We determined the existence of a sensing domain in the reductase formed by residues located from K110 to M126 (Figure 4a), which correlated with a loop that contributes to most of the interactions between the FAD-binding domain of the reductase and the adenine dinucleotide moiety of FAD, as previously indicated (Figure 1, yellow background). This area was described as important since the oxygen atoms of the phosphate group of NADH form H bonds with K110 and Y112 and, also, with some of the amino acid residues that are part of the Cb_5 interacting domain of Cb_5R (F120, G123, K125) [37]. Moreover, several motifs were also detected as effectors, which are susceptible to respond to changes in perturbations associated with sensing areas.

These motifs are centered on L46, F61, I69, L80, R91, V105, K110, S127, F141, G177, H204, T237, L261, L268, A282, and N286, which are amino acid residues labeled in red Figure 4b. The allocation in the structural model of Cb_5R of the sensing and effector residues is shown in Figure 4c. Our analysis indicates that some of these residues form part of the effector motifs of Cb_5R, which also participate in the interaction with Cb_5. These amino acid residues are Y79, R91, and R142, from which R91 was proposed to participate in the tethering of the FAD cofactor by Cb_5R [57]. Additionally, through ENM 1.0, a spectrum of motions near their physiological conditions, which often assist in adapting to intermolecular interactions or accomplishing their biological function, can be obtained for Cb_5R. In the obtained animated structural model (Figure 4d and Supplementary Figure), those residues implicated in sensing to deliver an effect translated into a fast response are highlighted in red and correlated to those residues which response is slow (blue). By comparing the animation obtained in this figure with that of Figure 3a, which shows the interaction of

Cb_5R with Cb_5 in the same orientation, it can be noted that the binding domain of Cb_5 could become altered upon a perturbation acting in the sensing domain to hang the Cb_5 molecule. Furthermore, when an NADH molecule interacts with the Cb_5R molecule at the NADH-binding lobe K110-M126, a closer interaction between Cb_5 and the FAD moiety is promoted. Additionally, these simulations of the Cb_5R dynamics allow the comprehension of some mutations effect, such as those found in residues V105M and M126V, which induce methemoglobinemia since they form part of the sensing motif that will cause a substantial alteration in the protein dynamics.

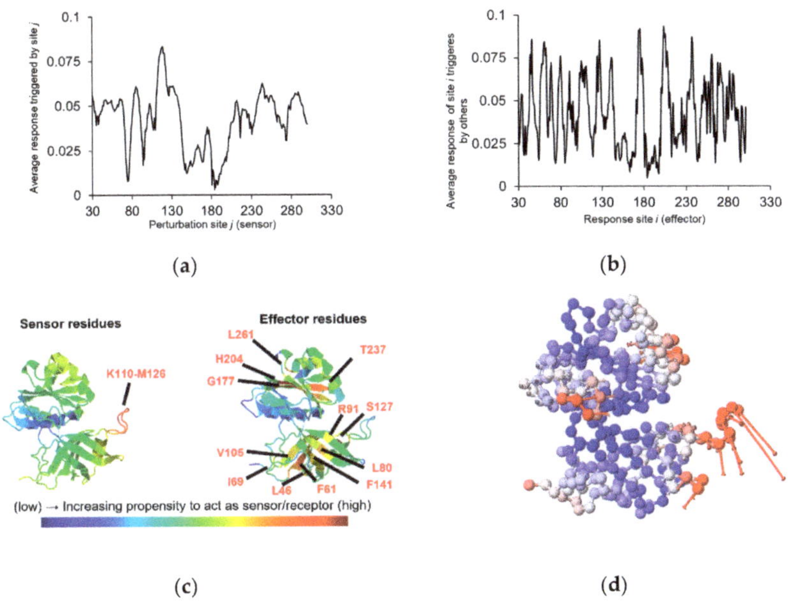

Figure 4. High sensor/effector residues of Cb_5R and molecular motions based on compiled information obtained through analysis of the intrinsic dynamics. Correlation between the average response by a site triggered with a perturbation in the Cb_5R sequence acting as a sensor is shown in panel (**a**). Correlation between the average response of a site triggered by others with a responsive site in the Cb_5R sequence acting as an effector is shown in panel (**b**). Representation of the Cb_5R areas, which have an increasing propensity to act as sensor/receptor and label residues located in hot spotted areas, is shown in panel (**c**). Animation obtained with the second-highest ranked model based on the previously analyzed information, where the areas are color coded based on the size of motions where the color indicate the type of fluctuations in a red-white-blue scale (red colors correspond to large fluctuations and blue colors to small fluctuations.), is shown in panel (**d**). Also in this panel, vectors (red lines) represent the direction of motion of each residue.

Other suggestions were also proposed for the interaction of Cb_5R with ligands modulating the Cb_5R function. For example, two reductase conformations were characterized by experiments performed at alkaline pHs, a strategy that we also used to describe the novel enzymatic activities of Cb_5 [71,72]. Additionally, a second inactive swollen form in which at least a cysteine and a tyrosine residue are exposed can be stabilized by the addition of mersalyl, a Cb_5R inhibitor [28]. These changes support the existence of altered conformations associated with the binding of certain ligands in vivo and by changes in the ionization of the lateral side chain of some amino acid residues.

Supplementary Materials: The following are available online at https://www.mdpi.com/article/10.3390/ijms23010118/s1.

Author Contributions: Conceptualization, C.G.-M. and A.K.S.-A.; methodology, C.G.-M. and A.K.S.-A.; software, A.K.S.-A.; validation, C.G.-M. and A.K.S.-A.; formal analysis, A.K.S.-A.; investigation, C.G.-M., O.H.M.-C., M.M., and A.K.S.-A.; resources, A.K.S.-A.; data curation, C.G.-M. and A.K.S.-A.; writing—original draft preparation, M.M., O.H.M.-C., C.G.-M., and A.K.S.-A.; writing—review and editing, All authors; visualization, A.K.S.-A.; supervision, A.K.S.-A. All authors have read and agreed to the published version of the manuscript.

Funding: This research received no external funding.

Institutional Review Board Statement: Not applicable.

Informed Consent Statement: Not applicable.

Data Availability Statement: The Cb_5:Cb_5R complex model has been previously published [37]. The data regarding Cb_5R Molecular dynamics were obtained by submitting the PDB file of the protein (1UMK) to the dynOmics portal 1.0 server (http://dynomics.pitt.edu/ (accessed on 21 June 2021)).

Acknowledgments: This research was funded by grant RTI2018-093864-B-I00, from the Spanish "Ministry of Science, Innovation, and Universities" and ERDF/FEDER funds to M.M.

Conflicts of Interest: The authors declare no conflict of interest.

References

1. Müller, F. Flavin radicals: Chemistry and biochemistry. *Free Radic. Biol. Med.* **1987**, *3*, 215–230. [CrossRef]
2. Strittmatter, P. The Reaction Sequence in Electron Transfer in the Reduced Nicotinamide Adenine Dinucleotide-Cytochrome B5 Reductase System. *J. Biol. Chem.* **1965**, *240*, 4481–4487. [CrossRef]
3. Iyanagi, T. Redox Properties of Microsomal Reduced Nicotinamide Adenine Dinucleotide-Cytochrome B5 Reductase and Cytochrome B5. *Biochemistry* **1977**, *16*, 2725–2730. [CrossRef]
4. Meyer, T.E.; Shirabe, K.; Yubisui, T.; Takeshita, M.; Bes, M.T.; Cusanovich, M.A.; Tollin, G. Transient Kinetics of Intracomplex Electron-Transfer in the Human Cytochrome B5 Reductase-Cytochrome B5 System: NAD+ Modulates Protein-Protein Binding and Electron Transfer. *Arch. Biochem. Biophys.* **1995**, *318*, 457–464. [CrossRef]
5. Kobayashi, K.; Iyanagi, T.; Ohara, H.; Hayashi, K. One-Electron Reduction of Hepatic NADH-Cytochrome B5 Reductase as Studied by Pulse Radiolysis. *J. Biol. Chem.* **1988**, *263*, 7493–7499. [CrossRef]
6. Anderson, R.F. Pulse radiolysis studies on the equilibria between reduced and oxidized free flavin species and the effect of molecular oxygen. In *Pulse Radiolysis Studies on the Equilibria between Reduced and Oxidized Free Flavin Species and the Effect of Molecular Oxygen*; De Gruyter: Berlin, Germany, 2019; pp. 57–60. ISBN 978-3-11-152135-0.
7. Deng, B.; Parthasarathy, S.; Wang, W.; Gibney, B.R.; Battaile, K.P.; Lovell, S.; Benson, D.R.; Zhu, H. Study of the Individual Cytochrome B5 and Cytochrome B5 Reductase Domains of Ncb5or Reveals a Unique Heme Pocket and a Possible Role of the CS Domain. *J. Biol. Chem.* **2010**, *285*, 30181–30191. [CrossRef] [PubMed]
8. Siendones, E.; Ballesteros, M.; Navas, P. Cellular and Molecular Mechanisms of Recessive Hereditary Methaemoglobinaemia Type II. *J. Clin. Med.* **2018**, *7*, 341. [CrossRef] [PubMed]
9. Samhan-Arias, A.K.; López-Sánchez, C.; Marques-da-Silva, D.; Lagoa, R.; Garcia-Lopez, V.; García-Martínez, V.; Gutierrez-Merino, C. Biochemical and Anatomical Basis of Brain Dysfunctions Caused by Cytochrome B5 Reductase Deficiency or Dysregulation. *J. Neurol. Neuromed.* **2016**, *1*, 61–65. [CrossRef]
10. Ozols, J.; Carr, S.A.; Strittmatter, P. Identification of the NH2-Terminal Blocking Group of NADH-Cytochrome B5 Reductase as Myristic Acid and the Complete Amino Acid Sequence of the Membrane-Binding Domain. *J. Biol. Chem.* **1984**, *259*, 13349–13354. [CrossRef]
11. Borgese, N.; D'Arrigo, A.; De Silvestris, M.; Pietrini, G. NADH-Cytochrome b 5 Reductase and Cytochrome b 5 Isoforms as Models for the Study of Post-Translational Targeting to the Endoplasmic Reticulum. *FEBS Lett.* **1993**, *325*, 70–75. [CrossRef]
12. Wu, F.F.; Vergeres, G.; Waskell, L. Kinetics of the Reduction of Cytochrome B5 with Mutations in Its Membrane-Binding Domain. *Arch. Biochem. Biophys.* **1994**, *308*, 380–386. [CrossRef] [PubMed]
13. Borgese, N.; Longhi, R. Both the Outer Mitochondrial Membrane and the Microsomal Forms of Cytochrome B5 Reductase Contain Covalently Bound Myristic Acid. Quantitative Analysis on the Polyvinylidene Difluoride-Immobilized Proteins. *Biochem. J.* **1990**, *266*, 341–347. [CrossRef] [PubMed]
14. Shirabe, K.; Landi, M.T.; Takeshita, M.; Uziel, G.; Fedrizzi, E.; Borgese, N. A Novel Point Mutation in a 3′ Splice Site of the NADH-Cytochrome B5 Reductase Gene Results in Immunologically Undetectable Enzyme and Impaired NADH-Dependent Ascorbate Regeneration in Cultured Fibroblasts of a Patient with Type II Hereditary Methemoglobinemia. *Am. J. Hum. Genet.* **1995**, *57*, 302–310.
15. Samhan-Arias, A.K.; Gutierrez-Merino, C. Cytochrome B5 as a pleitropic metabolic modulator in mammalian cells. In *Cytochromes b and c: Biochemical Properties, Biological Functions and Electrochemical Analysis*; Rurik, T., Ed.; Nova Publishers: New York, NY, USA, 2014; ISBN 978-1-63117-467-4.

16. Borgese, N.; Aggujaro, D.; Carrera, P.; Pietrini, G.; Bassetti, M. A Role for N-Myristoylation in Protein Targeting: NADH-Cytochrome B5 Reductase Requires Myristic Acid for Association with Outer Mitochondrial but Not ER Membranes. *J. Cell Biol.* **1996**, *135*, 1501–1513. [CrossRef] [PubMed]
17. Gordon, J.I.; Duronio, R.J.; Rudnick, D.A.; Adams, S.P.; Gokel, G.W. Protein N-Myristoylation. *J. Biol. Chem.* **1991**, *266*, 8647–8650. [CrossRef]
18. Gómez-Tabales, J.; García-Martín, E.; Agúndez, J.A.G.; Gutierrez-Merino, C. Modulation of CYP2C9 Activity and Hydrogen Peroxide Production by Cytochrome B5. *Sci. Rep.* **2020**, *10*, 15571. [CrossRef] [PubMed]
19. Samhan-Arias, A.K.; Garcia-Bereguiain, M.A.; Martin-Romero, F.J.; Gutierrez-Merino, C. Clustering of Plasma Membrane-Bound Cytochrome B5 Reductase within "lipid Raft" Microdomains of the Neuronal Plasma Membrane. *Mol. Cell. Neurosci.* **2009**, *40*, 14–26. [CrossRef] [PubMed]
20. Samhan-Arias, A.K.; Gutierrez-Merino, C. Purified NADH-Cytochrome B5 Reductase Is a Novel Superoxide Anion Source Inhibited by Apocynin: Sensitivity to Nitric Oxide and Peroxynitrite. *Free Radic. Biol. Med.* **2014**, *73*, 174–189. [CrossRef] [PubMed]
21. Pearl, N.M.; Wilcoxen, J.; Im, S.; Kunz, R.; Darty, J.; Britt, R.D.; Ragsdale, S.W.; Waskell, L. Protonation of the Hydroperoxo Intermediate of Cytochrome P450 2B4 Is Slower in the Presence of Cytochrome P450 Reductase Than in the Presence of Cytochrome B5. *Biochemistry* **2016**, *55*, 6558–6567. [CrossRef]
22. Hahne, K.; Haucke, V.; Ramage, L.; Schatz, G. Incomplete Arrest in the Outer Membrane Sorts NADH-Cytochrome B5 Reductase to Two Different Submitochondrial Compartments. *Cell* **1994**, *79*, 829–839. [CrossRef]
23. The Oxidation of External NADH by an Intermembrane Electron Transfer in Mitochondria from the Ubiquinone-Deficient Mutant E3-24 of *Saccharomyces cerevisiae*—ScienceDirect. Available online: https://www.sciencedirect.com/science/article/abs/pii/0003986184905514?via%3Dihub (accessed on 9 November 2021).
24. Ott, G.; Havemeyer, A.; Clement, B. The Mammalian Molybdenum Enzymes of MARC. *J. Biol. Inorg. Chem.* **2015**, *20*, 265–275. [CrossRef]
25. Yuan, S.; Hahn, S.A.; Miller, M.P.; Sanker, S.; Calderon, M.J.; Sullivan, M.; Dosunmu-Ogunbi, A.M.; Fazzari, M.; Li, Y.; Reynolds, M.; et al. Cooperation between CYB5R3 and NOX4 via Coenzyme Q Mitigates Endothelial Inflammation. *Redox Biol.* **2021**, *47*, 102166. [CrossRef] [PubMed]
26. Takac, I.; Schröder, K.; Zhang, L.; Lardy, B.; Anilkumar, N.; Lambeth, J.D.; Shah, A.M.; Morel, F.; Brandes, R.P. The E-Loop Is Involved in Hydrogen Peroxide Formation by the NADPH Oxidase Nox4. *J. Biol. Chem.* **2011**, *286*, 13304–13313. [CrossRef]
27. Powis, G.; Svingen, B.A.; Appel, P. Factors Affecting the Intracellular Generation of Free Radicals from Quinones. *Adv. Exp. Med. Biol.* **1981**, *136 Pt A*, 349–358. [CrossRef]
28. Valério, G.N.; Gutiérrez-Merino, C.; Nogueira, F.; Moura, I.; Moura, J.J.G.; Samhan-Arias, A.K. Human Erythrocytes Exposure to Juglone Leads to an Increase of Superoxide Anion Production Associated with Cytochrome B5 Reductase Uncoupling. *Biochim. Biophys. Acta—Bioenerg.* **2020**, *1861*, 148134. [CrossRef] [PubMed]
29. Samoilova, R.I.; Crofts, A.R.; Dikanov, S.A. Reaction of Superoxide Radical with Quinone Molecules. *J. Phys. Chem. A* **2011**, *115*, 11589–11593. [CrossRef]
30. Linnane, A.W.; Kios, M.; Vitetta, L. Coenzyme Q(10)—Its Role as a Prooxidant in the Formation of Superoxide Anion/Hydrogen Peroxide and the Regulation of the Metabolome. *Mitochondrion* **2007**, *7*, S51–S61. [CrossRef]
31. Samhan-Arias, A.K.; Marques-da-Silva, D.; Yanamala, N.; Gutierrez-Merino, C. Stimulation and Clustering of Cytochrome B5 Reductase in Caveolin-Rich Lipid Microdomains Is an Early Event in Oxidative Stress-Mediated Apoptosis of Cerebellar Granule Neurons. *J. Proteom.* **2012**, *75*, 2934–2949. [CrossRef]
32. Fortalezas, S.; Poejo, J. Cholesterol-Rich Plasma Membrane Submicrodomains Can Be a Major Extramitochondrial Source of Reactive Oxygen Species in Partially Depolarized Mature Cerebellar Granule Neurons in Culture. *J. Neurophysiol. Neurol. Disord.* **2019**, *1*, 1.
33. Samhan-Arias, A.K.; Fortalezas, S.; Cordas, C.M.; Moura, I.; Moura, J.J.G.; Gutierrez-Merino, C. Cytochrome B5 Reductase Is the Component from Neuronal Synaptic Plasma Membrane Vesicles That Generates Superoxide Anion upon Stimulation by Cytochrome c. *Redox Biol.* **2018**, *15*, 109–114. [CrossRef]
34. Pietrini, G.; Aggujaro, D.; Carrera, P.; Malyszko, J.; Vitale, A.; Borgese, N. A Single MRNA, Transcribed from an Alternative, Erythroid-Specific, Promoter, Codes for Two Non-Myristylated Forms of NADH-Cytochrome B5 Reductase. *J. Cell Biol.* **1992**, *117*, 975–986. [CrossRef] [PubMed]
35. Borgese, N.; Macconi, D.; Parola, L.; Pietrini, G. Rat Erythrocyte NADH-Cytochrome B5 Reductase. Quantitation and Comparison between the Membrane-Bound and Soluble Forms Using an Antibody against the Rat Liver Enzyme. *J. Biol. Chem.* **1982**, *257*, 13854–13861. [CrossRef]
36. Martín-Romero, F.J.; Gutiérrez-Martín, Y.; Henao, F.; Gutiérrez-Merino, C. The NADH Oxidase Activity of the Plasma Membrane of Synaptosomes Is a Major Source of Superoxide Anion and Is Inhibited by Peroxynitrite. *J. Neurochem.* **2002**, *82*, 604–614. [CrossRef] [PubMed]
37. Samhan-Arias, A.K.; Almeida, R.M.; Ramos, S.; Cordas, C.M.; Moura, I.; Gutierrez-Merino, C.; Moura, J.J.G. Topography of Human Cytochrome B5/Cytochrome B5 Reductase Interacting Domain and Redox Alterations upon Complex Formation. *Biochim. Biophys. Acta Bioenerg.* **2018**, *1859*, 78–87. [CrossRef]
38. Hampton, M.B.; Zhivotovsky, B.; Slater, A.F.; Burgess, D.H.; Orrenius, S. Importance of the Redox State of Cytochrome c during Caspase Activation in Cytosolic Extracts. *Biochem. J.* **1998**, *329 Pt 1*, 95–99. [CrossRef]

39. Constantinescu, A.; Han, D.; Packer, L. Vitamin E Recycling in Human Erythrocyte Membranes. *J. Biol. Chem.* **1993**, *268*, 10906–10913. [CrossRef]
40. Lagoa, R.; Samhan-Arias, A.K.; Gutierrez-Merino, C. Correlation between the Potency of Flavonoids for Cytochrome c Reduction and Inhibition of Cardiolipin-Induced Peroxidase Activity. *Biofactors* **2017**, *43*, 451–468. [CrossRef]
41. Kobayashi, K.; Harada, Y.; Hayashi, K. Kinetic Behavior of the Monodehydroascorbate Radical Studied by Pulse Radiolysis. *Biochemistry* **1991**, *30*, 8310–8315. [CrossRef]
42. Hara, T.; Minakami, S. On Functional Role of Cytochrome B5. II. NADH-Linked Ascorbate Radical Reductase Activity in Microsomes. *J. Biochem.* **1971**, *69*, 325–330. [CrossRef]
43. Villalba, J.M.; Navarro, F.; Gómez-Díaz, C.; Arroyo, A.; Bello, R.I.; Navas, P. Role of Cytochrome B5 Reductase on the Antioxidant Function of Coenzyme Q in the Plasma Membrane. *Mol. Asp. Med.* **1997**, *18*, 7–13. [CrossRef]
44. Xiaoyuan Wang, P.J.Q. The Location and Function of Vitamin E in Membranes (Review). *Mol. Membr. Biol.* **2000**, *17*, 143–156. [CrossRef] [PubMed]
45. Wefers, H.; Sies, H. The Protection by Ascorbate and Glutathione against Microsomal Lipid Peroxidation Is Dependent on Vitamin E. *Eur. J. Biochem.* **1988**, *174*, 353–357. Available online: https://febs.onlinelibrary.wiley.com/doi/full/10.1111/j.1432-1033.1988.tb14105.x?sid=nlm%3Apubmed (accessed on 9 November 2021). [CrossRef] [PubMed]
46. Frei, B.; Kim, M.C.; Ames, B.N. Ubiquinol-10 Is an Effective Lipid-Soluble Antioxidant at Physiological Concentrations. *Proc. Natl. Acad. Sci. USA* **1990**, *87*, 4879–4883. [CrossRef]
47. Clemente, S.M.; Martínez-Costa, O.H.; Monsalve, M.; Samhan-Arias, A.K. Targeting Lipid Peroxidation for Cancer Treatment. *Molecules* **2020**, *25*, 5144. [CrossRef] [PubMed]
48. Ross, D.; Siegel, D. Functions of NQO1 in Cellular Protection and CoQ10 Metabolism and Its Potential Role as a Redox Sensitive Molecular Switch. *Front. Physiol.* **2017**, *8*, 595. [CrossRef]
49. Maroz, A.; Anderson, R.F.; Smith, R.A.J.; Murphy, M.P. Reactivity of Ubiquinone and Ubiquinol with Superoxide and the Hydroperoxyl Radical: Implications for in Vivo Antioxidant Activity. *Free Radic. Biol. Med.* **2009**, *46*, 105–109. [CrossRef]
50. Samhan-Arias, A.K.; Duarte, R.O.; Martín-Romero, F.J.; Moura, J.J.G.; Gutiérrez-Merino, C. Reduction of Ascorbate Free Radical by the Plasma Membrane of Synaptic Terminals from Rat Brain. *Arch. Biochem. Biophys.* **2008**, *469*, 243–254. [CrossRef]
51. Cabelli, D.E.; Bielski, B.H.J. Kinetics and Mechanism for the Oxidation of Ascorbic Acid/Ascorbate by HO2/O2- (Hydroperoxyl/Superoxide) Radicals. A Pulse Radiolysis and Stopped-Flow Photolysis Study. *J. Phys. Chem.* **1983**, *87*, 1809–1812. [CrossRef]
52. Fridovich, I. Superoxide Dismutases. *Annu. Rev. Biochem.* **1975**, *44*, 147–159. [CrossRef]
53. Dym, O.; Eisenberg, D. Sequence-Structure Analysis of FAD-Containing Proteins. *Protein Sci.* **2001**, *10*, 1712–1728. [CrossRef]
54. Bewley, M.C.; Marohnic, C.C.; Barber, M.J. The Structure and Biochemistry of NADH-Dependent Cytochrome B5 Reductase Are Now Consistent. *Biochemistry* **2001**, *40*, 13574–13582. [CrossRef] [PubMed]
55. Marohnic, C.C.; Crowley, L.J.; Davis, C.A.; Smith, E.T.; Barber, M.J. Cytochrome B5 Reductase: Role of the Si-Face Residues, Proline 92 and Tyrosine 93, in Structure and Catalysis. *Biochemistry* **2005**, *44*, 2449–2461. [CrossRef]
56. Kimura, S.; Emi, Y.; Ikushiro, S.; Iyanagi, T. Systematic Mutations of Highly Conserved His49 and Carboxyl-Terminal of Recombinant Porcine Liver NADH-Cytochrome B5 Reductase Solubilized Domain. *Biochim. Biophys. Acta—Protein Struct. Mol. Enzymol.* **1999**, *1430*, 290–301. [CrossRef]
57. Marohnic, C.C.; Barber, M.J. Arginine 91 Is Not Essential for Flavin Incorporation in Hepatic Cytochrome B5 Reductase. *Arch. Biochem. Biophys.* **2001**, *389*, 223–233. [CrossRef] [PubMed]
58. Kimura, S.; Kawamura, M.; Iyanagi, T. Role of Thr66 in Porcine NADH-Cytochromeb 5 Reductase in Catalysis and Control of the Rate-Limiting Step in Electron Transfer. *J. Biol. Chem.* **2003**, *278*, 3580–3589. [CrossRef] [PubMed]
59. Kugler, W.; Pekrun, A.; Laspe, P.; Erdlenbruch, B.; Lakomek, M. Molecular Basis of Recessive Congenital Methemoglobinemia, Types I and II: Exon Skipping and Three Novel Missense Mutations in the NADH-Cytochrome B5 Reductase (Diaphorase 1) Gene. *Hum. Mutat.* **2001**, *17*, 348. [CrossRef]
60. Yamada, M.; Tamada, T.; Takeda, K.; Matsumoto, F.; Ohno, H.; Kosugi, M.; Takaba, K.; Shoyama, Y.; Kimura, S.; Kuroki, R.; et al. Elucidations of the Catalytic Cycle of NADH-Cytochrome B5 Reductase by X-Ray Crystallography: New Insights into Regulation of Efficient Electron Transfer. *J. Mol. Biol.* **2013**, *425*, 4295–4306. [CrossRef]
61. Ainsley Davis, C.; Barber, M.J. Cytochrome B5 Oxidoreductase: Expression and Characterization of the Original Familial Ideopathic Methemoglobinemia Mutations E255- and G291D. *Arch. Biochem. Biophys.* **2004**, *425*, 123–132. [CrossRef]
62. Roma, G.W.; Crowley, L.J.; Davis, C.A.; Barber, M.J. Mutagenesis of Glycine 179 Modulates Both Catalytic Efficiency and Reduced Pyridine Nucleotide Specificity in Cytochrome B5 Reductase. *Biochemistry* **2005**, *44*, 13467–13476. [CrossRef]
63. Iyanagi, T. Molecular Mechanism of Metabolic NAD(P)H-Dependent Electron-Transfer Systems: The Role of Redox Cofactors. *Biochim. Biophys. Acta Bioenerg.* **2019**, *1860*, 233–258. [CrossRef]
64. Baymann, F.; Schoepp-Cothenet, B.; Duval, S.; Guiral, M.; Brugna, M.; Baffert, C.; Russell, M.J.; Nitschke, W. On the Natural History of Flavin-Based Electron Bifurcation. *Front. Microbiol.* **2018**, *9*, 1357. [CrossRef]
65. Christgen, S.L.; Becker, S.M.; Becker, D.F. Chapter one—Methods for determining the reduction potentials of flavin enzymes. In *Methods in Enzymology*; New Approaches for Flavin Catalysis; Palfey, B.A., Ed.; Academic Press: Cambridge, MA, USA, 2019; Volume 620, pp. 1–25.
66. Takaba, K.; Takeda, K.; Kosugi, M.; Tamada, T.; Miki, K. Distribution of Valence Electrons of the Flavin Cofactor in NADH-Cytochrome B5 Reductase. *Sci. Rep.* **2017**, *7*, 43162. [CrossRef] [PubMed]

67. Bonomi, F.; Iametti, S. Redox titration of flavoproteins: An overview. In *Flavins and Flavoproteins: Methods and Protocols;* Methods in Molecular Biology; Barile, M., Ed.; Springer: New York, NY, USA, 2021; pp. 119–133. ISBN 978-1-07-161286-6.
68. Ishikita, H.; Saito, K. Proton Transfer Reactions and Hydrogen-Bond Networks in Protein Environments. *J. R. Soc. Interface* **2014**, *11*, 20130518. [CrossRef] [PubMed]
69. Alizadeh-Pasdar, N.; Li-Chan, E.C. Comparison of Protein Surface Hydrophobicity Measured at Various PH Values Using Three Different Fluorescent Probes. *J. Agric. Food Chem.* **2000**, *48*, 328–334. Available online: https://pubs.acs.org/doi/10.1021/jf990393p (accessed on 27 June 2021). [CrossRef] [PubMed]
70. Li, H.; Chang, Y.-Y.; Lee, J.Y.; Bahar, I.; Yang, L.-W. DynOmics: Dynamics of Structural Proteome and Beyond. *Nucleic Acids Res.* **2017**, *45*, W374–W380. [CrossRef] [PubMed]
71. Samhan-Arias, A.K.; Cordas, C.M.; Carepo, M.S.; Maia, L.B.; Gutierrez-Merino, C.; Moura, I.; Moura, J.J.G. Ligand Accessibility to Heme Cytochrome B5 Coordinating Sphere and Enzymatic Activity Enhancement upon Tyrosine Ionization. *J. Biol. Inorg. Chem.* **2019**, *24*, 317–330. [CrossRef]
72. Samhan-Arias, A.K.; Maia, L.B.; Cordas, C.M.; Moura, I.; Gutierrez-Merino, C.; Moura, J.J.G. Peroxidase-like Activity of Cytochrome B5 Is Triggered upon Hemichrome Formation in Alkaline PH. *Biochim. Biophys. Acta Proteins Proteom.* **2018**, *1866*, 373–378. [CrossRef]

International Journal of *Molecular Sciences*

MDPI

Article

A Mechanistic Model of NMDA and AMPA Receptor-Mediated Synaptic Transmission in Individual Hippocampal CA3-CA1 Synapses: A Computational Multiscale Approach

Pietro Micheli [1], Rui Ribeiro [1,2,*] and Alejandro Giorgetti [1,2,*]

[1] Department of Biotechnology, University of Verona, 37134 Verona, Italy; pietro.micheli@studenti.univr.it
[2] Institute for Neuroscience and Medicine (INM-9) and Institute for Advanced Simulations (IAS-5), "Computational Biomedicine", Forschungszentrum Jülich, 52428 Jülich, Germany
* Correspondence: rui.ribeiro@univr.it (R.R.); alejandro.giorgetti@univr.it (A.G.)

Abstract: Inside hippocampal circuits, neuroplasticity events that individual cells may undergo during synaptic transmissions occur in the form of Long-Term Potentiation (LTP) and Long-Term Depression (LTD). The high density of NMDA receptors expressed on the surface of the dendritic CA1 spines confers to hippocampal CA3-CA1 synapses the ability to easily undergo NMDA-mediated LTP and LTD, which is essential for some forms of explicit learning in mammals. Providing a comprehensive kinetic model that can be used for running computer simulations of the synaptic transmission process is currently a major challenge. Here, we propose a compartmentalized kinetic model for CA3-CA1 synaptic transmission. Our major goal was to tune our model in order to predict the functional impact caused by disease associated variants of NMDA receptors related to severe cognitive impairment. Indeed, for variants Glu413Gly and Cys461Phe, our model predicts negative shifts in the glutamate affinity and changes in the kinetic behavior, consistent with experimental data. These results point to the predictive power of this multiscale viewpoint, which aims to integrate the quantitative kinetic description of large interaction networks typical of system biology approaches with a focus on the quality of a few, key, molecular interactions typical of structural biology ones.

Keywords: CA3-CA1 synapses; NMDA; AMPA; systems biology; multiscale modeling; Schaffer collateral-CA1 synapses

Citation: Micheli, P.; Ribeiro, R.; Giorgetti, A. A Mechanistic Model of NMDA and AMPA Receptor-Mediated Synaptic Transmission in Individual Hippocampal CA3-CA1 Synapses: A Computational Multiscale Approach. *Int. J. Mol. Sci.* **2021**, *22*, 1536. https://doi.org/10.3390/ijms22041536

Academic Editor: Masoud Jelokhani-Niaraki

Received: 1 December 2020
Accepted: 1 February 2021
Published: 3 February 2021

Publisher's Note: MDPI stays neutral with regard to jurisdictional claims in published maps and institutional affiliations.

1. Introduction

Ionotropic glutamatergic receptors are a class of membrane receptors divided into three main subtypes, classified according to their activation to the selective agonists: NMDA (N-Methyl-D-aspartic acid), AMPA (α-amino-3-hydroxy-5-methyl-4-isoxazolepropionic acid), and Kainato. They play a key role in the process of synaptic transmission, which takes place in excitatory glutamatergic synapses, and dysregulations in their normal activities have been widely linked to numerous neurological disorders and synaptopathies [1–5]. Particularly, NMDA and AMPA receptors have been identified as crucial in the molecular mechanism underlying the process of synaptic plasticity, a process that leads to the modulation in the strength of the neuronal response to stimulation, linked to learning and memory [6–8].

Complex cognitive functions such as learning and multiple forms of memory are carried out by the hippocampal formation, which can dynamically sample, encode, store, and retrieve information coming from the sensory experience [9–11]. The constant encoding and integration of new information is possible thanks to the ability of a neural circuit to continuously reshape its topology and modulate the strength of its connections. In the hippocampal circuits, synaptic plasticity events that individual cells may undergo during synaptic transmissions occur in the form of Long Term Potentiation (LTP) and Long Term Depression (LTD). The trisynaptic circuit, particularly, has been extensively studied because of its apparently simple connectivity and the experimental accessibility of its structures. Inside this pathway, CA3 Shaffer collateral axons innervate CA1 pyramidal cells, forming excitatory glutamatergic synapses. The high density of NMDA receptors expressed on the surface of the dendritic CA1 spines confers to this synapse the ability to easily undergo NMDA receptor-mediated LTP and LTD, which has been substantially evidenced to be essential for some forms of explicit learning in mammals [12,13].

In Schaffer collateral-CA1 synapses, AMPA and NMDA receptors populate the membrane of the CA1 spine, actively participating in synaptic transmission. AMPA receptors are GluR1-GluR4 containing homo/hetero-tetrameric receptors that mediate fast excitatory neurotransmission in glutamatergic synapses. The early phase of synaptic plasticity events that occur in Schaffer collateral-CA1 synapses are associated with alterations in the number of AMPA receptors expressed on the spine membrane through activation of exocytosis or endocytosis mechanisms, as well as changes in AMPA receptors conductance through phosphorylation modifications [14,15]. Together, these molecular mechanisms lead to fine modulations in the strength of the synaptic transmission. The reactions underlying such modulation are controlled by the transient variations in the Ca^{2+} concentration that occur in the post-synaptic spine, especially, due to the activation of NMDA receptors. NMDA receptors are hetero-tetrameric glutamatergic ionotropic receptors permeable to Na^{2+}, K^{+}, Ca^{2+}, and Mg^{2+} ions [16,17]. The permeability to Mg^{2+} ions gives to NMDA receptors a pronounced voltage-dependent behavior. At resting membrane potentials, external Mg^{2+} ions enter into the receptor's pore but, unlike the other permeating ions, they bind tightly to the pore, blocking it and impairing further ion permeation [18,19]. One of the most accepted physiological mechanisms needed to efficiently unblock NMDA receptors, thus generating an inward Ca^{2+} flux, is a temporal coincidence between the release of pre-synaptic neurotransmitter and a depolarization of the post-synaptic spine (of sufficient amplitude and duration) elicited by post-synaptic activity. This synchronicity is taken into account in the *Spike Timing Dependent Plasticity (STDP)* paradigm that also includes the post-synaptic dendritic activity expressed in the form of *back-propagating action potentials* (bAPs) [20,21]. The transient post-synaptic Ca^{2+} inward current generated by the activation and unblocking of NMDA receptors critically acts on the kinetic equilibrium of the different calcium-binding proteins involved in LTP/LTD-inducing pathways, such as Ca^{2+}/Calmodulin-dependent Kinase II (CaMKII) [22–24].

Dysfunctions on LTP/LTD-mediated synaptic plasticity have been associated with many neurological disorders such as epilepsy and Alzheimers, Hughtington, and Parkinson's diseases [4,25–30].

A comprehensive and detailed understanding of the molecular mechanisms underlying synaptic transmission and neuroplasticity is then crucial for the physio-pathological characterization of many cognitive functions. However, even if LTP/LTD-mediated synaptic plasticity has been extensively studied, providing a substantial description of a full integration of the interaction networks underlying the whole synaptic transmission, deeply characterized at the molecular level, is currently a major challenge. This could be the starting point for the identification of new therapeutic strategies, aimed at re-tuning the global behavior of the intricate network of molecular interactions underlying synaptic plasticity, thus restoring its functional integrity.

Systems biology models have been shown to be key in approaching the complexity of this type of interaction networks. These models use a holistic approach to unveil the complexity of the molecular pathways and to catalogue all the biological complexes and the relationships between them [31]. They have evolved from empirical descriptions to fundamental mathematical equations applied by computational methods, allowing us to envision how such systems change over time under different conditions. In this way, one can infer qualitative features of the whole system, such as the downstream consequences of a single altered interaction, and consequently identify, for example, pharmacological targets or even predict the severity of a structural variant of a molecular species.

Here, we present, and render available to the scientific community (see Data Availability section), a mathematical model of the CA3 Schaffer collateral-CA1 transmission. Although other integrated and detailed models of glutamatergic synapses have been proposed recently [32,33], a clinical-oriented application of such models, able to also take into account the molecular characterization of particular disease-associated variants, is lacking. The rationale of our work was to provide a synaptic model that can be easily reproduced, run, and be integrated into larger analytical pipelines, proposing a novel viewpoint on the possible applications of comprehensive and detailed system biology models.

Our model allows us to simulate several features of the CA3-CA1 synaptic transmission process. These include (1) glutamate release inside the synaptic cleft as a result of a pre-synaptic stimulation, (2) bAP in the post-synaptic dendritic spine, (3) kinetic description of the gating mechanism of both NMDA and AMPA receptors, (4) estimation of the excitatory post-synaptic currents (EPSCs) and excitatory post-synaptic potentials (EPSPs), including the explicit calculation of the NMDA-mediated inward Ca^{2+} current, and (5) kinetic descriptions of the Ca^{2+}-dependent molecular reactions that take place inside the post-synaptic spine and lead to the activation of CaMKII. Here we report some of the qualitative features observed in the receptors-specific contributions to synaptic transmission, as well as in the timing of pre/post-synaptic stimulation. Finally, we offer a further integration of our systems biology approach with a molecular level modeling of disease associated variants. This approach may pave the way to novel multiscale approaches to be used in the pharmacology or structural systems biology field. Because complex biological systems do not rely on individual metabolic networks, having a fully integrated description of metabolic networks allows us to envision the system as a whole instead of a sum of its parts [34]. It follows that the combination of integrated pathways with molecular detail observations, as the one we are presenting here, may bring to light new therapeutic strategies and bring us closer to the new era of personalized medicine.

2. Results and Discussions

This section is divided in two main subsections. In the first part, we present the implementation of the mechanistic model, providing an overview on the structure of the pipeline through the description of the individual modules, implemented to describe different fragments of the system. The second part contains the simulation of the model under different parameter configurations. This allows us to infer some qualitative features of the system, with a particular focus on the timing between pre and post-synaptic stimuli, and finally to assess shifts in the global system behavior given by the introduction of rare variants in the NMDA receptors associated with diseases.

2.1. An Integrative, Python-Based Pipeline for Simulating Glutamatergic Synaptic Transmission

We developed an integrative mathematical pipeline for the easy running of numerical simulations of synaptic transmission in individual CA3 Schaffer collateral-CA1 synapses, driven by both pre- and post-synaptic stimulation. The pipeline is composed of four different main modules, each one aimed at modelling a different part of the whole transmission process. Starting from the definition of a stimulation pattern, new modules were progressively implemented and added on top of each other, defining a linear pipeline for simulating the synaptic transmission (Figure 1).

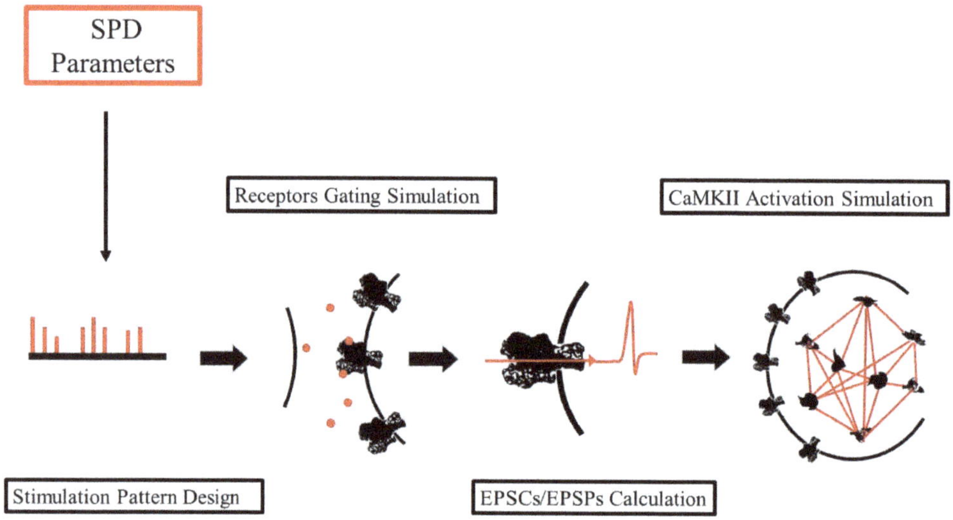

Figure 1. Conceptual scheme of the pipeline to simulate our synaptic transmission model. This scheme illustrates all of the four modules of our framework: (i) Stimulation Pattern Design (SPD module); (ii) Receptors Gating Simulation (RGS module); (iii) Excitatory Post-Synaptic Currents and Potentials (EPSCs/EPSPs) Calculation (CPC module); (iv) Ca^{2+}/Calmodulin-dependent Kinase II (CaMKII) Activation Simulation (CAS module).

2.1.1. Stimulation Pattern Design (SPD)

This module implements a series of functions that easily allow us to define a stimulation pattern that will drive the synaptic transmission. Such stimulation patterns can be composed of both pre- and post-synaptic stimuli, organized as trains of bursts. Here, highly customizable patterns can be designed by setting the number of stimuli composing each burst, the intra-burst, and the inter-burst frequencies for both pre- and post-synaptic stimuli (Figure 2).

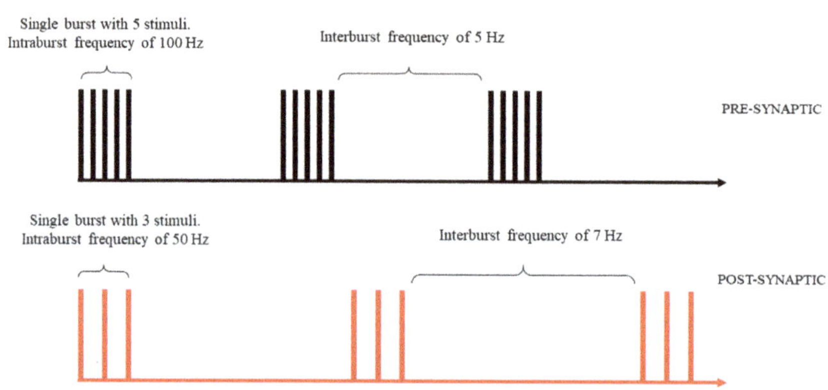

Figure 2. Example scheme of a stimulation pattern. Pre- and post-synaptic stimuli are organized as trains or bursts. Each burst is composed of a sequence of stimuli, delivered at an intra-burst frequency. Inter-burst frequency defines the interval between each burst. Number of stimuli per burst, intra-burst, and inter-burst frequencies can be defined during the stimulation pattern design, for both pre- and post-synaptic patterns.

Pre-synaptic stimuli are idealized and modeled as the instantaneous rise and fall of the free glutamate concentration in the synaptic cleft, assuming a square pulse-like shape. In this article, we will refer to a pre-synaptic stimulus as a "glutamate pulse". The quantity of released glutamate (i.e., the pulse amplitude, expressed in μM) and the glutamate exposure time inside the cleft (i.e., the pulse width, expressed in ms) of each pre-synaptic stimulus can be independently parametrized. Post-synaptic stimuli are modeled as dendritic back-propagating action potentials, consisting of transient depolarization potentials of the post-synaptic spine membrane. The shape of such stimuli has been defined using a two-component exponential function (see methods, Section 3.1.1 for further details), as proposed by Shouval et al. [35]. The stimulation pattern defined in this module will constitute the input of the following modules.

2.1.2. Receptors Gating Simulation (RGS)

Pre-synaptic stimuli, defined during the design of the stimulation pattern, are used as input to a second module, which is used to simulate the interactions between the neurotransmitter and the AMPA and NMDA receptors. This module contains the compartmental kinetic description of both the receptor-neurotransmitter binding reactions and the gating mechanisms that lead to the opening of the channels. Particularly, the latter consists of state-transition models (including closed, pre-open, open, and desensitized states) that statistically represent the stochastic distribution of the current traces recorded by electrophysiological experiments. We selected and integrated one kinetic model for both AMPA and NMDA receptors, proposed by Koike et al. and Amico-Ruvio and Popescu [36,37], respectively. Then, we translated both models into systems of first-order differential equations, implemented in a single larger kinetic model using the python PySB package (see methods for further details). Finally, a numerical integration was performed, allowing the simulation of the receptor's behavior with a high temporal resolution (integration step of 1 μs). We tested the reliability of these *ex-novo* implementations by comparing the behaviors predicted by our model, for both AMPA and NMDA receptors, with the behaviors reported in the works by Koike et al. and by Amico-Ruvio and Popescu [36,37] (Table S1). We observed a strong consistency between the kinetic features of both AMPA and NMDA receptors predicted by our PySB-based model and the respective original models, pointing to a high reliability of our implementation.

2.1.3. EPSCs/EPSPs Calculation (CPC)

The third module of our framework consists of a system of equations used to explicitly calculate the EPSCs and the respective EPSPs generated during the simulation of the synaptic transmission. The EPSCs are estimated by calculating, over the simulation, the ion fluxes that permeate each open channel (predicted with the RGS module described in Section 2.1.2). This estimation is made according to the channel-specific conductance, the channel-specific reversal potential, and the depolarization level of the post-synaptic membrane. The EPSPs are then derived from the EPSCs (see methods Section 3.1.3 for further details). All the depolarization potentials, which include the EPSPs and, eventually, the bAPs arising from the post-synaptic stimulation, are summed together to assess the global changes in the membrane depolarization value. In this module, the equation for the explicit estimation of the NMDA-mediated Ca^{2+} current is used to assess the post-synaptic changes in the Ca^{2+} concentration according to a simple model proposed by Shouval et al. [35] (see methods, Section 3.1.3 for further details).

2.1.4. CaMKII Activation Simulation (CAS)

The last module of our pipeline aims to simulate a kinetic description of the post-synaptic molecular interactions that controls the CaMKII kinase autophosphorylation events. For this purpose, as previously described for the RGS module (Section 2.1.2), we selected from the literature a detailed kinetic model based on its reproducibility, and we transcribed all its reactions into a second PySB model as a system of first-order differential equation. We chose to implement a model for the CaMKII activation proposed by Pepke et al. [38], and we integrated it into the simulation pipeline. This kinetic model includes a large number of reactions, mainly characterizing the interactions between free Ca^{2+} ions, calcium-binding messenger CaM, and the CaMKII enzyme. Particularly, the Ca^2-CaM mediated autophosphorylation of CaMKII enzyme, which leads to its own activation, directly plays a pivotal role in inducing the early phase of synaptic plasticity [22–24]. Although the changes in the synaptic strength are currently not explicitly assessed in our model, the variations in the activated CaMKII accumulation allows one to assess the relative efficiency of the simulated synaptic transmission.

2.2. Kinetic Behavior Analysis of AMPA and NMDA Receptors under Different Pre-Synaptic Stimulation Conditions

We explored how AMPA and NMDA receptors kinetically behave under different stimulations patterns, exploiting the RGS module (Section 2.1.2). For this purpose, we simulated the model using different pre-synaptic stimulation patterns, consisting of either a single glutamate pulse or bursts of multiple glutamate pulses, delivered at different frequencies (ranging from 10 to 100 Hz). The amplitude of the glutamate pulses was set into a physiological range of 1–2 mM [39,40], while the time width was varied in a range between 1 ms and 1.5 s.

We first focused on the kinetic behavior of AMPA receptors under a single glutamate pulse of 1 mM, simulated with 1, 5, and 10 ms width. The desensitization kinetics of AMPA receptors predicted by the gating model shows a much slower time course (τ = ~25 ms, fitted with single exponential function) compared to the deactivation kinetics (τ = ~0.55 ms, fitted with single exponential function) after the end of a single glutamate pulse (Figure 3A). Moreover, both the exposure time of the glutamate (defined by the pulse width) and the stimulation frequency seem to strongly affect the number of desensitized receptors reached after a single pre-synaptic event [36] (Figure 3B). The faster deactivation, compared to the desensitization predicted by the model, points to the property of AMPA receptors to preferentially undergo a temporal accumulation of desensitized states instead of the open states.

We then analyzed how the variation of the glutamate pulses duration affects the summation of desensitized states under a single pre-synaptic burst stimulation. The latter was simulated by a single burst composed of 5 glutamate pulses of 1 mM amplitude and 1,5, and 10 ms width, with an intra-burst frequency of 100 Hz. We observed a significant increase in the temporal summation of desensitized AMPA receptors as the glutamate exposure values increased (Figure 3C–E, respectively).

Next, we analyzed the predicted kinetic behavior of NMDA receptors. By simulating a single glutamate pulse of 1 mM amplitude and 1 ms, 500 ms, and 1.5 s width, we observed a significatively slower deactivation and desensitization kinetics compared to AMPA receptors (Figure 4). Fitting the curves with a single exponential function, we found time constants of 163, 195, and 210 ms for the deactivation kinetics after 1 ms, 500 ms, and 1.5 s of glutamate exposure, respectively, and a time constant of 1.95 s for the desensitization kinetics (Figure 4A,B). From these results, we got a ratio between the desensitization and the deactivation time constant ($\tau_{desens}/\tau_{deact}$) of ~12 for the NMDA receptors and ~45 for the AMPA receptors. The lower value found for the NMDA receptors leads to a more efficient temporal summation of its open states. In fact, when we simulated the model with a single pre-synaptic burst of 5 glutamate pulses of 1 mM amplitude and 1, 5, and 10 ms width, with intra-burst frequencies of 10, 50, and 100 Hz, we observed, effectively, summation of the open NMDAs (Figure 4C–E).

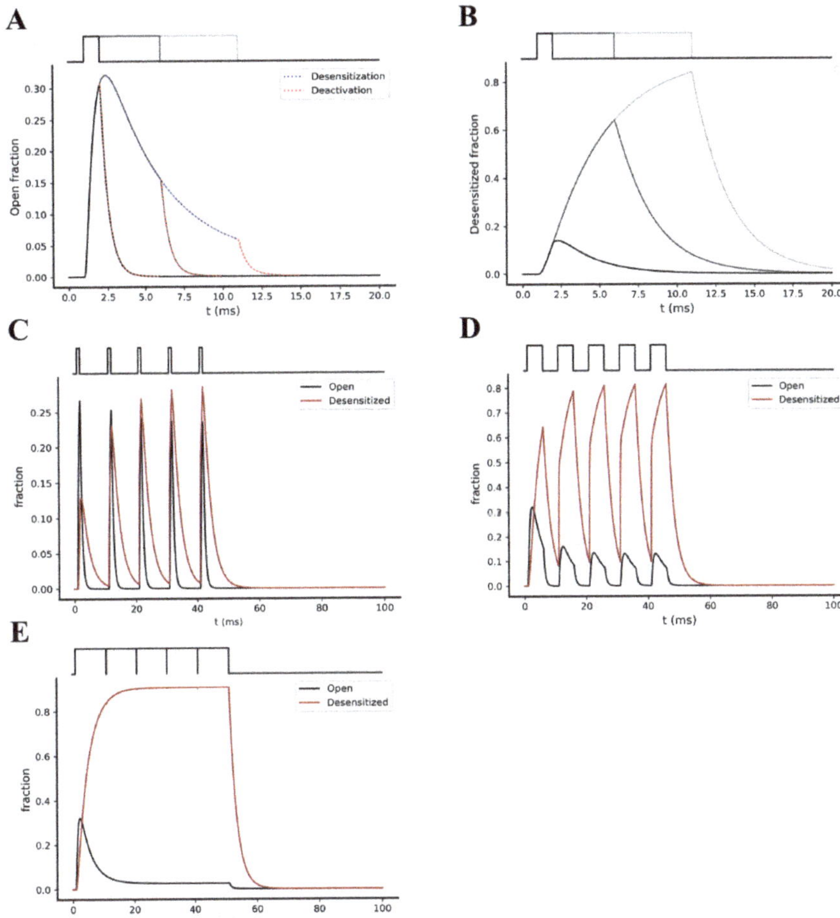

Figure 3. Opening and desensitization kinetics of the AMPA receptors. (**A**) Open fraction kinetics following a stimulation with a single glutamate pulse of 1 mM amplitude and width of 1 ms (black), 5 ms (dark grey) and 10 ms (light grey). Blue dotted trace shows the desensitization kinetics, while red dotted traces show the deactivation kinetics following glutamate removal from the synaptic cleft. (**B**) Desensitized fraction kinetics following a stimulation with a single glutamate pulse of 1 mM amplitude and width of 1 ms (black), 5 ms (dark grey), and 10 ms (light grey). (**C–E**) Kinetics of open and desensitized fractions following pre-synaptic stimulations with a burst composed of 5 glutamate pulses, with glutamate pulse amplitude of 1 mM, an intra-burst frequency of 100 Hz, and a pulses width of 1 ms (**C**), 5 ms (**D**), and 10 ms (**E**).

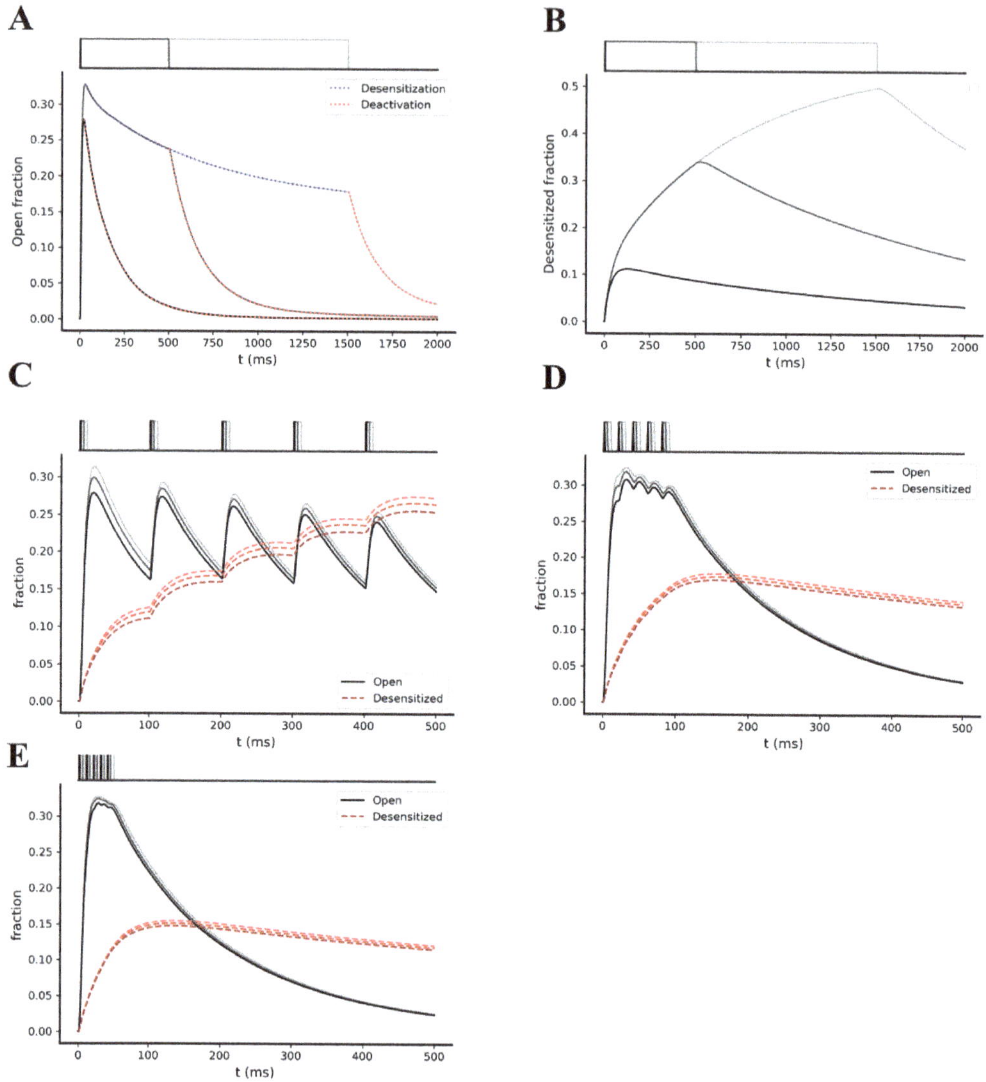

Figure 4. Opening and desensitization kinetics of the NMDA receptors. (**A**) Open fraction kinetics following a stimulation with a single glutamate pulse of 1mM amplitude and width of 1 ms (black), 500 ms (dark grey), and 1.5 s (light grey). Blue dotted trace shows the desensitization kinetics, while red dotted traces show the deactivation kinetics following glutamate removal from the virtual synaptic cleft. (**B**) Desensitized fraction kinetics following a stimulation with a single glutamate pulse of 1mM amplitude and width of 1 ms (light grey), 5 ms (dark grey), and 10 ms (black). (**C–E**) Kinetics of open and desensitized fractions following pre-synaptic stimulations with a burst composed of 5 glutamate pulses, with glutamate pulse amplitude of 1 mM, pulse width of 1 ms (black), 5 ms (dark grey), and 10 ms (light grey) and an intra-burst frequency of 10 Hz (**C**), 50 Hz (**D**), and 100 Hz (**E**).

To have a better insight into the difference between the kinetic behavior of AMPA and NMDA receptors, we simulated our model with a single pre-synaptic burst of 5 glutamate pulses of 1 mM amplitude and 1, 5, and 10 ms width, varying the intra-burst frequency between 10 and 100 Hz. For each intra-burst frequency, we calculated the ratio between the total number of desensitized and open receptors. According to our model, these simulations pointed out that the desensitized/open ratio of AMPA receptors depends more on the stimulation frequencies and on the glutamate pulses durations compared to the desensitized/open ratio of the NMDA receptors (Figure 5).

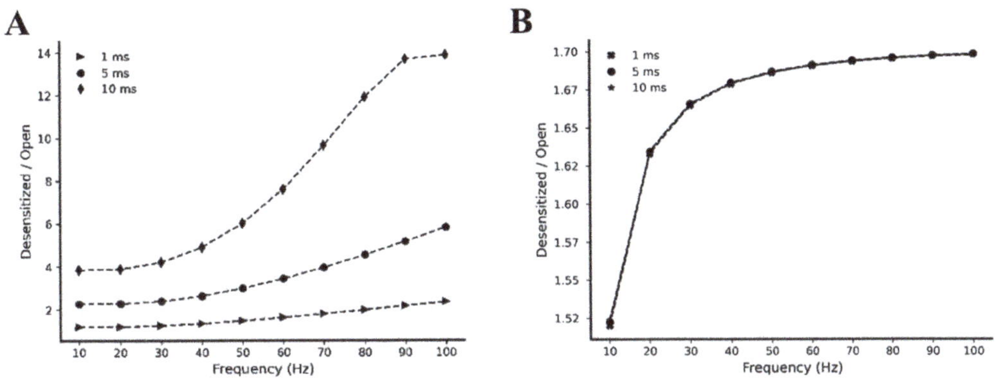

Figure 5. Desensitized/Open ratio expressed as a function of stimulation frequency. Simulations were performed using single pre-synaptic glutamate pulses of 1 mM amplitude and 1, 5, and 10 ms width. For each simulation, the ratio between the desensitized and the open fraction has been calculated for (**A**) AMPA (triangles, circles and diamonds symbols refer, respectively, to 1, 5 and 10 ms pulse width) and (**B**) NMDA receptors (crosses, circles and stars symbols refer, respectively, to 1, 5 and 10 ms pulse width).

2.3. Temporal Relationship between Pre- and Post-Synaptic Stimuli Strongly Impacts Synaptic Transmission Efficiency

During the stimulation of the synapse, the equations implemented in the CPC module (Section 2.1.3) allow us to explicitly assess the individual contribution of both AMPA and NMDA receptors to the global electrical transmission. Pre-synaptic-induced excitatory potentials and post-synaptic dendritic back-propagation events, programed during the stimulation pattern design, are integrated together to continuously estimate the variations in the NMDA permeability, as well as in the Ca^{2+} flux driving force (see methods Section 3.1.3 for further details). We explored, through several simulations, how the temporal relationship between pre- and post-synaptic stimuli can shape the efficiency of the electro-chemical transmission.

2.3.1. AMPA-Mediated EPSPs Are Not Sufficient to Efficiently Relieve the Mg^{2+} Block from NMDA Receptors

The pronounced voltage-dependent affinity of NMDA receptor for the extracellular Mg^{2+} ions causes the actual permeation of the channel to be strongly modulated by the depolarization level of the membrane [19]. We have previously observed that the kinetic equations implemented in the RGS module predict no effective temporal summations of open AMPA receptors because of their fast desensitization and deactivation kinetics, as observed in other studies [36,41]. Analyzing the output of the RGS module using the equations implemented in the CPC module (Sections 2.1.2 and 2.1.3), we then observed that, coherently, the AMPA-mediated responses also tend not to summate (Figure S1).

This observation prompted us to investigate if the amplitude of an AMPA-mediated EPSP evoked by a single pre-synaptic event was high enough to relieve the Mg^{2+} block from NMDA receptors. Because the EPSPs amplitudes of AMPA and NMDA receptors are influenced by their levels of expression on the post-synaptic spine surface, we performed multiple simulations of a single glutamate pulse of 1 mM amplitude and 1, 5, and 10 ms width, varying the level of available AMPA receptors in a range between 20 and 200 [42]. Simulation results reported that the maximum AMPA-mediated EPSPs peaks elicited by single-pulse pre-synaptic stimulations reach −40 mV with 200 units of AMPA receptors (Figure 6A). According to the Mg^{2+} unblocking probability function that we have incorporated into the model (see methods Section 3.1.3 for further details), such a depolarization level can effectively release the Mg^{2+} ion from NMDA receptors only if the extracellular Mg^{2+} concentration is very low compared to the physiological concentration (Figure 6B), which is near to 1 mM [19].

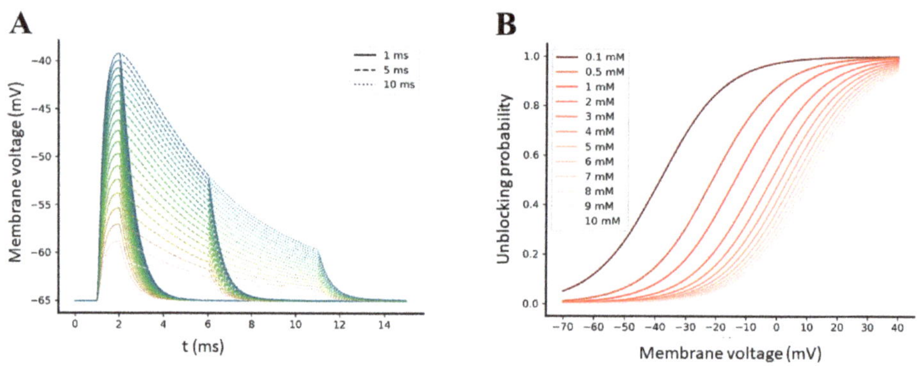

Figure 6. (**A**) Simulated AMPA-mediated EPSPs evoked by different numbers of available AMPA receptors, ranging from 20 (lower trace) to 200 (upper trace). Solid, dashed, and dotted traces refer to single pulse stimulation performed with a glutamate pulse width of, respectively, 1, 5, and 10 ms. (**B**) Sigmoidal unblocking probability function for Mg^{2+} block, expressed as a function of membrane voltage. Each trace corresponds to a different value of extracellular Mg^{2+} concentration.

These results emphasize the fact that pre-synaptic events on their own may not be enough to ensure an effective Ca^{2+} permeation. As supported by the STDP paradigm, temporal coordination between pre- and post-synaptic events must occur in order to allow a significant Ca^{2+} influx that can effectively trigger plasticity [43].

2.3.2. Synchronization between Pre- and Post-Synaptic Stimulation Significantly Increases the NMDA Receptor Contribution to Synaptic Transmission

We further investigated how the synchrony between pre- and post-synaptic activity can affect the efficiency of synaptic transmission, particularly by increasing the amplitude of the NMDA receptors-mediated EPSCs and EPSPs. For this purpose, we compared the individual responses of the AMPA and NMDA receptors obtained from two different stimulation patterns, one including only pre-synaptic stimulation and one including coupled pre- and post-synaptic stimulations. In both stimulation pattern, the pre-synaptic stimulation consisted of a single theta burst composed of 5 glutamate pulses of 1 mM amplitude and 1, 5, and 10 ms width, with an intra-burst frequency of 100 Hz [44]; post-synaptic stimulation was designed as a single dendritic back-propagation event, which occurs in the post-synaptic spine 1 ms after the first pre-synaptic stimuli was delivered. Simulations were performed in the presence of 20 AMPA and 15 NMDA receptors [42,45], with extracellular Mg^{2+} concentration set to 1 mM. As expected, significant increases in the total NMDA receptor-mediated current peak (~2.5 fold), as well as in the Ca^{2+} that permeated the channel (~4.5 fold), were observed during the coupled pre- and post-synaptic stimulation compared to the pre-synaptic stimulation alone, showing the impact of bAP-mediated synaptic facilitation on the NMDA receptors conductance (Figure 7).

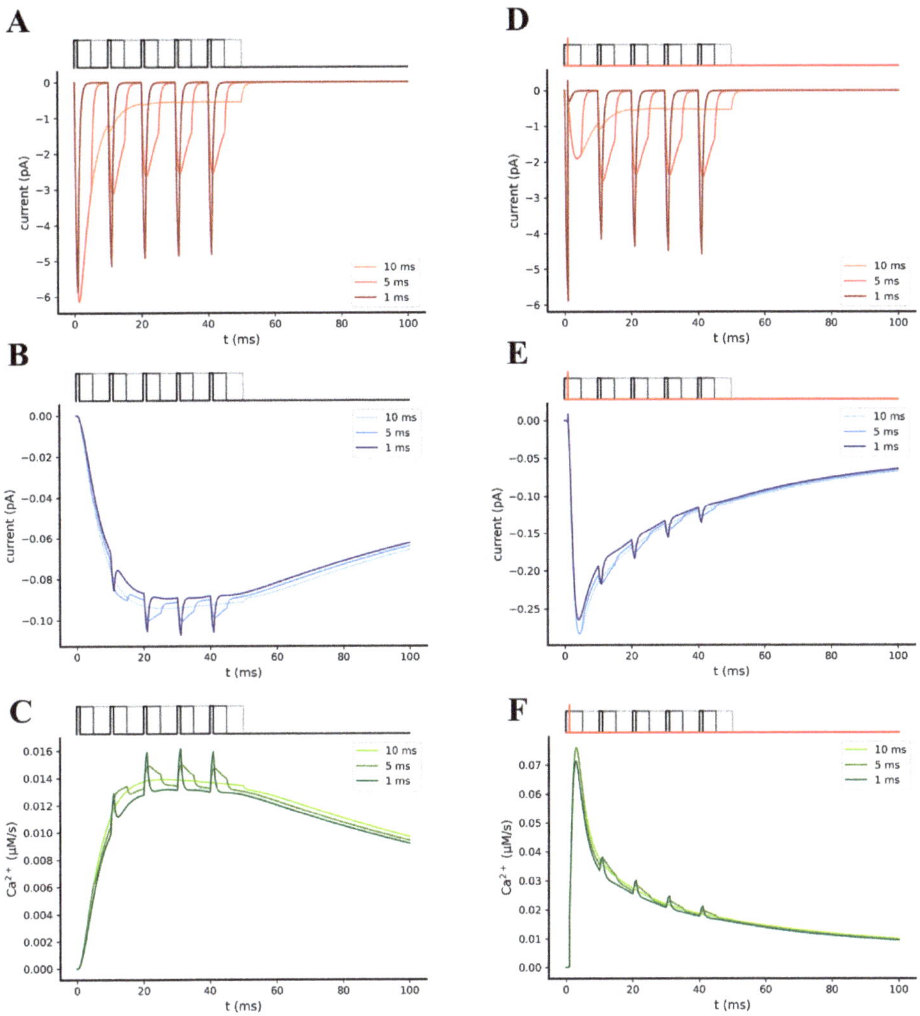

Figure 7. Simulation of synaptic transmission elicited by a single pre-synaptic burst of 5 glutamate pulses, in the absence of (**A–C**) or in the presence of (**D–F**) a single post-synaptic back-propagating action potential (bAP). (**A**,**D**) Time course of the individual AMPA-mediated EPSC. (**B**,**E**) Time course of the individual NMDA-mediated EPSC. (**C**,**F**) Time course of the Ca^{2+} molar flowrate that permeate NMDA receptors during the simulations. Pre-synaptic bursts were composed of 5 glutamate pulses of 1 mM amplitude and 1 ms (black pulses), 5 ms (dark grey pulses), and 10 ms (light grey pulses) width; in each plot, the responses elicited by 1 to 10 ms widths are represented by different colors, respectively, from the darkest to the brightest. Post-synaptic activity (red trace) was programmed as a single dendritic back-propagation event that occurs 1 ms after the first pulse of the pre-synaptic burst began. Both simulations were performed in the presence of 20 AMPA, 15 NMDA, and 1 mM of extracellular Mg^{2+}.

Because we had observed that the presence of a bAP during stimulation significantly increases the NMDA receptor mediated EPSC, we analyzed how variations in temporal coordination level between pre- and post-synaptic stimuli impacts the amplitude of the elicited Ca^{2+} influx. For this purpose, we performed multiple simulations varying the time interval between pre- and post-synaptic stimuli ($\Delta t = t_{post} - t_{pre}$). For each simulation, we then evaluated the effect of the bAP-induced synaptic facilitation by calculating the maximum Ca^{2+} concentration reached in the post-synaptic spine. Simulating a single pre-synaptic glutamate pulse of 1 mM amplitude and 1 ms width, together with a single post-synaptic bAP, we found that post-synaptic Ca^{2+} rises from a value of ~200 nM (the post-synaptic Ca^{2+} concentration elicited by a single pre-synaptic event alone) to a maximum of ~1.4 μM (Figure 8). This value is obtained when the pre-synaptic event precedes the post-synaptic event (positive Δt) of ~20 ms, in agreement with the Hebbian STDP paradigm for synaptic plasticity (see *Feldman 2012* [20] for a review).

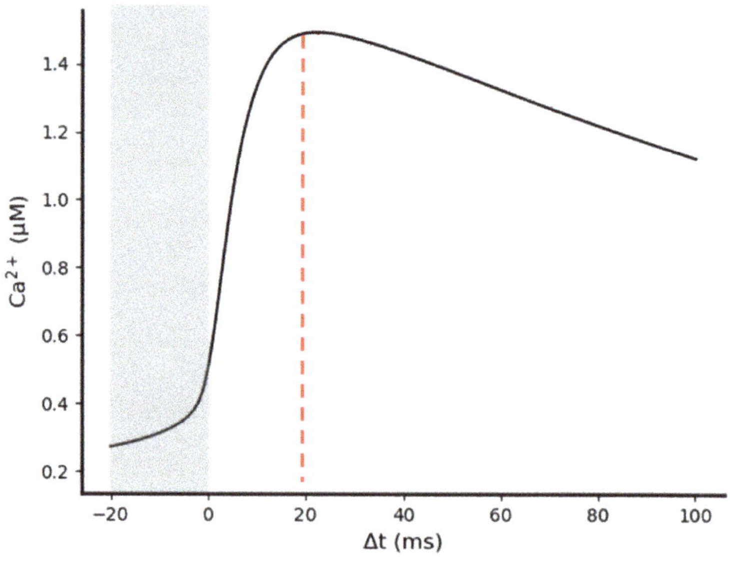

Figure 8. Relationship between pre/post-synaptic stimulation timing and the Ca^{2+} concentration peaks reached in the post-synaptic spine. Simulations were performed in the presence of 20 AMPA, 15 NMDA, and 1 mM of extracellular Mg^{2+}. Maximum post-synaptic Ca^{2+} concentration was reached with $\Delta t \approx 20$ ms. Gray rectangle highlights negative Δt values in which post-synaptic stimuli precede pre-synaptic stimuli.

2.4. Kinetic and Pharmacological Analysis of NMDA Variants: Multiscale Integration

Deactivation time course defines the time required by the receptor-mediated current to decay after the removal of the agonist from the synaptic cleft. This kinetic feature, together with EC_{50} values of the agonist, constitute a prominent quantitative feature used to perform functional analysis of ion channels [46]. Many published studies on rare NMDA receptor variants have tried to assess the severity of a certain mutation, considering its impact on both glutamate potency and deactivation time constant [25,47–49].

We used our model to predict the glutamate affinity (K_d) and the weighted deactivation time constant (τ_w) in NMDA receptor variants, based on the EC_{50} values that have been reported in different experimental and computational studies [25,47,48,50]. In particular, we focused on two rare variants: Glu413Gly and Cys461Phe that fall inside the GluN2B binding pocket (Figure 9). These variants have been shown to decrease the glutamate potency, which may result from a decrease in the glutamate affinity [47,48,50,51].

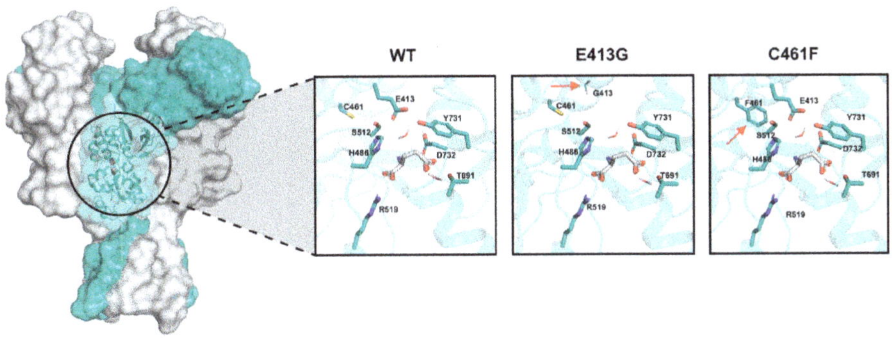

Figure 9. Structure of human GluN1/GluN2A NMDA receptor (PDB accession code: 4TLM). The GluN2B subunit is colored in light blue. The insights show the glutamate binding domain of the wild-type (WT) receptor and the two structural variants Glu413Gly (E413G) and Cys461Phe (C461F). Each window focuses on the docked glutamate (white molecule) and the crucial residues that directly participate to the interaction. Red arrows point to the residue substitution of each of the two structural variants.

Therefore, we tuned the NMDA kinetic model to reproduce the same concentration-response behaviors experimentally observed for both the Glu413Gly and Cys461Phe variants.

Exploiting our kinetic model, we were able to computationally assess the NMDA-glutamate concentration-response relationship by using the following approach: firstly, we sampled concentration values in a range between 0.01 and 1000 mM; next, for each value, we ran the RGS module, simulating a single glutamate pulse, with amplitude corresponding to the current glutamate concentration value and width of 1.5 s, as reported by experiments [47], setting the number of AMPA receptors to 0 (because we were interested in isolating the NMDA response). Finally, calculating from each simulation the peak of the evoked current, EC_{50} value was obtained by fitting the concentration-response data with the logistic function.

To predict the shifts in the NMDA receptor-glutamate affinity associated with the rare variants Glu413Gly and Cys461Phe, knowing their experimental EC_{50} values (75–79 μM for Glu413Gly [47,48] and 169 μM for Cys461Phe [47]), we progressively increase, during a sequence of multiple simulations, the ratio between the rate constants k_{off} and k_{on} (i.e., the K_d) of the equations describing the interaction between the NMDA receptor and the glutamate. For each simulation, we computed the EC_{50} value, and at the end of all the simulations we selected the K_d that rendered the EC_{50} values closest to the experimental ones.

As a result, we found that the NMDA receptor kinetic behavior generated by predicted K_d values shows a current deactivation time constantly very close to the experimental ones (Table 1).

Table 1. Predicted K_d and deactivation time constant for NMDA Wt and variants. Deactivation decay was fitted with a two-component exponential function, and the weighted Tau was then calculated (see methods Section 3.2.2).

	Predicted K_d (µM)	Weighted Tau (ms)	
		Predicted	Exp.
Wt	2.5	328	314–570 [47,48,52]
Glu413Gly	190.5	29	20–34 [47,48]
Cys461Phe	446.5	27	28 [48]

The kinetic model of the NMDA receptor was tuned by only increasing the k_{off} rate constant of the glutamate binding reactions. Therefore, we reasoned that the coherence between our results and the experimental data points to the fact that the analyzed variants are likely to affect the affinity of the receptor (thus causing an EC_{50} shifting) by negatively altering the glutamate residence time inside the binding pocket of the receptor.

For the wild-type NMDA receptor, we found a K_d value of 2.5 µM and a deactivation constant of 328 ms, whereas, for the Glu413Gly and Cys461Phe variants, we found K_d values of 190.5 and 446.5 µM and deactivation constants of 29 and 27 ms, respectively (Figure 10 and Table 1). As these results imply, the Glu413Gly and Cys461Phe variants increase the K_d of the glutamate ~75 and ~180-fold (Table 1).

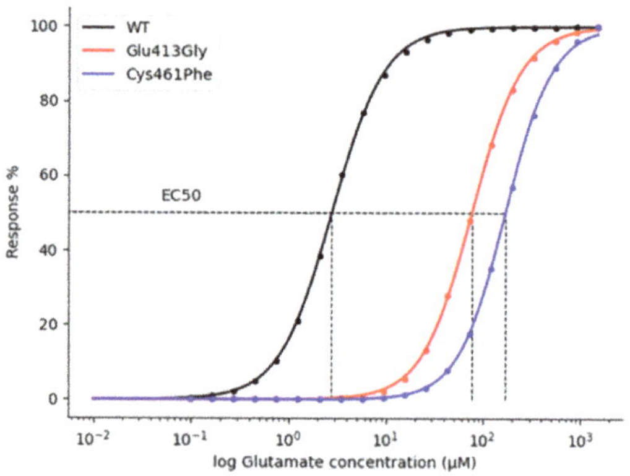

Figure 10. Dose-response curves of the effect of glutamate on wild-type (WT) and variant-NMDA receptors. Simulation data was fitted with logistic regression. EC_{50} values of 2.7, 76, and 169 µM for WT, Glu413Gly, and Cys461Phe-NMDA receptors, respectively.

The next step in our multiscale analysis of NMDA Glu413Gly and Cys461Phe receptor variants consisted in further investigating if the calculated affinity alterations can impact the synaptic plasticity mechanism. To address this question, we simulated the effects of the structural variants on the amplitude of the rise in the post-synaptic Ca^{2+} concentration and on the amount of activated CaMKII, an enzyme that directly plays a pivotal role in triggering synaptic plasticity events in CA3-CA1 synapses. This latter estimation was done by exploiting the CAS module (Section 2.1.4). This module contains a mathematical description of the Ca^{2+}-dependent CaM-CaMKII transduction pathway, which, starting from Ca^{2+} transients, leads to activation of CaMKII kinase (see methods Section 3.1.4 for further details). We stimulated our virtual synapse with a pair of single pre- and post-synaptic stimuli (glutamate pulse of 1 mM amplitude and 1ms width, time interval between pre- and post-synaptic stimuli of 20 ms). As expected, we found that the predicted decrease in the NMDA glutamate affinity significatively attenuates the amplitude of the elicited post-synaptic Ca^{2+} variation ~5 and ~8.5 fold for the Glu413Gly and Cys461Phe variants, respectively (Figure 11A). Moreover, the kinetic model for the Ca^{2+}-mediated activation of the CaMKII enzyme predicted much lower amounts of activated CaMKII for Glu413Gly (~13 fold) and Cys461Phe (~23 fold) variants compared to the wild type (Figure 11B). Considering the key role that the CaMKII enzyme plays in the molecular mechanism underlying the synaptic plasticity process, the predicted drastic decrease in the activation efficiency of such enzyme points out the severity of these rare structural variants. In fact, because CaMKII-driven neuroplasticity seems to be negatively affected in a significant way, severe neuropathological phenotypes, including learning and memory impairment, are likely to arise.

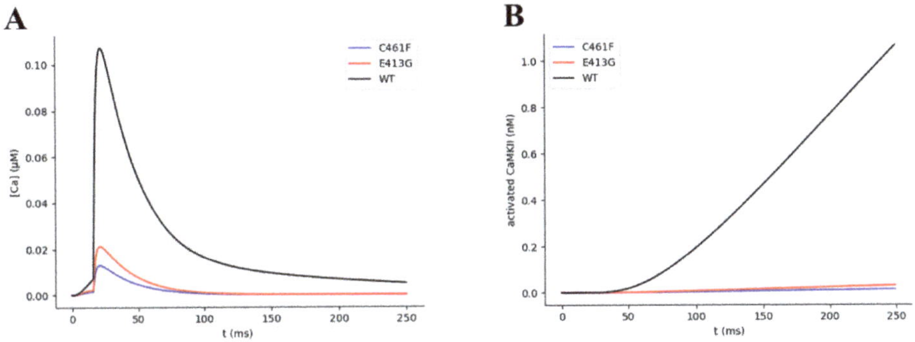

Figure 11. Variation of (**A**) Ca2+ concentration and (**B**) activated CaMKII over time for the WT and variant NMDA receptors. All simulations were performed under one pair of single pre- and post-synaptic pulses, with a pre-synaptic pulse of 1 ms of glutamate exposure, a delay between the pre- and the post-synaptic stimuli of 20 ms, and 1 mM of Mg^{2+}.

In the last part of our in silico experiment, we were interested in reporting a more general representation of the relationship between NMDA-glutamate affinity and CaMKII enzyme activation efficiency. Here, our rationale was to search for an NMDA-glutamate affinity threshold that can be used for discriminating between high and low-severity variants, knowing their respective K_d. We proceeded, for this purpose, to simulate the whole synaptic model with the same basic stimulation pattern previously adopted for the analysis of the Glu413Gly and Cys461Phe variants but varying the K_d affinity value in a range between 1 and 1000 μM. For each simulation (i.e., for each K_d value), we selected the maximum amount of activated CaMKII observed. Data were first normalized to the maximum response observed across all the simulations and then fitted with the four-parameter logistic function (see methods Section 3.2.3 for further details) (Figure 12). Finally, the threshold was calculated by finding the bending point of the fitted curve, which corresponded to a K_d value of ~19 μM (Figure 12).

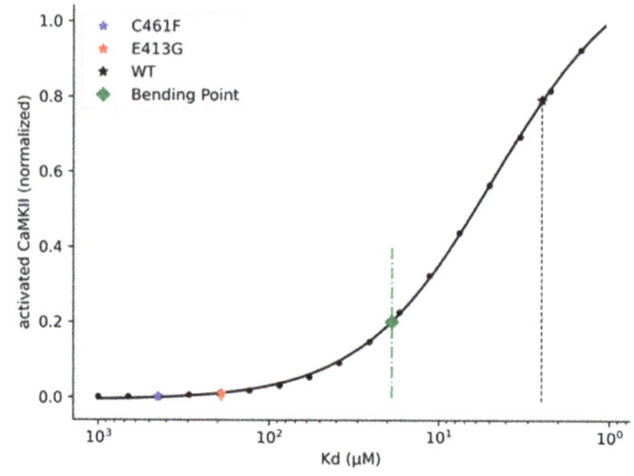

Figure 12. Variation of activated CaMKII as a function of the NMDA-glutamate K_d values. All simulations were performed under one pair of single pre- and post-synaptic pulses, with a pre-synaptic pulse of 1 ms of glutamate exposure, a delay between the pre- and the post-synaptic stimuli of 20 ms, and 1 mM of Mg^{2+}.

The identification of this type of thresholds can be very useful for a rapid assessment of the downstream effects of variants and can be easily integrated into larger analytical pipelines. We are currently working on a further implementation of this synaptic model that also integrates a detailed kinetic description of the reactions controlling the phosphorylation of AMPA receptors by the CaMKII enzyme, an event which is known to directly control synaptic strength modulations (LTP and LTD) by altering the conductance and trafficking of these receptors. With this further extension, we aim to explicitly quantify synaptic plasticity events that can occur during the stimulations.

3. Methods

In this section, we provide a full and detailed description of all the individual modules that compose the proposed mathematical model, each of which implements a different fragment of the whole synaptic transmission process.

This modular rationale at the base of the framework implementation guarantees an easy customization of the simulation pipeline, as well as the further extensibility of the system.

The current build of the framework includes:

- Stimulation Pattern Design (SPD) module, where both pre- and post-synaptic stimuli can be programmed independently. This module allows us to define the inputs of the virtual synapse.
- Receptors Gating Simulation (RGS) module. This module performs a compartmentalized kinetic simulation of the virtual synaptic cleft, where a neurotransmitter released from pre-synaptic stimuli interacts with ionotropic membrane receptors expressed on the post-synaptic spine.
- EPSCs/EPSPs Calculation (CPC) module. This module analyzes the data coming from the RGS module and, calculating synaptic currents and their respective potentials, integrates pre- and post-synaptic stimuli. It constitutes a "bridge" between the extracellular and the intracellular compartments.
- CaM/CaMKII Activation Simulation (CAS) module. This module performs a compartmentalized kinetic simulation of a set of molecular reactions that takes place in the virtual post-synaptic spine, which includes the interactions between Ca^{2+}, Calmodulin (CaM), and Ca^{2+}/CaM-dependent Kinase II (CaMKII).

The kinetic equations used to describe the reactions contained in both the RGS and CCS modules are implemented, exploiting the PySB python package [53] as systems of first-order differential equations. Numerical integration is performed using the SciPy ODE integrator [54]. All of the data analysis and fittings were performed using SciPy and Numpy packages [54,55]. Finally, all the plots were generated using the Matplotlib library [56].

All the code is stored in a publicly available github repository (https://github.com/pietromicheli/CA3-CA1_SynapticModel), where a jupyter notebook file for running simulations and performing basic analysis can also be found.

3.1. Mathematical Model Implementation

3.1.1. SPD Module

In this module, the stimulation pattern of the virtual synapse can be designed. Bidirectionality is a crucial feature of neuronal communication. The functional and topological properties of the brain neural network can be significantly shaped by the temporal relationship between forward and backward signals, as the STDP paradigm for the synaptic plasticity claims [20,21,57]. Therefore, integration of pre- and post-synaptic stimuli constitute a logic core of our implementation. For this purpose, patterns of pre- and post-synaptic stimuli can be programmed and simulated independently in order to analyze how the system behaves under different levels of synchronization between pre- and post-synaptic activities. Each pattern is modeled as a train of bursts. Numbers of stimuli per burst, intra-burst, and inter-bursts frequencies can be specified to design custom stimulation patterns (Figure 1).

In our model, pre-synaptic stimuli have been idealized as glutamate pulses, representing the instantaneous rise and fall in the free neurotransmitter concentration available inside the cleft compartment following pre-synaptic action potentials. Amplitude (i.e., the amount of available free glutamate) and width (i.e., the exposure time of the free glutamate) of the pre-synaptic glutamate pulses can be defined during the stimulation design.

On the other hand, post-synaptic stimuli have been modeled as transient depolarizations of the post-synaptic spine generated by dendritic back-propagating action potentials (bAP). Each bAP is shaped using a two-component exponential function, taken from the work by Shouval et al. [35]:

$$\mathrm{bAP}(t) = V_{max} * \left(\left(I_{fast} * exp\left(\frac{-t}{\tau_{fast}} \right) \right) + \left(I_{slow} * exp\left(\frac{-t}{\tau_{fast}} \right) \right) \right) \quad (1)$$

where V_{max} is the maximum depolarization value for bAP value set to +67 mV [58], I_{fast} and I_{slow} are the relative magnitudes of the fast and slow components of the bAP that sum to one, and τ_{fast} and τ_{slow} are the relative time constants that describe the exponential decays of the two components.

3.1.2. RGS Module

This module contains a system of kinetic equations describing the interactions between glutamate and AMPA/NMDA receptors, which take place inside the cleft compartment. The aim of this module is to accurately simulate both the receptors-neurotransmitter binding reactions and the gating mechanism that lead to opening or desensitization of the receptors.

Individual models describing the kinetic behavior of both AMPA and NMDA receptors have been selected from the literature based on their reproducibility and subsequently implemented as systems of first-order differential equations inside the PySB framework. To reproduce the kinetic behavior of AMPA receptors, we chose a model proposed by Koike et al. [36] for homomeric GluR2(flip) receptors. The model assumes two glutamate binding steps, one pre-open transient state, three desensitized states, and one open state of the receptor (Figure S2B). For the kinetic description of the gating mechanism of NMDA receptors, we used the model for GluN1/GluN2B NMDA receptor proposed by Amico-Ruvio and Pospescu [37]. This kinetic scheme includes two sequential glutamate binding steps, three pre-open transient states, two desensitization states, and one open state of the receptor (Figure S2A). Because we assume a saturating concentration of glycine inside the clef compartment, the binding steps with this molecule are not included in the kinetic model. Thus, all the resting NMDA receptors are considered glycine bound.

3.1.3. CPC Module

In this module, we implemented a set of equations with the aim of assessing the EPSCs and the respective EPSPs generated by the open fractions of both AMPA and NMDA receptors. EPSPs are then integrated with the back-propagating action potentials programmed during the stimulation design. Finally, the sum of all the depolarizing contributions is used to assess the variations of the post-synaptic membrane potential.

Many synaptic models that have been proposed in the past estimated the EPSCs and/or the EPSPs simply by using two-component exponential functions fitted on electrophysiological recordings [35,58–60]. On the contrary, in our model, the open probabilities of the receptors vary according to a system of deterministic rate equations that represent mass-action kinetics of receptors-neurotransmitter interactions [53]. For this reason, the rising and decay phases of both receptor-mediated EPSCs and EPSPs responses are shaped by the complex receptor-specific interaction kinetics with the neurotransmitter. This confers more flexibility to our model, allowing us, for example, to explore the responses generated by mutant forms of the receptors by tuning the rate constants of some of the kinetic equations.

We defined the EPSCs of AMPA and NMDA receptors as follows:

$$EPSC_{AMPA}(t) = O_{AMPA}(t) * G_{AMPA} * \left(V_m(t - \Delta t) - V_{E_{AMPA}}\right) \tag{2}$$

$$EPSC_{NMDA}(t) = O_{NMDA}(t) * G_{NMDA} * \left(V_m(t - \Delta t) - V_{E_{NMDA}}\right) * B(V_m(t - \Delta t)) \tag{3}$$

where $O_{AMPA}(t)$ and $O_{NMDA}(t)$ are the number of open NMDA and AMPA receptors at each time step; G_{NMDA} and G_{AMPA} are the single channel conductance set to 40 pS and 15 pS, respectively [61–63]; $V_m(t - \Delta t)$ is the membrane potential at time $(t - \Delta t)$; V_E is the channel-specific equilibrium reversal potential and defines the value of the membrane potential for which the electrochemical equilibrium is reached and, thus, the net flux through the channel is 0 (we assume that $V_{EAMPA} = V_{ENMDA} = 0$) [64]; and $B(V_m)$ describes the voltage dependence of the NMDA current given by the Mg^{2+} blocks defined by [35]:

$$B(V_m) = \frac{1}{1 - exp(-K_M V_m) * \left(\frac{[Mg^{2+}]}{3.27}\right)} \tag{4}$$

Once the EPSCs have been calculated, the relative EPSPs are determined simply by applying the law of Ohm:

$$EPSP_{AMPA}(t) = EPSC_{AMPA}(t) * R_s \tag{5}$$

$$EPSP_{NMDA}(t) = EPSC_{NMDA}(t) * R_s \tag{6}$$

where R_s is the spine's resistance set to 500 MΩ [65].

Finally, the total membrane potential, defined as the sum of the partial depolarization contributions, is calculated according to the equation:

$$V_m(t) = V_r + EPSP_{AMPA}(t) + EPSP_{NMDA}(t) + bAP(t) \tag{7}$$

where V_r is the resting membrane potential of the spine (−65 mV).

In CA3 Schaffer collateral-CA1 synapses, the key mediator of the post-synaptic response is the elicited intracellular Ca^{2+} variation. Because NMDA receptors are the major source of Ca^{2+} during spine stimulation [66], we explicitly calculate the NMDA receptor-mediated Ca^{2+} molar flowrate as follows:

$$I_{Ca^{2+}}(t) = O_{NMDA}(t) * G_{Ca^{2+}} * \left(V_M(t-\Delta t) - V_{E_{Ca^{2+}}}\right) * B(V_m(t-\Delta t)) \quad (8)$$

where $G_{Ca^{2+}}$ expresses the permeability of the NMDA receptor to Ca^{2+} ions, set to 2 $nM \cdot ms^{-1} \cdot mV^{-1}$ [58] and $V_{E_{Ca^{2+}}}$ is the reversal equilibrium potential for Ca^{2+} set to +130 mV [58].

Finally, the calcium dynamics in the post-synaptic cell is integrated by a simple first-order differential equation [35,58]:

$$\frac{d\left[Ca^{2+}(t)\right]}{dt} = I_{Ca^{2+}}(t) - \frac{\left[Ca^{2+}(t)\right]}{\tau_{Ca^{2+}}} \quad (9)$$

where $\tau_{Ca^{2+}}$ is the passive decay time constant of post-synaptic Ca^{2+} concentration, set to 20 ms [35].

A full list of all the parameters used in the equations described above is provided in the supplementary data (Table S2).

3.1.4. CAS Module

The last module of the pipeline contains a compartmentalized kinetic description of a reaction network that takes place inside the post-synaptic spine. Here, our rationale was to assess the variability in the amount of activated CaMKII enzyme upon different stimulation conditions. Because CaMKII plays a crucial role in the positive regulation of the early phase of LTP in CA3 Schaffer collateral-CA1 synapses [22–24], this estimation allows us to qualitatively infer the strength and the efficiency of the synaptic transmission. As previously described for the RGS module, we selected from the literature a kinetic model based on its reproducibility; we translated it inside the PySB framework and, finally, we appended the new block to the pipeline. For this purpose, we selected from the BioModels database [67] a model describing a set of interactions that, starting from post-synaptic rise in Ca^{2+} concentration, leads to the autophosphorylation (i.e., the activation) of monomeric CaMKII [38]. Particularly, the set of reactions implemented includes:

- Binding reactions between Ca^{2+} ions and CaM and CaM-CaMKII species;
- Dimerization reactions between Ca^{2+}-CaM and monomeric CaMKII;
- Dimerization reactions between two Ca^{2+}-CaM-CaMKII complexes;
- Autophosphorylation reactions of CaMKII monomers inside the 2 (Ca^{2+}-CaM-CaMKII) complexes.

3.2. *Data Fitting*

3.2.1. Concentration-Response Curves

We computed the glutamate concentration-response curves for NMDA receptors by stimulating the system with a glutamate pulse of 1.5 s in the absence of Mg^{2+} [47]. We run multiple simulations varying the amplitude of the glutamate pulse, with a concentration range between 0.01 and 1500 μM, and calculating the NMDA receptor-mediated current peak values. The EC_{50} values were then calculated by fitting the concentration-response data with the following equation:

$$Response\ \% = \frac{100}{\left(1 + \frac{EC_{50}}{[glutamate]}\right)^n} \quad (10)$$

where n is the Hill slope.

3.2.2. Two-Component Exponential Function Fitting

The deactivation time constant for the NMDA wild-type receptor and Glu413Gly and Cys461Phe variants were estimated as weighted time constants of the double exponential fit of the NMDA receptor current decay after the exposure of 1 mM glutamate for 1.5 s. The two-component exponential function used for the fitting takes the form:

$$I(t) = I_{fast} * exp\left(\frac{-t}{\tau_{slow}}\right) + I_{slow} * exp\left(\frac{-t}{\tau_{slow}}\right) \tag{11}$$

where I is the current, I_{fast} and I_{slow} are the amplitudes of the fast and slow components, respectively, and τ_{fast} and τ_{slow} are the respective decay time constants. The weighted time constant of decay (τ_w) was calculated using the following equation:

$$\tau_w = \frac{I_{fast}}{I_{fast} + I_{slow}} * \tau_{fast} + \frac{I_{slow}}{I_{slow} + I_{fast}} * \tau_{slow} \tag{12}$$

3.2.3. Four-Parameter Logistic Function and Bending Points

The data generated by the simulation of the relationship between different glutamate-NMDA K_d values and the concentration peaks of activated CaMKII enzyme (see results Section 2.4) were fitted with the four-parameter logistic function:

$$Y = \frac{a - d}{1 + \left(\frac{X}{c}\right)^b} + d \tag{13}$$

where Y represents the activated CaMKII response, X represents the affinity value K_d (expressed in µM), a is the lower asymptote, d is the upper asymptote, c represents the K_d that generates a mid-way response between the estimated a and d, and b is a slope factor. The bending point of the curve was then computed as follow:

$$X_{bend} = \frac{a - d}{1 + k} + d \tag{14}$$

$$Y_{bend} = c * \left(\frac{a - Y_{bend}}{Y_{bend} - d}\right)^{\frac{1}{b}} \tag{15}$$

where k is a constant value, set to 4.6805 [68].

4. Conclusions

We proposed a compartmental model for the hippocampal synapse CA3-CA1. Our goal was to provide a simple and portable, python-based program to run kinetics simulations of the synaptic transmission, which embodied both pre- and post-synaptic activity. The rationale that drove us through the implementation, as well as the application, of this model was to focus on the integration between system biology and structural biology viewpoints. Exploiting this hybrid multiscale approach, we analyzed the impact that single disease associated variants of NMDA receptors, related to neurological disorders and cognitive impairments, may have on the whole synaptic transmission process. We were able to consistently reproduce experimental data and to quantitatively infer molecular-level causality of a variant-related functional impairment. Therefore, these results show the predictive power of such a multiscale approach, aimed at observing behavioral shifts of a complex system that emerge from amplification of small, quantifiable, molecular-level alterations.

A future improvement of our model will allow us to explicitly quantify synaptic plasticity events by adding further biological details, e.g., AMPA receptors conductance modulation and translocation by CaMKII-mediated phosphorylation. The next step will be to extend the structural analysis to the multiple molecular entities involved in the transmission and modulation processes, recalibrating the kinetic constants of the interactions according to the conformational rearrangements caused by specific mutations. This will allow us to explicitly simulate the molecular effects, as well as the impact on the single-neuron functionality, of mutational signatures linked to neurological and cognitive impairments, which affect one or multiple entities of the modeled interactome. Finally, this approach may be extended to post-synaptic receptors belonging to other families, such as G-protein coupled receptors.

Supplementary Materials: Supplementary Materials can be found at https://www.mdpi.com/1422-0067/22/4/1536/s1, Figure S1: AMPA-mediated EPSPs generated by pre-synaptic stimulations composed of a single glutamate pulse or a burst of 5 glutamate pulses delivered at 100 Hz, Figure S2: Kinetic schemes used for simulating the gating mechanisms of NMDA and AMPA receptors, Table S1: Comparison between peak open probability and deactivation time constants values obtained from our implementation and the ones reported by the original models, Table S2: List of the parameters used in the equations of the CPC module.

Author Contributions: Conceptualization, P.M., R.R., and A.G.; methodology, P.M.; software, P.M.; validation, P.M. and R.R.; formal analysis, P.M.; investigation, P.M. and R.R.; resources, A.G.; data curation, P.M. and R.R.; writing—original draft preparation, P.M. and R.R.; writing—review and editing, P.M., R.R., and A.G.; supervision, R.R. and A.G.; project administration, R.R. and A.G.; funding acquisition, A.G. All authors have read and agreed to the published version of the manuscript.

Funding: This research was funded by the ordinary research fund form the Dept. of Biotechnology of the University of Verona.

Data Availability Statement: All the code used for implementing and simulating the compartmentalized model is stored in a publicly available github repository (https://github.com/pietromicheli/CA3-CA1_SynapticModel) where a jupyter notebook file for running simulations and performing basic analysis can be also found.

Conflicts of Interest: The authors declare no conflict of interest. The funders had no role in the design of the study; in the collection, analyses, or interpretation of data; in the writing of the manuscript, or in the decision to publish the results.

Abbreviations

AMPA	α-amino-3-hydroxy-5-methyl-4-isoxazolepropionic acid
bAP	back-propagating action potential
CaM	Calmodulin
CaMKII	Ca^{2+}/CAM-dependent Kinase II
CAS	CaMKII Activation Simulation
CPC	Current/Potential Calculation
LTD	Long Term Depression
LTP	Long Term Potentiation
mCaMKII	Individual subunits of Calmodulin Kinase II
NMDA	N-Methyl-D-aspartic acid
RGS	Receptors Gating Simulation
STDP	Spike Timing Dependent Plasticity

References

1. Mattson, M.P. Pathways Towards and Away from Alzheimer's Disease. *Nature* **2004**, *430*, 631–639. [CrossRef] [PubMed]
2. Grosjean, B.; Tsai, G.E. NMDA neurotransmission as a critical mediator of borderline personality disorder. *J. Psychiatry Neurosci.* **2007**, *32*, 103–115.
3. Hsieh, H.; Boehm, J.; Sato, C.; Iwatsubo, T.; Tomita, T.; Sisodia, S.; Malinow, R. AMPA-R Removal Underlies Aβ-induced Synaptic Depression and Dendritic Spine Loss. *Neuron* **2006**, *52*, 831–843. [CrossRef] [PubMed]
4. Snyder, E.M.; Nong, Y.; Almeida, C.G.; Paul, S.; Moran, T.; Choi, E.Y.; Nairn, A.C.; Salter, M.W.; Lombroso, P.J.; Gouras, G.K.; et al. Regulation of NMDA receptor trafficking by amyloid-beta. *Nat. Neurosci.* **2005**, *8*, 1051–1058. [CrossRef] [PubMed]
5. Lerma, J.; Marques, J.M. Kainate receptors in health and disease. *Neuron* **2013**, *80*, 292–311. [CrossRef]
6. Martin, S.J.; Grimwood, P.D.; Morris, R.G. Synaptic plasticity and memory: An evaluation of the hypothesis. *Annu. Rev. Neurosci.* **2000**, *23*, 649–711. [CrossRef]
7. Mansvelder, H.D.; Verhoog, M.B.; Goriounova, N.A. Synaptic plasticity in human cortical circuits: Cellular mechanisms of learning and memory in the human brain? *Curr. Opin. Neurobiol.* **2019**, *54*, 186–193. [CrossRef]
8. Baez, M.V.; Cercato, M.C.; Jerusalinsky, D.A. NMDA Receptor Subunits Change after Synaptic Plasticity Induction and Learning and Memory Acquisition. *Neural Plast.* **2018**, *2018*. [CrossRef]
9. Kandel, E.R.; Dudai, Y.; Mayford, M.R. The Molecular and Systems Biology of Memory. *Cell* **2014**, *157*, 163–186. [CrossRef]
10. Voss, J.L.; Bridge, D.J.; Cohen, N.J.; Walker, J.A. A closer look at the hippocampus and memory. *Trends Cogn. Sci.* **2017**, *21*, 577–588. [CrossRef]
11. Lisman, J.; Buzsáki, G.; Eichenbaum, H.; Nadel, L.; Ranganath, C.; Redish, A.D. Viewpoints: How the hippocampus contributes to memory, navigation and cognition. *Nat. Neurosci.* **2017**, *20*, 1434–1447. [CrossRef] [PubMed]
12. Kumar, A. Long-Term Potentiation at CA3–CA1 Hippocampal Synapses with Special Emphasis on Aging, Disease, and Stress. *Front. Aging Neurosci.* **2011**, *3*, 7. [CrossRef] [PubMed]
13. Tsien, J.Z.; Huerta, P.T.; Tonegawa, S. The Essential Role of Hippocampal CA1 NMDA Receptor–Dependent Synaptic Plasticity in Spatial Memory. *Cell* **1996**, *87*, 1327–1338. [CrossRef]
14. Henley, J.M.; Wilkinson, K.A. Synaptic AMPA receptor composition in development, plasticity and disease. *Nat. Rev. Neurosci.* **2016**, *17*, 337–350. [CrossRef] [PubMed]
15. Chater, T.E.; Goda, Y. The role of AMPA receptors in postsynaptic mechanisms of synaptic plasticity. *Front. Cell. Neurosci.* **2014**, *8*, 401. [CrossRef] [PubMed]
16. Blanke, M.L.; VanDongen, A.M.J. Activation Mechanisms of the NMDA Receptor. In *Biology of the NMDA Receptor*; Van Dongen, A.M., Ed.; Frontiers in Neuroscience; CRC Press/Taylor & Francis: Boca Raton, FL, USA, 2009; ISBN 978-1-4200-4414-0.
17. Vyklicky, V.; Korinek, M.; Smejkalova, T.; Balik, A.; Krausova, B.; Kaniakova, M.; Lichnerova, K.; Cerny, J.; Krusek, J.; Dittert, I.; et al. Structure, function, and pharmacology of NMDA receptor channels. *Physiol. Res.* **2014**, *63*, S191–S203. [CrossRef] [PubMed]
18. Mayer, M.L.; Westbrook, G.L.; Guthrie, P.B. Voltage-dependent block by Mg^{2+} of NMDA responses in spinal cord neurones. *Nature* **1984**, *309*, 261–263. [CrossRef]
19. Jahr, C.E.; Stevens, C.F. A quantitative description of NMDA receptor-channel kinetic behavior. *J. Neurosci.* **1990**, *10*, 1830–1837. [CrossRef]
20. Feldman, D.E. The spike timing dependence of plasticity. *Neuron* **2012**, *75*, 556–571. [CrossRef]
21. Markram, H.; Gerstner, W.; Sjöström, P.J. A History of Spike-Timing-Dependent Plasticity. *Front. Synaptic Neurosci.* **2011**, *3*, 4. [CrossRef]
22. Lee, S.-J.R.; Escobedo-Lozoya, Y.; Szatmari, E.M.; Yasuda, R. Activation of CaMKII in single dendritic spines during long-term potentiation. *Nature* **2009**, *458*, 299–304. [CrossRef] [PubMed]
23. Lisman, J.; Yasuda, R.; Raghavachari, S. Mechanisms of CaMKII action in long-term potentiation. *Nat. Rev. Neurosci.* **2012**, *13*, 169–182. [CrossRef] [PubMed]
24. Barria, A.; Muller, D.; Derkach, V.; Griffith, L.C.; Soderling, T.R. Regulatory Phosphorylation of AMPA-Type Glutamate Receptors by CaM-KII during Long-Term Potentiation. *Science* **1997**, *276*, 2042–2045. [CrossRef] [PubMed]
25. Sibarov, D.A.; Bruneau, N.; Antonov, S.M.; Szepetowski, P.; Burnashev, N.; Giniatullin, R. Functional Properties of Human NMDA Receptors Associated with Epilepsy-Related Mutations of GluN2A Subunit. *Front. Cell Neurosci.* **2017**, *11*, 155. [CrossRef] [PubMed]
26. Shankar, G.M.; Li, S.; Mehta, T.H.; Garcia-Munoz, A.; Shepardson, N.E.; Smith, I.; Brett, F.M.; Farrell, M.A.; Rowan, M.J.; Lemere, C.A.; et al. Amyloid-β protein dimers isolated directly from Alzheimer's brains impair synaptic plasticity and memory. *Nat. Med.* **2008**, *14*, 837–842. [CrossRef] [PubMed]
27. Shipton, O.A.; Leitz, J.R.; Dworzak, J.; Acton, C.E.J.; Tunbridge, E.M.; Denk, F.; Dawson, H.N.; Vitek, M.P.; Wade-Martins, R.; Paulsen, O.; et al. Tau Protein Is Required for Amyloid β-Induced Impairment of Hippocampal Long-Term Potentiation. *J. Neurosci.* **2011**, *31*, 1688–1692. [CrossRef]
28. Bagetta, V.; Ghiglieri, V.; Sgobio, C.; Calabresi, P.; Picconi, B. Synaptic dysfunction in Parkinson's disease. *Biochem. Soc. Trans.* **2010**, *38*, 493–497. [CrossRef]
29. Usdin, M.T.; Shelbourne, P.F.; Myers, R.M.; Madison, D.V. Impaired Synaptic Plasticity in Mice Carrying the Huntington's Disease Mutation. *Hum. Mol. Genet.* **1999**, *8*, 839–846. [CrossRef]
30. Murphy, K.P.S.J.; Carter, R.J.; Lione, L.A.; Mangiarini, L.; Mahal, A.; Bates, G.P.; Dunnett, S.B.; Morton, A.J. Abnormal Synaptic Plasticity and Impaired Spatial Cognition in Mice Transgenic for Exon 1 of the Human Huntington's Disease Mutation. *J. Neurosci.* **2000**, *20*, 5115–5123. [CrossRef]

31. Kaplan, U.; Türkay, M.; Biegler, L.; Karasözen, B. Modeling and simulation of metabolic networks for estimation of biomass accumulation parameters. *Discret. Appl. Math.* **2009**, *157*, 2483–2493. [CrossRef]
32. Bartol, T.M.; Keller, D.X.; Kinney, J.P.; Bajaj, C.L.; Harris, K.M.; Sejnowski, T.J.; Kennedy, M.B. Computational reconstitution of spine calcium transients from individual proteins. *Front. Synaptic Neurosci.* **2015**, *7*, 17. [CrossRef] [PubMed]
33. Hu, E.; Mergenthal, A.; Bingham, C.S.; Song, D.; Bouteiller, J.-M.; Berger, T.W. A Glutamatergic Spine Model to Enable Multi-Scale Modeling of Nonlinear Calcium Dynamics. *Front. Comput. Neurosci.* **2018**, *12*, 58. [CrossRef] [PubMed]
34. Pujol, A.; Mosca, R.; Farrés, J.; Aloy, P. Unveiling the role of network and systems biology in drug discovery. *Trends Pharmacol. Sci.* **2010**, *31*, 115–123. [CrossRef] [PubMed]
35. Shouval, H.Z.; Bear, M.F.; Cooper, L.N. A unified model of NMDA receptor-dependent bidirectional synaptic plasticity. *Proc. Natl. Acad. Sci. USA* **2002**, *99*, 10831–10836. [CrossRef] [PubMed]
36. Koike, M.; Tsukada, S.; Tsuzuki, K.; Kijima, H.; Ozawa, S. Regulation of Kinetic Properties of GluR2 AMPA Receptor Channels by Alternative Splicing. *J. Neurosci.* **2000**, *20*, 2166–2174. [CrossRef] [PubMed]
37. Amico-Ruvio, S.A.; Popescu, G.K. Stationary Gating of GluN1/GluN2B Receptors in Intact Membrane Patches. *Biophys. J.* **2010**, *98*, 1160–1169. [CrossRef]
38. Pepke, S.; Kinzer-Ursem, T.; Mihalas, S.; Kennedy, M.B. A Dynamic Model of Interactions of Ca^{2+}, Calmodulin, and Catalytic Subunits of Ca^{2+}/Calmodulin-Dependent Protein Kinase II. *PLoS Comput. Biol.* **2010**, *6*, e1000675. [CrossRef]
39. Dzubay, J.A.; Jahr, C.E. The Concentration of Synaptically Released Glutamate Outside of the Climbing Fiber–Purkinje Cell Synaptic Cleft. *J. Neurosci.* **1999**, *19*, 5265–5274. [CrossRef]
40. Moussawi, K.; Riegel, A.; Nair, S.; Kalivas, P.W. Extracellular Glutamate: Functional Compartments Operate in Different Concentration Ranges. *Front. Syst. Neurosci.* **2011**, *5*, 94. [CrossRef]
41. Robert, A.; Howe, J.R. How AMPA Receptor Desensitization Depends on Receptor Occupancy. *J. Neurosci.* **2003**, *23*, 847–858. [CrossRef]
42. Nusser, Z.; Lujan, R.; Laube, G.; Roberts, J.D.B.; Molnar, E.; Somogyi, P. Cell Type and Pathway Dependence of Synaptic AMPA Receptor Number and Variability in the Hippocampus. *Neuron* **1998**, *21*, 545–559. [CrossRef]
43. Markram, H.; Lübke, J.; Frotscher, M.; Sakmann, B. Regulation of Synaptic Efficacy by Coincidence of Postsynaptic APs and EPSPs. *Science* **1997**, *275*, 213–215. [CrossRef] [PubMed]
44. Albensi, B.C.; Oliver, D.R.; Toupin, J.; Odero, G. Electrical stimulation protocols for hippocampal synaptic plasticity and neuronal hyper-excitability: Are they effective or relevant? *Exp. Neurol.* **2007**, *204*, 1–13. [CrossRef] [PubMed]
45. Racca, C.; Stephenson, F.A.; Streit, P.; Roberts, J.D.B.; Somogyi, P. NMDA Receptor Content of Synapses in Stratum Radiatum of the Hippocampal CA1 Area. *J. Neurosci.* **2000**, *20*, 2512–2522. [CrossRef] [PubMed]
46. Lester, R.A.J.; Clements, J.D.; Westbrook, G.L.; Jahr, C.E. Channel kinetics determine the time course of NMDA receptor-mediated synaptic currents. *Nature* **1990**, *346*, 565–567. [CrossRef] [PubMed]
47. Swanger, S.A.; Chen, W.; Wells, G.; Burger, P.B.; Tankovic, A.; Bhattacharya, S.; Strong, K.L.; Hu, C.; Kusumoto, H.; Zhang, J.; et al. Mechanistic Insight into NMDA Receptor Dysregulation by Rare Variants in the GluN2A and GluN2B Agonist Binding Domains. *Am. J. Hum. Genet.* **2016**, *99*, 1261–1280. [CrossRef] [PubMed]
48. Wells, G.; Yuan, H.; McDaniel, M.J.; Kusumoto, H.; Snyder, J.P.; Liotta, D.C.; Traynelis, S.F. The GluN2B-Glu413Gly NMDA receptor variant arising from a de novo GRIN2B mutation promotes ligand-unbinding and domain opening. *Proteins* **2018**, *86*, 1265–1276. [CrossRef]
49. Yuan, H.; Hansen, K.B.; Vance, K.M.; Ogden, K.K.; Traynelis, S.F. Control of NMDA receptor function by the NR2 subunit amino-terminal domain. *J. Neurosci.* **2009**, *29*, 12045–12058. [CrossRef]
50. Hu, C.; Chen, W.; Myers, S.J.; Yuan, H.; Traynelis, S.F. Human GRIN2B variants in neurodevelopmental disorders. *J. Pharm. Sci.* **2016**, *132*, 115–121. [CrossRef]
51. Adams, D.R.; Yuan, H.; Holyoak, T.; Arajs, K.H.; Hakimi, P.; Markello, T.C.; Wolfe, L.A.; Vilboux, T.; Burton, B.K.; Fajardo, K.F.; et al. Three rare diseases in one Sib pair: RAI1, PCK1, GRIN2B mutations associated with Smith–Magenis Syndrome, cytosolic PEPCK deficiency and NMDA receptor glutamate insensitivity. *Mol. Genet. Metab.* **2014**, *113*, 161–170. [CrossRef]
52. Hansen, K.B.; Ogden, K.K.; Yuan, H.; Traynelis, S.F. Distinct functional and pharmacological properties of triheteromeric GluN1/GluN2A/GluN2B NMDA receptors. *Neuron* **2014**, *81*, 1084–1096. [CrossRef] [PubMed]
53. Lopez, C.F.; Muhlich, J.L.; Bachman, J.A.; Sorger, P.K. Programming biological models in Python using PySB. *Mol. Syst. Biol.* **2013**, *9*, 646. [CrossRef] [PubMed]
54. Virtanen, P.; Gommers, R.; Oliphant, T.E.; Haberland, M.; Reddy, T.; Cournapeau, D.; Burovski, E.; Peterson, P.; Weckesser, W.; Bright, J.; et al. SciPy 1.0: Fundamental algorithms for scientific computing in Python. *Nat. Methods* **2020**, *17*, 261–272. [CrossRef] [PubMed]
55. Harris, C.R.; Millman, K.J.; van der Walt, S.J.; Gommers, R.; Virtanen, P.; Cournapeau, D.; Wieser, E.; Taylor, J.; Berg, S.; Smith, N.J.; et al. Array programming with NumPy. *Nature* **2020**, *585*, 357–362. [CrossRef] [PubMed]
56. Hunter, J.D. Matplotlib: A 2D Graphics Environment. *Comput. Sci. Eng.* **2007**, *9*, 90–95. [CrossRef]
57. Stone, D.B.; Tesche, C.D. Topological dynamics in spike-timing dependent plastic model neural networks. *Front. Neural Circuits* **2013**, *7*, 70. [CrossRef]
58. Rackham, O.J.L.; Tsaneva-Atanasova, K.; Ganesh, A.; Mellor, J.R. A Ca^{2+}-Based Computational Model for NMDA Receptor-Dependent Synaptic Plasticity at Individual Post-Synaptic Spines in the Hippocampus. *Front. Synaptic Neurosci.* **2010**, *2*, 31. [CrossRef]

59. Lajeunesse, F.; Kröger, H.; Timofeev, I. Regulation of AMPA and NMDA receptor-mediated EPSPs in dendritic trees of thalamocortical cells. *J. Neurophysiol.* **2013**, *109*, 13–30. [CrossRef]
60. Pongrácz, F.; Poolos, N.P.; Kocsis, J.D.; Shepherd, G.M. A Model of NMDA Receptor-Mediated Activity in Dendrites of Hippocampal CA1 Pyramidal Neurons. *J. Neurophysiol.* **1992**, *68*, 2248–2259. [CrossRef]
61. Traynelis, S.F.; Wollmuth, L.P.; McBain, C.J.; Menniti, F.S.; Vance, K.M.; Ogden, K.K.; Hansen, K.B.; Yuan, H.; Myers, S.J.; Dingledine, R. Glutamate Receptor Ion Channels: Structure, Regulation, and Function. *Pharm. Rev.* **2010**, *62*, 405–496. [CrossRef]
62. Momiyama, A.; Feldmeyer, D.; Cull-Candy, S.G. Identification of a native low-conductance NMDA channel with reduced sensitivity to Mg^{2+} in rat central neurones. *J. Physiol.* **1996**, *494*, 479–492. [CrossRef] [PubMed]
63. Benke, T.; Traynelis, S.F. AMPA-type glutamate receptor conductance changes and plasticity: Still a lot of noise. *Neurochem. Res.* **2019**, *44*, 539–548. [CrossRef] [PubMed]
64. Higley, M.J.; Sabatini, B.L. Calcium Signaling in Dendritic Spines. *Cold Spring Harb. Perspect. Biol.* **2012**, *4*, a005686. [CrossRef] [PubMed]
65. Di Maio, V.; Ventriglia, F.; Santillo, S. A model of cooperative effect of AMPA and NMDA receptors in glutamatergic synapses. *Cogn. Neurodyn.* **2016**, *10*, 315–325. [CrossRef] [PubMed]
66. Bloodgood, B.L.; Sabatini, B.L. Nonlinear Regulation of Unitary Synaptic Signals by CaV2.3 Voltage-Sensitive Calcium Channels Located in Dendritic Spines. *Neuron* **2007**, *53*, 249–260. [CrossRef]
67. Malik-Sheriff, R.S.; Glont, M.; Nguyen, T.V.N.; Tiwari, K.; Roberts, M.G.; Xavier, A.; Vu, M.T.; Men, J.; Maire, M.; Kananathan, S.; et al. BioModels—15 years of sharing computational models in life science. *Nucleic Acids Res.* **2020**, *48*, D407–D415. [CrossRef] [PubMed]
68. Sebaugh, J.L.; McCray, P.D. Defining the linear portion of a sigmoid-shaped curve: Bend points. *Pharm. Stat.* **2003**, *2*, 167–174. [CrossRef]

International Journal of
Molecular Sciences

MDPI

Review

The SARS-Coronavirus Infection Cycle: A Survey of Viral Membrane Proteins, Their Functional Interactions and Pathogenesis

Nicholas A. Wong * and Milton H. Saier, Jr. *

Department of Molecular Biology, Division of Biological Sciences, University of California at San Diego, La Jolla, CA 92093-0116, USA
* Correspondence: naw016@ucsd.edu (N.A.W.); msaier@ucsd.edu (M.H.S.J.); Tel.: +1-650-763-6784 (N.A.W.); +1-858-534-4084 (M.H.S.J.)

Abstract: Severe Acute Respiratory Syndrome Coronavirus-2 (SARS-CoV-2) is a novel epidemic strain of *Betacoronavirus* that is responsible for the current viral pandemic, coronavirus disease 2019 (COVID-19), a global health crisis. Other epidemic *Betacoronaviruses* include the 2003 SARS-CoV-1 and the 2009 Middle East Respiratory Syndrome Coronavirus (MERS-CoV), the genomes of which, particularly that of SARS-CoV-1, are similar to that of the 2019 SARS-CoV-2. In this extensive review, we document the most recent information on Coronavirus proteins, with emphasis on the membrane proteins in the Coronaviridae family. We include information on their structures, functions, and participation in pathogenesis. While the shared proteins among the different coronaviruses may vary in structure and function, they all seem to be multifunctional, a common theme interconnecting these viruses. Many transmembrane proteins encoded within the SARS-CoV-2 genome play important roles in the infection cycle while others have functions yet to be understood. We compare the various structural and nonstructural proteins within the Coronaviridae family to elucidate potential overlaps and parallels in function, focusing primarily on the transmembrane proteins and their influences on host membrane arrangements, secretory pathways, cellular growth inhibition, cell death and immune responses during the viral replication cycle. We also offer bioinformatic analyses of potential viroporin activities of the membrane proteins and their sequence similarities to the Envelope (E) protein. In the last major part of the review, we discuss complement, stimulation of inflammation, and immune evasion/suppression that leads to CoV-derived severe disease and mortality. The overall pathogenesis and disease progression of CoVs is put into perspective by indicating several stages in the resulting infection process in which both host and antiviral therapies could be targeted to block the viral cycle. Lastly, we discuss the development of adaptive immunity against various structural proteins, indicating specific vulnerable regions in the proteins. We discuss current CoV vaccine development approaches with purified proteins, attenuated viruses and DNA vaccines.

Keywords: SARS-CoV-2; Betacoronavirus; Coronaviridae; transmembrane proteins; pathogenesis; inflammation; immunity; vaccines

Citation: Wong, N.A.; Saier, M.H., Jr. The SARS-Coronavirus Infection Cycle: A Survey of Viral Membrane Proteins, Their Functional Interactions and Pathogenesis. *Int. J. Mol. Sci.* **2021**, *22*, 1308. https://doi.org/10.3390/ijms22031308

Academic Editor: Masoud Jelokhani-Niaraki

Received: 1 December 2020
Accepted: 22 January 2021
Published: 28 January 2021

1. Introduction

The rapid spread of the recently emerging coronavirus disease, COVID-19, caused by the severe acute respiratory syndrome coronavirus-2 (SARS-CoV-2), has given rise to the current pneumonia pandemic, which has a moderate (5%) fatality rate compared to SARS-CoV-1 (10% fatality rate) and Middle East Respiratory Coronavirus (MERS-CoV) (34% fatality rate), due to sepsis/acute respiratory distress syndrome (SARS) [1,2]. Rapid transmission can be hindered by hand washing, distance maintenance between people and the use of masks by infected individuals to trap virus-carrying breath droplets [3]. As of 20 November 2020, there have been over 55 million cases, worldwide, with over 12 million (~22%) being within the USA. The pandemic shows doubling of cases every 5 days and an estimated incubation period of 5 days, both in the USA and world-wide [3]. Well over 1.3 million deaths have resulted, with over 250,000 of them occurring in the USA. The rapid inter-human disease transmission has caught global health professionals by surprise, and as of 20 November 2020, we have no clinically approved anti-viral drug for treatment, and vaccines, recently developed for prevention, are not yet available for distribution [4,5].

The disease arose in the city of Wuhan (Hubei Province, China), and seems to have been caused by the sale of live wild animals as a food source in the Wuhan wholefood market [6]. These animals were thought to be the intermediate carrier of the virus, originating in *Rhinolophus* bats [1]. Although the primary epidemiological factors contributing to the rapid spread of the disease [7] and the immune responses to the virus [8] are reasonably well understood, the animal intermediate carrier has not yet been identified. A greater incidence of the disease in older people, particularly those with co-morbidities such as hypertension and cardiovascular disease, and disproportionate distributions of cases and fatalities among people of different ethnic groups (among African, Hispanic and Native Americans as compared with European Americans) are now well established [9,10].

2. Viral Proteins and Their Roles in the Infection Cycle

While the widespread availability of a vaccine is foreseeable, the development of antivirals, targeting specific proteins involved in SARS-CoV-2's pathogenesis and the infection cycle, may prove to be feasible in a longer time frame. The subsequent progression of disease conditions produced by the 2003 SARS-CoV-1 and the 2020 SARS-CoV-2 include a deadly viral pneumonia and Severe Acute Respiratory Syndrome (SARS), caused in part by the overstimulation of the host's innate immune system. Patients infected with SARS-CoV-1 or SARS-CoV-2 exhibit similar lung pathological symptoms, again reflecting the similarities between these two viral types. However, additional symptoms affecting other bodily organs have been documented [11,12].

A number of anti-viral agents, known to inhibit or otherwise negatively influence the infectivity or transmission of related viruses, the closest being SARS-CoV-1, have been considered, and some of these have even been tested, either in vitro or in vivo. For example, micronutrients such as vitamins C, D and E have been suggested to be of value for prophylaxis or treatment [7]. Controversial therapies include treatment with inhibitors of the angiotensin-converting enzyme, ACE, the receptor for SARS-CoV-2 [13], non-steroidal and steroidal anti-inflammatory drugs such as ibuprofen and corticosteroids, and various Chinese herbal medicines. These have all been considered and recommended in published articles [7,14,15]. However, several antiviral agents, effective against other viruses, are being tested, and in some cases, they are being used in clinical trials. One of the most promising drugs is remdesivir and its derivatives, broad spectrum antiviral drugs; current limited evidence suggests that they have been effective as a therapeutic option in combating COVID-19 [16,17]. Chloroquine (CG) and hydroxychloroquine (HCG), anti-malarial drugs, have been used, but caution is required because of the toxicity of these drugs and because higher doses seem to be required to combat the virus than are used for malaria. At present, there is insufficient evidence to suggest that CQ/HCQ are safe and effective treatments for COVID-19, and most of the evidence suggests that at safe doses, they are not effective [18,19].

Potential drug targets of SARS-CoV-2 include several viral structural and non-structural proteins. The effective development of novel drugs specifically for this virus depends on an understanding of the structures and functions of the constituent proteins [20]. The genome sequences of several *Betacoronaviruses* (β-CoVs) have been available for several years, and that of the new CoV-2 has been available almost since the initial outbreak in the city of Wuhan [21]. A typical betacoronavirus genome includes at least six genes (open reading frames; ORFs), some or all of which can be post-transcriptionally/translationally proteolytically cleaved into smaller proteins or peptides with importance to the viral infection cycle, and these may prove to be suitable drug targets [22]. These proteins include the structural Spike (S), Envelope (E), Membrane (M), Nucleocapsid (N), Hemagglutinin esterase (HE) and Helicase (H) proteins, and nonstructural proteins (nsp) which include Proteases, papain-like proteases (PLP or PL^{pro}) and 3C-like protease (3CL^{pro}), and Replicase proteins [22]. The Spike (S) protein, found as protrusions on the surface of the viral particle, includes the fusion peptide while the envelope (E) protein has viroporin activity, important for normal completion of the viral infection cycle. The nucleocapsid (N) protein forms a nucleoprotein with the single (+) stranded RNA genome, and this nucleoprotein complex is maintained within the capsid comprised in part of the matrix (M) protein. The protease proteins, which lie in ORF1a/ORF1ab, have been a target of anti-viral action by lopinavir/ritonavir, with and without arbidol, an anti-envelope viral indole derivative [16,23]. Four main structural proteins are encoded by ORFs 10 and 11 near the 3′ terminus of the viral genome (Figure 1) [21]. Among the potential targets of drug action are E-protein viroporins [24] and 'Spike' membrane fusion proteins [25]. In this review, we discuss several of the SARS-CoV-2 proteins but focus on these two proteins as potential targets of novel potential drugs to combat COVID-19, SARS and MERS.

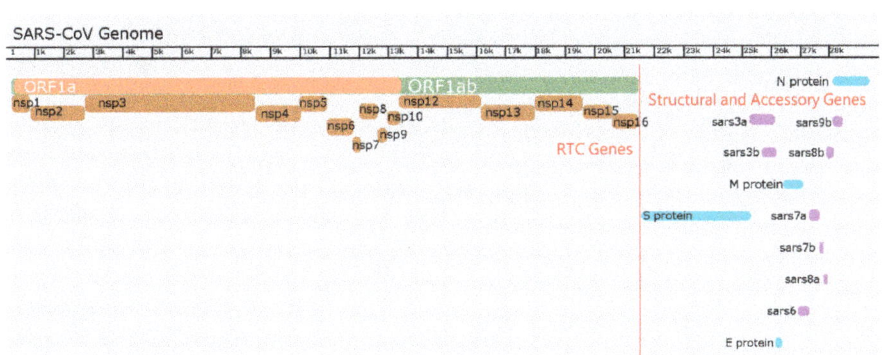

Figure 1. Genome graphic of SARS-CoV, showing where the genes encoding the recognized viral proteins reside.

3. The CoV Replication Cycle (from Entry to Exit)

The coronaviruses comprise a family of enveloped RNA viruses with positive sense strand genomes, meaning that expression of most genes, with the exception of the late structural genes, are transcribed directly from the genome. The first obstacle of any virus is entry into its host cell, utilizing common and often highly expressed viral receptors present on the host membranes. These receptors not only play a role in adherence to the host surface but also facilitate membrane fusion [26]. Attachment to host surface proteins often leads to a chain reaction that activates entry mechanisms in either the host, the virus, or both [27]. At some point during or after viral entry into the host cell, the virus may release its genome to begin hijacking the cellular machinery, allowing replication of its genome with the eventual production of viral products (Figure 2). These receptors may vary widely amongst viruses; they may be carbohydrates as is the case for influenza virus which uses sialic acid in respiratory epithelial cells, or they may be specific for Clusters of Differentiation (CD) ligands on immune cells. For example, HIV targets CD4 and CCR5/CXCR4 of human T-cells. Viruses may also utilize coreceptors that do not activate entry mechanisms but increase their affinity for their host to improve the probability of eventually engaging the correct viral receptor for entry.

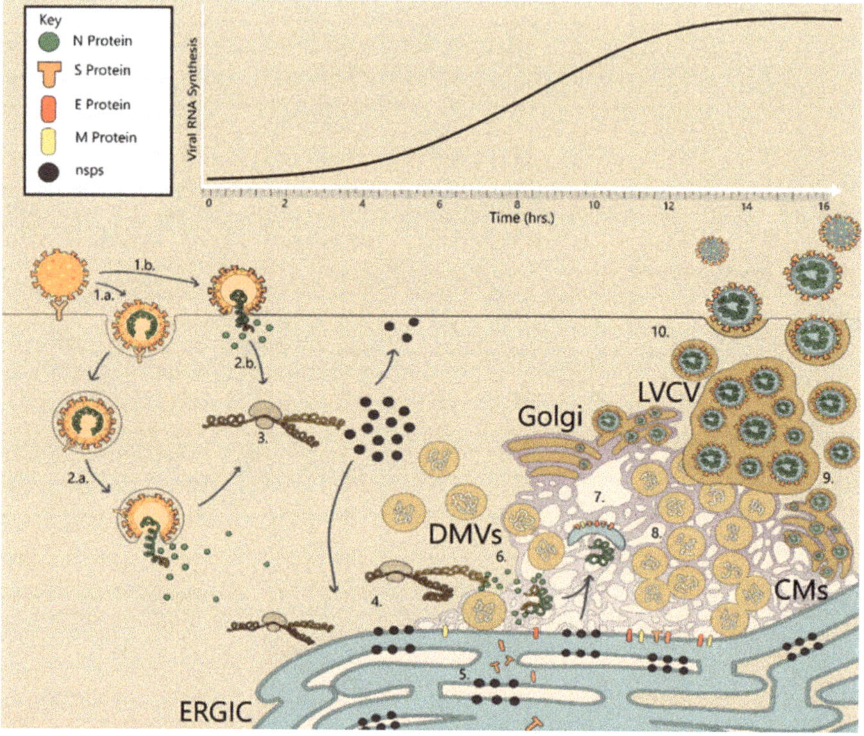

Figure 2. The infection cycle of coronaviruses. 1.a. Endocytic entry of virus particles after attaching to receptor proteins. 1.b. Immediate fusion at the surface of the cell upon attachment to receptor protein. 2.a. Endocytic fusion of viral and host membranes after endosomal maturation. 3. After shedding N protein from the genome, viral RNA is translated, expressing nsps from pp1a and pp1ab. Nsps migrate to the ER and perinuclear spaces, and RdRp complexes form to reverse transcribe additional viral transcripts and genomes. 4. Nsp3, nsp4 and nsp6 begin membrane rearrangements and form CMs and DMVs. 5. Viral structural proteins are produced and post-translationally modified in the ER/Golgi. 6. Viral genomes are encapsulated by N proteins. 7. Encapsulated genomes are wrapped by lipid envelopes assembled by structural proteins S, M and E. 8. Late-stage membrane rearrangements, many interconnected DMVs spread through perinuclear space with dsRNA within them. The Golgi has many virions budding from it with LVCVs attached to its membranes. 9. Mature CoV virions bud from ERGIC. 10. Exit of progeny CoV viruses via lysosomal exocytosis.

For enveloped viruses, entry into the host involves fusion of the host membrane with its own viral envelope to release its genome into the cell (Figure 2). All human infecting coronaviruses are capable of infecting the respiratory tract, and the three epidemic strains of β coronaviruses, SARS-CoV-1, SARS-CoV-2 and MERS, take advantage of prevalent ectopeptidases, surface glycoproteins found not only in respiratory organs, but also in epithelial cells of the gastrointestinal (GI) tract and excretory system as well as immune cells [28,29]. Dipeptidyl peptidase-4 (DPP4) is utilized by MERS, and both SARS-CoV-1 and SARS-CoV-2 use human angiotensin-converting enzyme 2 (ACE2) [26,30–32]. ACE2 is commonly found throughout human epithelial tissues including the oral and nasal mucosa, nasopharynx, lungs, stomach, small intestine, colon, lymph nodes, thymus, bone marrow, spleen, liver, kidney and even the brain which sometimes leads to systemic infection in late pathogenesis of Covid diseases [29]. For both SARS-CoV-1 and SARS-CoV-2, The S1 subunit of the Spike (S) protein engages ACE2 and primes activation of the S fusion core complex by host proteases. The S protein fusion complex of β-CoVs is promiscuous and can be cleaved by a variety of proteases such as furin, trypsins, trypsin-like proteases, cathepsins,

PC1, transmembrane serine protease-2 (TMPRSS-2), TMPRSS-4, type II transmembrane serine protease (TTSP) matriptase, and human airway trypsin-like (HAT) protease [32,33]. Membrane fusion, and release of the viral genome into the host cell may occur through a non-endosomal pathway at the membrane surface of the cell, although this is a minor pathway for entry in SARS-CoV-1 and MERS [34]. Studies have confirmed that SARS-CoV-1 and MERS prefer to enter the host cell through endocytic pathways, activating the fusion complex and releasing their genomes after maturation of the endosome following initial entry, similar to the endosomal pH-dependence of influenza viral fusion complexes [35]. However, it is the availability of these host proteases, present at extracellular binding sites, that dictates if the CoV will fuse its membrane at the surface, or after endocytosis, which may explain the varying infection dynamics amongst different cell types [32]. For instance, colocalization of ACE-2 and TMPRSS-2, possibly on lipid rafts of cell surfaces, may optimize cell to cell entry of SARS-CoVs and syncitia formation in respiratory epithelial cells [36].

Induction of endocytic pathways is a common theme among enveloped viruses, often utilizing commonly characterized methods such as clathrin- or caveolae-coated pits, pinocytosis, or macro-pinocytosis [37]. Drugs inhibiting endocytic pathways have proven to be effective antivirals for various viruses such as influenza virus, simian virus, herpes simplex virus, and coronaviruses [35,38–40]. Human infecting coronaviruses have been shown to use clathrin- and caveolae-mediated endocytosis; however, alternative SARS-CoV-1 entry mechanisms, independent of clathrin and caveolae, have also been confirmed, distinguishing SARS-CoV-1 from the other coronaviruses and enveloped viruses in general [34,41–43].

Clinically available chlorpromazine (CPZ), a clathrin inhibiting drug, significantly reduced MERS entry and infection rates, but not those of SARS-CoV-1 [34,44]. Additionally, sequestering cholesterol lipid raft-dependent caveolae-mediated endocytosis with filipin or nystatin did not disrupt SARS-CoV-1 entry [34]. The lack of reliance on clathrin and caveolae for cell entry of SARS-CoV-1 is similar to influenza viral entry, which can also infect cells via clathrin- and caveolin-independent endocytic pathways [35]. Interestingly, Ou et al. reported that SARS-CoV-2 endosomal entry can be inhibited using apilimod, a potent phosphoinositide 3-kinase (PIKfyve) inhibitor. PIKfyve is the only mammalian producer of phosphatidylinositol-3,5-bisphosphate (PI(3,5)P_2) in early endosomes, and it regulates endosomal sorting [32,45]. Although Ou et al. reported no in vitro cytotoxicity of apilimod at any of the tested concentrations on HEK293 and HeLa cell lines, despite its low abundance in endosomes, the PIKfyve-PI(3,5)P_2 complex is crucially important to cellular homeostasis and signal pathways [45,46]. Disruption of this complex has been linked to a variety of neurological diseases and neurodegeneration in humans with links to Alzheimer's disease and amyotrophic lateral sclerosis (ALS) [45,46].

Downstream of PI(3,5)P_2 is the NAADP-PI(3,5)P_2-activated cation-sensitive Two Pore Calcium Channels (TPC) [47]. TPC1 (TCID 1.A.1.11.25) and TPC2 (TCID: 1.A.1.11.19) have been shown to play roles in calcium signaling in endosomes and facilitate the fusion of acidic protease-rich lysosomes with endosomes to form late endo-lysosomes, providing a versatile array of cell type-specific responses [48–50]. Generally, both TPC1 and TPC2, but primarily TPC2, efflux calcium from endosomes into the cytosol of the cell to signal endo-lysosomal fusion, with TPC2 localizing to both early and late endosomes as well as lysosomes [51,52]. Mature endo-lysosomes may continue to use TPC2-mediated calcium signaling to regulate endo-lysosome interactions with the ribosomal rich ER membrane, a region of importance for enveloped RNA viruses [51].

Interest in TPC2 has grown after confirmation of its role in NAADP-induced calcium signaling for endosomal sorting and maturation, relevant to virus entry. Inhibition of TPC2 has shown positive results in preventing endosomal viral entry into cells for both enveloped and nonenveloped viruses. The enveloped virulent viruses, Ebola [53], MERS [54], SARS-CoV-2 [32], and henipaviruses [55] all show weakened or abrogated abilities to release viral contents into the host cell cytosol after either deleting or inhibiting TPC2, indicating that TPC2 is a necessary and common entry participant for various viruses. Disruption of TPC2 activity also blocks translocation of virus-containing endosomes through the endocytic pathway in MERS infected cells [54]. Unfortunately, none of the studies pinpointed a more complete reason as to why TPC2 is crucial for viral entry, but an understanding of the requirements of viral fusion proteins may provide insight.

Another overlap among the mentioned viruses is the dependency on the host endolysosomal protease, cathepsin L, which cleaves and activates the various viral fusion proteins [56]. While other proteases may be used by various viruses, SARS-CoV-2 showed a specific dependency on cathepsin L; deletion of this lysosomal protease drastically decreased viral release [32]. TPC2's calcium signaling could be a crucial step in (1) endolysosome formation, (2) incorporation of lysosomal proteases, (3) acidification of the virus-containing endosome as a prelude to membrane fusion/viral release, and (4) navigation to the ER-golgi membrane [51] where membrane-associated protein synthesis occurs. Indeed, TPC2 inhibition with fangchinoline inhibited endosomal furin protease activity and mobility of endosomal structures [54]. It remains elusive if TPC2-mediated signaling is required for cathepsin L protease activity. To the best of our knowledge, no such study exists as of yet connecting the full nature of endosomal maturation pathways, protease activity and viral membrane fusion.

4. Coronavirus Genome Structure, Replication and Expression

The coronavirus genomes are the largest of any positive stranded RNA viruses, ranging from 26 to 32 kb [57]. Each of these genomes is a single continuous strand of RNA that mimics host mRNAs with the existence of a 5′ cap and poly A tail [58]. Coronaviruses share a common structure, but in the interest of the epidemic strains of β-CoVs, we will primarily discuss the genomes of MERS, SARS-CoV-1, SARS-CoV-2 and other analogous β-CoVs. The 5′ end of the genome contains ORF1a and ORF1ab which contain the early nonstructural genes that compose the CoV replication-transcription complex (RTC) (Figure 1).

While the genes in the RTC region are directly expressed, replication and transcription of the genome is more complicated. Like other RNA viruses, Coronaviruses encode their own RNA-dependent RNA polymerases (RdRp) for genome and gene amplification. Both genome replication and expression of sub-genomic RNAs towards the 3′ end of the genome require reverse transcription and production of negative strand mRNA intermediates. Further complicating amplification and expression of late genes is the presence of transcription-regulating sequences (TRSs) that exist at the beginning of each structural or accessory gene [59]. These TRSs are composed of clusters of weak affinity oligonucleotides; for instance, SARS-CoV-1 TRSs contain the core sequence (5′-AAACGAAC-3′) [60]. These unique TRSs act as RdRp pause sites, giving the viral replication complex opportunity to dissociate and randomly walk around the genomic template until it finds another TRS to resume transcription. Transcription or random recombination of these sub-genomic RNA strands does not stop until the RdRp reaches a final TRS at the leader sequence on the 5′ end of the genome [59]. Alternatively, the RdRp can continuously transcribe through the genome, creating complete negative strand templates for full genome replication. This novel ability to discontinuously transcribe negative strand templates and recombine transcripts can provide insight into how CoVs are so readily able to mutate and recombine into novel strains. Coinfection of a single cell by two different CoVs can experience an RdRp jump from the TRS of one genome to that of another CoV genome, potentially creating a recombined genomic transcript [61]. Altogether, negative strand RNA templates comprise only up to 1–2% of the total viral RNA products in an infected cell, implying that only a few negative strand templates are required for amplification of full genomes and sub-genomic late genes [59,62]. All negative strand templates must then be reverse transcribed, again to create positive strand mRNA templates for protein synthesis or production of viable genomes for packaging.

Like other positive sense single stranded RNA viruses, early expression genes are transcribed and translated directly off its genome. These early genes exist in two nested open reading frames, ORF1a and ORF1ab, and include the nsps (Figure 1). These proteins are translated by cellular ribosomes into polyproteins, ORF1a polyprotein (pp1a) and ORF1ab polyprotein (pp1ab) [63]. pp1a is cleaved into 10-11 nsps, and pp1ab is cleaved into as many as 16 mature viral nsps that participate in immediate catalysis of viral RNA synthesis (Table 1) [64]. Cleavage of these polyproteins is accomplished by virally derived papain-like proteases PL1^{pro} and PL2^{pro} in nsp3 (but only PL1^{pro} designated as PL^{pro} in SARS-CoV) [65] and 3CL^{pro} in nsp5 [66]. nsp3 cleaves off itself, nsp1 and nsp2 from pp1a or pp1ab while 3CL^{pro} cleaves itself from nsp4 and nsp6, and it additionally cleaves nsp7-16 [67]. Expression of both pp1a and pp1ab is made possible by a heptameric 'slippery sequence' (5′-UUUAAAC-3′) and a 'pseudoknot' that exists at the end of ORF1a before ORF1ab. In SARS-CoV-1, and likely other betacoronaviruses, the pseudoknot is composed of 3 highly conserved, thermodynamically stable stem loops (free energy ~ −21.12 kcal/mol). These stem loops are digestible by dsRNAses, and disruption of any of these stem loops reduces frame shifting and expression of pp1ab by up to 94% in vero E6 cells [68]. The pseudoknot occasionally restricts the ribosome from proceeding farther than the ORF1a stop codon, halting expression at this point and allowing for the ribosome to 'slide' on the slippery sequence. This ribosomal sliding may shift the reading frame backwards by 1 nucleotide from ORF1a to ORF1ab, extending translation to express pp1ab [59]. However, most of the time, the ribosome is able to melt the pseudoknot, progressing through the slippery sequence uninterrupted until it reaches the ORF1a stop codon, producing more pp1a than pp1ab [68,69].

Table 1. Functions and properties of corona virus structural (S, E, M, N) and non-structural proteins (Nsps1–16) as well as ORFs 3a & b, 6a, 7a & b, 8a & b, and 9b.

Nsp1	Likely induces cell arrest in the G_0/G_1 phase and interferes with type I IFNs [70].	Nsp15	Endoribonuclease activity; Immune evasion; degrades viral polyuridine sequences to prevent host antiviral detection [71].
Nsp2	Potential nonessential role in pathogenesis [72].	Nsp16	2′O-methyl-transferase activity [73].
Nsp3	Membrane rearrangements for replication organelle formation. Viral proteolytic activity. Membrane anchoring of other viral proteins to perinuclear membranes; potential role in genome packaging. IFN antagonism.	S	Spike protein for binding/fusion/entry. Role in ER stress and syncytium formation.
Nsp4	Membrane rearrangements for replication organelle formation.	E	Envelope protein essential for virion formation and exit. Viroporin and membrane rearranging activity. Role in NLRP3 inflammasome activation.
Nsp5	Viral proteolytic activity.	M	Membrane/Matrix protein essential for virion formation. Binds to S, E and lipids to form the virial envelope-protein capsule.

Table 1. *Cont.*

Nsp6	Essential for membrane rearrangements for SARS-CoV. May induce autophagy.	N	Nucleocapsid protein, binds to viral RNAs. Necessary for packaging genome and protection from host RNAases.
Nsp7	Cofactor for RdRp complex [74].	(ORF)3a	Viroporin activity similar to E. Interacts with S, E, and M. Activates NLRP3 inflammasomes [75,76].
Nsp8	RNA binding, RNA polymerase activity and essential RdRp complex cofactor protein [74].	(ORF)3b	Putative function in upregulating cytokine secretion [75].
Nsp9	Novel ssRNA binding protein. May participate in RNA processing [77].	(ORF)6a	Colocalizes with nsp3, nsp8 and RO-associated membranes. Upstream and downstream Type I IFN antagonist [75].
Nsp10	Replicative cofactor to nsp14 [78].	(ORF)7a	Interacts with structural proteins M, E and S. May form a complex with 3a. Possibly essential for viral replication [75].
Nsp11	Unknown.	(ORF)7b	Possibly essential for viral replication. [75].
Nsp12	RNA-dependent RNA polymerase (RdRp) [79].	(ORF)8a	Viroporin activity similar to 3a and E and activates NLRP3 inflammasome [80].
Nsp13	Cofactor for the RdRp complex. Viral helicase. Unwinding duplex RNA [81].	(ORF)8b	Contributor to lysosomal stress, autophagy and inflammation; activation of NLRP3 inflammasome [82].
Nsp14	S-adenosyl methionine-dependent (N7-guanine)-methyl transferase, assembling cap1 structure at 5′ end of viral mRNA to promote translation and avoid antiviral detection. Proofreading of viral RNA transcripts [78].	(ORF)9b	Suppresses innate immunity by usurping poly-C-binding protein 2 and HECT domain E3 ligase AIP4 to degrade MAVS/TRAF3/TRAF6 signalosome [83].

ORF1a encodes nsp1–11, which include several membrane spanning proteins associated with viral replication. Within ORF1ab are the genes for the important nsp12–16 which all directly participate in viral RNA replication [63]. Collectively, the nsps derived from ORF1a and ORF1ab are called the RTC [62]. The important nsp12 functions as the viral RdRp for reverse transcription and is associated with nascent viral RNA synthesis [62]. To emulate host mRNAs, provide stability, ensure proper translation, and avoid degradation by host nucleases, the viral genomic and nascent RNAs are all capped and methylated at the 5′ end via nsp14 N7 cap methyltransferase [84]. With expression of the RTC, the virus is equipped with the proper arsenal to hijack the cell, requisitioning and modifying important host organelles and the machinery for its own replication.

5. CoV Replication Organelle Formation: A Dance of Membrane Rearrangements

In order to review the nature of coronavirus membrane reorganizations, it is important to note that all nidovirales rearrange host membranes in similar manners, thanks to nsps derived from pp1a and pp1ab. In this section, we present collected information from various studies of early membrane rearrangements induced by members of the Nidovirales (Coronaviruses and Arteviruses) with an emphasis on β-CoVs (SARS-CoV-1/2, MERS-CoV and MHV). Throughout the replication cycle of Nidovirales viruses, dramatic rearrangements of host membranes are utilized by the viruses for replication, protein expression, assembly and exit. Virally induced zippered ER, production of double membrane vesicles (DMVs), convoluted membranes (CMs), and later Vesicle Packets (VPs), Double Membrane Spherules (DMSs), the occasional membrane whorls (MWs), and giant vesiculations (GVs) are hallmarks of coronavirus and other nidovirales infections. Other more elusive structures include cubic membrane structures (CMSs) and tubular bodies (TBs) [84]. All membrane rearrangements are visible through electron microscopy (EM) and electron tomography (ET) and occupy generous portions of the cell space. While minute details will obviously cause slight differentiations in the dynamics of early infection, the nsps responsible for membrane rearrangements are astoundingly similar amongst the Nidovirales, allowing for a comparative analysis within the virus order. See Table 2 for comparisons between important membrane rearrangements amongst different Coronaviruses.

Table 2. Membrane structures found in various coronaviruses. Bold checkmarks indicate dominating membrane structures.

Membrane Structures	SARS	MERS	MHV	HCoV-229E	PEDV	PDCov	IBV
DMV	✓	✓	✓	✓	✓	✓	✓
DMS	✓	✓	✓	✓		✓	✓
CM	✓	✓	✓	✓	✓		
VP	✓	✓	✓				
Zippered ER	✓	✓	✓	✓		✓	✓
LVCV	✓	✓	✓	✓	✓	✓	
GVP	✓	✓	✓				
TB	✓	✓	✓		✓		
Interconnections	✓	✓	✓	✓	✓	✓ (Perinuclear DMVs)	✓ (Perinuclear DMVs)

In all (+) stranded RNA viruses, viral-induced reorganization of cellular membranes has been documented and associated with viral RNA synthesis [85]. Benefits of associating viral RNA synthesis with modified membranes include more efficient catalysis of viral products with association of macromolecule-rich membranes such as the ER and Golgi. Viruses may reorganize these organelle membranes into 'viral organelles' to separate specific parts of viral replication stages and allow induction and confinement of RNA synthesis and dsRNA intermediates to microenvironments shielded from host innate antivirals [86]. Viral organelles specific for replication and nucleotide synthesis are appropriately called replication organelles (ROs). Nidovirales, and more specifically, the subfamilies Coronaviruses and their related subfamily Arteriviruses appear to have a novel rearrangement of host membranes not seen in any other (+) sense RNA viruses [85,87]. In β-CoVs, early expression and cleavage of nsp2–6 TM proteins participate in membrane rearrangement of the ER into DMVs which are completely sealed from the cytosol [MERS, SARS, MHV, HCoV-229E, PEDV, IBV], with cleavage of nsp3 and nsp4 being crucial for SARS-CoV-1 and MERS. Plasmid expression of only MHV or MERS-CoV-1 nsp3 and nsp4 was sufficient to replicate clustered DMV membrane rearrangements reminiscent of MERS-infected cells [88,89]. Similarly, SARS-CoV-1 nsp3 and nsp4 were sufficient to induce phenotypic membrane rearrangements reminiscent of SARS-CoV-1 infected cells [89], but nsp6 may additionally promote these rearrangements [57]. Together, these findings imply that late membrane-bound structural proteins are not required to modify host membranes in the early infection, and this process is primarily associated with the RTC.

In the following text, we attempt to provide an approximate timeline of CoV RO formations with an emphasis on β-CoVs. Nearly immediately after the early synthesis and expression of RTC RNAs, viral replication complexes begin to form within cytoplasmic membranes. Abundant transmembrane nsps begin to accumulate in the membranes of the ER and induce unusual modifications, localizing in the perinuclear region of the cell. With β-CoVs, as early as 1–2 hpi, isolated DMVs can be seen forming in the cytoplasm with a slight proximity to the ER, which may exhibit a slight zippering shape [90,91]. Low levels of viral RNA synthesis are also detectable in MERS, SARS, MHV, and IBV [62]. Shortly after, in β-CoVs, large CMs (0.2–2 μm) and reticular inclusions form with connections to DMVs and the ER [90,91]. By 4 hpi, DMVs have drastically increased in amount and cluster throughout the cell with increasing localization in the perinuclear space [90,91]. At some point, the golgi membrane is also incorporated in the DMV-CM-ER network [91,92]. By 5–7 hpi, newly assembled and budding virions appear in the Golgi cisternae as do virion-containing golgi-derived LCVCs [90,91,93]. α-, β-, δ- and γ-CoV ER morphologies adopt a zipper form [92–94]. γ- and δ-CoVs also contain DMSs sprouting from the ER with small 'neck like' openings to the cytosol [93,94]. At 7 hpi and beyond, there can be as many as 200–300 DMVs clustered within the cells infected with SARS [90]. After this point in time, the outer membranes of several DMVs have fused together to create an interconnected system of single membrane vesicles contained within one communal outer membrane forming large VPs. At least 95% of DMVs have one or multiple thin 'neck like' connections (~8 nm) to one or several other DMVs, CM or the ER, implying that late stage modified membranes form one large uninterrupted network with the ERGIC membranes [90,91]. Late infection membrane rearrangements in α- and β-CoVs may also include DMSs embedded in CMs and have interconnections with CMs, CM-ER, and ER membranes, although the exact time frame when these are first formed in β-CoVs is not clear [92,93,95]. This network is appropriately called the reticulovesicular network (RVN). [64]. Finally, at ≥10 hpi, assembly of new virions can be seen budding into and from the heavily modified ERGIC lumen, and the virus has completely repurposed the cell for viral replication [90,92].

There can be up to 1000 vesiculations in a single SARS-infected cell [90], and separate VPs fuse together to form GVPs which may be interconnected with LVCVs and Golgi membranes containing both vesicles and significant numbers of budding complete virions [90,95]. In all CoVs, at every recorded stage, no DMVs exhibit openings to the cytosol [64,90–96]. DMVs close to the ER exhibit tightly apposed inner and outer membranes, but perinuclear residing DMVs have looser organization between the two membranes [90].

While the patterns of membrane rearrangements are generally common amongst all coronaviruses studied thus far, slight differences do exist among the families. γ-CoV IBV and δ-CoV PDCoV did not show visible membrane rearrangements until after 6 hpi, but the phenotypes were remarkably similar to those of the observed β-CoVs and remained consistent through to 24 hpi [92–94]. Another stark difference between β-CoVs and γ-CoVs is the organization of CMs and DMVs. In β-CoVs, DMVs likely form first and cluster in a net of CMs [90,91], while in γ-CoVs, zippered ERs and DMSs are the primary structures, and DMVs tend to appear as free-floating vesicles away from the ER [94]. α-CoVs, HCoV-229E and PEDV, have membrane rearrangement patterns similar to those of β-CoVs, but they take between 24–60 hpi to develop [92,95]. PEDV also exhibits unique late infection membrane rearrangements called endoplasmic reticular bodies (ERB) which occur in a minority of cells and virion-positive endolysosomal compartments [95].

As mentioned before, β-CoVs require nsp3 and nps4 for modification of cell membranes. Exactly how these nsps are capable of modifying the ER membrane to produce zippering and vesiculations remains uncertain. Disruption of nsp3 and nsp4 expression greatly inhibits RVN formation and viral genomic replication [97,98]. Co-expression of nsp3, nsp4 [88,89] and nsp6 (for SARS) [57] induces clustered DMVs and disorganized double membraned CM-like structures sprouting from the ER. It is thus proposed that accessory structures other than DMVs are induced by other viral proteins or viral replication.

Initially, it was thought that host vesicular or secretory pathways might associate with these nsps to induce membrane changes. Autophagosomes and endo-lysosomes have been known to be induced by coronavirus infections and appear during mid-late infection, but they are not required for viral replication [95,99]. Additionally, during mid infection, virion positive Golgi born LVCVs form in conjunction with ERGIC and DMV constructs [91]. However, no matter which CoV, there exists a high statistical correlation between the abundance of different membrane rearrangements. In MHV, the abundance of DMVs with CMs in a single cell is comparable to the abundance between DMVs and DMSs in an IBV infected cell [100]. These findings provide evidence that membrane rearrangements for a given CoV are highly related to each other in the formation of an RVN. Perhaps a dominating structure in a CoV may be able to compensate in function for another missing membrane structure.

Picornavirus utilizes some parts of the secretory pathway for membrane reorganization (DMVs) and replication [101]. Treatment of infected cells with brefeldin A, an antiviral drug that inhibits secretory protein transport from the ER to Golgi, suppresses host secretory pathways completely and prevents picornavirus replication, but it only partially inhibits coronavirus replication [64]. In fact, upon suppression of native host secretory pathways, early induction of DMVs was hindered merely to 20% of normal activity, and late-stage infection exhibited a similar phenotype [64]. Interestingly, this experiment revealed that host ER secretory protein Sec61α (TCID: 3.A.5.9.1), a transport protein subunit that anchors ribosomes to ER membranes and shuttles unfolded polypeptides into the lumen of the ER, was redistributed to RVNs upon SARS infection. However, of interest, Brefeldin A treatment of SARS-infected cells (1–7 hpi) accelerated RVN formation, while slightly inhibiting the expression of nsp3 and N protein causing small differences in RVN morphology. DMV formation seemed to be accelerated, and aggregation into LVCVs occurred during mid infection (7 hpi) as opposed to late infection times. Additionally, the luminal space between inner and outer membranes of DMVs appeared more open, but not on the sides facing CMs. Intracellular virions could still form, but brefeldin A treatment prevented secretion of these particles, due to downstream inhibition of excretory pathways necessary for virion budding [64]. Hence, suppression of secretory pathways slows early CoV infection and lowers virion productivity, but it does not stop RNA synthesis and virion production altogether.

Protein disulfide-isomerase (PDI), another luminal ER-associated protein that participates in the formation of disulfide bridges of unfolded polypeptides, migrated to MHV- and SARS-induced RVNs and also partially localized with nsp3 [64,102]. Interestingly, when nsp3 or nsp4 is singly expressed within a cell, it localizes with PDI, but not when they are co-expressed [88]. Unfortunately, what the relationships between CoV RVNs and host PDIs are remains elusive. Any antiviral responses involving suppression of host ER protein interactions with nsps must be further investigated [102]. Autophagosomes have also been proposed to be involved in membrane rearrangements, but several studies seem to have argued against this hypothesis, as deletion of autophagy-related-genes (encoding ATGs) did not prevent DVM formation, despite a SARS nsp6 association with the ATG pathways [103–105]. Additionally, microtubules are not required, despite LC3 decorating the outer membranes of DMVs [85].

This departure from common viral replicase themes in early membrane rearrangements has brought Nidovirales research to the currently supported hypothesis that the nsps themselves are sufficient to induce these dramatic membrane rearrangements. Associated host pathways discussed above merely support mechanisms that promote membrane rearrangement without being required. Bioinformatic analyses of the TM domains as well as NMR and X-ray crystallography analyses of these nsps, support the theory that simply the shapes and protein-protein interactions amongst ER spanning nsps is enough to scaffold membrane rearrangements. Both nsp3 and nsp4 span the membrane multiple times and induce the ER early membrane pairing necessary for RVN formation. Intriguingly, nsp3 and nsp4 of β-CoVs and their equivalents in other CoVs retain only modest primary sequence homology, in spite of strong structural similarity, specifically, in the proximal location of the TMDs and large luminal loops [106]. This could explain why early infection membranes are characterized by a zippered ER with minimal vesiculation in some CoVs. Over time, as RNA synthesis increases, expression of pp1a and pp1ab allows for sufficient accumulation of membrane spanning nsps in the ER, causing dramatic vesiculation and reconstruction of host membranes. Differences in dominating structures may be attributed to small differences in integral membrane nsp topologies, or to other infection dynamics. Lone expression of nsp3 or nsp4 individually only localizes the protein to the ER, but co-expression of nsp3 and nsp4, or polyprotein nsp3–6, gave rise to perinuclear localization of these nsps and formation of clustered DMVs [57,88]. Structural inspection of these nsps revealed that a truncated nsp3, where the deletion spans from the first TMS to the C-terminus, can still localize with nsp4 to perinuclear regions [88]. Even more interesting is a possible direct involvement of nsp3 and nsp4 in viral replication in addition to RO formation.

6. Structures of nsp3 and nsp4

Initially, pp1a and pp1ab are cleaved by the PL1^{pro} or PL2^{pro} papaine-like proteases that are encoded in the region adjacent to nsp3. In SARS and MERS-CoVs, only the PL2^{pro} protease is present, but it performs the same function and cleaves the pp1a or pp1ab into the nsp1, nsp2, and nsp3-polyproteins. Further proteolysis of the nsp3-polyprotein is fulfilled by nsp5 to generate the individual nsp3, and remaining nsp proteins. Also, to the N-terminal side of nsp3 is a ubiquitin-like domain 1 (Ubl1) which, while retaining poor primary sequence conservation amongst CoVs, has well conserved secondary folding and is present in all CoVs [98]. Ubl1 is associated with ssRNA, interacts with the N protein, and is essential for viral RNA synthesis [106]. Ubl1 weakly binds to N, but the exact regions that associate with each other may vary among CoVs [98]. However, deletion of Ubl1 in MHV prevents viral replication [98]. It is possible that Ubl1 serves as a dock for viral genomes to early RTCs, bringing N and the nascent viral genomes close together for packaging. This would suggest that nsp3 not only facilitates the remodeling of host membranes, but also serves as an active protein in genome packaging. Additionally, the Ubl1 of SARS-CoV-1 might disrupt Ras regulated cell-cycle progression. The Ubl1 is similar to the Ras-interacting domain of the Ral guanine nucleotide dissociation stimulator (RalGDS). Since SARS and MHV infections are known to induce cellular arrest in the G_0/G_1 phase, it could be that nsp3/Ubl1 disrupts the interaction between Ras and its downstream effectors [98].

Following the Ubl1 domain are the macrodomains (Mac1, Mac2 and Mac3), with Mac2–3 forming a portion of what was once thought of as a SARS Unique Domain (SUD). It was thought the SUD was unique to SARS due to the high variability in CoV genomes, but secondary structure analysis has revealed strong structural similarity in these regions amongst other CoVs, and Mac2-3 may not be unique to SARS [106]. The Mac domains are all similar in structure and are hypothesized to be gene duplicates of Mac1. However, only Mac3 was shown to be necessary for SARS replication in a cDNA study [107]. Following Mac3 on the C-terminal end is Domain Preceding Ubl2 and PL2^{pro} (DPUP), which forms antiparallel β-sheets. The Mac2-Mac3-DPUP complex has an affinity for nucleic acids and binds to RNA. Specifically, Mac3 was shown to bind oligo(G) [107] and oligo(A) [98]. Thus, Mac3 may bind to the poly(A) tails of RNA molecules present in viral genomic, subgenomic and host RNAs. Also present in Mac3, is a unique antiparallel β sheet that exists in MHV [108]. Following the DPUP and Ubl2 domains is PL2^{pro}, previously mentioned to cleave the nascent viral polyprotein into nsp1/2, nsp2/3, and nsp3/4 [109]. In addition to autocleavage activity, PL2^{pro} has deubiquitinating and deISGylating activities, which may participate in suppression and evasion of intracellular innate immune pathways [98,109].

Note: ISGylation is the process of IFN-induced gene ISG15 ubiquitin-like protein associating with targets (ISG15 targets) [110]. The SARS-CoV-1 Mac2-3 and PL2^{pro} domains together may elicit innate immune suppression by competitively binding to host E3 ubiquitin ligase RCHY1, leading to down-regulation of antiviral pro-apoptotic transcription factor p53 [111,112]. In contrast to the enzymatic nature of the N-terminal side of nsp3, the C-terminal one third end of the nsp3 protein (nsp3C) interacts with nsp4 and other nsps (e.g., nsp8) [57,88,97,98]. The specific interaction of nsp3 with nsp4 may induce the hallmark membrane curvature in the ER.

The nsp4 protein has four TMSs, one large luminal loop between the 1st and 2nd TMSs, and a small luminal loop between the 3rd and 4th TMSs. The 3rd and 4th TMSs are dispensable for SARS and MHV nsp4, but deletion of TMSs 2–4 affects localization with nsp3. Moreover, deletion of either TMS 1 + the large loop, or TMSs 2–4 + this large loop, completely prevents localization with nsp3 and nsp3c [88]. Changing the luminal loop of nsp4 of one CoV for another also prevents localization with nsp3, suggesting that the exact structure of the N-terminal large luminal loop is specific per CoV nsp3c binding site [88]. Within the large luminal loop resides 4–10, cysteine residues as well as glycosylation sites are conserved amongst CoVs [57,88]. Deletion of glycosylated regions produces aberrant DMVs with large luminal spaces and increased levels of CMs [57]. Replacement of the cysteine residues results in low localization with nsp3, suggesting possible disulfide bridge formation amongst nsps during membrane pairing [88], as well as a possible reliance on PDI and Sec61α. In SARS, the region responsible for localizing with nsp3C are within the regions 112–164 aas and 220–234 aas. Deleting those two regions, or only the specific residues H120 and F121 within nsp4, prevents localization with nsp3C, replicon formation and thus, viral replication, but the phenotype can be rescued when wild type nsp4 is reintroduced via an encoding cDNA [97].

In SARS-infected cells, nsp6 is required to induce an RVN phenotype akin to wild-type. Without nsp6, DMVs migrate farther away from the RVN. Nsp6 may break up elongated stretches of membrane pairings, causing DMV vesiculation to cluster near CMs, causing RVN formation for SARS. The existence of polyprotein nsp4–6 following nsp3 cleavage may be critical for efficient SARS DMV formation [57]. Overall, nsp3, nsp4 and nsp6 are essential for normal SARS replication. In the pursuit of antivirals, inhibition of the papain-like proteases that cleave the early polyproteins would be an excellent upstream target to block early viral replication.

7. The Replicon Conundrum and a Putative Nucleopore

Based on the themes of other (+) sense RNA viral replicases, it would be expected that coronaviruses may modify membranes for similar reasons as for the existing canonical replication themes of many RNA viruses, using DMVs as RNA factories. In other RNA viruses, the existence of dsRNAs is often a marker for nascent viral RNA synthesis since dsRNA molecules are a necessary intermediate. Labelling for dsRNA in CoVs revealed that their presence within DMVs could be observed throughout infection, with the signal growing stronger over time. Following the logic of other RNA viruses, perhaps DMVs might provide an encapsulated environment free of host innate immune mechanisms to protect viral RNA synthesis. Utilizing membrane rearrangements sourced from the ER would also benefit CoVs as their proteomes contain many essential structural and nsp TMD-containing proteins. However, in all previously recorded CoVs and Arteviruses, the DMVs are sealed from the cytosol, and nsps delocalize from DMVs during mid to late infection [90,92]. Spatially, DMVs migrate away from the ER-CM constructs and cluster in the perinuclear regions forming the VPs in conjunction with modified Golgi and LVCVs during late infection. Electron micrographs of some VPs and virion budding LVCVs documented that the connected membranes seem to compartmentalize, keeping former DMVs on one side while mature and budding virions form on the other [90,92].

Studies using either BrU or Click chemistry and labelled RdRp and an nsp (nsp8 or nsp12) to detect nascent RNA synthesis revealed that CoV viral synthesis first localizes with dsRNAs and TM nsps, but migrates to perinuclear regions later in infection, eventually spreading throughout the cell [62]. On the other hand, nsps primarily accumulate in the ER and CMs [62,64]. Additionally, dsRNAs did not incorporate BrU or clickU, implying that these intermediates were catalytically inactive and did not participate in RNA synthesis at the time of labelling, despite RdRp nsps also co-occurring in the lumen of DMVs [90]. Similar findings in studies with γ-CoVs showed that less than 1.5% of RdRp nsps colocalized with dsRNAs [94]. These results suggested that dsRNAs and minor localizations of RdRp nsps are not bona-fide markers for viral RNA synthesis. Rather, encapsulating dsRNAs within DMVs may be used to protect the replicase from activating immune pathways.

Another proposed structure for RNA synthesis is the DMS-ER network. DMSs have been detailed to also occur in α-virus Semliki Forest Virus (SFV) of the Togaviridae, and they are reported to be sites of RNA synthesis [113]. Due to the small openings in DMSs that connect them to the cytoplasm, it was hypothesized that DMSs would be probable sources of RNA synthesis in CoVs since they also occur in abundance in α-CoVs [94], δ-CoVs [PDCov], γ-CoVs [IBV], and recently also, in low quantities, in β-CoVs [92]. However, actual labelling of nascent RNAs with radioactive ^{3}H-U and EM imaging revealed little to no localization mid-late infection near zippered ER/CMs or DMSs in IBV, MERS, and SARS [92]. Intriguingly, both ^{3}H-U and indirect immunogold-BrU labelling still supported membrane structures in proximity to DMVs as the primary regions of RNA synthesis [114], and not DMSs or CMs [92]. However, these conclusions were drawn only from images taken in mid to late infection and do not represent the dynamic findings of previous CoV RNA synthesis. Although DMSs tend to occur abundantly in IBV, RNA synthesis and virion production were not hindered in the pathogenic M41 strain of IBV that produced significantly less DMSs [100]. This could mean that very few DMS structures are necessary for subsequent RNA synthesis and virion production. Since (1) DMSs occur in low abundance with non-γ-CoVs, (2) they are difficult to discern among other membrane rearrangements, and (3) their time of production remains in question, it is possible that very few DMSs are required in early stages of infection. As the infection progresses, late RNA synthesis could migrate from DMSs or CMs to DMV-associated structures. Since IBV DMVs have repeatedly been reported to occur as lone vesiculations with only a few connected to the ER, it may not be a requirement for them to be connected to the ER as previously stated. Still, RNA synthesis has also been reported to have low background activity in the cytosol of infected MHV cells [114], and ^{3}H-U or immunogold-BrU labelling with EM may not be sensitive enough to indicate minor regions of RNA synthesis. Following similar reasoning, the existence of CMs that do not occur in some CoVs may perform functions similar to those of DMSs. Alternatively, CMs may simply be consequential constructs caused by an overaccumulation of nsps despite their close relationship to DMVs. To elucidate the nature of RO-RVN formations and RNA synthesis, it would be best if a future study combines labels for TM nsps, RdRp complexes (nsp7 + 8 and nsp12), nascent RNA labelling, dsRNA labelling and EM all together throughout several time points spanning early to late infection with high resolution imaging. Of course, all results and findings are at the mercy of sample preparation and handling to carefully preserve the delicate nature of these RVNs.

This unpairing between dsRNAs and RdRp complexes seems to be a common theme among all of the CoVs examined in this review. Even more curious is the observation that MHV nascent RNAs delocalize with dsRNAs between 4.15–5 hpi to 8–9 hpi, with a Pearson correlation coefficient dropping from ~0.6 to ~0.35 [62]. In the perspective of virion synthesis, statistical correlations between membrane rearrangements, CM-DMV-virion abundance in MHV and DMV-DMS-virion abundance were low [100]. However, these findings still do not exclude DMVs from being involved in RNA synthesis, as EM imaging revealed that nascent RNAs still localize near DMVs and the ER in early and late infections [90,92]. Determining how dsRNAs collect inside the lumen of DMVs has also been troubling since no imaging provided has been able to detail DMV intermediates so far.

Returning to the pore hypothesis, recent cryo-EM tomograms of DMVs revealed a pore complex formed by nsps, sparsely scattered on DMV surfaces, opening the lumen to the cytosol in MHV and SARS-CoV-2 infected cells [115]. The pore has 6-fold symmetry, and the channel begins with a 6 nm wide opening facing the cytosol, surrounded by a protruding crown with 6 prongs extending 13 nm outward and 14 nm away from the central axis of the opening. The pore is stated to be analogous to the reoviridae genome packaging pore [115]. RNA export function has yet to be confirmed, but the structure and its 6-fold symmetry is primarily composed of nsp3, which has RNA binding capacities [106]. The luminal side of the pore complex appeared denser, and Wolff et al. speculated that other luminal DMV-associated proteins such as N and/or nsp12 RdRp associate with this pore. Since nsp7, nsp8 and nsp12 form a reverse transcription tunnel, it is possible that the viral transcriptome associates with the luminal nsp3s of the pore. This suggestion is additionally supported by nsp3′s known function as a scaffold protein for other nsps in the RTC [116]. Such a complex could then transcribe RNAs and export them upon synthesis in an efficient manner. Despite the pore being composed of nsps with catalytic activity, the pore itself has no confirmed catalytic activity, characteristic of other viral portals, such as those in bacteriophages, Reoviridae and Herpes [117–119]. We thus suggest the RdRp could be used as a motor to feed transcripts into the pore. Meanwhile, aborted, or malfunctioned transcripts are not exported, but instead are left within the DMVs, due to the size limitations of the pore. Confirmation of this structure's existence is a major step forward to completing the coronavirus RO puzzle. It is also yet to be confirmed if equivalent pores exist in CoVs outside of the β-CoVs.

It is possible that after significant suppression of host innate antiviral pathways due to the accumulation of nsps or other ORF proteins, that RNA replication no longer needs to reside in perinuclear membranes and may migrate throughout the cytosol during mid-late infections. This pore has been confirmed only in the β-CoVs, MHV and SARS-CoV-2, but due to the phenotypic similarity between many β- and γ-CoVs, it is likely that pores exist beyond the findings of Wolff et al. [115].

8. Viral Proteins–Structures, Expression and Assembly

8.1. The Nucleocapsid (N) Protein: Genome Packaging

The nucleocapsid (N) proteins of coronaviruses are reasonably well conserved proteins, although the SARS-CoV-1 and SARS-CoV-2 N-proteins, nearly 90% identical to each other, show only about 25% sequence identity with those from other members of the *Coronaviridae* family [120]. Nevertheless, most of them exhibit only moderate variation in size, usually being between just below 400 amino acyl residues (aas) (e.g., Porcine transmissible gastrointestinal CoV, TEGV, of 382 aas), to just over 450 aas (e.g., Murine CoV-3, MHV3, of 454 aas) [121]. They have three domains, an N-terminal domain (NTD) nearly 200 residues in size which is the dominant RNA-binding domain, a central Ser/Arg-rich flexible linker domain with a striated box of about 50 residues, and a C-terminal domain (CTD), which like the NTD, is roughly 200 residues in length, but functions in dimer/oligomer formation [122]. The N protein has primary functions in self dimerization/oligomerization and RNA binding, yet although the NTD serves a primary function in RNA binding, all three domains have affinity for nucleic acids [123]. In addition, there are intrinsically disordered regions near the N- and C-termini of these N-proteins, each about 50 residues in length [124].

N proteins have multiple functions including but not limited to: (1) forming stable but dynamic complexes with the genomic RNA for compaction of the nucleic acid in the viral particle, (2) interacting with the structural membrane (M) protein to promote membrane envelop folding and virion assembly, (3) interacting via two distinct regions of N with the non-structural protein, nsp3, to allow proper recruitment of N to the replication/transcription complex, (4) playing an essential role in enhancing the transcription of genomic RNA and viral mRNA, (5) increasing RNA replication efficiency, in part, by facilitating separation of the two RNA strands, (6) interfering with host cellular defense processes such as interferon production, and (7) promoting host cell death (apoptosis) [125–128]. The N-protein can be phosphorylated to facilitate condensation with RNA and the M-protein and to modulate the liquid-liquid phase separation [129,130]. As noted above, its recruitment to the RTC plays a role in the coronavirus life cycle [131].

The details of many of these functions have been elucidated to a considerable degree, and several of them are clearly interrelated [123]. Its recruitment to RTC plays a crucial role in the overall coronavirus infection cycle [131]. Self-association of the N-protein, which also depends on its RNA binding capacity [132], is required for formation of the viral capsid, which occurs at intracellular membranes of the ER-Golgi intermediate compartment. Unfortunately, many details of the molecular packaging inside the virion have not been fully elucidated. Early electron microscopy revealed that the ribonucleoproteins (RNPs) are helical, consisting of coils of 9–16 nm in diameter with a hollow interior of about 3–4 nm [121]. In the mature virus particle, the capsid protects the viral genome from caustic chemicals and extreme physical conditions [133]. In this regard, it is important to note that N has protective RNA folding/chaperone activity, due in part, to the central disordered domain (the LKR domain), reducing the free energy barrier for dissociation of the nascent minus RNA chain from the genomic RNA template during discontinuous RNA transcription. N also promotes template switching, which may be a primary cause of its acceleration of transcription [134].

The 3-D structure of N together with NMR analyses revealed that the basic region between aas 248 and 280 in the SARS-CoV-1 N protein binds RNA, while the region just C-terminal to this sequence promotes octamerization of the CTD [135–138]. The former region forms a positively charged groove, being able to accommodate either single stranded or double stranded negatively charged nucleic acids [139]. Such interactions allow formation of a compact ribonucleoprotein complex, the nucleocapsid, that ensures timely replication, reliable transmission and proper regulation of translation while in the cell, before formation of the filamentous nucleocapsid of about 12 nm in diameter and up to several hundred nm in length, that will be incorporated into the viral particle during assembly [123]. The assembly process also depends on the M protein, which together with the E protein, is a primary core constituent in the final virion. By using 3D cryo-electron tomography with MHV particles, it was possible to see that the viral membrane was nearly twice the thickness of a typical cell membrane, possibly due to the C-terminal domain of the M protein [140]. It should be clear that the ribonucleoprotein complex, together with the closely associated M-protein, plays a major role in envelope formation and viral budding within intracellular ER-Golgi complexes (see the section on the M-protein). For this reason, the assembly of the N-protein oligomer with its associated RNA has been considered to be an appropriate target of drug action [141] (see below).

The N-protein has proven to be a successful target for antiviral drugs and may be useful for the potential development of vaccines. This topic has been extensively reviewed recently [142], and only a couple of examples will be provided here. Cyclosporin A and its non-immunosuppressive derivatives are effective antiviral agents for coronaviruses and many other viruses. They normally bind to cellular cyclophilins, thus inactivating the cis-trans peptidyl-prolyl isomerase activities of the latter. Cyclosporin A (but not cyclosporin B) binds to and blocks the interaction between the N-proteins of various CoVs and cyclophilin to prevent viral RNA replication. Thus, cyclophilin inhibitors such as cyclosporin A block this protein-protein interaction, inhibit replication, and thus prevent infectivity [143]. Examining several cyclophilin inhibitors revealed that even non-immunosuppressive cyclosporin derivatives can block replication, showing that they could be effective antiviral agents with minimal side-affects [143]. Clearly these compounds might prove effective at blocking diseases such as Covid-19.

In another recent study, Lin et al. [138] examined the structure-based stabilization of N-protein-protein interactions for the purpose of designing antiviral drugs. This unique approach for the discovery of novel drugs was based on the high resolution 3-dimensional structure of the N-terminal domain of the MERS-CoV nucleocapsid protein (N-NTD).

Non-native interacting interfaces of the dimeric N-protein surface proved to form a conserved hydrophobic cavity that could be used for targeted drug screening. The authors evaluated the complementary surface as a potential binding pocket for drugs and identified 5-benzyloxygramine as an ortho-steric stabilizer that exhibits both antiviral and N-NTD protein-stabilizing activities. X-ray analyses revealed that 5-benzyloxygramine stabilizes the N-NTD dimer through hydrophobic interactions between the protein and the compound. This causes abnormal oligomerization of the protein. Thus, novel approaches can be used to identify potential drugs that can be used to fight viral infections [138].

In case the antiviral approaches discussed above do not prove successful in combating the current or any future pandemic, there may be a need for novel antiviral approaches that can target emerging viruses, particularly when no effective vaccine or pharmaceutical is available, as is currently the case for Covid-19. Abbott et al. [144] showed that a CRISPR-Cas13-based strategy, which they called PAC-MAN (prophylactic antiviral CRISPR in human cells), can be used for viral inhibition by effectively degrading viral RNA in intact cells. The approach was tried against SARS-CoV-1 and live influenza A virus in human lung epithelial cells. CRISPR RNAs targeted conserved regions of the target proteins and proved to reduce viral load. The authors concluded that a set of only six CRISPR RNAs could target more than 90% of all coronaviruses, thus being potentially applicable to diseases caused by both human and animal coronaviruses. This technique could be developed for safe and effective delivery into the respiratory tracts of intact animals [144].

8.2. The Envelope (E) Protein: Viral Assembly

Among the essential conserved transmembrane proteins in the Coronaviridae family, the Envelope (E) proteins are multifunctional viroporins. The genomes of CoVs may encode up to 2 additional viroporins, 3a and 8a, making CoVs among the most viroporin-rich RNA viruses. E proteins are 74–109 aas long [145] with multiple domains and cellular associations. The N-terminal end is a short hydrophilic region followed by a hydrophobic region containing the α-helical trans-membrane-spanning segment (TMS). Following this TMS is a C-terminal hydrophilic region. E proteins contain an unusually short, palindromic transmembrane helical hairpin around a pseudo-center of symmetry, a structural feature which seems to be unique to CoVs [146]. The hairpin deforms lipid bilayers by way of increasing their curvature, providing a molecular explanation for E protein's pivotal role in viral budding [147]. Depending on the CoV, E protein may be glycosylated [148], palmitoylated [149] and ubiquinated [150]. These conclusions have been extensively confirmed [151–153], although unfortunately, a high-resolution X-ray or cryoEM structure is not yet available. Deletions in various parts of the E protein throughout its length produces an attenuated virus.

Expression of the E and M proteins together in transfected cells is sufficient for VLP formation in MHV, TGEV, BCoV, IBV and SARS-CoV-1 [149]. In some CoVs, their expression is not essential to produce intracellular particles; however, their loss may result in a severe reduction of the number of released virions from the Golgi as for SARS, MERS and MHV. In many CoVs, deletion of or mutations within the E protein gene attenuates the virus both in vivo and in vitro and reduces the progression of disease and mortality in animal models. For SARS, ΔE mutants are able to replicate viable particles albeit at a lower efficiency [154]. HCoV-OC43 ΔE mutants have a dramatic deficiency in viral replication but are still able to replicate at a much lower efficiency than wildtype in CNS tissue culture and in mice with decreased pathogenicity [155]. TGEV ΔE and MERS ΔE mutants completely lose their ability to bud from host cells, making intracellular virions that are unable to infect new cells [156,157]. Deleting E from IBV results in lethality for the virus [24].

The ion channel activity of an E protein is a major contributor to the hallmark inflammatory response [158], leading to the cytokine storm and acute respiratory distress syndrome (ARDS) associated with respiratory CoV infections [159]. The two other recognized ion channels, 3a and 8a, have also been shown to illicit inflammatory stress in a similar manner; however, targeted changes to E have the strongest attenuation of viral infections in SARS [160] and MERS [157]. Some CoVs, such as γ-CoV IBV do not have other viroporins [24], and studies on IBV E have shed some light on the various responsibilities of E proteins during infection without the noise of other accessory viroporins.

E proteins are largely interchangeable between β- and γ-CoVs, but not between E proteins of these CoVs and the α-CoVs [161]. Structurally, β- and γ-CoV Es are more similar to each other than they are to α-CoVs E proteins, containing predicted β-hairpin structural motifs in their C-terminal cytoplasmic facing tails (see preceding paragraph), responsible for localizing the protein to the membrane. In all CoVs, E localizes to the ER/ERGIC/Golgi perinuclear membranes, consistent with the CoV-induced reticulovesicular network. Specifically, the cytoplasmic tail of IBV-E targets and binds to the golgi tag GM130 and trans golgi tag p115, while SARS E localizes with GM130, ERGIC tag ERGIC53 and trans-Golgi tag p230 [162,163]. Surprisingly, neither the TMD nor the β-hairpin motif is necessary for Golgi localization as shown with truncated SARS E, suggesting the presence of a second Golgi localization tag within the N-terminal region of E. A truncated SARS-E, containing its N-terminus attached to the C-terminus of the VSV G protein still localized with GM130, ERGIC53 and p230 [163].

All CoVs have a well conserved proline residue in the β-strands of E tails. Mutating Pro54 to alanine (P54A) in IBV E disrupts its localization with the Golgi [163]. While the E protein is produced in abundance during infection, very few copies are actually incorporated into mature virus particles [148,158]. Despite this fact, E deletion mutants of SARS, MERS, MHV, TGEV and IBV produced weakened viruses that could either not escape cells (MERS, IBV, TGEV) [24,156,157], or had difficulty budding (SARS, MHV) [154,164]. For SARS, deleting the E protein leads to 100–1000-fold lower viral titers in lungs and nasal turbinates of infected hamsters [154] and lower NFκB (a major immune transcription factor) activation [158]. SARS with both 3a and E proteins deleted were non-viable but were rescuable if either 3a or E was reintroduced [160], thus suggesting marginal flexibility and exchangeability in viroporin roles.

Like many of the other CoV proteins, E may perform multiple roles during the infection process. Since it is only incorporated into virions in small numbers, it has been proposed that E helps scaffold newly forming virions, adding to their structural integrity. EM scans of intracellular virions revealed no change in SARS morphology upon deletion of the E gene, but upon viral purification, many of the ΔE mutant viruses had aberrant or misshaped morphologies, suggesting that an E deficiency makes CoV particles susceptible to shearing forces [154]. In all E deleted CoVs, smaller in vitro plaques are observed, and viral titers are reduced [148].

When purifying IBV E protein, two distinct molecular weight pools for the protein were extracted, suggesting oligomerization properties [165]. The lower molecular weight pool was predicted to consist of monomers and/or homodimers, while the higher molecular weight pool was predicted to consist of homopentamers, or possibly, hetero-oligomers associated with host proteins [165]. Wild type IBV E tended to favor the higher molecular weight pool, suggesting that a majority of the resulting conformations were homopentameric. Disrupting the hydrophobic domain composing the TMD of IBV E produced no high molecular weight pool, implying that the pentamers are formed through α-helical interactions [165]. Homopenatmers of E had already been suggested from earlier studies [166]. Indeed, it was predicted that the E protein requires homopentamerization for ion channel (IC) activity, where the amphipathic/hydrophobic α-helical TMSs form a continuous channel just large enough (4–5 A for SARS) to fit a dehydrated cation (H^+, Na^+, K^+, or Ca^{2+}) through it [167,168].

β-CoV Es tend to be selective for Na^+ over K^+ ions, but conflicting evidence suggests that SARS E is slightly selective for K^+ and Ca^{2+} ions and is dependent on the slight negative charge of ER membranes [168,169]. However, because the ER and Golgi are large Ca^{2+} stores in cells, it is more probable that E functions as a Ca^{2+} efflux channel, while minimal amounts of Na^+ and K^+ are imported into the ER/Golgi lumen [168]. For SARS, the predicted residues responsible for conferring IC activity to E are N15 and

V25 [167]. Solution NMR analyses in dodecyl-phosphatidylcholine micelles revealed dynamic conformational changes in the homopentamer that could accompany cation translocation [164].

The SARS E N15 residue and its polar equivalents in other CoVs may provide a cation selectivity filter, while V25 and V28 form a 2.0–2.3 Å constriction, predicted to correspond to the closed state of the IC [167]. In fact, creating a recombinant SARS E virus with a mutated residue at position 15 (N15A), but not at position 25, consistently eliminated IC activity and reduced pathogenicity in mice [170]. V25F mutants reverted back to the IC^{+} phenotype either by directly mutating F25 to C, or by mutating neighboring residues: L19A, F20L, F26L, L27S, T30I, and L37R 2 dpi in mice. Modifying the equivalent IC residues (T16A or A26F) in IBV E in vitro gave similar results, with E-A26F unable to form VLPs in vitro. However, no reversion mutations were recorded [24]. The reversion of V25F to various other residues just after 2 dpi in mice revealed an obvious danger in generating attenuated point mutant viral vaccines. However, analysis of the N15A mutant, which did not mutate back within the time interval of the study, suggests that mutations in the predicted cation selectivity filter are more lethal and specific than the structural V25/V28 residues. Perhaps leaving the filter intact causes the pore to retain selectivity, and opening the channel, obstructed by V25F, is easier than reverting the N15A filter residue.

The drug, hexamethylene amiloride (HMA), shown to abolish IC activity in MHV E and HCoV-299 E, also inhibits the IC activity of SARS E, likely by associating with the N15 residue and the equivalent residues in other CoVs [167]. Further investigations into SARS N15A mutants revealed 80–100% survivability rates in mice, despite similar disease progression during the first 2 dpi [170]. Overall, lung autopsies of infected mice revealed less swollen alveolar walls and airways free of pulmonary edema, as opposed to the typical Acute Respiratory Distress Syndrome (ARDS) phenotype induced by SARS [170]. Furthermore, neutrophil recruitment was lower in N15A mice due to the reduced amount of secreted IL-1β, TNF and IL-6 proinflammatory cytokines. IC activity also promotes the fitness and release of IBV viral particles [171,172]. HMA treated MHV or HCoV-299 infected cell lines exhibited much smaller plaques as opposed to the HMA-free infected cells with plaques roughly 3–4 mm in diameter [173].

In addition to inducing pro-inflammatory responses, the IC activity of E may confer major modifications to secretory or apoptotic pathways. IBV infections are associated with p53-independent, caspase-dependent, CHOP transcription factor and IRE sensor-mediated unfolded protein response (UPR) pathway-regulated apoptosis. This pathway is stimulated by ER stress, marked by the cleavage of downstream poly ADP-ribose polymerase (PARP) [24]. Similarly, SARS induces apoptosis in cell cultures via protein kinase R (PKR) [174], caspase-3-mediated ER stress, JNK-dependent pathways [175], and PERK and eIF2α-mediated UPR activation [176]. However, CoVs prefer nonapoptotic budding of virions and regulate the apoptotic pathways, likely through E protein IC activity to optimize virus release. Specifically, CoV-induced apoptosis is related to ER stress, induced by the viral replication in the ER-derived RVN and the expression of unfolded, unprocessed accessory and structural proteins [24,177]. In SARS, S induced the greatest ER stress [176], although the E IC activity may also induce stress to a lesser degree late in infection [24,178].

E IC deficiency in CoVs, induced by mutations or drugs, leads to smaller plaques in in vitro tissue cultures and reduced pathogenicity in vivo. When eliminating E IC activity by generating recombinant IBV E-T16A or E-A26F mutants, levels of cleaved PARP and pro-inflammatory mRNAs for IL-6 and IL-8 were reduced [24]. SARS E protein alone reduced ER stress of vero-E6 and MA-104 cells when the stress was induced externally by adding either respiratory syncytial virus or an ER stress inducing drug, tunicamycin or thapsigargin. In comparison to wildtype, SARS-CoV-ΔE underwent higher rates of apoptosis and increased the expression of double specificity phosphatases, DUSP-1 and DUSP-10, despite expressing lower levels of proinflammatory chemokines CXCL2 and CCL2 [179]. DUSP-1 and DUSP-10 negatively regulate mitogen-activated protein kinase (MAPK) signaling, reduced the viral induced inflammatory response, and reduced the synthesis and secretion of TNF, IL-6, CCL2/MCP-1, CCL3, CCL4 and CXCL2/MIP-2 [179]. Thus, deleting E in SARS or IBV leads to a weakened proinflammatory response while also attenuating virus production and infectivity. Possibly, E allows host tolerance to the virus during early-mid infection, preventing apoptosis for a long enough period to allow production of more viral particles through budding.

Meanwhile, wild type SARS had reduced expression of ER stress induced GRP78, GRP94 and MHCI antigen-presenting facilitator HSPs on the surfaces of infected cells compared to the mutant [179], and as noted above, SARS-CoV-ΔE deletion mutants induce apoptosis in infected cells at a greater rate than their wild type counterpart [179]. By contrast, late infection IBV IC activity may induce apoptosis by destabilizing the ion gradients between the Golgi lumen and the cytosol [24]. PEDV E protein was also reported to induce ER stress through the UPR, but the results were attained through a transfected plasmid encoding only PEDV E [178]. The IC activity of E protein [168] as well as those of the other accessory viroporins, 3a and 8a [160], activate the NLRP3 inflammasome by effluxing Ca^{2+} from the lumen of the ER/ERGIC/Golgi, altering the homeostatic levels of cytosolic Ca^{2+} [168,180] and resulting in upregulation and secretion of pro-inflammatory TNF-α, IL-1β, IL-6 and IL-18 [160,181]. ER stress [182] and ROS production [183] are also activators of the NLRP3 inflammasome, and due to E protein's regulation of ER stress, they may also activate NLRP3 through an alternative mechanism.

Release of assembled CoV virions requires secretory pathways during early-mid infection, and lysosomal pathways for egress in late infection [62,182,184]. While secretory pathways are necessary for production of CoV structural proteins and processing, the Arl8b-dependent lysosome exocytosis pathway has recently been shown to be the exit pathway of mature MHV and SARS-CoV-2 virions [184]. Hence, E protein's role in viral release may be most important during assembly, before egress of mature virions to the cellular membrane. It is possible that E protein modifies secretory pathways to facilitate the release of intracellular particles since E mutants of several CoVs have difficulty leaving the cell. IBV infected cells, or expression of E alone, induces the neutralization of the Golgi pH, suggesting a role for E in altering secretion [185]. Replacing the hydrophobic domain in IBV E with VSV Glycoprotein HD led to a decrease in viral shedding, increase in damaged particles and accumulation of prematurely cleaved S protein [185], all suggesting a protective role for E in the maturation of other structural proteins. On the other hand, merely replacing the residues in the HD responsible for IC activity in IBV did not affect glycosylation or proteolytic processing of S [24]. Hence, the HD domain in its entirety, but not IC activity of E alone, may contribute to some of the purported functions of E. This suggestion is further supported by the fact that monomers were more strongly correlated with IBV-induced secretory modifications than pentamers although IC activity supports virion assembly [165]. Thus, additional conformations of E may equip CoVs with multifunctional molecular tools.

E proteins also contain a C-terminal class-II hydrophobic PDZ binding motif (PBM) [158] that anchors them to lipid membranes and participates in the relocalization of syntenin-1, a multifunctional adaptor protein that is involved in trafficking of membrane proteins to perinuclear regions [186], thus, activating p38 MAP kinase-mediated inflammation [160]. This PBM has also been found to interact with the host PALS1 protein, an epithelial cell polarization protein [187], disrupting the tight junctions between epithelial cells. It may perform a role in non-apoptotic virus release through a cell to cell exit mechanism [188]. Deletion of the PBM in SARS, either by truncating the E protein at the C-terminus or by replacing the residues within the PBM to produce a mutant E of the same length, did not affect viral replication efficiency in vero-E6 and DBT-mACE2 cells [189]. However, mice infected with virus possessing E, but lacking the PBM, did show a decrease in expression of inflammatory cytokines and active p38 MAPK in their lungs, reducing the pathogenic response and mortality [160,189]. Additionally, SARS transfected vero-E6 cell lines in which the full length PBM of E was disrupted, and deleting 3a (Δ3a, E-PBM$^-$), or its inverse (3a-PBM$^-$, ΔE) resulted in an infectious virus [160]. Introducing a stop codon to truncate the E protein missing the PBM (3a, E-ΔPBM) reverted back to wildtype [160], showing an ability of different CoV PBMs to substitute for each other. These studies revealed the essentiality of PBMs in SARS-CoV-1 infections. HCoV-OC43 E PBM greatly improves propagation in human and mouse neuronal cells and infectivity in the brain and spinal cords of mice, and its removal attenuates the virus [155].

Other mutations introduced into the E protein also promoted attenuation of the virus, implying that other important regions of the E protein contribute to viral pathogenicity. Deleting regions along the hydrophilic C-terminus of SARS E led to reduced pathogenicity, although deletions at the very end of the C-terminal tail had no such effect in a mouse-adapted model [190]. Clearance of viral infections, typical of CoV survivors, is associated with elevated levels of T cell production [191]. Infection by SARS is in part attributed to reduced numbers of T cells, primarily CD4+ T cells, leading to the host's inability to clear the infection [190,191]. Attenuating the SARS E protein by deleting these regions leads to less lung tissue damage and higher T cell counts, likely disrupting the β-hairpin Golgi localization motif or the PBM [163]. In MHV, replacing clustered charged residues within the C-terminal end of the E protein with alanine (E-K63A/K67A or E-D60A/R61A) resulted in thermally unstable virus particles with much smaller plaque morphologies [192].

8.3. The Spike (S) Protein: The Primary Receptor and Membrane Fusion Mediator

Spike (S) proteins of coronaviruses are the receptor binding glycoproteins and class I fusion proteins of CoVs. S proteins are large [1162–1376 aas], are synthesized in the ER/ERGIC, and are post translationally modified in the Golgi, undergoing proteolysis and extensive O- and N-linked glycosylation as well as palmitoylation. SARS S proteins are relatively unique among CoVs, sharing little sequence similarity with their relatives despite strongly conserved structures and functions [193]. SARS-CoV-2 S is similar to its 'predecessor', SARS-CoV-1 S, with a 76% aa identity with SARS-Urbani S and 80% identity with bat SARS-CoV ZXC21 S and ZC45 S [194], and 98% identity with bat RaTG13 [32,195], conserving several N-linked glycosylation sites [194]. During synthesis, the protein may be cleaved into the S1 (head and receptor binding) and S2 (membrane embedded stalk and fusion) subunits by either host or viral proteases [196], or it can be left as a full-length S protein, requiring cleavage at S1/S2 upon receptor binding [197]. If cleaved, these subunits then remain noncovalently bound to each other [193]. The S1 subunit can be further divided into the N-terminal domain (NTD) and C terminal domain (CTD), both of which participate in receptor binding [198]. The S1/S2 structure then trimerizes with 2 other S1/S2 molecules to form the complete S protein [196,198].

There are multiple important domains among the S1 and S2 subunits that contribute to the binding and fusion functionalities of the protein. The S1 subunit (N-terminal residues 14–685) for SARS-CoV-2) [199] contains the receptor binding domain (RBD) that associates with the host receptor (DPP4 for MERS or hACE2 for SARS). Comprising the RBD is either the S^A domain of CoV-HKU1, HCoV-OC43 and MERS or the S^B domain of SARS-CoV-1 and SARS-CoV-2, which interacts with the host receptors [194]. S^B directly binds to ACE2 to allow viral entry of target cells [194]. The S^B domain operates like a lock and key, existing in an open or closed conformation which possibly induces differential folding at the S1/S2 junction [194,200]. The open and closed conformations of S^B are transient states, stochastically revealing and sheltering the RBD [195]. For SARS-CoV-2, the closed conformation is indicated by the RBD bound in trans in a pocket provided by the NTD and RBD of the neighboring S1 monomer [201]. SARS-CoV-2 S has been reported to have an even higher binding affinity [202], about 20-fold higher than SARS-CoV-1 S, for the ACE2 receptor [195]. SARS-CoV-1 S also binds to the C-type lectin DC-SIGN (dendritic cell specific intercellular adhesion molecule grabbing nonintegrin) as well as DC-SIGNR of dendritic cells without engaging the fusion complex [203]. Since dendritic cells migrate to lymphatic tissues, SARS may utilize dendritic cells as 'ferries', traversing blood and lymphatic vessels to new $ACE2^+$ tissues, leading to systemic infections [203].

Upon binding to ACE2, the S^B domain goes into the open configuration, releasing constraints at the S1/S2 site [194]. SARS-CoV-2 is much more susceptible to fusion activation than is SARS-CoV-1, indicating the presence of an additional furin cleavage site, confirmed to exist between S1 and S2 (residues 677–687) [194]. Depending on the mode of entry and the CoV strain, the furin cleavage site can be cleaved by host furin, transmembrane protease serine protease-2 (TMPRSS-2), TMPRSS-4, trypsin, lysosomal cathepsins or airway trypsin like protease (HAT), priming the class I fusion complex during synthesis, either at the cell surface or within an endosome [32]. The additional furin cleavage site in SARS-CoV-2 S may expand its tropism or propensity to fuse with host cells. Many SARS-CoV-1 or SARS-CoV-2 pseudo-virions/virions contain pre-cleaved S1/S2, indicating that cleavage can occur during S synthesis [197,199].

Fusion is likely pH-independent [204] but may be regulated by endosomal maturation and Ca^{2+} [51,52,205]. After binding to the receptor and cleavage of S1/S2, a final cleavage must occur at the S2′ site (residue R797 in SARS-CoV-1) [206]. Once cleaved, steric bulk is released from two amphipathic α-helical, 4-3 coiled-coil heptad repeats, HR1 (residues 910–988 in SARS-CoV-2) and HR2 (1162–1206 in SARS-CoV-2), in the S2 subunit, releasing stiffness from this joint. X-ray crystallography revealed that the post fusion conformation of these HRs is characterized by a hip-knee-ankle style folding, where three HR2 helices collapse onto the hydrophobic grooves in an antiparallel manner of the central coiled-coil of the HR1 helices [194,207]. This folding reduces the distance between the viral envelope and the host surface/endosomal membrane, allowing insertion of a fusion peptides (FP) into the host membrane before membrane fusion [206].

Several efforts have been made to identify the FPs and regions in S2 that contribute to membrane fusion. Three regions in SARS S2, termed R1, R2 and R3, were found to have membrane-associating properties. R1 (858–886) is upstream of HR1, R2 (1077–1092) is situated between HR1 and HR2, and R3 (1190–1202) is proximal to the TM portion of S2. Of these regions, mutations in R1 led to a decrease in syncytia formation [206]. Upstream and overlapping with R1 are regions discovered through Wimley and White interfacial hydrophobicity analysis. They are called WW-I (770–778) and WW-II (864–886) and are strongly associated with membranes [206]. An exposed FP (FP1 798–818) is likely to occur in the WW-I N-terminal side of HR1, containing Ca^{2+} salt bridge-forming residues, D830 and L831. Immediately following FP1 is FP2 (816–835) which contains two disulfide bridge-forming cysteine residues C822 and C833. FP1 is highly conserved among CoVs with a crucial invariant LLF motif that when mutated to alanines causes defective fusion [208]. Additional putative FPs are the Alt-FP (770–788), overlapping with the WW-I, downstream of the S1/S2 cleavage site and internal FP (IFP) (873–888), coinciding with R1 [209]. Additionally, the region upstream of the TMS (1185–1202) seems to have membrane association properties [209]. ESR analyses conducted on each of these lone peptide segments, exposing multilamellar vesicles (MLVs) to these putative FP segments, induced an increase of membrane ordering, an indicator of viral fusion peptide activity. In all FPs, a requirement for Ca^{2+} was observed, and activity occurred from pH 5 to pH 7, consistent with the ability of S to fuse membranes at neutral pH values. FP1 and FP2 likely work in concert with each other, embedding themselves into membranes, forming disulfide and Ca^{2+} salt bridges to stabilize the fusion complex [206,209]. Fusion activity of FP2 was undetectable when Ca^{2+} was not present or if the complex was treated with disulfide bond reducing dithiothreitol [209]. Understanding their conformations may provide additional drug targets for the inhibition of SARS entry.

S protein also induces cell to cell fusion with production of syncytia with tissue damage [210]. New intracellular CoV particles can exit a cell and enter directly into adjacent cells of epithelial tissues, leading to disruption of cellular barriers and production of multinucleated cells. SARS-CoV-1, MHV, IBV, MERS and SARS-CoV-2 have all been reported to induce formation of syncytia in vitro in cell lines, likely due to S protein's ability to engage its fusion complex at neutral pH [32,211–214]. However, SARS-CoV-2 has an unprecedented capacity to form syncytia, producing multinucleated cells with hundreds of nuclei per 293T or Huh-7 cell [199]. Host proteases on membrane surfaces may be required for immediate fusion and entry of the virus into neighboring cells. TMPRSS-2, present on the opposite side of the membrane as S, seems to be required for syncytium formation in SARS-CoV-1 and SARS-CoV-2-infected veroE6 cells [214,215]. Lone SARS S, expressed via a cDNA in one of several cell lines, expressed S on the surface of transfected cells, and for some CoVs, syncytia have been reported to be formed by S alone [212]. These results suggest that SARS S binds to ACE2 and is readily cleaved by membrane bound proteases [215]. An accumulation of secreted S protein on the surface of cells may induce fusion between neighboring cell membranes.

Further supporting the requirement of a membrane protease to generate syncytia is work conducted with MHV-2, a strain of MHV with S that can only be cleaved by cathepsins and cannot generate syncytia [211]. TMPRSS-2 was identified as a potent entry factor for SARS-CoV-2 in nasal epithelial cells [216], and TMPRSS-2 is abundantly distributed in respiratory epithelia [217]. Hence, developmental therapeutics, suppressing the interaction of SARS with TMPRSS-2, may reduce viral replication in tissue. Additionally, mutating all 9 palmitoylated residues in S, which does not disrupt folding, trafficking or core functions, does disrupt syncytium formation [218]. The increased capacity of SARS-CoV-2 to form syncytia could be due to the additional furin cleavage site and the increased occurrence of pre-primed S protein, so that only the S2′ cleavage site is a prerequisite for fusion. Recently, it was discovered that pan-coronavirus fusion inhibitor, EK1 peptide variant EK1C4, could inhibit SARS-CoV-2 fusion in a dose-dependent manner by binding to HR1, but the exact mechanism was not revealed [199].

Tropism for ACE2 may be mediated by cholesterol. SARS and other enveloped viruses such as HIV have been reported to be dependent on lipid rafts for entry [219]. ACE2 associates with detergent-resistant cholesterol-rich microdomains in membranes, but treating cells with methyl-β-cyclodextrins (MβCDs) did not affect expression of ACE2 [220]. Cholesterol depletion in several cell lines treated with MβCD inhibited binding of S to ACE2 and SARS entry [220,221]. ACE2 receptor binding may be promoted by S palmitoylation, since palmitoylation was found to promote S association with lipid rafts and detergent-resistant membranes [218]. However, due to the virus' ability to infect new cells through cell-to-cell fusion, depletion of cholesterol and lipid rafts could not completely suppress viral replication [219]. Thus, lipid rafts are promoters of viral entry, but not necessarily of viral replication within tissues.

It was noted earlier that S can bind to DC-SIGN, a C-type lectin on the surfaces of dendritic cells. However, SARS has also been known to infect monocytes through both ACE2-independent and ACE2-dependent mechanisms in the lungs of SARS patients [222,223]. Infection of T cells by SARS-CoV-2 has also been reported [224]. This may contribute to the severe inflammation and depletion of T cells. A preprint study detailed tropism for white blood cells which may be mediated through CD147. However, further evidence is required for confirmation [225]. While abundant IFN-γ, present in macrophages, may suppress viral replication [222], SARS has multiple IFN-suppressing strategies that allow it to evade and silence innate antiviral activity in monocytes [223]. For instance, the virus can avoid detection from intracellular pattern recognition receptors (PRRs) such as MDA5 and RIG-I. These PRRs, which illicit a specific antiviral response upon detection of a virus, are either never activated or silenced by suppression of IRF-3 [223]. Rather than an antiviral response, cytokine secretions may be dominated by nonspecific inflammatory mediators that contribute to SARS-CoV-1 and SARS-CoV-2-associated diseases [223].

S protein may activate complement in early infections, which could explain the early onslaught of cytokines circulating in SARS patients. IgM, IgG, mannose binding lectin (MBL) or an alternative pathway may allow recognition of and binding to S, thereby activating complement. Subsequent activation of complement downstream pathways results in a flush of proinflammatory cytokines, possibly leading to a cytokine storm. Since complement can be activated directly by the presence of antigen, S protruding from viral particles may be a major contributor in the development of disease. Since S is abundantly glycosylated with N-linked mannosyl oligosaccharides, sequestering of SARS could occur early by binding MBL to S. MBL has been shown to bind to S in SARS-CoV-1 bearing pseudo-viruses, specifically at an N-linked oligomannosyl glycosylation site in the RBD. This critical localization within the RBD prevents S from binding to DC-SIGN, but not to ACE2 [226]. MBL was also found to bind to SARS-CoV-1 infected FRhK-4 cells and immobilize actual SARS-CoV-1 particles, inhibiting their infectivity [227]. Thus, low serum levels of MBL may be a susceptibility factor for the acquisition of SARS [227]. Despite these findings, it remains unclear what role MBL and C3b may have in activating complement early in the infection, even though complement activation has been confirmed in SARS-CoV-1, MERS and SARS-CoV-2 [228,229].

8.4. *The Membrane Matrix (M) Protein, the Virion Scaffold*

The homologous M proteins of CoVs and many other envelope viruses have been called the membrane proteins, the matrix proteins, the M proteins, or simply "M". M is the most abundant protein in any one coronavirus virion, and it is among the most conserved and constrained of all the viral structural proteins [230]. This may be attributed to its many functions in the viral infection cycle as well as in interferon antagonism (see Table 3) [230]. All of the major structural proteins of these viruses are derivatized and/or hydrolyzed at specific positions by post translational modification (PTM) reactions (see Fung & Liu, 2018 [231] for a review). These derivatization reactions involve (1) protease-mediated hydrolysis by both virus- and host cell-encoded proteases, (2) either O (serine or threonine)- or N (asparagine)-glycosylation, and often both, (3) palmitoylation of the spike (S) and envelope (E) proteins, (4) protein phosphorylation by ATP-dependent protein kinases, and (5) ADP-ribosylation of the nucleocapsid (N)-protein. Other PTMs of nonstructural "accessory" proteins have also been documented [231].

Table 3. Potential Functions of Coronavirus Matrix (M) proteins.

	Association with Itself and All Other Structural Proteins to Assemble Virions
1.	Interference with the host immunological response by interferon (IFN) antagonism
2.	Involvement of M in host cell cycle arrest
3.	Induction of endoplasmic reticulum (ER) stress and the unfolded protein response (UPR)
4.	Coronavirus-induced autophagy and abortive apoptosis
5.	Functioning of M as a protective antigen
6.	Viroporin activities: The E and 3a proteins, and the ability of M to substitute for E

The most complicated of these PTM reactions is glycosylation. M proteins of SARS-CoVs are O-glycosylated (on seryl and/or threonyl residues), and not N-glycosylated (on asparaginyl residues) [232,233]. The structures of the O-linkages are known and include O-linked N-acetylgalactosamine to which galactosyl and sialyl residues are glycosidically linked [234]. O-Glycosylation occurs in the Golgi and has been used as a marker for proper M protein intracellular trafficking, membrane insertion and maturation [235]. It seems that glycosylation is non-essential for assembly of some CoV virions, but it greatly facilitates the formation of active virus particles, and it also regulates interferon production (see below). The established or probable functions of M proteins are presented in Table 3.

The primary function of M is assembly of newly formed viral particles. As noted above, it is the most prevalent protein component of the virion. It provides a homo-dimeric scaffold for virion assembly and has affinity not only for itself, but also for all of the constituent structural proteins found in the virion. Thus, to provide the "master assembly function", it has both homotypic and heterotypic associative properties. In one study, based on cryoEM, tomography and statistical analyses, Neuman et al., 2011 [236] suggested that M can assume two distinct conformations. One, they suggested, is elongated, being associated with rigidity, spike clustering and a narrow range of membrane curvature, while the other is more compact and is associated with greater flexibility and a lower spike density. Presumably, the proper ratio of these two forms determines the final virion construction. As noted above, M associates with itself to form dimers, but also with the nucleocapsid (N)-protein, the spike (S)-protein and the envelope (E)-protein as well as the genomic RNA. Thus, with M as the 'glue', holding the complex together, these primary constituents of the viral particle determine virion size and shape with M playing the dominant role.

A subsequent study led to the suggestion that initial self-assembly and ultimate release of the membrane-enveloped vesicle/particle (the virion) depends most importantly on the association of M with N and the viral RNA [237]. Assembly seems to be a multi-step process as illustrated in Figure 2: First, M self-associates, creating an M-protein homodimer, and this self-association process involves several distinct regions of M, explaining why this occurs with high affinity. Its heterotrophic interactions, then, may be largely responsible for the order of the protein associations. Because of the TMSs in M, this early intermediate is likely to already be membrane associated. Second, although M is made in the ER, it acts either in the trans-Golgi network, or the ER-Golgi intermediate compartment (ERGIC) during assembly, depending on the specific coronavirus under study, clearly requiring specific host-catalyzed trafficking of M through the endo-membrane network [238]. In this regard, it is important to note that two motifs in the C-terminal domain (DxEER and KxGxYR in the MERS CoV M protein) are ER export and trans-Golgi network retention sequences, respectively [238]. Third, M associates with N, the nucleocapsid protein, again probably via multiple sites in M, although a particularly important sequence for this association is the di-leucine motif in the C-terminal tail of the protein [237]. Based on mutational analyses, the N-terminal exo-domain or the central TMSs appear to be of lesser importance for the association of N with M. However, a central cysteyl residue (C158) also plays a role [237]. The C-terminal domain of N is largely responsible for the association with M [239]. Fourth, the M-N association allows the genomic RNA to become part of the developing particle complex because of the high affinity of N for this nucleic acid. However, the inclusion of the genomic RNA within the complex may occur simultaneously with step 4 because of the high affinity of N for the genomic RNA. Fifth, the endo-membrane-M association allows recruitment of the Spike (S)-protein to the particle. In fact, M has affinity for ALL of the other structural proteins that end up in the virion. Sixth, several M-protein residues seem to be involved in the final secretion and budding processes, and these residues are scattered throughout the protein, probably playing specific roles [237]. Finally, the E-protein, together with M, with which it interacts, plays a significant but less well-defined role in the assembly process [239]. The stage(s) of its involvement in the temporal scheme outlined here are not as well defined as the general scheme itself. As will become apparent, the viroporin functions of E and 3a are assumed to play a role (see below).

In addition to the associations with its own viral proteins, Gordon et al., 2020 [240] cloned, tagged and expressed 26 of the 29 SARS-CoV-2 proteins in human cells and identified the human proteins physically associated with each of the 26 viral proteins using affinity-purification/mass spectrometry (AP-MS). They identified 332 high-confidence SARS-CoV-2-human protein-protein interactions (PPIs). Among these, were 66 human potential drug targets (host factors), and these were targeted by 69 compounds. This work therefore provides a guide for the development of anti-viral drugs that may act against SARS-CoV-2 to block different aspects of the viral infection cycle.

9. Viroporin Activities: The E, 3a and 4a Proteins, and the Ability of Mutated M to Substitute for E

Three distantly related proteins in SARS-CoV-1 and other related coronaviruses display very similar hydropathy plots. The first of these is M (TC# 1.A.117), the second is protein 4a (TC# 1.A.89), and the third is protein 3a (which also can assume other designations, depending on the virus) (TC# 1.A.57). The similarities of their topologies can be viewed in Figure 3 as hydropathy plots. As shown in this figure, variations within each family occur, but they are similar in all three families (Figure 3). Moreover, surprisingly similar hydropathy plots with sequence similarity of borderline significance can be observed. As noted above, viroporin activities have been demonstrated for the E, 4a and 3a proteins as well as a fourth family of apparent viroporins classified under TC family # 1.C.99; however, porin activity has not been demonstrated for M [80,241,242]. Nevertheless, a most interesting study, suggesting a functional relationship between the M and E proteins, was conducted by Kuo and Masters [243]. E was eliminated by deletion of its structural gene in the mouse hepatitis virus. The virus was found to still be infective, but it showed poor assembly with altered virion morphology, and it gave rise to tiny plaques. The authors then selected for "suppressor" mutations with at least partially restored viral growth and virion production, giving rise to much larger plaques. The secondary mutations were found to be in M, and these mutants arose in a sequential process involving M-gene duplication, where one copy retained the native M gene while the second M-gene encoded an altered M protein (M*) with a truncated C-terminus. Both M and M* were incorporated into the virion. It seems that M* served as a surrogate for E, providing a new gene function through recombination. Since E is known to have viroporin activity [244,245], it seems plausible that M* had recovered (at least partly) the viroporin activity of the deleted E protein. Although these authors had a different interpretation of their observations, we suggest that the N-terminal transmembrane domain of M may be capable amino acid substitution that allows it to form transmembrane pores, a suggestion that needs to be confirmed or refuted. In Figure 4, we provide comparative hydropathy plots between the SARS 3a viroporin, HCoV 299E 4a viroporin and M protein.

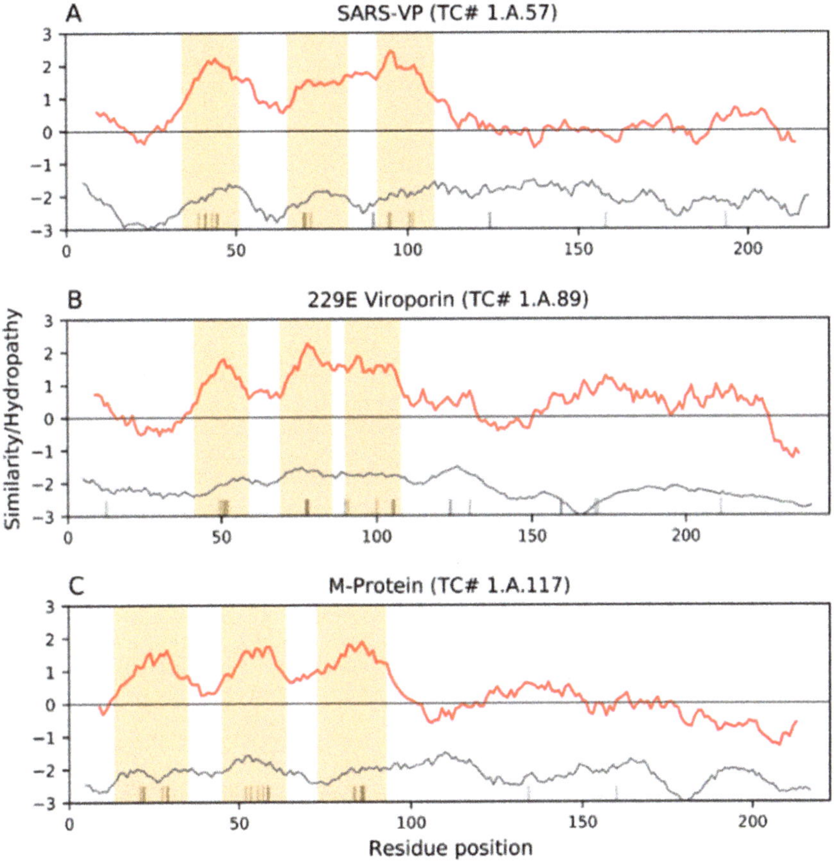

Figure 3. Average hydropathy and similarity within three families. (**A**) Family of SARS-3a cation-selective viroporins. (**B**) Family of HCoV-229E-4a cation-selective viroporins. (**C**). Family of M (matrix)-proteins. Red curves indicate average hydropathy, gray curves indicate average similarity per position, and vertical thin black bars on the x-axis indicate regions predicted to be part of TMSs. Conserved hydrophobic peaks (inferred TMSs) are highlighted with moccasin-colored bars. Proteins within each family were aligned with MAFFT [246] using the L-INS-i algorithm and then edited with trimAL [247] to keep positions with less than 30% gaps. Plots were generated with the program AveHAS [248]. Notice the high topological similarity among the three families, despite their poor sequence similarity.

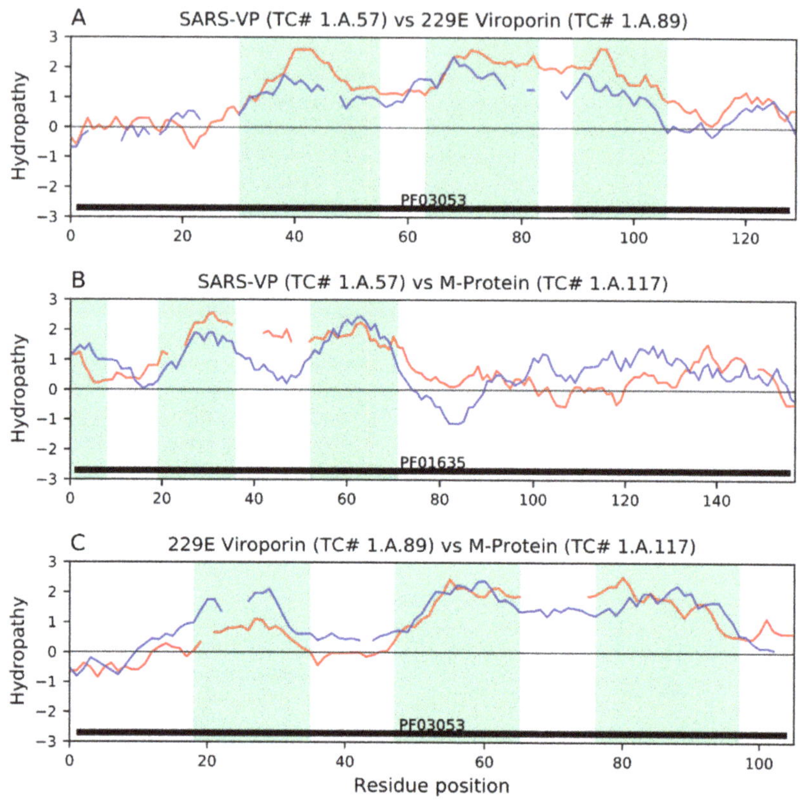

Figure 4. Topological relationships between (A) SARS-VP (3a) viroporins, 229E (4a) viroporins, and (C) M-proteins. Families were compared using our methodological pipeline based on the transitivity property of homology [249,250]. Hydrophobic peaks (inferred TMSs) are highlighted as green bars. Pfam domains were projected with the program GetDomainTopology [250] and drawn as solid black bars above the x-axis. (**A**) Hydropathy plots of representative alignments (E-value: 1.3×10^{-5}) between a SARS-VP (3a) viroporin homolog AWV67041 (red) and a 229E (4a) viroporin homolog ADX59489 (blue). The characteristic Pfam domain of family 229E viroporins (PF03053) was projected to the SARS-VP homolog ADX59489 (E-value: 8.7×10^{-4}). (**B**) Hydropathy plots of the representative alignments (E-value: 6.1×10^{-7}) between a SARS-VP (3a) homolog ADX59475 (red) and an M-protein homolog ARI44791 (blue). The characteristic Pfam domain of the M-protein family (PF01635) was projected to the SARS-VP homolog ADX59475 (E-value: 4.8×10^{-3}). (**C**) Hydropathy plots of the representative alignments (E-value: 1.4×10^{-6}) between a 229E (3a) viroporin homolog ABQ57217 (red) and an M-Protein homolog YP_003858587 (blue). The characteristic Pfam domain of family 229E viroporin (PF03053) was projected to the M-protein homolog YP_003858587 (E-value: 1.4×10^{-3}). Notice how the projected domains cover the entire length of the alignments in panels A-C. Altogether, the compatibility of TMS topologies (Figure 3) and the similarity of sequence characteristics between these three families suggest that they form a superfamily.

10. Post Translational Modifications (PTMs) to Coronaviral Structural Proteins

10.1. Glycosylation

Glycosylation of viral proteins is a common theme among enveloped viruses due to their hydrophobic natures [251]. Glycosylation facilitates proper folding of the nascent polypeptide through recruitment of chaperone proteins and can play a role in cellular trafficking [251]. Cellular glycosylation also plays important roles in homeostasis and receptor signaling [252], which can be hijacked by viruses [251]. Viruses may utilize glycosylation to mimic or complement host proteins for receptor binding and entry, viral assembly/release, and/or immune evasion [251]. Several CoV E proteins and all S proteins contain N-linked and O-linked glycosylation sites. N-linked glycosylation is characterized by covalent en block binding of an N-linked 14-unit glycan precursor [253] onto an asparagine residue located within a recognition sequence Asn-X-Ser/Thr [254], where X is any amino acid except proline [251]. Attachment of the N-linked glycan to the Asn-X-Ser/Thr is performed by the enzyme, oligosaccharyltransferase, followed by further modifications by glycosidases and glycosyltransferases [231]. O-linked glycosylation involves the attachment of an oligosaccharide to the side chain oxygen atom of a serine or threonine residue, initially through the activity of N-acetylgalactosamine (GalNAc) transferase, and this reaction does not require a consensus sequence as does N-linked glycosylation. Rather than occurring en block, simple sugar residues are sequentially added after the initial O-linked addition. N- and O-linked glycosylations can occur simultaneously and do not compete with each other.

10.2. Palmitoylation

Palmitoylation is a common post-translational modification that occurs in the Golgi where palmitoyl transferases add a fatty acid, palmitate, to a cysteine residue via a thioester linkage [255]. Palmitoylation enhances the hydrophobicity of proteins and plays an important role in subcellular trafficking of proteins between membrane compartments [256]. Proteins can be readily palmitoylated, and the modification is typically reversible. Addition and removal of palmitate can play roles in protein and membrane regulation [256]. Palmitoylated proteins can be modified by either a single palmitoyl group, or dually modified with one or more palmitoyl groups and one or more prenyl or myristoyl groups. Palmitoylated proteins can be divided into four types, (1) single palmitoyl modifications, often at the end of a protein, (2) palmitoylation near a transmembrane domain, (3) dual palmitoylation and prenylation and (4) dual palmitoylation and myristylation [255]. Viral proteins may arrange on membranes in accordance to their palmitoylation status, and it affects assembly and localization of viral oligomers [257].

10.3. Ubiquitination

Ubiquitination is the process of adding the peptide, ubiquitin, to a lysine, serine or threonine residue in Ub or Ubl domains of a protein. Proteins can be mono or poly ubiquitinated, and the ubiquitination process is reversible through the action of deubiquitinating enzymes (DUBs). The process is performed by a sequential cascade of ubiquitin-activating enzymes (E1s), ubiquitin-conjugating enzymes (E2s) and ubiquitin ligases (E3s). Ubiquitination can regulate endocytic trafficking, influence inflammation, and target proteins to the proteasome for degradation. Viruses can take advantage of ubiquitination, often to avoid host cell defense mechanisms such as apoptosis, the type I IFN response, and MHC I antigen presentation. Many viruses including CoVs also have deubiquitinating proteins (SARS nsp3), although the biochemical functions of these proteins are not well defined [258].

10.4. A Focus on S Protein PTMs

As noted above, glycosylation of receptor binding viral glycoproteins is common among enveloped viruses. Spike (S) of Coronaviruses, Hemagglutinin (HA)/Neuraminidase (NA) of influenza, and glycoprotein 120 (gp120) of HIV are all receptor binding glycosylated proteins that protrude from the viral envelope to allow association with their corresponding cellular receptors. When S is expressed in the ERGIC, it is likely transferred to the Golgi and is co-translationally N- and O-glycosylated, and it then trimerizes if properly folded. Glycosylation plays a key role in successful folding, trimerization and secretion of the protein.

Due to their sizes, geometries, and extracellular natures, receptor binding glycoproteins are crucial targets in adaptive immune responses. Typically, humoral immunity, provided by serum antibodies secreted by B cells, sequester the released virus, preventing it from infecting new cells by binding to the viral glycoproteins or marking them for ingestion by patrolling phagocytes. Antibodies can be extremely specific and mimic the shape of the viral receptors, attacking the glycoprotein receptor binding domain head. They can also assume specific shapes along the glycoprotein neck and pre-fusion complexes, preventing conformational changes required for membrane fusion preceding entry [259]. Since the S protein of SARS is both a receptor binding (S1) and class I fusion protein (S1/2 and S2), there is potential for multiple antibodies to form and prevent SARS entry. 2003 SARS S-specific human monoclonal antibody CR3022 provides protection against the virus, and recent experiments revealed in vitro potent cross-neutralization against the SARS-CoV-2 S RBD [260]. Donated antibody serum against SARS-CoV-2 from survivors is currently being researched, but at the time of writing, data meant to establish its effectiveness are inconclusive. Regardless, donated plasma must come from healthy blood donors with known medical history having no evidence for past blood-borne infectious diseases [261].

Humoral immunity against 2003 SARS was dominated by IgG antibodies specific to S (residues 669–1255) and N [262]. In non-intensive care unit (ICU) patients, an increase of S-IgG positively correlated with a decrease in C-reactivity, a marker for patient recovery; it was the longest and most secreted antibody [263]. Despite these vulnerabilities, the S protein provides structural defense against potential antibodies through glycosylation. Anti-SARS-CoV-2-IgG antibodies appeared weeks before clearance of infection, suggesting that these antibodies were not neutralizing [264], and that patients must survive the infection long enough for true neutralizing antibodies to develop. It is possible that S glycosylation provides a glycan shield against antibodies, such as those made against the fusion protein of HIV-1 [265]. Glycan shields are characterized by densely clustered oligomannose glycans that extensively interact with each other as well as intricate structures within the protein to shield it from antibodies [266].

Glycan analysis revealed a preference for oligo-mannose type glycans in SARS and MERS as well as α- and δ-CoVs [266,267]. SARS-CoV-1 S contains 22 N-linked glycan sites, while SARS-CoV-2 S has 23, sharing 18 of its glycan sites with its "predecessor" [266]. However, Cryo-EM and further glycan analyses revealed that SARS and MERS mannooligosaccharides are more loosely scattered around the S1 head and S2 subunits and lack the characteristic dense organization of oligo-mannose when it serves as a glycan shield [266,268]. The S2 subunit, which forms a class one fusion complex, is much less susceptible to mutation than S1, due to less selective pressure and its conserved mechanical nature. In fact, bioinformatic analysis of SARS-CoV-2 and SARS-CoV-1 revealed that S2 glycan sites were completely conserved while S1 glycan sites experienced deletions and additions of other residues [266]. It is difficult to say whether the loosely clustered glycans in SARS and MERS provide a glycan shield, but extensive glycosylation of the protein has been confirmed.

Since antibodies preferentially target the S1 subunit [266], it is reasonable that movement of glycans, due to a higher mutational frequency in the S1 subunit, can provide structural differences sufficient to prevent cross immunity between previous and novel versions of the virus as the virus transits host reservoirs over time. Indeed, stripping the SARS S of N-glycans with peptide N-glycosidase-F abolished neutralization of the protein by purified antisera developed against purified virions [269], thus indicating the specificity of anti-S antibodies.

O-linked glycosylation of the SARS-CoV-2 S protein exists on Ser68, Thr323, Ser325, Ser673 and Thr678, with the last 3 being conserved O-glycosylation sites among CoVs [268,270]. Similar to the N-linked mannooligosaccharide shield, O-linked glycans can form glycan shields by forming mucin-like domains [270]. Although only a few O-linked glycan sites have been confirmed in SARS-CoV-2 S, data on this S protein are still preliminary because methodologies for the extraction of the monomeric/trimeric proteins, imaging, and computational predictions can affect results. Thus, the importance of glycosylation of SARS-CoV-2 S, and its role in immune evasion, have yet to be fully elucidated. Since glycans are important for the development of vaccines as antigens and adjuvants [271], the densely glycosylated S protein will likely play a substantial role in the search for vaccines.

Palmitoylation of S protein has both accessory and replicative functions. The S protein is palmitoylated on cytoplasmic cysteine clusters within the endo-domain [272]. Removal of all cytoplasmic palmitoylated cysteine residues does not affect folding, trafficking or association with M protein [273]. Intriguingly, deletion of the two cysteine residues C1234 and C1235 in SARS S prevented its incorporation into VLPs, despite still being able to associate with M [274]. Similar results were found for TGEV [275]. During receptor binding, palmitoylation promotes association with detergent-resistant membrane microdomains associated with ACE2 on cell membranes [218]. This hydrophobic nature of palmitoylated S improves binding with receptors associated with lipid rafts as well as syncytia formation [218]. A preprint study of SARS-CoV-2 S revealed 9 putative palmitoylation sites similar to SARS-CoV-1. SARS-CoV-2 S palmitoylation plays a key role in infecting cells with high cholesterol density in their membranes. During inflammation, certain cytokines induce production of cholesterol on cell surfaces, which subsequently promotes additional infections, thus upregulating further inflammation. Hence, cholesterol may be a determinant of pathogenicity in SARS-CoV-2 [276].

10.5. A Focus on M Protein PTMs

The M proteins of all known CoVs are O- and N-glycosylated which contribute to folding, structure, stability, trafficking and immune responses [277]. Glycosylation sites and hydropathy patterns are remarkably well conserved in CoV M proteins [278], suggesting an importance for function. SARS-CoV-1 M protein is glycosylated on residue N4, although the consequences of this carbohydrate derivative are elusive. Suppression of this N-glycosylation site does not impair its accumulation in the Golgi or the assembly and infectivity of SARS virions [279]. O-glycosylation in MHV and TGEV M proteins was found to induce antiviral cytokine IFN-α, and mutating O-glycan sites to N-glycan sites in MHV induced higher levels of IFNs [278,280]. Strangely, changing the glycosylation state of M to (O^-/N^+) in recombinant MHV improved infectivity in vivo in mice [280]. Despite these pathological observations, other roles for O- or N-glycosylation in M for CoVs remain unknown, as suppressing glycosylation did not hamper recombinant viral production [278].

10.6. A Focus on E Protein PTMs

The role of glycosylation in E protein structure, localization and stability is relatively under-studied [148]. SARS E is reported to have two putative glycan sites, N48 and N66, which may or may not be glycosylated in the fully processed protein [281]. Typically, SARS E has its HD facing the membranes, but whether the C terminal end faces the cytosol while N faces the ER lumen, or both terminal ends face the cytosol, is uncertain [281,282]. However, in at least one minor form, SARS E N66 is glycosylated with the C-terminal tail exposed to the lumen of the ER/Golgi [281]. This additional minor conformation with glycosylated N66 may contribute to alternative dimers and trimers. Since it is minimally glycosylated, it is unfortunately difficult to establish a role for singly glycosylated proteins such as E [251].

E protein has a cluster of 2–3 cysteine residues (SARS-CoV-1 C40, C43 and C44) on the carboxy side of the HD that are all palmitoylated in IBV and SARS [149,283]. While these cysteines may participate in disulfide bridges to form homodimer/trimers [284] and other possible hetero-oligomers, these residues are not important for homopentamer formation or IC activity [283]. Mutating these residues to alanine prevents E from oligomerizing with M, but not with N [149,283]. MHV E palmitoylation on C40, C44 and C47 likely promotes association with membranes, possibly embedding part of its α-helical HD into the membrane while the palmitoylated cysteine cluster stabilizes its association with membrane lipids [257]. This interaction may contribute to the production of viral particles, as mutating the cysteines dramatically reduces the production of VLPs [257]. These findings support the hypothesis that multiple conformations of E play different roles of CoV replication and pathogenesis. In addition, these different conformations and palmitoyl-assisted membrane anchoring may contribute to viral particle structural integrity.

PL^{pro} contains secondary deubiquitinating activity, suggesting a role in host and viral protein modulatory function. Despite this property, the E protein is ubiquitinated. Following the theme of viral protein ubiquitination, E ubiquitination may allow avoidance of host cell defense mechanisms. The N-terminal Ubl1 domain of nsp3 interacts with E, and the complex localizes in the cytoplasm of infected cells [150]. As stated earlier, the nsp3 Ubl1 domain can act as an anchor for other viral proteins such as N. Since nsp3 and E are involved in viral replication, this association could be important for the synthesis of new virions, bringing E close to the RVN. Additionally, nsp3 deubiquitinating PL^{pro} may dynamically alter the ubiquitination state of E, regulating potentially different protein-protein interactions, protection from the proteasome, and sorting of the protein. Cellular ion channels are also known to be regulated by ubiquitination, where misfolded membrane-bound protein is marked for degradation [285]. E has many conformations and oligomerization states, and perhaps, the concentrations of each may be influenced by ubiquitination. While it is not clear what the role of E ubiquitination is, suppressing ubiquitination in MHV interferes with viral RNA synthesis and may inhibit proteasome and viral nsp proteolytic activities [286].

11. Viral Responses to and Interference with Normal Cellular Function

11.1. Interference with Host Immunological Responses by Interferon (IFN) Antagonism

The type-I IFN system is an important first line of defense against viral infections, participating specifically in antiviral responses. IFN is an effective inhibitor of coronavirus replication and is detected in significant amounts in CoV infected animals and cell lines, but its expression is delayed both in vivo and in vitro [287,288]. In fact, SARS-CoV-2 is more susceptible to IFN treatment than its predecessor, despite its strong ability to suppress IFN pathways [289]. Activation of IFN pathways can occur through detection of dsRNAs through cytosolic RIG-I and RIG-I-like receptors (RLRs) such as melanoma differentiation gene 5 (MDA5) [290]. Toll-like receptors (TLRs), TLR3, TLR7 and TLR8, also detect viral single or double stranded RNAs [291]. Upon activation, RIG-I and MDA5 caspase activation recruitment domains (CARD) are modified with ubiquitin [292] and bind with adaptor mitochondrial antiviral signaling (MAVS) protein, also known as IFN-β promoter stimulator 1 (IPS-1), to form the IPS-1 signalosome [293]. The IPS-1 signalosome then interacts with IKK-related kinases, TANK-binding kinase 1 (TBK1) and IκB kinase (IKKε) [293]. Both kinases can phosphorylate interferon regulatory factors 3 and 7 (IRF3/7) [290]. Phosphorylated IRF3 and IRF7 form homo- and heterodimers and translocate to the nucleus to activate expression of IFN-α/β [290]. The IPS-1 signalosome can also recruit IKKα and IKKβ kinases which activate NFκB. NFκB then translocates to the nucleus to activate expression of proinflammatory cytokines, TNFα and IL-1β, and it upregulates type-I IFN expression [294]. Similarly, TLRs recruit TRIF and/or MyD88 which activates IKKε/IKKi kinases which phosphorylate IRF-3 [291]. TLRs can also activate NFκB through MyD88-IRAK-TRAF6 signaling which activates IKKα/IKKβ [291]. Once IFNs are secreted, they behave as autocrine and paracrine factors to stimulate the expression of IFN-stimulated genes (ISGs) through Janus activated kinase (JAK)-signal transducers and STAT signaling pathways [295]. Binding of IFN to IFN receptors on cell surfaces stimulates the JAK-STAT pathway, which utilizes JAK1 and Tyk2 kinases to phosphorylate STAT1 and STAT2 which triggers their dimerization and translocation to the nucleus where they activate ISGs [296]. Downstream effects of ISGs include upregulation of chemokines (including additional IFNs) and chemokine receptors [297], induced resistance to viral replication in cells [298], activation of monocytes/macrophages [299], activation of Natural Killer cells to kill virus-infected cells [300], and regulation of adaptive T and B cell responses [301,302]. The antiviral innate immune responses of animal hosts commonly interfere with essential viral processes such as the formation of replication-associated membrane structures [303]. In response to the anti-viral activities of these host proteins, many viruses combat the interferon-mediated anti-viral activities of the host by a number of mechanisms [304]. Coronaviruses are equipped with a large array of viral proteins that have secondary functions in IFN suppression or evasion including nsps 1, 3, 7, 12, 13, 14, 15, 16 [305–308], structural proteins M [309], N [292] and E, and accessory proteins ORF3b, 4a [310], 4b [311], 5 [292], 6 [306,307] and 9b [292].

11.2. The M Protein

M-proteins have been reported to be potent interferon (IFN) antagonists in MERS and SARS-CoV-1/SARS-CoV-2 [306,308,309]. In these experiments, the genes were individually cloned into plasmids, transfected into cells, and expressed. While all three of these proteins were effective, ORF4a seemed to be the most potent at counteracting the antiviral effects of IFN via the inhibition of IFN-β promoter activity and NF-κB activation as well as the ISRE (interferon-stimulated response element) promoter signaling pathways [309]. These studies were continued with SARS-CoV-1 M protein, showing that M suppresses type I interferon production by impeding the formation of a functional TRAF3-containing complex. This IFN antagonizing activity is mediated by the first TMS (TMS1) at the N-terminus of the protein. Some specificity was surprisingly noted, since the human Coronavirus HKU1 M protein lacked the inhibitory activity observed for the SARS-CoV-1 M. TMS1 of SARS-CoV-1 M targets the protein to the Golgi apparatus, and Golgi localization seems to play a role in its action as an IFN antagonist. Using the MERS-CoV M protein, the authors suggested that TMS1 prevents the interaction of TRAF3 with its downstream effectors [312,313], confirming its ability to help evade the host innate antiviral response by suppressing type I IFN expression in response to various agents and RNAs. They reported that M interacted with TRAF3, blocking the TRAF3-TBK1 association, which in turn reduced activation of the INF regulatory factor 3 (IRF3). Liu et al. also found that the N-terminal hydrophobic TMS, but not the C-terminal hydrophilic region of M, was important for the response, confirming the results of Siu et al. [312]. The M-mediated interferon antagonism noted here seems to be a common characteristic of a large number of (but definitely not all) viruses from different viral classifications [314,315].

Surprisingly, M may also be able to promote IFN-β induction via a Toll-like Receptor (TLR)-related, TRAF3-independent mechanism [316]. In this case, M itself (rather than its mRNA) seemed to function as the cytosolic pathogen-associated molecular pattern (PAMP) to stimulate type I interferon production. In fact, both NF-κB and TBK1-IRF3 signaling cascades were reported to be activated by M-gene products. Activation of IFN-β production seemed to be generated from within the cell, and the wild type M-protein induced production of both IFN-β and NFκB through a TLR-related signaling cascade. Interestingly, a V68A mutant of M had the opposite effect, markedly inhibiting SARS-CoV-promoted INF-β production [316]. These observations illustrate the complexity of virus-host cell interactions and reveal the high degree of specificity observed for different envelop viruses.

11.3. The N Protein

The N-protein of the MERS coronavirus suppresses Type I and Type III interferon (IFN) induction (virus-induced IFN-β and IFN-lambda1) by targeting RIG-I signaling. This is accomplished by reducing the IFN gene promoter activities and therefore their mRNA levels, thereby blocking production of the bioactive IFNs. The C-terminal domain of the N-protein plays a pivotal role in this antagonistic activity, and it is particularly important, as these interferons are at the frontline of the larger antiviral defense that triggers the activation of hundreds of downstream antiviral genes [317].

Details of the transcriptional signaling pathway have been elucidated [317], and in an earlier study, Likai et al. [318] found that the porcine δ-coronavirus N-protein suppressed IFN-β production in piglets. These observations suggest that in many, if not all coronaviruses, the N-protein functions to allow the virus to escape the immune surveillance of the host. In all studied cases, the mechanism of suppression involves the N-protein targeting the promoters of interferon genes. This is accomplished by targeting the retinoic acid-inducible gene 1 (pRIG-1) and the TNF receptor by direct interaction. The two studies, using very different coronaviruses, indicate that similar mechanisms of action are involved in both cases. In fact, even earlier studies had provided evidence for such a mechanism [319].

SARS N protein also interferes with TRIM25-mediated RIG-I ubiquitination [292]. Rather than binding to RIG-I or MDA5 [320], N protein associates with the RIG-I effector molecule, TRIM25 [292]. Upon detecting a PAMP, RIG-I is ubiquitinated by TRIM25 [321] to begin the essential antiviral signal cascade. SARS N protein C-terminal residues 364–422 competitively bind to the TRIM25 SPRY domain and interferes with its binding to RIG-I, disrupting the necessary ubiquitination [292]. Such inhibition would blind the cell from ever detecting the presence of non-host RNAs accumulating during viral replication through RIG-I. Downstream IFN signaling is also disrupted by the SARS-CoV-2 N protein by inhibiting the phosphorylation of STAT1 and STAT2 through direct binding to STATs. Coimmunoprecipitation assays revealed that truncated N residues 1–361 are sufficient to prevent STAT signaling, with region 319–422 aas being indispensable for STAT binding [322].

12. Nonstructural Protein Interference with IFN Gene Expression

12.1. nsp1

Nsp1 suppresses IFN activation in an unprecedented manner, unique to all other enveloped RNA viruses considered thus far. Rather than inhibiting protein interactions involved in IFN cascades, nsp1 promotes the degradation of host mRNAs. To assess the contribution nsp1 has to IFN suppression through host RNA degradation, SARS-CoV-1 carrying mutant nsp1 had higher levels of IFN-β coupled with higher levels of host mRNAs. Specifically, residues 160–173 in the C-terminal end of nsp1 participate in mRNA degradation [323]. SARS-CoV-1 nsp1 localizes to translation complexes and has been shown to directly bind to the 40S ribosomal subunit to access the mRNAs, abrogating translation [324]. In MERS, an endonuclease was confirmed to exist within nsp1 despite not being able to bind to the 40S ribosome, indicating that RNA degradation activity may vary even within the β-CoVs [325]. SARS-CoV-2 nsp1 was shown to bind to both 40S and 80S ribosomal subunits through its C-terminal region, physically blocking RNAs from entering the entrance region of ribosomes. It was proposed that SARS-CoV-1 nsp1 may degrade host mRNAs in a two-pronged manner, where it first binds to 40S ribosome subunits, and then applies modifications to host RNAs at the 5′ caps, rendering them translationally incompetent [326]. It was suspected that nsp1 triggers template-dependent endonucleolytic RNA cleavage in the 5′ region of RNAs [327], which is then completed by exonucleolytic activity from host Xrn1 [324]. Viral transcripts have been reported to possibly escape nsp1-induced degradation due to the differences in 5′ caps in viral and host transcripts [327]. When viral proteins are expressed on clonal plasmids, nsp1 can promote the degradation of its own transcripts in transfected cells [323]. Nsp1 does not prevent IRF3 dimerization but does prevent the expression of IFN transcripts. In addition to its endonucleolytic activity, nsp1 can disrupt downstream IFN signaling, where SARS-CoV-1 nsp1 inhibits STAT1, but not STAT2 phosphorylation [328].

12.2. nsp3

As noted previously, nsp3 is a large nonstructural protein containing multiple domains, and it participates in a wide array of functions. Its PL^{pro} domains have deubiquitinating and deISGylating activities and are speculated to participate in immunomodulation. Indeed, SARS and the HCoV-NL63 $PL2^{pro}$ domains can interfere with IRF-3 phosphorylation without affecting respective kinases and stimulation of NFκB dependent genes [329–331]. Interestingly, the enzymatic activities of nsp3 are not solely responsible for IFN-β suppression. Deleting the catalytic residue, C1678 in SARS-CoV-1, and H1836 in HCoV-NL63

PL2^{pro}, to eliminate proteolytic and deubiquitinating activities of the protein, only slightly decreased IFN suppression [331]. Treating PL2^{pro} transfected cells with protease inhibitor GRL-0617S had no effect on IFN suppression but did abrogate NFκB stimulated gene suppression [331]. Since nsp3 is a membrane-spanning protein, the authors of this study also examined if TM forms of the PL2^{pro} domain could inhibit IFN expression. Truncating nsp3 to only include the PL2^{pro} domain attached to a TMS was shown to be a potent IFN antagonist, and it could suppress N-RIG stimulated IFN-β production [331].

To investigate the possible role deubiquitination has on IFN suppression, IRF-3(5D), a phosphomimetic of IRF3, was shown to be deubiquitinated by PL2^{pro}. Despite being deubiquitinated, IRF-3(5D) was still able to dimerize, translocate to the nucleus and bind to DNA, but it could not induce IFN expression. The authors proposed that its interaction with other transcriptional machinery is altered so that IFN expression cannot be achieved [332]. These results suggest that nsp3 is a potent inhibitor of IFN expression both upstream and downstream of IRF-3 phosphorylation. It has yet to be detailed how nsp3 can prevent the phosphorylation of IRF-3. Enzymatic activity may also be required for NFκB stimulated gene suppression [331].

13. Accessory Protein Interference with IFN Expression

13.1. ORF6

The SARS-CoV-1 ORF6 protein has been shown to have IFN-inhibiting abilities, suppressing both upstream and downstream effectors of IFN pathways. Expression of the protein suppressed Sendai virus-induced IFN expression by inhibiting phosphorylation and subsequent translocation of IRF3 to the nucleus [308,333]. ORF6 was also shown to inhibit STAT1 nuclear translocation, despite not preventing STAT1 phosphorylation [334]. Similarly, SARS-CoV-2 ORF6 is able to broadly suppress type-I IFN expression in vitro. Clonal expression of SARS-CoV-2 ORF6 and C-terminally truncated ORF6 inhibited multiple stages of IFN activation as well as downstream pathways of IFN signaling. Specifically, residues 53–61 of the protein overexpressed in HEK293T cells suppressed IRF3 activation by interfering with RIG-I, MDA5, and MAVS complex assembly [308]. Interestingly, the same region was also able to inhibit STAT1 nuclear translocation in IFN-stimulated HEK293T cells [308]. A possible explanation is that SARS-CoV-1 ORF6 localizes in the ER/Golgi membranes. Its C-terminus binds to nuclear import factors karyopherin-α2 and karyopherin-β1, disrupting the formation of nuclear import complexes. Phosphorylated STAT1 is then unable to enter the nucleus. Deletion of ORF6, or removal of the C-terminus, restored STAT1 nuclear translocation [335]. It therefore seems that ORF6 can prevent the expression and secretion of IFNs, thus preventing the downstream upregulation of ISGs.

13.2. ORF3b

Lone transfection of clonal SARS-CoV-1 ORF3b in A549 cells co-infected with recombinant Newcastle disease virus (NDV) prevented replication in the presence of type-I IFN-rescued NDV replication. ORF3b is able to prevent IRF-3 phosphorylation, and thus its translocation to the nucleus. Interestingly, ORF3b was found to localize to the nucleus and nucleolus of cells, associating with B23, C23, and fibrillarin through a nuclear localization signal (NLS) in its C-terminal end [336]. Despite its nuclear localization, it is the cytosolic ORF3b that participates in IFN antagonism, as recently shown for SARS-CoV-2. In fact, deletion of the NLS improves the IFN antagonism of SARS-CoV-2 variants, making this region of the protein an indicator for coronaviral pathogenesis [337]. Similar to ORF6, ORF3b also prevents stimulation of downstream IFN pathways, inhibiting expression from an IRSE promoter [334]. The function of nuclear localization of ORF3b has yet to be detailed.

14. Complement Activation by CoV Structural Proteins

S, E, N and a few nsps likely play roles in the activation of complement, the immediate innate immune response and the bridge between innate and adaptive immune systems.

Complement is a double-edged sword and has only recently undergone more thorough investigation as a major contributor to over-inflammation and pathology. Progression of disease in many pathogenic infections are often the result of hyperactive innate immune responses, inducing severe inflammation. Complement activation is a multistage process involving a large array of activation products. It is a crucial driver of early inflammation and provides protection from infections, stimulating proinflammatory and cytotoxic cytokine secretion, activation and proliferation of leukocytes, vascular constriction, and stimulation of adaptive immune cells (B and T cells) [338,339]. If complement is overstimulated, a cytokine storm may ensue, and disease can be characterized by intense fever, immense vasoconstriction, plasma coagulation, necrosis of infected cells and severe tissue damage. SARS-CoV-1 and other respiratory viruses such as Influenza induce intense fever, severe pulmonary tissue damage, vasoconstriction, and thrombosis in alignment with symptoms of overactivated complement. Evidence exists supporting the suggestion that MERS, SARS-CoV-1 and SARS-CoV-2 all induce complement. Mice infected with either SARS-CoV-1 or MERS have elevated levels of complement proteins in sera [228], and preprint studies on SARS-CoV-2 patients revealed elevated complement-associated proteins in alveolar spaces and blood vessels [229]. Newer proposed treatments for viral infections involve suppressing complement to increase the host tolerance for the pathogen, allowing the virus to proliferate while reducing the severity of pathogenicity.

Complement can be activated via three routes, first, the classical pathway, mediated by natural IgM or antigen-specific IgG, second, the mannose binding lectin (MBL) pathway, mediated by MBL binding to antigen, and third, the alternative pathway, activated by plasma. In all three pathways, production of Complement (C)3 cleavage products, C3a and C3b, are required to begin the downstream effects. In the classical pathway, pentameric IgM or at least 2 IgGs bind(s) to antigen and associate(s) with complement C1 proteins, C1q, C1r and C1s, to form the C1-complex. The C1 complex activates the C1r subunit, a serine protease which splits C4 and C2 into C4a plus C4b and C2a plus C2b, respectively. C4b and C2a associate to form C4bC2a, the C3-convertase which cleaves C3 into C3a and C3b. Similarly, the MBL pathway utilizes opsonin, MBL and ficolins to activate MBL-associated serine proteases (MASP-1 and MASP-2) which cleave C2 and C4 into C2a plus C2b, and C4a plus C4b, respectively. The alternative pathway differs the most and is independent of the C4 derived protease. Rather, it requires the spontaneous hydrolysis of C3 in plasma to form $C3(H_2O)$. $C3(H_2O)$ binds to factor B (fB) to form $C3(H_2O)fB$, which is cleaved by factor D (fD) to form the alternative fluid phase C3 convertase $C3(H_2O)Bb$ which can cleave C3 into C3a and C3b. This spontaneous production of $C3(H_2O)Bb$ ensures a stable and abundant level of C3b in plasma. C3b deposits on pathogens or infected cell membrane surfaces. Free C3b can induce the alternative pathway if it directly binds to the surface of a pathogen. Membrane bound C3b is still able to associate with fB, and in the presence of factor D, it will produce membrane-bound C3bBb, the alternative pathway C3 convertase. All complement pathways converge on the C3bBb C3 convertase to promote the cleavage of C3 in a positive feedback loop. C3a acts as a proinflammatory chemokine. Downstream, C3b becomes a C5 convertase by associating with other C cleavage products, the classical/MBL (C4b2b3b or C4b2a3b), or an alternative (C3bBbC3b). Terminal C5 cleavage products, C5a and C5b, result in a final form of complement. C5a, like C3a acts as a chemoattractant for leukocytes. C5b can bind to cell surfaces and oligomerizes with C6, C7, C8 and poly C9 to produce the C5b9 membrane attack complex (MAC), the innate immune system's cytotoxic warhead. MAC breaches a hole in bacteria, virus-infected (recognized non-self) cells, and even viral envelopes, causing extracellular fluids to rush into the cell/virus, inducing lysis. Other roles of C5b and C5b9 promote chemokine secretion and inflammation [338,339] (see Figure 5 for an illustration of the complement pathways). A majority of complement proteins are produced in the liver and secreted into the blood [340]. Damaged liver tissue seems to require activation of complement for regeneration, relying specifically on C3 and C5 [340], complicating the balance between suppressing and activating complement in diseases that affect the liver. Extensive damage in the liver has been linked to severe disease in SARS-CoV-2 infections [341].

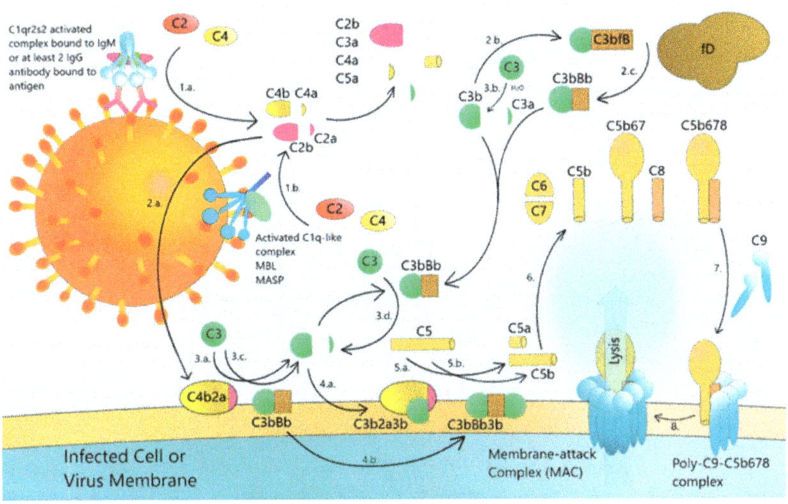

Figure 5. Schematic diagram of how coronavirus stimulates the complement cascade. 1.a. C1qr2s2 bound to IgM or at least 2 IgG antibodies binds to antigen. C1qr2s2 is activated and cleaves C2 and C4 into C2a, C2b, C4a, and C4b. 1.b. Activated C1q-like complex bound to MBL-MASP binds to mannose-glycosylated antigen and cleaves C2 and C4 into C2a, C2b, C4a, and C4b. 2.a. C4b and C2a bind to form C4b2a C3 convertase. 2.b. C3b binds to fB. 2.c. C3bfB is cleaved into C3bBb. 3.a. C3 is cleaved into C3a and C3b by C4b2a. 3.b. C3 is hydrolytically cleaved spontaneously into C3a and C3b. 3.c. and 3.d. C3 is alternatively cleaved by C3bBb either freely or on membrane surface. 4. C3b binds to C4b2a to form C5 convertase C4b2a3b. 4.b. C3b binds to C3Bb to form C3bBbC3b alternative C5 convertase. 5.a. C4b2a3b cleaves C5 into C5a and C5b. 5.b. C3bBbC3b cleaves C5 into C5a and C5b. 6. C5b binds to C6, C7, and C8 to form C5b678. 7. C9 is recruited to antigen presenting membranes to form poly-C9 and binds to C5b678 8. MAC is formed, and lysis occurs.

Inhibition of complement pathways attenuate disease progression despite continued replication of the pathogen in the host. Inhibiting C5 reduces intravascular coagulation and prevents organ failure, cytokine storms and sepsis in *E. coli*-infected Baboons [342]. For influenza virus, C5 induces over-recruitment of neutrophils and CD8+ T cells as well as cytokine secretion, inducing acute lung injury in H1N1 or H5N1 infected mice. Treatment with a C5 inhibitor significantly attenuates respiratory inflammation and tissue damage [343]. Along with septic shock, typically, multi-organ failure and kidney damage are associated with complement overactivation [344,345]. Since severe SARS-CoV-1 infections accompanying kidney damage, while rarer, are linked to systemic over inflammation rather than viral tropism for this tissue [346]. It is not unlikely that complement plays a role in the multiple organ failure seen in SARS-CoV-2. Investigations into the role of complement in CoV-induced disease revealed that SARS-CoV-MA15 infected C3-/- mice resisted severe disease progression. In comparison, control mice, having elevated complement proteins in mouse sera, exhibited 15% weight loss with lung tissue damage [228]. C3-/- infected mice did not lose weight 2–4 dpi, had reduced cytokine proinflammatory secretions (IL-6, TNF-α, and IL-1β), had reduced monocyte infiltration and exhibited little pulmonary tissue damage. It is likely that multiple branches of complement activation are required for infection in SARS-infected mice, as neither C4-/- nor fB -/- mice were protected from weight loss [228].

Complete suppression of C3 in humans is probably not a valid strategy for combating SARS-CoV-2 due to the necessity of C3 in other immune pathways. Despite having a seemingly beneficial effect in SARS infected mice, C3-/- mice infected with H5N1 or

H1N1 actually had more inflammation and tissue damage due to failure to activate adaptive humoral and cell immunity [347–349]. Rather, downstream C5 products are responsible for severe and lethal infections similar to the *E. coli* and influenza studies [229]. Endothelial injury from C5 activation products was detected in infected and damaged $ACE2^+$ tissues. Together with the formation of C5 products, over recruitment of neutrophils and macrophages was observed. C5a interacts with membrane C5aR on endothelial cells, inducing downregulation of thrombomodulin and activation of coagulation with secretion of P-selecting promoting platelet adhesion, aggregation and recruitment of white blood cells. Besides forming MAC, C5b9 induces endothelial activation and dysfunction, upregulating tissue factors and adhesion molecules for migrating white blood cells. Additional inflammatory chemokines are secreted along with increased vascular permeability and coagulation. In unpublished observations, abundant C5b9 was observed in microvasculature of interalveolar septa, large caliber vessels of the lung parenchyma and microvasculature in occluded arteries of SARS-CoV-2 patients [229]. C5b9 deposits were also found in septal capillaries colocalized with the S and E proteins, indicating that CoV structural proteins are involved in the induction of complement. Downstream suppression of C5a and C5b activities would be reasonable as all complement pathways result in C5 cleavage products. This would prevent the most severe effects of complement from occurring without affecting other peripheral pathways, such as stimulation of adaptive immunity [339]. Anti C5aR antibodies prevented MERS-induced upregulation of proinflammatory cytokines in serum, thus reducing leukocyte infiltration and tissue damage [350]. Suppression of C5 products could be achieved by the use of the approved drug, eculizumab, which inhibits C5, preventing its cleavage [351], or by candidate C5aR inhibitor, CCX168, currently in phase III clinical trials [352].

15. Induction of Endoplasmic Reticulum (ER) Stress and the Unfolded Protein Response (UPR)

Perturbation of the ER, for example by pore-formation, causes ER stress, leading to the activation of cell signaling pathways including the unfolded protein response (UPR). As noted above, SARS-CoV-1 uses the ER/Golgi apparatus for synthesis and processing of viral proteins, and for this purpose, it uses the UPR. Although several viral proteins may contribute to the UPR, the Spike (S)-protein appears to be the primary inducer of several UPR effectors, including glucose-regulated protein 78 (GRP78), GRP94, and the C/EBP homologous protein. However, the expression of S exerts different effects on the three major signaling pathways of the UPR. Thus, it induces GRP78/94 through the PKR-like ER kinase, PERK, but it has no effect on activating transcription factor 6 or X box-binding protein 1. The S-protein appears to specifically modulate the UPR to facilitate viral replication [176,353]. However, overexpression of ORF3a, ORF3b, ORF6 or ORF7a can also induce apoptosis. Interestingly, inhibitors of Caspase-3 and JNK block ORF-6 induced apoptosis. Thus, ORF-6 induces apoptosis via Caspase-3-mediated ER stress and JNK-dependent pathways [175]. ORF3a also down regulates the type 1 interferon receptor [354], while Nsp6 activates omegasome and autophagosome formation [355]. Interestingly, the E-protein of SARS-CoV-1 seems to decrease the stress responses while increasing inflammation [179], yet the same protein, as well as N of Porcine Epidemic Diarrhea Virus (PEDV), can cause ER stress. However, both proteins also up-regulate interleukin-8 expression [178,356,357], while overexpression of Nsp7 down-regulates interleukin 8 [356]. In fact, many of the CoV proteins, including 3, 8b, and the ion channel activity of the IBV E-protein, influence ER stress and the translation apparatus [24,82,356,358]. Interestingly, although not essential for replication, glycosylation of the IBV M protein ectodomain plays important roles in activating ER stress, apoptosis and the pro-inflammatory response, thereby contributing to the pathogenesis of IBV [359]. All of these analyses reveal (1) how complicated the viral induction of ER stress is, and (2) the large number of viral proteins that influence this process.

16. Coronavirus-Induced Host Cell Cycle Arrest

Several early studies demonstrated that various proteins encoded within coronaviral genomes can cause cell cycle arrest in the infected cells in various growth phases. One of these is 3a of SARS-CoV-1 which is mainly localized to the Golgi apparatus together with M in co-transfected cells. Expression of 3a inhibited cell growth and prevented 5-bromodeoxyuridine incorporation, suggesting that 3a deregulates cell cycle progression [360]. 3a expression blocked cell cycle progression at the G1 phase in various tissue cells 24–60 h after transfection. Mutational analysis of 3a revealed that the C-terminal region, from residue 176, which includes a potential calcium ATPase motif, was essential for cell cycle arrest. As noted above, like the M-protein, 3a predominantly localized to the Golgi apparatus, with its N-terminus residing in the lumen and its C-terminus in the cytosol. In the relevant experiments, 3a expression correlated with a reduction of the cyclin D3 level. Increases in p53 phosphorylation on Ser-15 were observed in both SARS-CoV-1 M and 3a transfected cells, suggesting that this phosphorylation activity might not be responsible for the 3a-induced G0/G1 phase arrest. Thus, there was evidence that 3a and M might function independently to inhibit cell cycle progression, but that their detailed mechanisms might be different. ORF7a expression may also block cell cycle progression in the G0/G1 phase, and it apparently can induce apoptosis via a caspase-dependent pathway [361]. ORF7a expression is associated with blockage of cell cycle progression in several cell lines after 24 to 60 h post-transfection. Mutational analysis of ORF7a revealed that the domain spanning amino acyl residues 44–82 was essential for its induction of cell cycle arrest. Since ORF7a expression correlated with a reduction of cyclin D3 mRNA levels and phosphorylation of the retinoblastoma (Rb) protein on serine residues, it was suggested that the insufficient expression of cyclin D3 might have caused the decreased activity of cyclin D/cdk4/6, resulting in the inhibition of Rb phosphorylation. Accumulation of hypo- or non-phosphorylated Rb thus may have prevented cell cycle progression during the G0/G1 phase.

Virulent strains of porcine epidemic diarrhea virus (PEDV), an enteropathogenic α-coronavirus, cause a highly contagious enteric disease in swine, characterized by severe enteritis, vomiting, and watery diarrhea. Xu et al. [362] investigated the subcellular localization and function of the PEDV M-protein through examination of its effects on cell growth, cell cycle progression, and interleukin 8 (IL-8) production. Their results revealed that after infection, the M-protein seemed to localize throughout the cell cytoplasm. M altered porcine intestinal epithelial cell line (IEC) growth, and it induced cell cycle arrest at the S-phase via the cyclin A pathway. S-phase arrest proved to be associated with a decreased level of cyclin A, but M did not induce endoplasmic reticulum (ER) stress (see the next section). Moreover, it did not activate NF-κB which is important for IL-8 and Bcl-2 expression. Thus, the PEDV M-protein induces cell cycle arrest when cells are in the S-phase. Sun et al., 2018 [363] confirmed many of the observations of Xu et al. (2015) [362], and further showed that the p53-dependent pathway plays an important role in PEDV-induced cell cycle arrest. In fact, inhibition of p53 signaling reversed arrest. They additionally showed that cell cycle arrest contributes to viral infection and involves down-regulation of the Cyclin E protein gene.

17. Coronavirus-Induced Autophagy and Abortive Apoptosis

Macro-autophagy (hereafter referred to as autophagy) is an evolutionarily conserved intracellular catabolic transport route that generally allows the lysosomal degradation of cytoplasmic components, including bulk cytosol, protein aggregates, damaged or superfluous organelles and invading microbes [356]. Notably, autophagy participates in both innate and adaptive immune pathways. The innate role is through an autophagy subroutine called xenophagy for the elimination of intracellular parasites and viruses. The adaptive immune system utilizes autophagy for the purpose of antigen presentation. Autophagy allows for cells to cross-present antigens between the MHC class I and MHC class II molecules. Typically, MHC class I present antigens of endogenous sources, while MHC class II presents antigens from extracellular spaces. Autophagy permits endogenous antigens to enter the MHC class II presentation pathway. It is then no surprise that several viruses have evolved mechanisms to inhibit or hijack autophagy pathways and associated proteins. It remains a debated question if Coronaviruses can also utilize autophagy for their own replication, or if autophagy is an effective antiviral response to Coronavirus infection.

Porcine hemagglutinating encephalomyelitis virus (PHEV) infection induces atypical autophagy and causes the appearance of autophagosomes, but it blocks fusion with lysosomes [360]. In addition, transmissible gastroenteritis virus (TGEV) infection induces autophagy of mitochondria (mitophagy) to promote cell survival and possibly viral infection while counteracting oxidative stress and apoptosis [361]. In fact, non-canonical autophagy is believed to converge with the infection cycles of many DNA and RNA viruses that utilize membranes from the ER and cis-Golgi [362]. $PL2^{pro}$ may act as a novel autophagy-inducing protein, but it induces incomplete autophagy by increasing the accumulation of autophagosomes while blocking the fusion of autophagosomes with lysosomes. Furthermore, $PL2^{pro}$ interacts with the key host cell autophagy regulators, LC3 and Beclin1 to promote a Beclin1 interaction with STING, the key regulator for antiviral IFN signaling. Finally, knockdown of Beclin1 partially reversed the $PL2^{pro}$ inhibitory effect on innate immunity while resulting in decreased coronaviral replication [364]. Nsp6 of β-CoVs MERS, SARS-1 and SARS-2, and γ-CoV IBV have also been documented to restrict autophagosome expansion, ultimately preventing the delivery of viral components to lysosomes for degradation [99]. While coronavirus replication complex formation requires constituents of the host autophagy system [365], it does not require the autophagy protein, ATG5, that normally completes autophagy and promotes fusion of the autophagocytic vesicles with lysosomes [102]. So far, there is little or no evidence that the M-protein plays more than an indirect role in autophagy.

Abortive apoptosis is a last resort mechanism of cells in response to intracellular stress, and detection of DNA damage. Many viruses can induce apoptosis in cells either "intentionally" for replicative purposes, or "unintentionally" due to consequences of hijacking cellular machinery. Whether or not apoptosis is beneficial or not to CoV replication remains questionable. While all CoVs induce ER stress, and can induce apoptosis, there are many proteins that suppress the UPR mediated abortive apoptosis pathway. Coronaviruses have been indicated to induce said intracellular stress as mentioned in earlier sections of this review, in particular the production of CoV proteins in the ER activates the UPR pathway. Specifically, the extensive post translational modifications of the various membrane spanning proteins in the CoV proteome rely heavily on the limited protein chaperons inside the ER [358]. Prolonged UPR and failure to reattain homeostasis leads to ER stress-associated abortive apoptosis [358]. The protein, ER-resident transmembrane kinase-endoribonuclease inositol-requiring enzyme 1 (IRE1), a UPR signal transduction molecule that behaves as a timer for heavy ER stress, indicates the cell to switch from cytoprotective phase to apoptosis. IRE1 functions as a RNase, splicing the mRNA of the X box binding protein 1 (XBP1) gene, producing XBP1s mRNAs which encode a potent activator of many UPR genes. Conversely, unspliced XBP1 confers an inhibitor of UPR genes. Thus, prolonged IRE1 signaling and splicing of XBP1 results in overactivation of UPR and decreased cell survival over time. MHV and IBV were shown to activate the IRE1-XBP1, but XBP1s protein expression is suppressed in MHV possibly by persistent phosphorylation of eIF2α, suppressing host translation [366]. Interestingly, SARS seems to prevent splicing of XBP1 altogether through some unknown mechanism related to the E protein [179]. Since CoVs rely on budding of virions from the host, as opposed to lysis, apoptosis would appear to be an inhibitory mechanism to optimal CoV replication.

18. Structural Proteins as Protective Antigens in Survivors, and Vaccine Development

18.1. S Protein as a Protective Antigen

Antisera of SARS survivors have shown representative IgG antigen recognition against the S1 subunit of the S protein [262,367]. Specifically, the RBD of the S1 subunit has been a prime target for adaptive humoral immunity against the virus [367]. Cryo-EM of the highly potent anti-RBD S230 antigen-binding fragment (Fab), purified from a SARS survivor antiserum, bound to the S protein, and displayed specific localization with the S^B domain existing in 2 states. The state 1 complex showed multiple orientations of each of the S230 Fabs associated with intermediate and open conformations of the S^B domain. State 2 complexes had all three S^B domains in the open conformation but lacked 3-fold symmetry [367], suggesting that S230 can bind to S^B domains in varying degrees of openness. The residues involved in the Fab-S^B complex were as follows: S230:CDRH2 F59 and S230:CDRH3 Y106, F107 and Y110, localized near SARS Y408, Y492, F460 and Y475 centered around L443 [367]. S230 potency may derive from its ability to mimic the ACE2 receptor and bind to the RBD with even higher affinity. Its mimicry of ACE2 also allows the molecule to trick S protein fusion activation, locking all S^B domains into the open conformation upon binding, leading to the relaxation of S2 subunit folds and subsequent proteolysis and premature activation [367]. Thus, S230 not only sequesters the S protein, but also deactivates its ability to fuse viral and host membranes. Because of the similarities between SARS-CoV-1 S and SARS-CoV-2 S, it is possible that cross immunity from polyclonal antibodies may exist between the two viruses. Consistent data have yet to confirm this suggestion, but both human and rabbit mono/polyclonal anti-SARS-CoV-1 S antibodies unfortunately had weak to no neutralizing capacity against either SARS-CoV-2 S pseudo-virions [32] or SARS-CoV-2 S itself [195], suggesting limited cross-immunity.

Since natural immunity against SARS-CoV-1 S is characterized by antibodies targeting the RBD, vaccination efforts have homed in on methods to develop anti-RBD vaccines. Successful and potent anti-RBD vaccines can be produced through recombinant IgG1-Fc-$RBD_{(318–510)}$ in 293T cells [364] or a truncated S $RBD_{(318–510)}$ fragment in mammalian 293T cells, insect Sf9 cells, or *E. coli* [365]. In these studies, potent SARS neutralizing antibodies were produced in rodent models, preventing infections both in vitro and in vivo. The strongest SARS neutralizing antibodies were IgGs from mice vaccinated with truncated $RBD_{(318–510)}$, originating from transfected mammalian 293T cells, neutralizing 100% of SARS virions upon the first boost in veroE6 cells [365]. No viral RNA was detected in the lungs of mice 5 dpi, vaccinated with any of the truncated RBD vaccines, while unvaccinated mice suffered infection [365]. Regardless of the source, all RBD vaccines elicited strong anti-SARS activity, although it is not clear if the mechanism of protection is similar to that of S230.

T cell responses, which are essential for the clearance of any viral infection, are also targets for activation by vaccines. Patients recovering from SARS have elevated levels of activated T cells. In a study measuring the adaptive immunity against SARS, at least 50% of SARS survivors tested positive for the T cell response a year after infection [368]. Patients who experienced severe illness had many memory T cells (CD26+/CD45RO+) and polyfunctional CD4+ T producing IFN-γ, TNF-α and CD107a degranulation. Many of the CD4+ T cells were largely specific for S protein [368]. In nearly all patients, the elevated T cell response was coupled with anti-S IgG antibodies, indicating that clearance of SARS is both humoral and cell-mediated, and centered around structural proteins, specifically S, although N is also targeted extensively. Thus, the development of vaccines that can also induce T cell responses would provide stronger protection, similar to immunity gained through infection.

DNA (or RNA) vaccination is a radically new method of vaccination. Viral DNA is cloned as cDNA plasmids and injected directly into a tissue of the subject in order to induce an immune response. DNA vaccines were shown to be effective against HIV, hepatitis B, hepatitis C, influenza and rabies [369]. Several studies involving developing cDNA vaccines based on the SARS S protein were able to rapidly mount humoral and cell-mediated immunity against the virus in rodent models. In a study by Huang et al., a full-length cDNA S plasmid was used as a vaccine in BALB/c mice. After 1-week post vaccination, elevated secretions of IFN-γ were detected in the spleens of mice after challenging them with S antigen, and the response was increased by 3–30-fold if the mouse was given a vaccination boost, indicating a specific response to S. IFN-γ producing CD4+ and CD8+ T cells were also detected in lymph nodes, spleen and lungs post immunization. However, CD8+ T cells were preferential for IFN-γ while CD4+ T cells preferentially produce IL-2. Eight weeks after immunization, T cells specific to SARS S remained in the lymph nodes, spleen and lungs [370]. Additionally, a majority of memory CD4+ and CD8+ T cells were found to be effector memory cells in lungs of mice [370]. Many memory T cells also expressed IL-7Rα, which plays a role in managing the homeostasis of memory CD8+ T cells. In another study, pcDNA vaccines of the SARS-CoV-1 structural proteins S, M and N revealed that S can induce a stronger and more lasting humoral immunity compared to the other structural proteins tested.

The researchers separated S into overlapping C- and N-terminal subunits, denoted as pcDNASa and pcDNASb in a 1:1 ratio in BALB/c mice. While the humoral immunity was strongest for the S vaccine, it induced a weaker cytotoxic T cell response in comparison to M and N pcDNA [369]. Additionally, purified lymphocytes from the pcDNASa-pcDNASb vaccinated BALB/c mice hardly proliferated when restimulated with S protein [369]. Truncated S cDNA is also effective in mounting immunity. Either deleting the TM domain (SΔTM) or the cytoplasmic domain (SΔCD) produced an effective T cell response with neutralizing antibodies in BALB/c mice [371]. Surprisingly, the role of T cells in providing immune protection appears minor. Depleting T cells from the spleen and liver of vaccinated mice still resulted in protection. On the other hand, infecting mice with SARS, and then donating T cells from vaccinated mice, did not prevent infection, although donor IgG antisera did [371]. Despite this, the T cells generated produced either IFN-γ or TFN-α in response to S antigen, indicative of S specificity. All mice vaccinated with any of the cDNA vaccines were protected from SARS infection 30 days after immunization [371]. Finally, cDNA vaccination against S can be strengthened with pcDNA-IL-2 as an adjuvant [372]. Mice vaccinated with pcDNA-S + pcDNA-IL-2 had the strongest conferred cellular and humoral immunity.

In the same study, different vaccination methods were compared using injection, oral administration and electroporation. The authors noted the preference of IgG subclasses in the different vaccines tested. 10 days after immunization, the IgG1 subclass was detected primarily in pcDNA-S + pcDNA-IL-2, while pcDNA-S vaccinated mice produced primarily IgG2α [372]. Conventional intramuscular immunization produced a better antigen-specific T cell response than electroporation, but electroporation produced better humoral immunity. Additionally, specific subsets of cytokine secreting CD4+ T cells, Th1 and Th2, were discerned and measured. In all groups, Th1(IFN-γ secreting) and Th2 (IL-4 secreting) were present, but Th1 composed the majority of Th cells, consistent with an inflammatory response associated with SARS. The addition of IL-2 as an adjuvant indicates that immunization against structural proteins alone is not enough to activate the immune system to its greatest potential. IL-2 is a modulatory cytokine for both innate and adaptive immune cells, activating Th cells, cytotoxic T cells, B cells, macrophages and Natural Killer cells.

Overall, DNA vaccines against S are probably effective, due to the production of humoral immunity followed by a T cell response, even if an infection were to still occur, possibly from a closely related virus. Upon detecting the S antigen, Th1 cells release IFN-γ, recruiting and activating phagocytes to regions where SARS is present. IgG antibodies sequester the S protein while activated white blood cells consume the viral particles. The seeming ineffectiveness of T cells in protecting against SARS as reported by Yang et al. could be attributed to the role the T cell response has to a SARS infection. T cells were detected in patients with mild to severe infections, coupled with elevated IgG. If an abundance of cells in tissues are to be infected, CD8+ cytotoxic T cells would be required to kill cells to prevent further replication of the virus, consistent with the measured higher level CD8+ T cell response over the CD4+ T cell response in SARS survivors [373].

Meanwhile, Th1 cells may modulate and enhance secretion of IgG2a to continue sequestering viral particles, thereby increasing inflammation in and chemotaxis to infected tissues. The danger of the SARS-induced T cell response resides in the cytokine storm characteristic of severe infections. In combination with the innate immune response (complement) to SARS infection, and the inflammatory response due to the virus' pathogenicity, an imbalance of Th1 and Th2 cells could be a major contributor to the disease progression.

These vaccination studies revealed a preference for IFN-γ and IL-2 secreting CD4+ T cells, indicators of Th1 cells [370,372]. Excessive accumulation of IFN-γ in the host without the anti-inflammatory secretion of Th2 could result in hyperinflammation, overactivity of white blood cells and extreme pulmonary tissue damage in mid-late infection. Hence, it would be crucial to also mount a Th2 response when developing a vaccine.

Some SARS patients had elevated Th1 cytokines, IFN-γ, IL-1β, IL-6, and IL-12, with limited to elevated anti-inflammatory Th2 IL-10 in the blood plasma [368,374]. While additional data are needed, similar results were obtained for SARS-CoV-2 patients, indicating a preference for Th1 over Th2 cells [375]. On the other hand, prolonged overproduction of Th2 IL-10 along with elevated CD8+ T cells was associated with fatal infections [368,376], suggesting that an imbalance towards Th2 may also be lethal. Some SARS-CoV-2 patients with worsening disease displayed elevated IL-10 with decreased CD4+ and CD8+ T cells [377], suggesting that IL-10 may be secreted by monocytes rather than Th2 cells, and that the T cell suppressing role from overexpressed IL-10 is detrimental. Regardless, the balance of secreted cytokines can be easily disturbed, but it is crucial for the determination of severe disease progression in both SARS and SARS-CoV-2. While additional information is necessary to determine the nature of the T cell response during SARS-CoV-2 infection, vaccines should be able to mount a full immune response.

18.2. N-Protein as a Protective Antigen

The N-protein has been considered by several groups for use in vaccine design. For example, Yong et al. described in 2019 recent advances in the development of vaccines against the MERS coronavirus, and N is one of several viral structural proteins used in this endeavor, others being the S, E and nsp16 CoV proteins. These authors emphasized immune responses and potential antibody-dependent enhancement of infection, but they also discussed animal models to evaluate vaccine candidates. In another study, Jiang et al., 2020 [378], using a SARS-CoV-2 proteomic microarray, characterized the IgG and IgM antibody responses to sera from 29 convalescent Covid-19 patients to most of the viral proteins. All patients produced antibodies most abundantly to the N and S1 proteins. Moreover, Basu and Brown [379] and Lee and Koohy [380] analyzed immunogenic peptides from nucleocapsid and surface proteins of several CoVs, identifying areas of the N-proteins that are conserved and therefore of interest for vaccine development. Ahmed et al. [381] conducted similar analyses, finding regions in the N-protein that were identical between CoV-1 and CoV-2 and therefore would likely prove appropriate for cross reactive vaccine development. It is encouraging that memory T-cell responses targeting SARS-CoV-1 persisted up to eleven years post-infection [382]. Additionally, several novel approaches are now being used, such as reverse vaccinology and machine learning, to develop a vaccine against CoV-2 [383].

18.3. M-Protein as a Protective Antigen

The M-proteins of several coronaviruses have been shown to act as dominant protective immunogens, being antigens for the humoral response [384,385]. Specifically, the N-terminal transmembrane region of M contains a T-cell epitope cluster, and this provides a major fraction of the immunogenicity of the virus [386]. The M-protein therefore serves as one possible candidate for the development of a vaccine against one or several of the human respiratory coronaviruses. However, early studies with AIBV suggested that the M glycoprotein elicited antibodies in low titers and of limited cross-reactivity. Moreover, immunization of chickens with the purified M protein did not induce protection against virulent challenge [387,388]. However, in the same year, Saif [389] reported that the M proteins of several animal coronaviruses can induce antibodies that neutralize the viruses in the presence of complement.

Twelve years later, Okada et al. [390] showed that M DNA from SARS-CoV-1, using the pcDNA 3.1(+) plasmid vector, evoked T cell immune responses (CTL induction and proliferation) in mice against this M protein. These observations were confirmed and extended by Liu et al. [385] who showed that the M-protein of SARS-CoV-1 acts as a dominant immunogen as revealed by a clustering region of novel functionally and structurally defined cytotoxic T-lymphocyte epitopes. Soon thereafter, Zhang et al. [391] concluded that a conserved linear B-cell epitope was present in the M-protein of PEDV, and Yan et al. [392] identified a similar epitope in this M-protein. Similar developments were reported by Takano et al. [393] for the Feline Infectious Peritonitis Virus (FIPV). Immune responses to pcDNA vaccines against M protein elicited stronger lymphocyte proliferation and cytotoxic T cell lysis activity than pcDNASa and pcDNASb by week 12 post vaccination. Humoral immune responses followed an interesting trend, with M-specific antibodies reaching higher levels than pcDNASa-pcDNASb and pcDNAN within 6 weeks, but they rapidly declined over the following weeks, while pcDNASa-pcDNASb retained stable levels after week 8. These results give hope that vaccines directed against the M-proteins of human pathogenic coronaviruses, including that of SARS-CoV-2, will be forthcoming in the future.

19. Conclusions

The novel SARS-CoV-2 virus is projected to remain a threat to global public health for at least another two years since its first occurrence in late 2019. While vaccine and antiviral research is underway, the rapid spread and fatality due to the virus indicates that pharmaceuticals will not be enough to stop this disease. Global health policy and coordination between local and national governments will be essential in order to slow the spread. The effectivity of a future vaccine must be coupled with proper social distancing, public health practices and education [394]. In this review, we detail the extensive coronavirus genome, its proteins, and their roles in viral replication and pathogenesis. The virus is notoriously capable of evading host innate immune systems, while still inducing severe disease and inflammation. Long term adaptive immunity against the virus remains in question, placing a larger pressure on effective vaccine research [395]. Optimistically, there are many targets within the coronavirus proteome, especially in the transmembrane proteins, to counteract the severe inflammation in the interest of antivirals detailed in this review. Research efforts over the past 20 years have revealed several targetable sites in most CoV proteins with identified immunoregulatory functions.

Novel human infecting pathogenic viruses resulting from zoonotic jumping are not uncommon as virologists and epidemiologists fervently study pig and bird influenzas that may have jumped to humans. Other viruses that have caused widespread epidemics that sparked a wave of research into antivirals are HIV and Ebola [396,397]. Until recently, coronaviruses have gone largely underrepresented as growing threats to civilization, despite being responsible for two epidemics in China 2003 (SARS) and the Middle East 2012-present (MERS). Additionally, minor outbreaks of HKU1-CoV occurred in Hong Kong and the USA causing mild to severe pneumonia [398,399]. Coronaviruses are also a costly agricultural nuisance with Porcine Epidemic Diarrhea Virus and Infectious Bovine Virus, largely affecting pork and cattle supply and economy, respectively [400,401]. Despite this history, antivirals against CoVs are virtually nonexistent, making humanity pharmaceutically defenseless against the Covid-19 pandemic. Since CoVs have repeatedly challenged civilization for the past 20 years, and with the widespread dispersion of SARS-CoV-2 infections, we can expect CoVs to be a major contributor to future diseases along with influenza and antibiotic resistant bacteria. The capacitance for CoVs to genetically recombine and zoonotically jump also reveals a growing vulnerability in disease prevention strategies, especially in rural areas. Increasing globalization and rapid development into rural areas allows infectious diseases to spread into cities and other countries, leading more easily to pandemics [402].

Predicting the next epidemic disease is computationally impractical and requires massive surveillance. Nevertheless, hindering acquisition of infectious diseases can be achieved through widespread education and distribution of sanitary equipment. Such practices were performed in Africa to limit the spread of diarrheal diseases and attenuate the spread of Ebola [403,404]. In the meantime, research for antivirals and a vaccine against the SARS-CoV-2 may lead to solutions against the Covid-19 pandemic, and possibly future CoVs.

Author Contributions: Both authors contributed significantly to the conceptualization of this review, but N.A.W. initially conceived of the project. For the methodology, validation, formal analysis and data curation, both authors contributed equally, but N.A.W. designed the format and verified the primary conclusions. Both authors contributed to the writing of the original drafts and the editing. M.H.S.J. conducted the project administration and was responsible for funding acquisition. Both authors have read and agreed to the published version of the manuscript. Authorship is limited to those who have contributed substantially to the work reported. All authors have read and agreed to the published version of the manuscript.

Funding: This research was funded by the U.S. National Institutes of Health, grant number GM077402.

Institutional Review Board Statement: Not applicable.

Informed Consent Statement: Not applicable.

Acknowledgments: We would like to acknowledge the contribution of Zhongge Zhang and Luis Arturo Medrano-Soto for useful discussion, and the latter for the analyses necessary to generate Figures 3 and 4. We are thankful for the artistic representation of the Coronavirus replication cycle in Figure 2, illustrated by Yatavee Vajaphattana.

Conflicts of Interest: The authors declare no conflict of interest.

Abbreviations

Viruses and associated abb'ns	
α β γ δ	Coronavirus (CoV) types
ACoV	avian CoV
BCoV	bovine respiratory CoV
CoV	coronavirus
CCoV	canine CoV
CoV HKU1	a species of CoV that infects humans
HCoV-OC43	another species of CoV that infects humans
Covid-19	the disease caused by CoV-2
FCoV	feline CoV
FIPV	feline infectious peritonitis virus
HAV, HBV, HCV	hepatitis A, B or C coronavirus
HCoV	human CoV
H1N1	an influenza A viral subtype; swine flu
H5N1	an influenza A viral subtype; asian avian flu
IBV or AIBV	avian infectious bronchitis coronavirus (avian CoV)
MERS	middle east respiratory syndrome
MHV	murine (mouse) hepatitis coronavirus
NDV	Newcastle disease virus
PEDV or PDCoV	porcine epidemic diarrhea coronavirus
PHEV	porcine hemagglutinating encephalomyelitis virus
RSV	respiratory syncytial virus

RV	Rhino virus
SARS, SARS-CoV or SARS-CoV-1	severe acute respiratory syndrome coronavirus-1
SARS-CoV-2 or CoV-2	severe acute respiratory syndrome coronavirus-2
SFV	semliki forest virus
TGEV	transmissible gastroenteritis coronavirus
VSV	vesicular stomatitis virus
VIRAL PROTEINS and RNAs	
E	envelope protein
H	helicase
HE	hemagglutinin esterase
M	membrane matrix protein
N	nucleocapsid protein
nsp or NSP	(viral) non-structural protein (e.g., nsp3a 3b, 4a, 5, 6, 7a, etc.)
P	protease
PLP	PL^{pro}, papain-like protease
Pp	polyprotein (i.e., pp1a (Orf1a) is spliced to give nsp1–11 pp1ab (Orf1b) is spliced to give nsp12 - 16.
RdRp	RNA-dependent RNA polymerase
S	spike protein (the S1 domain binds ACE2, receptor for several CoVs the S2 domain induces membrane fusion)
gRNA	guide RNA
gsRNA	single guide RNA
ssRNA	single stranded RNA
dsRNA	double stranded RNA
TRS	transcription regulating sequence
HOST ENZYMES AND OTHER PROTEINS	
ACE2	angiotensin-converting enzyme-2, receptor for several CoVs
ATG5	an autophagy (ATG) protein - promotes fusion of phagocytic vesicles with lysosomes
Bcl-2	an apoptosis regulator it contains tandemly repeated PDZ domains that bind the cytoplasmic, C-terminal domains of a variety of transmembrane proteins.
Beclin1	a regulator of autophagy, ATG.
CCAAT	enhancer-binding proteins transcription factors
CCL1, 2, 3, 4	CxCL2, proinflammatory chemokines
CDRH1, 2, 3	the first, second, third complementarity-determining variable region of antibody heavy chains
Cdk	PKR-like endoplasmic reticulum (ER) kinase
C/EBP	CCAAT-enhancer binding protein
CHOP	C/EBP homologous protein (transcription factor)
Cyclin	protein involved in the cell cycle (e.g., cyclin D3, cyclin E)
DC-SIGN	dendritic cell-specific intercellular adhesion molecule grabbing nonintegrin also, CD209
DUB	deubiquitinating enzyme
DPP4	dipeptidyl peptidase 4 (adenosine deaminase complexing protein 2)
DUSP	dual specificity phosphatase
eIF2	eukaryotic initiation factor 2
GRP	glucose regulated protein (e.g., GPR78, GRP94)
HSP	heat shock protein
IFN	interferon
IL	interleukin (i.e., IL-1β)
IPS-1	INFβ promoter stimulator
IRE1	inositol-requiring enzyme (protein kinase a sensor)
IRF3	interferon regulatory factor 3
ISG	INF stimulated gene

JNK	cJUN amino terminal kinase
LC3	cytosolic ubiquitin-like protein (regulates macroautophagy (autophagy, ATG))
MAPK	MAP kinase, Mitogen-activated protein kinase
MBL	mannose binding lectin
MHCI and II	major histocompatibility complex, classes I and II
NF-κB or NFκB	a protein complex that controls transcription of DNA, cytokine production and cell survival.
NLRP3	NLR family pyrin domain-containing protein-3
PALS1	protein associated with Lin seven 1, a Maguk PDZ guanylate kinase
PARP	poly ADP-ribose polymerase
PDI	protein disulfide isomerase
PERK	protein kinase RNA (PKR)-like ER kinase
p53 or TP53	protein that regulates the cell cycle and hence functions as a tumor suppressor.
PIKfyve	phosphoinositol 3-kinase (makes PI-3,5-bisphosphate - a regulator of endosome sorting)
PKR	protein kinase R
Proteases	
CTP	cathepsin protease
DPP4	dipeptidyl peptidase-4
Furin	a host protease
HAT	human airway trypsin-like protease
MASP1	MASP2, mannose-binding lectin (MBL)-associated serine proteases 1,2
PLP or PL^{pro}	papain-like protease
TMPRSS-2 or -4	transmembrane serine protease-2 or -4
PRR	pattern recognition receptor
Rb or pRb	retinoblastoma protein
STING	a key regulator of antiviral interferon (IFN)
Syntenin-1	contains tandemly repeated PDZ domains that bind the cytoplasmic C-terminal domains of a variety of transmembrane proteins.
TBK1	Tank-binding kinase-1
TLR	Toll-like receptor
TNF	tumor necrosis factor
TPC	two pore calcium (cation) channel
TPC1 and 2 are in endosomes	
TRAF3	TNF receptor-associated protein 3
XBP	X-box binding protein - a transcription factor involved in endoplasmic reticulum (ER) stress
(UPR)	regulates the unfolded protein response
INTRACELLULAR MEMBRANE STRUCTURES AND PROCESSES	
CM	convoluted intracellular membrane
CTL	cytotoxic T lymphocyte involved in the T-cell immune response
DMS	double membrane spherule
DMV	double membrane vesicle
EM	electron microscopy
ER	endoplasmic reticulum
ERB	ER body
ERGIC	ER-golgi intermediate compartment (a mobile complex that delivers cargo from the ER to the golgi)
GV	giant vesicle; giant vesiculation
IRF3	interferon (INF) regulatory factor 3
ISR	interferon-stimulated response;
ISRE	ISR element
JNK	the C-Jun N-terminal kinase signaling pathway
LCVC	lysosomal virion-containing cisternae (golgi-derived)

MLB	multilamellar (maze-like) body
NMR	nuclear magnetic resonance
Omegasome	a cell membrane-enclosed compartment enriched for phosphatidylinositol-3-P
PPI	protein-protein interaction
PTM	post translational modification
RO	(viral) replication organelle
RTC	replication-transcription complex
RVN	reticulovesicular network
TRS	transcription regulating sequence (at the beginning of a gene)
UPR	unfolded protein response
VLP	vesicle-like particle
VP,	vesicle packet or vesicle particle
zipped membranes	ER membranes that fold to form spherules
TERMS	
aa	amino acyl residue
AP-MS	affinity purification-mass spectrometry
ARDS	acute respiratory distress syndrome
BrU	bromouridine (U uridine)
CDRH	complementarity-determining region of heavy chain antibody (variable region)
Click chemistry	simple chemistry involving the joining of two molecules
CPZ	chlorpromazine (a clathrin inhibitor)
Cryo EM	cryogenic electron microscopy
CTD	C-terminal domain
dpi	days post infection
DPUP	domain preceding Ubi2 and PL2^{pro}
ESR	electron spin resonance
Fab	antigen-binding fragment of an antibody
FP	fusion peptide
GI (tract)	gastrointestinal (tract)
HD	hydrophobic domain (in a protein)
usually forms TMSs	
HMA	hexamethylene amiloride
hpi	hours post infection
HR1 HR2	heptad repeat 1, 2 . . .
IC	ion channel across a membrane
ISRE	interferon-stimulated response element
Mac	macrodomain
MβCD	methyl-β-cyclodextrins
Mitophagy	autophagy of mitochondria
NTD	N-terminal domain
ORF	open reading frame (a protein encoding gene)
PAMP	pathogen-associated molecular pattern
PBM	PDZ binding motif
PDZ	a protein domain in proteins that recognizes motifs in other proteins and is therefore a protein-protein interaction (PPI) domain
perinuclear	the cytoplasm immediately surrounding the nucleus
PPI	protein-protein interaction
PTM	post translational modification
RBD	receptor binding domain
RNP	ribonuclear protein
ROS	reactive oxygen species
RTC	replication-transcription complex
SUD	SARS unique domain
TM	transmembrane

TMD	transmembrane domain
TMS	transmembrane (α-helical) segment
TRS	transcription-regulating sequence
Ubl1	ubiquitin-like (domain) 1
VLP	virus-like particle

References

1. Gabutti, G.; d'Anchera, E.; Sandri, F.; Savio, M.; Stefanati, A. Coronavirus: Update related to the current outbreak of covid-19. *Infect. Dis. Ther.* **2020**, *9*, 1–13. [CrossRef]
2. Wang, N.; Shang, J.; Jiang, S.; Du, L. Subunit vaccines against emerging pathogenic human coronaviruses. *Front. Microbiol.* **2020**, *11*, 298. [CrossRef]
3. Park, M.; Cook, A.R.; Lim, J.T.; Sun, Y.; Dickens, B.L. A systematic review of covid-19 epidemiology based on current evidence. *J. Clin. Med.* **2020**, *9*, 967. [CrossRef]
4. Shereen, M.A.; Khan, S.; Kazmi, A.; Bashir, N.; Siddique, R. Covid-19 infection: Origin, transmission, and characteristics of human coronaviruses. *J. Adv. Res.* **2020**, *24*, 91–98. [CrossRef]
5. Amanat, F.; Krammer, F. Sars-cov-2 vaccines: Status report. *Immunity* **2020**, *52*, 583–589. [CrossRef]
6. Rabi, F.A.; Al Zoubi, M.S.; Kasasbeh, G.A.; Salameh, D.M.; Al-Nasser, A.D. Sars-cov-2 and coronavirus disease 2019: What we know so far. *Pathogens* **2020**, *9*, 231. [CrossRef]
7. Kakodkar, P.; Kaka, N.; Baig, M.N. A comprehensive literature review on the clinical presentation, and management of the pandemic coronavirus disease 2019 (covid-19). *Cureus* **2020**, *12*, e7560. [CrossRef]
8. Rokni, M.; Ghasemi, V.; Tavakoli, Z. Immune responses and pathogenesis of sars-cov-2 during an outbreak in iran: Comparison with sars and mers. *Rev. Med. Virol.* **2020**, *30*, e2107. [CrossRef]
9. Emami, A.; Javanmardi, F.; Pirbonyeh, N.; Akbari, A. Prevalence of underlying diseases in hospitalized patients with covid-19: A systematic review and meta-analysis. *Arch. Acad. Emerg. Med.* **2020**, *8*, e35.
10. Shahid, Z.; Kalayanamitra, R.; McClafferty, B.; Kepko, D.; Ramgobin, D.; Patel, R.; Aggarwal, C.S.; Vunnam, R.; Sahu, N.; Bhatt, D.; et al. Covid-19 and older adults: What we know. *J. Am. Geriatr. Soc.* **2020**, *68*, 926–929. [CrossRef]
11. Smyk, W.; Janik, M.K.; Portincasa, P.; Milkiewicz, P.; Lammert, F.; Krawczyk, M. Covid-19: Focus on the lungs but do not forget the gastrointestinal tract. *Eur. J. Clin. Investig.* **2020**, *50*, e13276. [CrossRef]
12. Tariku, M.; Hajure, M. Available evidence and ongoing hypothesis on corona virus (covid-19) and psychosis: Is corona virus and psychosis related? A narrative review. *Psychol. Res. Behav. Manag.* **2020**, *13*, 701–704. [CrossRef]
13. Gheblawi, M.; Wang, K.; Viveiros, A.; Nguyen, Q.; Zhong, J.C.; Turner, A.J.; Raizada, M.K.; Grant, M.B.; Oudit, G.Y. Angiotensin-converting enzyme 2: Sars-cov-2 receptor and regulator of the renin-angiotensin system: Celebrating the 20th anniversary of the discovery of ace2. *Circ. Res.* **2020**, *126*, 1456–1474. [CrossRef]
14. Md Insiat Islam, R. Current drugs with potential for treatment of covid-19: A literature review. *J. Pharm. Pharm. Sci.* **2020**, *23*, 58–64. [CrossRef]
15. Luo, H.; Tang, Q.L.; Shang, Y.X.; Liang, S.B.; Yang, M.; Robinson, N.; Liu, J.P. Can chinese medicine be used for prevention of corona virus disease 2019 (covid-19)? A review of historical classics, research evidence and current prevention programs. *Chin. J. Integr. Med.* **2020**, *26*, 243–250. [CrossRef]
16. Martinez, M.A. Compounds with therapeutic potential against novel respiratory 2019 coronavirus. *Antimicrob. Agents Chemother.* **2020**, *64*, e00399-20. [CrossRef]
17. Amirian, E.S.; Levy, J.K. Current knowledge about the antivirals remdesivir (gs-5734) and gs-441524 as therapeutic options for coronaviruses. *One Health* **2020**, *9*, 100128. [CrossRef]
18. Gbinigie, K.; Frie, K. Should chloroquine and hydroxychloroquine be used to treat covid-19? A rapid review. *BJGP Open* **2020**, *4*, bjgpopen20X101069. [CrossRef]
19. Li, R.; Yin, K.; Zhang, K.; Wang, Y.Y.; Wu, Q.P.; Tang, S.B.; Cheng, J.D. Application prospects of virtual autopsy in forensic pathological investigations on covid-19. *Fa Yi Xue Za Zhi* **2020**, *36*, 149–156.
20. Barlow, A.; Landolf, K.M.; Barlow, B.; Yeung, S.Y.A.; Heavner, J.J.; Claassen, C.W.; Heavner, M.S. Review of emerging pharmacotherapy for the treatment of coronavirus disease 2019. *Pharmacotherapy* **2020**, *40*, 416–437. [CrossRef]
21. Mousavizadeh, L.; Ghasemi, S. Genotype and phenotype of covid-19: Their roles in pathogenesis. *J. Microbiol. Immunol. Infect.* **2020**. [CrossRef]
22. Prajapat, M.; Sarma, P.; Shekhar, N.; Avti, P.; Sinha, S.; Kaur, H.; Kumar, S.; Bhattacharyya, A.; Kumar, H.; Bansal, S.; et al. Drug targets for corona virus: A systematic review. *Indian J. Pharmacol.* **2020**, *52*, 56–65.
23. Deng, L.; Li, C.; Zeng, Q.; Liu, X.; Li, X.; Zhang, H.; Hong, Z.; Xia, J. Arbidol combined with lpv/r versus lpv/r alone against corona virus disease 2019: A retrospective cohort study. *J. Infect.* **2020**, *81*, e1–e5. [CrossRef]
24. Li, S.; Yuan, L.; Dai, G.; Chen, R.A.; Liu, D.X.; Fung, T.S. Regulation of the er stress response by the ion channel activity of the infectious bronchitis coronavirus envelope protein modulates virion release, apoptosis, viral fitness, and pathogenesis. *Front. Microbiol.* **2019**, *10*, 3022. [CrossRef]
25. Tang, T.; Bidon, M.; Jaimes, J.A.; Whittaker, G.R.; Daniel, S. Coronavirus membrane fusion mechanism offers a potential target for antiviral development. *Antivir. Res.* **2020**, *178*, 104792. [CrossRef]

26. Hofmann, H.; Pohlmann, S. Cellular entry of the sars coronavirus. *Trends Microbiol.* **2004**, *12*, 466–472. [CrossRef]
27. Perrotta, F.; Matera, M.G.; Cazzola, M.; Bianco, A. Severe respiratory sars-cov2 infection: Does ace2 receptor matter? *Respir. Med.* **2020**, *168*, 105996. [CrossRef]
28. Fagerberg, L.; Hallstrom, B.M.; Oksvold, P.; Kampf, C.; Djureinovic, D.; Odeberg, J.; Habuka, M.; Tahmasebpoor, S.; Danielsson, A.; Edlund, K.; et al. Analysis of the human tissue-specific expression by genome-wide integration of transcriptomics and antibody-based proteomics. *Mol. Cell Proteom.* **2014**, *13*, 397–406. [CrossRef]
29. Hamming, I.; Timens, W.; Bulthuis, M.L.; Lely, A.T.; Navis, G.; van Goor, H. Tissue distribution of ace2 protein, the functional receptor for sars coronavirus. A first step in understanding sars pathogenesis. *J. Pathol.* **2004**, *203*, 631–637. [CrossRef]
30. Barlan, A.; Zhao, J.; Sarkar, M.K.; Li, K.; McCray, P.B., Jr.; Perlman, S.; Gallagher, T. Receptor variation and susceptibility to middle east respiratory syndrome coronavirus infection. *J. Virol.* **2014**, *88*, 4953–4961. [CrossRef]
31. Hoffmann, M.; Kleine-Weber, H.; Schroeder, S.; Kruger, N.; Herrler, T.; Erichsen, S.; Schiergens, T.S.; Herrler, G.; Wu, N.H.; Nitsche, A.; et al. Sars-cov-2 cell entry depends on ace2 and tmprss2 and is blocked by a clinically proven protease inhibitor. *Cell* **2020**, *181*, 271–280.e8. [CrossRef]
32. Ou, X.; Liu, Y.; Lei, X.; Li, P.; Mi, D.; Ren, L.; Guo, L.; Guo, R.; Chen, T.; Hu, J.; et al. Characterization of spike glycoprotein of sars-cov-2 on virus entry and its immune cross-reactivity with sars-cov. *Nat. Commun.* **2020**, *11*, 1620. [CrossRef]
33. Jaimes, J.A.; Millet, J.K.; Whittaker, G.R. Proteolytic cleavage of the sars-cov-2 spike protein and the role of the novel s1/s2 site. *iScience* **2020**, *23*, 101212. [CrossRef]
34. Wang, H.; Yang, P.; Liu, K.; Guo, F.; Zhang, Y.; Zhang, G.; Jiang, C. Sars coronavirus entry into host cells through a novel clathrin- and caveolae-independent endocytic pathway. *Cell Res.* **2008**, *18*, 290–301. [CrossRef]
35. Lakadamyali, M.; Rust, M.J.; Zhuang, X. Endocytosis of influenza viruses. *Microbes Infect.* **2004**, *6*, 929–936. [CrossRef]
36. Shulla, A.; Heald-Sargent, T.; Subramanya, G.; Zhao, J.; Perlman, S.; Gallagher, T. A transmembrane serine protease is linked to the severe acute respiratory syndrome coronavirus receptor and activates virus entry. *J. Virol.* **2011**, *85*, 873–882. [CrossRef]
37. Barrow, E.; Nicola, A.V.; Liu, J. Multiscale perspectives of virus entry via endocytosis. *Virol. J.* **2013**, *10*, 177. [CrossRef]
38. Anderson, H.A.; Chen, Y.; Norkin, L.C. Bound simian virus 40 translocates to caveolin-enriched membrane domains, and its entry is inhibited by drugs that selectively disrupt caveolae. *Mol. Biol. Cell* **1996**, *7*, 1825–1834. [CrossRef]
39. Devadas, D.; Koithan, T.; Diestel, R.; Prank, U.; Sodeik, B.; Dohner, K. Herpes simplex virus internalization into epithelial cells requires na+/h+ exchangers and p21-activated kinases but neither clathrin- nor caveolin-mediated endocytosis. *J. Virol.* **2014**, *88*, 13378–13395. [CrossRef]
40. Yang, N.; Shen, H.M. Targeting the endocytic pathway and autophagy process as a novel therapeutic strategy in covid-19. *Int. J. Biol. Sci.* **2020**, *16*, 1724–1731. [CrossRef]
41. Milewska, A.; Nowak, P.; Owczarek, K.; Szczepanski, A.; Zarebski, M.; Hoang, A.; Berniak, K.; Wojarski, J.; Zeglen, S.; Baster, Z.; et al. Entry of human coronavirus nl63 into the cell. *J. Virol.* **2018**, *92*, e01933-17. [CrossRef] [PubMed]
42. Liu, D.X.; Liang, J.Q.; Fung, T.S. Human coronavirus-229e, -oc43, -nl63, and -hku1. *Ref. Modul. Life Sci.* **2020**, *77*, 4435–4438.
43. Liu, H.; Liu, Y.; Liu, S.; Pang, D.W.; Xiao, G. Clathrin-mediated endocytosis in living host cells visualized through quantum dot labeling of infectious hematopoietic necrosis virus. *J. Virol.* **2011**, *85*, 6252–6262. [CrossRef]
44. Cong, Y.; Hart, B.J.; Gross, R.; Zhou, H.; Frieman, M.; Bollinger, L.; Wada, J.; Hensley, L.E.; Jahrling, P.B.; Dyall, J.; et al. Mers-cov pathogenesis and antiviral efficacy of licensed drugs in human monocyte-derived antigen-presenting cells. *PLoS ONE* **2018**, *13*, e0194868. [CrossRef]
45. Currinn, H.; Guscott, B.; Balklava, Z.; Rothnie, A.; Wassmer, T. App controls the formation of pi(3,5)p(2) vesicles through its binding of the pikfyve complex. *Cell Mol. Life Sci.* **2016**, *73*, 393–408. [CrossRef]
46. McCartney, A.J.; Zhang, Y.; Weisman, L.S. Phosphatidylinositol 3,5-bisphosphate: Low abundance, high significance. *Bioessays* **2014**, *36*, 52–64. [CrossRef]
47. Pitt, S.J.; Reilly-O'Donnell, B.; Sitsapesan, R. Exploring the biophysical evidence that mammalian two-pore channels are naadp-activated calcium-permeable channels. *J. Physiol.* **2016**, *594*, 4171–4179. [CrossRef]
48. Chao, Y.K.; Schludi, V.; Chen, C.C.; Butz, E.; Nguyen, O.N.P.; Muller, M.; Kruger, J.; Kammerbauer, C.; Ben-Johny, M.; Vollmar, A.M.; et al. Tpc2 polymorphisms associated with a hair pigmentation phenotype in humans result in gain of channel function by independent mechanisms. *Proc. Natl. Acad. Sci. USA* **2017**, *114*, E8595–E8602. [CrossRef]
49. Zhang, Z.H.; Lu, Y.Y.; Yue, J. Two pore channel 2 differentially modulates neural differentiation of mouse embryonic stem cells. *PLoS ONE* **2013**, *8*, e66077. [CrossRef]
50. Zhu, M.X.; Ma, J.; Parrington, J.; Galione, A.; Evans, A.M. Tpcs: Endolysosomal channels for ca2+ mobilization from acidic organelles triggered by naadp. *FEBS Lett.* **2010**, *584*, 1966–1974. [CrossRef]
51. Kilpatrick, B.S.; Eden, E.R.; Hockey, L.N.; Yates, E.; Futter, C.E.; Patel, S. An endosomal naadp-sensitive two-pore ca(2+) channel regulates er-endosome membrane contact sites to control growth factor signaling. *Cell Rep.* **2017**, *18*, 1636–1645. [CrossRef]
52. Grimm, C.; Chen, C.C.; Wahl-Schott, C.; Biel, M. Two-pore channels: Catalyzers of endolysosomal transport and function. *Front. Pharmacol.* **2017**, *8*, 45. [CrossRef] [PubMed]
53. Sakurai, Y.; Kolokoltsov, A.A.; Chen, C.C.; Tidwell, M.W.; Bauta, W.E.; Klugbauer, N.; Grimm, C.; Wahl-Schott, C.; Biel, M.; Davey, R.A. Ebola virus. Two-pore channels control ebola virus host cell entry and are drug targets for disease treatment. *Science* **2015**, *347*, 995–998.

54. Gunaratne, G.S.; Yang, Y.; Li, F.; Walseth, T.F.; Marchant, J.S. Naadp-dependent ca(2+) signaling regulates middle east respiratory syndrome-coronavirus pseudovirus translocation through the endolysosomal system. *Cell Calcium* **2018**, *75*, 30–41. [CrossRef] [PubMed]
55. Pager, C.T.; Dutch, R.E. Cathepsin l is involved in proteolytic processing of the hendra virus fusion protein. *J. Virol.* **2005**, *79*, 12714–12720. [CrossRef]
56. Simmons, G.; Gosalia, D.N.; Rennekamp, A.J.; Reeves, J.D.; Diamond, S.L.; Bates, P. Inhibitors of cathepsin l prevent severe acute respiratory syndrome coronavirus entry. *Proc. Natl. Acad. Sci. USA* **2005**, *102*, 11876–11881. [CrossRef] [PubMed]
57. Angelini, M.M.; Akhlaghpour, M.; Neuman, B.W.; Buchmeier, M.J. Severe acute respiratory syndrome coronavirus nonstructural proteins 3, 4, and 6 induce double-membrane vesicles. *mBio* **2013**, *4*, e00524-13. [CrossRef]
58. Te Velthuis, A.J.; van den Worm, S.H.; Snijder, E.J. The sars-coronavirus nsp7+nsp8 complex is a unique multimeric rna polymerase capable of both de novo initiation and primer extension. *Nucleic Acids Res.* **2012**, *40*, 1737–1747. [CrossRef]
59. Fehr, A.R.; Perlman, S. Coronaviruses: An overview of their replication and pathogenesis. *Methods Mol. Biol.* **2015**, *1282*, 1–23.
60. Rota, P.A.; Oberste, M.S.; Monroe, S.S.; Nix, W.A.; Campagnoli, R.; Icenogle, J.P.; Penaranda, S.; Bankamp, B.; Maher, K.; Chen, M.H.; et al. Characterization of a novel coronavirus associated with severe acute respiratory syndrome. *Science* **2003**, *300*, 1394–1399. [CrossRef]
61. Yount, B.; Roberts, R.S.; Lindesmith, L.; Baric, R.S. Rewiring the severe acute respiratory syndrome coronavirus (sars-cov) transcription circuit: Engineering a recombination-resistant genome. *Proc. Natl. Acad. Sci. USA* **2006**, *103*, 12546–12551. [CrossRef]
62. Hagemeijer, M.C.; Vonk, A.M.; Monastyrska, I.; Rottier, P.J.; de Haan, C.A. Visualizing coronavirus rna synthesis in time by using click chemistry. *J. Virol.* **2012**, *86*, 5808–5816. [CrossRef] [PubMed]
63. Sawicki, S.G.; Sawicki, D.L.; Siddell, S.G. A contemporary view of coronavirus transcription. *J. Virol.* **2007**, *81*, 20–29. [CrossRef] [PubMed]
64. Knoops, K.; Swett-Tapia, C.; van den Worm, S.H.; Velthuis, A.J.T.; Koster, A.J.; Mommaas, A.M.; Snijder, E.J.; Kikkert, M. Integrity of the early secretory pathway promotes, but is not required for, severe acute respiratory syndrome coronavirus rna synthesis and virus-induced remodeling of endoplasmic reticulum membranes. *J. Virol.* **2010**, *84*, 833–846. [CrossRef] [PubMed]
65. Graham, R.L.; Denison, M.R. Replication of murine hepatitis virus is regulated by papain-like proteinase 1 processing of nonstructural proteins 1, 2, and 3. *J. Virol.* **2006**, *80*, 11610–11620. [CrossRef]
66. Muramatsu, T.; Takemoto, C.; Kim, Y.T.; Wang, H.; Nishii, W.; Terada, T.; Shirouzu, M.; Yokoyama, S. Sars-cov 3cl protease cleaves its c-terminal autoprocessing site by novel subsite cooperativity. *Proc. Natl. Acad. Sci. USA* **2016**, *113*, 12997–13002. [CrossRef]
67. Stobart, C.C.; Sexton, N.R.; Munjal, H.; Lu, X.; Molland, K.L.; Tomar, S.; Mesecar, A.D.; Denison, M.R. Chimeric exchange of coronavirus nsp5 proteases (3clpro) identifies common and divergent regulatory determinants of protease activity. *J. Virol.* **2013**, *87*, 12611–12618. [CrossRef]
68. Plant, E.P.; Perez-Alvarado, G.C.; Jacobs, J.L.; Mukhopadhyay, B.; Hennig, M.; Dinman, J.D. A three-stemmed mrna pseudoknot in the sars coronavirus frameshift signal. *PLoS Biol.* **2005**, *3*, e172. [CrossRef]
69. Irigoyen, N.; Firth, A.E.; Jones, J.D.; Chung, B.Y.; Siddell, S.G.; Brierley, I. High-resolution analysis of coronavirus gene expression by rna sequencing and ribosome profiling. *PLoS Pathog.* **2016**, *12*, e1005473. [CrossRef]
70. Narayanan, K.; Ramirez, S.I.; Lokugamage, K.G.; Makino, S. Coronavirus nonstructural protein 1: Common and distinct functions in the regulation of host and viral gene expression. *Virus Res.* **2015**, *202*, 89–100. [CrossRef]
71. Hackbart, M.; Deng, X.; Baker, S.C. Coronavirus endoribonuclease targets viral polyuridine sequences to evade activating host sensors. *Proc. Natl. Acad. Sci. USA* **2020**, *117*, 8094–8103. [CrossRef]
72. Angeletti, S.; Benvenuto, D.; Bianchi, M.; Giovanetti, M.; Pascarella, S.; Ciccozzi, M. Covid-2019: The role of the nsp2 and nsp3 in its pathogenesis. *J. Med. Virol.* **2020**, *92*, 584–588. [CrossRef] [PubMed]
73. Menachery, V.D.; Debbink, K.; Baric, R.S. Coronavirus non-structural protein 16: Evasion, attenuation, and possible treatments. *Virus Res.* **2014**, *194*, 191–199. [CrossRef] [PubMed]
74. Kirchdoerfer, R.N.; Ward, A.B. Structure of the sars-cov nsp12 polymerase bound to nsp7 and nsp8 co-factors. *Nat. Commun.* **2019**, *10*, 2342. [CrossRef] [PubMed]
75. McBride, R.; Fielding, B.C. The role of severe acute respiratory syndrome (sars)-coronavirus accessory proteins in virus pathogenesis. *Viruses* **2012**, *4*, 2902–2923. [CrossRef] [PubMed]
76. Tang, X.; Li, G.; Vasilakis, N.; Zhang, Y.; Shi, Z.; Zhong, Y.; Wang, L.F.; Zhang, S. Differential stepwise evolution of sars coronavirus functional proteins in different host species. *BMC Evol. Biol.* **2009**, *9*, 52. [CrossRef]
77. Egloff, M.P.; Ferron, F.; Campanacci, V.; Longhi, S.; Rancurel, C.; Dutartre, H.; Snijder, E.J.; Gorbalenya, A.E.; Cambillau, C.; Canard, B. The severe acute respiratory syndrome-coronavirus replicative protein nsp9 is a single-stranded rna-binding subunit unique in the rna virus world. *Proc. Natl. Acad. Sci. USA* **2004**, *101*, 3792–3796. [CrossRef]
78. Ma, Y.; Wu, L.; Shaw, N.; Gao, Y.; Wang, J.; Sun, Y.; Lou, Z.; Yan, L.; Zhang, R.; Rao, Z. Structural basis and functional analysis of the sars coronavirus nsp14-nsp10 complex. *Proc. Natl. Acad. Sci. USA* **2015**, *112*, 9436–9441. [CrossRef]
79. Te Velthuis, A.J.; Arnold, J.J.; Cameron, C.E.; van den Worm, S.H.; Snijder, E.J. The rna polymerase activity of sars-coronavirus nsp12 is primer dependent. *Nucleic Acids Res.* **2010**, *38*, 203–214. [CrossRef]
80. Chen, C.C.; Kruger, J.; Sramala, I.; Hsu, H.J.; Henklein, P.; Chen, Y.M.; Fischer, W.B. Orf8a of sars-cov forms an ion channel: Experiments and molecular dynamics simulations. *Biochim. Biophys. Acta* **2011**, *1808*, 572–579. [CrossRef]

81. Ivanov, K.A.; Thiel, V.; Dobbe, J.C.; van der Meer, Y.; Snijder, E.J.; Ziebuhr, J. Multiple enzymatic activities associated with severe acute respiratory syndrome coronavirus helicase. *J. Virol.* **2004**, *78*, 5619–5632. [CrossRef]
82. Shi, C.S.; Nabar, N.R.; Huang, N.N.; Kehrl, J.H. Sars-coronavirus open reading frame-8b triggers intracellular stress pathways and activates nlrp3 inflammasomes. *Cell Death Discov.* **2019**, *5*, 101. [CrossRef] [PubMed]
83. Shi, C.S.; Qi, H.Y.; Boularan, C.; Huang, N.N.; Abu-Asab, M.; Shelhamer, J.H.; Kehrl, J.H. Sars-coronavirus open reading frame-9b suppresses innate immunity by targeting mitochondria and the mavs/traf3/traf6 signalosome. *J. Immunol.* **2014**, *193*, 3080–3089. [CrossRef] [PubMed]
84. Chen, Y.; Guo, D. Molecular mechanisms of coronavirus rna capping and methylation. *Virol. Sin.* **2016**, *31*, 3–11. [CrossRef] [PubMed]
85. Angelini, M.M.; Neuman, B.W.; Buchmeier, M.J. Untangling membrane rearrangement in the nidovirales. *DNA Cell Biol.* **2014**, *33*, 122–127. [CrossRef] [PubMed]
86. Nagy, P.D.; Strating, J.R.; van Kuppeveld, F.J. Building viral replication organelles: Close encounters of the membrane types. *PLoS Pathog.* **2016**, *12*, e1005912. [CrossRef] [PubMed]
87. van der Hoeven, B.D.; Oudshoorn, D.; Koster, A.J.; Snijder, E.J.; Kikkert, M.; Barcena, M. Biogenesis and architecture of arterivirus replication organelles. *Virus Res.* **2016**, *220*, 70–90. [CrossRef] [PubMed]
88. Hagemeijer, M.C.; Monastyrska, I.; Griffith, J.; van der Sluijs, P.; Voortman, J.; van Bergen en Henegouwen, P.M.; Vonk, A.M.; Rottier, P.J.; Reggiori, F.; de Haan, C.A. Membrane rearrangements mediated by coronavirus nonstructural proteins 3 and 4. *Virology* **2014**, *458–459*, 125–135. [CrossRef]
89. Oudshoorn, D.; Rijs, K.; Limpens, R.; Groen, K.; Koster, A.J.; Snijder, E.J.; Kikkert, M.; Barcena, M. Expression and cleavage of middle east respiratory syndrome coronavirus nsp3-4 polyprotein induce the formation of double-membrane vesicles that mimic those associated with coronaviral rna replication. *mBio* **2017**, *8*, e01658-17. [CrossRef]
90. Knoops, K.; Kikkert, M.; Worm, S.H.; Zevenhoven-Dobbe, J.C.; van der Meer, Y.; Koster, A.J.; Mommaas, A.M.; Snijder, E.J. Sars-coronavirus replication is supported by a reticulovesicular network of modified endoplasmic reticulum. *PLoS Biol.* **2008**, *6*, e226. [CrossRef]
91. Ulasli, M.; Verheije, M.H.; de Haan, C.A.; Reggiori, F. Qualitative and quantitative ultrastructural analysis of the membrane rearrangements induced by coronavirus. *Cell Microbiol.* **2010**, *12*, 844–861. [CrossRef]
92. Snijder, E.J.; Limpens, R.; de Wilde, A.H.; de Jong, A.W.M.; Zevenhoven-Dobbe, J.C.; Maier, H.J.; Faas, F.; Koster, A.J.; Barcena, M. A unifying structural and functional model of the coronavirus replication organelle: Tracking down rna synthesis. *PLoS Biol.* **2020**, *18*, e3000715. [CrossRef] [PubMed]
93. Doyle, N.; Hawes, P.C.; Simpson, J.; Adams, L.H.; Maier, H.J. The porcine deltacoronavirus replication organelle comprises double-membrane vesicles and zippered endoplasmic reticulum with double-membrane spherules. *Viruses* **2019**, *11*, 1030. [CrossRef] [PubMed]
94. Maier, H.J.; Hawes, P.C.; Cottam, E.M.; Mantell, J.; Verkade, P.; Monaghan, P.; Wileman, T.; Britton, P. Infectious bronchitis virus generates spherules from zippered endoplasmic reticulum membranes. *mBio* **2013**, *4*, e00801-13. [CrossRef]
95. Zhou, X.; Cong, Y.; Veenendaal, T.; Klumperman, J.; Shi, D.; Mari, M.; Reggiori, F. Ultrastructural characterization of membrane rearrangements induced by porcine epidemic diarrhea virus infection. *Viruses* **2017**, *9*, 251. [CrossRef] [PubMed]
96. Zhang, W.; Chen, K.; Zhang, X.; Guo, C.; Chen, Y.; Liu, X. An integrated analysis of membrane remodeling during porcine reproductive and respiratory syndrome virus replication and assembly. *PLoS ONE* **2018**, *13*, e0200919. [CrossRef]
97. Sakai, Y.; Kawachi, K.; Terada, Y.; Omori, H.; Matsuura, Y.; Kamitani, W. Two-amino acids change in the nsp4 of sars coronavirus abolishes viral replication. *Virology* **2017**, *510*, 165–174. [CrossRef]
98. Lei, J.; Kusov, Y.; Hilgenfeld, R. Nsp3 of coronaviruses: Structures and functions of a large multi-domain protein. *Antivir. Res.* **2018**, *149*, 58–74. [CrossRef]
99. Carmona-Gutierrez, D.; Bauer, M.A.; Zimmermann, A.; Kainz, K.; Hofer, S.J.; Kroemer, G.; Madeo, F. Digesting the crisis: Autophagy and coronaviruses. *Microb. Cell* **2020**, *7*, 119–128. [CrossRef]
100. Maier, H.J.; Neuman, B.W.; Bickerton, E.; Keep, S.M.; Alrashedi, H.; Hall, R.; Britton, P. Extensive coronavirus-induced membrane rearrangements are not a determinant of pathogenicity. *Sci. Rep.* **2016**, *6*, 27126. [CrossRef]
101. Belov, G.A.; Sztul, E. Rewiring of cellular membrane homeostasis by picornaviruses. *J. Virol.* **2014**, *88*, 9478–9489. [CrossRef]
102. Chamberlain, N.; Anathy, V. Pathological consequences of the unfolded protein response and downstream protein disulphide isomerases in pulmonary viral infection and disease. *J. Biochem.* **2020**, *167*, 173–184. [CrossRef] [PubMed]
103. Reggiori, F.; Monastyrska, I.; Verheije, M.H.; Cali, T.; Ulasli, M.; Bianchi, S.; Bernasconi, R.; de Haan, C.A.; Molinari, M. Coronaviruses hijack the lc3-i-positive edemosomes, er-derived vesicles exporting short-lived erad regulators, for replication. *Cell Host Microbe* **2010**, *7*, 500–508. [CrossRef] [PubMed]
104. Zhao, Z.; Thackray, L.B.; Miller, B.C.; Lynn, T.M.; Becker, M.M.; Ward, E.; Mizushima, N.N.; Denison, M.R.; Virgin, H.W., IV. Coronavirus replication does not require the autophagy gene atg5. *Autophagy* **2007**, *3*, 581–585. [CrossRef] [PubMed]
105. Schneider, M.; Ackermann, K.; Stuart, M.; Wex, C.; Protzer, U.; Schatzl, H.M.; Gilch, S. Severe acute respiratory syndrome coronavirus replication is severely impaired by mg132 due to proteasome-independent inhibition of m-calpain. *J. Virol.* **2012**, *86*, 10112–10122. [CrossRef] [PubMed]
106. Neuman, B.W. Bioinformatics and functional analyses of coronavirus nonstructural proteins involved in the formation of replicative organelles. *Antivir. Res.* **2016**, *135*, 97–107. [CrossRef]

107. Kusov, Y.; Tan, J.; Alvarez, E.; Enjuanes, L.; Hilgenfeld, R. A g-quadruplex-binding macrodomain within the "sars-unique domain" is essential for the activity of the sars-coronavirus replication-transcription complex. *Virology* **2015**, *484*, 313–322. [CrossRef]
108. Johnson, M.A.; Chatterjee, A.; Neuman, B.W.; Wuthrich, K. Sars coronavirus unique domain: Three-domain molecular architecture in solution and rna binding. *J. Mol. Biol.* **2010**, *400*, 724–742. [CrossRef]
109. Baez-Santos, Y.M.; John, S.E.S.; Mesecar, A.D. The sars-coronavirus papain-like protease: Structure, function and inhibition by designed antiviral compounds. *Antivir. Res.* **2015**, *115*, 21–38. [CrossRef]
110. Nakashima, H.; Nguyen, T.; Goins, W.F.; Chiocca, E.A. Interferon-stimulated gene 15 (isg15) and isg15-linked proteins can associate with members of the selective autophagic process, histone deacetylase 6 (hdac6) and sqstm1/p62. *J. Biol. Chem.* **2015**, *290*, 1485–1495. [CrossRef]
111. Ma-Lauer, Y.; Carbajo-Lozoya, J.; Hein, M.Y.; Muller, M.A.; Deng, W.; Lei, J.; Meyer, B.; Kusov, Y.; von Brunn, B.; Bairad, D.R.; et al. P53 down-regulates sars coronavirus replication and is targeted by the sars-unique domain and plpro via e3 ubiquitin ligase rchy1. *Proc. Natl. Acad. Sci. USA* **2016**, *113*, E5192–E5201. [CrossRef]
112. Ozaki, T.; Nakagawara, A. Role of p53 in cell death and human cancers. *Cancers* **2011**, *3*, 994–1013. [CrossRef] [PubMed]
113. Grimley, P.M.; Berezesky, I.K.; Friedman, R.M. Cytoplasmic structures associated with an arbovirus infection: Loci of viral ribonucleic acid synthesis. *J. Virol.* **1968**, *2*, 1326–1338. [CrossRef] [PubMed]
114. Gosert, R.; Kanjanahaluethai, A.; Egger, D.; Bienz, K.; Baker, S.C. Rna replication of mouse hepatitis virus takes place at double-membrane vesicles. *J. Virol.* **2002**, *76*, 3697–3708. [CrossRef] [PubMed]
115. Wolff, G.; Limpens, R.; Zevenhoven-Dobbe, J.C.; Laugks, U.; Zheng, S.; de Jong, A.W.M.; Koning, R.I.; Agard, D.A.; Grunewald, K.; Koster, A.J.; et al. A molecular pore spans the double membrane of the coronavirus replication organelle. *Science* **2020**, *369*, 1395–1398. [CrossRef] [PubMed]
116. Imbert, I.; Snijder, E.J.; Dimitrova, M.; Guillemot, J.C.; Lecine, P.; Canard, B. The sars-coronavirus plnc domain of nsp3 as a replication/transcription scaffolding protein. *Virus Res.* **2008**, *133*, 136–148. [CrossRef]
117. Hilbert, B.J.; Hayes, J.A.; Stone, N.P.; Duffy, C.M.; Sankaran, B.; Kelch, B.A. Structure and mechanism of the atpase that powers viral genome packaging. *Proc. Natl. Acad. Sci USA* **2015**, *112*, E3792–E3799. [CrossRef]
118. Taraporewala, Z.F.; Patton, J.T. Nonstructural proteins involved in genome packaging and replication of rotaviruses and other members of the reoviridae. *Virus Res.* **2004**, *101*, 57–66. [CrossRef]
119. Trus, B.L.; Cheng, N.; Newcomb, W.W.; Homa, F.L.; Brown, J.C.; Steven, A.C. Structure and polymorphism of the ul6 portal protein of herpes simplex virus type 1. *J. Virol.* **2004**, *78*, 12668–12671. [CrossRef]
120. Kannan, S.; Ali, P.S.S.; Sheeza, A.; Hemalatha, K. Covid-19 (novel coronavirus 2019)—Recent trends. *Eur. Rev. Med. Pharm. Sci.* **2020**, *24*, 2006–2011.
121. Chang, C.K.; Hou, M.H.; Chang, C.F.; Hsiao, C.D.; Huang, T.H. The sars coronavirus nucleocapsid protein—Forms and functions. *Antivir. Res.* **2014**, *103*, 39–50. [CrossRef]
122. Satarker, S.; Nampoothiri, M. Structural proteins in severe acute respiratory syndrome coronavirus-2. *Arch. Med. Res.* **2020**, *51*, 482–491. [CrossRef] [PubMed]
123. McBride, R.; van Zyl, M.; Fielding, B.C. The coronavirus nucleocapsid is a multifunctional protein. *Viruses* **2014**, *6*, 2991–3018. [CrossRef] [PubMed]
124. Malik, Y.A. Properties of coronavirus and sars-cov-2. *Malays. J. Pathol.* **2020**, *42*, 3–11. [PubMed]
125. Zhang, X.; Shi, H.; Chen, J.; Shi, D.; Dong, H.; Feng, L. Identification of the interaction between vimentin and nucleocapsid protein of transmissible gastroenteritis virus. *Virus Res.* **2015**, *200*, 56–63. [CrossRef] [PubMed]
126. Woo, J.; Lee, E.Y.; Lee, M.; Kim, T.; Cho, Y.E. An in vivo cell-based assay for investigating the specific interaction between the sars-cov n-protein and its viral rna packaging sequence. *Biochem. Biophys. Res. Commun.* **2019**, *520*, 499–506. [CrossRef] [PubMed]
127. Barik, S. Genus-specific pattern of intrinsically disordered central regions in the nucleocapsid protein of coronaviruses. *Comput. Struct. Biotechnol. J.* **2020**, *18*, 1884–1890. [CrossRef]
128. Ye, Q.; West, A.M.V.; Silletti, S.; Corbett, K.D. Architecture and self-assembly of the sars-cov-2 nucleocapsid protein. *Protein Sci.* **2020**. [CrossRef]
129. Lu, S.; Ye, Q.; Singh, D.; Villa, E.; Cleveland, D.W.; Corbett, K.D. The sars-cov-2 nucleocapsid phosphoprotein forms mutually exclusive condensates with rna and the membrane-associated m protein. *bioRxiv* **2020**. [CrossRef]
130. Carlson, C.R.; Asfaha, J.B.; Ghent, C.M.; Howard, C.J.; Hartooni, N.; Morgan, D.O. Phosphorylation modulates liquid-liquid phase separation of the sars-cov-2 n protein. *bioRxiv* **2020**. [CrossRef]
131. Cong, Y.; Ulasli, M.; Schepers, H.; Mauthe, M.; V'Kovski, P.; Kriegenburg, F.; Thiel, V.; de Haan, C.A.M.; Reggiori, F. Nucleocapsid protein recruitment to replication-transcription complexes plays a crucial role in coronaviral life cycle. *J. Virol.* **2020**, *94*. [CrossRef]
132. Zlotnick, A. Theoretical aspects of virus capsid assembly. *J. Mol. Recognit.* **2005**, *18*, 479–490. [CrossRef] [PubMed]
133. Surjit, M.; Liu, B.; Kumar, P.; Chow, V.T.; Lal, S.K. The nucleocapsid protein of the sars coronavirus is capable of self-association through a c-terminal 209 amino acid interaction domain. *Biochem. Biophys. Res. Commun.* **2004**, *317*, 1030–1036. [CrossRef]
134. Zuniga, S.; Cruz, J.L.; Sola, I.; Mateos-Gomez, P.A.; Palacio, L.; Enjuanes, L. Coronavirus nucleocapsid protein facilitates template switching and is required for efficient transcription. *J. Virol.* **2010**, *84*, 2169–2175. [CrossRef] [PubMed]
135. Jayaram, H.; Fan, H.; Bowman, B.R.; Ooi, A.; Jayaram, J.; Collisson, E.W.; Lescar, J.; Prasad, B.V. X-ray structures of the n- and c-terminal domains of a coronavirus nucleocapsid protein: Implications for nucleocapsid formation. *J. Virol.* **2006**, *80*, 6612–6620. [CrossRef] [PubMed]

136. Takeda, M.; Chang, C.K.; Ikeya, T.; Guntert, P.; Chang, Y.H.; Hsu, Y.L.; Huang, T.H.; Kainosho, M. Solution structure of the c-terminal dimerization domain of sars coronavirus nucleocapsid protein solved by the sail-nmr method. *J. Mol. Biol.* **2008**, *380*, 608–622. [CrossRef] [PubMed]
137. Chen, I.J.; Yuann, J.M.; Chang, Y.M.; Lin, S.Y.; Zhao, J.; Perlman, S.; Shen, Y.Y.; Huang, T.H.; Hou, M.H. Crystal structure-based exploration of the important role of arg106 in the rna-binding domain of human coronavirus oc43 nucleocapsid protein. *Biochim. Biophys. Acta* **2013**, *1834*, 1054–1062. [CrossRef] [PubMed]
138. Lin, S.M.; Lin, S.C.; Hsu, J.N.; Chang, C.K.; Chien, C.M.; Wang, Y.S.; Wu, H.Y.; Jeng, U.S.; Kehn-Hall, K.; Hou, M.H. Structure-based stabilization of non-native protein-protein interactions of coronavirus nucleocapsid proteins in antiviral drug design. *J. Med. Chem.* **2020**, *63*, 3131–3141. [CrossRef] [PubMed]
139. Chen, C.Y.; Chang, C.K.; Chang, Y.W.; Sue, S.C.; Bai, H.I.; Riang, L.; Hsiao, C.D.; Huang, T.H. Structure of the sars coronavirus nucleocapsid protein rna-binding dimerization domain suggests a mechanism for helical packaging of viral rna. *J. Mol. Biol.* **2007**, *368*, 1075–1086. [CrossRef]
140. Barcena, M.; Oostergetel, G.T.; Bartelink, W.; Faas, F.G.; Verkleij, A.; Rottier, P.J.; Koster, A.J.; Bosch, B.J. Cryo-electron tomography of mouse hepatitis virus: Insights into the structure of the coronavirion. *Proc. Natl. Acad. Sci. USA* **2009**, *106*, 582–587. [CrossRef]
141. Yadav, R.; Imran, M.; Dhamija, P.; Suchal, K.; Handu, S. Virtual screening and dynamics of potential inhibitors targeting rna binding domain of nucleocapsid phosphoprotein from sars-cov-2. *J. Biomol. Struct. Dyn.* **2020**, 1–16. [CrossRef]
142. Chang, C.K.; Lo, S.C.; Wang, Y.S.; Hou, M.H. Recent insights into the development of therapeutics against coronavirus diseases by targeting n protein. *Drug Discov. Today* **2016**, *21*, 562–572. [CrossRef] [PubMed]
143. Ma-Lauer, Y.; Zheng, Y.; Malesevic, M.; von Brunn, B.; Fischer, G.; von Brunn, A. Influences of cyclosporin a and non-immunosuppressive derivatives on cellular cyclophilins and viral nucleocapsid protein during human coronavirus 229e replication. *Antivir. Res.* **2020**, *173*, 104620. [CrossRef] [PubMed]
144. Abbott, T.R.; Dhamdhere, G.; Liu, Y.; Lin, X.; Goudy, L.; Zeng, L.; Chemparathy, A.; Chmura, S.; Heaton, N.S.; Debs, R.; et al. Development of crispr as an antiviral strategy to combat sars-cov-2 and influenza. *Cell* **2020**, *181*, 865–876.e12. [CrossRef] [PubMed]
145. Venkatagopalan, P.; Daskalova, S.M.; Lopez, L.A.; Dolezal, K.A.; Hogue, B.G. Coronavirus envelope (e) protein remains at the site of assembly. *Virology* **2015**, *478*, 75–85. [CrossRef] [PubMed]
146. Bianchi, M.; Benvenuto, D.; Giovanetti, M.; Angeletti, S.; Ciccozzi, M.; Pascarella, S. Sars-cov-2 envelope and membrane proteins: Structural differences linked to virus characteristics? *Biomed. Res. Int.* **2020**, *2020*, 4389089. [CrossRef] [PubMed]
147. Arbely, E.; Khattari, Z.; Brotons, G.; Akkawi, M.; Salditt, T.; Arkin, I.T. A highly unusual palindromic transmembrane helical hairpin formed by sars coronavirus e protein. *J. Mol. Biol.* **2004**, *341*, 769–779. [CrossRef] [PubMed]
148. Schoeman, D.; Fielding, B.C. Coronavirus envelope protein: Current knowledge. *Virol. J.* **2019**, *16*, 69. [CrossRef] [PubMed]
149. Tseng, Y.T.; Wang, S.M.; Huang, K.J.; Wang, C.T. Sars-cov envelope protein palmitoylation or nucleocapid association is not required for promoting virus-like particle production. *J. Biomed. Sci.* **2014**, *21*, 34. [CrossRef]
150. Alvarez, E.; DeDiego, M.L.; Nieto-Torres, J.L.; Jimenez-Guardeno, J.M.; Marcos-Villar, L.; Enjuanes, L. The envelope protein of severe acute respiratory syndrome coronavirus interacts with the non-structural protein 3 and is ubiquitinated. *Virology* **2010**, *402*, 281–291. [CrossRef]
151. Khattari, Z.; Brotons, G.; Akkawi, M.; Arbely, E.; Arkin, I.T.; Salditt, T. Sars coronavirus e protein in phospholipid bilayers: An X-ray study. *Biophys. J.* **2006**, *90*, 2038–2050. [CrossRef]
152. Manor, J.; Arbely, E.; Beerlink, A.; Akkawi, M.; Arkin, I.T. Use of isotope-edited ftir to derive a backbone structure of a transmembrane protein. *J. Phys. Chem. Lett.* **2014**, *5*, 2573–2579. [CrossRef] [PubMed]
153. Xie, Q.; He, X.; Yang, F.; Liu, X.; Li, Y.; Liu, Y.; Yang, Z.; Yu, J.; Zhang, B.; Zhao, W. Analysis of the genome sequence and prediction of b-cell epitopes of the envelope protein of middle east respiratory syndrome-coronavirus. *IEEE/ACM Trans. Comput. Biol. Bioinform.* **2018**, *15*, 1344–1350. [CrossRef] [PubMed]
154. DeDiego, M.L.; Alvarez, E.; Almazan, F.; Rejas, M.T.; Lamirande, E.; Roberts, A.; Shieh, W.J.; Zaki, S.R.; Subbarao, K.; Enjuanes, L. A severe acute respiratory syndrome coronavirus that lacks the e gene is attenuated in vitro and in vivo. *J. Virol.* **2007**, *81*, 1701–1713. [CrossRef] [PubMed]
155. Stodola, J.K.; Dubois, G.; le Coupanec, A.; Desforges, M.; Talbot, P.J. The oc43 human coronavirus envelope protein is critical for infectious virus production and propagation in neuronal cells and is a determinant of neurovirulence and cns pathology. *Virology* **2018**, *515*, 134–149. [CrossRef]
156. Ortego, J.; Ceriani, J.E.; Patino, C.; Plana, J.; Enjuanes, L. Absence of e protein arrests transmissible gastroenteritis coronavirus maturation in the secretory pathway. *Virology* **2007**, *368*, 296–308. [CrossRef]
157. Almazan, F.; DeDiego, M.L.; Sola, I.; Zuniga, S.; Nieto-Torres, J.L.; Marquez-Jurado, S.; Andres, G.; Enjuanes, L. Engineering a replication-competent, propagation-defective middle east respiratory syndrome coronavirus as a vaccine candidate. *mBio* **2013**, *4*, e00650-13. [CrossRef]
158. DeDiego, M.L.; Nieto-Torres, J.L.; Jimenez-Guardeno, J.M.; Regla-Nava, J.A.; Castano-Rodriguez, C.; Fernandez-Delgado, R.; Usera, F.; Enjuanes, L. Coronavirus virulence genes with main focus on sars-cov envelope gene. *Virus Res.* **2014**, *194*, 124–137. [CrossRef]
159. Ye, Q.; Wang, B.; Mao, J. The pathogenesis and treatment of the 'cytokine storm' in covid-19. *J. Infect.* **2020**, *80*, 607–613. [CrossRef]

160. Castano-Rodriguez, C.; Honrubia, J.M.; Gutierrez-Alvarez, J.; DeDiego, M.L.; Nieto-Torres, J.L.; Jimenez-Guardeno, J.M.; Regla-Nava, J.A.; Fernandez-Delgado, R.; Verdia-Baguena, C.; Queralt-Martin, M.; et al. Role of severe acute respiratory syndrome coronavirus viroporins e, 3a, and 8a in replication and pathogenesis. *mBio* **2018**, *9*, e02325-17. [CrossRef]
161. Kuo, L.; Hurst, K.R.; Masters, P.S. Exceptional flexibility in the sequence requirements for coronavirus small envelope protein function. *J. Virol.* **2007**, *81*, 2249–2262. [CrossRef]
162. Corse, E.; Machamer, C.E. The cytoplasmic tail of infectious bronchitis virus e protein directs golgi targeting. *J. Virol.* **2002**, *76*, 1273–1284. [CrossRef] [PubMed]
163. Cohen, J.R.; Lin, L.D.; Machamer, C.E. Identification of a golgi complex-targeting signal in the cytoplasmic tail of the severe acute respiratory syndrome coronavirus envelope protein. *J. Virol.* **2011**, *85*, 5794–5803. [CrossRef] [PubMed]
164. Kuo, L.; Masters, P.S. The small envelope protein e is not essential for murine coronavirus replication. *J. Virol.* **2003**, *77*, 4597–4608. [CrossRef] [PubMed]
165. Westerbeck, J.W.; Machamer, C.E. A coronavirus e protein is present in two distinct pools with different effects on assembly and the secretory pathway. *J. Virol.* **2015**, *89*, 9313–9323. [CrossRef]
166. Parthasarathy, K.; Lu, H.; Surya, W.; Vararattanavech, A.; Pervushin, K.; Torres, J. Expression and purification of coronavirus envelope proteins using a modified beta-barrel construct. *Protein Expr. Purif.* **2012**, *85*, 133–141. [CrossRef]
167. Pervushin, K.; Tan, E.; Parthasarathy, K.; Lin, X.; Jiang, F.L.; Yu, D.; Vararattanavech, A.; Soong, T.W.; Liu, D.X.; Torres, J. Structure and inhibition of the sars coronavirus envelope protein ion channel. *PLoS Pathog.* **2009**, *5*, e1000511. [CrossRef]
168. Nieto-Torres, J.L.; Verdia-Baguena, C.; Jimenez-Guardeno, J.M.; Regla-Nava, J.A.; Castano-Rodriguez, C.; Fernandez-Delgado, R.; Torres, J.; Aguilella, V.M.; Enjuanes, L. Severe acute respiratory syndrome coronavirus e protein transports calcium ions and activates the nlrp3 inflammasome. *Virology* **2015**, *485*, 330–339. [CrossRef]
169. Verdia-Baguena, C.; Nieto-Torres, J.L.; Alcaraz, A.; DeDiego, M.L.; Torres, J.; Aguilella, V.M.; Enjuanes, L. Coronavirus e protein forms ion channels with functionally and structurally-involved membrane lipids. *Virology* **2012**, *432*, 485–494. [CrossRef]
170. Nieto-Torres, J.L.; DeDiego, M.L.; Verdia-Baguena, C.; Jimenez-Guardeno, J.M.; Regla-Nava, J.A.; Fernandez-Delgado, R.; Castano-Rodriguez, C.; Alcaraz, A.; Torres, J.; Aguilella, V.M.; et al. Severe acute respiratory syndrome coronavirus envelope protein ion channel activity promotes virus fitness and pathogenesis. *PLoS Pathog.* **2014**, *10*, e1004077. [CrossRef]
171. Ruch, T.R.; Machamer, C.E. A single polar residue and distinct membrane topologies impact the function of the infectious bronchitis coronavirus e protein. *PLoS Pathog.* **2012**, *8*, e1002674. [CrossRef]
172. To, J.; Surya, W.; Fung, T.S.; Li, Y.; Verdia-Baguena, C.; Queralt-Martin, M.; Aguilella, V.M.; Liu, D.X.; Torres, J. Channel-inactivating mutations and their revertant mutants in the envelope protein of infectious bronchitis virus. *J. Virol.* **2017**, *91*. [CrossRef] [PubMed]
173. Wilson, L.; Gage, P.; Ewart, G. Hexamethylene amiloride blocks e protein ion channels and inhibits coronavirus replication. *Virology* **2006**, *353*, 294–306. [CrossRef]
174. Krahling, V.; Stein, D.A.; Spiegel, M.; Weber, F.; Muhlberger, E. Severe acute respiratory syndrome coronavirus triggers apoptosis via protein kinase r but is resistant to its antiviral activity. *J. Virol.* **2009**, *83*, 2298–2309. [CrossRef] [PubMed]
175. Ye, Z.; Wong, C.K.; Li, P.; Xie, Y. A sars-cov protein, orf-6, induces caspase-3 mediated, er stress and jnk-dependent apoptosis. *Biochim. Biophys. Acta* **2008**, *1780*, 1383–1387. [CrossRef] [PubMed]
176. Chan, C.P.; Siu, K.L.; Chin, K.T.; Yuen, K.Y.; Zheng, B.; Jin, D.Y. Modulation of the unfolded protein response by the severe acute respiratory syndrome coronavirus spike protein. *J. Virol.* **2006**, *80*, 9279–9287. [CrossRef] [PubMed]
177. Fung, T.S.; Huang, M.; Liu, D.X. Coronavirus-induced er stress response and its involvement in regulation of coronavirus-host interactions. *Virus Res.* **2014**, *194*, 110–123. [CrossRef] [PubMed]
178. Xu, X.; Zhang, H.; Zhang, Q.; Dong, J.; Liang, Y.; Huang, Y.; Liu, H.J.; Tong, D. Porcine epidemic diarrhea virus e protein causes endoplasmic reticulum stress and up-regulates interleukin-8 expression. *Virol. J.* **2013**, *10*, 26. [CrossRef]
179. DeDiego, M.L.; Nieto-Torres, J.L.; Jimenez-Guardeno, J.M.; Regla-Nava, J.A.; Alvarez, E.; Oliveros, J.C.; Zhao, J.; Fett, C.; Perlman, S.; Enjuanes, L. Severe acute respiratory syndrome coronavirus envelope protein regulates cell stress response and apoptosis. *PLoS Pathog.* **2011**, *7*, e1002315. [CrossRef]
180. Murakami, T.; Ockinger, J.; Yu, J.; Byles, V.; McColl, A.; Hofer, A.M.; Horng, T. Critical role for calcium mobilization in activation of the nlrp3 inflammasome. *Proc. Natl. Acad. Sci. USA* **2012**, *109*, 11282–11287. [CrossRef]
181. Swanson, K.V.; Deng, M.; Ting, J.P. The nlrp3 inflammasome: Molecular activation and regulation to therapeutics. *Nat. Rev. Immunol.* **2019**, *19*, 477–489. [CrossRef]
182. Chen, X.; Guo, X.; Ge, Q.; Zhao, Y.; Mu, H.; Zhang, J. Er stress activates the nlrp3 inflammasome: A novel mechanism of atherosclerosis. *Oxid Med. Cell Longev.* **2019**, *2019*, 3462530. [CrossRef] [PubMed]
183. Harijith, A.; Ebenezer, D.L.; Natarajan, V. Reactive oxygen species at the crossroads of inflammasome and inflammation. *Front. Physiol.* **2014**, *5*, 352. [CrossRef] [PubMed]
184. Ghosh, S.; Dellibovi-Ragheb, T.A.; Kerviel, A.; Pak, E.; Qiu, Q.; Fisher, M.; Takvorian, P.M.; Bleck, C.; Hsu, V.W.; Fehr, A.R.; et al. Beta-coronaviruses use lysosomes for egress instead of the biosynthetic secretory pathway. *Cell* **2020**, *183*, 1520–1535.e14. [CrossRef] [PubMed]
185. Westerbeck, J.W.; Machamer, C.E. The infectious bronchitis coronavirus envelope protein alters golgi ph to protect the spike protein and promote the release of infectious virus. *J. Virol.* **2019**, *93*. [CrossRef]

186. Gordon-Alonso, M.; Rocha-Perugini, V.; Alvarez, S.; Moreno-Gonzalo, O.; Ursa, A.; Lopez-Martin, S.; Izquierdo-Useros, N.; Martinez-Picado, J.; Munoz-Fernandez, M.A.; Yanez-Mo, M.; et al. The pdz-adaptor protein syntenin-1 regulates hiv-1 entry. *Mol. Biol. Cell* **2012**, *23*, 2253–2263. [CrossRef]
187. Pieczynski, J.; Margolis, B. Protein complexes that control renal epithelial polarity. *Am. J. Physiol. Renal Physiol.* **2011**, *300*, F589–F601. [CrossRef]
188. Teoh, K.T.; Siu, Y.L.; Chan, W.L.; Schluter, M.A.; Liu, C.J.; Peiris, J.S.; Bruzzone, R.; Margolis, B.; Nal, B. The sars coronavirus e protein interacts with pals1 and alters tight junction formation and epithelial morphogenesis. *Mol. Biol. Cell* **2010**, *21*, 3838–3852. [CrossRef]
189. Jimenez-Guardeno, J.M.; Nieto-Torres, J.L.; DeDiego, M.L.; Regla-Nava, J.A.; Fernandez-Delgado, R.; Castano-Rodriguez, C.; Enjuanes, L. The pdz-binding motif of severe acute respiratory syndrome coronavirus envelope protein is a determinant of viral pathogenesis. *PLoS Pathog.* **2014**, *10*, e1004320. [CrossRef]
190. Regla-Nava, J.A.; Nieto-Torres, J.L.; Jimenez-Guardeno, J.M.; Fernandez-Delgado, R.; Fett, C.; Castano-Rodriguez, C.; Perlman, S.; Enjuanes, L.; DeDiego, M.L. Severe acute respiratory syndrome coronaviruses with mutations in the e protein are attenuated and promising vaccine candidates. *J. Virol.* **2015**, *89*, 3870–3887. [CrossRef]
191. Channappanavar, R.; Zhao, J.; Perlman, S. T cell-mediated immune response to respiratory coronaviruses. *Immunol. Res.* **2014**, *59*, 118–128. [CrossRef]
192. Fischer, F.; Stegen, C.F.; Masters, P.S.; Samsonoff, W.A. Analysis of constructed e gene mutants of mouse hepatitis virus confirms a pivotal role for e protein in coronavirus assembly. *J. Virol.* **1998**, *72*, 7885–7894. [CrossRef] [PubMed]
193. Chan, W.E.; Chuang, C.K.; Yeh, S.H.; Chang, M.S.; Chen, S.S. Functional characterization of heptad repeat 1 and 2 mutants of the spike protein of severe acute respiratory syndrome coronavirus. *J. Virol.* **2006**, *80*, 3225–3237. [CrossRef] [PubMed]
194. Walls, A.C.; Park, Y.J.; Tortorici, M.A.; Wall, A.; McGuire, A.T.; Veesler, D. Structure, function, and antigenicity of the sars-cov-2 spike glycoprotein. *Cell* **2020**, *181*, 281–292.e6. [CrossRef] [PubMed]
195. Wrapp, D.; Wang, N.; Corbett, K.S.; Goldsmith, J.A.; Hsieh, C.L.; Abiona, O.; Graham, B.S.; McLellan, J.S. Cryo-em structure of the 2019-ncov spike in the prefusion conformation. *Science* **2020**, *367*, 1260–1263. [CrossRef] [PubMed]
196. Spiga, O.; Bernini, A.; Ciutti, A.; Chiellini, S.; Menciassi, N.; Finetti, F.; Causarono, V.; Anselmi, F.; Prischi, F.; Niccolai, N. Molecular modelling of s1 and s2 subunits of sars coronavirus spike glycoprotein. *Biochem. Biophys. Res. Commun.* **2003**, *310*, 78–83. [CrossRef]
197. Watanabe, R.; Matsuyama, S.; Shirato, K.; Maejima, M.; Fukushi, S.; Morikawa, S.; Taguchi, F. Entry from the cell surface of severe acute respiratory syndrome coronavirus with cleaved s protein as revealed by pseudotype virus bearing cleaved s protein. *J. Virol.* **2008**, *82*, 11985–11991. [CrossRef]
198. Wang, Q.; Zhang, Y.; Wu, L.; Niu, S.; Song, C.; Zhang, Z.; Lu, G.; Qiao, C.; Hu, Y.; Yuen, K.Y.; et al. Structural and functional basis of sars-cov-2 entry by using human ace2. *Cell* **2020**, *181*, 894–904.e9. [CrossRef]
199. Xia, S.; Liu, M.; Wang, C.; Xu, W.; Lan, Q.; Feng, S.; Qi, F.; Bao, L.; Du, L.; Liu, S.; et al. Inhibition of sars-cov-2 (previously 2019-ncov) infection by a highly potent pan-coronavirus fusion inhibitor targeting its spike protein that harbors a high capacity to mediate membrane fusion. *Cell Res.* **2020**, *30*, 343–355. [CrossRef]
200. Hoffmann, M.; Hofmann-Winkler, H.; Pöhlmann, S. Priming time: How cellular proteases arm coronavirus spike proteins. In *Activation of Viruses by Host Proteases*; Böttcher-Friebertshäuser, E., Garten, W., Klenk, H.D., Eds.; Springer International Publishing: Cham, Switzerland, 2018; pp. 71–98.
201. Xiong, X.; Qu, K.; Ciazynska, K.A.; Hosmillo, M.; Carter, A.P.; Ebrahimi, S.; Ke, Z.; Scheres, S.H.W.; Bergamaschi, L.; Grice, G.L.; et al. A thermostable, closed sars-cov-2 spike protein trimer. *Nat. Struct. Mol. Biol.* **2020**, *27*, 934–941. [CrossRef]
202. Shang, J.; Wan, Y.; Luo, C.; Ye, G.; Geng, Q.; Auerbach, A.; Li, F. Cell entry mechanisms of sars-cov-2. *Proc. Natl. Acad. Sci. USA* **2020**, *117*, 11727–11734. [CrossRef]
203. Marzi, A.; Gramberg, T.; Simmons, G.; Moller, P.; Rennekamp, A.J.; Krumbiegel, M.; Geier, M.; Eisemann, J.; Turza, N.; Saunier, B.; et al. Dc-sign and dc-signr interact with the glycoprotein of marburg virus and the s protein of severe acute respiratory syndrome coronavirus. *J. Virol.* **2004**, *78*, 12090–12095. [CrossRef] [PubMed]
204. Walls, A.C.; Tortorici, M.A.; Snijder, J.; Xiong, X.; Bosch, B.J.; Rey, F.A.; Veesler, D. Tectonic conformational changes of a coronavirus spike glycoprotein promote membrane fusion. *Proc. Natl. Acad. Sci. USA* **2017**, *114*, 11157–11162. [CrossRef] [PubMed]
205. Benesova, O.; Pavlik, A. Perinatal treatment with glucocorticoids and the risk of maldevelopment of the brain. *Neuropharmacology* **1989**, *28*, 89–97. [CrossRef]
206. Millet, J.K.; Whittaker, G.R. Physiological and molecular triggers for sars-cov membrane fusion and entry into host cells. *Virology* **2018**, *517*, 3–8. [CrossRef] [PubMed]
207. Bosch, B.J.; Martina, B.E.; van der Zee, R.; Lepault, J.; Haijema, B.J.; Versluis, C.; Heck, A.J.; de Groot, R.; Osterhaus, A.D.; Rottier, P.J. Severe acute respiratory syndrome coronavirus (sars-cov) infection inhibition using spike protein heptad repeat-derived peptides. *Proc. Natl. Acad. Sci. USA* **2004**, *101*, 8455–8460. [CrossRef] [PubMed]
208. Madu, I.G.; Roth, S.L.; Belouzard, S.; Whittaker, G.R. Characterization of a highly conserved domain within the severe acute respiratory syndrome coronavirus spike protein s2 domain with characteristics of a viral fusion peptide. *J. Virol.* **2009**, *83*, 7411–7421. [CrossRef] [PubMed]

209. Lai, A.L.; Millet, J.K.; Daniel, S.; Freed, J.H.; Whittaker, G.R. The sars-cov fusion peptide forms an extended bipartite fusion platform that perturbs membrane order in a calcium-dependent manner. *J. Mol. Biol.* **2017**, *429*, 3875–3892. [CrossRef]
210. Follis, K.E.; York, J.; Nunberg, J.H. Furin cleavage of the sars coronavirus spike glycoprotein enhances cell-cell fusion but does not affect virion entry. *Virology* **2006**, *350*, 358–369. [CrossRef]
211. Freeman, M.C.; Peek, C.T.; Becker, M.M.; Smith, E.C.; Denison, M.R. Coronaviruses induce entry-independent, continuous macropinocytosis. *mBio* **2014**, *5*, e01340-14. [CrossRef]
212. McBride, C.E.; Li, J.; Machamer, C.E. The cytoplasmic tail of the severe acute respiratory syndrome coronavirus spike protein contains a novel endoplasmic reticulum retrieval signal that binds copi and promotes interaction with membrane protein. *J. Virol.* **2007**, *81*, 2418–2428. [CrossRef]
213. de Wilde, A.H.; Raj, V.S.; Oudshoorn, D.; Bestebroer, T.M.; van Nieuwkoop, S.; Limpens, R.; Posthuma, C.C.; van der Meer, Y.; Barcena, M.; Haagmans, B.L.; et al. Mers-coronavirus replication induces severe in vitro cytopathology and is strongly inhibited by cyclosporin a or interferon-alpha treatment. *J. Gen. Virol.* **2013**, *94*, 1749–1760. [CrossRef] [PubMed]
214. Matsuyama, S.; Nao, N.; Shirato, K.; Kawase, M.; Saito, S.; Takayama, I.; Nagata, N.; Sekizuka, T.; Katoh, H.; Kato, F.; et al. Enhanced isolation of sars-cov-2 by tmprss2-expressing cells. *Proc. Natl. Acad. Sci. USA* **2020**, *117*, 7001–7003. [CrossRef] [PubMed]
215. Matsuyama, S.; Nagata, N.; Shirato, K.; Kawase, M.; Takeda, M.; Taguchi, F. Efficient activation of the severe acute respiratory syndrome coronavirus spike protein by the transmembrane protease tmprss2. *J. Virol.* **2010**, *84*, 12658–12664. [CrossRef] [PubMed]
216. Sungnak, W.; Huang, N.; Becavin, C.; Berg, M.; Queen, R.; Litvinukova, M.; Talavera-Lopez, C.; Maatz, H.; Reichart, D.; Sampaziotis, F.; et al. Sars-cov-2 entry factors are highly expressed in nasal epithelial cells together with innate immune genes. *Nat. Med.* **2020**, *26*, 681–687. [CrossRef] [PubMed]
217. Bertram, S.; Dijkman, R.; Habjan, M.; Heurich, A.; Gierer, S.; Glowacka, I.; Welsch, K.; Winkler, M.; Schneider, H.; Hofmann-Winkler, H.; et al. Tmprss2 activates the human coronavirus 229e for cathepsin-independent host cell entry and is expressed in viral target cells in the respiratory epithelium. *J. Virol.* **2013**, *87*, 6150–6160. [CrossRef] [PubMed]
218. McBride, C.E.; Machamer, C.E. Palmitoylation of sars-cov s protein is necessary for partitioning into detergent-resistant membranes and cell-cell fusion but not interaction with m protein. *Virology* **2010**, *405*, 139–148. [CrossRef] [PubMed]
219. Li, G.M.; Li, Y.G.; Yamate, M.; Li, S.M.; Ikuta, K. Lipid rafts play an important role in the early stage of severe acute respiratory syndrome-coronavirus life cycle. *Microbes Infect.* **2007**, *9*, 96–102. [CrossRef]
220. Lu, Y.; Liu, D.X.; Tam, J.P. Lipid rafts are involved in sars-cov entry into vero e6 cells. *Biochem. Biophys. Res. Commun.* **2008**, *369*, 344–349. [CrossRef]
221. Glende, J.; Schwegmann-Wessels, C.; Al-Falah, M.; Pfefferle, S.; Qu, X.; Deng, H.; Drosten, C.; Naim, H.Y.; Herrler, G. Importance of cholesterol-rich membrane microdomains in the interaction of the s protein of sars-coronavirus with the cellular receptor angiotensin-converting enzyme 2. *Virology* **2008**, *381*, 215–221. [CrossRef]
222. Yilla, M.; Harcourt, B.H.; Hickman, C.J.; McGrew, M.; Tamin, A.; Goldsmith, C.S.; Bellini, W.J.; Anderson, L.J. Sars-coronavirus replication in human peripheral monocytes/macrophages. *Virus Res.* **2005**, *107*, 93–101. [CrossRef]
223. Jafarzadeh, A.; Chauhan, P.; Saha, B.; Jafarzadeh, S.; Nemati, M. Contribution of monocytes and macrophages to the local tissue inflammation and cytokine storm in covid-19: Lessons from sars and mers, and potential therapeutic interventions. *Life Sci.* **2020**, *257*, 118102. [CrossRef] [PubMed]
224. Liang, Y.; Wang, M.L.; Chien, C.S.; Yarmishyn, A.A.; Yang, Y.P.; Lai, W.Y.; Luo, Y.H.; Lin, Y.T.; Chen, Y.J.; Chang, P.C.; et al. Highlight of immune pathogenic response and hematopathologic effect in sars-cov, mers-cov, and sars-cov-2 infection. *Front. Immunol.* **2020**, *11*, 1022. [CrossRef] [PubMed]
225. Wang, K.; Chen, W.; Zhou, Y.-S.; Lian, J.-Q.; Zhang, Z.; Du, P.; Gong, L.; Zhang, Y.; Cui, H.-Y.; Geng, J.-J.; et al. Sars-cov-2 invades host cells via a novel route: Cd147-spike protein. *bioRxiv* **2020**. [CrossRef]
226. Zhou, Y.; Lu, K.; Pfefferle, S.; Bertram, S.; Glowacka, I.; Drosten, C.; Pohlmann, S.; Simmons, G. A single asparagine-linked glycosylation site of the severe acute respiratory syndrome coronavirus spike glycoprotein facilitates inhibition by mannose-binding lectin through multiple mechanisms. *J. Virol.* **2010**, *84*, 8753–8764. [CrossRef] [PubMed]
227. Ip, W.K.; Chan, K.H.; Law, H.K.; Tso, G.H.; Kong, E.K.; Wong, W.H.; To, Y.F.; Yung, R.W.; Chow, E.Y.; Au, K.L.; et al. Mannose-binding lectin in severe acute respiratory syndrome coronavirus infection. *J. Infect. Dis.* **2005**, *191*, 1697–1704. [CrossRef]
228. Gralinski, L.E.; Sheahan, T.P.; Morrison, T.E.; Menachery, V.D.; Jensen, K.; Leist, S.R.; Whitmore, A.; Heise, M.T.; Baric, R.S. Complement activation contributes to severe acute respiratory syndrome coronavirus pathogenesis. *mBio* **2018**, *9*, e01753-18. [CrossRef]
229. Noris, M.; Benigni, A.; Remuzzi, G. The case of complement activation in covid-19 multiorgan impact. *Kidney Int.* **2020**, *98*, 314–322. [CrossRef]
230. Cagliani, R.; Forni, D.; Clerici, M.; Sironi, M. Computational inference of selection underlying the evolution of the novel coronavirus, severe acute respiratory syndrome coronavirus 2. *J. Virol.* **2020**, *94*. [CrossRef]
231. Fung, T.S.; Liu, D.X. Post-translational modifications of coronavirus proteins: Roles and function. *Future Virol.* **2018**, *13*, 405–430. [CrossRef]
232. Holmes, K.V.; Doller, E.W.; Sturman, L.S. Tunicamycin resistant glycosylation of coronavirus glycoprotein: Demonstration of a novel type of viral glycoprotein. *Virology* **1981**, *115*, 334–344. [CrossRef]

233. Stern, D.F.; Sefton, B.M. Coronavirus proteins: Biogenesis of avian infectious bronchitis virus virion proteins. *J. Virol.* **1982**, *44*, 794–803. [CrossRef] [PubMed]
234. Niemann, H.; Geyer, R.; Klenk, H.D.; Linder, D.; Stirm, S.; Wirth, M. The carbohydrates of mouse hepatitis virus (mhv) a59: Structures of the o-glycosidically linked oligosaccharides of glycoprotein e1. *EMBO J.* **1984**, *3*, 665–670. [CrossRef] [PubMed]
235. Locker, J.K.; Rose, J.K.; Horzinek, M.C.; Rottier, P.J. Membrane assembly of the triple-spanning coronavirus m protein. Individual transmembrane domains show preferred orientation. *J. Biol. Chem.* **1992**, *267*, 21911–21918. [CrossRef]
236. Neuman, B.W.; Kiss, G.; Kunding, A.H.; Bhella, D.; Baksh, M.F.; Connelly, S.; Droese, B.; Klaus, J.P.; Makino, S.; Sawicki, S.G.; et al. A structural analysis of m protein in coronavirus assembly and morphology. *J. Struct. Biol.* **2011**, *174*, 11–22. [CrossRef] [PubMed]
237. Tseng, Y.T.; Chang, C.H.; Wang, S.M.; Huang, K.J.; Wang, C.T. Identifying sars-cov membrane protein amino acid residues linked to virus-like particle assembly. *PLoS ONE* **2013**, *8*, e64013. [CrossRef]
238. Perrier, A.; Bonnin, A.; Desmarets, L.; Danneels, A.; Goffard, A.; Rouille, Y.; Dubuisson, J.; Belouzard, S. The c-terminal domain of the mers coronavirus m protein contains a trans-golgi network localization signal. *J. Biol. Chem.* **2019**, *294*, 14406–14421. [CrossRef]
239. Kuo, L.; Hurst-Hess, K.R.; Koetzner, C.A.; Masters, P.S. "Analyses of coronavirus assembly interactions with interspecies membrane and nucleocapsid protein chimeras. *J. Virol.* **2016**, *90*, 4357–4368. [CrossRef]
240. Gordon, D.E.; Jang, G.M.; Bouhaddou, M.; Xu, J.; Obernier, K.; White, K.M.; O'Meara, M.J.; Rezelj, V.V.; Guo, J.Z.; Swaney, D.L.; et al. A sars-cov-2 protein interaction map reveals targets for drug repurposing. *Nature* **2020**, *583*, 459–468. [CrossRef]
241. Lu, W.; Zheng, B.J.; Xu, K.; Schwarz, W.; Du, L.; Wong, C.K.; Chen, J.; Duan, S.; Deubel, V.; Sun, B. Severe acute respiratory syndrome-associated coronavirus 3a protein forms an ion channel and modulates virus release. *Proc. Natl. Acad. Sci. USA* **2006**, *103*, 12540–12545. [CrossRef]
242. Zhang, R.; Wang, K.; Lv, W.; Yu, W.; Xie, S.; Xu, K.; Schwarz, W.; Xiong, S.; Sun, B. The orf4a protein of human coronavirus 229e functions as a viroporin that regulates viral production. *Biochim. Biophys. Acta* **2014**, *1838*, 1088–1095. [CrossRef]
243. Kuo, L.; Masters, P.S. Evolved variants of the membrane protein can partially replace the envelope protein in murine coronavirus assembly. *J. Virol.* **2010**, *84*, 12872–12885. [CrossRef] [PubMed]
244. Madan, V.; Mde, J.G.; Sanz, M.A.; Carrasco, L. Viroporin activity of murine hepatitis virus e protein. *FEBS Lett.* **2005**, *579*, 3607–3612. [CrossRef]
245. Verdia-Baguena, C.; Nieto-Torres, J.L.; Alcaraz, A.; Dediego, M.L.; Enjuanes, L.; Aguilella, V.M. Analysis of sars-cov e protein ion channel activity by tuning the protein and lipid charge. *Biochim. Biophys. Acta* **2013**, *1828*, 2026–2031. [CrossRef] [PubMed]
246. Katoh, K.; Standley, D.M. Mafft multiple sequence alignment software version 7: Improvements in performance and usability. *Mol. Biol. Evol.* **2013**, *30*, 772–780. [CrossRef] [PubMed]
247. Capella-Gutierrez, S.; Silla-Martinez, J.M.; Gabaldon, T. Trimal: A tool for automated alignment trimming in large-scale phylogenetic analyses. *Bioinformatics* **2009**, *25*, 1972–1973. [CrossRef] [PubMed]
248. Zhai, Y.; Saier, M.H., Jr. A web-based program for the prediction of average hydropathy, average amphipathicity and average similarity of multiply aligned homologous proteins. *J. Mol. Microbiol. Biotechnol.* **2001**, *3*, 285–286. [PubMed]
249. Wang, S.C.; Davejan, P.; Hendargo, K.J.; Javadi-Razaz, I.; Chou, A.; Yee, D.C.; Ghazi, F.; Lam, K.J.K.; Conn, A.M.; Madrigal, A.; et al. Expansion of the major facilitator superfamily (mfs) to include novel transporters as well as transmembrane-acting enzymes. *Biochim. Biophys. Acta Biomembr.* **2020**, *1862*, 183277. [CrossRef]
250. Medrano-Soto, A.; Ghazi, F.; Hendargo, K.J.; Moreno-Hagelsieb, G.; Myers, S.; Saier, M.H., Jr. Expansion of the transporter-opsin-g protein-coupled receptor superfamily with five new protein families. *PLoS ONE* **2020**, *15*, e0231085. [CrossRef]
251. Watanabe, Y.; Bowden, T.A.; Wilson, I.A.; Crispin, M. Exploitation of glycosylation in enveloped virus pathobiology. *Biochim. Biophys. Acta Gen. Subj.* **2019**, *1863*, 1480–1497. [CrossRef]
252. Ferreira, I.G.; Pucci, M.; Venturi, G.; Malagolini, N.; Chiricolo, M.; Dall'Olio, F. Glycosylation as a main regulator of growth and death factor receptors signaling. *Int. J. Mol. Sci.* **2018**, *19*, 580. [CrossRef]
253. Berg, J.M.; Tymoczko, J.L.; Stryer, L. *Carbohydrates Can Be Attached to Proteins to Form Glycoproteins*; W. H. Freeman and Company: New York, NY, USA, 2002.
254. Reily, C.; Stewart, T.J.; Renfrow, M.B.; Novak, J. Glycosylation in health and disease. *Nat. Rev. Nephrol.* **2019**, *15*, 346–366. [CrossRef] [PubMed]
255. Guan, X.; Fierke, C.A. Understanding protein palmitoylation: Biological significance and enzymology. *Sci. China Chem.* **2011**, *54*, 1888–1897. [CrossRef] [PubMed]
256. Greaves, J.; Salaun, C.; Fukata, Y.; Fukata, M.; Chamberlain, L.H. Palmitoylation and membrane interactions of the neuroprotective chaperone cysteine-string protein. *J. Biol. Chem.* **2008**, *283*, 25014–25026. [CrossRef] [PubMed]
257. Boscarino, J.A.; Logan, H.L.; Lacny, J.J.; Gallagher, T.M. Envelope protein palmitoylations are crucial for murine coronavirus assembly. *J. Virol.* **2008**, *82*, 2989–2999. [CrossRef] [PubMed]
258. Lindner, H.A. Deubiquitination in virus infection. *Virology* **2007**, *362*, 245–256. [CrossRef]
259. Murin, C.D.; Wilson, I.A.; Ward, A.B. Antibody responses to viral infections: A structural perspective across three different enveloped viruses. *Nat. Microbiol.* **2019**, *4*, 734–747. [CrossRef]
260. Tian, X.; Li, C.; Huang, A.; Xia, S.; Lu, S.; Shi, Z.; Lu, L.; Jiang, S.; Yang, Z.; Wu, Y.; et al. Potent binding of 2019 novel coronavirus spike protein by a sars coronavirus-specific human monoclonal antibody. *Emerg Microbes Infect.* **2020**, *9*, 382–385. [CrossRef]

261. Radosevich, M.; Burnouf, T. Intravenous immunoglobulin g: Trends in production methods, quality control and quality assurance. *Vox Sang.* **2010**, *98*, 12–28. [CrossRef]
262. Qiu, M.; Shi, Y.; Guo, Z.; Chen, Z.; He, R.; Chen, R.; Zhou, D.; Dai, E.; Wang, X.; Si, B.; et al. Antibody responses to individual proteins of sars coronavirus and their neutralization activities. *Microbes Infect.* **2005**, *7*, 882–889. [CrossRef]
263. Sun, B.; Feng, Y.; Mo, X.; Zheng, P.; Wang, Q.; Li, P.; Peng, P.; Liu, X.; Chen, Z.; Huang, H.; et al. Kinetics of sars-cov-2 specific igm and igg responses in covid-19 patients. *Emerg. Microbes Infect.* **2020**, *9*, 940–948. [CrossRef]
264. De Vriese, A.S.; Reynders, M. Igg antibody response to sars-cov-2 infection and viral rna persistence in patients on maintenance hemodialysis. *Am. J. Kidney Dis.* **2020**, *76*, 440–441. [CrossRef] [PubMed]
265. Wagh, K.; Kreider, E.F.; Li, Y.; Barbian, H.J.; Learn, G.H.; Giorgi, E.; Hraber, P.T.; Decker, T.G.; Smith, A.G.; Gondim, M.V.; et al. Completeness of hiv-1 envelope glycan shield at transmission determines neutralization breadth. *Cell Rep.* **2018**, *25*, 893–908.e7. [CrossRef] [PubMed]
266. Watanabe, Y.; Berndsen, Z.T.; Raghwani, J.; Seabright, G.E.; Allen, J.D.; Pybus, O.G.; McLellan, J.S.; Wilson, I.A.; Bowden, T.A.; Ward, A.B.; et al. Vulnerabilities in coronavirus glycan shields despite extensive glycosylation. *Nat. Commun.* **2020**, *11*, 2688. [CrossRef] [PubMed]
267. Ritchie, G.; Harvey, D.J.; Feldmann, F.; Stroeher, U.; Feldmann, H.; Royle, L.; Dwek, R.A.; Rudd, P.M. Identification of n-linked carbohydrates from severe acute respiratory syndrome (sars) spike glycoprotein. *Virology* **2010**, *399*, 257–269. [CrossRef] [PubMed]
268. Watanabe, Y.; Allen, J.D.; Wrapp, D.; McLellan, J.S.; Crispin, M. Site-specific glycan analysis of the sars-cov-2 spike. *Science* **2020**, *369*, 330–333. [CrossRef] [PubMed]
269. Song, H.C.; Seo, M.Y.; Stadler, K.; Yoo, B.J.; Choo, Q.L.; Coates, S.R.; Uematsu, Y.; Harada, T.; Greer, C.E.; Polo, J.M.; et al. Synthesis and characterization of a native, oligomeric form of recombinant severe acute respiratory syndrome coronavirus spike glycoprotein. *J. Virol.* **2004**, *78*, 10328–10335. [CrossRef]
270. Shajahan, A.; Supekar, N.T.; Gleinich, A.S.; Azadi, P. Deducing the n- and o- glycosylation profile of the spike protein of novel coronavirus sars-cov-2. *Glycobiology* **2020**, *30*, 981–988. [CrossRef]
271. Zimmermann, S.; Lepenies, B. Glycans as vaccine antigens and adjuvants: Immunological considerations. *Methods Mol. Biol.* **2015**, *1331*, 11–26.
272. Petit, C.M.; Chouljenko, V.N.; Iyer, A.; Colgrove, R.; Farzan, M.; Knipe, D.M.; Kousoulas, K.G. Palmitoylation of the cysteine-rich endodomain of the sars-coronavirus spike glycoprotein is important for spike-mediated cell fusion. *Virology* **2007**, *360*, 264–274. [CrossRef]
273. Vuoksimaa, E.; Eriksson, C.J.; Pulkkinen, L.; Rose, R.J.; Kaprio, J. Decreased prevalence of left-handedness among females with male co-twins: Evidence suggesting prenatal testosterone transfer in humans? *Psychoneuroendocrinology* **2010**, *35*, 1462–1472. [CrossRef]
274. Ujike, M.; Huang, C.; Shirato, K.; Matsuyama, S.; Makino, S.; Taguchi, F. Two palmitylated cysteine residues of the severe acute respiratory syndrome coronavirus spike (s) protein are critical for s incorporation into virus-like particles, but not for m-s co-localization. *J. Gen. Virol.* **2012**, *93*, 823–828. [CrossRef] [PubMed]
275. Gelhaus, S.; Thaa, B.; Eschke, K.; Veit, M.; Schwegmann-Wessels, C. Palmitoylation of the alphacoronavirus tgev spike protein s is essential for incorporation into virus-like particles but dispensable for s-m interaction. *Virology* **2014**, *464–465*, 397–405. [CrossRef] [PubMed]
276. Wang, H.; Yuan, Z.; Pavel, M.A.; Hansen, S.B. The role of high cholesterol in age-related covid19 lethality. *bioRxiv* **2020**. [CrossRef]
277. Oostra, M.; de Haan, C.A.; de Groot, R.J.; Rottier, P.J. Glycosylation of the severe acute respiratory syndrome coronavirus triple-spanning membrane proteins 3a and m. *J. Virol.* **2006**, *80*, 2326–2336. [CrossRef] [PubMed]
278. de Haan, C.A.; de Wit, M.; Kuo, L.; Montalto, C.; Masters, P.S.; Weiss, S.R.; Rottier, P.J. O-glycosylation of the mouse hepatitis coronavirus membrane protein. *Virus Res.* **2002**, *82*, 77–81. [CrossRef]
279. Voss, D.; Pfefferle, S.; Drosten, C.; Stevermann, L.; Traggiai, E.; Lanzavecchia, A.; Becker, S. Studies on membrane topology, n-glycosylation and functionality of sars-cov membrane protein. *Virol. J.* **2009**, *6*, 79. [CrossRef]
280. de Haan, C.A.; de Wit, M.; Kuo, L.; Montalto-Morrison, C.; Haagmans, B.L.; Weiss, S.R.; Masters, P.S.; Rottier, P.J. The glycosylation status of the murine hepatitis coronavirus m protein affects the interferogenic capacity of the virus in vitro and its ability to replicate in the liver but not the brain. *Virology* **2003**, *312*, 395–406. [CrossRef]
281. Yuan, Q.; Liao, Y.; Torres, J.; Tam, J.P.; Liu, D.X. Biochemical evidence for the presence of mixed membrane topologies of the severe acute respiratory syndrome coronavirus envelope protein expressed in mammalian cells. *FEBS Lett.* **2006**, *580*, 3192–3200. [CrossRef]
282. Nieto-Torres, J.L.; Dediego, M.L.; Alvarez, E.; Jimenez-Guardeno, J.M.; Regla-Nava, J.A.; Llorente, M.; Kremer, L.; Shuo, S.; Enjuanes, L. Subcellular location and topology of severe acute respiratory syndrome coronavirus envelope protein. *Virology* **2011**, *415*, 69–82. [CrossRef]
283. Liao, Y.; Yuan, Q.; Torres, J.; Tam, J.P.; Liu, D.X. Biochemical and functional characterization of the membrane association and membrane permeabilizing activity of the severe acute respiratory syndrome coronavirus envelope protein. *Virology* **2006**, *349*, 264–275. [CrossRef]
284. Liao, Y.; Lescar, J.; Tam, J.P.; Liu, D.X. Expression of sars-coronavirus envelope protein in escherichia coli cells alters membrane permeability. *Biochem. Biophys. Res. Commun.* **2004**, *325*, 374–380. [CrossRef] [PubMed]
285. Abriel, H.; Staub, O. Ubiquitylation of ion channels. *Physiology* **2005**, *20*, 398–407. [CrossRef] [PubMed]

286. Raaben, M.; Posthuma, C.C.; Verheije, M.H.; te Lintelo, E.G.; Kikkert, M.; Drijfhout, J.W.; Snijder, E.J.; Rottier, P.J.; de Haan, C.A. The ubiquitin-proteasome system plays an important role during various stages of the coronavirus infection cycle. *J. Virol* **2010**, *84*, 7869–7879. [CrossRef] [PubMed]
287. Kindler, E.; Thiel, V. Sars-cov and ifn: Too little, too late. *Cell Host Microbe* **2016**, *19*, 139–141. [CrossRef] [PubMed]
288. Rose, K.M.; Elliott, R.; Martinez-Sobrido, L.; Garcia-Sastre, A.; Weiss, S.R. Murine coronavirus delays expression of a subset of interferon-stimulated genes. *J. Virol.* **2010**, *84*, 5656–5669. [CrossRef]
289. Mantlo, E.; Bukreyeva, N.; Maruyama, J.; Paessler, S.; Huang, C. Antiviral activities of type i interferons to sars-cov-2 infection. *Antivir. Res.* **2020**, *179*, 104811. [CrossRef]
290. Loo, Y.M.; Gale, M., Jr. Immune signaling by rig-i-like receptors. *Immunity* **2011**, *34*, 680–692. [CrossRef]
291. Kawasaki, T.; Kawai, T. Toll-like receptor signaling pathways. *Front. Immunol.* **2014**, *5*, 461. [CrossRef]
292. Hu, Y.; Li, W.; Gao, T.; Cui, Y.; Jin, Y.; Li, P.; Ma, Q.; Liu, X.; Cao, C. The severe acute respiratory syndrome coronavirus nucleocapsid inhibits type i interferon production by interfering with trim25-mediated rig-i ubiquitination. *J. Virol.* **2017**, *91*, e02143-16. [CrossRef]
293. Ramos, H.J.; Gale, M., Jr. Rig-i like receptors and their signaling crosstalk in the regulation of antiviral immunity. *Curr. Opin. Virol.* **2011**, *1*, 167–176. [CrossRef]
294. Liu, T.; Zhang, L.; Joo, D.; Sun, S.C. Nf-kappab signaling in inflammation. *Signal Transduct. Target. Ther.* **2017**, *2*, 17023. [CrossRef] [PubMed]
295. Crosse, K.M.; Monson, E.A.; Beard, M.R.; Helbig, K.J. Interferon-stimulated genes as enhancers of antiviral innate immune signaling. *J. Innate Immun.* **2018**, *10*, 85–93. [CrossRef] [PubMed]
296. Au-Yeung, N.; Mandhana, R.; Horvath, C.M. Transcriptional regulation by stat1 and stat2 in the interferon jak-stat pathway. *JAKSTAT* **2013**, *2*, e23931. [CrossRef] [PubMed]
297. Schneider, W.M.; Chevillotte, M.D.; Rice, C.M. Interferon-stimulated genes: A complex web of host defenses. *Annu. Rev. Immunol.* **2014**, *32*, 513–545. [CrossRef] [PubMed]
298. Schoggins, J.W.; Rice, C.M. Interferon-stimulated genes and their antiviral effector functions. *Curr. Opin. Virol.* **2011**, *1*, 519–525. [CrossRef]
299. Dallagi, A.; Girouard, J.; Hamelin-Morrissette, J.; Dadzie, R.; Laurent, L.; Vaillancourt, C.; Lafond, J.; Carrier, C.; Reyes-Moreno, C. The activating effect of ifn-gamma on monocytes/macrophages is regulated by the lif-trophoblast-il-10 axis via stat1 inhibition and stat3 activation. *Cell Mol. Immunol.* **2015**, *12*, 326–341. [CrossRef]
300. Paolini, R.; Bernardini, G.; Molfetta, R.; Santoni, A. Nk cells and interferons. *Cytokine Growth Factor Rev.* **2015**, *26*, 113–120. [CrossRef]
301. Crouse, J.; Kalinke, U.; Oxenius, A. Regulation of antiviral t cell responses by type i interferons. *Nat. Rev. Immunol.* **2015**, *15*, 231–242. [CrossRef]
302. Swann, J.B.; Hayakawa, Y.; Zerafa, N.; Sheehan, K.C.; Scott, B.; Schreiber, R.D.; Hertzog, P.; Smyth, M.J. Type i ifn contributes to nk cell homeostasis, activation, and antitumor function. *J. Immunol.* **2007**, *178*, 7540–7549. [CrossRef]
303. Oudshoorn, D.; van der Hoeven, B.; Limpens, R.W.; Beugeling, C.; Snijder, E.J.; Barcena, M.; Kikkert, M. Antiviral innate immune response interferes with the formation of replication-associated membrane structures induced by a positive-strand rna virus. *mBio* **2016**, *7*, e01991-16. [CrossRef]
304. Fung, S.Y.; Yuen, K.S.; Ye, Z.W.; Chan, C.P.; Jin, D.Y. A tug-of-war between severe acute respiratory syndrome coronavirus 2 and host antiviral defence: Lessons from other pathogenic viruses. *Emerg. Microbes Infect.* **2020**, *9*, 558–570. [CrossRef] [PubMed]
305. Kindler, E.; Thiel, V.; Weber, F. Interaction of sars and mers coronaviruses with the antiviral interferon response. *Adv. Virus Res.* **2016**, *96*, 219–243. [PubMed]
306. Totura, A.L.; Baric, R.S. Sars coronavirus pathogenesis: Host innate immune responses and viral antagonism of interferon. *Curr. Opin. Virol.* **2012**, *2*, 264–275. [CrossRef] [PubMed]
307. Yuen, C.K.; Lam, J.Y.; Wong, W.M.; Mak, L.F.; Wang, X.; Chu, H.; Cai, J.P.; Jin, D.Y.; To, K.K.; Chan, J.F.; et al. Sars-cov-2 nsp13, nsp14, nsp15 and orf6 function as potent interferon antagonists. *Emerg. Microbes Infect.* **2020**, *9*, 1418–1428. [CrossRef]
308. Lei, X.; Dong, X.; Ma, R.; Wang, W.; Xiao, X.; Tian, Z.; Wang, C.; Wang, Y.; Li, L.; Ren, L.; et al. Activation and evasion of type i interferon responses by sars-cov-2. *Nat. Commun.* **2020**, *11*, 3810. [CrossRef]
309. Yang, Y.; Zhang, L.; Geng, H.; Deng, Y.; Huang, B.; Guo, Y.; Zhao, Z.; Tan, W. The structural and accessory proteins m, orf 4a, orf 4b, and orf 5 of middle east respiratory syndrome coronavirus (mers-cov) are potent interferon antagonists. *Protein Cell* **2013**, *4*, 951–961. [CrossRef]
310. Comar, C.E.; Goldstein, S.A.; Li, Y.; Yount, B.; Baric, R.S.; Weiss, S.R. Antagonism of dsrna-induced innate immune pathways by ns4a and ns4b accessory proteins during mers coronavirus infection. *mBio* **2019**, *10*, e00319-19. [CrossRef]
311. Yang, Y.; Ye, F.; Zhu, N.; Wang, W.; Deng, Y.; Zhao, Z.; Tan, W. Middle east respiratory syndrome coronavirus orf4b protein inhibits type i interferon production through both cytoplasmic and nuclear targets. *Sci. Rep.* **2015**, *5*, 17554. [CrossRef]
312. Siu, K.L.; Chan, C.P.; Kok, K.H.; Woo, P.C.; Jin, D.Y. Suppression of innate antiviral response by severe acute respiratory syndrome coronavirus m protein is mediated through the first transmembrane domain. *Cell Mol. Immunol.* **2014**, *11*, 141–149. [CrossRef]
313. Lui, P.Y.; Wong, L.Y.; Fung, C.L.; Siu, K.L.; Yeung, M.L.; Yuen, K.S.; Chan, C.P.; Woo, P.C.; Yuen, K.Y.; Jin, D.Y. Middle east respiratory syndrome coronavirus m protein suppresses type i interferon expression through the inhibition of tbk1-dependent phosphorylation of irf3. *Emerg. Microbes Infect.* **2016**, *5*, e39. [CrossRef]

314. Di Palma, F.; Daino, G.L.; Ramaswamy, V.K.; Corona, A.; Frau, A.; Fanunza, E.; Vargiu, A.V.; Tramontano, E.; Ruggerone, P. Relevance of ebola virus vp35 homo-dimerization on the type i interferon cascade inhibition. *Antivir. Chem. Chemother.* **2019**, *27*, 2040206619889220. [CrossRef] [PubMed]
315. Schwerk, J.; Negash, A.; Savan, R.; Gale, M., Jr. Innate immunity in hepatitis c virus infection. *Cold Spring Harb. Perspect. Med.* **2020**, a036988. [CrossRef] [PubMed]
316. Wang, Y.; Liu, L. The membrane protein of severe acute respiratory syndrome coronavirus functions as a novel cytosolic pathogen-associated molecular pattern to promote beta interferon induction via a toll-like-receptor-related traf3-independent mechanism. *mBio* **2016**, *7*, e01872-15. [CrossRef] [PubMed]
317. Chang, C.Y.; Liu, H.M.; Chang, M.F.; Chang, S.C. Middle east respiratory syndrome coronavirus nucleocapsid protein suppresses type i and type iii interferon induction by targeting rig-i signaling. *J. Virol.* **2020**, *94*, e00099-20. [CrossRef] [PubMed]
318. Likai, J.; Shasha, L.; Wenxian, Z.; Jingjiao, M.; Jianhe, S.; Hengan, W.; Yaxian, Y. Porcine deltacoronavirus nucleocapsid protein suppressed ifn-beta production by interfering porcine rig-i dsrna-binding and k63-linked polyubiquitination. *Front. Immunol.* **2019**, *10*, 1024. [CrossRef]
319. Zhang, Q.; Shi, K.; Yoo, D. Suppression of type i interferon production by porcine epidemic diarrhea virus and degradation of creb-binding protein by nsp1. *Virology* **2016**, *489*, 252–268. [CrossRef]
320. Lu, X.; Pan, J.; Tao, J.; Guo, D. Sars-cov nucleocapsid protein antagonizes ifn-beta response by targeting initial step of ifn-beta induction pathway, and its c-terminal region is critical for the antagonism. *Virus Genes* **2011**, *42*, 37–45. [CrossRef]
321. Martin-Vicente, M.; Medrano, L.M.; Resino, S.; Garcia-Sastre, A.; Martinez, I. Trim25 in the regulation of the antiviral innate immunity. *Front. Immunol.* **2017**, *8*, 1187. [CrossRef]
322. Mu, J.; Fang, Y.; Yang, Q.; Shu, T.; Wang, A.; Huang, M.; Jin, L.; Deng, F.; Qiu, Y.; Zhou, X. Sars-cov-2 n protein antagonizes type i interferon signaling by suppressing phosphorylation and nuclear translocation of stat1 and stat2. *Cell Discov.* **2020**, *6*, 65. [CrossRef]
323. Narayanan, K.; Huang, C.; Lokugamage, K.; Kamitani, W.; Ikegami, T.; Tseng, C.T.; Makino, S. Severe acute respiratory syndrome coronavirus nsp1 suppresses host gene expression, including that of type i interferon, in infected cells. *J. Virol.* **2008**, *82*, 4471–4479. [CrossRef]
324. Gaglia, M.M.; Covarrubias, S.; Wong, W.; Glaunsinger, B.A. A common strategy for host rna degradation by divergent viruses. *J. Virol.* **2012**, *86*, 9527–9530. [CrossRef] [PubMed]
325. Nakagawa, K.; Narayanan, K.; Wada, M.; Popov, V.L.; Cajimat, M.; Baric, R.S.; Makino, S. The endonucleolytic rna cleavage function of nsp1 of middle east respiratory syndrome coronavirus promotes the production of infectious virus particles in specific human cell lines. *J. Virol.* **2018**, *92*, e01157-18. [CrossRef] [PubMed]
326. Kamitani, W.; Huang, C.; Narayanan, K.; Lokugamage, K.G.; Makino, S. A two-pronged strategy to suppress host protein synthesis by sars coronavirus nsp1 protein. *Nat. Struct. Mol. Biol.* **2009**, *16*, 1134–1140. [CrossRef] [PubMed]
327. Huang, C.; Lokugamage, K.G.; Rozovics, J.M.; Narayanan, K.; Semler, B.L.; Makino, S. Sars coronavirus nsp1 protein induces template-dependent endonucleolytic cleavage of mrnas: Viral mrnas are resistant to nsp1-induced rna cleavage. *PLoS Pathog.* **2011**, *7*, e1002433. [CrossRef] [PubMed]
328. Wathelet, M.G.; Orr, M.; Frieman, M.B.; Baric, R.S. Severe acute respiratory syndrome coronavirus evades antiviral signaling: Role of nsp1 and rational design of an attenuated strain. *J. Virol.* **2007**, *81*, 11620–11633. [CrossRef] [PubMed]
329. Frieman, M.; Ratia, K.; Johnston, R.E.; Mesecar, A.D.; Baric, R.S. Severe acute respiratory syndrome coronavirus papain-like protease ubiquitin-like domain and catalytic domain regulate antagonism of irf3 and nf-kappab signaling. *J. Virol.* **2009**, *83*, 6689–6705. [CrossRef]
330. Devaraj, S.G.; Wang, N.; Chen, Z.; Chen, Z.; Tseng, M.; Barretto, N.; Lin, R.; Peters, C.J.; Tseng, C.T.; Baker, S.C.; et al. Regulation of irf-3-dependent innate immunity by the papain-like protease domain of the severe acute respiratory syndrome coronavirus. *J. Biol. Chem.* **2007**, *282*, 32208–32221. [CrossRef]
331. Clementz, M.A.; Chen, Z.; Banach, B.S.; Wang, Y.; Sun, L.; Ratia, K.; Baez-Santos, Y.M.; Wang, J.; Takayama, J.; Ghosh, A.K.; et al. Deubiquitinating and interferon antagonism activities of coronavirus papain-like proteases. *J. Virol.* **2010**, *84*, 4619–4629. [CrossRef]
332. Matthews, K.; Schafer, A.; Pham, A.; Frieman, M. The sars coronavirus papain like protease can inhibit irf3 at a post activation step that requires deubiquitination activity. *Virol. J.* **2014**, *11*, 209. [CrossRef]
333. Li, J.Y.; Liao, C.H.; Wang, Q.; Tan, Y.J.; Luo, R.; Qiu, Y.; Ge, X.Y. The orf6, orf8 and nucleocapsid proteins of sars-cov-2 inhibit type i interferon signaling pathway. *Virus Res.* **2020**, *286*, 198074. [CrossRef]
334. Kopecky-Bromberg, S.A.; Martinez-Sobrido, L.; Frieman, M.; Baric, R.A.; Palese, P. Severe acute respiratory syndrome coronavirus open reading frame (orf) 3b, orf 6, and nucleocapsid proteins function as interferon antagonists. *J. Virol.* **2007**, *81*, 548–557. [CrossRef] [PubMed]
335. Frieman, M.; Yount, B.; Heise, M.; Kopecky-Bromberg, S.A.; Palese, P.; Baric, R.S. Severe acute respiratory syndrome coronavirus orf6 antagonizes stat1 function by sequestering nuclear import factors on the rough endoplasmic reticulum/golgi membrane. *J. Virol.* **2007**, *81*, 9812–9824. [CrossRef] [PubMed]
336. Yuan, X.; Yao, Z.; Shan, Y.; Chen, B.; Yang, Z.; Wu, J.; Zhao, Z.; Chen, J.; Cong, Y. Nucleolar localization of non-structural protein 3b, a protein specifically encoded by the severe acute respiratory syndrome coronavirus. *Virus Res.* **2005**, *114*, 70–79. [CrossRef] [PubMed]

337. Konno, Y.; Kimura, I.; Uriu, K.; Fukushi, M.; Irie, T.; Koyanagi, Y.; Sauter, D.; Gifford, R.; Consortium, U.-C.; Nakagawa, S.; et al. Sars-cov-2 orf3b is a Potent Interferon Antagonist Whose Activity is Increased by a Naturally Occurring Elongation Variant. *Cell Rep.* **2020**, *32*, 108185. [CrossRef] [PubMed]
338. Stoermer, K.A.; Morrison, T.E. Complement and viral pathogenesis. *Virology* **2011**, *411*, 362–373. [CrossRef]
339. Dunkelberger, J.R.; Song, W.C. Complement and its role in innate and adaptive immune responses. *Cell Res.* **2010**, *20*, 34–50. [CrossRef]
340. Thorgersen, E.B.; Barratt-Due, A.; Haugaa, H.; Harboe, M.; Pischke, S.E.; Nilsson, P.H.; Mollnes, T.E. The role of complement in liver injury, regeneration, and transplantation. *Hepatology* **2019**, *70*, 725–736. [CrossRef]
341. Abenavoli, L.; Gentile, I.; Maraolo, A.E.; Negro, F. Sars-cov-2 and liver damage: A possible pathogenetic link. *Hepatobiliary Surg. Nutr.* **2020**, *9*, 322–324. [CrossRef]
342. Keshari, R.S.; Silasi, R.; Popescu, N.I.; Patel, M.M.; Chaaban, H.; Lupu, C.; Coggeshall, K.M.; Mollnes, T.E.; DeMarco, S.J.; Lupu, F. Inhibition of complement c5 protects against organ failure and reduces mortality in a baboon model of escherichia coli sepsis. *Proc. Natl. Acad. Sci. USA* **2017**, *114*, E6390–E6399. [CrossRef]
343. Garcia, C.C.; Weston-Davies, W.; Russo, R.C.; Tavares, L.P.; Rachid, M.A.; Alves-Filho, J.C.; Machado, A.V.; Ryffel, B.; Nunn, M.A.; Teixeira, M.M. Complement c5 activation during influenza a infection in mice contributes to neutrophil recruitment and lung injury. *PLoS ONE* **2013**, *8*, e64443. [CrossRef]
344. Thurman, J.M.; Le Quintrec, M. Targeting the complement cascade: Novel treatments coming down the pike. *Kidney Int.* **2016**, *90*, 746–752. [CrossRef]
345. Rittirsch, D.; Redl, H.; Huber-Lang, M. Role of complement in multiorgan failure. *Clin. Dev. Immunol.* **2012**, *2012*, 962927. [CrossRef] [PubMed]
346. Chu, K.H.; Tsang, W.K.; Tang, C.S.; Lam, M.F.; Lai, F.M.; To, K.F.; Fung, K.S.; Tang, H.L.; Yan, W.W.; Chan, H.W.; et al. Acute renal impairment in coronavirus-associated severe acute respiratory syndrome. *Kidney Int.* **2005**, *67*, 698–705. [CrossRef] [PubMed]
347. Kopf, M.; Abel, B.; Gallimore, A.; Carroll, M.; Bachmann, M.F. Complement component c3 promotes t-cell priming and lung migration to control acute influenza virus infection. *Nat. Med.* **2002**, *8*, 373–378. [CrossRef] [PubMed]
348. Kim, Y.J.; Kim, K.H.; Ko, E.J.; Kim, M.C.; Lee, Y.N.; Jung, Y.J.; Lee, Y.T.; Kwon, Y.M.; Song, J.M.; Kang, S.M. Complement c3 plays a key role in inducing humoral and cellular immune responses to influenza virus strain-specific hemagglutinin-based or cross-protective m2 extracellular domain-based vaccination. *J. Virol.* **2018**, *92*, e00969-18. [CrossRef] [PubMed]
349. Rattan, A.; Pawar, S.D.; Nawadkar, R.; Kulkarni, N.; Lal, G.; Mullick, J.; Sahu, A. Synergy between the classical and alternative pathways of complement is essential for conferring effective protection against the pandemic influenza a(h1n1) 2009 virus infection. *PLoS Pathog.* **2017**, *13*, e1006248. [CrossRef]
350. Jiang, Y.; Zhao, G.; Song, N.; Li, P.; Chen, Y.; Guo, Y.; Li, J.; Du, L.; Jiang, S.; Guo, R.; et al. Blockade of the c5a-c5ar axis alleviates lung damage in hdpp4-transgenic mice infected with mers-cov. *Emerg. Microbes Infect.* **2018**, *7*, 77. [CrossRef]
351. Horiuchi, T.; Tsukamoto, H. Complement-targeted therapy: Development of c5- and c5a-targeted inhibition. *Inflamm. Regen.* **2016**, *36*, 11. [CrossRef]
352. Bekker, P.; Dairaghi, D.; Seitz, L.; Leleti, M.; Wang, Y.; Ertl, L.; Baumgart, T.; Shugarts, S.; Lohr, L.; Dang, T.; et al. Characterization of pharmacologic and pharmacokinetic properties of ccx168, a potent and selective orally administered complement 5a receptor inhibitor, based on preclinical evaluation and randomized phase 1 clinical study. *PLoS ONE* **2016**, *11*, e0164646. [CrossRef]
353. Versteeg, G.A.; van de Nes, P.S.; Bredenbeek, P.J.; Spaan, W.J. The coronavirus spike protein induces endoplasmic reticulum stress and upregulation of intracellular chemokine mrna concentrations. *J. Virol.* **2007**, *81*, 10981–10990. [CrossRef]
354. Minakshi, R.; Padhan, K.; Rani, M.; Khan, N.; Ahmad, F.; Jameel, S. The sars coronavirus 3a protein causes endoplasmic reticulum stress and induces ligand-independent downregulation of the type 1 interferon receptor. *PLoS ONE* **2009**, *4*, e8342. [CrossRef] [PubMed]
355. Cottam, E.M.; Maier, H.J.; Manifava, M.; Vaux, L.C.; Chandra-Schoenfelder, P.; Gerner, W.; Britton, P.; Ktistakis, N.T.; Wileman, T. Coronavirus nsp6 proteins generate autophagosomes from the endoplasmic reticulum via an omegasome intermediate. *Autophagy* **2011**, *7*, 1335–1347. [CrossRef] [PubMed]
356. Xu, X.; Zhang, H.; Zhang, Q.; Huang, Y.; Dong, J.; Liang, Y.; Liu, H.J.; Tong, D. Porcine epidemic diarrhea virus n protein prolongs s-phase cell cycle, induces endoplasmic reticulum stress, and up-regulates interleukin-8 expression. *Vet. Microbiol.* **2013**, *164*, 212–221. [CrossRef]
357. Zhang, Q.; Xu, Y.; Chang, R.; Tong, D.; Xu, X. Transmissible gastroenteritis virus n protein causes endoplasmic reticulum stress, up-regulates interleukin-8 expression and its subcellular localization in the porcine intestinal epithelial cell. *Res. Vet. Sci.* **2018**, *119*, 109–115. [CrossRef] [PubMed]
358. Zou, D.; Xu, J.; Duan, X.; Xu, X.; Li, P.; Cheng, L.; Zheng, L.; Li, X.; Zhang, Y.; Wang, X.; et al. Porcine epidemic diarrhea virus orf3 protein causes endoplasmic reticulum stress to facilitate autophagy. *Vet. Microbiol.* **2019**, *235*, 209–219. [CrossRef]
359. Liang, J.Q.; Fang, S.; Yuan, Q.; Huang, M.; Chen, R.A.; Fung, T.S.; Liu, D.X. N-linked glycosylation of the membrane protein ectodomain regulates infectious bronchitis virus-induced er stress response, apoptosis and pathogenesis. *Virology* **2019**, *531*, 48–56. [CrossRef]
360. Yuan, X.; Yao, Z.; Wu, J.; Zhou, Y.; Shan, Y.; Dong, B.; Zhao, Z.; Hua, P.; Chen, J.; Cong, Y. G1 phase cell cycle arrest induced by sars-cov 3a protein via the cyclin d3/prb pathway. *Am. J. Respir. Cell Mol. Biol.* **2007**, *37*, 9–19. [CrossRef]

361. Yuan, X.; Wu, J.; Shan, Y.; Yao, Z.; Dong, B.; Chen, B.; Zhao, Z.; Wang, S.; Chen, J.; Cong, Y. Sars coronavirus 7a protein blocks cell cycle progression at g0/g1 phase via the cyclin d3/prb pathway. *Virology* **2006**, *346*, 74–85. [CrossRef]
362. Xu, X.G.; Zhang, H.L.; Zhang, Q.; Dong, J.; Huang, Y.; Tong, D.W. Porcine epidemic diarrhea virus m protein blocks cell cycle progression at s-phase and its subcellular localization in the porcine intestinal epithelial cells. *Acta Virol.* **2015**, *59*, 265–275. [CrossRef]
363. Sun, P.; Wu, H.; Huang, J.; Xu, Y.; Yang, F.; Zhang, Q.; Xu, X. Porcine epidemic diarrhea virus through p53-dependent pathway causes cell cycle arrest in the g0/g1 phase. *Virus Res.* **2018**, *253*, 1–11. [CrossRef]
364. He, Y.; Zhou, Y.; Liu, S.; Kou, Z.; Li, W.; Farzan, M.; Jiang, S. Receptor-binding domain of sars-cov spike protein induces highly potent neutralizing antibodies: Implication for developing subunit vaccine. *Biochem. Biophys. Res. Commun.* **2004**, *324*, 773–781. [CrossRef] [PubMed]
365. Du, L.; Zhao, G.; Chan, C.C.; Sun, S.; Chen, M.; Liu, Z.; Guo, H.; He, Y.; Zhou, Y.; Zheng, B.J.; et al. Recombinant receptor-binding domain of sars-cov spike protein expressed in mammalian, insect and e. Coli cells elicits potent neutralizing antibody and protective immunity. *Virology* **2009**, *393*, 144–150. [CrossRef] [PubMed]
366. Fung, T.S.; Liu, D.X. Coronavirus infection, er stress, apoptosis and innate immunity. *Front. Microbiol.* **2014**, *5*, 296. [CrossRef] [PubMed]
367. Walls, A.C.; Xiong, X.; Park, Y.J.; Tortorici, M.A.; Snijder, J.; Quispe, J.; Cameroni, E.; Gopal, R.; Dai, M.; Lanzavecchia, A.; et al. Unexpected receptor functional mimicry elucidates activation of coronavirus fusion. *Cell* **2019**, *176*, 1026–1039.e15. [CrossRef]
368. Li, C.K.; Wu, H.; Yan, H.; Ma, S.; Wang, L.; Zhang, M.; Tang, X.; Temperton, N.J.; Weiss, R.A.; Brenchley, J.M.; et al. T cell responses to whole sars coronavirus in humans. *J. Immunol.* **2008**, *181*, 5490–5500. [CrossRef]
369. Wang, Z.; Yuan, Z.; Matsumoto, M.; Hengge, U.R.; Chang, Y.F. Immune responses with DNA vaccines encoded different gene fragments of severe acute respiratory syndrome coronavirus in balb/c mice. *Biochem. Biophys. Res. Commun.* **2005**, *327*, 130–135. [CrossRef]
370. Huang, J.; Ma, R.; Wu, C.Y. Immunization with sars-cov s DNA vaccine generates memory cd4+ and cd8+ t cell immune responses. *Vaccine* **2006**, *24*, 4905–4913. [CrossRef]
371. Yang, Z.Y.; Kong, W.P.; Huang, Y.; Roberts, A.; Murphy, B.R.; Subbarao, K.; Nabel, G.J. A DNA vaccine induces sars coronavirus neutralization and protective immunity in mice. *Nature* **2004**, *428*, 561–564. [CrossRef]
372. Hu, H.; Lu, X.; Tao, L.; Bai, B.; Zhang, Z.; Chen, Y.; Zheng, F.; Chen, J.; Chen, Z.; Wang, H. Induction of specific immune responses by severe acute respiratory syndrome coronavirus spike DNA vaccine with or without interleukin-2 immunization using different vaccination routes in mice. *Clin. Vaccine Immunol* **2007**, *14*, 894–901. [CrossRef]
373. Janice Oh, H.L.; Gan, S.K.; Bertoletti, A.; Tan, Y.J. Understanding the t cell immune response in sars coronavirus infection. *Emerg. Microbes Infect.* **2012**, *1*, e23. [CrossRef]
374. Wong, C.K.; Lam, C.W.; Wu, A.K.; Ip, W.K.; Lee, N.L.; Chan, I.H.; Lit, L.C.; Hui, D.S.; Chan, M.H.; Chung, S.S.; et al. Plasma inflammatory cytokines and chemokines in severe acute respiratory syndrome. *Clin. Exp. Immunol.* **2004**, *136*, 95–103. [CrossRef]
375. Grifoni, A.; Weiskopf, D.; Ramirez, S.I.; Mateus, J.; Dan, J.M.; Moderbacher, C.R.; Rawlings, S.A.; Sutherland, A.; Premkumar, L.; Jadi, R.S.; et al. Targets of t cell responses to sars-cov-2 coronavirus in humans with covid-19 disease and unexposed individuals. *Cell* **2020**, *181*, 1489–1501.e15. [CrossRef]
376. Huang, J.L.; Huang, J.; Duan, Z.H.; Wei, J.; Min, J.; Luo, X.H.; Li, J.G.; Tan, W.P.; Wu, L.Z.; Liu, R.Y.; et al. Th2 predominance and cd8+ memory t cell depletion in patients with severe acute respiratory syndrome. *Microbes Infect.* **2005**, *7*, 427–436. [CrossRef] [PubMed]
377. Costela-Ruiz, V.J.; Illescas-Montes, R.; Puerta-Puerta, J.M.; Ruiz, C.; Melguizo-Rodriguez, L. Sars-cov-2 infection: The role of cytokines in covid-19 disease. *Cytokine Growth Factor Rev.* **2020**, *54*, 62–75. [CrossRef] [PubMed]
378. Jiang, H.W.; Li, Y.; Zhang, H.N.; Wang, W.; Yang, X.; Qi, H.; Li, H.; Men, D.; Zhou, J.; Tao, S.C. Sars-cov-2 proteome microarray for global profiling of covid-19 specific igg and igm responses. *Nat. Commun.* **2020**, *11*, 3581. [CrossRef] [PubMed]
379. Basu, B.V.; Brown, O.R. Comparative analysis of coronaviridae nucleocapsid and surface glycoprotein sequences. *Front. Biosci. (Landmark Ed.)* **2020**, *25*, 1894–1900. [CrossRef] [PubMed]
380. Lee, C.H.; Koohy, H. In silico identification of vaccine targets for 2019-ncov. *F1000Res* **2020**, *9*, 145. [CrossRef]
381. Ahmed, S.F.; Quadeer, A.A.; McKay, M.R. Preliminary identification of potential vaccine targets for the covid-19 coronavirus (sars-cov-2) based on sars-cov immunological studies. *Viruses* **2020**, *12*, 254. [CrossRef]
382. Ng, O.W.; Chia, A.; Tan, A.T.; Jadi, R.S.; Leong, H.N.; Bertoletti, A.; Tan, Y.J. Memory t cell responses targeting the sars coronavirus persist up to 11 years post-infection. *Vaccine* **2016**, *34*, 2008–2014. [CrossRef]
383. Ong, E.; Wong, M.U.; Huffman, A.; He, Y. Covid-19 coronavirus vaccine design using reverse vaccinology and machine learning. *Front. Immunol.* **2020**, *11*, 1581. [CrossRef]
384. de Haan, C.A.; Rottier, P.J. Molecular interactions in the assembly of coronaviruses. *Adv. Virus Res.* **2005**, *64*, 165–230. [PubMed]
385. Liu, J.; Sun, Y.; Qi, J.; Chu, F.; Wu, H.; Gao, F.; Li, T.; Yan, J.; Gao, G.F. The membrane protein of severe acute respiratory syndrome coronavirus acts as a dominant immunogen revealed by a clustering region of novel functionally and structurally defined cytotoxic t-lymphocyte epitopes. *J. Infect. Dis.* **2010**, *202*, 1171–1180. [CrossRef] [PubMed]
386. Li, Y.H.; Hu, C.Y.; Wu, N.P.; Yao, H.P.; Li, L.J. Molecular characteristics, functions, and related pathogenicity of mers-cov proteins. *Engineering* **2019**, *5*, 940–947. [CrossRef] [PubMed]

387. Ignjatovic, J.; Galli, L. Structural proteins of avian infectious bronchitis virus: Role in immunity and protection. *Adv. Exp. Med. Biol.* **1993**, *342*, 449–453. [PubMed]
388. Ignjatovic, J.; Galli, L. The s1 glycoprotein but not the n or m proteins of avian infectious bronchitis virus induces protection in vaccinated chickens. *Arch. Virol.* **1994**, *138*, 117–134. [CrossRef]
389. Saif, L.J. Coronavirus immunogens. *Vet. Microbiol.* **1993**, *37*, 285–297. [CrossRef]
390. Okada, M.; Takemoto, Y.; Okuno, Y.; Hashimoto, S.; Yoshida, S.; Fukunaga, Y.; Tanaka, T.; Kita, Y.; Kuwayama, S.; Muraki, Y.; et al. The development of vaccines against sars corona virus in mice and scid-pbl/hu mice. *Vaccine* **2005**, *23*, 2269–2272. [CrossRef]
391. Zhang, Z.; Chen, J.; Shi, H.; Chen, X.; Shi, D.; Feng, L.; Yang, B. Identification of a conserved linear b-cell epitope in the m protein of porcine epidemic diarrhea virus. *Virol. J.* **2012**, *9*, 225. [CrossRef]
392. Yan, F.; Zhao, Y.; Hu, Y.; Qiu, J.; Lei, W.; Ji, W.; Li, X.; Wu, Q.; Shi, X.; Li, Z. Protection of chickens against infectious bronchitis virus with a multivalent DNA vaccine and boosting with an inactivated vaccine. *J. Vet. Sci.* **2013**, *14*, 53–60. [CrossRef]
393. Takano, T.; Morioka, H.; Gomi, K.; Tomizawa, K.; Doki, T.; Hohdatsu, T. Screening and identification of t helper 1 and linear immunodominant antibody-binding epitopes in spike 1 domain and membrane protein of feline infectious peritonitis virus. *Vaccine* **2014**, *32*, 1834–1840. [CrossRef]
394. Ventola, C.L. Immunization in the united states: Recommendations, barriers, and measures to improve compliance: Part 1: Childhood vaccinations. *P T* **2016**, *41*, 426–436. [PubMed]
395. Duggan, N.M.; Ludy, S.M.; Shannon, B.C.; Reisner, A.T.; Wilcox, S.R. Is novel coronavirus 2019 reinfection possible? Interpreting dynamic sars-cov-2 test results through a case report. *Am. J. Emerg. Med.* **2020**. [CrossRef] [PubMed]
396. Gunthard, H.F.; Saag, M.S.; Benson, C.A.; del Rio, C.; Eron, J.J.; Gallant, J.E.; Hoy, J.F.; Mugavero, M.J.; Sax, P.E.; Thompson, M.A.; et al. Antiretroviral drugs for treatment and prevention of hiv infection in adults: 2016 recommendations of the international antiviral society-USA panel. *JAMA* **2016**, *316*, 191–210. [CrossRef] [PubMed]
397. Mirza, M.U.; Vanmeert, M.; Ali, A.; Iman, K.; Froeyen, M.; Idrees, M. Perspectives towards antiviral drug discovery against ebola virus. *J. Med. Virol.* **2019**, *91*, 2029–2048. [CrossRef] [PubMed]
398. Lau, S.K.; Woo, P.C.; Yip, C.C.; Tse, H.; Tsoi, H.W.; Cheng, V.C.; Lee, P.; Tang, B.S.; Cheung, C.H.; Lee, R.A.; et al. Coronavirus hku1 and other coronavirus infections in hong kong. *J. Clin. Microbiol.* **2006**, *44*, 2063–2071. [CrossRef]
399. Kanwar, A.; Selvaraju, S.; Esper, F. Human coronavirus-hku1 infection among adults in cleveland, ohio. *Open Forum. Infect. Dis.* **2017**, *4*, ofx052. [CrossRef]
400. Schulz, L.L.; Tonsor, G.T. Assessment of the economic impacts of porcine epidemic diarrhea virus in the united states. *J. Anim. Sci.* **2015**, *93*, 5111–5118. [CrossRef]
401. Legnardi, M.; Tucciarone, C.M.; Franzo, G.; Cecchinato, M. Infectious bronchitis virus evolution, diagnosis and control. *Vet. Sci.* **2020**, *7*, 79. [CrossRef]
402. Wu, T.; Perrings, C.; Kinzig, A.; Collins, J.P.; Minteer, B.A.; Daszak, P. Economic growth, urbanization, globalization, and the risks of emerging infectious diseases in china: A review. *Ambio* **2017**, *46*, 18–29. [CrossRef]
403. Njume, C.; Goduka, N.I. Treatment of diarrhoea in rural african communities: An overview of measures to maximise the medicinal potentials of indigenous plants. *Int. J. Environ. Res. Public Health* **2012**, *9*, 3911–3933. [CrossRef]
404. Levy, B.; Edholm, C.; Gaoue, O.; Kaondera-Shava, R.; Kgosimore, M.; Lenhart, S.; Lephodisa, B.; Lungu, E.; Marijani, T.; Nyabadza, F. Modeling the role of public health education in ebola virus disease outbreaks in sudan. *Infect. Dis. Model.* **2017**, *2*, 323–340. [CrossRef] [PubMed]

International Journal of
Molecular Sciences

MDPI

Article

Syk Inhibitors: New Computational Insights into Their Intraerythrocytic Action in *Plasmodium falciparum* Malaria

Giuseppe Marchetti [1,*], Alessandro Dessì [2,*], Roberto Dallocchio [2], Ioannis Tsamesidis [1,3], Maria Carmina Pau [1], Francesco Michelangelo Turrini [4] and Antonella Pantaleo [1]

1 Department of Biomedical Sciences, University of Sassari, 07100 Sassari, Italy; johntsame@gmail.com (I.T.); paumc81@tiscali.it (M.C.P.); apantaleo@uniss.it (A.P.)

2 National Research Council (CNR) Institute of Biomolecular Chemistry, 07100 Sassari, Italy; roberto.dallocchio@cnr.it

3 UMR 152 Pharma-Dev, Université de Toulouse III, IRD, UPS, 31000 Toulouse, France

4 Department of Oncology, University of Turin, 10126 Turin, Italy; francesco.turrini@unito.it

* Correspondence: gmarchetti@uniss.it (G.M.); alessandro.dessi@cnr.it (A.D.)

Received: 17 July 2020; Accepted: 18 September 2020; Published: 23 September 2020

Abstract: Resistance to antimalarial drugs has spread rapidly over the past few decades. The WHO recommends artemisinin-based combination therapies for the treatment of uncomplicated malaria, but unfortunately these approaches are losing their efficacy in large areas of Southeast Asia. In 2016, artemisinin resistance was confirmed in 5 countries of the Greater Mekong subregion. We focused our study on Syk inhibitors as antimalarial drugs. The Syk protein is present in human erythrocytes, and the membrane of protein band 3 is its major target following activation by oxidant stress. Tyr phosphorylation of band 3 occurs during *P. falciparum* growth, leading to the release of microparticles containing hemicromes and structural weakening of the host cell membrane, simplifying merozoite reinfection. Syk inhibitors block these events by interacting with the Syk protein's catalytic site. We performed in vitro proteomics and in silico studies and compared the results. In vitro studies were based on treatment of the parasite's cellular cultures with different concentrations of Syk inhibitors, while proteomics studies were focused on the Tyr phosphorylation of band 3 by Syk protein with the same concentrations of drugs. In silico studies were based on different molecular modeling approaches in order to analyze and optimize the ligand–protein interactions and obtain the highest efficacy in vitro. In the presence of Syk inhibitors, we observed a marked decrease of band 3 Tyr phosphorylation according to the increase of the drug's concentration. Our studies could be useful for the structural optimization of these compounds and for the design of novel Syk inhibitors in the future.

Keywords: band 3; red blood cells; antimalarial drugs; molecular docking; molecular dynamics

1. Introduction

Spleen tyrosine kinase (Syk) is a cytosolic non-receptor tyrosine kinase that functions downstream of antigen receptors in immune cells such as mast cells, B lymphocytes, and macrophages. Syk is a crucial signal transducer of activated immunoreceptors in multiple downstream events, which differ depending on the cell type, including proliferation, differentiation, and phagocytosis [1–3]. Syk function might, therefore, be an attractive target for therapeutic interventions for autoimmune or inflammation diseases [4]. Syk is composed of two Src homology tandem domains, defined as N-SH2 and C-SH2 [5]; these are important for activity regulation and for localizing this kinase in the cell membrane. The tandem SH2 (tSH2) module is also separated by an inter-SH2 linker of 50 amino acids. This is the most conserved region in the kinase family, with 65% sequence homology. Additionally, tSH2 presents an α-helix, which has an important role in protein–protein interactions and serves as a docking platform for tyrosine-based immune receptor activating motifs (ITAMs), which are displayed on the cytosolic side of the plasma membrane [6–8]. An interdomain linker of 80–100 amino acids is located between C-SH2 and the catalytic domain. This interdomain is important in regulating kinase activity because it contains phosphotyrosine residues. A catalytic domain or SH1 containing 300 amino acids follows the interdomain linker. It contains the binding sites for ATP and two autophosphorylation sites (Tyr525 and Tyr526) [9]. Syk protein ends with a C-terminal tail, the function of which is currently unidentified.

During previous studies of human erythrocyte membranes, we observed that the erythrocytes possess a mechanism that is involved in the expulsion of denatured hemoglobin, requiring the activation of Syk [10–14]. This function could play a role in the process of asexual *P. falciparum* growth, as malaria parasites exert oxidative stress in erythrocytes, causing denaturation of hemoglobin, the oxidation of band 3, and its subsequent phosphorylation by Syk [15,16]. The protein band 3 (also known as the anion exchanger, AE1) constitutes the major attachment site of the spectrin-based cytoskeleton to the erythrocyte's lipid bilayer, and thereby contributes critically to the stability of the red cell membrane [13,17]. Under steady conditions, this linkage confers to the membrane the required elasticity and mechanical stability. The oxidation of band 3 and its subsequent phosphorylation by Syk kinase cause its detachment from the cytoskeleton and the destabilization of the membrane with the release of microvesicles [10,14]. Scheme 1 shows a schematic model representing the proposed mechanism of action.

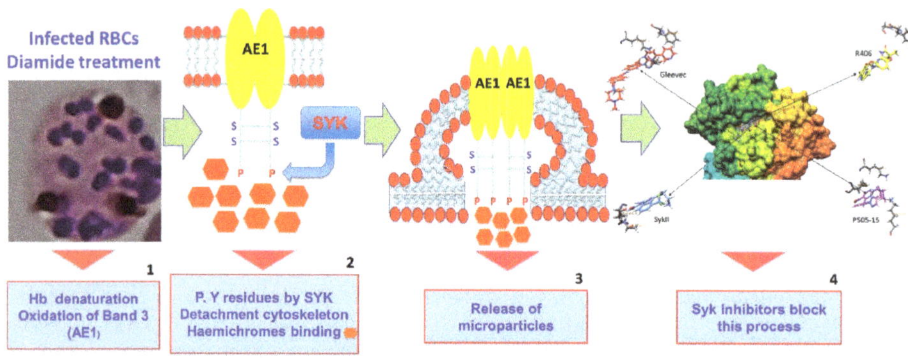

Scheme 1. Schematic model of the proposed mechanism of Syk inhibitor action. (**1**) Diamide treatment induces band 3 oxidation and hemoglobin denaturation. (**2**) Syk binds to band 3 and catalyzes band 3 cytoplasmic domain tyrosine phosphorylation, causing its detachment from the cytoskeleton. (**3**) Oxidized and phosphorylated band 3 forms large clusters, which are released in microparticles through the vesiculation. (**4**) Syk inhibitors block this process.

Taking in consideration that Syk kinase inhibitors block the expulsion of denatured hemoglobin and its accumulation inside the parasitized erythrocytes [15–17], more knowledge of the mechanisms responsible for the protein-ligand recognition and binding will facilitate the design, development, and discovery of a new promising class of antimalarial drugs.

We first characterized the ligands with the aim of screening new drugs more efficiently in order to cure malaria, using quantum mechanics (QM) and molecular descriptors to assess the electronic density, molecular electrostatic potential (MEP), and the charge distribution. Molecular modelling approaches, such as docking and molecular dynamics (MD) simulation [18,19], were performed to analyze the binding mode of Syk/ligands. In an effort to pursue a more unbiased approach towards identifying protein tyrosine kinase (PTK) inhibitors with antimalaria activity, we screened some ATP-competitive inhibitors of Syk characterized by different IC_{50s} (Half maximal inhibitory concentrations) for the Syk catalytic subunits [17]. Figure 1 shows the chemical structures and the IC_{50s} values of Imatinib, R406, Syk inhibitor II, and P505-15.

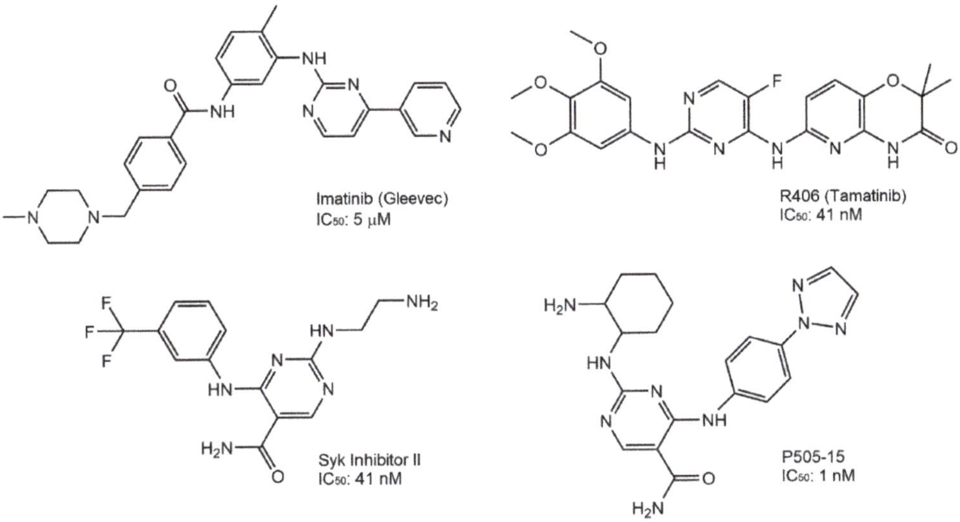

Figure 1. Chemical structures and IC_{50s} for the Syk catalytics subunits of Imatinib, R406, Syk inhibitor II, and P505-15.

Imatinib, sold under the brand names Gleevec, is a well-tolerated tyrosine kinase inhibitor. It is FDA-approved for use in children [20–22] and prevents parasite-induced tyrosine phosphorylation of band 3 and terminates *P. falciparum* parasitemia in vitro by blocking parasite egress at clinically relevant concentrations [15]. This drug is also used for the treatment of chronic myeloid leukemia (CML), acute lymphoblastic leukemia (ALL), and gastrointestinal stromal tumor (GIST). R406 (tamatinib) is an active metabolite of prodrug R788 (fostamatinib), which has already been used in clinical trials for rheumatoid arthritis [23], autoimmune thrombocytopenia [24], autoimmune hemolytic anemia, IgA nephropathy, and lymphoma [25,26].

P505-15 is a candidate drug that has already been used in in vivo studies in mice for treatment of rheumatoid arthritis, non-Hodgkin lymphoma (NHL), and chronic lymphocytic leukemia (CLL) [27,28]. Syk inhibitor II is already used for inhibition of serotonin (5-HT) release in rat basophilic leukemia (RBL) cells and to treat allergic diseases [29,30].

Here, we document the interactions between Syk and its ligands in order to understand the biology at the molecular level, with the aim of improving and modifying the compounds' structures and discovering new drugs to inhibit infected RBCs.

2. Results and Discussion

2.1. Densitometric Analysis of Band 3 Tyrosine Phosphorylation in Diamide-Treated Erythrocytes after Treatment with Syk Inhibitors

In vitro studies have shown that diamide treatment increases the band 3 tyrosine phosphorylation state, suggesting a possible functional connection between membrane oxidative damage and modulation of signal transduction pathways involving kinases or phosphatases [31,32]. To better understand the relationship between the activity of Syk inhibitors and their abilities to reduce band 3 tyrosine phosphorylation, we carried out experiments using 2 mM diamide as the control (Figure 2A–D, lane 2) and 2 mM diamide in presence of increasing Syk inhibitor concentration ranging from 0.2 μM to 10 μM (Figure 2A–D, lanes 3–9). As we expected, in accordance with previous reports, diamide caused a rapid band 3 Tyr phosphorylation, without any other phosphorylative changes occurring in erythrocyte membrane proteins [11,33], while Syk inhibitors led to a substantial decrease of band 3 phosphorylation (95–250 KDa) related to the increasing amounts of inhibitors. This trend is clearly evident at 95 KDa bands (phosphorylated band 3).

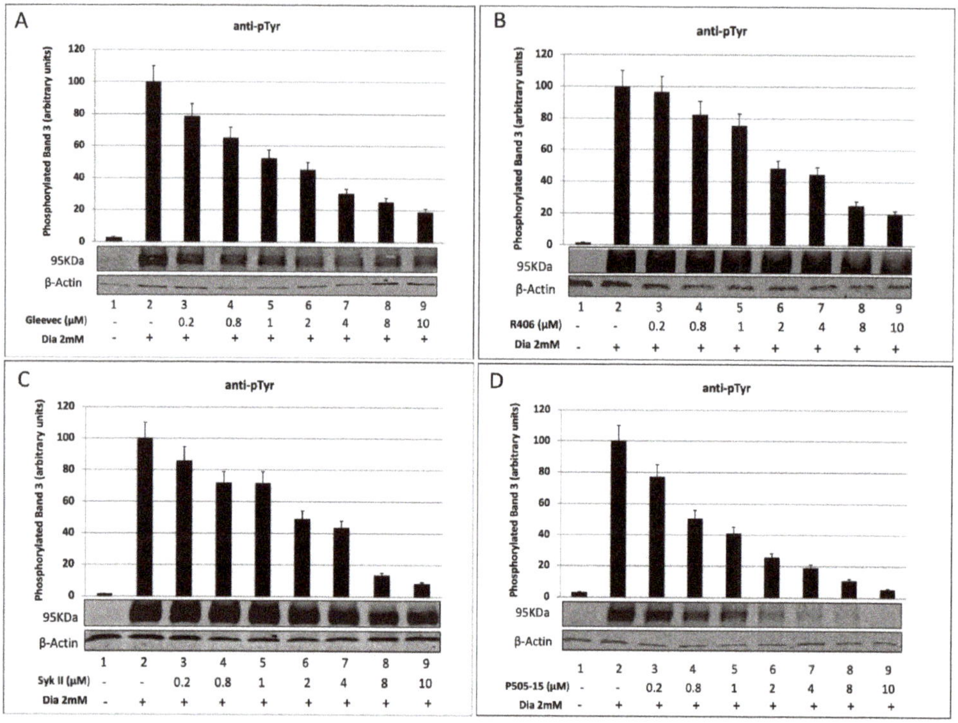

Figure 2. Dose–response course of erythrocyte membrane proteins treated with an oxidant agent and Syk inhibitors related to quantitative analysis of band 3 tyrosine phosphorylation. Erythrocytes were treated with 2 mM diamide (Dia) and different concentrations of the Syk inhibitors (0–10 μM) (**A**) Gleevec, (**B**) R406, (**C**) Syk II, and (**D**) P505-15. **Lane 1** shows the untreateded control sample. Erythrocytes were separated by 8% SDS-PAGE, blotted on a nitrocellulose membrane, and stained with antiphosphotyrosine (apTyr) antibodies. Images were acquired using a laser IR fluorescence detector (Odyssey, Licor, USA). Band 3 Tyr phosphorylation was quantified using Image J software. Values are the means ± for four independent experiments, normalized to total beta-actin levels. All graphs show relative phosphorylation, expressed as a percentage of the maximum observed in each experiment (100%). The error bars represent the standard deviation (SD) of the data.

All densitometry analyses performed on membranes scanned on the Odyssey CLx were done using Odyssey 3.0 software, confirming the large decreases of Tyr phosphorylation levels in band 3 residues caused by Syk protein. Diamide-treated erythrocytes (Figure 2, lane 2) showed high levels of phosphorylation due to the oxidative stress conditions. All the tested Syk inhibitors efficiently suppressed band 3 phosphorylation (Figure 2A–D, lanes 3–9). Table 1 shows the IC_{50} values obtained by densitometric analysis; these results were in agreement with those obtained in previous published studies treating RBCs with different concentrations of each drug and quantitating residual parasitemia 24 and 48 h later (Table 2) [15,17]. P505-15 was the most potent Syk inhibitor, with an IC_{50} of 0.64 μM, while Syk II was the least efficient, with an IC_{50} of 1.72 μM.

Table 1. IC_{50} values obtained from the densitometry analysis performed on anti-p-Tyr Western blotting membranes.

Syk Inhibitors	IC50 (μM)
P505-15	0.64
Gleevec	0.77
R406	0.83
Syk II	1.72

Table 2. Approximate IC_{50} values for each drug were determined by treating ring-stage cultures of *P. falciparum*, Palo Alto strain, with different concentrations of each drug and by quantitating residual parasitemia 24 and 48 h later.

Palo Alto Strain	24 h		48 h	
	IC50 (μM)	Range (μM)	IC50 (μM)	Range (μM)
P505-15	0.83±0.06	0.78−0.90	0.49±0.07	0.42−0.61
R406	2.62±0.83	1.42−3.85	0.55±0.19	0.25−0.92
Gleevec	3.81±0.55	3.24−4.95	1.55±0.13	1.32−1.74
SYK II	5.01±0.44	4.65−5.87	0.90±0.16	0.82−1.25

2.2. Assessment of Syk Inhibitor Efficacy through Computational Studies

To contribute to the understanding of the mechanism of Syk kinase inhibitors in the treatment of malaria and the possible role of Syk inhibition in parasite growth via suppression of band 3 phosphorylation, we also performed computational studies. This kind of approach has been used by many research groups around the world, e.g., by the "Global Online Fight Against Malaria" project of The Scripps Research Institute (TSRI) in La Jolla, CA, U.S.A. In this study, we used the same techniques and the same software, with the aim of speeding up the knowledge acquisition and the process of antimalarial drug discovery. Antimalarial drugs are as effective as artemisinin derivatives, thus providing new hope for the control of malaria.

Previous studies showing the conformational research on R406 [34], Gleevec, Syk inhibitor II, and P505-15 inhibitors [35] were taken into consideration to select the conformation level with the lowest energies and best stability. The highest occupied molecular orbital–lowest unoccupied molecular orbital (HOMO-LUMO) distribution values, energy values, and energy gaps for the hit molecules were computed in order to understand the biological activity [36] (Figure 3).

Figure 4 shows the molecular electrostatic potential (MEP) computed using GaussView 5.0, which allowed us to visualize several sites with abundant electrons by analyzing the charge distributions within a molecule in three dimensions. These maps were used to predict how molecules interact with the binding site of Syk.

The X-ray crystal structure for 4FL2.pdb (2.19 resolution) was available in the RCSB PDB database; this structure was the most complete, although the active site compared to other structures found in the same database was forced by an activation loop.

The docking results obtained from the crystal structure of Syk (4FL2) in the complex with ligands showed conserved H-bond interactions (Table 3); however, a different disposition of ligands in the pocket was observed compared to the poses of the crystal structure of reference.

Although the ligands have a common pattern of interaction with Syk, which is known for all tyrosine kinases [37], our molecules presented different rotations and torsions in the site binding, with the exception of P505-15 [38], which showed an RMSD of 1.39 Å related to its reference (PDB accession number 4RX9), (Figure 5).

Table 3. The data shown represent the H-bond binding found in docking analyses from the different inhibitors (Gleevec, R406, Syk II, and P505-15) interacting with Syk protein. Abbreviations: hydrogen donor, HD; oxygen acceptor, OA; nitrogen acceptor, NA; Debie, D; and Angstrom, Å. The cross-bridge H-bond interactions with the same aa are listed in bold".

H-Bond Interaction								
Ligands	**Pose**	**%**	**H-Bond**	**Ligands Atom**	**Protein Atom**	**Distance (Å)**	**LogP**	**D. M. (Debye)**
Gleevec	1	73	4	N33(NA)	Lys402:HZ1(HD)	2.217	3.83	5.1101
				O10(OA)	Ala451:HN(HD)	2.393		
				N38(NA)	Arg498:HE(HD)	2.147		
				H27(HD)	Asp512:OD2(OA)	1.995		
R406	11	16	3	H9(HD)	Leu377:O(OA)	1.875	3.37	3.4870
				H25(HD)	Ala451:O(OA)	**2.262**		
				O36(OA)	Ala451:HN(HD)	**2.317**		
SykII	1	56	7	H23(HD)	Glu449:O(OA)	**2.075**	1.22	4.4175
				H24(HD)	Glu449:O(OA)	**2.233**		
				O25(OA)	Ala451:HN(HD)	2.299		
				H29(HD)	Arg498:O(OA)	2.063		
				H30(HD)	Ser511:HG(HD)	2.558		
				H8(HD)	Asp512:OD2(OA)	**2.191**		
				H30(HD)	Asp512:OD2(OA)	**2.025**		
P505-15	8	37	4	H8(HD)	Ala451:O(OA)	**2.301**	1.02	2.4285
				O21(OA)	Ala451:HN(HD)	**1.888**		
				H23(HD)	Glu449:O(OA)	**1.940**		
				H24(HD)	Glu449:O(OA)	**2.448**		

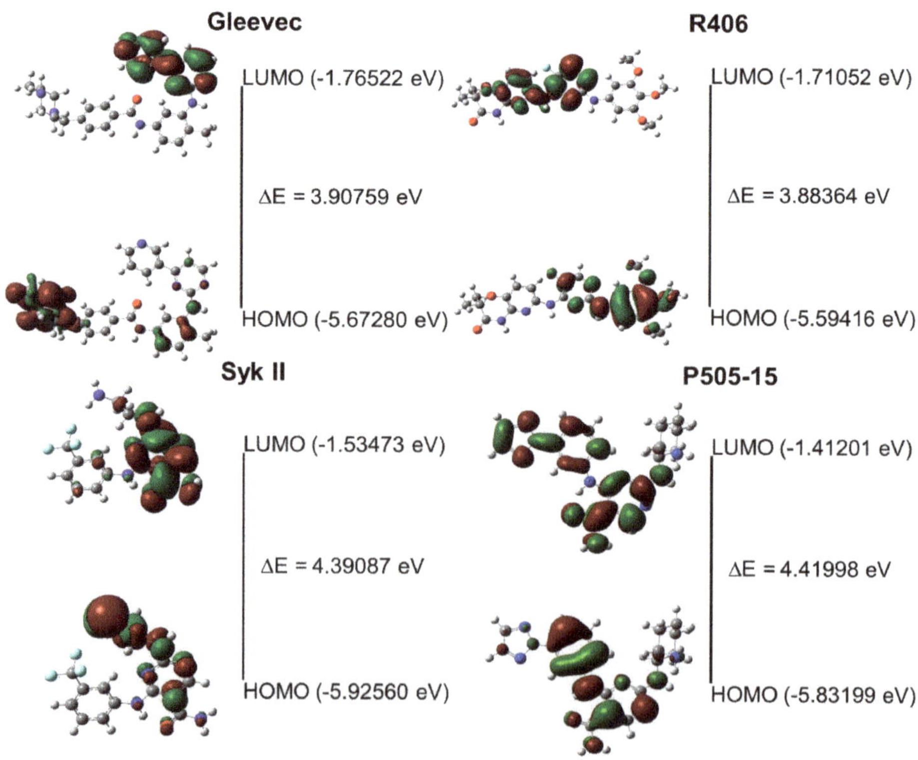

Figure 3. The most probable statistical positions in the highest occupied molecular orbital–lowest unoccupied molecular orbital (HOMO-LUMO) descriptor values.

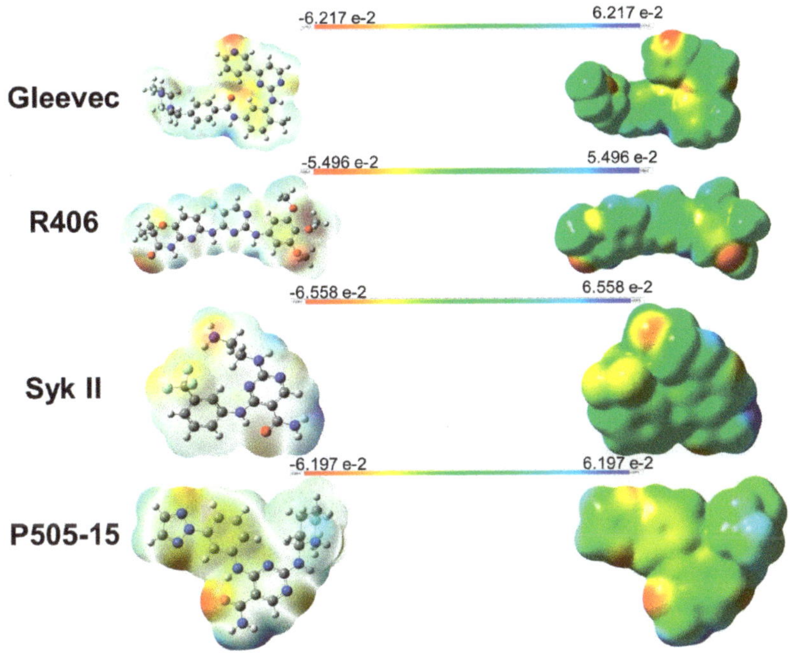

Figure 4. Molecular electrostatic potential (MEP) of surface Syk inhibitors. The color scheme ranges from red (negative potential) via green (zero potential) to blue (positive potential). The unit of isosurface potential is electrostatic potential (eV).

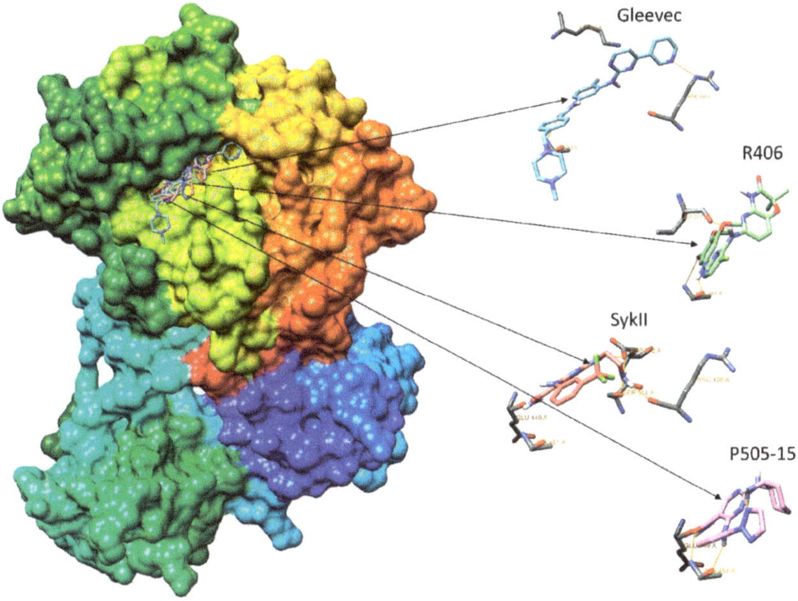

Figure 5. The 3D surface structure of Syk protein, showing Syk inhibitors interacting in the catalytic site. The compounds, with the main H-bond interactions, were reported alone, outside of the pocket of binding.

Further analysis of the docking demonstrated that the amino acids Leu377, Val385, Ala400, Met448, Met450, Ala451, Gly454, and Leu501 present in the active site of the protein interacted with all tested Syk inhibitors. Table 4 shows an interesting hydrophobic interaction of Tyr525 with Gleevec, representing an autophosphorylation site of Syk protein that was previously found [39]. Gleevec interacts through H-bonds with Lys402, Arg498, and Asp512. R406 shows H-bonds with Leu377 and Ala451. Syk II interacts with Glu449, Ala451, Arg498, Ser511, and Asp512, while P505-15 shows H-bonds with Glu449 and Ala451.

Table 4. Hydrophobic interactions and H-bonds between Syk inhibitors and proteins. The bonds are shown highlighted with * (1 H-bond) and with ** (2 H-bonds). Amino acids that are underlined have a high level of Syk specificity. The amino acid Tyr525 is double underlined, which is an autophosphorylation site of Syk protein that forms an interesting interaction with Gleevec.

R406	Syk II	Gleevec	P505-15
Leu377 *	Leu377	Arg338	Leu377
Gly378	Gly378	Leu377	Val385
Ser379	Val385	Ser379	Ala400
Gly380	Ala400	Gly380	Met448
Val385	Lys402	Asn381	Glu449 **
Ala400	Met448	Phe382	Met450
Lys402	Glu449 **	Val385	Ala451 **
Val433	Met450	Ala400	Glu452
Met448	Ala451 *	Lys402 *	Gly454
Glu449	Gly454	Met448	Pro455
Met450	Pro455	Met450	Lys458
Ala451 **	Lys458	Ala451 *	Leu501
Glu452	Arg498 *	Leu453	
Gly454	Leu501	Gly454	
Pro455	Ser511 *	Arg498 *	
Lys458	Asp512 **	Asn499	
Leu501		Leu501	
Asp512		Ser511	
		Asp512 *	
		Tyr525	
		His531	

The different interactions between ligands and proteins observed through their docking and different conformations might be due to the fact that the pocket is capped by an activation loop, causing a steric hindrance. Due to the presence of this buried active site, we considered the 4FL2 crystal's structure in its active conformation compared to the other proteins that we have analyzed in the RCSB database, where their binding pockets were opened.

The main differences among these structures are in the activation loops (a.a. 520–534) [40], which translate into the active conformation of Syk, closing the ligands in the pocket. The active conformation is characterized for containing the Syk activation loop that closes the ATP-binding pocket once a ligand interacts with the amino acids in the site.

The estimation of the inhibition constants (Ki) of all tested compounds obtained by docking simulation showed that the lowest value was for Gleevec (13.65 nM). This result was unexpected; we assume that it could be due to the increasing number of protein–ligand interactions, which enhance the stability inside the pocket. Table 5 shows the binding energy and Ki values obtained by computational analysis. These data, which were obtained by combining the results of both tests, are useful in understanding the role of Syk inhibitors and their ability to enable the implementation of in vitro testing.

Table 5. Data showing the binding energy and inhibition constant (Ki) values obtained from the docking analysis of Syk inhibitors with 4FL2. Abbreviations: mean binding energy, M.B.E; estimated free energy of binding, E.F.E.B; estimated inhibition constant, E.I.C.

Ligand	M.B.E	E.F.E.B	E.I.C, Ki
Gleevec	−10.57	−10.73	13.65 nM
P505-15	−7.69	−8.81	346.88 nM
R406	−7.28	−7.94	1.50 μM
SykII	−6.80	−6.95	8.03 μM

Dynamic molecular analyses were performed to compute the ligands' trajectory and their interactions with the protein. P505-15 and R406, both alone or in combination with the crystal structure of Syk (4FL2), showed higher chemical stability if compared to Gleevec and Syk II, showing a lower conformation change in the pocket.

Further evidence is shown in Figure 6, where the high variability of the conformations of Gleevec and Syk II, both alone or in combination, is evident from the peaks, as compared to P505-15 and R406. Gleevec shows significant variation of the structure disposition, with a value range of 1.5 to 3.8 Å.

Figure 6. Root mean square deviation (RMSD) values (values on the *y*-axis reported in Å) for all deviation atoms calculated over 10 ns, with the values on the *x*-axis reported as steps (1 step = 2 ps) of 4FL2, the compounds alone, and the four complexes—4FL2-Gleevec, 4FL2-SykII, 4FL2-R406, 4FL2-P505-15.

Based on these data, we could establish the potential mechanism of the interactions between the competitive ATP inhibitors and the binding site of the protein, which is involved in the inhibition process.

It should be noted that the autophosphorylation site Tyr 525 interacts only with Gleevec, which could explain the different inhibition levels observed in in vitro (IC_{50} 3.81 μM) and in silico (Ki 13.65 nM) studies. Furthermore, the data demonstrated that most of the amino acid interactions established by Gleevec could be relevant in providing better chemical stability and lower binding energy in silico, although this ligand is not specific to Syk. In this paper, we also report that all Syk inhibitors interact with Met450, Leu453, and Pro455; these amino acids are of great importance because of their high levels of Syk specificity [41–43].

We postulate that all of the discrepancies between the different methodologies used in this study could be due to numerous biological variables that occur in cultures of infected RBCs (i.e., RBC membrane transport, infection by parasites, ATP consumption) and to the difficulties founded in in silico studies caused by the inability to use fixed concentrations for the tested compounds.

A good correlation was found for the IC_{50} values among in vitro and in silico experiments; R406 and P505-15 followed the same trend and were more effective. Gleevec demonstrated major interactions and had the lowest Ki value in the docking analysis. The free binding energy values confirmed the high affinity of Syk inhibitors toward the catalytic site. The computed energy gap between the HOMO and LUMO was used to determine the chemical stability and molecular features of all tested compounds. All of this information will be useful in facilitating the selection of similar Syk inhibitors and antimalarial compounds. Based on these conclusions and the fact that the parasite cannot mutate an erythrocyte tyrosine kinase, we can speculate that Syk inhibitors could contribute to the potency levels of ACTs. It must also be noted that none of the currently used antimalaria drugs prevent the rupture of infected erythrocytes and reinvasion or inhibit host targets that cannot be mutated by the parasite in order to develop drug resistance.

Future computational and proteomic analyses will be necessary to better understand the importance of some amino acids in the pattern of interaction, basing the research of new compounds on the catalytic site features in order to improve their in vitro efficacy.

3. Materials and Methods

In this work, we performed a series of in vitro experiments in infected RBCs in order to investigate the biological activity of different Syk inhibitors on *P. falciparum* cultures and evaluate their IC_{50} concentrations. We measured their activity at varying concentrations, for various durations, and at different parasite stages. Furthermore, *in proteomics* studies, the levels of Tyr phosphorylation in oxidized RBCs were quantified using diamide (a reagent that oxidizes sulphydryl groups to the disulfide form). These analyses were followed by the identification of amino acid interactions in the catalytic site of Syk protein through in silico studies.

3.1. In Vitro Experiments

Freshly drawn blood (R+) samples from healthy adults were used to sustain the parasites in in vitro cultures. Healthy adults provided written, informed consent in ASL.1-Sassari. The in vitro studies were conducted using *P. falciparum* (Palo Alto strain), as previously reported for the Palo Alto (PA) strain (mycoplasma-free) according to standard protocols [44,45]. The Palo Alto (PA) strain is a reference parasite strain that is used to study various antimalarial drugs in *P. falciparum*. The PA strain was isolated from a Ugandan patient and is considered a reference strain due to its high genetic stability [46].

Parasite cultures were synchronized as described by Lambros and Vanderberg [47]. Throughout this procedure, *P. falciparum* cultures maintained synchronicity for 2–3 cycles. For all experiments, mature parasites (shizonts and segmenters) after Percoll separation [48] were added to washed RBCs; 12 h after the infection (occurring within 6 h), the cultures were ready for the experimental procedures.

All experiments were carried out by starting with 2% hematocrit and 2% parasitemia. Each well of a 24-multiwell plate contained 500 µl of growth medium treated with different concentrations of the drugs (0.2, 0.8, 1, 2, 4, 8, 10 µM) R406, Syk II, Gleevec, and P505-15 for 24 and 48 h. The parasitemia was evaluated by optic microscopy and the IC_{50} value of each compound was calculated using ICEstimator 1.2 [49,50].

3.2. Treatment of Red Blood Cells

Venous blood was drawn from healthy volunteers following informed consent and pelleted at 1000 g for 10 min at room temperature. After removal of the buffy coat, RBCs were again pelleted and washed 3 times with phosphate-buffered saline (127 mM NaCl, 2.7 mM KCl, 8.1 mM Na2HPO4, 1.5 mM KH2PO4, 20 mM HEPES, 1 mM MgCl2, and pH 7.4) in 5 mM glucose (PBS glucose) to obtain packed cells. RBCs were suspended at a hematocrit level of 30% in PBS glucose and pretreated in different experiments with Syk inhibitor II (Merck, Darmstadt, Germany.), R406 (Selleckchem, Darmstadt, Germany), and Gleevec at different concentrations (0.2, 0.8, 1, 2, 4, 8, 10 μM) for 1 h at 37 °C in the dark, then in the presence of the oxidant diamide at a 2 mM concentration for 45 min. For all the protocols described, untreated controls and controls treated with only 2 mM of diamide were identically processed. To prevent further phosphorylation of band 3, after incubation we washed the cells with cold buffer and the membranes were immediately prepared.

3.3. RBC Membrane Preparation

Membrane proteins were prepared at 4 °C on ice as previously described [16]. Briefly, 150 μL of packed RBCs was diluted into 1.5 mL of cold hemolysis buffer (HB) (5 mM disodium phosphate, 1 mM EDTA, pH 8), containing a protease and a phosphatase inhibitor cocktail, then washed up to 4 more times in the same buffer (until membranes became white) in a refrigerated Eppendorf microfuge at 25,000× *g*. The samples were stored frozen at −20 °C until use. The membrane protein content was quantified using the CD Protein Assay (Bio-Rad).

3.4. SDS-PAGE

To perform one-dimensional electrophoresis, membrane proteins were solubilized in Laemmli buffer [51] at a volume ratio of 1:1.30 μg of protein-to-antiphosphotyrosine. Then, the samples were put in a thermomixer at 1400 rpm and 28 °C for 30 min. Next, the samples were stored at 95 °C for 5 min, then separated on 8% polyacrylamide gel under reducing and non-reducing conditions. The electrophoretic run was performed on the Bio-Rad Mini-Protean 3 setup.

3.5. Western Blot Analysis

Proteins separated by SDS-PAGE were transferred to nitrocellulose membranes as previously described with Trans-Blot Turbo Bio-Rad and then probed with antiphosphotyrosine antibody (sc7020, Santa Cruz, CA, USA). This was produced in mice in Santa Cruz, CA, USA, and was diluted to 1:2000. Secondary antibodies conjugated with infrared fluorescent dyes excitable at 680 nm or 800 nm (IRDye, Antimouse 800 CW 926-32210, Li-COR, Lincoln, NE, USA) were then used to visualize the desired antigens with a laser scanner (Odyssey, Licor, Lincoln, NE, USA). Quantitative densitometry analyses of tyrosine phosphorylation levels were carried out by analyzing Western blot images using Image J software. The values were expressed as arbitrary units. The rate of band 3 phosphorylation was expressed as the PTP activity and as a percentage of the maximal activity in RBCs treated with 2 mM diamide. All graphs precisely show the relative phosphorylation, expressed as a percentage of the maximum observed in each experiment (100%). The results show the average of four experiments, normalized to total beta-actin levels. The error bars represent the standard deviations (SDs) of the data. The IC_{50} values of different drugs were calculated using ICEstimator 1.2 software.

3.6. Molecular Mechanics (MM) and Quantum Chemicals (QC)

Computational modelling was performed on IBM Blade Center HS22 7870 multiprocessor machines, using OS Ubuntu 16.04 or Windows 10. The small molecules were constructed with standard bond lengths and angles from the fragment database with MacroModel 5.5 [52]. Minimization of structures by conformational search was performed with the MacroModel/BachMin 6.0 program using the AMBER force field.

An extensive conformational search was further carried out using Monte Carlo energy minimization [53] (Ei-E min < 5 Kcal/mole, the energy difference between the generated conformation and the current minimum).

The atomic charges were assigned using the Gasteiger–Marsili method [54]. Representative minimum energy conformations of each compound were optimized using the quantum chemistry program Gaussian 09W with the DFT B3LYP/6-311G method basis set. Visual quantum chemical calculation analysis was performed with GaussView version 5.0 [55,56].

3.7. Molecular Electrostatic Potential (MEP)

The molecular electrostatic potentials (MEP) related to the dipole moment, electronegativity, and partial charges, and showing the reactivity of a molecule were computed. Positive potential values reflect nucleus predominance, while negative values represent rearrangements of electronic charges and lone pairs of electrons.

The analyzed MEP were expressed as different colors depending on the densities of organic molecules and electrophilic electrons, with red representing a negative charge and blue representing a positive charge.

3.8. Molecular Docking

All docking tests were performed by considering a $60 \times 60 \times 60$ grid and adopting the default grid spacing (0.375 Å), treating the docking active site as rigid and the ligands as flexible, i.e., all non-ring torsions were considered active (free to rotate).

Binding of the compounds was analyzed using MGLTools 1.5.7rc1 [57] and AutoDock 4.2 docking programs [58,59].

From the estimated free energy values of ligand binding (E.F.E.B., ΔG), the inhibition constant (Ki) for each ligand was evaluated. Ki was calculated using the equation: $Ki = \exp((\Delta G \times 1000)/(R \times T))$, where ΔG is the docking energy, R (gas constant) is 1.98719 cal K−1 mol−1, and T (temperature) is 298.15 K. The protein target Syk complex with the AMP-PNP ligand (PDB ID: 4FL2; resolution of 2.19 Å) was chosen, which is deposited in RCSB Protein Data Bank [60]. The structure was the most defined and complete, except for the first part of the N-terminus (a.a. 1–8) and the interdomain linker region (a.a. 265–336). The crystallographic water molecules were stripped and hydrogen atoms were added using the AutoDockTools (ADT) module.

3.9. Molecular Dynamics (MD)

Molecular dynamics (MD) calculations were performed to simulate the interactions with the active site of Syk protein, with the best conformation scores predicted by Autodock for all ligands. The MD protocol) for the production simulations were carried out using the Particle Mesh Ewald Molecular Dynamics (PMEMD) version included in the AMBER14 program [61], after careful relaxation of the system using minimization and equilibration protocols.

10 nanoseconds of molecular dynamics production trajectory was saved (5000 frames of 0.002 nanoseconds). The ionizable residues were set to their normal ionization states at pH 7, while the protein atoms and all water molecules of the crystal structure were surrounded by a periodic box of TIP3P32 water molecules that extended 10 Å from the protein. Counterions (Cl^- or Na^+) were placed by xleap to neutralize the system with a Ewald force field and TIP3P water [62].

The ff10 version of the AMBER force field was used to model the protein, and the General AMBER Force Field (GAFF) was used for the organic ligand using the Austin Model 1–Bond Charge Corrections (AM1-BCC) partial charges derived from the antechamber program of the AMBER suite. In the MD simulation protocol, the SHAKE algorithm was used to constrain all bonds involving hydrogen atoms. A non-bonded cutoff of 8.0 Å was used. Langevin dynamics were used to control the temperature (300 K) using a collision frequency of 1.0 ps-1, along with isotropic position scaling to maintain the pressure (1 atm). Periodic boundary conditions were applied to simulate a continuous system. To include the contributions of long-range interactions, the particle mesh Ewald (PME) method was used with a grid spacing of 1 Å combined with a fourth-order B-spline interpolation to compute the potentials and forces in between grid points. The trajectories were analyzed using the Processing Trajectory program (PTRAJ) module of AMBER.

Molecular mechanics and Poisson–Boltzmann (or generalized Born) surface area (MM/PB(GB)SA) calculations and analyses were done with the MM-PBSA program in Amber14 suite.

Graphical representation of the hypothetical positions derived from the docking calculations and the trajectory analysis of molecular dynamics calculations was performed using Chimera [63] and Visual Molecular Dynamics (VMD) [64] software.

Calculations and energy comparisons were conducted using the method proposed by Ross Walker [65].

3.10. Validation of Molecular Docking Protocol

The reliability of the docking approach was further verified with two methods, with one test performed by extracting the phosphoaminophosphonic acid adenylate ester (ANP) from the catalytic site of 4FL2, with a 2.19 Å resolution crystal structure, leading the re-docking. After repositioning the ANP into the protein, the new ANP location was the same as in the original X-ray structure, with RMSD 1.54 Å, which was repositioned with only minimal conformational changes, hence confirming the reliability of the system (RMSD 1.54 Å). The second test was conducted by extraction of the N-{6-[3-(piperazin-1-yl)phenyl]pyridin-2-yl}-4-(trifluoromethyl)pyridin-2-amine (0SB) from X-ray images of 4F4P.pdb, with a 2.37 Å resolution, which was docked with the same macromolecule derived from 4FL2.pdb. After the docking, the new 0SB location was the same as in the original X-ray structure, with RMSD 0.99 Å, which was repositioned with only minimal conformational changes, hence further confirming the reliability of the system (Figure 7). The multiple ligand–protein interactions were analyzed through a 2D diagram in LigPlot+ [66] and the RMSD values of 1.54 Å and 0.99 Å were consequently evaluated in ANP and 0SB experiments, respectively. We could assume the validation of the docking protocol to be satisfactory, considering the values obtained were lower than 2 Å.

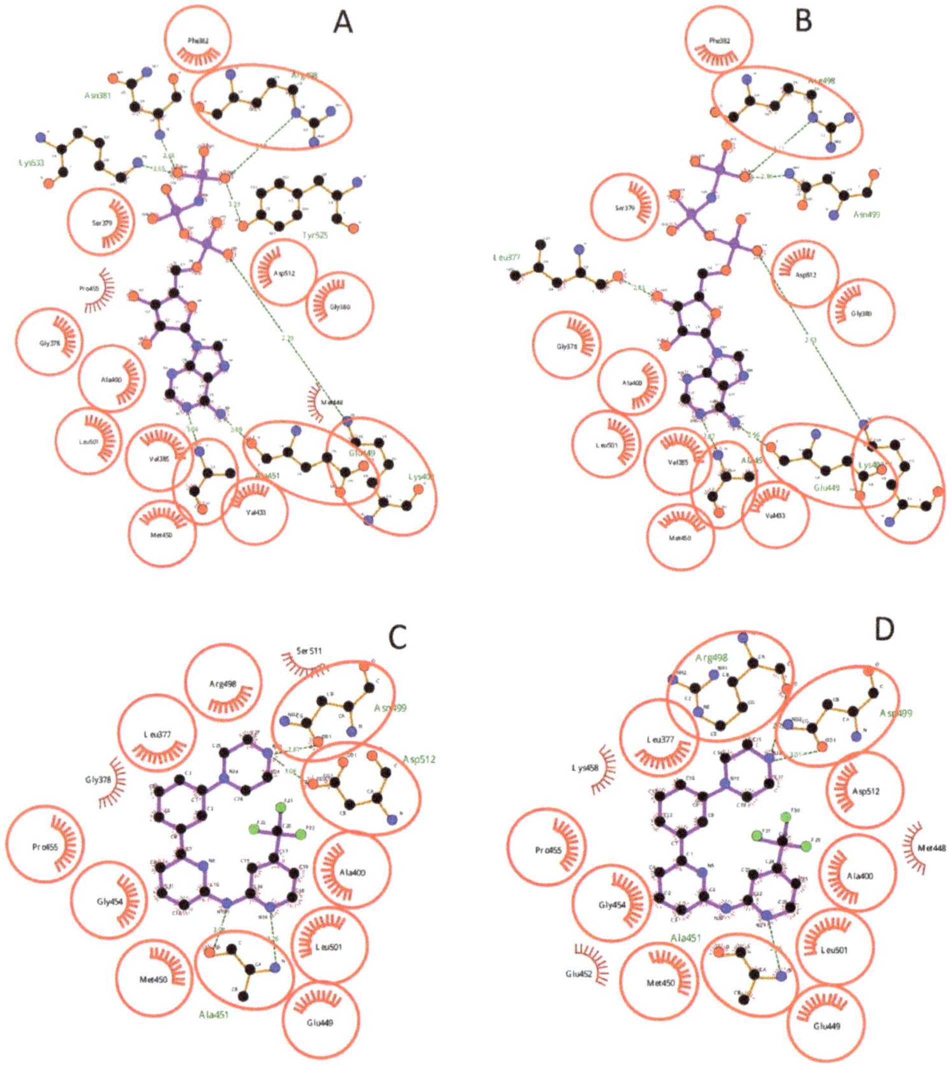

Figure 7. Two different approaches used to validate the docking analysis protocol related to X-ray images of 4FL2 (**A**) and 4F4P (**C**) crystals. The first test consisted of re-docking among 4FL2 and its ligand ANP (**B**), while the second test involved the docking of 4FL2 with the ligand 0SB (**D**). The 2D diagrams were achieved through the LigPlot+ tool.

4. Conclusions

Molecular docking is one of the most widely used computational approaches in the design of structure-based drugs [67,68]. This approach can be used both to identify the correct conformation of the ligand within the target binding pocket and to estimate the interaction energy between a target and a ligand [69,70]. In the present study, different Syk inhibitors were investigated using different methodologies to test their efficacy in vitro and in silico as promising new antimalarial drugs to block Syk phosphorylation events in RBCs infected by parasites. Thorough the computational studies, we gathered useful information about the tested inhibitors, analyzing their chemical features and the patterns of interaction in the Syk protein pocket, proving their inhibitory activity in in vitro experiments. Based on the above data and other publications from our labs, we demonstrated that these compounds actually block the phosphorylation of Tyr residues in protein band 3. The in vitro experiments using a proteomic approach allowed us to notice that Syk inhibitors terminate the life cycle during the process of merozoite egress from the infected RBC, leading to a decrease of band 3 phosphorylation and avoiding the destabilization and weakening of the erythrocyte membrane. The biological data obtained by evaluating the IC_{50} values using Western blotting for all tested compounds confirmed the results achieved in in silico studies. The development of new drugs that can effectively kill malaria parasites is of increasing importance as the threat of drug resistance continues to grow. However, as highlighted above, the findings of this study require further exploration and testing to elucidate or confirm the role of Syk inhibitors as potential drug targets.

Author Contributions: A.P., G.M., A.D., and R.D. conceived of the study and design and supervised the study, contributing to interpretation of data and writing of the paper. G.M., together with I.T., A.D., and R.D., performed experiments. M.C.P. and I.T. contributed to the formulation of results. A.P., F.M.T., A.D., and R.D. have given approval to the final version of the manuscript. All authors have read and agreed to the published version of the manuscript.

Funding: This study was supported by grants from Fondo di Ateneo per la Ricerca 2020.

Acknowledgments: The authors would like to thank Claudio Fozza for his collaboration.

Conflicts of Interest: The authors declare no conflict of interest.

Abbreviations

ACTs	Artemisinin-based combination therapies
Syk	Spleen tyrosine kinase
ANP	Phosphoaminophosphonic acid adenylate ester
OSB	N-{6-[3-(piperazin-1-yl)phenyl]pyridin-2-yl}-4-(trifluoromethyl)pyridin-2-amine
MD	Molecular dynamics
QM	Quantum mechanics
QC	Quantum chemistry
RMSD	Root mean square deviation
RBCs	Red blood cells
HOMO	Highest occupied molecular orbital
LUMO	Lowest unoccupied molecular orbital
MEP	Molecular electrostatic potential

References

1. Berton, G.; Mocsai, A.; Lowell, C.A. Src and Syk kinases: Key regulators of phagocytic cell activation. *Trends Immunol.* **2005**, *26*, 208–214. [CrossRef] [PubMed]
2. Sada, K.; Takano, T.; Yanagi, S.; Yamamura, H. Structure and function of Syk protein-tyrosine kinase. *J. Biochem.* **2001**, *130*, 177–186. [CrossRef] [PubMed]
3. Siraganian, R.P.; Zhang, J.; Suzuki, K.; Sada, K. Protein tyrosine kinase Syk in mast cell signaling. *Mol. Immunol.* **2002**, *38*, 1229–1233. [CrossRef]
4. Singh, R.; Masuda, E.S.; Payan, D.G. Discovery and development of spleen tyrosine kinase (SYK) inhibitors. *J. Med. Chem.* **2012**, *55*, 3614–3643. [CrossRef]
5. Chan, A.C.; Shaw, A.S. Regulation of. antigen receptor signal transduction by protein tyrosine kinases. *Curr. Opin. Immunol.* **1995**, *8*, 394–401. [CrossRef]
6. Van Oers, N.S.; Weiss, A. The Syk/ZAP-70 protein tyrosine kinase connection to antigen receptor signalling processes. *Semin. Immunol.* **1995**, *7*, 227–236. [CrossRef]
7. Filippakopoulos, P.; Müller, S.; Knapp, S. SH2 domains: Modulators of nonreceptor tyrosine kinase activity. *Curr. Opin. Struct. Biol.* **2009**, *19*, 643–649. [CrossRef]
8. Bond, P.J.; Faraldo-Gómez, J.D. Molecular mechanism of selective recruitment of Syk kinases by the membrane antigen-receptor complex. *J. Biol. Chem.* **2011**, *286*, 25872–25881. [CrossRef]

9. Hall, J.; Aulabaugh, A.; Rajamohan, F.; Liu, S.; Kaila, N.; Wan, Z.K.; Ryan, M.; Magyar, R.; Qiu, X. Biophysical and mechanistic insights into novel allosteric inhibitor of spleen tyrosine kinase. *J. Biol. Chem.* **2012**, *10*, 7717–7727. [CrossRef]
10. Pantaleo, A.; Ferru, E.; Carta, F.; Mannu, F.; Simula, L.F.; Khadjavi, A.; Turrini, F. Irreversible AE1 Tyrosine Phosphorylation Leads to Membrane Vesiculation in G6PD Deficient Red Cells. *PLoS ONE* **2011**, *6*. [CrossRef]
11. Pantaleo, A.; Ferru, E.; Giribaldi, G.; Mannu, F.; Carta, F.; Matte, A.; De Franceschi, L.; Turrini, F. Oxidized and poorly glycosylated band 3 is selectively phosphorylated by Syk kinase to form large membrane clusters in normal and G6PD-deficient red blood cells. *Biochem. J.* **2009**, *2*, 359–367. [CrossRef] [PubMed]
12. Pantaleo, A.; Ferru, E.; Carta, F.; Valente, E.; Pippia, P.; Turrini, F. Effect of heterozygous beta thalassemia on the phosphorylative response to Plasmodium falciparum infection. *J. Proteom.* **2012**, *76*, 251–258. [CrossRef] [PubMed]
13. Ferru, E.; Giger, K.; Pantaleo, A.; Campanella, E.; Grey, J.; Ritchie, K.; Vono, R.; Turrini, F.; Low, P.S. Regulation of membrane-cytoskeletal interactions by tyrosine phosphorylation of erythrocyte band 3. *Blood* **2011**, *22*, 5998–6006. [CrossRef]
14. Ferru, E.; Pantaleo, A.; Carta, F.; Mannu, F.; Khadjavi, A.; Gallo, V.; Turrini, F. Thalassemic erythrocytes release microparticles loaded with hemichromes by redox activation of p72Syk kinase. *Haematologica* **2014**, *99*, 570–578. [CrossRef] [PubMed]
15. Kesely, K.R.; Pantaleo, A.; Turrini, F.M.; Olupot-olupot, P.; Low, P.S. Inhibition of an Erythrocyte Tyrosine Kinase with Imatinib Prevents Plasmodium falciparum Egress and Terminates Parasitemia. *PLoS ONE* **2016**, *11*, e0164895. [CrossRef] [PubMed]
16. Pantaleo, A.; Kesely, K.R.; Pau, M.C.; Tsamesidis, I.; Schwarzer, E.; Skorokhod, O.A.; Chien, H.D.; Ponzi, M.; Bertuccini, L.; Low, P.S.; et al. Syk inhibitors interfere with erythrocyte membrane modification during *P falciparum* growth and suppress parasite egress. *Blood* **2017**, *130*, 1031–1040. [CrossRef] [PubMed]
17. Tsamesidis, I.; Reybier, K.; Marchetti, G.; Pau, M.C.; Virdis, P.; Fozza, C.; Nepveu, F.; Low, P.S.; Turrini, F.M.; Pantaleo, A. Syk kinase inhibitors synergize with artemisinins by enhancing oxidative stress in *Plasmodium falciparum*-parasitized erythrocytes. *Antioxidants* **2020**, *9*, 753. [CrossRef]
18. Sechi, M.; Derudas, M.; Dallocchio, R.; Dessì, A.; Bacchi, A.; Sannia, L.; Carta, F.; Palomba, M.; Ragab, O.; Chan, C.; et al. Design and Synthesis of Novel Indole -Diketo Acid Derivatives as HIV-1 Integrase Inhibitors. *J. Med. Chem.* **2004**, *47*, 5298–5310. [CrossRef]
19. Pani, G.; Dessì, A.; Dallocchio, R.; Scherm, B.; Azara, E.; Delogu, G.; Migheli, Q. Natural Phenolic Inhibitors of Trichothecene Biosynthesis by the Wheat Fungal Pathogen Fusarium culmorum: A Computational Insight into the Structure-Activity Relationship. *PLoS ONE* **2016**, *11*, e0157316. [CrossRef]
20. Marangon, E.; Citterio, M.; Sala, F.; Barisone, E.; Lippi, A.A.; Rizzari, C.; Biondi, A.; D'Incalci, M.; Zucchetti, M. Pharmacokinetic profile of imatinib mesylate and N-desmethyl-imatinib (CGP 74588) in children with newly diagnosed Ph+ acute leukemias. *Cancer Chemother. Pharmacol.* **2009**, *3*, 563–566. [CrossRef]
21. Champagne, M.A.; Capdeville, R.; Krailo, M.; Qu, W.; Peng, B.; Rosamilia, M.; Therrien, M.; Zoellner, U.; Blaney, S.M.; Bernstein, M. Imatinib mesylate (STI571) for treatment of children with Philadelphia chromosome-positive leukemia: Results from a Children's Oncology Group phase 1 study. *Blood* **2004**, *9*, 2655–2660. [CrossRef] [PubMed]
22. Annesley, C.E.; Brown, P. Novel agents for the treatment of childhood acute leukemia. *Ther. Adv. Hematol.* **2015**, *6*, 61–79. [CrossRef] [PubMed]
23. Weinblatt, M.E.; Genovese, M.C.; Ho, M.; Rosiak-Jedrychowicz, S.H.K.; Kavanaugh, A.; Millson, D.S.; Leon, G.; van der Heijde, D. Effects of Fostamatinib, an Oral Spleen Tyrosine Kinase Inhibitor, in Rheumatoid Arthritis Patients with an Inadequate Response to Methotrexate: Results from a Phase III, Multicenter, Randomized, Double-Blind, Placebo-Controlled, Parallel-Group Study. *Arthritis Rheumatol.* **2014**, *66*, 3255–3264. [CrossRef] [PubMed]
24. Podolanczuk, A.; Lazarus, A.H.; Crow, A.R.; Grossbard, E.; Bussel, J.B. Of mice and men: An open-label pilot study for treatment of immune thrombocytopenic purpura by an inhibitor of Syk. *Blood* **2009**, *113*, 3154–3160. [CrossRef] [PubMed]
25. Tauzin, S.; Ding, H.; Burdevet, D.; Borisch, B.; Hoessli, D.C. Membrane-associated signaling in human B-lymphoma lines. *Exp. Cell. Res.* **2011**, *2*, 151–162. [CrossRef]

26. Friedberg, J.W.; Sharman, J.; Sweetenham, J.; Johnston, P.B.; Vose, J.M.; LaCasce, A.; Schaefer-Cutillo, J.; De Vos, S.; Sinha, R.; Leonard, J.P.; et al. Inhibition of Syk with fostamatinib disodium has significant clinical activity in non-Hodgkin Lymphoma and chronic lymphocytic leukemia. *Blood* **2010**, *115*, 2578–2585. [CrossRef]
27. Coffey, G.; DeGuzman, F.; Inagaki, M.; Pak, Y.; MDelaney, S.; Ives, D.; Betz, A.; Jia, Z.J.; Pandey, A.; Baker, D.; et al. Specific Inhibition of Spleen Tyrosine Kinase Suppresses Leukocyte Immune Function and Inflammation in Animal Models of Rheumatoid Arthritis. *J. Pharmacol. Exp. Ther.* **2011**, *340*, 350–359. [CrossRef]
28. Spurgeon, S.E.; Coffey, G.; Fletcher, L.B.; Burke, R.; Tyner, J.W.; Druker, B.J.; Loriaux, M.M. The Selective Syk Inhibitor P505-15 (PRT062607) Inhibits B Cell Signaling and Function In Vitro and In Vivo and Augments the Activity of Fludarabine in Chronic Lymphocytic Leukemia. *J. Pharmacol. Exp. Ther.* **2013**, *344*, 378–387. [CrossRef]
29. Hisamichi, H.; Naito, R.; Toyoshima, A.; Kawano, N.; Ichikawa, A.; Orita, A.; Tsukamoto, S. Synthetic studies on novel Syk inhibitors. Part 1: Synthesis and structure-activity relationships of pyrimidine-5-carboxamide derivatives. *Bioorg. Med. Chem.* **2005**, *13*, 4936–4951. [CrossRef]
30. Yi, Y.; Son, Y.; Ryou, C.; Sung, G.; Kim, J.; Cho, J.Y. Functional Roles of Syk in Macrophage-Mediated Inflammatory Responses. *Mediat. Inflamm.* **2014**, *2014*, 1–12. [CrossRef]
31. Bordin, L.; Brunati, A.M.; Donella-Deana, A.; Baggio, B.; Toninello, A.; Clari, G. Band 3 is an anchor protein and a target for SHP-2 tyrosine phosphatase in human erythrocytes. *Blood* **2002**, *100*, 276–282. [CrossRef] [PubMed]
32. Harrison, M.L.; Isaacson, C.C.; Burg, D.L.; Geahlen, R.L.; Low, P.S. Phosphorylation of human erythrocyte band 3 by endogenous p72syk. *J. Biol. Chem.* **1994**, *269*, 955–959. [PubMed]
33. Bordin, L.; Zen, F.; Ion-Popa, F.; Barbetta, M.; Baggio, B.; Clari, G. Band 3 tyr-phosphorylation in normal and glucose-6-phospate dehydrogenase-deficient human erythrocytes. *Mol. Membr. Biol.* **2005**, *5*, 411–420. [CrossRef]
34. Villaseñor, A.G.; Kondru, R.; Hoangdung, H.; Wang, S.; Papp, E.; Shaw, D.; Barnett, J.W.; Browner, M.F.; Kuglstatter, A. Structural Insights for Design of Potent Spleen Tyrosine Kinase Inhibitors from Crystallographic Analysis of Three Inhibitor Complexes. *Chem. Biol. Drug Des.* **2009**, *73*, 466–470. [CrossRef] [PubMed]
35. Castillo, M.; Forns, P.; Erra, M.; Lopez, M.; Maldonado, M.; Orellana, A.; Carreno, C.; Ramis, I.; Miralpeix, M.; Vidal, B. Highly potent aminopyridines as Syk kinase inhibitors. *Bioorg. Med. Chem. Lett.* **2012**, *22*, 5419–5423. [CrossRef] [PubMed]
36. Al-Sabagh, A.M.; Nasser, N.M.; Farag, A.A.; Migahed, M.A.; Eissa AM, F.; Mahmoud, T. Structure effect of some amine derivatives on corrosion inhibition efficiency for carbon steel in acidic media using electrochemical and Quantum Theory Methods. *Egypt. J. Pet.* **2013**, *22*, 101–116. [CrossRef]
37. Traxler, P.; Furet, P. Strategies toward the Design of Novel and Selective Protein Tyrosine Kinase Inhibitors. *Pharmacol. Ther.* **1999**, *82*, 195–206. [CrossRef]
38. Thoma, G.; Smith, A.B.; Eis, M.J.; Van Vangrevelinghe, E.; Blanz, J.; Aichholz, R.; Zerwes, H. Discovery and Profiling of a Selective and Efficacious Syk Inhibitor. *J. Med. Chem.* **2015**, *58*, 1950–1963. [CrossRef]
39. Miah SM, S.; Sada, K.; Tuazon, P.T.; Ling, J.; Maeno, K.; Kyo, S.; Yamamura, H. Activation of Syk Protein Tyrosine Kinase in Response to Osmotic Stress Requires Interaction with p21-Activated Protein Kinase Pak2/γ-PAK. *Mol. Cell. Biol.* **2004**, *24*, 71–83. [CrossRef]
40. Grädler, U.; Schwarz, D.; Dresing, V.; Musil, D.; Bomke, J.; Frech, M.; Wegener, A. Structural and biophysical characterization of the Syk activation switch. *J. Mol. Biol.* **2013**, *425*, 309–333. [CrossRef]
41. Zeifman, A.A.; Titov, Y.; Svitanko, I.V.; Rakitina, T.V.; Lipkin, A.V.; Stroylov, V.S.; Chilov, G.G. Rational design and synthesis of novel Syk-kinase inhibitors. *Mendeleev Commun.* **2012**, *22*, 73–74. [CrossRef]
42. Padilla, F.; Bhagirath, N.; Chen, S.; Chiao, E.; Goldstein, D.M.; Hermann, J.C.; Lucas, M.C. Pyrrolopyrazines as selective spleen tyrosine kinase inhibitors. *J. Med. Chem.* **2013**, *56*, 1677–1692. [CrossRef]
43. Jiang, M.; Liu, R.; Chen, Y.; Zheng, Q.; Fan, S.; Liu, P. A combined experimental and computational study of Vam3, a derivative of resveratrol, and syk interaction. *Int. J. Mol. Sci.* **2014**, *15*, 17188–17203. [CrossRef] [PubMed]
44. Trager, W.; Jensen, J.B. Human malaria parasites in continuous culture. *Science* **1976**, *193*, 673–675. [CrossRef] [PubMed]

45. De Lucia, S.; Tsamesidis, I.; Pau, M.C.; Kesely, K.R.; Pantaleo, A.; Turrini, F. Induction of high tolerance to artemisinin by sub-lethal administration: A new in vitro model of P. falciparum. *PLoS ONE* **2018**, *13*. [CrossRef] [PubMed]
46. Fandeur, T.; Bonnefoy, S.; Mercereau-Puijalon, O. In vivo and in vitro derived Palo Alto lines of Plasmodium falciparum are genetically unrelated. *Mol. Biochem. Parasitol.* **1991**, *47*, 167–178. [CrossRef]
47. Lambros, C.; Vanderberg, J.P. Synchronization of Plasmodium falciparum erythrocytic stages in culture. *J. Parasitol.* **1979**, *65*, 418–420. [CrossRef]
48. Rivadeneira, E.M.; Wasserman, M.; Espinal, C.T. Separation and concentration of schizonts of Plasmodium falciparum by Percoll gradients. *J. Protozool.* **1983**, *30*, 367–370. [CrossRef]
49. Le Nagard, H.; Vincent, C.; Mentré, F.; Le Bras, J. Online analysis of in vitro resistance to antimalarial drugs through nonlinear regression. *Comput. Methods Programs Biomed.* **2010**, *104*, 10–18. [CrossRef]
50. Kaddouri, H.; Nakache, S.; Houzé, S.; Mentré, F.; Le Bras, J. Drug Susceptibility of Plasmodium falciparum Clinical Isolates from Africa using Plasmodium Lactate Dehydrogenase Immunodetection Assay and inhibitory Emax model for precise IC50 measurement. *Antimicrob. Agents Chemother.* **2006**, *50*, 3343–3349. [CrossRef]
51. Laemmli, U.K. Cleavage of structural proteins during the assembly of the head of bacteriophage T4. *Nature* **1970**, *227*, 680–685. [CrossRef]
52. Mohamadi, F.; Richards, N.G.J.; Guida, W.C.; Liskamp, R.; Lipton, M.; Caufield, C.; Chang, G.; Hendrickson, T.; Still, W.C. Macromodel-an integrated software system for modeling organic and bioorganic molecules using molecular mechanics. *J. Comput. Chem.* **1990**, *11*, 440–467. [CrossRef]
53. Chang, G.; Guida, W.C.; Still, W.C. An internal-coordinate Monte Carlo method for searching conformational space. *J. Am. Chem. Soc.* **1989**, *111*, 4379–4386. [CrossRef]
54. Gasteiger, J.; Marsili, M. Iterative partial equalization of orbital electronegativity—A rapid access to atomic charges. *Tetrahedron* **1980**, *36*, 3219–3228. [CrossRef]
55. Frisch, M.J.; Trucks, G.W.; Schlegel, H.B.; Scuseria, G.E.; Robb, M.A.; Cheeseman, J.R.; Scalmani, G.; Barone, V.; Petersson, G.A.; Nakatsuji, H.; et al. *Gaussian 09*; Gaussian Inc.: Wallingford, CT, USA, 2009.
56. Dennington, R.; Keith, T.; Millam, J. *Gauss View, Version 5*; Semichem Inc.: Shawnee Mission, KS, USA, 2009.
57. Sanner, M.F. Python: A programming language for software integration and development. *J. Mol. Graph.* **1999**, *17*, 57–61.
58. Morris, G.M.; Goodsell, D.S.; Halliday, R.S.; Huey, R.; Hart, W.E.; Belew, R.K.; Olson, A.J. Automated docking using a Lamarckian genetic algorithm and an empirical binding free energy function. *J. Comput. Chem.* **1998**, *19*, 1639–1662. [CrossRef]
59. Huey, R.; Morris, G.M.; Olson, A.J.; Goodsell, D.S. A semiempirical free energy force field with charge-based desolvation. *J. Comput. Chem.* **2007**, *28*, 1145–1152. [CrossRef]
60. Berman, H.M.; Westbrook, J.; Feng, Z.; Gilliland, G.; Bhat, T.N.; Weissig, H.; Shindyalov, I.N.; Bourne, P.E. The Protein Data Bank. *Nucleic Acids Res.* **2000**, *28*, 235–242. [CrossRef]
61. Case, D.A. *AMBER 14*; University of California: San Francisco, CA, USA, 2014.
62. Joung, I.S.; Cheatham, T.E. Determination of Alkali and Halide Monovalent Ion Parameters for Use in Explicitly Solvated Biomolecular Simulations. *J. Phys. Chem. B* **2008**, *112*, 9020–9041. [CrossRef]
63. Pettersen, E.F.; Goddard, T.D.; Huang, C.C.; Greenblatt, D.M.; Meng, E.C.; Ferrin, T.E. UCSF Chimera—A visualization system for exploratory research and analysis. *J. Comput. Chem.* **2004**, *25*, 1605–1612. [CrossRef]
64. Humphrey, W.; Dalke, A.; Schulten, K. VMD: Visual molecular dynamics. *J. Mol. Graph.* **1996**, *14*, 33–38. [CrossRef] [PubMed]
65. Kaus, J.W.; Pierce, L.T.; Walker, R.C.; McCammon, J.A. Improving the Efficiency of Free Energy Calculations in the Amber Molecular Dynamics Package. *J. Chem. Theory Comput.* **2013**, *9*, 4131–4139. [CrossRef]
66. Laskowski, R.A.; Swindells, M.B. LigPlot+: Multiple Ligand–Protein Interaction Diagrams for Drug Discovery. *J. Chem. Inf. Model.* **2011**, *51*, 2778–2786. [CrossRef] [PubMed]
67. Dessì, A.; Peluso, P.; Dallocchio, R.; Weiss, R.; Andreotti, G.; Allocca, M.; Aubert, E.; Pale, P.; Mamane, V.; Cossu, S. Rational design, synthesis, characterization and evaluation of iodinated 4,4′-bipyridines as new transthyretin fibrillogenesis inhibitors. *Molecules* **2020**, *25*, 2213. [CrossRef]
68. Wang, Z.; Sun, H.; Yao, X.; Li, D.; Xu, L.; Li, Y.; Tian, S.; Hou, T. Comprehensive evaluation of ten docking programs on a diverse set of protein–ligand complexes: The prediction accuracy of sampling power and scoring power. *Phys. Chem. Chem. Phys.* **2016**, *18*, 12964–12975. [CrossRef]

69. Irwin, J.J.; Shoichet, B.K. Docking screens for novel ligands conferring new biology. *J. Med. Chem.* **2016**, *59*, 4103–4120. [CrossRef]
70. Ferreira, L.G.; dos Santos, R.N.; Oliva, G.; Andricopulo, A.D. Molecular docking and structure-based drug design strategies. *Molecules* **2015**, *20*, 13384–13421. [CrossRef]

International Journal of
Molecular Sciences

MDPI

Article

Transient Excursions to Membrane Core as Determinants of Influenza Virus Fusion Peptide Activity

Remigiusz Worch [1], Anita Dudek [2], Paulina Borkowska [1] and Piotr Setny [2,*]

[1] Institute of Physics, Polish Academy of Sciences, Aleja Lotnikow 32/46, 02-668 Warsaw, Poland; remiwo@ifpan.edu.pl (R.W.); borkowska@ifpan.edu.pl (P.B.)
[2] Centre of New Technologies, University of Warsaw, Banacha 2c, 02-097 Warsaw, Poland; a.dudek@cent.uw.edu.pl
* Correspondence: p.setny@cent.uw.edu.pl

Abstract: Fusion of viral and host cell membranes is a critical step in the life cycle of enveloped viruses. In the case of influenza virus, it is mediated by subunit 2 of hemagglutinin (HA) glycoprotein whose N-terminal fragments insert into the target membrane and initiate lipid exchange. These isolated fragments, known as fusion peptides (HAfp), already possess own fusogenic activity towards liposomes. Although they have long been studied with the hope to uncover the details of HA-mediated fusion, their actual mechanism of action remains elusive. Here, we use extensive molecular dynamics simulations combined with experimental studies of three HAfp variants to fully characterize their free energy landscape and interaction with lipid bilayer. In addition to customary assumed peptides localization at lipid–water interface, we characterize membrane-spanning configurations, which turn out to be metastable for active HAfps and unstable for the fusion inactive W14A mutant. We show that, while the degree of membrane perturbation by surface peptide configurations is relatively low and does not show any mutation-related differences, the effect of deeply inserted configurations is significant and correlates with insertion depth of the N-terminal amino group which is the highest for the wild type HAfp. Finally, we demonstrate the feasibility of spontaneous peptide transition to intramembrane location and the critical role of strictly conserved tryptofan residue 14 in this process.

Keywords: influenza virus fusion peptides; peptide-membrane interactions; membrane fusion

Citation: Worch, R.; Dudek, A.; Borkowska, P.; Setny, P. Transient Excursions to Membrane Core as Determinants of Influenza Virus Fusion Peptide Activity. *Int. J. Mol. Sci.* **2021**, *22*, 5301. https://doi.org/10.3390/ijms22105301

Academic Editor: Masoud Jelokhani-Niaraki

Received: 15 April 2021
Accepted: 13 May 2021
Published: 18 May 2021

1. Introduction

Influenza virus entry to the cells requires fusion of viral and cellular membranes. It is mediated by a homotrimeric viral protein hemagglutinin (HA), whose HA1 subunits bind to cellular receptors and HA2 subunits are responsible for membrane remodeling [1,2]. A critical, yet still not exactly understood role in this latter respect is played by amphipatic N-terminal HA2 fragments, so-called fusion peptides (HAfp), which are directly inserted into target lipid bilayer and initiate the fusion process [3]. Their high sequence conservation and the existence of nonfunctional mutants, which nevertheless were demonstrated to embed within lipid bilayers, indicate that their role is not merely limited to act as inert anchors, but must also involve specific interplay with bilayer structure [4,5]. Synthetic peptides corresponding to only the first 20 N-terminal amino acids of HA2 have long been known to initiate the fusion of lipid vesicles and content mixing in vitro [6,7], without the mechanical action provided by the rest of the protein. Since the effect of fusion-abolishing mutations within HAfp is generally reproduced by the loss of function in complete fusion proteins [5], they have been extensively used as model systems to study basic aspects of viral fusion [8].

Experimental data concerning synthetic HAfp indicate that, whereas disordered in aqueous solution, they adopt mostly helical forms once bound to lipid structures such as membranes or micelles [9]. The N-terminal, strictly conserved part (residues 1–11) folds into a stable α-helix, followed by a kink region (residues 12–14) and a second, malleable C-terminal α-helix that starts from residue 15. Early studies, focussed on possibly shortest fusiogenic peptides composed of only 20 residues ($HAfp_{1-20}$), suggested that both helices form a boomerang like structure, with the N-terminal arm partially inserted into hydrophobic lipid core, followed by a solvent-exposed kink, and the C-terminal segment located at lipid–water interface [10]. This view has been challenged by subsequent, nuclear magnetic resonance (NMR)-based investigation of a longer and more active 23 amino acid peptide ($HAfp_{1-23}$), which was shown to adopt a tight helical hairpin structure in micelles, supposedly remaining at the lipid–water interface [11]. The hairpin structure is apparently not affected by further sequence elongation. According to NMR data, residues 24–28 form solvent exposed random coil [12], which suggests that they belong to a flexible tether that links the fusion peptide to the rigid coiled coil stem in a fusion ready, complete HA [13].

A sharply bent hairpin seems to be the actual fusion-active form of HAfp. Indeed, this explains well a strict conservation of several glycine residues aligned in a ridge that enables tight packing of helical arms as well as the fact that many amino acid substitutions that are known to abrogate fusion activity [8] are sterically inconsistent with the hairpin structure [11]. The activity of boomerang-like $HAfp_{1-20}$ has been attributed to the fact that it actually exists in an ensemble of states with residual ($\sim$11%) closed hairpin population [12], thus also explaining its lower activity compared to $HAfp_{1-23}$. Intriguingly, however, fusion-inactive W14A mutant, with complete helices and no steric hindrance that would preclude the hairpin formation, appears to exist as an overly flexible boomerang structure [14]. It indicates that hairpin stabilization may depend not only on intra-peptide contacts but also on its interaction with environment, since the W14 side chain, apparently needed to maintain the hairpin conformation, is exposed to solvent and does not participate in interhelical interactions.

Unambiguous interpretation of the above observations is hampered by the lack of clear evidence concerning the actual peptide placement within lipid bilayer and its impact on membrane structure. The length of the N-terminal helix (11 residues) is much shorter compared to typical transmembrane helical protein domains (around 20 residues [15]). Given this notion, several NMR-based studies conclude that fusion peptides orient parallel to membrane surface and stay at the lipid–water interface [11,16]. On the other hand, measurements based on attenuated total reflection Fourier-transform infrared spectroscopy (ATR-FTIR) or spin-label electron paramagnetic resonance (EPR) suggest that peptides insert obliquely into the external membrane leaflet with the N-helix tilted on average $\sim$50 degrees with respect to bilayer normal [9,10,17,18], which results in buried N-terminal group and solvent-facing kink region. Moreover, recent molecular dynamics (MD) simulations indicated the possibility of fully transmembrane peptide placement which can be achieved owing to local membrane thinning caused by aqueous solvent attraction towards hydrophilic residues on both hairpin poles [19–22].

Such mixed results regarding HAfp structure and location within membrane entail the lack of consensus concerning its actual fusogenic mechanism. In this respect, multiple modes of action have been proposed, such as the promotion of lipid tails protrusions [23], lipid heads intrusions [24], membrane thinning [25,26], stabilization of positive membrane curvature [27], stabilization of negative membrane curvature [16], local membrane dehydration [28], or membrane rupture by transmembrane HAfp bundle with subsequent formation of π shaped intermediates [29]. Of the above, the likelihood of lipid tails protrusions [30] has been demonstrated via MD simulations as a viable predictor of membrane fusogenic propensity, and the extent to which it was increased by various peptide mutants was shown to correlate with their experimental activity [23,24]. In a recent simulation study of complete HA-mediated fusion [31], a particularly important role in inducing lipid tails protrusions was demonstrated for peptide configurations in which the charged HAfp N-terminus was buried within membrane core in agreement with oblique insertion mode suggested by spectroscopic measurements.

In this study, to thoroughly characterize $HAfp_{1-23}$ configuration landscape within POPC membranes and to assess the determinants of its fusogenic activity, we combine MD simulations with spectroscopic methods. To this end, we consider wild type (wt) peptide and two mutants, E11A and W14A, that are known to have reduced or none activity, respectively [14]. We demonstrate that the peptides are rather dynamic in membrane environment. While they preferentially remain on bilayer surface, as previously indicated, it seems that the key to explain their function is to assume transient excursions into the membrane core. According to our analysis, in contrast to generally similar characteristics of surface conformations among the considered mutants, the ability to adopt deep configuration is affected by amino acid substitutions and significantly contributes to explaining the observed, mutation-related differences in activity.

2. Materials and Methods

2.1. Computational Methods

2.1.1. Simulated Systems

We considered a 23-residue long HAfps: wt GLFGAIAGFIEGGWQGMVDGWYG and E11A and and W14A mutants. The N-terminus was modeled as a charged amino group, and the C-terminus was amidated. For most cases, we simulated E11 in protonated (neutral) state; however, a peptide with its charged version, denoted as wt^-, was considered as well for membrane-spanning configurations. Simulated systems included one peptide, 162 POPC molecules (81 per leaflet) and 9337 TIP3P [32] water molecules together with sodium and chloride ions necessary to construct a neutral system consisting of a membrane slab with $\sim$20 Å of 0.15 mol/L NaCl solution margins on both sides. Peptides and lipids were modeled with Amber99SB-ILDNP* [33] and Amber Lipid14 [34] force fields, respectively. In addition, we considered wt HAfp simulations in transmembrane hairpin configuration using Charmm36 force field [35]. Starting geometries for surface bound and transmembrane peptides were taken from our previous runs [20]. Necessary mutations were introduced with Discovery Studio Visualiser (Biovia).

2.1.2. MD Simulations

MD simulations were carried out with Gromacs software [36]. They were constructed using periodic boundary conditions, with a time step of 1 fs (2 fs for CHARMM simulations), interatomic bonds constrained using LINCS algorithm [37], van der Waals interactions smoothly shifted to 0 at 10 Å (or cut off at 12 Å, with force-switch for CHARMM simulations), and electrostatic interactions calculated using particle mesh Ewald method [38] with 1.2 Å mesh spacing. Desired temperature and pressure of 1 bar were maintained by velocity-rescale thermostat and seimiisotropic ParinelloRahman barostat [39], respectively. Temperature replica exchange runs (tREMD) [40] were conducted using 24 or 40 replicas, at temperatures, T, ranging from 310 to 350 or 377 K, respectively (detailed list in the Supplementary Materials, Table S2). Exchanges were attempted every 1 ps. The diffusion of trajectories in temperature space was monitored to assure that each trajectory was able to reversibly sample the entire temperature spectrum. The summary of conducted runs is given in the Supplementary Materials (Table S3).

2.1.3. Kinetic Analysis

Opening and closing of HAfp structures was considered as a two state process. The assignment of peptide configurations visited during tREMD simulations to hairpin or boomerang states was based on root mean square deviation (RMSD) of backbone heavy atoms with respect to NMR hairpin structure (pdb 2kxa, model 1) [11], with dividing threshold of 2.5 Å that corresponded to a minimum in bimodal RMSD distributions obtained at T = 310 K. A set of temperature dependent kinetic equations was fitted to time evolution of open state fraction in tREMD trajectories and extrapolated to infinite time to give an estimate of equilibrium populations at T = 310 K, according to method introduced by van der Spoel and Seibert [41]. All calculations were carried out with the use of g_kinetics Gromacs module. Final fractions of hairpin structures were evaluated as an average of two asymptotic infinite time estimates based on two tREMD runs that started from fully closed or fully opened conformations.

2.1.4. Free Energy Calculations

Potentials of mean force for peptide translocation along the axis perpendicular to membrane plane (z axis) were obtained using umbrella sampling simulations based on tREMD runs with 24 replicas. The peptides were restrained to hairpin geometry (reference NMR structure, pdb 2kxa, model 1) with harmonic potential, U_h, acting on pairs of Cα atoms that were closer than 7 Å in the reference structure, with a force constant of 2.39 kcal/mol/Å^2. Umbrella sampling was performed using biasing harmonic potentials with a force constant of 2.39 kcal/mol/Å^2 that acted in the z direction between the center of mass of peptide 1–20 Cα atoms (PCOM) and the membrane center (MCOM) defined based on the positions of three terminal carbon atoms of each lipid acyl chain, located within a cylinder of 30 Å radius and z axis passing through PCOM. PCOM relative positions, $z_p = z_{PCOM} - z_{MCOM}$, were gathered for the replica run at T = 310 K. Window spacing along the reaction coordinate was 1 Å for $z \in [-5, 16]$ Å range and 2 Å intervals for $z \in [18, 34]$ Å.

To assess necessary equilibration length and the final time of production phase, t_{end}, for each umbrella window, we determined the time, t_{OK}, at which z_p distributions gathered for $t \in [t_{OK}, \frac{1}{2}(t_{end} + t_{OK})]$ and for $t \in [\frac{1}{2}(t_{end} + t_{OK}), t_{end}]$ were similar according to Kolomogorov–Smirnov test with $p > 0.1$. Then, z_p distributions gathered for $t \in [t_{OK}, t_{end}]$, with the requirement that $t_{end} - t_{OK} > 15$ ns, were analyzed with weighted histogram analysis method [42] as implemented in Gromacs WHAM module, with a standard bootstraping error analysis.

The free energy cost of hairpin restraining with harmonic potential in membrane environment, U_h, was evaluated based on unrestrained runs using free energy perturbation formula [43]: $\Delta G_{0\to h} = -k_B T \langle \exp(-\beta U_h) \rangle_{\{F\}}$, where k_B is Boltzmann constant, $T = 310$ K. As a source of simulation frames, we used the last 500 ns of two tREMD runs (250 ns for E11A) that were started from closed and open structures (see above). A set $\{F\}$ of 10,000 frames was randomly drawn from this pool such that to fulfil the proportion of open and closed structures as determined based on the kinetic analysis. The procedure was repeated 1000 times to obtain an average $\Delta G_{0\to h}$ and its error as a standard deviation. Analogous process in aqueous environment was split into simulation parts, in which the force constant of the restraining potential was gradually increased in steps 0, 0.001, 0.01, 0.1, and 1.0 to its full value used in U_h.

Free energy of unrestrained peptides as a function of PCOM position along the z axis, z_p, defined as above was evaluated based on the probability distributions of $p(z_p)$ gathered during unrestrained tREMD simulations, according to $G(z_p) = -k_B T \ln(p(z_p)) + G_0$ relation where G_0 is an arbitrary constant. In the case of transmembrane configurations, an additional biasing potential was introduced: $U_b(z_p) = \frac{1}{2}k(z_p - z_0)^2$ for $z_p > z_0$ and 0 otherwise, that prevented peptide from surfacing, with $z_0 = 6$ Å for wt and E11A and $z_0 = 4.5$ Å for W14A peptides. The resulting biased probability distribution, $p'(z_p)$, was subsequently reweighted to obtain the unbiased $p(z_p)$, according to the following formula: $p(z_p) = p'(z_p) \exp(\beta U_b(z_p))c$, with c being a normalization constant [44].

2.1.5. Tryptophan Fluorescence Calculations

To estimate depth-dependent fluorescence quenching by brominated lipids based on our simulations, we adopted the model proposed by A. Ladokhin [45]. For a given average depth of lipid carbon-bound Br probe, h_m, and its dispersion σ_m (both established based on pure POPC tREMD runs, see Table S1 for values), we calculated simulation averages of depth-dependent fluorescence profiles:

$$\frac{F_m}{F_0} = \sum_{W \in \{14,21\}} \left\langle \exp\left(- G(h_W - h_m, \sigma_m, S) - G(h_W + h_m, \sigma_m, S)\right) \right\rangle_{MD} \quad (1)$$

where $\frac{F_m}{F_0}$ is the ratio of tryptophan fluorescence quenching by lipids specifically brominated at position m, to fluorescence without the quencher, h_W is the position of tryptophan indole ring center along the z axis in simulation frames, and the two Gaussian terms, $G(\Delta h, \sigma, S) = \frac{S}{\sigma\sqrt{2\pi}} \exp\left(\frac{(\Delta h)^2}{\sigma^2}\right)$, describe contributions from both membrane leaflets with S being the assumed quenching intensity [45] (see the Supplementary Materials, Figure S7). Given sets of F_m/F_0 for Br probes at $(4,5)$, $(6,7)$, $(9,10)$, or $(11,12)$ lipid acyl carbon atoms experimentally determined for each peptide, we used our simulation data to check what configuration of the respective peptide provided for the lowest root mean square error (RMSE) with respect to these values. To this end, we evaluated Equation (1) for the corresponding set of h_m and σ_m values based on simulation frames representing the considered peptide configuration, and determined RMSE subject to S minimization. To assess the fraction f and $1-f$ of transmembrane (TM) and surface (SURF) configurations, respectively, we assumed $F_m/F_0 = f(F_m/F_0)_{TM} + (1-f)(F_m/F_0)_{SURF}$ and minimized the RMSE subject to $f \in [0,1]$ and S.

2.1.6. Membrane Perturbation

Lipid splays were defined as events when any of carbon atoms within lipid acyl chain was at least 2 Å further from membrane midplane than the phosphate atom of the same lipid. Lipids proximity to peptide was assessed based on the closest distance of their phosphate atoms to any heavy peptide atom. Lipids closer than 7 Å were considered as "close" and lipids further than 30 Å from peptides were considered for reference calculations. All results were block averaged, with block length of 50 ns, and the analysis was conducted for replicas simulated at 310 K.

Water membrane permeability, P, was estimated assuming inhomogeneous solubility-diffusion mechanism [46], based on water density profile across the membrane, $\rho(z)$, and position dependent water diffusion coefficient in z direction, $D(z)$:

$$\frac{1}{P} = \int_{z_1}^{z_2} \frac{\rho_0}{\rho(z)D(z)} \mathrm{d}z \quad (2)$$

with ρ_0 denoting bulk water density, and integral extending between ± 40 Å from membrane center (we note that P was rather insensitive to integration interval as long as it encompassed membrane interior; hence, we resorted to such simple choice). Water density profiles were evaluated from tREMD runs for replica at $T = 310$ K with `gmx density` tool. For the calculation of diffusion profiles tREMD trajectories were demuxed, and continuous trajectory fragments that remained at $T = 310$ K for at least 10 ps were used for analysis. The z axis was discretized into bins, z_i of 2 Å width and if a water molecule was found within a given bin at time t, i.e., $z(t) \in z_i$, its displacement within $\Delta t = 10$ ps contributed to $D(z_i) = \langle (z(t+\Delta t) - z(t))^2 / 2\Delta t \rangle$, where the averaging includes all such instances. To obtain final diffusion profiles for integration, $D(z_i)$ were interpolated with a Gaussian kernel, with $\sigma_D = 1$ Å dispersion. Water permeability obtained for pure POPC was $(9 \pm 2) \times 10^{-3}$ cm/s, with error based on three simulation blocks (30 ns each), which is in fair agreement with experimental value of $(13.0 \pm 0.4) \times 10^{-3}$ cm/s [47]. Water permeability in the presence of a peptide was calculated for the entire membrane patch and should be interpreted as corresponding to experimental conditions with 1:162 peptide to lipid ratio.

2.1.7. Peptide Supervised Insertion

The procedure of peptide supervised insertion comprised 10 rounds of 50 simulations, 5 ns each. Starting structures included two random peptide placements and two selected from unconstrained simulations as already deeply inserted, all in hairpin conformation.

After each round, the distance of W14 Cα atom from membrane surface (center of mass of phosphate atoms within given leaflet) was evaluated within the 4 last ns of each simulation, and a frame with maximum insertion depth was selected as a seed for subsequent round. The probability of reaching particular position within i rounds was evaluated as $p = \prod_i \frac{n_i}{50}$ with n_i being the number of rounds in which the maximum insertion depth was within 0.7 Å from the global maximum.

2.2. Experimental Methods

2.2.1. Materials

Materials were custom ordered with purity > 95% (Lipopharm, Gdańsk, Poland). Sequences were as follows: wt GLFGAIAGFIEGGWQGMVDGWYGSGKKKKD and its E11A and W14A mutants. In all cases, the N-terminus was unmodified and the C-terminus was an amide. The -SGKKKD sequence was introduced to increase peptide solubility. Stocks were prepared from weighted amounts dissolved in water as 300–500 µM solutions. Concentrations were checked by UV spectroscopy using the extinction coefficient at 280 nm of 12490 $M^{-1}cm^{-1}$ for wt and E11A peptides and 6990 $M^{-1}cm^{-1}$ for W14A. N-(7-Nitrobenz-2-Oxa-1,3-Diazol-4-yl)-1,2-Dihexadecanoyl-*sn*-Glycero-3-Phosphoethanolamine (NBD-PE) and Lissamine™ Rhodamine B 1,2-Dihexadecanoyl-*sn*-Glycero-3-Phosphoethanolamine (N-Rh-PE) used in fusion assays were from ThermoFisher Scientific. POPC (1-palmitoyl-2-oleoyl-*sn*-glycero-3-phosphocholine), $Br_{4,5}$ (1-palmitoyl-2-stearoyl-(4,5)dibromo-*sn*-glycero-3-phosphocholine), $Br_{6,7}$ (1-palmitoyl-2-(6,7-di-bromo)stearoyl-*sn*-glycero-3-phosphocholine), $Br_{9,10}$ (1-palmitoyl-2-(9,10-dibro-mo)stearoyl-*sn*-glycero-3-phosphocholine), $Br_{11,12}$ (1-palmitoyl-2-(11-,12-di-bromo)ste-aroyl-*sn*-glycero-3-phos-phocholine), and all other chemicals were from Sigma Aldrich (Merck, St. Louis, MO, USA). All experiments were performed in buffer pH 5.0 (10 mM citric acid/NaOH, 150 mM NaCl).

2.2.2. Liposome Preparation

Desired amounts of lipids in chloroform were dried under a stream of nitrogen overnight under vacuum, followed by rehydration with appropriate buffer to 2–10 mg/mL concentration. For LUV preparation, the dispersion was frozen and thawed in liquid nitrogen and room temperature at least 5 times, followed by extrusion (15–21 times) through polycarbonate filters with 100 nm pores (Whatman) using Avanti Mini Extruder. For SUV preparation, the dispersion was sonicated with a tip sonicator (VibraCell VCX130) in 7–20 pulses lasting 10 s separated by 10 s breaks until the solution was clear.

2.2.3. Lipid Mixing

Lipid mixing of membrane fusion was measured by FRET using a Cary Eclipse (Varian) spectrofluorometer. For each lipid composition, unlabeled and labeled LUVs were prepared. To prepare the labeled LUVs, we included 1 mol % NBD-PE and N-Rh-PE in the lipid mixture before drying the lipids in the liposome preparation procedure. Unlabeled and labeled LUVs were mixed at a 9:1 ratio in pH 5.0 buffer. The total lipids concentration was 0.2 mM. After the equilibration of the vesicles, an appropriate amount of peptides from a stock solution was added to give final concentrations of 4 µM. Then, 10% Triton X-100 was added to achieve a final concentration of 1% after fusion had been completed. Fluorescence intensity of the acceptor (excitation with 463 nm and emission at 590 nm) before the addition of peptides and after the addition of Triton X-100 was defined as 0% and 100% fusion, respectively. Experiments were performed in triplicates and averaged signal is shown.

2.2.4. Tryptophan Fluorescence

Fluorescence measurements were made with a Carry Eclipse (Varian) spectrofluorometer with an excitation wavelength of 280 nm, excitation/emission (2.4/4 nm), and photomultiplier voltage of 800 V. Spectra were measured using a 4 × 10-mm cuvettes (Hellma) in of 295–500 nm emission region with an increment of 1 nm. Peptide solutions were used in 2–10 μM concentrations in 1000 μL volume, titrated with increasing portions of SUV suspension up to 1 mM in 13–20 steps. Normally, for each lipid concentration, three spectra were averaged to achieve an adequate signal-to-noise ratio. The cuvette was in the contact with a thermostat, assuring constant temperature of 22.0 ± 0.5 °C. From each spectrum background was subtracted (by measuring blank titration). The titration curves were constructed as normalized intensity values for the wavelength for which the maximum spectral shift was observed between free and liposome-bound peptide (∼328 nm). Such procedure was shown to govern a linear response of the signal in respect to the titrated peptide [48]. The titration curves were further corrected for SUV scattering [48] by using the tryptophan (N-acetyl-L-tryptophanamide) fluorescence registered under the same conditions in the presence of SUV solution at concentration [L] according to the equation:

$$F_{pept}^{corr}([L]) = F_{pept}([L])(F_{Trp}^{buffer})/(F_{Trp}([L])) \quad (3)$$

To corrected data point, non-linear hyberbolic curve was fitted according to the equation:

$$F = 1 + (I-1)(K_x[L])/([W] + K_x[L]) \quad (4)$$

where I denotes asymptotic intensity value, [W] is the molar water concentration (55.55 M), and K_x is the molar partition coefficient. Gibbs free energies were calculated as: $\Delta G_x^{\circ} = -RT \ln K_x$.

2.2.5. Tryptophan Quenching

We measured the quenching of tryptophan residues inside lipid bilayers using lipids labeled with bromine at carbons 4–5, 6–7, 9–10 and 11–12 in the acyl chain (denoted as $Br_{4,5}$, $Br_{6,7}$, $Br_{9,10}$, and $Br_{11,12}$, respectively), with 20 mol% brominated PCs in place of POPC. The same spectral conditions and spectrofluorometer settings were used as in binding experiments. Aliquots of peptide solution were added to vesicle suspensions to achieve a final peptide/lipid ratio of 1:500 and incubated for 2 min before recording tryptophan fluorescence intensity. The signal from an identical sample without peptide was used to determine the quenching (F_m/F_0).

2.3. Leakage Assay

Leakage assay was performed by monitoring fluorescence increase of calceinrealeased from LUV interior. Calcein was encapsulated in LUVs by hydrating lipid film by dye solution (112 mM of calcein in 0.27 M NaOH). Free dye and liposomes were separated to fraction by size exclusion chromatography (PD-10 columns, GE Healthcare). The most concentrated fractions with liposomes were pooled. Final concentrations of lipids were 0.625 mg/mL. Fluorescence was recorded with Carry Eclips (Varian) using single excitation/emission wavelengths (495/518 nm). Fluorescence baseline of 3.65 μM liposomes was observed for 2 min (subsequently rescaled to 0% leakage). Then, the peptides were added to the solution at 1:50 peptide/lipid ratio and the leakage was measured over 4000 s. The remaining intact liposomes were disrupted by adding 10 μL of 10% Triton X-100 to the solution (100% leakage).

3. Results

3.1. Peptide Configurations

To investigate HAfp configuration space and its alteration by two known function-affecting mutations, we conducted a set of tREMD simulations for surface and deeply inserted peptides. Taking into account structures based on experimental data, the simulations of surface bound peptides started from all hairpin or all boomerang structures. In turn, all deeply inserted peptides were initially modeled in hairpin conformation, based on our previous results obtained using the self assembly technique [20].

3.1.1. Surface Placement

Regardless of initial configurations, surface bound peptides started to interconvert between tight helical hairpin and open boomerang forms, as is apparent based on the gradual formation of bimodal RMSD distributions (Figure 1a). Representative structures, obtained in each case as the centroids of the most populated conformation clusters, correspond well to experimentally determined boomerang and hairpin geometries, with minor differences between the considered peptide versions (Figure 1b and Figure S1). The insertion depth of surface configurations turns out to be the same for each variant (Figure 1e and Figure S1) and displays only slight conformation-dependent alterations. In general, the N-helices do not change their position relative to membrane surface, irrespective of hairpin opening or closing, while the C-helices of boomerang configurations are slightly more deeply inserted.

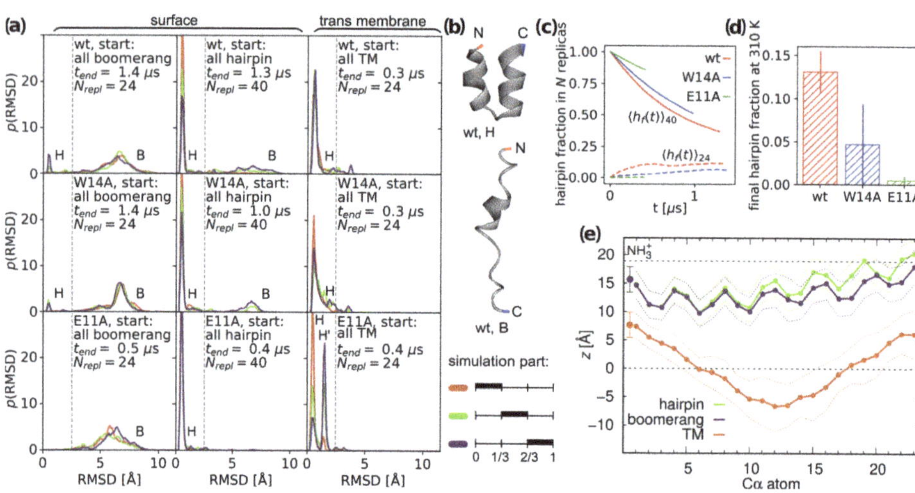

Figure 1. (**a**) Time evolution for distributions of backbone atoms RMSD with respect to tight helical hairpin (NMR structure, PDB id 2kxa, model 1) at 310 K from tREMD simulations. (**b**) Representative conformations for two most populated states. (**c**) Time evolution of hairpin fraction in tREMD simulations based on fitted kinetic equations. (**d**) Equilibrium hairpin fractions at 310 K. (**e**) Insertion depths of N-terminal amino group nitrogen and Cα atoms of wt HAfp ± one standard deviation (dotted lines). Dashed lines are the membrane surface (the maximum of phosphate atoms density).

Given the adopted tREMD protocol and relatively small size of the peptides under study, we aimed to determine equilibrium populations of surface conformations in each case. Unfortunately, despite the long accumulated sampling time (85, 73, and 28 μs for wt, W14A, and E11A, respectively), the RMSD distributions were still far from equilibria, indicating rather slow conformational dynamics on membrane surface. To estimate the final proportion of closed and open configurations, we assumed that the transition between them is a two-state process and considered a set of temperature-dependent kinetic equations operating on stochastic tREMD trajectories (Figure 1c) [41]. The extrapolation of fitted solutions to infinite time for tREMD simulations starting from all hairpin, and, likewise, all boomerang structures should, in principle, lead to the same equilibrium hairpin population at $T = 310$ K. Accordingly, we estimated this population as an average from both kinds of tREMD runs and obtained 0.13 ± 0.02, 0.05 ± 0.05, and 0.005 ± 0.005 hairpin fractions for wt HAfp, W14A, and E11A mutants, respectively, at $T = 310$ K (Figure 1d). In qualitative agreement with experimental data, this result indicates greater preference towards hairpin structure for wt HAfp compared to W14A mutant [11,14]. We did not find direct experimental evidence for the nature of the dominant E11A conformation, but, based on its reported thermodynamic binding signature with moderately unfavorable entropy component similar to W14A and rather different from highly entropy penalized wt HAfp [14], the dominance of less constrained boomerang structure is likely also in its case.

Our predicted population of 0.13 closed structures for wt HAfp23 is considerably lower compared to estimates based on NMR in DPC micelles that suggest exclusively hairpin conformation at neutral pH, exemplified by 2kxa PDB structure [11]. In the same system, but at pH = 4, to which our simulation conditions are adjusted (protonated E11), closed conformations were found to constitute 0.8 of the entire population and were estimated to interchange with the open ones on ∼25 µs time scale [49]. Furthermore, as compared to detergent micelles, the closed form is apparently less stable in membranes, where its population of 0.7 was reported already at neutral pH [50]. Taken this into account, as well as the fact that sole W14A substitution is enough to permanently open up the hairpin, it is plausible that the stabilization of the closed structure is not very strong, and that its interconversion with non-negligible boomerang fraction indeed occurs in POPC membranes for wt HAfp.

3.1.2. Intramembrane Placement

Although shorter than typical transmembrane helical fragments, HAfp in hairpin conformation may be capable of adopting intramembrane position [20]. It can do so by inducing local membrane thinning that provides hydration of both hydrophilic hairpin poles necessary for the stabilization of the membrane-spanning orientation (Figure 2a). In the case of wt HAfp, the deeply inserted hairpin is located centrally within the membrane, with residues 6 and 18 positioned at the bilayer midplane (Figure 1e). Notably, the nuclear Overhauser effect between peptide amide groups and terminal methyl groups of lipid acyl chains was observed exactly and exclusively for these two residues in the study of wt HAfp23 in DPC micelles [11]. Given relatively shallow insertion depths observed in simulations of open structures (Figure 1e), this would be not possible at all without assuming intramembrane configurations.

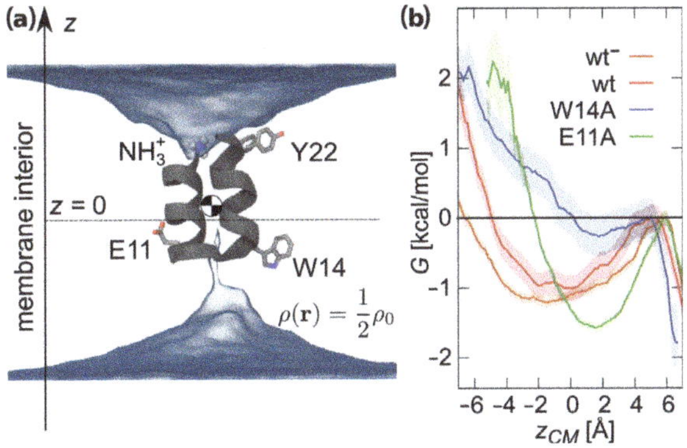

Figure 2. (**a**) Representative membrane-spanning configuration of wt HAfp with average water density isosurface at half of bulk density. (**b**) Free energy profiles for unrestrained peptides center of mass in membrane-spanning configurations.

Hairpin conformation appears to be well preserved for deeply inserted peptides (Figure 1a), although we observed solitary unfolding events that lead to surfacing of the C-terminal arm at the opposite membrane side to the N-terminus—we note, however, that this would be impossible, if the peptide was attached via a linker to the remaining part of the HA2 subunit. Our simulations indicate that the intramembrane configuration is generally metastable, and can be abandoned by peptide escaping to the surface in the direction of hairpin opening. We observed no successful, complete egress in the opposite direction (i.e., towards the hairpin apex) in any of our simulations, likely due to firm anchoring of the positively charged N-terminal amino group within the aqueous compartment.

The stability of intramembrane localization appears to be affected by amino acid composition of the peptide. Wt HAfp remains most deeply inserted among the considered variants, and is separated from surface configuration by ∼1 kcal/mol free energy barrier (Figure 2b). Intriguingly, the deprotonation of E11, which may be plausible owing to the exposure of hairpin apex to the neutral pH of cellular environment on the outer side of the endosomal membrane, further deepens and stabilizes peptide insertion. The W14A mutant is least stable, with practically no free energy barrier for surfacing, and roughly 3 Å shallower insertion depth compared to wt HAfp. We attribute this behavior to the decrease of hairpin apex hydrophilicity upon the removal of W14 side chain. In turn, E11A substitution leads to intermediate insertion depth, albeit relatively well stabilized within the membrane core. In this case, in contrast to other variants, a partial remodeling of the C-terminal hairpin arm is also observed (Figure S1).

To check whether the stability of intramembrane configurations is not merely a spurious effect of the selected Amber force field family, we carried out an independent set of tREMD simulations for wt using unrelated Charmm force field (peptide and lipids). We observed a stable, membrane-spanning hairpin, with even greater free energy barrier for surfacing of ∼2.5 kcal/mol than in the Amber force field (Figure S5).

3.2. Potentials of Mean Force

To investigate peptides partitioning between membrane surface and interior, we evaluated free energy differences between the respective configurations (Figure 3). To this end, we took into account the free energy cost of peptide restraining to hairpin conformation at membrane surface assuming hairpin to boomerang ratios summarized in Figure 1d, the potential of mean force (PMF) for restrained hairpin translocation into the membrane core, and free energy gain upon the release of the restraining potential thereof. To further complement the calculations, we assessed the cost of hairpin unbinding from surface and unfolding in bulk solvent (for details, see the Supplementary Materials).

The calculated free energies of peptides binding to membrane surface (Table 1, $\Delta G^{B \to S}$) are in the order of 11 kcal/mol. They do not include, however, the effect of peptide liberation from a fixed position in the bulk, which should generally decrease the calculated affinities by 1–2 kcal/mol, depending on the assumed free peptide concentration. Our estimate of the wt HAfp binding strength is ∼1 kcal/mol above that of the W14A mutant, which is in fair agreement with our own measurements (Table 1, ΔG_{exp}, Figure S9) and other published data [14]. Free energy profiles for further peptide transition from membrane surface towards the core rise steeply in all three considered cases, and barriers leading to relatively shallow intramembrane minima are achieved when peptides centers of mass are around 6 Å from bilayer midplane (Figure 3). The estimated free energy cost of peptide transitioning from membrane surface to the core, $\Delta G^{S \to D}$, is 10 kcal/mol for the wt HAfp and 11 kcal/mol for both mutants. The overall transition cost from surface to deep minimum is a combination of three effects. The first is the cost of hairpin restraining on membrane surface, $-\Delta G^{S}_{h \to 0}$. Here, naturally closed peptides have an advantage over those assuming predominantly open conformation, as is reflected by almost 3 kcal/mol difference between wt HAfp and E11A.

The second, is the cost of restrained hairpin translocation to intramembrane position. Somewhat surprisingly, the results for wt HAfp and W14A mutant are quite similar in this respect (see Figure S3 for full, detailed PMF). It indicates that, instead of changing intrinsic peptide propensity to move from membrane surface to the core, W14A mutation rather affects peptide behavior within both free energy wells by promoting the open surface form and weakening the stability of the membrane-spanning configuration. In turn, membrane penetration of the restrained E11A mutant is least energetically costly, most likely owing to increased hydrophobicity of the hairpin apex, that is the peptide part that actually penetrates through the core. The third free energy contribution, $\Delta G^{D}_{h \to 0}$, comes from the removal of the restraining potential for membrane-spanning configurations. We do not see large differences here between the considered HAfp versions. In all cases, free energy gain is smaller than the corresponding effect on the surface, indicating greater hairpin rigidity within the membrane core.

Table 1. Experimental, ΔG_{exp}, and calculated free energy differences in kcal/mol (error estimates in subscript): $\Delta G^{B\to S}$ binding from bulk solvent to membrane surface, $\Delta G^{S\to D}$ transition from surface to deep configuration, $\Delta G^{B}_{0\to h}$ the cost of peptide restraining to hairpin in bulk solvent, $\Delta G^{B\to S}_{h}$ transition of restrained peptide from bulk solvent to membrane surface, $\Delta G^{S}_{h\to 0}$ restraints removal for surface configuration, $\Delta G^{S\to D}_{h}$ transition of restrained peptide from surface to deep configuration, $\Delta G^{D}_{h\to 0}$ restraints removal for deep configuragion. * experimental energies are for peptides with solubility tags, and † value for E11A based on experimental difference wrt. wt.

Peptide	ΔG^{*}_{exp}	$\Delta G^{B\to S}$	$\Delta G^{S\to D}$	$\Delta G^{B}_{0\to h}$	$\Delta G^{B\to S}_{h}$	$\Delta G^{S}_{h\to 0}$	$\Delta G^{S\to D}_{h}$	$\Delta G^{D}_{h\to 0}$
wt	$-10.2_{0.1}$	$-11.7_{0.9}$	$10.1_{0.7}$	$18.3_{0.3}$	$-20.0_{0.7}$	$-9.9_{0.2}$	$8.6_{0.3}$	$-8.5_{0.2}$
W14A	$-8.6_{0.1}$	$-10.9_{1.1}$	$11.3_{0.8}$	$17.9_{0.6}$	$-18.6_{0.6}$	$-10.3_{0.2}$	$8.5_{0.3}$	$-7.7_{0.1}$
E11A	$-10.3_{0.1}$	$-11.8^{\dagger}_{0.9}$	$11.0_{1.4}$	–	–	$-12.6_{1.2}$	$6.5_{0.2}$	$-8.5_{0.2}$

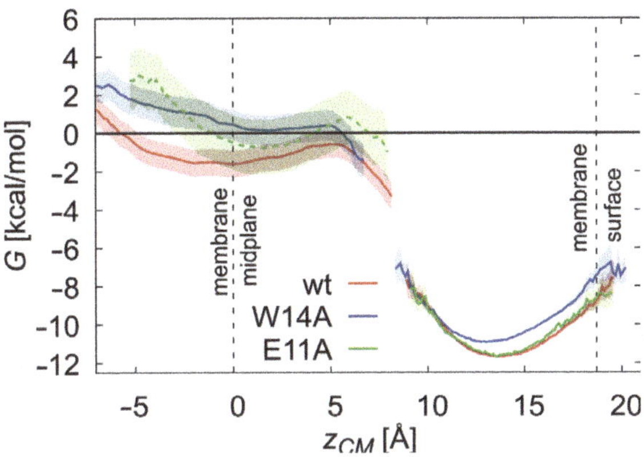

Figure 3. PMF for unrestrained peptides in deep and surface free energy minima. $G = 0$ corresponds to free peptides in the aqueous phase. Relative depths of free energy wells for surface and deep configurations were obtained based on: continuous PMF for restrained peptides, (un)restraining free energy calculations, and free peptides simulations (Table 1 and Figure S4). Dashed line for E11A reflects the fact that its surface minimum was estimated based on binding free energy for wt.

Taken together, the overall cost of peptide translocation into the membrane core appears to be rather high and, based on the considered simulation setup, indicates that the population of deeply bound peptides is very small. Nevertheless, the order wt > E11A > W14A observed both for the availability and stability of membrane-spanning configurations correlates well with peptides fusogenic activities (Figure S8), suggesting a hypothesis that deep insertion may be relevant to their mechanism of action.

3.3. N-Helix Orientation

Experimental evidence concerning peptide orientation within the membrane often comes from ATR-FTIR spectroscopy [51]. Assuming the dominance of helical structure, at least within the N-helix, the obtained order parameter, S_h, allows determination of the average angle between helix axis and membrane normal, z. In this respect, S_h for an at least 23 residue wt HAfp is reported to be in the range from 0.32 to 0.59, depending on pH and membrane composition [17,18,52], which is interpreted as an oblique helix insertion, with 30–50 degrees average tilt with respect to the z axis. A similar oblique insertion angle of 50 degrees was suggested based on EPR measurements [10]. Our simulations do not support this view. In all surface configurations, the N-helix remains parallel to membrane plane (Figures 1e and 4), and its capping NH_3^+ group does not reveal any tendency for deep membrane penetration. On the contrary, if the peptides are forced to move into the membrane during umbrella sampling runs or supervised insertions (see below), they penetrate the hydrophobic core exclusively with the hairpin apex. In doing so, however, they do not assume any intermediate, oblique orientations, but rather shift sharply from parallel to almost perpendicular orientation upon passing the free energy barrier between surface and deep configurations (Figure 4).

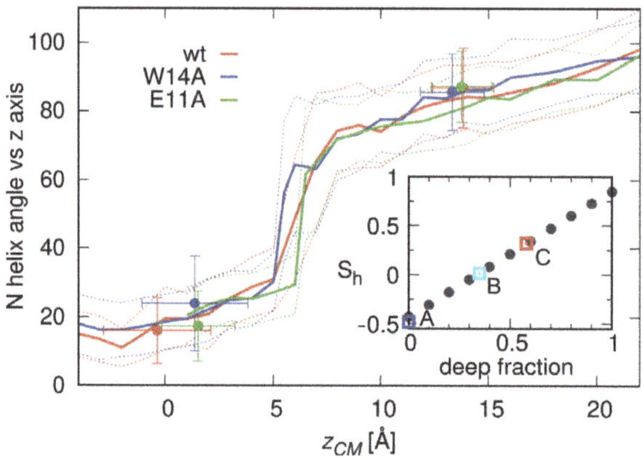

Figure 4. N-helix angle with respect to membrane normal in degrees as a function of insertion depth. Lines represent data from umbrella sampling simulations for restrained hairpin conformations, while points are data from unrestrained simulations. Error bars correspond to one standard deviation. Inset: The estimated N-helix order parameter as a function of deep configurations fraction. Squares are the experimental data for S_h: (A) W14A [14]; (B) wt HAfp20 [53]; and (C) wt HAfp23 [17,18,52].

This simulation based picture suggests a two states contribution to the order parameter. If we calculate S_h with the assumption of bimodal tilt angle distribution taken from simulation data, the experimental values for wt HAfp23 are recovered for 0.5–0.6 fraction of deep configurations (Figure 4, inset). This fraction would be lower for HAfp20 (reported $S_h = 0.18$ [53], deep fraction 0.4) and nonexistent for the W14A mutant (reported $S_h = -0.48$ [14]), in agreement with surface only configurations in this case.

3.4. TRP Fluorescence

To further confront our simulations with experimental findings, we analyzed depth-dependent tryptophan fluorescence quenching by brominated lipids [54]. Here, the decrease in TRP fluorescence intensity with respect to a reference value, F/F_0, caused by Br atoms located at known positions along lipid acyl chains reveals an approximate insertion depth of TRP side chains. Assuming distance-dependent model of fluorescence quenching (see Section 2.1.5 for details) [54] and applying it to peptide-lipids configurations from our tREMD simulations, we assessed what combination of major macrostate populations: surface hairpin, surface boomerang, and deeply inserted hairpin, best reproduces experimental data.

For all three considered peptides, the measured quenching effect is strongest for Br probes that are closest to bilayer surface ($Br_{4,5}$) and generally decreases towards membrane interior (Figure 5a). Importantly, however, the readouts differ between peptide mutants, indicating higher fluorescence quenching by deeply buried probes in the case of the E11A mutant and wt HAfp, compared to the W14A mutant. In all cases root mean square errors (RMSE) between experimental F/F_0 ratios and predictions based on MD simulations clearly disfavor standalone surface hairpin configurations (Figure 5b). Since for the wt HAfp the hairpin is indicated as a dominant structure by NMR [11], it suggests the need to include deep configurations in order to explain fluorescence data. Indeed, 20% of deep configurations in addition to surface hairpin leads to the overall lowest RMSE in this case. For the W14A mutant, the hairpin structure is unlikely [14], which leaves surface boomerang in best agreement with fluorescence data. Notably, in this case, the inclusion of deep configurations does not improve RMSE. On the contrary, favorable agreement for the

E11A mutant clearly requires the assumption of ≳30% of intramembrane locations, but the results do not discriminate between boomerang and hairpin configurations.

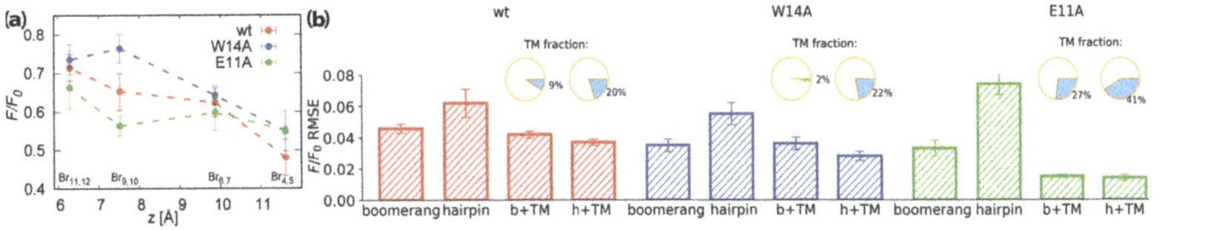

Figure 5. (**a**) Experimentally determined fluorescence quenching ratio, F/F_0, for peptides in brominated lipids; and (**b**) RMSE values for best fit between experimental and calculated F/F_0. Circle plots indicate the fraction of transmembrane configuration providing the lowest RMSE.

The results unanimously suggest a greater tendency to penetrate membrane interior for the wt HAfp and E11A mutant compared to the W14A mutant. While there is certainly no quantitive match between population fractions assessed from the PMF calculations and the estimates based on TRP fluorescence quenching or ATR-FTIR spectroscopy discussed above, we note that the simulation model most likely overestimates the free energy difference between surface and deep minima due to limited system size that enhances the compactness of the considered membrane slab, but it still captures qualitative differences between the considered peptide variants.

3.5. Membrane Perturbation

3.5.1. Lipid Tail Protrusions

The extent of peptide-promoted lipid tail protrusions [30] has often been invoked as a predictor of their ability to induce lipid mixing between membranes, as a proxy for fusogenic activity [23]. According to our simulations, the presence of surface bound peptides indeed increases the likelihood of lipid tail protrusions, but only roughly sixfold compared to pure POPC membrane (Figure 6a). Aside from being relatively weak and showing little dependence on peptide conformations, the effect observed for surface configurations is equally strong for fusion active wt HAfp and inactive W14A mutant, and, hence, does not explain the loss of function in this latter case.

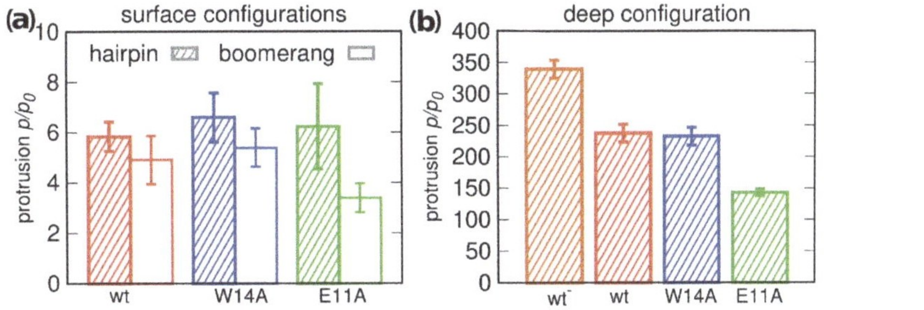

Figure 6. Ratios of lipid tail protrusions within 7 Å distance from peptides to those in pure POPC membrane for: (**a**) surface; and (**b**) deep configurations.

The situation changes, if membrane-spanning configurations are considered. Firstly, they appear to be capable of increasing the intensity of lipid tail protrusions roughly 250 fold compared to pure membrane (Figure 6b). It is most likely a consequence of deeper insertion of the positively charged N-terminal amino group (Figure 1e and Figure S2), which allows more effective dragging of phosphate groups below membrane surface, thus promoting lipid tilting and splaying. Secondly, the assumption of only surface bound configurations for the W14A mutant, inferred from previous results, would explain its inactivity. Thirdly, comparably lower protrusion promoting capability of the E11A mutant compared to the wt HAfp corresponds well to their relative fusogenic activity. The difference is possibly related to shallower insertion depth the E11A mutant, and consequently lower burial of the N-terminal peptide charge (Figure S2). Finally, of note is the highest rate of lipid protrusions generated by the wt$^-$ HAfb with E11 deprotonated, whose occurrence in real life scenarios cannot be excluded given the exposure of the hairpin apex to cellular environment.

3.5.2. Membrane Water Permeability

To further assess membrane perturbation induced by the peptides, we calculated relative membrane water permeability and compared it with experimental leakage assays of LUV encapsulated calcein.

According to simulations, peptides located at the membrane surface do not lead to any increase in its water permeability compared to pure POPC membrane (Figure 7a). However, a sharp onset of permeability is observed upon transition to deep minimum. This effect is somewhat less pronounced for the E11A mutant, possibly due to its greater hydrophobicity and smaller insertion depth. The highest permeation was observed for the most deeply inserted wt$^-$ variant. The simulated results are in fair agreement with the experimental leakage assays (Figure 7b). Apparently lower leakage caused by W14A mutant as compared to other peptides may be explained by its less stable deep configuration (Figure 3). Notably, however, relative magnitudes of leakage for the wt HAfp and E11A mutant observed in experiments are opposite to those suggested by simulations. In agreement with the interpretation of fluorescence quenching experiment (Figure 5b), this may indicate that the fraction of deeply inserted E11A configurations may be actually somewhat greater than in the case of wt. Taken together with greater stability but shallower depth of E11A transmembrane free energy minimum (Figure 2b), this would imply that comparably greater fusogenic activity of the wt is predominantly the effect of its ability to localize deeper within the membrane core. To check whether the difference between W14A and wt/E11A peptides was not driven by a diminished surface concentration resulting from an order of magnitude lower partition coefficient for W14A (~1.6 kcal/mol difference in Gibbs binding free energy, Table 1), we also performed the experiment at lower (1:500) peptide/lipid ratio. All three leakage curves looked more similar to each other; however, wt and E11A peptide still led to more efficient permeation (Figure S10).

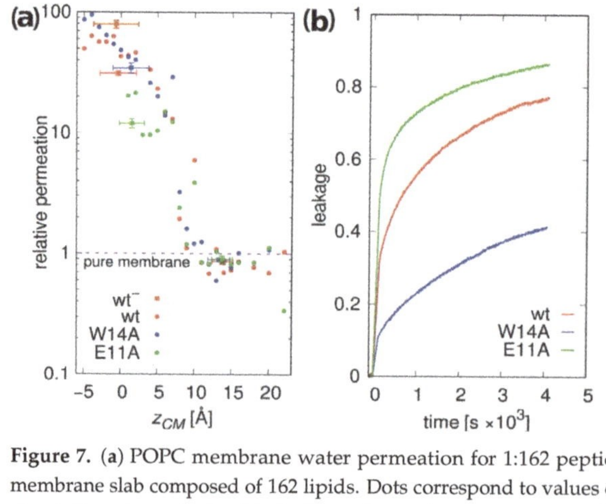

Figure 7. (**a**) POPC membrane water permeation for 1:162 peptide to lipid ratio, relative to pure membrane slab composed of 162 lipids. Dots correspond to values obtained from umbrella sampling simulations for restrained hairpin conformations, squares to values obtained from unrestrained simulations. (**b**) Peptide-induced calcein leakage from POPC LUVs.

3.6. Supervised Insertion

As indicated by our PMF calculations, the peptides need to traverse a substantial free energy barrier of roughly 10 kcal/mol in order to reach membrane interior from surface configurations. To check whether such putative transitions are feasible for unconstrained peptides and to assess their time scale, we devised a supervised insertion scheme. It is based on consecutive simulation rounds, each including relatively short (5 ns) multiple (50) unconstrained MD runs. After each round, a structure that is most advanced along the assumed pathway is selected among all runs and is used as a seed for a subsequent round of simulations. We focussed on wt HAfp and considered the position of W14 Cα atom along the z axis as a measure of insertion progression.

Of four independent supervised insertion procedures initiated from diverse starting configurations two turned out to be successful in peptide reaching the deep free energy well within 10 simulation rounds (Figure 8a, Top). In each such case, prior to achieving membrane-spanning configuration, the peptide induced the formation of a thin water wire between W14 side chain and aqueous compartment on the opposite membrane side (Figure 8b). The apparent role in this process was played by W14 indole nitrogen atom. It was effective, however, only in one of three possible side chain conformers that provided cross-membrane facing nitrogen orientation (Figure 8c), and without assuming this conformer peptide progression was stalled. Intriguingly, this conformer appears more frequently among closed rather than open surface configurations (Figure 8d), thus contributing further to hairpin insertion readiness, in addition to its already more favorable shape. To test an alternative hypothesis of N-terminus driven insertion, we also carried out simulations based on supervised terminal amino group nitrogen atom position, but none of the four trials for hairpin structure was successful (Figure S6).

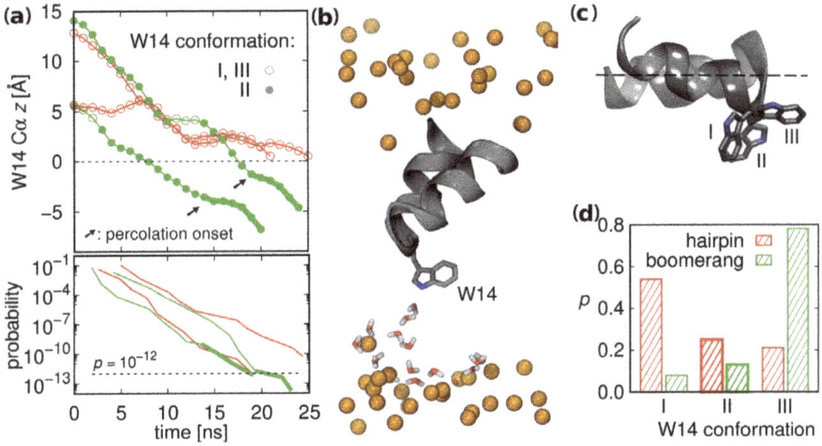

Figure 8. (**a**) (Top) Supervised insertion trajectories for two successful (green) and two unsuccessful trials (red); and (Bottom) the probability of insertion advancing along z axis. (**b**) Water wire established by interaction with W14 at percolation onset (trajectory snapshot). The water molecules within 10 Å of protein atoms and z coordinates restricted to protein-distal membrane leaflet are shown. The orange spheres represent membrane phosphate atoms. (**c**) Three main W14 conformations. (**d**) Probability distribution for finding W14A conformers in surface configurations.

The time scale of complete peptide insertion was roughly 20 ns. By simply evaluating the probability of completing the entire route as a product of ratios of productive to unproductive runs in each of 10 rounds (see methods for details), we obtain the probability of success in a single attempt, $p \sim 10^{-12}$ (Figure 8a, bottom). Following a coarse reasoning, for a GUV with 10 µm diameter and 160 lipids to peptide ratio, this gives an estimate of

>150 insertions per second, per vesicle, thus approaching the time scales of experimental fusion observation.

3.7. Possible Insertion Modes

The dominant view of wt HAfp23 configurations based on experimental insights to date corresponds to a tightly closed hairpin [11] which remains partially inserted into membrane, with buried N-terminus and solvent-exposed kink, such that the overall tilt angle of the N-helix with respect to membrane normal is ~40 degrees (Figure 9c) [10,14]. We do not find support for this view in our simulations. We also note that such configuration would entail at least partial burial of the C-terminal hairpin fragment (together with a few residues from subsequent linker region, if complete HA structure was considered) whereas its strictly conserved sequence, ^{21}WYG^{23}, suggests rather a preference to the membrane–water interface region. Instead, we propose a more complex model, in which the peptides fluctuate between closed and open surface conformations (Figure 9a,b) and occasionally dive into the membrane core, adopting more or less stable transmembrane configuration (Figure 9d,e). In this view, the fusogenic HAfp activity would be related either to this latter configuration alone or to membrane-perturbing transitions of buried hairpins back to membrane surface.

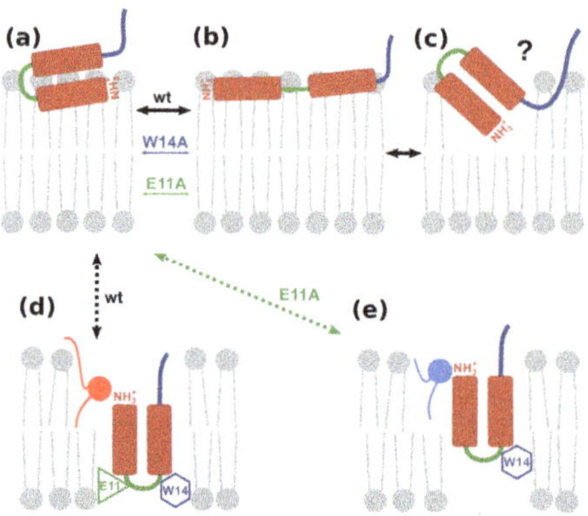

Figure 9. Possible insertion modes of HAfp into lipid membrane: (**a**,**b**) surface hairpin and boomerang, respectively; (**c**) obliquely inserted hairpin; (**d**) deeply inserted transmembrane hairpin (wt); and (**e**) shallowly inserted transmembrane hairpin (E11A).

4. Discussion

Our simulations indicate the highly dynamic nature of HAfp, which includes interconversions between open and closed surface conformations, as well as transient excursions into the membrane core. The lack of notable differences in the level of membrane perturbation by surface configurations of active and inactive mutants combined with apparently greater ability of the active ones to adopt transmembrane placement with highly membrane disruptive potential indicate that peptides' ability to dive into the membrane core may be the key element of their fusogenic function.

Indeed, experimentally observed activity of the three considered HAfp versions appears to correlate with the stability and insertion depth of their membrane-spanning configurations. According to our results, the actual HAfp fusogenic mechanism is related to partial burial of positively charged N-terminal amino group whose interaction with membrane phosphates promotes lipids tilting and, eventually, acyl tails protrusions. While such role of the N-terminal charge has already been postulated, its burial was ascribed to oblique peptide insertion whose existence is not supported by our study. Instead, we advocate that the experimental measurements consistent with such oblique insertion reflected an average over two, distinct states corresponding to in-plane and transmembrane orientations. While at first sight the relatively short length of HAfp helical hairpin arms appears incompatible with transmembrane configurations, in fact, it is well suited to provide for efficient burial of the N-terminal amino group within bilayer core. An extension of the N-helix would lead to surfacing of the N-terminus and a stable, yet benign membrane-spanning structure, typical for integral membrane proteins. Instead, the necessity to maintain a metastable deep configuration by keeping balanced interaction with aqueous compartment on both membrane sides imposes strict requirements for the kink region that likely explain the conservation of residues that support tight packing of both helices, such as the glycine ridge, as well as the existence of the essential W14, whose role in membrane penetration and water attraction was highlighted by our simulations.

The free energy difference between surface and deep configurations resulting from our simulations is large and suggests that the fraction of deeply inserted peptides may be rather small. We note, however, that our free energy calculations likely provide an upper limit to the actual value due to relatively small size of the membrane patch considered in our model. In reality, deep peptide insertions may be happening opportunistically supported by local membrane fluctuations. However, even rare such events yet of relatively high fusogenic potential are enough to explain the course of experimentally observed vesicle fusion, whose time scale extends to minutes.

5. Conclusions

Our study leads to the following conclusions:

- A model assuming the presence of HAfp exclusively at lipid–water interface is not sufficient to explain mutation-related differences in peptides activity, experimentally estimated helix tilting angle, or peptide-induced membrane water permeability.
- Simulations together with tryptophan fluorescence quenching experiments and the observation of peptide-induced membrane water permeability suggest the existence of membrane-spanning configurations with high membrane-perturbing potential.
- The effect of amino acids mutations on fusogenic activity correlates with peptides ability to achieve and maintain deeply inserted configurations and with the insertion depth of the N-terminal amino group.
- Although surface HAfp binding seems to be the dominant mode, simulations demonstrate the feasibility of spontaneous deep peptide insertions at sufficient rate to promote fusion in experimentally observed time scale.

Supplementary Materials: The following are available online at https://www.mdpi.com/1422-0067/22/10/5301/s1, Figure S1: Representative conformations from unrestrained simulations for three peptide variants and the distribution of root mean square deviation from tight hairpin structure (PDB. 2kxa, model 1) for wild type, E11 deprotonated hemagglutin fusion peptide (wt$^-$ HAfp) simulation in trans-membrane orientation. Figure S2: Insertion depths of N-terminal amino group nitrogen atom and Cα atoms of wt HAfp in considered configurations. Figure S3: Potentials of mean force for translocation between bulk water and membrane center for restrained peptides. Figure S4: Relations between contributions to free energy considered for obtaining binding free energies of unrestrained peptides at surface and deep minima, relative to unrestrained peptides in water. Figure S5: Results for Charmm36 force field. Figure S6: Supervised insertion attempts based on N-terminus progression. Figure S7: Scheme for Trp quenching calculations. Figure S8: Experimental results for lipid mixing. Figure S9: Experimental results for peptide binding to POPC SUVs. Figure S10: Experimental results for calcein leakage. Table S1: Insertion depths of Br atoms. Table S2: List of tREMD temperatures and list of conducted simulations. Table S3: List of umbrella sampling runs.

Author Contributions: Conceptualization, P.S. and R.W.; investigation, P.S., A.D., P.B. and R.W.; writing—original draft preparation, P.S.; and writing—review and editing, R.W. and P.S. All authors have read and agreed to the published version of the manuscript.

Funding: This research was funded by EMBO Installation Grant 3051/2015 and the Opus grant (National Science Centre UMO-2018/29/B/NZ1/02434) to P.S. R.W. and P.B. were supported by the Sonata Bis 8 grant (National Science Centre UMO-2018/30/E/NZ1/00257). Computer simulations were carried out with the support of the Interdisciplinary Centre for Mathematical and Computational Modelling (ICM), University of Warsaw, under grant No. GA76-7, and by PLGrid Infrastructure.

Institutional Review Board Statement: Not applicable.

Informed Consent Statement: Not applicable.

Data Availability Statement: MD configuration files, system topologies, system structures, trajectories, and source experimental data are available on figshare (DOI 10.6084/m9.figshare.14365508).

Acknowledgments: We acknowledge Anna Wegrzyn (Faculty of Biology, University of Warsaw) for her assistance in binding experiments.

Conflicts of Interest: The authors declare no conflict of interest.

References

1. Wilson, I.A.; Skehel, J.J.; Wiley, D.C. Structure of the haemagglutinin membrane glycoprotein of influenza virus at 3 Å resolution. *Nature* **1981**, *289*, 366–373. [CrossRef] [PubMed]
2. Skehel, J.J.; Wiley, D.C. Receptor Binding and Membrane Fusion in Virus Entry: The Influenza Hemagglutinin. *Annu. Rev. Biochem.* **2000**, *69*, 531–569. [CrossRef] [PubMed]
3. Durrer, P.; Galli, C.; Hoenke, S.; Corti, C.; Glück, R.; Vorherr, T.; Brunner, J. H+-induced Membrane Insertion of Influenza Virus Hemagglutinin Involves the HA2 Amino-terminal Fusion Peptide but Not the Coiled Coil Region. *J. Biol. Chem.* **1996**, *271*, 13417–13421. [CrossRef] [PubMed]
4. Longo, M.L.; Waring, A.J.; Gordon, L.M.; Hammer, D.A. Area Expansion and Permeation of Phospholipid Membrane Bilayers by Influenza Fusion Peptides and Melittin. *Langmuir* **1998**, *14*, 2385–2395. [CrossRef]
5. Nieva, J.L.; Agirre, A. Are fusion peptides a good model to study viral cell fusion? *Biochim. Biophys. Acta Biomembr.* **2003**, *1614*, 104–115. [CrossRef]
6. Lear, J.D.; DeGrado, W.F. Membrane binding and conformational properties of peptides representing the NH2 terminus of influenza HA-2. *J. Biol. Chem.* **1987**, *262*, 6500–6505. [CrossRef]
7. Rafalski, M.; Rockwell, A.; Lear, J.D.; Grado, W.F.; Wilschut, J.; Ortiz, A.; Ginkel, L.C.; Ortiz, A.; Ginkel, L.C. Membrane Fusion Activity of the Influenza Virus Hemagglutinin: Interaction of HA2 N-Terminal Peptides with Phospholipid Vesicles. *Biochemistry* **1991**, *30*, 10211–10220. [CrossRef]
8. Cross, K.J.; Langley, W.A.; Russell, R.J.; Skehel, J.J.; Steinhauer, D.A. Composition and functions of the influenza fusion peptide. *Protein Pept. Lett.* **2009**, *16*, 766–778. [CrossRef]
9. Martin, I. Structure and Topology of the Influenza Virus Fusion Peptide in Lipid Bilayers. *J. Biol. Chem.* **1995**, *270*, 27606–27614. [CrossRef]
10. Han, X.; Bushweller, J.H.; Cafiso, D.S.; Tamm, L.K. Membrane structure and fusion-triggering conformational change of the fusion domain from influenza hemagglutinin. *Nat. Struct. Biol.* **2001**, *8*, 715–720. [CrossRef]
11. Lorieau, J.L.; Louis, J.M.; Bax, A. The complete influenza hemagglutinin fusion domain adopts a tight helical hairpin arrangement at the lipid:water interface. *Proc. Natl. Acad. Sci. USA* **2010**, *107*, 11341–11346. [CrossRef]
12. Lorieau, J.L.; Louis, J.M.; Bax, A. The impact of influenza hemagglutinin fusion peptide length and viral subtype on its structure and dynamics. *Biopolymers* **2013**, *99*, 189–195. [CrossRef]
13. Tamm, L.K.; Han, X.; Li, Y.; Lai, A.L. Structure and function of membrane fusion peptides. *Biopolym. Pept. Sci. Sect.* **2002**, *66*, 249–260. [CrossRef] [PubMed]
14. Lai, A.L.; Park, H.; White, J.M.; Tamm, L.K. Fusion Peptide of Influenza Hemagglutinin Requires a Fixed Angle Boomerang Structure for Activity. *J. Biol. Chem.* **2006**, *281*, 5760–5770. [CrossRef]
15. Sharpe, H.J.; Stevens, T.J.; Munro, S. A Comprehensive Comparison of Transmembrane Domains Reveals Organelle-Specific Properties. *Cell* **2010**, *142*, 158–169. [CrossRef]
16. Smrt, S.T.; Draney, A.W.; Lorieau, J.L. The influenza hemagglutinin fusion domain is an amphipathic helical hairpin that functions by inducing membrane curvature. *J. Biol. Chem.* **2015**, *290*, 228–238. [CrossRef]
17. Gray, C.; Tatulian, S.A.; Wharton, S.A.; Tamm, L.K. Effect of the N-terminal glycine on the secondary structure, orientation, and interaction of the influenza hemagglutinin fusion peptide with lipid bilayers. *Biophys. J.* **1996**, *70*, 2275–2286. [CrossRef]
18. Han, X.; Steinhauer, D.A.; Wharton, S.A.; Tamm, L.K. Interaction of mutant influenza virus hemagglutinin fusion peptides with lipid bilayers: Probing the role of hydrophobic residue size in the central region of the fusion peptide. *Biochemistry* **1999**, *38*, 15052–15059. [CrossRef]
19. Victor, B.L.; Lousa, D.; Antunes, J.M.; Soares, C.M. Self-Assembly Molecular Dynamics Simulations Shed Light into the Interaction of the Influenza Fusion Peptide with a Membrane Bilayer. *J. Chem. Inf. Model.* **2015**, *55*, 795–805. [CrossRef]
20. Worch, R.; Krupa, J.; Filipek, A.; Szymaniec, A.; Setny, P. Three conserved C-terminal residues of influenza fusion peptide alter its behavior at the membrane interface. *Biochim. Biophys. Acta Gen. Subj.* **2017**, *1861*, 97–105. [CrossRef]
21. Worch, R.; Dudek, A.; Krupa, J.; Szymaniec, A.; Setny, P. Charged n-terminus of influenza fusion peptide facilitates membrane fusion. *Int. J. Mol. Sci.* **2018**, *19*, 578. [CrossRef] [PubMed]
22. Lousa, D.; Pinto, A.R.; Campos, S.R.; Baptista, A.M.; Veiga, A.S.; Castanho, M.A.; Soares, C.M. Effect of pH on the influenza fusion peptide properties unveiled by constant-pH molecular dynamics simulations combined with experiment. *Sci. Rep.* **2020**, *10*, 1–18. [CrossRef]

23. Larsson, P.; Kasson, P.M. Lipid Tail Protrusion in Simulations Predicts Fusogenic Activity of Influenza Fusion Peptide Mutants and Conformational Models. *PLoS Comput. Biol.* **2013**, *9*, e1002950. [CrossRef] [PubMed]
24. Légaré, S.; Lagüe, P. The influenza fusion peptide promotes lipid polar head intrusion through hydrogen bonding with phosphates and N-terminal membrane insertion depth. *Proteins* **2014**, *82*, 2118–2127. [CrossRef]
25. Huang, Q.; Chen, C.L.; Herrmann, A. Bilayer conformation of fusion peptide of influenza virus hemagglutinin: A molecular dynamics simulation study. *Biophys. J.* **2004**, *87*, 14–22. [CrossRef]
26. Lagüe, P.; Roux, B.; Pastor, R.W. Molecular dynamics simulations of the influenza hemagglutinin fusion peptide in micelles and bilayers: Conformational analysis of peptide and lipids. *J. Mol. Biol.* **2005**, *354*, 1129–1141. [CrossRef]
27. Fuhrmans, M.; Marrink, S.J. Molecular View of the Role of Fusion Peptides in Promoting Positive Membrane Curvature. *J. Am. Chem. Soc.* **2012**, *134*, 1543–1552. [CrossRef]
28. Chakraborty, H.; Lentz, B.R.; Kombrabail, M.; Krishnamoorthy, G.; Chattopadhyay, A. Depth-Dependent Membrane Ordering by Hemagglutinin Fusion Peptide Promotes Fusion. *J. Phys. Chem. B* **2017**, *121*, 1640–1648. [CrossRef]
29. Risselada, H.J.; Marelli, G.; Fuhrmans, M.; Smirnova, Y.G.; Grubmüller, H.; Marrink, S.j.; Müller, M. Line-tension controlled mechanism for influenza fusion. *PLoS ONE* **2012**, *7*, 16–20. [CrossRef]
30. Tahir, M.A.; Van Lehn, R.C.; Choi, S.H.; Alexander-Katz, A. Solvent-exposed lipid tail protrusions depend on lipid membrane composition and curvature. *Biochim. Biophys. Acta Biomembr.* **2016**, *1858*, 1207–1215. [CrossRef]
31. Pabis, A.; Rawle, R.J.; Kasson, P.M. Influenza hemagglutinin drives viral entry via two sequential intramembrane mechanisms. *Proc. Natl. Acad. Sci. USA* **2020**, *117*, 7200–7207. [CrossRef] [PubMed]
32. Jorgensen, W.L.; Chandrasekhar, J.; Madura, J.D.; Impey, R.W.; Klein, M.L. Comparison of simple potential functions for simulating liquid water. *J. Chem. Phys.* **1983**, *79*, 926–935. [CrossRef]
33. Lindorff-Larsen, K.; Piana, S.; Palmo, K.; Maragakis, P.; Klepeis, J.L.; Dror, R.O.; Shaw, D.E. Improved side-chain torsion potentials for the Amber ff99SB protein force field. *Proteins Struct. Funct. Bioinform.* **2010**, *78*, 1950–1958. [CrossRef] [PubMed]
34. Dickson, C.J.; Madej, B.D.; Skjevik, Å.A.; Betz, R.M.; Teigen, K.; Gould, I.R.; Walker, R.C. Lipid14: The amber lipid force field. *J. Chem. Theory Comput.* **2014**, *10*, 865–879. [CrossRef]
35. Huang, J.; MacKerell, A.D. CHARMM36 all-atom additive protein force field: Validation based on comparison to NMR data. *J. Comput. Chem.* **2013**, *34*, 2135–2145. [CrossRef]
36. Abraham, M.J.; Murtola, T.; Schulz, R.; Páall, S.; Smith, J.C.; Hess, B.; Lindahl, E. Gromacs: High performance molecular simulations through multi-level parallelism from laptops to supercomputers. *SoftwareX* **2015**, *1–2*, 19–25. [CrossRef]
37. Hess, B.; Bekker, H.; Berendsen, H.J.C.; Fraaije, J.G.E.M. LINCS: A linear constraint solver for molecular simulations. *J. Comput. Chem.* **1997**, *18*, 1463–1472. [CrossRef]
38. Essmann, U.; Perera, L.; Berkowitz, M.L.; Darden, T.; Lee, H.; Pedersen, L.G. A smooth particle mesh Ewald method. *J. Chem. Phys.* **1995**, *103*, 8577–8593. [CrossRef]
39. Parrinello, M.; Rahman, A. Polymorphic transitions in single crystals: A new molecular dynamics method. *J. Appl. Phys.* **1981**, *52*, 7182–7190. [CrossRef]
40. Sugita, Y.; Okamoto, Y. Replica-exchange molecular dynamics method for protein folding. *Chem. Phys. Lett.* **1999**, *314*, 141–151. [CrossRef]
41. Van der Spoel, D.; Seibert, M.M. Protein Folding Kinetics and Thermodynamics from Atomistic Simulations. *Phys. Rev. Lett.* **2006**, *96*, 238102–238104. [CrossRef] [PubMed]
42. Souaille, M.; Roux, B. Extension to the weighted histogram analysis method: Combining umbrella sampling with free energy calculations. *Comput. Phys. Commun.* **2001**, *135*, 40–57. [CrossRef]
43. Zwanzig, R.W. High-Temperature Equation of State by a Perturbation Method. I. Nonpolar Gases. *J. Chem. Phys.* **1954**, *22*, 1420–1426. [CrossRef]
44. Bussi, G.; Tribello, G.A. Analyzing and Biasing Simulations with PLUMED. *Methods Mol. Biol.* **2019**, *2022*, 529–578. [CrossRef]
45. Ladokhin, A.S. Distribution analysis of depth-dependent fluorescence quenching in membranes: A practical guide. *Methods Enzymol.* **1997**, *278*, 462–473. [CrossRef]
46. Marrink, S.J.; Berendsen, H.J.C. Simulation of water transport through a lipid membrane. *J. Phys. Chem.* **1994**, *98*, 4155–4168. [CrossRef]
47. Mathai, J.C.; Tristram-Nagle, S.; Nagle, J.F.; Zeidel, M.L. Structural Determinants of Water Permeability through the Lipid Membrane. *J. Gen. Physiol.* **2008**, *131*, 69–76. [CrossRef]
48. Ladokhin, A.S.; Jayasinghe, S.; White, S.H. How to Measure and Analyze Tryptophan Fluorescence in Membranes Properly, and Why Bother? *Anal. Biochem.* **2000**, *285*, 235–245. [CrossRef]
49. Lorieau, J.L.; Louis, J.M.; Schwieters, C.D.; Bax, A. pH-triggered, activated-state conformations of the influenza hemagglutinin fusion peptide revealed by NMR. *Proc. Natl. Acad. Sci. USA* **2012**, *109*, 19994–19999. [CrossRef]
50. Ghosh, U.; Xie, L.; Jia, L.; Liang, S.; Weliky, D.P. Closed and Semiclosed Interhelical Structures in Membrane vs Closed and Open Structures in Detergent for the Influenza Virus Hemagglutinin Fusion Peptide and Correlation of Hydrophobic Surface Area with Fusion Catalysis. *J. Am. Chem. Soc.* **2015**, *137*, 7548–7551. [CrossRef]
51. Tamm, L.K.; Tatulian, S.A. Infrared spectroscopy of proteins and peptides in lipid bilayers. *Q. Rev. Biophys.* **1997**, *30*, 365–429. [CrossRef] [PubMed]

52. Wu, C.W.; Cheng, S.F.; Huang, W.N.; Trivedi, V.D.; Veeramuthu, B.; Kantchev, A.B.; Wu, W.G.; Chang, D.K. Effects of alterations of the amino-terminal glycine of influenza hemagglutinin fusion peptide on its structure, organization and membrane interactions. *Biochim. Biophys. Acta Biomembr.* **2003**, *1612*, 41–51. [CrossRef]
53. Han, X.; Tamm, L.K. A host-guest system to study structure-function relationships of membrane fusion peptides. *Proc. Natl. Acad. Sci. USA* **2000**, *97*, 13097–13102. [CrossRef] [PubMed]
54. Ladokhin, A.S.; Holloway, P.W. Fluorescence of membrane-bound tryptophan octyl ester: A model for studying intrinsic fluorescence of protein-membrane interactions. *Biophys. J.* **1995**, *69*, 506–517. [CrossRef]

International Journal of
Molecular Sciences

MDPI

Review

The FtsHi Enzymes of *Arabidopsis thaliana*: Pseudo-Proteases with an Important Function

Laxmi S. Mishra and Christiane Funk *

Department of Chemistry, Umeå University, SE-901 87 Umeå, Sweden; laxmi.mishra@umu.se
* Correspondence: christiane.funk@umu.se; Tel.: +46-(0)-907-867-633

Abstract: FtsH metalloproteases found in eubacteria, animals, and plants are well-known for their vital role in the maintenance and proteolysis of membrane proteins. Their location is restricted to organelles of endosymbiotic origin, the chloroplasts, and mitochondria. In the model organism *Arabidopsis thaliana*, there are 17 membrane-bound FtsH proteases containing an AAA^+ (ATPase associated with various cellular activities) and a Zn^{2+} metalloprotease domain. However, in five of those, the zinc-binding motif HEXXH is either mutated (FtsHi1, 2, 4, 5) or completely missing (FtsHi3), rendering these enzymes presumably inactive in proteolysis. Still, homozygous null mutants of the pseudo-proteases FtsHi1, 2, 4, 5 are embryo-lethal. Homozygous *ftshi3* or a weak point mutant in *FTSHi1* are affected in overall plant growth and development. This review will focus on the findings concerning the FtsHi pseudo-proteases and their involvement in protein import, leading to consequences in embryogenesis, seed growth, chloroplast, and leaf development and oxidative stress management.

Keywords: AAA-type protease; *Arabidopsis thaliana*; FtsH metalloprotease; chloroplast; embryo lethal; leaf variegation; plastid biogenesis; protein import; oxidative stress

Citation: Mishra, L.S.; Funk, C. The FtsHi Enzymes of *Arabidopsis thaliana*: Pseudo-Proteases with an Important Function. *Int. J. Mol. Sci.* **2021**, *22*, 5917. https://doi.org/10.3390/ijms22115917

Academic Editor: Masoud Jelokhani-Niaraki

Received: 5 May 2021
Accepted: 29 May 2021
Published: 31 May 2021

Publisher's Note: MDPI stays neutral with regard to jurisdictional claims in published maps and institutional affiliations.

1. Introduction

Cells have evolved an extensive system of molecular chaperones, folding catalysts, and proteases that control protein quality and prevent damage. In addition to the well-studied degradative removal of damaged or superfluous proteins [1], proteolysis is highly important in regulating protein preprocessing, maturation, post-translational protein modifications, and signaling [2,3]. Therefore, it is is no overstatement that proteolysis is directly or indirectly involved in most cellular processes [4].

1.1. Proteases in the Plant Chloroplast

Plant proteases associated with a particular proteolytic activity are present in various cellular compartments and organelles constituting up to 3% of the plant proteome [5,6]. Proteases are classified according to their catalytic types. Except for glutamic acid proteases, representatives from all protease classes (threonine, cysteine, serine, aspartic, metalloproteases) have been detected in the plant *Arabidopsis thaliana* [7]. The chloroplast is a unique organelle of the plant cell; absorption and conversion of light energy in the photosynthetic reaction lead to a permanent need for protein turnover (processing and degradation) to adapt to different light conditions. Excess light adsorption further can cause the formation of reactive oxygen species and damage proteins. Therefore, protein quality and quantity controls are essential [8,9]. In addition to photosynthesis, several metabolic reactions happen in the chloroplast, including the biosynthesis of lipids, amino acids, chlorophylls, and carotenoids; therefore, plastidic proteases are vital regulators [9]. More than 20 different families of chloroplast proteases have been detected, with members localized in specific sub-organellar compartments [3,9].

1.2. Plant Pseudo-Proteases

In addition to the proteolytically active proteases, members with mutations in their active site attracted the attention of researchers. Despite their putative proteolytically inactivity, many of these pseudo-proteases have essential roles in the cell, ranging from structural proteins via chaperones [10,11] to enzymes with a new function [12–20]. In the chloroplast, pseudo-proteases are found in the families of serine Clp-proteases (Cl-pRs, [21–23]) and FtsH metalloproteases (FtsHis [18,24–26]). ClpRs of the Clp protease family lack their catalytic triad. These proteolytically inactive subunits are of structural importance to form a tetradecameric proteolytic core together with the catalytically active ClpP subunits [21–23]. Most ClpR proteins are essential for the proteolytic function of the Clp core function.

This review will focus on pseudo-proteases belonging to the family of the membrane-bound ATP-dependent FtsH metalloproteases, which are termed FtsHi (*i*-inactive) [23]. These presumably proteolytically inactive FtsHi enzymes are restricted to the plant chloroplast and, similar to the ClpR, affect chloroplast and overall plant development [17,23–25].

2. Filamentation Temperature-Sensitive Protein H (FtsH) Protease Family

The name FtsH (filamentation temperature-sensitive) erroneously originated from the growth behavior of a Y16 *Escherichia coli* strain deficient in its *ftsh* gene. However, later, a second, independent mutation was found to be responsible for the observed temperature-sensitive phenotype [23].

FtsH proteases, known as zincins, belong to the MEROPS peptidase family M41, which in turn belongs to a larger family of zinc metalloproteases [27]. Within the M41 peptidase domain, the Zn^{2+} ion is ligated by two histidine residues, forming the HEXXH motif (where X is any uncharged residue) as well as a glutamic acid residue [28]. Functional homo- or hetero-hexameric complexes are inserted into the membrane by one or more N-terminal transmembrane domains per subunit [29–32]. The highly conserved AAA^+ domain, a cassette of about 200–250 residues that contains the ATP-binding motif (Walker A and Walker B) and the second region of homology (SRH), is situated between the trans-membrane region and the active site (Figure 1). Unlike other well-studied ATP-dependent proteases, FtsHs lack robust unfoldase activity [33]. Instead, ATP hydrolysis by FtsH is used to translocate unfolded substrates sequentially into the hexameric pore [33,34]. The AAA^+ domain is required for nucleotide binding and hydrolysis [29,35] and responsible for alternating between a closed and open state of the FGV pore motif, which is a con-served hydrophobic area at the proteolytic chamber [29]. The substrate is pulled into the degradation chamber via a narrow pore [36,37]. Recent cryo-electron microscopy studies enabled the study of substrate processing of AAA^+ proteins in detail (reviewed in [38]) and revealed a conserved spiraling organization of ATPase hexamers around the translocating protein substrate.

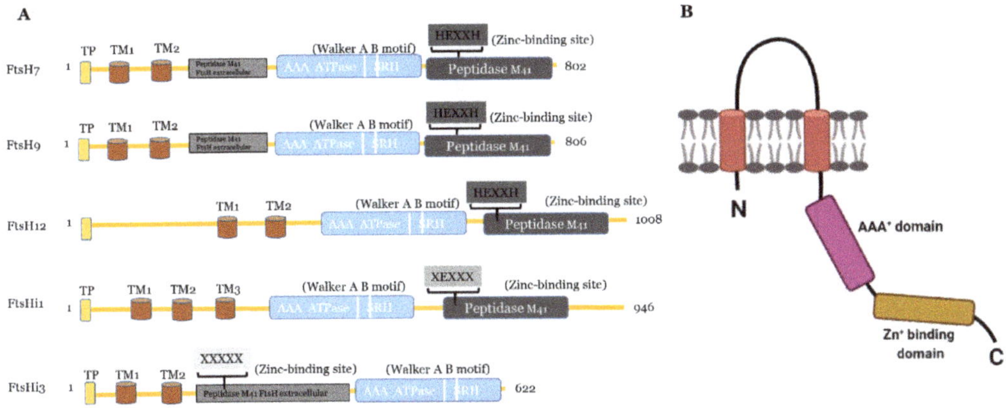

Figure 1. (**A**) Predicted domains and motifs of AtFtsHis in comparison to the presumably active AtFtsH7, 9, and 12. TP, transit peptide; transmembrane domains, TM1-3; Walker A B motifs are indicated as two white lines between the AAA$^+$ ATPase and SRH; SRH, second region of homology. Active FtsH proteases contain the Zn^{2+}-binding motif (HEXXH) in the their peptidase M41 domain, which is substituted or absent in presumably inactive FtsHis (substitution of both histidines indicated as XEXXX). In FtsHi3, the peptidase M41 domain is annotated as "FtsH extracellular" and is located at the N-terminal to the ATPase; its HEXXH motif is completely missing (XXXXX). FtsH7 and FtsH9 contain an FtsH extracellular" peptidase domain additionally to their protease domain. AtFtsHis are predicted to contain three (FtsHi1, 5), two (FtsHi2, 3), or one (FtsHi4) transmembrane domains (http://www.cbs.dtu.dk/services/TMHMM/, accessed on 3 April 2021) (Supplementary Figure S1 [26]). (**B**) Schematic drawing of the structure of a monomeric FtsH protease with two membrane-spanning regions (shown in red), the proteolytic domain (in range) and the AAA$^+$-domain (in pink). Created with BioRender.com (accessed on 27 May 2021).

2.1. The FtsH Protease Family of *Arabidopsis thaliana*

The annual plant *Arabidopsis thaliana* contains 17 different FtsH proteases. Gene comparison studies showed that of the 12 *FTSH* genes potentially coding for fully functional proteases, ten are found in highly homologous pairs. While the pairs AtFtsH1/5, AtFtsH2/8, and AtFtsH7/9 are targeted to the chloroplast, AtFtsH3/10 and AtFtsH4 have been identified in mitochondria. AtFtsH11, the pair partner of AtFtsH4, was initially reported to be dual targeted to mitochondria and the chloroplast [39]. However, Wagner and coworkers confirmed its location to be only in chloroplasts [40]. AtFtsH3 and AtFtsH10 were shown not to be crucial for growth under optimal conditions [41]. Loss of AtFtsH4 leads to oxidative stress and the accumulation of oxidized proteins [42,43]. FtsH10 is involved in the assembly and/or stability of complexes I and V of the mitochondrial oxidative phosphorylation system [43].

2.1.1. FtsH Proteases Located in the Thylakoid Membrane

Of the plastidic FtsHs, FtsH1, 2, 5, and 8 are localized in the thylakoid membrane. These members are the most abundant FtsH proteases and extensively studied [44]; they form hetero-hexameric complexes, in which FtsH1 and FtsH5 (Type A) and FtsH2 and FtsH8 (Type B) can partially substitute for each other [45]. A threshold in the amount of type A and B subunits was postulated to determine the proper function and development of chloroplasts and thylakoid membrane [46–49]. This thylakoid located protease complex plays a vital role in the degradation and assembly of the Photosystem II reaction center protein D1 and other transmembrane subunits of the photosynthetic machinery. Mutants lacking FtsH2/VAR2 or FtsH5/VAR1 show strongly or slightly variegated leaves, respectively [50,51]. Functional loss of, e.g., FtsH2 results in upregulation of other FtsH proteins in the green leaf sectors to maintain proper function and development of the chloroplasts [47,50,52]. FtsH6 is also localized in the thylakoid membrane. It is essential for thermotolerance and thermomemory in seedlings [53], while no phenotype was observed in adult plants when grown in semi-natural outdoor conditions [54].

2.1.2. FtsH Proteases Located in the Chloroplast Envelope

The other plastidic FtsH enzymes are believed to be localized in the chloroplast envelope [19]. Deleting FtsH7 and 9 does not result in any obvious phenotype [53], and the proteases are not required for PSII repair [55]. FtsH11 is crucial for growth in long photoperiods [40] and thermotolerance [56,57]. FtsH12 was co-immuno-precipitated in a complex with FtsHi1, 2, 4, 5 and NAD-dependent malate dehydrogenase (MDH) and shown to be involved in protein import [58,59]. In addition to FtsHi1, 2, 4, and 5, even FtsHi3 belongs to the five plastidic FtsH homologues, which are incapable of proteolysis in *Arabidopsis*. The FtsHi enzymes either have a mutation in their HEXXH motif (FtsHi1, 2, 4, and 5), or the entire motif is missing (FtsHi3) [18,26]. Compared to AtFtsHi1, 2, 4, and 5, FtsHi3 contains a very short C-terminal domain. Interestingly, FtsHi3 also has undergone a domain swap: the whole M41 domain is located at the N-terminal instead of at the C-terminal to the AAA^+ domain [26]. Comparing the domain organization of AtFtsHi3 with AtFtsHi1, AtFtsH7, 9, 12 (Figure 1), also AtFtsH7/9 contain this "peptidase M41 FtsH extracellular" domain N-terminal to the AAA^+ domain, which is additional to their protease domain located in the C-terminal to the AAA^+ domain. This additional domain is not present in other FtsHs or FtsHis. Whether the N-terminal "peptidase M41 FtsH extracellular" domain of FtsH7, FtsH9, and FtsHi3 enables these enzymes to form a common complex with a specific function remains to be shown. Three independent pre-protein translocating models (pSSU-TEV-protein A, pL11Flag-TEV-Protein A, pLHCP-TEV-protein A) suggested FtsHi3 to form a 1-MD complex separate from the FtsH12/FtsHi1,2,4,5/MDH complex [58] and different to the 1-MD TIC complex [60]. The identity of other components in this complex is unknown. *FTSHi3* is not co-expressed with the tight cluster of *FTSH12/FTSHi1, 2, 4, 5*, but instead with a gene encoding OTP51 [26,58,59]. This pentatricopeptide repeat protein is required for the splicing of group IIa introns and impacts photosystem I and II assembly [61]. The tight co-expression with *FTSHi3* indicates a common function of OTP51 and FtsHi3; therefore, OTP51 is another hypothetical complex partner.

3. Pseudo-Proteases with an Important Enzymatic Activity

In addition to being pseudo-proteases, the AAA^+ domain of FtsHis is intact and—based on the seed lethality of many FtsHi mutants—highly important for their activity, plastid, and overall plant development. Four out of the five AtFtsHi members have been demonstrated to form an inner envelope-bound heteromeric AAA^+ (ATPase associated with diverse cellular activities) ATPase complex. This complex consisting of FtsH12/FtsHi1,2,4,5/pdNAD-MDH was found to be involved in ATP-driven protein import across the chloroplast envelope [58,59,62,63]. Even Ycf2 was observed being part of this 2 MDa complex using transgenic lines overexpressing *FTSH12* [58]. Still, the protein could not be detected in complexes isolated from wild-type and tic56-3 plastids using a combination of native gel electrophoresis and protein quantification [64] and neither after pull-down of FtsH12 using its native promoter [65]. Kikuchi and coworkers [57] determined the super-complex Ycf2-FtsH12-FtsHi1,2,4,5-pdNAD-MDH physically to interact with TIC components such as Tic214 (Ycf1), Tic100, and Tic56 as well as with the pre-protein translocation components Toc75 and Toc159 [59]. In this complex, neither pdNAD-MDH activity [59] nor FtsH12 proteolytic activity [57] are required; an FtsH12 (H769Y) mutant developed normal chloroplasts with functional pre-protein import abilities.

The finding of ATP-driven protein import across the chloroplast membrane by an FtsH12/FtsHi1,2,4,5/pdNAD-MDH complex has steered an intense debate [66,67] on the importance of the long-accepted chloroplast protein import machinery, i.e., Tic110 and Tic40 forming a common translocon in the inner chloroplast membrane and recruiting the stromal chaperones Hsp93/ClpC1, cpHsp70, and Hsp90C [67]. If indeed the FtsH12/FtsHi complex plays the main role in protein import into the chloroplast [67] remains to be shown. The absence of this complex in most monocots (see Section 4) as well as its lower impact in adult plants (see Section 5) rather point to a specific function during chloroplast development in dicotyledons. We refer the reader to [66,67] and references therein for a detailed description of the subunits involved in protein import.

4. Evolution of FtsHi Pseudo-Proteases

Orthologues of the *Escherichia coli* FtsH protease exist in eubacteria and eukaryotic organelles of prokaryotic origin (chloroplasts and mitochondria) but not in Archae. While bacterial genomes contain only one *ftsh* gene, multiple *FTSH* genes are present in photosynthetic organisms, ranging from cyanobacteria to higher plants (reviewed in [18,68]). Even FtsHi enzymes are found already in prokaryotes: the chlorophyll b containing cyanobacterium *Prochlorococcus marinus SS120*, which has a highly reduced genome, contains one gene encoding a presumably inactive version [26]. The evolutionary origin of *ftshi* genes is not resolved yet [15,26,55,58]. Kikuchi and coworkers [58] proposed a progenitor of the chloroplast-encoded Ycf2 to be the predecessor of FtsHi members. Ycf2 has a putative AAA$^+$ ATP-binding domain [69] and is essential for cell survival [70]. FtsH1 and FtsH2 were found to associate with Ycf2 in cyanobacteria (*Synechocystis* sp. *PCC 6803*) and Rhodophyta (*Cyanidioschyzon* and *Pyropia*) [58]. Alternatively, bacteria belonging to the Firmicutes were found to have paralogs to the FtsH/FtsHi enzymes acquired via horizontal gene transfer [55].

Based on the finding of an FtsH12/FtsHi1,2,4,5/MDH import complex, we hypothesize a common evolution of multiple FtsHis together with FtsH12 [26]. The protein sequence of AtFtsH12 was blasted against the translated genomes of various species of cyanobacteria, green algae, and higher plants; the BLAST scores, percentage of identity to AtFtsH12, as well as the number of FtsH and FtsHi proteins identified in each species are shown in Figure 2. In *Spirodela* and eudicots, FtsH12 orthologues were identified with an identity of more than 70%. This phylogenetic group also shows the highest copy number of presumably proteolytically inactive FtsH enzymes. The unusually high number of FtsH/FtsHi orthologues found in *Glycine max* is most likely caused by the two duplication events of its genome and several chromosome rearrangements resulting in a palaeopolyploid genome with up to 75% of the genes present in multiple copies [71].

In eudicots, an envelope-located FtsH12–FtsHi complex seems to be necessary for viability. However, two phylogenetic groups, gymnosperms and grasses, have replaced the function of an FtsH12–FtsHi complex. Apart from keeping a high copy number of FtsHs for maintenance of the photosystems and the respiratory chain, these groups contain neither genes with high sequence similarity to FtsH12 nor multiple copies of FtsHi (Figure 2) [62]. In grasses, the FtsH12/FtsHi complex most likely was replaced by an energetically more efficient protein import system that involves Hsp70-type molecular chaperones [66] similar to the mitochondrial protein import system [63].

In early evolutionary plants and cyanobacteria, BLAST scores and percent of amino acid identities to FtsH12 are low (Figure 2). In addition, FtsHi proteins in these organisms are either absent or present in low amounts. Three of the nine FtsH enzymes found in the green alga *Chlamydomonas reinhardtii* are FtsHi pseudo-enzymes [15]. Green algae might have already utilized multiple FtsHs for maintenance and/or import through their inner envelope (Figure 2).

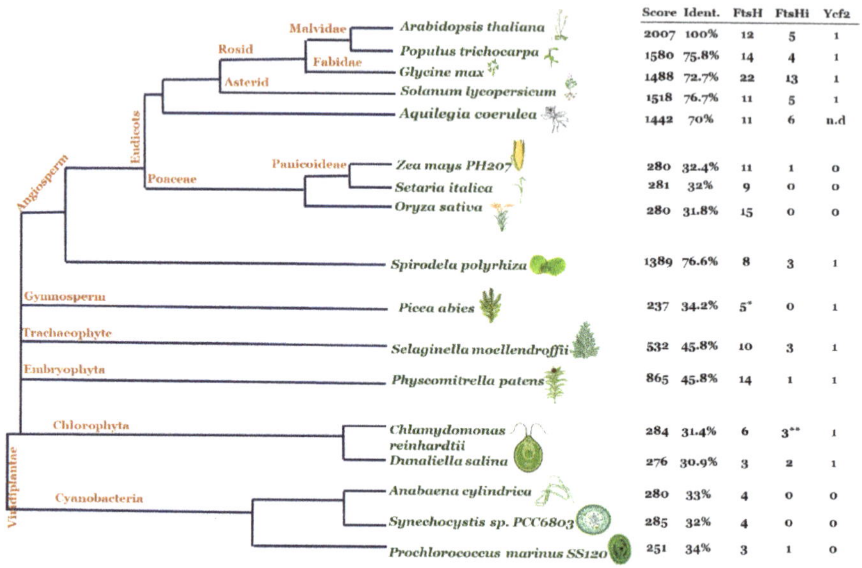

Figure 2. Co-evolution of FtsH12 and FtsHi. The AtFtsH12 amino acid sequence was blasted via phytozome (https://phytozome.jgi.doe.gov/, accessed on 16 August 2018). Hits with the highest score and identity to Arabidopsis FtsH12 are shown at the right hand of the species tree. Furthermore, the number of proteolytic active FtsH and inactive FtsHi in these species were determined by manually investigating the presence of an AAA-like domain and peptidase M41 domain. Sequences containing a zinc-binding motif (HEXXH) within M41 were assumed to correspond to active proteolytic proteases. Data for *Picea abies* were additionally retrieved from congenie.org (marked by *). Data for *Prochlorococcus*, *Synechocystis*, and *Anabaena* were retrieved from https://www.ncbi.nlm.nih.gov/ (accessed on 16 August 2018). Data for *Chlamydomonas reinhardtii* and *Oryza sativa* were supplemented with literature data ([15,68] marked by **). The search for the plastid-encoded Ycf2 was performed in the NCBI organelle genome resources, the chloroplast genome of *Aquilegia coerulea* is not available yet (marked as n.d.). Adapted from [26], created with BioRender.com.

5. Phenotypic Consequences

While chloroplast import is a fundamental process throughout the plant´s lifespan, it is interesting to note that reduced levels of FtsHis have a more substantial impact on seedlings than on adult plants. *FTSHi* knock-down mutants display a leaf variegated or pale seedling phenotype, while adult plants look similar to WT [26,72–74]. Therefore, import via the FtsH12/FtsHi/MDH complex might be highly important during chloroplast development. These phenotypic consequences further point toward a dose-dependent threshold of single subunits in the FtsH12/FtsHi complex, as observed for FtsHi5 [75]. Studies on the strong and weak *ftshi* mutants may provide insights into the regulatory processes and possible thresholds accountable for a 'gradient' of compromised and normal chloroplasts during leaf and chloroplast development.

5.1. FtsHi1

Arabidopsis ARC1 (accumulation and replication of chloroplasts 1) was isolated by map-based cloning and was found to encode FtsHi1 (At4g23940) [72]. Homozygous *ftshi1-2* knock out mutants are embryo lethal, while the missense mutant *ftshi1-1/arc1* displays a pale-seedling phenotype [72]. Albino seeds in developing siliques of *ftshi1-1/arc1* showed arrest at three different stages of embryo development, namely late globular, early heart, and late heart stage [72]. Seeds harvested from field-grown *ftshi1-1/arc* plants displayed a significant delay in germination compared to their respective WT [26]. The *ftshi1-1/arc1* mutant shows an increased number of chloroplasts, and the plants have smaller rosette sizes throughout their life span [72]. The chloroplast ultrastructure of *ftshi1-1/arc1* showed wavy, swollen, and less organized thylakoids with starch grains accumulating, indicating the chloroplasts still being metabolic active. While chloroplasts of the *ftshi1-1/arc1* mutant contain assembled Ycf2/FtsH12/FtsHi complexes, these plants are impaired in in vitro protein import, which is most likely caused by miss-folding of the AAA-ATPase domain [58]. Mutated FtsHi1 or FtsH12 might partially substitute FtsHi1 in the complex. Gene expression of *FTSH12* and the other *FTSHis* was significantly lower in the *ftshi1-1/arc1* mutant than in WT [26]. Vice versa, the expression of *FTSHi1* was down-regulated in single *FTSHi* mutants [26]. The overexpression of *FTSHi1 (35S: FTSHi1–YFP)* generated few recovered transgenic plants, which were mildly variegated in appearance [72]. These overexpression plants further showed a substantial increase in chloroplast size, but fewer chloroplasts than WT. *FTSHi1* transcript levels were similar in the white and green sectors of the variegated leaves, but in variegated tissue, the level of FtsHi1–YFP was low, while green tissue accumulated FtsHi1–YFP similar to WT [72].

The FtsHi enzymes of the chloroplast envelope have been suggested to respond to photo-oxidative stress [72]. During de-etiolation in white light at intensities of 15, 100, or 300 μmol m^{-2} s^{-1}, *ftshi1-1/arc1* mutant seedlings accumulated only half of the chlorophyll amount compared to WT controls. Low light (1 μmol m^{-2} s^{-1}) irradiance with 40% blue or 60% red (BR) light led to a significant increase in the mutant's chlorophyll accumulation. The authors concluded that *ftshi1-1/arc1* at low irradiance still can attenuate chloroplast biogenesis without causing photo-oxidative stress. At normal growth conditions, *ftshi1-1/arc1* displayed lower non-photochemical quenching (NPQ) values than WT. However, during exposure to various stresses (continuous light, long day/high light, short day/4 °C, and long day/30 °C), the mutant sustained similar to WT [26], protection mechanisms might have been triggered [76,77]

5.2. FtsHi2

During embryogenesis in plants, a fertilized ovule develops into a plant embryo [78]. Large-scale screens for mutants with altered gametophyte development [75] displayed EMBRYO-DEFECTIVE (EMB) genes that are essential for the growth and overall development of *Arabidopsis* [75,79]. *FTSHi2*, along with *FTSHi4* and *5*, are listed at http://seedgenes.org/index.html (accessed on 16 August 2020) for being EMBRYO-DEFECTIVE (EMB) genes [75,79]. *emb2083-1*, *emb2083-2*, and *emb2083-4* (or *ftshi2-1*, *ftshi2-2*, and *ftshi2-3*, respectively) were investigated by Lu and coworkers [73], and 25% of the ovules in those heterozygous lines were found to be albinos with an embryo-lethal phenotype. Heterozygous *ftshi2* mutants exhibited no evident defects before the globular stage, but then, ≈80% ($n \geq 60$) were arrested at the globular stage, and the remaining 20% reached the heart-shaped stage with an abnormal division pattern [73]. Consistent with the finding of the FtsH12/FtsHi complex [58,59], FtsHi2 and FtsHi4 were found to interact with each other in vitro [73] and in silico co-expression and qPCR analysis [26]. *FTSH12* and all *FTSHis* co-express with genes encoding enzymes involved in plastid translation, division, and positioning and—with the exception of *FTSHi3*—involved in amino acid metabolism [80].

5.3. FtsHi3

FtsHi3 is not associated with the FtsH12/FtsHi1,2,4,5 import complex but instead seems to form a 1-MD complex with unknown partners [58,64]. Further investigations are required to identify its role in the chloroplast envelope. *ftshi3*-KO plants showed residual albino growth in young leaves [58]. The homozygous *ftshi3-2* mutant displays a pale-seedling phenotype when grown for eight days on agar plates, indicating delayed chloroplast and thylakoid membrane development. Six-week-old *ftshi3-2* mutant plants are pale compared to WT when grown in cold stress under short-day conditions [26]. *ftshi3-2* plants displayed reduced Darwinian fitness in comparison to WT [26]. After stress exposure for some days, *ftshi3-2* mutants showed a significant drop in NPQ values than WT but later recovered. After exposure to stress for 6 weeks, the mutant displayed higher NPQ values than the control. Therefore, loss of FtsHi3 might enhance tolerance to photo-oxidative stress and photo-protection [81]. Another homozygous *AtFTSHi3* knock-down mutant *(ftshi3-1(kd))* displayed a significant delay in seed germination compared to WT (Mishra and Funk, unpublished results). This phenotype was attributed to over-accumulation of ABA, while *ftshi3-1 (kd)* seedlings showed partial sensitivity to exogenous ABA. *ftshi3-1 (kd)* plants were drought-tolerant up to 20 days after the irrigation was stopped, whereas wild-type plants wilted after 12 days. Although *ftshi3-1(kd)* displayed a drought-tolerant phenotype in aboveground tissue, its root-associated bacterial community responded to drought (Mishra and Funk, unpublished results).

5.4. *FtsHi4*

In addition to its suborganellar location in the chloroplast envelope [19,58,59], FtsHi4 was also identified as a thylakoid membrane-associated protein [73]. If FtsHi4 indeed is dual-targeted to the envelope and the thylakoid membrane or if this result is due to antibody cross-reaction (as shown for FtsH11 [40]) or impurity of the preparation (in the study, no envelope marker protein was used to examine the purity of the thylakoid fraction [73]) remains to be shown. Ubiquitous transcript levels of *FTSHi4* were detected in all organs of 40-day-old wild-type plants. The lowest *FTSHi4* transcripts were present in roots; transcripts were most abundant in young leaves [73]. *A. thaliana* mutants depleted of FtsHi4 display embryo lethality and disrupted thylakoid formation. Heterozygous *ftshi4/FTSHi4* plants exhibit abnormal division pattern within the same silique, with 80% wild-type embryos reaching maturity and 20% arresting at the heart-shaped stage. Then, albino and green seeds are distributed in developing siliques [73]. These results imply that even FtsHi4 can be substituted in the FtsH12/FtsHi complex. Gene expression of *FTSHi2* and *FTSHi3* was enhanced in homozygous *ftshi4-2* mutants compared to WT [26], while on the protein level, the amount of FtsH12 was slightly diminished in the mutant [65].

Significantly lower numbers of seeds per siliques were observed in *ftshi4/FTSHi4-1* mutants grown under semi-natural conditions. These heterozygous *ftshi4/FTSHi4-1* plants [26] and RNAi-FtsHi4 mutant plants [69] are smaller than WT; their cotyledons have white and yellowish leaves. Six-week-old *ftshi4-2* mutant plants exhibited pale phenotypes compared to WT when exposed to cold stress under short-day conditions [26].

5.5. *FtsHi5*

Similar to the other *FTSHi* mutants, *ftshi5* has a chlorotic seedling phenotype. Under ambient CO_2 conditions 14-day-old *ftshi5* mutant plants displayed partially impaired thylakoid morphology with reduced density, while the chloroplasts developed normally in high CO_2 conditions [74]. Using a dexamethasone (DEX)-inducible RNA-interference transgene in *FTSHi5* (DEX: RNAi-*FtsHi5*), Wang and coworkers could induce a dose-dependent albino phenotype in new leaves of *A. thaliana* [74]. Thylakoids in *DEX: RNAi-FtsHi5* plants looked wavy, swollen, and less organized upon DEX induction [74]. RNAi-*FtsHi5* mutant plants also exhibited pale-green leaves upon DEX induction [74].

Transcripts of *FTSHi5* were detected in pre-mature seeds, inflorescences, and young leaves [74]. *FTSHi5pro::GUS* transgenic reporter lines showed the highest *FTSHi5* expression in developing seeds, leaves, and pistils. *FTSHi5* transcription exhibits a circadian rhythm with elevated transcript levels at midday and lower levels at night. Transcription also increased after exposure to high light, while high CO_2 concentrations had no noticeable effect [74]. Lowered *FTSHi5* expression altered the expression of senescence-related genes and genes encoding enzymes of the oxidation-reduction process. Additionally, *ftshi5-1* plants produced higher levels of H_2O_2 and higher amounts of antioxidants to maintain the cellular redox balance [74].

6. Conclusions

Early plastid differentiation occurs at the globular-to-heart transition stage during plant embryogenesis [82,83]. The accumulation of chlorophyll in embryos begins at the heart-shaped stage. Therefore, chloroplast biogenesis is associated with embryo development and seedling growth [82,83]. Lack of the plastidic FtsHi proteins affects embryogenesis at the globular–heart transition [72,73,75,79]; therefore, the role of these enzymes is critical during chloroplast biogenesis. While most of the phenotypic characteristics observed in *FTSHi* single mutants can be explained by impaired protein import into the chloroplast, the strong impact during early development is intriguing and should be studied further. Chloroplast development is known to proceed differently in the cotyledons and true leaves in dicotyledonous plants [84,85]. The FtsH12/FtsHi1,2,4,5 complex is absent in grasses, and chloroplast development also proceeds differently in monocotyledonous and dicotyledonous species [84,85]. Whether the role of FtsHi is restricted to import via the FtsH12/Ftshi1,2,4,5 complex or has broader impact remains to be shown. Critical evaluation of all available data is necessary to review or extend our current models. Modern techniques, e.g., cryo-EM, should elucidate the comprehensive molecular structures and underlying mechanisms of the TOC-TIC-Ycf2/FtsHi motor complexes. The various weak and strong FtsHi protease mutants might be perfect tools to answer open questions.

Author Contributions: Writing—original draft preparation, L.S.M.; writing—review and editing, C.F. funding acquisition, C.F. All authors have read and agreed to the published version of the manuscript.

Funding: We acknowledge financial support by the Swedish Research Council VR (grant number 2019-04472) and Umeå University.

Institutional Review Board Statement: Not applicable.

Informed Consent Statement: Not applicable.

Data Availability Statement: Not applicable.

Conflicts of Interest: The authors declare no conflict of interest.

References

1. Ehrmann, M.; Clausen, T. Proteolysis as a Regulatory Mechanism. *Annu. Rev. Genet.* **2004**, *38*, 709–724. [CrossRef] [PubMed]
2. Stael, S.; Van Breusegem, F.; Gevaert, K.; Nowack, M.K. *Plant Proteases and Programmed Cell Death*; Oxford University Press: Oxford, UK, 2019.
3. van Wijk, K.J. Protein Maturation and Proteolysis in Plant Plastids, Mitochondria, and Peroxisomes. *Annu. Rev. Plant Biol.* **2015**, *66*, 75–111. [CrossRef] [PubMed]
4. López-Otín, C.; Bond, J.S. Proteases: Multifunctional enzymes in life and disease. *J. Biol. Chem.* **2008**, *283*, 30433–30437. [CrossRef]
5. Bonner, P.L. *Peptidases in Plant Tissue*; John Wiley & Sons, Inc.: Hoboken, NJ, USA, 2021; pp. 1–12.
6. García-Lorenzo, M.; Sjödin, A.; Jansson, S.; Funk, C. Protease gene families in Populus and Arabidopsis. *BMC Plant Biol.* **2006**, *6*, 30. [CrossRef] [PubMed]
7. Schaller, A. A cut above the rest: The regulatory function of plant proteases. *Planta* **2004**, *220*, 183–197. [CrossRef]
8. Adam, Z. Chloroplast proteases and their role in photosynthesis regulation. In *Regulation of Photosynthesis*; Springer: Berlin/Heidelberg, Germany, 2001; pp. 265–276.
9. Nishimura, K.; Kato, Y.; Sakamoto, W. Chloroplast proteases: Updates on proteolysis within and across suborganellar compartments. *Plant Physiol.* **2016**, *171*, 2280–2293. [CrossRef]
10. Rep, M.; van Dijl, J.M.; Suda, K.; Schatz, G.; Grivell, L.A.; Suzuki, C.K. Promotion of mitochondrial membrane complex assembly by a proteolytically inactive yeast Lon. *Science* **1996**, *274*, 103–106. [CrossRef]
11. Voos, W.; Pollecker, K. The mitochondrial Lon protease: Novel functions off the beaten track? *Biomolecules* **2020**, *10*, 253. [CrossRef] [PubMed]
12. Zhou, J.W.; Fang, L.R.; Yang, Z.X.; Xu, S.G.; Lv, M.T.; Sun, Z.; Chen, J.Y.; Wang, D.; Gao, J.; Xiao, S.B. Identification of novel proteolytically inactive mutations in coronavirus 3C-like protease using a combined approach. *FASEB J.* **2019**, *33*, 14575–14587. [CrossRef]
13. Klemenčič, M.; Asplund-Samuelsson, J.; Dolinar, M.; Funk, C. Phylogenetic distribution and diversity of bacterial pseudo-orthocaspases underline their putative role in photosynthesis. *Front. Plant Sci.* **2019**, *10*, 293. [CrossRef]
14. Reynolds, S.L.; Fischer, K. Pseudoproteases: Mechanisms and function. *Biochem. J.* **2015**, *468*, 17–24. [CrossRef]

15. Malnoe, A.; Wang, F.; Girard-Bascou, J.; Wollman, F.A.; de Vitry, C. Thylakoid FtsH protease contributes to photosystem II and cytochrome b6f remodeling in Chlamydomonas reinhardtii under stress conditions. *Plant Cell* **2014**, *26*, 373–390. [CrossRef] [PubMed]
16. Pulido, P.; Toledo-Ortiz, G.; Phillips, M.A.; Wright, L.P.; Rodriguez-Concepcion, M. Arabidopsis J-protein J20 delivers the first enzyme of the plastidial isoprenoid pathway to protein quality control. *Plant Cell* **2013**, *25*, 4183–4194. [CrossRef]
17. Pulido, P.; Llamas, E.; Llorente, B.; Ventura, S.; Wright, L.P.; Rodriguez-Concepcion, M. Specific Hsp100 Chaperones Determine the Fate of the First Enzyme of the Plastidial Isoprenoid Pathway for Either Refolding or Degradation by the Stromal Clp Protease in Arabidopsis. *PLoS Genet.* **2016**, *12*, e1005824. [CrossRef]
18. Wagner, R.; Aigner, H.; Funk, C. FtsH proteases located in the plant chloroplast. *Physiol. Plant.* **2012**, *145*, 203–214. [CrossRef] [PubMed]
19. Ferro, M.; Brugiere, S.; Salvi, D.; Seigneurin-Berny, D.; Court, M.; Moyet, L.; Ramus, C.; Miras, S.; Mellal, M.; Le Gall, S.; et al. AT_CHLORO, a comprehensive chloroplast proteome database with subplastidial localization and curated information on envelope proteins. *Mol. Cell Proteom.* **2010**, *9*, 1063–1084. [CrossRef]
20. Wu, J.; Jin, Y.; Zhong, S.; Chen, R.; Zhu, S.; Wang, W.; Lu, Q.; Xiong, Y. A unique group of inactive serine protease homologues from snake venom. *Toxicon* **2008**, *52*, 277–284. [CrossRef]
21. Nishimura, K.; Apitz, J.; Friso, G.; Kim, J.; Ponnala, L.; Grimm, B.; van Wijk, K.J. Discovery of a Unique Clp Component, ClpF, in Chloroplasts: A Proposed Binary ClpF-ClpS1 Adaptor Complex Functions in Substrate Recognition and Delivery. *Plant Cell* **2015**, *27*, 2677–2691. [CrossRef]
22. Nishimura, K.; van Wijk, K.J. Organization, function and substrates of the essential Clp protease system in plastids. *Biochim. Biophys. Acta* **2015**, *1847*, 915–930. [CrossRef]
23. Andersson, F.I.; Tryggvesson, A.; Sharon, M.; Diemand, A.V.; Classen, M.; Best, C.; Schmidt, R.; Schelin, J.; Stanne, T.M.; Bukau, B.; et al. Structure and function of a novel type of ATP-dependent Clp protease. *J. Biol. Chem.* **2009**, *284*, 13519–13532. [CrossRef]
24. Sokolenko, A.; Pojidaeva, E.; Zinchenko, V.; Panichkin, V.; Glaser, V.M.; Herrmann, R.G.; Shestakov, S.V. The gene complement for proteolysis in the cyanobacterium Synechocystis sp. PCC 6803 and Arabidopsis thaliana chloroplasts. *Curr. Genet.* **2002**, *41*, 291–310. [CrossRef] [PubMed]
25. Janska, H.; Kwasniak, M.; Szczepanowska, J. Protein quality control in organelles—AAA/FtsH story. *Biochim. Biophys. Acta Mol. Cell Res.* **2013**, *1833*, 381–387. [CrossRef]
26. Mishra, L.S.; Mielke, K.; Wagner, R.; Funk, C. Reduced expression of the proteolytically inactive FtsH members has impacts on the Darwinian fitness of Arabidopsis thaliana. *J. Exp. Bot.* **2019**, *70*, 2173–2184. [CrossRef] [PubMed]
27. Rawlings, N.D.; Waller, M.; Barrett, A.J.; Bateman, A. MEROPS: The database of proteolytic enzymes, their substrates and inhibitors. *Nucleic Acids Res.* **2013**, *42*, D503–D509. [CrossRef]
28. Saikawa, N.; Ito, K.; Akiyama, Y. Identification of glutamic acid 479 as the gluzincin coordinator of zinc in FtsH (HflB). *Biochemistry* **2002**, *41*, 1861–1868. [CrossRef]
29. Bieniossek, C.; Niederhauser, B.; Baumann, U.M. The crystal structure of apo-FtsH reveals domain movements necessary for substrate unfolding and translocation. *Proc. Natl. Acad. Sci. USA* **2009**, *106*, 21579–21584. [CrossRef]
30. Bieniossek, C.; Schalch, T.; Bumann, M.; Meister, M.; Meier, R.; Baumann, U. The molecular architecture of the metalloprotease FtsH. *Proc. Natl. Acad. Sci. USA* **2006**, *103*, 3066–3071. [CrossRef] [PubMed]
31. Ogura, T.; Wilkinson, A.J. AAA+ superfamily ATPases: Common structure—Diverse function. *Genes Cells* **2001**, *6*, 575–597. [CrossRef]
32. Tomoyasu, T.; Gamer, J.; Bukau, B.; Kanemori, M.; Mori, H.; Rutman, A.J.; Oppenheim, A.B.; Yura, T.; Yamanaka, K.; Niki, H.; et al. Escherichia-Coli Ftsh Is a Membrane-Bound, Atp-Dependent Protease Which Degrades the Heat-Shock Transcription Factor Sigma(32). *EMBO J.* **1995**, *14*, 2551–2560. [CrossRef]
33. Herman, C.; Prakash, S.; Lu, C.Z.; Matouschek, A.; Gross, C.A. Lack of a robust unfoldase activity confers a unique level of substrate specificity to the universal AAA protease FtsH. *Mol. Cell* **2003**, *11*, 659–669. [CrossRef]
34. Asahara, Y.; Atsuta, K.; Motohashi, K.; Taguchi, H.; Yohda, M.; Yoshida, M. FtsH recognizes proteins with unfolded structure and hydrolyzes the carboxyl side of hydrophobic residues. *J. Biochem.* **2000**, *127*, 931–937. [CrossRef]
35. Karata, K.; Inagawa, T.; Wilkinson, A.J.; Tatsuta, T.; Ogura, T. Dissecting the role of a conserved motif (the second region of homology) in the AAA family of ATPases. Site-directed mutagenesis of the ATP-dependent protease FtsH. *J. Biol. Chem.* **1999**, *274*, 26225–26232. [CrossRef]
36. Hinnerwisch, J.; Fenton, W.A.; Furtak, K.J.; Farr, G.W.; Horwich, A.L. Loops in the central channel of ClpA chaperone mediate protein binding, unfolding, and translocation. *Cell* **2005**, *121*, 1029–1041. [CrossRef]
37. Park, E.; Rho, Y.M.; Koh, O.J.; Ahn, S.W.; Seong, I.S.; Song, J.J.; Bang, O.; Seol, J.H.; Wang, J.; Eom, S.H.; et al. Role of the GYVG pore motif of HslU ATPase in protein unfolding and translocation for degradation by HslV peptidase. *J. Biol. Chem.* **2005**, *280*, 22892–22898. [CrossRef] [PubMed]
38. Puchades, C.; Sandate, C.R.; Lander, G.C. The molecular principles governing the activity and functional diversity of AAA+ proteins. *Nat. Rev. Mol. Cell Biol.* **2020**, *21*, 43–58. [CrossRef]
39. Urantowka, A.; Knorpp, C.; Olczak, T.; Kolodziejczak, M.; Janska, H. Plant mitochondria contain at least two i-AAA-like complexes. *Plant Mol. Biol.* **2005**, *59*, 239–252. [CrossRef] [PubMed]

40. Wagner, R.; Von Sydow, L.; Aigner, H.; Netotea, S.; Brugière, S.; Sjögren, L.; Ferro, M.; Clarke, A.; Funk, C. Deletion of FtsH11 protease has impact on chloroplast structure and function in Arabidopsis thaliana when grown under continuous light. *Plant Cell Environ.* **2016**, *39*, 2530–2544. [CrossRef]
41. Kolodziejczak, M.; Skibior-Blaszczyk, R.; Janska, H. m-AAA Complexes Are Not Crucial for the Survival of Arabidopsis under Optimal Growth Conditions Despite Their Importance for Mitochondrial Translation. *Plant Cell Physiol.* **2018**, *59*, 1006–1016. [CrossRef]
42. Gibala, M.; Kicia, M.; Sakamoto, W.; Gola, E.M.; Kubrakiewicz, J.; Smakowska, E.; Janska, H. The lack of mitochondrial AtFtsH4 protease alters Arabidopsis leaf morphology at the late stage of rosette development under short-day photoperiod. *Plant J.* **2009**, *59*, 685–699. [CrossRef]
43. Marta, K.; Marta, G.; Adam, U.; Hanna, J. The significance of Arabidopsis AAA proteases for activity and assembly/stability of mitochondrial OXPHOS complexes. *Physiol. Plant.* **2007**, *129*, 135–142. [CrossRef]
44. Kato, Y.; Sakamoto, W. FtsH Protease in the Thylakoid Membrane: Physiological Functions and the Regulation of Protease Activity. *Front. Plant Sci.* **2018**, *9*, 855. [CrossRef]
45. Zaltsman, A.; Ori, N.; Adam, Z. Two types of FtsH protease subunits are required for chloroplast biogenesis and Photosystem II repair in Arabidopsis. *Plant Cell* **2005**, *17*, 2782–2790. [CrossRef]
46. Chen, M.; Choi, Y.D.; Voytas, D.F.; Rodermel, S. Mutations in the Arabidopsis VAR2 locus cause leaf variegation due to the loss of a chloroplast FtsH protease. *Plant J.* **2000**, *22*, 303–313. [CrossRef]
47. Miura, E.; Kato, Y.; Matsushima, R.; Albrecht, V.; Laalami, S.; Sakamoto, W. The balance between protein synthesis and degradation in chloroplasts determines leaf variegation in Arabidopsis yellow variegated mutants. *Plant Cell* **2007**, *19*, 1313–1328. [CrossRef]
48. Takechi, K.; Sodmergen; Murata, M.; Motoyoshi, F.; Sakamoto, W. The Yellow Variegated (VAR2) locus encodes a homologue of FtsH, an ATP-dependent protease in Arabidopsis. *Plant Cell Physiol.* **2000**, *41*, 1334–1346. [CrossRef]
49. Yu, F.; Liu, X.; Alsheikh, M.; Park, S.; Rodermel, S. Mutations in Suppressor of Variegation1, a factor required for normal chloroplast translation, suppress var2-mediated leaf variegation in Arabidopsis. *Plant Cell* **2008**, *20*, 1786–1804. [CrossRef]
50. Yu, F.; Park, S.; Rodermel, S.R. The Arabidopsis FtsH metalloprotease gene family: Interchangeability of subunits in chloroplast oligomeric complexes. *Plant J.* **2004**, *37*, 864–876. [CrossRef]
51. Sakamoto, W.; Zaltsman, A.; Adam, Z.; Takahashi, Y. Coordinated regulation and complex formation of yellow variegated1 and yellow variegated2, chloroplastic FtsH metalloproteases involved in the repair cycle of photosystem II in Arabidopsis thylakoid membranes. *Plant Cell* **2003**, *15*, 2843–2855. [CrossRef]
52. Zaltsman, A.; Feder, A.; Adam, Z. Developmental and light effects on the accumulation of FtsH protease in Arabidopsis chloroplasts—Implications for thylakoid formation and photosystem II maintenance. *Plant J.* **2005**, *42*, 609–617. [CrossRef]
53. Sedaghatmehr, M.; Mueller-Roeber, B.; Balazadeh, S. The plastid metalloprotease FtsH6 and small heat shock protein HSP21 jointly regulate thermomemory in Arabidopsis. *Nat. Commun.* **2016**, *7*, 12439. [CrossRef] [PubMed]
54. Wagner, R.; Aigner, H.; Pruzinska, A.; Jankanpaa, H.J.; Jansson, S.; Funk, C. Fitness analyses of Arabidopsis thaliana mutants depleted of FtsH metalloproteases and characterization of three FtsH6 deletion mutants exposed to high light stress, senescence and chilling. *New Phytol.* **2011**, *191*, 449–458. [CrossRef]
55. Shao, S.; Cardona, T.; Nixon, P.J. Early emergence of the FtsH proteases involved in photosystem II repair. *Photosynthetica* **2018**, *56*, 163–177. [CrossRef]
56. Adam, Z.; Aviv-Sharon, E.; Keren-Paz, A.; Naveh, L.; Rozenberg, M.; Savidor, A.; Chen, J.P. The Chloroplast Envelope Protease FTSH11-Interaction With CPN60 and Identification of Potential Substrates. *Front. Plant Sci.* **2019**, *10*, 428. [CrossRef]
57. Chen, J.; Burke, J.J.; Velten, J.; Xin, Z. FtsH11 protease plays a critical role in Arabidopsis thermotolerance. *Plant J.* **2006**, *48*, 73–84. [CrossRef]
58. Kikuchi, S.; Asakura, Y.; Imai, M.; Nakahira, Y.; Kotani, Y.; Hashiguchi, Y.; Nakai, Y.; Takafuji, K.; Bedard, J.; Hirabayashi-Ishioka, Y.; et al. A Ycf2-FtsHi Heteromeric AAA-ATPase Complex Is Required for Chloroplast Protein Import. *Plant Cell* **2018**, *30*, 2677–2703. [CrossRef]
59. Schreier, T.B.; Clery, A.; Schlafli, M.; Galbier, F.; Stadler, M.; Demarsy, E.; Albertini, D.; Maier, B.A.; Kessler, F.; Hortensteiner, S.; et al. Plastidial NAD-Dependent Malate Dehydrogenase: A Moonlighting Protein Involved in Early Chloroplast Development through Its Interaction with an FtsH12-FtsHi Protease Complex. *Plant Cell* **2018**, *30*, 1745–1769. [CrossRef]
60. Kikuchi, S.; Bedard, J.; Hirano, M.; Hirabayashi, Y.; Oishi, M.; Imai, M.; Takase, M.; Ide, T.; Nakai, M. Uncovering the Protein Translocon at the Chloroplast Inner Envelope Membrane. *Science* **2013**, *339*, 571–574. [CrossRef]
61. De Longevialle, A.F.; Hendrickson, L.; Taylor, N.L.; Delannoy, E.; Lurin, C.; Badger, M.; Millar, A.H.; Small, I. The pentatricopeptide repeat gene OTP51 with two LAGLIDADG motifs is required for the cis-splicing of plastid ycf3 intron 2 in Arabidopsis thaliana. *Plant J.* **2008**, *56*, 157–168. [CrossRef]
62. Nakai, M. New Perspectives on Chloroplast Protein Import. *Plant Cell Physiol.* **2018**, *59*, 1111–1119. [CrossRef]
63. Herrmann, J.M. A Force-Generating Machine in the Plant's Powerhouse: A Pulling AAA-ATPase Motor Drives Protein Translocation into Chloroplasts. *Plant Cell* **2018**, *30*, 2646–2647. [CrossRef]
64. Schafer, P.; Helm, S.; Kohler, D.; Agne, B.; Baginsky, S. Consequences of impaired 1-MDa TIC complex assembly for the abundance and composition of chloroplast high-molecular mass protein complexes. *PLoS ONE* **2019**, *14*, e0213364. [CrossRef]

65. Mielke, K.; Wagner, R.; Mishra, L.S.; Demir, F.; Perrar, A.; Huesgen, P.F.; Funk, C. Abundance of metalloprotease FtsH12 modulates chloroplast development in Arabidopsis thaliana. *J. Exp. Bot.* **2020**, *72*, 3455–3473. [CrossRef] [PubMed]
66. Nakai, M. Reply: The Revised Model for Chloroplast Protein Import. *Plant Cell* **2020**, *32*, 543–546. [CrossRef]
67. Li, H.M.; Schnell, D.; Theg, S.M. Protein Import Motors in Chloroplasts: On the Role of Chaperones. *Plant Cell* **2020**, *32*, 536–542. [CrossRef]
68. García-Lorenzo, M.; Pružinská, A.; Funk, C. ATP-dependent proteases in the chloroplast. In *ATP-Dependent Proteases;* Kutejova, E., Ed.; Research Signpost: Kerala, India, 2008; pp. 145–176.
69. Wolfe, K.H. Similarity between Putative Atp-Binding Sites in Land Plant Plastid Orf2280 Proteins and the Ftsh/Cdc48 Family of Atpases. *Curr. Genet.* **1994**, *25*, 379–383. [CrossRef]
70. Drescher, A.; Ruf, S.; Calsa, T., Jr.; Carrer, H.; Bock, R. The two largest chloroplast genome-encoded open reading frames of higher plants are essential genes. *Plant J.* **2000**, *22*, 97–104. [CrossRef] [PubMed]
71. Schmutz, J.; Cannon, S.B.; Schlueter, J.; Ma, J.; Mitros, T.; Nelson, W.; Hyten, D.L.; Song, Q.; Thelen, J.J.; Cheng, J.; et al. Genome sequence of the palaeopolyploid soybean. *Nature* **2010**, *463*, 178–183. [CrossRef]
72. Kadirjan-Kalbach, D.K.; Yoder, D.W.; Ruckle, M.E.; Larkin, R.M.; Osteryoung, K.W. FtsHi1/ARC1 is an essential gene in Arabidopsis that links chloroplast biogenesis and division. *Plant J.* **2012**, *72*, 856–867. [CrossRef] [PubMed]
73. Lu, X.D.; Zhang, D.Y.; Li, S.P.; Su, Y.P.; Liang, Q.J.; Meng, H.Y.; Shen, S.D.; Fan, Y.L.; Liu, C.M.; Zhang, C.Y. FtsHi4 Is Essential for Embryogenesis Due to Its Influence on Chloroplast Development in Arabidopsis. *PLoS ONE* **2014**, *9*, e99741. [CrossRef]
74. Wang, T.; Li, S.; Chen, D.; Xi, Y.; Xu, X.; Ye, N.; Zhang, J.; Peng, X.; Zhu, G. Impairment of FtsHi5 function affects cellular redox balance and photorespiratory metabolism in Arabidopsis. *Plant Cell Physiol.* **2018**, *59*, 2526–2535. [CrossRef]
75. Meinke, D.; Muralla, R.; Sweeney, C.; Dickerman, A. Identifying essential genes in Arabidopsis thaliana. *Trends Plant Sci.* **2008**, *13*, 483–491. [CrossRef]
76. Juvany, M.; Muller, M.; Munne-Bosch, S. Photo-oxidative stress in emerging and senescing leaves: A mirror image? *J. Exp. Bot.* **2013**, *64*, 3087–3098. [CrossRef]
77. Matsubara, S.; Schneider, T.; Maurino, V.G. Dissecting Long-Term Adjustments of Photoprotective and Photo-Oxidative Stress Acclimation Occurring in Dynamic Light Environments. *Front. Plant Sci.* **2016**, *7*, 1690. [CrossRef]
78. de Vries, S.C.; Weijers, D. Plant embryogenesis. *Curr. Biol.* **2017**, *27*, R870–R873. [CrossRef] [PubMed]
79. Meinke, D.W. Genome-wide identification of Embryo-Defective (EMB) genes required for growth and development in Arabidopsis. *New Phytol.* **2020**, *226*, 306–325. [CrossRef]
80. Majsec, K.; Bhuiyan, N.H.; Sun, Q.; Kumari, S.; Kumar, V.; Ware, D.; van Wijk, K.J. The Plastid and Mitochondrial Peptidase Network in Arabidopsis thaliana: A Foundation for Testing Genetic Interactions and Functions in Organellar Proteostasis. *Plant Cell* **2017**, *29*, 2687–2710. [CrossRef]
81. Muller, P.; Li, X.P.; Niyogi, K.K. Non-photochemical quenching. A response to excess light energy. *Plant Physiol.* **2001**, *125*, 1558–1566. [CrossRef]
82. Mansfield, S.; Briarty, L. Early embryogenesis in Arabidopsis thaliana. II. The developing embryo. *Can. J. Bot.* **1991**, *69*, 461–476. [CrossRef]
83. Tejos, R.I.; Mercado, A.V.; Meisel, L.A. Analysis of chlorophyll fluorescence reveals stage specific patterns of chloroplast-containing cells during Arabidopsis embryogenesis. *Biol. Res.* **2010**, *43*, 99–111. [CrossRef]
84. Pogson, B.J.; Albrecht, V. Genetic Dissection of Chloroplast Biogenesis and Development: An Overview. *Plant Physiol.* **2011**, *155*, 1545–1551. [CrossRef] [PubMed]
85. Pogson, B.J.; Ganguly, D.; Albrecht-Borth, V. Insights into chloroplast biogenesis and development. *Biochim. Biophys. Acta* **2015**, *1847*, 1017–1024. [CrossRef] [PubMed]

International Journal of *Molecular Sciences*

Article

Direct Measurement of the Affinity between tBid and Bax in a Mitochondria-Like Membrane

Markus Rose [1], Martin Kurylowicz [1], Mohammad Mahmood [1], Sheldon Winkel [1], Jose M. Moran-Mirabal [2] and Cécile Fradin [1,3,*]

[1] Department of Physics and Astronomy, McMaster University, Hamilton, ON L8S 4M1, Canada; rosemm2@mcmaster.ca (M.R.); marty@scopesys.ca (M.K.); mahmooma@mcmaster.ca (M.M.); sheldon.winkel@gmail.com (S.W.)

[2] Department of Chemistry and Chemical Biology, McMaster University, Hamilton, ON L8S 4M1, Canada; mirabj@mcmaster.ca

[3] Department of Biochemistry and Biomedical Sciences, McMaster University, Hamilton, ON L8S 4K1, Canada

* Correspondence: fradin@physics.mcmaster.ca

Abstract: The execution step in apoptosis is the permeabilization of the outer mitochondrial membrane, controlled by Bcl-2 family proteins. The physical interactions between the different proteins in this family and their relative abundance literally determine the fate of the cells. These interactions, however, are difficult to quantify, as they occur in a lipid membrane and involve proteins with multiple conformations and stoichiometries which can exist both in soluble and membrane. Here we focus on the interaction between two core Bcl-2 family members, the executor pore-forming protein Bax and the truncated form of the activator protein Bid (tBid), which we imaged at the single particle level in a mitochondria-like planar supported lipid bilayer. We inferred the conformation of the proteins from their mobility, and detected their transient interactions using a novel single particle cross-correlation analysis. We show that both tBid and Bax have at least two different conformations at the membrane, and that their affinity for one another increases by one order of magnitude (with a 2D-K_D decreasing from $\simeq$1.6 μm^{-2} to $\simeq$0.1 μm^{-2}) when they pass from their loosely membrane-associated to their transmembrane form. We conclude by proposing an updated molecular model for the activation of Bax by tBid.

Keywords: apoptosis; mitochondria; Bcl-2 family; Bax; Bid; membrane protein; protein–protein interaction; protein oligomerization; fluorescence; single particle detection

Citation: Rose, M.; Kurylowicz, M.; Mahmood, M.; Winkel, S.; Moran- Mirabal, J.M.; Fradin, C. Direct Measurement of the Affinity between tBid and Bax in a Mitochondria-Like Membrane. *Int. J. Mol. Sci.* **2021**, *22*, 8240. https://doi.org/10.3390/ijms22158240

Academic Editor: Masoud Jelokhani-Niaraki

Received: 8 June 2021
Accepted: 22 July 2021
Published: 31 July 2021

1. Introduction

The regulation step in mitochondrial apoptosis is the permeabilization of the Mitochondrial Outer Membrane (MOM) [1]. It is under the control of the Bcl-2 family of proteins, whose members carry between 1 and 4 Bcl-2 protein homology (BH) regions [2]. The BH3 region, present in all Bcl-2 family proteins, is thought to act as a "death ligand" mediating heterodimerization between family members [3]. Members of the Bcl-2 family have been classified into four different subgroups [2,4,5]. Pro-apoptotic multidomain family members, Bax and Bak, are the executors of the permeabilization process. They oligomerize and form pores in the MOM in response to apoptotic signals. They can be activated by pro-apoptotic BH3-only family members (such as truncated Bid—known as tBid, Bim and Puma [6–8]), through a process thought to involve a direct physical interaction [9–11]. The Bcl-2 family also includes anti-apoptotic multidomain members (e.g., Bcl-2, Bcl-XL, Mcl-1), which work by either directly inhibiting Bax and Bak, or sequestering BH3-only proteins. The last subgroup in the family is made of the so-called sensitizer proteins (e.g., Bad), which regulates MOM permeabilization by inhibiting anti-apoptotic Bcl-2 family members.

The fate of cells is dictated by the intricate balance of interactions between the different groups of Bcl-2 family proteins, a balance often found to be dysregulated in drug-resistant cancer cells and therefore a sought-after target for pharmacological intervention against cancer [12–15]. At the centre of the Bcl-2 family interaction network are several competing interactions: the activator-effector interaction (e.g., between tBid and Bax), the repressor-effector interaction (e.g., between Bcl-XL and Bax) and the activator-repressor interaction (e.g., between between tBid and Bcl-XL), each of them able to tip the balance either towards death or survival. Several models have been put forward that try and capture the main features of the activation of Bax, each giving more or less importance to these different interactions, and with the most recent models incorporating the influence of the lipid membrane, increasingly recognized as a main player in the process [16–19]. One way to discriminate between these different models would be to measure the affinity between different pairs of Bcl-2 family proteins in conditions as close as possible as those encountered in the MOM.

Quantification of the affinity between different pairs of Bcl-2 family proteins has been achieved using ensemble fluorescence resonance energy transfer [10,20,21]. However, it has since become clear that a number of Bcl-2 family proteins (e.g., Bax, tBid, Bcl-XL) exist in a dynamic equilibrium between solution and membrane [22–24]. Ensemble measurements, which do not distinguish between the soluble and membrane forms of the proteins, can thus only return an apparent affinity, which depends on the amount and type of lipids present in the system. Interactions between different pairs of Bcl-2 family proteins embedded in the membrane of giant unilamellar vesicles have also been detected using linear scan cross-correlation spectroscopy [25–27]. Yet these experiments cannot distinguish between the different conformations or stoichiometries these proteins are known to adopt in lipid membranes [28,29]. In contrast, single particle experiments allow measuring affinities between pairs of proteins with specific conformations and stoichiometries, and have been successfully used by us and others to characterize the oligomerization of tBid [29] and Bax [30] in planar membranes. It is therefore the approach we have chosen here to characterize Bcl-2 family protein interactions. One challenge associated with such measurements is that a low surface density of proteins, typically below the 2D-K_D, must be used for single particle detection, and therefore interacting protein pairs are rare events—a needle in a haystack. A second issue is that coincidental co-localization events may be observed, leading to an underestimate of the 2D-K_D [31]. We solved both issues by combining the ideas of single particle detection and image correlation spectroscopy (building on previous ideas by others [32]), and calculating a single particle cross-correlation coefficient for each detected pair of particles to identify true binding interactions [33].

Here we focus on the activator-effector interaction, using tBid and Bax as a model system. Bid is a BH3-only direct activator of Bax [10,34,35]. It is unique amongst the BH3-only proteins because in solution it adopts a alpha-helical structure packed into a globular fold which is structurally homologous to that of multidomain family members Bcl-2 and Bax [36,37]. Other BH3-only proteins, on the other hand, are known or predicted to be unstructured or to have a structure distinct from that of other Bcl-2 family proteins [38–41]. Bid gets cleaved by caspase-8 in response to ligand binding to death receptors. The larger of the two resulting Bid fragments, called truncated Bid (tBid), then inserts into the MOM [42]. Upon contact with the membrane, tBid changes conformation and increases its affinity for Bax, which in turn inserts into the MOM [10,43]. Bax then forms oligomers that increase the membrane permeability [6,7]. To fully characterize the interaction between the two proteins at the membrane, we used a reconstituted system capturing all facets of this interaction—full-length fluorescently-labelled purified proteins and a Supported Lipid Bilayer (SLB) with mitochondria-like lipid composition. We used single particle detection on confocal images to sort them into different categories according to their stoichiometry and their mobility, as we have done previously for tBid alone [29], and to detect associations between particles. This allowed us to measure the 2D-K_D for each protein subpopulation, which shows the tBid-Bax interaction becomes stronger as the proteins insert deeper into the membrane.

2. Results

In order to quantify the affinity between tBid and Bax, we incubated SLB of mitochondria-like composition with purified full-length tBid (labelled with Alexa647) and purified full-length Bax (labelled with HyLite488), at concentrations low enough to achieve single-particle detection. Separate controls were also carried out where the membrane was either incubated with tBid alone or Bax alone. Dual-color confocal image stacks were recorded for each sample, and analyzed in order to extract two-dimensional dissociation constants for different categories of proteins. In the next three sections, we explain how the proteins were detected and classified according to their mobility, how they were further classified according to their stoichiometry, and finally how interactions between tBid and Bax were identified and characterized.

2.1. Different Membrane Conformations Are Detected for tBid and Bax Based on Their Mobility

It is well established that tBid and Bax can each adopt different conformations when interacting with a lipid membrane [29]. Proteins with different conformations may have different mobilities, resulting in distinct signatures in both widefield [44] and confocal images [29,45]. We thus looked to classify particles detected in our confocal images according to their mobility.

2.1.1. Signature of Stationary and Mobile Particles in Confocal Images

We performed simulations to establish the link between particles' apparent shape in confocal images and their mobility. We simulated the two-dimensional diffusion of particles with varying mobilities and molecular brightness, and the acquisition of confocal images using conditions similar to those typically used in our experiments (confocal detection volume radius $w_0 = 300$ nm, pixel size $d = 100$ nm, pixel dwell time $\delta = 1$ ms). Examples of simulated images for both completely stationary particles and diffusing particles ($D = 2\ \mu\mathrm{m}^2\,\mathrm{s}^{-1}$) can be seen in Figure 1. Events can easily be detected in both cases, but their appearance is strikingly different: diffraction limited spots for stationary particles and single line streaks along the scanning direction for diffusing particles (as previously observed for both proteins and lipids in SLB [29,46]).

The single particle detection procedure described in the Methods section was applied to multiple series of simulated images of either stationary or diffusing particles (noise images were also analyzed for comparison). For each detected event, this procedure returns the radii of the 2D-Gaussian spot that best fit the particle (Equation (4)), $w_{x,p}$ along the scanning direction and $w_{y,p}$ perpendicular to it, as well as the intensity of the particle, I_p. By inspecting the eccentricity maps (two-dimensional distributions of particle radii, shown in Figure 1), multiple clusters can be identified. In images containing only background photon noise (Figure 1a), only single pixel events concentrated in a small region centred around $w_{x,p} \approx w_{y,p} \approx d = 100$ nm are detected, as expected since photon noise is not spatially correlated. (There is both a horizontal gap and a vertical gap around $\approx$50 nm, the result of the instability of the Gaussian fitting of a signal detected in a single pixel.) Stationary particles (Figure 1b) appear as a cluster of diffraction-limited events around $w_{x,p} \approx w_{y,p} \approx w_0 = 300$ nm and $I_p \approx B = 20$ kHz. They are well separated, both by their dimensions and intensity, from events due to noise, also detected in this case. We thus decided to classify events as stationary particles if they were found within a distance $w_0/2$ of their expected position (w_0, w_0) on the eccentricity map. For diffusing particles (Figure 1c), the eccentricity map shows an extended cluster of events with $w_{y,p} < w_0/2 = 150$ nm and I_p around or below $B = 20$ kHz. The distinction between diffusing and stationary particles, based on their shape, is therefore straightforward in this case. However, the distinction between noise and diffusing particles is not, as the distributions of these two types of events overlap, both on the eccentricity map and on the intensity map. As a compromise between false positive and false negative detection, we decided to classify events as mobile particles only if $w_{y,p} < w_0/2$ and $w_{x,p} > 1.2\,d = 120$ nm. In these conditions, noise events only contribute a maximum of 10% of the mobile particle detection events, which was judged acceptable.

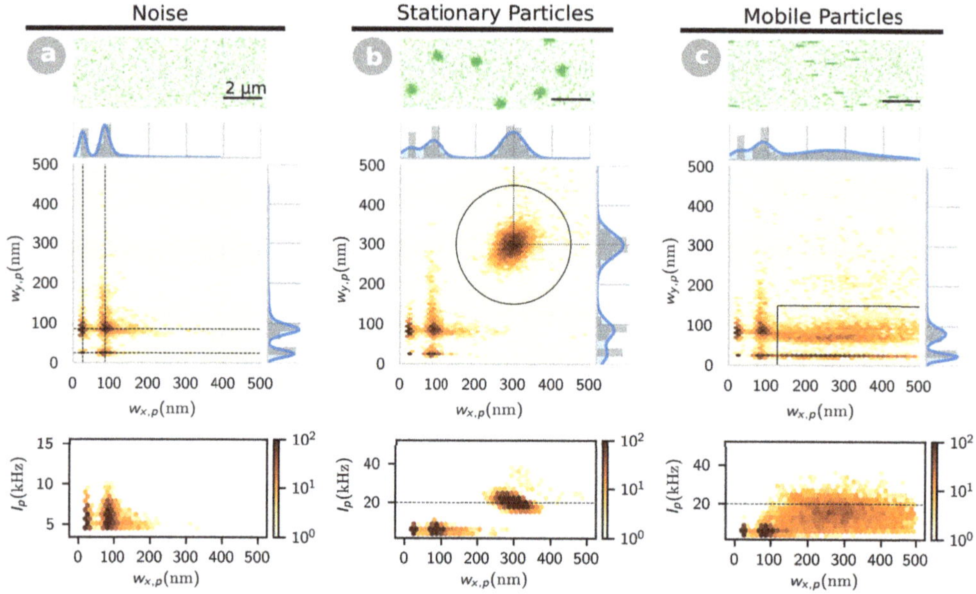

Figure 1. Confocal image simulations for (**a**) background Poisson noise ($i_B = 1.3$ kHz), (**b**) stationary particles ($B = 20$ kHz) and (**c**) mobile particles ($D = 2\ \mu\mathrm{m}^2\,\mathrm{s}^{-1}$, $B = 20$ kHz). For each condition, a single image section is shown as an example (top panel), as well as the particle size distribution (eccentricity map) generated from the analysis of a large number of such images (middle panel), and the relationship between particle width and intensity (bottom panel). In the eccentricity map in (**a**), the dashed lines indicate the position of the two peaks corresponding to spurious photon noise events. In (**b**), the dashed lines show the expected dimensions of an immobile particle (w_0 both along and perpendicular to the scanning direction), and the circle delimitates the area in which particles are considered "spots". In (**c**), the rectangle delimitates the area in which the particles are considered "streaks". In the bottom panels in (**b**,**c**), the dashed line indicates the value of the molecular brightness, B.

To further characterize the mobility of the diffusing particles, we considered the length and eccentricity of the streaks, which depend on the particles' trajectory during the imaging, and therefore on their mobility [47]. Simulations (Figure 2) show that the average streak length, $\langle w_{x,p} \rangle$, decreases as D increases, while the average eccentricity, $\langle w_{x,p}/w_{y,p} \rangle$, first increases from 1 to $\simeq 2w_0/d = 6$, then decreases. For $D > 20\ \mu\mathrm{m}^2\,\mathrm{s}^{-1}$, particle events become difficult to separate from noise events, as the lengths of the streaks shorten. The time needed to image a full line in the confocal image is $n_x\delta = 0.1$ s in the conditions of our experiments, thus the time necessary to capture a stationary particle is $\simeq n_x\delta \times w_0/d = 0.3$ s. The cut-off between particles appearing as "spots" and "streaks" should occur when they are able to diffuse over a distance w_0 during that time, in other words when $D > D_c = w_0^2/(4n_x\delta \times w_0/d) = 0.08\ \mu\mathrm{m}^2\,\mathrm{s}^{-1}$. This is indeed what is observed in our simulations (Figure 2a). Thus, the classification method introduced here allows, in the conditions of our experiments, sorting events into "stationary" proteins ($D \lesssim 0.1\ \mu\mathrm{m}^2\,\mathrm{s}^{-1}$) and "mobile" proteins ($D \simeq 0.1 - 20\ \mu\mathrm{m}^2\,\mathrm{s}^{-1}$).

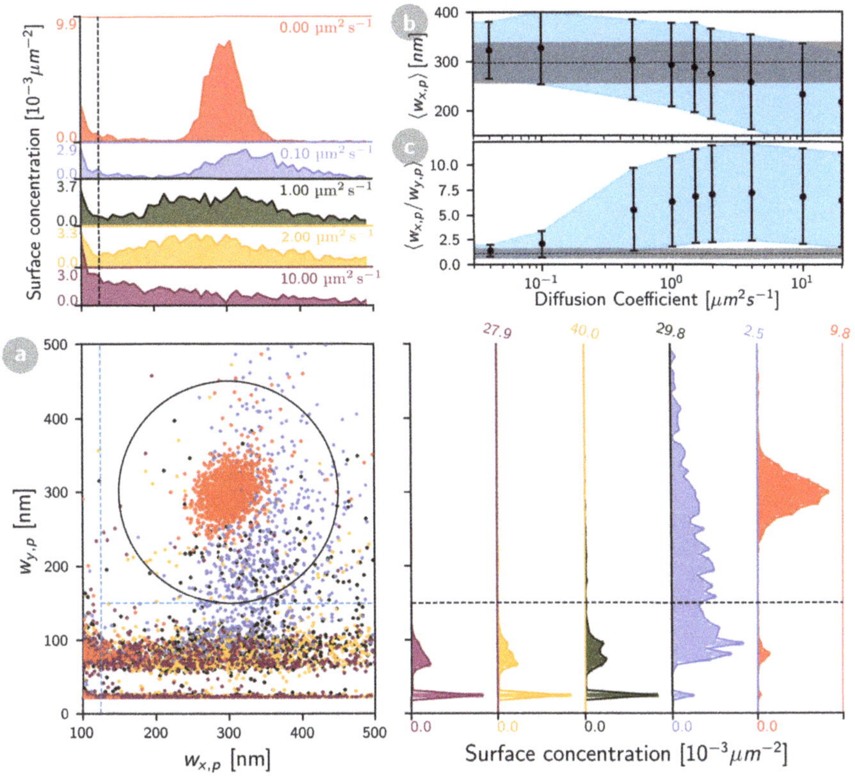

Figure 2. Influence of particle diffusion coefficient on the appearance of events detected in simulated confocal images. Simulation parameters were otherwise the same as in Figure 1. (**a**) Distribution of $w_{x,p}$ and $w_{y,p}$ values for particles with different diffusion coefficients (red: 0 µm^2 s^{-1}, blue: 0.1 µm^2 s^{-1}, green: 1 µm^2 s^{-1}, yellow: 2 µm^2 s^{-1}, purple: 10 µm^2 s^{-1}). (**b**) Average streak length and (**c**) average eccentricity of all events with $w_{x,p} > 120$ nm (error bars are the standard deviation) as a function of diffusion coefficient. The dashed line and the grey area represents the values obtained for stationary particles.

2.1.2. tBid and Bax Mobility in the Mitochondria-Like SLB

We applied the single particle detection and classification method outlined in the previous section to a set of 258 dual-color confocal images of SLB incubated successively with 0.2 nM tBid-Alexa647 and Bax-HyLite488 at concentrations ranging from 0.1 to 2 nM (see Methods for experimental details). An example of a pair of such images (green and red detection channels) can be seen in Figure 3a, where detected events are marked with a box (a number of detection events are rejected, either because they are too close to the border or to another already detected event, because they cannot be correctly fitted by a Gaussian, or because they have an intensity below the set threshold intensity). In these images, both spots and streaks are observed, indicative of the presence of both stationary and mobile proteins. Accordingly, the corresponding eccentricity maps show particles both in the mobile and stationary particle regions (Figure 3b). The number of particles detected per frame varied only slightly across conditions, and was roughly the same for both proteins (SI Figure S2). In each channel, we detected on average $\simeq$5 to 10 mobile particles per frame, significantly more than stationary particles, found at a rate of $\simeq$1 to 3 particles per frame. Even including poorly defined particles rejected from the final analysis, the total number of detected particles remained on average below 30 per frame and per channel, i.e., below a concentration of 0.3 particles/µm^2. For both Bax and tBid, our data thus indicates the presence of at least two different types of membrane conformation—an abundant highly mobile form and a rarer stationary form. When the same experiment was repeated with tBid alone, both mobile and stationary particles were also observed in the membrane, as we already reported in Ref. [29]. In contrast, when the experiment was repeated with Bax alone, no particles could be detected at the membrane (even at the highest 2 nM Bax concentration used for incubation, see SI Figure S3), highlighting the important role played by tBid in retaining Bax at the membrane in this reconstituted system.

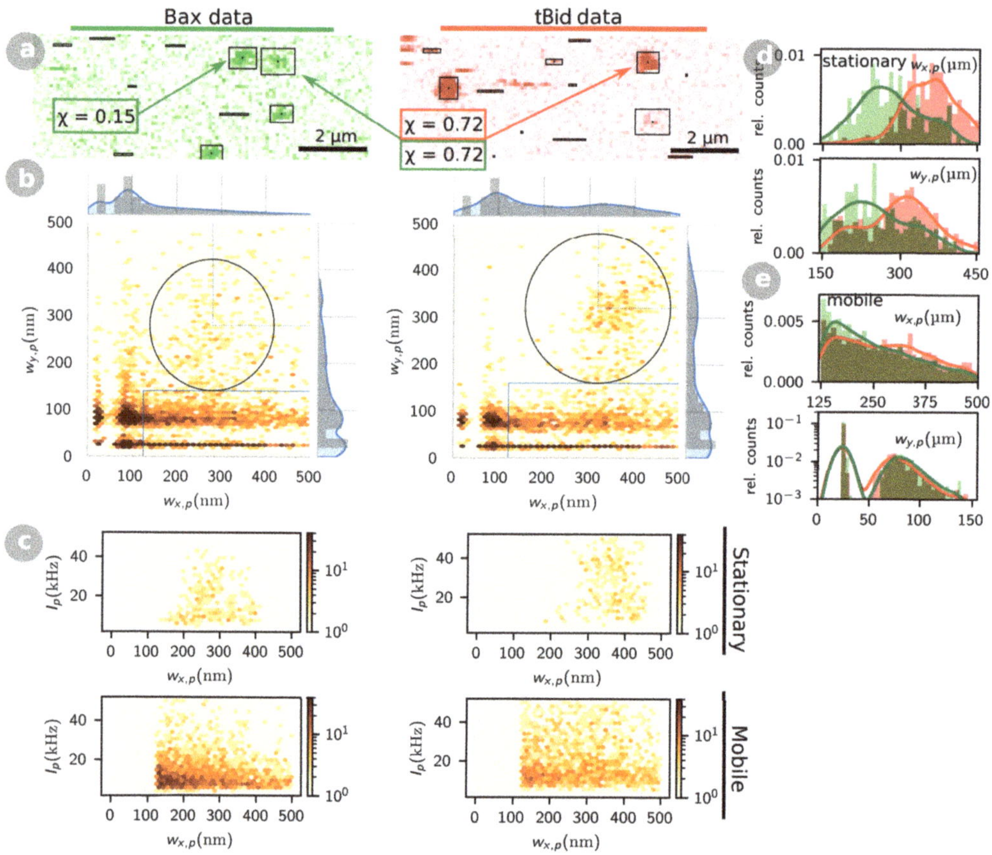

Figure 3. Event detection for confocal images of SLB incubated with tBid-Alexa647 and Bax-HyLite488. (**a**) Representative example of a section of a pair of confocal images (left: green detection channel, right: red detection channel) acquired for a SLB incubated successively with 0.2 nM tBid-Alexa647 and 1.0 nM Bax-HyLite488. Black boxes highlight all the events detected in these images. (**b**) Eccentricity maps and (**c**) intensity maps obtained for the entire set of 258 images acquired (regardless of Bax concentration). Solid lines delimitate the regions used to classify the particles as stationary or mobile, considering $w_0 = 280\,\text{nm}$ (green channel) or $w_0 = 320\,\text{nm}$ (red channel). The distributions of observed $w_{x,p}$ and $w_{y,p}$ values are given in (**d**) for stationary particles and in (**e**) for mobile particles (green lines and bars: Bax, red lines and bars: tBid).

Stationary Bax-HiLyte488 have an average apparent radius $\langle w_{x,p} \rangle = (280 \pm 60)$ nm (mean ± stdv), 10 to 20% smaller than that of stationary tBid-Alexa647 ($\langle w_{x,p} \rangle = (360 \pm 50)$ nm), as expected given the difference in excitation and detection wavelengths between the green and red channels, and in each case close to the radius of the confocal observation volume measured by FCS. Of note, the distributions of values for $\langle w_{x,p} \rangle$ and $\langle w_{y,p} \rangle$ obtained for stationary Bax and tBid (Figure 3d), while clearly belonging to a well-defined population, are more spread out than that obtained in simulations (Figure 1b), and the average eccentricity of these particles is slightly above 1 ($\langle w_{x,p}/w_{y,p} \rangle = 1.1$ for Bax and 1.2

for tBid). Thus, "stationary" tBid and Bax particles might include particles with a small mobility, on the order of 0 to 0.1 µm^2 s^{-1} (other effects, such as photobleaching, could also explain this slight eccentricity).

For mobile particles, the distribution of detected shapes, as seen in the eccentricity maps (Figure 3b), and the distributions of $w_{x,p}$ and $w_{y,p}$ values (Figure 3e), are also as expected from the simulations. From the average length of these streaks ($\langle w_{x,p} \rangle = 250$ nm for Bax and 280 nm for tBid) and their average eccentricity ($\langle w_{x,p}/w_{y,p} \rangle = 4.4$ for Bax and 4.9 for tBid), which can be compared to that obtained for simulated particles with different mobilities (Figure 2), we conclude that the mobile membrane fractions of both tBid and Bax have diffusion coefficients on the order of 10 to 20 µm^2 s^{-1}, corresponding to protein configurations only loosely associated with the membrane (since they are faster than lipids, which in SLB have diffusion coefficients in the 1 to 5 µm^2 s^{-1} range [48–50]).

2.2. tBid and Bax form Oligomers at the Membrane

Because of photon noise, and because of eventual motions during imaging, the intensity of an event may not be equal to the particle molecular brightness (see Figure 1b,c, bottom panels). Simulations of stationary particles for different values of the brightness B (and at a concentration of 20 particles per image, comparable to those encountered in our experiments), do show that the distribution of intensities for detected particles is usually quite narrowly peaked around B (Figure 4a). For low intensity particles, for which the fixed detection threshold becomes on the order of B, the intensity distribution becomes broader and is centred slightly above B (Figure 4a, left panel). However, as the mobility of the particles increases, the centre of the distribution shifts from B towards lower values (Figure 4a, right panel). In stark contrast, the distribution of intensities for recorded images of tBid and Bax is very broad, in some cases with visible peaks at intensities that are much larger than B (Figures 3c and 4b), as was previously observed for both proteins separately [29,30]. These extended distributions are consistent with the presence of oligomers, with intensities representative of the number of monomers it contains (barring issues with incomplete labelling or photobleaching). When comparing the size distribution of the tBid oligomers observed in the presence of Bax (Figure 4b, lower panels) with that observed in the absence of Bax (as reported in Ref. [29]), the only visible difference is a slight shift of the distribution towards smaller size oligomers in the presence of Bax.

We first concentrate on the distribution of intensities for stationary tBid and Bax (Figure 4b, left panels, in which the results of all our experiments are congregated regardless of Bax concentration). Detected event intensities have been normalized by the known intensity of the monomer, which was separately measured before each experiment using fluorescence correlation spectroscopy (FCS), and found to be on average $\langle B \rangle = 13$ kHz for cBid-Alexa647 and $\langle B \rangle = 7.6$ kHz for Bax-HyLite488. For both tBid and Bax, we see a peak around a normalized intensity of 1 (corresponding to particles with brightness B) followed by a series of harder to distinguish peaks going up to a normalized intensity of about 10. Intensity bins of width 1 and centred on integer values (corresponding to $B = 1, 2,$ etc.) were used to sort the data into putative monomers, dimers, etc. Although this classification is not very precise (because of day-to-day variation in the value of B, incomplete labelling and photobleaching), it allows us to conclude that, for both proteins, there is a broad distribution of stationary oligomers up to the tetramers, with rarer oligomers as large as decamers.

The distribution of intensities for mobile particles is noticeably different than that of stationary particles, with a shorter tail at higher intensities (Figure 4b, right panels). Since these particles have a diffusion coefficient around 10 to 20 µm^2 s^{-1}, our simulations show that an oligomer with brightness nB should have an intensity around 0.7 nB (Figure 4a, right panel). The abundance of different mobile oligomers was therefore estimated by binning the distributions of normalized intensities into bins of width 0.7 centred around multiples of 0.7. For both tBid and Bax, the proportion of low stoichiometry detections is higher for mobile particles than for stationary particles. This is true even if, for mobile particles, the intensity threshold ($\simeq B/2$) is close to the centre of the monomer bin, meaning that a significant number of mobile monomers must go undetected. This suggests that oligomer immobilization in the SLB, which is probably due to deeper insertion into the lipid bilayer, tends to be associated with larger stoichiometries (in agreement with previous observations for tBid [29]).

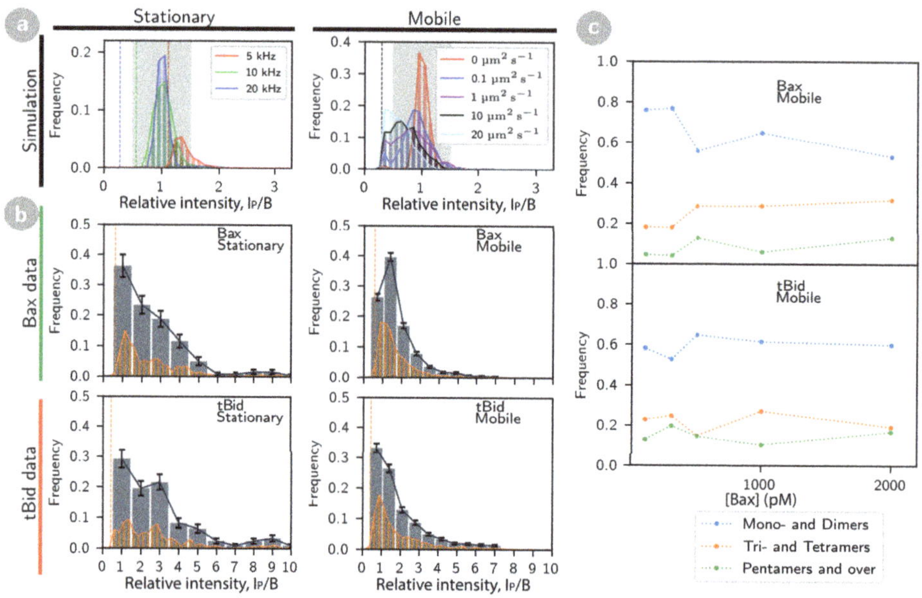

Figure 4. Intensity distribution and stoichiometry of detected events. (**a**) Intensity distributions for events detected in simulated confocal images for either stationary particles with different brightness (left panel) or particles with brightness $B = 20$ kHz and different diffusion coefficients. (**b**) Intensity distributions for events detected in confocal images of SLB incubated with tBid-Alexa647 and Bax-HyLite488, classified for each protein into stationary and mobile particles (regardless of Bax concentration). All intensities have been normalized by the known particle brightness of the monomer. The experimental distributions are binned in two different ways, with a small bin width of $B/4$ for visualization (orange bars) and with a larger bin width of B or $0.7B$ for stationary and mobile particles, respectively, representing the expected apparent brightness of a monomer for these two types of particles (blue bars). The coarse binning thus provides an estimate of the relative abundance of different protein stoichiometry. In (**a**,**b**), the dashed vertical bars indicate the threshold intensity used for particle detection. (**c**) Oligomer frequency versus Bax-HyLite488 concentration (for a constant 0.2 nM tBid-Alexa647 concentration).

Looking at relative oligomer abundance as a function of the concentration of Bax incubated with the SLB (Figure 4c), and focussing on the mobile particles which are more abundant and for which we have better statistics, shows that in all the explored experimental conditions, Bax was active enough and concentrated enough to form oligomers with a size (tetramer) generally considered large enough to constitute a pore [51]. A slight decrease in the concentration of low stoichiometry Bax oligomers (monomers and dimers), at the profit of larger size oligomers, can be seen when Bax concentration is increased (no discernible change in oligomer composition can be detected for tBid).

2.3. Quantification of the Interaction between tBid and Bax

Two-channel confocal images hold information about potential molecular interactions via the colocalization of events detected in both channels. A difficulty in positively identifying interactions is that particles might coincidentally be located within the same pixel. To avoid this issue, we have previously proposed the use of a quantity called the particle cross-correlation coefficient, χ, calculated as the spatial cross-correlation between the fluorescence intensity fluctuations recorded in both channels over a small region around the particle (see the Methods section for details) [33]. For perfectly co-localized particles, we expect $\chi = 1$, while for distant particles $\chi = 0$. Here we explore the sensitivity of χ on the distance between particles through simulations, before using it to detect interactions between tBid and Bax particles.

2.3.1. The Particle Cross-Correlation Coefficient Decreases with the Distance between Particles

Particle tracks and dual-channel confocal images were simulated for pairs of particles (one in the green detection channel, with $w_{0,g} = 320$ nm, and one in the red detection channel, with $w_{0,r} = 370$ nm). The position of the two particles was either completely independent or offset by a fixed value Δr, and the particles were either mobile or stationary. For each condition, 100 pairs of confocal images (each containing 14 to 20 particles) were generated. Images were analyzed as described above, and for each detection event in the first channel, χ was calculated (Equation (5)). For exactly co-localized particles ($\Delta r = 0$), the obtained values of χ are peaked around 1, but when Δr is increased, the average value of χ decreases continuously, down to $\bar{\chi} \simeq 0$ as Δr reaches $w_0 \simeq 0.35\,\mu$m, and below 0 to a minimum of $\bar{\chi} \simeq -0.2$ for $\Delta r \simeq 2w_0$ (Figure 5a). For larger separations $\bar{\chi} \simeq$ goes back to 0. In addition, the distribution of values for χ broadens when Δr increases. A decrease in molecular brightness (corresponding to an increase in photon noise), also results in a broadening of the peak, but not in a shift of the peak position, as expected since the systematic effect of photon noise is corrected for when calculating χ (data not shown). Thus, the average value of χ reflects the distance between the particles when it is below the diffraction limit. Similar conclusions are obtained from the simulation of mobile particles (data not shown).

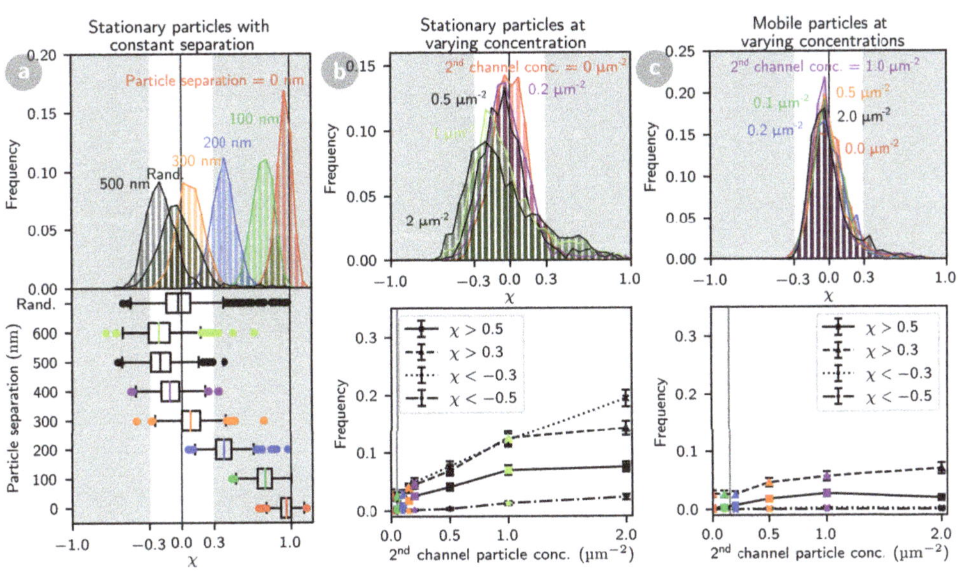

Figure 5. Distribution of particle cross-correlation coefficients obtained for simulated two-channel confocal images. (**a**) Particle cross-correlation distribution for particles separated by a fixed distance Δr. (**b**,**c**) Effect of increasing the particle concentration in the second channel for randomly placed particles, either stationary (**b**) or mobile (**c**). The lower panels in (**b**,**c**) give the frequency of accidentally correlated ($\chi > 0.3$ and $\chi > 0.5$) and anti-correlated ($\chi < -0.3$ and $\chi < -0.5$) events. All distributions show the cross-correlation of the particles detected in the green channel. Grey vertical lines in (**b**,**c**) lower panels indicate the experimental conditions.

As the issue of accidental co-localization becomes more preponderant at high surface concentration of particles, we also performed simulations where the concentration of the particles in the second channel was increased up to 200 particles per image (2 particles per μm^2). When particles in both channels are stationary (Figure 5b), the fraction of particles for which $\chi > 0.3$ increases up to $\simeq$14%, as more particles in the second channel accidentally co-localize with detected particles in the first channel. Interestingly, this is accompanied by a comparable increase in the fraction of particles with $\chi < -0.3$, as some of the particles in the second channel happen to be within a distance w_0 to $2w_0$ of the detected particles. Similar results are obtained when the particles in the second channel are mobile, except that the frequency of accidental co-localizations is slightly lower (SI Figure S4). Thus, accidental co-localizations, when the detected particles are immobile (and regardless of whether the particles in the second channel are mobile or immobile), have a characteristic signature which is different from that of true colocalization events (which never result in $\chi < -0.3$).

The situation is different when the detected particles are mobile (Figure 5c), since in that case accidental co-localizations (i.e., events with high χ) are much less likely to occur, even at high surface concentration of particles in the second channel. This is both because events have a smaller footprint and because each mobile particle has a unique trajectory and therefore a unique shape in the confocal image, which is unlikely to be accidentally matched by that of another particle. Calculating the particle cross-correlation coefficient is therefore especially interesting in this case. In our experimental conditions (mobile particle concentration of 0.15 particles/μm^2), we expect accidental co-localizations to occur for only 4% of events (if choosing $\chi = 0.3$ as the detection threshold) or 1.7% of events (if choosing $\chi = 0.5$ as the detection threshold).

2.3.2. Co-Localization of tBid and Bax

The result of the particle cross-correlation analysis applied to the set of confocal images acquired for SLB incubated with tBid-Alexa647 and Bax-HiLyte488 is shown in Figure 6, where the distributions shown are aggregated data for all the Bax concentrations studied. All these distributions (whether the particles are detected in the tBid or the Bax channel, and whether particles are stationary or mobile) show a dominant narrow peak exactly centred at $\chi = 0$ (corresponding to a large population of non-interacting particles), a second smaller peak centred at χ = 0.3–0.5 (corresponding to a population of particles with significant cross-correlation with a particle in the other channel), and a few rare events with $\chi > 0.5$ (corresponding to particles positively bound to a particle in the other channel). Almost no events with clear anti-correlation ($\chi < -0.3$) are detected.

Concentrating first on stationary particles (Figure 6, two left panels), we see that the number of positively correlated events is above the expected maximum number of false positives (lower panels), both when considering all events together, and when separating them into different stoichiometries. In addition, since these positively correlated events are not accompanied by any negatively correlated events, they do not fit the pattern expected for accidental co-localizations. We therefore considered all stationary events with $\chi > 0.3$ as corresponding to tBid-Bax complexes.

In the case of mobile tBid oligomers, the frequency of positively correlated events (whether they are defined as $\chi > 0.3$ or $\chi > 0.5$) is significantly larger than what would be expected for accidental co-localizations (Figure 6, right panels), and becomes very substantial for larger oligomers. The case is not as clear cut for mobile Bax oligomers, as the levels of detected positively correlated events are very close to the maximum expected level for accidental co-localizations, except for larger Bax oligomers where an above-background positive correlation is clearly seen. We note that when separating the data according to the Bax concentration used for incubation, no obvious trend was observed (SI Figure S5). For mobile particles, we conservatively estimated the number of tBid-Bax complexes by subtracting the maximum number of accidental co-localizations from the number of mobile events with $\chi > 0.3$.

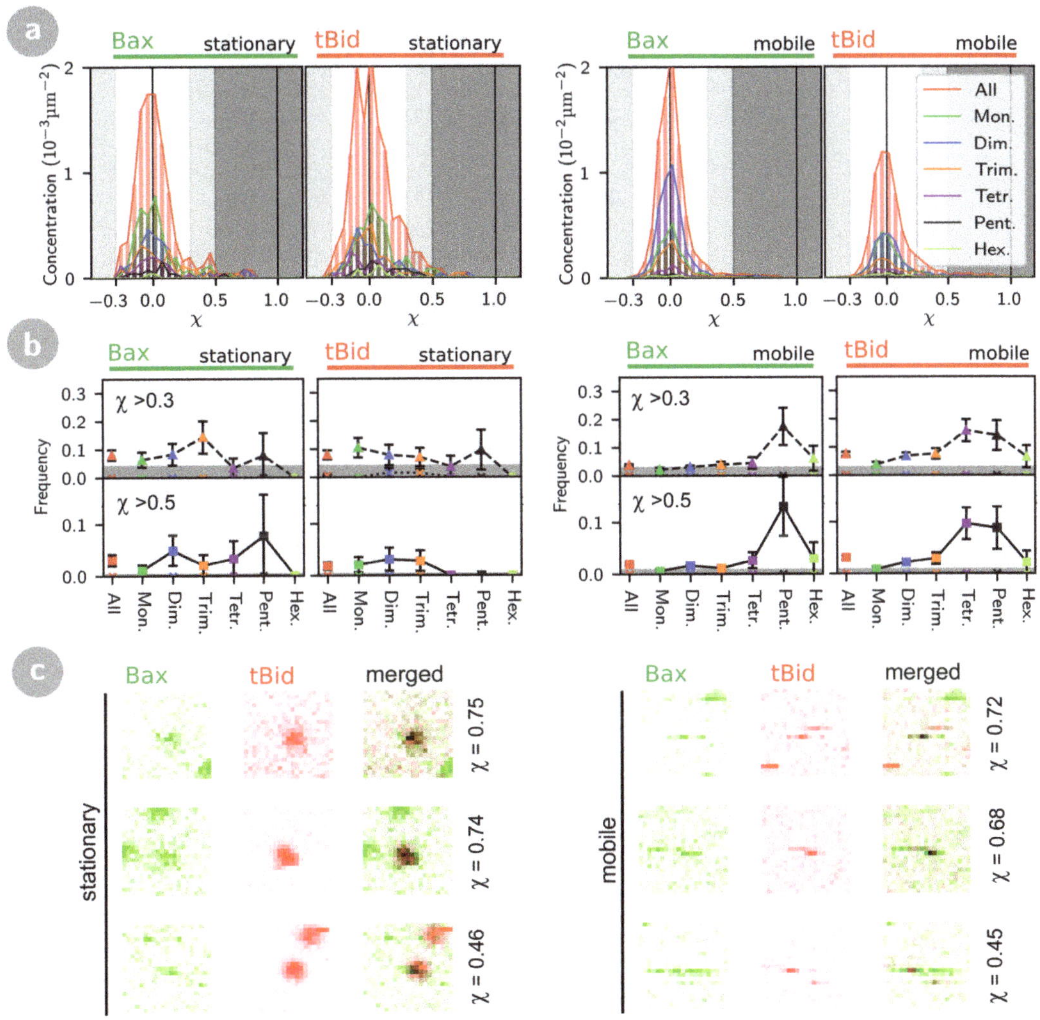

Figure 6. Particle cross-correlation coefficients measured for stationary and mobile tBid and Bax oligomers. (**a**) Cross-correlation coefficient distributions for all oligomers. The light grey areas correspond to slight cross- or anti-correlation ($0.3 < \chi < 0.5$ or $-0.5 < \chi < -0.3$) and the dark grey area highlights strong cross-correlation ($\chi > 0.5$). (**b**) Fraction of correlated particles ($\chi > 0.3$, upper panels, triangles and dashed lines) and strongly correlated particles ($\chi > 0.5$, lower panels, squares and continuous lines) in each oligomer class. The highlighted areas show the absolute maximum amount of accidental positive correlations to be expected given a particle concentration similar to that observed in experiments (0.05 particles/µm^2 for stationary particles and 0.15 particles/µm^2 for mobile particles), as obtained from the simulations shown in Figure 5b,c. In each panel, the fraction of detected anti-correlated particles ($\chi < -0.3$ or $\chi < -0.5$) are also shown (X symbols and dotted lines). (**c**) Representative examples of detected particle pairs with either slight or strong correlation. The shown particle cross-correlation coefficient is the one calculated for the green particle.

2.3.3. Dissociation Constant of tBid-Bax Complexes

The two-dimensional dissociation constant for tBid and Bax ($2D - K_D$), was obtained using either the tBid or the Bax channel to detect complexes, and for particles with different mobilities or stoichiometries. The $2D - K_D$ was calculated from the observed surface concentrations of non-interacting particles (c_{Bax} and c_{tBid} with $\chi < \chi_{\text{tresh}}$) and of interacting complexes ($c_{\text{Bax-tBid}}$, calculated from the number of events with $\chi > 0.3$ in the considered channel, and in the case of mobile tBid subtracting the maximum number of accidental co-localizations predicted from simulations), using Equation (7). Dissociation constants have the dimension of a concentration, thus the $2D - K_D$ has the dimension of a surface concentration. When considering all membrane species of tBid and Bax (mobile and stationary, regardless of stoichiometry and initial Bax concentration), we find an apparent dissociation constant $2D - K_D = 1.1\,\mu\text{m}^{-2}$. However, separating the data into mobile particles (using the tBid channel to detect complexes) and stationary particles (using either the tBid or the Bax channel to detect complexes) show that the nature of that equilibrium changes significantly when the proteins change conformation: the complex is much more stable for stationary (membrane-inserted) proteins ($2D - K_{D,\text{stationary}} = 0.1\,\mu\text{m}^{-2}$) than for mobile (loosely-bound) proteins ($2D - K_{D,\text{mobile}} = 1.6\,\mu\text{m}^{-2}$).

In contrast, we see no significant change in the dissociation constant when considering samples incubated with different Bax concentrations, or populations of either tBid or Bax with different apparent stoichiometries. This is illustrated in Figure 7, which shows the concentration of protein complexes, $c_{\text{Bax-tBid}}$, as a function of the product $c_{\text{Bax}}\ c_{\text{tBid}}$, for all these different populations. Populations with the same $2D - K_D$ should lay on the same line of slope 1. All subpopulations of mobile particles (regardless of the initial Bax concentration or particle stoichiometry) more or less fall on a line corresponding to $2D - K_{D,\text{m}} \simeq 1.6\,\mu\text{m}^{-2}$, while all subpopulations of stationary particles fall on a $2D - K_{D,\text{s}} \simeq 0.1\,\mu\text{m}^{-2}$ line. Thus, the main determinant of the $2D - K_D$ is the conformation of the protein and its degree of insertion in the membrane.

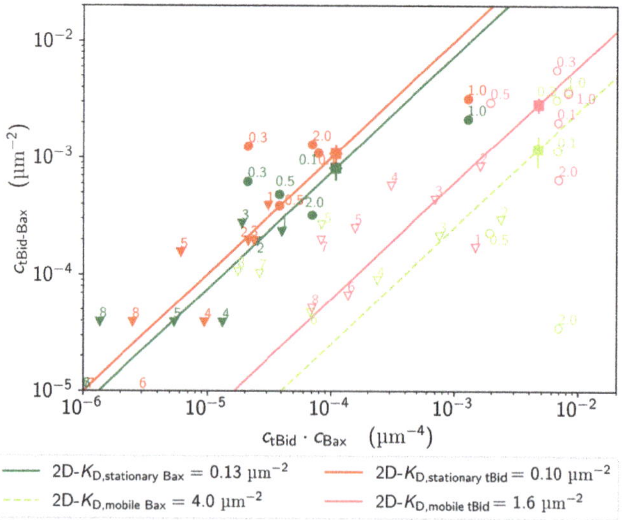

Figure 7. Surface concentration of the tBid-Bax complexes (defined as particles with $\chi_{\text{thresh}} > 0.3$) detected either in the Bax channel (green symbols) or the tBid channel red symbols) as a function of the product of the reactants. Data are shown both for stationary particles (dark red and dark green symbols) or mobile particles (light red and light green symbols). Each circle represents the aggregated data for a specific Bax concentration used for incubation (indicated on the figure, in nM), while each triangle represent the aggregated data for a specific oligomer size (also indicated on the figure). Squares represent the value obtained for the whole data set. The solid lines are fit of the data with Equation (7) for stationary proteins on the one hand and mobile proteins on the other hand, assuming a constant $2D - K_D$ (on this log-log plot the slope of these lines is 1 and their vertical offset, $\log[2D - K_D^{-1}]$ is linked to the inverse of the dissociation constant).

3. Discussion

In the present work, by combining the ideas of single particle detection (which let us identify and characterize specific protein complexes) and image cross-correlation spectroscopy (which let us identify real protein–protein interactions) we were able to directly measure the two-dimensional dissociation coefficient of the membrane-associated tBid-Bax complex. This was achieved in spite of the extremely low fraction of interacting proteins found at the low surface concentrations (below the $2D - K_D$) necessary to reach single particle detection conditions. The reconstituted system we have used differs from in vivo cellular systems in several ways, amongst which: the lipid membrane in our reconstituted system is planar, supported by a solid substrate and devoid of proteins other than tBid and Bax; we chose to use human Bax and murine tBid, for direct comparison with previous studies that looked at quantifying the tBid/Bax interaction (and had shown these two proteins had a direct interaction and that together they were able to permeabilize membranes [10,34,52–54]); the purified tBid and Bax proteins added to the system have each been labelled with a small organic fluorophore, and Bax has been subjected to three point mutations; for convenience the experiments were performed at room temperature rather than physiological temperature. Because of these differences, the dissociation constants we report here can be expected to be slightly different from those encountered in vivo in human cells. However, our measurements: (*i*) bring a proof-of-principle that two-dimensional dissociation constants for membrane proteins can be measured from confocal images using the concept of particle cross-correlation coefficient; and (*ii*) allow comparing the dissociation constants for several membrane conformations of tBid and Bax. Indeed, our most important finding is that the stability of the tBid/Bax complex depends not only on whether tBid and Bax are in a soluble form or in a membrane form, but also on their membrane conformation, a refinement of our current understanding of Bax activation which is summarized in Figure 8.

If one does not distinguish between the soluble and membrane forms of the proteins, the equilibrium of the system can simply be described by an apparent dissociation constant (Figure 8a). This dissociation constant was previously measured, using ensemble fluorescence resonance energy transfer, and found to be $K_D^{app} = 20$ nM in a liposome solution with total lipid concentration $[L] = 125\,\mu\text{M}$ [10]. However, this apparent dissociation constant is insufficient to describe the tBid/Bax system. First, tBid and Bax hardly interact in their soluble form [10,26]. This is true of many other pairs of Bcl-2 family proteins and the tenet of the "embedded together" model which emphasizes the importance of the membrane in enabling interactions between Bcl-2 family proteins [16,55]. Second, most Bcl-2 family proteins are in a dynamic equilibrium between their soluble and membrane forms, a phenomenon known as "retro-translocation" [22–24,56]. These considerations have led to models such as the one illustrated in Figure 8b, where the proteins exists both in soluble and membrane forms, and where regulation of the equilibrium between the two forms of Bax becomes another way of controlling its activity [2,5,57]. The equilibrium of the system is then characterized by two partition coefficients (P_{tBid} and P_{Bax}) describing the equilibrium between the soluble form and membrane form of tBid and Bax, and one two-dimensional dissociation constant for the membrane tBid/Bax complex ($2D - K_D$). The value of this dissociation constant can be estimated from the measured value of K_D^{app}, since comparing these two models shows that $2D - K_D \approx K_D^{\text{app}} \times d[L]v_L P_{tBid} P_{Bax}$, where $d = 4\,\text{nm}$ is the membrane thickness and $v_L = 7.6 \times 10^{-4}\,\text{L}\,\text{mol}^{-1}$ is the molar volume of the lipids (see supplemental material and Figure S1 for a derivation and discussion of this relationship). Considering that $P_{tBid} = 7000$ [29] and assuming $P_{Bax} = 3500$, leads to the order of magnitude estimate $2D - K_D \sim 0.1\,\mu\text{m}^{-2}$.

Our single-particle detection confocal experiments, however, show that the model shown in Figure 8b is oversimplified, by confirming the multiplicity and complexity of the tBid and Bax membrane forms glimpsed in previous studies, with both proteins in equilibrium between at least two different types of membrane conformations. In the first ensemble of conformations, the proteins have a very low mobility ($D = 0$ to $0.5\,\mu\text{m}^2\,\text{s}^{-1}$ in the SLB), signalling that they must be inserted in the bilayer with one or more helices going all the way through. The existence of transmembrane conformations linked to membrane permeabilization was suggested for both tBid and Bax by single residues fluorescence labelling experiments [43,58], and for tBid by single particle tracking experiments [29]. Solving the membrane structure of these proteins has proven challenging, yet models for transmembrane forms of Bax have been proposed based on x-ray, electron paramagnetic resonance and fluorescence experiments [58–61] or simulations [62]. Models also exist for tBid [43,63,64], although the consensus in this case is that tBid is only partially inserted into the membrane. In contrast to these membrane-embedded forms, the second detected membrane conformational ensemble is characterized by a very high mobility ($D \simeq 10$ to $20\ \mu m^2\ s^{-1}$), considerably larger than that of the lipids ($D_{lipid} \simeq 2$ to $3\ \mu m^2\ s^{-1}$ in mitochondria-like SLB [29,46]). This second fraction thus corresponds to loosely bound proteins, probably associated to the membrane via electrostatic interactions, and probably retaining a structure close to that of the soluble form (as suggested for Bax by simulations [62] and small-angle scattering experiments [65]). This population has been observed before in the same system for tBid using single-point FCS, and its average diffusion coefficient estimated to be $D = 22\ \mu m^2\ s^{-1}$ [29]. The loosely-bound form, appearing as streaks in our confocal images, is too fast to be detected in camera-based single-particle tracking experiments, thus its existence has not often been acknowledged. Despite its elusiveness in most experiments, this form is probably crucial in vivo, both by allowing a dynamic equilibrium with the soluble form of the proteins (retro-translocation) and by allowing the protein to perform a quick two-dimensional search for binding partners on the membrane (electrostatic scanning [66]).

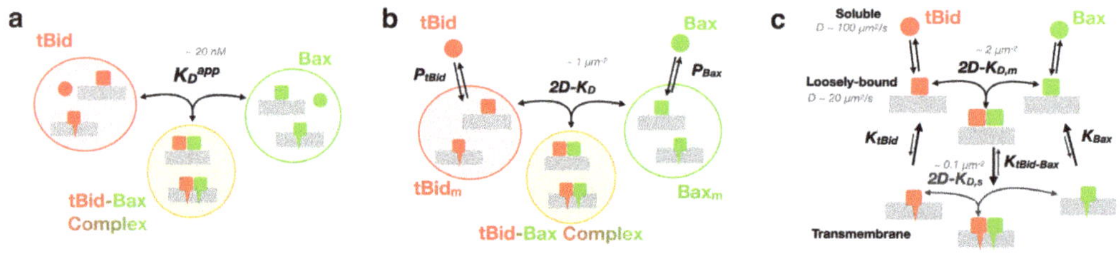

Figure 8. Three different levels of models for the tBid-Bax interaction. (**a**) In the simplest model, all the forms of tBid, Bax and the tBid-Bax complex are each considered as a separate conformational ensemble. The interaction between the tBid and Bax ensembles is described by a simple apparent dissociation constant, K_D^{app}. (**b**) An improved version of the simple model in (a) considers the solution and membrane forms of each protein as separate conformational ensembles (the tBid-Bax complexes exist only as membrane-associated complexes). At equilibrium the system is described with two partition coefficients, P_{tBid}, P_{Bax}, characterizing the equilibrium (retro-translocation) between the solution and membrane forms of each protein, and a two-dimensional dissociation constant, $2D - K_D$. (**c**) A more accurate model, where the membrane forms of each protein are separated into two different conformational ensembles based on their degree of insertion in the membrane (but regardless of stoichiometry), is necessary to explain the results of our single particle experiments. The equilibrium of this system is described by four independent partition coefficients, and two two-dimensional dissociation constants, $2D - K_{D,m}$ for the mobile loosely-bound proteins and $2D - K_{D,s}$ for the stationary deeply inserted proteins.

Our results show that there is one dissociation constant associated with the mobile species superficially bound to the membrane, $2D - K_{D,m} \simeq 1.6\ \mu m^{-2}$, and a second significantly lower one associated with the stationary transmembrane species, $2D - K_{D,s} \simeq 0.1\ \mu m^{-2}$. This leads us to propose the model shown in Figure 8c, where, for each protein, the membrane conformational ensemble is explicitly separated into loosely-bound (mobile) and transmembrane (stationary) species, and within each of these ensembles the tBid and Bax binding equilibrium is characterized by a different dissociation constant. It is striking that, whereas the dissociation constant we measure for the superficially bound tBid/Bax complex is comparable to that measured in vivo for membrane receptor dimerization [31,67], the dissociation constant measured for the deeply inserted tBid/Bax complex is one order of magnitude lower, signalling a much tighter interaction. The introduction of a loosely bound state for tBid and Bax with a detectable but weak affinity for each other is consistent with the notion that interactions between Bcl-2 family proteins are tighter when the proteins are interacting with the membrane, with the added intricacy that the affinity between the two proteins progressively increases as they get inserted deeper into the membrane. This model is also consistent with the finding that Bax oligomers can form before Bax membrane insertion [68].

The model shown in Figure 8 is still likely a simplification, since loosely-bound transmembrane species might be able to directly interact, and stoichiometry of complexes are not taken into account. It also says nothing of the actual pore formation process which comes after Bax activation due to the formation of large Bax oligomers. Large complexes were detected in our system, both for tBid and Bax, and a fraction of these large complexes were mixed complexes that contained both Bax and tBid proteins (an observation that is somewhat reminiscent of the recent finding that oligomers of Bcl-XL can bind both tBid and Bad at the same time [69]). This is surprising, because the tBid/Bax interaction is often thought of as a hit-and-run interaction, where the tBid/Bax complex dissociates before the activated Bax can bind other Bax molecules (allowing tBid to activate more Bax)—instead, it seems that either tBid and Bax can remain associated as the Bax oligomers grow or that tBid is able to re-bind to already active oligomeric Bax. The capacity of tBid to participate in large oligomeric complexes, usually thought to be specific to the multidomain effector proteins (Bax and Bak), might be related to the fact that it is structurally related to Bax [70,71]. It might thus not be representative of the behavior of other BH3-only activator proteins, which do not share this structural homology.

Our work also informs on the dynamics of the tBid/Bax interaction. A stable and short range interaction (on the timescale of the confocal imaging performed here, i.e., the time to image a single particle, 0.1 to 0.3 s in the conditions of our experiments) would result in a particle cross-correlation coefficient close to $\chi \simeq 1$ (as shown by our simulations, see Figure 5a). Instead, the experimental values of χ we measured for interacting particles were lower, around 0.3 to 0.8. This is not an artefact due to imperfect channel overlap, as values $\chi \simeq 1$ are obtained for doubly-labelled liposomes [33]. One intriguing possibility would be that instead of being the result of a direct molecular interaction between tBid and Bax, the complexes that we detect are in fact accumulation of proteins in small $\simeq$100 nm lipid domains. This would explain the very varied stoichiometries observed for both proteins separately as well as for the mixed tBid/Bax complexes, but it would not be totally consistent with the fact that direct contact can be detected between both proteins using FRAP [10]. Perhaps a more likely possibility is that this imperfect correlation is due to the transient nature of the tBid/Bax interaction, which is known to be dynamic [10] and was suggested to be short lived [26]. A tBid/Bax interaction with a lifetime on the order of 0.1 s (similar to the time necessary to record the few lines in the confocal image in which the complex appears) would result in a value of χ that is positive but less than 1, as we observed.

Recent protein knockout experiments have shown that Bax can be activated without the help of any of the eight canonical BH3-only proteins [72,73]. This calls into question the biological relevance of the direct activation model, which is centred around the idea that BH3-only activators (such as tBid and Bim) activate effector proteins (Bax and Bak) through direct physical interaction [19]. It is true that ever since Bid was first identified in a screen of proteins binding to Bax [34], a direct interaction between tBid and Bax has remained hard to detect in live cells (which is maybe not so surprising since this interaction is so strongly correlated with cell death). On the other hand, as shown in multiple reconstituted lipid systems, there is no question that tBid and Bax interact as long as a lipid membrane is present, and that tBid triggers the membrane localization and the pore-forming activity of Bax [7,10]. Even so, as far as we know, our experiments represent the first time the tBid/Bax interaction could be observed at the single particle level (Figure 6).

Following this challenging of the direct activation model, it has been proposed that the mitochondrial membrane (and not BH3-only activators such as tBid) played the role of Bax activator [19]. This proposal is consistent with observations that a cardiolipin-rich membrane can lead to Bax auto-activation [74]. However, cardiolipin appears in the outer mitochondrial membrane only after apoptosis is triggered, so in cells this activation mechanism might be used for amplification rather than initiation. In any case, in order to be activated by or with the help of the membrane, Bax needs to be brought and held at the membrane. This might be precisely the role of activator proteins such as tBid and Bim, which have strong affinities for the membrane and can associate with it quickly and almost irreversibly. Indeed what stands out in the tBid/Bax interaction model inspired by our experiments (Figure 8c) is that the interaction between tBid and Bax seems designed to pull Bax down and anchors it in the membrane, maybe just long enough for Bax to become activated by the membrane, and maybe greatly accelerating a process which may happen in the absence of tBid, but would be much slower. This formalizes an already existing narrative for the tBid/Bax interaction, which is that tBid "recruits" Bax to the membrane, and can shift the equilibrium between the soluble and membrane forms of Bax [22–24,75].

Eventually, a strong test of the direct interaction model will be to directly compare the affinity of tBid for Bax to that of tBid for Bcl-XL. Values are available for the K_D of tBid/Bax (20 nM) [10] and tBid/Bcl-XL (1 nM) [76]) in the presence of membranes. However, these values cannot be compared, because they have been measured in different reconstituted systems (liposomes vs. mitochondria) and without knowing exactly how much of the proteins was at the membrane. Considering only the membrane fraction of the proteins and measuring the $2D - K_D$ like we have done here resolves this issue. If indeed the direct interaction model is irrelevant for tBid, then it should be possible to show that the $2D - K_D$ for the Bcl-XL/Bid interaction is much smaller than that for the tBid/Bax interaction. In the future, the method employed here can also be applied to compare the $2D - K_D$ of different tBid and Bax mutants (e.g., using mutations in the BH3 motif of tBid in order to modify the affinity of tBid for Bax, Bcl-2 and Bcl-XL [34,77]).

4. Materials and Methods

4.1. Protein Purification and Labelling

4.1.1. tBid

Full-length murine Bid with a N-terminal His-tag and a cysteine to serine mutation at position 30 (leaving a single cysteine residue at position 126, and thus referred to as Bid 126C), was purified as previously described [10,28,43,76]. Briefly, a plasmid encoding the Bid 126C construct was transformed into *E. coli* BL21 arabinose inducible cells by electroporation. Single colonie large scale 5 L cultures in lysogeny broth (LB) were induced with 0.2% arabinose for 3 h at 37 °C. Cells were then spun down at 5000 g for 30 min at 4 °C and the pellet was resuspended in Bid lysis buffer: 10 mM HEPES pH 7, 100 mM NaCl, 10 mM imidazole, 1 mM PMSF, 10 µg mL^{-1} DNase and 1× Halt Proteinase Inhibitor Cocktail (Thermo Fischer, Waltham, MA, USA). Cells were then vortexed, passed through a 16 gauge needle and lysed by at least 2 passes through a French Press. Cell debris were removed by centrifugation at 15,000 *g* for 30 min at 4 °C. The supernatant was incubated with 1.5 mL Ni-NTA agarose resin (Qiagen, Hilden, Germany) for 1.5 h at 4 °C with mixing. The slurry was added to a 20 mL column (Bio-Rad, Hercules, CA, USA) that retained the resin, after which the column was washed with 50 mL Bid wash buffer: 10 mM HEPES pH 7, 300 mM NaCl, 1% (*w*/*v*) CHAPS (Bioshop Canada, Burlington, ON, Canada) and 10 mM imidazole. The protein was eluted in Bid elution buffer (10 mM HEPES pH 7, 100 mM NaCl, 0.1% (*w*/*v*) CHAPS, 200 mM imidazole and 10% glycerol).

The purified Bid 126C was then fluorescently labelled with Alexa647 (Molecular Probes, now Thermo Fisher), which formed a covalent bond via maleimide chemistry to the cysteine residue at position 126 [43]. Briefly, proteins in Bid storage buffer (10 mM HEPES pH 7, 100 mM NaCl, 0.2 mM EDTA, 10% glycerol) supplemented with 0.5% *w*/*v* CHAPS were incubated with a 10-fold molar excess of dye for 2 h at room temperature and in the dark while under constant rotation. Unreacted free dye was then first quenched by addition of 1 mM DTT, then removed by gel filtration in a G-25 fine Sephadex column (GE Healthcare, Chicago, IL, USA). The labelling efficiency was determined from absorption measurements that gave both protein and dye concentrations, and found to be 70% in this case.

The fluorescently labelled Bid was then cleaved with Caspase-8 to obtain cleaved Bid (cBid), as described in Ref. [43]. The fluorescent label (at residue 126) is on the larger C-terminal fragment of the protein, referred to as truncated Bid (tBid). Since both the His-tag and the C30S mutation are on the small N-terminal fragment of Bid, the only difference between purified tBid-Alexa47 and wild type tBid is the presence of the fluorescent Alexa647 tag.

4.1.2. Bax

Full-length human Bax, with the two endogenous cysteine residues (at positions 62 and 126) replaced with alanine, and with an additional leucine to cysteine mutation at position 47 (Bax 47C) was purified as previously described [10,58,76,78]. Briefly, a plasmid encoding the Bax 47C construct fused to a C-terminal chitin binding domain followed by a self-cleaving intein sequence (intein-chitin binding domain or CBD, New England Biolabs, Ipswich, MA, USA) was transformed into *E. coli* BL21 arabinose inducible cells by electroporation. Single colonies were selected for large scale 5 L cultures in lysogeny broth (LB), and induced with 0.2% arabinose for 5 h at 30 °C. Cells were then spun down at 5000 g for 30 min at 4 °C. The pellet was resuspended in Bax lysis buffer: 10 mM HEPES pH 7, 100 mM NaCl, 0.2% (*w*/*v*) CHAPS, 1 mM PMSF, 10 µg mL^{-1} DNase and 1× Halt Proteinase Inhibitor Cocktail (Thermo Fischer). Cells were then vortexed, passed through a 16 gauge needle and lysed by at least 2 passes through a French Press. Cell debris were removed by centrifugation at 15,000× *g* for 30 min at 4 °C. The supernatant was incubated with 2 mL chitin resin (New England Biolabs) for 1.5 h at 4 °C with mixing. The slurry was added to a 20 mL column (Bio-Rad) that retained the resin, after which the column was washed with 50 mL Bax wash buffer (10 mM HEPES pH 7, 500 mM NaCl, 0.5% (*w*/*v*) CHAPS), followed by flushing with 60 mL Bax cleavage buffer (10 mM HEPES pH 7, 200 mM NaCl, 0.2 mM EDTA, 0.1% (*w*/*v*) CHAPS, 100 mM hydroxylamine). The column was then incubated in Bax cleavage buffer for 36 to 48 h at 4 °C, after which the protein was eluted in Bax cleavage buffer. Hydroxylamine was removed by dialysis in Bax storage buffer (10 mM HEPES pH 7, 200 mM NaCl, 0.2 mM EDTA, 10% (*w*/*v*) glycerol).

Purified Bax was labelled with the small neutral dye HyLite488 (Anaspec, Fremont, CA, USA) through maleimide chemistry, as described above for Bid, but at pH 7.5 in presence of a 2-fold molar excess of TCEP to promote reduction of the thiol group, and the labelling reaction was allowed to proceed for a total of 6 h. Free dye was removed and labelling efficiency was assessed as for Bid. The labelling efficiency in that case was found to be 81%. We note that labelling was also attempted for several different dyes (Alexa488, Atto488, Atto495) and one other position (residue 126), but these resulted in very low labelling efficiencies, as did the protocol without TCEP. To confirm that labeling occurred only at the cysteine residue, control experiments were done with a cysteine-less mutant of Bax and wild-type Bax. As expected, labelling was very low in the first case (11%) and about double that of Bax 47C in the second case (155%).

4.1.3. Purified Protein Activity

Since the purified and labeled proteins differ from their wild-type equivalent due to the presence of a fluorescence tag (and, in the case of Bax, due to the presence of three point mutations) it was important to test their activity. This was done for each batch of purified and fluorescently labelled protein using a standard ANTS release assay [78]. The pore-forming activity of cBid-Alexa647 (assessed in conjunction with purified unlabeled wild-type Bax), was found to be slightly (10 to 20%) higher than that of its wild-type equivalent. In contrast, the pore-forming activity of Bax-HyLite488 (assessed in conjunction with purified unlabeled cBid) was found to be lower (30 to 40%) than that of its wild-type equivalent. When using both labelled proteins together in the assay, the pore-forming activity was reduced by only 20 to 30% when compared to the corresponding wild-type proteins, which was judged acceptable.

4.2. Mitochondria-Like Supported Lipid Bilayers

Mitochondria-like solid-supported lipid bilayers were prepared by liposome fusion [29]. The lipid composition was chosen to represent that of mitochondrial membranes in yeast and vertebrates [79,80], with phosphatidylcholine (PC), phosphatidylethanolamine (PE), phosphatidylinositol (PI), phosphatidylserine (PS) and cardiolipin (CL) in a molar ratio of 48:28:10:10:4. All lipids were acquired from Avanti Polar Lipids (Alabaster, AL, USA) as either egg extracts (PC, PE), liver extracts (PI) or monounsaturated synthetic lipids ($(18:1)_2$ PS, $(18:1)_4$ CL). The different lipids, dissolved in chloroform, were mixed in the right proportion to reach a total lipid mass of 1 mg, after which the chloroform was allowed to evaporate first under a stream of argon and later in a vacuum at 25 inHg for at least 2 h. The lipid film was rehydrated with 1 ml of an assay buffer solution (10 mM HEPES pH 7, 200 mM KCl, 1 mM $MgCl_2$, and 0.2 mM EDTA), followed by 10 freeze-thaw cycles to obtain unilamellar liposomes, and extrusion through a 100 nm pore filter. The liposome solution was then injected into a 400 µL perfusion chamber (Bioptechs FCS2, Butler, PA, USA), fitted with a mica-coated glass coverslip, the manufacture of which is important both for proper SLB formation and confocal image quality. In order to ensure that the total thickness of the coverslip was less than 220 µm (the maximum thickness acceptable given the objective used for imaging), glass coverslips with thickness 170 µm were used, and the mica (25 mm diameter circular sheet, V-1 grade, purchased from SPI Supplies, West Chester, PA, USA) was first cleaved to a thickness of 6 to 12 µm (as determined from its weight). The mica sheet was gently pressed down onto the glass coverslip coated with an optical adhesive (Norland Products, Cranbury, NJ, USA) pre-heated to a temperature of 50 °C, and the assembly was exposed to UV illumination to fix the glue. Just before perfusion chamber assembly and lipid solution injection, the top layers of the mica were removed using clear packing tape to expose a clean mica surface. The sample was incubated at 37 °C for 1 h in order to give time to the liposome to fuse onto the mica surface and form a single SLB. The lipid membrane was washed by slow injection of assay buffer in the chamber, after which a 0.2 nM solution of cBid-Alexa647 in assay buffer was injected, followed by a 15 min incubation period at 37 °C. A Bax-HyLite488 assay buffer solution (with Bax concentration varying between 0.1 and 2 nM) was then injected in the chamber, again followed by an incubation period (2 h at 37 °C).

4.3. Confocal Imaging

Fluorescent proteins associated with the SLB were imaged using a dual-colodr confocal fluorescence fluctuation microscope (Insight, Evotec Technologies, Hamburg, Germany, now Perkin-Elmer, Waltham, MA, USA). Excitation was provided simultaneously at $\lambda_{ex} =$ 488 nm and $\lambda_{ex} = 635$ nm by two lasers (Sapphire 488-20 CDRH and Radius 635-25, both from Coherent, Santa Clara, CA, USA), whose beams were combined into a single optical fiber, before being focused into the sample using a water immersion objective (U-Apo, 40×, 1.15 NA, Olympus, Tokyo, Japan). The fluorescence signal was refocused through a 40 µm confocal pinhole, separated into green and red detection channels by a dichroic mirror and detected by two avalanche photodiodes (SPCM-CD3017, Perkin-Elmer). Before each experiment, the confocal pinhole was aligned and the observation volume was calibrated in both channels by performing a FCS experiment using nanomolar solutions of Alexa488 (diffusion coefficient $D = 390\,\mu m^2\,s^{-1}$ at 20 °C [81], $\lambda_{em} = 525$ nm) and Alexa647 (diffusion coefficient $D = 290\,\mu m^2\,s^{-1}$ at 20 °C [82], $\lambda_{em} = 665$ nm), both purchased from Molecular Probes (now Thermo Fisher). Calibration experiments were performed in a sample chamber prepared with a microscope coverslip coated with mica, exactly as the SLB sample to be imaged. Typically, the $1/e^2$ radius of the detection volume was estimated to be $w_g = 320\,\text{nm}$ in the green detection channel and $w_r = 370\,\text{nm}$ in the red detection channel, a bit larger than the expected values in ideal conditions $w_g = 0.51(\lambda_{\text{ex}} + \lambda_{\text{em}})/2/(2\ln 2)^{1/2}/\text{NA} = 190\,\text{nm}$ and $w_r = 245\,\text{nm}$. A second calibration step was then carried out using nanomolar solutions of Bax-HyLite488 and cBid-Alexa647, also through a mica-coated coverslip, to determine their molecular brightness. At the 20 µW excitation laser power used in each channel for these experiments, the molecular brightness of Bax-HyLite488 was found to be $B_g = 6(1)$ kHz and that of cBid-Alexa647 to be $B_r = 11(2)$ kHz. Finally, dual-color z-stacks of 10 µm by 10 µm (100 × 100 pixels) images were acquired with an oversampling pixel size $d = 100\,\text{nm}$ and a pixel dwell time $\delta = 1\,\text{ms}$ for multiple regions of interest in each sample. Typically, each stack contained 5 or 10 pairs of images, spaced by 0.5 or 1 µm, centred around the focal plane. For each stack, the pair of images with the highest maximum intensity was considered to correspond to the images for which the membrane plane was in focus, and selected for single particle analysis.

4.4. Simulations

To validate the particle detection and classification scheme used on confocal images, a program was written in Python (using the NumPy and Pandas libraries) to simulate particle diffusion in a plane and generate confocal images of this plane (with dimensions $n_x \times n_y = 100 \times 100$ pxl, pixel size $d = 100\,\text{nm}$ and pixel dwell time $\delta = 1\,\text{ms}$). To emulate the behaviour of particles which may move during the recording of a pixel intensity, the time step for the simulation τ was chosen to be a fraction of the exposure time, $\tau = \delta/m$, with $m = 10$. Initial positions for N different particles, $\vec{r}_0$, were randomly selected inside the image area. Tracks for mobile particles were simulated with the constant diffusion coefficient D, by generating at each time step n a displacement $\delta\vec{r}(n) = (\delta x, \delta y)$ for each particle, whose components were drawn from a Gaussian distribution:

$$\Psi[\delta x] = \frac{1}{\sqrt{4\pi D\tau}} \exp\left\{-\frac{\delta x^2}{4D\tau}\right\}. \tag{1}$$

With the knowledge of all particle positions at all times ($\vec{r}_p(t)$), the confocal image was generated on a pixel-by-pixel basis, were the intensity at pixel (i,j) is recorded starting at time $t_{ij} = (n_x j + i)\delta$. Assuming a Gaussian beam profile with $1/e^2$ radius w_0, a particle found at a distance $\Delta\vec{r}$ from the center of the pixel contributes on average the following number of photons during the time interval τ:

$$g[\Delta\vec{r}] = B\tau \exp\left\{-\frac{2(\Delta\vec{r})^2}{w_0^2}\right\}, \tag{2}$$

where B is the molecular brightness of the particle. For the simulations we used $w_0 = 0.3\,\mu\text{m}$ and $B = 5$ to 20 kHz. The average pixel photon intensity recorded at pixel (i,j) between t and $t+\tau$ can then be calculated by summing over all particles and over the whole dwell time:

$$I(i,j) = \sum_{n=0}^{m-1} \sum_{p=1}^{N} g\big[\vec{r}_{ij} - \vec{r}_p(t_{ij} + n\tau)\big] + i_B\delta, \tag{3}$$

where the constant background intensity of $i_B = 1.3$ kHz was added at each pixel, to match the typical background intensity measured in the experimental confocal images. Finally, photon noise was added to the image, by replacing the average photon count calculated using Equation (3) by a photon count drawn from a Poisson distribution with the average pixel intensity as the expectation value.

The Python code used to simulate confocal images is available in the publicly accessible repository: https://github.com/cecilefradin/BidBax_Simulation_and_Analysis (accessed on 30 July 2021).

4.5. Image Analysis

The confocal images were analyzed using an in-house algorithm, written as a Java plugin for ImageJ (https://github.com/cecilefradin/BidBax_Simulation_and_Analysis, accessed on 30 July 2021). This algorithm detects single particles in one channel, determines their size, shape and fluorescence intensity, then calculate for each one a single particle cross-correlation coefficient, which allows establishing whether it is bound to another particle in the second channel. The intensity distribution for a population of particles inform about the distribution of stoichiometries in that population, while the surface concentrations of interacting and non-interacting particles allow calculating two-dimensional dissociation constants.

4.5.1. Single Particle Detection

Particle detection and identification was done as detailed in [29]. Images were first searched for local maxima. Each maximum with intensity above the threshold value $I_T = \langle I_B \rangle + 0.2B$ (with B the value of the molecular brightness measured for the solubleprotein by FCS on the same confocal instrument, and $\langle I_B \rangle$ initially chosen as the average intensity in a region of 5×5 pixels centred around the pixel with lowest intensity in the image) was considered in turn, starting with the brightest one, and fitted with a two-dimensional Gaussian function:

$$H(x,y) = I_p \exp\left\{ -\frac{2(x-\bar{x}_p)^2}{w_{x,p}^2} \right\} \exp\left\{ -\frac{2(y-\bar{y}_p)^2}{w_{y,p}^2} \right\} + I_{B,p}. \quad (4)$$

This fit returned the particle position $(\bar{x}_p, \bar{y}_p)$ with sub-pixel precision, the particle fluorescence intensity (I_p), the $1/e^2$ radii of the image of the particle along the x and y direction ($w_{x,p}$ and $w_{y,p}$) an estimate of the local background intensity $(I_{B,p})$, and a normalized χ-squared value $(\chi_{N,p})$. Once the fit was over, a square region of $0.5\,\mu\text{m} \times 0.5\,\mu\text{m}$ around the particle was erased from the image and the next most intense remaining local maximum was considered, using an updated threshold value (with $\langle I_B \rangle$ now calculated as the average of all the local background intensities estimated from previous fits). Only particles for which the fit was judged acceptable ($\chi_{N,p} < 2,\ I_p + I_{B,p} > I_T + \sqrt{I_T}$) were selected for further analysis.

4.5.2. Single Particle Cross-Correlation Coefficient

To determine whether a particle detected in a given channel interacts with another particle in the other channel, a single particle cross-correlation coefficient was calculated for that particle. First, an area of size 7×7 pixel was delimited around the detected particle. Then the intensities I_{ch1} and I_{ch2} recorded at each pixel within that area for either channels were cross-correlated via:

$$\chi = \frac{\langle (I_{ch1} - \langle I_{ch1} \rangle) \cdot (I_{ch2} - \langle I_{ch2} \rangle) \rangle}{\sqrt{\sigma_{ch1}^2 - \langle I_{ch1} \rangle}\sqrt{\sigma_{ch2}^2 - \langle I_{ch2} \rangle}} \quad (5)$$

with

$$\sigma_{ch(k)}^2 = \left\langle (I_{ch(k)} - \left\langle I_{ch(k)} \right\rangle)^2 \right\rangle, \quad (6)$$

and where averages are taken over the box drawn around the particle. Equation (5) includes a correction for the photon noise, via subtraction of the variance of the Poisson distributed photon noise ($\left\langle I_{ch(k)} \right\rangle$) from the total variance ($\sigma_{ch(k)}^2$), leaving only the contribution due to the average signal of the particle, as discussed in Friaa and Fradin [33].

4.5.3. Dissociation Constant

A value of the two-dimensional dissociation constant, 2D-K_D, was obtained by comparing surface concentrations of uncorrelated particles with the channel-averaged surface concentration of the correlated particles. It can be written as:

$$2D - K_D = \frac{c_{\text{Bax}}\, c_{\text{tBid}}}{c_{\text{Bax-tBid}}}, \tag{7}$$

where c_{Bax} and c_{tBid} are the respective sums of particles per area with $\chi < 0.3$ (no correlated signal in the other channel) and $c_{\text{tBid-Bax}}$ is the sum of particles per area with $\chi > 0.3$ (correlation of the signals in both channels) corrected for the expected number of incidental correlation predicted by the simulations.

Supplementary Materials: The following are available online at https://www.mdpi.com/1422-0067/22/15/8240/s1.

Author Contributions: Conceptualization, C.F.; methodology, M.R., M.K., M.M. and C.F.; software, M.R., M.M., S.W. and J.M.M.-M.; formal analysis, M.R. and C.F.; investigation, M.R., M.K. and M.M.; writing—original draft preparation, M.R.; writing—review and editing, C.F.; visualization, M.R.; funding acquisition, C.F. All authors have read and agreed to the published version of the manuscript.

Funding: This work was supported by grants FRN86657 from the Canadian Institutes of Health Research (CIHR) and RGPIN-2015-06362 from the Natural Sciences and Engineering Research Council of Canada (NSERC) to C.F.

Institutional Review Board Statement: Not applicable.

Informed Consent Statement: Not applicable.

Acknowledgments: We thank D.W. Andrews and B. Leber for their help and guidance with this project.

Conflicts of Interest: The authors declare no conflict of interest.

References

1. Kalkavan, H.; Green, D.R. MOMP, cell suicide as a BCL-2 family business. *Cell Death Differ.* **2018**, *25*, 46. [CrossRef]
2. Kale, J.; Osterlund, E.J.; Andrews, D.W. BCL-2 family proteins: Changing partners in the dance towards death. *Cell Death Differ.* **2018**, *25*, 65. [CrossRef]
3. Kelekar, A.; Thompson, C.B. Bcl-2-family proteins: The role of the BH3 domain in apoptosis. *Trends Cell Biol.* **1998**, *8*, 324–330. [CrossRef]
4. Moldoveanu, T.; Follis, A.V.; Kriwacki, R.W.; Green, D.R. Many players in BCL-2 family affairs. *Trends Biochem. Sci.* **2014**, *39*, 101–111. [CrossRef]
5. Westphal, D.; Kluck, R.; Dewson, G. Building blocks of the apoptotic pore: How Bax and Bak are activated and oligomerize during apoptosis. *Cell Death Differ.* **2014**, *21*, 196–205. [CrossRef] [PubMed]
6. Eskes, R.; Desagher, S.; Antonsson, B.; Martinou, J.C. Bid induces the oligomerization and insertion of Bax into the outer mitochondrial membrane. *Mol. Cell. Biol.* **2000**, *20*, 929–935. [CrossRef] [PubMed]
7. Kuwana, T.; Mackey, M.R.; Perkins, G.; Ellisman, M.H.; Latterich, M.; Schneiter, R.; Green, D.R.; Newmeyer, D.D. Bid, Bax, and lipids cooperate to form supramolecular openings in the outer mitochondrial membrane. *Cell* **2002**, *111*, 331–342. [CrossRef]
8. Putcha, G.V.; Le, S.; Frank, S.; Besirli, C.G.; Clark, K.; Chu, B.; Alix, S.; Youle, R.J.; LaMarche, A.; Maroney, A.C.; et al. JNK-mediated BIM phosphorylation potentiates BAX-dependent apoptosis. *Neuron* **2003**, *38*, 899–914. [CrossRef]
9. Walensky, L.D.; Pitter, K.; Morash, J.; Oh, K.J.; Barbuto, S.; Fisher, J.; Smith, E.; Verdine, G.L.; Korsmeyer, S.J. A stapled BID BH3 helix directly binds and activates BAX. *Mol. Cell* **2006**, *24*, 199–210. [CrossRef]
10. Lovell, J.F.; Billen, L.P.; Bindner, S.; Shamas-Din, A.; Fradin, C.; Leber, B.; Andrews, D.W. Membrane binding by tBid initiates an ordered series of events culminating in membrane permeabilization by Bax. *Cell* **2008**, *135*, 1074–1084. [CrossRef]
11. Chi, X.; Pemberton, J.; Nguyen, D.; Osterlund, E.J.; Liu, Q.; Brahmbhatt, H.; Zhang, Z.; Lin, J.; Leber, B.; Andrews, D. W. The carboxyl-terminal sequence of bim enables bax activation and killing of unprimed cells. *eLife* **2020**, *9*, e44525. [CrossRef]
12. Singh, R.; Letai, A.; Sarosiek, K. Regulation of apoptosis in health and disease: The balancing act of BCL-2 family proteins. *Nat. Rev. Mol. Cell Biol.* **2019**, *20*, 175–193. [CrossRef]
13. Knight, T.; Luedtke, D.; Edwards, H.; Taub, J.W.; Ge, Y. A delicate balance-The BCL-2 family and its role in apoptosis, oncogenesis, and cancer therapeutics. *Biochem. Pharmacol.* **2019**, *162*, 250–260. [CrossRef]
14. Hafezi, S.; Rahmani, M. Targeting BCL-2 in Cancer: Advances, Challenges, and Perspectives. *Cancers* **2021**, *13*, 1292. [CrossRef]

15. Pogmore, J.P.; Uehling, D.; Andrews, D.W. Pharmacological Targeting of Executioner Proteins: Controlling Life and Death. *J. Med. Chem.* **2021**, *64*, 5276–5290. [CrossRef] [PubMed]
16. Leber, B.; Lin, J.; Andrews, D.W. Embedded together: The life and death consequences of interaction of the Bcl-2 family with membranes. *Apoptosis* **2007**, *12*, 897–911. [CrossRef]
17. Bogner, C.; Leber, B.; Andrews, D.W. Apoptosis: Embedded in membranes. *Curr. Opin. Cell Biol.* **2010**, *22*, 845–851. [CrossRef] [PubMed]
18. Westphal, D.; Dewson, G.; Czabotar, P.E.; Kluck, R.M. Molecular biology of Bax and Bak activation and action. *Biochim. Biophys. Acta BBA Mol. Cell Res.* **2011**, *1813*, 521–531. [CrossRef]
19. Luo, X.; O'Neill, K.L.; Huang, K. The third model of Bax/Bak activation: A Bcl-2 family feud finally resolved? *F1000Research* **2020**, *9*. [CrossRef] [PubMed]
20. Shamas-Din, A.; Bindner, S.; Chi, X.; Leber, B.; Andrews, D.W.; Fradin, C. Distinct lipid effects on tBid and Bim activation of membrane permeabilization by pro-apoptotic Bax. *Biochem. J.* **2015**, *467*, 495–505. [CrossRef]
21. Yang, F.; Qu, W.; Du, M.; Mai, Z.; Wang, B.; Ma, Y.; Wang, X.; Chen, T. Stoichiometry and regulation network of Bcl-2 family complexes quantified by live-cell FRET assay. *Cell. Mol. Life Sci.* **2020**, *77*, 2387–2406. [CrossRef] [PubMed]
22. Edlich, F.; Banerjee, S.; Suzuki, M.; Cleland, M.M.; Arnoult, D.; Wang, C.; Neutzner, A.; Tjandra, N.; Youle, R.J. Bcl-xL retrotranslocates Bax from the mitochondria into the cytosol. *Cell* **2011**, *145*, 104–116. [CrossRef] [PubMed]
23. Todt, F.; Cakir, Z.; Reichenbach, F.; Youle, R.; Edlich, F. The C-terminal helix of Bcl-x L mediates Bax retrotranslocation from the mitochondria. *Cell Death Differ.* **2013**, *20*, 333. [CrossRef] [PubMed]
24. Shamas-Din, A.; Satsoura, D.; Khan, O.; Zhu, W.; Leber, B.; Fradin, C.; Andrews, D. Multiple partners can kiss-and-run: Bax transfers between multiple membranes and permeabilizes those primed by tBid. *Cell Death Dis.* **2014**, *5*, e1277. [CrossRef]
25. García-Sáez, A.J.; Ries, J.; Orzáez, M.; Pérez-Payà, E.; Schwille, P. Membrane promotes tBID interaction with BCL XL. *Nat. Struct. Mol. Biol.* **2009**, *16*, 1178. [CrossRef]
26. Bleicken, S.; Hantusch, A.; Das, K.K.; Frickey, T.; Garcia-Saez, A.J. Quantitative interactome of a membrane Bcl-2 network identifies a hierarchy of complexes for apoptosis regulation. *Nat. Commun.* **2017**, *8*, 73. [CrossRef]
27. Murad, F.; Garcia-Saez, A.J. Quantification of the Interactions Between BCL-2 Proteins by Fluorescence Correlation Spectroscopy. In *BCL-2 Family Proteins*; Springer: Berlin/Heidelberg, Germany, 2019; pp. 337–350.
28. Dlugosz, P.J.; Billen, L.P.; Annis, M.G.; Zhu, W.; Zhang, Z.; Lin, J.; Leber, B.; Andrews, D.W. Bcl-2 changes conformation to inhibit Bax oligomerization. *EMBO J.* **2006**, *25*, 2287–2296. [CrossRef] [PubMed]
29. Shivakumar, S.; Kurylowicz, M.; Hirmiz, N.; Manan, Y.; Friaa, O.; Shamas-Din, A.; Masoudian, P.; Leber, B.; Andrews, D.W.; Fradin, C. The proapoptotic protein tBid forms both superficially bound and membrane-inserted oligomers. *Biophys. J.* **2014**, *106*, 2085–2095. [CrossRef]
30. Subburaj, Y.; Cosentino, K.; Axmann, M.; Pedrueza-Villalmanzo, E.; Hermann, E.; Bleicken, S.; Spatz, J.; García-Sáez, A.J. Bax monomers form dimer units in the membrane that further self-assemble into multiple oligomeric species. *Nat. Commun.* **2015**, *6*, 8042. [CrossRef]
31. Kasai, R.S.; Suzuki, K.G.; Prossnitz, E.R.; Koyama-Honda, I.; Nakada, C.; Fujiwara, T.K.; Kusumi, A. Full characterization of GPCR monomer–dimer dynamic equilibrium by single molecule imaging. *J. Cell Biol.* **2011**, *192*, 463–480. [CrossRef]
32. Dupont, A.; Stirnnagel, K.; Lindemann, D.; Lamb, D. Tracking image correlation: Combining single-particle tracking and image correlation. *Biophys. J.* **2013**, *104*, 2373–2382. [CrossRef]
33. Friaa, O.; Fradin, C. Coincidence Measurements in Dual-Color Confocal Microscopy: A Combined Single-Particle and Fluorescence Correlation Approach. *Biophys. Rev. Lett.* **2014**, *9*, 249–271. [CrossRef]
34. Wang, K.; Yin, X.M.; Chao, D.T.; Milliman, C.L.; Korsmeyer, S.J. BID: A novel BH3 domain-only death agonist. *Genes Dev.* **1996**, *10*, 2859–2869. [CrossRef]
35. Shamas-Din, A.; Brahmbhatt, H.; Leber, B.; Andrews, D.W. BH3-only proteins: Orchestrators of apoptosis. *Biochim. Biophys. Acta BBA Mol. Cell Res.* **2011**, *1813*, 508–520. [CrossRef] [PubMed]
36. Chou, J.J.; Li, H.; Salvesen, G.S.; Yuan, J.; Wagner, G. Solution structure of BID, an intracellular amplifier of apoptotic signaling. *Cell* **1999**, *96*, 615–624. [CrossRef]
37. McDonnell, J.M.; Fushman, D.; Milliman, C.L.; Korsmeyer, S.J.; Cowburn, D. Solution structure of the proapoptotic molecule BID: A structural basis for apoptotic agonists and antagonists. *Cell* **1999**, *96*, 625–634. [CrossRef]
38. Hinds, M.; Smits, C.; Fredericks-Short, R.; Risk, J.; Bailey, M.; Huang, D.; Day, C. Bim, Bad and Bmf: Intrinsically unstructured BH3-only proteins that undergo a localized conformational change upon binding to prosurvival Bcl-2 targets. *Cell Death Differ.* **2007**, *14*, 128–136. [CrossRef]
39. Rautureau, G.J.; Day, C.L.; Hinds, M.G. Intrinsically disordered proteins in bcl-2 regulated apoptosis. *Int. J. Mol. Sci.* **2010**, *11*, 1808–1824. [CrossRef]
40. Peng, Z.; Xue, B.; Kurgan, L.; Uversky, V. Resilience of death: Intrinsic disorder in proteins involved in the programmed cell death. *Cell Death Differ.* **2013**, *20*, 1257–1267. [CrossRef] [PubMed]
41. Aouacheria, A.; Combet, C.; Tompa, P.; Hardwick, J.M. Redefining the BH3 death domain as a 'short linear motif'. *Trends Biochem. Sci.* **2015**, *40*, 736–748. [CrossRef]

42. Gross, A.; Yin, X.M.; Wang, K.; Wei, M.C.; Jockel, J.; Milliman, C.; Erdjument-Bromage, H.; Tempst, P.; Korsmeyer, S.J. Caspase cleaved BID targets mitochondria and is required for cytochrome c release, while BCL-XL prevents this release but not tumor necrosis factor-R1/Fas death. *J. Biol. Chem.* **1999**, *274*, 1156–1163. [CrossRef]
43. Shamas-Din, A.; Bindner, S.; Zhu, W.; Zaltsman, Y.; Campbell, C.; Gross, A.; Leber, B.; Andrews, D.W.; Fradin, C. tBid undergoes multiple conformational changes at the membrane required for Bax activation. *J. Biol. Chem.* **2013**, *288*, 22111–22127. [CrossRef]
44. Schuster, J.; Cichos, F.; Von Borczyskowski, C. Diffusion measurements by single-molecule spot-size analysis. *J. Phys. Chem. A* **2002**, *106*, 5403–5406. [CrossRef]
45. Brown, C.; Dalal, R.; Hebert, B.; Digman, M.; Horwitz, A.; Gratton, E. Raster image correlation spectroscopy (RICS) for measuring fast protein dynamics and concentrations with a commercial laser scanning confocal microscope. *J. Microsc.* **2008**, *229*, 78–91. [CrossRef]
46. Rose, M.; Hirmiz, N.; Moran-Mirabal, J.; Fradin, C. Lipid diffusion in supported lipid bilayers: A comparison between line-scanning fluorescence correlation spectroscopy and single-particle tracking. *Membranes* **2015**, *5*, 702–721. [CrossRef] [PubMed]
47. Friaa, O.; Furukawa, M.; Shamas-Din, A.; Leber, B.; Andrews, D.W.; Fradin, C. Optimizing the Acquisition and Analysis of Confocal Images for Quantitative Single-Mobile-Particle Detection. *Chemphyschem* **2013**, *14*, 2476–2490. [CrossRef]
48. Guo, L.; Har, J.Y.; Sankaran, J.; Hong, Y.; Kannan, B.; Wohland, T. Molecular diffusion measurement in lipid bilayers over wide concentration ranges: A comparative study. *ChemPhysChem* **2008**, *9*, 721–728. [CrossRef]
49. Macháň, R.; Hof, M. Recent developments in fluorescence correlation spectroscopy for diffusion measurements in planar lipid membranes. *Int. J. Mol. Sci.* **2010**, *11*, 427–457. [CrossRef] [PubMed]
50. Mulligan, K.; Jakubek, Z.J.; Johnston, L.J. Supported lipid bilayers on biocompatible polysaccharide multilayers. *Langmuir* **2011**, *27*, 14352–14359. [CrossRef]
51. Saito, M.; Korsmeyer, S.J.; Schlesinger, P.H. BAX-dependent transport of cytochrome c reconstituted in pure liposomes. *Nat. Cell Biol.* **2000**, *2*, 553. [CrossRef] [PubMed]
52. Gavathiotis, E.; Suzuki, M.; Davis, M.L.; Pitter, K.; Bird, G.H.; Katz, S.G.; Tu, H.C.; Kim, H.; Cheng, E.H.Y.; Tjandra, N.; et al. BAX activation is initiated at a novel interaction site. *Nature* **2008**, *455*, 1076. [CrossRef]
53. Czabotar, P.E.; Westphal, D.; Dewson, G.; Ma, S.; Hockings, C.; Fairlie, W.D.; Lee, E.F.; Yao, S.; Robin, A.Y.; Smith, B.J.; et al. Bax crystal structures reveal how BH3 domains activate Bax and nucleate its oligomerization to induce apoptosis. *Cell* **2013**, *152*, 519–531. [CrossRef]
54. Moldoveanu, T.; Grace, C.R.; Llambi, F.; Nourse, A.; Fitzgerald, P.; Gehring, K.; Kriwacki, R.W.; Green, D.R. BID-induced structural changes in BAK promote apoptosis. *Nat. Struct. Mol. Biol.* **2013**, *20*, 589–597. [CrossRef] [PubMed]
55. Leber, B.; Lin, J.; Andrews, D.W. Still embedded together binding to membranes regulates Bcl-2 protein interactions. *Oncogene* **2010**, *29*, 5221. [CrossRef] [PubMed]
56. Schellenberg, B.; Wang, P.; Keeble, J.A.; Rodriguez-Enriquez, R.; Walker, S.; Owens, T.W.; Foster, F.; Tanianis-Hughes, J.; Brennan, K.; Streuli, C.H.; et al. Bax exists in a dynamic equilibrium between the cytosol and mitochondria to control apoptotic priming. *Mol. Cell* **2013**, *49*, 959–971. [CrossRef] [PubMed]
57. Edlich, F. The great migration of Bax and Bak. *Mol. Cell. Oncol.* **2015**, *2*, e995029. [CrossRef] [PubMed]
58. Annis, M.G.; Soucie, E.L.; Dlugosz, P.J.; Cruz-Aguado, J.A.; Penn, L.Z.; Leber, B.; Andrews, D.W. Bax forms multispanning monomers that oligomerize to permeabilize membranes during apoptosis. *EMBO J.* **2005**, *24*, 2096–2103. [CrossRef]
59. Westphal, D.; Dewson, G.; Menard, M.; Frederick, P.; Iyer, S.; Bartolo, R.; Gibson, L.; Czabotar, P.E.; Smith, B.J.; Adams, J.M.; et al. Apoptotic pore formation is associated with in-plane insertion of Bak or Bax central helices into the mitochondrial outer membrane. *Proc. Natl. Acad. Sci. USA* **2014**, *111*, E4076–E4085. [CrossRef]
60. Bleicken, S.; Jeschke, G.; Stegmueller, C.; Salvador-Gallego, R.; García-Sáez, A.J.; Bordignon, E. Structural model of active Bax at the membrane. *Mol. Cell* **2014**, *56*, 496–505. [CrossRef]
61. Lv, F.; Qi, F.; Zhang, Z.; Wen, M.; Kale, J.; Piai, A.; Du, L.; Wang, S.; Zhou, L.; Yang, Y.; et al. An amphipathic Bax core dimer forms part of the apoptotic pore wall in the mitochondrial membrane. *EMBO J.* **2021**, *40*, e106438. [CrossRef] [PubMed]
62. Veresov, V.G.; Davidovskii, A.I. Activation of Bax by joint action of tBid and mitochondrial outer membrane: Monte Carlo simulations. *Eur. Biophys. J.* **2009**, *38*, 941–960. [CrossRef]
63. Oh, K.J.; Barbuto, S.; Meyer, N.; Kim, R.S.; Collier, R.J.; Korsmeyer, S.J. Conformational changes in BID, a pro-apoptotic BCL-2 family member, upon membrane binding: A site-directed spin labeling study. *J. Biol. Chem.* **2005**, *280*, 753–767. [CrossRef]
64. Veresov, V.G.; Davidovskii, A.I. Monte Carlo simulations of tBid association with the mitochondrial outer membrane. *Eur. Biophys. J.* **2007**, *37*, 19–33. [CrossRef]
65. Satsoura, D.; Kučerka, N.; Shivakumar, S.; Pencer, J.; Griffiths, C.; Leber, B.; Andrews, D.W.; Katsaras, J.; Fradin, C. Interaction of the full-length Bax protein with biomimetic mitochondrial liposomes: A small-angle neutron scattering and fluorescence study. *Biochim. Biophys. Acta BBA Biomembr.* **2012**, *1818*, 384–401. [CrossRef]
66. Zhou, H.X.; Pang, X. Electrostatic interactions in protein structure, folding, binding, and condensation. *Chem. Rev.* **2018**, *118*, 1691–1741. [CrossRef] [PubMed]
67. King, C.; Hristova, K. Direct measurements of VEGF–VEGFR2 binding affinities reveal the coupling between ligand binding and receptor dimerization. *J. Biol. Chem.* **2019**, *294*, 9064–9075. [CrossRef]
68. Sung, T.C.; Li, C.Y.; Lai, Y.C.; Hung, C.L.; Shih, O.; Yeh, Y.Q.; Jeng, U.S.; Chiang, Y.W. Solution structure of apoptotic BAX oligomer: Oligomerization likely precedes membrane insertion. *Structure* **2015**, *23*, 1878–1888. [CrossRef] [PubMed]

69. Bogner, C.; Kale, J.; Pogmore, J.; Chi, X.; Shamas-Din, A.; Fradin, C.; Leber, B.; Andrews, D.W. Allosteric regulation of BH3 proteins in Bcl-xL complexes enables switch-like activation of bax. *Mol. Cell* **2020**, *77*, 901–912. [CrossRef]
70. Youle, R.J.; Strasser, A. The BCL-2 protein family: Opposing activities that mediate cell death. *Nat. Rev. Mol. Cell Biol.* **2008**, *9*, 47–59. [CrossRef] [PubMed]
71. Billen, L.; Shamas-Din, A.; Andrews, D. Bid: A Bax-like BH3 protein. *Oncogene* **2008**, *27*, S93–S104. [CrossRef]
72. Greaves, G.; Milani, M.; Butterworth, M.; Carter, R.J.; Byrne, D.P.; Eyers, P.A.; Luo, X.; Cohen, G.M.; Varadarajan, S. BH3-only proteins are dispensable for apoptosis induced by pharmacological inhibition of both MCL-1 and BCL-X L. *Cell Death Differ.* **2019**, *26*, 1037–1047. [CrossRef] [PubMed]
73. Huang, K.; O'Neill, K.L.; Li, J.; Zhou, W.; Han, N.; Pang, X.; Wu, W.; Struble, L.; Borgstahl, G.; Liu, Z.; et al. BH3-only proteins target BCL-xL/MCL-1, not BAX/BAK, to initiate apoptosis. *Cell Res.* **2019**, *29*, 942–952. [CrossRef] [PubMed]
74. Lai, Y.C.; Li, C.C.; Sung, T.C.; Chang, C.W.; Lan, Y.J.; Chiang, Y.W. The role of cardiolipin in promoting the membrane pore-forming activity of BAX oligomers. *Biochim. Biophys. Acta BBA Biomembr.* **2019**, *1861*, 268–280. [CrossRef]
75. Garcia, A.; Murad, F. Bcl-xL inhibits tBid and Bax via distinct mechanisms. *Faraday Discuss.* **2020**. [CrossRef]
76. Pogmore, J.P.; Pemberton, J.M.; Chi, X.; Andrews, D.W. Using Förster-Resonance energy transfer to measure protein interactions between Bcl-2 family proteins on mitochondrial membranes. In *Programmed Cell Death*; Springer: Berlin/Heidelberg, Germany, 2016; pp. 197–212.
77. Cartron, P.F.; Gallenne, T.; Bougras, G.; Gautier, F.; Manero, F.; Vusio, P.; Meflah, K.; Vallette, F.M.; Juin, P. The first α helix of Bax plays a necessary role in its ligand-induced activation by the BH3-only proteins Bid and PUMA. *Mol. Cell* **2004**, *16*, 807–818. [CrossRef]
78. Yethon, J.A.; Epand, R.F.; Leber, B.; Epand, R.M.; Andrews, D.W. Interaction with a membrane surface triggers a reversible conformational change in Bax normally associated with induction of apoptosis. *J. Biol. Chem.* **2003**, *278*, 48935–48941. [CrossRef]
79. Van Meer, G.; Voelker, D.R.; Feigenson, G.W. Membrane lipids: Where they are and how they behave. *Nat. Rev. Mol. Cell Biol.* **2008**, *9*, 112. [CrossRef] [PubMed]
80. Horvath, S.E.; Daum, G. Lipids of mitochondria. *Prog. Lipid Res.* **2013**, *52*, 590–614. [CrossRef]
81. Petrášek, Z.; Schwille, P. Precise measurement of diffusion coefficients using scanning fluorescence correlation spectroscopy. *Biophys. J.* **2008**, *94*, 1437–1448. [CrossRef]
82. Kapusta, P. *Absolute Diffusion Coefficients: Compilation of Reference Data for FCS Calibration*; Picoquant Application Note; 2010. Available online: https://www.picoquant.com/images/uploads/page/files/7353/appnote_diffusioncoefficients.pdf (accessed on 30 July 2021)

International Journal of
Molecular Sciences

MDPI

Article

NO• Represses the Oxygenation of Arachidonoyl PE by 15LOX/PEBP1: Mechanism and Role in Ferroptosis

Karolina Mikulska-Ruminska [1,2,*], Tamil S. Anthonymuthu [3], Anastasia Levkina [4,5], Indira H. Shrivastava [1,4], Alexandr A. Kapralov [4], Hülya Bayır [3,4], Valerian E. Kagan [4,6,7,8,9,*,†] and Ivet Bahar [1,*,†]

1 Department of Computational and Systems Biology, School of Medicine, University of Pittsburgh, Pittsburgh, PA 15260, USA; ihs2@pitt.edu
2 Institute of Physics, Faculty of Physics, Astronomy and Informatics, Nicolaus Copernicus University in Toruń, Grudziadzka 5, 87-100 Toruń, Poland
3 Department of Critical Care Medicine, Safar Center for Resuscitation Research, Children's Neuroscience Institute, Children's Hospital of Pittsburgh, University of Pittsburgh, Pittsburgh, PA 15260, USA; ATAMIL@pitt.edu (T.S.A.); bayihx@ccm.upmc.edu (H.B.)
4 Department of Environmental and Occupational Health and Center for Free Radical and Antioxidant Health, University of Pittsburgh, Pittsburgh, PA 15260, USA; levkina.a.anastasia@gmail.com (A.L.); olk6@pitt.edu (A.A.K.)
5 Institute of Translational Medicine, Pirogov Russian National Research Medical University, Ostrovityanova 1, 117997 Moscow, Russia
6 Department of Radiation Oncology, University of Pittsburgh, Pittsburgh, PA 15260, USA
7 Department of Chemistry, University of Pittsburgh, Pittsburgh, PA 15260, USA
8 Department of Pharmacology and Chemical Biology, University of Pittsburgh, Pittsburgh, PA 15260, USA
9 Institute of Regenerative Medicine, IM Sechenov Moscow State Medical University, 119048 Moscow, Russia
* Correspondence: karolamik@fizyka.umk.pl (K.M.-R.); kagan@pitt.edu (V.E.K.); bahar@pitt.edu (I.B.)
† I.B. and V.E.K. contributed equally to this work.

Citation: Mikulska-Ruminska, K.; Anthonymuthu, T.S.; Levkina, A.; Shrivastava, I.H.; Kapralov, A.A.; Bayır, H.; Kagan, V.E.; Bahar, I. NO• Represses the Oxygenation of Arachidonoyl PE by 15LOX/PEBP1: Mechanism and Role in Ferroptosis. *Int. J. Mol. Sci.* **2021**, *22*, 5253. https://doi.org/10.3390/ijms22105253

Academic Editor: Masoud Jelokhani-Niaraki

Received: 3 April 2021
Accepted: 12 May 2021
Published: 17 May 2021

Publisher's Note: MDPI stays neutral with regard to jurisdictional claims in published maps and institutional affiliations.

Abstract: We recently discovered an anti-ferroptotic mechanism inherent to M1 macrophages whereby high levels of NO• suppressed ferroptosis via inhibition of hydroperoxy-eicosatetraenoyl-phosphatidylethanolamine (HpETE-PE) production by 15-lipoxygenase (15LOX) complexed with PE-binding protein 1 (PEBP1). However, the mechanism of NO• interference with 15LOX/PEBP1 activity remained unclear. Here, we use a biochemical model of recombinant 15LOX-2 complexed with PEBP1, LC-MS redox lipidomics, and structure-based modeling and simulations to uncover the mechanism through which NO• suppresses ETE-PE oxidation. Our study reveals that O_2 and NO• use the same entry pores and channels connecting to 15LOX-2 catalytic site, resulting in a competition for the catalytic site. We identified residues that direct O_2 and NO• to the catalytic site, as well as those stabilizing the esterified ETE-PE phospholipid tail. The functional significance of these residues is supported by in silico saturation mutagenesis. We detected nitrosylated PE species in a biochemical system consisting of 15LOX-2/PEBP1 and NO• donor and in RAW264.7 M2 macrophages treated with ferroptosis-inducer RSL3 in the presence of NO•, in further support of the ability of NO• to diffuse to, and react at, the 15LOX-2 catalytic site. The results provide first insights into the molecular mechanism of repression of the ferroptotic Hp-ETE-PE production by NO•.

Keywords: nitric oxide; ferroptosis; lipid peroxidation; lipoxygenase structure; O_2 and NO• binding mechanisms; 1-stearoyl-2-arachidonoyl phosphatidylethanolamine (1-SA-2-ETE-PE or SAPE); lipidomics; MD simulations

1. Introduction

Recent years have brought attention to ferroptosis, an iron-dependent form of regulated cell death implicated in a broad range of diseases [1,2], and to the selective peroxidation of eicosatetraenoyl-phosphatidylethanolamine (ETE-PE) by 15-lipoxygenase (15LOX) complexed with PE-binding protein 1 (PEBP1) [3–5] as a ferroptotic mechanism. ETE-PE peroxidation produces hydroperoxy-ETE-PE (HpETE-PE) molecules that, in turn, serve as pro-ferroptotic signals. 15LOX catalytic action includes the Fe-driven abstraction of a hydrogen atom from the carbon in bis-allylic position, formation of a carbon-centered radical followed by addition of molecular oxygen, culminating in the production of 15-HpETE.

The role of 15LOX in the onset of ferroptosis is now well-established [3–6]. Under physiological conditions, arachidonic acid (AA), a polyunsaturated fatty acid (PUFA) also termed *cis*-5,8,11,14-ETE or shortly ETE, is the preferred substrate of these non-heme iron-containing enzymes; but under inflammatory conditions, such as those stimulated by interleukin-13/4, the formation of the complex 15LOX/PEBP1 shifts the substrate-specificity of 15LOX from free AA to AA esterified into PE (or ETE-PE). The AA-esterification of PE mostly occurs at the sn-2 acyl chain leaving the sn-1 chain for saturated or mono-unsaturated acyls, predominantly palmitic (C16:0), stearic (S) (C18:0), and oleic (C18:1). Of interest below will be 1-SA-2-ETE-PE, i.e., PE with stearic acid (SA) and arachidonic acid (AA or ETE) at the respective sn-1 and sn-2 chains, which will be shortly called SAPE. 15LOX-catalyzed peroxidation of SAPE at its sn-2 chain results in the production of 15-HpHETE-PE [3]—one of the major players in the ensuing induction of ferroptosis. As such, the 15LOX-2/PEBP1 complex emerged as a master promoter of ferroptotic cell-death [3]. This involvement of 15LOX-2/PEBP1 has been supported by the ability of ferrostatin-1, the most common ferroptosis inhibitor, to bind the complex [4] and inhibit the formation of 15-HpHETE-PE.

As a multistage process, ferroptosis is regulated by several enzymatic events occurring independently or concertedly: (i) biosynthesis of PUFA-PE peroxidation substrates by acyl Co-A synthetase 4 (ACSL4) and lysophosphatidylcholine acyltransferase (LPCAT) [7]; (ii) prevention of the formation of 15-HpETE-PE by flavin-containing oxidoreductase, ferroptosis suppressor protein 1 (FSP1) [8,9]; and (iii) reduction in 15-HpETE-PE into an innocuous alcohol, 15-HETE-PE, by glutathione peroxidase 4 (GPX4) [10,11]. Lately, this list of ferroptosis regulators has been extended to include new members, such as the Ca^{2+}-independent phospholipase A2β (iPLA2β) [12], which can hydrolyze peroxidized phospholipids and, therefore, can eliminate the ferroptotic cell death signal, and the inducible isoform of nitric oxide synthase, iNOS, that generates $NO^{\bullet}$ [13]. Notably, $NO^{\bullet}$'s reactivity towards different radical intermediates of lipid peroxidation has been discovered more than three decades ago [14]. Furthermore, $NO^{\bullet}$ inhibits lipid peroxidation by lipoxygenase and cyclooxygenase [15].

In the present study, we first performed a lipidomics analysis to observe that $NO^{\bullet}$ indeed suppresses the production of HpETE-PE by 15LOX-2/PEBP1, while promoting the formation of nitrosylated ETE-PE. As the 15LOX-2/PEBP1 complex, rather than 15LOX-2 alone, catalyzes the oxidation of 15-ETE-PE (rather than free ETE), we focused here on the interactions of $NO^{\bullet}$ with SAPE and15LOX-2/PEBP1, and on its competition with the lipoxygenase substrate O_2. Notably, the precise delivery of molecular oxygen to form a peroxyl radical from the carbon-centered radical of ETE at the 15LOX-2 catalytic site occurs via a specialized channel, whose structure and accessibility is altered upon binding of ETE to the enzyme [16]. With this in mind, we studied the effects of $NO^{\bullet}$ peroxidation of SAPE by 15LOX-2/PEBP1 and the structural dynamics of the oxygen delivery process to the catalytic Fe^{3+} in the 15LOX-2/PEBP1 complex and the competition with $NO^{\bullet}$, in the presence and absence of the ETE-PE substrate compared to free ETE in 15LOX-2. Our study provides a mechanistic description of the time-evolved interactions between 15LOX-2 (bound to PEBP1) and its ligands, the ETE-PE- and PEBP1-induced changes in 15LOX-2 conformation or catalytic site accessibility which alter the delivery of O_2 and $NO^{\bullet}$, and the competition between O_2 and $NO^{\bullet}$ for binding the active site. Several key residues mediating the events prior to peroxidation and the interference of $NO^{\bullet}$ to repress them are identified, the critical roles of which are consolidated by in silico saturation mutagenesis experiments and comparisons with recent work.

2. Results

2.1. Lipidomics Analysis of the Effect of NO• on 15LOX-2/PEBP1 Peroxidation Activity

To examine the effects of NO• on the production of the pro-ferroptotic signal, 15-HpETE-PE, by 15LOX-2/PEBP1, we incubated one of the lipoxygenase isoforms, 15LOX-2, and PEBP1 in the presence of an NO•-donor, propylamine (PAPA) NONOate [17]. This donor releases NO• at a constant rate with a decay half-time of 15 min. We found that NO• strongly suppressed the production of hydroperoxide of 1-SA-2-ETE-PE (HpETE-PE) (Figure 1a). Detailed analysis of redox lipidomics data led to the identification of a nitrosylated product, nitroso-ETE-PE (1-SA-2-ETE(-NO)-PE) generated by 15LOX-2/PEBP1 only in the presence of NO• (Figure 1b–d).

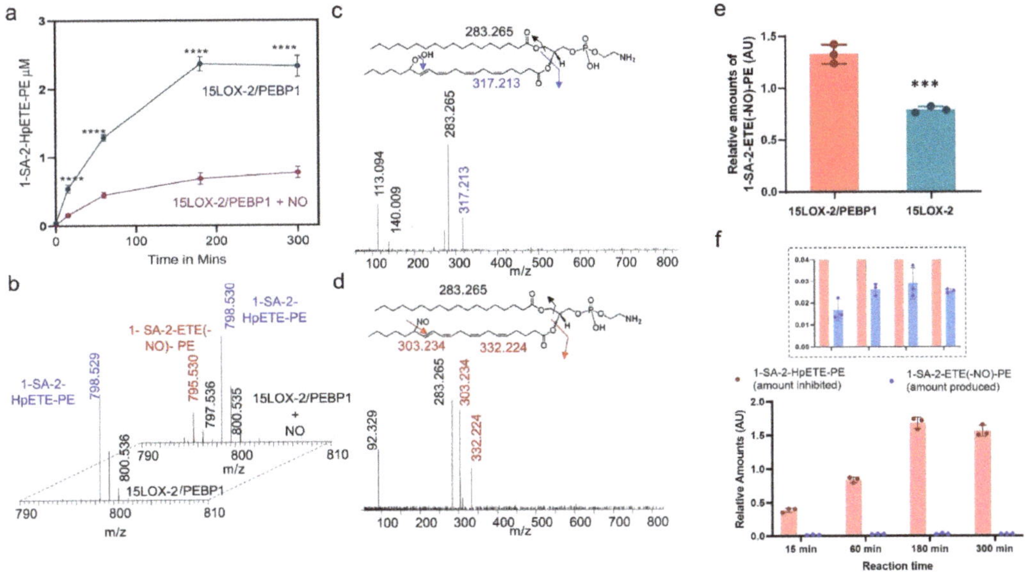

Figure 1. Effect of NO• on peroxidation of 1-SA-2-ETE-PE by the 15LOX-2/PEBP1 complex. (**a**) Kinetics of the production of 1-SA-2-HpETE-PE by 15LOX-2/PEBP1 in the presence and absence of NO• donor. Data are mean ± SD, $n = 3$, *** definition; **** $p < 0.0001$ vs. 15LOX-2/PEBP1 + NO, two way ANOVA, Sidak post-hoc analysis. (**b**) Mass spectrum showing the nitrosylated product formed upon 1-SA-2-ETE-PE incubation with 15LOX-2/PEBP1+PAPA NONOate (back). Nitroso-1-SA-2-ETE-PE (1-SA-2-ETE(-NO•)-PE) was not detected in the system without PAPA NONOate (front). Fragmentation analysis of 1-SA-2-HpETE-PE (**c**) and 1-SA-2-ETE(-NO)-PE (**d**) showing representative fragments. The insets at the top show the structures of 1-SA-2-HpETE-PE, 1-SA-2-ETE(-NO)-PE and their possible fragments. (**e**) Bar plot comparing 1-SA-2-ETE(-NO)-PE formation by 15LOX-2 and by 15LOX-2/PEBP1, both in the presence of NO•. Data are mean + SEM., $n = 3$, *** $p = 0.0007$, student's *t*-test. (**f**) Bar plot showing that the amount of 1-SA-2-HpETE-PE inhibited by NO• is significantly larger than the amount of 1-SA-2-ETE(-NO)-PE produced, as a function of reaction time. The inset bar plot shows the enlarged view of the lower plot, between the ordinate values $0.0 < y < 0.04$ (AU).

These observations suggest that a direct reaction occurs between the ETE-PE carbon-centered radical, an intermediate of the 15LOX-catalyzed dioxygenase cycle [18], and the NO• radical. No nitroxygenated products formed from the reaction of O_2 with the reactive intermediates of NO• [19] could be found. Notably, the 15LOX-2/PEBP1 complex produced a greater amount of NO•-ETE-PE than 15LOX-2 alone (Figure 1e). Lack of appropriate standards of nitrosylated ETE-PE precluded the absolute quantification of nitrosylated ETE-PE and its comparison with 1-SA-2-HpETE-PE levels. However, upon normalization with the same internal standard, the relative amounts of the nitrosylated ETE-PE were about 50 folds lower than the inhibited amounts of 1-SA-2-HpETE-PE (Figure 1f).

Combined, these results suggest that at least two mechanisms contribute to the observed suppression of ETE-PE peroxidation by NO• in the 15LOX-2/PEBP1 complex: (i) direct reaction of NO• with the ETE-PE carbon-centered intermediate, and (ii) potential presence or occupancy of the O_2-binding site and channeling path by NO• or a competition between the two small molecules—O_2 and NO•—interfering with the peroxidation of the oxidizable substrates. To explore the latter mechanism, we analyzed the changes in the conformational state and dynamics of 15LOX-2/PEBP1 invoked upon binding of the oxidizable substrate, as well as O_2 and NO•.

2.2. *In silico Analysis of the Effect of NO• on the Structure, Dynamics, and Interactions of 15LOX-2/PEBP1*

We performed structure-based modeling and simulations to investigate the molecular basis of the experimentally observed reduced activity of 15LOX-2/PEBP1 in the presence of NO•. First, we explored how the accessibility of the catalytic site of 15LOX-2 to O_2 and NO• molecules is altered upon complexation with PEBP1. Next, we investigated how the presence of NO• affects O_2-binding and channeling to the catalytic pocket. We further analyzed the effect of oxidizable substrates (1-SA-2-ETE-PE, shortly called SAPE, vs. free ETE (AA)) on the accessibility of the 15LOX-2 catalytic site to O_2 and NO•, in the presence and absence of PEBP1. Overall, we conducted 10 sets of molecular dynamics (MD) simulations in duplicate (or triplicate) runs of 150 nanoseconds (ns) under different conditions, summing up to total simulation duration of 3.6 microseconds in explicit water, as summarized in the Supplementary Table S1.

2.3. *Specific Porous Regions on 15LOX-2 Surface Enable Access of O_2 and NO• to the Catalytic Cavity*

Lipoxygenases are non-heme iron-containing enzymes. The iron at the catalytic site in the central part of the catalytic domain of 15LOX-2 is coordinated by four highly conserved residues including three histidines: H373, H378 (on helix α12-14), H553 (on helix α27), and the C-terminal isoleucine, I676 (Supplementary Figure S1). Simulations revealed three porous regions on 15LOX-2 surface, denoted here as Entrances 1–3 (*E1–E3*), which could potentially allow for the entry of the radicals, O_2 or NO•, to potentially access the catalytic cavity of the enzyme (Figure 2a): *E1*, near the loop Y154-P159 where Y154 and W158 side chains appear to be gating the entrance (Figure 2b and Supplementary Figure S1a,c); *E2*, at the 15LOX-2/PEBP1 binding interface (Figure 2a); and *E3*, near S573, P595, and A599 (Figure 2d). Although all three entrances could potentially provide access to the catalytic site [13], *E2* was largely obstructed by the oxidizable substrate (SAPE or AA in our simulations) and, therefore, inaccessible to O_2 and NO•. In contrast, *E1* served as the main access point to a tunnel leading to the catalytic site, for both O_2 and NO• (Figure 2a,b, orange trace), while *E3* provided an alternative path leading to the catalytic site (Figure 2a, blue trace and Figure 2d). Supplementary Figure S1 provides close-up views of *E1* and *E3* from different perspectives, to facilitate the visualization of their connection to the catalytic site; and Supplementary Figure S2 describes the secondary structure of 15LOX-2 [20].

The entrance *E1* was lined by Y154, N155, G157, W158, I421, I435, F438, and S439 (Figure 2b). It was transiently occupied by NO• or O_2 at the initial stage of their interaction with the enzyme. After this first recognition event (at site *S0*), the small molecules moved deeper towards two attractive sites, binding sites *S1* and *S2*, that arrested them for extended durations (Figure 2c): *S1*, composed of N413, A416, R417, L374, and L379, was occupied by these ligands for more than 70 ns; and *S2*, containing I433, T431, V427, F365, and E369, for more than 35 ns, during 150 ns runs. Both sites made close contacts (atom-atom distance less than 3.5 Å) with NO• or O_2 for sufficiently long durations (Figure 2e) meeting the criteria (see Methods) to qualify as binding sites.

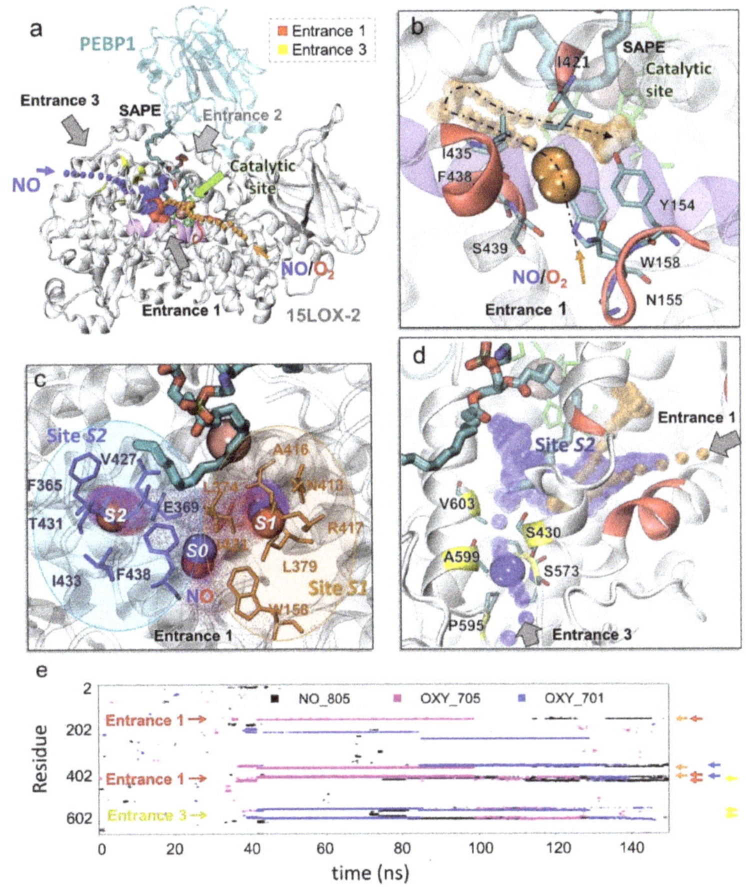

Figure 2. Ligand ($NO^{\bullet}$ and O_2) binding and channeling to the catalytic site of 15LOX-2 and critical interactions mediating this process in the presence of SAPE. Results from MD simulations for 15LOX-2/PEBP1 in the presence of SAPE, five $NO^{\bullet}$ and five O_2 molecules are presented (Supplementary Table S1; last row). (**a**) 15LOX-2/PEBP1/SAPE complex stabilized by the end of a 150 ns MD run. 15LOX-2 and PEBP1 are displayed in white and cyan ribbon diagrams, respectively; and SAPE in cyan sticks (with atoms in CPK colors). Catalytic residues (H373, H378, H553, I676) are in green sticks, and the iron ion in pink sphere. Entrances 1 (red), 2 (inaccessible in the presence of PEBP1 and substrate), and 3 (yellow) are indicated by black/gray arrows. The series of orange dots represents the entry or diffusion path of O_2 and $NO^{\bullet}$ molecules through Entrance 1, and the blue dots those through Entrance 3. The region between helices α12 and α14 (K365-L380, UniProt ID: O15296), colored violet, contains two of the catalytic histidines. (**b**) A close-up view of Entrance 1, from a different angle. An O_2 molecule (orange) approaches it around 37 ns and translocate to the catalytic site (orange dots in panels b and d) (**c**) A close-up view of the catalytic site with three $NO^{\bullet}$ molecules occupying the recognition site *S0* (Entrance 1), and the binding sites *S1* (orange) and *S2* (blue). The cloud of small dots shows where and how long (more dots) $NO^{\bullet}$ travelled or remained bound during simulations. $NO^{\bullet}$ molecules are in blue-red spheres corresponding to their nitrogen and oxygen atoms, respectively. (**d**) A close-up view of the Entrance 3 providing access to a $NO^{\bullet}$ molecule (in blue). Key residues lining the path are displayed in yellow. The series of orange dots represents the entry or diffusion path for O_2 and $NO^{\bullet}$ through Entrance 1 and the blue dots show the path through Entrance 3. (**e**) Illustration of the time evolution of contacts between 15LOX-2 residues (ordinate) and $O_2/NO^{\bullet}$ molecules (labeled NO_805, OXY_701 (shown in panel b) and OXY_705; shown by the respective black, magenta, and blue traces) during a typical MD run. Red, yellow, orange, and blue arrows along the ordinate correspond to the Entrance 1 (*E1*), Entrance 3 (*E3*), Binding Sites *S1* and *S2*, respectively. All three molecules were arrested for extended durations at *S1* or *S2*, inserting to the catalytic site through Entrance 1 or 3.

The entrance *E3*, on the other hand, included (in addition to S573, P595, and A599), S430 and V603 near the surface (Figure 2d). These residues assist in directing the $O_2/NO^\bullet$ molecules from *E3* to the catalytic site. Entrances 1 and 2 are predominantly composed of hydrophobic residues: (i) the cluster I216, I604, F561, C564, and A565 that initially bound $O_2/NO^\bullet$ for ~40 ns; and (ii) the cluster L607, L610, L420, V426, and V427 that retained the molecules for ~11 ns. The third site that serves as a bridge between *E3* and catalytic site partially overlaps with *S2*: it contains the residues L246, E364, F365, H368, E369, and L570 that bound $O_2/NO^\bullet$ for ~40 ns. Thus, *S2* serves as an attractor for O_2 or $NO^\bullet$ binding entering through either *E1* or *E3*. Two highly conserved residues therein, F365 and E369, will be shown below to play an important role in binding or redirecting the O_2 and $NO^\bullet$ molecules.

A summary of all these residues involved in various roles, entry (*E1* and *E3*), binding to *S1* and *S2*, or insertion towards the catalytic site (mainly hydrophobic clusters), is given in Supplementary Table S2, along with corresponding secondary structures (based on [21]).

These simulations also indicated an additional pore that connected from the surface to the catalytic site through a short tunnel with a wide entrance. This path, available in both unbound- and PEBP1-bound-15LOX-2 (Supplementary Figure S3a,b, in green), was not selected by either $NO^\bullet$ or O_2. This tunnel included five residues highly conserved across LOX family members [22], L610, Q560, and the three catalytic residues, H373, H553, I676, along with E613 and S557. It is conceivable that this tunnel plays a role in 15LOX-2 dioxygenase activity, yet to be explored.

2.4. O_2 and $NO^\bullet$ Compete for the Catalytic Site

Simulations with the substrate revealed an interesting effect which arises upon the complexation 15LOX-2 with PEBP1, namely a competition between O_2 and $NO^\bullet$ for a position near the C15 carbon in the arachidonoyl chain of SAPE within the catalytic site (Figure 3, Movie 1). It is known that 15LOX-2/PEBP1 complex converts SAPE substrate to its peroxidized form by facilitating the addition of a hydroperoxyl- group at the C15 position of the ETE sn-2 chain at the catalytic site. Thus, the presence of an O_2 molecule close to C15 is necessary to initiate the process. Here, we observed that the positioning of O_2 at C15 was disrupted by the interference of $NO^\bullet$ radicals. Figure 3 illustrates such an observation. Panels a–i display a series of snapshots showing how $NO^\bullet$ approaches the region originally occupied by an O_2 molecule, to displace O_2, and position itself near the C15 atom for extended durations (>50 ns), until the departure of O_2 from the catalytic site, consistent with the experimentally observed ability of $NO^\bullet$ to inhibit the peroxidation of SAPE.

The panel j in Figure 3 indicates the 15LOX-2 residues (ordinate) that make successive contacts with $NO^\bullet$ and O_2 molecules, starting from first encounters with O_2 (left panel, $40 < t < 75$ ns, shaded in green), followed by the arrival of $NO^\bullet$ to the catalytic site and the competition between $NO^\bullet$ and O_2 for binding the catalytic pocket residues (both panels, $80 < t < 140$ ns; shaded in orange (left) and yellow (right)), finally ending with the dislocation of O_2, while $NO^\bullet$ remained stabilized near C15 and Fe^{3+} ion. Among residues ligating (successively or competitively) both O_2 and $NO^\bullet$ during this trajectory, we note the highly conserved residues P365 and E369 in site S2, and the hydrophobic cluster V603, A606, L609, and L610, revealing that O_2 and $NO^\bullet$ channeled to the catalytic site through Entrance 3 in this case.

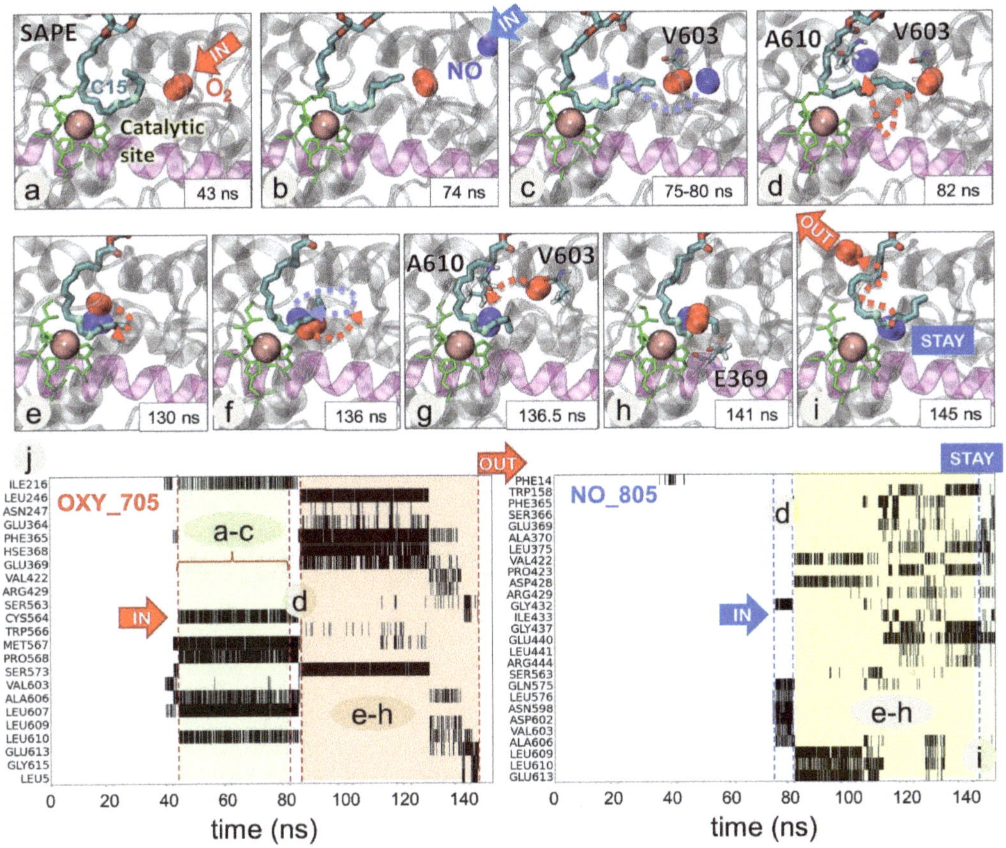

Figure 3. Competition between $NO^{\bullet}$ and O_2 molecules near the C15 atom of the AA-chain of SAPE at the catalytic site of 15LOX-2 complexed with PEBP1. A sequence of MD snapshots (**a**–**i**) illustrates the time evolution of the positions of a $NO^{\bullet}$ radical (NO_805) and an O_2 molecule (OXY_705) during the period 43 < t < 150 ns of the simulations. A first O_2 (**a**) reaches the catalytic site at t = 43 ns, followed by a $NO^{\bullet}$ at t = 74 ns (**b**,**c**) which disrupts the binding of O_2. After a competition process during which the two molecules undergo multiple fluctuations and dislocations in their positions, the O_2 molecule ends up leaving the catalytic cavity at 145 ns (**i**). Catalytic site residues (H373, H378, H553, and I676) are displayed in green and sn-2 chain of SAPE as cyan-red-blue sticks. Two highlighted atoms of SAPE (light cyan spheres) denote C13 and C15 atoms as reference points. $NO^{\bullet}$ and O_2 are shown in blue and red spheres, respectively. The region between the helices α12 and α14 is colored in light violet. Lower panels (**j**) display the detailed time evolution of contacts between 15LOX-2 residues and O_2 (**left panel**) and $NO^{\bullet}$ (**right panel**) molecules. This panel includes only those residues that undergo frequent contacts (cumulative contact time of 2.5 ns) with O_2 (**left**) or $NO^{\bullet}$ (**right**) during the simulation period of 150 ns. Contacts are shown by black shades/bars for each residue. A603 and A606 are involved in initial contacts with O_2 which are broken upon the arrival and interference of $NO^{\bullet}$. See Movie 1 for a visualization of the interplay between O_2 and $NO^{\bullet}$ at the catalytic site.

2.5. Selected Residues Stabilize O_2 and $NO^{\bullet}$ Near 15LOX-2 Catalytic Site in a Substrate-Dependent Manner

We further carried out a statistical analysis of four independent runs to identify the 15LOX-2 residues which made frequent and long-lived contacts with $O_2/NO^{\bullet}$ molecules (see Methods), among those located within 20 Å of the Fe^{3+} ion. Supplementary Figure S4a displays such residues. Panels b–d list the PEBP1-bound 15LOX-2, free 15LOX-2 and PEBP1 residues, respectively, that make contacts for extended durations with O_2 and $NO^{\bullet}$, and the corresponding number of runs in which such contacts are observed. Clearly 15LOX-2 residues at Entrances 1 and 3 (*E1* and *E3*) are detected among them, along with sites *S1* and *S2* residues, as indicated by the bars between panels b and c. Of particular importance is the long helix α12-14 (F365-L379) comprising both *S1* (L374, L379) and *S2* (F365, E369) residues also noted [13] in the absence of the phospholipid substrate. Interestingly, our recent computational analysis of a dataset of 88 crystal structures resolved for lipoxygenase showed that this particular region acts as a strong effector of allosteric signals [22]. The highly conserved WxxAK motif (W353-K357) shared by other LOXs [22] also takes part in the same region. Comparison of panels b and c shows that certain residues (E168, L172-A179 (except for I174), and F219-P223) interact with O_2 and NO• in the absence of PEBP1 but become inaccessible upon PEBP1 binding.

2.6. The Change in 15LOX-2 Structure Upon PEBP1 Binding Renders the Catalytic Site Accessible to Both O_2 and NO• Molecules

We observed that in the presence of PEBP1, both O_2 and NO• co-localize within the catalytic site (Figure 4a) where they compete for a position near the C15 atom of SAPE (Figure 3, Movie 1). Remarkably, in the absence of PEBP1 and at the small concentration of NO•, the NO• molecules were unable to access the catalytic site (Figure 4b, red oval). This behavior consistently reproduced in independent runs (see Supplementary Figure S5) was due to a conformational change (opening or exposure of a binding site) between the α2 helix and the T166-A179 region stabilized upon PEBP1 binding (black arrows Figure 4c). When this site was not exposed (in the absence of PEBP1), O_2 and NO• molecules were attracted to the hydrophobic sn-1 (stearoyl) chain of SAPE and to the 15LOX-2 residues L172-A179. At a higher NO• concentration, however, this effect was partially suppressed and a few NO• molecules took a place near C15 atom of SAPE in the catalytic site (black dotted oval Supplementary Figure S6). Simulations repeated with AA bound to 15LOX-2 (Supplementary Figure S7) showed that NO• molecules were able to access the catalytic site in the absence of the sn-1 chain of SAPE that otherwise sequestered the NO• molecules.

These simulations therefore lead to the following conclusions: (i) PEBP1 increases the affinity of 15LOX-2 to bind both O_2 and NO• molecules to the catalytic site; whereas in the absence of PEBP1 and without higher NO• concentration, only O_2 molecules access the active site, while NO• molecules preferentially bind to the phospholipid tail of SAPE; (ii) competitive binding of NO• to the catalytic site in the presence of PEBP1 (Figure 4a) is expected to interfere with the peroxidation activity of 15LOX-2 resulting in lipid nitrosylation, as well as decreased conversion of SAPE to SAPE-OOH; and (iii) in absence of PEBP1, the NO• molecules co-localize at the exposed sn-1 stearoyl chain and do not compete with the O_2 molecules at the catalytic site (Supplementary Figure S5b).

2.7. The Precise Positioning of SAPE for Peroxidation by 15LOX-2 Is Assisted by α2 Residues N181, Y185, and G189, and by E3/S1 Residues A416 and A606

Lipid peroxidation by 15LOX-2 requires O_2 molecule to be positioned in proximity (<7.5 Å) of the carbon C15 in the arachidonoyl chain of SAPE, while the C13 in the arachidonoyl chain of SAPE would approach (by <7.5 Å) the iron ion at the catalytic site. We analyzed the MD trajectories to examine whether such poses were sampled. We identified a total of 121 and 538 poses, respectively, for PEBP1-bound and -unbound 15LOX-2/SAPE (Figure 5a) that satisfied this requirement. The 15LOX-2 residues, G189, A416 (at *S1*) and A606 (at *E3*), were distinguished by their high tendency (75% of MD snapshots) to coordinate SAPE, in both systems. Furthermore, α2-helix residues N181, Y185, and G189 exhibited a high propensity to coordinate the substrate, Y185 playing a major role in the presence of PEBP1; and N181 in the absence. Additionally, PEBP1 D144 and P186 contributed to stabilizing the SAPE.

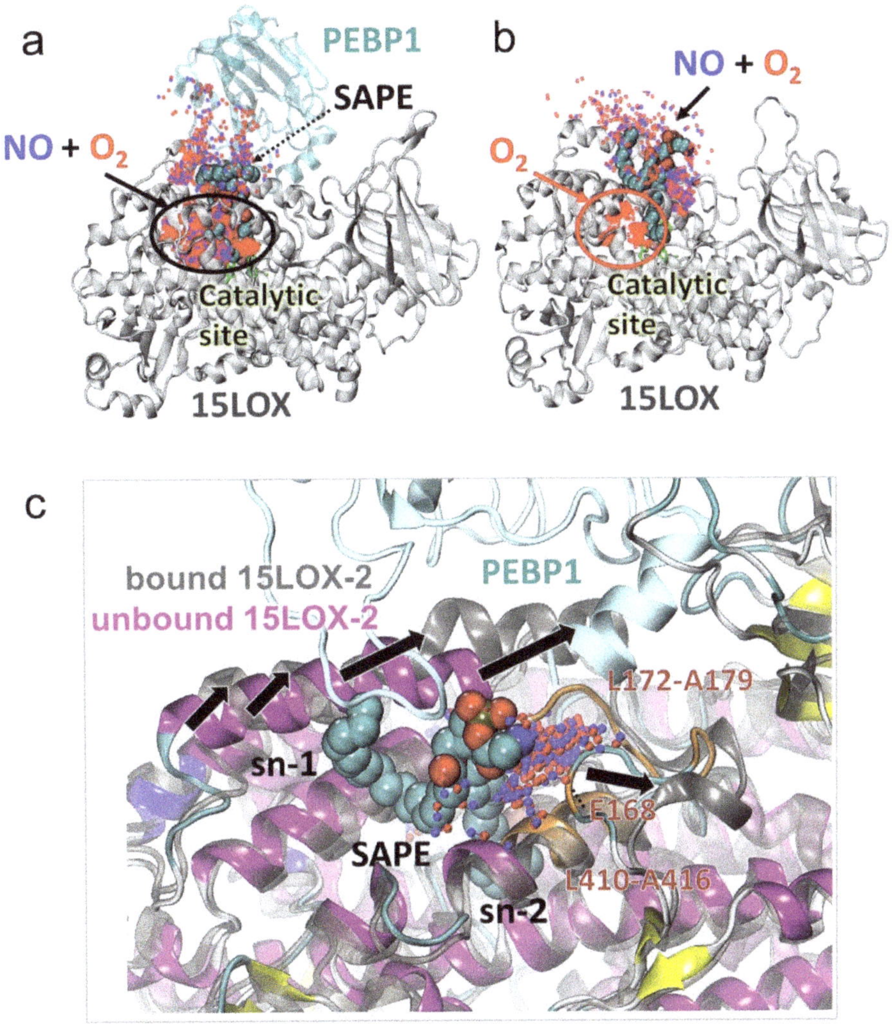

Figure 4. Conformational change in 15LOX-2 induced upon PEBP1 binding allows $NO^{\bullet}$ binding to the catalytic sites in the presence of SAPE. Panels **a** and **b** compare the binding patterns of $O_2/NO^{\bullet}$ to 15LOX-2 in the presence (**a**) and absence (**b**) of complexation with PEBP1. The positions of $NO^{\bullet}$ (blue dots) and O_2 (red dots) sampled during MD snapshots are displayed. These refer to contacts (within 3.5 Å) between $O_2/NO^{\bullet}$ and (**a**) SAPE-bound 15LOX-2/PEBP1 complex and (**b**) SAPE-bound 15LOX-2 (with $O_2/NO^{\bullet}$ molecules within 7 Å from any SAPE atom). Both O_2 and $NO^{\bullet}$ molecules sample the catalytic site in the presence of PEBP1 (panel a, black oval). In the absence of PEBP1 the catalytic site exclusively harbors the O_2 molecules (red oval). Snapshots from another run (Supplementary Figure S5) illustrate the reproducibility of the results. (**c**) Structural change induced by PEBP1 binding. Structural alignment of 15LOX-2 structure (after 150 ns simulation) in PEBP1-bound (dark grey) and unbound (magenta) forms with SAPE substrate (spheres) and accumulation of $NO^{\bullet}$ molecules near the loop L172-A179 (orange) on the surface, are shown. Black arrows show the direction of conformational change in the $\alpha 2$ helix and the T166-A179 loop, providing access to both $NO^{\bullet}$ (blue dots) and O_2 (red dots). Other $NO^{\bullet}$ molecules attracted by the stearic acid sn-1 chain of SAPE are hidden for better visualization.

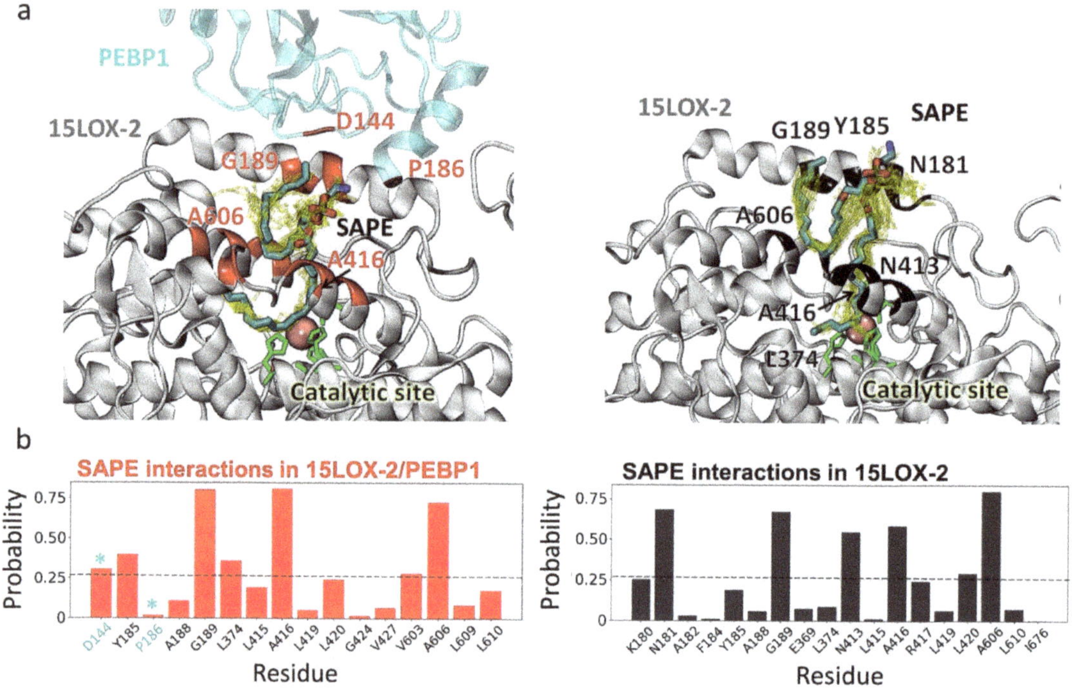

Figure 5. Substrate-binding residues of 15LOX-2 in the presence and absence of PEBP1. (**a**) Interfacial contacts (within 3.5 Å) between the substrate SAPE and PEBP1/15LOX-2 (left panel) and 15LOX-2 (right panel) observed in MD simulations. (**b**) 15LOX-2 residues exhibiting the highest probabilities of contacts with SAPE. The ordinate shows the probabilistic occurrence (the number of counts divided by the total number of selected frames that satisfy the contact requirement). Two PEBP1 residues, D144 and P186, also observed to make frequent contacts are also included (indicated by cyan stars). The fluctuations in the conformation of SAPE during simulations are indicated by yellow sticks.

2.8. In Silico Saturation Mutagenesis Analysis Confirms the Critical Role of Selected Residues Amongst Those Identified to Mediate $O_2/NO^{\bullet}$ Entry and Translocation to the Catalytic Site

The present study points to several residues implicated in initial entry, channeling, and binding of $O_2/NO^{\bullet}$ taking part in *E1* or *E3*, or associated sites *S1*, *S2*, and other clusters (of mostly hydrophobic residues) paving the way to the catalytic site (Table S2). As a test of the potential impact of substitutions at those sites, we performed an in silico saturation mutagenesis analysis, using *Rhapsody* [23]. This machine learning tool scans all possible 19 substitutions at all the *N* amino acid positions of the protein to predict the so-called pathogenicity probabilities, a measure of the expected impact of specific substitutions on the protein function, varying from zero (neutral) to 1 (most damaging or pathogenic). The predictions are based on the evaluation of the sequence-, structure- and dynamics-properties of 15LOX family proteins in comparison to the features observed in more than 20,000 missense variants used as learning dataset [23,24].

The results for residue segments of interest are presented as a heatmap in Figure 6a, color-coded from blue (neutral) to red (deleterious). The wild-type (WT) residues are shown in white. The ribbon diagram in panel b is also colored by the expected 'pathogenicity' from blue to red, mainly using the average values plotted under the heat map for each residue. The three curves therein represent the residue-averaged pathogenicity values predicted by Rhapsody (red), EVmutation [25] (green), and PolyPhen-2 [26] (blue). EVmutation takes rigorous account of residue (co)evolutionary properties and has proven to provide highly accurate results, while Polyphen-2 is broadly used for evaluating the effect of mutations due to its applicability in the absence of structural data or sufficiently large multiple-sequence-alignments. These curves provide a consolidated view of the sensitivity of 15LOX-2 amino acids to mutations, the peaks describing the sites that would be most resistant to substitutions in general.

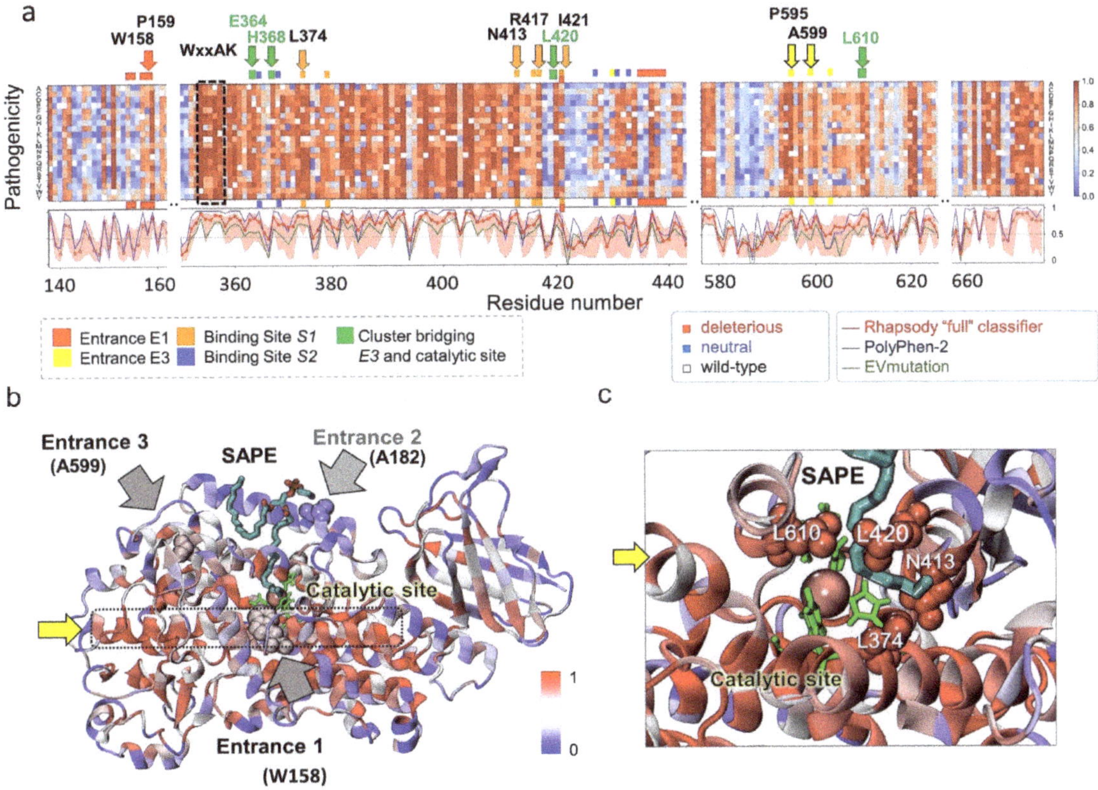

Figure 6. In silico saturation mutagenesis results for human 15LOX-2. (**a**) Pathogenicity probabilities for all substitutions, plotted as a function of residue number (abscissa) for all possible amino acid substitutions (ordinate). The probabilities are represented by a heatmap color-coded from blue (neutral) to red (deleterious). Colored horizontal bars (red, yellow, orange, and blue) along the upper abscissa denote the sites linked to specific functions (see labels under the map, Figure 2 and Table S2). Those distinguished by highly deleterious response to substitution are labeled. Black dashed box highlights the conserved WxxAK motif. The curves in the panel under the map indicate the sensitivity of a given residue to any mutation, as predicted by Rhapsody (red curve), EVmutation (green), and PolyPhen-2 (blue). Entrance 2 (on α2 helix) is broadly neutral to substitutions and not included in the heatmap. (**b**) Ribbon diagram of 15LOX-2/SAPE color-coded by pathogenicity probabilities (if mutated). The regions shown in space-filling representation and labeled (pointed by the black/grey arrows) are the entrances E1–E3. Catalytic residues are displayed in green and SAPE as cyan-red-blue sticks. (**c**) Close-up view of the catalytic site with bound substrate coordinated by four residues, L374, L420, L610, and N413, whose substitutions would be highly damaging to function. Yellow arrow points to the α12-14 scaffolding helix.

As expected, substitutions at the catalytic residues H373, H378, H553, and I676, and at the WxxAK motif were deleterious, irrespective of the type of amino acid substitution. Of interest is, however, to see whether (or which of) the residues involved in mediating the interactions with the ligand (O_2/$NO^{\bullet}$) or substrate (SAPE) are predicted to be critical to function. Our analysis in fact revealed that the residues labeled along the upper abscissa, written in boldface in Table S2, to be intolerant to mutations. These residues include P595 and A599 lining *E3*, W158-P159 and I421 at *E1*, and N413, R417, L421, and L374 at site *S1*. Among them we note that some play a dual role of coordinating the substrate SAPE too (L374, N413, L420, and L610; see Figure 6c). In contrast, the α2-helix residues A177-G199 (*E2* residues and binding interface for PEBP1) are found to be tolerant to mutations (Figure 6b), as well as the surface-exposed G189, even though it participates in 70% of the interactions with SAPE.

Residues located along the helix α12-14 (K350-Q391) are particularly sensitive to substitutions. Examination of the structure shows that this region, composed of a long helix (with disruptions at two positions, hence the labeling as α12, α13, and α14) spans the entire structure at the center, making contacts with both *E1* and *E3* residues and lining the catalytic pocket (Figure 6b, dotted black box; see also Supplementary Figure S1). Its scaffolding role and multiple contacts appear to be critical for maintaining the stability and functionality of the enzyme.

2.9. Identification of Nitrosylated PE Species in Cells Treated with NO• Donors

Encouraged by these results, we next examined whether the nitrosylated PE products are formed in cells in which ferroptosis is inhibited by NO•. We had previously shown that RAW 264.7 M2 macrophages are susceptible to ferroptosis when their phospholipid hydroperoxide-specific glutathione peroxidase is inhibited by RSL3 [17]. In these cells, ferroptosis induced by RSL3 is suppressed by two inhibitors, Ferrostatin-1 and DTPA NONOate. Ferrostatin-1 suppresses ferroptosis by inhibiting the 15LOX-2/PEBP1 complex and through radical trapping antioxidant action. DTPA NONOate is a donor of NO• [3]. We had earlier shown that rescue from ferroptosis by DTPA NONOate is associated with a reduction in HPETE-PE contents [17]. Both Fer-1 and DTPA NONOate rescued from RSL3 induced ferroptosis with approximately similar effectiveness (Figure 7a). We identified two nitrosylated PE products 1-SA-2-ETE(-NO)-PE and 1-OA-2-ETE(-NO)-PE with m/z values of 795.541 and 793.530, respectively (Figure 7b). The precursors for these nitrosylated lipids, 1-SA-ETE-PE and 1-OA-ETE-PE, are the two most abundant ETE-containing PE species in cells. There was about 10-fold excess of these nitosylated PE species in cells treated with DTPA NONOate compared to Fer-1 treated cells (Figure 7c). Though these amounts are very low and we do not have an appropriate standard to quantify these nitrosylated PE species, our findings point to the presence of NO• in the close proximity of carbon-centered radicals formed by lipoxygenase.

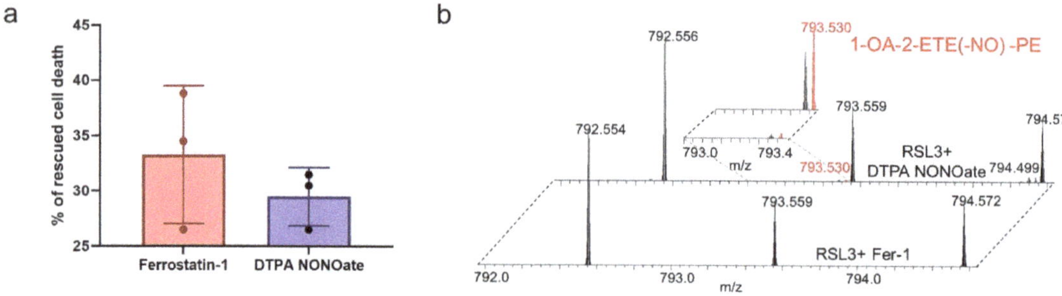

Figure 7. *Cont.*

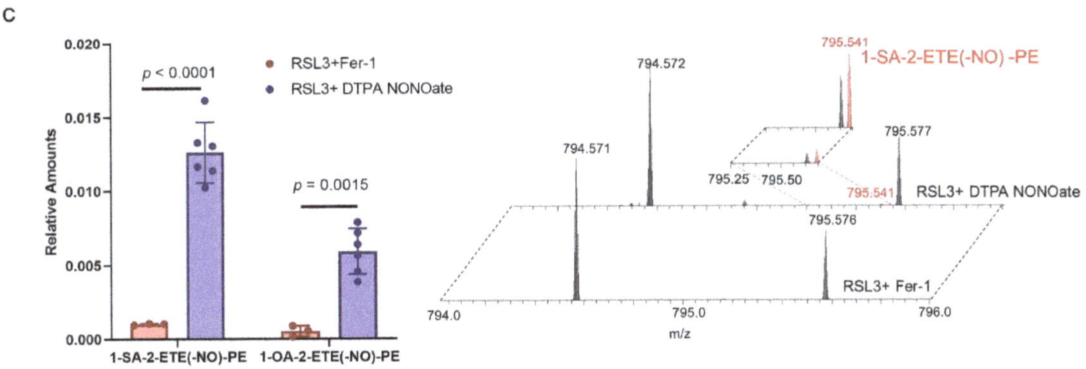

Figure 7. Formation of nitrosylated PE during the inhibition of ferroptosis by NO• (**a**) Bar graph showing the decrease in cell death upon addition of ferroptosis inhibitor Ferrostatin-1 and DTPA NONOate in RAW 264.7 M2 macrophages treated with RSL3. Data are mean ± SD, $n = 3$ and no statistical significance was observed. (**b**) Mass spectrum showing the presence of two nitrosylated PE species, 1-OA-2-ETE(-NO)-PE (top panel) and 1-SA-2-ETE(-NO)-PE (bottom panel) in RAW 264.7 M2 macrophages treated with RSL3+Fer-1 (bottom section) and RSL3+DTPA NONOate (top section). Inset shows the magnified spectrum. (**c**) Quantities of 1-OA-2-ETE(-NO)-PE and 1-SA-2-ETE(-NO)-PE in RAW 264.7 M2 macrophages treated with RSL3+Fer-1 and RSL3+DTPA NONOate. Data are mean ± SD, $n = 3$ for RSL3+Fer-1 cells and 6 for RSL3+DTPA NONOate cells. p values are calculated using two-way ANOVA followed by Sidak post-hoc test.

3. Discussion

Among different aspects of ferroptosis, its physiological regulation by mechanisms other than the well-established effect of GPX4, attracted particular attention in recent years [27]. Among the most recent developments is a demonstration of iNOS/NO• potency to directly control ferroptosis in macrophages and microglia, and distantly in several neighboring (e.g., epithelial) cells [13]. Interest in the role of NO• in ferroptosis further increased by the possibility of using NO• donors for attenuating, suppressing, or delaying ferroptotic death [13]. Although several mechanisms, including direct reaction of NO• with lipid radical intermediates (L•, LO•, $LO^{\bullet}{}_2$), have been considered [19], the exact nature of NO•'s inhibitory effect remained elusive.

Given the reported participation of 15LOX-2/PEBP1 complex in the generation of 15HpETE-PE that serve as a pro-apoptotic agent, NO•'s potential regulation of this catalytic complex has emerged as an important consideration. Our experiments in a model biochemical system with purified recombinant proteins directly demonstrate the ability of NO• to inhibit the catalysis of ETE-PE oxidation to 15HpETE-PE by the complex but not by 15LOX-2 alone. This suggests that structural features specific to the 15LOX-2/PEBP1 complex could account for this inhibitory effect. Given that O_2 is delivered to the 15LOX-2 catalytic site via a channel regulated by the binding of the oxidizable substrate, the role of structural changes in the 15LOX-2/PEBP1/SAPE in controlling O_2 delivery and possible competition with NO• became and even more fascinating yet unsolved puzzle. The present LC-MS data showing nitrosylation of ETE-PE, i.e., direct interaction of NO• with a carbon-centered radical, rather than with the peroxyl radical intermediate, LO2•, provide further evidence for the direct interference of NO•.

The lipidomics experiments showed that the 15LOX-2/PEBP1 complex produced greater amount of thc nitrosylated product (NO•-SAPE) than 15LOX-2 alone suggesting that a larger number of NO• molecules were able to enter the catalytic pocket of PEBP1-bound 15LOX-2 compared to free 15LOX-2. In accord with these observations, our simulations showed the higher propensity of NO• molecules to access the catalytic site in the presence of PEBP1 bound to 15LOX-2. In contrast, in the absence of PEBP1 and at small concentration of NO•, NO• were attracted to the sn-1 tail of SAPE or grouped in a new cavity formed after loop L172-A179 reorganization in 15LOX-2 alone simulations.

The experimentally observed formation of nitrosylated-ETE-PE catalyzed (albeit at lower levels) by 15LOX-2 alone [28], is possible due to presence of $NO^{\bullet}$ at the catalytic site. The in silico studies at similar concentrations of both molecules, could not detect any $NO^{\bullet}$ at the catalytic site. However, an increase in the number of $NO^{\bullet}$ molecules in the MD system lead to a few $NO^{\bullet}$ accessing the active site, explaining the nitrosylated-ETE-PE formation in 15LOX-2 alone. We also note that the turnover rate for 15LOX-2 is rather slow (~8.5–25/s). Thus, the current simulations mainly provide insights into the mechanistic aspects of $NO^{\bullet}$ actions, rather than a quantitative description of the overall kinetics.

Concentrations of both gases in physiological conditions may vary. Several studies indicate that the intra-cellular concentrations of O_2 may be as high as 30 μM and the $NO^{\bullet}$ concentration can only go up to 5 μM [29]. In aerobically incubated cell culture, O_2 can go up to 200 μM whereas $NO^{\bullet}$ concentration can be manipulated within the wide range dependently on the type of the $NO^{\bullet}$ donors added [30,31]. These concentrations may differ significantly within the micro-environment of 15LOX-2/PEBP1 complex. O_2 is the substrate for many biologically relevant systems, such as cellular respiration [32], NADPH oxidase [33], dehydrogenases of the TCA cycle [34], while $NO^{\bullet}$ can be consumed in nitroxygenation and S-nitrosylation reactions of proteins and their thiols [35] and avidly react with $O_2^{\bullet -}$ to yield peroxynitrite [36]. Furthermore, the ratios of $O_2/NO^{\bullet}$ may vary significantly dependently on the cell types. For example, in macrophages polarized to M1 phenotype, high expression of iNOS leads to a sharp increase in the $NO^{\bullet}$ production [37]. At the same time, NADPH oxidase consumes O_2 to generate $O_2^{\bullet -}$ hence depleting O_2 required for the 15LOX-catalyzed reactions [38]. Notably, iNOS generated $NO^{\bullet}$ can diffuse intra- and extra-cellularly to reach high levels sufficient for quenching the production of pro-ferroptotic signals by the 15LOX-2/PEBP1 complexes [13] (Dar et al., Manuscript under revision). In contrast M2 macrophages express negligible amounts of iNOS and $NO^{\bullet}$ likely resulting in preponderance of O_2 vs. $NO^{\bullet}$ amounts [39]. Evidently, even more dramatic variations in intracellular contents of $NO^{\bullet}$ and O_2 may occur in disease conditions related to inflammation and sepsis [40–42].

Our simulations unambiguously showed that this effect, occurring in the presence of PEBP1 is indeed due to the ability of $NO^{\bullet}$ molecules to bind and diffuse to the catalytic site of 15LOX-2, favored by a conformational change in 15LOX-2 induced upon complexation with PEBP1. The 15LOX-2/PEBP1/SAPE simulations revealed that the accessibility paths of both O_2 and $NO^{\bullet}$ are similar, and $NO^{\bullet}$ often out-competes O_2 and occupies the catalytic site in the 15LOX-2/PEBP1 complex, leading to the production of the nitrosylated product, $NO^{\bullet}$-SAPE observed in the biochemical and in vitro experiments. Notably, during free AA peroxidation, the $NO^{\bullet}$ molecules get access to the catalytic pocket. This suggests that at low concentration of $NO^{\bullet}$ molecules the exposed sn-1-acyl chain of SAPE in 15LOX-2 alone sequesters $NO^{\bullet}$ thereby preventing its access to the catalytic site. The lipidomics data further showed that lipid peroxidation by 15LOX-2 alone (ETE oxidation into 15-HpETE) was not altered significantly by $NO^{\bullet}$, in line with the lower accessibility to the 15LOX-2 catalytic site (in the absence of PEBP1) shown in MD simulations.

Finally, our work provides a detailed mechanistic description of the interactions of $O_2/NO^{\bullet}$ molecules with 15LOX-2 residues during their journey towards the catalytic site. There are two available entrances to the catalytic cavity. The first (*E1*) is near Y154-W158 loop the flexibility of which is presumably limited by a proline, P159, highly resistant to substitutions. The second (*E2*) is near S430, S573, P595, A599, and L603. Both entrances provide access to the catalytic site through intra-protein channels that mediate the translocation of the small molecules. Our analysis highlighted the importance of L374, N413, L420, I421, and L610 in enabling efficient communication with the catalytic site. Detailed analysis of the substrate interaction site with 15LOX-2 and PEBP1 at the most favorable geometric positioning for lipids peroxidation also revealed crucial interactions with selected residues (L374, L420, L610, and N413) that formed a substrate-binding epitope.

Comparison of the critical residues with prior observations made for lipoxygenase family variants lends support to the functional significance of several sites identified here.

For example, loss of 15LOX-2 activity has been observed in the variant A416D [43], a residue taking part in the binding site *S1* and in the cluster of hydrophobic residues that connect the entrance *E3* to the catalytic site. Likewise, the mutation N426M in 5LOX, which is the homologous counterpart of the site *S2* T431 in 15LOX-2, gives rise to a loss of activity when associated with F360W and A425I [44]. Likewise, the mutation A417A in LOX12 (counterpart of 15LOX-2 S430 at *E3*) reduced catalytic activity and altered the stereo-selectivity of the oxygenation reaction [45]. It is also worth noting that many of these identified residues are highly conserved across lipoxygenase family members as indicated in Supplementary Table S2.

The overall analysis identified not only key sites, but their intricate couplings to enable the catalytic activity of 15LOX-2 complexed with PEBP1. In the presence of $NO^{\bullet}$, the enzymatic machinery is still in place, but it cannot effectively produce HpETE-PE. Even though the direct reaction of $NO^{\bullet}$ with the ETE-PE carbon-centered intermediate takes place, albeit at a very low level, our analysis strongly suggests that the observed suppression of ETE-PE peroxidation by $NO^{\bullet}$ in the 15LOX-2/PEBP1 complex is mostly due to the occupancy of the O_2-binding site or channeling path by $NO^{\bullet}$, and a competition between the two radicals (O_2 and $NO^{\bullet}$) thus resulting in the reduced, if at low physiological $NO^{\bullet}$ amounts not completely abrogated, ETE-PE peroxidation. Interestingly, our highly sensitive redox lipidomics analysis did not reveal the formation of nitroxygenated ETE-PE derivatives. This suggests that the direct chemical reaction of $NO^{\bullet}$ with O_2—very effective in gas and liquid phases [46]—is strongly suppressed within the structural confinements of the channel. It remains to be seen if this type of interference of $NO^{\bullet}$ to repress the peroxidation of free and esterified PUFAs by 15LOX-2/PEBP1 could be exploited in designing anti-ferroptotic therapies.

4. Materials and Methods

4.1. Molecular Dynamics (MD) Simulations

Conventional full-atomic MD simulations [47–49] were performed for human 15LOX-2 (PDB id: 4NRE [20]) and 15LOX-2/PEBP1 complex (PDB ids: 4NRE and 1BEH [50]), with a bound substrate (SAPE), in the presence of randomly distributed nitric oxide and oxygen molecules with different ratios, such as 1:1 (five of each), 1:3, and 3:1. Multiple MD runs (see Supplementary Table S1) of 150 ns with different initial spatial distributions of $NO^{\bullet}$ and O_2 molecules were performed for each structure (15LOX-2 and its complex with PEBP1) using the NAMD [51] software with the CHARMM [52] force field and 2 fs time steps. The proteins were solvated with explicit water models (TIP3P [53]) at physiological salt concentrations. Docking simulations generated structural models for the 15LOX-2/PEBP1 complex, using the protocols described previously [3]. The binding site and pose of SAPE were predicted using SMINA [54] ligand-protein docking package derived from AutoDock Vina [55]. CHARMM force field parameters for $NO^{\bullet}$, O_2, and covalently bonded Fe^{3+} were created based on bound O_2 and heme group using Gaussian [56] package. Prior to productive runs, the following protocol was adopted: 0.2 ns of water equilibration, 10,000 steps of minimization, 0.35 ns of heating from 0 to 300 K, and 0.15 ns equilibration of the whole system. Simulations were performed with a cutoff of 12 Å for non-bonded interactions and Langevin piston algorithm to maintain the temperature at 300K and pressure at 1 atm. We used VMD [57] for visualization and ProDy [58–60] for trajectory analysis with in-house scripts. CAVER [61] 3.0 with PyMOL Molecular Graphics System, Version 1.8, Schrödinger, LLC was used to represent and display cavities and interior surfaces.

4.2. Rhapsody and in Silico Saturation Mutagenesis Analysis

The Rhapsody tool [24] was used for automated scanning of all residue substitutions in 15LOX-2 to predict the functional consequences of single amino acid variants (SAV). A random forest-based classifier was trained on an integrated dataset of 20,854 missense mutations functionally characterized to date. For each training sample, eight features incorporating the effects of structural dynamics and sequence-based (co)evolution properties were calculated using ProDy [58], Evol [58], and PolyPhen-2 [62]. The 15LOX-2 structure (PDB id: 4NRE [20]) was used as input. Residue-averaged scores evaluated by Rhapsody, PolyPhen-2, and EVmutation [25] were examined for consolidation of the results.

4.3. Lipoxygenase Assay

Human recombinant 15LOX-2 and PEBP1 were recombinantly expressed an purified as described previously [3,63]. 15LOX-2 was pre-activated with 13-hydroperoxy-octadecadienoic acid (HpODE) (5 μM) for 30 min at 37 °C. The pre-activated 15LOX-2 (3 pmols) was added to a reaction mixture containing 100 μM lipid in Tris-HCl (50 mM, pH7.4), 25 μM PAPA NONOAte and DTPA (100 μM) for a total volume of 50 μL. DTPA was added 15 min prior to the start of the reaction. For reactions with the 15LOX-2/PEBP1 complex, equal quantity of PEBP1 and 15LOX-2 were mixed prior to pre-activation. The reaction mixture was incubated on a shaker mixer at 37 °C for 30 min. For time course, the reaction was allowed to continue up to a specific time. The reaction was stopped with addition of 9 volumes (450 μL) of 100% acetonitrile, and the samples were centrifuged at 10,000× *g* for 15 min at 4 °C. Then, 20 μL of the supernatant was transferred to an auto sampler vial, and 5 μL was injected into the LC-MS/MS system. LC-MS/MS analysis was performed as described previously.

4.4. Cell Culture

RAW 264.7 cells were obtained from the American Type Culture Collection (ATCC). Cultured at 37 °C and 5% CO_2 in DMEM or RPMI (ATCC) supplemented with 10% heat-inactivated fetal calf serum (FCS; Sigma–Aldrich, St. Louis, MO, USA) and 50 U ml^{-1} penicillin–streptomycin (Thermo Fisher Scientific, Waltham, MA, USA). RAW 264.7 macrophages were polarized by to M2 stage by incubating in DMEM containing 10% FBS, 50 U mL^{-1} penicillin–streptomycin, and IL-4 (20 ng mL^{-1}) for 48 h. For ferroptotis experiments, cells were incubated with RSL3 (0.5 μM), 2 μM), +Fer-1 (~400 nM), or ±DPTA NONOate (25 μM), for 5 h. Cell death was determined by flow cytometry.

4.5. LC–MS Analysis of Phospholipids

MS analysis of phospholipids was performed on an Orbitrap mass spectrometer (Thermo Fisher). Phospholipids were separated on a normal phase column (Luna 3 μm Silica (2) 100 Å, 150 × 2.0 mm, Phenomenex, Torrance, CA, USA) at a flow rate of 0.2 mL min^{-1} on a Dionex Ultimate 3000 HPLC system (Dionex, Idstein, Germany). The column was maintained at 35 °C. Analysis was performed using gradient solvents (A and B) containing 10 mM ammonium acetate. Solvent A contained propanol:hexane:water (285:215:5, vol/vol/vol) and solvent B contained propanol:hexane:water (285:215:40, vol/vol/vol). All solvents were LC–MS grade. The column was eluted for 0–23 min with a linear gradient of 10–32% B; 23–32 min using a linear gradient of 32–65% B; 32–35 min with a linear gradient of 65–100% B; 35–62 min held at 100% B; 62–64 min with a linear gradient from 100% to 10% B followed by and equilibration from 64 to 80 min at 10% B. The instrument was operated with the electrospray ionization probe in negative polarity mode. Data was analyzed using Quan Browser of xcalibur software (Thermo Fisher).

Supplementary Materials: The following are available online at https://www.mdpi.com/1422-0067/22/10/5253/s1, Figure S1: Location of the two entrances E1 and E2 of 15LOX-2 that enable access of O_2 and $NO^•$ to the catalytic site, Figure S2: Secondary structure of 15LOX-2, Figure S3: Additional pores or tunnels leading to the catalytic site, Figure S4: Close-up view of contacts between 15LOX-2 residues and O_2/$NO^•$ molecules, Figure S5: Same results as Figure 4a–b, reproduced by an independent run, Figure S6: 15LOX-2/SAPE complex and its interactions with O_2 and $NO^•$ molecules observed in MD simulations with higher concentration of $NO^•$ molecules, Figure S7: 15LOX-2/AA complex and its interactions with O_2 and $NO^•$ molecules observed in MD simulations, Table S1: Summary of the simulated systems, compositions, and durations, Table S2: Key residues in 15LOX-2 that play a role in regulation of lipid peroxidation, Movie 1: Competition between $NO^•$ and O_2 molecules near the SAPE at the catalytic site of 15LOX-2 complexed with PEBP1.

Author Contributions: V.E.K., K.M.-R., and I.B. conceived the study. T.S.A. designed lipidomics experiments. V.E.K. and H.B. designed experiments; T.S.A. and A.L. performed lipidomics experiments and analyzed the experimental data. A.A.K. performed RAW M2 macrophages experiments. K.M.-R. and I.H.S. designed the computational experiments. K.M.-R. performed the computational modeling and simulations. I.B., K.M.-R., and I.H.S. interpreted the MD data. K.M.-R., I.B., and V.E.K. wrote the manuscript, which was edited by all authors. All authors have read and agreed to the published version of the manuscript.

Funding: This research was funded by NIH (HL114453, U01AI156924, U01AI156923, CA165065, NS076511, NS061817, P41 GM103712, P01 DK096990), and by Polish National Science Centre no. 2019/35/D/ST4/02203.

Institutional Review Board Statement: Not applicable.

Informed Consent Statement: Not applicable.

Data Availability Statement: Data and codes generated during the study and included in this article are available from the corresponding authors upon request.

Acknowledgments: This work was supported by NIH (HL114453, U01AI156924, U01AI156923, CA165065, NS076511, NS061817, P41 GM103712, P01 DK096990), by Polish National Science Centre no. 2019/35/D/ST4/02203. Karolina Mikulska-Ruminska is thankful for the facilities and computer time allocated by the Interdisciplinary Center for Modern Technologies, NCU. The authors thank Ted Holman (University of California at Santa Cruz) and Andrew VanDemark (University of Pittsburgh) for their generous gift of recombinant 15LOX-2 and PEBP1, respectively.

Conflicts of Interest: The authors have declared that no conflicts of interest exists.

References

1. Stockwell, B.R.; Angeli, J.P.F.; Bayir, H.; Bush, A.I.; Conrad, M.; Dixon, S.J.; Fulda, S.; Gascón, S.; Hatzios, S.K.; Kagan, V.E. Ferroptosis: A regulated cell death nexus linking metabolism, redox biology, and disease. *Cell* **2017**, *171*, 273–285. [CrossRef] [PubMed]
2. Jiang, X.; Stockwell, B.R.; Conrad, M. Ferroptosis: Mechanisms, biology and role in disease. *Nat. Rev. Mol. Cell Biol.* **2021**, *22*, 1–17. [CrossRef] [PubMed]
3. Wenzel, S.E.; Tyurina, Y.Y.; Zhao, J.; Croix, C.M.S.; Dar, H.H.; Mao, G.; Tyurin, V.A.; Anthonymuthu, T.S.; Kapralov, A.A.; Amoscato, A.A. PEBP1 Wardens Ferroptosis by Enabling Lipoxygenase Generation of Lipid Death Signals. *Cell* **2017**, *171*, 628–641.e26. [CrossRef] [PubMed]
4. Anthonymuthu, T.S.; Tyurina, Y.Y.; Sun, W.-Y.; Mikulska-Ruminska, K.; Shrivastava, I.H.; Tyurin, V.A.; Cinemre, F.B.; Dar, H.H.; VanDemark, A.P.; Holman, T.R.; et al. Resolving the paradox of ferroptotic cell death: Ferrostatin-1 binds to 15LOX/PEBP1 complex, suppresses generation of peroxidized ETE-PE, and protects against ferroptosis. *Redox Biol.* **2020**, *38*, 101744. [CrossRef]
5. Anthonymuthu, T.S.; Kenny, E.M.; Shrivastava, I.; Tyurina, Y.Y.; Hier, Z.E.; Ting, H.-C.; Dar, H.H.; Tyurin, V.A.; Nesterova, A.; Amoscato, A.A.; et al. Empowerment of 15-Lipoxygenase Catalytic Competence in Selective Oxidation of Membrane ETE-PE to Ferroptotic Death Signals, HpETE-PE. *J. Am. Chem. Soc.* **2018**, *140*, 17835–17839. [CrossRef] [PubMed]
6. Shah, R.; Shchepinov, M.S.; Pratt, D.A. Resolving the role of lipoxygenases in the initiation and execution of ferroptosis. *ACS Cent. Sci.* **2018**, *4*, 387–396. [CrossRef]
7. Doll, S.; Proneth, B.; Tyurina, Y.Y.; Panzilius, E.; Kobayashi, S.; Ingold, I.; Irmler, M.; Beckers, J.; Aichler, M.; Walch, A. ACSL4 dictates ferroptosis sensitivity by shaping cellular lipid composition. *Nat. Chem. Biol.* **2017**, *13*, 91. [CrossRef]
8. Bersuker, K.; Hendricks, J.; Li, Z.; Magtanong, L.; Ford, B.; Tang, P.H.; Roberts, M.A.; Tong, B.; Maimone, T.J.; Zoncu, R. The CoQ oxidoreductase FSP1 acts parallel to GPX4 to inhibit ferroptosis. *Nature* **2019**, *575*, 688–692. [CrossRef]
9. Doll, S.; Freitas, F.P.; Shah, R.; Aldrovandi, M.; da Silva, M.C.; Ingold, I.; Grocin, A.G.; da Silva, T.N.X.; Panzilius, E.; Scheel, C. FSP1 is a glutathione-independent ferroptosis suppressor. *Nature* **2019**, *575*, 693–698. [CrossRef]
10. Imai, H. Biological significance of lipid hydroperoxide and its reducing enzyme, phospholipid hydroperoxide glutathione peroxidase, in mammalian cells. *Yakugaku Zasshi J. Pharm. Soc. Jpn.* **2004**, *124*, 937–957. [CrossRef]
11. Björnstedt, M.; Hamberg, M.; Kumar, S.; Xue, J.; Holmgren, A. Human thioredoxin reductase directly reduces lipid hydroperoxides by NADPH and selenocystine strongly stimulates the reaction via catalytically generated selenols. *J. Biol. Chem.* **1995**, *270*, 11761–11764. [CrossRef] [PubMed]
12. Sun, W.-Y.; Tyurin, V.A.; Mikulska-Ruminska, K.; Shrivastava, I.H.; Anthonymuthu, T.S.; Zhai, Y.-J.; Pan, M.-H.; Gong, H.-B.; Lu, D.-H.; Sun, J. Phospholipase iPLA 2 β averts ferroptosis by eliminating a redox lipid death signal. *Nat. Chem. Biol.* **2021**, *17*, 465–476. [CrossRef] [PubMed]
13. Kapralov, A.A.; Yang, Q.; Dar, H.H.; Tyurina, Y.Y.; Anthonymuthu, T.S.; Kim, R.; Croix, C.M.S.; Mikulska-Ruminska, K.; Liu, B.; Shrivastava, I.H.; et al. Redox lipid reprogramming commands susceptibility of macrophages and microglia to ferroptotic death. *Nat. Chem. Biol.* **2020**, *16*, 278–290. [CrossRef] [PubMed]

14. Hogg, N.; Kalyanaraman, B. Nitric oxide and lipid peroxidation. *BBA Bioenerg.* **1999**, *1411*, 378–384. [CrossRef]
15. Kanner, J.; Harel, S.; Granit, R. Nitric oxide, an inhibitor of lipid oxidation by lipoxygenase, cyclooxygenase and hemoglobin. *Lipids* **1992**, *27*, 46. [CrossRef] [PubMed]
16. Ivanov, I.; Heydeck, D.; Hofheinz, K.; Roffeis, J.; O'Donnell, V.B.; Kuhn, H.; Walther, M. Molecular enzymology of lipoxygenases. *Arch. Biochem. Biophys.* **2010**, *503*, 161–174. [CrossRef] [PubMed]
17. Keefer, L.K.; Nims, R.W.; Davies, K.M.; Wink, D.A. "NONOates"(1-substituted diazen-1-ium-1, 2-diolates) as nitric oxide donors: Convenient nitric oxide dosage forms. In *Methods in Enzymology*; Elsevier: Amsterdam, The Netherlands, 1996; Volume 268, pp. 281–293.
18. Anthonymuthu, T.S.; Kenny, E.M.; Bayır, H. Therapies targeting lipid peroxidation in traumatic brain injury. *Brain Res.* **2016**, *1640*, 57–76. [CrossRef]
19. Wood, I.; Trostchansky, A.; Rubbo, H. Structural considerations on lipoxygenase function, inhibition and crosstalk with nitric oxide pathways. *Biochimie* **2020**, *178*, 170–180. [CrossRef]
20. Kobe, M.J.; Neau, D.B.; Mitchell, C.E.; Bartlett, S.G.; Newcomer, M.E. The structure of human 15-lipoxygenase-2 with a substrate mimic. *J. Biol. Chem.* **2014**, *289*, 8562–8569. [CrossRef]
21. Laskowski, R.A. PDBsum: Summaries and analyses of PDB structures. *Nucleic Acids Res.* **2001**, *29*, 221–222. [CrossRef]
22. Mikulska-Ruminska, K.; Shrivastava, I.; Krieger, J.; Zhang, S.; Li, H.; Bayır, H.; Wenzel, S.E.; VanDemark, A.P.; Kagan, V.E.; Bahar, I. Characterization of Differential Dynamics, Specificity, and Allostery of Lipoxygenase Family Members. *J. Chem. Inf. Model.* **2019**, *59*, 2496–2508. [CrossRef] [PubMed]
23. Ponzoni, L.; Peñaherrera, D.A.; Oltvai, Z.N.; Bahar, I. Rhapsody: Predicting the pathogenicity of human missense variants. *Bioinformatics* **2020**, *36*, 3084–3309. [CrossRef] [PubMed]
24. Ponzoni, L.; Bahar, I. Structural dynamics is a determinant of the functional significance of missense variants. *Proc. Natl. Acad. Sci. USA* **2018**, *115*, 4164–4169. [CrossRef] [PubMed]
25. Hopf, T.A.; Ingraham, J.B.; Poelwijk, F.J.; Scharfe, C.P.; Springer, M.; Sander, C.; Marks, D.S. Mutation effects predicted from sequence co-variation. *Nat. Biotechnol.* **2017**, *35*, 128–135. [CrossRef] [PubMed]
26. Adzhubei, I.; Jordan, D.M.; Sunyaev, S.R. Predicting functional effect of human missense mutations using PolyPhen-2. *Curr. Protoc. Hum. Genet.* **2013**, *76*, 7.20.1–7.20.41. [CrossRef] [PubMed]
27. Maiorino, M.; Conrad, M.; Ursini, F. GPx4, lipid peroxidation, and cell death: Discoveries, rediscoveries, and open issues. *Antioxid. Redox Signal.* **2018**, *29*, 61–74. [CrossRef]
28. Kühn, H.; Barnett, J.; Grunberger, D.; Baecker, P.; Chow, J.; Nguyen, B.; Bursztyn-Pettegrew, H.; Chan, H.; Sigal, E. Overexpression, purification and characterization of human recombinant 15-lipoxygenase. *BBA Lipids Lipid Metab.* **1993**, *1169*, 80–89. [CrossRef]
29. Brown, G.C. Nitric oxide regulates mitochondrial respiration and cell functions by inhibiting cytochrome oxidase. *FEBS Lett.* **1995**, *369*, 136–139. [CrossRef]
30. Jarazo Dietrich, S.; Fass, M.I.; Jacobo, P.V.; Sobarzo, C.M.A.; Lustig, L.; Theas, M.S. Inhibition of NOS-NO system prevents autoimmune orchitis development in rats: Relevance of NO released by testicular macrophages in germ cell apoptosis and testosterone secretion. *PLoS ONE* **2015**, *10*, e0128709. [CrossRef]
31. Stuart, J.A.; Fonseca, J.; Moradi, F.; Cunningham, C.; Seliman, B.; Worsfold, C.R.; Dolan, S.; Abando, J.; Maddalena, L.A. How supraphysiological oxygen levels in standard cell culture affect oxygen-consuming reactions. *Oxid. Med. Cell. Longev.* **2018**, *2018*, 1–13. [CrossRef]
32. Schmidt-Rohr, K. Oxygen is the high-energy molecule powering complex multicellular life: Fundamental corrections to traditional bioenergetics. *ACS Omega* **2020**, *5*, 2221–2233. [CrossRef] [PubMed]
33. Clancy, R.; Leszczynska-Piziak, J.; Abramson, S. Nitric oxide, an endothelial cell relaxation factor, inhibits neutrophil superoxide anion production via a direct action on the NADPH oxidase. *J. Clin. Investig.* **1992**, *90*, 1116–1121. [CrossRef] [PubMed]
34. Palmieri, E.M.; Gonzalez-Cotto, M.; Baseler, W.A.; Davies, L.C.; Ghesquière, B.; Maio, N.; Rice, C.M.; Rouault, T.A.; Cassel, T.; Higashi, R.M. Nitric oxide orchestrates metabolic rewiring in M1 macrophages by targeting aconitase 2 and pyruvate dehydrogenase. *Nat. Commun.* **2020**, *11*, 1–17. [CrossRef] [PubMed]
35. Jaffrey, S.R.; Erdjument-Bromage, H.; Ferris, C.D.; Tempst, P.; Snyder, S.H. Protein S-nitrosylation: A physiological signal for neuronal nitric oxide. *Nat. Cell Biol.* **2001**, *3*, 193–197. [CrossRef] [PubMed]
36. Ischiropoulos, H.; Zhu, L.; Beckman, J.S. Peroxynitrite formation from macrophage-derived nitric oxide. *Arch. Biochem. Biophys.* **1992**, *298*, 446–451. [CrossRef]
37. MacMicking, J.; Xie, Q.-W.; Nathan, C. Nitric oxide and macrophage function. *Ann. Rev. Immun.* **1997**, *15*, 323–350. [CrossRef]
38. Babior, B.; Lambeth, J.; Nauseef, W. The neutrophil NADPH oxidase. *Arch. Biochem. Biophys.* **2002**, *397*, 342–344. [CrossRef]
39. Gordon, S.; Martinez, F.O. Alternative activation of macrophages: Mechanism and functions. *Immunity* **2010**, *32*, 593–604. [CrossRef]
40. Victor, V.M.; Rocha, M.; Esplugues, J.V. Role of free radicals in sepsis: Antioxidant therapy. *Curr. Pharm. Des.* **2005**, *11*, 3141–3158. [CrossRef]
41. Darley-Usmar, V.; Wiseman, H.; Halliwell, B. Nitric oxide and oxygen radicals: A question of balance. *FEBS Lett.* **1995**, *369*, 131–135. [CrossRef]
42. Swindle, E.J.; Metcalfe, D.D. The role of reactive oxygen species and nitric oxide in mast cell-dependent inflammatory processes. *Immun. Rev.* **2007**, *217*, 186–205. [CrossRef]

43. Horn, T.; Kakularam, K.R.; Anton, M.; Richter, C.; Reddanna, P.; Kuhn, H. Functional characterization of genetic enzyme variations in human lipoxygenases. *Redox Biol.* **2013**, *1*, 566–577. [CrossRef] [PubMed]
44. Hofheinz, K.; Kakularam, K.R.; Adel, S.; Anton, M.; Polymarasetty, A.; Reddanna, P.; Kuhn, H.; Horn, T. Conversion of pro-inflammatory murine Alox5 into an anti-inflammatory 15S-lipoxygenating enzyme by multiple mutations of sequence determinants. *Arch. Biochem. Biophys.* **2013**, *530*, 40–47. [CrossRef] [PubMed]
45. Chen, X.S.; Funk, C.D. Structure-function properties of human platelet 12-lipoxygenase: Chimeric enzyme and in vitro mutagenesis studies. *FASEB J.* **1993**, *7*, 694–701. [CrossRef] [PubMed]
46. Thomas, D.D.; Liu, X.; Kantrow, S.P.; Lancaster, J.R. The biological lifetime of nitric oxide: Implications for the perivascular dynamics of NO and O_2. *Proc. Natl. Acad. Sci. USA* **2001**, *98*, 355–360. [CrossRef] [PubMed]
47. Perilla, J.R.; Schulten, K. Physical properties of the HIV-1 capsid from all-atom molecular dynamics simulations. *Nat. Commun.* **2017**, *8*, 1–10. [CrossRef]
48. Mikulska-Ruminska, K.; Kulik, A.; Benadiba, C.; Bahar, I.; Dietler, G.; Nowak, W. Nanomechanics of multidomain neuronal cell adhesion protein contactin revealed by single molecule AFM and SMD. *Sci. Rep.* **2017**, *7*, 8852. [CrossRef]
49. Cheng, Z.; Lan, Y.; Guo, J.; Ma, D.; Jiang, S.; Lai, Q.; Zhou, Z.; Peplowski, L. Computational Design of Nitrile Hydratase from Pseudonocardia thermophila JCM3095 for Improved Thermostability. *Molecules* **2020**, *25*, 4806. [CrossRef]
50. Banfield, M.J.; Barker, J.J.; Perry, A.C.; Brady, R.L. Function from structure? The crystal structure of human phosphatidylethanolamine-binding protein suggests a role in membrane signal transduction. *Structure* **1998**, *6*, 1245–1254. [CrossRef]
51. Phillips, J.; Braun, R.; Wang, W.; Gumbart, J.; Tajkhorshid, E.; Villa, E.; Chipot, C.; Skeel, R.; Kale, L.; Schulten, K. Scalable molecular dynamics with NAMD. *J. Comput. Chem.* **2005**, *26*, 1781–1802. [CrossRef] [PubMed]
52. MacKerell, A., Jr.; Bashford, D.; Bellott, M.; Dunbrack, R., Jr.; Evanseck, J.; Field, M.; Fischer, S.; Gao, J.; Guo, H.; Ha, S. All-Atom Empirical Potential for Molecular Modeling and Dynamics Studies of Proteins. *J. Phys. Chem. B* **1998**, *102*, 3586–3616. [CrossRef] [PubMed]
53. Jorgensen, W.; Chandrasekhar, J.; Madura, J.; Impey, R.; Klein, M. Comparison of simple potential functions for simulating liquid water. *J. Chem. Phys.* **1983**, *79*, 926. [CrossRef]
54. Koes, D.R.; Baumgartner, M.P.; Camacho, C.J. Lessons learned in empirical scoring with smina from the CSAR 2011 benchmarking exercise. *J. Chem. Inf. Mod.* **2013**, *53*, 1893–1904. [CrossRef] [PubMed]
55. Trott, O.; Olson, A.J. AutoDock Vina: Improving the speed and accuracy of docking with a new scoring function, efficient optimization, and multithreading. *J. Comput. Chem.* **2010**, *31*, 455–461. [CrossRef] [PubMed]
56. Frisch, M.; Trucks, G.; Schlegel, H.; Scuseria, G.; Robb, M.; Cheeseman, J.; Montgomery, J., Jr.; Vreven, T.; Kudin, K.; Burant, J. *Gaussian 03, Revision B. 05*; Gaussian Inc.: Pittsburgh, PA, USA, 2003; p. 12478.
57. Humphrey, W.; Dalke, A.; Schulten, K. VMD: Visual molecular dynamics. *J. Mol. Graph.* **1996**, *14*, 33–38. [CrossRef]
58. Bakan, A.; Dutta, A.; Mao, W.; Liu, Y.; Chennubhotla, C.; Lezon, T.R.; Bahar, I. Evol and ProDy for bridging protein sequence evolution and structural dynamics. *Bioinformatics* **2014**, *30*, 2681–2683. [CrossRef]
59. Bakan, A.; Meireles, L.M.; Bahar, I. ProDy: Protein dynamics inferred from theory and experiments. *Bioinformatics* **2011**, *27*, 1575–1577. [CrossRef] [PubMed]
60. Zhang, S.; Krieger, J.M.; Zhang, Y.; Kaya, C.; Kaynak, B.; Mikulska-Ruminska, K.; Doruker, P.; Li, H.; Bahar, I. ProDy 2.0: Increased Scale and Scope after 10 Years of Protein Dynamics Modelling with Python. *Bioinformatics* **2021**, btab187. [CrossRef] [PubMed]
61. Chovancova, E.; Pavelka, A.; Benes, P.; Strnad, O.; Brezovsky, J.; Kozlikova, B.; Gora, A.; Sustr, V.; Klvana, M.; Medek, P. CAVER 3.0: A tool for the analysis of transport pathways in dynamic protein structures. *PLoS Comput. Biol.* **2012**, *8*, e1002708. [CrossRef]
62. Adzhubei, I.A.; Schmidt, S.; Peshkin, L.; Ramensky, V.E.; Gerasimova, A.; Bork, P.; Kondrashov, A.S.; Sunyaev, S.R. A method and server for predicting damaging missense mutations. *Nat. Methods* **2010**, *7*, 248–249. [CrossRef]
63. Jameson, J.B., II; Kenyon, V.; Holman, T.R. A high-throughput mass spectrometric assay for discovery of human lipoxygenase inhibitors and allosteric effectors. *Anal. Biochem.* **2015**, *476*, 45–50. [CrossRef] [PubMed]

International Journal of
Molecular Sciences

MDPI

Article

Loose Morphology and High Dynamism of OSER Structures Induced by the Membrane Domain of HMG-CoA Reductase

Ricardo Enrique Grados-Torrez [1,†], Carmen López-Iglesias [2,3], Joan Carles Ferrer [4] and Narciso Campos [1,4,*]

1 Centre for Research in Agricultural Genomics (CRAG) CSIC-IRTA-UAB-UB, Department of Molecular Genetics, Campus UAB, Bellaterra (Cerdanyola del Vallès), 08193 Barcelona, Spain; regrados.old@umsa.bo
2 Scientific and Technological Centers, University of Barcelona, 08028 Barcelona, Spain; c.lopeziglesias@maastrichtuniversity.nl
3 Microscopy CORE Lab, Maastricht Multimodal Molecular Imaging Institute, Maastricht University, 6229 ER Maastricht, The Netherlands
4 Department of Biochemistry and Molecular Biomedicine, Faculty of Biology, University of Barcelona, 08028 Barcelona, Spain; jcferrer@ub.edu
* Correspondence: ncampos@ub.edu
† Present address: Facultad de Ciencias Farmacéuticas y Bioquímicas, Universidad Mayor de San Andrés, Zona Miraflores 22-24, La Paz, Bolivia.

Abstract: The membrane domain of eukaryotic HMG-CoA reductase (HMGR) has the conserved capacity to induce endoplasmic reticulum (ER) proliferation and membrane association into Organized Smooth Endoplasmic Reticulum (OSER) structures. These formations develop in response to overexpression of particular proteins, but also occur naturally in cells of the three eukaryotic kingdoms. Here, we characterize OSER structures induced by the membrane domain of *Arabidopsis* HMGR (1S domain). Immunochemical confocal and electron microscopy studies demonstrate that the 1S:GFP chimera co-localizes with high levels of endogenous HMGR in several ER compartments, such as the ER network, the nuclear envelope, the outer and internal membranes of HMGR vesicles and the OSER structures, which we name ER-HMGR domains. After high-pressure freezing, ER-HMGR domains show typical crystalloid, whorled and lamellar ultrastructural patterns, but with wide heterogeneous luminal spaces, indicating that the native OSER is looser and more flexible than previously reported. The formation of ER-HMGR domains is reversible. OSER structures grow by incorporation of ER membranes on their periphery and progressive compaction to the inside. The ER-HMGR domains are highly dynamic in their formation versus their disassembly, their variable spherical-ovoid shape, their fluctuating borders and their rapid intracellular movement, indicating that they are not mere ER membrane aggregates, but active components of the eukaryotic cell.

Keywords: HMG-CoA reductase; HMGR; HMGR vesicle; ER-HMGR domain; mevalonate; endoplasmic reticulum; OSER; high-pressure freezing; chemical fixation

Citation: Grados-Torrez, R.E.; López-Iglesias, C.; Ferrer, J.C.; Campos, N. Loose Morphology and High Dynamism of OSER Structures Induced by the Membrane Domain of HMG-CoA Reductase. *Int. J. Mol. Sci.* **2021**, *22*, 9132. https://doi.org/10.3390/ijms22179132

Academic Editor: Masoud Jelokhani-Niaraki

Received: 18 July 2021
Accepted: 21 August 2021
Published: 24 August 2021

1. Introduction

Eukaryotic HMG-CoA reductase (HMGR) has a key regulatory role in the mevalonate pathway for isoprenoid biosynthesis [1,2]. Isoprenoid products derived from this pathway are required for many diverse essential functions, including membrane biogenesis (sterols), control of growth and development (steroid hormones and plant cytokinins), protein prenylation (farnesyl and geranyl groups), protein glycosylation (dolichols) and respiration (ubiquinones) [3]. In plants, the mevalonate pathway also provides a wide variety of secondary metabolites required for defence against herbivores and pathogens or for the attraction of beneficial organisms [4]. In all plant species, HMGR is encoded by a multigene family. This was first proposed after analysis of few model plants [5], but has been confirmed by high throughput sequencing of an ever-increasing number of genomes [6]. In *Arabidopsis thaliana*, two genes (*HMG1* and *HMG2*) encode three HMGR isoforms (HMGR1S, HMGR1L and HMGR2) [5,7]. It has been suggested that different variants of plant HMGR are physically associated with other enzymes, forming metabolons for the synthesis of particular isoprenoid products, and that these metabolons would be located at particular sites of the endomembrane system [8]. The association of sterol biosynthetic enzymes at the ER membrane (one of the branches of the isoprenoid pathway) has been shown in yeast, mammals and plants [9–11], but no proof of the existence of metabolons involving HMGR has yet been provided.

HMGR is composed of an N-terminal membrane domain, with low or no sequence similarity among eukaryotic kingdoms, and a highly conserved catalytic domain [12–14]. In plant HMGR, the membrane domain has only two hydrophobic segments, whereas in yeast and animal HMGR eight membrane-spanning regions have been predicted [12,15,16]. The three *Arabidopsis* HMGR isoforms are primarily targeted to the ER by the two hydrophobic sequences of the membrane domain that interact specifically with the Signal Recognition Particle [12,17]. However, immunolocalization whole-mount studies in *Arabidopsis* cotyledon suggested that endogenous HMGR mostly localizes within spherical vesicular structures, which were therefore named HMGR vesicles [18,19]. It is not known how the integral membrane protein HMGR reaches the inside of vesicular structures, nor what relationships exist between these vesicles and the ER.

Despite diverging evolution, the membrane domain of HMGR from the three eukaryotic kingdoms has the common capacity to induce massive proliferation of ER membranes that subsequently constitute Organized Smooth Endoplasmic Reticulum (OSER) structures [19–21]. When examined by transmission electron microscopy (EM), OSER structures contain tightly associated ER membranes according to three different patterns: ordered arrays with hexagonal or cubic symmetry (crystalloid ER), concentric layers (whorled ER) or simply stacked (lamellae or perinuclear karmellae) [21–23]. A highly conserved N-terminal motif of plant HMGR is required for OSER biogenesis [19], but no equivalent sequence has been identified in yeast or animal HMGR nor has the morphogenic mechanism been described. In *Arabidopsis*, OSER structures induced by the membrane domain of HMGR1S fused to GFP (1S:GFP chimera) also accumulate high amounts of endogenous HMGR and, therefore, have been named ER-HMGR domains [19].

Highly proliferated ER with ordered repetitive patterns was first described in the 1960s, as naturally occurring in diverse cell types from animals and plants [24–30] and as readily developing upon exposure to drugs [31,32]. Since then, diverse forms of hypertrophied ER have been identified in many natural and induced systems and referred to with a variety of terms, such as *cotte de mailles* [33], *paracrystalline arrays* [34], *elaborate rings of granular ER* [35], *double membrane arrays* [36], *tubuloreticular structures* [37], *undulating membranes* [38], *membrane lattice* [39], *stacks of flattened smooth ER* [40], *interlaced smooth surfaced tubules* [41], *compact areas of smooth membranes* [42], *paracrystalline ER* [43], *crystalloid membranes* [44], *organized smooth endoplasmic reticulum* [45] or *cubic membranes* [46]. An exhaustive review [47], with about 200 examples, reported that *cubic membranes* (OSER structures) are broadly distributed in the three eukaryotic kingdoms. These structures are found in numerous cell types under certain physiological conditions or appear in response to stress or disease [47]. However, in most of the aforementioned studies, images were obtained by transmission EM after chemical fixation [47]. Alternative preparation and observation techniques are necessary to further expand our knowledge on OSER ultrastructure.

In this work we first study in more depth the subcellular location of *Arabidopsis* HMGR and, particularly, the HMGR vesicles. We also characterize ER-HMGR domains in *Arabidopsis* and *Nicotiana* cells, focusing on their biogenesis, ultrastructure and dynamism. Our EM analyses uncover differences in OSER ultrastructure because of the fixation method. We find that the ER-HMGR domains are flexible live entities, fully integrated in ER architecture and dynamism.

2. Results

2.1. Subcellular Location of *Arabidopsis* HMGR in WT and 1S:GFP Transgenic Plants

Immunolocalization whole-mount studies in *Arabidopsis* cotyledon indicated that endogenous HMGR mostly localized inside HMGR vesicles ranging from 0.2 to 0.6 μm in diameter [18]. These studies were done with a crude rabbit polyclonal antibody raised against the catalytic domain of *Arabidopsis* HMGR1 (Ab-CD1), but it was later reported that this serum cross-reacts with *E. coli* proteins [48]. Before proceeding with a deeper localization analysis of HMGR, we wanted to confirm the whole-mount studies with an immunopurified fraction of the antibody (Ab-CD1-i) [48]. In this improved assay, we confirmed that *Arabidopsis* HMGR mostly localizes in vesicular structures of parenchymal cells, in close proximity with chloroplast (Figure 1a–c). Our observations suggest that the HMGR vesicles can measure up to 2 μm in diameter, somewhat more than previously reported.

We subsequently performed immunochemical transmission EM studies of the HMGR vesicles, both in wild type (WT) and 1S:GFP-overexpressing *Arabidopsis* plants. We used the immunopurified serum Ab-CD1-i to detect endogenous HMGR, and a commercial antibody against GFP (Ab-5450) to detect the chimeric 1S:GFP. We found selective deposition of gold particles on the surface and the inside of the HMGR vesicles, denoting the presence of both endogenous HMGR and the 1S:GFP chimera (Figure 1d–f,l). The HMGR vesicles were associated in small groups connected by ER strands (Figure 1e,f,l). This connecting ER was also immunolabeled with the antibodies against HMGR and 1S:GFP (Figure 1e,l). The ultrastructural analysis uncovered that the HMGR vesicles were delimited by an ER membrane (Figure 1e). In addition, they possessed internal ER membranes (Figure 1d,f). Both the outer and internal membranes were recognized by the Ab-CD1-i (Figure 1d,e,l) and Ab-5450 (Figure 1f,l) antibodies. These results provide a rational explanation for the whole-mount detection of HMGR protein within HMGR vesicles.

As previously reported, overexpression of 1S:GFP in transgenic *Arabidopsis* plants induces ER proliferation and OSER structure biogenesis [19]. Whole-mount and immunochemical transmission EM analyses demonstrate colocalization of the 1S:GFP chimera and high levels of endogenous HMGR in the OSER formations (Figure 1g,l). Because of the presence of HMGR protein, we name them ER-HMGR domains. They have a disordered and heterogeneous crystalloid pattern, but with a characteristic layer of large loops in their external face and more compressed structures in the inside (Figure 1h,i,l). Precise deposition of immunogold particles indicates an abundance of 1S:GFP chimera and endogenous HMGR in the ER strands of ER-HMGR domains, both in the distended external loops and the internal membrane aggregates. High levels of the 1S:GFP chimera and endogenous HMGR were also detected in the ER network (Figure 1j,l) and nuclear envelope (Figure 1j,m). Few immunogold particles were observed inside the nucleus (Figure 1j), whereas no immunolabelling was found in the Golgi apparatus (Figure 1j), mitochondria (Figure 1h) or chloroplast (Figure 1d,h,j). In the negative control, no labelling was obtained without Ab-CD1-i and Ab-5450 primary antibodies (Figure 1k).

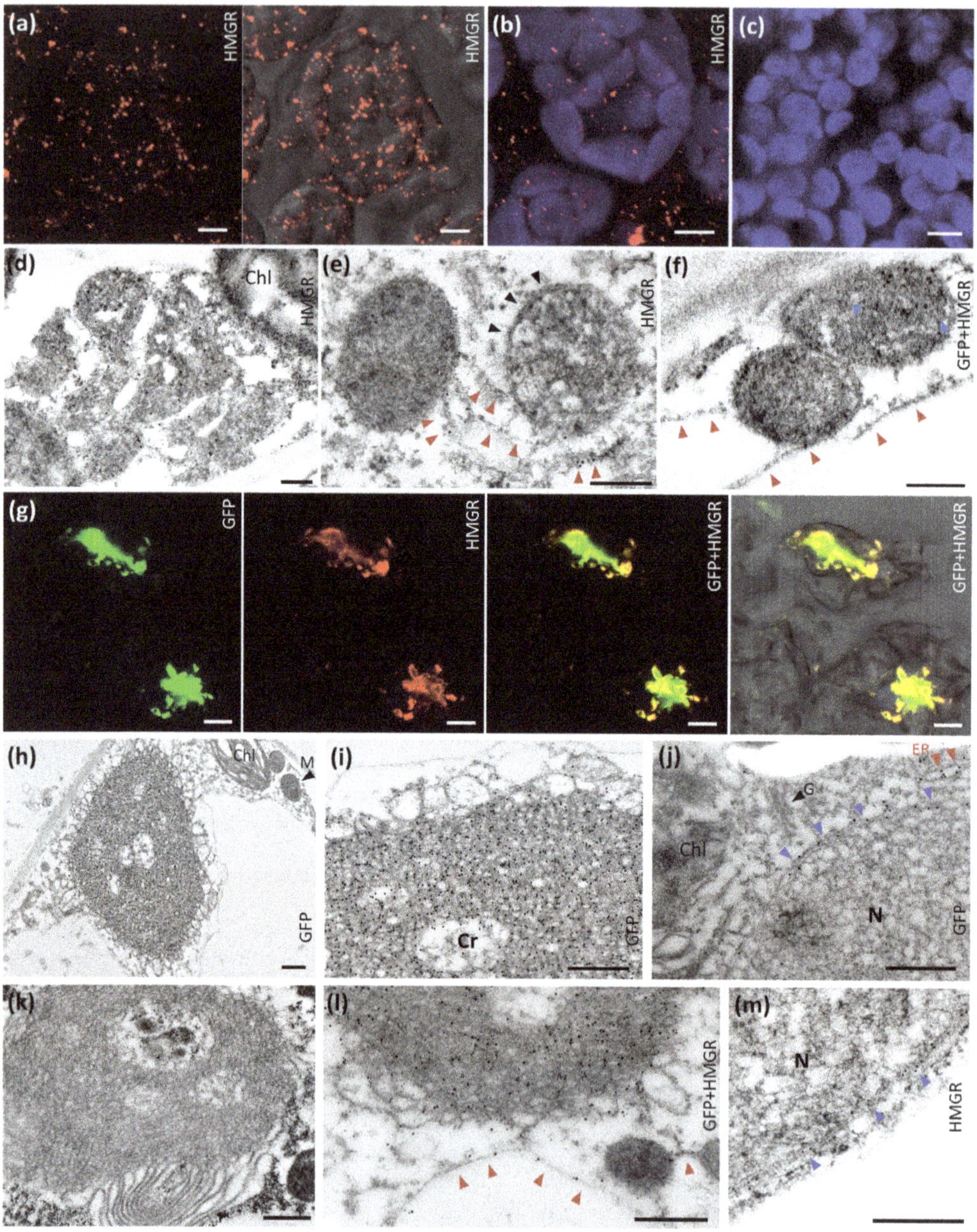

Figure 1. *Arabidopsis* HMGR localizes in the ER network, nuclear envelope, HMGR vesicles and ER-HMGR domains. (**a**–**c**) Whole-mount immunohistochemical analysis of cotyledon parenchymal cells from 6-day-old *Arabidopsis* WT seedlings (**a**) Immunodetection of HMGR with Ab-CD1-i and anti-rabbit IgG secondary antibody (Alexa Fluor 555, in red), visualized by confocal microscopy under dark (left) or bright fields (right). (**b**) Immunodetection of HMGR with Ab-CD1-i and anti-rabbit IgG secondary antibody (AlexaFluor 555, in red), and simultaneous detection of chlorophyll (in blue). (**c**) Negative control without the Ab-CD1-i antibody. The irregular corpuscles, 0.2 to 2 μm in length, correspond to HMGR vesicles. The elliptic bodies, 6 to 8 μm in diameter, correspond to chloroplasts. Images were obtained by Z-projection encompassing 3 (**a**), 10 (**b**) or 4 (**c**) μm in the Z-axis. Bars, 5 μm. (**d**–**f**) Immunochemical study of HMGR vesicles by transmission EM. Leaf samples from 10-day-old *Arabidopsis* WT or 1S:GFP seedlings were processed by HPF and embedded with Lowicryl

HM20. (**d**) HMGR was detected in cotyledon from WT seedlings with Ab-CD1-i and anti-rabbit-IgG (18 nm particle). (**e**) HMGR was detected in true leaf from 1S:GFP seedlings with Ab-CD1-i and anti-rabbit-IgG (12 nm particle). (**f**) Double immunolocalization of HMGR and 1S:GFP in true leaf from 1S:GFP seedlings. HMGR was detected with Ab-CD1-i and anti-rabbit-IgG (12 nm particle) and 1S:GFP was detected with Ab-5450 and anti-goat-IgG (18 nm particle). Black and blue arrowheads indicate, respectively, the external and internal membranes from HMGR vesicles. Red arrowheads indicate ER strands. Bars, 250 nm. (**g**) Whole-mount immunohistochemical analysis of cotyledon parenchymal cells from 6-day-old *Arabidopsis* 1S:GFP transgenic seedlings. 1S:GFP was detected with Ab-5450 and secondary antibody Alexa fluor 488 (green). HMGR was detected with Ab-CD1-i and secondary antibody Alexa fluor 594 (red). Images were obtained by Z-projection encompassing 12 μm in the Z-axis. Bar, 5 μm. (**h–m**) Immunolocalization of HMGR and 1S:GFP by transmission EM. True leaves from 10-day-old *Arabidopsis* WT seedlings were processed by HPF and embedded with Lowicryl HM20. (**h–j**) 1S:GFP was detected with Ab-5450 and anti-goat-IgG (18 nm particle). (**k**) Negative control without Ab-CD1-i and Ab-5450. (**l**) Double immunolocalization of 1S:GFP (18 nm particle) and HMGR (12 nm particle). (**m**) HMGR was detected with Ab-CD1-i and anti-rabbit-IgG (12 nm particle). Chloroplast (Chl). ER strands (red arrowheads). Golgi apparatus (G). Mitochondria (M). Nuclear envelope (blue arrowheads). Nucleus (N). Bars, 500 nm.

2.2. Reversible Formation of ER-HMGR Domains

To further characterize the biogenesis of ER-HMGR domains, we induced the transient expression of 1S:GFP in *Nicotiana benthamiana* leaves. The agroinfiltration approach allowed generalized and abundant expression of the 1S:GFP construct in leaf epidermis (Figure 2a). At day two after transfection, massive ER proliferation led to formation of OSER structures in the transfected tissue that were detectable even at low magnification (Figure 2a). Many small OSER structures appeared at ER network junctions and a single large OSER was formed around the nuclear envelope (Figure 2c,d). The 1S:GFPm construct, containing monomeric GFP, similarly induced small OSER structures at the network junctions and a large OSER aggregate around the nucleus (Figure 2b). As previously reported [19], this indicates that the membrane domain of *Arabidopsis* HMGR, and not its dimerizing GFP partner, induces ER proliferation and membrane association into OSER. In contrast to 1S:GFP, the 1S:GFPm chimera also accumulated in hypertrophied ER strands. Thick ER strands are usually present in epidermal cells and can be visualized with the ER-GFP luminal marker (Figure 2e), but become more prominent in the case of 1S:GFPm (Figure 2b). At day six after transfection, the expression of 1S:GFP was severely reduced. Concomitant with that, OSER structures virtually disappeared, with only some remnants around the nucleus and in the cytosol (Figure 2f). The resulting ER had the usual thick strands, although broad cisternae replaced the fine network (Figure 2f). ER cisternae are common in epidermal cells transfected with the luminal ER-GFP marker, although they have a smaller size (Figure 2e). Our observations in *Nicotiana* epidermal cells indicate that OSER biogenesis is reversible. The OSER structures are not a permanent consequence of transfection with 1S:GFP or 1S:GFPm, but can be replaced by quite normal ER when the levels of the chimeric protein decrease.

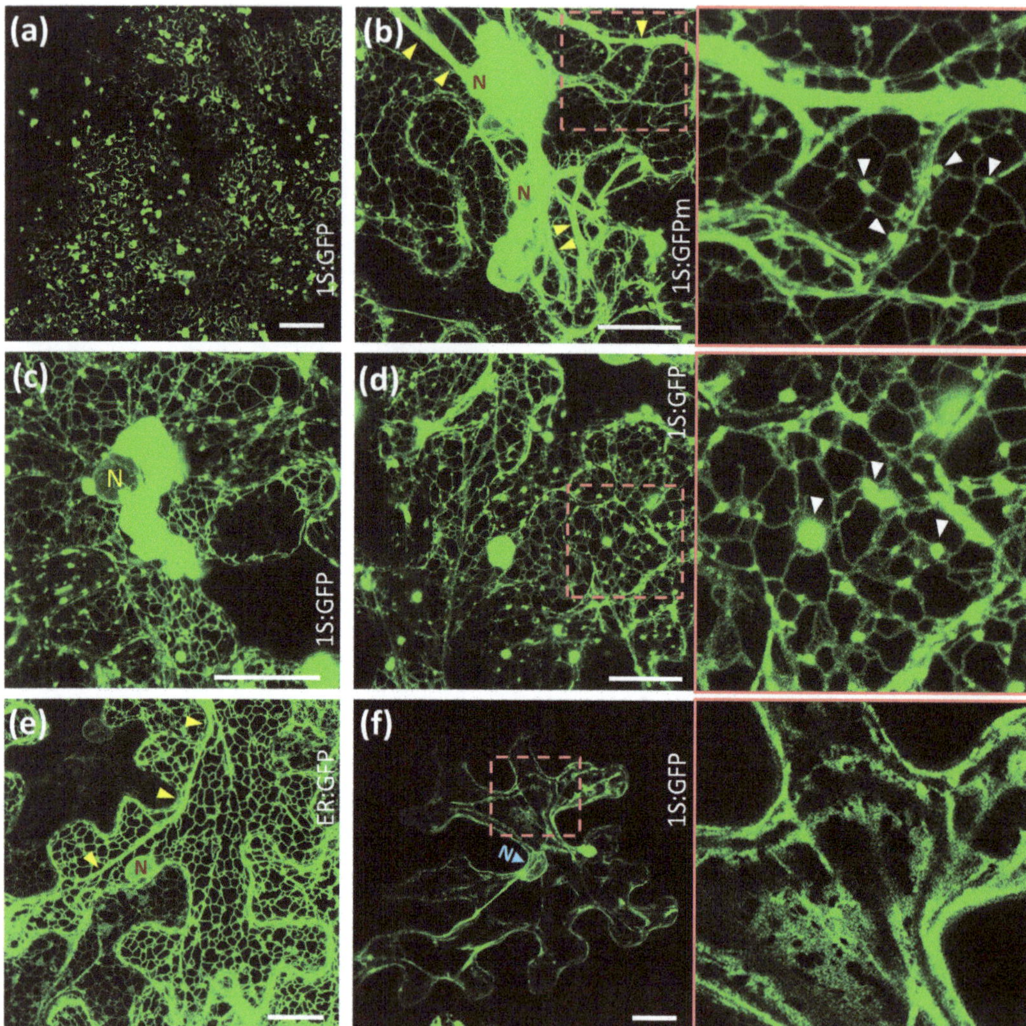

Figure 2. Reversible biogenesis of ER-HMGR domains in transfected *Nicotiana* cells. *Nicotiana* leaves were transfected with constructs encoding 1S:GFP (**a**,**c**,**d**,**f**), 1S:GFPm (**b**) or the control ER lumen marker ER:GFP (**e**) and epidermal cells were visualized by confocal laser microscopy. The expression time was 2 days (**a**), 3 days (**b**–**e**) or 6 days (**f**). The square regions indicated in (**b**,**d**,**f**) are shown enlarged on the right. Images are a single section (**a**) or Z-projections encompassing 10 (**b**), 21 (**c**), 7 (**d**), 12 (**e**) or 14 (**f**) µm in the Z-axis. Nucleus (N). OSER structures (white arrowheads). Thick ER strands (yellow arrowheads). Bars, 100 µm (**a**), 20 µm (**b**–**f**).

2.3. ER-HMGR Domains Are Highly Dynamic

To further inspect OSER structure morphology and dynamism, we obtained transgenic *Arabidopsis* plants stably expressing the 1S:GFP construct. A panoramic view of seedling root epidermis showed high expression of the 1S:GFP chimera accumulating at the ER (Figure 3a). The transgenic construct induced large OSER structures (up to 10 µm in diameter) around the nuclei and smaller OSER formations at ER network junctions (Figure 3a and Supplementary Movie 1). As previously reported [49,50], the ER network is highly dynamic with continuous strand movement and fusion or fission events. We found that

OSER structures are connected to the ER network and participate in its dynamism. Many strands of the ER network associate with OSER formations (Figure 3b). The ER strands rapidly connect to, slide along or separate from the OSER surface (Supplementary Movie 2). Small OSER structures migrate along fine or thick ER strands, whereas the large OSER formations have a more limited motion (Supplementary Movies 1 and 2). In the nuclear OSER, this movement may imply a brief separation from the nuclear envelope (Supplementary Movie 1). OSER structures have spherical-ovoid shapes with slight continuous variation (Supplementary Movies 2 and 3). The OSER borders are not sharp, but have a fluctuating blurry aspect (Figure 3b,c), suggesting the incorporation or emergence of GFP-labelled material (likely membranes) in the OSER surface (Supplementary Movies 2 and 3). We conclude that, in *Arabidopsis* cells, OSER structures are highly dynamic entities. They change in shape, have a moving surface and migrate intracellularly.

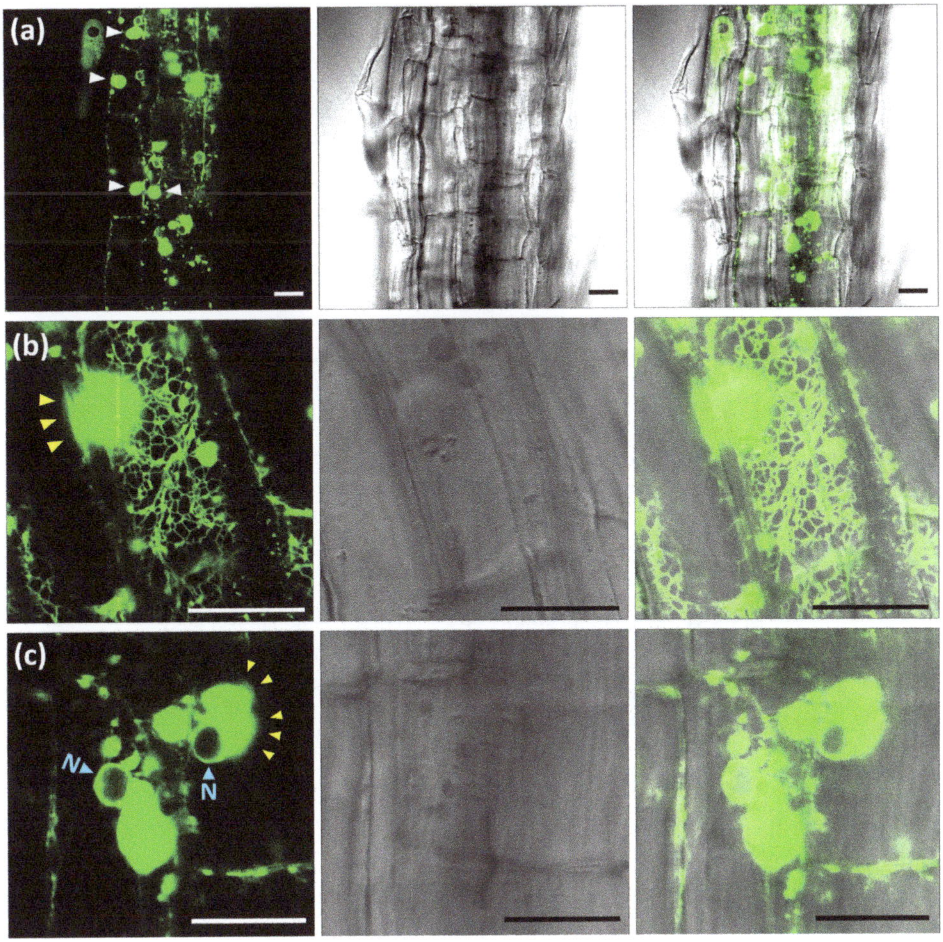

Figure 3. Characterization of ER-HMGR domains in *Arabidopsis* 1S:GFP transgenic plants. *Arabidopsis* 1S:GFP 9-day-old seedlings were analysed by confocal laser microscopy. The pictures show three channels (GFP, bright field and merge) of single sections from root epidermal cells that were subsequently characterized by live imaging: (**a**) panoramic view of root epidermis with large ER-HMGR domains indicated by white arrowheads (Supplementary Movie 1); (**b**) large ER-HMGR domain with dynamic connections to the ER network (Supplementary Movie 2); (**c**) nuclear ER-HMGR domains with changing spherical-ovoid shape (Supplementary Movie 3). Blurry fluctuating borders (yellow arrowheads). Nucleus (N). OSER structures (white arrowheads). Bars, 20 μm.

2.4. The Fixation and Dehydration Method Severely Affects OSER Ultrastructure

Our above confocal microscope observations of ER-HMGR domains do not fit the concept of OSER structures as rigid entities. Their tight, repetitive pattern, obtained after chemical fixation, contrasts with the flexibility and dynamism of OSER structures. To capture single states of OSER change at the ultrastructure level, we expressed 1S:GFP in *Nicotiana* leaf and submitted samples to high-pressure freezing (HPF) followed by freeze-substitution, to finally observe epidermal cells by transmission EM. We compared the HPF results with those of chemical fixation and subsequent dehydration at room temperature. With either method, OSER structures show a combination of crystalloid, lamellar and whorled membrane patterns (Figure 4a,b,g,h,j). In both cases, there is also a coincidence in cytosolic and luminal spaces. Luminal spaces correspond to the continuous internal cavity of the ER and have an electron-lucent aspect at transmission EM (Figure 4b,f,i,k). Cytosolic spaces result from the apposition of adjacent ER membranes and have a quite constant width, which is about 10–15 nm in both chemical and HPF fixation (Figure 4b,f,i,k). Cytosolic spaces have a darker aspect than luminal spaces, probably reflecting the presence of proteins that mediate the intermembrane attachment.

In spite of the above-mentioned coincidences, OSER structures have quite different overall morphology depending on the fixation method. The most remarkable feature of OSER formations after chemical fixation is the presence of highly repetitive convoluted membranes, which have a very different smooth and turgid aspect in HPF images (Figure 4a,c,e,h,i). The convoluted pattern is exclusive to chemically fixed crystalloid domains, whereas aligned membranes are present in whorled domains, both with HPF and chemical fixation (Figure 4b,j,k). However, the crystalloid (Figure 4a,c,d) and whorled (Figure 4b) domains are far looser after HPF than after chemical fixation (Figure 4h,j). The morphological heterogeneity generated by fixation is not due to the cytosolic spaces, which are always narrow and uniform, but occurs in the luminal spaces (Figure 4b,f,i,k). The small size and regularity of luminal spaces obtained after chemical fixation contributes to the repetitive pattern (Figure 4h,i). In contrast, HPF results in larger turgid luminal spaces, which are very variable in size and morphology (Figure 4e,f). After HPF and subsequent embedding by two alternative methods, the OSER morphology is disordered and heterogeneous, but has the above-mentioned row of large loops at the periphery and is internally more compact (see similarities of Figure 4a,c,d with Figure 1h,i,l).

To confirm the OSER ultrastructure resulting from chemical fixation, we expressed the membrane domain of *Arabidopsis* HMGR1S fused to monomeric GFP (chimera 1S:GFPm) in *Nicotiana* cells. Constructs 1S:GFP and 1S:GFPm differ in just one amino acid residue [19]. The 1S:GFPm chimera generated crystalloid, lamellar and whorled patterns with constant cytosolic distance between membranes and also regular luminal spaces (Figure 4l–o), similar to those obtained with 1S:GFP (Figure 4h–j). However, in OSER structures derived from 1S:GFPm the core of crystalloid domains was usually looser than the peripheral parts (Figure 4m). This was not observed in OSER structures derived from 1S:GFP, even when they were much larger (Figure 5e). We conclude that the dimerizing capacity of GFP may influence OSER membrane compaction during the chemical fixation process.

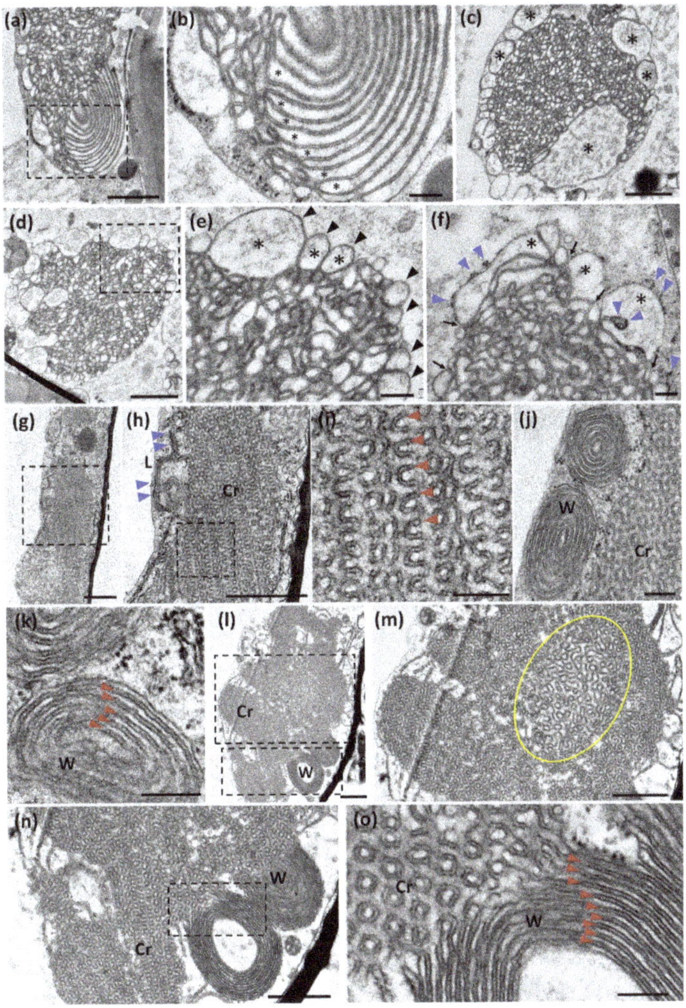

Figure 4. Ultrastructural analysis of ER-HMGR domains in *Nicotiana* epidermal cells. Three days after agroinfiltration with 1S:GFP (**a–k**) or 1S:GFPm (**l–o**), *Nicotiana* leaves were processed for transmission EM. (**a–f**) 1S:GFP-transfected leaves were submitted to HPF and embedded in Epon resin. (**a**) Panoramic view ER-HMGR domain with a crystalloid region in sagittal section and a whorled region in transversal section. (**b**) Magnified region of panel (**a**), to show the clear luminal spaces alternating with darker (denser for electrons) cytosolic spaces. (**c**,**d**) Panoramic view of crystalloid ER-HMGR domains with large loops surrounding the internal, more compact structure. (**e**) Magnified region of (**d**) with large loops. (**f**) Crystalloid structure with interspersed clear-luminal and dark-cytosolic spaces. The OSER cytosolic spaces are continuous with the cytoplasm (black arrows). Ribosomes are visible at the external face of the OSER structure. Occasionally, they are trapped in internal cytosolic spaces, near the periphery of the ER aggregate. (**g–k**) 1S:GFP-transfected leaves were submitted to chemical fixation and embedded in Spurr resin. (**g**) Panoramic view of crystalloid ER-HMGR domain. (**h**) Magnified region of (**g**) to show the repetitive smooth ER and a thin lamellar region on the left side. Ribosomes are excluded from the ER-HMGR domain. (**i**) Magnified regions of (**h**). (**j**) Whorled and crystalloid regions. (**k**) Whorled region with regular cytosolic and luminal spaces. (**l–o**) 1S:GFPm-transfected leaves were submitted to chemical fixation and embedded in Spurr resin. (**l**) Panoramic view of ER-HMGR domain with crystalloid, lamellar and whorled patterns. (**m**) Magnified region of (**l**) to show the looser region (circled) in the core of the ER-domain. (**n**) Magnified region of (**l**) to show the transition between whorled and crystalloid regions. (**o**) Magnified region of (**n**). Large membrane loops (black arrowheads). Crystalloid region (Cr). Cytosolic spaces (red arrowheads). Lamellar region (L). Luminal spaces (asterisks). Ribosomes (blue arrowheads). Whorled region (W). Bars, 1 μm (**a**,**c**,**d**,**g**,**h**,**l–n**), 200 nm (**b**,**e**,**f**,**i–k**,**o**).

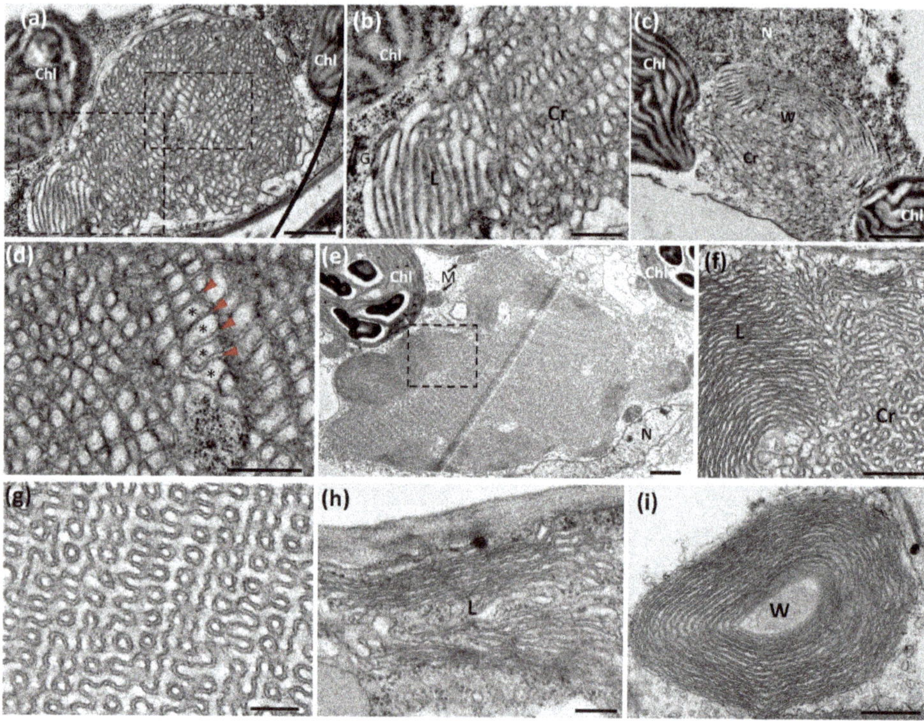

Figure 5. Ultrastructural analysis of ER-HMGR domains in *Arabidopsis* parenchymal cells. Emerging true leaves from *Arabidopsis* 10-day-old seedlings transgenic for 1S:GFP were submitted to HPF or chemical fixation and embedded in Spurr resin. (**a**–**d**) Samples obtained by HPF. (**a**) Panoramic view of a crystalloid-lamellar ER-HMGR domain located between chloroplasts. (**b**) Magnified region of panel (**a**), to show the transition between crystalloid and lamellar regions. (**c**) Nuclear ER-HMGR domain with crystalloid and whorled patterns. (**d**) Magnified region of (**a**), to show alternation of cytosolic (dark) and luminal (clear) spaces. Notice the narrow homogenous width of the cytosolic spaces and the broader and more variable size of the luminal spaces. (**e**–**i**) Samples obtained by chemical fixation. (**e**) Panoramic view of ER-HMGR domain with crystalloid, lamellar and whorled patterns. Notice the large size of the OSER structure (12 μm long). (**f**) Magnified region of panel (**e**), to show the transition between lamellar and crystalloid regions. (**g**) Detail of a crystalloid structure to show the repetitive pattern of dark cytosolic spaces and lighter luminal spaces. (**h**) Lamellar ER-HMGR domain. (**i**) Whorled ER-HMGR domain. Chloroplast (Chl). Crystalloid region (Cr). Cytosolic spaces (red arrowheads). Golgi apparatus (G). Lamellar region (L). Luminal spaces (asterisks). Mitochondria (M). Nucleus (N). Whorled region (W). Bars, 1 μm (**a**,**c**,**e**), 500 nm (**b**,**d**,**f**,**i**), 200 nm (**g**,**h**).

We also examined OSER ultrastructure in the emerging true leaves of the 1S:GFP *Arabidopsis* seedlings (10-day-old transgenic plants). We observed OSER structures with crystalloid, lamellar and whorled membrane patterns in parenchymal cells of both HPF and chemically fixed samples (Figure 5a,c,f). As mentioned above for *Nicotiana*, in *Arabidopsis* cells the three OSER ultrastructural patterns had a more relaxed morphology after HPF than after chemical fixation (Figure 5a,b,e,f). This is due to larger and more variable luminal spaces in HPF than in chemical fixation samples, whereas the dark cytosolic spaces had a similar width in both techniques (Figure 5b,d,g–i). The convoluted patterns observed in crystalloid domains after chemical fixation were absent in samples obtained by HPF (Figure 5d,g). Since HPF and chemical fixation were performed in parallel from the same samples, we conclude that the morphological differences of 1S:GFP-containing OSER were generated during the fixation process. Equivalent results were obtained in the two assayed systems, transfected *Nicotiana* leaves and *Arabidopsis* seedlings. As HPF immobilizes water and prevents its loss from the sample, the resulting transmission EM images are likely more similar to native OSER than those obtained by chemical fixation. In addition, the results obtained by HPF are more consistent with a dynamic view of the ER-HMGR domains observed with the confocal microscope.

3. Discussion

We found that *Arabidopsis* HMGR has several subcellular locations, such as the ER network, the nuclear envelope, HMGR vesicles, and the hypertrophied ER-HMGR domains. However, all these compartments are morphological variations of the ER. Therefore, our immunochemical results uncover that the primary targeting site of *Arabidopsis* HMGR is also its final subcellular destination. This protein does not migrate through the endomembrane system. Our observations underline the importance of the membrane domain of HMGR, which not only determines its primary and final destination sites, but also induces ER proliferation and OSER biogenesis. The resulting OSER structures (ER-HMGR domains) accumulate high levels of the chimeric 1S:GFP and endogenous HMGR ([19] and this work). Our transient expression assays indicate that the ER-HMGR domains disappear when the expression of 1S:GFP decreases, thus re-establishing normal ER morphology. A similar observation was made in transgenic *Arabidopsis* plants in which 1S:GFP expression suffered silencing after a few weeks of development [19]. Hence, in non-transfected cells natural OSER could also form or disappear, depending on the expression level of endogenous OSER-inducing proteins.

The HMGR vesicles have an outer membrane enclosing membrane material and the whole structure remains connected to the ER network. Both the internal and surrounding membranes are recognized by antibodies against HMGR and 1S:GFP, which denote ER identity. We therefore conclude that the HMGR vesicles are membrane aggregates that derive directly from the ER. The HMGR localized in the outer and the internal membranes might be metabolically active. Thus, HMGR vesicles could be a differentiated organelle of the ER involved in the synthesis of particular isoprenoid products. Additional experiments, such as metabolic labelling assays or the localization of other enzymes of the isoprenoid pathway, are required to determine whether the HMGR vesicles contain functional metabolon-like assemblies.

The HMGR domains are not a mere consequence of the accumulation of proliferated membranes in the cytosol, but behave as live structures. The OSER aggregates are an integral part of the ER network and participate in its dynamism. Small ER-HMGR domains migrate along ER strands, whereas the large ones have a more limited motion. The large OSER formations reversibly establish dynamic connections to the ER network. The ER-HMGR domains show a slow but steady change in shape, alternating between ovoid and spherical forms. Particularly intriguing are the fluctuating borders of ER-domains, which suggest continuous incorporation or release of 1S:GFP-containing material. This interpretation is reinforced by ultrastructural and immunolabelling EM studies, showing that the ER-HMGR domains have a layer of large membrane loops with 1S:GFP that surround the more compressed core. Such a morphology suggests that the OSER periphery is a site for membrane compaction or unfolding. Since samples were prepared from growing *Arabidopsis* and *Nicotiana* leaves engaged in high expression of 1S:GFP and ongoing OSER biogenesis, the moving and static images likely indicate incorporation of new membranes to the OSER aggregate. Therefore, our results suggest that crystalloid OSER structures incorporate ER membranes in their external face with subsequent compression (reduction of luminal spaces) to the inside.

Our finding that OSER structures are loose, dynamic and flexible entities could explain previous observations. The OSER-inducing chimera cytochrome b(5)-GFP was highly mobile within OSER structures and diffused rapidly between these formations and the ER network [45]. The stacked membrane associations (karmellae) produced by HMGR overexpression in yeast did not interfere with protein transit from the ER to the Golgi apparatus [51]. Overexpression of rat liver aldehyde dehydrogenase in monkey COS-1 cells induced OSER formations, but the resulting crystalloid ER did not disturb protein transport from the ER to plasma membrane or lysosomes [44]. In hamster UT-1 cells, the G protein from vesicular stomatitis virus entered and egressed from OSER aggregates freely, without apparent restriction [52]. This fast transit of proteins across OSER formations seems difficult to reconcile with the small luminal spaces and apparently rigid architecture deduced at EM after chemical fixation. The above observations are more consistent with the broad luminal spaces of OSER structures obtained by HPF and the high flexibility and dynamism seen by live imaging with the confocal microscope.

We have analysed OSER ultrastructure by transmission EM after either chemical fixation or HPF. In both cases, OSER formations are similarly composed of crystalloid, lamellar and whorled patterns, indicating that this structural diversity is not generated during fixation, but is present in the original aggregates. The two fixation methods are also reproducible in the narrow cytosolic spaces between OSER membranes. The cytosolic spaces are dense for electrons in both methods and have a constant width. These two features may suggest that membrane apposition at a fixed distance is mediated by specific proteins. In morphological free-fracture EM analysis of animal crystalloid OSER structures, a high density of homogenous particles were protruding from the cytosolic side of the internal membranes [53]. These particles may correspond to intermembrane protein bridges, the components of which should be highly abundant in the crystalloid domain.

Cryoimmobilization is regarded as the most reliable fixation strategy for ultrastructural analysis by transmission EM [54]. Diverse artefacts, particularly affecting cellular membranes, are produced by chemical fixation. It was believed that gram-positive bacteria possessed mesosomes in their plasma membrane, but it was later found that such folded invaginations are not natural but produced by chemical fixation [55–57]. Similarly, in sea-urchin eggs, glutardehyde fixation caused the formation of large membrane vesicles at the initial site of fusion between plasma and granule membranes [58]. More recently, in a human cell-line transformed with Epstein-Barr virus, aldehyde fixation notably reduced endosomal volume, without affecting the length of its outer membrane [59]. Thus, chemically fixed early endosomes became irregular ovoid bodies with broad protruding tubules, whereas after cryofixation the same organelles were round and turgid with an incidental short tubule [59]. It was proposed that this endosome shrinkage was produced by dehydration during chemical fixation [59]. Chemical fixation could similarly dehydrate the luminal spaces of OSER structures, without affecting membrane length. This would produce an overall shrinkage in the OSER formation, concomitant with membrane accommodation in convoluted shapes. We reproducibly obtained the convoluted membranes after chemical fixation, but never after HPF, although both methods were applied to equivalent *Arabidopsis* or *Nicotiana* samples. It is somehow amazing that a crystal-like pattern can develop from disordered and turgid membranes by the mere application of chemical fixative agents. However, it would be even more surprising that OSER morphologies having smooth turgid membranes, heterogeneous luminal spaces and large peripheral loops were generated in HPF treatment from a hypothetical native convoluted pattern. We therefore propose that the OSER ultrastructure observed after HPF is closest to the original morphology existing in vivo.

OSER structure depictions have been obtained by transmission EM from a large variety of chemically fixed eukaryotic cells and tissues, always with similar convoluted shapes in crystalloid domains [47]. Therefore, we believe that a common feature of OSER membranes may guide their compaction during chemical fixation. The aggregated lipid bilayers are highly enriched in the OSER-inducing protein and few other uncharacterized polypeptides [60,61]. For instance, not only 1S:GFP but also endogenous HMGR accumulate at high levels in ER-HMGR domains ([19] and this work). Any of these proteins could self-associate, GFP as a dimer or HMGR as a tetramer [62,63]. Highly abundant self-associating proteins could then provide a structural frame to build the crystal-like convoluted pattern that would be established during slow dehydration and chemical cross-linking.

Crystalloid ultrastructural patterns generated by 1S:GFP were homogeneous and compact, whereas those generated by 1S:GFPm had a looser core (Figure 4g,h,m), indicating that the presence of monomeric GFP instead of the original dimerizing GFP was responsible for the difference. Since chemical fixatives rely on diffusion to penetrate the sample and are consumed during fixation, their effects are discontinuous with slower cross-linking in the inside [54]. We speculate that this temporal discontinuity, together with a lower association capacity of 1S:GFPm with respect to 1S:GFP, could reduce membrane compaction at the core of the crystalloid domain. Independently of the molecular mechanism, the heterogeneous membrane compaction with 1S:GFPm reinforces the notion that chemical fixation may illegitimately affect OSER ultrastructure.

4. Materials and Methods

4.1. Plant Material

Experiments were performed with *Arabidopsis* (*Arabidopsis thaliana*) WT Col 0 or a transgenic line overexpressing 1S:GFP in the same genetic background. The preparation of the 1S:GFP transgenic line was described previously [19]. In this line, 1S:GFP is under control of the cauliflower mosaic virus 35S promoter, which confers high expression in the whole plant. Seeds were surface sterilized by washing three times with 70% (v/v) ethanol and three times with 100% (v/v) ethanol, sowed in petri dishes with half concentrated Murashige and Skoog (MS) medium [64] and vernalized at 4 °C for 3 d. Growth was at 22–24 °C and long day conditions (16 h light/8 h darkness).

4.2. Constructs

The identity and source of protein constructs was as follows: 1S:GFP and 1SGFPm, N-terminal fragment 1-178 from *Arabidopsis* HMGR1S (accession AAF16652) fused, respectively, to GFP or monomeric GFP [19]. ER:GFP, GFP sequence with appended ER signal peptide at the N-terminus and ER retention signal at the C-terminus, encoded by plasmid pVKH-GFP-HDEL [65].

4.3. Transient Expression in Nicotiana Benthamiana Leaves

Transient expression of 1S:GFP and 1S:GFPm in *N. benthamiana* leaves was achieved by agroinfiltration as previously described [66]. 1S:GFP and 1S:GFPm was previously cloned into plasmid pPCV002 under control of the cauliflower mosaic virus 35S promoter, which confers high constitutive expression [19]. The pPCV002 derivatives were transformed into *Agrobacterium tumefaciens* strain GV3101 pMP90RK [67]. The transformed bacteria were grown in YEB medium (per litre: 5 g beef extract, 1 g yeast extract, 5 g bacteriological peptone, 5 g sucrose, 2 mmol $MgSO_4$) containing 100 µg/mL each of kanamycin, rifampicin and carbenicillin at an OD_{600} of 0.5 to 1.0. Cells were harvested by centrifugation and suspended in 10 mM *N*-morpholino ethanesulfonic acid (MES, pH 5.7), 10 mM $MgCl_2$ and 0.2 mM acetosyringone to an OD_{600} of 1.0. Bacteria were infiltrated into leaves with a 1-mL disposable syringe without a needle. Expression was examined daily, until day 6 after agroinfiltration.

4.4. Source and Use of Antibodies

The catalytic domain of *Arabidopsis* HMGR1S (CD1) produced in *Escherichia coli* was used as immunogen to produce a polyclonal antibody in rabbit and the resulting serum was immunosubstracted to remove IgG reacting against the bacterial proteins [48]. The immunopurified serum (Ab-CD1-i) was used as primary antibody at 1:500 for whole mount and 1:1000 for transmission EM. Anti-rabbit IgG secondary antibodies for HMGR detection were code Ab150066 (Abcam, Toronto, ON, Canada) coupled to Alexa Fluor-555 at 1:1000 for whole mount (Figure 1a,b), code Ab150068 (Abcam, Toronto, ON, Canada) coupled to Alexa Fluor-594 at 1:1000 for whole mount (Figure 1g), code 111-215-144 (Jackson Immunoresearch, Cambridge, UK) coupled to an 18-nm gold particle at 1:30 for transmission EM (Figure 1d) and code 111-205-144 (Jackson Immunoresearch, Cambridge, UK) coupled to a 12-nm gold particle at 1:30 for transmission EM (Figure 1e,f,l,m).

GFP was detected with Ab-5450 (Abcam, Toronto, ON, Canada) as the primary antibody at 1:1000 for whole mount and transmission EM. Anti-goat IgG secondary antibodies for GFP detection were code Ab150133 (Abcam, Toronto, ON, Canada) coupled to Alexa Fluor-488 at 1:1000 for whole mount, and code 705-215-147 (Jackson Immunoresearch, Cambridge, UK) coupled with an 18-nm gold particle at 1:15 for transmission EM.

4.5. Immunolocalization in Whole Mount

Whole-mount in situ immunolocalization was done as described [68] with modifications. After fixation in 4% (*w*/*v*) paraformaldehyde, seedlings were incubated in methanol to remove chlorophylls. Five cycles of seedling freezing and thawing on glass slides were performed to permeate tissue, followed by incubation with 2% (*w*/*v*) Driselase (D8037; Sigma Aldrich-Merk, Madrid, Spain) to allow for the penetration of antibodies through the plant cell wall. After blocking with 3% (*w*/*v*) bovine serum albumin (BSA) solution, simultaneous incubation with one or two primary antibodies (Ab-CD1-i at 1:500, Ab-5450 at 1:1000) was performed directly on the slides. Samples were then washed with phosphate-buffered saline (PBS) and incubated with the corresponding secondary antibody at 1:1000. The secondary antibody to detect HMGR was Abcam Ab150066 (Alexa Fluor-555) or Ab150068 (Alexa Fluor-594). The secondary antibody to detect GFP was Abcam Ab150133 (Alexa Fluor-488). After washing with PBS, samples were sealed for observation at the confocal microscope.

4.6. Confocal Microscopy

Confocal laser microscopy was performed with spectral microscope Olympus FV1000 (objectives UPLSAPO 60x O, numerical aperture: 1.35 and UPLSAPO 60x W, numerical aperture: 1.20) or Leica TCS SP5 (objectives HCX PL APO CS 40x/1.25 Oil UV, HCX PL APO 63x/1.20 W corrected for UV, and HCX PL APO CS 63x/1.20 water UV) at room temperature. Fluorophores were detected with the following excitation and emission wavelengths: GFP (excitation = 488 nm, emission = 500–545 nm), Alexa Fluor-555 (excitation = 559 nm, emission = 570–610 nm) and Alexa Fluor-594 (excitation = 559 nm, emission = 575–620 nm) and chlorophyll (excitation = 559 nm, emission = 640–680 nm). Images were acquired with the software FV10-ASW 4.1 (Olympus, Hamburg, Germany) and LAS AF 2.7.3.9723 (Leica Microsystems, Wetzlar, Germany) and processed with the software ImageJ 1.50i (http://imagej.nih.gov/ij, accessed on 28 May 2020).

4.7. Chemical Fixation for Ultrastructural Studies

Explants from *Nicotiana* leaves or *Arabidopsis* seedlings were excised under a stereomicroscope and transferred to glass vials filled with 1.5% (*v*/*v*) paraformaldehyde and 1.5% (*v*/*v*) glutaraldehyde in 0.1 M cacodylate buffer (pH 7.4) containing 2 mM $CaCl_2$. The vials were degassed briefly to allow penetration of the fixative into tissue and incubated at 4 °C for 24 h. After washing with the cacodylate-$CaCl_2$ buffer without fixative, samples were post-fixed for 3 h at 4 °C with 1% (*w*/*v*) osmium tetroxide and 0.8% (*w*/*v*) $K_3Fe(CN)_6$ in the same buffer. Samples were subsequently dehydrated in acetone, infiltrated with Spurr resin for 2 d, embedded in the same resin and polymerized at 60 °C for 48 h.

4.8. High-Presure Freezing (HPF) for Ultrastructural Studies

Explants from *Nicotiana* leaves or *Arabidopsis* seedlings were excised under a stereomicroscope and transferred to aluminium planchette with a 200 µm-deep cavity that was subsequently filled with yeast (*Saccharomyces cerevisiae*) paste and cryoimmobilized immediately with a high pressure freezer (EM Pact (Leica)). Freeze substitution of frozen samples was performed in an automatic freeze substitution system (EM AFS (Leica)) with acetone containing 2% (*w*/*v*) osmium tetroxide and 0.1% (*w*/*v*) uranyl acetate, for 3 d at

−90 °C. On the fourth day, the temperature was raised by 5 °C per hour to room temperature. At this temperature, samples were rinsed in acetone, infiltrated with Epon (*Nicotiana* samples, Figure 4) or Spurr (*Arabidopsis* samples, Figure 5) resin for 2 d, embedded in a thin layer of the same resin and polymerized at 60 °C for 48 h.

4.9. *Ultrastructural Analysis*

Embedded blocks for ultrastructural analysis were submitted to thin sectioning and cell integrity was confirmed at the light microscope. Ultrathin sections were obtained using an Ultracut UC6 Ultramicrotome (Leica) and mounted on Formvar-coated copper grids. Samples were stained with 2% (*w*/*v*) uranyl acetate in water and lead citrate. Samples were observed in a JEM-1010 Electron Microscope (JEOL) equipped with a CCD Camera SIS Megaview III and the AnalySIS software. After capture, image brightness and contrast were adjusted with ImageJ 1.50i for a better visualization.

4.10. *Immunochemical Ultrastructural Analysis*

Explants from *Arabidopsis* seedlings were cryoimmobilized by high-pressure freezing using an EM Pact (Leica) with yeast paste as the filler. Freeze substitution of frozen samples was performed in an automatic freeze substitution system (EM AFS (Leica)) with methanol containing 0.5% (*w*/*v*) uranyl acetate at −90 °C for 3 d. On the fourth day, the temperature was raised by 5 °C per hour to −50 °C. At this temperature, samples were rinsed in acetone, infiltrated and flat embedded in Lowicryl HM20 for 4 d. Ultrathin sections were picked up on Formvar-coated nickel grids. Sample-containing grids were incubated on drops of PBS with 5% (*w*/*v*) BSA for 20 min at room temperature. After removal of the washing solution, drops of PBS with the primary antibody (Ab-CD1-i or Ab-5450 at 1:1000) and 1% (*w*/*v*) BSA were added and incubated for 2 h. Grids were washed three times for 30 min with a drop of PBS with 0.25% (*v*/*v*) Tween 20 and incubated for 1 h in drops of PBS with the secondary antibody and 1% (*w*/*v*) BSA. Secondary antibodies (at 1:30) for HMGR detection were codes 111-205-144 (12-nm gold particle) or 111-215-144 (18-nm particle) from Jackson Immunoresearch (Cambridge, UK). The secondary antibody for GFP detection (code 705-215-147; 18-nm particle; Jackson Immunoresearch, Cambridge, UK) was used at 1:15. The grids were washed three times with a drop of PBS for 5 min and two times with distilled water and air-dried. In control assays for the nonspecific binding of the gold-conjugated antibody, the primary antibody was omitted. Sections were stained with 2% (*w*/*v*) uranyl acetate in water and lead citrate and observed in a JEM-1010 electron microscope (JEOL) with an SIS Mega View III CCD camera.

Supplementary Materials: The following are available online at https://www.mdpi.com/1422-0067/22/17/9132/s1.

Author Contributions: Conceptualization, N.C. and R.E.G.-T.; methodology, R.E.G.-T., C.L.-I. and N.C.; validation, R.E.G.-T. and N.C.; investigation, R.E.G.-T. and N.C.; resources, N.C.; data curation, N.C.; writing—original draft preparation, N.C.; writing—review and editing, N.C. and C.L.-I.; visualization, R.E.G.-T. and N.C.; supervision, J.C.F. and N.C.; project administration, N.C.; funding acquisition, J.C.F. and N.C. All authors have read and agreed to the published version of the manuscript.

Funding: This research was funded by the Generalitat de Catalunya (grant 2017SGR710), Centres de Recerca de Catalunya (CERCA), the Ministerio de Asuntos Exteriores y de Cooperación-Agencia Española de Cooperación Internacional para el Desarrollo (predoctoral scholarship to R.E.G.-T.) and the Ministerio de Economía y Competitividad through the Severo Ochoa Program for Centers of Excellence in R&D 2016–2019 (SEV-2015-0533) and 2020–2023 (CEX2019-000902-S). The Article Processing Charge was financed by Ajuts de la Universitat de Barcelona per publicar en accés obert.

Acknowledgments: We thank Montse Amenós for help in image capturing and processing, Katharina Göttmann and Víctor Campos for the revision of English usage, the Confocal Microscopy Service of the Centre for Research in Agricultural Genomics (CRAG), the Advanced Optic Microscopy Unit of the Scientific and Technological Centers of the University of Barcelona (Faculty of Biology) and the Servei de Camps Experimentals de la Universitat de Barcelona and the Plant Growth Facility of CRAG.

Conflicts of Interest: The authors declare no conflict of interest. The funders had no role in the design of the study; in the collection, analyses, or interpretation of data; in the writing of the manuscript, or in the decision to publish the results.

References

1. Burg, J.S.; Espenshade, P.J. Regulation of HMG-CoA reductase in mammals and yeast. *Prog. Lipid Res.* **2011**, *50*, 403–410. [CrossRef] [PubMed]
2. Gutensohn, M.; Nagegowda, D.A.; Dudareva, N. Involvement of compartmentalization in monoterpene and sesquiterpene biosynthesis in plants. In *Isoprenoid Synthesis in Plants and Microorganisms: New Concepts and Experimental Approaches*; Springer: New York, NY, USA, 2013; pp. 155–169. ISBN 9781461440635.
3. Rodríguez-Concepción, M.; Boronat, A. Breaking new ground in the regulation of the early steps of plant isoprenoid biosynthesis. *Curr. Opin. Plant Biol.* **2015**, *25*, 17–22. [CrossRef] [PubMed]
4. Kutchan, T.M.; Gershenzon, J.; Moller, B.L.; Gang, D.R. Natural products. In *Biochemistry and Molecular Biology of Plants*; Buchanan, B.B., Gruissem, W., Jones, R.L., Eds.; Wiley Blackwell: Chichester, UK; Hoboken, NJ, USA, 2015; pp. 1132–1206. ISBN 9780470714218.
5. Lumbreras, V.; Campos, N.; Boronat, A. The use of an alternative promoter in the *Arabidopsis thaliana HMG1* gene generates an mRNA that encodes a novel 3-hydroxy-3-methylglutaryl coenzyme A reductase isoform with an extended N-terminal region. *Plant J.* **1995**, *8*, 541–549. [CrossRef] [PubMed]
6. NCBI Genome Browser. Available online: https://www.ncbi.nlm.nih.gov/genome/browse (accessed on 9 July 2021).
7. Enjuto, M.; Balcells, L.; Campos, N.; Caelles, C.; Arro, M.; Boronat, A. *Arabidopsis thaliana* contains two differentially expressed 3-hydroxy-3-methylglutaryl-CoA reductase genes, which encode microsomal forms of the enzyme. *Proc. Natl. Acad. Sci USA* **1994**, *91*, 927–931. [CrossRef] [PubMed]
8. Chappell, J. The Biochemistry and Molecular Biology of Isoprenoid Metabolism. *Plant Physiol.* **1995**, *107*, 1–6. [CrossRef]
9. Mo, C.; Bard, M. A systematic study of yeast sterol biosynthetic protein–protein interactions using the split-ubiquitin system. *Biochim. Biophys. Acta* **2005**, *1737*, 152–160. [CrossRef]
10. Mialoundama, A.S.; Jadid, N.; Brunel, J.; Di Pascoli, T.; Heintz, D.; Erhardt, M.; Mutterer, J.; Bergdoll, M.; Ayoub, D.; Van Dorsselaer, A.; et al. Arabidopsis ERG28 tethers the sterol C4-demethylation complex to prevent accumulation of a biosynthetic intermediate that interferes with polar auxin transport. *Plant Cell* **2013**, *25*, 4879–4893. [CrossRef]
11. Gachotte, D.; Eckstein, J.; Barbuch, R.; Hughes, T.; Roberts, C.; Bard, M. A novel gene conserved from yeast to humans is involved in sterol biosynthesis. *J. Lipid Res.* **2001**, *42*, 150–154. [CrossRef]
12. Campos, N.; Boronat, A. Targeting and topology in the membrane of plant 3-hydroxy-3-methylglutaryl coenzyme A reductase. *Plant Cell* **1995**, *7*, 2163–2174. [CrossRef]
13. Basson, M.E.; Thorsness, M.; Finer-Moore, J.; Stroud, R.M.; Rine, J. Structural and functional conservation between yeast and human 3-hydroxy-3-methylglutaryl coenzyme A reductases, the rate-limiting enzyme of sterol biosynthesis. *Mol. Cell. Biol.* **1988**, *8*, 3797–3808. [CrossRef]
14. Liscum, L.; Finer-Moore, J.; Stroud, R.M.; Luskey, K.L.; Brown, M.S.; Goldstein, J.L. Domain structure of 3-hydroxy-3-methylglutaryl coenzyme A reductase, a glycoprotein of the endoplasmic reticulum. *J. Biol. Chem.* **1985**, *260*, 522–530. [CrossRef]
15. Olender, E.; Simon, R. The intracellular targeting and membrane topology of 3-hydroxy-3-methylglutaryl-CoA reductase. *J. Biol. Chem.* **1992**, *267*, 4223–4235. [CrossRef]
16. Profant, D.A.; Roberts, C.J.; Wright, R.L. Mutational analysis of the karmellae-inducing signal in Hmg1p, a yeast HMG-CoA reductase isozyme. *Yeast* **2000**, *16*, 811–827. [CrossRef]
17. Campos, N.; Palau, J.; Zwieb, C. Diversity of 7 SL RNA from the signal recognition particle of maize endosperm. *Nucleic Acids Res.* **1989**, *17*, 1573–1588. [CrossRef]
18. Leivar, P.; González, V.M.; Castel, S.; Trelease, R.N.; López-Iglesias, C.; Arró, M.; Boronat, A.; Campos, N.; Ferrer, A.; Fernàndez-Busquets, X. Subcellular localization of *Arabidopsis* 3-hydroxy-3-methylglutaryl-coenzyme A reductase. *Plant Physiol.* **2005**, *137*, 57–69. [CrossRef] [PubMed]
19. Ferrero, S.; Grados-Torrez, R.E.; Leivar, P.; Antolín-Llovera, M.; López-Iglesias, C.; Cortadellas, N.; Ferrer, J.C.; Campos, N. Proliferation and morphogenesis of the endoplasmic reticulum driven by the membrane domain of 3-hydroxy-3-methylglutaryl coenzyme A reductase in plant cells. *Plant Physiol.* **2015**, *168*, 899–914. [CrossRef] [PubMed]
20. Jingami, H.; Brown, M.S.; Goldstein, J.L.; Anderson, R.G.; Luskey, K.L. Partial deletion of membrane-bound domain of 3-hydroxy-3-methylglutaryl coenzyme A reductase eliminates sterol-enhanced degradation and prevents formation of crystalloid endoplasmic reticulum. *J. Cell Biol.* **1987**, *104*, 1693–1704. [CrossRef] [PubMed]
21. Wright, R.G.; Basson, M.D.; D'Ari, L.; Rine, J. Increased amounts of HMG-CoA reductase induce "karmellae": A proliferation of stacked membrane pairs surrounding the yeast nucleus. *J. Cell Biol.* **1988**, *107*, 101–114. [CrossRef] [PubMed]
22. Sandor, A.; Fricker, M.D.; Kriechbaumer, V.; Sweetlove, L.J. IntEResting structures: Formation and applications of organized smooth endoplasmic reticulum in plant cells. *Plant Physiol.* **2020**, *185*, 550–561. [CrossRef]

23. Chin, D.J.; Luskey, K.L.; Anderson, R.G.; Faust, J.R.; Goldstein, J.L.; Brown, M.S. Appearance of crystalloid endoplasmic reticulum in compactin-resistant Chinese hamster cells with a 500-fold increase in 3-hydroxy-3-methylglutaryl-coenzyme A reductase. *Proc. Natl. Acad. Sci.* **1982**, *79*, 1185–1189. [CrossRef]
24. Eymé, M.J. Infrastructure des cellules nectarigènes de *Diplotaxis erucoides D. C., Helleborus niger* L. et *H. fœtidus* L. *Comptes Rendus des Séances l'Académie des Sci. Ser. D* **1996**, *262*, 1629–1632.
25. Eymé, M.J.; Le Blanc, M. Contribution à l'étude inframicroscopique d'inclusions cytoplasmatiques présentes dans les ovules de *Ficaria* et dans les nectaries d'*Helleborus*. *Comptes Rendus des Séances l'Académie des Sci. Ser. D* **1963**, *256*, 4958–4959.
26. Eymé, J. Nouvelles observations sur l'infrastructure des tissus nectarigènes floraux. *Le Bot. (serie L)* **1967**, *50*, 169–183.
27. Bassot, J.M. Une forme microtubulaire et paracristalline de reticulum endoplasmique dans les photocytes des annelides polynoinæ. *J. Cell Biol.* **1966**, *31*, 135–158. [CrossRef] [PubMed]
28. Wooding, F.B.P. Endoplasmic reticulum aggregates of ordered structure. *Planta* **1967**, *76*, 205–208. [CrossRef] [PubMed]
29. Christensen, A.K. The fine structure of testicular interstitial cells in guinea pigs. *J. Cell Biol.* **1965**, *26*, 911–935. [CrossRef] [PubMed]
30. Christensen, A.K.; Fawcett, D.W. The normal fine structure of opossum testicular interstitial cells. *J. Cell Biol.* **1961**, *9*, 653–670. [CrossRef]
31. Orrenius, S.; Ericsson, J.L.E.; Ernster, L. Phenobarbital-induced synthesis of the microsomal drug-metabolizing enzyme system and its relationship to the proliferation of endoplasmic membranes. *J. Cell Biol.* **1965**, *25*, 627–639. [CrossRef]
32. Jones, A.L.; Fawcett, D.W. Hypertrophy of the agranular endoplasmic reticulum in hamster liver induced by phenobarbital (with a review on the functions of this organelle in liver). *J. Histochem. Cytochem.* **1966**, *14*, 215–232. [CrossRef]
33. Franke, W.W.; Scheer, U. Some structural differentiations in the HeLa cell: Heavy bodies annulate lamellae and cotte de maillet endoplasmic reticulum. *Cytobiologie* **1971**, *4*, 317–329.
34. Black, V.H. The development of smooth-surfaced endoplasmic reticulum in adrenal cortical cells of fetal guinea pigs. *Am. J. Anat.* **1972**, *135*, 381–417. [CrossRef]
35. Barton, B.R.; Hertig, A.T. Ultrastructure of quiescent oocytes of *Cebus albifrons*. *J. Anat.* **1975**, *120*, 227–238. [PubMed]
36. Davidowitz, J.; Philips, G.H.; Pachter, B.R.; Breinin, G.M. Particle-free and glycogen-bearing double membrane arrays in extraocular muscle of rabbit. *Am. J. Pathol.* **1975**, *78*, 191–198.
37. Grimley, P.M.; Schaff, Z. Significance of tubuloreticular inclusions in the pathobiology of human diseases. *Pathobiol. Annu.* **1976**, *6*, 221–257.
38. Schaff, Z.; Lapis, K.; Grimley, P.M. Undulating membraneous structures associated with the endoplasmic reticulum in tumour cells. *Int. J. Cancer* **1976**, *18*, 697–702. [CrossRef] [PubMed]
39. Linder, J.C.; Staehelin, A. The membrane lattice: A novel organelle of the trypanosomatid flagellate *Leptomonas collosoma*. *J. Ultrastruct. Res.* **1980**, *72*, 200–205. [CrossRef]
40. Yamamoto, A.; Otsu, H.; Yoshimori, T.; Maeda, N.; Mikoshiba, K.; Tashiro, Y. Stacks of flattened smooth endoplasmic reticulum highly enriched in inositol 1,4,5-trisphosphate($InsP_3$)receptor in mouse cerebellar Purkinje cells. *Cell Struct. Funct.* **1991**, *16*, 419–432. [CrossRef]
41. Karnaky, K.J.; Lau, K.R.; Garretson, L.T.; Schultz, S.G. Seasonal variations in the fine structure of the *Necturus maculosus* urinary bladder epithelium: Low transporters and high transporters. *Am. J. Anat.* **1984**, *171*, 227–242. [CrossRef] [PubMed]
42. Kerr, J.B.; Weiss, M. Spontaneous or experimentally induced formation of a special zone in the adrenal cortex of the adult brush-tailed possum(*Trichosurus vulpecula*). *Am. J. Anat.* **1991**, *190*, 101–117. [CrossRef] [PubMed]
43. Wolf, K.W.; Motzko, D. Paracrystalline endoplasmic reticulum is typical of gametogenesis in Hemiptera species. *J. Struct. Biol.* **1995**, *114*, 105–114. [CrossRef]
44. Yamamoto, A.; Masaki, R.; Tashiro, Y. Formation of crystalloid endoplasmic reticulum in COS cells upon overexpression of microsomal aldehyde dehydrogenase by cDNA transfection. *J. Cell Sci.* **1996**, *109*, 1727–1738. [CrossRef] [PubMed]
45. Snapp, E.; Hegde, R.; Francolini, M.; Lombardo, F.; Colombo, S.; Pedrazzini, E.; Borgese, N.; Lippincott-Schwartz, J. Formation of stacked ER cisternae by low affinity protein interactions. *J. Cell Biol.* **2003**, *163*, 257–269. [CrossRef] [PubMed]
46. Almsherqi, Z.A.; Kohlwein, S.D.; Deng, Y. Cubic membranes: A legend beyond the Flatland* of cell membrane organization. *J. Cell Biol.* **2006**, *173*, 839–844. [CrossRef]
47. Almsherqi, Z.A.; Landh, T.; Kohlwein, S.D.; Deng, Y. Chapter 6: Cubic membranes the missing dimension of cell membrane organization. *Int. Rev. cell Mol. Biol.* **2009**, *274*, 275–342. [CrossRef] [PubMed]
48. Leivar, P.; Antolín-Llovera, M.; Ferrero, S.; Closa, M.; Arró, M.; Ferrer, A.; Boronat, A.; Campos, N. Multilevel control of *Arabidopsis* 3-hydroxy-3-methylglutaryl coenzyme A reductase by protein phosphatase 2A. *Plant Cell* **2011**, *23*, 1494–1511. [CrossRef]
49. Westrate, L.; Lee, J.; Prinz, W.; Voeltz, G. Form follows function: The importance of endoplasmic reticulum shape. *Annu. Rev. Biochem.* **2015**, *84*, 791–811. [CrossRef] [PubMed]
50. Hawes, C.; Kiviniemi, P.; Kriechbaumer, V. The endoplasmic reticulum: A dynamic and well-connected organelle. *J. Integr. Plant Biol.* **2015**, *57*, 50–62. [CrossRef]
51. Nishikawa, S.; Hirata, A.; Nakano, A. Inhibition of endoplasmic reticulum (ER)-to-Golgi transport induces relocalization of binding protein (BiP) within the ER to form the BiP bodies. *Mol. Biol. Cell* **1994**, *5*, 1129–1143. [CrossRef]
52. Bergmann, J.E.; Fusco, P.J. The G protein of vesicular stomatitis virus has free access into and egress from the smooth endoplasmic reticulum of UT-1 cells. *J. Cell Biol.* **1990**, *110*, 625–635. [CrossRef]

53. Anderson, R.; Orci, L.; Brown, M.; Garcia-Segura, L.; Goldstein, J. Ultrastructural analysis of crystalloid endoplasmic reticulum in UT-1 cells and its disappearance in response to cholesterol. *J. Cell Sci.* **1983**, *63*, 1–20. [CrossRef] [PubMed]
54. Hurbain, I.; Sachse, M. The future is cold: Cryo-preparation methods for transmission electron microscopy of cells. *Biol. Cell* **2011**, *103*, 405–420. [CrossRef]
55. Ebersold, H.R.; Cordier, J.-L. Bacterial mesosomes: Method dependent artifacts. *Arch. Microbiol.* **1981**, *130*, 19–22. [CrossRef]
56. Higgins, M.L.; Tsien, H.C.; Daneo-Moore, L. Organization of mesosomes in fixed and unfixed cells. *J. Bacteriol.* **1976**, *127*, 1519–1523. [CrossRef]
57. Silva, M.; Sousa, J.; Polónia, J.; Macedo, M.; Parente, A.M. Bacterial mesosomes: Real structures of artifacts? *Biochim. et Biophys. Acta (BBA)-Biomembr.* **1976**, *443*, 92–105. [CrossRef]
58. Chandler, D. Comparison of quick-frozen and chemically fixed sea-urchin eggs: Structural evidence that cortical granule exocytosis is preceded by a local increase in membrane mobility. *J. Cell Sci.* **1984**, *72*, 23–36. [CrossRef] [PubMed]
59. Murk, J.L.A.N.; Posthuma, G.; Koster, A.; Geuze, H.J.; Verkleij, A.J.; Kleijmeer, M.J.; Humbel, B.M. Influence of aldehyde fixation on the morphology of endosomes and lysosomes: Quantitative analysis and electron tomography. *J. Microsc.* **2003**, *212*, 81–90. [CrossRef] [PubMed]
60. Orci, L.; Brown, M.S.; Goldstein, J.L.; Garcia-Segura, L.; Anderson, R.G. Increase in membrane cholesterol: A possible trigger for degradation of HMG-CoA reductase and crystalloid endoplasmic reticulum in UT-1 cells. *Cell* **1984**, *36*, 835–845. [CrossRef]
61. Kochevar, D.; Anderson, R. Purified crystalloid endoplasmic reticulum from UT-1 cells contains multiple proteins in addition to 3-hydroxy-3-methylglutaryl coenzyme A reductase. *J. Biol. Chem.* **1987**, *262*, 10321–10326. [CrossRef]
62. Zacharias, D.A.; Violin, J.D.; Newton, A.C.; Tsien, R.Y. Partitioning of lipid-modified monomeric GFPs into membrane microdomains of live cells. *Science* **2002**, *296*, 913–916. [CrossRef] [PubMed]
63. Istvan, E.S.; Palnitkar, M.; Buchanan, S.K.; Deisenhofer, J. Crystal structure of the catalytic portion of human HMG-CoA reductase: Insights into regulation of activity and catalysis. *EMBO J.* **2000**, *19*, 819–830. [CrossRef] [PubMed]
64. Murashige, T.; Skoog, F. A revised medium for rapid growth and bio-assays with tobacco tissue cultures. *Physiol. Plant.* **1962**, *15*, 473–497. [CrossRef]
65. Batoko, H.; Zheng, H.-Q.; Hawes, C.; Moore, I. A rab1 GTPase is required for transport between the endoplasmic reticulum and Golgi apparatus and for normal Golgi movement in plants. *Plant Cell* **2000**, *12*, 2201–2217. [CrossRef] [PubMed]
66. Van Der Hoorn, R.A.L.; Laurent, F.; Roth, R.; de Wit, P. Agroinfiltration is a versatile tool that facilitates comparative analyses of Avr9/Cf-9-induced and Avr4/Cf-4-induced necrosis. *Mol. Plant-Microbe Interactions* **2000**, *13*, 439–446. [CrossRef] [PubMed]
67. Koncz, C.; Schell, J. The promoter of TL-DNA gene 5 controls the tissue-specific expression of chimaeric genes carried by a novel type of *Agrobacterium* binary vector. *Mol. Genet. Genom.* **1986**, *204*, 383–396. [CrossRef]
68. Sauer, M.; Paciorek, T.; Benková, E.; Friml, J. Immunocytochemical techniques for whole-mount in situ protein localization in plants. *Nat. Protoc.* **2006**, *1*, 98–103. [CrossRef] [PubMed]

International Journal of *Molecular Sciences*

MDPI

Article

Hot Spot Mutagenesis Improves the Functional Expression of Unique Mammalian Odorant Receptors

Yosuke Fukutani [1,*], Yuko Nakamura [1], Nonoko Muto [1], Shunta Miyanaga [1], Reina Kanemaki [1], Kentaro Ikegami [1], Keiichi Noguchi [2], Ikuroh Ohsawa [3], Hiroaki Matsunami [4] and Masafumi Yohda [1]

1 Department of Biotechnology and Life Science, Tokyo University of Agriculture and Technology, Tokyo 184-8588, Japan; yuko.nakamura@yohda.net (Y.N.); nonoko.muto@yohda.net (N.M.); shunta.miyanaga@yohda.net (S.M.); reina.kanemaki@yohda.net (R.K.); kentaro.ikegami@yohda.net (K.I.); yohda@cc.tuat.ac.jp (M.Y.)
2 Instrumentation Analysis Center, Tokyo University of Agriculture and Technology, Tokyo 184-8588, Japan; knoguchi@cc.tuat.ac.jp
3 Biological Process of Aging, Tokyo Metropolitan Institute of Gerontology, Tokyo 173-0015, Japan; iohsawa@tmig.or.jp
4 Department of Molecular Genetics and Microbiology, Duke University Medical Center, Durham, NC 27710, USA; matsu004@mc.duke.edu
* Correspondence: fukutani@cc.tuat.ac.jp; Tel.: +81-42-388-7479

Abstract: Vertebrate animals detect odors through olfactory receptors (ORs), members of the G protein-coupled receptor (GPCR) family. Due to the difficulty in the heterologous expression of ORs, studies of their odor molecule recognition mechanisms have progressed poorly. Functional expression of most ORs in heterologous cells requires the co-expression of their chaperone proteins, receptor transporting proteins (RTPs). Yet, some ORs were found to be functionally expressed without the support of RTP (RTP-independent ORs). In this study, we investigated whether amino acid residues highly conserved among RTP-independent ORs improve the functional expression of ORs in heterologous cells. We found that a single amino acid substitution at one of two sites (N_{BW}3.39 and 3.43) in their conserved residues (E and L, respectively) significantly improved the functional expression of ORs in heterologous cells. $E^{3.39}$ and $L^{3.43}$ also enhanced the membrane expression of RTP-dependent ORs in the absence of RTP. These changes did not alter the odorant responsiveness of the tested ORs. Our results showed that specific sites within transmembrane domains regulate the membrane expression of some ORs.

Keywords: odorant receptor; chemosensory; membrane traffic; heterologous expression

Citation: Fukutani, Y.; Nakamura, Y.; Muto, N.; Miyanaga, S.; Kanemaki, R.; Ikegami, K.; Noguchi, K.; Ohsawa, I.; Matsunami, H.; Yohda, M. Hot Spot Mutagenesis Improves the Functional Expression of Unique Mammalian Odorant Receptors. *Int. J. Mol. Sci.* **2022**, *23*, 277. https://doi.org/10.3390/ijms23010277

Academic Editors: Masoud Jelokhani-Niaraki and Edgar Soria-Gomez

Received: 19 November 2021
Accepted: 24 December 2021
Published: 28 December 2021

Publisher's Note: MDPI stays neutral with regard to jurisdictional claims in published maps and institutional affiliations.

1. Introduction

Olfaction is an essential sensation for animals to continue carrying out their daily life activities while experiencing changes in the surrounding environment. Mammalian olfactory receptors (ORs) are expressed on the cell surface of olfactory sensory neurons (OSNs) to detect and discriminate the vast number of odors in the environment [1]. ORs form the largest and most diverse family of G protein-coupled receptors (GPCRs) [2]. There are more than 400 and 1200 functional ORs in the olfactory sensory neurons of humans and mice, respectively [3–5]. It is thought that this diverse odor recognition of ORs has made it possible for olfaction to discriminate odors, which are a mixture of various volatile molecules, with high sensitivity [6,7]. A major difficulty in characterizing the function of these mammalian ORs is that most ORs are not exported to the plasma membrane and are instead retained in the endoplasmic reticulum (ER) in nonolfactory cells, making in vitro characterization extremely challenging [2]. Consequently, functional analysis of ORs is far behind that of other canonical Class A GPCRs. Therefore, the flexible and selective odor molecule recognition mechanism of ORs is almost unknown. To fill this gap in knowledge, it is necessary to develop a method for improving OR expression without changing the function of ORs. The use of a Rho tag or IL6 tag attached to the N-terminus of ORs has been suggested as a method due to the technology's ability to promote the enhancement of surface expression of ORs [8–10]. Additionally, co-expression of chaperone proteins, Hsp70t or Ric-8B, is effective for the functional expression of some ORs [11,12]. However, both modifications only enhance the expression of a limited number of ORs.

Heterologous expression of many ORs is achieved with the assistance of receptor transporting protein (RTP) 1S and RTP2 [13–17]. These are single transmembrane proteins, a class of chaperones, that mediate the transport of ORs into the plasma membrane. However, the detailed mechanism of the membrane trafficking of ORs by RTPs is unknown. Interestingly, a small subset of ORs can localize to the cell membrane without RTP1 and RTP2 in heterologous cells and function even in double knockout mice lacking RTP1S and RTP2 [18]. We previously demonstrated the importance of a specific amino acid residue within transmembrane domain 4 ($G^{4.53}$) for the functional expression of OR in heterologous cells. Poor cell surface expression seems to correlate with structural instability [19].

This study investigated whether mammalian OR expression could be improved by replacing an amino acid with one that is highly conserved in a well-expressed OR (RTP-independent OR). We also tried to identify important amino acid residues that contribute to the functional expression of mammalian ORs.

2. Results

2.1. Screening of the Amino Acid Mutations That Enhance OR Expression

We hypothesized that highly conserved amino acid residues in RTP-independent ORs play essential roles in OR trafficking in heterologous cells. Our previous study showed that amino acid residues at 66 sites are associated with RTP independence [19]. Among them, the conserved amino acid residues at 26 sites appeared more frequently in RTP-independent ORs than in all intact 1093 mouse ORs. Their positions were dispersed throughout the OR sequence (Figure 1A and Supplementary Data S1). We hypothesized that these ORs that have non-conserved amino acid residues at these positions might show higher levels of expression when these residues are changed to conserved ones.

We tested this hypothesis by constructing 40 mutant ORs, transiently expressing them in HEK293T cells and analyzing their cell surface expression by FACS. We first tested 14 sites out of the 26 in 14 RTP-independent ORs (Figure 1A Red circle). Among them, Olfr544 (also known as OR-S6), D115E ($N_{BW}3.39$), Olfr194_128H (IC2), and Olfr78_Q297E (C-term) exhibited a significant increase in surface expression (Figures 1B, S1 and S2).

We next tested whether the changes to the conserved residues in these sites improve the expression of RTP-dependent ORs. In the same manner, we changed a non-conserved amino acid residue to a conserved residue in 11 RTP-dependent ORs. Ninety-seven single amino acid mutants covering all 26 sites were constructed and expressed in HEK293T cells without RTP1S coexpression. Two mutants (Olfr992D111E ($N_{BW}3.39$) and Olfr982Y120L ($N_{BW}3.43$)) showed enhanced plasma membrane localization by more than 30% (Figures 1C, S1 and S3). Since $N_{BW}3.39$ and $N_{BW}3.43$ are in proximity in the same transmembrane helix, interactions between them may contribute to the expression of ORs. Based on these screenings with both RTP-dependent and RTP-independent ORs, we decided to focus on our study on residues within transmembrane domain 3.

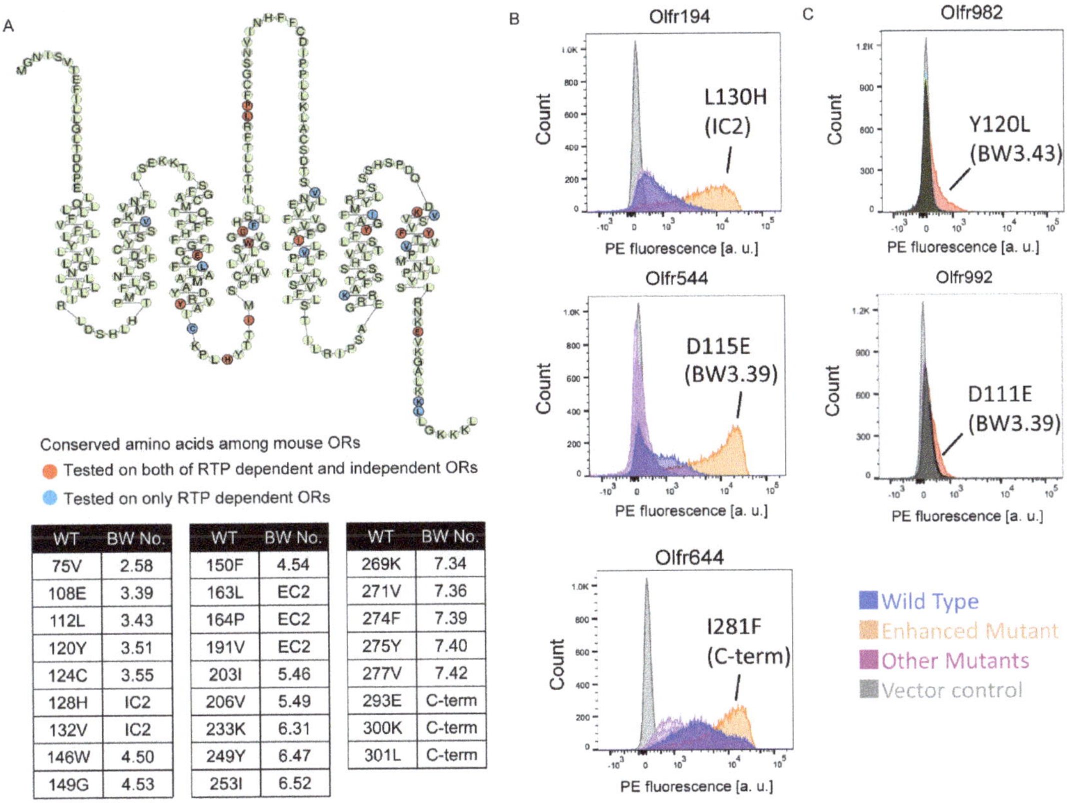

WT	BW No.	WT	BW No.	WT	BW No.
75V	2.58	150F	4.54	269K	7.34
108E	3.39	163L	EC2	271V	7.36
112L	3.43	164P	EC2	274F	7.39
120Y	3.51	191V	EC2	275Y	7.40
124C	3.55	203I	5.46	277V	7.42
128H	IC2	206V	5.49	293E	C-term
132V	IC2	233K	6.31	300K	C-term
146W	4.50	249Y	6.47	301L	C-term
149G	4.53	253I	6.52		

Figure 1. Identification of mutation residues enhancing OR expression. (**A**) Snake plot showing the consensus sequence of mouse ORs. Circle: Amino acid site that constructed the mutant. Red: tested on both of RTP dependent and independent ORs, Blue: tested on RTP dependent ORs only. The table shows the mutation sites and BW (Ballesteros-Weinstein) number [20] tested in this study. (**B**,**C**) The cell-surface expression of the RTP independent OR (Olfr194, Olfr544 and Olfr644) and RTP dependent ORs (Olfr982 and Olfr992), respectively. Each mutant was transfected into HEK293T cells, and the cell surface expression level was measured. X-axis: PE fluorescence, Y-axis: cell number.

2.2. Improvement of Surface Expression by the 3.39E and 3.43L Mutations in Many ORs

In 1093 intact mouse ORs registered in the HORDE database [21], E(glutamic acid) accounted for 85.4% (933/1093) of N_{BW} 3.39, and L(leucine) accounted for 93.3% (1020/1093) of N_{BW} 3.43 (Figure 2A,B). To examine whether mutations of nonconserved N_{BW}3.39 and N_{BW} 3.43 residues to E (3.39E mutation) and L (3.43L mutation) are effective for expressing various ORs, we applied the mutations to mouse ORs with known ligands. We made 3.39E mutants for six ORs (Olfr339, Olfr992, Olfr1352, Olfr1353, Olfr1377, and Olfr960) and 3.43L mutants for five ORs (Olfr62, Olfr982, Olfr978, Ofr979, and Olfr960) and measured the mutant expression level on the cell surface (Figure 2A–C). Surface expression on heterologous cells was improved for all single mutants except Olfe1352 D3.39E in the co-expression of RTP1S. Furthermore, 6 ORs were localized on cell membranes without co-expression of RTP1S after introducing mutations of 3.39E or 3.43L (Figure 2D). The cell surface expression of Olfr960, which has both L3.39 and Y3.43, was not improved by either mutation (L3.39E and/or Y3.43L). However, the effect of RTP1S co-expression on the mutants was higher than that on the wild type (Figure 2C,D)

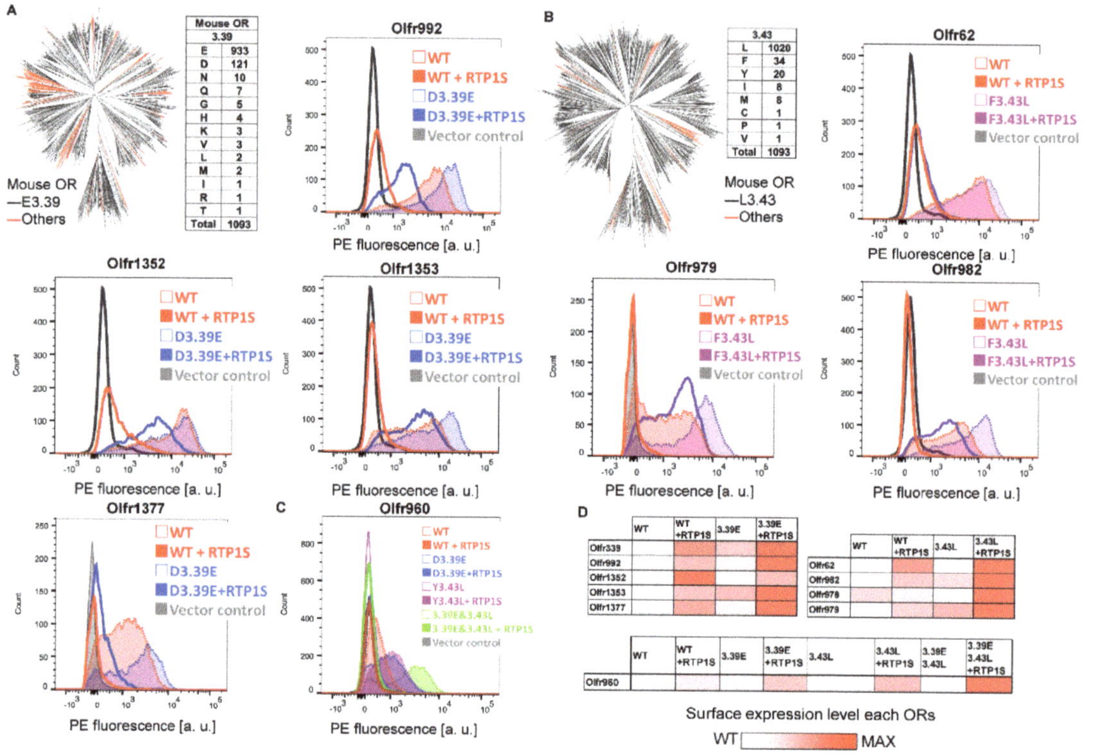

Figure 2. Effect of the 3.39E and 3.43L mutants on mouse ORs. (**A**) Phylogenic tree of mouse ORs showing the E3.39 or others. Red indicates an OR in which the residue of $N_{BW}3.39$ is not E (Glu) in 1093 intact mouse ORs. Cell-surface expression of the Rho-tagged ORs and 3.39E mutants. Each mutant was transfected into HEK293T cells, and the cell surface expression level was measured. X-axis: PE fluorescence, Y-axis: cell number. (**B**) Phylogenic tree of mouse ORs showing the L3.43 or others. Red indicates an OR in which the residue of $N_{BW}3.43$ is not L (Leu). Cell-surface expression of the Rho-tagged ORs and 3.43L mutants. (**C**) Cell-surface expression of Rho-tagged Ofr960 and its mutants. (**D**) The values of the effect of mutations of 3.39E and 3.43L were used as the max intensity (red) and wild type (white) in each OR, respectively.

2.3. *Olfr544 D3.39E Showed Significantly High Expression without a Change in Ligand Selectivity*

The single amino acid mutation D3.39E significantly improved the membrane expression of Olfr544. The expression level of Olfr544D115E$^{BW3.39}$ alone was significantly higher than with the co-expression of RTP1S (Figure 3A,B). Olfr544 responds to dicarboxylic acids, including nonanedioic acid (azelaic acid) [22]. We tested Olfr544 WT and D3.39E mutant responses to dicarboxylic acids with various carbon chain lengths (Figure 3C). The responses of the D3.39E mutant were higher than those of the WT, correlating with the higher expression level in the mutant. The selectivity for dicarboxylic acids with different carbon chains was not affected by the mutation. The D3.39E mutant responded to concentrations of nonanedioic acid 100 times lower than the WT, with statistical significance (Figure 3D). The ligand selectivity did not change for the tested ligands.

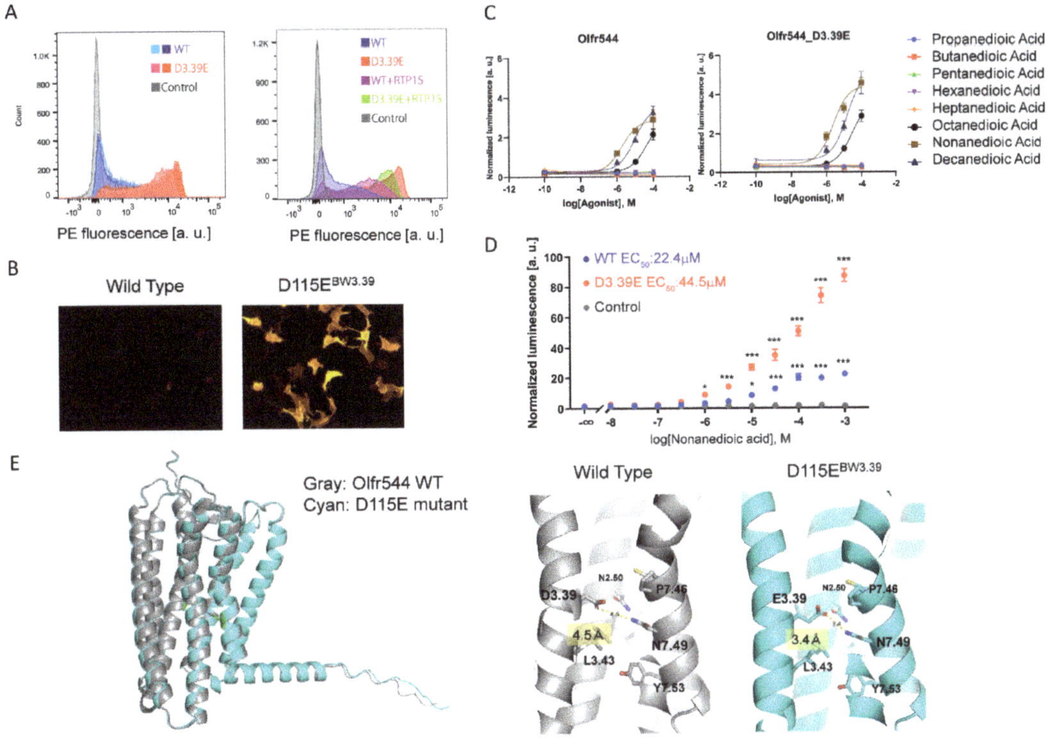

Figure 3. Detailed analysis of Olfr544 D115E$^{BW3.39}$ (**A**) Cell-surface expression of the Rho-tagged ORs and D3.39E mutants. Each mutant was transfected into HEK293T cells, and the cell surface expression level was measured. X-axis: PE fluorescence, Y-axis: cell number. Left: Comparison between the WT and the D3.39E mutant. The results from two independent experiments are shown in the graph. Right: The effect of RTP1S co-expression of Olfr544 and D3.39E mutant. (**B**) Immunocytochemical image of Olfr544 and the D3.39E mutant stained with anti-Rho4D2 mouse antibody and PE-conjugated anti-mouse IgG goat antibody. (**C**) Agonist selectivity of Olfr544 and the D3.39E mutant for various dicarboxylic acids. Error bars indicate s.e.m (n = 3). (**D**) Dose response curve of Olfr544 and the D3.39E mutant to nonanedioic acid. Error bars indicate s.e.m (n = 3). Multiple comparisons were performed using one-way ANOVA followed by Dunnett's test (* $p < 0.05$, *** $p < 0.001$). (**E**) Structural model of Olfr544 and the D3.39E mutant using Alphafold 2 prediction. It was predicted that the distance between the residue of 3.39 and N7.49 was 4.5 Å in the WT (D3.39) and 3.4 Å in the E3.39 mutant.

Olfr544 belongs to the Class I OR family, and D3.39 is widely conserved in Olfr544 homologs (Figure S4), suggesting that Olfr544D3.39 appeared relatively early and was subsequently maintained during evolution. When D3.39 of Olfr544 was mutated to various amino acids, the D3.39E mutant showed the highest cell surface expression level among the mutants (Figure S5). The D3.39R and D3.39N mutants showed increased expression but lost responsiveness to their agonist. Since $N_{BW}3.39$ of ORs is located in the sodium ion binding site in Class A GPCRs [23,24], it is plausible that the interaction of glutamate with sodium ions is important for the stability and function of ORs.

To gain insights into the mechanism of the improvement of stability by the $N_{BW}3.39E$ mutation, we constructed structural models of Olfr544 variants, wild type, D3.39E (D115E) mutants, and D3.39R (D115R) mutants using AlphaFold 2 [25]. The overall structures of the models were almost identical. The distance between the side chain of N7.49 and the side chain of D3.39 (D115) is 4.5 Å in the wild type. In the D3.39E mutant, the distance between E3.39 and N7.49 was shortened to 3.4 Å to form a hydrogen bond (Figure 3E).

In the model of the D3.39R mutant, the arginine residue of N_{BW}3.39 was located in the center of the lumen of the GPCR, similar to other reported Class A GPCRs (Figure S5) [26]. The effects of the D3.39E and D3.39R mutations are related to the coordination and allosteric action of the sodium ion. In other words, this suggests that N_{BW}3.39 of ORs also functions in the allosteric movement of ORs as a sodium ion binding site, which is similar to other Class A GPCRs.

2.4. Agonist Response of N_{BW}3.39E and N_{BW}3.43L Mutant ORs

Then, we examined the effects of the N_{BW}3.39E or N_{BW}3.43L mutation on the ligand responses of various ORs. Most OR mutants showed an improved odor response compared to WT (Figure 4A–C and Supplementary Data S2), correlating with the increase in surface expression. However, the Olfr992D3.39E and Olfr62F3.43L mutants did not respond to the original ligands of octanoic acid and 2-coumaranon, respectively (Figure 4D). Since there is a correlation between amounts of OR expression and ligand responsiveness in heterologous cells [17], these results suggest that the mutation sites Olfr992D3.39 and Olfr62F3.43 are likely to play a role in ligand binding.

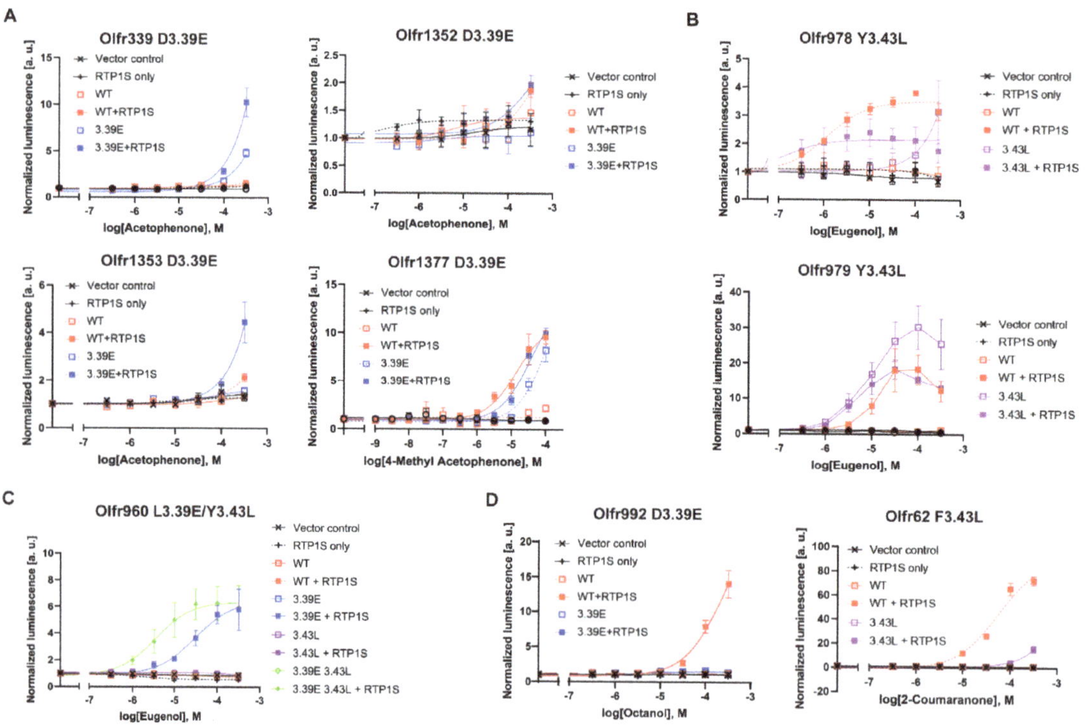

Figure 4. Dose-response curve of each mouse OR mutant. (**A**) D3.39E mutants of Olfr339, Olfr1352, Olfr1353 and Olfr1377, (**B**) Y3.43L mutants of Olfr978 and Olfr979, (**C**) Olfr960 L3.39E, Y3.43L and double mutants, (**D**) ORs with reduced agonist response due to mutations. Olfr992D3.39E against octanol and Olfr62F3.43L against 2-coumaranone. All constructs, including the vector control, were also tested under RTP1S co-expression conditions. Error bars indicate s.e.m (n = 3).

2.5. The Change in Conserved E3.39 and L3.43 Caused a Loss of Function

To investigate whether mutation of the conserved N_{BW}3.39E and N_{BW}3.43L alters the functional expression of ORs, we constructed E3.39D and L3.43Y mutants. We selected Olfr1508 (agonist: Cu^{2+} ion) and Olfr733 (agonist: hexanal), which are highly expressed in the cell membrane without the assistance of RTPs [19,27]. The E3.39D and L3.43Y mutations significantly reduced their cell surface expression and ligand response ability (Figure 5).

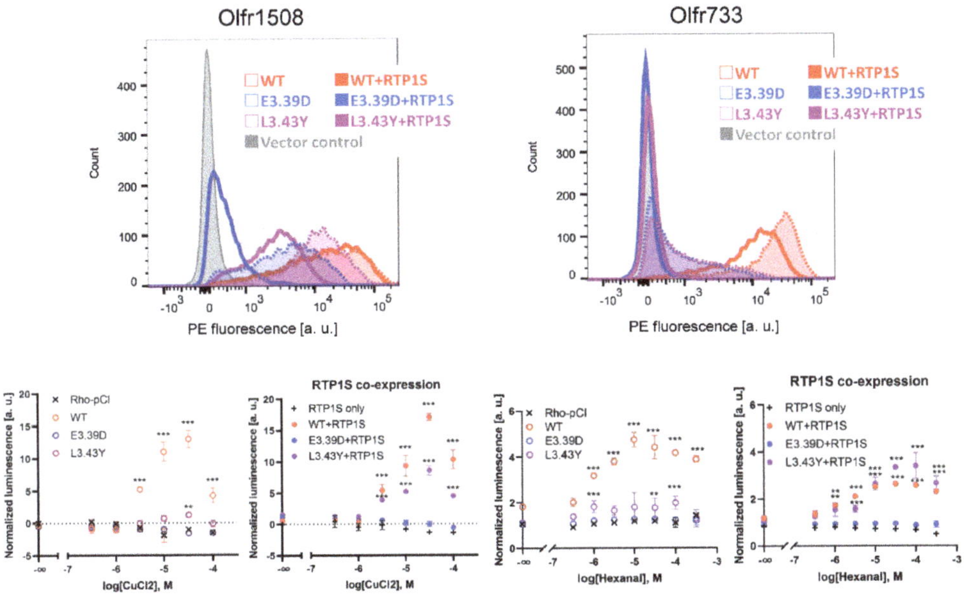

Figure 5. The change in conserved E3.39 and L3.43 led to a loss of function. (**Top**) The cell-surface expression of the Rho-tagged wild-type, E3.39D and L3.43Y mutants of Olfr1508 and Olfr733. Each OR was transfected into HEK293T cells, and the cell surface expression level was measured. X-axis: PE fluorescence, Y-axis: cell number. Left graph: Comparison between the WT and D3.39E mutant. The results from two independent experiments are shown in the graph. Right: The effect of RTP1S co-expression on them. (**Bottom**) Dose-response curve of Olfr1508 and Olfr733 mutants to CuCl2 (Cu2+) and hexanal, respectively. Error bars indicate s.e.m (n = 3). Multiple comparisons were performed using one-way ANOVA followed by Dunnett's test (** $p < 0.01$, *** $p < 0.001$).

Although co-expression of RTP1S could rescue the functional expression of L3.43Y mutants, the E3.39D mutant was not recovered. These results suggest that both N_{BW}3.39 and N_{BW}3.43 of ORs are important residues for the appropriate expression of Ors.

2.6. N_{BW} 3.39E and N_{BW} 3.43L Mutations Are also Effective in Human Ors

We examined the effects of the 3.39E and 3.43L mutations on human Ors. In human Ors, E(Glutaminc acid) accounted for 81% (315/390) of N_{Bw}3.39 and L(Leucine) accounted for 92.5% (361/390) of N_{Bw}3.43. (Figure 6A,B). Ors without conserved E3.39 tend to be clustered in a few families, suggesting evolutionary advantages in these families. On the other hand, the Ors whose 3.43 was not L were dispersed throughout the phylogenetic tree, and many of them were pseudogenes. Then, we constructed 3.39E and 3.43L mutants and examined their surface expression and ligand response (Figures 6 and S6). OR7D4 is one of the characteristic Ors in which human perception of androstenone changes dramatically depending on genetic variants [28]. The D3.39E mutation significantly increased the cell surface expression of OR7D4, and the detection limit of androstenone (5α-androst-16-en-3-one) was improved to less than 1 μM (Figure 6C). A significant increase was also observed in OR10G7 Y3.43L, which was expressed without co-expression of RTP1S (Figure 6D). This mutant also detected a significant response against Eugenol even at lower stimulus concentrations than the wild type [29]. Mutations in the N_{BW}3.39 and 3.43 amino acids can be effective ways to improve the functional expression of human ORs. Taken together, the amino acids located at N_{BW}3.39 and 3.43 are involved in regulating the expression of mammalian ORs.

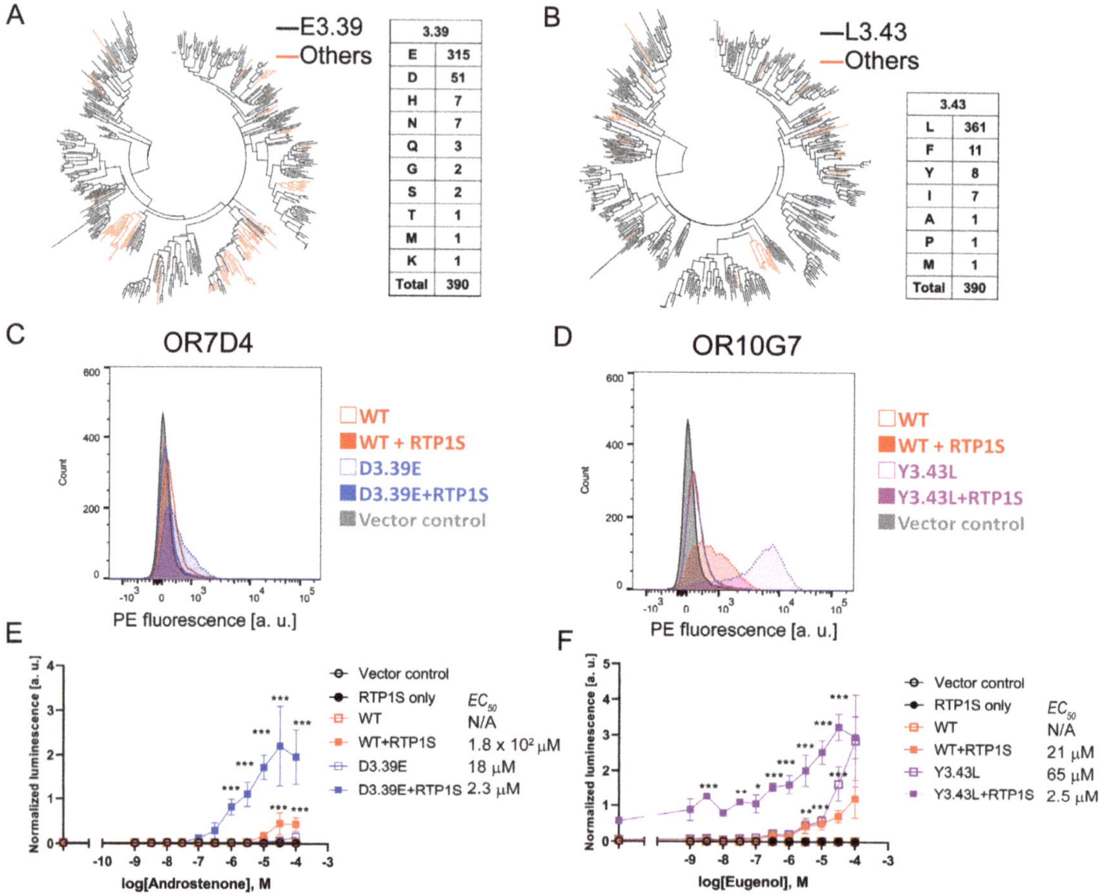

Figure 6. Effects of 3.39E and 3.43L mutants on human ORs (**A**,**B**) Phylogenic tree of mouse ORs showing the E3.39 or others, or L3.43 or others in 390 intact human ORs registered in the HORDE database. (**C**,**D**) Cell-surface expression of the Rho-tagged ORs and 3.39E mutants. Each mutant was transfected into HEK293T cells, and the cell surface expression level was measured. X-axis: PE fluorescence, Y-axis: cell number. (**E**,**F**) Dose response curves of OR7D4 (WT and muant) and OR10G7 (WT and mutant) to androstenone and eugenol, respectively. Error bars indicate s.e.m (n = 3). Multiple comparisons were performed using one-way ANOVA followed by Dunnett's test (* $p < 0.05$, ** $p < 0.01$, *** $p < 0.001$).

3. Discussion

Animals can detect various odorants in the environment through pattern recognition of the response by hundreds of ORs [7,30]. We presented a hypothesis that olfactory neurons acquired a high level of evolutionally capacitance to allow rapid evolution of ORs [19]. This rapid evolution allows ORs to respond to structurally diverse environmental odor molecules but compromises conformational stability, resulting in difficulties in functional expression of ORs in non-olfactory neurons [19]. In this study, from the amino acid residues conserved among the RTP-independent ORs with high conformational stability, we found that N_{BW}3.39 and 3.43 in the third transmembrane domain are related to OR expression. The 3.39E and 3.43L substitutions were effective in improving the membrane localization of multiple ORs in mice and humans. Curiously, conservation of N_{BW}3.39E and N_{BW}3.43L in human ORs is lower than that in mouse ORs. This may, in part, explain why expression levels of the human ORs are generally lower than those of mouse ORs in heterologous cell expression systems. 3.39E mutations of the ORs containing D at N_{BW}3.39 often enhanced membrane expression (7/10 in tested ORs), while the 3.39E mutation was less effective for ORs with other amino acids at N_{BW}3.39, such as Q (1/4 in tested ORs). Both N_{BW}3.39 and 3.43 are in the transmembrane helices and face the same direction. Interestingly, N_{BW}4.53, which affects the RTP dependency of Olfr539, also faces towards N_{BW}3.39 and N_{BW}3.43 [19]. Together, the luminal environment of each transmembrane helix may be important for the stability of ORs and their associated membrane localization. This may occur via the stabilization of membrane packing. Residues in both TM3 and TM4 of ORs may act in combination to stabilize membrane packing. This work successfully identified two specific residues to facilitate the expression of ORs. The expression levels of the human ORs were lower than those of mouse ORs in heterologous cell expression systems. By introducing these mutations, detailed analysis of orphan ORs, which could not be analyzed due to its low expression level, may be achieved.

We previously report that the introduction of arginine to specific residues predicted to improve thermal stability also improved the functional expression of consensus ORs [19]. It has also been reported that the functional expression of OR was improved by introducing mutations to transmembrane domains other than TM3 and TM4 and the C-terminal region [31,32]. These methods that have been effective to some ORs appears to be ineffective with other ORs. As a consequence, no single method can be adapted to improve the expression of all mammalian ORs. There may be different solutions to the OR trafficking problem depending on the individual ORs. However, the solutions can be broadly categorized into a small number of methods, one of which would be to stabilize membrane packing by adding appropriate mutations.

N_{BW}3.39 and N_{BW}3.43 are located in the vicinity of the sodium ion binding site in other Class A GPCRs. N_{BW}3.39 is frequently mutated to increase the production yield for crystallization of GPCRs [26,33]. The 8 amino acids out of the 15 sodium-binding sites reported are different from those of ORs to the canonical Class A GPCR (Figure S7), and the relationship between the functional activity of the ORs and the sodium ion has not been reported thus far. This difference in amino acid conservation of the sodium ion-binding site may mean that the origin of the mammalian ORs is fundamentally different from other GPCRs.

It is very likely that sodium ions are coordinated in the vicinity of N_{BW}3.39 and act allosterically, the same as in other GPCRs. While the three-dimensional structures of many Class A GPCRs have been reported, none of the ORs have been solved. Reports of ligand docking analysis and MD simulation of ORs using a model structure are increasing [34–37]. The progress of structure prediction technologies including AlphaFold 2 has been remarkable [25]. Our study identifying important residues that contribute to the functional expression of ORs will help lead to the elucidation of the functional mechanism of ORs and the development of high-sensitivity odor sensing technology using mammalian ORs.

The introduction of the 3.39E or 3.43 L mutation significantly affected the functional expression of ORs. Most mutant ORs retained odorant selectivity except for Olfr544 D3.39R, which caused a loss of ligand response while improving membrane localization. Although the mutations (N_{BW}3.39E and N_{BW}3.43L) we found in this study are applicable to a small number of ORs (approximately 19% and 7.5% of human ORs, respectively), the ORs may have evolved to have unique characteristics in odor detection in some cases. Future experiments can address this hypothesis by expanding the number of ORs and odorants tested using N_{BW}3.39E and N_{BW}3.43L mutations. Currently, however, the majority of mammalian ORs are still poorly expressed even with the aid of RTPs, suggesting that pieces are still missing for the stable expression of ORs in heterologous cells.

4. Materials and Methods

4.1. DNA and Vector Preparation

Open reading frames of OR genes were subcloned into pCI (Promega, Madison, WI, USA) with a rhodopsin tag at the N-terminus. To generate mutants of ORs, DNA fragments of OR genes were amplified by Phusion polymerase (Thermo Fisher Scientific, Waltham, MA, USA) and PrimeStar MAX polymerase (Takara bio, Shiga, Japan). The fragments were mixed and amplified by PCR to obtain full sequences. All plasmid DNA was purified by NucleoSpin plasmid transfection grade (MACHEREY-NAGEL GmbH & Co, Düren, Deutschland). Other expression vectors were the same as used in a previous study [17]. All plasmid sequences were verified using Sanger sequencing.

4.2. Cell Culture

HEK293T and Hana 3A [13] cells were grown in Minimal Essential Medium (MEM) containing 10% FBS (vol/vol) with penicillin-streptomycin and amphotericin B. Hana 3A cells were authenticated using polymorphic short tandem repeats (STRs) at the Duke DNA Analysis Facility using GenePrint 10 (Promega) and shown to share profiles with the reference (ATCC). All cell lines were incubated at 37 °C, saturating humidity and 5% CO_2. No mycoplasma infection was detected in any cell culture.

4.3. Flow Cytometry Analyses

HEK293T cells were grown to confluence, resuspended, and seeded onto 35 mm plates at 25% confluency. The cells were cultured overnight. The Rho-tagged OR in the plasmid pCI and GFP expression vector were transfected using Viafect transfection reagent (Promega). After 18–24 h, the cells were resuspended in a cell stripper (Corning, NY, USA) and then kept on ice. The cells were spun down at 4 °C and resuspended in PBS containing 15 mM NaN_3 and 2% FBS to wash away the cell stripper. They were then incubated with a primary antibody mouse anti-Rho4D2 mouse antibody (Merck-Millipore, Burlington, MA, USA), washed, and then stained with phycoerythrin (PE)-conjugated donkey anti-mouse $F(ab')_2$ fragment antibody (Abcam, Cambridge, UK) in the dark. To stain dead cells, 7-amino-actinomycin D (Merck-Millipore, Burlington, MA, USA) was added. The cells were analyzed using BD FACSCanto II FACS and BD LSRFortessa with gating allowing for GFP-positive, single, spherical, viable cells, and the measured PE fluorescence intensities were analyzed and visualized using FlowJo.

4.4. Immunocytochemistry

Live-cell surface staining was performed as described previously [38]. The primary antibody used was mouse anti-rhodopsin 4D2 (Merck-Millipore). The secondary antibodies used were Cy3-conjugated anti-mouse IgG antibody (Thermo Fisher Scientific, Waltham, MA, USA). After antibody staining, the cells were fixed in 4% paraformaldehyde, and then the slides were mounted with Mowiol and visualized by a fluorescence microscopy M1 imager (Carl Zeiss, Oberkochen, Deutschland).

4.5. Dual Luciferase Reporter Gene Assay

All odorants were purchased from FUJIFILM Wako Chemicals (Osaka, Japan) and TCI chemicals (Tokyo, Japan). The Dual-Glo Luciferase Assay (Promega) was used to determine the activities of firefly and Renilla luciferase in Hana3A cells as previously described [38]. Briefly, firefly luciferase, driven by a cAMP response element promoter (CRE-Luc; Stratagene), was used to measure the OR activation levels. For each well of a 96-well plate, 5 ng SV40-RL, 10 ng CRE-Luc, 5 ng mouse RTP1, 2.5 ng M3 receptor 3, and 5 ng of Rho-tagged receptor plasmid DNA were transfected. Normalized activity for each well was further calculated as (Luc)/(Rluc). Luc and Rluc are the luminescence of firefly luciferase and Renilla luminescence, respectively. The basal activity was averaged from six wells in the absence of odorants and further corrected by subtracting that of the control empty vector. Odorant-induced activity was averaged from at least three wells and further corrected by subtracting the basal activity of that receptor. Odorant-induced responses were normalized to that of WT. EC50 value was calculated using GraphPad Prism software and data shown in Supplementary Data S2.

4.6. Statistical Analysis

Multiple comparisons were performed using one-way or two-way analysis of variance (ANOVA) using GraphPad Prism. Student's t tests were performed using the built-in function in Microsoft Excel. The average is shown as the mean $\pm$ standard errors.

Supplementary Materials: The following are available online at https://www.mdpi.com/article/10.3390/ijms23010277/s1.

Author Contributions: Conceptualization, Y.F. and H.M.; methodology, Y.F.; validation, Y.F.: formal analysis, Y.F.; investigation, Y.F., Y.N., N.M., S.M., R.K. and K.I.; resources, K.N. and I.O.; data curation, K.I.; writing—original draft preparation, Y.F., H.M. and M.Y.; writing—review and editing, Y.F., H.M. and M.Y.; visualization, Y.F.; supervision, Y.F. and M.Y.; project administration, Y.F.; funding acquisition, Y.F. and M.Y. All authors have read and agreed to the published version of the manuscript.

Funding: This work was supported by grants from JSPS-KAKENHI (18K14060 and 20K15745 to Y.F.; 20H02532 to M.Y.) and JST, ACT-X Grant Number JPMJAX201C to Y.F.

Acknowledgments: We thank Priya Meesa at Duke University for reading and editing the manuscript. YF stayed at Duke University with financial support from the JSPS Program for Advancing Strategic International Networks to Accelerate the Circulation of Talented Researchers (R2801). KI stayed at Duke University with financial support from Tokyo University of Agriculture and Technology as a student of the program for leading graduate schools in Japan.

Conflicts of Interest: The authors declare no conflict of interest.

Abbreviations

OR	Odorant receptor
GPCR	G-protein coupled receptor
RTP	Receptor transporting protein
BW	Ballesteros-Weinstein number [20]
ER	endoplasmic reticulum
HEK293T	Human Embryonic Kidney 293T
FACS	fluorescence-activated cell sorting

References

1. Buck, L.; Axel, R. A novel multigene family may encode odorant receptors: A molecular basis for odor recognition. *Cell* **1991**, *65*, 175–187. [CrossRef]
2. Lu, M.; Echeverri, F.; Moyer, B.D. Endoplasmic reticulum retention, degradation, and aggregation of olfactory G-protein coupled receptors. *Traffic* **2003**, *4*, 416–433. [CrossRef] [PubMed]
3. Malnic, B.; Godfrey, P.A.; Buck, L.B. The human olfactory receptor gene family. *Proc. Natl. Acad. Sci. USA* **2004**, *101*, 2584–2589. [CrossRef] [PubMed]
4. Saraiva, L.R.; Kondoh, K.; Ye, X.; Yoon, K.H.; Hernandez, M.; Buck, L.B. Combinatorial effects of odorants on mouse behavior. *Proc. Natl. Acad. Sci. USA* **2016**, *113*, E3300–E3306. [CrossRef] [PubMed]
5. Barnes, I.H.A.; Ibarra-Soria, X.; Fitzgerald, S.; Gonzalez, J.M.; Davidson, C.; Hardy, M.P.; Manthravadi, D.; Van Gerven, L.; Jorissen, M.; Zeng, Z.; et al. Expert curation of the human and mouse olfactory receptor gene repertoires identifies conserved coding regions split across two exons. *BMC Genom.* **2020**, *21*, 196. [CrossRef] [PubMed]
6. Su, C.Y.; Menuz, K.; Carlson, J.R. Olfactory perception: Receptors, cells, and circuits. *Cell* **2009**, *139*, 45–59. [CrossRef]
7. Kida, H.; Fukutani, Y.; Mainland, J.D.; de March, C.A.; Vihani, A.; Li, Y.R.; Chi, Q.; Toyama, A.; Liu, L.; Kameda, M.; et al. Vapor detection and discrimination with a panel of odorant receptors. *Nat. Commun.* **2018**, *9*, 4556. [CrossRef]

8. Krautwurst, D.; Yau, K.W.; Reed, R.R. Identification of ligands for olfactory receptors by functional expression of a receptor library. *Cell* **1998**, *95*, 917–926. [CrossRef]
9. Noe, F.; Geithe, C.; Fiedler, J.; Krautwurst, D. A bi-functional IL-6-HaloTag((R)) as a tool to measure the cell-surface expression of recombinant odorant receptors and to facilitate their activity quantification. *J. Biol. Methods* **2017**, *4*, e82. [CrossRef]
10. Noe, F.; Frey, T.; Fiedler, J.; Geithe, C.; Nowak, B.; Krautwurst, D. IL-6-HaloTag((R)) enables live-cell plasma membrane staining, flow cytometry, functional expression, and de-orphaning of recombinant odorant receptors. *J. Biol. Methods* **2017**, *4*, e81. [CrossRef]
11. Neuhaus, E.M.; Mashukova, A.; Zhang, W.; Barbour, J.; Hatt, H. A specific heat shock protein enhances the expression of mammalian olfactory receptor proteins. *Chem. Senses* **2006**, *31*, 445–452. [CrossRef] [PubMed]
12. Von Dannecker, L.E.; Mercadante, A.F.; Malnic, B. Ric-8B promotes functional expression of odorant receptors. *Proc. Natl. Acad. Sci. USA* **2006**, *103*, 9310–9314. [CrossRef]
13. Saito, H.; Kubota, M.; Roberts, R.W.; Chi, Q.; Matsunami, H. RTP family members induce functional expression of mammalian odorant receptors. *Cell* **2004**, *119*, 679–691. [CrossRef] [PubMed]
14. Zhuang, H.; Matsunami, H. Synergism of accessory factors in functional expression of mammalian odorant receptors. *J. Biol. Chem.* **2007**, *282*, 15284–15293. [CrossRef] [PubMed]
15. Wu, L.; Pan, Y.; Chen, G.Q.; Matsunami, H.; Zhuang, H. Receptor-transporting protein 1 short (RTP1S) mediates translocation and activation of odorant receptors by acting through multiple steps. *J. Biol. Chem.* **2012**, *287*, 22287–22294. [CrossRef] [PubMed]
16. Yu, T.; Su, X.; Pan, Y.; Zhuang, H. Receptor-transporting protein (RTP) family members play divergent roles in the functional expression of odorant receptors. *PLoS ONE* **2017**, *12*, e0179067. [CrossRef]
17. Fukutani, Y.; Tamaki, R.; Inoue, R.; Koshizawa, T.; Sakashita, S.; Ikegami, K.; Ohsawa, I.; Matsunami, H.; Yohda, M. The N-terminal region of RTP1S plays important roles in dimer formation and odorant receptor-trafficking. *J. Biol. Chem.* **2019**, *294*, 14661–14673. [CrossRef]
18. Sharma, R.; Ishimaru, Y.; Davison, I.; Ikegami, K.; Chien, M.S.; You, H.; Chi, Q.; Kubota, M.; Yohda, M.; Ehlers, M.; et al. Olfactory receptor accessory proteins play crucial roles in receptor function and gene choice. *Elife* **2017**, *6*, e21895. [CrossRef]
19. Ikegami, K.; de March, C.A.; Nagai, M.H.; Ghosh, S.; Do, M.; Sharma, R.; Bruguera, E.S.; Lu, Y.E.; Fukutani, Y.; Vaidehi, N.; et al. Structural instability and divergence from conserved residues underlie intracellular retention of mammalian odorant receptors. *Proc. Natl. Acad. Sci. USA* **2020**, *117*, 2957–2967. [CrossRef]
20. Juan, A.; Ballesteros, H.W. Integrated methods for the construction of three-dimensional models and computational probing of structure-function relations in G protein-coupled receptors. *Methods Neurosci.* **1995**, *25*, 366–428.
21. Olender, T.; Nativ, N.; Lancet, D. HORDE: Comprehensive resource for olfactory receptor genomics. *Methods Mol. Biol.* **2013**, *1003*, 23–38. [PubMed]
22. Malnic, B.; Hirono, J.; Sato, T.; Buck, L.B. Combinatorial receptor codes for odors. *Cell* **1999**, *96*, 713–723. [CrossRef]
23. Gutierrez-de-Teran, H.; Massink, A.; Rodriguez, D.; Liu, W.; Han, G.W.; Joseph, J.S.; Katritch, I.; Heitman, L.H.; Xia, L.; Ijzerman, A.P.; et al. The role of a sodium ion binding site in the allosteric modulation of the A(2A) adenosine G protein-coupled receptor. *Structure* **2013**, *21*, 2175–2185. [CrossRef] [PubMed]
24. Katritch, V.; Fenalti, G.; Abola, E.E.; Roth, B.L.; Cherezov, V.; Stevens, R.C. Allosteric sodium in class A GPCR signaling. *Trends Biochem. Sci.* **2014**, *39*, 233–244. [CrossRef] [PubMed]
25. Tunyasuvunakool, K.; Adler, J.; Wu, Z.; Green, T.; Zielinski, M.; Zidek, A.; Bridgland, A.; Cowie, A.; Meyer, C.; Laydon, A.; et al. Highly accurate protein structure prediction for the human proteome. *Nature* **2021**, *596*, 590–596. [CrossRef]
26. Yasuda, S.; Kajiwara, Y.; Toyoda, Y.; Morimoto, K.; Suno, R.; Iwata, S.; Kobayashi, T.; Murata, T.; Kinoshita, M. Hot-Spot Residues to be Mutated Common in G Protein-Coupled Receptors of Class A: Identification of Thermostabilizing Mutations Followed by Determination of Three-Dimensional Structures for Two Example Receptors. *J. Phys. Chem. B* **2017**, *121*, 6341–6350. [CrossRef]
27. Li, S.; Ahmed, L.; Zhang, R.; Pan, Y.; Matsunami, H.; Burger, J.L.; Block, E.; Batista, V.S.; Zhuang, H. Smelling Sulfur: Copper and Silver Regulate the Response of Human Odorant Receptor OR2T11 to Low-Molecular-Weight Thiols. *J. Am. Chem. Soc.* **2016**, *138*, 13281–13288. [CrossRef]
28. Keller, A.; Zhuang, H.; Chi, Q.; Vosshall, L.B.; Matsunami, H. Genetic variation in a human odorant receptor alters odour perception. *Nature* **2007**, *449*, 468–472. [CrossRef]
29. Mainland, J.D.; Li, Y.R.; Zhou, T.; Liu, W.L.; Matsunami, H. Human olfactory receptor responses to odorants. *Sci. Data* **2015**, *2*, 150002. [CrossRef]
30. Saito, H.; Chi, Q.; Zhuang, H.; Matsunami, H.; Mainland, J.D. Odor coding by a Mammalian receptor repertoire. *Sci. Signal* **2009**, *2*, ra9. [CrossRef]
31. Bubnell, J.; Jamet, S.; Tomoiaga, D.; D'Hulst, C.; Krampis, K.; Feinstein, P. In Vitro Mutational and Bioinformatics Analysis of the M71 Odorant Receptor and Its Superfamily. *PLoS ONE* **2015**, *10*, e0141712. [CrossRef] [PubMed]
32. Kotthoff, M.; Bauer, J.; Haag, F.; Krautwurst, D. Conserved C-terminal motifs in odorant receptors instruct their cell surface expression and cAMP signaling. *FASEB J.* **2021**, *35*, e21274. [CrossRef] [PubMed]
33. Liu, W.; Chun, E.; Thompson, A.A.; Chubukov, P.; Xu, F.; Katritch, V.; Han, G.W.; Roth, C.B.; Heitman, L.H.; Ap, I.J.; et al. Structural basis for allosteric regulation of GPCRs by sodium ions. *Science* **2012**, *337*, 232–236. [CrossRef] [PubMed]
34. Abaffy, T.; Bain, J.R.; Muehlbauer, M.J.; Spasojevic, I.; Lodha, S.; Bruguera, E.; O'Neal, S.K.; Kim, S.Y.; Matsunami, H. A Testosterone Metabolite 19-Hydroxyandrostenedione Induces Neuroendocrine Trans-Differentiation of Prostate Cancer Cells via an Ectopic Olfactory Receptor. *Front. Oncol.* **2018**, *8*, 162. [CrossRef] [PubMed]

35. Bushdid, C.; de March, C.A.; Topin, J.; Do, M.; Matsunami, H.; Golebiowski, J. Mammalian class I odorant receptors exhibit a conserved vestibular-binding pocket. *Cell. Mol. Life Sci.* **2019**, *76*, 995–1004. [CrossRef]
36. de March, C.A.; Topin, J.; Bruguera, E.; Novikov, G.; Ikegami, K.; Matsunami, H.; Golebiowski, J. Odorant Receptor 7D4 Activation Dynamics. *Angew. Chem. Int. Ed. Engl.* **2018**, *57*, 4554–4558. [CrossRef]
37. Charlier, L.; Topin, J.; de March, C.A.; Lai, P.C.; Crasto, C.J.; Golebiowski, J. Molecular modelling of odorant/olfactory receptor complexes. *Methods Mol. Biol.* **2013**, *1003*, 53–65.
38. Zhuang, H.; Matsunami, H. Evaluating cell-surface expression and measuring activation of mammalian odorant receptors in heterologous cells. *Nat. Protoc.* **2008**, *3*, 1402–1413. [CrossRef]

MDPI
St. Alban-Anlage 66
4052 Basel
Switzerland
Tel. +41 61 683 77 34
Fax +41 61 302 89 18
www.mdpi.com

International Journal of Molecular Sciences Editorial Office
E-mail: ijms@mdpi.com
www.mdpi.com/journal/ijms

www.ingramcontent.com/pod-product-compliance
Lightning Source LLC
Chambersburg PA
CBHW061502220926
51850CB00004B/150
* 9 7 8 3 0 3 6 5 6 5 3 7 8 *